Glossary of Symbols *(continued)*

Symbol	Meaning
μ_1, μ_2	means of two populations
$\mu_1 - \mu_2$	difference between two population means
μ_i	ith treatment mean, ANOVA
μ_p	mean, sampling distribution of proportion
μ_T	mean of distribution of T statistic, large sample
μ_U	mean of distribution of U statistic, large sample
$\mu_{\hat{\theta}}$	mean, sampling distribution of the statistic $\hat{\theta}$
$\mu_{\bar{x}}$	mean, sampling distribution of the mean
m	mean of a Poisson random variable
Md	median
Mo	mode
Md_0	hypothetical population median
Md_1, Md_2	medians of two populations
MSA	mean square due to factor A, ANOVA
MSAB	mean square due to interaction, ANOVA
MSB	mean square due to factor B; mean square due to blocks, ANOVA
MSE	mean square for error, ANOVA, regression analysis
MSR	mean square for regression, regression analysis
MSTR	mean square for treatments, ANOVA
ν	number of degrees of freedom
n	sample size
n_i	size of ith sample, ANOVA
$\tilde{n}$	harmonic mean of sample sizes, ANOVA
N	population size; total number of observations
N(0, 1)	normally distributed variable with mean 0, variance 1
N(μ, σ^2)	normally distributed variable with mean μ, variance σ^2
π	population proportion
π_0	[illegible]
$\pi_1, \pi_2, \ldots$	[illegible] proportions in [illegible] populations
$\pi_1 - \pi_2$	difference between two population proportions
$\hat{\pi}$	pooled proportion for testing $\pi_1 = \pi_2$
p	probability of success on one trial of a binomial random variable; sample proportion
$p_1, p_2, \ldots$	proportions in two or more samples
$p_1 - p_2$	difference between two sample proportions
P, P value	observed significance level for a statistical test
$P(\ )$	probability of ( )
$P(A)$	probability of event A
$P(\sim A)$	probability of the complement of A
$P(A \text{ and } B)$	joint probability that A and B both occur
$P(A \text{ or } B)$	probability that at least one of A or B occurs
$P(A\|B)$	conditional probability of A, given B
P^n_r	number of permutations of n objects taken r at a time
PSD	pseudo-standard deviation
q	Studentized range variable
$q_\alpha\{k, N-k\}$	critical value of Studentized range statistic
Q_1	first quartile
Q_3	third quartile
ρ	coefficient of correlation for a population
ρ_s	Spearman correlation coefficient for a population
ρ^2	coefficient of determination for a population
r	coefficient of correlation for a sample
r_s	Spearman correlation coefficient for a sample
r^2	coefficient of determination for a sample

STATISTICS TODAY
A Comprehensive Introduction

STATISTICS TODAY

A Comprehensive Introduction

Donald R. Byrkit
The University of West Florida

With contributions from Robert L. Schaefer & John H. Skillings,
Miami University of Ohio

The Benjamin/Cummings Publishing Company, Inc.
Menlo Park, California • Reading, Massachusetts
Don Mills, Ontario • Wokingham, U.K. • Amsterdam • Sydney
Singapore • Tokyo • Madrid • Bogota • Santiago • San Juan

To Steven and Linda

Sponsoring Editor: Craig Bartholomew
Developmental Editor: Martine Westermann
Production Editor: Richard Mason, Bookman Productions
Production Supervisor: Mary Picklum
Copy Editor: Don Yoder
Interior and Consulting Designer: Hal Lockwood
Illustrations: Carl Brown
Composition: Graphic Typesetting Services

The basic text of this book was designed using the Modular Design System, as developed by Wendy Earl and Design Office Bruce Kortebein.

Library of Congress Cataloging-in-Publication Data

Byrkit, Donald R.
  Statistics today.
  Includes index.
  1. Statistics.   I. Title.
QA276.12.B96   1987   519.5   86-20710
ISBN 0-8053-0740-0

DEFGHIJ-DO-89

The Benjamin/Cummings Publishing Company, Inc.
2727 Sand Hill Road
Menlo Park, California 94025

Brief Contents

Detailed Contents

10 Nonparametric Methods 541

11 Single-Factor Analysis of Variance 588

14 Multiple Regression 791

Preface

Statistics is a dynamic field of study today. The computer has augmented the ability of statisticians to analyze large data sets thoroughly. New statistical techniques such as robust methods and exploratory data analysis are making it easier to evaluate data sets that were once considered "messy." The student of statistics is now entering a field that is on the cutting edge of our information-oriented economy. To keep up with the current developments in this exciting field, I have written *Statistics Today*. Rather than a patched-together book from another era, this is a new text that reflects the position of statistics. . .*today*.

What's Different About This Book?

Statistics Today goes beyond previous books on elementary statistics. A few key elements of this book set it apart:

- Coverage of Minitab and the SAS software system——Minitab and SAS procedures are thoroughly covered in the book. Each chapter concludes with an optional section that presents the appropriate commands for both Minitab and the SAS software system. Later in the book SAS software systems' usage is integrated throughout each chapter. This is a step beyond the convention of merely inserting computer output as an afterthought.
- Modern Methods——Many contemporary statistical methods are included in this book. If these methods cannot be covered in class due to time

constraints, it is hoped that the student will refer to them for further statistical study. Such topics include stem and leaf plots, box plots, normal probability plots, the pseudo-standard deviation, using trimmed and Winsorized data sets, and the Fisher-Behrens t′.

- Thinking "statistically"——An effort has been made in this book to encourage the student to approach problems like a real statistician. In addition, a number of historical comments and notes are presented in the margins in the interest of building an appreciation for statistics as a continually developing science.
- Motivation——A common misconception amongst uninitiates is that statistics is a boring field of study. This is not true! Every effort has been made in this book to make the text enjoyable reading, by showing the utility and real-life application of statistical methods. Each chapter is motivated by an intriguing chapter-opening case that should pique the reader's interest. This case is solved at the end of the chapter by using the methods presented in that chapter. Real-data problems and examples are also sprinkled throughout each chapter.
- Nonparametric Methods——Most books "tack on" nonparametric tests as an apparent afterthought. In this book some nonparametric tests are introduced in Chapter 10. Appropriate nonparametric techniques are then presented side-by-side with their counterparts in Chapters 11, 12, and 13.
- Foundations for further study——Rather than presenting *only* a brief overview of statistics suitable for a general education survey course, this book also lays a complete foundation for further courses requiring statistics. Such courses could include experimental psychology, biometrics, econometrics, decision theory, marketing research, and many others. This text also provides alternative procedures to fit most situations encountered in real life.

Courses That Might Use This Book

Statistics Today provides a complete first course in general statistics at the algebra-based level. It is suitable for either a one-term or two-term course, and for students with a wide range of mathematical maturity. A first course in high-school algebra and some familiarity with the use of a calculator is recommended. Examples and problems have been chosen from a variety of disciplines including business, economics, accounting, management, biology, agriculture, psychology, sociology, anthropology, and education.

Flexibility and Organization

The book has been organized in a logical sequence to allow for great flexibility in a course. The first eight chapters contain the core of a beginning statistics course, although much additional material has been included. The accompanying dependency chart (pp. xxiv–xxv) shows which topics are considered essential to the logical development of the course. A few options are:

- For a two-course sequence, the first course might include the "core" of Chapters 1–8 (see dependency chart), along with additional topics from these chapters, or with some of Chapter 9. The second course could then cover most of the remainder of the material from Chapters 9–14.
- A one-term introductory course should include all of those topics listed as the "core" of the course, together with a selection of additional topics chosen by the instructor.
- For a course minimizing the coverage of probability, much of Chapter 3 can be omitted. The fundamental ideas of probability can be covered in Section 3.1 along with the idea of independence (in Section 3.2), which may be presented intuitively.
- An introduction to categorical data analysis, extension of two-sample methods to more than two populations, and an introduction to prediction via regression analysis might include Chapter 9, Sections 11.1 and 11.2, and Sections 13.1 and 13.2.
- An investigation of the methods for checking assumptions and using alternate approaches might include Sections 2.4, 6.4.3–6.4.5, 7.2.3–7.2.4, 8.2.3–8.2.5, and Chapter 10.

Pedagogical Aids

As mentioned previously, a concerted attempt has been made to motivate the student to study and enjoy the material in this book. The chapter opening Cases should stimulate the student's interest in each chapter. In addition to these Cases the following study aids are woven into the text:

- Proficiency Checks——Most sections contain proficiency checks. These are short exercises designed to test to what extent students have mastered the concepts in that section, and to reinforce the retention of those concepts. Complete solutions of the proficiency checks are provided for easy reference at the conclusion of each chapter.

- Subsections——Each section has been divided into smaller "bite-size" subsections. Each subsection presents a single topic for students to master. It is hoped that these subsections will be helpful for students when reviewing the material for tests or for reference.
- Examples——Nearly every new idea is illustrated by at least one example showing its application. These examples are meant to amplify and clarify the theory behind an idea.
- Problems——The book contains a large number of both end-of-section and chapter-review problems. All problems are divided into two categories. Those problems labeled *Practice* are designed to provide exercise in using the mechanics covered in that section or chapter. Those problems marked *Applications* are realistic scenarios in which students can apply these methods. Many of these application problems are based on studies or situations that have appeared in newspapers, magazines, or journals.
- Symbols——Symbols that appear for the first time in each chapter are listed at the end of the chapter. All symbols that appear in the book are listed with their meaning on the book's endpapers.
- Key Terms——Key terms are printed in boldface type where they first occur. An index to key terms has been provided at both the end of each chapter and at the back of the book.
- References——References for additional reading are provided where appropriate in each chapter. These are repeated in a list at the end of the chapter along with additional references.

Use of Computer Packages

Minitab and the SAS software system have been chosen for extensive use in this book. These systems were selected both because they are widely used and available, and because they are sufficiently different to be of interest to different users. Both packages are now available for mainframes and microcomputers.

To give the student something more useful than the standard "printout followed by interpretation" approach, emphasis has been placed on hands-on use of the computer. Each chapter includes basic explanations of how to use Minitab and the SAS software system, with examples and illustrations chosen from that chapter. Thus, very little explanation or supervision should be required by the instructor in order for the student to work problems using either a SAS software system or Minitab installation, once the local system has been mastered.

Minitab is introduced in Chapter 1, together with the data-entry, storage, and retrieval methods used throughout the book. A separate section in each chapter is devoted to using Minitab to perform some or all of the procedures introduced in that chapter. In some chapters the computer coverage is used to amplify or illustrate some of the concepts presented in that chapter. We suggest that these sections be read even if a computer is not used in the course. A complete listing of Minitab commands is presented in Appendix A.1.

Coverage of the SAS software system parallels that of Minitab in Chapters 1–10. Beginning in Chapter 11, SAS software system coverage is integrated within each section, since the computer is an important adjunct to the material presented in this chapter and succeeding chapters. A selected listing of SAS statements is presented in Appendix A.2.

Appendix B contains two data sets for use with the computer. Data set B.1 contains data used in a study of 100 female students at Miami University, Oxford, Ohio; data set B.2 reproduces a survey of 166 elderly people conducted by the U.S. Department of Agriculture. Additional data sets are included in the Instructor's Guide.

Testing of Assumptions

One of the most important aspects of statistical inference is understanding the meaning of the statistical procedures being used. To enhance student understanding, all procedures in this book are carefully examined, yet without using intimidating mathematical proofs or notation. A knowledge of the underlying assumptions for each procedure is essential if these procedures are not to be misused. Therefore, those assumptions are frequently emphasized. Unlike most other books, methods are given to test the assumptions. For example, several methods are given to test the ubiquitous assumption of normality, including normal probability plots, a goodness-of-fit test, and some modern, exploratory data-analysis procedures. Hartley's test is applied to the assumption of homogeneity of variance in ANOVA.

Supplements

A number of supplements to this book are provided to assist the instructor and the student in presenting and learning the material:

- *Instructor's Guide*——The Instructor's Guide contains additional optional material not included in the book. This includes derivations, the use of

Chebyshev's theorem on statistical inference, additional data sets, and additional SAS statements and procedures. Transparency masters and the answers to even-numbered problems are also included here. All data sets are available on disk or tape.

- *Student Solutions Manual*——Step-by-step worked-out solutions to the odd-numbered problems (answers to most of which are in the back of the text) are available in a separate Solutions Manual. This supplement also includes comprehensive chapter summaries.
- *Testbank*——A Testbank for instructors includes a large number of problems for use in tests.
- *STATDISK*——STATDISK is a set of statistical programs available for the IBM-PC or the Apple II. It is available from the publisher for adopters of this book. The accompanying *Statdisk Manual,* a tutorial self-study student manual, is available at a reasonable cost for student purchase.

Acknowledgments

Thanks are due to many people who helped in the preparation of this book, especially to Craig Bartholomew, who believed in the project, and Martine Westermann, who made many, many valuable suggestions for improvements. Particular thanks are due to John Skillings and Robert Schaefer of Miami University of Ohio for supplying a draft of the Minitab material. Specials thanks also go to my wife, Marnette, who encouraged me through many 80- and 90-hour work weeks over the past year and a half, and gave up our weekends to work on the book. Thanks also to the many reviewers who made valid criticisms and valuable suggestions: Barnard Bissinger, Penn State University; John W. Dirksey, California State University, Bakersfield; Shirley Dowdy, West Virginia University; Richard G. Driskell, Ball State University; Judith M. Ekstrand, San Francisco State University; Dale Everson, University of Idaho; Iris Ibrahim, Clemson University; William Koellner, Montclair State College; Purushottam Laud, Northern Illinois University; Bruce Lind, University of Puget Sound; Michael Shing, Stanford University; John Skillings, Miami University; Leonard Sweet, University of Akron; John Wasik, North Carolina State University; and Ann Watkins, Los Angeles Pierce College.

I am indebted to the SAS Institute and Minitab, Inc. for permission to document their packages, and to Barbara F. Ryan of Minitab who checked all Minitab commands and printouts. Output from SAS procedures are printed with permission of SAS Institute Inc., Cary, NC. 27511-8000, Copyright © 1985. I am also indebted to the Biometrika Trustees for permission to use Tables 8, 12, 18, 29, and 31 from *Biometrika Tables for Statisticians,*

Volume I, Third Edition, by Pearson and Hartley; to American Cyanamid for permission to use Tables 1 and 2 from *Some Rapid Approximate Statistical Procedures* by Wilcoxon and Wilcox; to the authors and publishers of *Statistical Tables* by R. R. Sokal and F. J. Rohlf (W. H. Freeman and Company) for permission to reprint part of Table O; and to the staff of the Institute for Statistical and Mathematical Modeling of the University of West Florida for compiling Tables 2, 3, and 4.

Donald R. Byrkit
Pensacola, Florida

Dependency Chart

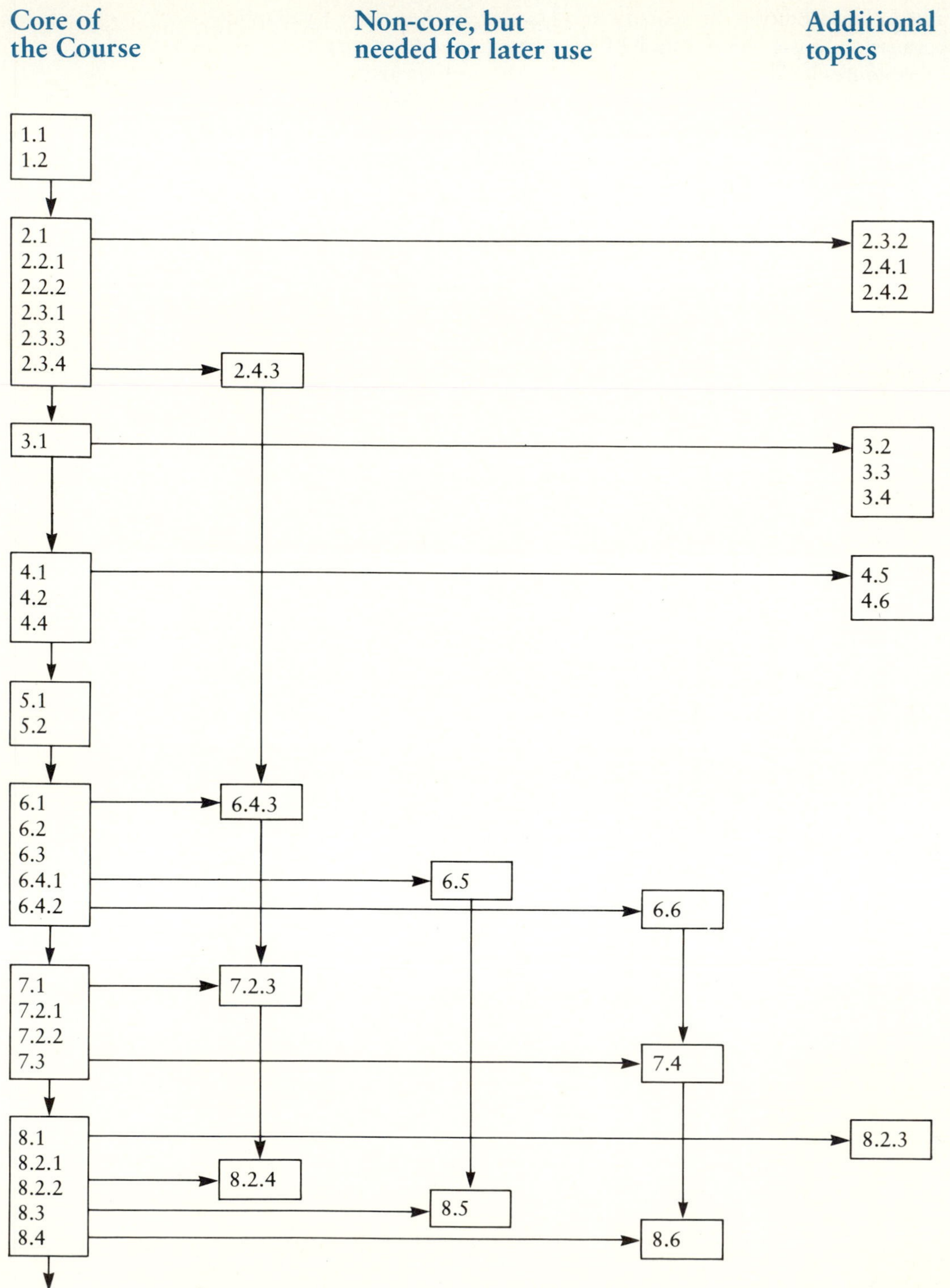

Dependency Chart

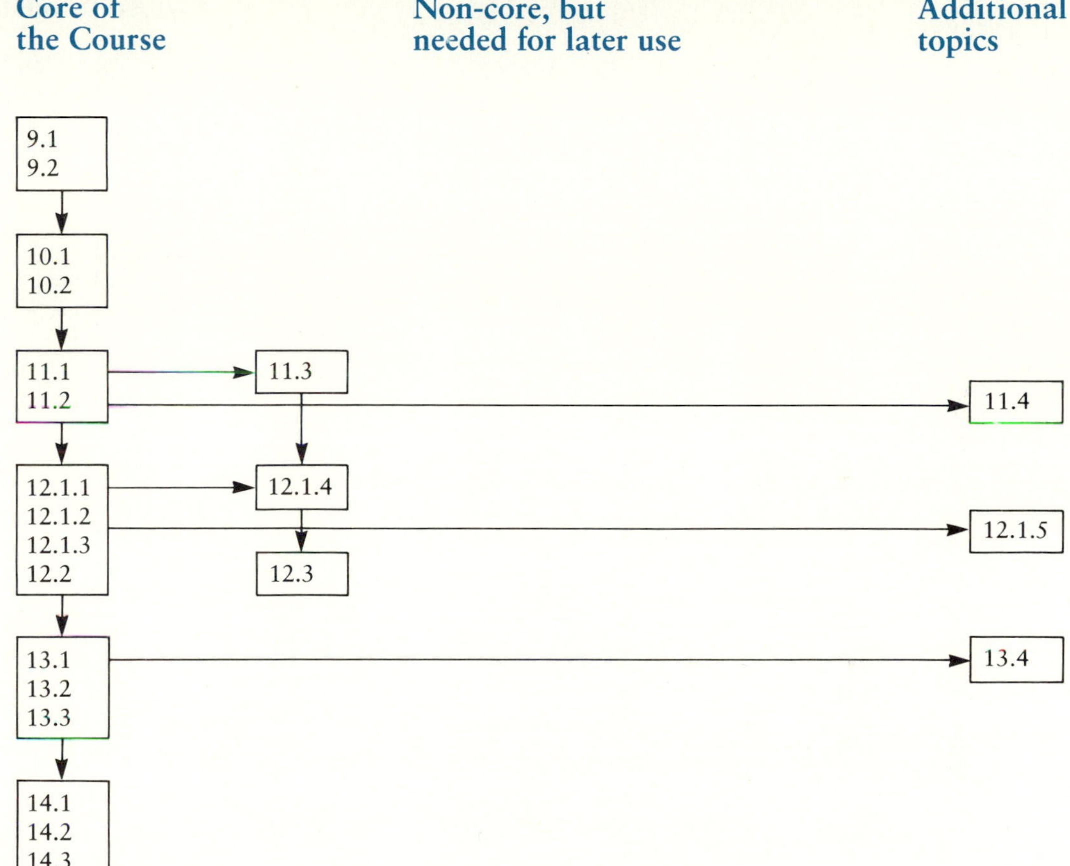

1

Organizing and Presenting Data

The study of statistics is important to understanding and interpreting our world. If it were not for the methods of statistical analysis, for instance, the work of Jonas Salk in developing a vaccine for poliomyelitis (1954) would not have been accepted as quickly as it was. Statistical methods presented in later chapters were among those Salk used to show that his vaccine actually worked, and thus statistics helped prevent the unnecessary crippling of countless thousands of people.

In this chapter we learn how to organize and present facts in such a way that they can be more easily understood and interpreted. The summarization techniques presented in this chapter are the most widely used methods. As a result of studying this chapter you should be able to organize, summarize, and give visual presentations of data that are quickly comprehended even by those who have not studied statistics.

1.0 Case 1: What Grapes To Grow?

Giovanni DiPalma of Tuscany, Italy, was left a farm in California by his uncle. Giovanni grew wine grapes in Italy, and he knew that California had a very good reputation for wine, so he decided to continue his profession there. When he arrived, he found that his uncle had not been growing grapes but instead had grown other crops. Giovanni set about converting the farm to wine grapes and had to decide what variety to grow. There are hundreds of varieties of wine grapes, each growing best in a particular climate and type of soil. In Tuscany, Giovanni had grown the Sangiovese grape, from which Chianti is made. In Europe, climatic conditions are fairly stable throughout a region so that the same varieties of grape can be grown throughout. In California, however, conditions can vary widely from one valley to the next and even within the same valley. Giovanni found that the farm next door on the east grew Chardonnay and Riesling grapes, which require a cool climate. The farm to the south, however, grew Sylvaner and Muscat grapes, which grow in a slightly warmer climate. An important decision needed to be made because planting the wrong grape could have been an economic disaster. A neighbor suggested that Giovanni use statistics published by the University of California, Davis, to make his decision.

1.1 Frequency Distributions

1.1.1 What Is Statistics?

Most people tend to think of the word *statistics* in terms of yards gained in a football game, the consumer price index, selling price of houses, polls on voter preferences, Nielsen ratings, and other such items that may or may not be of interest to you. What does it mean when the TV news reports that a family with median income can afford 85% of a median price house (that is, that the "affordability index" is 0.85)? What does it mean when the business news section of your newspaper reports that "the rate of increase in inflation is decreasing"—is this a contradiction?

Statistics consists of more than mere numbers or sets of figures. Statistics can be described as the science, and art, of classifying and organizing data in order to draw inferences. When we want to find out something, we usually gather information. In statistics this information generally takes the form of numbers. After we obtain the information, we must organize it into understandable segments and then examine it closely in order to determine what conclusions we may reach. The collection and organization of data is usually called *descriptive* statistics; the process of deciding what conclusions may be reached is called *inferential* statistics. The studies which culminated in the conclusion that cigarette smoking is harmful to health relied greatly on statistical inference. To reach this conclusion it was necessary to gather and analyze a great amount of data and show that cancer of the respiratory passages and other medical problems were attributable to cigarette smoke and not to chance or some external cause. In the face of concerted opposition by the tobacco industry, a great deal of clear evidence was needed to reach this conclusion. Modern statistical methods provided the means to demonstrate that the conclusion was inescapable.

The science of inferential statistics is very young. Calculus was invented in the seventeenth century (Isaac Newton and Gottfried Leibniz); algebra goes back to the third century (Diophantus of Alexandria); geometry flourished in the fifth century B.C. (the Pythagoreans) and even before (Thales and the Egyptians). In contrast, although statistical methods were used by John Graunt in the seventeenth century, the *science* of statistics may date only from the nineteenth century (Karl Gauss and John Galton) and has developed into the twentieth century (Karl Pearson and Ronald Fisher).

John Graunt (1620–1674) is generally considered to be the father of the science of demography, the statistical study of human populations. While active as a haberdasher he studied death records and classified death rates according to cause of death. His work (published in 1662) paved the way for mortality tables used in the insurance industry. It also influenced a number of other statistical pioneers, including Edmund Halley, the astronomer royal. After his business was destroyed in the London Fire of 1666, Graunt held municipal offices and a militia command. He was a charter member of the Royal Society, the oldest scientific society in England, founded in 1660.

The introduction and rapid development of the computer have opened the door for more widespread use of statistical methods, due to the ease of computation, and new methods are continually being devised and investigated. The field of exploratory data analysis, advanced by John Tukey and Frederick Mosteller, is a new development in statistics, and some of its methods are gaining wide acceptance.

Statistics is used in practically every human endeavor. *Statistics and Public Policy,* by William B. Fairley and Frederick Mosteller (1977), contains a number of cases in which statistical analysis was used. For example, data on graduate admissions to the University of California, Berkeley, for fall 1973 appear to show a pattern of bias against female applicants. A statistical analysis of the data, however, shows a small but significant bias in *favor* of women. Other articles explore such issues as whether motor vehicle inspection reduces accident mortality rates (it does), when hurricanes should be seeded with silver iodide crystals to reduce their destructive force, applications of statistics in legal proceedings, and many other interesting topics.

1.1.2 The Basics

An important aspect of classifying data is the effective organization and presentation of data. An unorganized mass of figures is more often confusing than clarifying. This chapter is concerned with methods of deriving meaning from numerical data.

*Since the Latin **data** is plural, the word is used in the plural form throughout this book.*

As a first step toward learning statistics, we must agree on the meanings of certain basic terms that will be used throughout the book. We call attention to important terms by printing them in boldface type. The word *data,* for example, was used in the preceding paragraphs without explanation, but now we wish to define it. The term **data** refers to the set of values, elements, or numbers under consideration. The complete set of all observations pertaining to a single characteristic of interest is called a statistical **population,** while anything less than the complete set is called a *sample.* Thus a **sample** is a set of data selected or otherwise obtained from a statistical population. A **random sample** is a sample obtained in such a way that all samples of the same size are equally likely. Such a sample is considered to be representative of the population. Examples of statistical populations include the set of ages of all students, the set of their grade-point averages, and the set of makes of their automobiles as well as the set of weights of the automobiles, the numbers of miles driven, and the numbers of scratches on each one.

Each element of a statistical population is called a **data point, piece of data,** or **observation.** These terms are used interchangeably. A listing of all the data points in a population is called a **census.** The amount of money spent by a customer in a store on a particular day, for example, is a piece of data, while the collection of all expenditures by all customers on that day would comprise a complete set of data. This set, in turn, could be considered a sample of the population of all expenditures of all customers on all days in the store.

Note that a census is a listing of the elements of a population. The set of incomes of all people living in a city is a population, but obtaining a census for this population might be difficult. A population is abstract; a census is concrete.

Data can be *qualitative* or *quantitative.* **Qualitative data** (sometimes called attribute or categorical data) result from information that has been sorted into categories. Each piece of data clearly belongs to one **classification** or category. The classification by make of automobiles on a parking lot is one example of qualitative data. In general, the terms *qualitative, attribute, categorical,* and *classification* are used interchangeably when applied to types of data. **Quantitative** or **numerical data** result from counting or measuring. We might count the number of nicks and scratches in the paint of each car and then list the number of cars with zero scratches, one scratch, two scratches, and so forth. A car has a whole number of scratches (it cannot have 3.7 scratches, for instance), so there are clear divisions between the values. This type of quantitative data is called **discrete** or countable. We say

PROFICIENCY CHECKS

1. Which of the following could be a statistical population?
   a. The number of children in each family in a city
   b. The ages of owners of automobiles with four-wheel drive
   c. The number of insurance policies sold by Fred Jones in a month
   d. The number of insurance policies sold by each of a company's insurance agents in a single month
   e. The number of each different type of policy sold in a month
   f. The number of whole life policies of each insured value sold in a month
   g. The number of whole life policies sold in a month
2. Classify each of the following as quantitative or qualitative data:
   a. The manufacturer's brand name of your new refrigerator
   b. The cost of your new refrigerator
   c. The color of your new refrigerator
   d. Your weight
   e. The number of different courses you are taking
   f. The make of the tires on your car
3. Classify each of the following variables as discrete or continuous. Which, if any, could be considered either one in different circumstances?
   a. Distance
   b. Weight
   c. Number of beans in a jar
   d. Time elapsed since the start of a race
   e. Incomes (monetary value)
   f. Number of students in a classroom

that a set of numbers is countable if the numbers can be arranged in such a way that they can be counted. If we weighed the cars, we could get the weight to the nearest pound, but it would be possible for a car to weigh slightly more or less than a whole number of pounds. In fact, it would be theoretically impossible ever to give the *exact* weight of the car, no matter how many decimal places we used, due to limitations of our measuring devices. A reported weight of 2,877 pounds, for example, would probably indicate a weight somewhere between 2,876.5 and 2,877.5 pounds. This is an example of **continuous** measurement yielding continuous data. Generally, data arising from measuring are continuous whereas data arising from counting are discrete.

Table 1.1 HEIGHTS OF ADULT MALES IN CITY.

HEIGHT (IN.)	FREQUENCY
61	28
62	94
63	206
64	411
65	604
66	712
67	848
68	931
69	817
70	808
71	783
72	731
73	547
74	388
75	316
76	284
77	157
78	132
79	116
80	44
81	11
Total	8968

Often we have an option regarding the system of classification to use. The characteristic being observed is usually called a *variable*. A **variable** is a quantity or classification whose value is not fixed. In the example of the customer in the store, we could classify each customer according to whether he or she bought anything. The variable would simply have the values "yes" and "no," which would be a qualitative classification. We could let the variable be the number of items bought, which would be quantitative and discrete. We could let the variable be the amount of time spent in the store, which would be quantitative and continuous. Several other variables could be used as well. A likely variable would be the amount of money spent. Although this variable is technically discrete, since the customer does not spend fractional parts of a penny, it is common practice to consider as "practically" continuous those variables whose unit is quite small in relation to the amounts involved. If we were talking about the different numbers of pennies several children had, the variable would obviously be discrete. In terms of the national debt, the variable could be considered continuous for all practical purposes. A matter of judgment is involved. Some variables seemingly pose the opposite problem. One common variable is age. Since a person's age is a measure of the time that the person has been living, the variable age would seem to be continuous. The common practice with age, however, is to give one's age as of the last birthday. Reported in this way the variable is discrete.

1.1.3 Frequency Tables

Suppose you asked someone for data on the height of adult males in a city of some 25,000 people, and the person responded by giving you a list of 8,968 heights. Unless the data were organized in some fashion, this list in its raw form would not be of much use. One way of organizing data is called a *frequency distribution*. In its simplest form, a **frequency distribution** consists of listing each value the data could have and enumerating the total number, called the **frequency,** of the pieces of data that have each value. If height is reported to the nearest inch in the height example, such a frequency distribution might appear as in Table 1.1. Such a table tells you that most of the data points have values from 63 to 76 inches and that the numbers above 76 inches and below 63 inches are relatively quite small. In short, the table gives you a very good and accurate profile of this set of data.

Table 1.2 MAKES OF CARS ON FACULTY PARKING LOT.

MAKE	NUMBER OF CARS
Chevrolet	33
Ford	28
Pontiac	11
Datsun	9
Toyota	7
Plymouth	4
Others*	6
Total	98

*Two or fewer each.

A frequency table may also be used to present qualitative data. For example, Table 1.2 lists the number of automobiles of each of several makes in the faculty parking lot. Note that it does not matter whether tables are arranged in ascending or descending order. The guiding principles should be ease and clarity of understanding of what is presented.

It is sometimes useful to determine the proportion of cases for each value of the variable. This proportion, called the **relative frequency,** is equal to the number of observations (frequency) for a given value divided by the total number of observations (total frequency). Relative frequency can also be reported as a percentage. In Table 1.1, we see that 931 men were 68 inches tall (to the nearest inch) out of the total of 8968, so the relative frequency for 68 inches is 931/8968 or about 0.104, or 10.4%.

There is an easy procedure for constructing a frequency table. For small amounts of data we can rewrite the data given into ascending (or descending) order. We then have the data arranged in order and the construction of the table is generally quite easy. Another plan, more useful in cases where a great deal of data are involved, is to find the highest and lowest values and then list these values and all values between them. In a second column we **tally** each observation by putting a slash (tally) mark for each one as we come to it, usually crossing out the number in the original list, and we then summarize the results in a frequency table. If one has access to a computer, the data may be copied into a data set and, depending upon the software for the computer, a sorting procedure may be used to arrange the data in order. The final table, if intended for presentation, should be self-contained unless there is accompanying explanatory material; that is, it should have a title and sufficient information attached to make it self-explanatory. The *stem and leaf plot,* presented in Section 1.2.4, may be used instead of a tally chart.

Example 1.1

A biologist is keeping careful records of the time required, measured to the nearest minute, for a spore culture to double in a petri dish. The number of minutes required for each of 25 determinations is listed here. Construct a frequency table for these data:

57	59	58	61	56	63	57	69	
59	59	57	63	72	59	62	60	
59	63	59	65	57	61	58	59	63

Solution

Since there are few numbers here, rewriting them in order would be logical. An example of the tally method would be helpful at this point, however, so it is presented in Table 1.3. Using the tally chart we can construct the frequency table in Table 1.4.

Table 1.3 TALLY CHART FOR EXAMPLE 1.1.

TIME (MIN)	TALLY	TIME (MIN)	TALLY
72	/	63	////
71		62	/
70		61	//
69	/	60	/
68		59	~~////~~ //
67		58	//
66		57	////
65	/	56	/
64			

Table 1.4 TIME REQUIRED FOR A SPORE CULTURE TO DOUBLE.

TIME (MIN)	FREQUENCY	RELATIVE FREQUENCY
72	1	0.04
69	1	0.04
65	1	0.04
63	4	0.16
62	1	0.04
61	2	0.08
60	1	0.04
59	7	0.28
58	2	0.08
57	4	0.16
56	1	0.04
Total	25	1.00

Note: Experiment conducted 8 April 1985.

Sometimes it is convenient or perhaps even more informative to combine a few of the values when the frequencies for several of them are zero. This is often done when such values fall at one end of the distribution. In Table 1.4 we note that there are only three observations with values greater than 63. We could combine them into one set of values, namely 64–72. The frequency would be 3 and the relative frequency 0.12. This arrangement does not significantly impair our understanding of the data presented in the table, but in this case we do not know the actual values of the three observations.

PROFICIENCY CHECK

4. The following are ages (as of last birthday) of 25 randomly selected college students. Construct a table showing frequency and relative frequency for each age.

   19 18 22 20 20 21 23 20 19 20 22 21 20 19 19
   21 19 20 18 20 22 19 20 21 20

1.1.4 Grouping Data

Suppose that a set of data lists incomes (to the nearest dollar) of families in a city. It is obvious that trying to convert such a listing to a frequency table involves the listing of possibly thousands of numbers. This produces a table that would probably not be meaningful, since many income amounts would have a frequency of only 1. One way out of such a difficulty is the method of **grouping** data. We group together observations that have values close to each other so that instead of having a large number of values, each with a small number of observations, we have a smaller set of intervals, each containing a larger number of observations. We group these data by dividing the interval of numbers that contains all the values of the data into smaller intervals, most frequently called *classes*. If the incomes range from very small to upward of $50,000, for example, we may classify together all incomes from, say, $0 to $3999, all incomes from $4000 to $7999, and so on. The interval of values in each case is called a **class.** These classes should not overlap—to eliminate the possibility of counting the same measurement more than once—and there should be enough classes to include all the data.

A few general observations govern the grouping of data. First, we do not want too many or too few classes, since this might result in a distortion of the picture we want to convey. Second, the classes should all be the same size (although this rule may be bent in certain circumstances). Third, the classes should be easy to work with. It has been found that the number of classes that will achieve these ends depends, to a certain extent, on the number of observations in the data set. The larger the data set, the greater the number of classes we should use. Trial and error is often needed to get just the right number of classes. The need for judgment in the selection of classes is an example of the *art* of statistics. One rule of thumb often used as a gauge is that five classes will be about right for 10 to 22 observations and that the number of classes should be increased by one for each doubling of the number of data.

Once we have determined the approximate number of classes we wish to have, we must determine the approximate width of each class. To do this we must first determine the *range* of the data. The **range** of a set of numerical data is defined to be the largest value in the set minus the smallest value in the set. We might measure the amount of time taken to burn a specified amount of rocket fuel, for example, and then repeat the experiment for a total of 30 observations. If the longest time was 38.446 seconds and the shortest time was 34.923 seconds, the range is 3.523 seconds.

The height data presented in Table 1.1 are an example of continuous data being reported as if they were discrete. In this kind of arrangement, a height of 72 inches represents all heights from 71.5 to nearly 72.5 inches. We need to keep this in mind when we group the data into classes larger than 1 inch in width. (To avoid confusion, we agree that any heights that fall on the border—in this case, 71.5 or 72.5 inches—will be rounded to the next *higher* class.) Since we have already rounded the data to whole numbers, we can now define *class limits* and *class boundaries*. First we note the units in which the data are reported—in this case, integral number of inches. The **class limits** are the highest and lowest values specified for each class; they are given in the same units in which the data are reported. The **class boundaries** are the actual dividing lines between the classes and are midway between successive class limits. The distance between the class boundaries is called the **class width.**

Now suppose we wish to group the height data in order to make the presentation more compact. We must determine the number of classes and the size of each class before we can proceed any further. Since the data are continuous, the smallest height could be 60.5 inches and the largest almost 81.5 inches, so the range would be 81.5–60.5 or 21 inches. A class width of 3 inches would give us seven or eight classes (depending on where we start), while a class width of 2 inches would require about eleven classes. We arbitrarily decide to have seven classes with a class width of 3 inches. The lowest height in the data set is 61 inches, so we let the lower limit of the first class be 61. The class will contain the values 61, 62, 63, so we say that the first class is 61–63. The upper class limit is 63. The lower limit of the next class is 64, and the boundary between the two classes is 63.5. This is the upper boundary of the class 61–63 and the lower boundary of the class 64–66. The center of the first class is 62 and the center of the second class is 65. The center of an interval, called the **class mark,** is equal to half the sum of the class limits. When the ungrouped data are not available, the class marks are often used to perform arithmetic on grouped data.

Using the general guidelines discussed here, we could construct a frequency table for the data of Table 1.1. In Table 1.5 we present this set of data grouped into classes as shown, together with the class boundaries and class marks (neither of which are generally presented in a table).

In another example, suppose that income data are collected for 9311 employees of a company, nationwide, at the level of lower middle management and below, including part-time and temporary employees. The lowest income is \$343; the highest is \$43,764. A frequency distribution for the data might be constructed as follows. The range is \$43,764 − \$343, or \$43,421. The number of employees and the range are both large, so we may need a fairly large number of classes, say 12 to 15. Dividing the range by 12, we obtain \$3318; dividing by 15, we obtain about \$2895. A convenient class width, then, would be about \$3000. We could start with \$343, but

Table 1.5 CLASS INFORMATION FOR HEIGHT DATA.

HEIGHT (IN.)			
CLASS LIMITS	CLASS BOUNDARIES	CLASS MARK	FREQUENCY
61–63	60.5–63.5	62	328
64–66	63.5–66.5	65	1727
67–69	66.5–69.5	68	2596
70–72	69.5–72.5	71	2322
73–75	72.5–75.5	74	1251
76–78	75.5–78.5	77	573
79–81	78.5–81.5	80	171

Table 1.6 INCOME DATA FOR COMPANY EMPLOYEES.

INCOME ($)	NUMBER OF EMPLOYEES
42,000–44,999	3
39,000–41,999	7
36,000–38,999	12
33,000–35,999	17
30,000–32,999	28
27,000–29,999	60
24,000–26,999	216
21,000–23,999	268
18,000–20,999	448
15,000–17,999	621
12,000–14,999	949
9000–11,999	1421
6000–8999	2844
3000–5999	2123
0–2999	294

it is more usual to start with a round number such as zero or $1000. Since $343 is less than $1000, we might obtain the distribution shown in Table 1.6.

Example 1.2

A random sample of 50 football players currently playing in the NFL shows the weights listed here. Group the data into ten classes and construct a frequency table; also list relative frequencies.

193	240	217	283	268	212	251	263
275	208	230	288	259	225	252	236
243	247	280	234	250	236	277	218
243	268	231	269	224	239	258	231
255	228	202	245	246	271	249	255
265	235	243	219	255	245	238	257
254	284						

Solution

The weights are continuous, so the smallest number in the set could be 192.5, the largest could be nearly 288.5, and the range is 96. Dividing 96 by 10, we obtain 9.6, so that the class width would be 10. We could start with any number that is less than or equal to 193 and at least 184 (so that 193 would be included), but it would be most natural to start with either 190 or 191. Since it does not matter, let us start with 190. Then the classes will be 190–199, 200–209, 210–219, 220–229, 230–239, 240–249, 250–259, 260–269, 270–279, and 280–289. We may obtain the number in each class by means of a tally chart or by arranging the numbers in order. The result will be as shown in Table 1.7.

Table 1.7 WEIGHTS OF 50 NFL PLAYERS.

CLASS	FREQUENCY	RELATIVE FREQUENCY
190–199	1	0.02
200–209	2	0.04
210–219	4	0.08
220–229	3	0.06
230–239	9	0.16
240–249	9	0.18
250–259	10	0.22
260–269	5	0.10
270–279	3	0.06
280–289	4	0.08

PROFICIENCY CHECKS

5. Several thousand IQ scores were found to have a high of 168 and a low of 72.
   a. What is the range of the scores?
   b. What class width should be selected to divide the data into approximately 12 classes?
   c. Make two lists of classes, both using 71 as the lower limit of the first class—one using the class width from part (*b*) and the other using a class width of 10.

6. Suppose that a set of data has been grouped into eight classes as shown here:

CLASS NUMBER	CLASS
1	6–10
2	11–15
3	16–20
4	21–25
5	26–30
6	31–35
7	36–40
8	41–45

   a. What is the class mark for class number 8?
   b. What is the class width for these classes?
   c. What are the limits for class number 5?
   d. What are the boundaries for 3?
   e. If the data had been rounded off from tenths, in what class would a class number value of 25.7 be placed?

Table 1.8 TIME REQUIRED FOR A SPORE CULTURE TO DOUBLE.

TIME (MIN)	FREQUENCY	CUMULATIVE FREQUENCY	RELATIVE CUMULATIVE FREQUENCY
56	1	1	0.04
57	4	5	0.20
58	2	7	0.28
59	7	14	0.56
60	1	15	0.60
61	2	17	0.68
62	1	18	0.72
63	4	22	0.88
64–72	3	25	1.00

Note: Experiment conducted 8 April 1985.

40.5 · 45.5

1.1.5 Cumulative Frequency Tables

For some purposes we may wish to know the number of observations either *less than or equal to* or *greater than or equal to* each value of the variable. These numbers are called **cumulative frequencies.** If we show the number of observations less than or equal to each value of the variable, the table is said to be *cumulative in ascending order.* If we show the number greater than or equal to each value of the variable, the table is said to be *cumulative in descending order.* Although it does not matter whether the accumulation is done in ascending or descending order when a table is being used for presentation only, most statistical purposes call for a table to be in ascending order. The cumulative frequency for each value is the sum of the frequencies for values less than or equal to the value. The **relative cumulative frequency** (or **cumulative relative frequency**) for a value is the proportion of the data that is equal to or less than the value. For the data of Table 1.4, such a table could be as shown in Table 1.8. In general, only the columns that are needed would be listed.

Problems

Practice

1.1 Determine the limits, the width, the mark, and the boundaries for each of the following classes: 20–26, 27–33, 34–40, 41–47, 48–54.

1.2 You are given the eight classes 50–59, 60–69, 70–79, 80–89, 90–99, 100–109, 110–119, 120–129.

a. What is the width of these classes?
b. What is the class mark for the class 50–59?
c. What are the boundaries for the class 100–109?
d. To what class would 79.4 be assigned?
e. To what class would 119.5 be assigned?

1.3 Construct a table showing frequency, cumulative frequency, relative frequency, and relative cumulative frequency for the following data.

114	111	98	124	110	103	112	125	122	107
95	102	110	116	122	111	102	110	98	116
100	124	111	121	110	102	120	112	107	106

1.4 Consider the following data set:

82	51	75	114	81	67	121	55	91	73
76	93	82	65	83	77	103	76	85	72
91	73	104	122	76	113	55	61	81	92
100	92	81	69	113	81	123	85	107	84
74	95	68	83	89	62	91	84	56	77

a. What is the range of the data?
b. Group the data into classes, starting at 50 and using a class width of 10.
c. Construct a table showing frequencies and cumulative frequencies for these classes.

Applications

1.5 In a contest to see who can drive a golf ball the farthest, several hundred balls are hit. The shortest drive (to the nearest yard) is 167 yards and the longest is 314 yards.
a. What is the range of these scores?
b. Make a list of the classes for the data using 20 as class width, beginning with 160 yards as the first lower limit.
c. What class width will yield approximately ten classes?

1.6 The following are the weekly sales, in cases, of Roger's Cat Food at each of 80 retail outlets:

14	27	81	36	92	60	17	34
83	54	37	40	27	30	26	36
29	71	23	37	31	37	12	36
61	17	70	36	35	77	83	13
39	61	48	54	23	37	35	23
61	97	31	46	13	24	30	19
26	73	70	17	23	10	38	11
65	67	14	45	70	55	24	27
45	64	24	86	28	16	27	25
11	15	53	65	12	58	62	53

a. Group these numbers into a frequency table with eight classes.
b. Make a relative frequency table for the data.
c. Make a cumulative frequency table for the data.
d. Make a relative cumulative frequency table for the data.

1.7 The following are percentages (to the nearest tenth of a percent) of beginning inventory remaining after 45 days at each of 30 warehouses owned by a large concern:

56.1	54.4	47.6	58.1	49.7	54.3	50.9	52.8
56.5	54.8	48.4	57.3	54.9	51.0	56.8	51.5
52.9	55.4	52.1	58.5	58.1	49.0	54.3	50.2
59.8	53.4	62.5	55.7	50.4	52.8		

a. Construct a frequency table for the data.
b. What class width will yield eight classes? Make a list of these classes, beginning with 47.0.
c. Make a frequency and cumulative frequency table for the grouped data.
d. It has been determined that the most efficient operation is obtained when at least 50% but no more than 60% of beginning inventory is left each month. What proportion of the data fit these guidelines?

1.8 A psychology student conducts an experiment dealing with reaction times and records the following times to the nearest hundredth of a second:

0.48	0.96	0.52	0.36	0.49	0.58	0.64	0.44
0.39	0.59	0.37	0.66	0.57	0.68	0.64	0.81
1.02	0.54	0.38	0.67	0.74	0.77	0.52	0.68
0.55	0.67	0.34	0.67	0.74	0.81	0.91	0.66
0.54	0.68	0.83	0.52	0.37	0.56	0.90	0.47
0.54	0.69	0.58	0.55	0.52	0.94	0.39	0.57
0.66	0.43	0.68	0.71	0.42	0.60	0.58	0.74
0.55	0.49	0.58	0.64				

a. Construct a frequency table for the data with 12 classes.
b. What is the class mark for the lowest class?
c. What are the boundaries for the second lowest class?
d. Subjects with reaction times that are too slow (greater than 0.9 second) or too fast (less than 0.4 second) are unsuitable for further research. What proportion of the subjects are suitable for further research?

1.9 The following data, obtained from the *1980 Census of Population and Housing* (U. S. Bureau of the Census), give the percentage of single-family houses built before 1940 in 40 selected metropolitan areas:

Abilene, TX	16.3	Albuquerque, NM	6.7
Anchorage, AK	1.1	Atlanta, GA	10.3
Baltimore, MD	27.7	Baton Rouge, LA	8.3
Bellingham, WA	27.4	Birmingham, AL	19.0
Boston, MA	50.6	Bradenton, FL	7.0
Buffalo, NY	43.4	Charleston, SC	11.9
Chicago, IL	34.7	Dallas, TX	8.6
Denver, CO	14.4	Detroit, MI	24.1
El Paso, TX	11.0	Flint, MI	21.7
Fort Myers, FL	3.9	Honolulu, HI	8.4
Houston, TX	6.9	Lafayette, IN	26.0
Las Vegas, NV	1.5	Los Angeles–Long Beach, CA	17.7
Miami–Hialeah, FL	7.3	Milwaukee, WI	32.1
Nashville, TN	14.6	New York, NY	47.8
Orlando, FL	6.4	Pensacola, FL	8.9
Portland, ME	47.0	Reno, NV	7.7
Roanoke, VA	21.8	St. Louis, MO	26.8
San Francisco, CA	34.7	Seattle, WA	20.5
Syracuse, NY	40.6	Tulsa, OK	15.9
Wheeling, WV	49.9	York, PA	38.4

a. Construct a relative frequency distribution for the data. Use a class width of 5.0 percentage points, beginning with 0.0.

b. What proportion of the sample had at least 20% of the single-family houses built before 1940?

c. What proportion of the sample had less than 30% of the single-family houses built before 1940?

d. What proportion of the sample had at least 20% but less than 30% of the single-family houses built before 1940?

1.10 According to data obtained from the College Entrance Examination Board (unpublished data, 1984), the following were average SAT scores made by college-bound seniors who took the test, reported by state:

Alabama	970	Alaska	914
Arizona	678	Arkansas	1003
California	897	Colorado	982
Connecticut	904	Delaware	902
Florida	890	Georgia	822
Hawaii	869	Idaho	992
Illinois	981	Indiana	864
Iowa	1089	Kansas	1051
Kentucky	997	Louisiana	980
Maine	892	Maryland	898
Massachusetts	896	Michigan	976
Minnesota	1020	Mississippi	992
Missouri	981	Montana	1034
Nebraska	1041	Nevada	931
New Hampshire	931	New Jersey	876
New Mexico	1014	New York	894
North Carolina	827	North Dakota	1054
Ohio	968	Oklahoma	1009
Oregon	907	Pennsylvania	887
Rhode Island	885	South Carolina	803
South Dakota	1086	Tennessee	1009
Texas	866	Utah	1045
Vermont	907	Virginia	894
Washington	968	West Virginia	976
Wisconsin	1007	Wyoming	1034

a. Using a class width of 50, beginning with 650, construct a relative cumulative frequency distribution (ascending) for the data.

b. What proportion of the states had average SAT scores of at least 1000?

c. What proportion of the states had average SAT scores of less than 850?

d. What proportion of the states had average SAT scores of at least 800 but less than 900?

1.2 Graphical Methods

One disadvantage of frequency distribution tables is that their presentation lacks visual appeal. They do not call attention to the outstanding features of the data as well as a pictorial presentation does. Since there are many different ways to present data in pictorial form, only a few of the important methods are presented here. For a complete discussion of the presentation of data in visual form, as well as many excellent examples, see *The Visual Display of Quantitative Information* by Edward R. Tufte (1983).

1.2.1 Bar Graphs

Certain types of data lend themselves well to certain types of graphs. Discrete data are usually represented well by means of **bar graphs.** Each grouping is represented by a bar, or some variation of a bar, whose height represents the frequency for the group. An example of a bar graph is shown in Figure 1.1.

It is not necessary to list the frequencies at the top of the bar, but it is sometimes an aid to understanding. Bar graphs may also be drawn with horizontal bars. A common representation uses little pictures of the items represented in place of bars. An example of this type of bar graph (also called a *pictogram* or *pictograph*) is shown in Figure 1.2.

Age is often represented by bar graphs as well. As mentioned earlier, when age is measured to last birthday, the variable age is actually discrete. Under these circumstances a bar graph is a reasonable way of graphing ages, as shown in Figure 1.3.

The bars on a bar graph are usually separated as in Figures 1.1 and 1.3. The width and amount of separation of the bars have no significance at all, but all bars are usually made the same width and same distance apart in order to avoid distortion. A line works just as well as a wide bar, but usually it has less visual appeal.

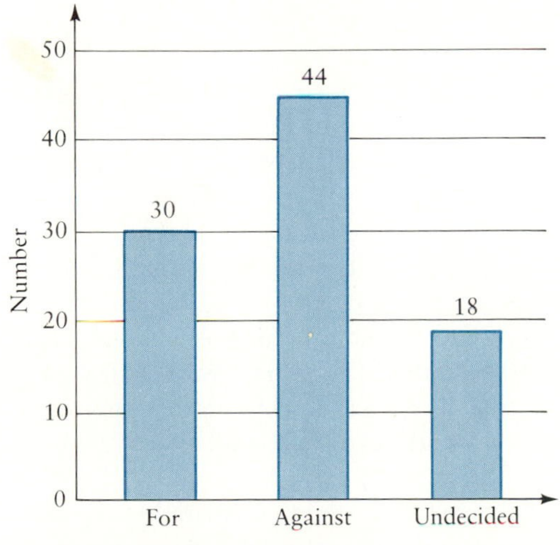

Figure 1.1
Bar graph of results of a poll of voters on an upcoming bond election.

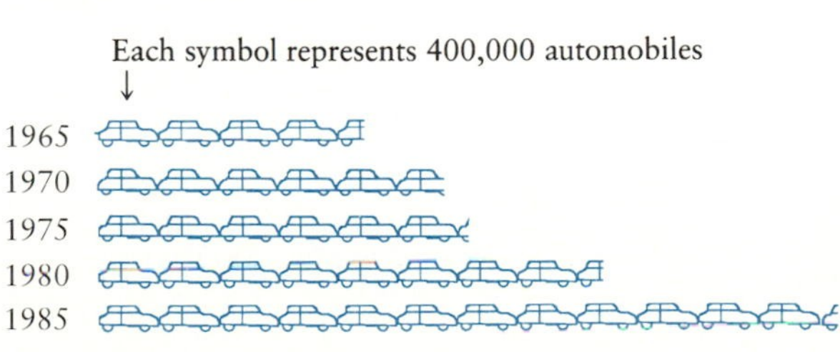

Figure 1.2
Pictogram of U.S. imports of automobiles.

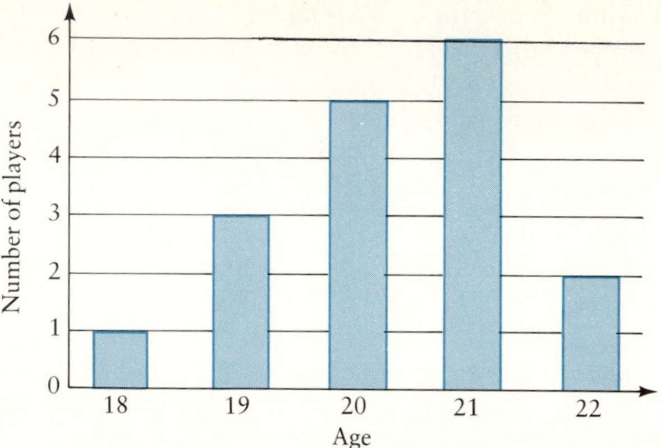

Figure 1.3
Bar graph of distribution of ages of players on the Meridian College basketball team.

PROFICIENCY CHECK

7. The following data are the numbers of bull's-eyes in 25 shots of 25 sharpshooters. Make a bar graph for the data.

18	20	22	22	20	20	21
21	20	23	19	21	20	19
19	21	22	19	20	19	20
19	20	18	20			

1.2.2 Histograms

A graph used to display continuous data, usually reported in classes, is called a **histogram.** A histogram is similar to a bar graph, but the area of the rectangle is very important. To construct a histogram, we first divide the horizontal axis into intervals corresponding to the classes. Then we construct a rectangle with each class interval as a base and the height of the rectangle equal to the number of observations in the class. An example of a histogram is shown in Figure 1.4.

In this example we can see that most of the shipments were about 40, 50, or 70 pounds. The class width is 10 pounds for each class with multiples of 10 falling at or near the center of the classes at the class mark. Judging from the graph, the classes begin at 14.5 and have class widths of 10 and thus class marks of 19.5, 29.5, 39.5, and so on.

There are two important points to remember when constructing a histogram. First of all, if a class has no observations—that is, a frequency of zero—a gap will appear in the histogram. If, for example, the diamond shipments graphed in Figure 1.4 included no shipments in the class centered near 120 pounds, but had one or more shipments in the class centered near

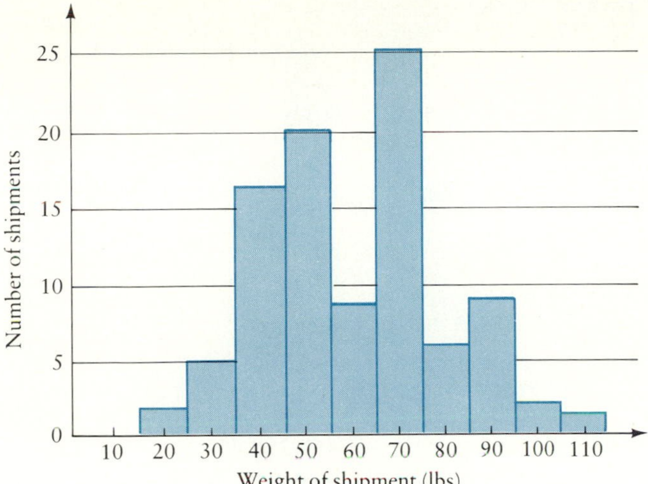

Figure 1.4
Histogram for distribution of diamond shipment weight from the Abercrombie Mine in 1985.

130 pounds, there would be no rectangle centered above 120. The second important thing to remember is that the *areas,* and not simply the heights, of the rectangles will represent the frequencies. For this reason all classes should have the same width. If we combine classes, the height of the rectangle for the combined class should equal the total of the frequencies for the combined classes divided by the number of classes combined. In Figure 1.4, for example, it appears that the first two classes, centered near 20 and 30, respectively, contain frequencies of 2 and 5. If we wished to combine them into one class, the class would have boundaries of 14.5 and 34.5, be centered near 25, and have a frequency of 7. The area of the new rectangle should equal the sum of the areas of the two rectangles combined. Thus the height of the rectangle should be 3.5. Similarly, if we combined the classes centered near 90, 100, and 110, the combined frequency would appear to be about 12, so the height of the rectangle representing these three classes would be 4. Use of this procedure keeps the relationships in proper perspective. Figure 1.5 shows the results of combining the classes discussed and adding a few shipments near 130 pounds.

Another type of histogram, more useful for most statistical purposes, is called a **relative frequency histogram.** In this histogram the vertical scale represents relative frequency. The data used for Figure 1.4 can also be used to draw a relative frequency histogram, as presented in Figure 1.6.

Histograms can give other information about samples as well. We may note the *shape* of the graph of the distribution, commonly called the **shape of the distribution.** Some frequently encountered types of distributions are shown in Figure 1.7. By far the most important of these types is the mound-shaped distribution. A special form of mound-shaped distribution is called the *normal* distribution, which is introduced in Chapter 2, discussed in great detail in Chapter 5, and used extensively thereafter.

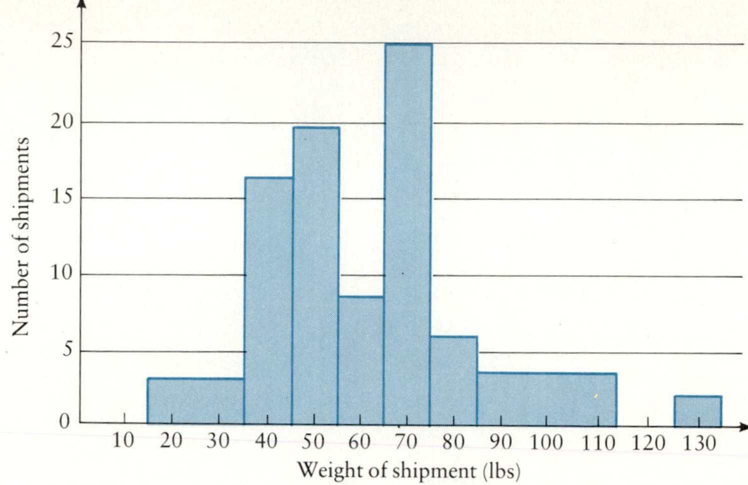

Figure 1.5
Histogram for diamond shipment data with classes combined.

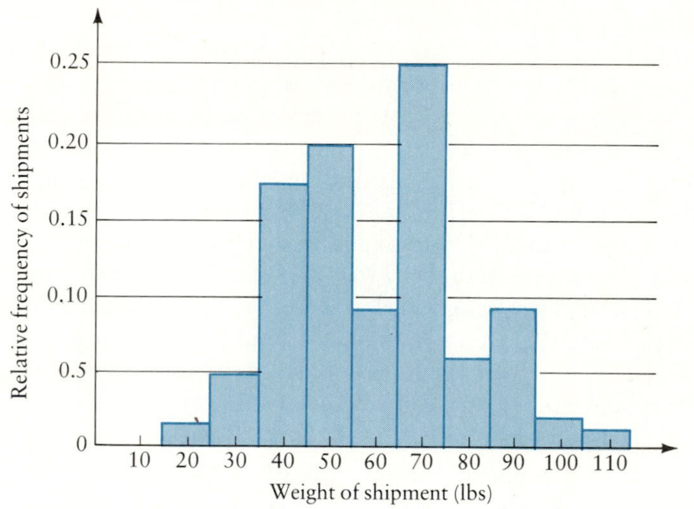

Figure 1.6
Relative frequency histogram for the diamond shipment data.

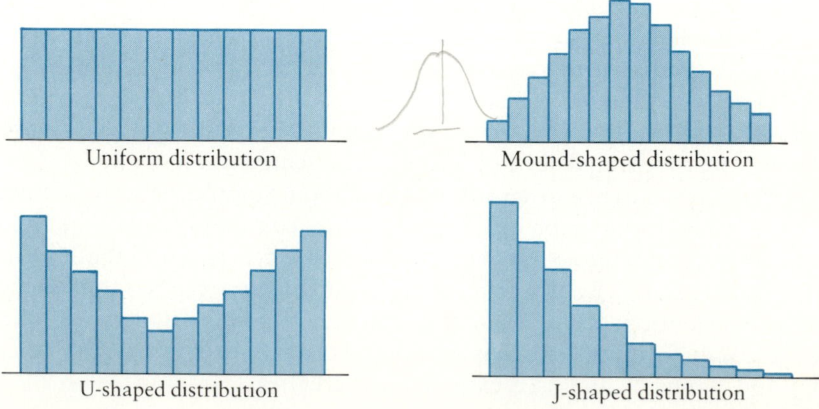

Figure 1.7
Histograms for four common distributions.

PROFICIENCY CHECKS

8. Make a histogram for the following data. Use a class width of 1.

55	48	53	55	49	57	51	52
49	56	47	54	49	50	53	55
48	49	51	56	50	48	57	48
47	54	55	53	52	47	49	49
52	51	50	55	54	56	57	59

9. Make a histogram for the following data:

CLASS	FREQUENCY
1–3	4
4–6	12
7–9	8
10–12	20
13–15	18
16–18	15
19–21	7
22–24	12
25–27	4

1.2.3 Frequency Polygons

For presentation purposes a *frequency polygon* is sometimes more useful than a histogram. A frequency polygon seems to have more visual appeal, but it lacks the comparative power of the histogram since areas are no longer totally representative of the data. To construct a **frequency polygon,** we use the same vertical and horizontal axes we would use for a histogram. We use the class mark to represent each class, and we place a dot above the class mark at the height (on the vertical scale) equal to the frequency for that class. (This procedure is equivalent to placing a dot in the center of the upper edge of the rectangle in a histogram, although there is no need to draw the histogram first.) We then connect the dots in the proper order. To give the final graph a finished look, we may add a class at the beginning and a class at the end, each with zero frequency. For the diamond shipment example we would add a class centered near 10 and a class centered near 120, each with frequency zero. The resulting frequency polygon is shown in Figure 1.8. A **relative frequency polygon** would look the same, but the vertical axis would represent relative frequency rather than the actual frequency.

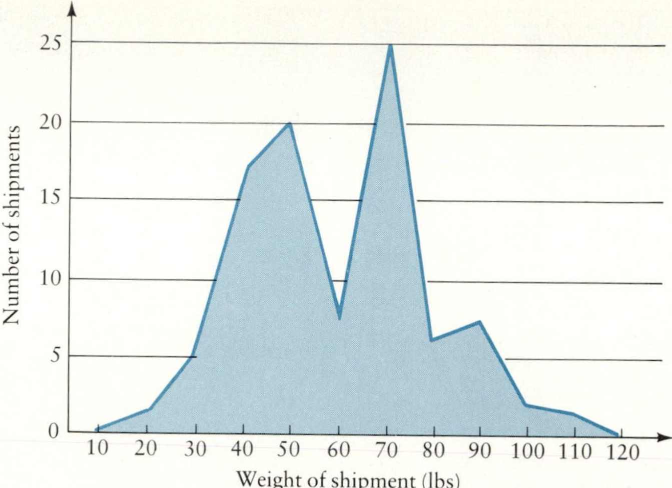

Figure 1.8
Frequency polygon for the diamond shipment data.

PROFICIENCY CHECKS

10. Make a frequency polygon for the data of Proficiency Check 8.
11. Make a frequency polygon for the data of Proficiency Check 9.

1.2.4 Stem and Leaf Plots

A procedure that simultaneously sorts the data as in a tally chart and provides a graph of the distribution is the **stem and leaf plot,** introduced by John Tukey (1977). Consider the data of Example 1.1:

57	59	58	61	56	63	57	69	
59	59	57	63	72	59	62	60	
59	63	59	65	57	61	58	59	63

The data range from 57 to 72. We first note the units—ones—and then list the tens of units vertically:

5
6
7

These are the *stems*. The units' digits are the *leaves*. We list the leaves next to the stems as we encounter them in the data set, using vertical bars to separate the leaves from the stems. The result looks like this:

5|79867997999789
6|1393203513
7|2

If we wish, we can divide each ten into two parts, putting two stems for each. The first of each pair will be used for digits 0 through 4, the second for digits 5 through 9. The revised plot looks like this:

5|
5|79867997999789
6|13320313
6|95
7|2
7|

It is a simple task for us to construct a frequency table and a graph from this display, but the plot already gives a good picture of the shape of the distribution.

PROFICIENCY CHECKS

12. In Example 1.2, the weights of a random sample of 50 football players currently playing in the NFL were listed. Make a stem and leaf plot for the data.
13. Make a stem and leaf plot for the data of Problem 1.6.

1.2.5 Other Graphical Methods

Methods of graphical representation are limited only by the imagination. Particularly useful methods include *pie charts* and *pictograms*. **Pie charts** are graphs made by dividing a circle into parts depicting how much of a total amount is represented by each classification. They are particularly useful in presenting percentages or proportions and are familiar to many because of their widespread use in showing sources of the federal government's dollar income and expenditure. Figure 1.9 shows a pie chart for sources of revenue for American movies.

Pictograms are graphs in which a picture (usually stylized) of the items being presented is used in the graph. Pictograms are difficult to use properly, however, because they often convey a false impression. Figure 1.10, for example, presents a somewhat deceptive picture. The fish on top is just about three times as long as the one on the bottom, since $31{,}353/9566 \doteq 3.28$, but because $(3.28)^2 \doteq 10.8$, its area is more than ten times as great and this is the impression the eye receives.

This impression is further compounded by adding the illusion of depth as in Figure 1.11. Tufte (1983, pp. 53–77) has many similar examples. A fairer impression is shown by using all symbols of the same size, as is done in Figure 1.12. We saw in Figure 1.2 that this type of pictogram is effectively a bar graph.

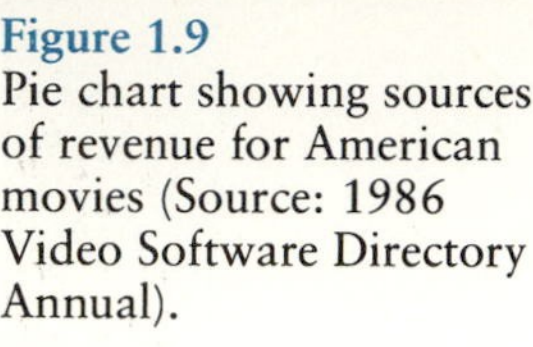

Figure 1.9
Pie chart showing sources of revenue for American movies (Source: 1986 Video Software Directory Annual).

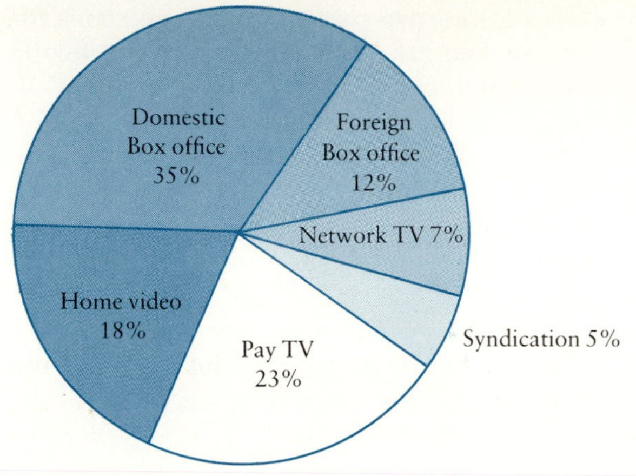

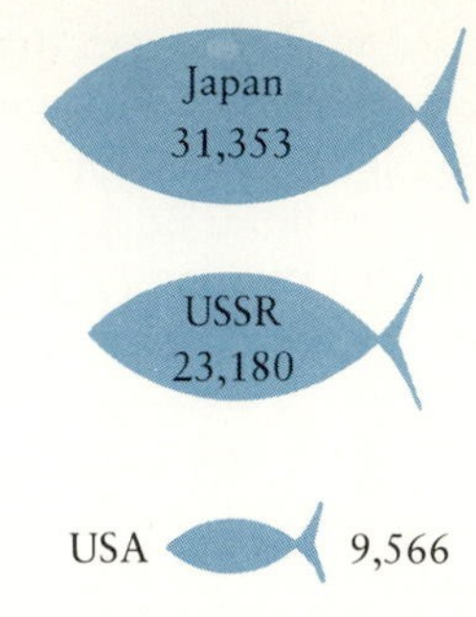

Figure 1.10
Pictogram showing catch of fish and shellfish (millions of pounds).

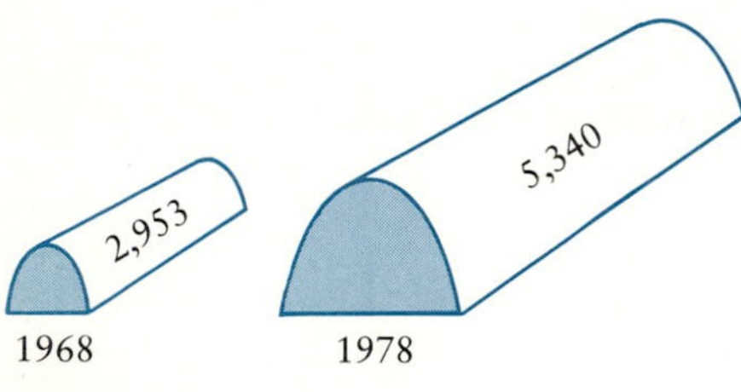

Figure 1.11
U.S. aluminum production (thousands of tons) (Source: *Encyclopedia Britannica*).

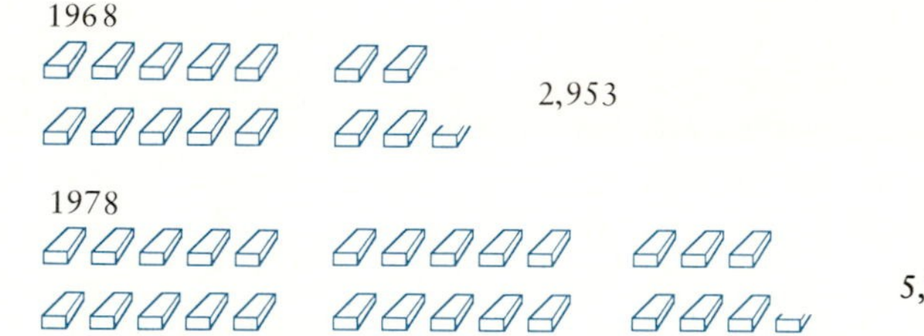

Figure 1.12
U.S. aluminum production (thousands of tons) (Source: *Encyclopedia Britannica*).

1.2.6 A Word About Distortion

Graphs are often highly useful in presenting information visually, but they can easily be manipulated to distort the picture being presented. Pictograms are not the only culprit. Most graphs are subject to distortion. A pie chart, for example, can be shown at an angle, as in Figure 1.13, emphasizing the portions at the top and bottom and deemphasizing those at the sides.

Figure 1.13
Pie chart at an angle showing hypothetical budget breakdown.

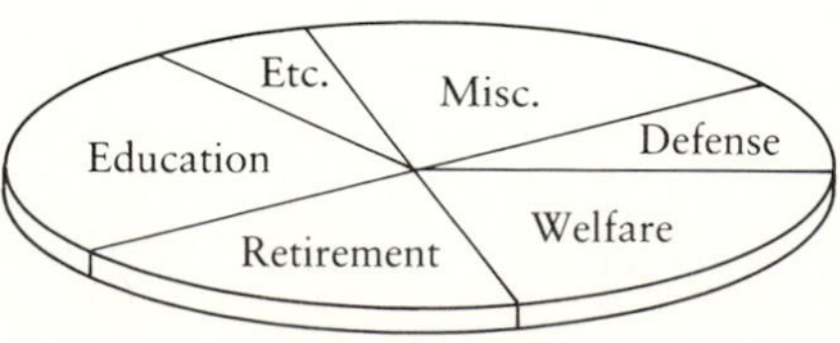

One of the ways to distort a bar graph, frequency polygon, or histogram is to compress or stretch either the horizontal or vertical scale unequally. Using a logarithmic scale, for example, creates a great deal of distortion. In a logarithmic scale, the scale starts at 1 and then goes to 10; the distance from 10 to 100 is the same as from 1 to 10; the distance from 100 to 1000 is the same; and so on. What are the guidelines for constructing a fair and accurate graph? Most graphs are more pleasing if they are wider than they are tall. Units used on each scale should be uniform—that is, distances numerically equal should be equal in length on each scale (but not usually the same on both scales). Vertical scales should always begin at zero unless they are clearly marked to show that they do not, and horizontal scales should be chosen to avoid distortion.

Even more distortion can creep in if the vertical scale does not start at zero. Unions asking for a wage increase might use a graph such as Figure 1.14 to support their claim that the company is making money "hand over fist." In proper perspective, the graph should look like Figure 1.15. In the first graph we get an impression of sharply increasing profits. In the second it appears that profits are approximately stable throughout this period. A similarly misleading graph could probably be used by the company to show increases in worker income.

Occasionally we may prefer not to start the scale (either vertical or horizontal) at zero because the values or the frequencies (or both) are compressed onto a fairly small segment of values. To avoid the impression that we are doing so deliberately to distort perspective, we usually indicate that the scale does not start at zero by means of a broken line or other such "flag." This should be done whenever the vertical axis does not start at zero. If the horizontal axis does not start at zero, the use of a flag is often helpful but not absolutely necessary. An example is shown in Figure 1.16.

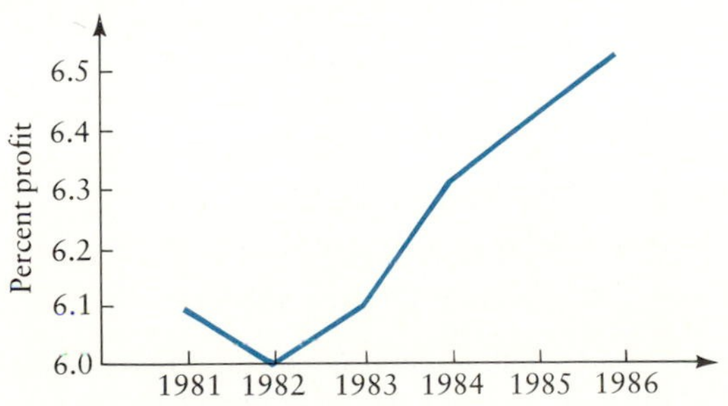

Figure 1.14
Company profits 1981–1986 (Distorted graph).

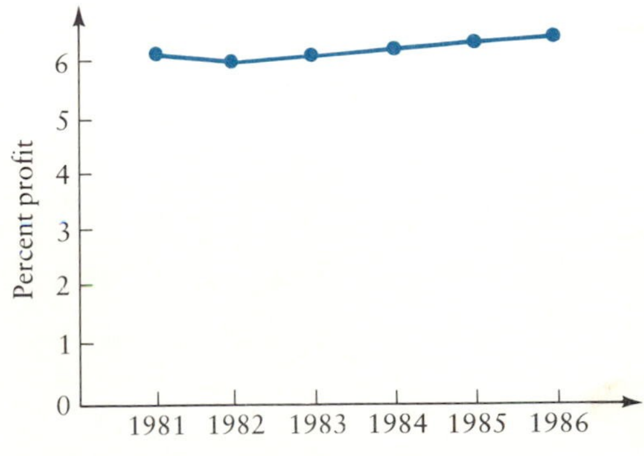

Figure 1.15
Company profits 1981–1986 (Correct graph).

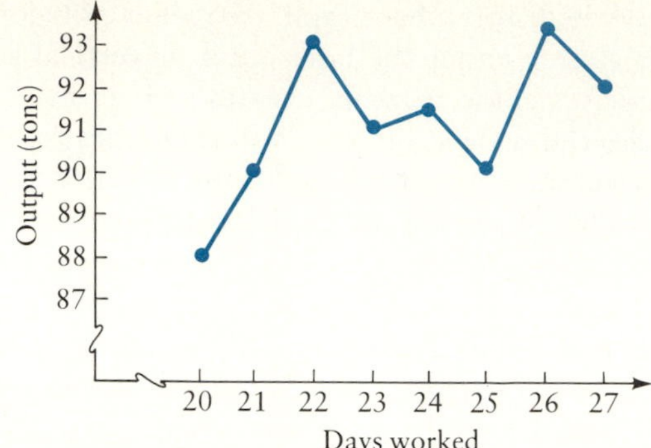

Figure 1.16
Graph using broken lines to indicate that axes do not start at zero.

Problems

Practice

1.11 Construct a bar graph for the data of Problem 1.3.

1.12 Refer to Problem 1.4.
a. Construct a histogram for the data using the groupings obtained there.
b. Construct a frequency polygon for the grouped data.

Applications

1.13 Refer to the height data of 50 incoming freshmen in the table below.
a. Construct a histogram for the data.
b. Construct a relative frequency polygon for the data.

HEIGHT (IN.)	FREQUENCY	RELATIVE FREQUENCY	CUMULATIVE RELATIVE FREQUENCY
61–63	3	0.06	0.06
64–66	5	0.10	0.16
67–69	12	0.24	0.40
70–72	15	0.30	0.70
73–75	9	0.18	0.88
76–78	4	0.08	0.96
79–81	2	0.04	1.00

1.14 Draw a pictogram for the basketball team data of Figure 1.3.

1.15 A family's monthly expenses are listed as follows: housing \$528; utilities \$194; medical expenses \$21; food \$181; transportation \$187; clothing \$35; savings \$35; miscellaneous \$69. Draw a pie chart showing this information graphically.

1.16 Problem 1.7 gave the percentages of beginning inventory remaining after 45 days at each of 30 warehouses.
a. Construct a stem and leaf plot for the data.
b. Using eight classes, beginning with 47.0, construct a relative frequency histogram for the data.

c. Using the results of part (*b*), construct a frequency polygon for the data.

1.17 Refer to the income data of Table 1.6. Using the data as presented, draw:
a. a histogram
b. a relative frequency histogram
c. a frequency polygon
d. a relative frequency polygon

1.18 The Institute for Socioeconomic Studies (Feb. 4, 1986) issued a survey of 139 winners of $1 million or more in state lotteries: 52 continued working, 33 quit their jobs, 22 retired, 1 worked longer hours, 4 quit their second jobs, 21 worked reduced hours, and 6 changed jobs. Construct a pie chart showing this information in percentages.

1.19 Refer to the housing data of Problem 1.9. Using the results there, construct a histogram for the grouped data.

1.20 The average salaries paid by major league baseball teams in 1985 and 1984 are shown below. (Source: Major League Players Association.)

AMERICAN LEAGUE			NATIONAL LEAGUE		
TEAM	1985	1984	TEAM	1985	1984
Yankees	$546,364	$458,544	Atlanta	$540,988	$402,689
Baltimore	438,256	360,204	Los Angeles	424,273	316,250
California	433,818	431,431	Cubs	413,765	422,193
Milwaukee	430,843	385,215	San Diego	400,497	311,199
Detroit	406,755	371,332	Philadelphia	399,728	401,476
Boston	386,597	297,878	Pittsburgh	392,271	330,661
Toronto	385,995	295,563	Mets	389,365	282,952
Kansas City	368,469	291,160	St. Louis	386,505	290,886
Oakland	352,004	384,027	Houston	366,250	382,991
White Sox	348,488	447,281	Cincinnati	336,786	269,019
Minnesota	258,039	172,024	San Francisco	320,370	282,132
Texas	257,573	247,081	Montreal	315,328	368,557
Cleveland	219,579	159,774			
Seattle	169,694	168,505			

Use the data to construct two bar graphs, one for each league, showing the average salaries for each team in such a way that the two years may be easily compared for each team.

1.3 Computer Usage

Computers play a very important role in statistics today. The availability of high-speed computers has changed the way statistics is now applied. Computers have enabled people to use statistical procedures more easily, since calculations that once had to be done by hand can now be done by machines in a matter of seconds with no computational errors. Computers also have

The mathematician William Shanks (1812–1882) spent many years of his life laboriously calculating the value of π (the ratio of the circumference of a circle to its diameter) to 707 decimal places. The first 607 were published in 1853, the last hundred in 1873. In 1946 it was found that Shanks had made an error in the 528th decimal place, invalidating over twenty years' worth of labor. In 1959, a computer calculated the value of π to 16,167 decimal places in a matter of hours (it can now be done in seconds), and in 1967 another computer calculated the value of π to 500,000 decimal places.

enabled researchers to analyze much larger sets of data and to perform sophisticated statistical analyses that were once too complicated for hand calculations.

One must be careful, however, when using computers for statistical analysis. The calculations can be carried out so simply on a computer that even people untrained in statistics can perform them. A statistical beginner may perform an incorrect analysis or unknowingly make an incorrect interpretation of the results on a computer output. Keep in mind that computers do not check to determine whether the correct analysis is being performed; only people can do this. It is important, therefore, that people using computers to perform statistical analysis know what is being calculated and what the results mean. Students must therefore learn and understand what statistics is all about before using a computer to perform the calculations.

The objective of this section on computer usage is to illustrate with examples how computers can be used to perform statistical calculations or graphs. Similar sections are provided in subsequent chapters of the text. These sections include sample computer printouts that illustrate some of the statistical techniques presented in the chapter. A few exercises are included as well for the student with access to a computer.

1.3.1 Computer Software

A statistical software package is a group of computer programs written to perform tasks that are frequently used in analyzing data. These tasks include entering data, manipulating data, graphing data, and performing statistical calculations. There are many statistical software packages available today. In this book two of these software packages are illustrated: Minitab®* and the SAS®† System. Minitab can be run on personal computers, minicomputers, and large mainframe computers. The SAS System is in wide use on mainframe systems and is also available in a version for personal computers.

Minitab is an easy-to-use yet quite versatile collection of programs developed at Pennsylvania State University in 1972 for use in elementary statistics classes. Although particularly well suited for use in elementary statistics courses, it can also be used to perform advanced statistical techniques. Because of this versatility the use of Minitab has become widespread and it is now available at many colleges and universities as well as in many businesses and industries.

The SAS System has a much greater scope than Minitab, and is just as easy to use, but it is not presently available in as many locations as Minitab. It is available, however, in many industries and university systems. In Florida, for example, SAS software is installed in the main computer at the

---

* Minitab is the registered trademark of Minitab Inc. For further information write to Minitab Inc., 215 Pond Laboratory, University Park, PA 16802.

† SAS is the registered trademark of SAS Institute Inc., Cary, NC, USA.

University of Florida and available at all locations connected to the university's computer. This network includes all eight other members of the Florida State University system and 28 public community colleges. Most of the SAS statements and procedures discussed in this book are available for personal computers on the SAS/STAT™ Software. This software requires Base SAS® Software as well. The SAS System runs on IBM PC XT, PC AT, and compatible machines under PC DOS. It can also be run on the IBM PC XT/370 and AT/370 under VM/PC and on some other systems.

The rest of this section presents preliminary information on the use of Minitab and SAS software. Included is information on how to enter data into the computer, ways to correct typing errors in Minitab, and ways to use Minitab and SAS software to obtain elementary graphs and calculations. Illustrations show how you can use Minitab and SAS software to perform some of the statistical techniques discussed in this chapter.

1.3.2 Introduction to Minitab

Minitab is known as an *interactive software package*. In effect, the user engages in a dialogue with the computer. Errors are often detected at the time of entry and may be corrected immediately.

Signing on the Computer and Minitab The process of signing on the computer and accessing the Minitab system depends on the installation, so general rules cannot be provided here. This information may be obtained from those who are familiar with the system. In many installations, a booklet or printout is provided to users.

Once you sign on the computer and access Minitab, you will see the Minitab "prompt"—MTB>. The computer will prompt you with the MTB> when it is ready for you to type your commands. In the remainder of this book all commands or data that you type are in color; the computer's response is in black. Whenever a Minitab example is presented, the minimum that must be typed is capitalized. Everything in lowercase type is ignored by Minitab and is provided only for clarification of the command's execution to the user. In fact Minitab recognizes only the first four letters in a command; for clarity, however, the entire command is printed here in uppercase letters.

Data or commands that you type are not entered into the computer until you press the RETURN or ENTER key after a command or line of data has been written. Mistakes can be corrected before you press the key; after you do so, what you have typed has been entered into the computer and, if incorrect, must be corrected by a subsequent command. Remember always to press the RETURN or ENTER key to enter a line of data or send a command to the computer. The Minitab system is exited by typing STOP. For more information on Minitab's capabilities and commands consult the *Minitab Student Handbook* (2nd ed.) by B. F. Ryan, B. L. Joiner, and T. A. Ryan.

Data Storage in Minitab Data in Minitab are stored in a row-and-column format. Each data value or number is stored in a particular row and column. The columns represent the different variables (age, weight, sex, and so on); the rows refer to the different members of the sample or population (person 1, person 2, and so on). Thus the entries in a particular row stored in the computer could give the age, weight, sex, and so forth of a particular person.

The typical analysis is done on a single variable or column of numbers in Minitab. To refer to different columns in Minitab we will use C1, C2, C3, and so on to represent column 1, column 2, column 3, and so on. In Minitab you do not have to start with column 1 and progress in order; rather, you can start with any column number and proceed in any order, forward or backward.

Data Input in Minitab There are two ways of entering the data into Minitab by using the computer keyboard. (Data may also be entered from a disk or tape file. Details can probably be obtained at your installation.) The first is by using the SET command. The SET command is used to enter an entire column of numbers at one time. To put the five data values 2, 8, 5, 4, and 10 into column 1 and the three data points 12.1, 9.3, and 8.7 into column 3 using the SET command, you would type the following highlighted material. Minitab "prompts" (MTB> and DATA>) are not highlighted. (Remember to press RETURN or ENTER following each line.)

```
MTB > SET the following data into column C1
DATA> 2 8 5 4 10
DATA> END of data
MTB > SET the following data into column C3
DATA> 12.1 9.3 8.7
DATA> END of data
```

Note that the data values are separated by at least one space. The data entered into Minitab now looks like this:

```
          C1         C2         C3
ROW 1      2                   12.1
    2      8                    9.3
    3      5                    8.7
    4      4
    5     10
```

Data can also be entered by using the READ command. The READ command can be used to enter more than one variable or column of data at a time into Minitab. This method is very useful if each observation consists of data recorded on several variables. Suppose that measurements on three variables have been recorded on each of four people and that these values are to be entered into columns 2, 1, and 4 as recorded. The READ command would be used as follows:

```
MTB >   READ the following data into columns C2, C1, C4
DATA>   8 5 12
DATA>   7 6 15
DATA>   5 5 10
DATA>   7 7 11
DATA>   END of data
```

The data entered into Minitab look like this:

	C1	C2	C3	
ROW 1	5	8		12
2	6	7		15
3	5	5		10
4	7	7		11

SAVE and RETRIEVE Commands in Minitab An advantage of statistical computer packages is their ability to store a set of data on the computer and retrieve the data at a later time. This eliminates the need to re-enter the data manually each time a different data analysis is performed on the same data set.

The ability to store data on a computer is dependent upon the nature of the system at your computer installation. To store a current version of the rows and columns of data in Minitab simply type

```
MTB >   SAVE 'filename'
```

at the end of your Minitab program, but before you type STOP.

Once the data set has been "saved" by Minitab it is recorded in a format that only Minitab can retrieve and understand. To have Minitab read a previously saved data set use the RETRIEVE command as follows:

```
MTB >   RETRIEVE 'filename'
```

where filename is the name given to the data set when it was saved. Minitab then creates a copy of the data of rows and columns identical to how they looked when they were saved. The data are also in the same rows and columns with the same name as when the data set was saved. You should note that only the current version of the data is saved. If you make any changes to the data after retrieving them these changes are not made in the saved data set unless that set is saved again.

Correcting Errors If you make a mistake in typing a command or data value and notice the mistake before you press the RETURN or ENTER key, simply backspace over the error and correct it. If you enter a mistyped command, Minitab will recognize the error and give an error message. You must then retype the command. If you enter incorrect data, you can correct them by using the LET command. To change the 5 in column 2 in the preceding data set to a 4, for example, type

```
MTB >  LET C2(3) = 4
```

This command will replace the third value in column 2 with the value 4.

Printing Data Once you have entered your data into Minitab it is a good practice to print the data in order to check for any errors you might have missed. To do so, you use the PRINT command followed by the list of columns to be printed. To print the data in columns 1, 2, and 8, you would type

```
MTB >  PRINT C1 C2 C8
```

Naming Variables Once you have entered data into Minitab, you use the columns to refer to the different variables. Often we wish to use a more meaningful name to refer to the column as well so that we will be reminded what variable is represented by the data in the column. This is done by using the NAME command. For example:

```
MTB >  NAME C1 'AGE' C21 'WEIGHT'
```

would name column 1 "age" and column 21 "weight." From then on we can refer to column 1 by either C1 or 'AGE.' The variable name *must* be enclosed between two apostrophe symbols (sometimes called "single quote" symbols). There are two restrictions on the names you can use: they must contain letters only (no numbers or symbols), and they cannot be longer than eight letters.

1.3.3 Minitab Commands for This Chapter

Minitab has three commands that allow you to perform some of the techniques illustrated in this chapter. In Minitab you can produce bar graphs, frequency tables, and stem-and-leaf plots of columns of data.

You can obtain an ungrouped frequency table by using the TABLE command. This command will produce a table with the frequency of each different value in the column. You may obtain the relative frequencies in addition to or instead of the frequency table by using subcommands. If you type

```
MTB >  TABLE the data in column C3;
SUBC>  COUNTS;
SUBC>  TOTPERCENTS.
```

you will obtain a table of the data in column 3 with frequencies for each value (specified by the subcommand COUNTS) and the relative frequencies as percentage (specified by the subcommand TOTPERCENTS). The semicolons are used to indicate a continuation of the original command (TABLE); the period indicates the completion of the last subcommand. If there is no subcommand, the TABLE command will give the frequencies.

The HISTOGRAM command will produce a bar graph with the bars horizontal and a grouped frequency table for each column specified. You can also specify the midpoint of the first class and the width of the classes. For example:

```
MTB >   HISTOGRAM C4;
SUBC>   INCREMENT or class width = 10;
SUBC>   START midpoint = 8.
```

specifies that a histogram is to be produced for the data of column 4, that the first class has midpoint 8, and that all classes have a width of 10. If you do not specify these values, Minitab will choose them. The printout will include the frequencies for each class as well as the bar graph.

You can obtain a stem and leaf plot for a column of data by using the STEM-AND-LEAF command:

```
MTB >   STEM-AND-LEAF the data in list of columns
```

This command will produce one plot for each column of data specified. Minitab will determine the number and value of the stems; these cannot be changed.

Example 1.3

Using the weights of the 50 football players (Example 1.2), obtain a grouped frequency table, a bar graph, and a stem and leaf plot.

Solution

Figure 1.17 show the complete Minitab printout including commmands, data entry, and results. Notice that the grouped frequency table agrees with the one found earlier in Table 1.7. The stem and leaf plot agrees with the one shown in the answer to Proficiency Check 12 at the end of this chapter.

Example 1.4

Use the data of Example 1.1 (the time, in minutes, required for a spore culture to double) to obtain a frequency and relative frequency table.

Solution

The program and results are shown in Figure 1.18. The printout gives the frequencies (COUNT) and relative frequencies (% OF TBL), which agree with those given in Table 1.4.

Figure 1.17
Minitab Printout for weights of football players.

```
MTB > SET the following data into C1
DATA> 193 240 217 283 268 212 251 263 275 208 230
DATA> 288 259 225 252 236 243 247 280 234 250 236
DATA> 277 218 243 268 231 269 224 259 258 231 255
DATA> 228 202 245 246 271 249 255 265 235 243 219
DATA> 245 245 238 257 254 284
DATA> END of data
MTB > LET C1(30) = 239
MTB > LET C1(45) = 255
MTB > NAME for C1 is 'WEIGHT'
MTB > PRINT the data in C1
      WEIGHT
         193    240    217    283    268    212    251    263    275    208    230
         288    259    225    252    236    243    247    280    234    250    236
         277    218    243    268    231    269    224    239    258    231    255
         228    202    245    246    271    249    255    265    235    243    219
         255    245    238    257    254    284
```

```
MTB >
MTB > HISTOGRAM the data in C1;
SUBC> INCREMENT or class width = 10;
SUBC> STARTING midpoint = 194.5.

      Histogram of WEIGHT N = 50

      Midpoint    Count
        194.5       1     *
        204.5       2     **
        214.5       4     ****
        224.5       3     ***
        234.5       9     *********
        244.5       9     *********
        254.5      10     **********
        264.5       5     *****
        274.5       3     ***
        284.5       4     ****
          ↑         ↑
        Class   Frequencies
        marks
```

```
MTB> STEM-AND-LEAF the data 'WEIGHT'

  STEM-AND-LEAF DISPLAY OF WEIGHT    N = 50
  LEAF UNIT =     1.0

     1   19   3
     3   20   28
     7   21   2789
    10   22   458
    19   23   011456689
   (9)   24   033355679  ←Class containing the median (9 is the frequency for this class)
    22   25   0124555789
    12   26   35889
     7   27   157
     4   28   0348
          ↑     ↑
       Stems  Leaves
```

```
MTB > SET the following data in C1
DATA> 57 59 58 61 56 63 57 69 59 59 57 63 72 59
DATA> 62 60 59 63 59 65 57 61 58 59 63
DATA> END of data
MTB > TABLE the data in C1;
SUBC> COUNTS;
SUBC> TOTPERCENTS.

     ROWS: C1  ┌── Frequency
               │        ┌── Relative Frequency as a percent
               ↓        ↓
          COUNT  % OF TBL
   56         1      4.00
   57         4     16.00
   58         2      8.00
   59         7     28.00
   60         1      4.00
   61         2      8.00
   62         1      4.00
   63         4     16.00
   65         1      4.00
   69         1      4.00
   72         1      4.00
   ALL       25    100.00
              ↑
              └── Sample size
```

Figure 1.18
Minitab Printout for spore culture data.

Miscellaneous We would like to point out that the examples and illustrations of Minitab in this book refer to the 85 Minitab version. If you do not have the 85 version available, some of your commands will operate differently and the output format will vary slightly for some commands. (For a complete listing of the 85 Minitab commands and their format, see Appendix A.1.) The version of Minitab that you have access to is given when you first enter Minitab. If you are using a previous version and a command, as given in one of our examples, does not work, then try the HELP command to find the correct format for that command. For example, to obtain an overview of the HELP command, simply type

```
MTB > HELP.
```

To obtain information about a specific Minitab command, for example the HISTOGRAM command, type

```
MTB > HELP HISTOGRAM.
```

Minitab will respond with information about the correct usage of that command.

1.3.4 Introduction to the SAS System

The SAS System is a large collection of programs, called procedures, which may be used on a set of data to obtain the desired results. In contrast to

Minitab, SAS software is not interactive in all installations. In many cases the user writes a program creating a data set and specifying what procedures are to be used and then sends the program to the computer, indicating that the SAS System is to be used. This type of program is called a "batch" job. The program may be written on cards or floppy disks—or, more often, the user sits in front of a terminal and creates a file that will be submitted to the computer. The specific procedure varies from installation to installation, and the user must first learn how to use the system at hand. In the subsections covering the SAS System, we will write the programs as if they were to be submitted as batch jobs. We use the SAS System Version 5 throughout the book. There are minor differences in earlier versions. Consult a guide for your version if you have problems. A selected list of SAS statements is given in Appendix A.2.

Creating a SAS Data Set To use the SAS System, we create a data set and then manipulate it with procedures that are stored in the computer. The first step in creating a SAS data set is to write the word DATA. Every SAS statement must be followed by a semicolon. Thus we must write

```
DATA;
```

which tells the computer "I am creating a data set." You may give the data set a name—for example, DATA ONE; specifies that the data set is named ONE. The *SAS User's Guide* (1985) recommends naming data sets. Naming a data set is particularly important if several data sets are created in a program. If you do not specify a name, SAS assigns a temporary name to the data set.

The next statement is the INPUT statement, which specifies how the data are to be entered. Data may be entered in several different ways, depending on their complexity. If only one or two variables are to be entered, the simplest method is to specify the name of the variable or variables being entered. If we wish to enter the sex, age, and weight of several people, for example, we would use the statement

```
INPUT SEX AGE WEIGHT @@;
```

The symbol @@ allows us to enter the data continuously rather than starting a new line with each entry of sex. If we plan to do any calculations with the data, we would be wise to specify variables that are not numbers with the symbol $ following the variable name (after a space). Since arithmetic cannot be performed on nonnumerical variables, this symbol keeps us from getting an error message if we forget to limit our arithmetic to numbers. Before we enter the data, we use the statement CARDS; to indicate that the data are to be entered.

As many statements may be entered on a line as there is room, and statements may be broken between lines so long as words are not broken.

For ease of correction and interpretation, however, it is probably better to write statements on separate lines. Data, however, cannot occupy the same line as a statement. After writing the statement CARDS; you must go to the next line to begin entering the data. After you have entered all the data, go to the next line before writing any additional statement.*

Printing Data To obtain a printout of sex, age, and weight of a number of persons, we could write

```
DATA EXAMPLE;
   INPUT SEX $ AGE WEIGHT @@;
CARDS;
M 18 165 M 21 172 F 19 118 M 22 155 F 37 123 M
24 188 F 18 135
PROC PRINT;
```

The result of this program would be the following printout:

```
OBS    SEX    AGE    WEIGHT
 1      M      18     165
 2      M      21     172
 3      F      19     118
 4      M      22     155
 5      F      37     123
 6      M      24     188
 7      F      18     135
```

If we wished, we could write the first three statements on the same line:

```
DATA; INPUT SEX $ AGE WEIGHT @@; CARDS;
```

Structuring the program as originally shown is the preferred approach, although it really does not make any difference.

SAS Procedures The statement PROC PRINT; is a command to print the data exactly as entered. The column OBS gives the number of the observation; each observation consists of values of three variables. Each SAS procedure is always obtained by a statement in which the first word is PROC. Additional statements may be used with most procedures. If we wish, we can have the data printed by sex, for example. In order to do this, we must first sort the data by sex, using the statement PROC SORT;—which must be followed by a

---

*In some earlier versions of the SAS System it was necessary to insert a semicolon on a line below the data and above the next statement.

statement indicating how we wish the data sorted. We may obtain a printout of the data, sorted by sex, if we write

```
PROC SORT;
   BY SEX;
PROC PRINT;
   BY SEX;
```

By convention, statements that are subsidiary to an initial statement are indented.

The data will be printed as follows:

```
OBS      AGE      WEIGHT
       SEX = F
 1        19        118
 2        37        123
 3        18        135
       SEX = M
 4        18        165
 5        21        172
 6        22        155
 7        24        188
```

If desired, we may specify only one or more of the variables to be printed. If we wished only age and wanted it to be printed by sex, we would write

```
PROC SORT;
   BY SEX;
PROC PRINT;
   VAR AGE;
   BY SEX;
```

The resulting printout would be the following:

```
OBS    AGE
 SEX = F
1       19
2       37
3       18
 SEX = M
4       18
5       21
6       22
7       24
```

1.3.5 SAS Statements for This Chapter

The SAS procedure FREQ can be used to obtain frequency tables. The frequency tables can be for either grouped or ungrouped data and will give the frequency, percentage, cumulative frequency, and cumulative percentage. Several different sets of statements will achieve the same ends but we will illustrate by using the football player weights. The data will be entered as in the Minitab example above, but we will group the data by a series of statements. We write

```
DATA;
    INPUT WT @@;
```

so that the variable WT will consist of the data as entered. We then group the data by creating the new variable WEIGHT and saying that if a value of WT is between the class limits, then the variable WEIGHT has a specified value. For example, the statement

```
IF 190<=WT<200 THEN WEIGHT = '190-199';
```

says that if the value of the variable WT is greater than or equal to 190 and less than 200, the corresponding value of the variable WEIGHT is the class 190–199. We repeat this statement for each class. We then specify PROC FREQ; with the statement TABLES *variable*. (Note that we use italics for an unspecified value or variable.) If we write PROC FREQ; TABLES WT; we will get a table for the ungrouped data. The statements PROC FREQ; TABLES WEIGHT; will give us a table for the grouped data.

We can obtain a bar graph by using PROC CHART. We may specify either horizontal or vertical bars (HBAR or VBAR), the variable to be used, and whether we want a frequency, relative frequency (percentage), cumulative frequency, or cumulative percentage chart. If we do not specify, a preselected option is used. When there are several options possible and none is specified, the option to be used has been specified by the programmer and is called the *default* option.

The default option for the CHART procedure is frequency. To obtain a horizontal bar frequency chart of the grouped weights, we write

```
PROC CHART;
    HBAR WEIGHT;
```

Many SAS procedures have options allowing specification of ways of proceeding. Options are specific to a particular SAS statement and are written in a statement following a / mark. We may use the option TYPE with the statement HBAR *variable* if we want a different type of chart. To obtain a vertical relative frequency bar chart of the grouped weights, we specify relative frequency by using the option TYPE = PCT; as follows:

```
PROC CHART;
    VBAR WEIGHT / TYPE=PCT;
```

We can obtain a stem and leaf plot for a set of data by using the SAS procedure UNIVARIATE. The stem and leaf plot is only one of many items produced by the univariate procedure, so we will save this until the next chapter.

Example 1.5

Use SAS methods to obtain frequency tables for the grouped and ungrouped football player weights, a horizontal bar graph for the grouped data, and a vertical relative frequency bar graph for the ungrouped data.

Solution

The program used and the output are shown in Figure 1.19. For illustrative purposes, titles have been included for the two bar graphs.

The output of a SAS program includes a listing of the entire input file (omitting data), together with notes, as well as error messages and any output specified by the PUT command (see Section 3.5.2). This part of the output file is called the *SAS Log*. The SAS Log is included in Figure 1.19. In future figures showing the SAS output, the SAS Log will be omitted when the input file is included. When the SAS Log is included, the notes will be deleted.

Figure 1.19
SAS printout for the football player weights data.

```
INPUT FILE

DATA FOOTBALL;
  INPUT WT @@;
  IF 190<=WT<200 THEN WEIGHT = '190-199';
  IF 200<=WT<210 THEN WEIGHT = '200-209';
  IF 210<=WT<220 THEN WEIGHT = '210-219';
  IF 220<=WT<230 THEN WEIGHT = '220-229';
  IF 230<=WT<240 THEN WEIGHT = '230-239';
  IF 240<=WT<250 THEN WEIGHT = '240-249';
  IF 250<=WT<260 THEN WEIGHT = '250-259';
  IF 260<=WT<270 THEN WEIGHT = '260-269';
  IF 270<=WT<280 THEN WEIGHT = '270-279';
  IF 280<=WT<290 THEN WEIGHT = '280-289';
  CARDS;
193 202 208 210 212 217 219 224 225 228 230 231 231 234 235
236 236 238 239 240 243 243 243 245 245 246 247 249 250 251
252 254 255 255 255 257 258 259 263 265 268 268 269
271 275 277 280 283 284 288
PROC FREQ;
   TABLES WT;
PROC FREQ;
   TABLES WEIGHT;
PROC CHART;
   TITLE 'FREQUENCY CHART OF WEIGHTS';
   HBAR WEIGHT;
PROC CHART;
   TITLE 'RELATIVE FREQUENCY CHART OF WEIGHTS (IN PERCENT)';
   VBAR WEIGHT/TYPE=PCT;
```

OUTPUT FILE

```
          DATA FOOTBALL;
             INPUT WT @@;
             IF 190<=WT<200 THEN WEIGHT = '190-199';
             IF 200<=WT<210 THEN WEIGHT = '200-209';
             IF 210<=WT<220 THEN WEIGHT = '210-219';
             IF 220<=WT<230 THEN WEIGHT = '220-229';
             IF 230<=WT<240 THEN WEIGHT = '230-239';
             IF 240<=WT<250 THEN WEIGHT = '240-249';
             IF 250<=WT<260 THEN WEIGHT = '250-259';
             IF 260<=WT<270 THEN WEIGHT = '260-269';
             IF 270<=WT<280 THEN WEIGHT = '270-279';
             IF 280<=WT<290 THEN WEIGHT = '280-289';
          CARDS;

NOTE: SAS WENT TO A NEW LINE WHEN INPUT STATEMENT
      REACHED PAST THE END OF A LINE.
NOTE: DATA SET WORK. FOOTBALL HAS 50 OBSERVATIONS AND 2 VARIABLES. 2470 OBS/TRK.
NOTE: THE DATA STATEMENT USED 0.17 SECONDS AND 360K.

          PROC FREQ;
             TABLES WT;
NOTE: THE PROCEDURE FREQ USED 0.20 SECONDS AND 700K AND PRINTED PAGE 1.

          PROC FREQ;
             TABLES WEIGHT;
NOTE: THE PROCEDURE FREQ USED 0.19 SECONDS AND 700K AND PRINTED PAGE 2.

          PROC CHART;
             TITLE 'FREQUENCY CHART OF WEIGHTS';
             HBAR WEIGHT;
NOTE: THE PROCEDURE CHART USED 0.20 SECONDS AND 508K AND PRINTED PAGE 3.

             PROC CHART;
                TITLE 'RELATIVE FREQUENCY CHART OF WEIGHTS (IN PERCENT)';
                VBAR WEIGHT/TYPE=PCT;
NOTE: THE PROCEDURE CHART USED 0.21 SECONDS AND 508K AND PRINTED PAGE 4.
NOTE: SAS USED 700K MEMORY.
NOTE: SAS INSTITUTE INC.
      SAS CIRCLE
      PO BOX 8000
      CARY, N.C.  27511-8000
```

SAS

WT	FREQUENCY	PERCENT	CUMULATIVE FREQUENCY	CUMULATIVE PERCENT
193	1	2.0	1	2.0
202	1	2.0	2	4.0
208	1	2.0	3	6.0
210	1	2.0	4	8.0
212	1	2.0	5	10.0
217	1	2.0	6	12.0
219	1	2.0	7	14.0
224	1	2.0	8	16.0
225	1	2.0	9	18.0
228	1	2.0	10	20.0
230	1	2.0	11	22.0
231	2	4.0	13	26.0

234	1	2.0	14	28.0
235	1	2.0	15	30.0
236	2	4.0	17	34.0
238	1	2.0	18	36.0
239	1	2.0	19	38.0
240	1	2.0	20	40.0
243	3	6.0	23	46.0
245	2	4.0	25	50.0
246	1	2.0	26	52.0
247	1	2.0	27	54.0
249	1	2.0	28	56.0
250	1	2.0	29	58.0
251	1	2.0	30	60.0
252	1	2.0	31	62.0
254	1	2.0	32	64.0
255	3	6.0	35	70.0
257	1	2.0	36	72.0
258	1	2.0	37	74.0
259	1	2.0	38	76.0
263	1	2.0	39	78.0
265	1	2.0	40	80.0
268	2	4.0	42	84.0
269	1	2.0	43	86.0
271	1	2.0	44	88.0
275	1	2.0	45	90.0
277	1	2.0	46	92.0
280	1	2.0	47	94.0
283	1	2.0	48	96.0
284	1	2.0	49	98.0
288	1	2.0	50	100.0

SAS

WEIGHT	FREQUENCY	PERCENT	CUMULATIVE FREQUENCY	CUMULATIVE PERCENT
190-199	1	2.0	1	2.0
200-209	2	4.0	3	6.0
210-219	4	8.0	7	14.0
220-229	3	6.0	10	20.0
230-239	9	18.0	19	38.0
240-249	9	18.0	28	56.0
250-259	10	20.0	38	76.0
260-269	5	10.0	43	86.0
270-279	3	6.0	46	92.0
280-289	4	8.0	50	100.0

FREQUENCY CHART OF WEIGHTS

```
                                 SAS

              FREQUENCY BAR CHART

WEIGHT                                                          FREQ  CUM.   PERCENT     CUM.
                                                                      FREQ             PERCENT
190-199   *****                                                   1     1     2.00       2.00

200-209   **********                                              2     3     4.00       6.00

210-219   ********************                                    4     7     8.00      14.00

220-229   ***************                                         3    10     6.00      20.00

230-239   *********************************************           9    19    18.00      38.00

240-249   *********************************************           9    28    18.00      56.00

250-259   **************************************************     10    38    20.00      76.00

260-269   *************************                               5    43    10.00      86.00

270-279   ***************                                         3    46     6.00      92.00

280-289   ********************                                    4    50     8.00     100.00

--------------|----|----|----|----|----|----|----|----|----|---------------------
              1    2    3    4    5    6    7    8    9    10
                                 FREQUENCY
```

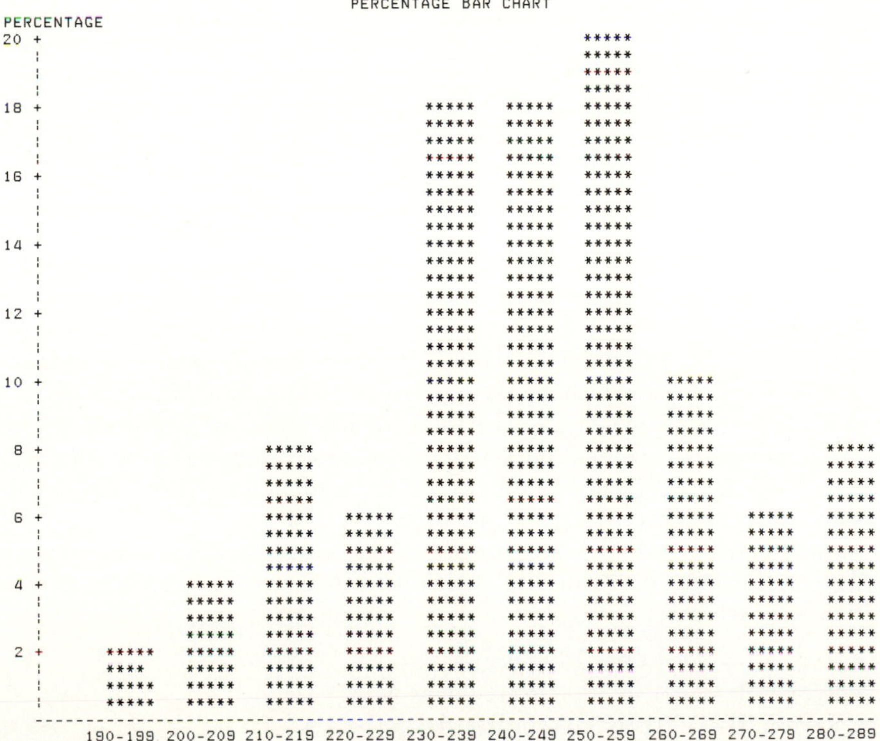

Problems

Applications

1.21 Use the computer to obtain a grouped frequency table, bar graph, and stem and leaf plot (if using Minitab) for the SAT scores given in Problem 1.10. Is the data set like a uniform, mound-shaped, U-shaped, or J-shaped distribution?

1.22 Use the computer to obtain an ungrouped frequency table for the weekly sales data given in Problem 1.6. Also obtain the relative frequencies.

1.23 Refer to the Miami Female Study data given in Appendix B.1. Use the computer to obtain frequency tables and relative frequency tables for the variable "hair color."

1.24 Graphs can provide a quick way of checking for differences between two populations. This can be done by obtaining an appropriate graph for a sample from each population and then inspecting the graphs to see if any differences are apparent. To illustrate, consider the Roger's Cat Food data of Problem 1.6. Suppose the first four columns of data are taken from retail outlets in the eastern United States, while the last four columns are taken from the western United States.

a. If using Minitab, obtain a stem and leaf plot for the outlets in the eastern United States. Then obtain a stem and leaf plot for the outlets in the western United States. After looking at the two graphs, comment on the sales for the two regions. Are the sales patterns about the same?

b. If using SAS software, obtain a bar graph of the data for the outlets in the eastern United States. Then obtain a bar graph of the data for the outlets in the western United States. After looking at the two graphs, comment on the sales for the two regions. Are the sales patterns about the same?

1.4 Case 1: What Grapes to Grow?—Solution

The University of California, Davis, has made a thorough study of the growing of grapes for wine. Grapes normally grow when the average daily temperature reaches 50°F. The sum of the average daily temperature above 50° for the growing season (about 1 April to sometime in October) is called the *degree-day*. A day in which the average daily temperature is 73°, for example, contributes 23 degree-days to the total for the year. The degree-days can differ substantially from location to location, so a reading has to be obtained for each property. To obtain the degree-day reading for any one day, 50 is subtracted from the average of the high and low reading for the day. A locality may be classified into one of five regions by the sum of the degree-days for the season. A locality is classified into regions according to the following scheme:

- Region I—fewer than 2500 degree-days
- Region II—2501 to 3000 degree-days
- Region III—3001 to 3500 degree-days

- Region IV—3501 to 4000 degree-days
- Region V—more than 4000 degree-days.

Most of France, except the southernmost regions, is classified as Region I; so is Germany. The major wine-growing regions of California, on the other hand, may be classified anywhere from Region I to Region IV.

The most favorable region for each grape variety has also been determined by the enologists (wine specialists) at Davis. For example, the grapes yielding the best wine (White Riesling, Chardonnay, Sauvignon Blanc, Cabernet Sauvignon, Pinot Noir, and Zinfandel) grow best in Region I, although all but Cabernet Sauvignon and Zinfandel also do reasonably well in Region II. A wine grower whose acreage lies in a warmer region (III–V) should grow grapes for wines that are of lesser quality but that are more profitable. (See Thompson (1980) and Muscatine and others (1984) for additional details.)

Armed with this information Giovanni will be able to determine the degree-days for his property and thus grow the grapes that flourish in his locale. Moreover, the Davis researchers have obtained much statistical information about the grapes, including best times to harvest, methods of crushing, and fermenting, and so on. The scientific study of grapes and wine making using statistical techniques has not only contributed to the improvement of wine in the last few decades but is also responsible to a large degree for keeping the price of wine relatively low in relation to its quality.

1.5 Summary

The organization and presentation of data is one of the primary purposes of descriptive statistics. Large amounts of data must be well organized in order to be understandable, and even small amounts of data may be more meaningful after organization.

One of the best ways to organize data is to use a frequency distribution, tallying the number of observations of each value of the variable or of certain classes of values of the variable. Variations of the basic frequency distribution are relative frequency, cumulative frequency, and relative cumulative frequency distributions.

Bar graphs, histograms, frequency polygons, pictograms, and pie charts are among the many methods used to present data in a visually appealing manner. Stem and leaf plots may also be used for visual presentation, although they are less flexible than other methods since the choice of classes is not completely open. Great care must be exercised in the construction of a graph to avoid giving a false impression.

An excellent collection of 46 essays describing important applications of statistics in many different fields is *Statistics: A Guide to the Unknown,* edited by Judith M. Tanur and others (1978). Another very interesting and witty book showing some of the uses and misuses of statistics is *How to Lie with Statistics* by Darrel Huff (1954).

Chapter Review Problems

For each of the problems presented here, use the computer as directed by your instructor.

Practice

1.25 Use the following data set to:

a. Construct a stem and leaf plot.
b. Construct a frequency distribution using ungrouped data.
c. Construct a bar graph for the data using the results of part (*b*).

11	7	13	17	9	6	16	22
3	19	18	13	11	18	13	11
4	10	9	16	12	10	17	23
20	14	8	13	14	17	10	6
14	17	22	16	13	15	9	11

1.26 A set of data has a high of 363 and a low of 107. Construct a table with 11 classes that contain the data. Give the class limits, boundaries, and marks.

1.27 The class marks of a distribution of sales (in dollars) are 125.5, 175.5, 225.5, 275.5, 325.5, 375.5, 425.5, 475.5, and 525.5. What are the class limits? The class boundaries?

1.28 Refer to the data of Problem 1.4.

a. Group the data into classes, starting at 50 and using a class width of 10.
b. Construct a table showing frequency and cumulative frequency for these classes.
c. Construct a histogram and relative frequency histogram for the data you constructed in part (*b*).
d. Construct a relative frequency polygon for the grouped data.

1.29 A frequency distribution is given here:

CLASS	FREQUENCY
11.00–12.99	2
13.00–14.99	7
15.00–16.99	13
17.00–18.99	24
19.00–20.99	11
21.00–22.99	4

a. What are the boundaries for the class 17.00–18.99?
b. What is the unit of measurement?
c. If a stem and leaf plot had been prepared, what would the stems have been?
d. What is the class mark for the class 13.00–14.99?
e. What is the class width?
f. What is the range if the highest score was 22.68 and the lowest 11.74?

Applications

1.30 Suppose that weights of shipments to the nearest pound are grouped into classes 0–75, 76–150, 151–225, 226–300, 301–375, 376–450, 451–525,

526–600, 601–675, 676–750, 751–825, 826–900, and 901 or more. If the frequency for each class is listed, is it possible to determine the number of shipments weighing (to the nearest pound):

a. 300 pounds or less?
b. less than 300 pounds?
c. more than 750 pounds?
d. 750 pounds or more?
e. more than 1000 pounds?
f. less than 700 pounds?

1.31 Sales of a certain candy bar each day for a period of 60 days are listed here:

48	34	36	40	26	29	25	30
22	32	34	41	47	24	42	37
37	30	26	37	42	41	41	23
31	22	34	41	24	42	35	41
40	29	32	32	22	31	28	41
27	34	37	24	33	33	24	28
29	43	46	32	32	28	34	34
28	27	43	34				

a. Construct a stem and leaf plot for the data.
b. Construct a frequency distribution with ten classes. Begin with 21.
c. Construct a bar graph for the data as presented in part (*b*).

1.32 The following table gives the frequency distribution of scores made by soldiers on an Officer Qualification Test:

SCORE	NUMBER
51–65	12
66–80	21
81–95	33
96–110	42
111–125	19
126–140	14
141–155	6

a. Construct a relative frequency histogram.
b. Construct a frequency polygon.

1.33 According to Runzheimer International (*USA Today*, Jan. 31, 1986), food and lodging costs for one night for one person average \$109.00 in the Pacific Region, \$84.59 in the Mountain Region, \$79.39 in the Plains Region, \$87.29 in the South Central Region, \$88.76 in the Great Lakes Region, \$85.19 in the Southeast Region, \$96.56 in the Middle Atlantic Region, and \$94.66 in the Northeast Region. Draw a bar graph to show this set of data.

1.34 During a period of one month, a total of 120 patients were admitted to a certain hospital. Their ages, in order of admittance, were as follows:

37	54	81	64	11	34	80	71
51	12	52	13	62	8	1	2
39	22	4	84	28	76	30	73
47	33	16	24	16	24	35	8

56	6	38	55	54	7	3	10
4	6	58	10	67	29	3	4
9	68	59	43	11	64	60	23
88	37	56	46	33	52	71	28
46	3	52	55	9	7	81	57
8	19	52	17	24	27	7	83
25	57	7	24	77	37	83	48
7	38	57	43	58	82	34	9
64	50	8	23	18	14	10	7
60	44	39	29	74	41	74	64
27	44	63	6	52	8	9	54

a. Construct a stem and leaf plot for the data.

b. Construct a frequency table, using a class interval of 10, beginning with age 1. Assume that ages are given to the nearest birthday rather than the last birthday.

c. Draw a histogram using the results of part (*b*).

d. Construct a histogram using the results of part (*b*).

1.35 The following are IQ scores of 110 randomly selected high school students:

154	131	122	100	113	119	121	128
128	112	133	119	115	117	110	104
125	85	120	135	93	103	103	122
109	147	103	113	107	98	128	93
90	105	118	134	89	113	108	122
85	108	106	116	151	117	116	80
111	127	81	96	114	103	126	119
122	102	99	106	105	111	127	108
104	91	123	122	97	110	150	103
87	98	108	117	94	96	111	101
118	104	127	94	115	101	125	109
91	110	97	115	108	109	133	107
115	83	102	116	101	113	112	82
114	112	113	102	115	123		

a. Construct a stem and leaf plot for the data.

b. Group the data into a frequency table with 15 classes.

c. Draw a histogram for the grouped data.

d. Construct a relative frequency table and a relative frequency polygon for the grouped data.

1.36 According to the U.S. League of Savings Institutions (*USA Today,* Jan. 20, 1986), the median housing expense (mortgage payment, taxes, utilities, and upkeep) was $770 in 1985. The amount differed from region to region and was $1020 in the west, $666 in the north central region, $838 in the east, and $719 in the south. Depict this information graphically.

Index of Terms

Answers to Proficiency Checks

1. **a.** Yes **b.** Yes **c.** No (only one piece of data)
d. Yes **e.** Yes **f.** Yes
g. No (only one piece of data; also unclear who sold the policies)

2. **a.** Qualitative **b.** Quantitative **c.** Qualitative
d. Quantitative **e.** Quantitative **f.** Qualitative

3. The answers to all but c, e, and f depend upon how the variable is measured. In a, b, and d, the underlying variables (distance, weight, and time) are continuous but are usually measured to the nearest convenient unit—inches, ounces, and seconds, for instance. Thus the data may be discrete but the variable continuous. Incomes (e) are discrete but nearly continuous, while count data (c and f) are discrete.

4.

AGE (YR)	TALLY	FREQUENCY	RELATIVE FREQUENCY
18	//	2	0.08
19	~~////~~ /	6	0.24
20	~~////~~ ////	9	0.36
21	////	4	0.16
22	///	3	0.12
23	/	1	0.04

5. **a.** 168 − 72 = 96 **b.** Actually, there is no integer class width that, beginning at 71, will yield 12 classes; a class width of 8 will require 13 classes, while 11 classes is enough for a class width of 9. The answer to the question, then, is "either 8 or 9" since either will give *approximately* 12 classes. **c.** For class widths of 8, 9 and 10, respectively, the classes are

71–78, 79–86, 87–94, 95–102, 103–110, 111–118, 119–126, 127–134, 135–142, 143–150, 151–158, 159–166, 167–174

71–79, 80–88, 89–97, 98–106, 107–115, 116–124, 125–133, 134–142, 143–151, 152–160, 161–169

71–80, 81–90, 91–100, 101–110, 111–120, 121–130, 131–140, 141–150, 151–160, 161–170

6. **a.** 43 **b.** 5 **c.** 26 and 30 **d.** 15.5 and 20.5 **e.** Class 5 (26–30)

7.

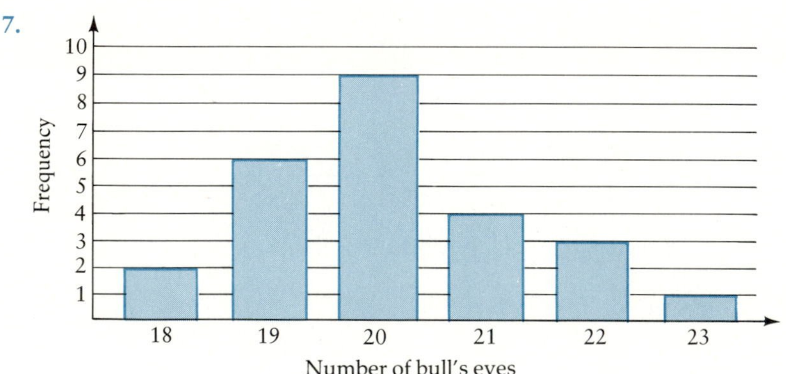

8.

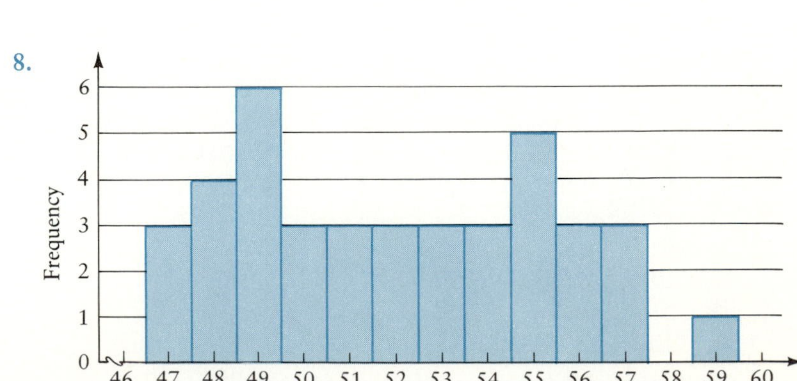

9.

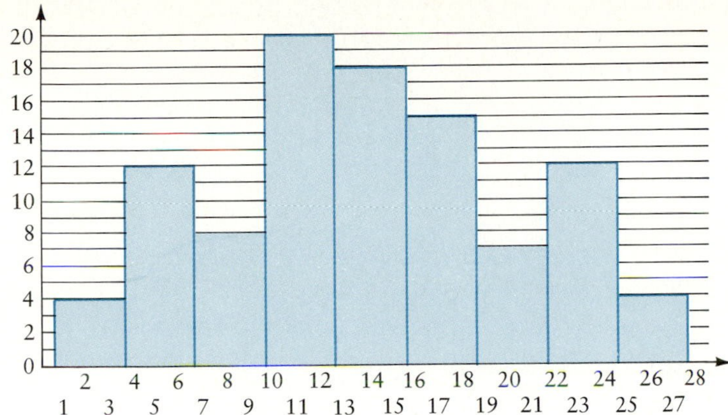
20
18
16
14
12
10
8
6
4
2
0
2 4 6 8 10 12 14 16 18 20 22 24 26 28
1 3 5 7 9 11 13 15 17 19 21 23 25 27

10.

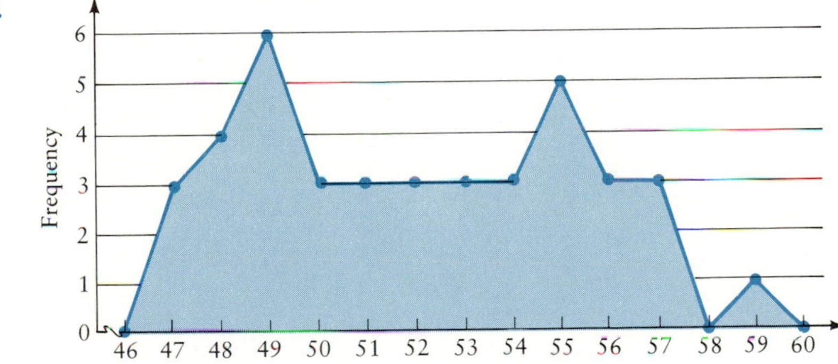
Frequency
6
5
4
3
2
1
0
46 47 48 49 50 51 52 53 54 55 56 57 58 59 60

11.

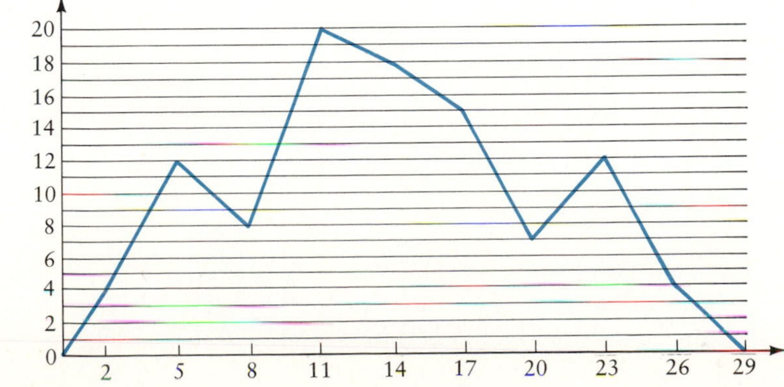
20
18
16
14
12
10
8
6
4
2
0
2 5 8 11 14 17 20 23 26 29

12. 19|3
19|
20|2
20|8
21|2
21|789
22|4
22|58
23|0114
23|5668
24|033
24|555669
25|0124
25|5557899
26|3
26|5889
27|1
27|57
28|034
28|8

13. 1|011223344
1|567779
2|3333444
2|566777789
3|00114
3|556666777789
4|0
4|5568
5|3344
5|58
6|011124
6|5557
7|00013
7|7
8|133
8|6
9|2
9|7

References

Fairley, William B., and Frederick Mosteller. *Statistics and Public Policy*. Reading, MA: Addison-Wesley, 1977.

Huff, Darrel. *How to Lie with Statistics*. New York: Norton, 1954.

Muscatine, Doris, Maynard A. Amerine, and Bob Thompson, eds. *The University of California/Sotheby Book of California Wine*. Berkeley: University of California Press/Sotheby Publications, 1984.

Ryan, B. F., B. L. Joiner, and T. A. Ryan. *Minitab Student Handbook*. 2nd ed. North Scituate, MA: Duxbury Press, 1985.

SAS Institute Inc. *SAS® User's Guide: Basics, Version 5 Edition*. Cary, NC: SAS Institute, Inc., 1985, 956 pp.

Tanur, Judith M., and others, eds. *Statistics: A Guide to the Unknown*. 2nd ed. San Francisco: Holden-Day, 1978.

Thompson, Bob. *The Pocket Encyclopedia of California Wines*. New York: Simon & Schuster, 1980.

Tufte, Edward R. *The Visual Display of Quantitative Information*. Cheshire, CT: Graphics Press, 1983.

Tukey, John W. *Exploratory Data Analysis*. Reading, MA: Addison-Wesley, 1977.

2

Summary Descriptive Measures

Organized data are more meaningful than unorganized data, but still more information can be gained from them by performing some arithmetic on those data. In this chapter we will discuss a few of the ways to *describe* a set of data to aid our understanding, and we will compare these descriptive measures.

In Chapter 1 we defined a statistical population as the set of all possible elements or pieces of data pertaining to a characteristic of interest. A listing of these elements is called a census. A sample consists of a limited set of data drawn from a population. Since the characteristics of a complete population are rarely known, except perhaps theoretically, most information comes from samples. We cannot know, for example, the exact distribution of heights of all the people in the world. For many purposes we wish to infer characteristics that describe a population from the characteristics of a sample. Such a procedure is generally called **statistical inference.** In this chapter we introduce several descriptive measurements of population and sample characteristics, together with some of their uses. This material will be essential to all later chapters on statistical inference.

2.0 Case 2: The Problem Promotion

Wilbur Wilson was vice president in charge of marketing for Smith, Incorporated, manufacturer and distributor of a wide array of items. Howard Sims, chief of marketing for the Midwestern Operations Division, was about to retire and had to be replaced. Company policy was to promote from within if possible. Mr. Wilson was responsible for making the decision. He had narrowed the field down to two candidates, Elizabeth Morton and Frederick Lasky, and simply could not decide between the two. Each had about the same amount and kind of experience, all reports from the field were glowing, and Sims had included their names on a list of five he thought were qualified to take his place.

One criterion to be used in making the selection was the result of a test of leadership and management ability the company administered to all personnel. One of the two had scored 143, the other 29. The company had changed the particular test it used, however, and the score of 29 was from the earlier test while the score of 143 was from the later one. Management decided that it would not be fair to give a retest since one of the candidates had already taken the latest test.

Wilson did note that about half the people taking the first test scored above 23, while about half the people taking the later test scored above 105. He could not decide, however, whether this information helped. With only a few days to go before the decision had to be made he turned to Marian Lee, the company's director of personnel, and asked her if there was any way of finding out which of the two candidates scored higher on the test. Ms. Lee said that she would investigate the information available about both tests and give him her findings within a day.

2.1 Measures of Central Tendency

A set of data has a number of important features. If the data set consists of numbers, we are most often interested in two characteristics of the data. The first is a number that is considered most representative of the set. Such a number is called a **measure of central tendency.** The second is a number that will tell us something about the spread of the data—that is, whether the data points are generally close to the center or spread out over a wider range. These numbers measure the **variability** or **dispersion** of the data set. In this section we will discuss several widely used measures of central tendency. Measures of variability are covered in Section 2.2.

Three measures of central tendency are often used, and each has its advantages. Consider the following episode. A campaigner knocks on your door asking for contributions to the County Fund. He stresses that he expects a generous gift, since the average income in your neighborhood is $52,000. A few days later, a campaigner asks you for a contribution for the poor of the neighborhood. After all, he says, most people around here make only $10,000 a year. Confused, you consult one of your neighbors, a university sociologist, who tells you that the middle income in the neighborhood is $24,000! Who is correct and who is lying, if anyone?

To sort out the truth, look at the actual annual incomes of the families in the neighborhood as presented in Table 2.1. A picture begins to unfold here. If we add all 13 incomes and divide by 13, we obtain $52,000. If we count down from the top (or up from the bottom) until we reach the middle (seventh) income, we obtain $24,000. The most frequently occurring income is $10,000. Thus none of the people you consulted was actually lying. Each of these numbers is a measure of central tendency.

Table 2.1 FAMILY INCOMES.

INCOME	FREQUENCY
$400,000	1
68,000	1
36,000	1
28,000	1
24,000	3
16,000	2
10,000	4

2.1.1 The Mean

The measure of central tendency most often used for purposes of statistical inference is called the *mean*. The **mean** of a set of data is found by adding up all the values of the data points and dividing by the total number of data points. From Table 2.1, we obtain a mean of $52,000. The term **average** generally refers to this mean, and it is sometimes called the arithmetic mean to distinguish it from other types of means that we need not discuss here.

We determine the mean of a population in the same way as the mean of a sample, but we use different symbols to indicate that they represent different concepts. The mean of a population describes the population and is one kind of population *parameter*. The word *parameter* means constant. Since a population has only one mean, that value is constant and so we use the word *parameter*.

We use the Greek letter μ (mu) to represent the mean of a population. We usually let x (or y) represent the value of a piece of data, and the symbol Σ (capital sigma) is used to indicate "the sum of." Thus Σx means the sum of the values of all observations in the population or sample. A brief overview of summation notation is given in the note at the end of this chapter. The total number of observations in a finite population is usually denoted by N. Thus we have the following definition of a population mean.

Mean of a Population

The mean of a population containing N pieces of data is given by the formula

$$\mu = \frac{\Sigma x}{N}$$

where Σx denotes the sum of the values of all data in the population.

The mean of a sample is similarly defined. We use the symbol $\bar{x}$ to represent the mean of a sample if x represents the value of the data points. The number of observations in a sample is usually denoted n. The following formula defines the mean of a sample.

Mean of a Sample

The mean of a sample containing n pieces of data is given by

$$\bar{x} = \frac{\Sigma x}{n}$$

where Σx denotes the sum of the values of all data in the sample.

Since many different samples can be drawn from a population, and since each of the samples has a mean, sample means from the same population usually vary for different samples. Each of the sample means is an estimate for the population mean, but since sample means are not constant from one sample to another, we refer to sample means as *sample statistics*. In general, a measure that describes a population is called a **parameter,** while the same measure describing a sample is called a **statistic.** In statistics we commonly use a Greek letter to denote a population parameter and the corresponding Roman letter (our alphabet) to denote the corresponding sample statistic. For this reason the mean of a sample is sometimes denoted by m. The use of $\bar{x}$ derives from physics, where it symbolizes the first moment of inertia, and it has become entrenched in scientific literature. Although the use of m is gaining acceptance in the social sciences, it is still not widespread.

Example 2.1

A random sample of five accounts in a department store shows the following balances at the end of the month: \$67.32; \$108.97; \$27.64; \$412.11; and \$81.96. Compute the mean balance.

Solution

Since the sum of the five amounts is \$698.00, the mean is given by $\bar{x}$ = \$698.00/5 = \$139.60.

Comment: At this point we need to say something about the precision of the mean. Precision of a number refers to how closely the numbers are measured. In Example 2.1, all the numbers are given to the cent. As a rule, this means that the mean should be given to the nearest cent. If, for example, the sum had been \$698.04, we would have 698.04/5 = 139.608, and we would have reported the mean to the nearest cent. When a number is approximated, not exact, we will use an equal sign with a dot over it. Thus we would have $\bar{x} \doteq$ = \$139.61. The $\doteq$ symbol will be used wherever appropriate throughout the book.

Example 2.2

The ages of 25 people in a certain income bracket are distributed as follows:

Age	29	33	37	38	39	40	42	43	45	47	50	59	66
Frequency	1	1	3	4	2	3	2	2	3	1	1	1	1

What is the mean age in this sample?

Solution

In the formula for the mean, Σx signifies the sum of the values of *all* observations. Therefore we must add up all 25 numbers. This sum is equal to

$$\begin{aligned} &29 + 33 + 3(37) + 4(38) + 2(39) + 3(40) + 2(42) + 2(43) \\ &\quad + 3(45) + 47 + 50 + 59 + 66 \\ &= 29 + 33 + 111 + 152 + 78 + 120 + 84 + 86 + 135 + 47 \\ &\quad + 50 + 59 + 66 \\ &= 1050 \end{aligned}$$

so that $\bar{x}$ = 1050/25 = 42.

Comment: A common error in finding the mean of data in a frequency distribution is overlooking the frequencies in finding the value of Σx or n. Each value of the variable should be counted as many times as it occurs in finding Σx. The simplest way is to multiply each value by its frequency in obtaining the total, as in Example 2.2. We obtain the value of n by taking the sum of the frequencies.

If the data are grouped, we lose a degree of accuracy because we do not know the exact value of each observation. The ease of presenting grouped data may compensate for the loss of accuracy, however. To calculate the mean from grouped data we use the class mark for each class to represent all the data in the class. The mean is then computed as in Example 2.2 but for the class marks rather than for the individual values; the latter have lost their identity.

Example 2.3

Find the mean of the data given in Table 2.2.

Table 2.2

CLASS	FREQUENCY
131–135	3
136–140	11
141–145	17
146–150	19
151–155	27
156–160	22
161–165	14
166–170	8
171–175	4

Solution

The sum of the frequencies is 125, so $n = 125$. The class marks are 133, 138, 143, 148, . . . , 173. Multiplying each class mark by its frequencies, we find

$$\begin{aligned}\Sigma x &= 133\cdot 3 + 138\cdot 11 + 143\cdot 17 + 148\cdot 19 + 153\cdot 27 + 158\cdot 22\\ &\quad + 163\cdot 14 + 168\cdot 8 + 173\cdot 4\\ &= 399 + 1518 + 2431 + 2812 + 4131 + 3476 + 2282 + 1344\\ &\quad + 692\\ &= 19{,}085\end{aligned}$$

Then $\overline{x} = 19{,}085/125 = 152.68 \doteq 152.7$.

Note that although the original data were probably given in whole numbers, we give the mean to the nearest tenth. The number of significant digits used in reporting the mean can be flexible. Strictly speaking, if we use the mean only to describe the approximate center of a data set, we should present the result in the same units the data are given. Here we would have $\overline{x} \doteq 153$. If the mean is going to enter into other calculations or be used for other statistical purposes, however, we commonly use more significant figures to avoid introducing error into subsequent calculations.

Example 2.4

A relative frequency table for a set of data is shown here. Compute the mean for the data.

x	RELATIVE FREQUENCY
15	0.3
16	0.2
17	0.2
18	0.2
19	0.1

Solution

The formula for the mean is $\overline{x} = (\Sigma x)/n$. This is equivalent to $\overline{x} = (\Sigma x)\cdot(1/n)$. Since the relative frequency of each observation is $1/n$, the mean of a set of data can be found by multiplying each value times its relative frequency and summing the results. Thus $\overline{x} = 15(0.3) + 16(0.2) + 17(0.2) + 18(0.2) + 19(0.1) = 16.6$.

Comment: In some cases it is possible to simplify the calculations greatly. If you add or subtract the same number to or from every piece of data in a data set, the mean is changed by the same amount. For example, suppose we want the mean of the following numbers: 234,567; 234,517; 234,544; 234,548; 234,561; 234,572. We can subtract 234,500 from each number and obtain the mean of the numbers—67, 17, 44, 48, 61, 72. For these numbers the sum is 309, so the mean is 309/6 = 51.5. Then the mean of the original data set is 234,500 + 51.5 = 234,51.5.

PROFICIENCY CHECKS

1. Determine the mean of each of the following data sets:
   a. 3, 5, 5, 6, 7
   b. 25, 31, 22, 16, 45
   c. 0.5, 0.7, 1.2, 2.2, 1.3, 2.1, 0.6, 1.6, 2.2, 1.0
   d. 44,445; 44,442; 44,443; 44,444; 44,449; 44,451; 44,448; 44,445; 44,443
2. A frequency distribution is given here. Determine the mean for the data. (Note that f represents the frequency for each value of x.)

x	74	75	76	78	79	80	81	83	88
f	3	6	8	12	21	13	9	4	1

3. A relative frequency distribution is given here. Determine the mean of the variable.

CLASS	RELATIVE FREQUENCY
111–115	0.02
116–120	0.04
121–125	0.07
126–130	0.18
131–135	0.16
136–140	0.28
141–145	0.14
146–150	0.05
151–155	0.04
156–160	0.02

2.1.2 The Median

In many cases involving small sets of data, or data sets in which most of the data are at one end with only a few extreme values, the mean is not truly representative because of the influence of these extreme values. In the income data of Table 2.1, for example, the mean is \$52,000, yet 12 of the 13 pieces of data are below the mean. Clearly the mean is not the measure of central tendency most representative of this set. The mean lacks a very desirable property—it is not **resistant** to changes. Resistance is a desirable property in summary statistics. (For a discussion see Mosteller and Tukey [1977, pp. 203ff].) If a change in a relatively small number of observations in a set of data can change the value of a statistic greatly, the statistic is not resistant to changes. In the income data, for example, if the family with the \$400,000 income were removed, the mean would drop from \$52,000 to \$23,000—a drastic change indeed. In such cases we want a more resistant measure of central tendency. Such a resistant measure is the central value of the data, called the **median** (symbol **Md**).

To find the median we arrange the data in ascending or descending order and count until we find the middle value—the median. If there is an odd number of observations, the median is the one in the middle. If there is an even number of observations, the median is defined as halfway between the two middle values. The income data of Table 2.1 have 13 observations, so the median is located at observation 7, or \$24,000. In general, to locate the median we obtain the number $(n + 1)/2$, where n is the total frequency. If n is odd, the median is the value of the observation at this location. If n is even, the location is between the two middle observations and the median is the average of the values of the two observations. If $n = 30$, for example, $(30 + 1)/2 = 15.5$; the median, then, is the average of the values of the fifteenth and sixteenth observations. The value of the median, like that of the mean, is usually expressed in the same units as the data or to one additional decimal place, whichever is most useful for clarity. In the case of small data sets it is usually helpful to arrange the numbers in numerical order and then simply count from the top or the bottom. A stem and leaf plot could be used for this procedure.

Example 2.5

Determine the median of each of the following data sets:

a. 25, 28, 13, 44, 17, 22, 19
b. 25, 28, 13, 44, 17, 22, 19, 32
c. 25, 28, 13, 44, 17, 22, 19, 32, 25, 25, 32, 39

Solution

We arrange the data sets in numerical order in each case.

a. 13, 17, 19, 22, 25, 28, 44; there are seven pieces of data, $(7 + 1)/2 = 4$, so the median is the fourth number. Counting, we obtain Md = 22.
b. 13, 17, 19, 22, 25, 28, 32, 44; there are eight pieces of data, $(8 + 1)/2 = 4.5$, so the median is between the fourth and fifth numbers. The fourth number from the bottom is 22, the fifth number is 25, so the median is the average of 22 and 25; Md = 23.5.
c. 13, 17, 19, 22, 25, 25, 25, 28, 32, 32, 39, 44; there are 12 pieces of data, $(12 + 1)/2 = 6.5$, so the median is between the sixth and seventh numbers. Both of these numbers are 25, so Md = 25.

Example 2.6

The ages of 25 people in a certain income bracket are distributed as follows:

Age	29	33	37	38	39	40	42	43	45	47	50	59	66
Frequency	1	1	3	4	2	3	2	2	3	1	1	1	1

Determine the median of these ages.

Solution

Since $n = 25$, the median is located at $(25 + 1)/2$, or 13 numbers from the bottom. Counting upward, we find the thirteenth number to be 40, so Md = 40.

PROFICIENCY CHECKS

4. Determine the median of each of the following data sets:
   a. 3, 5, 5, 6, 7
   b. 25, 31, 22, 16, 45, 38
   c. 0.5, 0.7, 1.2, 2.2, 1.3, 2.1, 0.6, 1.6, 2.2, 1.0
   d. 445, 442, 443, 444, 449, 451, 448, 445, 443
5. Each soldier in a platoon takes 25 shots at a target with the following number of bull's-eyes for each soldier:

11	7	13	17	9	6	16	22
3	19	18	13	11	18	13	11
4	10	9	16	12	10	17	23
20	14	8	13	14	17	10	6
14	17	22	16	13	15	9	11

   Determine the median number of bull's-eyes per soldier for the platoon.

2.1.3 The Mode

Occasionally it is useful to give a measure of the *most typical* value of a data set. For qualitative data, for example, the mean and the median would be meaningless, but we could determine the number of data points having a qualitative characteristic. The value of the variable with the largest frequency is called the **mode** (symbol **Mo**). In a grouped data frequency distribution, the class with the greatest frequency is called the **modal class.** If two values of the variable have the same highest frequency, the data set is said to be **bimodal.** If three or more have the same highest frequency, there is no mode. (The term *trimodal* could be used, but it has little application.) For grouped data, the term *bimodal* is sometimes used to describe a distribution in which two classes contain the highest frequencies (even if not exactly the same) and these classes are separated by other classes containing lesser frequencies (rather like two mountain peaks separated by a valley). A histogram of such a bimodal distribution is given in Figure 2.1.

In cases like this one, we may have collected data from two or more populations. A height distribution of the general adult population would most likely be bimodal with peaks at 5 feet 5 inches (the modal female height) and 5 feet 10 inches (the modal male height). Such bimodality is a signal for us to examine the possibility that the data are drawn from two or more different populations.

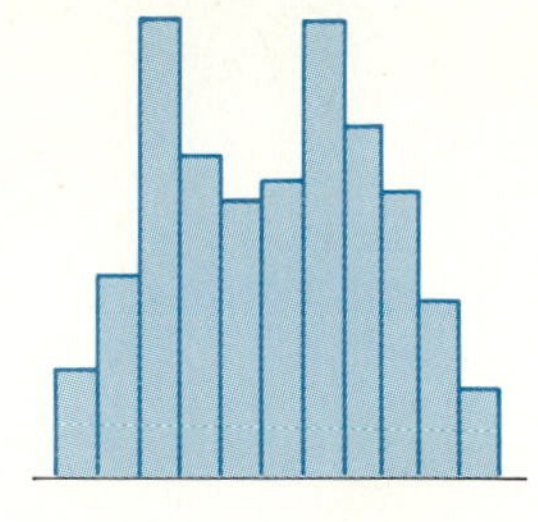

Figure 2.1
Histogram of a bimodal distribution.

To determine the mode, simply count the number of observations at each value; the one with the greatest frequency is the mode. In the data of Examples 2.2 and 2.6, for instance, the age 38 occurs four times, more than any other, so the mode is 38. In the data of Example 2.3, the class 151–155 contains 27 observations, more than any other class, so this is the modal class.

PROFICIENCY CHECKS

6. Determine the mode of each of the following data sets:
   a. 25, 28, 13, 44, 17, 22, 19, 22, 25, 28, 22
   b. 25, 28, 13, 44, 17, 22, 19, 22, 25, 28, 22, 25
   c. 25, 28, 13, 44, 17, 22, 19, 22, 25, 28, 22, 25, 28
7. The ages of 25 people in a certain income bracket are distributed as follows:

Age	29	33	37	38	39	40	42	43	45	47	50	59	66
Frequency	1	1	3	4	2	3	2	2	3	1	1	1	1

Determine the mode of these numbers.

Problems

Practice

Determine the mean, the median, and the mode of each of the data sets in Problems 2.1 to 2.4.

2.1 20, 22, 23, 26, 29, 30

2.2 223,728; 223,725; 223,720; 223,733; 223,727; 223,729; 223,723; 223,721; 223,724; 223,718; 223,730; 223,725

2.3 1.34, 1.69, 1.78, 1.89, 2.03, 2.27, 2.39, 2.88, 3.16, 3.34, 4.92, 5.57, 6.83, 7.44, 9.63, 11.82

2.4

x	17	18	19	20	21	22	23
Frequency	2	4	8	11	14	7	3

2.5 The heights of 50 men are collected and arranged in a frequency distribution as shown in the following table. Determine the mean height of this sample and the modal class.

HEIGHT (in.)	FREQUENCY
61–63	3
64–66	5
67–69	12
70–72	15
73–75	9
76–78	4
79–81	2

2.6 A frequency distribution is given in this table. Determine the mean for the data.

CLASS	FREQUENCY
11.00–12.99	2
13.00–14.99	7
15.00–16.99	13
17.00–18.99	24
19.00–20.99	11
21.00–22.99	4

Applications

2.7 According to *Platte's Oilgram Daily Report* and *Data Resources Inc.* (Associated Press, Feb. 15, 1986), U.S. oil prices dropped to nearly $15 per barrel, their lowest price since 1979. The report went on to say that some analysts were predicting a 10¢ to 15¢ per gallon drop in prices by mid or late spring. A local motorist conducts a telephone survey of 12 gas stations in town and asks how much the price of unleaded gas has dropped in the past month. He obtains the following figures (in cents). Determine the mean, median, and mode of these numbers.

5, 4, 4.5, 0, 2, 5, 5, 2, 3, 5, 5, 4

2.8 Determine the mean and the modal class for the income data of the following table.

INCOME ($)	NUMBER OF FAMILIES
42,000–44,999	3
39,000–41,999	7
36,000–38,999	12
33,000–35,999	17
30,000–32,999	28
27,000–29,999	60
24,000–26,999	216

21,000–23,999	268
18,000–20,999	448
15,000–17,999	621
12,000–14,999	949
9000–11,999	1421
6000–8999	2844
3000–5999	2123
0–2999	294

2.9 The following data, obtained from the *1980 Census of Population and Housing* (U.S. Bureau of the Census), give the percentage of single-family houses built before 1940 in 40 selected metropolitan areas:

Abilene, TX	16.3	Albuquerque, NM	6.7
Anchorage, AK	1.1	Atlanta, GA	10.3
Baltimore, MD	27.7	Baton Rouge, LA	8.3
Bellingham, WA	27.4	Birmingham, AL	19.0
Boston, MA	50.6	Bradenton, FL	7.0
Buffalo, NY	43.4	Charleston, SC	11.9
Chicago, IL	34.7	Dallas, TX	8.6
Denver, CO	14.4	Detroit, MI	24.1
El Paso, TX	11.0	Flint, MI	21.7
Fort Myers, FL	3.9	Honolulu, HI	8.4
Houston, TX	6.9	Lafayette, IN	26.0
Las Vegas, NV	1.5	Los Angeles–Long Beach, CA	17.7
Miami-Hialeah, FL	7.3	Milwaukee, WI	32.1
Nashville, TN	14.6	New York, NY	47.8
Orlando, FL	6.4	Pensacola, FL	8.9
Portland, ME	47.0	Reno, NV	7.7
Roanoke, VA	21.8	St. Louis, MO	26.8
San Francisco, CA	34.7	Seattle, WA	20.5
Syracuse, NY	40.6	Tulsa, OK	15.9
Wheeling, WV	49.9	York, PA	38.4

Determine the mean percentage for the data.

2.10 The average salaries paid by major league baseball teams in 1985 and 1984 are shown here. (Source: Major League Players Association.)

AMERICAN LEAGUE			NATIONAL LEAGUE		
TEAM	1985	1984	TEAM	1985	1984
Yankees	$546,364	$458,544	Atlanta	$540,988	$402,689
Baltimore	438,256	360,204	Los Angeles	424,273	316,250
California	433,818	431,431	Cubs	413,765	422,193
Milwaukee	430,843	385,215	San Diego	400,497	311,199
Detroit	406,755	371,332	Philadelphia	399,728	401,476
Boston	386,597	297,878	Pittsburgh	392,271	330,661
Toronto	385,995	295,563	Mets	389,365	282,952
Kansas City	368,469	291,160	St. Louis	386,505	290,886

Oakland	352,004	384,027	Houston	366,250	382,991
White Sox	348,488	447,281	Cincinnati	336,786	269,019
Minnesota	258,039	172,024	San Francisco	320,370	282,132
Texas	257,573	247,081	Montreal	315,328	368,557
Cleveland	219,579	159,774			
Seattle	169,694	168,505			

a. Determine the mean salary for each league for each year and for all major league teams for each year.

b. Determine the median salary for each league for each year and for all major league teams for each year.

c. Which measure—mean or median—do you think is a better measure of central tendency in this case?

2.2 Measures of Variability

When we determine the approximate center of a distribution, we have only a partial description of it. We can describe a distribution more fully by obtaining a measure of its spread. This kind of measure tells us whether the values in the distribution cluster closely about the center or stretch out in one or both directions. A measure of variability that is relatively small in terms of the unit of measurement indicates that the data cluster closely about the mean; a larger value indicates that the data set is spread out.

The simplest measure of variability, discussed in Chapter 1, is the range, which is the distance between the highest and lowest values of the data. The range is not a resistant measure, however. If the data contain a few extreme scores, or only one, the range can give a misleading impression of the spread of the scores. Furthermore, two data sets may have similar mean and range and be quite different in amount of variability. Each of the histograms in Figure 2.2 shows a distribution that has a mean of 3.5 and a range of 6, but they are quite different.

The distribution on the right is "heavier" in the vicinity of the mean. In fact, 23 of the 36 observations are 3 or 4, so 23/36 of the graph lies between 2.5 and 4.5, or within one unit of the mean. In contrast, only 4/36 of the

Figure 2.2
Histograms of distributions with a mean of 3.5 and a range of 6.

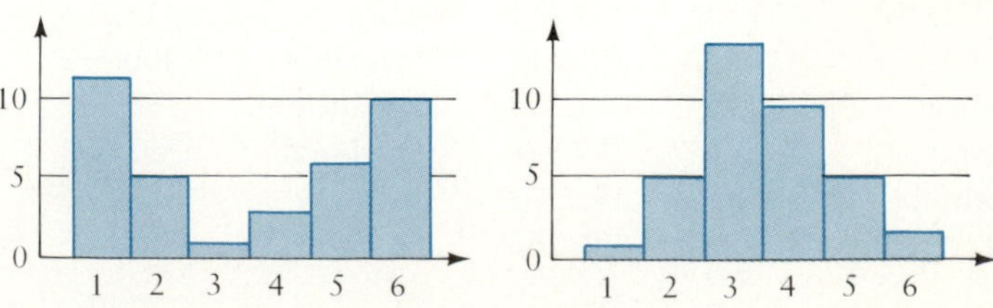

distribution on the left lies in the same interval. Thus the data whose distribution is on the left are more variable than the data whose distribution is on the right. This variability cannot be seen by reference to the mean or the range, however. We need a more specific measure of variability.

2.2.1 The Variance and Standard Deviation

In order to investigate the variability of a set of data about the mean, we need to discuss the *deviation* of each observation from the mean. If an observation has a value x and the mean of the population is μ, then the deviation of the observation from the mean is $x - \mu$. Because of the way we obtain the mean, the deviations above the mean and the deviations below the mean total to the same numbers but with opposite signs. Thus the sum of all the deviations from the mean is zero. For data that are clustered closely about the mean, most of the deviations have small absolute values; for data that are more variable, there are more deviations with large absolute values. The average of these absolute values, called the *mean deviation,* provides one measure of variability but it has limited applicability. A more useful measure is called the *median absolute deviation,* which is the median of the absolute values of deviations from the median. (See Mosteller [1977, p. 207].) The most widely used measures of variability are the *variance* and *standard deviation,* which we define here. The **variance** of a population is the mean of the squared deviations of each data point. The symbol used is σ^2, where σ is the lowercase Greek letter sigma. The variance of a population is defined mathematically as follows.

Variance of a Population

The variance of a population with mean μ and containing N pieces of data is given by

$$\sigma^2 = \frac{\Sigma(x - \mu)^2}{N}$$

If the variance is large, the variability of the data set is great; if the variance is small, the variability of the data set is small.

Most often we are interested in the variance of a sample because we have a sample of data rather than a population census. We denote the variance of a sample by s^2, and we measure the deviations from the sample mean rather than the population mean; that is, each deviation is $x - \bar{x}$. The sample variance is often used primarily to estimate the population variance. If we divide $\Sigma(x - \bar{x})^2$ by n to obtain a sample variance, we obtain a number that does not provide what is technically known as an *unbiased* estimate of σ^2, the population variance. An estimate for a population parameter is unbiased if the mean of all such estimates is equal to the parameter. We can show that a sample variance is an unbiased estimate of the variance of the population from which the sample was taken if we divide the sum of the squares of the deviations by $n - 1$ rather than n. We thus have the following definition of a sample variance.

Variance of a Sample

The variance of a sample with mean $\bar{x}$ and containing n pieces of data is defined by

$$s^2 = \frac{\Sigma(x - \bar{x})^2}{n - 1}$$

The formula given above is known as a *defining formula* for the sample variance. We will encounter defining formulas frequently in later chapters. Often the defining formula is used primarily to define a measure and rarely to calculate its value. In such a case we will give a *computational formula*—a mathematical shortcut formula generally used to calculate the value of the measure. We can show that the sum of the squares of the deviations from the mean can be calculated by using Σx^2 (the sum of the squares of the individual data points) and Σx.

Shortcut Formula for Sample Variance

The variance of a sample of n observations is given by

$$s^2 = \frac{\Sigma x^2 - \dfrac{(\Sigma x)^2}{n}}{n - 1}$$

Note that Σx^2 is the sum of the squares of the data values, while $(\Sigma x)^2$ is the square of the sum of the data values.

The numerator of the formula for the sample variance, defined as $\Sigma(x - \bar{x})^2$ and usually calculated as $\Sigma x^2 - (\Sigma x)^2/n$, is the *sum of squares of deviations from the sample mean for the variable x*. We usually shorten this term to the **sum of squares for x** and use the notation SSX. This term plays an important role in the later chapters on statistical inference.

Example 2.7

Find the variance of this sample data set: 1, 3, 4, 7, 9, 12.

Solution

The sum of the data is 36; the sum of the squares is 1 + 9 + 16 + 49 + 81 + 144, or 300. The number of observations is 6. Thus we have $\Sigma x^2 = 300$, $\Sigma x = 36$, and $n = 6$. Then

$$\Sigma x^2 - (\Sigma x)^2/n = 300 - (36)^2/6 = 300 - 216 = 84$$

and the sample variance is $84/(6 - 1) = 84/5 = 16.8$. If we use the defining formula, we get the same result.

The positive square root of the variance is called the **standard deviation.** It has an advantage over the variance in certain situations in that it is expressed in the same units as the data, since taking the square root of the variance "reverses" the prior squaring of the deviations. The standard deviation of a population is symbolized by σ, and the standard deviation of a sample is symbolized by s. For Example 2.7, $s = \sqrt{16.8}$ so that $s \doteq 4.1$.

Standard Deviation of a Sample

The standard deviation of a sample of n observations is given by

$$s = \sqrt{\frac{\Sigma(x - \bar{x})^2}{n - 1}} = \sqrt{\frac{\Sigma x^2 - \frac{(\Sigma x)^2}{n}}{n - 1}}$$

Example 2.8

Determine the variance and standard deviation of the following sample: 234,567; 234,517; 234,544; 234,548; 234,561; 234,572.

Solution

When we square these numbers to obtain Σx^2, the resulting numbers may overflow a calculator. But if we add or subtract the same number to or from every piece of data in a data set, the variance and standard deviation are *unchanged.* Thus the variance of the data given here is the same as the variance of the numbers 67, 17, 44, 48, 61, 72. For these numbers, $\Sigma x = 309$, $\Sigma x^2 = 17{,}923$, and $n = 6$, so

$$s^2 = \frac{17{,}923 - (309)^2/6}{5}$$

$$= \frac{2009.5}{5}$$

$$= 401.9$$

and $s = \sqrt{401.9} \doteq 20.05$.

Example 2.9

The ages of 25 people in a certain income bracket are distributed as follows:

Age	29	33	37	38	39	40	42	43	45	47	50	59	66
Frequency	1	1	3	4	2	3	2	2	3	1	1	1	1

Obtain the variance and standard deviation for the sample.

Solution

We apply the same formula as before, but we must remember to count each value of the variable as often as it occurs. In Example 2.2, we found that $\Sigma x = 1050$ for this set of data. We calculate Σx^2 as follows:

$$\Sigma x^2 = 29^2 + 33^2 + 3(37^2) + 4(38^2) + 2(39^2) + 3(40^2) + 2(42^2) + 2(43^2) + 3(45^2) + 47^2 + 50^2 + 59^2 + 66^2 = 45{,}502$$

Then

$$s^2 = \frac{45{,}502 - (1050)^2/25}{24} = \frac{45{,}502 - 44{,}100}{24} = \frac{1402}{24}$$

or

$$s^2 \doteq 58.42 \qquad \text{and} \qquad s \doteq 7.64$$

Comment: Most hand-held calculators have a function that determines the variance or standard deviation (or both) after the data have been entered. The user of such a calculator should check to make sure whether the sum of the squared deviations is divided by n or by $n - 1$. At one time, a Texas Instruments calculator had two functions, sd and var; sd gave the standard deviation using the formula for a sample standard deviation (dividing by $n - 1$), while var gave the variance using the formula for population variance (dividing by n). Several Sharp calculators had keys marked σ and s for population and sample standard deviation, respectively.

If a calculator does not have one of these functions, a simple calculator with one memory can be used to obtain the mean, variance, and standard deviation without writing any intermediate calculations on paper. Most calculators have square and square root keys. Here is a procedure you can often use to obtain the variance. First add the squares of the data values (multiplying each by its frequency if appropriate) and store the sum (Σx^2) in the memory. Then add the data values to obtain Σx. If you want to know $\bar{x}$, you can divide Σx by n to obtain $\bar{x}$ and then remultiply by n to display Σx. Square this number and divide by n; then change the sign and add the number in memory (Σx^2). This gives you SSX. Divide this by $n - 1$ to obtain s^2.

PROFICIENCY CHECKS

8. Determine the variance and standard deviation for the sample with values 3, 5, 5, 6, 8. Use the defining formula and the shortcut formula, and compare the efficiency of the methods.
9. Determine the standard deviation of the following sample of Nielsen ratings: 20.6, 11.3, 13.7, 9.2, 18.1, 7.2. Use the defining formula and the shortcut formula, and compare the efficiency of the methods.
10. A random sample of five accounts in a department store shows the following balances at the end of the month: \$67.32; \$108.97; \$27.64; \$412.11; \$81.96. Compute the standard deviation.
11. Determine the standard deviation of the following sample: 23,887; 23,889; 23,894; 23,913; 23,922
12. Refer to Proficiency Check 2. Determine the variance and standard deviation for the data.
13. Refer to Proficiency Check 5. Determine the variance and standard deviation for the data.

2.2.2 Chebyshev's Theorem and the Empirical Rule

The standard deviation and the mean give an amazingly accurate picture of the actual spread of the data. It has been proved that no matter what the distribution looks like, for any data set, population or sample, at least three-fourths of all pieces of data lie within two standard deviations of the mean, at least eight-ninths lie within three standard deviations of the mean, and, in general, at least $1 - 1/k^2$ of the data lie within k standard deviations of the mean, where k is any number equal to or greater than 1. This fact was proved by Pafnuty Chebyshev (1821–1894) and is called **Chebyshev's Theorem.** If the mean of a set of data is 24.8 and the standard deviation is 5.3, for example, then at least 75% of all the observations lie between 24.8 − 2(5.3) and 24.8 + 2(5.3); that is, at least 75% lie between 14.2 and 35.4. It is possible that a higher proportion of the data set lies between these limits, but Chebyshev's Theorem guarantees that *at least* 75% is between these extremes. Similarly, at least eight-ninths, or about 89% or more, of the data set lies between 8.9 and 40.7—that is, within three standard deviations of the mean. Furthermore, at least fifteen-sixteenths (93.75%) of the observations lie within four standard deviations of the mean, or between 3.6 and 46.0. Chebyshev's Theorem can be stated as follows.

Chebyshev's Theorem

Regardless of the shape of a data set's frequency distribution, the proportion of observations falling within k standard deviations of the mean is at least

$$1 - \frac{1}{k^2}$$

provided $k \geqslant 1$.

The proportions given by Chebyshev's Theorem are an *absolute minimum.* Most distributions have a higher percentage of the data included within these intervals. Often data collected for statistical analysis have a **mound-shaped distribution;** that is, they have a relatively symmetric distribution about the mean, tapering off in both directions. A certain type of mound-shaped distribution is called the **normal** (or *bell-shaped*) **distribution,** pictured in Figure 2.3.

Normal distributions have been investigated extensively, and it has been found that 68.26% of the area under a normal curve lies within one standard deviation from the mean, 95.44% of the area under the curve lies within two standard deviations from the mean, and 99.7% of the area under the curve lies within three standard deviations of the mean. These percentages apply to a distribution that is precisely normal in shape. Most sets of sample data do not have *exactly* normal distributions, but a mound-shaped distribution that is symmetric about the mean is said to be "approximately normal" in shape so long as the data do not extend too far. For such a distribution the *empirical rule* applies. Since the area under the curve corresponds to 100% of the data, the percentage of area under the curve between any two values on the horizontal axis corresponds to the percentage of the data between these two values. If 45% of the area under the graph of a set of

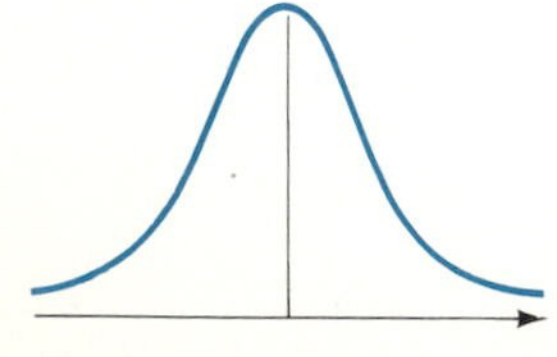

Figure 2.3
The normal or bell-shaped distribution.

data lies above 15.7 and below 28.3 then 45% of the observations have values between 15.7 and 28.3. The empirical rule is a generalization of this idea to normally distributed data.

According to the **empirical rule,** about 68% of the area under an approximately normal distribution lies within one standard deviation of the mean, about 95% of the area under the curve lies within two standard deviations of the mean, and nearly all of the distribution lies within three standard deviations of the mean. The empirical rule can be stated as follows.

The Empirical Rule

If a distribution of sample data is approximately normal, with mean $\bar{x}$ and standard deviation s, then approximately

68% of the data lie between $\bar{x} - s$ and $\bar{x} + s$.
95% of the data lie between $\bar{x} - 2s$ and $\bar{x} + 2s$.
Nearly all of the data lie between $\bar{x} - 3s$ and $\bar{x} + 3s$.

Although Chebyshev's Theorem applies to a distribution of any shape, the empirical rule is somewhat sensitive to deviations from normality. The empirical rule implies that the range of approximately normal data will be about six standard deviations, and this is accurate for very large, normally distributed data sets. Thus the standard deviation for such data sets will be about one-sixth of the range. For smaller data sets and nonnormal data, we still know that most of the data will fall within two standard deviations of the mean, so the range will be closer to four standard deviations, and the standard deviation will be about one-fourth of the range. As a general rule, we can say that the standard deviation of most data sets is between one-fourth and one-sixth of the range. If our calculations yield a standard deviation far outside these limits, we should check thoroughly for a possible error.

Some data sets appear to be approximately normal but have long tails—that is, extreme values on both ends of the distribution are farther from the mean than if the data were truly normal. For such data sets the mean is a good measure of central tendency, but the standard deviation is not a good measure of dispersion. See Section 2.4 for further discussion of long-tailed distributions. Short-tailed distributions do not pose a problem. Huber (1981, p. 4) points out that if a data set is short-tailed rather than normal, there is a negligible effect on the usefulness of the usual descriptive measures.

2.2.3 Pearson's Index of Skewness

Some data sets are *skewed;* that is, they have extreme values on one end only. If a distribution is highly asymmetric—if most of the values are bunched at one end with the distribution trailing off at the other end—we say that

PROFICIENCY CHECKS

14. A set of data has a mean of 40 and a standard deviation of 8. At least how much of the data will lie between:
    a. 24 and 56?
    b. 16 and 64?
    c. 32 and 48?
15. A set of data is approximately normally distributed with a mean of 40 and a standard deviation of 8. Approximately how much of the data will lie between:
    a. 24 and 56?
    b. 16 and 64?
    c. 32 and 48?
16. An approximately normal set of data has a mean of 40 and a standard deviation of 8. Approximately how much of the data will be:
    a. above 24?
    b. above 56?
    c. below 32?
    d. below 48?

the distribution is **skewed.** A distribution is *positively skewed* if the trailing end is off to the right and the mean is greater than the median; it is *negatively skewed* if the trailing end is off to the left and the mean is smaller than the median. Figure 2.4 shows two skewed distributions.

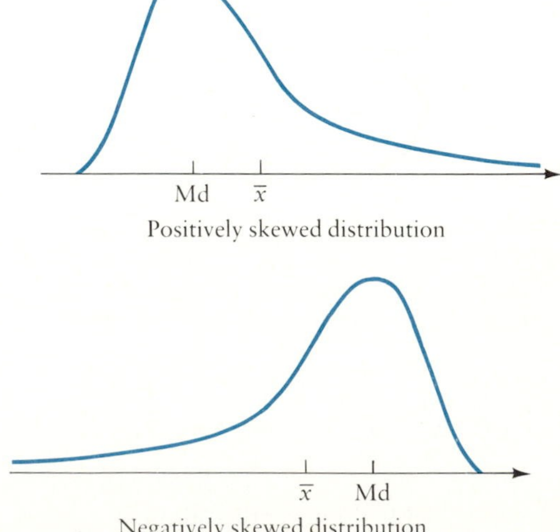

Figure 2.4
Graphs of positively and negatively skewed distributions.

Distributions that are bounded at one end and unbounded, or practically unbounded, at the other end are often skewed. Income is such an example. Income is bounded below at zero but generally has a very high upper end, far above the median.

Even with a skewed distribution, however, the mean and standard deviation are still good measures of central tendency and variability, respectively, unless the data are very greatly skewed. Karl Pearson (1857–1936) introduced an index of skewness to determine the amount of skewing in a distribution. He considered the amount of skewing to be significant if the difference between the mean and the mode is more than one-third of the standard deviation, although he stressed that this was not a hard-and-fast rule. A modification of his approach gives the following definition of **Pearson's Index of skewness.**

Pearson's Index of Skewness

If a distribution of data has mean $\overline{x}$, median Md, and standard deviation s, Pearson's Index of skewness is given by

$$I = \frac{3(\overline{x} - \text{Md})}{s}$$

If the index is 1.00 or greater or -1.00 or smaller, the data set is said to be *significantly skewed* and the mean and standard deviation are not valid measures of central tendency and variability, respectively. Using $+1$ and -1 as dividing lines for skewing was Pearson's choice, but the choice is quite arbitrary. If I is not zero, there is some skewing; the only question is how much is too much. The sign of the index gives the direction of skewing. If I is positive, the data set is positively skewed; if I is negative, the data set is negatively skewed. Pearson's Index should be used as one means of assessing whether or not a data set is skewed. As we will note in Section 2.4, the box plot and z-scores are also useful in determining whether or not a data set is skewed.

If the data set is badly skewed, the median should be used to measure the central tendency of the distribution. The appropriate measure of variability, called the *interquartile range,* is discussed in Section 2.3.

PROFICIENCY CHECK

17. The mean, median, and standard deviation for each of several data sets are given here. In each case determine whether the

data set is significantly skewed and, if so, whether it is negatively or positively skewed.

a. $\overline{x} = 31.7$, Md $= 32.8$, $s = 1.4$
b. $\overline{x} = 1437.2$, Md $= 1317.9$, $s = 415$
c. $\overline{x} = 1.877$, Md $= 1.912$, $s = 0.137$
d. $\overline{x} = 55.46$, Md $= 49.88$, $s = 16.10$
e. If $\overline{x} =$ Md, what shape does the distribution have?

Problems

Practice

Determine the variance and standard deviation of the data sets in Problems 2.11 to 2.13.

2.11 20, 22, 23, 26, 29, 30

2.12 1.34, 1.69, 1.78, 1.89, 2.03, 2.27, 2.39, 2.88, 3.16, 3.34, 4.92, 5.57, 6.83, 7.44, 9.63, 11.82

2.13

x	17	18	19	20	21	22	23
Frequency	2	4	8	11	14	7	3

2.14 Determine the mean and standard deviation of the data in this frequency distribution.

CLASS	FREQUENCY
15–24	9
25–34	14
35–44	22
45–54	28
55–64	17
65–74	8
75–84	3

2.15 A distribution of data has a mean of 114.7 and a standard deviation of 6.6. At least what proportion of the data lies between:
a. 101.5 and 127.9?
b. 94.9 and 134.5?
c. 88.3 and 141.1?

2.16 A normal distribution of data has a mean of 114.7 and a standard deviation of 6.6. About what proportion of the data lies:
a. between 101.5 and 127.9?
b. between 94.9 and 134.5?
c. between 108.1 and 121.3?
d. above 94.9?
e. below 108.1?
f. above 127.9?
g. below 134.5?

Applications

2.17 Intelligence tests report scores that have been standardized by being given to a large representative sample. A certain IQ test has scores that are normally distributed with a mean of 100 and a standard deviation of 15. If this test were given to a very large sample, what percentage of the IQ scores would be:

a. between 85 and 115?
b. between 70 and 130?
c. between 55 and 145?
d. above 70?
e. below 85?
f. above 115?
g. below 130?

2.18 A biologist was keeping careful records of the time required, measured to the nearest minute, for a spore culture to double in a petri dish. The number of minutes required for each of 25 determinations is listed here:

57	59	58	61	56	63	57	69	
59	59	57	63	72	59	62	60	
59	63	59	65	57	61	58	59	63

Determine the variance, standard deviation, and Pearson's Index of skewness for the data. Interpret the results.

2.19 Determine the mean and standard deviation of the following Nielsen ratings for "The Cosby Show" taken from several weekly ratings during the 1985–1986 season:

38.5 33.6 37.3 35.6 34.9 36.7

2.20 Refer to the income data of Problem 2.8.

a. Obtain the standard deviation of the data.
b. Assume that the data are spread out evenly within a subinterval. What proportion of the data lies between $\bar{x} - s$ and $\bar{x} + s$?
c. What proportion of the data lies between $\bar{x} - 2s$ and $\bar{x} + 2s$?
d. What proportion of the data lies between $\bar{x} - 3s$ and $\bar{x} + 3s$?
e. Based on these results, do you think the data set is normally distributed?

2.21 According to columnist Lew Sichelman (*Pensacola News Journal,* Feb. 15, 1986), the average contract price for new and existing houses was \$106,300 nationally as of Dec. 31, 1985. A real estate salesman had houses listing at the following prices: \$56,900, \$125,000, \$78,500, \$99,995, \$87,500, \$72,950, \$88,900, \$175,000, and \$71,400. Determine the mean and standard deviation of these prices.

2.22 Refer to the housing data of Problem 2.9.

a. Obtain the standard deviation of the data.
b. What proportion of the data lies between $\bar{x} - s$ and $\bar{x} + s$?
c. What proportion of the data lies between $\bar{x} - 2s$ and $\bar{x} + 2s$?
d. What proportion of the data lies between $\bar{x} - 3s$ and $\bar{x} + 3s$?
e. Based on these results, do you think the data set is normally distributed?

2.3 Other Descriptive Measures

The mean, median, and standard deviation give us a great deal of information about a data set, but they do not tell us anything about an individual observation. We frequently need to be able to determine the relation of a particular observation to the rest of the data; that is, we need to determine the position or standing of that observation relative to the rest of the data. We may also use measures of relative standing to compare observations in different data sets describing similar variables.

2.3.1 The z-Score

We can use the mean and standard deviation to determine the position of the value of an observation relative to the rest of the data. The measure we use to do this is called a **z-score,** which is usually given to two decimal places and is defined as follows.

The z-Score

The z-score of an observation with value x taken from a population with mean μ and standard deviation σ is given by

$$z = \frac{x - \mu}{\sigma}$$

The sample z-score of an observation with value x taken from a sample with mean $\bar{x}$ and standard deviation s is given by

$$z = \frac{x - \bar{x}}{s}$$

The value of z gives the position of the observation relative to the remainder of the data; it gives the number of standard deviations the observation is above the mean (or below the mean, if negative). The z-score provides a convenient way of interpreting some kinds of data. Suppose, for example, a man's blood test shows that he has a cholesterol level of 320. This piece of information is meaningless by itself. But suppose we know that the average level of blood cholesterol for adult men is 200 with a standard deviation of 60. This man's cholesterol level has a z-score of $z = (320 - 200)/60 = 2.00$. Thus this man's cholesterol level is two standard deviations above the mean. If cholesterol levels are approximately normally distributed (they are), then according to the empirical rule only 2.5% of the adult male population have cholesterol levels higher than his. Even if we did not know anything about the shape of the population, Chebyshev's Theorem would give us an approximation. We know that at least 75% of the population has cholesterol levels within two standard deviations of the mean—that is, with z-scores between -2.00 and $+2.00$—so 25% have z-scores above $+2.00$ or below -2.00. If the distribution is nearly symmetric, we can assume that about 12.5%, more or less, will have a standard score above $+2.00$, so this man's cholesterol level would be above about 87.5 percent of the adult male population. In either case, his level is extreme.

Another use for the z-score is to compare values measuring similar quantities obtained on different measuring instruments or from different samples. If two students take different IQ tests, for example, their scores are not directly comparable. Suppose John scores 112 on IQ test A while Jane scores 118 on IQ test B. Does this mean that Jane has a higher IQ than John? If IQ test A has a mean of 98 with a standard deviation of 16 and IQ test B has a mean of 104 with a standard deviation of 18, we can obtain z-scores for each one. If we denote John's z-score by z_1 and Jane's by z_2, then $z_1 = (112 - 98)/16 \doteq 0.88$ and $z_2 = (118 - 104)/18 \doteq 0.78$. Therefore John's score is higher relative to his test than Jane's is relative to hers. This result does not mean, however, that John's IQ is higher than Jane's. On a retest the results would probably not be the same. All we can do at this point is compare the z-scores.

We can also use the definition of z-score to determine the value of the observation with a given z-score. If a data set has a mean of 115.7 and a standard deviation of 13.6, for example, what value has a z-score of 2.30? Let x represent the value of the observation. Then

$$\frac{x - 115.7}{13.6} = 2.30$$

or

$$\begin{aligned} x - 115.7 &= 2.30(13.6) \\ &= 31.28 \\ &\doteq 31.3 \end{aligned}$$

so that

$$\begin{aligned} x &\doteq 31.3 + 115.7 \\ &\doteq 147.0 \end{aligned}$$

The z-score is an example of a *standard score*. A standard score is any score in which a set of values is transformed in order to obtain a set of scores with a predetermined mean and standard deviation. The z-score gives the number of standard deviations a certain value is above or below the mean; data transformed using the z-score formula have a mean of zero and a standard deviation of 1. A standard score widely used in educational and psychological testing is obtained by first computing the z-score and then multiplying by 10 and adding 50. The result is often rounded to the nearest whole number. This process results in a set of scores with mean 50 and standard deviation 10. Such scores are directly comparable. John's standard score for the IQ test would then be equal to $50 + 10(0.88) \doteq 59$, and Jane's would be $50 + 10(0.78) \doteq 58$. Data values can be standardized to any scale. Standard scores are not used in statistical inference, so they are not pursued in this text.

2.3.2 The Coefficient of Variation*

Suppose we want to compare the variability of two sets of data. One means of comparison is to compute the **coefficient of variation,** defined here.

Coefficient of Variation

The coefficient of variation for a population with mean μ and standard deviation σ is defined by

$$V = \frac{\sigma}{\mu} \cdot 100$$

The coefficient of variation for a sample with mean $\bar{x}$ and standard deviation s is defined by

$$V = \frac{s}{\bar{x}} \cdot 100$$

This measure expresses the standard deviation of a sample as a percentage of the mean. Highly variable data sets have relatively large values of V, while relatively small values of V indicate less variability. If the mean equals zero, V cannot be computed. We can use the coefficient of variation to compare the variability of two data sets, as it is a pure number without units.

Suppose, for example, we wish to purchase one of two stocks. Stock A is priced at \$80 per share and stock B is priced at \$20 per share. If we wish to minimize our risk we might wish to buy the stock that is less variable; on the other hand, we may wish to take a chance that the stock will go up quite a bit and would therefore want the stock that is more variable. We might take a sample of the price of each stock at the close of trading for the past few months and conclude that the mean price of stock A has been \$75.88 per share with a standard deviation of \$18.44, while the mean price of stock B has been \$21.87 per share with a standard deviation of \$11.55. The coefficient of variation for stock A, which we might express as V_A, is

*This section may be omitted without loss of continuity.

(18.44/75.88)(100) ≐ 24.30, while V_B = (11.55/21.87)(100) ≐ 52.81. Thus stock B is more than twice as variable as stock A.

PROFICIENCY CHECKS

18. Determine the z-scores for the given value of x if the mean of the data set is 30 and the standard deviation is 6.
    a. $x = 36$ b. $x = 39$ c. $x = 27$ d. $x = 24$ e. $x = 30$ f. $x = 19$ g. $x = 45$ h. $x = 4$ i. $x = 28$ j. $x = 61$

19. Determine the z-scores for the given value of x if the mean of the data set is 1235.7 and the standard deviation is 185.9.
    a. $x = 1431.5$ b. $x = 1000$ c. $x = 1143.5$ d. $x = 1655.4$ e. $x = 453.9$ f. $x = 1288.3$ g. $x = 2000$ h. $x = 750.5$

20. Which of the two distributions given in Proficiency Checks 18 and 19 is more variable?

21. Determine the value of x that has each of the following z-scores (to the nearest tenth) if the mean of the data set is 11.6 and the standard deviation is 1.8.
    a. $z = 1.15$ b. $z = -0.61$ c. $z = 3.16$ d. $z = -2.09$ e. $z = -0.13$ f. $z = -4.62$ g. $z = 1.51$ h. $z = 2.18$

2.3.3 Quartiles and the Interquartile Range

Yet another measure of relative standing is one that divides the data set into equal sections. The median is such a measure, as it divides the data into two equal parts. *Deciles* divide a data set into 10 equal parts, *percentiles* divide a data set into 100 equal parts, and *quartiles* divide the data set into 4 equal parts. Measures of this type are called **quantiles.** The most widely used quantiles, other than the median, are percentiles and quartiles. For purposes of statistical analysis the most useful and widely used of these are the quartiles, which we will discuss here.

The *first* **quartile,** Q_1, is the smallest value that exceeds one-fourth of the data; the *third quartile,* Q_3, is the smallest value that exceeds three-fourths of the data. The second quartile is the median, and there is no fourth quartile. Since the data are usually discrete, there is often no exact value for the quartiles. Koopmans (1981, p. 55) has suggested a simplified procedure that is exact enough for most purposes. First we obtain the position of the median; this is equal to $(n + 1)/2$, where n is the number of pieces of data. Then we can find the position of the quartiles by the formula

$$\text{Quartile position} = \frac{[(n + 1)/2] + 1}{2}$$

where [ ] is the **greatest integer function** (or bracket function); $[x]$ is the greatest integer not greater than x. That is, if x is positive, $[x]$ is the integer part of the number when written as a decimal number. If n is an integer, $(n + 1)/2$ is either an integer or an integer plus 0.5, so $[(n + 1)/2]$ is the integer without the 0.5. The resulting number gives us the position of the quartiles. We obtain the first quartile by counting in from the bottom of the data; we find the third quartile by counting in from the top of the data.

If the data set contains 54 pieces of data, for instance, the median position is (55/2) or 27.5; the median is between the twenty-seventh and twenty-eighth pieces of data. Thus $[27.5] = 27$, $27 + 1 = 28$; and half of 28 is 14, so Q_1 is the fourteenth piece of data from the bottom and Q_3 is the fourteenth piece of data from the top. If there are 55 pieces of data, the median position is (56/2) or 28; the median is the twenty-eighth piece of data. Half of 29 is 14.5, so Q_1 is halfway between the fourteenth and fifteenth pieces of data from the bottom and Q_3 is halfway between the fourteenth and fifteenth pieces of data from the top.

If the mean is the appropriate measure of central tendency for a set of data, the standard deviation is the appropriate measure of dispersion. If the median is the appropriate measure of central tendency for a data set (a significantly skewed data set, for example), the appropriate measure of dispersion is called the **interquartile range.** The interquartile range (**IQR**) is defined as follows.

The Interquartile Range

If Q_1 and Q_3 are the first and third quartiles, respectively, for a data set, the interquartile range is defined by

$$\text{IQR} = Q_3 - Q_1$$

The interquartile range, like the standard deviation, is a measure of dispersion. Relatively large values of IQR indicate relatively large variability; smaller values indicate less variability.

Example 2.10

A set of data is given here. Determine the values of Q_1, Q_3, and IQR.

3, 4, 5, 6, 6, 6, 7, 7, 7, 7, 7, 8, 8, 8, 8, 9, 9, 10, 11, 13, 15, 16, 16, 18, 19, 20, 21, 22, 22, 23, 23, 23, 23, 25, 26, 28, 29, 30, 31, 33, 33, 34, 35, 36, 39, 41, 44, 47, 49, 53, 58, 59, 63, 67, 75, 81

Solution

There are 56 pieces of data. One-half of 57 is 28.5, so the position of the median is between the twenty-eighth and twenty-ninth pieces of data. Half of 29 is 14.5, so Q_1 is between the fourteenth and fifteenth pieces of data from the bottom and Q_3 is between the fourteenth and fifteenth pieces of data from the top. Counting from the bottom, we find that $Q_1 = 8$. Counting from the top, we find that Q_3 is between 34 and 35, or 34.5. Then IQR $= 34.5 - 8 = 26.5$.

2.3.4 The Box Plot

The median and quartiles, together with the maximum and minimum of a distribution, give us five numbers that characterize a distribution reasonably well. A listing of these numbers is called a **five-number summary.** One visual representation of these numbers is called a **box plot** (or *box and whisker diagram*), which, like the stem and leaf plot, was introduced by Tukey (1977). The box plot may be drawn either vertically or horizontally. To draw a box plot vertically, we plot, on a vertical scale, the maximum and minimum values, the median, and the quartiles. Next we draw a short horizontal bar at the median and the quartiles, connecting them to form a box. We then draw a line from each end of the box to the extreme value. These lines are called **whiskers.** The result is a diagram like the one shown in Figure 2.5. A horizontal box plot may be drawn in a similar manner, exchanging the words "vertical" and "horizontal" in the foregoing discussion.

The box plot gives us a fairly clear picture of the data without constructing an elaborate graph. We see in Figure 2.5 that the median is about 180, the maximum about 197, and the minimum about 151. The quartiles are about 167 and 188. Thus about a quarter of the data are below 167, half are below 180, and a quarter are above 188. We see also that there is a slight negative skew to the data. If more data points are extreme, we usually draw a somewhat more elaborate version of the box plot (Section 2.4).

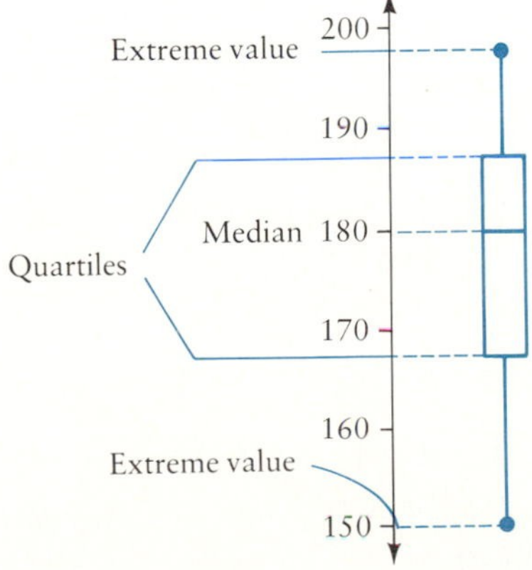

Figure 2.5
A vertical box plot.

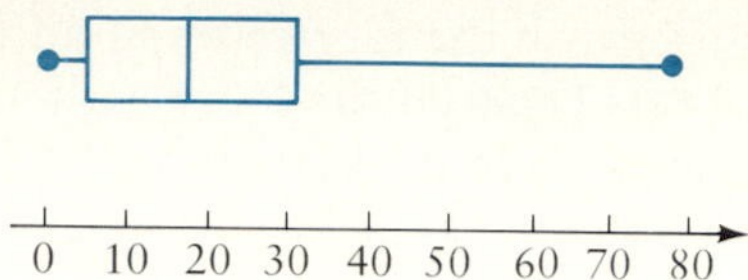

Figure 2.6
Box plot for Example 2.11.

Example 2.11

Draw a horizontal box plot for the data of Example 2.10.

Solution

The smallest value is 3, the largest value is 81, $Q_1 = 8$, and $Q_3 = 34.5$. We find the median to be 22, so we draw vertical bars at 8, 22, and 34.5 on a horizontal scale, connecting them to make a thin box. We then draw whiskers from the box to a dot at 81 and to a dot at 3 (see Figure 2.6). We can see in Figure 2.6 that the top fourth of the data is spread out much more than the remainder of the data, and we have a good picture of where each fourth of the data lies. This data set appears to be positively skewed.

We can obtain a box plot for grouped data by finding approximate values for the median and quartiles by assuming that the lowest value is the lowest class limit and the highest value is the highest class limit, and that the data in each class are evenly spaced.

PROFICIENCY CHECKS

Use the following data set for Proficiency Checks 22 to 24:

1.34, 1.69, 1.78, 1.89, 2.03, 2.27, 2.39, 2.88, 3.16, 3.34, 4.92, 5.57, 6.83, 7.44, 9.63, 11.82

22. Determine the first and third quartiles.
23. Determine the value of IQR, the interquartile range.
24. Construct a box plot for the data. What do you conclude?

Problems

Practice

2.23 A data set has a mean of 47 and a standard deviation of 12. Determine z-scores for data values of:
a. 17 b. 24 c. 33 d. 44 e. 53 f. 67 g. 81

2.24 A distribution has a mean of 156 and a standard deviation of 34. What is the value with a z-score of:
a. -1.63? b. -0.44? c. 0.76? d. 2.38?

2.25 Which of the two distributions in Problems 2.23 and 2.24 is more variable? Do you think the difference is important?

Use the data of Problem 2.3 for Problems 2.26 to 2.28.

2.26 Determine the first and third quartiles.

2.27 Determine the interquartile range.

2.28 Construct a box plot for the data. Does the data set appear symmetric?

Applications

2.29 Refer to the following IQ scores of 110 randomly selected high school students.
  a. Determine the median, quartiles, and IQR.
  b. Construct a box plot for the data.

154	131	122	100	113	119	121	128
128	112	133	119	115	117	110	104
125	85	120	135	93	103	103	122
109	147	103	113	107	98	128	93
90	105	118	134	89	113	108	122
85	108	106	116	151	117	110	80
111	127	81	96	114	103	126	119
122	102	99	106	105	111	127	108
104	91	123	122	97	110	150	103
87	98	108	117	94	96	111	101
118	104	127	94	115	101	125	109
91	110	97	115	108	109	133	107
115	83	102	116	101	113	112	82
114	112	113	102	115	123		

2.30 A subject was given three tests for potential leadership ability. The means and standard deviations of the scores for each test, determined by a reference sample, were as follows: test A, $\bar{x} = 4.20$, $s = 0.40$; test B, $\bar{x} = 160$, $s = 20$; test C, $\bar{x} = 36.2$, $s = 5.6$. The subject's scores for the three tests were as follows: test A, 4.38; test B, 168; test C, 37.6. Comparing these results to the reference samples, on which test was the score best? Worst?

2.31 Refer to the major league baseball player salaries of Problem 2.10.
  a. Obtain the mean and standard deviation for each of the four sets of data. (Since these are populations, we must obtain σ instead of s.)
  b. Construct the coefficient of variation for each of the data sets. Compare different leagues in the same year and different years in the same league. What do you conclude?
  c. Construct a box plot for each of the four data sets on the same axes.
  d. Compare the plots for the data sets. What do you conclude?

2.32 A 1985 survey of 534 videocassette recorder owners (*R. H. Bruskin Associates*) showed the following data about age of the owners:

Age	18–24	25–34	35–49	50–64	65+
Percent	18	29	31	14	8

Use the data to construct a box plot and determine the median and the interquartile range. Assume that the class 65 + is 65–85. Are any of your answers affected if this assumption is not accurate? What does this indicate about the resistance of these quantiles?

2.33 Two production lines are currently producing different items to be used in assembling a finished product. A certain amount of variability in the finished product is unavoidable, but products assembled from the most recent production runs are unacceptably variable. The problem has been traced to these two production lines. A sample of items from the most recent run of production line A is found to have a mean diameter of 0.55 millimeter with a standard deviation of 0.16 millimeter. A sample of items from the most recent run of production line B is found to have a mean diameter of 10.6 centimeters with a standard deviation of 0.8 centimeter. Which of the two lines is more likely to be the source of the problem?

2.4 Outliers and the Trimmed Mean

Most of the methods of traditional statistical inference assume that the data set is taken from a normal population. If this is not so, the mean and standard deviation no longer give a true picture of the data. Extreme values, called **outliers,** can cause problems. Outliers can either signal that the data set is not normal or that one or more of the observations do not belong with the rest of the data. The latter case can result from an error in measurement or recording, or the observation may have been placed in the data set in error. Since more often the outliers really do belong to the data set, this case causes the most concern. If the outliers are on only one end of the data, the data set is often skewed and the mean and the standard deviation are misleading measures of the set's central tendency and variability. If outliers are on both ends of the data set, the mean may well be a good measure of the center of the distribution, but the standard deviation may be so large that it overstates the variability of the data set. As mentioned before, the mean, variance, and standard deviation are not resistant measures and are therefore sensitive to the effect of outliers.

The first step in checking normality is to determine whether a data set has outliers and, if so, whether the outliers really belong to the data set. If the outliers cannot be attributed to recording errors, the distribution may be skewed (outliers on one end) or too long-tailed to be normal (outliers on both ends). This section presents some methods of detecting outliers and illustrates a way of compensating for lack of normality in calculating the mean and standard deviation of a data set.

2.4.1 Identifying Outliers

The most obvious outliers in a data set are those that lie more than three standard deviations from the mean. According to the empirical rule, virtually all the data in a normal distribution lie within three standard deviations of the mean. We can calculate the z-scores of the most extreme values in the data set; if the z-scores are greater than 3 or less than -3, those values are outliers. An observation with a z-scores of 8, for example, is an extreme value, and we should investigate further to make sure that value is correct. For data that are not normally distributed, Chebyshev's Theorem applies. According to this theorem, no more than 11 percent of the data in a data set can be more than three standard deviations from the mean. In such cases, too, observations with z-scores substantially greater than 3 or less than -3 can be considered outliers.

A second approach to detecting outliers utilizes the box plot. Most of the data in a normal distribution lie within one interquartile range (IQR) of the median. Thus if the median is, say, 80, and the IQR is 13, most of the data will lie between 67 and 93. There will be some variation, however, so we make the following distinctions introduced by Tukey (1977).

We first obtain the box plot for a set of data and then identify *fences*. The **inner fences** lie 1.5 IQR beyond the quartiles, and the **outer fences** lie 1.5 IQR beyond the inner fences. Data points lying between the inner and outer fences are said to be **mild outliers;** data points lying beyond the outer fences are considered **extreme outliers.**

When constructing a box plot for the purpose of detecting outliers, we make the diagram a bit differently. The box is constructed as before, but instead of placing a dot at the highest and lowest value, we place an × at the values inside the inner fences and nearest to them. These are called the **adjacent values.** Mild outliers are marked with open circles; extreme outliers are marked with solid circles. Such a box plot might look like the one in Figure 2.7. The fences are not usually marked on the plot but are included here for illustration. A plot such as this indicates that the data are probably not normal and may be skewed.

If there are outliers, but the distribution of data is reasonably symmetric, the data set is said to have **long tails.** This means that the graph of the population from which the data set is drawn is likely to be narrower around the center than the graph of a normal distribution, but with longer tails. Such a distribution has been called "peaked" in the past, but since this term calls attention to the center of the distribution, rather than to the tails (where the trouble lies), the term *long-tailed distribution* has come into general use. A graph of such a distribution is shown in Figure 2.8.

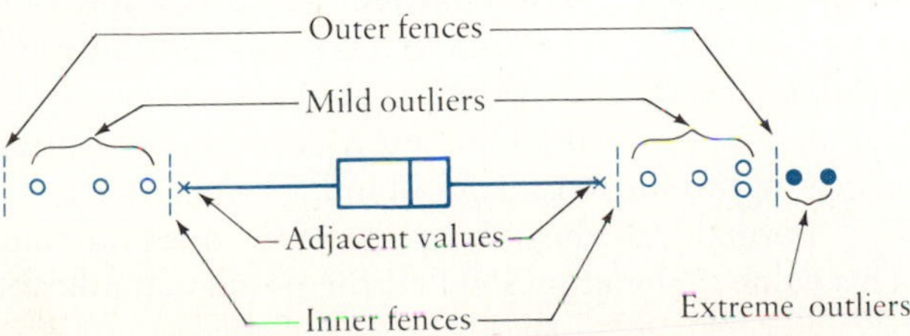

Figure 2.7
Procedure for detecting outliers by using the box plot.

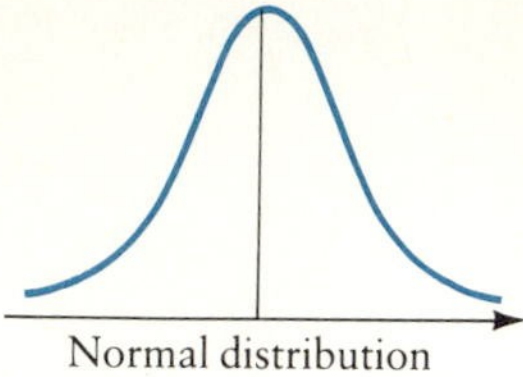

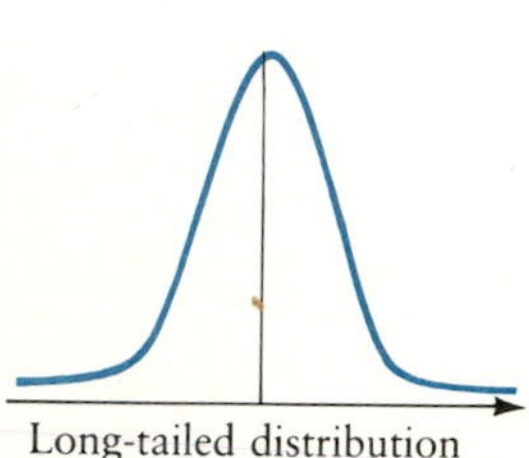

Figure 2.8
A comparison between normal and long-tailed distributions.

The matter of detecting outliers is crucial. The presence of a few outliers—or even one—can shift the mean or overstate the standard deviation to a marked degree. The procedure for detecting outliers is summarized here.

Using the Box Plot To Detect Outliers

To determine whether there are outliers (extreme values) in a data set, find the median, quartiles, and IQR. The inner fences lie at $Q_1 - 1.5 \cdot \text{IQR}$ and at $Q_3 + 1.5 \cdot \text{IQR}$; the outer fences lie at $Q_1 - 3 \cdot \text{IQR}$ and at $Q_3 + 3 \cdot \text{IQR}$. The adjacent values are those values between the quartiles and the inner fences and closest to the inner fences; mild outliers are those values between the inner and outer fences; extreme outliers are those beyond the outer fences. The box plot is constructed as usual except that the whiskers are drawn to the adjacent values (marked with an $\times$); the mild outliers are marked with open circles and the extreme outliers are marked with solid circles.

Example 2.12

Construct a box plot for the age data of Example 2.2 and test for outliers.

Solution

There are 25 pieces of data, so the median is located at observation number 13; that is, Md = 40. The quartiles are located at observation number 7 from each end; $Q_1 = 38$ and $Q_3 = 45$. Then IQR = 7; $1.5 \cdot \text{IQR} = 10.5$, so the inner fences are located at 27.5 and 55.5 and the outer fences are at 17 and 66. The adjacent values are 29 and 50, and the ages 59 and 66 are outliers. Since 66 is not beyond the outer fence, we will consider both observations to be mild outliers. A box plot for the data is shown in Figure 2.9. We can see that the data set is slightly positively skewed and is probably not normal.

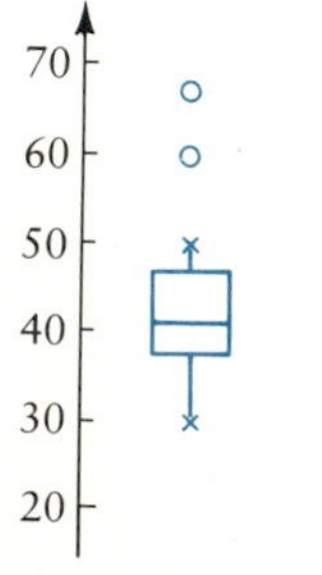

Figure 2.9
Box plot for the data of Example 2.13.

2.4.2 The Pseudo-Standard Deviation

There is a quick check we can use to determine whether the data set is normal. To do so, we must rely on a special extension of the empirical rule for normal data sets. In Chapter 5 we will learn how to determine the proportion of a normal distribution between any two values of the variable. It can be shown that 50% of the data in a normal distribution fall between approximately 0.675 standard deviation above the mean and 0.675 standard deviation below the mean. Thus for a normal distribution, the interquartile range is approximately 1.35 standard deviations. The standard deviation of a normal distribution is therefore approximately equal to IQR/1.35. This value is sometimes called the **pseudo-standard deviation.**

Pseudo-Standard Deviation of a Data Set

If a data set has interquartile range IQR, the pseudo-standard deviation of the data set is PSD, where

$$\text{PSD} = \frac{\text{IQR}}{1.35}$$

We can calculate the value of PSD quickly and compare it with s. If PSD is smaller than s, the data set probably has long tails; if it is larger than s, the data set may be short-tailed (bunched in the middle). The data set is normal only if PSD and s are fairly close and the data set is symmetric (not skewed).

Example 2.13

Using the data of Example 2.2, calculate PSD and s and compare the two. Draw conclusions and compare with the results of Example 2.12.

Solution

We found in Example 2.12 that IQR = 7. Then PSD = $7/1.35 \doteq 5.19$. In Example 2.9 we found that the standard deviation is approximately 7.64. Thus the standard deviation is considerably larger than PSD, and the data set is probably long-tailed.

2.4.3 Trimmed and Winsorized Data Sets

The power and comparative ease of using the mean and standard deviation have prompted statisticians to search for ways of adapting data sets so that they can use these two measures. C. P. Winsor once remarked that "all frequency distributions are normal in the middle" (Koopmans 1981, p. 58). This property is remarkably consistent, and since outliers are one cause of the problem of nonnormality, their removal may solve the problem. In the last 20 years or so, the process of removing outliers has been tested and found to lead to legitimate statistical results provided the distribution is symmetric. See, for instance, Tukey (1960) and Tukey and McLaughlin (1963). Koopmans (1981) has suggested removing 10% of the data from each tail, and the procedure seems to work well.

We can obtain the 10% **trimmed data set** by removing the smallest and largest 10% of the observations. If 10% of n is not an integer, we round up to the next integer to obtain the number to be removed. We can compute the mean and standard deviation of the trimmed sample as usual and denote them $\bar{x}(T)$ and $s(T)$. For some purposes a modification of this procedure works better. The **Winsorized data set** is obtained by replacing the trimmed values by the smallest and greatest remaining values; the mean and standard deviation of this sample are called the *Winsorized mean* and *Winsorized* *standard deviation*. Further uses for these measures will be found in later chapters.

The term *trimmed data set* means any data set that has had extreme values removed. There is no universal agreement on the percentage of values that should be removed; throughout this book we will use the term *trimmed* to mean that 10% of the data have been removed from each end unless it is specified otherwise.

Example 2.14

Use the data set of Example 2.2 and determine the trimmed mean and standard deviation. Compare with the PSD and decide whether the trimmed data set seems to be normal.

Solution

There are 25 pieces of data in the set. Since 10% of 25 is 2.5, we trim 3 data points off each end. A stem and leaf plot is useful to order the data if they are not already ordered. In this case we have the following stem and leaf plot:

Stem	Leaves
2	9
3	3
3	777888899
4	0002233
4	5557
5	0
5	9
6	
6	6

The underlined values are removed. The resulting trimmed data set has the following stem and leaf plot:

Stem	Leaves
3	77888899
4	0002233
4	5557

The trimmed mean and standard deviation are $\overline{x}(T) \doteq 40.84$ and $s(T) \doteq 3.10$. The quartiles are 38 and 43, so that IQR = 5; then PSD = $5/1.35 \doteq 3.70$, which agrees fairly well with $s(T)$.

Now in the original data set the inner fences are at 30.5 and 50.5 and the outer fences are at 23 and 58. In the original data set, then, 66 and 59 were mild outliers. The new data set has no outliers. The original data set had a mean of 42 and a standard deviation of 7.64. The median of both data sets is 40, so the mean and median are fairly close and the data set is reasonably symmetric, but a PSD of 5.19 compared to the standard deviation of 7.64 for the original data set indicates that this distribution had long tails. The revised data set is reasonably symmetric (although not quite) and has no outliers, so the procedure has achieved our goal.

PROFICIENCY CHECKS

Use the following data for Proficiency Checks 25 to 27:

1.34, 1.69, 1.78, 1.89, 2.03, 2.16, 2.27, 2.34, 2.39, 2.88, 2.92, 3.13, 3.36, 3.57, 3.67, 3.83, 5.44, 8.63, 11.82

25. Construct a box plot identifying outliers.
26. Obtain PSD and compare with s. What do you conclude?
27. Obtain the trimmed data set and $\bar{x}(T)$ and $s(T)$. Is the new data set more nearly normal?

Problems

Practice

Use the data of this table for Problems 2.34 to 2.36.

CLASS	FREQUENCY
111–115	1
116–120	2
121–125	4
126–130	11
131–135	9
136–140	14
141–145	7
146–150	6
151–155	3
156–160	1

2.34 Calculate the standard deviation and PSD. Do the data appear to be normal?

2.35 Obtain the trimmed data set, assuming the data points are located at the class mark. Calculate $\bar{x}(T)$ and $s(T)$.

2.36 Calculate PSD for the trimmed data set and compare with $s(T)$. Is the trimmed data set more nearly normal than the original data set?

Applications

A psychology student conducted an experiment dealing with reaction times and recorded the following times to the nearest hundredth of a second. Use the data for Problems 2.37 to 2.41.

0.48	0.96	0.52	0.36	0.49	0.58	0.64	0.44
0.39	0.59	0.37	0.66	0.57	0.68	0.64	0.81
1.02	0.54	0.38	0.67	0.74	0.77	0.52	0.68

0.55	0.67	0.34	0.67	0.74	0.81	0.91	0.66
0.54	0.68	0.83	0.52	0.37	0.56	0.90	0.47
0.54	0.69	0.58	0.55	0.52	0.94	0.39	0.57
0.66	0.43	0.68	0.71	0.42	0.60	0.58	0.74
0.55	0.49	0.58	0.64				

2.37 Make a stem and leaf plot for the data. Does the distribution appear to be approximately normal?

2.38 Make a box plot for the data and check for outliers.

2.39 Determine the variance and standard deviation. What is the PSD? Compare with s. What do you conclude?

2.40 Trim the upper and lower 10% off the data set. Obtain $\bar{x}(T)$ and $s(T)$. Compare with the results in Problem 2.39.

2.41 Obtain PSD for the trimmed data set and compare with $s(T)$. Is the trimmed data set more nearly normal than the original data set?

2.42 *Venture* magazine (November 1985) featured a story about a device called the Truant. The Truant contains an automatic telephone dialer that dials a parent in the evening, asks a series of questions, and records the answers or requests that the student bring a note to school. If the phone is busy it will call back. Absenteeism in school districts using the Truant has been reduced by 30% to 70%, according to the firm's president. To assess the need for such a device, a school district superintendent obtains a random sample of daily absentee figures for the past semester. The numbers of suspected truancies on each of 28 days are as follows:

132, 117, 143, 114, 125, 133, 197, 134, 113, 143, 121, 108, 131, 109, 117, 116, 84, 102, 153, 116, 98, 122, 127, 113, 111, 65, 122, 114

a. Make a stem and leaf diagram for the data.
b. Prepare a five-number summary of the data.
c. Construct a box plot and check for outliers. What do you conclude?

2.43 Refer to the housing data of Problem 2.9. Analyze them completely as in problems 2.37–2.41.

2.5 Computer Usage

We can use a computer to obtain many of the measurements presented in this chapter. Continuing our coverage of Minitab and the SAS System, we present some commands and printouts for this chapter's material.

2.5.1 Minitab Commands for This Chapter

We can calculate most of the measures of central tendency and dispersion discussed in this chapter using Minitab commands, particularly the DESCRIBE command. To obtain descriptive statistics for one or more variables entered as columns of data, for example, we may use the command

```
MTB > DESCRIBE the data in C, ..., C
```

The printed output we obtain by using this command will include the sample size, the mean, the median, a 5% trimmed mean, the standard deviation, the standard error of the mean ($s/\sqrt{n}$, discussed in Chapter 6), the maximum value, the minimum value, and the first and third quartiles. A 5% trimmed mean is similar to the 10% trimmed mean (Section 2.4) and is found by deleting the largest and smallest 5% of the data and calculating the mean for the remaining 90%.

We can also use Minitab to obtain a box plot for data stored in a certain column. The command

```
MTB > BOXPLOT the data in C1
```

will give a box plot for the data stored in column 1.

We cannot find the coefficient of variation automatically by using a Minitab command, but we can calculate it easily by using a sequence of commands. We are able to store constants such as the mean, median, and standard deviation of a set of data and then use them as necessary. To do so we use the symbol K; thus K1, K2, K3, . . . , represent different constants. The following three commands will calculate the mean, median, and standard deviation and store these values as constants K1, K2, and K3:

```
MTB > MEAN the data in C1 put the value in K1
MTB > MEDIAN the data in C1 put the value in K2
MTB > STDEV the data in C1 put the value in K3
```

After finding and storing the values for the mean and standard deviation, we can calculate Pearson's Index of skewness as follows:

```
MTB > LET K4 = 3 * (K1 - K2)/K3
```

The symbol "*" represents multiplication and the symbol "/" represents division. Exponentiation (raising to a power) is represented by the symbol "**." To determine the coefficient of variation we would write

```
MTB > LET K5 = (K3/K1)*100
```

We may print constants by using the PRINT command, as

```
MTB > PRINT K4 K5
```

This command will print the constants stored as constants K4 and K5. This command enables us to calculate and print the values for Pearson's Index of skewness and the coefficient of variation.

We can find z-scores for the data by using the CENTER command. The command has the form

```
MTB > CENTER the data in C, ..., C and store in C, ..., C
```

The CENTER command calculates z-scores for the values stored in the first list of columns and stores them in the second list of columns. To obtain z-scores for each entry in column 1 and store them in column 2, for example, we would write

```
MTB > CENTER the data in C1 store in C2
```

Example 2.15

Pollution indices for a city for 120 consecutive days are given here:

47	70	84	46	29	64	43	61
46	40	41	59	58	72	88	57
39	60	47	62	58	38	33	54
67	59	81	63	44	57	54	54
60	47	42	63	72	54	77	69
57	51	59	57	52	62	48	60
88	61	54	61	61	61	67	30
54	70	52	69	74	50	70	72
60	58	62	44	48	54	41	66
67	73	42	52	53	59	68	33
58	83	48	58	73	68	41	44
50	54	70	41	54	88	44	42
64	79	43	37	44	60	74	49
67	69	42	47	61	82	37	33
58	48	66	52	49	57	45	48

Using the data set, obtain descriptive statistics with the DESCRIBE command, obtain a bar graph using the HISTOGRAM command, and obtain a box plot, the coefficient of variation, the z-scores, and a stem and leaf display of the z-scores.

Solution

The computer commands and the printout are shown in Figure 2.10. We see from the bar graph that the data set is reasonably symmetric and probably normally distributed. This conclusion is further borne out by the box plot and the fact that all z-scores are between -3 and 3. Since there are no outliers and the data set is symmetric, an assumption of normality is reasonable. The coefficient of variation is 23.0137%, indicating a moderate amount of relative variation in the data.

```
MTB   > SET the following data in Column C1
DATA  >     47 70 84 46 29 64 43 61 46 40 41 59
DATA  >     58 72 88 57 39 60 47 62 58 38 33 54
DATA  >     67 59 81 63 44 57 54 54 60 47 42 63
DATA  >     72 54 77 69 57 51 59 57 52 62 48 60
DATA  >     88 61 54 61 61 61 67 30 54 70 52 69
DATA  >     74 50 70 72 60 58 62 44 48 54 41 66
DATA  >     67 73 42 52 53 59 68 33 58 83 48 58
DATA  >     73 68 41 44 50 54 70 41 54 88 44 42
DATA  >     64 79 43 37 44 60 74 49 67 69 42 47
DATA  >     61 82 37 33 58 48 66 52 49 57 45 48
DATA  >   END of data
MTB   >   NAME for C1 is 'P-INDEX'
MTB   > PRINT the data in C1

P-INDEX

   47 70 84 46 29 64 43 61 46 40 41 59 58
   72 88 57 39 60 47 62 58 38 33 54 67 59
   81 63 44 57 54 54 60 47 42 63 72 54 77
   69 57 51 59 57 52 62 48 60 88 61 54 61
   61 61 67 30 54 70 52 69 74 50 70 72 60
   58 62 44 48 54 41 66 67 73 42 52 53 59
   68 33 58 83 48 58 73 68 41 44 50 54 70
   41 54 88 44 42 64 79 43 37 44 60 74 49
   67 69 42 47 61 82 37 33 58 48 66 52 49
   57 45 48

      DESCRIBE the data in C1

           N   MEAN  MEDIAN  TRMEAN  STDEV  SEMEAN
P-INDEX  120   56.7   57.0    56.5   13.1    1.2
         MIN    MAX     Q1      Q3
P-INDEX 29.0   88.0   47.0    66.0


      HISTOGRAM the data in C1
Histogram of P-INDEX N = 120
Midpoint      Count
   30           2      **
   35           5      *****
   40          11      ***********
   45          14      **************
   50          14      **************
   55          15      ***************
   60          24      ************************
   65          10      **********
   70          12      ************
   75           5      *****
   80           3      ***
   85           2      **
   90           3      ***
```

Figure 2.10
Minitab printout for pollution data.

```
MTB > BOXPLOT the data in C1
                                    First quartile  Median  Third quartile

                                  -------------
          -------------I           +          I--------------
                                  -------------

          -+---------+---------+---------+--------
          30        45        60        75

      ONE HORIZONTAL SPACE = 0.15E + 01
      FIRST TICK AT   30.000

MTB > MEAN the data in C1 put the value in K1
         MEAN = 56.717

MTB > MEDIAN the data in C1 put the value in K2
         MEDIAN = 57.000

MTB > STDEV the data in C1 put the value in K3
         ST.DEV. = 13.053

MTB > LET K4 = 3 * (K1 - K2)/K3
MTB > LET K5 = (K3 / K1) * 100
MTB > PRINT K4 K5

      K4       -0.0651226 <---- Pearson's Index of Skewness
      K5       23.0137 <------- Coefficient of Variation

MTB > CENTER the data in C1 store the values in C2
MTB > NAME for C2 is 'Z-SCORES'

MTB > STEM-AND-LEAF the data in C2
        Stem-and-leaf of Z-scores   N = 120
        LEAF UNIT = 0.10

     2      -2* 10
     7      -1. 88855
    20      -1* 4322222111100
    41      -0. 999998887777666665555
    56      -0* 433332222222222
 -->(31)    +0* 0000000000011112222233333344444
    33      +0. 5577777788999
    20       1* 00001112233
     9       1. 5789
     5       2* 00333 <------------ Leaves
                 ^
                 Stems
Class containing the median
```

2.5.2 SAS Statements for This Chapter

The SAS procedure MEANS can be used to obtain the mean, the variance, the standard deviation, and the coefficient of variation, as well as the minimum and maximum values, the standard error of the mean, and the sum of the data. If more than one variable is entered, we may specify the variable or variables for which we want this information. We may also specify how the information is to be presented. The statement PROC MEANS, with no further options, will give us the mean, variance, and other descriptive information for each numerical variable for the entire data set.

Suppose we have entered the variables SEX, HEIGHT, AGE, and WEIGHT. To obtain the descriptive measures of height and age for each sex, we must first sort the data by sex (unless they were already sorted when entered), so we write

```
PROC SORT;
   BY SEX;
PROC MEANS;
   VAR HEIGHT AGE;
   BY SEX;
```

Suppose we have the scores on an achievement test for students in several schools and in several classes. We may wish to obtain the data sorted and printed out by school, by grade within school, and by sex within grade. We may also wish to obtain descriptive measures for each sex within grade and within school. Then the entire program (with a few data points entered) would be as follows:

```
DATA SCHOOLS;
   INPUT GRADE SCHOOL $ SEX $ SCORE @@;
CARDS;
11 A M 88 10 A F 84 12 B F 91 11 C M 76 ...
PROC SORT;
   BY SCHOOL GRADE SEX;
PROC PRINT;
   BY SCHOOL GRADE SEX;
PROC MEANS;
   BY SCHOOL GRADE SEX;
```

This program will result in a printout of the data as entered. Each combination of school, grade, and sex will be printed out separately, as well as the descriptive information for the variable score. If we wish the name of the school to be printed, rather than A, B, C, and so forth, we can use a slightly different entry statement by changing the name of the variable as shown in Section 1.3. We might enter the following:

```
DATA SCHOOLS;
 INPUT GRADE SCH $ SEX $ SCORE @@;
 IF SCH = 'A' THEN SCHOOL = 'LINCOLN';
 IF SCH = 'B' THEN SCHOOL = 'WILSON';
 IF SCH = 'C' THEN SCHOOL = 'JACKSON';
 IF SCH = 'D' THEN SCHOOL = 'FRANKLIN';
CARDS;
11 A M 88 10 A F 84 12 B F 91 11 C M 76 ...
```

The remainder of the statements will be the same as above. The single quotes are used for a nonnumerical variable. Figure 2.11 shows the input for data with the variables sex, height, age, and weight. (Part of this figure has been reduced in size. We will follow this practice when necessary to fit figures on the page width.) We obtained the height and weight for each combination of sex and age and obtained descriptive statistics on weight for each combination of sex and age.

Figure 2.11
SAS printout of MEANS procedure for sex, height, age, and weight data.

INPUT FILE

```
DATA EXAMPLE;
   INPUT SEX $ HEIGHT AGE WEIGHT @@;
   CARDS;
F 64 22 113 M 69 23 145 M 73 22 198 F 66 22 138 M 71 24 203
M 68 19 155 M 70 23 176 F 63 21 124 F 67 22 144 M 70 21 178
M 70 23 168 F 70 19 152 M 70 20 166 F 65 21 118 F 62 20 108
M 74 21 212 M 71 22 184 F 61 22 97
PROC SORT;
   BY SEX AGE;
PROC PRINT;
   BY SEX AGE;
PROC MEANS;
   VAR WEIGHT;
   BY SEX AGE;
```

OUTPUT FILE

```
------------------------------------SEX=F AGE=19------ ---------------------------------
                         OBS     HEIGHT     WEIGHT
                          1        70         152

------------------------------------SEX=F AGE=20------ ---------------------------------
                         OBS     HEIGHT     WEIGHT
                          2        62         108

------------------------------------SEX=F AGE=21------ ---------------------------------
                         OBS     HEIGHT     WEIGHT
                          3        63         124
                          4        65         118

------------------------------------SEX=F AGE=22------ ---------------------------------
                         OBS     HEIGHT     WEIGHT
                          5        64         113
                          6        66         138
                          7        67         144
                          8        61          97

------------------------------------SEX=M AGE=19------ ---------------------------------
                         OBS     HEIGHT     WEIGHT
                          9        68         155

------------------------------------SEX=M AGE=20------ ---------------------------------
                         OBS     HEIGHT     WEIGHT
                         10        70         166
```

(continued)

```
-----------------------------------SEX=M AGE=21------ -------------------------------
                        OBS      HEIGHT      WEIGHT
                        11         70          178
                        12         74          212

-----------------------------------SEX=M AGE=22------ -------------------------------
                        OBS      HEIGHT      WEIGHT
                        13         73          198
                        14         71          184

-----------------------------------SEX=M AGE=23------ -------------------------------
                        OBS      HEIGHT      WEIGHT
                        15         69          145
                        16         70          176
                        17         70          168

-----------------------------------SEX=M AGE=24------ -------------------------------
                        OBS      HEIGHT      WEIGHT
                        18         71          203
```

VARIABLE	N	MEAN	STANDARD DEVIATION	MINIMUM VALUE	MAXIMUM VALUE	STD ERROR OF MEAN	SUM	VARIANCE	C.V.
				SEX=F	AGE=19				
WEIGHT	1	152.00000000	.	152.00000000	152.00000000	.	152.00000000	.	.
				SEX=F	AGE=20				
WEIGHT	1	108.00000000	.	108.00000000	108.00000000	.	108.00000000	.	.
				SEX=F	AGE=21				
WEIGHT	2	121.00000000	4.24264069	118.00000000	124.00000000	3.00000000	242.00000000	18.00000000	3.506
				SEX=F	AGE=22				
WEIGHT	4	123.00000000	21.92411154	97.00000000	144.00000000	10.96205577	492.00000000	480.66666667	17.824
				SEX=M	AGE=19				
WEIGHT	1	155.00000000	.	155.00000000	155.00000000	.	155.00000000	.	.
				SEX=M	AGE=20				
WEIGHT	1	166.00000000	.	166.00000000	166.00000000	.	166.00000000	.	.
				SEX=M	AGE=21				
WEIGHT	2	195.00000000	24.04163056	178.00000000	212.00000000	17.00000000	390.00000000	578.00000000	12.329
				SEX=M	AGE=22				
WEIGHT	2	191.00000000	9.89949494	184.00000000	198.00000000	7.00000000	382.00000000	98.00000000	5.183
				SEX=M	AGE=23				
WEIGHT	3	163.00000000	16.09347694	145.00000000	176.00000000	9.29157324	489.00000000	259.00000000	9.873
				SEX=M	AGE=24				
WEIGHT	1	203.00000000	.	203.00000000	203.00000000	.	203.00000000	.	.

The other SAS procedure we use for this chapter is the UNIVARIATE procedure. The basic procedure produces a variety of information, including the mean, variance, and standard deviation, the coefficient of variation, the median and quartiles, the 1st, 5th, 10th, 90th, 95th, and 99th percentiles, the range, the IQR (given as Q3–Q1), and the mode. The UNIVARIATE procedure also has many options; the option PLOT, for example, will produce a stem and leaf plot and box plot, as well as a normal probability plot. If the normal probability plot of a set of data produces an approximately straight line, the original data set is normal. Figure 2.12 shows the program and output for both the MEANS and UNIVARIATE procedures for the data of Example 2.15.

Figure 2.12
SAS printout of UNIVARIATE procedure, including PLOT option, for pollution data of Example 2.15.

INPUT FILE

```
DATA POLLUTE;
   INPUT INDEX @@;
CARDS;
47 70 84 46 29 64 43 61 46 40 41 59 58 72 88 57 39 60 47 62 58 38 33 54
67 59 81 63 44 57 54 54 60 47 42 63 72 54 77 69 57 51 59 57 52 62 48 60
88 61 54 61 61 61 67 30 54 70 52 69 74 50 70 72 60 58 62 44 48 54 41 66
67 73 42 52 53 55 68 33 58 83 48 58 73 68 41 44 50 54 70 41 54 88 44 42
64 79 43 37 44 60 74 49 67 69 42 47 61 82 37 33 58 48 66 52 49 57 45 48
PROC MEANS;
PROC UNIVARIATE PLOT;
```

OUTPUT FILE

```
                                                      SAS

VARIABLE     N        MEAN        STANDARD      MINIMUM       MAXIMUM      STD ERROR         SUM           VARIANCE        C.V.
                                  DEVIATION      VALUE         VALUE        OF MEAN
INDEX       120   56.71666667   13.05269432   20.00000000   88.00000000   1.19154252   6806.0000000   170.37282913    23.014
                                                                SAS
                                                            UNIVARIATE

VARIABLE = INDEX

                 MOMENTS                                      QUANTILES (DEF = 4)                      EXTREMES
N               120  SUM WGTS          120           100% MAX    88     99%       88           LOWEST   HIGHEST
MEAN        56.7167  SUM              6806            75% Q3     66     95%    81.95               29        83
STD DEV     13.0527  VARIANCE      170.373            50% MED    57     90%       73               30        84
SKEWNESS   0.255218  KURTOSIS    -0.255761            25% Q1     47     10%       41               33        88
USS          406288  CSS           20274.4             0% MIN    29      5%       37               33        88
CV          23.0139  STD MEAN      1.19154                               1%    29.21               33        88
T:MEAN=0    47.5994  PROB>|T|       0.0001             RANGE     59
SGN RANK       3630  PROB>|S|       0.0001             Q3-Q1     19
NUM - = 0       120                                     MODE     54
```

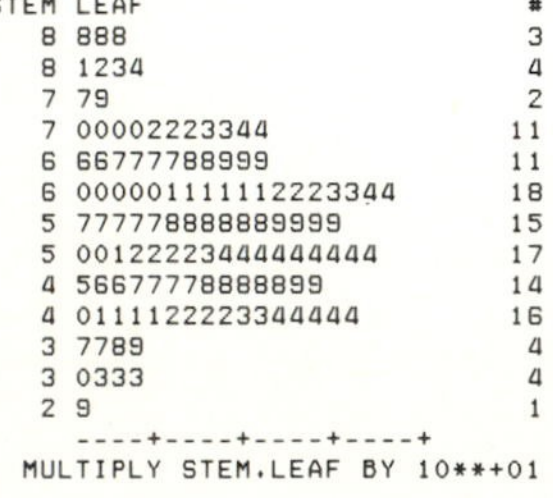

```
STEM LEAF                      #
   8 888                       3
   8 1234                      4
   7 79                        2
   7 00002223344              11
   6 66777788999              11
   6 000001111112223344       18
   5 777778888889999          15
   5 00122223444444444        17
   4 56677778888899           14
   4 0111122223344444         16
   3 7789                      4
   3 0333                      4
   2 9                         1
     ----+----+----+----+
  MULTIPLY STEM.LEAF BY 10**+01
```

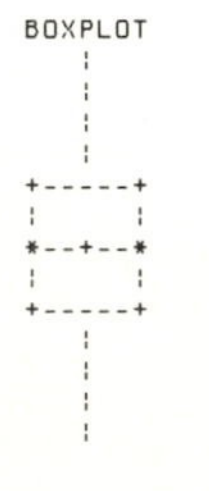

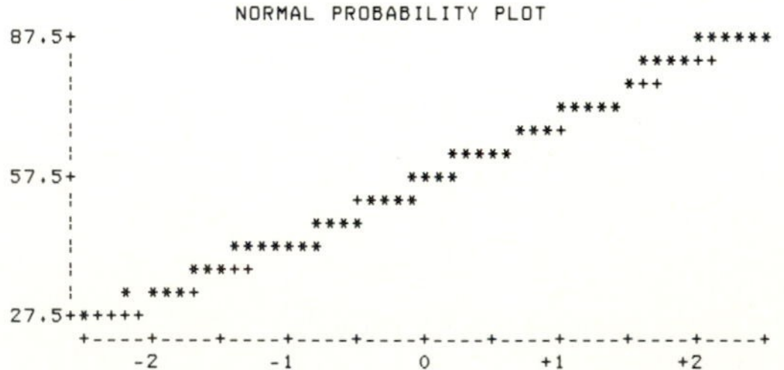

Problems

Applications

2.44 The following are lengths, in centimeters, of mullet taken from the bay:

31	29	35	37	39	35	40	32
37	35	36	41	42	33	39	35
41	36	37	37	34	41	37	31
38	43	37	38	37	34	36	35
36	32	39	37	34	36	37	35
37	35	33	36	30	34	33	37
38	36	37	36	36	33	40	39
38	37	37	37	35	37	40	33
37	38	38	36	37	35	33	36

Use either Minitab or appropriate SAS statements to:

a. Obtain the measures of central tendency and dispersion found by the DESCRIBE command or MEANS procedure.

b. Obtain the coefficient of variation.

c. Obtain a box plot and stem and leaf plot.

d. Identify any potential outliers. Does the data set appear normal? If you are using SAS statements, interpret the normal probability plot.

2.45 Refer to the Miami Female Study data given in Appendix B.1. Use Minitab or SAS statements to obtain information about the number of credit hours taken by the students.

a. Obtain the descriptive statistics found by the DESCRIBE command or MEANS procedure, as well as the coefficient of variation. Interpret the coefficient of variation.

b. Obtain a stem and leaf plot and a box plot for the data.

c. Identify any potential outliers. Does the data set appear normal? If you are using SAS statements, interpret the normal probability plot.

2.46 In Chapter 1 we discussed how graphs can be used to check for differences between two populations. Box plots are useful for this purpose. For the Miami Female Study data in Appendix B.1, obtain a box plot for the number of credit hours taken for students who have a job and for students who do not have a job. Compare the two plots and comment on any apparent differences between the two groups. Determine whether there are any outliers in either of the two groups.

2.6 Case 2: The Problem Promotion—Solution

After some research, Ms. Lee found that the earlier test had been given to several thousand employees and had had a mean score of 22.6, a median of

22.9, and a standard deviation of 2.8. The test now in use had been given to more than 600 employees and had a mean of 107.8, a median of 104.9, and a standard deviation of 17.4. She first determined that both sets of data were mound-shaped so that she could use the z-scores for comparison.

She then calculated the z-scores for the two candidates. The score of 29 on the first test had a z-score of 2.29, while the score of 143 on the second test had a z-score of 2.02. Thus the score of 29 was better relative to the set of scores made on that test, and she reported this information to Mr. Wilson. Wilson promoted the person with the higher z-score.

2.7 Summary

In this chapter we introduced many summary descriptive measures—numbers that are used to describe a population or sample—and showed procedures for their computation. To characterize a distribution of data we usually obtain a measure of central tendency (a value approximately at the center of the distribution) and a measure of variability (a number that gives information about the spread of the distribution).

Most often the primary measures of central tendency and variability are the mean and the standard deviation, respectively. The variance is the square of the standard deviation, and thus provides a measure of variability as well. The standard deviation by itself may tell us little about the sample or population, but it is extremely useful in connection with Chebyshev's Theorem and the Empirical Rule. The Empirical Rule governs proportions under the curve for approximately normal distributions, and Chebyshev's Theorem applies to any distribution.

For non-normal data sets, the median is often the best measure of central tendency, as it is resistant to the effects of extreme values. In that case, the interquartile range is probably the best measure of variability. The mode is a measure of central tendency that is particularly suitable for qualitative data, since it is the value of the variable with the greatest number of observations.

To be able to determine the appropriate measures of central tendency and variability for a data set, we must know its shape—normal, skewed, or long-tailed, for example. Non-normality and outliers can be detected using a box plot, while Pearson's Index of skewness can show skewing and the pseudo-standard deviation can indicate long tails. In certain cases, fairly symmetric, long-tailed data sets may be trimmed to allow for the computation of the trimmed mean and trimmed standard deviation.

Measures of comparison between values in a sample or population include the z-score and quantiles such as percentiles and quartiles. Finally, the coefficient of variation allows comparisons between the variabilities of different data sets that measure similar variables but not necessarily in the same units.

Chapter Review Problems

Practice

Use the frequency distribution in the following table for Problems 2.47 to 2.52.

CLASS	FREQUENCY
171–175	4
166–170	8
161–165	14
156–160	22
151–155	27
146–150	19
141–145	17
136–140	11
131–135	3

2.47 What is the mean of the data set?

2.48 What is the modal class?

2.49 Compute the z-scores for 150 and 165.

2.50 Compare s and PSD. What do you conclude?

2.51 Compute Pearson's Index of skewness. Is the data set significantly skewed?

2.52 Is this data set approximately normal? Why?

Applications

A store manager wishes to determine whether a product sells well enough to warrant continuing to carry it on the shelves. The number of units sold in each of the last 12 weeks is as follows:

61 44 51 32 76 44 38 52 43 56 18 67

Use this data set for Problems 2.53 to 2.57.

2.53 Determine the mean and median of the data.

2.54 Determine the standard deviation. Compare it with PSD.

2.55 Determine Pearson's Index of skewness.

2.56 Construct a box plot for the data and check for outliers. Find the z-scores for the highest and lowest values.

2.57 Does the distribution appear to be normal?

Twenty-five patients admitted to a hospital are tested for levels of blood sugar with the following results:

87	51	83	67	78	77	69	76	
68	85	84	85	70	68	80	74	
79	66	85	73	104	78	81	77	75

Use this data set for Problems 2.58 to 2.66.

2.58 Find the mean, median, and mode for the data.

2.59 Determine the standard deviation.

2.60 Obtain IQR and PSD.

2.61 Construct a box plot and check for outliers.

2.62 Obtain the 10% trimmed data set and compare $\overline{x}(T)$ to Md.

2.63 Compare $s(T)$ and PSD.

2.64 Is the data set skewed?

2.65 Which is the best measure of central tendency for the data—the mean, the median, or the trimmed mean?

2.66 Compute the coefficient of variation V.

2.67 A test of job-related stress was standardized and found to have a mean of 112.6 with a standard deviation of 13.8. A second test had a mean of 44.6 and a standard deviation of 6.3. Which of these tests is more highly variable?

A restaurant serves chicken and seafood dinners. The number of seafood dinners served each day for 60 days is given here. Use the data to answer Problems 2.68 to 2.72.

43	44	58	39	41	54	61	39	38
36	56	48	41	47	51	57	46	40
31	48	39	48	52	63	51	28	46
48	33	44	48	57	37	40	45	46
38	44	58	54	37	41	51	36	55
47	44	53	37	33	44	37	46	52
48	66	44	38	44	39			

2.68 What proportion of the data is within one, two, and three standard deviations of the mean? Compare with the empirical rule and with Chebyshev's Theorem. Is it likely that the data represent a normal distribution?

2.69 Assuming normal distribution, for what range of seafood dinners could the chef plan with about 95% assurance of having enough but not too many?

2.70 Construct a box plot and check for outliers.

2.71 Obtain Pearson's Index, compare PSD with s, and compare Md with $\overline{x}$. Given all the information gathered, is the data set probably normal?

2.72 Obtain the trimmed mean and standard deviation, and compare them with Md and PSD for the trimmed data set. Is the trimmed data set probably normal?

2.73 A hundred rats are fed a special diet and their weight gain is recorded after four weeks. The following represents the weight gain (in grams) for the rats in four weeks. Obtain the mean and standard deviation for the sample.

15	7	11	17	9	23	13	6
9	15	11	6	18	5	14	11
22	9	15	8	3	11	14	18
17	21	19	2	17	3	17	8
14	8	18	9	17	24	13	9
19	15	11	17	20	7	11	14
18	12	8	16	8	10	31	11

17	13	7	13	27	19	9	2
5	12	7	11	9	10	8	14
8	11	6	3	19	22	7	11
11	5	3	18	22	16	8	14
7	17	4	10	7	11	8	17
13	9	11	4				

2.74 A control group of rats, not fed the special diet, showed a mean weight gain over the four weeks of 5.1 grams with a standard deviation of 4.2 grams.

a. Comparing these data with the data of Problem 2.73, which data set is more highly variable?

b. Do you think the special group showed significantly higher gains than the control group? Why? (A procedure for answering this question will be given in Chapter 8. Just give a reasoned opinion here.)

2.75 A machine has been set to bore holes in aluminum extrusions. The holes will have a mean diameter of 0.15 millimeter with a standard deviation of 0.005 millimeter. Extrusions with holes smaller than 0.14 millimeter or greater than 0.16 millimeter are unusable. What proportion would you expect to be unusable if:

a. The distribution of hole sizes is normal?

b. Nothing is known about the distribution?

2.76 Bids for a project at site A (in thousands of dollars) are 150, 175, 150, 200, and 175. Bids for a similar project at site B are 250, 200, 175, 225, 200, 250, 200, and 220. The bids of the XYZ Corporation are 175 at site A and 200 at site B. Which bid is highest *relative to its group*?

2.77 Mr. Jones and Mr. Adams are given a physical examination for insurance. Men of Mr. Jones's age, height, and body structure are expected to have a mean weight of 144 pounds with a standard deviation of 12 pounds; for Mr. Adams's group, the mean weight should be 183 pounds with a standard deviation of 16 pounds. If Mr. Jones weighs 176 pounds and Mr. Adams weighs 212 pounds, which man is more overweight compared to his group? If the insurance company rejects applicants who are in the highest 5 or 6 percent of their reference groups, do you think either man should be rejected if the weight distributions are normal? What if nothing is known about the distribution of weights?

Index of Terms

(*continued*)

five-number summary, 83
greatest integer function, 82
inner fences, 87
interquartile range (IQR), 82
long-tailed distribution, 87
mean, 56
measure of central tendency, 56
median (Md), 61
modal class, 62
mode (Mo), 62
mound-shaped distribution, 72
normal distribution, 72
skewed, 74
standard deviation, 69
statistic, 58
statistical inference, 55
sum of squares, 68
trimmed data set, 89
variability, 56
variance, 67
whiskers, 83
Winsorized data set, 89
z-score, 78

Glossary of Symbols

IQR	interquartile range
μ	population mean
Md	median
Mo	mode
PSD	pseudo-standard deviation
Q_1	first quartile
Q_3	third quartile
σ	population standard deviation
σ^2	population variance
Σ	sum of
s	sample standard deviation
$s(T)$	trimmed standard deviation (sd of a trimmed data set)
s^2	sample variance
SSX	sum of squares of x
$\bar{x}$	sample mean
$\bar{x}(T)$	trimmed mean (mean of a trimmed data set)
z	z-score

Answers to Proficiency Checks

1. **a.** $26/5 = 5.2 \doteq 5$ **b.** $139/5 = 27.8 \doteq 28$ **c.** $13.4/10 = 1.34 \doteq 1.3$ **d.** The sum of 5, 2, 3, 4, 9, 11, 8, 5, and 3 is 50, and $50/9 \doteq 5.6$ or about 6; thus $\bar{x} \doteq$ 44,446.

2. $\bar{x} = 6064/77 \doteq 78.8$

3. $\bar{x} = 113(0.02) + 118(0.04) + 123(0.07) + 128(0.18) + 133(0.16) + 138(0.28) + 143(0.14) + 148(0.05) + 153(0.04) + 158(0.02) = 135.25$

4. **a.** Md = 5 **b.** Md = 28 **c.** Md = 1.25 **d.** Md = 445

5. Md = 13

6. **a.** 22 **b.** 22 and 25 **c.** No mode

7. 38

8. $\Sigma x = 27$, so the mean is 5.4. The deviations are -2.4, -0.4, -0.4, 0.6, and 2.6; $\Sigma(x - \bar{x})^2 = 13.2$, so $s^2 = 13.2/4 = 3.3$ and $s \doteq 1.82$. Using the shortcut formula, $\Sigma x^2 = 159$, so $s^2 = (159 - 27^2/5)/4 = 3.3$.

9. $s \doteq 5.2$

10. $s \doteq \$155.15$

11. The variance of the five numbers given is the same as the variance of -13, -11, -6, 13, and 22 (subtracting 23,900 from each number) or that of 7, 9, 14, 33, and 42 (subtracting 23,880 from each number). Using the latter approach, we find $\Sigma x = 105$ and $\Sigma x^2 = 3179$, so $\text{SSX} = 3179 - (105)^2/5 = 974$. Then the variance is $974/4 = 243.5$ and $s = \sqrt{243.5} \doteq 15.60$.

12. $s^2 = [478{,}004 - (6064)^2/77]/76 \doteq 5.85$; $s \doteq 2.42$.

13. $s^2 = [7158 - (468)^2/35]/34 \doteq 26$; $s \doteq 5.15$

14. **a.** 3/4 **b.** 8/9 **c.** No minimum amount

15. **a.** 95% **b.** Virtually all **c.** 68%

16. **a.** 97.5% **b.** 2.5% **c.** 16% **d.** 84%

17. **a.** $I = -2.36$, negatively skewed **b.** $I = 0.86$; not significantly skewed **c.** $I = -0.77$; not significantly skewed **d.** $I = 1.04$; positively skewed **e.** Symmetric

18. **a.** 1.00 **b.** 1.50 **c.** -0.50 **d.** -1.00 **e.** 0.00 **f.** -1.83 **g.** 2.50 **h.** -4.33 **i.** -0.33 **j.** 5.17

19. **a.** 1.05 **b.** -1.27 **c.** -0.50 **d.** 2.26 **e.** -4.21 **f.** 0.28 **g.** 4.11 **h.** -2.61

20. $V = 20.00$ for the distribution of Proficiency Check 18 and $V = 15.04$ for the distribution of Proficiency Check 19; therefore the data set of Proficiency Check 18 is more variable.

21. **a.** 13.7 **b.** 10.5 **c.** 17.3 **d.** 7.8 **e.** 11.4 **f.** 3.3 **g.** 14.3 **h.** 15.5

22. $Q_1 = 1.96$; $Q_3 = 6.20$

23. $\text{IQR} = 4.24$

24. Minimum = 1.34, maximum = 11.82, $Q_1 = 1.96$, Md = 3.02, $Q_3 = 6.20$. A plot is shown in the accompanying diagram. The data set is positively skewed.

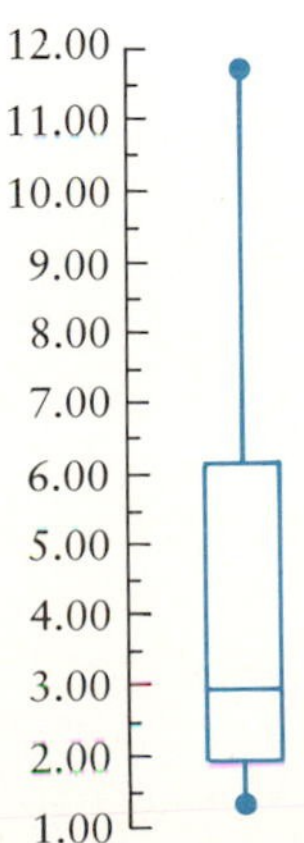

25. A stem and leaf plot for the data is shown here:

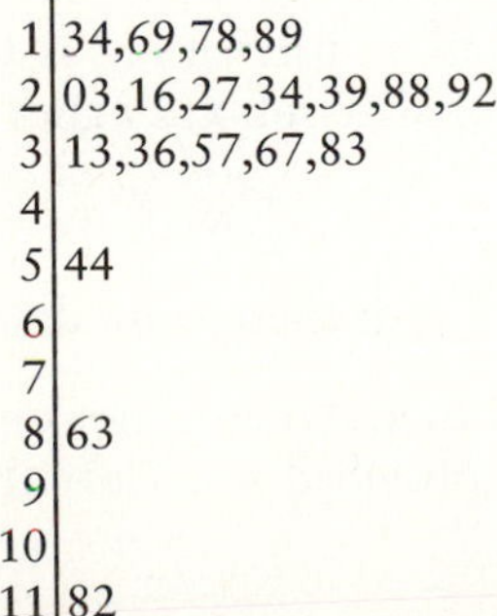

1	34,69,78,89
2	03,16,27,34,39,88,92
3	13,36,57,67,83
4	
5	44
6	
7	
8	63
9	
10	
11	82

There are 19 observations, so the median is located at observation number 10; that is, Md = 2.88. Each quartile, then, is 5.5 observations from the end; $Q_1 = 2.095$ and $Q_3 = 3.62$. Then IQR = 1.525 and 1.5 · IQR ≐ 2.29. The lower inner fence is below the data, so we need not worry about that; the upper inner fence is 5.91 and the outer fence is 8.20. Adjacent values are 1.34 and 5.44; 8.63 and 11.82 are extreme outliers. A box plot is shown in the accompanying diagram.

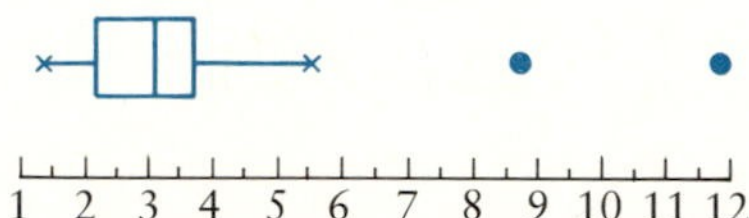

26. PSD = 1.525/1.35 ≐ 1.13; $s \doteq 2.60$. The two numbers do not agree. Moreover, $\bar{x} \doteq 3.53$, which is not very close to 2.88. Finally, the box plot shows a fair amount of skewness. In all, we may conclude that the data set is not normal.

27. Since there are 19 observations, we trim two data points off each end. The stem and leaf plot for the trimmed data set is shown here:

1	78,89
2	03,16,27,34,39,88,92
3	13,36,57,67,83
4	
5	44

Thus $\bar{x}(T) \doteq 2.91$, much closer to the median, while $s(T) \doteq 0.97$. For the trimmed data set $Q_1 = 2.215$ and $Q_3 = 3.465$, so IQR = 1.25. Hence PSD = 1.25/1.35 ≐ 0.93, so the trimmed data set is very close to a normal distribution.

Some Notes on Summation Notation

The summation notation used in most cases throughout this book is a shortcut notation. We write Σx to mean the sum of all the observations of the variable x and write Σx^2 to mean the sum of all the squares of the observations of the variable x. In many cases, however, we need more precision. We may represent a set of n observations by the symbols $x_1, x_2, x_3, \ldots, x_n$. We use the symbol x_i to represent any observation (we often say "the ith observation," or "x-sub-i"). Then the sum of all n observations is $x_1 + x_2 + x_3 + x_4 + \cdots + x_n$. This sum ("the sum of all x-sub-i as i goes from 1 to n") is symbolized as $\sum_{i=1}^{n} x_i$. Thus $\sum_{i=1}^{n} x_i = x_1 + x_2 + x_3 + x_4 + \cdots + x_n$. This notation is sometimes shortened to $\sum_i x_i$ if there is no doubt that the summation is to be for all values of i from 1 to n. If there is no confusion, we may further shorten this notation to Σx, provided it is clear that this means "the sum of all the data."

References

Huber, Peter J. *Robust Statistics*. New York: John Wiley & Sons, 1981.

Koopmans, Lambert H. *An Introduction to Contemporary Statistics*. Boston: Duxbury Press, 1981.

Mosteller, Frederick, and John W. Tukey. *Data Analysis and Regression*. Reading, MA: Addison-Wesley, 1977.

Tukey, John W. "A Survey of Sampling from Contaminated Distributions." In I. Olkin, ed., *Contributions to Probability and Statistics*. Stanford, CA: Stanford University Press, 1960.

Tukey, John W. *Exploratory Data Analysis*. Reading, MA: Addison-Wesley, 1977.

Tukey, John W., and D. H. McLaughlin. "Less Vulnerable Confidence and Significance Procedures for Location Based on a Single Sample: Trimming/Winsorization." *I. Sankyā,* Series A, 25 (1963): 334–352.

3

Probability

Although the primary emphasis in this book is on statistics and the use of statistical inference, you will need a knowledge of basic probability in order to use appropriate statistical techniques and understand the results of their application.

The term *probability* is difficult to define. In general, we use it to denote the relative likelihood of a certain event occurring, compared with the relative likelihood of all other possibilities. Thus probability deals with determining the chances that a certain outcome will occur in a known population of outcomes. If a lot of 80 parts contains 5 which are defective, for instance, we can determine the relative likelihood of the possible numbers of defectives in a sample of 6 parts drawn from this lot.

Statistical inference, on the other hand, provides a means for determining the approximate makeup of a population by investigating samples taken from the population. Suppose a company produces items on an assembly line and expects a certain number of defectives. Based on cost factors, the company has determined that if the number of defective items is more than 2% of the total, the assembly line should be shut down and refurbished. To control the quality of the items, they take a sample of 20 parts and examine them for defectives. If none are defective, the proportion of defectives may be minimal; on the other hand, if many of the sample items are defective, the assembly line should be shut down. A knowledge of probability and statistical techniques will enable management to make educated guesses about the proportion of defectives produced by the assembly line, to estimate the likelihood that these guesses are correct, and to make a reasoned decision whether or not the assembly line should be shut down. Thus probability computations and statistical inference techniques are complementary.

In short, probability helps us to understand the relationship between the known characteristics of a population and the possible characteristics of a sample taken from the population. Using probability we can determine the relative likelihood of the possible characteristics of a sample taken from a known population, and using statistical inference we can infer the likely characteristics of an unknown population from those of a sample taken from it.

This chapter presents an introduction to probability so that the topics of statistical inference covered in later chapters are easily understood. We will also consider counting principles and Bayes' Rule, which are not needed directly for this book, but may be needed in subsequent statistics courses.

3.0 Case 3: Where Did the Sponge Come From?

Probability is often used in legal cases to impress a jury or judge with the statistical likelihood—or unlikelihood—of a certain event. In one case a few years ago, a patient sued a hospital and a doctor for malpractice. The patient, a diabetic, had entered the hospital for implantation of a pacemaker. Several weeks after the operation, the incision still had not healed and was still somewhat open. The doctor decided to perform another operation to improve the patient's condition and, in the course of the second operation, it was found that the incision from the previous operation contained a surgical sponge (several layers of gauze designed to absorb fluids). The patient alleged that the sponge had been left during the earlier operation and sued the doctor and the hospital for a considerable sum of money.

Every hospital takes safeguards to prevent sponges being left in an incision. In this instance, sponges were counted three times before the operation and again three times after the operation. Moreover, the sponge discovered during the second operation was of a type the hospital had stopped using several months prior to the operation. The patient's lawyer contended that a box of the older sponges had been overlooked but was then found and used during the operation. No records had been kept regarding the type of sponges used in the operating room that day. The lawyer discussed a chain of circumstances that would have led to the sponge of that type being left in the patient. On the other hand, diabetics frequently fail to heal properly after an operation; indeed, several weeks after the second operation, the patient still had not begun to heal properly. Furthermore, the patient's wife was a nurse, and it had been her responsibility to clean and tend the incision. The hospital where she was employed did use sponges of the type found in the wound.

The lawyers for the hospital and doctor engaged the services of a statistician to determine the relative likelihoods of several possible courses of events. Although they knew that their case might not be helped by the results, they felt certain that the chain of events described by the opposing lawyer was relatively unlikely.

3.1 Probability

If you toss a fair coin, you know that it will land either head up or tail up. Each of these possibilities is called an *outcome*. We assume in this case that the two outcomes are equally likely. If this is true, the relative likelihood that a head will be showing is 1/2 and the relative likelihood that a tail will be showing is 1/2. We say that the *probability* of a head is 1/2 and the *probability* of a tail is 1/2. Thus the **probability** of an outcome in a situation is the relative likelihood that the outcome will occur.

3.1.1 Probabilities for Equally Likely Outcomes

If we shuffle together ten cards—four aces, three kings, two queens, and a jack— and then turn one over, the occurrence of a specific card is one outcome. Since we are no more likely to turn one card over than another, we say that all the outcomes are **equally likely**; therefore each has probability 1/10. On the other hand, if we consider only the denomination of a card and not which specific card is drawn, there are only four possibilities—the card will be an ace, a king, a queen, or a jack—and they are not equally likely. There are four possible ways to get an ace, three to get a king, two to get a queen, and only one to get a jack. For want of a better term, we call the act of drawing one card from these ten an *experiment*. In determining probabilities we say that an **experiment** consists of making one observation. We call each of the possible observations an **outcome**, and we define an **event** as a collection of one or more outcomes.

In the case of these ten cards, the *experiment* consists of drawing one card, each *outcome* is the specific card being drawn, and we may define some *events* as the face value of the card being drawn—that is, the events "an ace is drawn," "a king is drawn," and so forth. What then is the probability of each of these events? Intuitively we can see that if an experiment has s equally likely outcomes, r of which constitute an event, the probability of that event occurring is r/s.

Probability for Equally Likely Outcomes

If an experiment has s equally likely outcomes, r of which constitute an event, then the probability of the event is r/s.

Example 3.1

Suppose you shuffle together ten cards—four aces, three kings, two queens, and a jack—and then turn one over. Determine the probability of obtaining each denomination of card.

Solution

Since there are ten cards, each equally likely, and four of them are aces, the probability of obtaining an ace is 4/10; similarly, the probability of getting a king is 3/10, that of getting a queen is 2/10, and that of getting a jack is 1/10.

In many instances we need a special rule for determining the number of outcomes in order to apply our definition of probability. This rule is called the **multiplication rule.**

The Multiplication Rule

If an act requires n steps to complete, and these steps can be performed successively in $m_1, m_2, m_3, \ldots, m_n$ ways, then the total number of ways to perform the act is $m_1 \cdot m_2 \cdot m_3 \cdots m_n$.

Example 3.2 What is the probability of obtaining two heads on two tosses of a fair coin?

Solution

This act consists of two steps, each of which has two outcomes, head or tail, so there are $2 \cdot 2$ or 4 equally likely outcomes. One of these outcomes, head followed by head, makes up the desired event. The probability of this event, then, is 1/4 or 0.25.

PROFICIENCY CHECK

1. Four roads lead from East Siwash to Podunk; five roads lead from Podunk to High Town; three roads lead from High Town to Grand Burg. Determine the number of different routes:
   a. leading from East Siwash to High Town through Podunk
   b. leading from Podunk to Grand Burg through High Town
   c. leading from East Siwash to Grand Burg through Podunk and High Town

3.1.2 Sample Spaces

If we perform an experiment, the set of all possible outcomes is called the **sample space** for the experiment. The sample space is usually denoted by S. Each outcome is called a **point, sample point,** or **simple event.** Usually we show a sample space by listing each outcome. If a coin is to be tossed three times, the sample space of heads and tails corresponding to all the outcomes is the set S, where

$$S = \{(H,H,H), (H,H,T), (H,T,H), (H,T,T), (T,H,H), (T,H,T), (T,T,H), (T,T,T)\}$$

The first entry in each ordered triple is the result (head or tail) of the first toss, the second entry is the result of the second toss, and the third entry is the result of the third toss.

Any collection of points in the sample space is called an event. The set of no points is the empty set, denoted by $\emptyset$, and is included for completeness. In the experiment in which we toss a coin three times, the event "head appears twice" consists of the points (H,H,T), (H,T,H), and (T,H,H). If we denote this event by A, then

$$A = \{(H,H,T), (H,T,H), (T,H,H)\}$$

To facilitate further use of a sample space, we frequently use a shorthand notation omitting all set notation and commas. Thus, although the correct representation for the sample space is as shown above, we may represent this sample space in shorthand notation by

HHH HHT HTH HTT THH THT TTH TTT

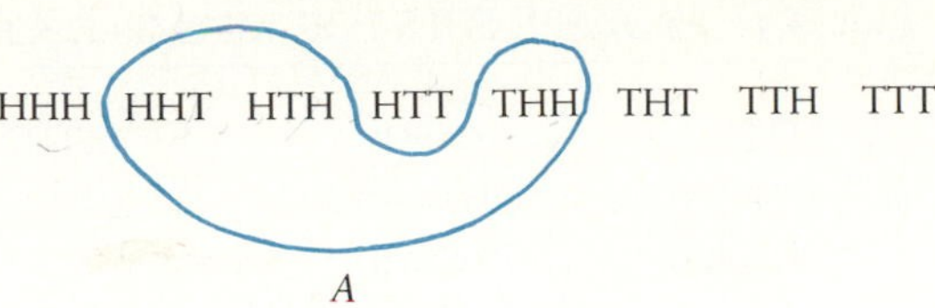

Figure 3.1
The event $A = \{(H,H,T), (H,T,H), (T,H,H)\}$.

This notation can facilitate the determination of events simply by enclosing the points which make up the event. We can indicate event A, as defined above, by circling the three points HHT, HTH, and THH as shown in Figure 3.1.

Since the coin is fair, all eight points are equally likely and the event A contains three points. The probability of A occurring is 3/8; that is, the probability is 3/8 that if we toss a coin three times, we will obtain exactly two heads. We denote the probability of an event by using the symbol $P(\quad)$; the event is denoted within the parentheses. In this case, we have $P(A) = 3/8 = 0.375$. We could also write $P(A) = 37.5\%$. All three notations, fractions, decimals, and percentages, are in common use for probabilities.

We could also define a variable x, which denotes the number of heads obtained on one experiment; the values that x can take on are 0, 1, 2, and 3. We would then denote the probability that $x = 2$ by the symbol $P(x = 2)$; then $P(x = 2) = 3/8$.

Suppose event B consists of the set of all points with at least two heads. We can see that event B consists of exactly four equally likely points, so that $P(B) = 4/8 = 0.5$. We can also use the same variable x, as before, denoting the number of heads obtained. In the latter case, $P(x \geqslant 2) = 0.5$. We can generalize this case to obtain the probability that an event will occur.

Probability of an Event

The probability that an event A will occur, denoted $P(A)$, is equal to the sum of the probabilities of the outcomes that make up the event.

Example 3.3

A balanced (fair) die is tossed once. List the sample space of outcomes and determine the probability of each of the following events:

a. obtaining at least a 3 4/6
b. at most a 4 4/6
c. more than 4 2/6
d. 5 or less 5/6
e. 2 or more
f. less than 5
g. at least 2 but less than 5
h. more than 2, and 5 or less

Table 3.1 PROBABILITIES FOR EXAMPLE 3.3.

	EVENT	SYMBOL	PROBABILITY
a.	{3,4,5,6}	$P(x \geq 3)$	2/3
b.	{1,2,3,4}	$P(x \leq 4)$	2/3
c.	{5,6}	$P(x > 4)$	1/3
d.	{1,2,3,4,5}	$P(x \leq 5)$	5/6
e.	{2,3,4,5,6}	$P(x \geq 2)$	5/6
f.	{1,2,3,4}	$P(x < 5)$	2/3
g.	{2,3,4}	$P(2 \leq x < 5)$	1/2
h.	{3,4,5}	$P(2 < x \leq 5)$	1/2

Solution

The sample space is the set {1, 2, 3, 4, 5, 6}, where each outcome has probability 1/6. The set corresponding to each event with its symbol and probability is shown in Table 3.1. The symbols correspond to the statement of the event desired. Note, for example, that the event in (*b*) and the event in (*f*) are the same but have two different symbols.

All the probabilities discussed thus far are **theoretical probabilities.** We may also determine probabilities in two other ways. **Empirical probability** is based on past events. On the assumption that past events are reasonably representative of future events, we use the relative frequency of past events to approximate the probabilities of future events. **Subjective probability** is an assignment of the relative likelihood of an event without any concrete evidence on which to make an assessment. A bettor at a racetrack may bet on a horse on a hunch. This hunch is merely based on a feeling that this horse has a better chance of winning than the others.

Many probability determinations are a combination of empirical and subjective probabilities. One such example is the weather forecaster's probability of rain tomorrow. This probability is based on a subjective assessment of the overall weather conditions and an objective, empirical probability based on the relative frequency of rain when these conditions were met in the past. Suppose that a patient is suffering from a disease, and the doctor has decided that it is probably one of two diseases that have similar symptoms. Based on his experience and other factors, the doctor concludes that one disease is twice as likely as the other and there is possibly one chance in ten that it is neither of these disorders. The doctor's assessment of the likelihoods provides the following probabilities: the probability that it is the first disease is 0.6, that it is the second disease is 0.3, and that it is neither is 0.1. These probabilities are based on a subjective assessment of empirical evidence. The following example uses empirical probabilities.

Example 3.4

Over the past 200 working days, the number of defective parts produced by an assembly line is shown in Table 3.2. We can use the relative frequency

Table 3.2 RELATIVE FREQUENCY TABLE FOR EXAMPLE 3.4.

NUMBER DEFECTIVE	DAYS	RELATIVE FREQUENCY
0	50	0.25
1	32	0.16
2	22	0.11
3	18	0.09
4	12	0.06
5	12	0.06
6	10	0.05
7	10	0.05
8	10	0.05
9	8	0.04
10	6	0.03
11	6	0.03
12	2	0.01
13	2	0.01

to estimate the probability of a certain number of defective parts on a given day. Determine the probability that tomorrow's output will have:

a. none defective
b. at least 1 defective
c. an odd number of defectives
d. no more than 5 defective
e. more than 5 defective
f. 13 or fewer defective
g. more than 13 defective

Solution

a. $P(0) = 0.25$

b. $P(x > 0) = P(x = 1,2,3,4,5,6,7, \ldots)$
$= P(1) + P(2) + P(3) + \cdots + P(13)$
$= 0.16 + 0.11 + 0.09 + \cdots + 0.01 = 0.75$

c. $P(x \text{ is an odd number}) = P(x = 1,3,5,7,9,11, \text{or } 13)$
$= P(1) + P(3) + P(5) + P(7) + P(9) + P(11) + P(13)$
$= 0.16 + 0.09 + 0.06 + 0.05 + 0.04 + 0.03 + 0.01$
$= 0.44$

d. $P(x \text{ is no more than } 5) = P(x \text{ is 5 or less})$
$= P(x \leq 5)$
$= P(0) + P(1) + P(2) + P(3) + P(4) + P(5)$
$= 0.25 + 0.16 + 0.11 + 0.09 + 0.06 + 0.06$
$= 0.73$

e. $P(x \text{ is more than } 5) = P(x > 5)$
$= P(x = 6,7,8,9,10,11,12, \text{ or } 13)$
$= 0.05 + 0.05 + 0.05 + 0.04 + 0.03 + 0.03 + 0.01$
$+ 0.01$
$= 0.27$

f. $P(x \leq 13)$ is the sum of all the probabilities listed, which is 1.00.

g. $P(x > 13) = 0$, since the event has no sample points.

PROFICIENCY CHECKS

2. Four poker chips are painted with the numbers 1, 2, 3, 4 and placed in a box. Two chips are drawn with the first one replaced before the second is drawn, and the numbers are added. List the sample space for the experiment. Then determine the probabilities for each of the following:
   a. The total is 2.
   b. The total is 4.
   c. The total is 5.
   d. The total is 6.
   e. The total is more than 2.
   f. The total is more than 4.
   g. The total is less than 8.
   h. The total is no more than 6.
3. Repeat Proficiency Check 2 but suppose the first chip is not replaced before the second is drawn.

3.1.3 Probability Rules

In the previous examples and proficiency checks, we have seen some of the rules by which we calculate probabilities. Clearly the probability of an event must be positive. An event that is impossible, such as $x > 13$ in Example 3.3, has probability zero. On the other hand, the sum of all the probabilities of the outcomes in a sample space is equal to 1. We thus conclude that the probability of an event cannot be negative or greater than 1.

Events that cannot both occur, such as those in parts (*c*) and (*g*) of Example 3.3, are **mutually exclusive** (or **disjoint**). Two mutually exclusive events that make up the entire sample space are called **complementary events,** or **complements** of each other. In Example 3.4, for instance, the events $x \leq 5$ and $x > 5$ in parts (*d*) and (*e*) are complementary events. We note that the sum of the probabilities of these two events is 1. Symbolically, the complement of an event A is written $\sim A$ (read "*not-A*"). The sum of the probabilities of A and $\sim A$ is equal to 1.

Finally, any combination of events that, taken all together, make up the entire sample space is said to be **exhaustive.** We note that complementary events are both mutually exclusive and exhaustive. A sample space for any experiment thus should consist of events that are both mutually exclusive and exhaustive.

We can put all these definitions together to obtain a few rules. We denote by $P(S)$ the probability that some event in the sample space will occur and by $P(\emptyset)$ the probability that none of the events will occur.

Probability Rules

1. For any event A, $0 \leq P(A) \leq 1$
2. For any sample space S, $P(S) = 1$ and $P(\emptyset) = 0$.
3. For any event A, $P(\sim A) = 1 - P(A)$.

Example 3.5

Suppose that you draw one marble at random from a box containing three white marbles, four red marbles, and one clear marble. Determine the probability that (a) the marble is red; (b) the marble is white; (c) the marble is not white.

Solution

Often we can illustrate a sample space schematically by using a **Venn diagram,** invented by John Venn (1834–1923) for use in set theory. The Venn diagram for this example is shown in Figure 3.2.

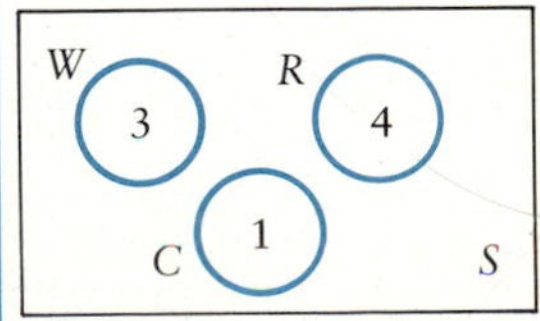

Figure 3.2
Venn diagram for Example 3.5.

We symbolize the event "a white marble is drawn" by W, the event "a red marble is drawn" by R, and the event "a clear marble is drawn" by C. These events are represented by the circles in the Venn diagram. The numbers inside the circles indicate the number of marbles of each color. The letter S represents the sample space. If there were marbles of other colors, but not of interest, the number of such marbles would appear outside the circles but inside the rectangle representing the sample space.

Since each marble is equally likely to be picked and there are eight marbles, the probability that a red marble is drawn, $P(R)$, is 4/8, or 0.5. Similarly, there are three white marbles, so $P(W) = 3/8$. The event "the marble drawn is not white" would be symbolized by $\sim W$; since five of the marbles are not white, $P(\sim W) = 5/8$.

We note that in this sample space the events R, W, and C are mutually exclusive. Since they account for all points of the sample space, they are also exhaustive. If events in a sample space are both mutually exclusive and exhaustive, we say that they form a **partition** of the sample space. Thus events R, W, and C form a partition of the sample space. The events W and $\sim W$ are complementary.

3.1.4 Compound Events

Suppose event A is the event "Fred gets a promotion" and event B is the event "Fred gets a raise." The event "Fred gets a promotion and a raise" is called a **compound event.** Compound events are of two types: *intersection* and *union.* The **intersection** of two events is the event in which both events occur. The event "Fred gets a promotion and a raise" is the intersection of events A and B. The **union** of two events is the event in which one or the other event, or both, occur. The union of A and B is the event "Fred gets a promotion, or a raise, or both." We symbolize the intersection of the events A and B *by "A and B"* and the union of the two events by "*A or B.*" The probability of the intersection of A and B is called the **joint probability** of A and B. (In set theory, the union is symbolized by $A \cup B$ and the intersection by $A \cap B$. These symbols are often used in statistics texts.) We will illustrate these concepts with an example.

Suppose that the Ellanjay Corporation has bid on ten contracts. Most of the projects will require the use of one or both of two machines—machine F and machine T. Three of the projects will require the use of machine F but not T, four of the projects will require the use of machine T but not F, and two of the projects will require both machines. One of the projects will require neither machine. Suppose that all contracts are equally likely to be awarded first and that the first contract to be obtained will be supervised by a new member of the managerial staff, Mr. Jones. Jones wishes to assess the various likelihoods of the two machines being used, as he is not familiar with either machine. He assigns F to the event "machine F will be used" and T to the event "machine T will be used." Then *F or T* represents the event "either machine F will be used, or machine T will be used, or both machines will be used," and *F and T* represents the event "both machine F and machine T will be used." We can illustrate the sample space by a Venn diagram as shown in Figure 3.3. The numbers indicate the number of potential contracts using each machine.

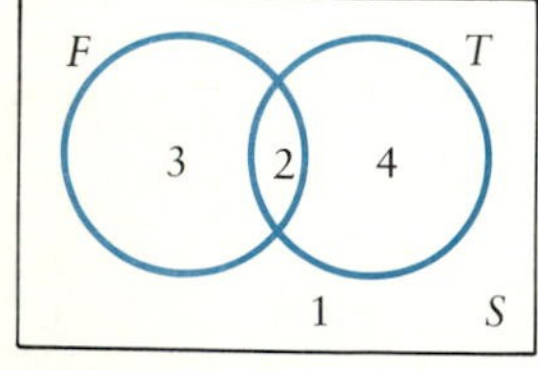

Figure 3.3
Venn diagram for sample space of the Ellanjay Corporation example.

Note that within circle F the number 2 is the number of contracts that require both machine F and machine T, while 3 is the number of contracts that require machine F and do not require machine T. The 4 in circle T that is outside circle F is the number of contracts that require machine T and do not require machine F. The number of contracts not requiring either of the machines is given by the 1 outside both circles. Note that it would be just as meaningful to list the probabilities rather than the number of contracts. In that case the numbers would be 0.3, 0.2, 0.4, and 0.1, instead of 3, 2, 4, and 1, respectively.

For the events F and T, we can easily see that $P(F) = 0.5$ since five of the ten projects require machine F and that $P(T) = 0.6$ since six of the ten projects require machine T. The overlapping portion of the two circles corresponds to the event F *and* T, so $P(F \text{ and } T) = 0.2$. Finally, there are nine projects that require at least one of the machines, so $P(F \text{ or } T) = 0.9$. Now note that $P(F) + P(T) = 0.5 + 0.6 = 1.1$. This result is not equal to $P(F \text{ or } T)$ since the overlapping portion has been counted twice in $P(F) + P(T)$. In order to obtain $P(F \text{ or } T)$ we must subtract $P(F \text{ and } T)$ from $P(F) + P(T)$ to compensate for counting it twice. This procedure leads to the **addition rule for probability.**

Addition Rule for Probability

If A and B are any events in a sample space,

$$P(A \text{ or } B) = P(A) + P(B) - P(A \text{ and } B)$$

If A and B are mutually exclusive,

$$P(A \text{ or } B) = P(A) + P(B)$$

(since in this case $P(A \text{ and } B) = 0$).

We can obtain other events and their probabilities from Figure 3.3. The probability that machine F will not be used is $P(\sim F)$, for example, and $P(\sim F) = 0.5$. Similarly, the probability that machine F will be used but machine T will not be used is $P(F \text{ and } \sim T)$, which can be seen to be 0.3.

In Example 3.5, the three events W, R, and C are mutually exclusive, so $P(W \text{ or } R) = 3/8 + 4/8 = 7/8$, $P(W \text{ or } C) = 3/8 + 1/8 = 4/8 = 1/2$, and $P(R \text{ or } C) = 4/8 + 1/8 = 5/8$.

We note in Figure 3.3 that the events F *and* T and F *and* $\sim T$ are mutually exclusive and their union is simply the event F. Thus $P(F \text{ and } T) + P(F \text{ and } \sim T) = 0.2 + 0.3 = 0.5 = P(F)$. This special rule has many applications and can be called the **rule of complementation.**

The Rule of Complementation

If A and B are any events in a sample space,

$$P(A) = P(A \text{ and } B) + P(A \text{ and } \sim B)$$

We can use a Venn diagram to good advantage in many instances. We fill in the known probabilities and then determine other probabilities from them.

Example 3.6

At a musical seminary, 10% of all students are studying the piano but no other instrument. Half the students are studying the piano, half are studying the cello, and 80% are studying the violin; 10% of the students are studying all three. Moreover, 40% of all the students are studying both violin and piano, and 30% are studying the violin and cello but not the piano. A student is selected at random and interviewed for a newspaper article. What is the probability that the student is studying:

a. none of the three instruments?
b. the violin or the cello?
c. the violin or the cello but not the piano?
d. the piano or violin but not the cello?

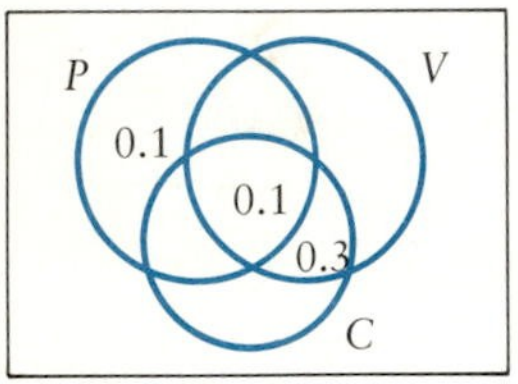

Figure 3.4
Known probabilities for Example 3.6.

Solution

We define the events "student is studying the piano," "student is studying the cello," and "student is studying the violin" as *P*, *C*, and *V*, respectively. We draw a Venn diagram with the three events shown as overlapping circles. Now 10% study the piano and no other instrument, 10% play all three, and 30% play the violin and cello but not the piano. Thus *P*(*P and* ~*V and* ~*C*) = 0.1, *P*(*P and V and C*) = 0.1, and *P*(~*P and V and C*) = 0.3, so we can fill in the spaces with the probabilities as shown in Figure 3.4.

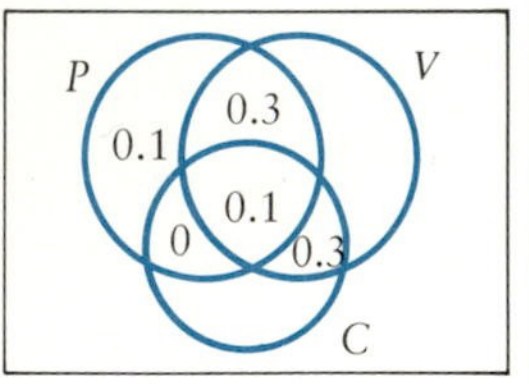

Figure 3.5
Further probabilities for Example 3.6.

Since *P*(*P and V*) = 0.4, we know that *P*(*P and V and* ~*C*) = 0.3; and since *P*(*P*) = 0.5, we have *P*(*P and* ~*V and C*) = 0. We now have the information shown in Figure 3.5.

Finally, since *P*(*V* = 0.8) and *P*(*C*) = 0.5, we can fill in the remaining spaces as shown in Figure 3.6. The sum of the numbers within the three circles is 1.0, so there are no students who are not studying at least one of the instruments. The probability that a student is studying the violin or the cello is equal to the sum of the numbers within both circles, or 0.9. The probability that a student is studying the violin or cello but not the piano is equal to the sum of the numbers within the circles *V* and *C* that are not in *P*; that is, 0.5. Similarly, 50% of the students study the piano or violin but not the cello. The answers to the questions, then, are

a. *P*(~*P and* ~*V and* ~*C*) = 0
b. *P*(*V or C*) = 0.9
c. *P*(~*P and V or C*) = 0.5
d. *P*(*P or V and* ~*C*) = 0.5

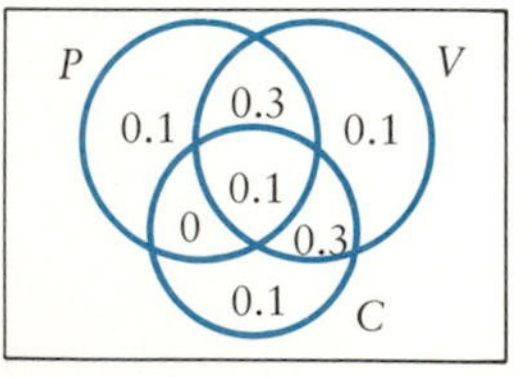

Figure 3.6
Completed Venn diagram for Example 3.6.

We can now summarize the probability rules given in this section.

Probability Rules

1. For any event A, $0 \leq P(A) \leq 1$.
2. For any sample space, $P(S) = 1$ and $P(\emptyset) = 0$.
3. For any event A, $P(\sim A) = 1 - P(A)$.
4. If A and B are any events in a sample space, $P(A \text{ or } B) = P(A) + P(B) - P(A \text{ and } B)$.
5. If A and B are mutually exclusive events in a sample space, $P(A \text{ or } B) = P(A) + P(B)$.
6. If A and B are any events in a sample space, $P(A) = P(A \text{ and } B) + P(A \text{ and } \sim B)$.

PROFICIENCY CHECKS

4. Use the Venn diagram of Figure 3.3 to determine the probability that the first project will require the machine or machines as listed.
   a. not machine F
   b. not machine T
   c. neither machine F nor machine T
   d. not machine F and not machine T
   e. not both machine F and machine T
   f. either not machine F or not machine T
   g. machine F but not machine T
   h. machine T but not machine F
5. The probability that June will get an A in College Algebra is about 0.4; the probability that Bill will get an A is about 0.35. Because they study together, the probability that they will both get an A is 0.2. Determine the following probabilities:
   a. the probability that June will get an A and Bill will not
   b. the probability that Bill will get an A and June will not
   c. the probability that at least one of them will get an A
   d. The probability that neither will get an A

Problems

Practice

3.1 An urn contains three white, two red, one blue, and four black balls. If one ball is drawn at random, what is the probability that it is:
a. black?

b. either black or white?
c. not white?
d. both red and blue?

3.2 In Problem 3.1, suppose that two balls will be drawn at random and the colors observed. List a sample space for the experiment:
a. if the first ball drawn is replaced before the second one is drawn
b. if the first ball is not replaced before the second one is drawn

3.3 In Problem 3.2(*a*), what is the probability of drawing:
a. two black balls?
b. one black and one white ball?
c. two blue balls?
d. at least one red ball?
e. at least one black ball?
f. at least one red ball and at least one white ball?
g. at least one red ball or at least one white ball?

3.4 In Problem 3.2(*b*), what is the probability of drawing:
a. two black balls?
b. one black and one white ball?
c. two blue balls?
d. at least one red ball?
e. at least one black ball?
f. at least one red ball and at least one white ball?
g. at least one red ball or at least one white ball?

3.5 A family is considering buying a dog. If the probability that they will buy a small dog is 0.1, that they will buy a medium-size dog is 0.3, that they will buy a large dog is 0.2, and that they will buy a very large dog is 0.1, what is the probability that the family will buy a dog?

3.6 If A and B are mutually exclusive and $P(A) = 0.4$ and $P(B) = 0.5$, what is $P(A \text{ or } B)$?

3.7 A psychologist claims that if a rat enters a maze, the probability that it will emerge at point A is 0.44; at point B it is 0.29, and at neither it is 0.17. Do you agree or disagree with this assertion? Why?

3.8 If $P(A) = 0.6$ and $P(B) = 0.7$, can A and B be mutually exclusive?

Applications

3.9 A student is taking five courses. In one course the instructor announces that the only grades to be given will be A, B, C, D, or F. Two courses will give grades A, B+, B, C+, C, D+, D, or F. One course, a seminar, will be graded only S or U. The student is taking the fifth course only for a grade of P or F.
a. How many different combinations of letter grades are possible?
b. Shortly before midterm the student drops the seminar, and near the end of the term he withdraws from one of the two courses that gives plus grades. The instructor will assign a WP or WF for that course. How many different combinations of grades are possible now?

3.10 Joe's Pizza Parlor is offering a special. For a given price you can purchase a large pizza and a drink or else a small pizza, a burger, and a drink. On the special, pizzas come plain or with a choice of either (but not both) of two toppings. Joe serves hamburgers, cheeseburgers, pizzaburgers, and onionburgers. For drinks the choices are cola, root beer, orange, grape, and cherry. How many different dinners can you purchase on the special?

3.11 A bingo player needs B-8, N-44, or G-53 for a BINGO. Only 27 counters remain in the container. What is the probability that the player will win on the next draw?

3.12 One of a group of 12 workers is to be chosen as shop foreman. If eight are men and four are women, what is the probability that a man will be chosen if all are equally likely to be picked? What is the probability that Fred Jones, one of the men, will be chosen?

3.13 An enthusiatic football fan claims that there are only 2 chances in 15 that his team will lose while the probability that it will win is 0.9. A second fan, more conservative, agrees that there are only 2 chances in 15 of losing, but he claims that the probability of winning is only 0.8. Comment on these two claims.

3.14 According to Internal Revenue Service data, 2.5 million returns were filed by January 25, 27.5 million were filed by February 22, 54.5 million were filed by March 29, and all 92.5 million were filed by April 26 with at least 9 million returns filed after April 15. If a return is selected at random from all those filed, what is the probability that it was filed after February 22 but by March 29?

3.15 According to a report in *Beverage Digest* (*San Francisco Examiner,* Feb. 2, 1986), Pepsi Cola is the leading cola on the market with 18.6% of soft drink sales compared to 15% for New Coke and 5.9% for Coke Classic. Suppose that a soft drink lover is asked his or her preference among all soft drinks, and that he or she has a preference. What is the probability that he or she

a. will prefer one of the cokes?

b. will prefer neither Pepsi nor one of the Cokes?

3.16 If a patient is admitted to a hospital with a certain disease, the probability that she will die of the disease is 0.12 and the probability that she will recover quickly is 0.24. The probability of recovering after a long convalescence without developing complications is 0.22, and the probability that she will develop complications is 0.42. Half the patients developing complications die from the complications, and half recover. If a patient suffering from this disease is admitted, what is the probability that she will:

a. recover from the disease without developing complications?

b. recover from the disease?

3.17 Among a group of children, 35% suffer from cognitive dissonance as a result of school experiences and 40% have a distorted sense of reality as measured by a standard measuring device. In a group of 200 children for which these percentages hold, a total of 32 have both cognitive dissonance and a distorted sense of reality. How many of these children are free from both problems?

3.18 A sociogram shows that 12% of the members of a group are social isolates. One-third of those are from low-income groups. If a member of this group is selected at random, what is the probability that this person is an isolate not from a low-income group?

3.19 Recently 1,000 adults were asked how many minutes of commercials are shown during each hour of prime time television (*USA Today*, Jan. 21, 1986). The actual number of minutes varies from 8 to 12. The results were as follows:

Estimate (min)	Less than 8	8–12	13–15	16 or more	Don't know
Number responding	140	270	290	220	80

Use these results to estimate the probability that:

a. A person selected at random who has an opinion thinks there are more than 12 minutes of commercials per hour.

b. A person selected at random will estimate the number of minutes of commercials per hour correctly.

3.20 Shuffle a deck of ordinary cards and deal them out, one at a time, until the first ace appears. Record the number of cards dealt, including the ace. Combine the results of five to ten experiments per student in the class. Compare the average number of cards dealt with the theoretical number, 10.6 per deal. (See the solutions manual for an explanation.)

3.21 Repeat problem 3.20 but count down until the ace of spades appears. Here the theoretical number is 26.5.

3.22 Repeat Problem 3.20 but count down until the first heart appears. What is the theoretical number this time?

3.23 Toss a coin 20 times and record the number of heads. Combine the results of the experiments of all students in the class. There should be a leveling effect, with approximately half the tosses showing a head. The probability of obtaining fewer than 268 or more than 332 heads in 600 tosses is only 1/100!

3.24 Select a column from a page in your telephone book and record the frequency of occurrence of last digits of telephone numbers. Combine your results in class and determine the proportions for each digit. If the telephone company assigns telephone numbers at random, how should the digits 0 through 9 be distributed?

3.2 Conditional Probability and Independent Events

In many cases we are interested in probabilities for events that have special conditions imposed upon them. The probability that an ace will be drawn from a deck of cards, for example, depends on whether the deck is a bridge deck or a pinochle deck; in the former case the probability is 1/13, while in the latter case it is 1/6. Decisions must often be made with reference to a certain order of events involving probabilities that may depend on the previous sequence of events. To determine relevant probabilities in such cases, we must consider the concept of *conditional probability*.

3.2.1 Conditional Probability

Suppose that the Arkay Corporation has let bids on a contract to two firms, Ellanjay and Zero. Based on past experience, Arkay management has determined that the subjective probability that Ellanjay will have the low bid, and therefore be awarded the contract, is 75%. One of two executives, Mr. Wells or Ms. Rogers, will be assigned as liaison for the project. Based on their experience, management expects Ms. Rogers to get the assignment if Ellanjay gets the contract and Mr. Wells to be the likely choice if the contract is awarded to Zero. Other factors must be considered, however, such as other projects the executives may be working on and the preferences of the low bidding firm.

Probabilities have been assigned subjectively to the various possibilities. If Ellanjay receives the contract, the probability is 70% that Ms. Rogers will get the liaison assignment; if it is Zero, there is only a 40% chance that she will get the assignment. The probabilities assigned to her chances are called *conditional probabilities*. The probability that she will get the assignment depends on which firm is awarded the contract. We define the **conditional probability** of A, given B, as the probability that event A will occur given that event B has occurred. This is symbolized as $P(A|B)$ (read "probability of A, given B"). If we symbolize by A the event "Ellanjay gets the contract," by B the event "Zero gets the contract," and by R the event "Ms. Rogers is assigned as liaison," we are given the following probabilities: $P(A) = 0.75$, $P(R|A) = 0.7$, and $P(R|B) = 0.4$.

We can infer several other probabilities. Since either Ellanjay or Zero will get the contract, the events A and B are complementary, so that $P(B) = 0.25$. If W symbolizes the event "Mr. Wells is assigned as liaison," we know that $P(W|A) = 0.3$ and $P(W|B) = 0.6$, since W and R are also complementary events. We could draw a Venn diagram for these various outcomes, but for sequential events a different type of drawing is generally more useful. Such a diagram of the various outcomes and their probabilities is called a **tree diagram**, shown in Figure 3.7.

The first event that will occur is awarding the contract, shown in Figure 3.7 by the first diverging lines—the **branches.** The second event will be the selection of the liaison executive. This occurrence is shown by the second set of branches. These events are actually the conditional events $W|A$, $R|A$, $W|B$, and $R|B$; since they are shown on the second set of branches, the conditional nature of the events is understood. The probabilities shown under the second set of branches are conditional probabilities. By following the branches, we see that there are actually four different final outcomes.

Figure 3.7
Tree diagram for awarding of Arkay contract.

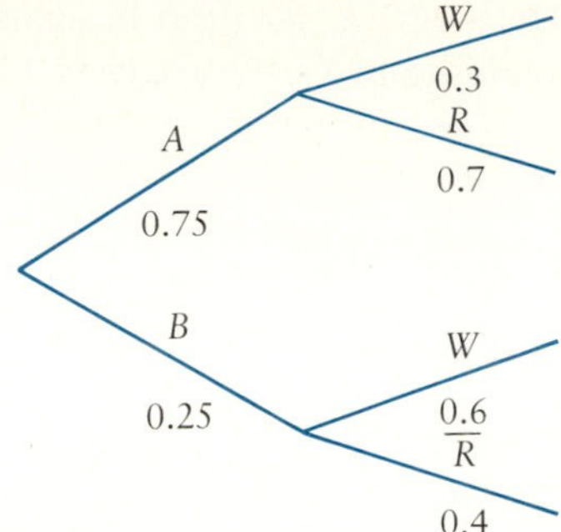

If Ellanjay gets the contract and Mr. Wells is appointed liaison, this is the event *A* and *W*. This outcome corresponds to the topmost branch on the tree. The other outcomes, in order, are *A and R, B and W*, and *B and R*, as shown in Figure 3.8.

Now the probability that Wells will be the liaison executive if Ellanjay is awarded the contract is 0.3; the probability that Ellanjay *will* be awarded the contract is 75%. There is a 30% chance of a 75% chance that both of these things will happen, so we can say that the probability that *A* will be awarded the contract *and* Wells will get the position is (0.75)(0.3) or 0.225. Similarly, the probability that Zero will be awarded the contract and Wells will get the position is (0.25)(0.6) or 0.150. Clearly we can find the probability of each of the compound events simply by multiplying the probabilities on the path leading to that event. Thus we see that

$$P(W|A) = 0.3 \text{ and } P(A) = 0.75, \text{ so } P(A \textit{ and } W) = 0.225.$$
$$P(R|A) = 0.7 \text{ and } P(A) = 0.75, \text{ so } P(A \textit{ and } R) = 0.525.$$
$$P(W|B) = 0.6 \text{ and } P(B) = 0.25, \text{ so } P(B \textit{ and } W) = 0.150.$$
$$P(R|B) = 0.4 \text{ and } P(B) = 0.25, \text{ so } P(B \textit{ and } R) = 0.100.$$

These probabilities are shown in Figure 3.9. We can generalize these results to the **multiplication rule for probability.**

Figure 3.8
Possible outcomes in the Arkay example.

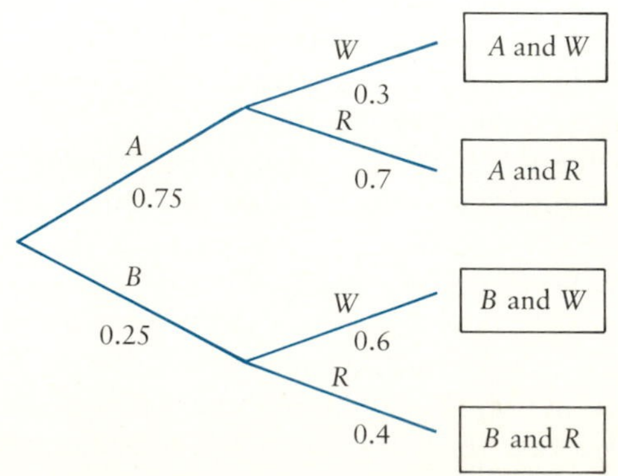

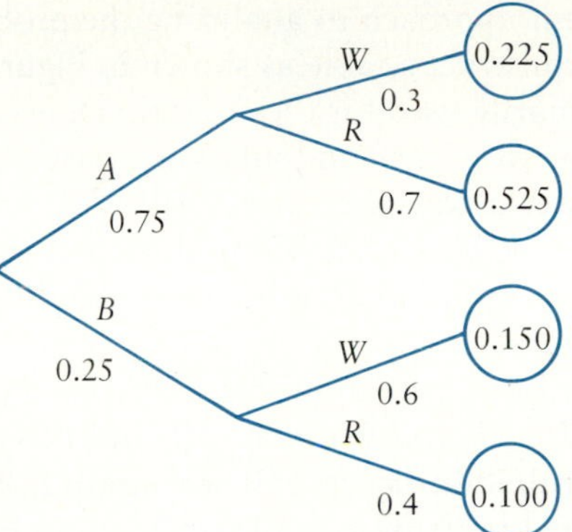

Figure 3.9
Completed tree diagram for the Arkay example.

Multiplication Rule for Probability

For any events A and B,

$$P(A \text{ and } B) = P(A|B) \cdot P(B)$$

and

$$P(A \text{ and } B) = P(B|A) \cdot P(A)$$

Who is more likely to get the assignment, Wells or Rogers? The probability that Wells gets the assignment is equal to the sum of all the mutually exclusive events in which he does get the assignment. In this case, $P(W) = P(A \text{ and } W) + P(B \text{ and } W) = 0.225 + 0.150 = 0.375$. Figure 3.10 shows these probabilities as unshaded in the circles. The shaded probabilities correspond to the events in which Rogers gets the assignment. We see that $P(R) = P(A \text{ and } R) + P(B \text{ and } R) = 0.525 + 0.100 = 0.625$. Ms. Rogers appears to have a greater chance of getting the assignment than Mr. Wells.

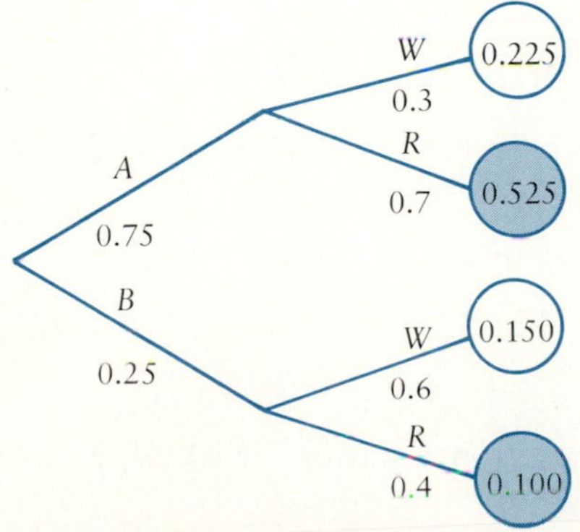

Figure 3.10
Probabilities of Rogers (shaded) or Wells (unshaded) getting the assignment.

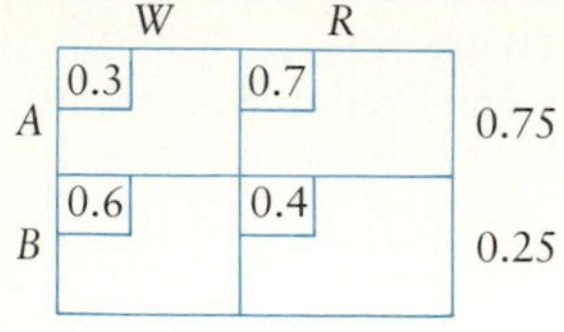

	W	R	
A	0.3	0.7	0.75
B	0.6	0.4	0.25

Figure 3.11
First step in probability table for Arkay example.

	W	R	
A	0.3 0.225	0.7 0.525	0.75
B	0.6 0.150	0.4 0.100	0.25

Figure 3.12
Joint probabilities added to table for Arkay example.

	W	R	
A	0.3 0.225	0.7 0.525	0.75
B	0.6 0.150	0.4 0.100	0.25
	0.375	0.625	

Figure 3.13
Finished probability table for Arkay example.

A slightly different approach to analyzing the problem involves the use of a table. We first construct a table as shown in Figure 3.11. The headings at the top are customarily used for the conditional events. We list the probabilities of A and B on the right-hand side while the conditional probabilities are placed in the inset boxes. We can read $P(W|A) = 0.3$, $P(R|A) = 0.7$, $P(W|B) = 0.6$, and $P(R|B) = 0.4$. We find the joint probabilities by multiplying the conditional probabilities by the probabilities on the right-hand side as shown in Figure 3.12.

We can read the joint probabilities as follows: $P(A \text{ and } W) = 0.225$, $P(A \text{ and } R) = 0.525$, $P(B \text{ and } W) = 0.150$, and $P(B \text{ and } R) = 0.100$. We then add the joint probabilities vertically to obtain $P(W) = 0.375$ and $P(R) = 0.625$. The finished probability table is shown in Figure 3.13.

Both a tree diagram and a probability table convey the same information. Which one should be used? The answer is often a matter of personal choice. We can use tree diagrams in cases where the table would be cumbersome, but in many cases we can use either format.

We can extend this method of finding the probabilities of the unconditional events (not dependent on other events) to any number of conditional events. Suppose that $\{B_1, B_2, B_3, \ldots, B_n\}$ is a partition of a sample space; that is, the events are mutually exclusive and exhaustive. Suppose also that A is any event in the same sample space. The probability of A equals the sum of the probabilities of each of the compound events $A \text{ and } B_1$, $A \text{ and } B_2$, and so forth. In Figure 3.14, the shaded portions represent the events $A \text{ and } B_1$, $A \text{ and } B_2$, and so on, through $A \text{ and } B_7$.

By the multiplication rule for probability, we can express the probability of each different shaded area in Figure 3.14 as the probability of A given a certain B_i times the probability of the B_i. Thus $P(A \text{ and } B_1) = P(A|B_1) \cdot P(B_1)$, $P(A \text{ and } B_2) = P(A|B_2) \cdot P(B_2)$, $P(A \text{ and } B_3) = P(A|B_3) \cdot P(B_3)$, and so on, through $P(A \text{ and } B_n) = P(A|B_n) \cdot P(B_n)$. The probability of A is the probability of the total shaded area; that is, it is the sum of the probabilities of each shaded area. We thus have the **rule of total probability.**

Rule of Total Probability

If $B_1, B_2, B_3, \ldots, B_n$ is a partition of a sample space and A is any event in the sample space, then

$$P(A) = P(A|B_1) \cdot P(B_1) + P(A|B_2) \cdot P(B_2) + \cdots + P(A|B_n) \cdot P(B_n)$$

We can use summation notation, if we wish, to write the rule more compactly. We can write

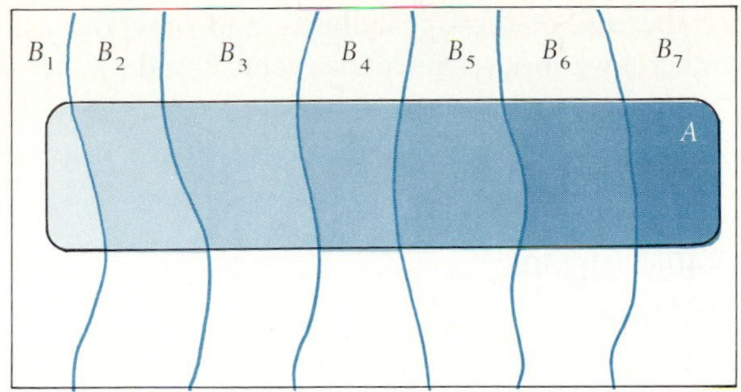

Figure 3.14 Joint probability diagram for partitioned sample space.

$$P(A) = \sum_{i=1}^{n} P(A|B_i) \cdot P(B_i)$$

or simply

$$\sum_i P(A|B_i) \cdot P(B_i)$$

(For a brief overview of this summation notation, see the note at the end of Chapter 2.) We can apply the rule of total probability to the Arkay example. Since A and B form a partition of the sample space, by the rule of total probability we have $P(W) = P(W|A) \cdot P(A) + P(W|B) \cdot P(B) = (0.3)(0.75) + (0.6)(0.25) = 0.225 + 0.150 = 0.375$. Similarly, $P(R)$ equals 0.625. Since W and R are complementary events, the sum of their probabilities must equal 1—and it does.

Example 3.7

A bill that has been passed by the legislature has been sent to the governor for signature. There is some opposition to the bill, however, and the governor has sent it back to a committee for further study and a clarification of several points. Representative Wilson, who sponsored the bill, notes that only one of the 14 members of the committee actually voted for its passage, so he believes that there is only 1 chance in 14 that the committee will write a report favorable to the bill. If it does, Wilson thinks there is about a 90% chance that the bill will be signed into law. If the committee report is not favorable, Wilson still feels there is about a 70% chance the governor will sign it anyway. If these probabilities are correct, what is the probability that the bill will be signed into law?

Solution

Let G represent the event "the governor signs the bill," and let C represent the event "the committee report is favorable." Then $P(C) = 1/14$, $P(\sim C) = 13/14$, $P(G|C) = 0.9$, and $P(G|\sim C) = 0.7$. We also know that $P(\sim G|C) = 0.1$ and $P(\sim G|\sim C) = 0.3$. A tree diagram for the problem is shown in Figure 3.15. The joint probabilities are given in the circles and boxes. The circles correspond to the events in which the governor signs the bill; the boxes correspond to the events in which the governor does not sign the bill. Then $P(G)$ is the sum of the numbers in the circles, and we see that $P(G) = 100/140 = 5/7$.

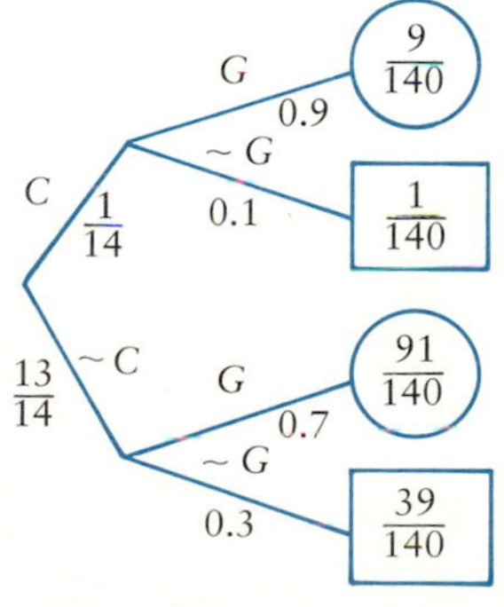

Figure 3.15 Tree diagram for Example 3.7.

We can use the rule of total probability and omit the use of the tree diagram altogether if we choose. Since the events C and $\sim C$ are the partition of the sample space we need to use, we have

$$\begin{aligned} P(G) &= P(G|C)\cdot P(C) + P(G|\sim C)\cdot P(\sim C) \\ &= (0.9)(1/14) + (0.7)(13/14) \\ &= 9/140 + 91/140 \\ &= 100/140 \\ &= 5/7 \end{aligned}$$

In many cases we do not know the conditional probabilities but need them. In such cases we may use the multiplication rule for probability to define conditional probability. If $P(A \text{ and } B) = P(A|B) \cdot P(B)$, then $P(A|B) = P(A \text{ and } B)/P(B)$.

Conditional Probability Rule

If both $P(A \text{ and } B)$ and $P(B)$ are given, then the conditional probability of A given B is defined by

$$P(A|B) = \frac{P(A \text{ and } B)}{P(B)}$$

provided $P(B) \neq 0$.

We may consider the conditional probability of A given B as the probability of event A *restricted to* B.

Example 3.8

In a group of 40 students, 20 are taking an English course, 16 are taking a business course, and 12 are taking both. Let E represent the event "student is taking an English course," and let B represent the event "student is taking a business course." Determine $P(B|E)$ and $P(E|B)$.

Solution

A Venn diagram may be helpful (Figure 3.16). We have $P(E) = 20/40 = 0.5$, $P(B) = 16/40 = 0.4$, and $P(E \text{ and } B) = 12/40 = 0.3$. If one student is selected and is known to be taking an English course, then the student is one of those 20. Thus we are restricted to the set E, which has exactly 20 elements. Since 12 of these elements are also in set B—that is, they also take a business course—the probability that our English-taking student also takes business is 12/20 or 0.6. By using the conditional probability rule,

$$P(B|E) = \frac{P(B\,and\,E)}{P(E)} = \frac{0.30}{0.50} = 0.60$$

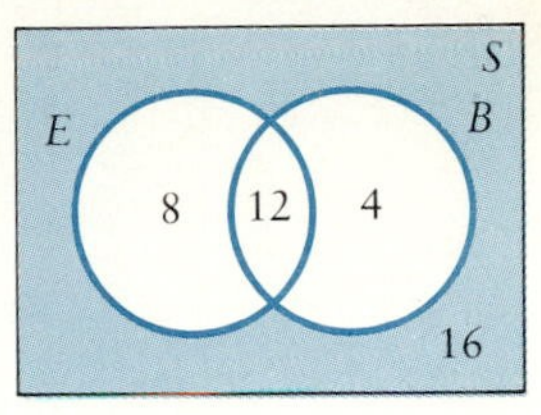

Figure 3.16
Venn diagram for Example 3.8.

The probability we obtain by the rule coincides with that shown in the Venn diagram, as it should. For more complex problems, we may not be able to draw a Venn diagram as easily understood as this one.

We may determine $P(E|B)$ either by reference to the Venn diagram (12/16 = 0.75) or by use of the conditional probability rule. In the latter case,

$$P(E|B) = \frac{P(E\,and\,B)}{P(B)} = \frac{0.30}{0.40} = 0.75$$

Often we are able to obtain raw data classified in a table called a *contingency table* (discussed further in Chapter 9). We can calculate joint, conditional, and total probabilities easily from such tables. Suppose that people in a random sample are classified as high or low in income and by party affiliation, with the following results:

	PARTY AFFILIATION			
INCOME LEVEL	REPUBLICAN	DEMOCRAT	OTHER (OR NONE)	TOTAL
High	44	62	18	124
Low	28	84	24	136
Total	72	146	42	260

With this table we can easily determine many probabilities. The probability that a person taken at random from this sample is a Republican is 72/260, or about 0.277. The joint probability that such a person is a Democrat with high income is 62/260, or about 0.238. The conditional probability that a

PROFICIENCY CHECKS

6. Suppose we know from experience that about 25 percent of the customers at a service station use super unleaded gasoline and 10% use super unleaded gasoline and pay with credit cards. If a customer drives in and buys super unleaded gasoline, what is the probability that the customer will use a credit card?

7. A doctor has a patient with a complaint that is probably one of two related conditions. The doctor thinks the probability that it is condition A is 0.35 and that it is condition B is 0.55. There is a 10% chance, she feels, that the patient's complaint is neither one of these conditions and is treatable by a simple palliative procedure. On the other hand, 90% of patients with condition A

(*continued*)

PROFICIENCY CHECKS *(continued)*

require surgery whereas only 30% of patients with condition B require surgery. If these probabilities apply, what is the probability that the patient will require surgery?

8. Among a group of patients at a mental health clinic, 35% suffer from cognitive dissonance as a result of their experiences and 40% have a distorted sense of reality as measured by a standard measuring device. A total of 28% of the group consists of patients suffering from both conditions. A staff member is assigned one of the patients at random and discovers that the patient has a distorted sense of reality. What is the probability that the patient suffers from cognitive dissonance?

Republican from this group has a high income is 44/72, about 0.611, and the conditional probability that a person with low income from this group is a Republican is 28/136, about 0.206.

3.2.2 Independent Events

Two events are said to be statistically *independent* if the occurrence or nonoccurrence of one event does not affect the probability that the other will occur. Suppose we toss a coin twice and the first toss comes up a head. Does this result affect the probability of getting a head on the second toss? Of course not. The outcome on the first toss does not affect the probability of the results of the second toss. Consider two events A and B. If the result of event B has no effect on the probability of event A, then $P(A|B)$ is no different from $P(A)$. The two events A and B are considered **independent.**

Independent Events

Two events A and B are said to be independent if and only if

$$P(A|B) = P(A)$$

We can determine several other facts about independent events. If A and B are independent, not only does $P(A|B) = P(A)$, but $P(B|A) = P(B)$; further, $P(A|{\sim}B) = P(A)$ and $P(B|{\sim}A) = P(B)$.

By the conditional probability rule, $P(A \text{ and } B) = P(A|B) \cdot P(B)$ for any events A and B. If A and B are independent events, $P(A|B) = P(A)$ and we have the **multiplication rule for independent events.**

Multiplication Rule for Independent Events

A and B are independent events if and only if

$$P(A \text{ and } B) = P(A) \cdot P(B)$$

The converse of this rule is also true. We can see this easily by observing that for any events A and B, we have $P(A \text{ and } B) = P(A|B) \cdot P(B)$. Therefore, if $P(A) \cdot P(B) = P(A \text{ and } B)$, then $P(A) \cdot P(B) = P(A|B) \cdot P(B)$. Dividing both sides by $P(B)$, provided $P(B) \neq 0$, we have $P(A) = P(A|B)$, and A and B are independent. Thus if $P(A) \cdot P(B) = P(A \text{ and } B)$, then A and B are independent.

Example 3.9

Determine whether events E and B in Example 3.8 are statistically independent.

Solution

We have $P(B) = 0.40$ and $P(B|E) = 0.60$, so the events are not independent; if B and E were independent, $P(B)$ and $P(B|E)$ would be equal. Another approach is to note that $P(E \text{ and } B) = 0.3$, while $P(E) \cdot P(B) = (0.5)(0.4) = 0.2$, which is not equal to $P(E \text{ and } B)$. Therefore the events are not independent.

Example 3.10

If two teams are approximately equally matched in baseball, assume that the probability the home team will win is 0.6. If this is true, and assuming independence of outcomes, what is the probability that team A will win the first four games of a World Series from team B? The first two games are played on one team's home field, and the next two games are played on the other team's home field.

Solution

If the first two games are played at team A's field and the next two games at team B's field, the probability that A will win all four games is $(0.6)(0.6)(0.4)(0.4) = 0.0576$. If the first two games are played at team B's field, the probability that A will win all four games is also $(0.4)(0.4)(0.6)(0.6) = 0.0576$. The probability that B will win all four games is 0.0576 as well. Thus the probability that the series will last only four games (without regard for which team wins) is 0.1152.

PROFICIENCY CHECKS

9. Successive rolls of a balanced die are independent. What is the probability of rolling three 6's in a row? What is the probability of rolling three different numbers in three rolls?

10. You are to roll a balanced die until a 6 appears. What is the probability that you will require exactly four rolls? What is the probability that you will need more than three rolls?

3.2.3 Bayes' Rule*

Let us return to Example 3.7 in which the legislature passed a bill and sent it back to a committee for recommendations. Suppose that the committee's recommendations are confidential, but the governor has signed the bill. What is the probability that the recommendation was favorable? The tree diagram, with all the probabilities shown, was given in Figure 3.15.

Before the bill is sent to the governor from the committee, the probability that the committee's recommendation will be favorable is 1/14 and the probability that it will be unfavorable is 13/14. These are called the **prior probabilities.** That is, they are the probabilities before (prior to) obtaining any other information. If we know that the governor has signed the bill, these probabilities no longer apply. We now want the conditional probability $P(C|G)$. We observe from Figure 3.15 that $P(C \textit{ and } G) = 9/140$ and that $P(\sim C \textit{ and } G) = 91/140$. These too are prior probabilities. If we know that the governor signed the bill, these probabilities no longer apply. The probability of the governor signing the bill is no longer 5/7, it is now 100%! The relative likelihoods of the committee's recommendation being favorable, however, remain the same—namely 9/140 to 91/140. Thus the chance that the committee report was favorable is (9/140)/(5/7), or 9%, while the likelihood that it was unfavorable is (91/140)/(5/7) or 91%. If the governor signed the bill, then the committee's recommendation was either favorable or unfavorable, so the probabilities of the two events are 0.09 and 0.91. These are called the **posterior probabilities.** That is, these are the probabilities after (posterior to) obtaining additional information. They can be written as $P(C|G)$ and $P(\sim C|G)$.

These probabilities are consistent with the conditional probability rule; $P(C|G) = P(C \textit{ and } G)/P(G)$ and $P(\sim C|G) = P(\sim C \textit{ and } G)/P(G)$. In terms of the prior probabilities, we could write $P(G)$ using the rule of total probability. Thus $P(G) = P(G|C) \cdot P(C) + P(G|\sim C) \cdot P(\sim C)$. Therefore, in terms

*This subsection may be omitted without loss of continuity.

of the prior probabilities, we can rewrite $P(C|G) = P(C \text{ and } G)/P(G)$ as

$$P(C|G) = \frac{P(G|C) \cdot P(C)}{P(G|C) \cdot P(C) + P(G|{\sim}C) \cdot P({\sim}C)}$$

By extension to any number of events B_i that form a partition of the sample space and a specific event B_j, we obtain **Bayes' Rule,** named for Clergyman Thomas Bayes (1702–1761).

Bayes' Rule

If $B_1, B_2, B_3, \ldots, B_n$ is a partition of a sample space and A is any event in the sample space, then

$$P(B_j|A) = \frac{P(A|B_j) \cdot P(B_j)}{P(A|B_1) \cdot P(B_1) + P(A|B_2) \cdot P(B_2) + \cdots + P(A|B_n) \cdot P(B_n)}$$

We can use summation notation, if we wish, to write the rule more compactly:

$$P(B_j|A) = \frac{P(A|B_j) \cdot P(B_j)}{\sum_{i=1}^{n} P(A|B_i) \cdot P(B_i)}$$

or simply

$$P(B_j|A) = \frac{P(A|B_j) \cdot P(B_j)}{\sum_{i} P(A|B_i) \cdot P(B_i)}$$

This rule is not as formidable as it appears. The denominator is simply the total probability of A, and the rule is just another way of writing $P(B_j|A) = P(A \text{ and } B_j)/P(A)$.

We may verify the result of this calculation of $P(C|G)$ by using Bayes' Rule as shown here:

$$\begin{aligned} P(C|G) &= \frac{P(G|C) \cdot P(C)}{P(G|C) \cdot P(C) + P(G|{\sim}C) \cdot P({\sim}C)} \\ &= \frac{(0.9)(1/14)}{(0.9)(1/14) + (0.7)(13/14)} \\ &= \frac{(9/140)}{(9/140) + (91/140)} \\ &= \frac{(9/140)}{(100/140)} \\ &= 0.09 \end{aligned}$$

Example 3.11

An oil company has selected a site to drill for a well. The site was chosen because the company thought there might be a dome structure, a shale formation deep in the ground that often presages the presence of oil. Based on geological survey information, the company geologist has assessed the probability of a dome formation at 0.4. Past data and geological information have led the geologist to estimate the probability of a major oil strike at 0.2 if there is a dome formation and 0.1 if there is not. If there is no major oil strike, there still could be a small oil strike; the probabilities of a small oil strike have been estimated at 0.5 if there is a dome and 0.2 if there is not. The only other possibility is no oil. Even after they have drilled they will not know whether there was a dome without more extensive tests. Given these prior probabilities, determine the posterior probability of there having been a dome if there was a) a major strike; b) a small strike; c) no oil.

Solution

Let D be the event "there is a dome" and $\sim D$ the event "there is no dome." Let M, S, N represent the events "major oil strike," "small oil strike," and "no oil," respectively. By Bayes' Rule:

a. $$P(D|M) = \frac{P(M|D)\cdot P(D)}{P(M|D)\cdot P(D) + P(M|\sim D)\cdot P(\sim D)}$$
$$= \frac{(0.2)\cdot(0.4)}{(0.2)\cdot(0.4) + (0.1)\cdot(0.6)}$$
$$= \frac{0.08}{(0.08) + (0.06)}$$
$$= \frac{0.08}{0.14}$$
$$\doteq 0.571$$

b. $$P(D|S) = \frac{P(S|D)\cdot P(D)}{P(S|D)\cdot P(D) + P(S|\sim D)\cdot P(\sim D)}$$
$$= \frac{(0.5)\cdot(0.4)}{(0.5)\cdot(0.4) + (0.2)\cdot(0.6)}$$
$$= \frac{0.20}{(0.20) + (0.12)}$$
$$= \frac{0.20}{0.32}$$
$$= 0.625$$

c. $$P(D|N) = \frac{P(N|D)\cdot P(D)}{P(N|D)\cdot P(D) + P(N|\sim D)\cdot P(\sim D)}$$
$$= \frac{(0.3)\cdot(0.4)}{(0.3)\cdot(0.4) + (0.7)\cdot(0.6)}$$

$$= \frac{0.12}{(0.12) + (0.42)}$$

$$= \frac{0.12}{0.54}$$

$$\doteq 0.222$$

Thus the probability of a dome structure if a major strike is made is 4/7, or about 0.571; if a small strike is made it is 5/8, or 0.625; and if no oil strike is made it is 2/9, or about 0.222. We also observe that the probability of a major strike is 0.14, that of a small strike is 0.32, and that of no strike is 0.54.

We can verify these probabilities by using the tree diagram shown in Figure 3.17. The probabilities in circles are for those compound events in which there was a major oil strike. Their sum is 0.14. Of these two, the event in which there was a dome has probability 0.08, so $P(D|M) = 0.08/0.14 \doteq 0.571$. The other probabilities can be similarly verified.

We could use a table just as easily. Figure 3.18 shows the completed table. The conditional probabilities of D given M, S, and N, respectively, can be found easily from the table. They are the joint probabilities in the first row divided by the total probabilities in the corresponding columns.

PROFICIENCY CHECK

11. Suppose that the patient in Proficiency Check 7 did not require surgery. What is the probability that the patient had:
    a. condition A?
    b. condition B?
    c. neither condition?

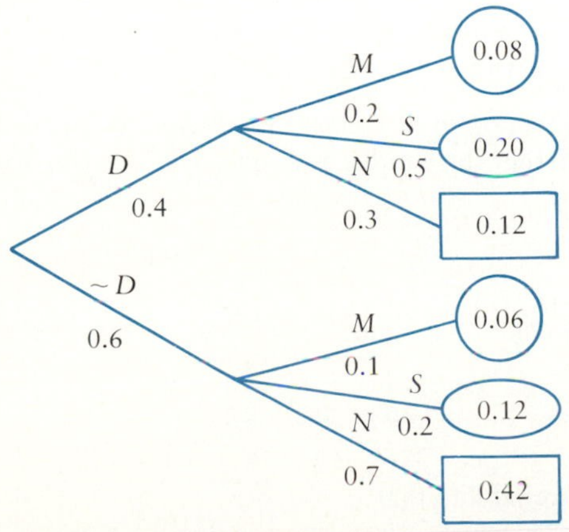

Figure 3.17
Tree diagram for the oil drilling data of Example 3.11.

Figure 3.18
Probability table for Example 3.11.

	M	*S*	*N*	
D	[0.2] 0.08	[0.5] 0.20	[0.3] 0.12	0.4
$\sim D$	[0.1] 0.06	[0.2] 0.12	[0.7] 0.42	0.6
	0.14	0.32	0.54	

Problems

Practice

3.25 Find $P(A \textit{ and } B)$ if $P(A) = 2/3$, $P(B) = 1/5$, and:
a. $P(A|B) = 0.1$ **b.** $P(B|A) = 0.1$
c. $P(A|B) = P(A)$ **d.** $P(B|A) = P(B)$

3.26 Fifteen white marbles and six red marbles are placed in a box. What is the probability of drawing three white marbles:
a. if the marbles are replaced after each drawing?
b. if the marbles are not replaced?

3.27 A pair of balanced dice is rolled until a total of 7 appears. If $P(x = 7) = 1/6$, what is the probability that more than three rolls will be necessary?

3.28 Three especially prepared cards are put into a hat. One is red on both sides, one is white on both sides, and the third is red on one side, white on the other side. A card is drawn at random from the hat and placed on the table. The side showing is red. What is the probability that the other side is red?

Applications

3.29 In a group of 100 students, 80 take an English course, 60 take a mathematics course, and 10 take neither. Are the two events "takes an English course" and "takes a mathematics course" independent? Why?

3.30 In a set of 60 teachers, 20 teach history and 6 are coaches. If coaching and teaching history are statistically independent for this set, how many of the coaches teach history?

3.31 An observer at a dog show counts the different numbers of dogs by breed and also notes whether the handler is a man or a woman. Some of the results are shown here:

SEX OF HANDLER	BREED: COLLIE	SPANIEL	POODLE	TERRIER
Male	12	34	21	36
Female	9	43	36	14

What is the probability that:
a. the handler is female, given that the dog is a terrier?

b. the dog is a collie, given that the handler is male?
c. the handler is male, given that the dog is a collie?

3.32 A bowler has a probability of 1/5 of getting a strike if he uses the proper weight ball. If he uses the improper weight ball, the probability drops to 1/8. There are six balls in the rack, but only two are the proper weight for him.
a. If he chooses a ball at random, what is the probability that he will get a strike?
b. If he gets a strike, what is the probability that he used the proper weight ball?

3.33 According to a survey reported in *Newsweek* (Feb. 17, 1986), 12% of all households own a telephone answering device. Suppose the probability that no one will be home to answer the phone between 9 AM and noon is 0.40 among households without an answering device and 0.80 among households with an answering device. Suppose that every household with an answering device turns on the device when they leave home. If you call a number at random between 9 AM and noon, what is the probability that if no one is home the call will be answered by an answering device?

3.34 Medical records of a doctor's patients show the following data:

	PATIENTS WHO WERE	
PATIENTS WHO HAVE	OVERWEIGHT	NOT OVERWEIGHT
High blood pressure	121	63
Normal blood pressure	52	88

What is the probability that one of the doctor's patients:
a. is overweight and has high blood pressure?
b. who is overweight has high blood pressure?
c. who has high blood pressure is overweight?
d. has high blood pressure?

3.35 A patient suffering from the disease described in Problem 3.16 is admitted to a hospital.
a. If the patient does not recover quickly, what is the probability that he will die?
b. If he does not die, what is the probability that he recovered quickly?

3.36 The owner of a department store is giving away a door prize by randomly selecting one of the entry forms filled out by the first 1000 customers to enter the store on a particular morning. The breakdown into those customers who have a charge account and those who do not and also into those who purchased a sale item and those who did not is as follows:

	CHARGE ACCOUNT	NO CHARGE ACCOUNT	TOTAL
Sale item purchasers	350	130	480
Nonpurchasers	210	310	520
Total	560	440	1000

a. What is the probability that the prize will be won by a charge account holder?
b. What is the probability that the prize will be won by a charge account holder who bought a sale item?
c. If the prize was won by a charge account holder, what is the probability that the prizewinner bought a sale item?
d. If the prize was won by a sale item purchaser, what is the probability that the prizewinner had a charge account?

3.37 Let R be the event "the stock market rises," K the event "Amalgamated Checkers stock rises," and T the event "the stock market falls." Suppose that $P(R) = 0.3$, $P(T) = 0.4$, and $P(K) = 0.2$. Assuming that K is independent of both R and T, which are obviously not independent of each other (why?), symbolize and determine each of the following probabilities.
a. the market rises and Amalgamated rises
b. the market falls and Amalgamated rises
c. the market does not rise, but Amalgamated rises
d. the market remains steady
e. Amalgamated does not rise, or the market falls
f. neither the market nor Amalgamated rises

3.38 It is highly unlikely, in Problem 3.37, that the performance of Amalgamated Checkers stock is independent of market fluctuations. Suppose $P(K|R) = 0.7$, $P(K|T) = 0.1$, and the probability that Amalgamated will rise if the market remains steady is 0.4. We still have $P(R) = 0.3$ and $P(T) = 0.4$.
a. Determine the probability that Amalgamated Checkers stock will rise.
b. Suppose Amalgamated does rise. What is the probability that there was a general rise in the stock market?

3.39 Two persons are to be selected from a group of people classified on the basis of social mobility and economic stability. The proportion of socially mobile people is 0.40, while that of economically stable people is 0.50. If the two factors are independent, what is the probability that at least one person selected from this group demonstrates both characteristics? What is the probability that at least one of those selected demonstrates at least one of the characteristics?

3.40 A certain species of toad has two interesting traits: resistance to a certain fungus infection varies widely among individuals, and skin variations known as warts also vary widely among individuals. A biologist thinks there is some relation between number of warts and resistance to the fungus. In a sample of toads, she classifies them into equal groups of high, average, and low incidence of warts. She then tests the toads for resistance to the fungus and finds one-fourth of them to have high resistance to the fungus. If number of warts and resistance to the fungus are actually independent, what proportion of the toads would be expected to have a large number of warts *and* high resistance to the fungus? (A means of testing such a situation will be given in Chapter 9.)

3.41 During the summer of 1983 a team of marine biologists took daily water samples from Bayou Chico to determine the proportion of plankton in bloom. The water sample was to be analyzed later at a university laboratory to be certain that it contained a minimum number of plankton. The sample was to be taken at 6:00 AM, but because of the possibility that a sample might not be acceptable, samples were also taken at 5:00 AM, 5:30 AM, 6:30 AM, and 7:00 AM. The analysis of the samples was begun at 8:00 AM. The 6:00 analysis was performed first. If the sample was acceptable, no other analyses were performed. If not, the 5:30, 6:30, 5:00, and 7:00 analyses were performed, in that order, until two of them were acceptable. The probability that an analysis would be acceptable was found to be 0.70. If the acceptability of two samples selected a half-hour apart were independent events, determine the probability that a given day's analysis would be successful—that is, either the 6:00 sample or two of the other samples would be acceptable.

3.3 Random Sampling

To make valid inferences about a population from a sample, we must be reasonably certain that the sample is representative of the population. There are many ways to obtain a sample that is representative of a population (see, for example, Sheaffer and others [1978] or Kohler [1985], pp. 8–53), and one of the simplest is called a *simple random sample*. Most statistical theory is based on the assumption that a sample being used to estimate a population parameter is a random sample. Although random samples are sometimes unavailable, many inferential results apply only to samples that are truly random.

A sample of size n is said to be a **random sample** of a population if every sample of size n has the same probability of being selected—that is, all the samples are equally likely. Many people think they can select a random sample fairly. Most people cannot. If you ask a hundred people to open a book at random, it is likely that every one of them will open the book somewhere in the middle third. There is almost no chance that a person will open the book near the front or the back. The systematic exclusion of most of the book is certainly not random. Similarly, if you were asked to write down 25 digits at random, you would very likely not write down a zero and you would probably not write down the same number twice in a row. Yet it can be shown that if 25 digits are selected truly at random, the probability that a zero will not appear is only about 0.07, and the probability that there will not be two consecutive digits alike at least once is only about 0.08. Systematic exclusion, as in the case of the person opening a book or writing down digits, means that the resulting "sample" is not random.

The problem with people attempting to achieve randomness without using a statistical procedure is illustrated well by the methods used by the federal government in draft lotteries. (For details see Stephen E. Fienberg [1971].) In 1940, capsules containing numbers of men eligible for the draft were drawn from a fishbowl. The capsules were stirred by means of a small paddle that did not reach very far into the bowl and also broke open some of the capsules. When the numbers were drawn, they were concentrated in certain hundreds, reflecting the fact that they had been placed in the bowl in lots of 100.

In 1970 a similar procedure was used, but there was no stirring. The order in which the birthdates were drawn reflected the fact that the end-of-the-year birthdates were put in last and drawn out first. The numbers drawn in both lotteries were anything but random. The 1971 draft lottery, however, made use of random number tables (described in Section 3.3.2) and was truly random.

3.3.1 Random Samples

Suppose we are asked to poll 10% of the student body. We may not be able to obtain a random sample by asking every tenth student. Consider the students standing in line to register. If we choose students 10, 20, 30, 40, and so on, we will have 10% of the students, but the sample will not have been chosen at random because a sample containing students 6 and 7, for example, will have had *no chance* of being chosen, no matter which student we began with. If the students were grouped in the line at random, however, without system or pattern, such a sample could be considered, for all practical purposes, to have been randomly obtained. This random grouping would be unlikely, though, for students often are assigned registration times based on classification, or alphabetically, or they could be grouped together by major or simply in friendship groups.

We could obtain a simple random sample in this case by using a spinner with numbers 1 through 10 on it, each having the same size of central angle so they will be equally likely. We could choose an indicator number, say 1, and spin the spinner each time a student shows up. If a 1 appeared, the student would be polled, otherwise not. There are many less clumsy methods that we could use. The most common method of selecting a random sample from a given population is to use random numbers either from a table or generated by a computer. Methods are generally applied beforehand to preselect the sample. If we wanted to ask one-twelfth of the students in a cafeteria line if they were satisfied with the food, we could use a pair of dice and call "10" a success, since the probability of obtaining a 10 is 1/12. Then we could toss the dice, counting the tosses, and record the number of the toss on which each 10 occurred. Such a list might look like this:

11, 17, 24, 42, 44, 50, 57, 78, 102, 104, 119, 143, . . .

We would continue until we had enough. Then, when the students were in the line, we would simply count and ask students 11, 17, 24, 42, 44, and so on, and be assured that our sample was randomly selected. If we could be absolutely sure that the students were randomly arranged in line, it would be all right to choose every twelfth one.

Sometimes there is no practical way to obtain a sample randomly. As a general rule, we can obtain a random sample from any population by a random procedure and from a randomly arranged population by a nonrandom procedure. If we obtain a sample from a nonrandomly arranged population by a nonrandom procedure, however, it cannot be considered random, and the statistical methods applicable to a random sample will not be valid. If you were to poll the faculty of a university by stopping in at every tenth office, the result would probably not be random. If the mathematics faculty were arranged in nine consecutive offices, they might have no chance of inclusion, and certainly there is no way more than one member of the mathematics faculty could be included.

There are many other ways to select a sample, including stratified samples, cluster samples, and sequential samples. Detailed methods of sampling can be found in books on sampling procedures.

3.3.2 Using a Random Number Table

A good procedure in selecting a sample is to use **random numbers.** Random numbers are digits selected in such a way that the probability of any digit being included at each selection is 0.1. There are specially prepared books containing random numbers, and there are computer procedures designed to generate numbers that are random or practically random (see Section 3.5 for examples).

A single page of a table of random numbers is reproduced in Table 1 in the back of the book. In such a table the digits 0, 1, 2, 3, 4, 5, 6, 7, 8, and 9 are arranged in sequence completely at random. The sequence is often generated by computer. The probability of a particular digit being selected as the "next" digit is 0.1 for each selection. The numbers can be used singly (0, 1, 2, . . . , 9), as two-digit numbers (00, 01, . . . , 99), or however we need them. In selecting the students in the cafeteria line, for example, we could let every twelfth number represent a selection; that is, if the number were 12, 24, 36, 48, 60, 72, 84, or 96, the student would be selected. One problem arises, however. We note that there are 8 such numbers. To preserve the ratio of 1 selection to 11 nonselections, we should have 88 numbers represent nonselections. Since $88 + 8 = 96$, we must exclude four numbers (any four) if the ratio of 1 selection in 12 is to hold. It might be easier to use 01 through 08 to represent a selection, 09 through 96 to represent nonselection, and ignore the remaining numbers (97, 98, 99, 00).

One method of using the table to select a sample would be to determine the size of the population to be sampled and then assign every member of the population a number. If we are to select 50 students from 5054, for instance, we must use four-digit numbers. The numbers 0001 to 5054 will represent the students. The numbers 5055 through 9999 and 0000 are ignored. If there were fewer than 5000 students, say 4500, it would be preferable not to ignore half the numbers. We could let the numbers 0001 through 4500 and 5001 through 9500 represent students, ignoring 4501 through 5000 and 9501 through 9999 and 0000. Numbers 3123 and 8123 would represent the same student. This method would effectively halve the amount of work necessary to select the sample.

There are often specific procedures for entering a table of random numbers, usually given in the book of tables. One way is to open the book anywhere and put your finger anywhere on the page. This is not your starting place, however, because the place you selected is not really at random. Rather it tells you where to start. Suppose the book contains more than 99 but fewer than 1000 pages. Then the first three digits specify the page on which you are to start. The next two digits identify the line, and the next two digits the column on which you are to begin. There are usually 50 lines and 50 columns on the pages, so you may decide beforehand to let two-digit numbers above 50 be reduced by 50 or to ignore them altogether. Now suppose the numbers you obtained were 1092778. You would then turn to page 109, line 27, column 28 (since $78 - 50 = 28$). Suppose the entries at that point are 30966005086192670496 . . . . In the example of the 5054 students, you would use these digits four at a time to obtain 3096, 6005, 0861, 9267, 0496. You would ignore numbers for which there are no students. If you have numbered the students from 1 to 5054, you would select the students numbered 3,096, 861, and 496 as your first selections. When you run out of numbers in the line at which you entered the table, you go on to the next line. You would stop when you had selected 50 different students.

Random numbers can also be used to simulate experiments. We can simulate a toss of 25 coins by calling odd digits (1, 3, 5, 7, 9) "head" and even digits (0, 2, 4, 6, 8) "tail." These outcomes are equally likely. If, in another case, we want the probability of success to be 5/14, we can let 01, 02, 03, 04, 05 represent success, 06, 07, 08, 09, 10, 11, 12, 13, 14 represent failure, and ignore all other two-digit numbers. This procedure would be quite time consuming, however, since we would be ignoring 86% of the numbers. A better method would be to use the greatest multiple of 14 less than 100, which is 98. Since $98 = 7(14)$, we would have 7(5) or 35 numbers representing success and 7(9) or 63 numbers representing failure. We would ignore the other two numbers. Thus 01 through 35 would represent success, 36 through 98 would represent failure, and 99 and 00 would be ignored. Selection of the actual numbers representing success and failure is arbitrary but unimportant, since all two-digit numbers have the same probability of being chosen.

Example 3.12

Use Table 1 in the back of the book to divide a group of 25 students randomly into three groups.

Solution

Supppose, for convenience, we let the students be represented by the letters *A* through *Y*. We let each number represent a group so that the three groups are equally likely. Suppose we close our eyes and stick a pin into Table 1 to determine where to enter it. Say that the pin hits line 20, column 5, third number. This number is a 1. The first four numbers are 1810. Thus we will enter the table in line 18, column 10. Our first few numbers, then, are 58174428882. We may use 1, 2, 3 for group I, use 4, 5, 6 for group II, use 7, 8, 9 for group III, and ignore 0. So we have the following pairings:

A	*B*	*C*	*D*	*E*	*F*	*G*	*H*	*I*	*J*	*K*	*L*	*M*	*N*	*O*	*P*	*Q*	*R*	*S*	*T*	*U*	*V*	*W*	*X*	*Y*
5	8	1	7	4	4	2	8	8	8	2	2	7	3	6	9	8	8	1	8	3	6	5	8	4

- Group I: *C G K L N S U*
- Group II: *A E F O V W Y*
- Group III: *B D H I J M P Q R T X*

Sometimes it is advantageous for the groups to be approximately or exactly (if possible) the same size. We can use a modification of the procedure described above to select samples of the same size. Since one-third of 25 is about 8, after the eighth assignment to group III we might ignore 7, 8, 9 until all other groups have 8 as well. Thus after *Q* is assigned to group III, we only use number 1 through 6, as follows:

R	*S*	*T*	*U*	*V*	*W*	*X*	*Y*
1	3	6	5	4	6	2	5

After *W* is assigned to group II, that group also has 8, so we can use only 1, 2, 3. After *X* is assigned to group I, all three groups have 8, so *Y* can have any integer 1 through 9. The procedure still produces a random grouping since the selection of the starting point in the table of random numbers was at random.

PROFICIENCY CHECK

12. Use Table 1 to select randomly a sample of five letters of the alphabet.

3.4 Counting Rules*

Often the number of possible outcomes of a statistical experiment is extremely large. It would be quite difficult and certainly tedious to list all these outcomes and count them. In many cases, however, rules can be developed for determining the number of outcomes without actually counting them. The multiplication rule, given in Section 3.1.1, is an example of such a rule.

*This section may be omitted without loss of continuity.

Several other rules can be developed from the multiplication rule and are presented in this section.

3.4.1 Permutations

Consider a set of three letters, A, B, and C. In how many ways can these three letters be arranged in order? We can answer this question by listing the arrangements as follows.

$ABC \quad ACB \quad BAC \quad BCA \quad CAB \quad CBA$

Counting these arrangements, we find that there are six different ones. Each of these arrangements is called a **permutation.** Suppose that we want to know how many ways there are to arrange three different letters of the alphabet in order. Writing the arrangements down and counting them is not a good idea in this case. Instead of writing them down we can apply our basic multiplication rule. In the case of the letters A, B, C, there are three choices for the first letter, two for the second, and only one for the third. By the multiplication rule, then, there are $3 \cdot 2 \cdot 1$ or 6 different arrangements. If we write down any three different letters of the alphabet, there would be $26 \cdot 25 \cdot 24$ or 15,600 different arrangements. (Note that if we were to write down three letters, allowing repetitions, then according to the multiplication rule the number of different ways to do this would be $26 \cdot 26 \cdot 26$ or 17,576.) We can easily extend this principle to obtain the **permutations rule.**

Permutations Rule

If r objects are to be selected from a set of n different objects in such a way that the order of selection is important, the number of permutations is P_r^n, where

$$P_r^n = n \cdot (n-1) \cdot (n-2) \cdots (n-r+1)$$

The rule follows directly from the multiplication rule, where $m_1 = n$, $m_2 = n-1$, $m_3 = n-2, \ldots, m_r = n-(r-1) = n-r+1$. The symbol for permutations can also be written ${}_nP_r$.

Example 3.13

In how many ways can a president, vice president, secretary, and treasurer be selected from an organization containing 30 members?

Solution

Since the order is important, we have four positions to be selected from 30 persons, so the number of permutations is $30 \cdot 29 \cdot 28 \cdot 27$ or 657,720.

As a consequence of the permutations rule, it can be seen that the total number of permutations of n different objects is equal to $n(n-1)(n-2)(n-3)\cdots 3\cdot 2\cdot 1$. This expression is usually written $\boldsymbol{n!}$ (read "n factorial") to facilitate writing it down. Using this notation we may rewrite the permutation rule in a more compact form.

Now $(n-r)!/(n-r)! = 1$, so if we multiply $n(n-1)(n-2)\cdots(n-r+1)$ by $(n-r)!/(n-r)!$ we obtain

$$n(n-1)(n-2)\cdots(n-r+1) = n(n-1)(n-2)\cdots(n-r+1)\frac{(n-r)!}{(n-r)!}$$

$$= \frac{n(n-1)(n-2)\cdots(n-r+1)(n-r)(n-r-1)(n-r-2)\cdots 3\cdot 2\cdot 1}{(n-r)!}$$

$$= \frac{n!}{(n-r)!}$$

Thus we have $P_r^n = n!/(n-r)!$. If $r = n$, we have the permutations of n objects taken n at a time, which is equal to $n!$. By the formula, however, we have $n!/(n-n)! = n!/0!$. In order for the formula to be valid in all cases, we must define $0! = 1$.

Example 3.14

In how many different ways can 11 people be seated in a row containing seats numbered from 1 to 11?

Solution

Since there are exactly 11 "objects" (the people), all of whom will be seated, there are 11! or 39,916,800 different orders in which they may be seated. Note that the numbering or the presence of the seats is immaterial. If the persons were seated in a row on the ground, the answer would be the same.

Example 3.15

In how many orders can 11 people be seated in a circle?

Solution

Unless there is some specified "first" position or numbered seats, it does not matter where the first person sits. If, for example, everyone sits down in a circle and then they all get up and move one position to the left or right around the circle, the order is not changed. Thus after one person is seated, all the rest may be seated in reference to that person, so the remaining ten persons may be seated in 10! or 3,628,800 different orders.

PROFICIENCY CHECKS

13. Suppose that three friends go to a movie and choose three vacant seats on an aisle. In how many different ways can they sit in the seats?

(continued)

PROFICIENCY CHECKS *(continued)*

14. Five friends order tickets for a rock concert by mail together. Unfortunately only three tickets are allowed to a customer, and they receive the three tickets and a refund for the rest. Further, the tickets are separated: one in the balcony, one in the center section, and one on the side. They draw lots for the tickets. In how many different ways can the tickets be distributed?
15. How many four-digit numbers can be formed from the six digits 1, 2, 3, 4, 5, 6, if no digit is to be repeated?

3.4.2 Partitions

In Section 3.1 we defined a partition as a division of a sample space into mutually exclusive and exhaustive events. In general, a partition of a set is a division of the entire set into distinct subsets. Earlier we used the term *permutation* to refer to arrangements of a set of objects, all of which were different. If some of the items are repetitions, such as the letters *e* in the word *tree,* the permutations rule obviously does not apply. If the two letters *e* were different, there would be 4! or 24 different arrangements of letters. Exchanging *e*'s, however, does not produce a different arrangement. Thus we must divide the 4! by 2!—the number of ways the repeated letters may be exchanged without changing the actual arrangement. Similarly, the letters in the word *bubble* can be rearranged in 6!/3! or 120 different ways. If more than one element is repeated, as in the word *classification,* each repetition is incorporated into the denominator. For *classification,* there are three occurrences of the letter *i* and two each of *a, c,* and *s*. Therefore the number of permutations of all the letters is 14!/(2!2!2!3!) or 1,816,214,400.

To determine the number of permutations of sets of elements in which there are repetitions, we regard the set as partitioned into groups of identical elements. The principle used with the word *classification* may be stated formally as the following **partitions rule.**

combine mult. rule w/ combinations

Partitions Rule

If a set contains n elements of which r_1 are of one kind, r_2 are of a second kind, and so on, through r_k, the number of permutations of all n elements is equal to

$$\frac{n!}{r_1!r_2!r_3!\cdots r_k!}$$

Example 3.16

Six screws are selected for a job, one 1 inch, two 1½ inch, and three 2 inch. The six holes are marked on the wall. In how many different orders can the screws be inserted?

Solution

By the partitions rule, the number of different ways is equal to

$$\frac{6!}{3!2!1!} = \frac{6\cdot5\cdot4\cdot3\cdot2\cdot1}{3\cdot2\cdot1\cdot2\cdot1\cdot1} = 6\cdot5\cdot2 = 60$$

This result can be checked by listing all the different possible ways, a few of which are shown here:

1 1½ 1½ 2 2 2	1 1½ 2 1½ 2 2	1 1½ 2 2 1½ 2
1 1½ 2 2 2 1½	1 2 1½ 1½ 2 2	1 2 1½ 2 1½ 2
1 2 1½ 2 2 1½	1 2 2 1½ 1½ 2	1 2 2 1½ 2 1½

If we completed the listing, we would count 60, confirming the result with the partitions rule. Note that if the screws should be inserted in a particular order but are instead put in at random, the probability that they are inserted correctly is only 1/60.

PROFICIENCY CHECK

16. How many different ten-digit numbers can we obtain by using the digits 2, 2, 2, 2, 6, 6, 6, 7, 8, 8?

3.4.3 Combinations

If we make a selection of elements in which the order is not important, the formulas for permutations do not apply. If we are to select three letters from the set $\{a, e, i, o, u\}$, for example, the number of permutations is $5 \cdot 4 \cdot 3$ or 60. If, however, we consider the set $\{a, e, o\}$ and the set $\{e, o, a\}$ to be the same—that is, if the order is not important—we must derive a new formula. Sets of this type are called **combinations.** To illustrate this concept, we list all 60 permutations of three letters selected from $\{a, e, i, o, u\}$.

aei	aie	eai	eia	iae	iea
aeo	aoe	eao	eoa	oae	oea
aeu	aue	eau	eua	uae	uea
aio	aoi	iao	ioa	oai	oia
aiu	aui	iau	iua	uai	uia
aou	auo	oau	oua	uao	uoa
eio	eoi	ieo	ioe	oei	oie
eiu	eui	ieu	iue	uei	uie
eou	euo	oeu	oue	ueo	uoe
iou	iuo	oiu	oui	uio	uoi

These 60 permutations have been grouped into ten sets of six each; each set of six consists of the permutations of the same three letters. There are, then, 60/6 or 10 different sets of letters if we disregard order. Thus to determine the total number of combinations of three different letters selected from a set of five, we first determine the total number of permutations, $5 \cdot 4 \cdot 3$, and then divide that number by the number of permutations of each distinct combination of three letters. This number is $3 \cdot 2 \cdot 1$, so the number of combinations of five objects taken three at a time is $(5 \cdot 4 \cdot 3)/(3 \cdot 2 \cdot 1)$, which may be written as $5!/(3!2!)$.

If we wish to select four different letters from the alphabet without regard to order, there are $26 \cdot 25 \cdot 24 \cdot 23$ different permutations of four letters selected from 26, and each different set of four can be arranged in $4 \cdot 3 \cdot 2 \cdot 1$ different orders, so the total number of combinations is

$$\frac{26 \cdot 25 \cdot 24 \cdot 23}{4 \cdot 3 \cdot 2 \cdot 1}$$

or 14,950. This result generalizes to the **combinations rule.** We write the symbol for combinations as C_r^n, as ${}_nC_r$, or as $\binom{n}{r}$, most often the latter.

Combinations Rule

The number of combinations of n different objects taken r at a time is equal to

$$\binom{n}{r} = \frac{n(n-1)(n-2)\cdots(n-r+1)}{r(r-1)(r-2)\cdots 3 \cdot 2 \cdot 1} = \frac{n!}{r!(n-r)!}$$

Example 3.17

How many committees of six can be chosen from an organization with 30 members?

Solution

Using the combination rule, we have

$$\binom{30}{6} = \frac{30 \cdot 29 \cdot 28 \cdot 27 \cdot 26 \cdot 25}{6 \cdot 5 \cdot 4 \cdot 3 \cdot 2 \cdot 1} = 593{,}775$$

Comment: In using a simple calculator to compute combinations, multiplication of the numbers in the numerator may yield an overflow. To avoid overflow, start out by entering 30, then divide by 6, multiply by 29, divide by 5, multiply by 27, divide by 4, and so on. An even better approach is to simplify first by canceling; $6 \cdot 5 = 30$, 4 goes into 28 seven times, 3 goes into 27 nine times, and 2 goes into 26 thirteen times. The cancellation is shown here:

$$\frac{\overset{7}{\cancel{30}} \cdot 29 \cdot \overset{9}{\cancel{28}} \cdot \overset{13}{\cancel{27}} \cdot \cancel{26} \cdot 25}{\cancel{6} \cdot \cancel{5} \cdot \cancel{4} \cdot \cancel{3} \cdot \cancel{2} \cdot 1}$$

Thus the quantity simplifies to $29 \cdot 7 \cdot 9 \cdot 13 \cdot 25$, which equals 593,775.

Example 3.18

A president, vice president, secretary, and treasurer are to be chosen in an organization with 30 members. These officers will serve on a steering committee with two additional members chosen from the remainder of the organization. How many different steering committees are there?

Solution

This problem does not precisely fit any of the situations we have encountered previously. A combination of approaches will work, however. In fact, there are two different approaches that work equally well. We may first select the officers; there are $30 \cdot 29 \cdot 28 \cdot 27$ different ways to do this. Then we may select two committee members from the remaining 26; using the combination rule, we obtain $26 \cdot 25/2$. Since we both select the officers and then the remaining members by the multiplication rule, the total number of different steering committees is $30 \cdot 29 \cdot 28 \cdot 27 \cdot 26 \cdot 25/2$, or 213,759,000. A second approach would be to select the steering committee first and then select the officers. By the combination rule (see Example 3.17), there are $30!/(24!6!)$ different committees. Then we select the officers. By the partitions rule, this can be done in $6!/2!$ different ways. By the multiplication rule, the number of ways to do both is equal to

$$\frac{30!}{24!6!} \cdot \frac{6!}{2!} = \frac{30 \cdot 29 \cdot 28 \cdot 27 \cdot 26 \cdot 25}{2} = 213{,}759{,}000$$

One other detail is of interest. We may observe that for every combination of r elements selected, there is a combination of $n - r$ elements not selected. It follows that

$$\binom{n}{r} = \binom{n}{n-r}$$

We can show this result easily by noting that, by the combinations rule, the number of combinations of n objects taken $n - r$ at a time is equal to $n!/\{[n - r]![n - (n - r)]!\} = n!/[(n - r)!r!] = n!/[n!(n - r)!]$, which is the number of combinations of n objects taken r at a time.

Thus selecting 10 birthday cards from a group of 15 different ones is equivalent to selecting a group of 5 from the 15. Instead of writing

$$\frac{15 \cdot 14 \cdot 13 \cdot 12 \cdot 11 \cdot 10 \cdot 9 \cdot 8 \cdot 7 \cdot 6}{10 \cdot 9 \cdot 8 \cdot 7 \cdot 6 \cdot 5 \cdot 4 \cdot 3 \cdot 2 \cdot 1}$$

we could just write

$$\frac{15 \cdot 14 \cdot 13 \cdot 12 \cdot 11}{5 \cdot 4 \cdot 3 \cdot 2 \cdot 1}$$

Notice that the product of 10, 9, 8, 7, 6 appears in both the numerator and the denominator of the former fraction, so we could cancel it to obtain the latter fraction.

PROFICIENCY CHECKS

17. In how many ways can 8 books be selected from a set of 20 different books?
18. Part 237KB22 can be purchased from 12 different suppliers. Seven of these parts must be purchased. To fulfill government regulations, each part must be purchased from a different supplier. In how many different ways can this be done?
19. In a modification of Proficiency Check 18, three of the seven parts will be purchased from one supplier, two will be purchased from a second supplier, and the other two parts will be purchased from two other suppliers. In how many ways may the orders be distributed among the twelve suppliers?

Problems

Practice

3.42 How many four-letter arrangements can be made from eight different letters if:

a. unlimited repetitions of a letter are allowed?
b. repetitions of a letter are not allowed?

3.43 How many different poker hands of 5 cards can be dealt from an ordinary deck of 52 different cards?

3.44 How many different permutations are there of the letters in each of the following words?

a. syzygy
b. bookkeeper
c. calamity
d. gorgeous
e. Mississippi
f. statistics

Applications

3.45 Jones Stables has entered three horses in a race, and Foster Stables has entered two horses in the same race. After the race, Mrs. Jones and Mrs. Foster talk about the race, which had only the five horses entered. Mrs. Jones notes that one of her horses finished first, but Mrs. Foster recalls that one of Mrs. Jones' horses finished last as well. If the Foster horses did not finish consecutively, how many different orders are possible based on this information?

3.46 A student is writing an examination consisting of two parts. Part A contains six problems; part B contains five problems. If the student is to omit two problems from each part, how many essentially different examinations can be written?

3.47 An advertising campaign will consist of a layout for magazines and a television campaign. There are six possible layouts for the magazine campaign and five possible television ads. If two layouts and three television ads will be used, how many essentially different advertising campaigns are there?

3.48 Among ten order slips, three are for model A, two for model B, one for model C, and four for model D. During a random check of billing procedures, one of the ten order slips is chosen at random. What is the probability that it is:

a. for model D?
b. for either model A or model D?
c. not for model A?
d. for both model B and model C?

3.49 A subcommittee of three representatives is to be selected from five men and three women members of the House Ways and Means Committee. In how many ways can the subcommittee be selected so that:

a. it consists of two men and one woman?
b. each sex is represented?
c. at least one man is on the subcommittee?
d. at least one woman is on the subcommittee?
e. there are no restrictions as to sex?

3.50 A perfect bridge hand consists of either 13 cards of the same suit or four aces, four kings, four queens, and a jack. There are four suits of 13 cards each. What is the probability of being dealt a perfect hand?

3.51 Three Ferraris, two Porsches, and five American cars of different makes are in a race. How many possible different ways can the cars finish by make?

3.52 Six friends arrive at a movie house and wish to sit together. The only row with six seats vacant, however, has people sitting in seats 1, 5, 9, and 10.

a. If a row has ten seats, in how many ways may the six friends sit in this row?
b. If all ten seats were vacant, in how many different ways could the six friends be seated in the row?

3.5 Computer Usage

In some problems it is very difficult or quite tedious to find exact probabilities for certain events. In these cases we can find an approximation for a probability by using a technique called **simulation** (also called *Monte Carlo methods*). In a simulation we repeat an experiment many times and record the outcome of the experiment each time. To approximate the probability of an event A, we simply use the ratio

$$P(A) \doteq \frac{\text{Number of experiments resulting in an outcome of } A}{\text{Number of repetitions of experiment}}$$

The accuracy of this approximation improves as the number of repetitions increases. Something like 100 repetitions is a small simulation and leads to fairly crude approximations. A simulation involving 5000 repetitions will lead to very accurate approximations.

Often we can use the computer to carry out a simulation for an experiment. Moreover, a computer can perform many repetitions of an experiment in a short period of time.

3.5.1 Minitab Commands for This Chapter

While some simulations are quite sophisticated, we can use Minitab to perform simple simulations. The RANDOM command is useful in simulations. When used along with the subcommand INTEGERS, this command randomly selects positive integers between two specified constants. The command format is as follows:

```
MTB > RANDOM K1 observations in each of the columns C, ... C ;
SUBC> INTEGERS selected randomly from K2 to K3 .
```

where the symbol K1 represents the quantity of random numbers desired and K2 to K3 is the range for the integers to be selected.

For example, to select 20 integers at random from 1 to 10, inclusive, and place the results in column 1 we use the command

```
MTB > RANDOM 20 observations in column C1 ;
SUBC> INTEGERS selected randomly from 1 to 10 .
```

The following example illustrates the simulation procedure.

Example 3.19

A game of chance is based on selecting at random, with replacement, three numbers from 1 to 50, inclusive, and adding the numbers together. A person wins the game if the resulting sum is divisible by 5 (that is, the sum is 5, 10, 15, 20, . . . , 150). Use simulation to determine the probability of winning the game.

Solution

To simulate this experiment we randomly select numbers from 1 through 50 and place them in columns 1, 2, and 3. The numbers in the same row of each column will represent the three randomly selected numbers for a single game. The sum of the three numbers in the same row is stored in C4, and each entry in C4 corresponds to the outcome for a single game. Figure 3.19 shows the results of this simulation for 200 repetitions of the game. (The numbers stored in C1 are shown for illustration, but the numbers stored in C2 and C3 are omitted.) Inspecting the results in column C4 on the printout, we see that the sum of the three randomly selected numbers is divisible by 5 thirty-five times. Our approximate probability for winning this game is about 35/200 or 0.175. Additional simulations may be used to improve the accuracy of the approximation.

Figure 3.19
Minitab printout for the simulation in Example 3.19.

```
MTB  > RANDOM 200 observations in C1 C2 C3 ;
SUBC> INTEGERS selected randomly between 1 and 50 .
MTB  > PRINT the values stored in C1

C1
   30.     2.    32.    21.    10.    44.    33.    29.    18.    49.
   45.    47.    11.     1.    24.    36.     2.    44.    42.    24.
    1.    22.    38.    44.    41.    22.    27.    11.    23.    19.
   40.    25.    16.    24.    19.    44.    21.    17.    25.    38.
    7.    46.    41.    16.    22.    11.    19.    19.    50.    37.
   49.    11.     6.    23.    15.    41.    11.    27.    15.    39.
   22.    43.    43.     7.    32.    46.    46.    26.    39.     4.
   30.    28.     2.    21.    23.    31.    41.     6.    22.     4.
   28.    36.    47.    25.    23.    40.    13.     4.    14.    48.
   37.    26.    48.     9.    46.    30.    45.    28.    42.    45.
   48.    36.    41.     8.    37.    31.     1.    36.    18.    14.
   38.    33.     1.    39.    13.    18.    30.    25.     5.    47.
   10.     7.     5.    50.    21.    24.    16.    42.    35.    36.
   45.    40.    29.    30.    13.    27.    49.    47.    24.    44.
   25.    30.    15.    39.     6.    46.    42.    30.    31.    23.
   27.    28.    21.    10.    46.    28.    50.    30.    10.    36.
   26.     4.    20.    37.    16.    16.    19.    50.    39.     5.
   40.    12.    43.    24.    45.    47.    13.    50.    15.    20.
    6.    41.    26.     2.     4.     5.    43.     4.    48.    17.
   17.    18.    25.    26.    43.    34.     8.    24.    22.    14.

MTB  > LET C4 = C1 + C2 + C3
MTB  > PRINT the values stored in C4

C4 Sum of the three randomly chosen numbers
   52     12     80*    81     56     78     65*    61     67     73    106
  121     59     87     60*    57     42     93    120*    76     83     90*
   89     90*   120*    40*    82     71     79     56     99     33     67
   46     79    121     74     81     54     91     81    113     98     42
   88     86     35*    91    109    119     56     70*    15*    66     63
  109     71     81     58    102     71     69     91     62    102     99
   73     86     90*    51     61     72     63     89     83     80*    89
   61     77     79    100*    77    107     43     82     56     51    100*
   24     71    110*    41     88     66    112     49    110*    45*    94
```

*denotes a win

```
128      72     79     78     51     89    105*    36     65*    66     39
102      77     67     75*    82     69    101     71     59    102     72
 87      14     84     86     78    107     78     77     96     85*    87
 86      50*    71     94    100*    80*    67     94     59     56     74
 64      75*    71    117    101     57     83     89     89     86     41
 93      82     92     98     60*    87     89     30*    97    102     84
 93      46    107    105*    25*    93     85*    85*   107    101     74
 44      88     26     77     58     72     88     30*    53     69     81
 51     119     78     60*   103     39     87     78    124     82     99
 65*     64
```

3.5.2 SAS Statements for This Chapter

We may use the SAS function RANUNI to obtain a result similar to that with Minitab in Example 3.19. A function is a command applied to a number (often called the *argument* of the function). For example, SQRT is the square root function, and SQRT(X) means "take the square root of x."

The SAS function RANUNI generates a uniform random number on the interval from 0 to 1—that is, a number from a set of numbers distributed uniformly over the interval. Each number in the interval thus has an equal chance of being selected. The numbers are expressed as decimals with six digits. Each random number is based on a "seed," a number that is operated on in a way to produce a number between 0 and 1. The number generated is not always the same for the same seed, since it is based on other factors as well (such as time of day). If we write

```
X = RANUNI(2345);
```

in the data step, for example, and print X, we might obtain 0.737226. We can find two-digit random numbers by multiplying x by 100 and dropping the fractional part. This result may be accomplished by using the SAS function INT. INT is the greatest integer function [ ] discussed in Section 2.3.3. We may bypass the PRINT command by using the PUT and RUN statements. The RUN statement causes the function to be executed immediately—before the next statement is read—and the PUT statement causes the output of the function to be printed in the SAS Log (that portion of the output consisting of a recounting of the input file). Thus if we had written

```
DATA RANNUMB;
  X = INT(100 * RANUNI(2345));
  PUT X;
RUN;
```

the number 73 would have been written on the SAS Log immediately following the RUN statement. Written in this way, the function will generate a random number from 0 to 99, inclusive. Since we want to generate random numbers from 1 through 50, inclusive, we multiply by 50 and add 1 before taking the greatest integer. Thus the statement

```
X = INT(50 * RANUNI(seed ) + 1);
```

will give us a random number from 1 through 50, inclusive. We may obtain 100 random numbers by using a "do loop." To obtain 100 different random numbers we may write

```
DO I = 1 TO 100;
   X = RANUNI(I);
   OUTPUT;
   END;
```

The statement OUTPUT says to store the resulting data, while the END statement is needed to stop the loop.

Here we wish to generate three different random numbers in each step. For variety (although it is not necessary) we use a different seed for each number in a step. We then add the numbers together to obtain X4. Our complete program for 100 repetitions of the simulated experiment is as follows:

```
DATA RANDOM:
 DO I = 1 TO 100;
    X1 = INT(50 * RANUNI(I) + 1);
    X2 = INT(50 * RANUNI(I+1) + 1);
    X3 = INT(50 * RANUNI(I+2) + 1);
    X4 = X1 + X2 + X3;
    OUTPUT;
    END;
```

The output of this program is printed in Figure 3.20. Nineteen of the hundred values of X4 are divisible by 5 (marked with *), so the probability of winning is about 0.19. This result compares favorably with the result obtained in Example 3.19.

```
INPUT FILE

DATA RANDOM;
   DO I = 1 TO 100;
      X1 = INT(50 * RANUNI(I) + 1);
      X2 = INT(50 * RANUNI(I + 1) + 1);
      X3 = INT(50 * RANUNI(I + 2) + 1);
      X4 = X1 + X2 + X3;
      OUTPUT;
      END;
PROC PRINT; VAR X1 X2 X3 X4;

OUTPUT FILE
```

Figure 3.20
SAS printout for simulation of 100 experiments.

SAS

OBS	X1	X2	X3	X4
1	10	49	20	79
2	13	47	49	109
3	28	27	3	58
4	4	41	27	72
5	43	4	48	95*
6	15	14	35	64
7	49	12	35	96
8	21	28	15	64
9	24	43	32	99
10	30	30	19	79
11	37	26	47	110*
12	47	30	15	92
13	20	24	34	78
14	9	9	44	62
15	15	47	46	108
16	29	3	7	39
17	26	22	9	57
18	34	21	7	62
19	23	10	29	62
20	37	22	3	62
21	27	18	2	47
22	36	47	23	106
23	48	36	6	90*
24	9	14	31	54
25	22	4	18	44
26	36	10	8	54
27	29	14	17	60*
28	29	3	22	54
29	46	27	37	110*
30	46	29	10	85*
31	17	35	7	59
32	10	14	33	57
33	22	2	14	38
34	22	42	43	107
35	44	14	16	74
36	20	18	39	77
37	28	31	28	87
38	37	19	33	89
39	28	44	29	101
40	38	8	2	48
41	41	33	2	76
42	43	19	19	81
43	26	36	19	81
44	12	35	28	75*
45	30	9	3	42
46	33	33	5	71
47	32	4	35	71
48	10	32	10	52
49	15	32	20	67
50	35	25	42	102
51	3	11	4	18
52	8	12	18	38
53	36	25	32	93
54	46	16	47	109
55	23	31	40	94
56	22	7	38	67
57	10	41	5	56

```
 58    10     17     36     63
 59    47     15     12     74
 60     8     30     24     62
 61    21     13     12     46
 62     5     22     34     61
 63    22     25     43     90*
 64    31     26     30     87
 65     3     26     35     64
 66    23     42     15     80*
 67    21     48     22     91
 68    40     17      9     66
 69    46     14     48    108
 70     7     14     28     49
 71    33     35      1     69
 72    39     31     27     97
 73     2     33     49     84
 74    10     31     49     90*
 75    18      8     34     60*
 76     4     30     36     70*
 77    22      3     25     50*
 78    47     33     27    107
 79    32     16     12     60*
 80    28     31     31     90*
 81     2     40     20     62
 82    40     14     11     65*
 83    38     23     19     80*
 84    21     39     16     76
 85    46     16     32     94
 86    38     26     22     86
 87    39     50     29    118
 88    41     26     11     78
 89    48     41     15    104
 90    16     26     40     82
 91    26     48     24     98
 92    48     32     25    105*
 93    33     26     12     71
 94    45     35     22    102
 95     8     12     10     30*
 96    39     50     18    107
 97    23     15      4     42
 98    28     10      3     41
 99    35     47     22    104
100    12     11      3     26
```

Problems

Applications

3.53 In a game of chance two integers are randomly selected from 1 to 10, inclusive. The same number may be selected twice. If both numbers agree then the person wins the game.

a. Find the exact probability of winning this game.

b. Use the computer to simulate the approximate probability of winning this game. Use 200 repetitions.

3.54 To simulate the sum obtained by tossing a pair of dice we can use Minitab to simulate integers from 1 to 6 in columns 1 and 2 and to store the sum of these columns in column 3. Using RANUNI in the SAS system, we could multiply RANUNI(*seed*) by 6, add 1, and then use INT to obtain numbers from 1 to 6. Use the computer and 200 repetitions to approximate probabilities for all different sums which may be obtained when tossing a pair of dice. Compare the results of the simulation with the theoretical probabilities.

3.55 Consider an experiment in which five dice are tossed at once. Use a simulation involving 200 repetitions to approximate the following probabilities.

a. the probability that the sum of the five dice is 20 or more

b. the probability that three or more dice show the same value

3.6 Case 3: Where Did the Sponge Come From?—Solution

The statistician obtained information on every aspect of the case. He also learned the procedure by which the sponges were ordered, received, stored, and used. He determined the average number of sponges used per day and the last day the earlier type of sponge was ordered so that he could determine the probability that one box of those sponges was overlooked. He also obtained information on the relative frequency of incorrect sponge counts at this hospital and determined the incidence of sponges left in incisions nationally.

When all these questions were taken into consideration, he assumed that the circumstances necessary for the sponge to have been left in the incision after the first operation consisted of independent events. For each of these events he assigned a range of probabilities from most likely to least likely, as well as a middle range. Note that these probabilities had to be both empirical and subjective. He then multiplied the probabilities in each set. The product of the least likely probabilities yielded an overall probability of 1 chance in 150,000,000. The product of the middle probabilities gave a probability of 9 in 25,000,000, while the product of the most likely probabilities gave an overall probability of 7 in 2,000,000. Thus, even using most favorable probabilities, the likelihood of the events happening as the patient's lawyer conjectured was about 0.0000035.

This is not to say it could not happen. The probability of each player in a bridge game being dealt a perfect hand (all one suit) is extremely small—practically zero—yet it does happen. The reason it does happen is that the number of rounds of bridge dealt in a single year is enormously large, so large that you would expect a perfect round perhaps once in 20 years, on the average.

In the case of the operation, however, we are dealing with one instance. The issue at hand was to determine the most likely set of circumstances that would have led to the sponge being left in the incision. Ultimately the judge found in favor of the hospital and doctor.

3.7 Summary

In this chapter we examined some of the basic principles of probability. To understand the material introduced in later chapters, it is essential to master the following definitions:

- *sample space:* the set of all possible outcomes of an experiment
- *point, sample point,* or *simple event:* one outcome of a sample space
- *event:* a collection of points of a sample space
- *complement of an event:* all points of the sample space not in the event
- *empty set:* the set of no points (symbolized $\emptyset$)
- *mutually exclusive events:* two or more events with no points in common
- *exhaustive events:* two or more events whose sample points make up the sample space
- *partition of a sample space:* a set of mutually exclusive and exhaustive events of a sample space
- *compound events:* a combination of two or more events
- *independent events:* two events such that neither has an effect on the probability of the other

The probabilities of the events listed above are governed by the basic probability rules, which are summarized throughout the chapter. Bayes' rule is used to obtain the *posterior probability* of an event—the conditional probability of an event given that a subsequent event has already occurred.

Counting rules, such as the multiplication rule and its "descendants" (the permutations, partitions, and combinations rules), greatly facilitate the determination of the number of outcomes of an experiment.

Since most statistical applications assume that a sample is randomly selected from a population, we must have methods of obtaining random samples. We can use a random number table for this purpose, or random numbers may be generated by a computer program. Both Minitab and the SAS system contain provisions for generating random numbers.

Chapter Review Problems

Practice

3.56 Find $P(A \text{ and } B)$ if $P(A) = 1/2$, $P(B) = 1/3$, and:
- **a.** $P(A \text{ or } B) = 4/5$
- **b.** $P(A|B) = 3/4$
- **c.** $P(B|A) = 2/5$
- **d.** A and B are independent
- **e.** A and B are mutually exclusive
- **f.** $P(\sim A \text{ and } B) = 1/10$

VENN DIAGRAM

3.57 Let A and B be events in sample space S. Suppose that if $P(B) = 0.30$, $P(A \text{ and } B) = 0.12$, and $P(\sim A \text{ and } \sim B) = 0.42$.
- **a.** Find $P(A)$.
- **b.** Find $P(A \text{ and } \sim B)$.
- **c.** Find $P(A|B)$.
- **d.** Are A and B independent events? Why?
- **e.** Are the events $A \text{ and } B$ and $\sim A \text{ and } \sim B$ mutually exclusive? Why?

3.58 Determine $P(A)$ if $P(A \text{ or } B) = 0.7$, $P(B) = 0.3$, and:
- **a.** $P(A \text{ and } B) = 0.2$
- **b.** $P(A|B) = 1/3$
- **c.** $P(\sim A) = 0.4$
- **d.** A and B are independent
- **e.** A and B are mutually exclusive
- **f.** $P(A \text{ or } \sim B) = 0.9$

3.59 One card is selected from a standard deck of 52. What is the probability of:
- **a.** selecting the 7 of hearts?
- **b.** selecting a spade?
- **c.** selecting a 7 or a spade?

3.60 Five different-colored marbles are placed in a jar. A marble is drawn and replaced, then another marble is drawn. List the points in the sample space, and determine the probability that the two marbles drawn are:
- **a.** the same color
- **b.** of different colors

3.61 An experiment consists of putting a dime, a nickel, and a penny in a cigar box, closing the lid, and shaking vigorously. The box is then opened and each coin observed to see whether it shows a head or a tail. The results are recorded as an ordered triple—the dime first, then the nickel, and finally the penny. Thus (H,T,H) shows that the dime and the penny came up heads while the nickel was a tail; (T,H,H) shows that the dime was a tail while the nickel and penny were heads.
- **a.** How many points are there in the sample space?
- **b.** List all points corresponding to the event where at least two coins showed heads.
- **c.** Let D be the event "dime shows a head." List all sample points corresponding to D.
- **d.** If N is the event "nickel shows a head," list all points in the event $\sim N$ *and* D.

3.62 The probability that Richard and his brother will go swimming this afternoon is 0.44. The probability that Richard will go swimming but his brother will not is 0.16. What is the probability that Richard will not go swimming this afternoon?

3.63 How many four-letter combinations may be formed from the letters of the word *stop* if:

a. repetitions of the letters are allowed?
b. no repetitions are allowed?

3.64 How many seven-letter combinations are possible using the letters *HHHTTTT*?

3.65 A box labeled A contains five red marbles and eight blue marbles, while box B contains three red and seven blue marbles.

a. Three marbles are drawn with replacement from box A. What is the probability that they are all red?
b. Three marbles are drawn without replacement from box B. What is the probability that they are all red?
c. A marble is drawn from A and placed in B. One marble is then drawn from B. What is the probability that it is blue?
d. One marble is drawn from B and placed in A. One marble is then drawn from A. What is the probability that it is blue?

3.66 Two urns are placed on a shelf. Urn A contains three white balls and two red balls; urn B contains three white balls and five red balls. A die is rolled. If 3 or 6 appears, urn A is chosen; otherwise urn B is selected. A ball is then drawn from the chosen urn.

a. What is the probability that a white ball is drawn?
b. What is the probability that a red ball is drawn?
c. If a red ball is drawn, what is the probability that it came from urn A?
d. If a white ball is drawn, what is the probability that it came from urn B?

3.67 The probability that a car parked at Pensacola Beach has a Florida license plate is 0.62; the probability that it is a foreign car (that is, foreign-made) is 0.28; the probability that it is a foreign car with Florida license plates is 0.14. Determine the following probabilities:

a. that a parked car is either a foreign car or has a Florida license plate
b. that such a car either does not have a Florida license plate or else is not a foreign car
c. that a car with a Florida license plate parked on the lot is a foreign car
d. that a foreign car parked on the lot has a Florida license plate

Applications

3.68 The probability that Mrs. Jones will be selected for jury duty is 0.40; the probability that Mr. Smith will be selected is 0.25. If the probability that neither will be selected is 0.45, what is the probability that both will be selected?

3.69 In how many ways can a committee of four be chosen from five married couples if:

a. all are equally eligible?
b. the committee must consist of three women and one man?
c. a husband and wife cannot serve together?

3.70 How many ways can a cross-country match finish by school if only three schools have entered and they have entered, respectively, six, seven, and eight runners?

3.71 How many six-person volleyball teams can be made from a group of seven men and three women if:
a. there is no sex discrimination?
b. the team must contain at least two women?

3.72 A multiple-choice exam consists of four questions. There are three choices for question 1, four choices for question 2, three choices for question 3, and five choices for question 4. If a student guesses on each question, what is the probability of getting:
a. at least one correct answer?
b. at least three correct answers?

3.73 How many seven-digit phone numbers are possible if the first three digits must be 478?

3.74 An airplane can complete its flight if at least half the engines are in working order. Suppose the probability of engine failure is 0.01 and engine failures are independent events. Calculate the probability that an airplane will complete its flight if it has:
a. two engines
b. three engines
c. four engines

3.75 Each machine in an assembly line has a probability of failure during a single day of 0.01. For each machine there is a backup machine that can be used in case of failure, and each of these backups has a 0.01 probability of failure. The assembly line will be shut down only if both a machine and its backup fail. Five machines are used on the assembly line. If machine failures are independent events, what is the probability that the assembly line will be shut down tomorrow?

3.76 Among a group of talented musicians, 10% play the piano but no other instrument. Half play the piano, half play the cello, 80% play the violin, and 10% play all three. Moreover, 40% play both the violin and the piano, and 30% play the violin and cello but not the piano.
a. Are any two events independent?
b. What percentage play none of the three instruments?
c. What is the probability that a pianist plays the violin?
d. What is the probability that a pianist plays the cello?
e. What is the probability that a pianist plays all three?

3.77 A ten-volume encyclopedia and a three-volume edition of *Tristram Shandy* are knocked onto the floor. If they are placed at random into the 13 adjacent vacant spaces, what is the probability that the volumes of *Tristram Shandy* are:
a. together?
b. together and in the proper order?

3.78 A utility tray is assembled from three pieces: a tray, a leg, and a stand. These pieces are manufactured separately and randomly assembled into a finished product. One in 20 of the trays, one in 100 of the legs, and one in 50 of the stands are flawed. The finished products are examined and rejected whenever at least two of the three pieces are flawed. What is the probability that a finished utility tray will be rejected?

3.79 A research study on the latest variety of influenza virus must be completed before November 1 so that a suitable vaccine can be developed and marketed. A vital piece of equipment breaks down on September 1 and a replacement is ordered immediately. Before it broke down, the project was on schedule, but now every day without the equipment reduces the probability of completing the project on time by 0.02. The supplier of the equipment estimates that the probability that the equipment will be delivered in 10 days after it is ordered is 0.50. If it is not delivered in 10 days, the probability is 0.80 that it will be delivered in 15. He guarantees that it will be delivered in 20 days.

a. What is the probability that the project will be completed on time?

b. If the project is completed on time, what is the probability that the equipment was delivered in 10 days?

3.80 Out of 100 applicants for a certain job, 70 have some experience, 28 are over 40 years old, and 65 are men. The distribution of applicants over these three factors is shown here:

	EXPERIENCE			NO EXPERIENCE	
	OVER 40	UNDER 40		OVER 40	UNDER 40
Male	15	40	Male	3	7
Female	5	10	Female	5	15

One person is chosen at random from the 100. Let F be the event "over 40," M the event "male," and E the event "has experience." Find each of the following probabilities:

a. $P(F)$ **b.** $P(\sim M)$ **c.** $P(M|F)$ **d.** $P(\sim M|E)$

3.81 Heredity is governed by laws of probability. In a certain species of guinea pig, three-fourths of each generation are short-haired and the remainder long-haired. An offspring of two long-haired guinea pigs is always long-haired, while eight-ninths of the offspring of two short-haired guinea pigs are short-haired. If the parents are of two different types, two-thirds of the offspring are short-haired. If a long-haired guinea pig is selected at random, what is the probability that both its parents are short-haired?

3.82 Closely akin to probabilities are *odds*. We define odds to be the ratio of favorable probability to unfavorable probability. If 2 of 12 eggs are spoiled, for example, the probability of selecting a good egg at random is 5/6, while the odds that an egg will be good are 5/6 to 1/6, or 5 to 1. The odds that an egg will be spoiled are 1 to 5. If an event has probability 0.15, the odds in favor of its occurrence are 3 to 17 (0.15 to 0.85) and the odds against its occurrence are 17 to 3.

a. If the probability of rain on a given day is said to be 0.4, what are the odds in favor of rain? Against rain?

b. A fuse box has 20 fuses in it, of which 4 are 20-ampere size. If a man takes a fuse out at random, what are the odds against getting a 20-ampere fuse?

c. At a racetrack, the odds that a horse will win are said to be 8 to 1; if this is so, what is the probability that the horse will win?

d. An American roulette wheel has slots numbered 00, 0, and 1 through 36. If you place your bet on a number, what are the odds that you will not get the number?

e. On an American roulette wheel, half the slots numbered 1 through 36 are red and half are black. The 0 and 00 are green. You can bet on red or black. If you bet on red, what are the odds that you will lose?

3.83 Approximately 1/15 of the male population is color-blind. Color-blindness is a sex-linked characteristic; this means that the female must have two color-blindness genes in order to be color-blind, while a male will be color-blind if he has one such gene. Fathers pass on the gene to daughters but not to sons. Mothers pass on the gene to either sons or daughters. Thus there is a 1/15 chance that a color-blindness gene will be passed on to a daughter by any given parent.

a. What is the probability that a female will be color-blind?

b. What proportion of the female population is color-blind?

c. If the ratio of males to females in the population is 103 to 105, what proportion of the entire population is color-blind?

3.84 Recall from Example 3.7 that a bill passed by the legislature has been sent back to a committee for further study. The probability of a favorable recommendation from the committee is 1/14. Suppose that the probability of a mildly unfavorable recommendation is 11/14 and the probability of a very unfavorable recommendation is 1/7. If the report is favorable, the probability the governor will sign the bill is 0.9; if the report is mildly unfavorable, the probability is 0.7; if the report is very unfavorable, however, the probability that the governor will sign the bill is only 0.3.

a. What is the probability that the governor will sign the bill?

b. If the governor signs the bill, what is the probability that the committee report was favorable? Mildly unfavorable? Very unfavorable?

3.85 G. R. Lindsey (1963) compiled a great deal of information on several hundred major league baseball games during 1959 and 1960. He found that if there were runners on first and second base with nobody out, the probability of scoring at least one run during the inning was 0.605. If there were runners on second and third with one out, the probability of scoring at least one run during the inning was 0.730.

If a bunt attempt is successful, the batter will be thrown out and the runners will advance one base each. Suppose that the probability of scoring a run with the bases loaded and none out is 0.922, and the probability of scoring a run with men on first and second and one out is 0.390. Suppose further that the probability that an attempted bunt will be successful is 0.545, the probability that all runners will be safe is 0.012, and the probability that the lead runner will be thrown out and the other runners safe is 0.443. (There are other possibilities, but for simplicity we will consider only these.) If a team is one run behind in the bottom of the ninth inning with men on first and second and nobody out, should the manager order a bunt attempt? That is, would a bunt improve the probability of scoring a run?

3.86 A mathematician is challenged to a game at her local bar. Two containers will be set in front of her together with ten black balls and ten red balls. She must use all the balls and distribute them as she pleases between the two containers, putting at least one in each container. The challenger will then select a container at random and draw one ball, also at random. If it is red, the challenger will buy the drinks; if black, the mathematician must pay. How should the mathematician distribute the balls to maximize her chances? If she does so, what is the probability that she will get a free drink?

3.87 The *Pensacola News-Journal* (February 24, 1986) reported that 1 adult male in 35 is either on probation, on parole, or in a correctional institution. Suppose that your class has 35 adult males in it. Does this mean that the expected number of adult males in one of these three categories in your class is 1? How does this proportion (1/35) apply to your class? What implication do your answers have for random sampling?

Index of Terms

Glossary of Symbols

C_r^n	number of combinations of n objects taken r at a time
$P(\ )$	probability of
$P(A)$	probability of event A
$P(\sim A)$	probability of the complement of A
$P(A \text{ and } B)$	joint probability that A and B both occur
$P(A \text{ or } B)$	probability that at least one of A and B occurs

$P(A\|B)$	conditional probability of A given B
P_r^n	number of permutations of n objects taken r at a time
$\emptyset$	empty set
$\binom{n}{r}$	number of combinations of n objects taken r at a time
!	factorial symbol

Answers to Proficiency Checks

1. **a.** 20 **b.** 15 **c.** 60

2. {(1,1), (1,2), (1,3), (1,4), (2,1), (2,2), (2,3), (2,4), (3,1), (3,2), (3,3), (3,4), (4,1), (4,2), (4,3), (4,4)}
a. 1/16 **b.** 3/16 **c.** 1/4 **d.** 3/16 **e.** 15/16
f. 5/8 **g.** 15/16 **h.** 13/16

3. {(1,2), (1,3), (1,4), (2,1), (2,3), (2,4), (3,1), (3,2), (3,4), (4,1), (4,2), (4,3)}
a. 0 **b.** 1/6 **c.** 1/3 **d.** 1/6 **e.** 1 **f.** 2/3 **g.** 1 **h.** 5/6

4. **a.** 0.5 **b.** 0.4 **c.** 0.1 **d.** 0.1 **e.** 0.8 **f.** 0.8 **g.** 0.3 **h.** 0.4

5. **a.** 0.2 **b.** 0.15 **c.** 0.55 **d.** 0.45

6. $P(U \textit{ and } C) = 0.10$ and $P(U) = 0.25$, where U and C represent the obvious events. Then $P(C|U) = 0.10/0.25 = 0.40$. Thus about 40% of super unleaded gasoline users pay with credit cards.

7. $(0.9)(0.35) + (0.3)(0.55) = 0.48$

8. 0.70

9. $(1/6)(1/6)(1/6) = 1/216$; $(1)(5/6)(4/6) = 5/9$

10. $(5/6)(5/6)(5/6)(1/6) = 125/1296$; $(5/6)(5/6)(5/6) = 125/216$

11. **a.** $(0.1)(0.35)/[(0.1)(0.35) + (0.7)(0.55) + (1.0)(0.10)] = (0.035)/(0.035 + 0.385 + 0.10) = 0.035/0.52 = 7/104 \doteq 0.0673$ **b.** $77/104 \doteq 0.740$ **c.** $5/26 \doteq 0.192$

12. There are many possible samples—in fact, 7,893,600. Let 26 two-digit numbers represent the letters, and ignore the rest. The first five such numbers will define the sample. For instance, suppose we use 01 through 26 to represent the letters, and enter the table in line 46, column 23. The numbers we obtain are 5711841564352906363803404909. Using two-digit numbers, but ignoring two-digit numbers above 26, we have 11, 15, 06, 03, 09. These represent the letters *K*, *O*, *F*, *C*, and *I*.

13. 6

14. 60

15. 360

16. 12,600

17. 125,970

18. 792

19. 5940

References

Fienberg, Stephen E. "Randomization and Social Affairs: The 1970 Draft Lottery." *Science,* January 22, 1971: 255–261.

Kohler, Heinz. *Statistics for Business and Economics.* Glenview, IL: Scott, Foresman & Co., 1985.

Lindsey, G. R. "An Investigation of Strategies in Baseball." *Operations Research* 11 (1963): 477–501.

Sheaffer, Richard L., William Mendenhall, and Lyman Ott. *Elementary Survey Sampling.* 2nd ed. North Scituate, MA: Duxbury Press, 1978.

4

Discrete Probability Distributions

The primary aim of statistical inference is to enable us to use the results of a sample to make generalizations about the characteristics of a population. These characteristics include population parameters—such as mean, median, and variance—and the relative frequency of occurence of the variable of interest. The relative frequency of each value of a variable provides us with the probabilities of these values. The set of all values of a variable and their probabilities is called a **probability distribution.** In this chapter we consider the probability distributions for several important discrete variables as we continue to pave the way for the study of statistical inference later in the book.

4.0 Case 4: The Parapsychological Puzzle

Psychological testing devices are frequently used to test for the presence of extrasensory perception (ESP)—the ability to know things that cannot be explained by the usual senses of sight, hearing, and so forth. A number of tests for ESP were devised by J. B. Rhine and other researchers at Duke University. In a test for mind-reading ability, subject A and subject B are in different rooms. Subject A has five cards, each with a different stylized picture on it. One card has a triangle, another a square, the third a cross, the fourth a circle, and the fifth wavy lines. Subject A randomly selects a card and concentrates on it while subject B writes down what he or she thinks subject A has selected. The process is repeated, usually for 20 trials, and the two lists are then compared. Subject B is rated for the presence or absence of mind-reading ability by the number of correct guesses.

A test for clairvoyance (the ability to see the future) is similar, except that a second subject is not necessary. A pack of the special cards is shuffled, and the subject tries to predict each card before it is turned over. Again, the number of correct guesses is scored and the subject's ability is rated by the number of correct guesses.

The classification of the subject's ability is based on the probability that the result obtained could be attributed to chance. One way to classify ESP ability would be the following. If the probability is less than 0.25 that the subject would guess that many cards or more correctly by chance, the subject is classified as having "some degree of ESP ability"; if the probability is less than 0.10, the subject may be classified as having "a high degree of ESP"; if the probability is less than 0.01, the subject would have "a very high degree of ESP." A subject who defied the laws of chance to the extent that there would be less than one chance in a thousand (0.001 probability) of guessing the result by chance would have "a remarkable degree of ESP." Four college friends took the tests. Listed here are the number of correct guesses in 20 trials for each. What do you suppose the classifications for each ability would be for each subject?

	NUMBER CORRECT	
SUBJECT	MIND READING	CLAIRVOYANCE
A	12	7
B	5	4
C	4	6
D	0	0

4.1 Random Variables and Probability Distributions

One of the central ideas in probability, and in statistics, is that of a probability distribution. We might guess that a probability distribution is something which describes how probabilities in a sample space are distributed. In Chapter 3 we studied probabilities for a number of different variables. The most important variables in statistics are known as *random variables*.

4.1.1 Random Variables

The outcome of a statistical experiment is always one of the points of a sample space, but the point itself is always determined by chance. In Chapter 1 we defined a variable as a quantity whose value is not fixed. A variable that takes on number values is a numerical variable. A numerical variable whose value is determined by chance—as by the result of an experiment—is called a **random variable.** A random variable can be defined in a sample space simply by assigning numbers to all the possible outcomes in the sample space generated by a particular experiment.

Random variables are of two types: discrete and continuous. Recall from Chapter 1 that discrete data are data with values that are finite, or countable, and continuous data are data obtained by measuring. Similarly we define a **discrete random variable** as a random variable whose values are countable; a **continuous random variable** is a random variable whose values arise from measurement. In this chapter we deal only with discrete random variables. Continuous random variables are discussed in Chapter 5 and later chapters.

Let us look at an example of a discrete random variable. If you roll two dice and observe the number of dots that come up on each die, you can examine the sample space generated by the experiment. To do this, list an array of ordered pairs, 36 in all, that gives all possible outcomes as points of the sample space. In each ordered pair, the first number indicates the number observed on the first die, and the second number indicates the number observed on the second die. Thus the ordered pairs (1,2) and (2,1) represent different points of the sample space. The sample space is shown here. (For clarity, the braces are omitted.)

(1,1) (2,1) (3,1) (4,1) (5,1) (6,1)
(1,2) (2,2) (3,2) (4,2) (5,2) (6,2)
(1,3) (2,3) (3,3) (4,3) (5,3) (6,3)
(1,4) (2,4) (3,4) (4,4) (5,4) (6,4)
(1,5) (2,5) (3,5) (4,5) (5,5) (6,5)
(1,6) (2,6) (3,6) (4,6) (5,6) (6,6)

A random variable x can be defined on this sample space by letting the value of the variable equal the sum of the number of dots showing on the two dice. We can see, then, that in this case x can take on all integer values from 2 through 12. If a coin is tossed four times, and we let the random variable be the number of heads that come up, this variable can take on all integer values from 0 through 4.

We can assign more than one random variable to a sample space. If x is the number of heads that come up in four tosses of a coin, we could let y represent the number of tails. We can define a random variable on a sample space in different ways. In each case we assign numbers to the mutually exclusive outcomes that make up a sample space.

In Chapter 3 we used $P(A)$ to indicate the probability of event A. In dealing with random variables, we use $P(x)$ to represent the probability of an unspecified value of the random variable x. For a specified value, or range of values, we will use the same symbolism—the probability that $x = 3$ is written $P(3)$, for example, and the probability that x is between 2 and 6 is written $P(2 < x < 6)$. Some books use the symbolism $P(x = 3)$ instead of $P(3)$; this notation is perhaps more precise, but if the variable is not in doubt there should be no confusion.

4.1.2 Discrete Probability Distributions

We cannot list all the values of a continuous random variable, but we may often list the values of a discrete random variable. The entire collection of values that a discrete random variable can take on, together with the probabilities for each of these values, is called a **discrete probability distribution.** From the probability rules in Chapter 3 we know that for any discrete random variable x, $P(x) \geqslant 0$ and the sum of all probabilities for each value of x is equal to 1. Since values of x for which $P(x) = 0$ are unimportant, they are usually not included as part of the distribution.

Example 4.1

Two fair dice are tossed. If x represents the total number of dots on the two faces, determine a probability distribution for this random variable.

Solution

Referring to the sample space mentioned previously, we find that we can group the sample points into distinct events as shown in Figure 4.1. We see that there are altogether 36 points, each equally likely, so the probability corresponding to each event is equal to the number of points in the event divided by 36, the total number of points. For example, the number of points

(1,1)	(1,2)	(3,1)	(4,1)	(5,1)	(6,1)
(1,2)	(2,2)	(3,2)	(4,2)	(5,2)	(6,2)
(1,3)	(2,3)	(3,3)	(4,3)	(5,3)	(6,3)
(1,4)	(2,4)	(3,4)	(4,4)	(5,4)	(6,4)
(1,5)	(2,5)	(3,5)	(4,5)	(5,5)	(6,5)
(1,6)	(2,6)	(3,6)	(4,6)	(5,6)	(6,6)

Figure 4.1
Events for Example 4.1.

corresponding to the value $x = 4$ is three—the points (1,3), (2,2), and (3,1), as illustrated—so $P(4) = 3/36$ (or 1/12). We can summarize the results with a tabulation such as Table 4.1.

Table 4.1 PROBABILITY DISTRIBUTION FOR EXAMPLE 4.1.

x	$P(x)$
2	1/36
3	2/36
4	3/36
5	4/36
6	5/36
7	6/36
8	5/36
9	4/36
10	3/36
11	2/36
12	1/36

Example 4.2

A fair coin is tossed three times or until a head appears, whichever comes first. Define the following random variables: x is the number of times the coin is tossed, and y is the number of tails which appear. Determine the probability distributions for x and y.

Solution

First we set up the sample space for the experiment:

$$S = \{(H), (T,H), (T,T,H), (T,T,T)\}$$

We list each point with its probability in Table 4.2. Note that the probability of a head (or a tail) on *each* toss is 1/2 and that successive tosses are independent, so we obtain the probabilities by simply applying the multiplication rule for probability.

Table 4.2 OUTCOMES AND PROBABILITIES FOR EXAMPLE 4.2.

Outcome	H	T,H	T,T,H	T,T,T
Probability	1/2	1/4	1/8	1/8

The number of tosses on one experiment can be one, two, or three, so the possible values for x are 1, 2, and 3. Only one point corresponds to $x = 1$, and its probability is 1/2; thus $P(1) = 1/2$. Similarly, $P(2) = 1/4$. Two points correspond to $x = 3$, however, each with probability 1/8, so $P(3) = 1/8 + 1/8 = 1/4$. We summarize these results in Table 4.3.

Table 4.3 PROBABILITY DISTRIBUTION OF NUMBER OF TOSSES.

x	$P(x)$
1	0.50
2	0.25
3	0.25

Now let us turn our attention to the random variable y, the number of tails in one experiment. Reference to the sample space (Table 4.2) gives us the probability distribution for y shown in Table 4.4. Note that for both x and y all probabilities are positive and that the sum of the probabilities for each variable is 1.

We can use various techniques to graph discrete probability distributions. The most useful and widely used technique is the *histogram* (Section

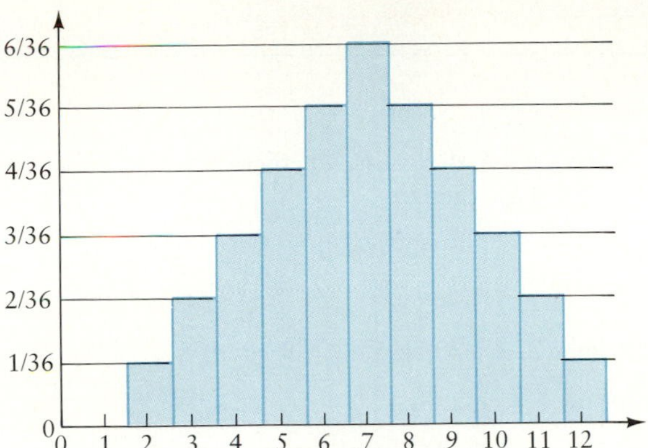

Figure 4.2
Histogram for probability distribution of dice in Example 4.1.

1.2). In the histogram, each value of the random variable is represented by a unit of length on the base and a rectangle drawn above it to indicate its probability relative to the other values of the variable. The probability for each value is then equal to the area of its rectangle divided by the sum of the areas of all the rectangles As an example, a histogram for the dice in Example 4.1 (see Table 4.1) is shown in Figure 4.2.

As in most histograms of this type, the numbers representing the values of the random variable, 2, 3, 4, and so on, lie in the center of the bases of the rectangles. That is, we represent 2 by the interval 1.5 to 2.5, we represent 3 by the interval 2.5 to 3.5, and so forth. The event "the sum of the dots is equal to 8" has the random variable value 8, which is represented by 7.5 to 8.5 on the horizontal axis. Thus the rectangle representing the probability for this value of the variable has a base equal to 1, a height equal to 5/36, and an area of 5/36 square unit. The sum of all the areas of the rectangles representing the complete distribution is 1.

Table 4.4
PROBABILITY DISTRIBUTION OF NUMBER OF TAILS.

y	$P(y)$
0	0.500
1	0.250
2	0.125
3	0.125

PROFICIENCY CHECKS

1. A jar contains 15 poker chips, each with a number painted on it. A "1" is painted on one chip, "2" on two chips, "3" on three chips, "4" on four chips, and "5" on five chips. An experiment consists of drawing one chip at random from the jar. Define the random variable x as the number painted on the chip drawn. Give the probability distribution for x and construct a histogram for the distribution.
2. A die is rolled three times or until a 6 appears, whichever comes first. Construct a probability distribution for the number of times the die is rolled.

Problems

Practice

In Problems 4.1 through 4.6, determine whether each is a probability distribution. Make a histogram for those that are.

4.1 $P(x) = x/25$ for $x = 1, 3, 5, 7, 9$

4.2 $P(3) = 5/7$, $P(5) = 1/7$, for $x = 3, 5$

4.3 $P(x) = 1/8$ for $x = 1, 2, 3, 4, 5, 6, 7, 8$

4.4 $P(x) = x/10$ for $x = 1, 2, 3, 4$

4.5 $P(x) = 1/5$ for $x = 0, 1, 2, 3, 4$

4.6 The data in the table at left.

x	$P(x)$
1	3/80
3	27/80
7	11/80
9	17/80
13	13/80
14	9/80

4.7 A probability distribution is defined by $P(x) = (x^2 + 1)/n$ for $x = 1, 2, 3, 4$. Determine n. Make a histogram for the distribution.

Applications

4.8 A box contains four good light bulbs and two defective ones. A bulb is selected at random and put in a socket. If it is good, it is left there; if defective, it is discarded and another bulb is selected. The process is continued until a good bulb is found. Give the probability distribution for the random variable x, the number of bulbs that will be tried.

4.9 A box contains four good light bulbs and two defective ones, as in Problem 4.8, but this time there are two sockets to fill with good bulbs. The same process is used. Give the probability distribution for the number of bulbs that will be tried.

4.10 A sociologist observes the social behavior of two young children over a period of several hours. His goal is to gather information concerning the socialization of this pair of children—whether they play together or independently. He has divided a play yard into four sections and observes the children every 5 minutes to see in which sections they are playing. If they are playing in the same section, a "0" is recorded; if they are in adjacent sections, a "1" is recorded. If they are separated by one section, a "2" is recorded; if they are at opposite ends of the play yard, a "3" is recorded. The random variable under consideration is the number that is recorded. Determine the probability distribution of this variable if the children behave independently and each pays no attention to the location of the other child. (In Chapter 9 we will consider a method for using this distribution to determine whether or not the children are in fact playing independently.)

4.11 In a study of the effect of oxygen deprivation, four guinea pigs are born in a litter that has been subjected to a 15% reduction in the amount of oxygen to the placenta. Theoretically, one-third of the group should show effects of this treatment. Let x represent the number of guinea pigs showing the effects.

Complete the following probability distribution:

x	0	1	2	3	4
$P(x)$	?	32/81	?	8/81	?

4.12 During the summer of 1983 a team of marine biologists took daily water samples from Bayou Chico to determine the proportion of plankton in bloom. The water sample was to be analyzed later at a university laboratory to be certain that it contained a minimum number of plankton. The sample was to be taken at 6:00 AM, but because of the possibility that a sample might not be acceptable, samples were also taken at 5:00 AM, 5:30 AM, 6:30 AM, and 7:00 AM. The analysis was begun at 8:00 AM. The 6:00 analysis was performed first. If the sample was acceptable, no other analyses were performed. If not, the 5:30, 6:30, 5:00, and 7:00 analyses were performed, in that order, until two of them were acceptable. The probability that an analysis would be acceptable was found to be 0.70. (See Problem 3.41.) If the acceptability of two samples selected a half-hour apart were independent events, construct a probability distribution for the number of samples that would have to be analyzed on a given day.

4.2 Mean and Variance of a Random Variable

A probability distribution describes the various relative likelihoods of the outcomes of an experiment. If we perform the experiment a very large number of times and record the value of the random variable each time, we can obtain the mean and standard deviation of this set of values. These statistics will be almost exactly equal to the mean and standard deviation of the theoretical population of outcomes represented by the probability distribution. We can obtain these population parameters from the probability distribution by using a mathematical concept known as *expected value*.

4.2.1 Mean of a Random Variable

The **expected value** of a random variable x, symbolized $E(x)$, is the long-term average value of the variable that would be expected if the experiment generating the variable were repeated a very great number of times. Thus it is the mean of the variable. If we were to toss a coin a few million times, for example, recording a "0" when a head appears and a "1" when a tail appears, we would expect, on the average, to have half the numbers equal to 0 and half equal to 1. Thus we would expect the average of all our numbers to be about 0.5. This is the expected value of the random variable x if x is the number of tails on one toss of a coin. We can define the expected value of a discrete random variable as follows.

Expected Value of a Discrete Random Variable

If a random variable x has values $x_1, x_2, x_3, \ldots, x_n$, with probabilities $P(x_1), P(x_2), P(x_3), \ldots, P(x_n)$, respectively, then the expected value of x is defined by

$$E(x) = x_1 \cdot P(x_1) + x_2 \cdot P(x_2) + \cdots x_n \cdot P(x_n)$$

We can write this equation more compactly as $E(x) = \Sigma[x \cdot P(x)]$. Note that the expected value is not something we "expect" in the usual sense. It need not be a possible value of x, nor should it be rounded to the nearest value of x if it is not an actual value of x.

Since the expected value of a random variable is the long-term average (or mean value) of the theoretical population of outcomes, it is identical to the population mean μ introduced in Chapter 2. We thus define the **mean of a random variable** as follows.

Mean of a Random Variable

For any random variable x,

$$\mu = E(x)$$

Since the concept of expected value has applicability beyond the mean of a random variable, we now consider some important properties that may be proved from the definition of expected value.

Properties of Expected Value

1. If c is any constant, $E(c) = c$.
2. If c is any constant and x is a random variable, $E(cx) = cE(x)$.
3. If x is a random variable and a and b are constants, $E(a + bx) = a + bE(x)$.

Example 4.3 A coin is tossed three times or until a head appears, whichever comes first. Determine the expected number of tosses and the expected number of tails. (See Example 4.2.)

Solution

The sample space for the experiment is shown here:

$S = \{(H), (T,H), (T,T,H), (T,T,T)\}$

If x is the random variable for the number of tosses, the probability distribution for x is as shown in Table 4.5. The expected value for x is given by $E(x) = 1(0.50) + 2(0.25) + 3(0.25) = 1.75$. Thus, on the average, we would expect 1.75 tosses per experiment. This is the mean of the random variable x.

We define the random variable y to be the number of tails in one experiment. The probability distribution for y is shown in Table 4.6. Then $E(y) = 0(0.500) + 1(0.250) + 2(0.125) + 3(0.125) = 0.875$. This is the mean of the random variable y. Since the expected number of tosses is 1.75 and the expected number of tails is 0.875, obviously the expected number of heads is $1.75 - 0.875$ or 0.875.

Table 4.5
PROBABILITY DISTRIBUTION FOR NUMBER OF TOSSES.

x	$P(x)$
1	0.50
2	0.25
3	0.25

Example 4.4

A promoter has scheduled an outdoor event that will be canceled in case of rain. If it rains, he will lose \$10,000. An insurance company agrees to cover his losses, if it rains, for a premium of \$430. If the premium includes \$30 in administrative costs and is fair to both parties, what probability does the insurance company assign to rain?

Solution

The insurance company is wagering \$9600 against the promoter's \$400 that it will not rain. That is, if it does rain the insurance company must pay \$10,000 (\$9600 of its own money and return the \$400 paid by the promoter); if it does not rain, the promoter will have paid the \$400 insurance fee, not including the administrative costs. In order for a bet or game to be fair, the expected value for all parties must be zero; that is, expected gain must equal expected loss. If we denote the probability of rain by p and the probability that it does not rain by $1 - p$, then the insurance company's expected gain is $(\$400)(1 - p)$ and their expected loss is $\$9600p$. Then $9600p = 400(1 - p)$, that is, $10{,}000p = 400$, so $p = 0.04$. Thus the insurance company believes that the probability of rain is 0.04.

Table 4.6
PROBABILITY DISTRIBUTION FOR NUMBER OF TAILS.

y	$P(y)$
0	0.500
1	0.250
2	0.125
3	0.125

PROFICIENCY CHECK

3. An ordinary die is rolled once. What is the expected value of the number of dots on the uppermost face?

Example 4.5

Example 3.4 gave the number of defective parts over the previous 200 working days. If x represents the number of defective parts on a given day, the probability distribution for x is shown in Table 4.7. Determine the mean of x.

Table 4.7 FREQUENCY AND PROBABILITY DISTRIBUTION FOR NUMBER OF DEFECTIVE PARTS.

x	FREQUENCY	$P(x)$
0	50	0.25
1	32	0.16
2	22	0.11
3	18	0.09
4	12	0.06
5	12	0.06
6	10	0.05
7	10	0.05
8	10	0.05
9	8	0.04
10	6	0.03
11	6	0.03
12	2	0.01
13	2	0.01

Solution

$\mu = 0(0.25) + 1(0.16) + 2(0.11) + 3(0.09) + 4(0.06) + 5(0.06) + 6(0.05) + 7(0.05) + 8(0.05) + 9(0.04) + 10(0.03) + 11(0.03) + 12(0.01) + 13(0.01) = 3.48$. Thus the mean number of defective parts that we would expect per day is 3.48.

To show that this is the same result obtained by using the formula for the population mean in Chapter 2, we observe that the total number of observations is 200. Since

$$0(50) + 1(32) + 2(22) + 3(18) + 4(12) + 5(12) + 6(10) + 7(10) + 8(10) + 9(8) + 10(6) + 11(6) + 12(2) + 13(2) = 696$$

the mean is $696/200 = 3.48$.

4.2.2 Variance of a Random Variable

Since random variables with the same mean can have quite different probability distributions, we must use a measure of variability or dispersion. We saw in Chapter 2 that the most common of these is the variance, symbolized σ^2. The **variance of a random variable** is defined as the expected value of the squared deviations from the mean of the random variable.

Variance of a Random Variable

For any random variable x, the variance is defined by

$$\sigma^2 = E(x - \mu)^2$$

The defining formula, however, is seldom used to compute the value of the variance. We can use the properties of expected value listed in the previous section to obtain a more practical shortcut formula. Since $(x - \mu)^2 = x^2 - 2\mu x + \mu^2$,

$$\begin{aligned} E(x - \mu)^2 &= E(x^2 - 2\mu x + \mu^2) \\ &= E(x^2) - E(2\mu x) + E(\mu^2) \end{aligned}$$

Since μ is a constant, $E(2\mu x) = 2\mu E(x)$ and $E(\mu^2) = \mu^2$. We know that $E(x) = \mu$, so $2\mu E(x) = 2\mu^2$. Finally,

$$\begin{aligned} E(x - \mu)^2 &= E(x^2) - 2\mu^2 + \mu^2 \\ &= E(x^2) - \mu^2 \end{aligned}$$

For a discrete random variable, $E(x^2)$ is simply the mean of the squares, so $E(x^2) = \Sigma[x^2 \cdot P(x)]$. Thus we have the computational formula for the variance of a discrete random variable.

Shortcut Formula for the Variance of a Discrete Random Variable

For a discrete random variable x, the variance is computed by using the formula

$$\sigma^2 = \Sigma[x^2 \cdot P(x)] - \mu^2$$

Table 4.8
PROBABILITY DISTRIBUTION OF THE RANDOM VARIABLE x.

x	$P(x)$
1	1/9
2	1/18
3	1/3
4	5/18
5	1/6
6	1/18

As before, the standard deviation is the square root of the variance. We symbolize the standard deviation of a random variable by σ, where σ is the square root of the variance σ^2. Note that the symbols for variance and standard deviation of a random variable are the same as those for the variance and standard deviation of a population.

Example 4.6

Determine the mean, variance, and standard deviation of the random variable x, with the probability distribution shown in Table 4.8. Draw a histogram for the distribution.

Solution

$\mu = 1(1/9) + 2(1/18) + 3(1/3) + 4(5/18) + 5(1/6) + 6(1/18) = 3.5$; $E(x^2) = 1(1/9) + 4(1/18) + 9(1/3) + 16(5/18) + 25(1/6) + 36(1/18) = 251/18 \doteq 13.9444$, so $\sigma^2 \doteq 13.9444 - (3.5)^2 = 13.9444 - 12.25 \doteq 1.69$. Then $\sigma \doteq \sqrt{1.69} \doteq 1.30$. The histogram is shown in Figure 4.3.

Example 4.7

Find the variance and standard deviation of the number of defective parts in Example 4.5 (Table 4.7).

Solution

The data are reproduced in Table 4.9 with appropriate calculations. Since we found μ to be 3.48, then $\sigma^2 = 24.32 - (3.48)^2 = 12.2096$ and $\sigma = \sqrt{12.2096} \doteq 3.49$.

Table 4.9 COMPUTATIONS USED IN SOLUTION TO EXAMPLE 4.7.

x	$P(x)$	x^2	$x^2 \cdot P(x)$
0	0.25	0	0
1	0.16	1	0.16
2	0.11	4	0.44
3	0.09	9	0.81
4	0.06	16	0.96
5	0.06	25	1.50
6	0.05	36	1.80
7	0.05	49	2.45
8	0.05	64	3.20
9	0.04	81	3.24
10	0.03	100	3.00
11	0.03	121	3.63
12	0.01	144	1.44
13	0.01	169	1.69
			24.32

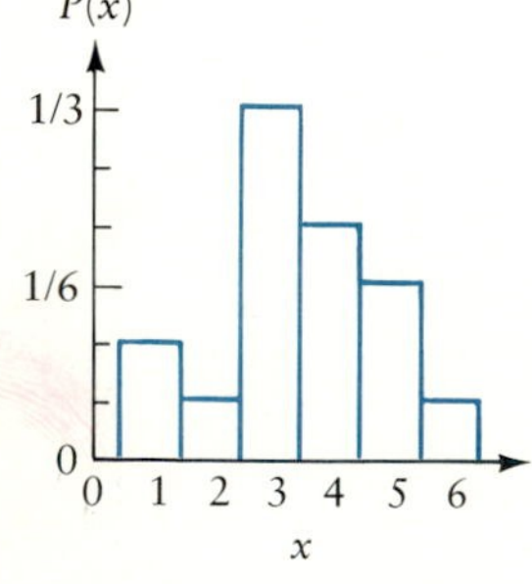

Figure 4.3 Histogram for the probability distribution of Example 4.6.

PROFICIENCY CHECKS

4. A random variable x is defined for $x = 0, 1, 2, 3$, and 4. Moreover, $P(0) = 0.1$, $P(2) = 0.4$, $P(3) = 0.2$, and $P(1)$ is twice $P(4)$. Determine a probability distribution for the variable, and determine its mean, variance, and standard deviation.

(continued)

PROFICIENCY CHECKS *(continued)*

5. A box contains two red and three blue balls. Two balls are drawn at random. Let x be the number of red balls drawn. Determine the mean and standard deviation of x:
   a. if the first ball is replaced before the second is drawn
   b. if the first ball is not replaced before the second is drawn
6. There are three light bulbs in a box, one of which is burnt out. A bulb is selected at random and tried. If it is no good, it is discarded and another one is selected. Let x be the number of light bulbs that must be tried to find a good one. Determine the mean and variance for x.

Problems

Practice

4.13 An urn contains six white and three red marbles. Four marbles are drawn at random without replacement. Determine the expected number of red marbles that will be drawn.

Determine the mean and standard deviation of each of the random variables defined in Problems 4.14 through 4.17.

4.14

x	13	17	21	24	25
$P(x)$	0.2	0.4	0.2	0.1	0.1

4.15

x	107	114	121	128	135
$P(x)$	0.22	0.27	0.16	0.21	0.14

4.16

x	5	11	13	18	22	27	33	34	39
$P(x)$	0.20	0.14	0.23	0.11	0.13	0.09	0.05	0.04	0.01

4.17

x	1000	1100	1200	1300
$P(x)$	0.1	0.2	0.3	0.4

4.18 Two dice are rolled and the sum of the numbers on the top faces is recorded (see Example 4.1). Determine the mean and variance of this random variable.

4.19 Suppose you are playing a game in which you may draw two cards at random, without replacement, from an ordinary deck of 52 cards, half black and half red. For each black card you draw, you receive $5. How much should you pay for the privilege of playing the game if it is to be fair?

Applications

4.20 Suppose the probabilities that zero, one, two, three, or four accidents will occur in the Holland Tunnel between 7:30 and 10:00 on a Monday morning are, respectively, 0.92, 0.04, 0.02, 0.01, and 0.01. How many accidents would you expect on the average on a Monday morning? How many accidents would you expect on 100 Monday mornings?

4.21 In the 1985 World Series between the Kansas City Royals and the St. Louis Cardinals, oddsmakers in Las Vegas were giving odds on the number of games the series would last. Using the odds, you could determine that the probabilities the series would last four, five, six, or seven games, were, respectively, 0.2, 0.4, 0.3, and 0.1. Before the series started, the number of games it would last was a random variable. If this random variable is denoted x, determine the mean, variance, and standard deviation of x.

4.22 The first prize in a lottery is \$10,000, and two second prizes of \$5000 each will also be awarded. Thirty thousand tickets have been sold for \$1 each. What is the expected gain for one ticket? What is the net expected value for one ticket? Is the lottery fair?

4.23 Mr. Jones is selling his business. A realtor promises that if Mr. Jones lists with her, the probability he will make \$200,000 is 20%, that he will make \$120,000 is 35%, that he will make \$40,000 is 10%, and that he will break even is 15%. She concedes, however, that market conditions may go sour—if so, Mr. Jones has a 15% chance of losing \$60,000 if he sells now—and there is even a chance he may lose \$120,000. She claims, however, that there are no other possibilities. If she is correct and Mr. Jones lists with her, what is his expected profit?

4.24 In the 1985 World Series, the oddsmakers considered St. Louis a 3–2 favorite. That is, they felt that the probability was 0.6 that St. Louis would win the series and 0.4 that Kansas City would win. (Kansas City won.) If the actual probability that a particular team will win is 0.6 per game, probabilities that the series will end in four, five, six, or seven games, are about 0.15, 0.27, 0.30, and 0.28, respectively. What is the expected number of games in such a series?. If Mrs. Black has tickets only to the fourth and fifth games, what is the expected number of games she will see?

4.25 On a roulette wheel, there are 38 slots, equally spaced, numbered 00, 0, 1, 2, 3, . . . , 36. If your number wins, you receive \$35 for a \$1 bet in addition to the return of your dollar; otherwise you lose your bet. Determine the expected net gain, on one bet, of a gambler who bets \$1 on a number.

4.26 On the roulette wheel described in Problem 4.25, 18 slots are red, 18 are black, and 00 and 0 are green. A person can place a bet on red or black. For a \$1 bet, a winner receives an additional \$1; a loser is out the \$1 bet. Calculate the expected value of a \$1 bet on either red or black. (The same principle applies to odd and even since 00 and 0 are considered neither by the house.)

4.27 Refer to Problem 4.10. Determine the mean and variance of this variable if the children behave independently and neither pays attention to the location of the other child.

4.28 Complete the following probability distribution for the guinea pigs described in Problem 4.11, and calculate the mean and variance of x.

x	0	1	2	3	4
$P(x)$	?	32/81	?	8/81	?

4.29 Refer to Problem 4.12. If the acceptabilities of two samples selected a half-hour apart are independent events, determine the mean and variance of the number of samples that will be taken on a given day.

4.30 Suppose that x is a random variable and c and k are constants. Define the random variable y, where $y = cx + k$ and $P(x) = P(y) = P(cx + k)$ for all values of x. Use the properties of expected value to show that if μ_1 and σ_1 are the mean and standard deviation of x and μ_2 and σ_2 are the mean and standard deviation of y, then $\mu_2 = c\mu_1 + k$ and $\sigma_2 = |c|\sigma_1$.

4.31 Obtain the mean and standard deviation of each of the following random variables. Compare them and show how they illustrate the result of Problem 4.30.

a.

x	-1	0	1	2
$P(x)$	0.3	0.2	0.3	0.2

b.

x	9	10	11	12
$P(x)$	0.3	0.2	0.3	0.2

c.

x	-3	0	3	6
$P(x)$	0.3	0.2	0.3	0.2

4.32 Show how the results of Problem 4.30 can be used to simplify the calculations in Problems 4.15 and 4.17.

4.3 Case Study: Repair, Replace, or Test?

A manufacturing plant makes parts for television picture tubes. Several years ago the plant was having trouble with one of its manufacturing processes and had to make a decision regarding which course of action to take. The problem described here is simplified by making the proportion of defective parts discrete, but we could easily extend the procedure to cover more complex situations.

A part used in the picture tubes was manufactured using several machines, and the process was showing signs of wear. The number of defective parts produced was running between 1% and 5% per production run. The number of defectives was especially critical since if a defective part was used in a picture tube, the tube would not work properly and had to be dismantled and the part replaced. Such a procedure cost an estimated \$5.48 in labor and lost time whenever it occurred. A second alternative was to identify the problem with the process itself and replace or repair whatever was causing the problem. Management had estimated the cost of this alternative to be about \$25,000 to \$30,000, taking into consideration tax allowances for depreciation and salvage values. A third possibility was to test each part before installation. This option would cost about 13¢ per part.

Normally, the best procedure would be to spend the capital for replacement or repairs. Only about 180,000 more of the parts were to be produced, however, so the total cost of inspection would be \$180,000(0.13) or \$23,400. Thus it would be cheaper to inspect each part than to provide the capital outlays.

The company's quality control expert (a statistician) suggested that it might be preferable simply to do nothing, absorbing the \$5.48 cost for each defective item. A study of past records since the problem began showed that about 43% of all lots contained 1% defective, 26% of the lots had 2% defective, 14% of the lots had 3% defective, 11% of the lots had 4% defective, and 6% of the lots had 5% defective. A further examination of the records showed no evidence that the proportion of defectives was increasing and also indicated that the proportion defective in a given lot appeared to be random. Thus the proportion defective was a random variable with values 0.01, 0.02, 0.03, 0.04, and 0.05. The probability distribution for this variable was empirically determined by past records to be as follows:

Proportion defective	0.01	0.02	0.03	0.04	0.05
Probability	0.43	0.26	0.14	0.11	0.06

The expected proportion defective was determined to be 0.0211, so the expected number defective in the 180,000 to be produced was estimated to be 180,000(0.0211) or 3798. At a cost of \$5.48 per defective unit, the cost of simply doing nothing was estimated to be about \$20,813. Thus the expected cost for this course of action was least of all, and the company continued its present policy.

4.4 Binomial Probability

Many decisions are of the either/or variety. A contract is signed, or it is not. A vaccine is effective, or it is not. Many populations have only two possibilities. A student passes a course or does not. A person speaks Spanish or does not. Many situations involve taking a sample from a population that has only two outcomes; in such a case the population is called a **dichotomous population.**

Two types of sampling from a dichotomous population arise naturally. If a coin is tossed, the outcome on the first toss has no effect on the outcome of the second toss; these outcomes are independent. We will turn our attention first to this type of sampling in which the successive outcomes are independent. In Section 4.6, we will discuss sampling in which successive outcomes are not independent.

4.4.1 The Binomial Distribution

When we take a sample from a dichotomous population in such a way that successive selections or trials are independent of each other, and if the number of trials is fixed, we may have an example of a **binomial experiment.**

Characteristics of a Binomial Experiment

A binomial experiment is an experiment consisting of a fixed number of identical trials for which:

1. There are exactly two outcomes for each trial, usually denoted S (success) and F (failure).
2. The probability of success, denoted p, remains constant from trial to trial.
3. Outcomes of successive trials are independent.

The probability of failure on each trial, then, is $1 - p$. This is sometimes denoted q, with the understanding that $p + q = 1$.

The random variable for a binomial experiment, called a **binomial random variable,** is the number of successes in the fixed number of trials. Each trial of a binomial experiment is called a *Bernoulli trial* in honor of Jakob Bernoulli (1654–1705). If, instead of a fixed number of trials, an experiment consists of repeating trials until a success is obtained, the number of trials is called a *geometric random variable*. The *geometric distribution* is discussed in Section 4.6. A binomial random variable is defined completely by two parameters (constants): the probability p of success on each trial and the number of trials.

Bernoulli was one of a family of eminent Swiss mathematicians and physicists. Jakob did a great deal of work in probability. His brother Johann (1667–1748) worked chiefly with calculus. Their nephew Nikolaus (1687–1759) was a professor of mathematics. Johann's son Nikolaus (1695–1726) became a professor of law in 1722 and accepted an appointment in mathematics in 1725. Daniel (1700–1782), son of Jakob, was a professor of mathematics, botany, anatomy, and natural philosophy, and he obtained important results in the field of hydrodynamics. Johann's son Johann II (1710–1790) received several prizes from the French Academy. Johann III (1744–1807), son of Johann II, was a doctor of law and director of the Berlin observatory; another son, Jakob (1759–1789), was a professor of mathematics.

We will need a formula for determining the probability for each value of a binomial random variable x. If there are n trials, clearly, the values that x can have are all the integers from zero through n. Let us use an example to derive this formula.

On a certain television show, a contestant has to choose one of 25 boxes. One box contains a check for a large sum of money. If the contestant chooses a box at random, he or she has exactly one chance in 25 of selecting the prize box. We would like to determine the probability distribution for the number of prize boxes that will be chosen in a week (five shows).

Now the probability of success on one trial—that is, of picking the right box—is 1/25 or 0.04, so the probability of failure is 0.96. The probability that there will be five winners in a week is clearly equal to (0.04)(0.04)

(0.04)(0.04)(0.04) or $(0.04)^5$, which is equal to 0.0000001024—not very likely. The probability that there will be no winners in a week is equally clearly $(0.96)^5$, or 0.8153726976. These calculations are easy. But what is the probability of exactly one winner in a week? Or two, or three, or four winners? Let us examine the events where there are two winners in a week.

If there are two winners in a week, there are three failures. If the successes occur on Monday and Tuesday and the failures on the remaining three days, the probability of that event is $(0.04)^2 \cdot (0.96)^3$. We can calculate this probability, but this is not the only way there could be two winners in a week. The successes could occur on any two of the five days and the failures on the other three. If you studied Section 3.4, you will know that the partitions rule can be used to determine the number of possible outcomes. We have a set of five objects—two successes and three failures—so there are $5!/(2!3!)$ ways of arranging them. This number is also equal to the number of combinations of five objects taken two at a time.

For those who did not study Section 3.4, we observe that if we wish to arrange the letters *SSFFF* in all possible orders, we can look at the problem as if we had five blanks in which to put these five letters. Once the *S*'s have been placed, the *F*'s will go in the remaining spots, so we need only to look at the ways in which we can place the *S*'s. There are five places to put the first *S* and four places to put the second *S*, so we have 5(4) or 20 ways to place the two *S*'s in a particular order. The order does not matter, however, so for each two orders we have only one arrangement. We may place the first *S* on Monday and the second on Wednesday, for example, or the first *S* on Wednesday and the second on Monday. In both cases the successes will be on Monday and Wednesday and the failures on the other days. Thus the total number of ways we can have two successes in five days is 20/2; that is, $(5 \cdot 4)/(2 \cdot 1)$, or 10. The probability of exactly two winners in one week, then, is $10(0.04)^2(0.96)^3$ or about 0.014.

We can generalize this result as follows. If we have a total of n trials and there are x successes, we have n choices for the first success, $n - 1$ for the second, $n - 2$ for the third, and so on, down to $(n - x + 1)$ for the xth success. Each positioning of x successes can be rearranged in many ways. There are x places to put the first success, $x - 1$ for the second, $x - 2$ for the third, and so on, down to the last success, which has only one possible spot. Thus there are $n(n - 1)(n - 2)(n - 3) \cdots (n - x + 1)$ ways to arrange x successes in n trials, and there are $x(x - 1)(x - 2)(x - 3) \cdots 3 \cdot 2 \cdot 1$ ways to rearrange any one such set so that the successes fall in the same positions. The total number of different ways that x successes and $(n - x)$ failures can be arranged is symbolized $\binom{n}{x}$. It is equal to

$$\frac{n(n-1)(n-2)(n-3)\cdots(n-x+1)}{x(x-1)(x-2)(x-3)\cdots 3\cdot 2\cdot 1}$$

If $x = 0$, this value is equal to 1.

In general, if we repeat a binomial experiment n times, the probability that x successes will occur is equal to the number of ways in which this can happen times the probability that each will happen in that particular order—namely $p^x(1 - p)^{n - x}$. This leads to the following rule.

Binomial Probability

If a binomial experiment has n trials and the probability of success on one trial is p, then the probability of exactly x successes in n trials is given by

$$P(x) = \binom{n}{x} p^x (1 - p)^{n - x}$$

where

$$\binom{n}{x} = \frac{n(n-1)(n-2)(n-3)\cdots(n-x+1)}{x(x-1)(x-2)(x-3)\cdots 3 \cdot 2 \cdot 1}$$

If x is a binomial random variable for a binomial experiment with n trials, then the set of all values of x, namely zero through n, together with all the probabilities, is called a **binomial distribution.** We can symbolize the statement "x is a binomial random variable of n trials with a probability of success on each trial equal to p" by writing $x = B(n, p)$.

Comment: Some hand calculators calculate $\binom{n}{x}$ automatically, using a key with the symbol ${}_nC_r$ or C_r^n. If it has to be calculated by hand, we usually cancel common factors before multiplying. For example:

$$\binom{15}{7} = \frac{(15)(\overset{2}{\cancel{14}})(13)(\overset{2}{\cancel{12}})(11)(\overset{2}{\cancel{10}})(\overset{3}{\cancel{9}})}{(7)(\cancel{6})(\cancel{5})(\cancel{4})(\cancel{3})(\cancel{2})(1)}$$
$$= (15)(13)(11)(3) = 6435$$

That is, 7 goes into 14 two times, 6 goes into 12 two times, 5 goes into 10 two times, $4 = 2 \times 2$ and so will cancel the 2's just obtained, 3 goes into 9 three times, and the last 2 divides the 2 above the 10.

Example 4.8

In a certain experiment five animals are given a drug. One-fifth of all animals given the drug develop certain symptoms. Determine the probability distribution for the number of animals that will develop the symptoms.

Solution

The number of animals that develop symptoms is 0, 1, 2, 3, 4, or 5. The outcome is dichotomous since an animal either develops symptoms or it does not. The other conditions are met, so this is an example of a binomial experiment. If x is the random variable for the number of animals that develop symptoms, we want the probability of x successes in five trials for $p = 0.2$, $1 - p = 0.8$, and $x = 0, 1, 2, 3, 4$, and 5. Table 4.10 shows how we calculate the probabilities for each value of the random variable x.

Table 4.10 CALCULATION OF PROBABILITIES FOR EXAMPLE 4.8.

x	$P(x)$	
0	$\binom{5}{0}(0.2)^0 \cdot (0.8)^5 = 1 \cdot 1 \cdot (0.32768)$	$\doteq 0.328$
1	$\binom{5}{1}(0.2)^1 \cdot (0.8)^4 = 5 \cdot (0.2) \cdot (0.4096) = 0.40960$	$\doteq 0.410$
2	$\binom{5}{2}(0.2)^2 \cdot (0.8)^3 = 10 \cdot (0.04) \cdot (0.512) = 0.20480$	$\doteq 0.205$
3	$\binom{5}{3}(0.2)^3 \cdot (0.8)^2 = 10 \cdot (0.008) \cdot (0.64) = 0.05120$	$\doteq 0.051$
4	$\binom{5}{4}(0.2)^4 \cdot (0.8)^1 = 5 \cdot (0.0016) \cdot (0.8) = 0.00640$	$\doteq 0.006$
5	$\binom{5}{5}(0.2)^5 \cdot (0.8)^0 = 1 \cdot (0.00032) \cdot 1 = 0.00032$	$\doteq 0+$

We rounded off the probabilities in Table 4.10 to three decimal places. To three decimal places 0.00032 must be rounded to zero. Since it is not really zero we use $0+$ to indicate that $P(5)$ is very small but not zero. A histogram of the distribution is shown in Figure 4.4.

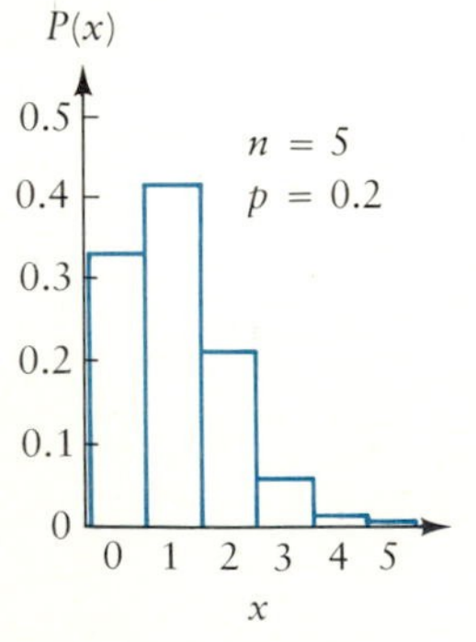

Figure 4.4
Histogram for the probability distribution of the number of animals developing symptoms in Example 4.8.

We can use the results of Table 4.10 to obtain a number of probabilities. The probability that no more than two animals will develop symptoms is $P(x \leqslant 2)$. Now $P(x \leqslant 2) = P(x = 0, 1, \text{or } 2) = P(0) + P(1) + P(2) \doteq 0.943$. The probability that at least three animals will develop symptoms is $P(x \geqslant 3)$. Now $P(x \geqslant 3) = P(3) + P(4) + P(5) \doteq 0.057 = 1 - P(x \leqslant 2)$. The probability that some animals develop symptoms is $P(x \geqslant 1) = P(x > 0) = 1 - P(0) \doteq 1 - 0.328 = 0.672$.

Histograms of some representative binomial distributions are shown in Figure 4.5. Note that for small values of n and p (Figure 4.4) the distribution is positively skewed. For small values of n and large values of p (Figure 4.5a) the distribution is negatively skewed. If p is close to 0.5 (Figure 4.5b) the distribution is symmetric, while for larger values of n (Figure 4.5c) the distribution appears nearly symmetric, even if p is small (or large). As a general rule, if both np and $n(1 - p)$ are at least 5 the distribution is symmetric.

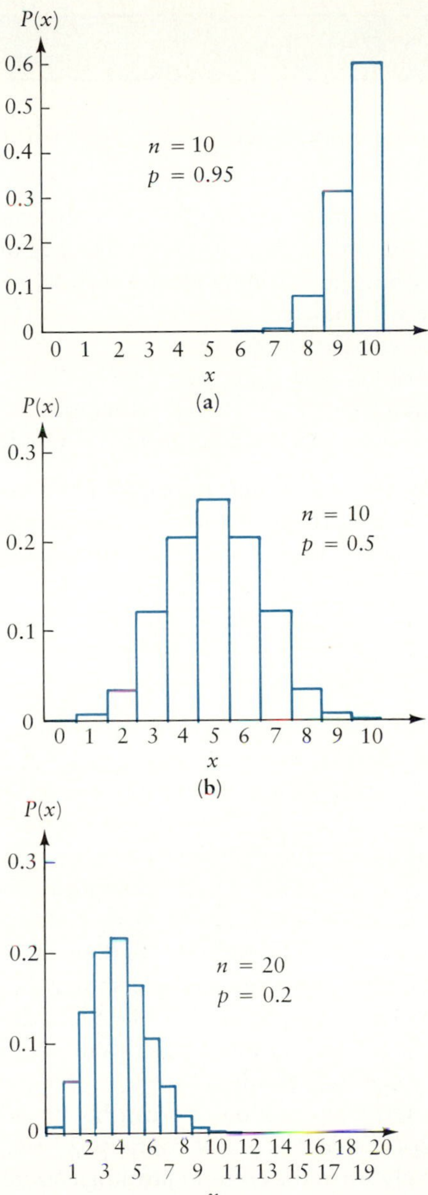

Figure 4.5
Histograms for three binomial probability distributions: (*a*) $n = 10$, $p = 0.95$; (*b*) $n = 10$, $p = 0.5$; (*c*) $n = 20$, $p = 0.2$.

4.4.2 Using Binomial Tables

Even the relatively small numbers of the preceding section make the task of computing the probabilities of a range of values quite tedious. To simplify the task, extensive tables for the probabilities of a binomial random variable are available. These tables generally come in two forms. One form gives individual probabilities—the probability of obtaining exactly x successes in n trials of a binomial experiment. The other form gives cumulative prob-

PROFICIENCY CHECKS

7. What is the probability of getting exactly five heads in 12 tosses of a fair coin?

8. A basketball player makes 80% of his free throws. Using this percentage as the probability of his making any one free throw and assuming that the binomial model applies, what is the probability that he will make:
   a. six of his next eight attempts?
   b. at least six of his next eight attempts?
   c. no more than six of his next eight attempts?
   d. at least four but fewer than six of his next eight attempts?

9. Approximately 10% of all items produced at a factory are defective and have to be returned for replacement. Suppose three items are purchased. Give the probability distribution for x, the number of defective items of the three purchased.

abilities—the probability of obtaining r *or fewer* successes in n trials of a binomial experiment, that is, $P(x \leq r)$. (Some tables accumulate the other way, that is, giving $P(x \geq r)$.) Since we can find the individual probabilities easily from cumulative tables but the converse is not generally true, we restrict our attention here to the cumulative tables.

A short table of cumulative binomial probabilities is given in Table 2 in the back of this book. Values are given for $n = 2, 3, 4, 5, \ldots, 25$ and for $p = 0.01, 0.05, 0.10, 0.20, 0.30, 0.40, 0.50, 0.60, 0.70, 0.80, 0.90, 0.95,$ and 0.99. Each three-digit entry in the table should be read as a decimal; that is, an entry of 983 means a probability of 0.983. The symbol 1 − means that the probability is less than 1 but greater than 0.9995, although for all practical purposes it can be considered to be 1. The symbol 0+ means a positive probability less than 0.0005 and can be considered equal to zero for practical purposes. Table 2 actually contains 24 such tables, which are presented consecutively since the column headings do not change. For illustrative purposes a small portion of Table 2 is reproduced in Table 4.11.

Each entry in Table 4.11 refers to a binomial experiment in which n, the number of trials, is equal to 10. The letter r stands for r or fewer successes. Thus if the probability of success on one trial is 0.3 and there are ten trials, the probability of four or fewer successes is about 0.850. The term $r = 4$ indicates that $x \leq 4$—that is, $x = 0, 1, 2, 3,$ or 4. Then $P(x \leq 4) = 0.850$ if $n = 10$ and $p = 0.3$. Note that this is an approximation correct to three decimal places. We can check that $P(x \leq 4) = 0.850$ by finding the individual probabilities and adding them:

Table 4.11 SELECTED BINOMIAL PROBABILITIES FOR $n = 10$.

		p			
n	r	0.05	0.30	0.60	0.70
10	0	599	028	0+	0+
	1	914	149	002	0+
	2	988	383	012	002
	3	999	650	055	011
	4	1−	850	166	047
	5	1−	953	367	150
	6	1−	989	618	350
	7	1−	998	833	617
	8	1−	1−	954	051
	9	1−	1−	994	972
	10	1	1	1	1

$$\begin{aligned}
P(x \leq 4) &= P(0) + P(1) + P(2) + P(3) + P(4) \\
&= (0.7)^{10} + \binom{10}{1}(0.3)^1(0.7)^9 + \binom{10}{2}(0.3)^2(0.7)^8 \\
&\quad + \binom{10}{3}(0.3)^3(0.7)^7 + \binom{10}{4}(0.3)^4(0.7)^6 \\
&= 0.0282475249 + 10(0.3)(0.040353607) \\
&\quad + 45(0.09)(0.05764801) + 120(0.027)(0.0823543) \\
&\quad + 210(0.0081)(0.117649) \\
&= 0.0282475249 + 0.121060821 + 0.2334744405 \\
&\quad + 0.266827932 + 0.200120949 \\
&= 0.8497316674 \\
&\doteq 0.850
\end{aligned}$$

Note how much simpler it is to look in the table.

To find the probability of more than r successes, note that $P(x > r) = 1 - P(x \leq r)$. Thus the probability of more than four successes in ten trials with $p = 0.3$ is about $1 - 0.850$ or 0.150. A second way to obtain the same result is to observe that more than four successes is the same as five or more failures. If the probability of success is 0.3, the probability of failure is 0.7, so we can look in Table 2 for $n = 10$, $p = 0.7$, $r = 5$ and directly read 0.150.

We can find the probability of exactly four successes in ten trials if $p = 0.30$ by observing that $P(x \leq 4) = P(0) + P(1) + P(2) + P(3) + P(4)$ and $P(x \leq 3) = P(0) + P(1) + P(2) + P(3)$. Then $P(x \leq 4) - P(x \leq 3) = P(4)$. We already have $P(x \leq 4) = 0.850$ and from Table 4.11 we see that, for $n = 10$ and $p = 0.3$, we have $P(x \leq 3) = 0.650$, so $P(4) = 0.200$. Similarly, if $n = 10$ and $p = 0.60$, then $P(x \leq 7) \doteq 0.833$ and $P(x \leq 8) \doteq 0.954$, so $P(8) \doteq 0.954 - 0.833 = 0.121$.

In general, if $a \leq b$ then $P(x \leq b) - P(x \leq a) = P(a < x \leq b)$. For instance, $P(x \leq 5) = P(0) + P(1) + P(2) + P(3) + P(4) + P(5)$ and $P(x \leq 8) = P(0) + P(1) + P(2) + P(3) + P(4) + P(5) + P(6) + P(7) + P(8)$, so $P(x \leq 8) - P(x \leq 5) = P(6) + P(7) + P(8) = P(5 < x \leq 8)$. We can find other ranges of values by converting to this same form. For example, $P(2 < x \leq 7)$ can be written $P(3 \leq x \leq 7)$, $P(2 < x < 8)$, or $P(3 \leq x < 8)$.

Here are a few more examples using $n = 10$ and $p = 0.30$:

$$P(2 < x \leq 7) = P(x \leq 7) - P(x \leq 2) \doteq 0.998 - 0.383 = 0.615$$

$$P(2 < x < 8) = P(2 < x \leq 7) \doteq 0.615$$

$$\begin{aligned} P(4 \leq x < 7) &= P(3 < x \leq 6) = P(x \leq 6) - P(x \leq 3) \\ &\doteq 0.989 - 0.650 = 0.339 \end{aligned}$$

$$\begin{aligned} P(1 \leq x \leq 5) &= P(0 < x \leq 5) = P(x \leq 5) - P(x \leq 0) \\ &\doteq 0.953 - 0.028 = 0.925 \end{aligned}$$

$$P(2 < x \leq 8) = P(x \leq 8) - P(x \leq 2) \doteq 1 - 0.383 \doteq 0.617$$

Example 4.9

A medical study showed that in a survey of more than 15,000 men who had a heart attack, recovered, and then subsequently died, 60% died of a second (or later) heart attack while 40% died of unrelated causes. The case histories of 20 men who have had a heart attack and recovered are under study. Using the empirically determined probability and the assumption that the binomial model is appropriate, determine the probability that at least 10 of these men will die of a later heart attack. Determine the probability that at least 8 but no more than 15 of the men will subsequently die of a heart attack.

Solution

$P(x \geq 10) = P(x > 9) = 1 - P(x \leq 9)$. Using Table 2, we have $p = 0.60$, $n = 20$, $r = 9$, so $P(x \leq 9) = 0.128$ and $P(x \geq 10) \doteq 0.872$. For $P(8 \leq x \leq 15)$, we write $P(7 < x \leq 15) \doteq 0.949 - 0.021 = 0.928$.

PROFICIENCY CHECKS

10. Use the binomial tables with $n = 10$ and $p = 0.6$ to calculate:
 a. $P(x \geq 6)$ b. $P(x > 4)$ c. $P(6)$
11. If $n = 10$ and $p = 0.7$, determine:
 a. $P(x < 5)$ b. $P(x \leq 7)$ c. $P(5)$
12. If $n = 10$ and $p = 0.4$, determine:
 a. $P(x \geq 4)$ b. $P(x > 7)$ c. $P(x < 5)$ d. $P(x \leq 3)$

(continued)

PROFICIENCY CHECKS (*continued*)

e. $P(5)$ f. $P(3 \leqslant x < 6)$ g. $P(4 < x < 9)$ h. $P(5 < x \leqslant 7)$
i. $P(2 \leqslant x \leqslant 8)$

13. If $n = 20$ and $p = 0.40$, determine:
a. $P(x \geqslant 12)$ b. $P(14)$ c. $P(x \leqslant 7)$ d. $P(5 < x \leqslant 10)$

14. From experience, a salesperson knows that about 60% of potential customers who spend more than 5 minutes examining a certain display will buy something. What is the probability that at least 9 of the first 15 customers who enter the store and spend more than 5 minutes examining the display will buy something?

15. From past experience, a bowler knows that she will get a strike (all pins down on the first ball) about 10% of the time. Using this assumption, what is the probability distribution (to three decimal places) for the number of strikes she will make on the first nine frames of a game? (The tenth frame is a bit more complicated since a bowler who makes a strike in the tenth frame is allowed two more balls.)

4.4.3 Mean and Variance of a Binomial Random Variable

We can calculate the mean and variance of a binomial random variable in the same way we calculate any other random variable, but there is a shortcut. First let us calculate the mean and variance of Example 4.8 the long way.

Example 4.10

Determine the mean and variance of the variable of Example 4.8.

Solution

We will need $E(x)$ and $E(x^2)$. The computations are summarized in Table 4.12. Here $E(x) = \Sigma xP(x) = 1$ and $E(x^2) = 1.8$, so $\mu = 1.00$. The variance, σ^2, is equal to $E(x^2) - \mu^2 = 1.8 - 1^2 = 0.8$. The standard deviation is the square root of the variance, so $\sigma = \sqrt{0.8} \doteq 0.8944$.

Table 4.12

x	$P(x)$	$xP(x)$	$x^2P(x)$
0	0.32768	0	0
1	0.40960	0.40960	0.40960
2	0.20480	0.40960	0.81920
3	0.05120	0.15360	0.46080
4	0.00640	0.02560	0.10240
5	0.00032	0.00160	0.00800
		1.00000	1.80000

We note that $n = 5$, $p = 0.2$, and $np = 5(0.2) = 1$. Further, $np(1 - p) = 5(0.2)(0.8) = 0.8$. For this example $\mu = np$ and $\sigma^2 = n(1 - p)$; these formulas are valid for any binomial random variable. We can derive them easily by using algebra and the properties of expected values.

Mean, Variance, and Standard Deviation of a Binomial Random Variable

If a binomial experiment is repeated n times and the probability of success on one trial is p, the mean and variance of the random variable denoting the number of successes are given by

$$\mu = np \qquad \sigma^2 = np(1 - p)$$

The standard deviation, then, is given by $\sigma = \sqrt{np(1 - p)}$.

PROFICIENCY CHECKS

16. On a multiple-choice test with five answers (one correct, four incorrect) to each question, a student randomly selects an answer to each of 225 questions. Determine the mean and standard deviation for the random variable denoting the number of correct answers.
17. The probability that a gun hits a certain target is 2/3. Five guns are fired simultaneously.
    a. What is the probability that the target is hit exactly once?
    b. What is the probability that the target is hit at least four times?
    c. Determine the mean and variance of the variable.
18. A fair coin is tossed six times. Give the probability distribution for the number of heads.
    a. Determine the mean and variance by using the special formulas.
    b. Verify these values by direct computation.

For the binomial distribution to apply, the probability p must be constant from trial to trial and the trials must be independent. If we have six light bulbs, four good and two defective, and we select two without replacement, the result of the second draw depends very much on the result of the first—

we say that the result of the second is *conditional* on the result of the first. We cannot use the binomial distribution to determine the various probabilities in this situation, however. The binomial probability distribution can be used only when the necessary conditions for a binomial experiment are met. Situations in which we are sampling from a population without replacement, such as taking a poll or doing a market survey, do not meet the technical definition of a binomial experiment.

A probability distribution appropriate for a given situation is called a **probability model.** If the requirements for a binomial experiment are met, we say that we may use the **binomial model.** For small samples without replacement, we need to use the *hypergeometric* model (Section 4.6). For large populations and for samples of a reasonably large size, we will see later that the discrepancy is so slight that the binomial model may often be used even in cases where it is not strictly appropriate.

Problems

Practice

In Problems 4.33 to 4.35, determine the probability distribution and the mean and variance of the variable.

4.33 Two dice are thrown. A success is 7 or 11. The experiment is repeated four times.

4.34 An urn contains four red and six blue balls. Five balls are drawn at random. The variable x denotes the number of blue balls drawn. Each ball is replaced before the next is drawn.

4.35 A coin is tossed ten times and a head is regarded as a success.

Determine the probabilities in Problems 4.36 to 4.43, where x is the number of successes in n trials of a binomial experiment with probability of success equal to p.

4.36 $P(x \geqslant 11)$ and $P(11)$ for $n = 20, p = 0.7$

4.37 $P(x \leqslant 5)$ and $P(5)$ for $n = 12, p = 0.2$

4.38 $P(x > 18)$ and $P(20)$ for $n = 22, p = 0.8$

4.39 $P(x > 14)$ and $P(x < 14)$ for $n = 25, p = 0.4$

4.40 $P(x \geqslant 7)$ and $P(3 \leqslant x < 8)$ for $n = 15, p = 0.5$

4.41 $P(x < 12)$ and $P(6 \leqslant x \leqslant 15)$ for $n = 25, p = 0.7$

4.42 $P(21 < x \leqslant 23)$ for $n = 24, p = 0.99$

4.43 $P(0 < x < 5)$ for $n = 23, p = 0.05$

4.44 For $n = 20$, find the value of p for which $P(x \geqslant 14) = 0.250$.

4.45 For $n = 25$ and $p = 0.4$, find the smallest value of r such that $P(x \geqslant r) < 0.05$.

Applications

4.46 It is known that 30 percent of a certain variety of flower bulb will not grow. If ten bulbs are planted, what is the probability that:

a. exactly six will bloom?
b. at least four will bloom?
c. no more than eight will bloom?
d. all of them will bloom?

4.47 The probability that a particular seed of red fescue grass will germinate is 0.75. A bag contains 10,000 seeds. Let x represent the number of seeds in the bag that will grow. What are the mean and standard deviation of x?

4.48 A multiple-choice quiz has nine questions with four responses (one correct) on each question. To pass, you must get at least six correct. If you guess on every question, what is the probability that you will pass?

4.49 The International Coffee Organization (*Pensacola News Journal,* Feb. 13, 1986) reported that the percentage of people over the age of 10 who drink decaffeinated coffee increased from 4% in 1960 to 17% in 1985. Suppose a random sample of five persons over 10 are asked if they drink decaffeinated coffee. What is the probability that at least two will answer yes?

4.50 Four students on campus are sampled at random and asked if they qualify for the college work-study program. If 10% of all students on campus qualify for work-study, give the probability distribution of x, the number of students qualifying for the program. Determine the mean and standard deviation of x. Use a binomial random variable.

4.51 A very erratic golfer has a probability of 3/4 that his tee shot will go at least 100 yards. If he plays five holes, what is the probability that his drive goes at least 100 yards on the majority of the five holes?

4.52 According to the National Fire Protection Association (*Associated Press,* Feb. 12, 1986), more than one-third of all hotel fires are caused by guest room furnishings. Suppose the probability that guest room furnishings cause a hotel fire is 35%. What is the probability that three of the next four hotel fires will be caused by guest furnishings, assuming independence of events?

4.53 A door-to-door salesman is allowed in the house on about two-thirds of his calls. He makes a sale about 30% of the time he is let in the house. What is the probability that he will make a sale on exactly 5 of his next 20 calls? On at least 8 of his next 25 calls?

4.54 A basketball player makes about 80% of his free throws. What is the probability that he will make at least 4 of his next 8 attempts? No more than 10 of his next 15 attempts?

4.55 If a penny is spun, rather than tossed, the probability of getting a tail is approximately 0.58. If a penny is spun five times, what is the probability of getting:

a. at least four tails?
b. exactly two tails?
c. more than one tail?
d. from one to four tails, inclusive?

4.56 If 30% of the viewers in a town watch the television program "Dallas," what is the probability that a majority of 25 persons sampled will not watch the show?

4.57 A couple is planning to have three children. The wife has been taking birth control pills for some time plus another medication to control the side effects. Because of this combination of medications her doctor informs her that the probability of having a girl on each birth is 0.44. Assuming that this is the case, that successive births are independent, that there are no multiple births, and that the couple has exactly three children, what is the probability that exactly two of the three children will be girls? Boys?

4.58 One-tenth of a group of subjects being studied at a psychiatric clinic have taken a hallucinogen at least once. If the population being studied is large enough to overlook the fact that sampling is done without replacement, what is the probability that exactly one-third of a sample of six subjects, chosen at random from the population, have taken a hallucinogen?

4.59 In a medical school, standard X-ray photographs are kept for use by students. Second-year students have a probability of 4/5 of diagnosing the X ray correctly if there is a pathological condition but only 3/5 of diagnosing it correctly if the X ray shows normal conditions. Three second-year students study the X rays and draw their conclusions. If we assume that the binomial model is applicable (which is a big assumption, since the abilities of the students will vary), what is the probability that at least two will make the correct diagnosis if:

a. a pathological condition is present?
b. the X ray shows normal conditions?

4.60 Refer to Problem 4.11. Determine the probability distribution of x, its mean, and its standard deviation.

4.61 A random sample is drawn from a group of Spanish-American people in a large city. If the group is large enough to use the binomial model as a good approximation, and if one-fifth of the group are unable to speak English, what is the probability that at least half the sample will be able to speak English in a sample of:

a. 6? **b.** 7? **c.** 8? **d.** 15? **e.** 20? **f.** 25?

4.62 Of every 1000 parts produced by a machine, on the average 10 are defective. What is the probability that some, but not all, of a sample of three parts are defective?

4.63 The sex ratio of humans at birth is 100 females to 105 males. What is the probability that, in six single births, at least half the babies born are females? Compare this probability with the result you would obtain if you used a sex ratio of 1 to 1.

4.64 A soft drink company claims that its cola tastes "unique." Five different brands of cola are set before a taster who is told to choose the one that is "different" (supposedly, of course, the company's brand). Now assume that there really is no difference in the taste of the colas but each of three tasters picks one of the five drinks anyway in a blindfold test. What is the probability that at least two of the tasters will select the "correct" drink by chance alone?

4.65 A sociologist plans on interviewing 20 family units of a certain socioeconomic status to assess their attitudes toward social changes. She thinks that the probability that any one of the family units she is interviewing will favor the change is 0.7. If this is so, what is the probability that more than 10 of the 20 will favor the change? Exactly 10 of the 20?

4.66 Suppose that the sociologist in Problem 4.65 interviews ten families. If her assumption about the probability is correct, give the probability distribution for the number of families favoring the social changes. Give the mean and standard deviation of the variable.

4.67 Suppose that the sociologist in Problem 4.65 interviews ten families and finds that only five of them favor the social changes. Determine the probability that five or fewer families would favor the changes if her hypothesis were correct. Do you think that this result refutes her hypothesis? (A method for testing the hypothesis that the probability is actually 0.7 is given in Section 7.3.)

4.68 Repeat Problem 4.67 but this time suppose that 10 of 20 families favor the changes.

4.5 The Poisson Random Variable

The next discrete distribution we will investigate in detail is the *Poisson distribution,* named for the French mathematician Siméon D. Poisson (1781–1840). Unlike the binomial random variable, a **Poisson random variable** focuses on the mean number of occurrences of some phenomenon per unit of time, distance, area, or volume. The random variable is the number of occurrences for some specified number of units.

4.5.1 The Poisson Distribution

The Poisson model applies when the occurrences are at random and independent of each other, independent of the starting point, and uniformly distributed over an interval of time or space. Independence of the starting point means that it does not matter where we start examining the occurrences. Uniform distribution over the interval means that if we expect 0.5 occurrence over some interval, we would expect 1.5 occurrences over an interval three times as large.

Poisson random variables include the number of telephone calls passing through a switchboard in a certain period of time, the number of electron tubes requiring replacement in an assembly per period of time, the number of specific parts needing replacement per job, the number of fish caught per person in a unit of time, the number of blemishes per square foot in the finish of an automobile, and of course many other examples. The formula for the **Poisson distribution** is as follows.

The Poisson Distribution

If x is a Poisson random variable with the expected number of occurrences per unit equal to m, the probability distribution of x over one unit is defined by

$$P(x) = \frac{e^{-m} \cdot m^x}{x!} \qquad \text{for } x = 0, 1, 2, 3, 4, \ldots$$

where

$$x! = x(x - 1)(x - 2)(x - 3) \cdots 3 \cdot 2 \cdot 1$$

The mean and variance of x are given by $\mu = m$ and $\sigma^2 = m$.

In this formula e is the base of natural logarithms, approximately 2.71828. For a Poisson distribution $\mu = E(x) = m$, and $\sigma^2 = m$ as well. Some histograms of Poisson distributions are shown in Figure 4.6. For small values of m the distribution is positively skewed; as m increases, however, the distribution becomes more nearly symmetric.

Suppose, for example, that an office is considering reducing its number of telephone lines as an economy measure. Operators have found that the average number of telephone calls during the busiest period, 10 AM to noon, and 1 PM to 4 PM (a period of 5 hours) is 75 and the average length of a call is 6 minutes. The office manager wants the probability to be small that all lines are busy. Since the average call lasts 6 minutes, we are interested in the number of calls that will come in during a given 6-minute period. There are fifty 6-minute periods in 5 hours, so the average number of calls

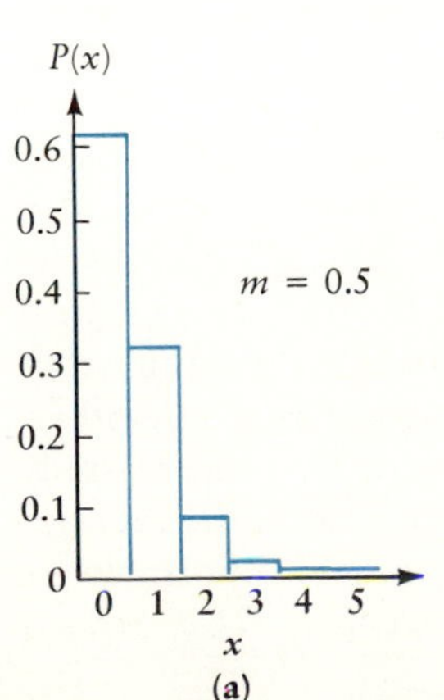

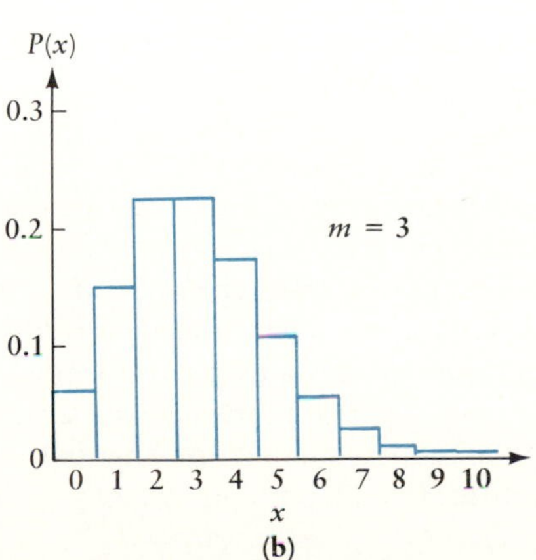

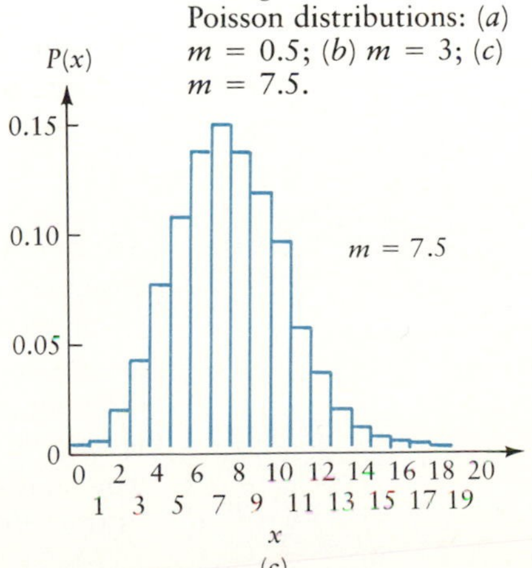

Figure 4.6 Histograms for three Poisson distributions: (a) $m = 0.5$; (b) $m = 3$; (c) $m = 7.5$.

Table 4.13 POISSON PROBABILITIES FOR $m = 1.5$.

r	$m = 1.5$
0	223
1	558
2	809
3	934
4	981
5	995
6	999
7	1
8	1

during a 6-minute unit of time is 75/50 or 1.5. Note that the calls can be considered to occur at random and are independent of each other. Moreover, because the time frame represents the busiest period, we may assume that any given 6-minute period will be like any other 6-minute period. (We may start at any point, and the distribution of calls is more or less uniform over the full 5 hours.) Thus the number of calls during a 6-minute period (x) can be considered to be a Poisson random variable. Table 3 in the back of the book lists cumulative Poisson probabilities for a selected set of values of m. Table 4.13 lists the portion of Table 3 relevant to this example.

We read Table 4.13 in exactly the same way as the tables of binomial probabilities. Thus $P(x \leq 3) \doteq 0.934$ and $P(3) \doteq 0.934 - 0.809 = 0.125$, for example. Since there is no upper bound for r, no cumulative probability can ever be equal to 1, although as r increases, the cumulative probability is practically indistinguishable from 1. In Table 3, all probabilities listed as 1 are actually $1-$.

Using Table 4.13, we find that the probability of more than four telephone calls in a 6-minute period is about $1 - 0.981$, or 0.019, less than one chance in 50, while the probability of more than three calls is 0.066. This means that if the office manager decides to have four lines, the probability that they will all be busy when a fifth call comes in is less than 2%. Thus there is less than a 2% chance that a caller will get a busy signal. If the office manager decides to have only three lines, the probability that a caller will get a busy signal is less than 7%. Note that a call is lost only if there are fewer lines than the number of calls. If there are five lines, the probability that a sixth call will be made during a 6-minute period (and thus get a busy signal) is only 0.005.

Ranges of probabilities and probabilities such as $P(x > r)$ can also be calculated as with the binomial tables. For instance, $P(x > 2) = 1 - P(x \leq 2) \doteq 0.191$; $P(2 \leq x < 4) = P(x \leq 3) - P(x \leq 1) \doteq 0.934 - 0.558 = 0.376$; $P(4) \doteq 0.981 - 0.934 = 0.047$.

Poisson probabilities occur in a great many practical situations—particularly in queuing situations such as telephone calls, lines at the tellers' windows in a bank, or lines at the checkout stands at a supermarket. Analysis of these situations can be helpful in determining the number of staff to have on duty during specific time periods.

Example 4.11

The number of tropical storms in a given area is approximated by a Poisson variable in which m is the mean number per period of time. The tropical storm period in this area lasts approximately 5 months. Assuming that storms are just as likely to occur at any part of this period as at any other, calculate the probability that a tropical storm will not occur while a family is staying in an area in which there are four storms per year on the average. The family will be staying there from August 1 to August 15 and August is in the storm season.

Solution

They will be staying for half a month; if the average number of storms for 5 months is four, then the average number of storms for half a month is 0.4. Using $m = 0.4$, we refer to Table 3 and find that the probability of no storms during that period is 0.670.

4.5.2 Approximating the Binomial Distribution

One particularly good use for the Poisson distribution is as an approximation to the binomial when p is small. Suppose that x is a binomial random variable, the probability of success on one trial is p, and there are n trials. As a rule of thumb, x can be approximated by a Poisson random variable with the same mean ($m = np$) when p is small (less than 0.10) and n is large (at least $1000p$). (If $1 - p$ is less than 0.10 and n is at least $1000(1 - p)$, the number of failures can be approximated by a Poisson random variable with $m = n(1 - p)$.) This approximation is particularly useful because the probabilities for a binomial variable depend upon both n and p, whereas the probabilities for a Poisson variable depend only upon m. Suppose, for example, that 20 people are selected at random from a population in which 1% suffer from a skin disorder. The cumulative binomial probabilities ($n = 20$, $p = 0.01$) and cumulative Poisson probabilities ($m = 20(0.01) = 0.2$) are shown in Table 4.14.

Table 4.14 COMPARISON OF BINOMIAL AND POISSON PROBABILITIES.

	BINOMIAL		POISSON
r	(TABLE2)	r	(TABLE 3)
0	818	0	819
1	983	1	982
2	999	2	999
3	1−	3	1
4	1−	4	1

The agreement in Table 4.14 is almost perfect. More important, we can use the Poisson tables to fill in gaps on the binomial tables—say for values of p such as 0.005, 0.02, and 0.03—and to extend the tables for larger values of n when p is small. We can use other approximation methods as well (see Section 5.3) so that tedious computations for the probabilities of a binomial random variable can practically be eliminated. For another example see Section 4.7.

Example 4.12

Determine the probability distribution for the number of new car buyers in a random sample of 25 college students taken from a population in which approximately 2 percent are new car buyers.

Solution

The binomial tables do not list the values for $p = 0.02$. However, $1000p = 20$ and $n = 25$, which is more than 20, so we may use the Poisson tables as an approximation with $m = 25(0.02) = 0.5$. From Table 3, we can convert the cumulative probability distribution to an *individual* probability distribution by subtracting successive terms, or we can calculate the Poisson probabilities directly from the definition. The individual Poisson probabilities with $m = 0.5$ are presented in Table 4.15.

We can calculate the individual binomial probabilities directly as a check. These are shown in Table 4.16. While not perfect, the agreement is quite good. If n is considerably larger than $1000p$, the agreement is nearly perfect.

Table 4.15 POISSON PROBABILITIES FOR $m = 0.5$.

x	$P(x)$
0	0.607
1	0.303
2	0.076
3	0.012
4	0.002
5+	0+

Table 4.16 BINOMIAL PROBABILITIES FOR $n = 25$ $p = 0.02$.

x	$P(x)$
0	0.603
1	0.308
2	0.075
3	0.012
4	0.002
5+	0+

PROFICIENCY CHECKS

19. A binomial experiment has $n = 70$, $p = 0.04$. Use the Poisson tables to obtain the following probabilities for x, the number of successes.
a. $P(x \geqslant 3)$ b. $P(x > 2)$ c. $P(x < 5)$ d. $P(x \leqslant 7)$
e. $P(6)$ f. $P(2 < x < 9)$ g. $P(3 \leqslant x \leqslant 7)$
h. $P(4 < x \leqslant 8)$

20. Of every 1000 parts produced by a machine, on the average 30 are defective. What is the probability that in a sample of 100 of these parts:
a. at least five are defective?
b. no more than two are defective?
c. the number defective is from two to six, inclusive?

Problems

Practice

4.69 A Poisson random variable has a mean of 2.2. List the probability distribution of the variable. What are the mean and standard deviation of the variable?

4.70 A binomial random variable has a probability of success equal to 0.005. The number of trials is eight. Determine the probability distribution of x:
a. using direct calculation b. using the Poisson tables

Applications

4.71 The number of "dimples" on the surface paint of an enameled refrigerator is a Poisson random variable. Suppose that the average 19-cubic-foot refrigerator, painted on the front, two sides, and the top, has 40 square feet of painted surface and that, on the average, there are 4.8 dimples per refrigerator. If the front contains 15 square feet of surface area, what is the probability that there will be at least two dimples on the front?

4.72 An oil company is planning on drilling three wells in an estuarine area of Central America. One part used in drilling is constantly under stress and has a breakage rate of once per five wells drilled. Breaks occur at random and without regard for the length of time the part has been in use. To avoid having to send back for additional parts, they wish to take along any replacement parts that will be needed. On the other hand, any parts not used will be discarded since they will be useless after having been exposed to weather conditions at the site. How many extra parts (in addition to the one already in the drill) should be taken so that the probability of having to send back for more parts is less than 0.05?

4.73 Customers arrive at a bank on Friday at an average rate of 60 during the period from 11:30 to 1:00. What is the probability that more than ten customers will arrive during a given 12-minute period?

4.74 Airlines deliberately overbook full flights on the (accurate) assumption that some passengers will not show up. (These passengers are called "no-shows.") According to Eastern Airlines, approximately 3% of all passengers are no-shows. If 200 passengers are booked for a flight that has 195 seats, what is the probability that there will be at least five no-shows?

4.75 The number of typographical errors on a page is a Poisson random variable. A typist has made an average of 2.5 errors per page over the past few weeks. He finds that if he makes more than four errors on a page, it is faster to retype the page than to go back and change the mistakes. What is the probability that he will retype a given page? What is the expected number of pages he will retype in a manuscript with 150 pages?

4.76 The number of crimes committed in a particular precinct of a city during the hours 11 PM to 2 AM is often a Poisson random variable. Suppose that the mean number of crimes committed in the Hull Street precinct during this period is 1.8.

a. What is the probability that more than five crimes will be committed during this period?

b. What is the probability that at least two crimes will be committed between midnight and 1 AM?

c. What is the probability than no more than two crimes will be committed between midnight and 2 AM?

4.77 Demand functions often have a Poisson probability distribution. Suppose that, according to the plant manager, demand at the factory for a particular part for the BANANA™ Personal Computer on Tuesday had a Poisson distribution with a mean of 2.3. The plant can manufacture five of these parts per day. What is the probability that the plant was unable to fill all the orders on Tuesday?

4.78 A branch store of Future Information Systems reported in July 1985 that the number of orders for a Macintosh 512K Computer was found to have a Poisson distribution with a mean of ten per month. The store received five Macintoshes from the factory twice per month.

a. How many Macintoshes should the shop keep on hand to ensure that the probability of running out during any particular half- month period is less than 0.05?

b. What is the probability that there will be no orders for the item in a particular month?

c. What is the probability that there will be no orders for the item during a particular half-month?

4.79 The number of automobiles passing through an intersection generally is a Poisson random variable for similar time periods on the same days of the week (holidays excluded). If the mean number of automobiles proceeding north through a particular intersection between 2 AM and 6 AM on a Monday morning is nine, what is the probability that there will be from six through ten, inclusive, proceeding north through the intersection between those hours on a particular Monday morning?

4.80 An appliance is sold with or without a service contract. Service contracts for this appliance cost $60 per year. The mean number of calls for service on this appliance is 1.5 per year and is a Poisson variable for the first 5 years of the life of the appliance. (Old age sets in after that, and the mean increases.) The cost of a service call is $30. Both service contracts and service calls cover labor only and parts are extra. Suppose Mrs. Jones buys an appliance.

a. What is the probability that she would be better off buying a service contract than paying for the calls as they occur?

b. What is the probability that she would be better off not buying the service contract?

c. What is the probability that it would not make any difference?

4.81 Accidents in a Monsanto textile plant occur at the rate of approximately one every 2 months and appear to be randomly occurring. Each accident costs the company about $2500 in work stoppage and slowdown. If more than two accidents occur in a month, the company suffers additional costs in employee discontent and slowdown estimated at about $8000. The company decides to take out insurance to pay $8000 in any month in which more than two accidents occur. Disregarding administrative costs, what would be a reasonable monthly premium for this protection? Why?

4.6 Other Discrete Random Variables*

The binomial distribution is appropriate only if the assumptions for a binomial experiment are satisfied. If sampling is done from a finite dichotomous

*This section may be omitted without loss of continuity.

population without replacement, the hypergeometric rather than the binomial model is applicable. If the number of trials is not fixed, the geometric rather than the binomial model is applicable. In this section we discuss these two additional models for a dichotomous population.

4.6.1 The Hypergeometric Distribution

If we take a sample of n observations at random from a dichotomous population without replacement, and if x is the random variable representing the number of "successes," the appropriate random variable is a **hypergeometric random variable.** Suppose, for example, we have six light bulbs, four good and two defective, and we select two without replacement. If x is the number of good bulbs, x is not a binomial random variable. The different outcomes, with their probabilities, are shown in the tree diagram of Figure 4.7. If the random variable x is the number of good bulbs on one draw, then the probability distribution of x is given by $P(0) = 1/15$, $P(1) = 8/15$, and $P(2) = 2/5$.

Sampling without replacement is commonplace. When we draw a sample without replacement from a finite population, the appropriate probabilities are those arising from application of conditional probabilities.

Suppose we draw a sample of 5 ball bearings from a shipment of 500 in which there are 200 defectives. We can find the probability that all 5 are good by multiplying the conditional probabilities that each successive ball bearing is good, given that the previous one is good. This probability is equal to the product

$$\frac{300}{500} \cdot \frac{299}{499} \cdot \frac{298}{498} \cdot \frac{297}{497} \cdot \frac{296}{496}$$

If the sample came from a shipment of 500 in which only 50 were defective, however, the probability that all 5 were good would be

$$\frac{450}{500} \cdot \frac{449}{499} \cdot \frac{448}{498} \cdot \frac{447}{497} \cdot \frac{446}{496}$$

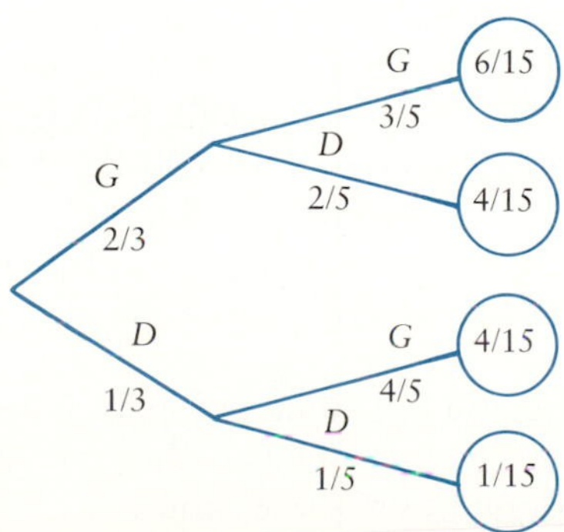

Figure 4.7
Tree diagram for hypergeometric probabilities of light bulb example.

Probabilities that 4, 3, 2, 1, and 0 are good can be calculated similarly, but the procedure is obviously tedious. Fortunately a simpler way exists.

We can find the total number of samples of five that can be chosen by using the combinations rule of Section 3.4; we give a brief discussion here. There are 500 different ways to choose the first ball bearing, then 499 ways to choose the second, 498 ways to choose the third, and so on. Thus there are $500 \cdot 499 \cdot 498 \cdot 497 \cdot 496$ ways to choose all five, in order. There are, however, $5 \cdot 4 \cdot 3 \cdot 2 \cdot 1$ ways to choose the same five, so the total number of different sets of five ball bearings that can be chosen from the 500 is equal to $(500 \cdot 499 \cdot 498 \cdot 497 \cdot 496)/(5 \cdot 4 \cdot 3 \cdot 2 \cdot 1)$. This is symbolized as $\binom{500}{5}$, the number of combinations of 500 objects taken 5 at a time, and is equal to 255,244,687,600.

In general, the number of combinations of n objects taken r at a time is given by

$$\binom{n}{r}$$

where

$$\binom{n}{r} = \frac{n \cdot (n-1) \cdot (n-2) \cdot (n-3) \cdots (n-r+1)}{r(r-1) \cdot (r-2) \cdot (r-3) \cdots 3 \cdot 2 \cdot 1}.$$

The number of ways in which we can choose 2 of 200 defective ball bearings is

$$\binom{200}{2}, \text{ or } \frac{200 \cdot 199}{2 \cdot 1},$$

which is equal to 19,900; similarly, the number of ways to choose 3 good ball bearings from among 300 is equal to

$$\binom{300}{3}, \text{ or } \frac{300 \cdot 299 \cdot 298}{3 \cdot 2 \cdot 1},$$

which is equal to 4,455,100. Thus the total number of ways to choose 2 defective *and* 3 good ball bearings from a shipment with 200 defective and 300 good is equal to $19{,}900 \cdot 4{,}455{,}100$, or 88,654,900,000. Since there are 255,244,687,600 ways to choose 5 ball bearings, the probability that a sample chosen at random from among 200 defective and 300 good ball bearings will have exactly 2 defective is 88,654,900,000/255,244,687,600, or about 0.347. We can easily generalize this procedure to obtain the **hypergeometric distribution**, shown at the top of page 211.

Example 4.13 A bag contains five red and four yellow apples. Five children each take an apple, at random, from the bag. What is the probability that the children took three of one color and two of the other?

The Hypergeometric Distribution

If a set consists of A objects of one kind and B objects of a second kind, and if n objects are selected at random from this set without replacement, the probability of selecting x objects of type A (and $n - x$ objects of type B) is given by

$$P(x) = \frac{\binom{A}{x}\binom{B}{n-x}}{\binom{A+B}{n}}$$

The mean and variance of the random variable x are as follows:

$$\mu = \frac{nA}{A+B} \quad \text{and} \quad \sigma^2 = \frac{nAB(A+B-n)}{(A+B)^2(A+B-1)}$$

Solution

We will define the random variable x to be the number of red apples. (The number of yellow would do as well, of course.) We thus want $P(3) + P(2)$. By the formula, with $A = 5$ and $B = 4$, we have

$$P(3) = \frac{\binom{5}{3}\binom{4}{2}}{\binom{9}{5}} = \frac{\frac{5\cdot4\cdot3}{3\cdot2\cdot1}\cdot\frac{4\cdot3}{2\cdot1}}{\frac{9\cdot8\cdot7\cdot6\cdot5}{5\cdot4\cdot3\cdot2\cdot1}} = \frac{10\cdot6}{126} = \frac{10}{21}$$

$$P(2) = \frac{\binom{5}{2}\binom{4}{3}}{\binom{9}{5}} = \frac{\frac{5\cdot4}{2\cdot1}\cdot\frac{4\cdot3\cdot2}{3\cdot2\cdot1}}{\frac{9\cdot8\cdot7\cdot6\cdot5}{5\cdot4\cdot3\cdot2\cdot1}} = \frac{10\cdot4}{126} = \frac{20}{63}$$

Thus $P(3) + P(2) = 100/126 \doteq 0.79$, so the probability that the children took three red and two yellow, or two red and three yellow, is about 0.79.

Example 4.14

Out of 50 people attending a conference, 20 speak Spanish and 30 do not. At the annual banquet, ten door prizes are given in a random drawing. The numbers to be drawn match those on the back of the name cards at the table. What is the probability that half of those who receive prizes will speak Spanish? If x is the number of Spanish-speaking people who receive prizes, what is the mean and variance of this random variable?

Solution

We have $A = 20$, $B = 30$, $n = 10$, and $x = 5$, so by the formula

$$P(5) = \frac{\binom{20}{5}\binom{30}{5}}{\binom{50}{10}} = \frac{\dfrac{\overset{\cancel{4}}{\cancel{20}}\cdot 19\cdot \overset{6}{\cancel{18}}\cdot 17\cdot \overset{8}{\cancel{16}}}{\cancel{5}\cdot\cancel{4}\cdot\cancel{3}\cdot\cancel{2}\cdot 1}\cdot\dfrac{\overset{6}{\cancel{30}}\cdot 29\cdot \overset{7}{\cancel{28}}\cdot \overset{9}{\cancel{27}}\cdot \overset{13}{\cancel{26}}}{\cancel{5}\cdot\cancel{4}\cdot\cancel{3}\cdot\cancel{2}\cdot 1}}{\dfrac{\overset{\cancel{5}}{\cancel{50}}\cdot \overset{7}{\cancel{49}}\cdot \overset{\cancel{6}}{\cancel{48}}\cdot 47\cdot 46\cdot \overset{5}{\cancel{45}}\cdot \overset{11}{\cancel{44}}\cdot 43\cdot \overset{\overset{7}{\cancel{14}}}{\cancel{42}}\cdot 41}{\cancel{10}\cdot\cancel{9}\cdot\cancel{8}\cdot\cancel{7}\cdot\cancel{6}\cdot\cancel{5}\cdot\cancel{4}\cdot\cancel{3}\cdot\cancel{2}\cdot 1}}$$

$$= \frac{19\cdot 3\cdot 17\cdot 8\cdot 6\cdot 29\cdot 9\cdot 13}{7\cdot 47\cdot 23\cdot 5\cdot 11\cdot 43\cdot 41}$$

$$= \frac{157{,}815{,}216}{733{,}734{,}155} \doteq 0.215$$

The mean and variance are equal to the following: $\mu = (10 \cdot 20)/50 = 4$; $\sigma^2 = [10 \cdot 20 \cdot 30 \cdot (50 - 10)]/[50^2(50 - 1)] = 240{,}000/122{,}500 \doteq 1.96$.

We observe that if the binomial model were applicable, we would have

$$P(5) = \binom{10}{5}(0.4)^5(0.6)^5 \doteq 0.201$$

$$\mu = 10\cdot 0.4 = 4$$

$$\sigma^2 = 10(0.4)(0.6) = 2.4$$

$$\sigma = 1.55$$

The probabilities are close, the means are the same, and although the variances are different the standard deviations (1.40 and 1.55, respectively) are fairly close. As the population increases, all the values become quite close, as the next example shows.

Example 4.15

A shipment containing 1000 phonograph cartridges, 950 of type IV and 50 of type V, arrived at a wholesale warehouse. Unfortunately the various cartridges became jumbled during shipment and it cannot be determined which is which without breaking the seals and opening the boxes. A retailer wants 5 type V and 15 type IV cartridges. The exasperated clerk simply takes 20 at random and gives them to the retailer.

a. What is the probability that the retailer actually received what was ordered?
b. What is the mean and variance of the random variable denoting the number of type V cartridges?
c. Repeat the problem using the binomial model as an approximation.

Solution

Since this problem involves sampling without replacement from a finite population, the hypergeometric model is appropriate. We find the probability of obtaining exactly five type V cartridges as follows:

$$P(5) = \frac{\binom{50}{5}\binom{950}{15}}{\binom{1000}{20}} \doteq 0.001978 \doteq 0.002$$

and $\mu = (20 \cdot 50)/1000 = 1$ and $\sigma^2 = (20 \cdot 50 \cdot 950 \cdot 980)/(1000^2 \cdot 999)$, or about 0.932. If we ignore the fact that the sampling is done without replacement and assume that other assumptions of the binomial are met, we have

$$P(5) = \binom{20}{5}(0.05)^5(0.95)^{15} \doteq 0.00224 \doteq 0.002$$

$$\mu = 20(0.05) = 1$$

$$\sigma^2 = 20(0.05)(0.95) = 0.95$$

These values are very close indeed.

If we draw a sample at random without replacement from a finite dichotomous population, the appropriate probability model is the hypergeometric. Calculation of hypergeometric probabilities is tedious at best, however, and practically impossible at worst. Thus if the sample is relatively small with respect to the population, we may use the binomial model as a very close approximation to the hypergeometric.

To justify the use of the binomial as an approximation to the hypergeometric model, let us take another example. Suppose a population consists of N observations and that one-fifth of these possess a characteristic of interest. If, as is usually the case in sampling, the first observation is not replaced before a second one is made, the outcomes are not independent, as is required for the binomial model, but *conditional*. That is, the outcome of the first observation has an effect on the probability of the outcome of the second observation. Thus the hypergeometric model, not the binomial, is appropriate. Let us examine the effect of population size on the probability of the outcome of the second observation. If $N = 10$, two observations have the characteristic (are "successes") and eight do not. Thus if the first observation is a success, the probability that the second one is a success is 1/9; if the first is not a success, the probability that the second one is a success is 2/9. Obviously the probabilities are quite different. As the population increases, however, the gap between the two probabilities lessens considerably. Table 4.17 shows the effect of population size on the probabilities of the outcome of the second observation. We symbolize success on the first observation by $S\{1\}$, failure on the first observation by $F\{1\}$, and success on the second observation by $S\{2\}$.

Table 4.17 CONDITIONAL PROBABILITIES OF $S\{2\}$ FOR DIFFERENT POPULATION SIZES.

N	$P(S\{2\}\|S\{1\})$	$P(S\{2\}\|F\{1\})$
10	0.1111	0.2222
50	0.1837	0.2041
100	0.1919	0.2020
500	0.1984	0.2004
1,000	0.1992	0.2002
5,000	0.1998	0.20004
10,000	0.1999	0.20002

We can see that for a sufficiently large population, the probabilities do not differ much. If we take a sample of 100 from a dichotomous population of 10,000, it can be shown that if none of the first 99 observations were successes, the probability that the 100th is a success is raised only to 0.202, which is not much different from 0.2. For similar reasons, if we draw a relatively small sample without replacement from a finite but relatively large population, we may use the binomial model as a very close approximation to the hypergeometric in certain cases. As a rule of thumb, suppose we draw a sample of size n from a population of size $N(= A + B)$ and define $p = A/N$. Then if $n < 0.05N$, we can ignore minor discrepancies and use the binomial model as a very good approximation.

PROFICIENCY CHECKS

21. Suppose that an organization contains ten members: six women and four men. Suppose further that a committee of three people is chosen at random from the ten. What is the probability distribution of women on the committee? What are the mean and variance of the variable?

22. An auditor selects 5 sales slips at random from 75 filled out by an employee. If 21 contain errors, what is the probability that more than one of the 5 selected will contain an error:
    a. Using the hypergeometric model?
    b. Using the binomial model?
    c. Which is appropriate in this case? Does the difference seem important?

4.6.2 The Geometric Distribution

Recall that a single trial of a binomial experiment is called a *Bernoulli trial.* If we perform a sequence of Bernoulli trials until we obtain a success, the random variable x representing the number of trials is called a **geometric random variable.** If a fair die is tossed until a 6 appears, for example, the number of tosses needed for one repetition of the experiment is a geometric random variable. Theoretically a 6 may *never* appear (although in all practicality we assume that it will or else we would stop tossing it). The probability that a 6 will appear on the first toss is 1/6; that it will appear for the first time on the second toss is $(5/6) \cdot (1/6)$; that it will appear for the first time on the tenth toss is $(5/6)^9 \cdot (1/6)$. Generalizing, we have the following definition of the **geometric distribution.**

The Geometric Distribution

An experiment will result in a geometric distribution if:

1. There are exactly two outcomes for each trial, usually denoted S (success) and F (failure).
2. The probability of success, denoted p, remains constant from trial to trial.
3. Outcomes of successive trials are independent.
4. The random variable x is the number of the trial on which the first success occurs.

The probability distribution for x is defined by

$$P(x) = (1 - p)^{x-1} \cdot p \qquad (x = 1, 2, 3, 4, \ldots)$$

The mean and variance of x are

$$\mu = \frac{1}{p} \qquad \text{and} \qquad \sigma^2 = \frac{1-p}{p^2}$$

We can derive the formulas for the mean and variance by using repeated applications of the formula for the sum of a geometric series from algebra.

Example 4.16

On the average, about one of ten persons passing a shop in a mall enters the shop. What is the probability that the first person to enter the shop after it opens in the morning is the fifth person who passes? The tenth? The twentieth?

Solution

Here $p = 0.1$ and the random variable is geometric. Then $P(5) = (0.9)^4(0.1) = 0.06561 \doteq 0.066$, $P(10) = (0.9)^9(0.1) \doteq 0.039$, and $P(20) = (0.9)^{19}(0.1) \doteq 0.014$.

Problems

Practice

4.82 An urn contains four red balls and six blue balls. Five balls are drawn at random. The variable x denotes the number of blue balls drawn. The balls are not replaced before the next drawing. Determine the probability distribution of x, and find its mean and variance.

4.83 What is the probability that a five-card poker hand contains exactly:
a. two aces?
b. one ace?
c. no aces?

4.84 A coin is tossed until a head appears. Give the probability distribution for the number of tosses (until the probability is less than 0.005), and make a histogram for the distribution. Compute the mean and variance using the formulas given in Section 4.2 and the special formulas. How do the results compare?

Applications

4.85 A certain mathematics department has ten members, of whom only four teach graduate statistics courses. Three members of the department are invited, at random, to a reception for the new dean. Let x represent the number of statistics teachers invited to the reception. Determine the probability distribution of x, and find its mean and variance.

4.86 Twenty X-ray plates are sent to a laboratory. Five are exposed on one patient. It turns out, however, that four of the twenty are defective. Let x represent the number of defective plates exposed on this patient. Determine the probability distribution of x, and find its mean and variance.

4.87 Suppose, in Problem 4.50, there are 1500 students on campus. Repeat Problem 4.50 using a hypergeometric random variable. Compare the results.

4.88 Refer to Problem 4.8. Give the probability distribution for the number of bulbs that will be tried. What are the mean and variance of the variable?

4.89 Refer to Problem 4.9. Give the probability distribution for the number of bulbs that will be tried. What are the mean and variance of the variable?

4.90 A random sample of five persons is selected from a group of N persons in which A favor capital punishment and B are opposed. Determine the probability that exactly two of those selected favor capital punishment:
a. if $N = 250$ and $A = 50$
b. if $N = 500$ and $A = 100$

c. if $N = 5000$ and $A = 1000$
d. if we use the binomial model with $p = 0.2$

4.91 When a vacancy occurs on the Supreme Court, prospective jurists are considered one at a time until one is chosen. Clearly all persons are not equally likely to be chosen, but suppose that the probability that any particular person is selected is 1/5.
a. What is the probability that the person selected will be the fifth one considered?
b. What is the probability that the person selected will be the sixth one considered or later?

4.92 A production run of 200 computers has 20 with a defective mother board. A random sample of 10 computers is to be selected from this run for testing. Determine the probability that at least 2 of the 10 will have defective mother boards:
a. using the hypergeometric distribution
b. using the binomial approximation to the hypergeometric
c. using the Poisson approximation to the binomial

4.93 Repeat Problem 4.92 but this time find the probability that exactly 4 of the 10 will have defective mother boards.

4.94 Repeat Problem 4.92 but suppose the run is 1000 computers with 20 defective.

4.95 Repeat Problem 4.94 but this time find the probability that exactly 4 of the 10 will be defective. Compare the results of Problems 4.92 to 4.95.

4.7 Computer Usage

You can use a computer to obtain probabilities for a variety of random variables. In this section we will learn how to use Minitab and SAS statements to find binomial and Poisson probabilities. This procedure can be very useful when the tables in this book cannot be used and the binomial probability formula of Section 4.4 is too tedious to calculate.

4.7.1 Minitab Commands for This Chapter

We can calculate individual binomial probabilities by using Minitab with the PDF command and the BINOMIAL subcommand. We can obtain $P(x)$ for $x = 0, 1, 2, 3, \ldots$ by using the command

```
MTB > PDF;
SUBC> BINOMIAL n = K1 p = K2 .
```

where K1 specifies the number of trials and K2 is the probability of success on any one trial. The computer will print the values of $P(x)$ for $x = 0, 1,$

2, 3, . . . until the individual probabilities are less than 0.00005. We can obtain cumulative probabilities, $P(x \leq r)$, by using the command

```
MTB > CDF;
SUBC> BINOMIAL n = K1 p = K2 .
```

The computer will print the values of $P(x \leq r)$ for $r = 0, 1, 2, 3, \ldots$ until the cumulative probability is greater than 0.99995.

Example 4.17

Automobile motors are assembled on an assembly line at a manufacturing plant. Past history shows that 93% of the motors are assembled correctly but 7% require some modification after assembly. A typical production run contains 50 motors. Use Minitab to answer the following questions.

a. What is the probability that exactly 2 of 50 motors require modification after assembly?
b. What is the probability that 4 or fewer of 50 motors require modification after assembly?
c. If a production run of 50 motors contains 11 that require modification, what can you conclude?

Solution

We assume a binomial model for this experiment and use Minitab to generate both individual and cumulative probabilities for a binomial random variable with $n = 50$ and $p = 0.07$. The program will give us a list of the probabilities for the number of motors that need modification after assembly on a typical run. The computer printout for this example is shown in Figure 4.8.

Figure 4.8
Minitab printout for motor assembly data of Example 4.17

```
MTB > PDF ;
SUBC> BINOMIAL n = 50 p = .07 .

 BINOMIAL PROBABILITIES FOR N = 50 AND P = 0.070000

      K          P(X = K)
      0           0.0266
      1           0.0999
      2           0.1843
      3           0.2219
      4           0.1963
      5           0.1359
      6           0.0767
      7           0.0363
      8           0.0147
      9           0.0052
     10           0.0016
     11           0.0004
     12           0.0001
```

Individual binomial probabilities

```
MTB > CDF ;
SUBC> BINOMIAL n = 50 p = .07 .

 BINOMIAL PROBABILITIES FOR N = 50 AND P = 0.070000

      K        P(X LESS OR = K)
      0             0.0266
      1             0.1265
      2             0.3108
      3             0.5327
      4             0.7290
      5             0.8650
      6             0.9417
      7             0.9780
      8             0.9927
      9             0.9978
     10             0.9994
     11             0.9999
     12             1.0000
```

Cumulative binomial probabilities

a. We see that the probability of getting exactly 2 motors that need modification is the binomial probability when $k = 2$, so $P(2) = 0.1843$. Thus we would expect to get exactly 2 motors needing modification on about 18.43% of the runs.

b. The probability of getting 4 or fewer motors needing modification in a run of 50 is $P(x \leq 4)$; from the second part of the table we find $P(x \leq 4) = 0.7290$ (that is, when $k = 4$). Therefore we should expect to have 4 or fewer motors needing modification on about 72.90% of the runs. Conversely, we would expect to have 5 or more motors needing modification on about $(100 - 72.90)\%$ or about 27.10% of the runs.

c. From the cumulative probability table we see that the probability of 10 or fewer motors needing modification is 0.9994; thus the probability that 11 or more would need modification is about 0.0006. This event is highly unlikely, so we can logically conclude one of two possibilities. Either this result is simply a chance occurrence, no matter how unlikely, or the distribution of motors needing modification is no longer binomial with $p = 0.07$. It is more likely that the probability of a motor needing modification is now higher than 0.07—particularly if we were to observe this many or more motors needing modification for several runs in a row.

We can obtain Poisson probabilities by using Minitab commands similar to those we used for the binomial probabilities. This procedure is particularly useful for finding cumulative Poisson probabilities which are not listed in this book. Individual probabilities are relatively easy to compute with a hand-held calculator which has an e^x function, but cumulative probabilities require a great deal of time. The appropriate Minitab command for obtaining individual Poisson probabilities is

```
MTB > PDF;
SUBC> POISSON mean = K1 .
```

where K1 is the expected number of occurrences (m) of a Poisson random variable per unit of time or space. We can find the cumulative probabilities $P(x \leq r)$ by using the command

```
MTB > CDF;
SUBC> POISSON mean = K1 .
```

The computer will print values for 0, 1, 2, 3, . . . until individual probabilities are less than 0.00005 or cumulative probability is greater than 0.99995, as before.

Example 4.18

At a large computing facility the average number of computer terminals that need repair during a 1-month period is 20. Suppose that the number of terminals needing repair in a given month is a Poisson random variable with a mean of 20.

a. What is the probability that exactly 10 terminals will need repair in a 1-month period?
b. What is the probability that more than 25 terminals will need repair in a 1-month period?

Solution

Figure 4.9 shows the Minitab printout of individual and cumulative Poisson probabilities for $m = 20$.

Figure 4.9 Minitab printout for Poisson variable with m = 20, Example 4.18.

```
MTB > PDF ;
SUBC> POISSON mean = 20 .

 POISSON PROBABILITIES FOR MEAN = 20.000

      K       P( X = K)
      0         0.0000
      1         0.0000
      2         0.0000
      3         0.0000
      4         0.0000
      5         0.0001
      6         0.0002
      7         0.0005
      8         0.0013
      9         0.0029
     10         0.0058
     11         0.0106
     12         0.0176
     13         0.0271
     14         0.0387
     15         0.0516
     16         0.0646
```

```
17          0.0760
18          0.0844
19          0.0888
20          0.0888
21          0.0846
22          0.0769
23          0.0669
24          0.0557
25          0.0446
26          0.0343
27          0.0254
28          0.0181
29          0.0125
30          0.0083
31          0.0054
32          0.0034
33          0.0020
34          0.0012
35          0.0007
36          0.0004
37          0.0002
38          0.0001
39          0.0001
40          0.0000
```

Individual Poisson probabilities

```
MTB > CDF ;
SUBC> POISSON mean = 20 .

 POISSON PROBABILITIES FOR MEAN = 20.000

    K        P(X LESS OR = K)
    0             0.0000
    1             0.0000
    2             0.0000
    3             0.0000
    4             0.0000
    5             0.0001
    6             0.0003
    7             0.0008
    8             0.0021
    9             0.0050
   10             0.0108
   11             0.0214
   12             0.0390
   13             0.0661
   14             0.1049
   15             0.1565
   16             0.2211
   17             0.2970
   18             0.3814
   19             0.4703
```

```
20    0.5591
21    0.6437
22    0.7206
23    0.7875
24    0.8432
25    0.8878
26    0.9221
27    0.9475
28    0.9657
29    0.9782
30    0.9865
31    0.9919
32    0.9953
33    0.9973
34    0.9985
35    0.9992
36    0.9996
37    0.9998
38    0.9999
39    0.9999
40    1.0000
```

Cumulative Poisson probabilities

a. By referring to this printout we see that the probability of exactly 10 terminals needing repair is 0.0058, a fairly unlikely event.

b. The probability that more than 25 terminals will need repair in a month, $P(x > 25)$, is equal to $1 - P(x \leq 25)$. From the printout, $P(x \leq 25) = 0.8878$, so $P(x > 25) = 1 - 0.8878 = 0.1122$. We can expect to have more than 25 terminals needing repair in about 11.22% of all months.

In Section 4.5.2 we observed that a Poisson distribution may be used to approximate a binomial distribution if p is small and $n > 1000p$. We use Minitab to show the accuracy of this approximation for $n = 1000$ and $p = 0.002$. Figure 4.10 shows the Minitab output for both individual and cumulative probabilities of a binomial random variable with $n = 1000$ and $p = 0.002$ and for both individual and cumulative probabilities of a Poisson random variable with $m = np = 1000(0.002) = 2$. We see that these distributions agree quite well. In fact, if we use the Poisson distribution as an approximation to the binomial, the error in the approximation occurs in the third place for the cumulative probabilities and in the fourth place for the individual probabilities.

Figure 4.10 Minitab printout on accuracy of Poisson approximation to the binomial distribution.

```
MTB > PDF ;
SUBC> BINOMIAL n = 1000 p = .002 .

 BINOMIAL PROBABILITIES FOR N = 1000 AND P = 0.002000

    K        P(X = K)
    0          0.1351
```

```
 1         0.2707
 2         0.2709
 3         0.1806
 4         0.0902
 5         0.0360
 6         0.0120
 7         0.0034
 8         0.0008
 9         0.0002
10         0.0000
11         0.0000
```

Individual probabilities

```
MTB > PDF ;
SUBC> POISSON mean = 2 .

 POISSON PROBABILITIES FOR MEAN = 2.000

    K        P(X = K)
    0          0.1353
    1          0.2707
    2          0.2707
    3          0.1804
    4          0.0902
    5          0.0361
    6          0.0120
    7          0.0034
    8          0.0009
    9          0.0002
   10          0.0000
```

Individual probabilities

```
MTB > CDF ;
SUBC> BINOMIAL n = 1000 p = .002 .

 BINOMIAL PROBABILITIES FOR N = 1000 AND P = 0.002000

    K        P(X LESS OR = K)
    0              0.1351
    1              0.4057
    2              0.6767
    3              0.8573
    4              0.9476
    5              0.9836
    6              0.9955
    7              0.9989
    8              0.9998
    9              1.0000
   10              1.0000
   11              1.0000
```

Cumulative probabilities

```
MTB > CDF ;
SUBC> POISSON mean = 2.

 POISSON PROBABILITIES FOR MEAN = 2.000

     K        P(X LESS OR = K)
     0              0.1353
     1              0.4060
     2              0.6767
     3              0.8571
     4              0.9473
     5              0.9834
     6              0.9955
     7              0.9989
     8              0.9998
     9              1.0000
    10              1.0000
```

Cumulative probabilities

4.7.2 SAS Statements for This Chapter

The SAS statement to generate probabilities of a binomial random variable is PROBBNML. We may write

```
P = PROBBNML(p, n, r );
```

to obtain cumulative probabilities for a binomial random variable with n trials and probability p of success on each trial. The output will be $P(x \leqslant r)$. To determine the probability that $x \leqslant 2$ if x is B(10, 0.4), we may write

```
DATA ONE;
  P = PROBBNML(.4, 10, 2);
  PUT P;
RUN;
```

The number 0.16729 will appear on the SAS Log. Thus $P(x \leqslant 2) \doteq 0.16729$ in this case. We can find individual binomial probabilities by subtracting consecutive cumulative probabilities. To find $P(6)$ if x is B(24, 0.26), we may write

```
DATA TWO;
  P1 = PROBBNML(.26, 24, 5);
  P2 = PROBBNML(.26, 24, 6);
  P = P2 - P1;
  PUT P;
RUN;
```

The number 0.1840955 will appear on the SAS Log; this is $P(6)$.

We can obtain a complete list of cumulative binomial probabilities by using a "do loop." We can obtain the cumulative probabilities for a binomial random variable with $n = 10$ and $p = 0.5$ by writing the statements

```
DATA BNML;
  DO R = 0 to 10;
  P = PROBBNML(.5, 10, R);
  OUTPUT;
  END;
PROC PRINT;
```

These statements will result in the values of r, from 0 to 10, being printed along with the cumulative probabilities $P(x \leq r)$.

To obtain all the individual probabilities, we need to calculate consecutive probabilities and then subtract. To obtain a printout with x (or r) corresponding to the correct numbers, we need, for example, $P(5) = P(x \leq 5) - P(x \leq 4)$. The following statements will give the desired probabilities:

```
DO X = 0 to 10;
P1 = PROBBNML(.5, 10, X);
IF X > 0 THEN P2 = PROBBNML(.5, 10, X-1);
IF X = 0 THEN P = P1;
IF X > 0 THEN P = P1 - P2;
OUTPUT;
END;
```

The statement "IF X > 0 THEN" is necessary since $P2$ will not exist if $x = 0$. Further, if $x = 0$ then $P(0) = P(x \leq 0)$. Thus we can use the SAS System to answer the question of Example 4.17. Figure 4.11 shows the output including the input program (shown on the SAS Log). We use CP to mean cumulative probability and IP to mean individual probability. The KEEP statement indicates which variables are to be kept in the data set. Alternatively (see Figure 4.12) we can print only those variables we wish by using the PRINT procedure. We performed only repetitions from 0 to 15 since we knew from Example 4.17 that we do not need more.

```
DATA BNML:
  DO X = 0 TO I5 ;
  CP = PROBBNML(.07, 50, X);
  IF X > 0 THEN P2 = PROBBNML(.07, 50, X-1);
  IF X = 0 THEN IP = CP;
  IF X > 0 THEN IP = CP - P2;
  OUTPUT;
  END;
  KEEP X CP IP;
PROC PRINT;
```

Figure 4.11
SAS printout for binomial probabilities of Example 4.17.

OBS	X	CP	IP
1	0	0.02656	0.026555
2	1	0.12649	0.099938
3	2	0.31079	0.184295
4	3	0.53274	0.221947
5	4	0.72903	0.196292
6	5	0.86495	0.135927
7	6	0.94169	0.076733
8	7	0.97799	0.036304
9	8	0.99268	0.014687
10	9	0.99784	0.005159
11	10	0.99943	0.001592
12	11	0.99986	0.000436
13	12	0.99997	0.000107
14	13	0.99999	0.000023
15	14	1.00000	0.000005
16	15	1.00000	0.000001

We can obtain cumulative Poisson probabilities by using the statement POISSON in the same way we used the PROBBNML statement. Since a Poisson variable has one parameter, m, we need to enter that value as well. We write

```
P = POISSON(m, r );
```

to obtain $P(x \leqslant r)$ for a Poisson random variable with mean m. We can obtain individual probabilities in the same way as with PROBBNML. The SAS output for Example 4.18 is shown in Figure 4.12.

Figure 4.12
SAS printout for Poisson probabilities of Example 4.18.

```
DATA POISSON;
   DO X = 0 TO 40;
   CP = POISSON (20, X);
   IF X > 0 THEN P2 = POISSON (20, X - 1);
   IF X = 0 THEN IP = CP;
   IF X > 0 THEN IP = CP - P2;
   OUTPUT;
   END;
PROC PRINT; VAR X CP IP;
```

OBS	X	CP	IP
1	0	0.000000	0.0000000
2	1	0.000000	0.0000000
3	2	0.000000	0.0000004
4	3	0.000003	0.0000027
5	4	0.000017	0.0000137
6	5	0.000072	0.0000550

7	6	0.000255	0.0001832
8	7	0.000779	0.0005235
9	8	0.002087	0.0013087
10	9	0.004995	0.0029082
11	10	0.010812	0.0058163
12	11	0.021387	0.0105751
13	12	0.039012	0.0176252
14	13	0.066128	0.0271156
15	14	0.104864	0.0387366
16	15	0.156513	0.0516489
17	16	0.221074	0.0645611
18	17	0.297028	0.0759542
19	18	0.381422	0.0843936
20	19	0.470257	0.0888353
21	20	0.559093	0.0888353
22	21	0.643698	0.0846051
23	22	0.720611	0.0769137
24	23	0.787493	0.0668815
25	24	0.843227	0.0557346
26	25	0.887815	0.0445876
27	26	0.922113	0.0342982
28	27	0.947519	0.0254061
29	28	0.965666	0.0181472
30	29	0.978182	0.0125153
31	30	0.986525	0.0083435
32	31	0.991908	0.0053829
33	32	0.995273	0.0033643
34	33	0.997312	0.0020390
35	34	0.998511	0.0011994
36	35	0.999196	0.0006854
37	36	0.999577	0.0003808
38	37	0.999783	0.0002058
39	38	0.999891	0.0001083
40	39	0.999947	0.0000556
41	40	0.999975	0.0000278

We can also use SAS statements to compare the Poisson and binomial probabilities. Figure 4.13 shows the SAS output analogous to the Minitab output of Figure 4.10. We use the MERGE statement to put together the two data sets using common values of x. We use BCP to indicate binomial cumulative probabilities, BIP for binomial individual probabilities, PCP for Poisson cumulative probabilities, and PIP for Poisson individual probabilities.

```
DATA BNML;
   DO X = 0 TO 11;
   BCP = PROBBNML(.002,1000,X);
   IF X > 0 THEN P2 = PROBBNML(.002,1000,X-1);
   IF X = 0 THEN BIP = BCP;
   IF X > 0 THEN BIP = BCP - P2;
   OUTPUT;
```

```
      END;
      KEEP X BCP BIP;
   DATA POISSON;
      DO X = 0 TO 11;
      PCP = POISSON(2,X);
      IF X > 0 THEN P2 = POISSON(2,X-1);
      IF X = 0 THEN PIP = PCP;
      IF X > 0 THEN PIP = PCP - P2;
      OUTPUT;
      END;
      KEEP X PCP PIP;
   DATA ALL;
   MERGE BNML POISSON;
   PROC PRINT;
   TITLE 'BINOMIAL PROBABILITIES (BCP AND BIP) FOR N=1000 P=0.002 AND
   POISSON PROBABILITIES (PCP AND PIP) FOR M = 2';
```

BINOMIAL PROBABILITIES (BCP AND BIP) FOR N = 1000 P = 0.002 AND
POISSON PROBABILITIES (PCP AND PIP) FOR M = 2

OBS	X	BCP	BIP	PCP	PIP
1	0	0.13506	0.135065	0.13534	0.135335
2	1	0.40573	0.270670	0.40601	0.270671
3	2	0.67668	0.270942	0.67668	0.270671
4	3	0.85730	0.180628	0.85712	0.180447
5	4	0.94753	0.090223	0.94735	0.090224
6	5	0.98354	0.036017	0.98344	0.036089
7	6	0.99551	0.011970	0.99547	0.012030
8	7	0.99897	0.003406	0.99890	0.003437
9	8	0.99977	0.000847	0.99976	0.000859
10	9	0.99995	0.000187	0.99995	0.000191
11	10	0.99999	0.000037	0.99999	0.000038
12	11	1.00000	0.000007	1.00000	0.000007

Problems

Applications

4.96 A speed reading program claims that it will double the reading speed for many people. In fact, it claims that the probability that a person's reading speed will double is 0.66. Use the computer to answer the following questions.

- **a.** Suppose the claim is true. If 38 people use this reading program, what is the probability that 30 or more will double their reading speeds?
- **b.** Find the probability that 20 or fewer people out of 38 will double their reading speeds if the claim is true.
- **c.** If you actually observe 15 or fewer out of the 38 people who double their reading speeds, would you accept the validity of the claim? Why or why not?

4.97 A manufacturer purchases transistors from another company. Whenever a shipment of transistors is received, some are tested for quality. If the quality is too low, the shipment is rejected. One inspection plan is to select 50 transistors at random and test them. If 2 or more defective transistors are found, the shipment is rejected. Use the computer to answer the following questions.

a. If the true probability that a transistor is defective is 0.02, what is the probability that a shipment will be accepted?

b. If the probability that a transistor is defective is 0.02, which of the following inspection plans has a higher probability of rejecting the shipment?
   i. Sample 50 and reject the shipment if 2 or more are defective.
   ii. Sample 25 and reject the shipment if 1 or more are defective.

4.98 The expected number of calls to a fire station during a 24-hour period is 4.56 calls. If we assume a Poisson model, use the computer to help answer the following questions.

a. What is the probability there will be more than eight calls to the station in a 24-hour period?

b. What is the most probable number of calls to the station in a 24-hour period?

c. What is the probability that there will be no calls during a particular 8-hour period?

4.99 Use the computer to determine the accuracy of the Poisson approximation to the binomial for the binomial experiments given here:

a. $n = 100, p = 0.15$

b. $n = 100, p = 0.001$

Do your results substantiate the rule that n should be at least $1000p$?

4.100 We can simulate outcomes of binomial experiments with specified values for n and p on the computer by using techniques shown in Section 3.5. To obtain K1 simulated values from a binomial experiment using Minitab, we use the following commands:

```
MTB > RANDOM K1 binomial outcomes and store in C;
SUBC> BINOMIAL n = K2 and p = K3.
```

Using the SAS System, we may use the statements

```
DO I = 1 TO N;
K = RANBIN(I, n, p );
OUTPUT;
END;
```

after the data step to generate N observations of a binomial random variable with n trials and probability p of success on one trial. Refer to Example 4.17 and perform the following:

a. Simulate the number of motors that need modification after assembly for each of 200 runs. Print out the results by using the Minitab command TABLE or the SAS statement PROC PRINT.

b. Find the mean and standard deviation for these simulated values by using the Minitab command DESCRIBE or the SAS statement PROC MEANS. Compare these values with the true mean and standard deviation for this binomial experiment.

4.101 We may simulate outcomes for a Poisson experiment by using the POISSON subcommand instead of the BINOMIAL subcommand in Minitab or using the RANPOI(*seed, m*) SAS statement instead of RANBIN(*seed, n, p*) and proceed as in Problem 4.100. Refer to Problem 4.98 and perform the following:

a. Simulate the number of fire calls for 200 time periods and list the results.

b. Find the mean of the simulated outcomes. Compare this result to the true Poisson mean.

4.8 Case 4: The Parapsychological Puzzle—Solution

When a subject guesses the card turned, the guess is either right or wrong. The number of trials is fixed at 20, and the probability of success on any trial is 0.2, one in five. If the cards are reshuffled before each trial, the number of correct guesses is a binomial random variable. The number of expected right guesses for any set of 20 trials is equal to 20(0.2), or 4. Deviations from this number in either direction need to be investigated.

Subject B got five right on the mind-reading trials and four right on the clairvoyance trials. Subject C got four right on the mind-reading trials and six right on the clairvoyance trials. Using Table 2, we find $P(x \geqslant 4) \doteq 0.589$, $P(x \geqslant 5) \doteq 0.370$, and $P(x \geqslant 6) \doteq 0.196$. Using the criteria described in Section 4.0, we conclude that subject C has some degree of clairvoyance. We note, however, that about one subject in five would do that well, strictly by chance.

Subject A, however, is a different story. We find that $P(x \geqslant 7) \doteq 0.087$, while $P(x \geqslant 12) = 0+$. The subject seems to have a high degree of clairvoyance and a remarkable degree of mind-reading ability.

What about subject D? Subject D got nothing right. Surely that result indicates no ESP ability, or does it? Researchers in this field are just as interested in negative results as positive ones. The probability that a subject would miss all 20 is only 0.012. This probability is so small, according to researchers, that this subject may well have a high degree of ESP, perhaps repressed, and further study of the subject is warranted.

4.9 Summary

In this chapter we have studied characteristics of the probability distributions of discrete random variables—in particular, the binomial, Poisson, hypergeometric, and geometric random variables. A random variable is a variable whose value is determined by chance. A *discrete* random variable is a random variable that has a countable number of values. For any discrete random variable x, the probability $P(x) \geq 0$, and the sum of the probabilities is 1. The expected value of any random variable x is the average value of the variable if the experiment generating the variable were repeated a great many times. Thus, the expected value equals the mean of the variable.

A binomial distribution is the set of probabilities of the binomial random variable denoting the number of successes in n trials of a binomial experiment. Any trial results either in a success or a failure; the probability of success is constant from trial to trial; and the outcomes of successive trials are independent. We can often obtain probabilities for a binomial random variable from tabulations such as Table 2.

A Poisson random variable denotes the number of occurrences of an event for which the mean number of occurrences per unit is m—if occurrences are at random and independent of each other, independent of the starting point, and uniformly distributed over any interval.

We can calculate Poisson probabilities directly or by using Table 3. We may also use Table 3 to estimate probabilities of a binomial random variable with mean $m = np$, provided that p is small (less than 0.10) and n is at least $1000p$.

A hypergeometric random variable denotes the number of successes when sampling is done without replacement from a finite dichotomous population. Its probabilities can be closely approximated by the probabilities of a binomial random variable if the sample is small relative to the population.

A geometric random variable denotes the number of the first success when a sequence of Bernoulli trials is performed.

Chapter Review Problems

Practice

4.102 From a box containing four red marbles and two white marbles, three are drawn in succession. The number of white marbles is observed. Find the probability distribution and the mean and standard deviation for this variable:

a. if each marble is replaced before the next is drawn

b. if the marbles are not replaced

4.103 A random variable is defined by $P(x) = x^2/30$ for $x = 1, 2, 3, 4$. Determine the mean and standard deviation of the variable.

4.104 Find the mean and standard deviation of the random variable with the probability distribution given in the table at right.

x	$P(x)$
1	3/80
3	27/80
7	11/80
9	17/80
14	13/80
16	9/80

4.105 Probabilities for the random variable x are defined by

$$P(x) = \frac{x^2 - 1}{n} \qquad \text{for } x = 2, 3, 4, 5$$

a. Determine n.

b. Make a histogram for the distribution.

c. Calculate the mean of the variable.

d. Calculate the standard deviation of the distribution.

4.106 A coin is tossed four times or until a head appears, whichever comes first. What is the expected number of:

a. tosses on one experiment?
b. tails on one experiment?

4.107 If you receive \$8 each time you roll a 1 on a fair die, how much should you pay each time you roll a number other than 1 to make this a fair game?

4.108 A bowling team consists of four members. The probability of each player bowling at least 150 is, respectively, 0.60, 0.80, 0.50, and 0.40.

a. Determine the probability distribution for the number of players on the team bowling at least 150.
b. Determine the mean and standard deviation of the variable.

4.109 In an experiment with three dice, tossing a 10 or 11 is considered a success. Suppose the experiment is repeated ten times.

a. Compute the mean and standard deviation of the variable.
b. What is the probability of obtaining at least one success?
c. What is the probability of obtaining fewer than three successes?

4.110 Two dice are tossed until a total of 7 on both dice is obtained. What is the probability that more than four tosses will be required?

4.111 Four aces are put into a hat—one ace from each suit. A card is drawn, observed, and returned to the hat. Assuming that consecutive trials are independent, how many drawings must you make so that the probability of drawing at least one ace of spades is greater than 4/5?

4.112 A regular pack of cards is shuffled, and the top and bottom cards are examined to see if at least one of them is an ace. If so, this is recorded as a success; otherwise it is a failure. The experiment is repeated 200 times. What are the mean and standard deviation of the variable?

Applications

4.113 A certain mathematics department has 12 members: 5 of senior rank and 7 of junior rank. If a committee of four is chosen at random, determine the probability that it will have more senior members than junior members.

4.114 Over the past 100 working days, the number of orders for a particular item per day has been as follows:

Items ordered	0	1	2	3	4	5	6
Number of days	11	16	21	24	15	9	4

Let x represent the number of orders in a day. We approximate $P(x)$ by the relative frequencies.

a. Determine the mean and variance of x.
b. What is the probability that more than four items will be ordered on a particular day?
c. The office manager hypothesizes that the distribution of orders is a Poisson random variable with mean 2.5. Determine the probability distribution under this assumption and compare it with the actual distribution of orders. Given these results and the results in part (*a*), do you think this assumption is reasonable? (A procedure to test this hypothesis will be given later in a discussion of *goodness of fit* in Chapter 9.)

4.115 Records on the unemployment rate in a certain region over the past 6 months show the following information:

UNEMPLOYMENT RATE (x)	RELATIVE FREQUENCY
$x \geq 10\%$	0.08
$8\% \leq x < 10\%$	0.40
$6\% \leq x < 8\%$	0.50
$x < 6\%$	?

Using these relative frequencies as estimates of future probabilities and assuming that successive observations are independent (in fact they are not), what is the probability that the unemployment rate next week will be:

a. less than 8%?
b. at least 6% but less than 10%?
c. at least 8%?

4.116 A man wishes to buy a 1-year-term policy insuring his life for $50,000. The insurance company determines that the probability that he will die during this period is 0.0042. Disregarding administrative costs, what should be the premium for the policy?

4.117 A retired teacher is considering purchasing a fast food franchise. Franchise A requires an investment of $70,000, of which $55,000 can be borrowed and repaid from the profits at the rate of $1763.32 per month for 3 years. Franchise B requires an investment of $60,000, of which $45,000 can be borrowed and repaid from the profits at the rate of $1442.70 per month for 3 years. The teacher has the needed $15,000. She examines the records of both franchises and estimates the approximate net weekly income for the two franchises to have the probability distributions shown here:

WEEKLY INCOME	PROBABILITY FRANCHISE A	PROBABILITY FRANCHISE B
$500	0.10	0.05
600	0.20	0.30
700	0.25	0.40
800	0.25	0.20
900	0.20	0.05

a. Which franchise has the higher expected weekly income?
b. Which franchise has the more variable income?
c. Which franchise has the greater expected monthly excess of income over the franchise payment? (One month = $4\frac{1}{3}$ weeks.)
d. What other factors might influence the teacher's decision on which franchise to purchase?

4.118 A public opinion poll is being taken on the issue of free housing for indigent laborers. If 30% of the population is in favor of the issue, what is the probability that, of 15 persons interviewed, 3, 4, 5, or 6 favor the free housing:

a. calculating the probabilities directly?
b. using Table 2?

4.119 The mortality rate for a certain disease is 40%. At a particular hospital, nine of the last ten patients admitted with the disease have recovered. Do you think these results are consistent with the known mortality rate, or should an investigation be made to see whether some new treatment is being used at the hospital?

4.120 Eight pints of blood are stored in a hospital laboratory. It is known that exactly 3 pints are type O, but it is not known which ones. Two pints of type O blood are needed. One pint at a time is removed from storage and typed. If it is type O, it is used; if not, it is labeled and the next pint is tested.
a. Make a probability distribution for the number of pints that must be tested in order to obtain 2 pints of type O.
b. Determine the mean and standard deviation of the variable.

4.121 Bill Olson pole-vaulted 19 feet $2\frac{3}{4}$ inches on December 28, 1985, for a new indoor world record. Suppose a college pole vaulter can vault 18 feet 3 inches on the average 20% of the time. The bar is set at 18 feet 3 inches and he gets three tries to complete the vault successfully. What is the probability that he will do so?

4.122 A biologist wishes to test whether the presence of a birth defect can be associated with a certain drug. On the average, the defect is present in one-tenth of the litters of newborn guinea pigs. He injects 25 pregnant guinea pigs with the drug and observes the number of litters in which the defect is present. If there is no association between the drug and the defect, what is the probability that at least five litters will contain the defect? Suppose that six of the litters contain the defect. Would you say that there is an association between the drug and the defect, or would you conclude that the result could be attributed to chance? (A method for testing this hypothesis is given in Section 7.3.)

4.123 Among the applicants for a certain position, one-tenth belong to racial minorities. Seven persons are asked for an interview. Assuming that the number of applicants is large and the qualifications of all individuals are about the same, what is the probability that at least one minority applicant will be called for an interview? Suppose that three or more minority members are called. Do you think the employer will be open to charges of reverse discrimination? Justify your conclusion.

4.124 AMA data show that among persons contracting leukemia, spontaneous remission occurs in about 1% of the cases. Among 25 persons given a new treatment at a local hospital, 2 experience complete remission. Do you think this result can be attributed to chance? Suppose 4 out of 50 experience complete remission. Do you think this result could be attributed to chance?

4.125 An efficiency expert believes that additional incentives offered to workers on an assembly line will lead to increased productivity. She selects 15 assembly lines at random and offers these workers additional coffee breaks without mentioning an increase in productivity. She then compares the productivity of these lines with their former productivity to see which, if any, did increase their productivity. If there is actually no difference, productivity for a line is just as likely to decrease as increase if a line showed a change in productivity, so that the probability of an increase (or decrease) is 0.5. Suppose that, of the 15 lines, 6 have an increase, 3 have a decrease, and the others have no change. What is the probability that at least 6 of the lines that changed would increase, by chance, if there really was no difference? Do you think this result supports the efficiency expert's belief? Support your conclusion.

4.126 At peak fishing times, according to the sport section of the newspaper, a fisherman will catch two fish per hour in Perdido Bay, on the average. If the number of catches is a Poisson random variable, what is the probability that the fisherman will catch:

a. three or more fish in an hour?
b. at least three fish in a half-hour?
c. more than ten fish in 4 hours?

4.127 We can easily extend the hypergeometric distribution to a multinomial population—that is, a population with several outcomes. A mother goes to the grocery store and takes six cans of baby food at random off a shelf containing six cans of applesauce, four cans of peaches, and three cans of pears. What is the probability that she has taken at least one of each kind?

4.128 A radiologist needs a sample of radioactive material. The retail price for a sample is \$1800. In an attempt to save her company money, she goes to a cut-rate supply house. The proprietor tells her that he has four samples on hand, but two of them have been around so long that they may be worthless. In fact, he thinks there is only one chance in five that they will be useful. The other two are definitely in excellent condition. Unfortunately, though, there was a mixup; the new samples were put in the same location as the old ones and there are no outward differences to be seen. As a result, he does not know which ones are which. She will be able to test them in her laboratory, but he insists she purchase any samples she takes with her. He will give her a good deal, however. He will sell any one or two of the samples for \$600 apiece, or any three for \$2100, and he will pay her \$900 for an extra one she does not need if she guarantees it to be good. Determine the possible courses of action and the probability distribution of the costs for each. Then determine the expected value for each course of action. If she wishes to minimize expected cost, what course of action should she take?

Index of Terms

Glossary of Symbols

$B(n,p)$	binomial random variable with n trials; p is the probability of success on one trial
$E(\ )$	expected value of
μ	mean of a random variable
σ	standard deviation of a random variable
σ^2	variance of a random variable
p	probability of success on one trial of a binomial random variable

Answers to Proficiency Checks

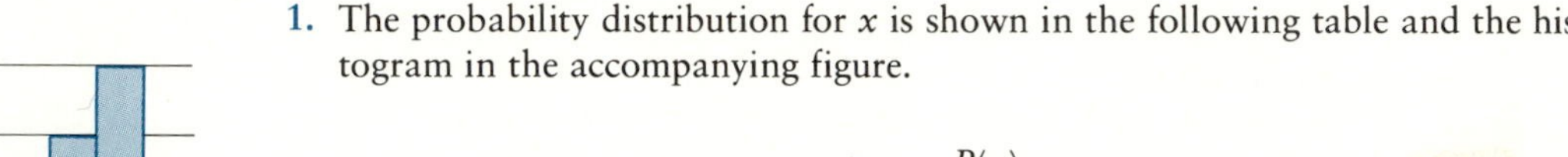

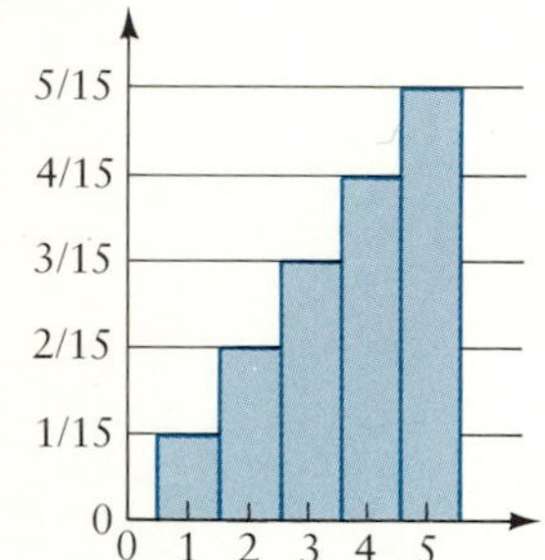

Figure 4.13
SAS printout comparing binomial and Poisson probabilities.

1. The probability distribution for x is shown in the following table and the histogram in the accompanying figure.

x	$P(x)$
1	1/15
2	2/15
3	3/15
4	4/15
5	5/15

x	$P(x)$
1	1/6
2	5/36
3	25/36

2. The sample space for the result can be written as {(6), (N,6), (N,N,6), (N,N,N)}, where N represents anything except a 6. Thus a 6 can be obtained on the first roll, the second roll, the third roll, or not at all. If a 6 is obtained on the first roll, the experiment ceases. The probability of this event is 1/6. If a 6 is not obtained on the first roll (the probability of this event is 5/6), we roll a second time. If a 6 is obtained, the experiment ceases. The conditional probability of obtaining a 6 on the second roll, given that a 6 was not obtained on the first roll, is 1/6. Thus the probability of not obtaining a 6 on the first roll *and* then obtaining a 6 on the second roll is $(5/6) \cdot (1/6)$, or 5/36. The remaining two possibilities both require that neither the first nor the second roll of the die shows a 6. The probability of this happening is $(5/6) \cdot (5/6)$, or 25/36. If we wish we can obtain the probabilities of the final two outcomes, but each outcome yields three rolls. We must add these probabilities together to obtain the probability that the number of rolls is three. As a check, we note that $1/6 + 5/36 + 25/36 = 1$. The probability distribution for x, the number of rolls that will be necessary, is given in the table at left.

3. 3.5

4. $\mu = 2.0$; $\sigma^2 = 5.2 - 4.0 = 1.2$; $\sigma \doteq 1.1$

5. a. $\mu = 0.80$; $\sigma^2 = 1.12 - 0.64 = 0.48$, $\sigma \doteq 0.69$
   b. $\mu = 0.80$; $\sigma^2 = 1.0 - 0.64 = 0.36$; $\sigma = 0.6$

6. $\mu = 4/3$; $\sigma^2 = 2/9$

7. $[(12 \cdot 11 \cdot 10 \cdot 9 \cdot 8)/(5 \cdot 4 \cdot 3 \cdot 2 \cdot 1)] \cdot (1/2)^5 \cdot (1/2)^7 = 99/512 \doteq 0.193$

8. **a.** 0.294 **b.** 0.797 **c.** 0.497 **d.** 0.193

9.

x	0	1	2	3
$P(x)$	0.729	0.243	0.027	0.001

10. **a.** 0.633 **b.** 0.834 **c.** 0.251

11. **a.** 0.047 **b.** 0.617 **c.** 0.103

12. **a.** 0.618 **b.** 0.012 **c.** 0.633 **d.** 0.382 **e.** 0.201
f. 0.667 **g.** 0.365 **h.** 0.154 **i.** 0.952

13. **a.** 0.057 **b.** 0.004 **c.** 0.416 **d.** 0.746

14. 0.610

15.

x	0	1	2	3	4	5	6,7,8,9
$P(x)$	0.387	0.388	0.172	0.045	0.007	0.001	0+ each

16. $\mu = 45, \sigma = 6$

17. **a.** 10/243, about 0.041 **b.** about 0.461 **c.** 10/3 and 10/9

18. $\mu = 3, \sigma^2 = 10.5 - 3^2 = 1.5$

19. $m = 2.8$ **a.** 0.531 **b.** 0.531 **c.** 0.848 **d.** 0.992
e. 0.041 **f.** 0.529 **g.** 0.523 **h.** 0.150

20. $m = 3$ **a.** 0.185 **b.** 0.423 **c.** 0.767

21.

x	0	1	2	3
$P(x)$	1/30	3/10	1/2	1/6

$\mu = 1.8, \sigma^2 = 0.56$

22. **a.** $P(0) \doteq 0.1832$, $P(1) \doteq 0.3848$, so $P(x \leq 1) \doteq 0.5680$; then $P(x > 1) \doteq 1 - 0.5680 = 0.4320$.
b. $P(0) \doteq 0.1935$, $P(1) \doteq 0.3762$, so $P(x > 1) \doteq 0.4303$.
c. The hypergeometric; difference does not seem important.

References

Freund, John E. *Mathematical Statistics, 2nd edition.* Englewood Cliffs, N.J.: Prentice-Hall, Inc., 1971.

Hogg, R. V. and A. J. Craig. *Introduction to Mathematical Statistics, 4th edition.* New York: Macmillan Publishing Co, 1978.

Huntsberger, David V. and Patrick Billingsley. *Elements of Statistical Inference, 3rd Edition.* Boston: Allyn and Bacon, Inc., 1973.

Mood, A. M., F. A. Graybill, and D. C. Boes. *Introduction to the Theory of Statistics, 3rd edition.* New York: McGraw-Hill Book Co., 1973.

Mosteller, Frederick, Robert E. K. Rourke, and George B. Thomas, Jr. *Probability with Statistical Applications, 2nd edition.* Reading, Mass: Addison-Wesley Publishing Co., Inc., 1970.

5

The Normal Distribution

The random variables we studied in Chapter 4 were discrete; that is, they could take on only a countable number of values. A set is countable if it contains the same number of elements as a set of positive integers. Another important type of random variable is a *continuous random variable.* **A continuous random variable** can take on any real number value in an interval or in a collection of two or more intervals. Examples of continuous random variables are those measuring time, weight, temperature, distance, area, and volume. Often a continuous variable is measured to the nearest whole number, such as height measured to the nearest inch. In such cases the variable can be treated as discrete.

This chapter covers an important class of continuous random variables—the *normal random variables*–and their applications in many typical situations. We will also see that many discrete random variables have distributions that are very close to that of a normal random variable. In these cases we may use approximate methods to simplify calculations. In particular, we may use these methods to approximate the probabilities in a binomial distribution.

5.0 Case 5: The Dubious Decisions

A large firm was accused of sex bias in its hiring practices. The firm's lawyer hired an industrial psychologist to determine whether or not the company's male personnel managers were in fact sexually biased when hiring people for vacant positions. Fifty personnel managers were each given the same five files and asked to select one applicant for a vacant position. As far as each manager knew, the request was routine and the decision would stand. They were told that the position was in another town and that they should make their decision without a personal interview.

The five files had been devised so that all candidates were about equally qualified for the position, but the names indicating sex were randomly varied among the files. Two of the names were female and three were male. When the selections were made, each file had been selected from 8 to 12 times, indicating that the files were about equally likely to have been selected, but a female name was selected only 17 times while a male name was chosen 33 times. Suppose you are the judge and know a little bit about statistics. Would you conclude that there was probably no sex bias on the part of the personnel managers, or does the evidence indicate that there probably was sex bias?

5.1 Normal Random Variables

5.1.1 Continuous Random Variables

Gauss is generally recognized as one of the greatest mathematicians of all time, along with Archimedes and Newton. He is often called the "Prince of Mathematicians." His doctoral thesis (1799) was the first proof of the fundamental theorem of algebra—that every algebraic equation has a solution. He was the first to develop noneuclidean geometry, but he did not publish this work because it ran counter to contemporary views. About 30 years later, when Bolyai and Lobachevsky published their works on noneuclidean geometry (around 1830) Gauss made his discoveries public. After his death, his unpublished work occupied mathematicians for several decades. In another important effort, he and Wilhelm Weber developed electric telegraphy, paving the way for worldwide communications.

We cannot define a probability distribution for a continuous random variable by listing all the outcomes and the probability of each. Such a definition is impossible because we cannot list all the values of the variable. Suppose we define a continuous random variable x over a sample space and let $f(x)$ define a function such that $f(x) \geqslant 0$ for all possible values of x. If all values of x lie between a and b, and the area under the graph of $y = f(x)$ between $x = a$ and $x = b$ is equal to 1, the function defined by $f(x)$ is called a **probability density function.** The distribution defined by a probability density function is called a **continuous probability distribution,** and in most cases the terms are used interchangeably. (In the Minitab section of Chapter 4, the command for determining individual probabilities was PDF—probability density function. The command for cumulative probabilities was CDF—cumulative density function.)

In effect, this definition means that any function $y = f(x)$ for which the area under the graph of the function is equal to 1 can be used to generate a probability distribution for a continuous random variable, provided that $y \geqslant 0$ for all values of x. The primary difference between probabilities for discrete and continuous variables is that while probabilities of a discrete random variable are defined for specific values of the variable, the probabilities of a continuous random variable are defined for a *range* of values of the variable. In the graph of a continuous probability distribution with probability density function $y = f(x)$, if x_1 and x_2 are values that the continuous random variable x can take, the probability that x will be between x_1 and x_2 is the area under the graph of the function between x_1 and x_2. That is, $P(x_1 < x < x_2)$ is equal to the shaded area in Figure 5.1. Technically there is no difference between $P(x_1 < x < x_2)$ and $P(x_1 \leqslant x \leqslant x_2)$ because the area under a point is zero ($P(x = x_2) = 0$).*

5.1.2 Normal Distributions

Among the most important and useful sets of continuous distributions in statistics is the set of **normal distributions** (discussed in Chapter 2). If the probability distribution of a random variable is a normal distribution, we say that the random variable is *normally distributed* or that it is a **normal random variable.** Reference is often made to *the* normal distribution. This term means the *standardized* normal distribution that is discussed in this section. Abraham DeMoivre (1667–1754) was the first to use a negative

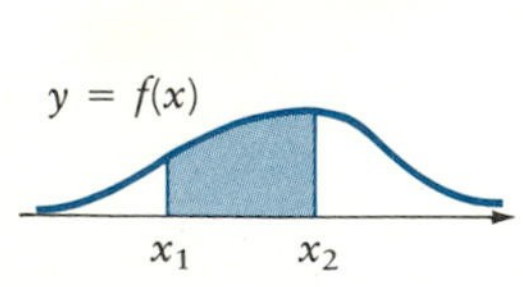

Figure 5.1
The shaded area under the curve represents $P(x_1 < x < x_2)$.

*The areas under a continuous function can be found by using the methods of calculus. If a probability density function is given by $y = f(x)$, and if a and b are values in the domain of the variable, then $P(a < x < b) = \int_a^b f(x)dx$.

exponential function in probability (in his *Miscellanea Analytica* of 1730), and he derived the normal curve in 1733 as an approximation to a binomial distribution (discussed in Section 5.2.4). The normal distribution was later used by Carl Friedrich Gauss (1777–1855) to deal with errors in experimental work and was found to have many other applications.

Many distributions of variables are very close to being normal—such as heights, weights, and IQ scores of people, many measurements in astronomy, and the useful lives of many manufactured items. For instance, items manufactured to meet certain specifications may have minor deviations from the specifications that affect performance. Measures of how these items perform nearly always follow a normal probability distribution.

Many of the procedures used in statistical inference require the assumption that a population is normal. A number of methods are available to determine the validity of that assumption. We have already encountered the use of the box plot and trimmed data set to detect and correct for lack of normality in a data set. The normal distribution is defined as follows.

The normal distribution occurs again and again in nature, but the reason for its prominence is elusive. Many other interesting mathematical phenomena have no clear explanation but nevertheless occur. The ancient Greeks, for example, obtained a number they called "the golden ratio." It is equal to $(1 + \sqrt{5})/2$, or about 1.618, and is symbolized by ϕ (the Greek letter phi). A rectangle whose sides were in this ratio was considered to have the most pleasing shape. Note that 3×5 and 5×8 reference cards have approximately this ratio. This number has been found to have some interesting properties—such as the fact that $1/\phi = \phi - 1$—and occurs in some unexpected places.

The set of integers 1, 1, 2, 3, 5, 8, 13, 21, 34, 55, . . . , in which each term is the sum of the preceding two, occurs in nature as well. It is called the Fibonacci sequence in honor of its discoverer Leonardo of Pisa (1170?–1240?), son of Bonaccio (Latin fils Bonaccio*), the most distinguished European mathematician of the Middle Ages. It has been conjectured that the number of leaves on a tree, seeds in a sunflower, and other such numbers are Fibonacci numbers. The ratio of each term in the sequence to the term preceding it approaches ϕ.*

The Normal Probability Distribution

A normal probability distribution is defined by the probability density function

$$f(x) = \frac{e^{-\frac{1}{2}\left(\frac{x-\mu}{\sigma}\right)^2}}{\sigma\sqrt{2\pi}} \qquad \text{for } (-\infty < x < \infty)$$

where μ = mean of the distribution

σ = standard deviation of the distribution

$e = 2.71828\ldots$

$\pi = 3.14159\ldots$

If a variable x is a normal random variable with mean μ and variance σ^2 we write "x is $N(\mu,\sigma^2)$."

Just as for the Poisson distribution, there are tables of normal probabilities that provide us with all the information we need, so it will not be necessary to make reference to the defining formula.*

*If x is a normal random variable, $\int_a^b f(x)dx$ cannot be determined by usual integration techniques. Values of $\int_0^a f(x)dx$ have been obtained by approximation techniques and are presented in Table 4 at the back of the book.

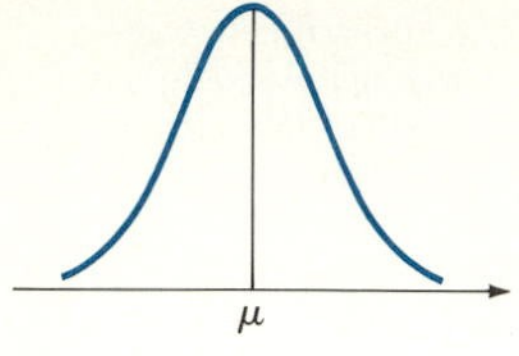

Figure 5.2
Graph of a normal distribution.

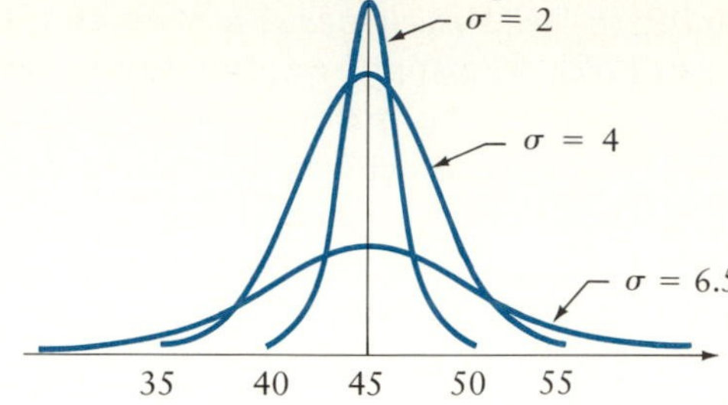

Figure 5.3
Three normal probability distributions with mean equal to 45 and standard deviations equal to 2, 4, and 6.5.

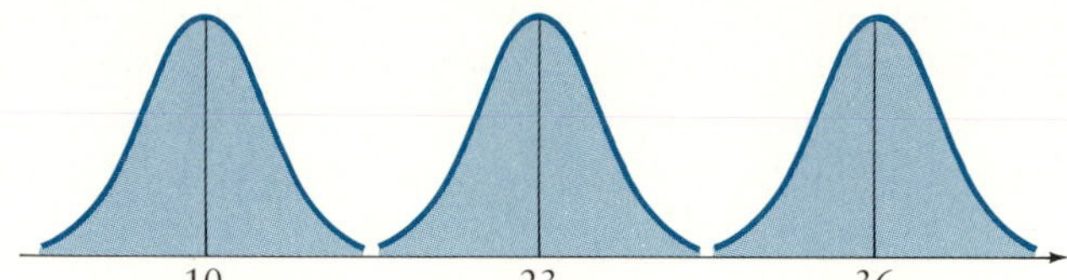

Figure 5.4
Graphs of normal probability distributions with equal standard deviations and means equal to 10, 23, and 36.

A graph of a normal distribution is shown in Figure 5.2. The graph of a normal distribution is called a **normal curve.** A normal curve is symmetric about the mean and trails off abruptly in both directions. It is the idealized shape of the "mound-shaped" curves mentioned in Chapter 2 and is often called the *bell-shaped curve* for obvious reasons. A normal distribution is completely determined by the mean and standard deviation of the variable. Thus graphs of the probability distributions of normal random variables with the same mean, but different standard deviations, differ only in amount of dispersion and height as shown in Figure 5.3. Normal random variables with the same standard deviation but different means have probability distributions whose graphs look identical in shape and differ only in their placement on the horizontal axis as shown in Figure 5.4.

Although it may not be apparent from these small drawings, theoretically the curve never touches the horizontal axis. When the value of the variable is about four standard deviations from the mean, however, the curve approaches the axis so closely that the area under the curve more than three standard deviations from the mean is, for all practical purposes, negligible. In terms of mean and standard deviation, the horizontal axis under any normal curve can be scaled as shown in Figure 5.5.

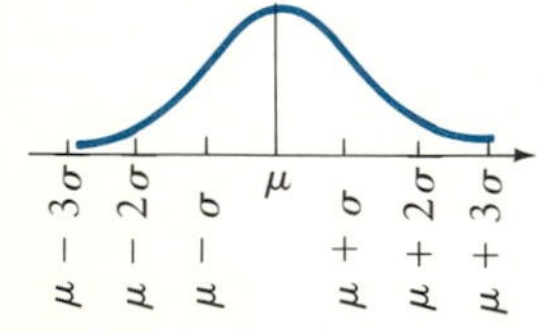

Figure 5.5
Normal curve with range of standard deviations.

5.1.3 The Standardized Normal Distribution

A normal random variable with mean zero and variance equal to 1 is called a **standardized normal variable** and is usually denoted by the symbol z. This is the same symbol z we used for the z-score in Chapter 2, and the choice of z for this notation is not accidental, as we will see. The probability distribution of the variable z is called the **standardized normal distribution.** Its graph, called the **standardized normal curve,** is shown in Figure 5.6.

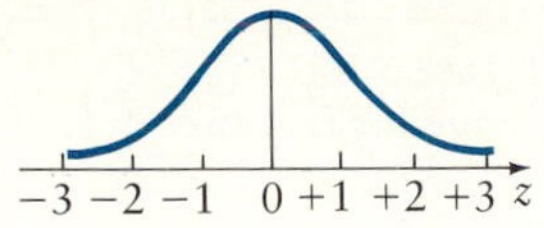

Figure 5.6
The standardized normal curve.

We observed in Chapter 2 that if x is a variable with mean μ and standard deviation σ, the z-score—that is, $z = (x - \mu)/\sigma$—gives the number of standard deviations that a value of x is above or below the mean. If x is a normal random variable with mean μ and standard deviation σ, then the variable z defined by $z = (x - \mu)/\sigma$ is also a normal random variable. The value of z for any particular value of x is called the **standard normal deviate** for that x. Since z gives the number of standard deviations that a value of x lies above or below the mean, this variable has mean zero and standard deviation 1. Thus any normal random variable can be related to the standardized normal variable and we can say that if x is $N(\mu,\sigma^2)$, then $z = (x - \mu)/\sigma$ is $N(0,1)$. For this reason the standardized normal distribution has been studied extensively. The important results are restated here.

Standardized Normal Distribution

If x is a normally distributed random variable with mean μ and standard deviation σ, the **standardized normal variable** z, defined by

$$z = \frac{x - \mu}{\sigma}$$

is a normally distributed random variable with mean zero and variance 1. Thus if x is $N(\mu,\sigma^2)$, then z as defined here is $N(0,1)$. The probability distribution for z is called the **standardized normal distribution,** and its graph is called the **standardized normal curve.** The value of z for a particular value of x is called the **standard normal deviate** for that value of x.

Example 5.1

Suppose the random variable x has a mean $\mu = 30.1$ and standard deviation $\sigma = 2.4$. Determine standard normal deviates for $x = 18.3, 27.9, 34.4, 39.3$.

Solution

We find the values of z for each value of x as follows:

If $x = 18.3$, then $z = (18.3 - 30.1)/2.4 \doteq -4.92$.

If $x = 27.9$, then $z = (27.9 - 30.1)/2.4 \doteq -0.92$.

If $x = 34.4$, then $z = (34.4 - 30.1)/2.4 \doteq 1.79$.

If $x = 39.3$, then $z = (39.3 - 30.1)/2.4 \doteq 3.83$.

A sketch of the curve is shown in Figure 5.7.

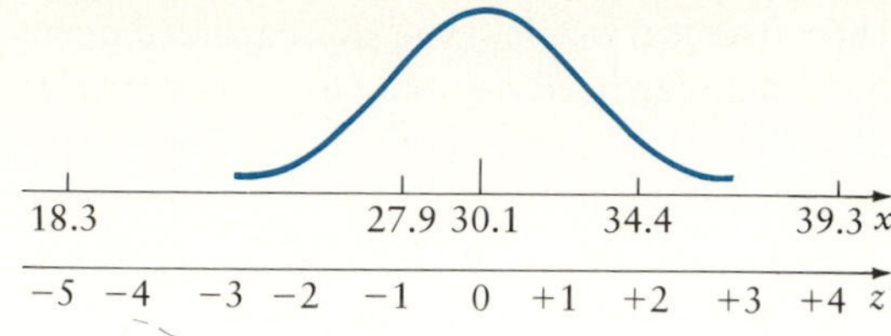

Figure 5.7
Curve for Example 5.1.

We can also use the formula for the standardized normal variable z to determine values of x if we know z. If $z = (x - \mu)/\sigma$, then $z\sigma = x - \mu$, so that $x = z\sigma + \mu$. We illustrate use of this formula in Example 5.2.

Example 5.2

If x is the same random variable as in Example 5.1, determine the values of x with the standard normal deviates -3.07, -1.04, 0.73, and 2.44.

Solution

A sketch of the curve is shown in Figure 5.8. Using $x = z\sigma + \mu$, we get the following:

If $z = -3.07$, then $x = (-3.07)(2.4) + 30.1 \doteq 22.7$.
If $z = -1.04$, then $x = (-1.04)(2.4) + 30.1 \doteq 27.6$.
If $z = 0.73$, then $x = (0.73)(2.4) + 30.1 \doteq 31.9$.
If $z = 2.44$, then $x = (2.44)(2.4) + 30.1 \doteq 36.0$.

PROFICIENCY CHECKS

1. The variable x is N(132.2, 129.96). That is, it is normally distributed with a mean of 132.2 and a variance of 129.96; thus its standard deviation is $\sqrt{129.96}$ or 11.4. Sketch a graph of the curve and determine standard normal deviates to two decimal places for the following values of x:
a. 155.6 b. 98.8 c. 119.6 d. 138.6
2. The variable x is N(0.67, 0.0144). Determine (to two decimal places) values of the variable with the following standard normal deviates:
a. −2.87 b. 1.85 c. −0.91 d. 3.32
3. The normal random variable x has mean 31.7. The standard normal deviate of 42.1 is 1.30. What is the standard deviation of x?
4. The normal random variable x has a variance equal to 136.89. A value of 11.3 has a standard normal deviate equal to 2.10. Sketch a graph and determine the mean of x.

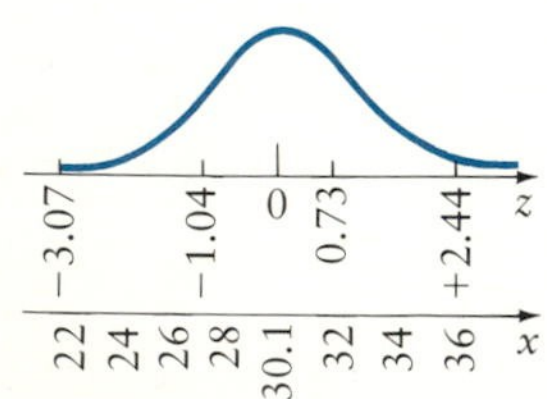

Figure 5.8
Curve for Example 5.2.

5.1.4 Using the Normal Area Table

The fact that we can relate any normal random variable to the standardized normal variable is very important. Because of this relationship, the standardized normal distribution has been studied in great detail and the results can be transferred to any normal distribution. Table 4 in the back of the book gives the area under a standardized normal curve between the mean (zero) and any particular value of z, where $0 \leq z \leq 3.59$.

Recall the empirical rule from Chapter 2. About 68% of the area under a normal curve lies within one standard deviation of the mean—that is, between $z = +1$ and $z = -1$. About 95% of the area lies within two standard deviations of the mean—actually between $z = +1.96$ and $z = -1.96$—and virtually all of the area under the curve lies within three standard deviations of the mean. We can show that 99.74% of the area under a normal curve lies between $z = +3$ and $z = -3$. When the pseudo-standard deviation was introduced, we stated that 50% of the area under a normal curve lies between $z = +0.675$ and $z = -0.675$. All these values can be verified from Table 4.

Table 5.1 shows two portions of Table 4. The entries on the left and top correspond to values of z. The integer value and first decimal value of z are given in the column at the left; the second decimal value is in the top row. The entries in the body of the table are the areas (correct to four decimal places) under the normal curve between the mean (zero) and the given value of z. Since most of the values of z we will be using are approximations, the results are approximations as well and should usually be viewed as being no more precise than the values of z used to get them.

Table 5.1 SOME AREAS UNDER THE NORMAL CURVE.

z	.00	.01	.02	.03
0.0	.0000	.0040	.0080	.0120
.	.	.	.	.
.	.	.	.	.
.	.	.	.	.
1.6	.4452	.4463	.4474	.4484

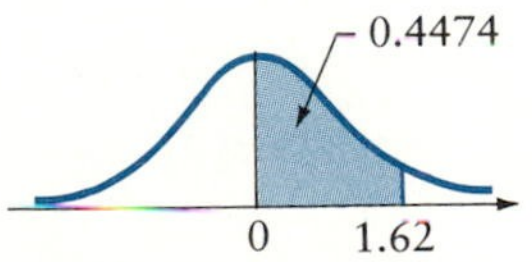

Figure 5.9 The shaded area under the standardized normal curve represents $P(0 < z < 1.62)$.

If $z = 1.62$, for instance, to find the corresponding area look down the left column to find 1.6; then look along the top row to find 0.02. The entry that is in both the row of 1.6 and the column of 0.02 is 0.4474 (see Table 5.1). Thus the area under the standardized normal curve between $z = 0$ and $z = 1.62$ is 0.4474, as shown in Figure 5.9. This means the probability that a standardized normal variable has a value between 0 and 1.62 is 0.4474.

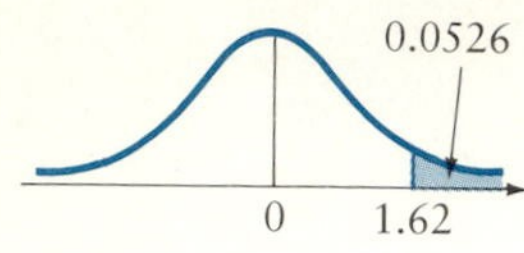

Figure 5.10
The shaded area under the standardized normal curve represents $P(z > 1.62)$.

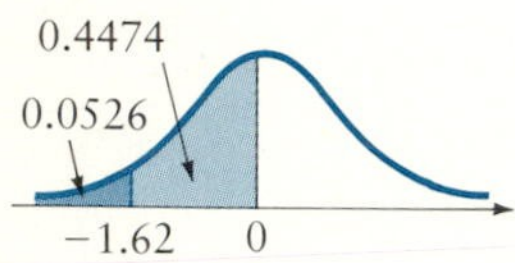

Figure 5.11
The darkly shaded area represents $P(z < -1.62)$; the lightly shaded area represents $P(-1.62 < z < 0)$.

Remember that the area under the curve must always be positive. Remember also that the entries on the edge of Table 4 (left and top) represent values of the standardized normal variable (values of z), whereas the entries in the *body* of Table 4 represent the area under the standardized normal curve between zero and the given value of z. Equivalently, they represent the probability that a value of z will lie between zero and the given value.

We can derive several other pieces of information from Figure 5.9. Since the normal curve is symmetric, each side contains half the area under the curve. As the area under the graph of any probability distribution is equal to 1, each half of the curve covers an area equal to 0.5000. Thus the area under the curve to the right of 1.62 is 0.5000 − 0.4474 or 0.0526, as illustrated in Figure 5.10.

Further, since the normal curve is symmetric, the area between $-z$ and zero is the same as the area between zero and z. Therefore in this case the area between −1.62 and zero is also 0.4474, and the area to the left of −1.62 is 0.0526, as shown in Figure 5.11.

Finally, the area between −1.62 and 1.62 is twice the area between zero and 1.62, or 0.8948. Thus the probability that a standardized normal variable z has a value between −1.62 and 1.62 is equal to 0.8948; that is, $P(-1.62 < z < 1.62) = 0.8948$, as illustrated in Figure 5.12.

Example 5.3

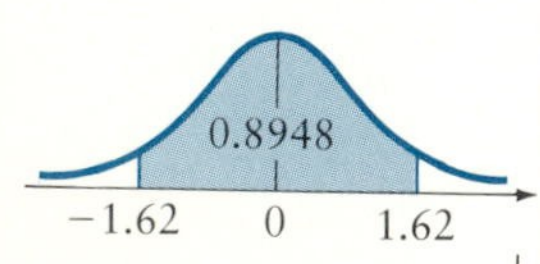

Figure 5.12
The shaded area represents $P(-1.62 < z < 1.62)$.

Find the area under the standardized normal curve between zero and z if z is equal to:

a. 0.07 b. 0.83 c. 1.70 d. 2.56 e. −0.24 f. −1.12 g. −3.01

Solution

From Table 4 we read the following values:

a. 0.0279 b. 0.2967 c. 0.4554 d. 0.4948 e. 0.0948
f. 0.3686 g. 0.4987

Note that for values of z off the table—that is, $|z| > 3.59$—the area between zero and z should for most purposes be considered to be equal to 0.4999 unless $|z|$ is very large, say greater than 5. In the latter case, assume for practical purposes that the area is 0.5000.

Example 5.4

Find the values of the standardized normal variable for which the area under the curve between zero and the value is:

a. 0.2019 b. 0.3621 c. 0.4345 d. 0.4599 e. 0.4908

Solution

We can read the values from Table 4 for all but part (*e*):

a. 0.53 b. 1.09 c. 1.51 d. 1.75

For part (*e*) we note that no entry in the table is precisely equal to 0.4908. The two entries closest to it are 0.4906 and 0.4909. Since 0.4908 is closer to 0.4909 than it is to 0.4906, we may use $z \doteq 2.36$, the value for which the area is equal to 0.4909. For most purposes two-decimal precision for values of z is satisfactory, so this result is sufficiently accurate.

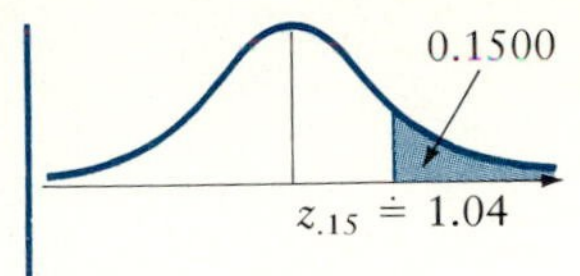

Figure 5.13
$z_{.15} \doteq 1.04$; $P(z > 1.04) = 0.15$.

In cases where the given area is midway between two table entries—such as 0.3953, which is halfway between 0.3944 ($z = 1.25$) and 0.3962 ($z = 1.26$)—it is common practice to round upward to the next unit. Then for an area of 0.3953 we have $z = 1.26$ to two decimal places. Sometimes it may be preferable to use a third decimal place, particularly for values of z that are used often. For example, we stated that 50% of the area under a normal curve lies between $z = +0.675$ and $z = -0.675$. This means that the area under the normal curve between $z = 0$ and $z = 0.675$ is 0.2500. Referring to Table 4, we see that 0.2500 lies almost exactly between 0.2486 and 0.2517, for which $z = 0.67$ and 0.68, respectively. Thus $z = 0.675$ is a reasonable approximation to use in this case.

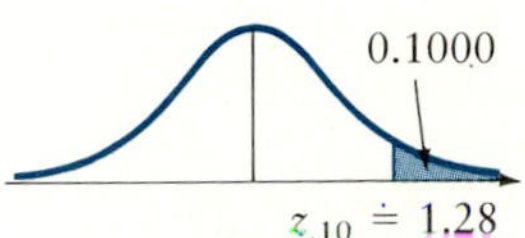

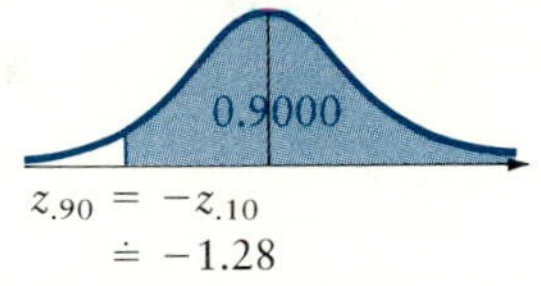

Figure 5.14
$P(z > 1.28) = 0.1$ (top); $P(z > -1.28) = 0.9$ (bottom).

One useful bit of notation is a subscript to indicate the area under the standardized normal curve to the right of a particular value of z. The notation $z_{.05}$, for instance, indicates the value of z for which 0.05 (that is, 5%) of the area under the curve is to its right. We can estimate this value from Table 4 to be almost exactly halfway between 1.64 and 1.65. Using the rule we stated before, to two decimal places we would use 1.65. This number, however, is used quite a bit, and to more decimal places it is equal to about 1.6446. Thus we will use $z_{.05} \doteq 1.645$. Similarly, $z_{.15} \doteq 1.04$ as shown in Figure 5.13.

Since the subscript indicates areas to the right of the given z, we have $P(z > z_a) = a$. If a is greater than 0.5, then z is negative. To determine $z_{.85}$, for example, we note that if 15% of the area lies to the right of $z = 1.04$, then 15% lies to the left of $z = -1.04$. Therefore 85% of the area lies to the right of $z = -1.04$, so that $z_{.85} = -1.04$. It follows, then, that for any value of a, we have $z_a = -z_{1-a}$. For example, $z_{.10} \doteq 1.28$ and $z_{.90} \doteq -1.28$. (See Figure 5.14.) Thus $P(z > 1.28) \doteq 0.10$, while $P(z > -1.28) \doteq 0.90$.

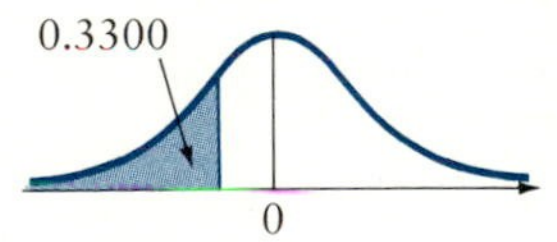

Figure 5.15
Standardized normal curve with all necessary information for Example 5.5.

Example 5.5

Determine the value of z for which 0.3300 of the area under the curve is to its left.

Solution

In solving problems such as this one, it is a good idea to sketch an appropriate curve and include all the necessary information. (See Figure 5.15.) Here we know that *less* than half the area is to the left of the desired value of z, so z must be negative. In terms of the foregoing discussion, we are looking for $z_{.67}$, and we know that this is equal to $-z_{.33}$. Table 4 gives areas between z and the mean, so the area under the curve between $-z_{.33}$ and the mean is 0.17. The value of z corresponding to an area of 0.1700 is 0.44; that is, $z_{.33} = 0.44$. Since we are looking for $-z_{.33}$, it follows that $-z_{.33} = -0.44$ as illustrated in Figure 5.16.

Example 5.6

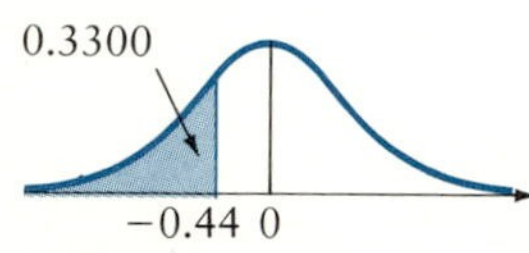

Figure 5.16 Standardized normal curve showing solution of Example 5.5.

Find the areas under the standardized normal curve that give the probabilities $P(-1.34 < z < 0.57)$ and $P(0.59 < z < 1.27)$.

Solution

For $z = -1.34$ and $z = 0.57$, the areas under the normal curve between each value of z and the mean are 0.4099 and 0.2157, respectively. Since $z = -1.34$ and $z = 0.57$ are on *opposite* sides of the mean, we must add the areas to obtain the area under the curve between -1.34 and 0.57 (Figure 5.17).

Thus $P(-1.34 < z < 0.57) = 0.6256$. For $z = 0.59$ and $z = 1.27$, the corresponding areas are 0.2224 and 0.3980. Since both values of z are positive, the corresponding areas overlap and their *difference* is the desired area (Figure 5.18). We then see that $P(0.59 < z < 1.27) = 0.1756$.

Example 5.7

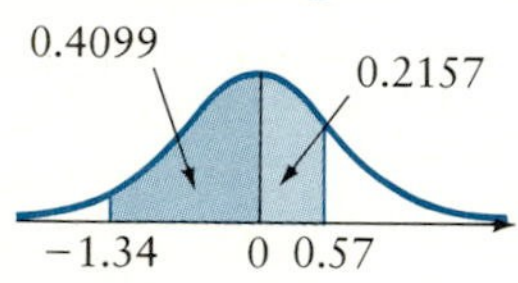

Figure 5.17 Standardized normal curve showing $P(-1.34 < z < 0.57)$.

The amount of time necessary for people to take a certain test is a normal random variable with mean 38.7 minutes and standard deviation 10.2 minutes. That is, if x is the amount of time as stated, x is N(38.7, 104.04). Estimate the probability that a person taking the test will require between 29.6 and 44.8 minutes; that is, $P(29.6 < x < 44.8)$.

Solution

We first find the standard normal deviates for 29.6 and 44.8. Since $(29.6 - 38.7)/10.2 \doteq -0.89$ and $(44.8 - 38.7)/10.2 \doteq 0.60$, we find the areas from Table 4 corresponding to $z = -0.89$ and $z = 0.60$. These are 0.3133 and 0.2257, as shown in Figure 5.19. Since 29.6 and 44.8 are on opposite sides of the mean, 38.7, we add these areas; because the values for z are rounded to two decimal places, we round to two decimal places to obtain $P(29.6 < x < 44.8) \doteq 0.54$.

Example 5.8

If x is N(133, 441), determine a number such that 80% of all values of x fall within that number from the mean.

Solution

We are looking for a value of n for which 40% of the area under the curve lies between 133 and $133 + n$, and 40% lies between 133 and $133 - n$, as shown in Figure 5.20. This means that 10% of the area under the curve lies above $133 + n$. (Figure 5.21).

The standard normal deviate we are looking for, then, is $z_{.10}$. We have already determined that $z_{.10}$ is about 1.28, so n is 1.28 standard deviations above 133. The standard deviation is $\sqrt{441}$ or 21, so $n \doteq 1.28(21)$ or about 27. Thus 80% of all values of x lie within 27 units of the mean; that is, $P(106 < x < 160) \doteq 0.80$. (Figure 5.22).

PROFICIENCY CHECKS

5. Find the area under the standardized normal curve between zero and z if z is:
   a. 1.33 b. -0.25 c. 2.87 d. -1.93
6. Find the values of the standardized normal variable for which the area under the curve between zero and the value is:
   a. 0.4131 b. 0.2486 c. 0.4862 d. 0.1790
7. Find the area between the given values of the standardized normal variable:
   a. 0 and 1.55 b. 0 and -0.93 c. -0.54 and 2.16
   d. 1.16 and 2.05
8. Find the area under the standardized normal curve:
   a. to the right of $z = 0.76$ b. to the left of $z = -1.80$
   c. to the right of $z = -1.25$ d. to the left of $z = 1.76$
9. The normal random variable x has a mean of 120 and a standard deviation of 10. What is the probability that a value of x selected at random will lie:
   a. between 120 and 145? b. between 100 and 130?
   c. above 125?
10. The variable z is N(60, 25). Determine:
   a. $P(x > 75)$ b. $P(55 < x < 72)$ c. $P(x < 65)$

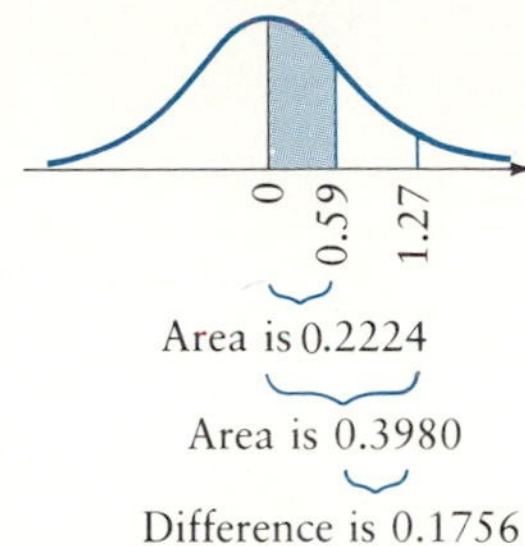

Figure 5.18
Standardized normal curve showing $P(0.59 < z < 1.27)$.

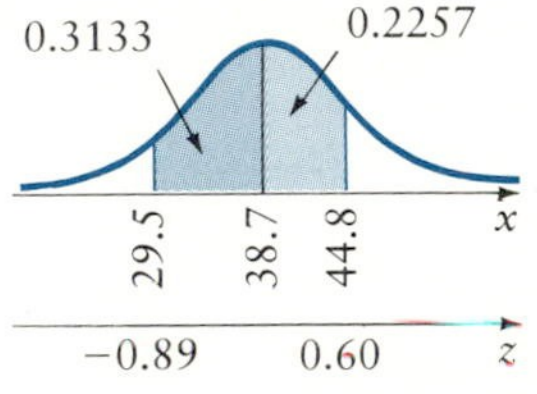

Figure 5.19
Standardized normal curve showing $P(29.6 < x < 44.8)$.

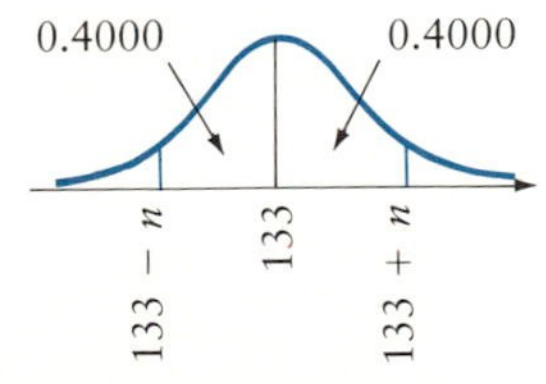

Figure 5.20
Curve for Example 5.8.

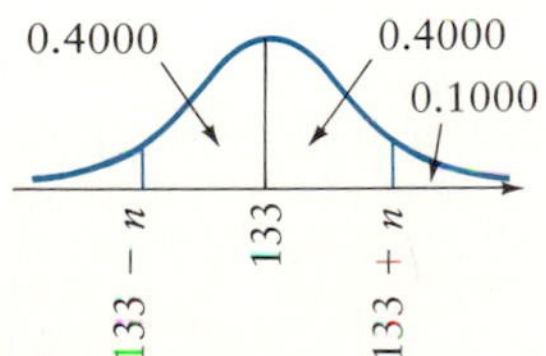

Figure 5.21
Curve for Example 5.8.

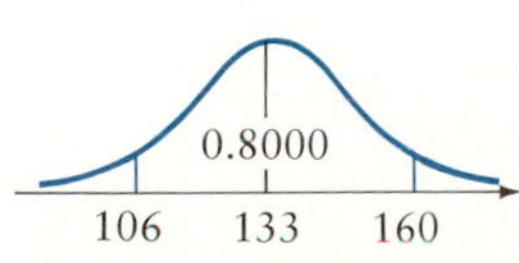

Figure 5.22
$P(106 < x < 160) \doteq 0.80$.

Problems

Practice

5.1 If x is a normal random variable with mean 284.7 and standard deviation 14.6, determine the standard normal deviates for each of the following values of x:
a. 261.4 b. 303.7 c. 259.0 d. 280.4 e. 293.9 f. 321.2

5.2 If x is N(10.4, 139.24), that is, x is a normal random variable with mean 10.4 and variance 139.24 (standard deviation 11.8)—determine the values of x with each of the following standard normal deviates:
a. 1.64 b. 2.07 c. −2.16 d. 0.50 e. −0.13 f. 1.14

5.3 Find the area under the standardized normal curve between the given values of z:
a. 0 and 2.18 b. −1.04 and 1.54 c. 1.56 and 2.93 d. −0.49 and −0.12 e. −3.04 and 1.63 f. −0.43 and 2.09

5.4 Find the area under the standardized normal curve:
a. to the right of $z = 1.43$ b. to the left of $z = -1.03$
c. to the right of $z = -0.77$ d. to the left of $z = 2.01$

5.5 Find the area under the standardized normal curve between z and $-z$ if z is equal to:
a. 1 b. 2 c. 3 d. 1.28 e. 1.645 f. 1.96 g. 2.33 h. 2.58

5.6 If x is N(193.4, 187.69), determine each of the following probabilities:
a. $P(x > 210.4)$ b. $P(x < 186.5)$ c. $P(x > 179.9)$ d. $P(x < 200.0)$ e. $P(173.4 < x < 186.8)$ f. $P(166.1 < x < 207.9)$ g. $P(200.0 < x < 210.0)$ h. $P(x > 200.0 \text{ or } x < 180.0)$ i. $P(x = 195)$

5.7 Determine $z_{.1230}$ and $z_{.8770}$.

5.8 Determine:
a. $z_{.01}$ b. $z_{.005}$ c. $z_{.025}$ d. $z_{.05}$ e. $z_{.10}$ f. $z_{.99}$ g. $z_{.995}$ h. $z_{.975}$ i. $z_{.95}$ j. $z_{.90}$

5.9 The mean of a normal random variable is 100.0. If the probability that the variable is greater than 121.0 is 0.1446, what is the variance of the variable?

5.10 A normal random variable has a standard deviation of 130. The probability that the variable is less than 1072 is 0.7734. What is the mean of the variable?

Applications

5.11 A Hertz Corporation report (*USA Today,* Dec. 17, 1985) stated that the average commuter spends 30.2 minutes commuting to work. Suppose this is the true mean commuting time for Los Angeles commuters and the standard deviation is 11.7 minutes. What proportion of all commuters will spend more than 45 minutes commuting?

5.12 Intelligence test scores tend to be normally distributed. A study once contended that the best drivers of trucks and automobiles are people with IQ scores between 90 and 100, slightly below average. The reason given was that people in this range are perfectly capable of driving machinery (more capable than those with lower scores) but must give all their attention to driving and thus will not let their minds wander as do people with higher scores. Suppose that IQ is scored on a continuous scale with a mean of 100 and a standard deviation of 16. What proportion of the population will then qualify as best drivers according to this contention?

5.13 A. M. Best and Company reported (*USA Today,* January 28, 1986) that the average amount paid for automobile insurance was \$343.42 per car in 1984, up 6.4% from 1983. Assuming that $\sigma = \$185$, determine the proportion of insurance policies costing between \$250 and \$600 annually.

5.14 Military artillery uses the concept of *probable error.* The probable error of a variable is a number such that 50% of the values of the variable lie within that number of the mean. If the probable error of a mortar at 1000 yards is 40 feet, for instance, this means that 50% of all the shells will fall within 40 feet of the target at 1000 yards, provided the mortar is aimed correctly. By this definition, then, the probable error is half the interquartile range.

a. Determine a number such that 50% of the area under the standardized normal curve lies within that number of the mean; that is, $(z_{.25} - z_{.75})/2$.
b. Suppose that several artillery shells are fired in a salvo at a target. If 50% of the shells will (on the average) fall within n yards of the target, n is the probable error of the salvo. Suppose an artillery piece is aimed at a target 2740 yards away, and the standard deviation of the distances the shells will fall from the target at that range is 40 yards. Determine the probable error of a salvo of shells fired at the target.

5.2 Applications of the Normal Distribution

5.2.1 Application to Normal Random Variables

Many sets of data have approximately normal distributions. For these sets we can use the methods of Section 5.1 to obtain a great deal of information. In particular, items manufactured to certain specifications may vary from item to item, but the distribution of the measurements is generally normally distributed with a mean equal to the specification. Measurements made on people, such as heights, IQ scores, and so on, generally have approximately normal distributions as well. A few examples will show the method.

Example 5.9

A certain brand of spaghetti sauce is packed in cans that are supposed to have a net weight of 15.5 ounces. Since the packing is done by machine and weight may vary from can to can, the machine is usually set to average a little more than the stated net weight. (Otherwise half the cans would have less than they are supposed to have.) Suppose that the machine is set to fill

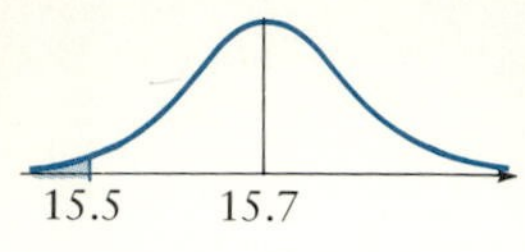

Figure 5.23
Graph of probability distribution of x, where x represents the number of ounces of spaghetti sauce in a can (Example 5.9).

the cans with 15.7 ounces, and the actual weight of the cans is normally distributed with a mean of 15.7 ounces and a standard deviation of 0.12 ounce. How many of a lot of 10,000 cans would you expect to find with less than 15.5 ounces?

Solution

If x represents the number of ounces in a can, then x is N(15.7, 0.0144). The graph of the probability distribution of x is shown in Figure 5.23.

The standard normal deviate for $x = 15.5$ is $(15.5 - 15.7)/0.12$ or about -1.67, so we can look in Table 4 and find that about 0.4525 of the area under the curve lies between 15.5 and 15.7. Therefore the area under the curve to the left of 15.5 (the shaded area in Figure 5.23) is about 0.0475. Thus $P(x < 15.5) \doteq 0.0475$. This is the probability that any single can contains less than 15.5 ounces. For a set of 10,000 cans. we would expect (on the average) 10,000(0.0475) or 475 cans to contain less than 15.5 ounces. On the other hand, about 475 cans would contain more than 15.9 ounces. From a practical standpoint, the probability that a Food and Drug Administration investigator would select, at random, a can weighing less than 15.5 ounces is about 0.0475, less than 1 chance in 20. If such a can were selected, the probability that a second can selected for confirmation would also have less than 15.5 ounces is also 0.0475. Thus the probability that it would happen twice in a row (assuming independent outcomes) is only $(0.0475)^2$, or about 0.0023, less than 1 chance in 400.

This is actually a very practical problem, since the standard deviation is relatively difficult to adjust on such a machine. Any changes would have to be made on the mean.

Example 5.10

Workers in an industrial plant assemble a delicate instrument. The time required for a skilled worker to assemble the instrument ranges from 34.25 to 55.10 minutes with a mean of 45.00 minutes. The times are normally distributed with a standard deviation of 4.00 minutes. One worker is observed assembling the instrument. What is the probability she will take between 48.00 and 50.00 minutes to assemble the instrument? What is the probability that she will take at least 50.00 minutes?

Solution

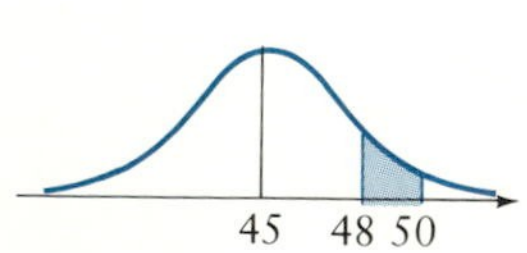

Figure 5.24
Graph of probability distribution for time required to assemble an instrument (Example 5.10).

First we assume that the probability distribution for all workers applies to this individual. If this is true, the graph of the probability distribution for the time this worker will take is shown in Figure 5.24. The area of the shaded portion gives the desired probability.

If z_1 and z_2 denote the standard normal deviates for 48.00 and 50.00, respectively, then

$$z_1 = (48 - 45)/4 = 0.75$$

$$z_2 = (50 - 45)/4 = 1.25$$

The areas under the normal curve for each are, respectively, 0.2734 and 0.3944. The area between 48.00 and 50.00, then, is 0.3944 − 0.2734 or 0.1210. Thus if x is the time required to assemble one instrument, $P(48.00 < x < 50.00) = 0.1210$.

The area above 50.00 is 0.5000 − 0.3944, or 0.1056, so $P(x > 50.00) \doteq 0.1056$. Note that the question was what is $P(x \geqslant 50)$? If the variable is continuous, $P(a < x < b)$ and $P(a \leqslant x \leqslant b)$ mean exactly the same since $P(x = a)$ and $P(x = b)$ are equal to zero for the continuous variable. Thus $P(x \geqslant 50) \doteq 0.1056$, as well.

PROFICIENCY CHECKS

11. A technician is testing a certain type of resistor and finds that the distribution of resistance is approximately normal with a mean of 15.08 ohms and a standard deviation of 1.75 ohms. What is the probability that a randomly chosen resistor will have:
    a. a resistance of more than 17 ohms?
    b. a resistance of 17 ohms or more?
    c. a resistance within 1 ohm of the mean?
    d. at least the minimum acceptable resistance—14.5 ohms?
12. A standardized test has a mean of 50.0 and a standard deviation of 10.0. Suppose the scores are normally distributed.
    a. What proportion of those taking the test should have scores between 35 and 70?
    b. How many of 1000 students who take the test should have scores between 35 and 70?

5.2.2 Approximating Discrete Distributions

One of the many uses for continuous probability distributions is as an approximation to a discrete distribution. The appropriate graph for the probability distribution of a discrete random variable is a histogram. Suppose, for instance, that Table 5.2 gives the probability distribution of the scores of 7000 randomly selected bowling games.

A histogram representing this distribution is shown in Figure 5.25. Except for the first interval, each interval on the horizontal scale (the base of the rectangle) represents the class width, 25 pins. The first interval is zero through 125.5. Since it is five times the size of the other intervals, its height is 162.4, or one-fifth of 812. The second class is represented by the interval 125.5 to 150.5; the third interval is 150.5 to 175.5; and so forth.

Table 5.2 PROBABILITY DISTRIBUTION OF BOWLING SCORES.

SCORE (PINS)	NUMBER OF GAMES	PROBABILITY
Under 126	812	0.1160
126–150	1764	0.2520
151–175	2433	0.3476
176–200	911	0.1301
201–225	646	0.0923
225–250	294	0.0420
251–275	103	0.0147
276–300	37	0.0053

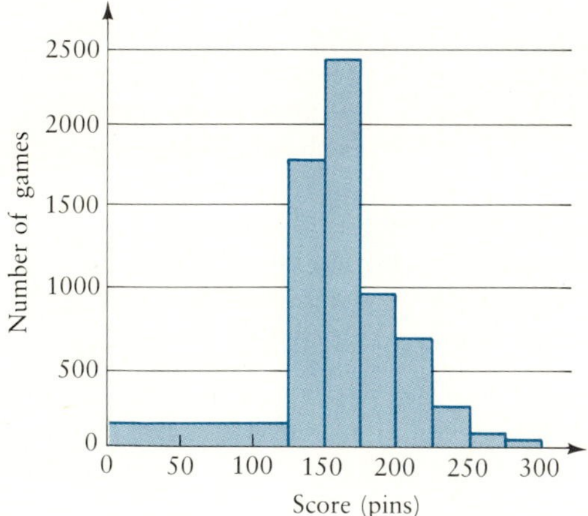

Figure 5.25
Histogram for distribution of scores of 7000 bowling games.

Now if we drew a histogram with each game given to the exact number of pins, the rectangles would be so narrow that the outline of the histogram would be very close to a continuous curve. We can use the histogram in Figure 5.25 to draw a continuous curve that approximates the histogram very closely. First we draw a frequency polygon and connect the midpoints of the top edges of the rectangles, beginning and ending at the extreme values of the variable. We then smooth out the curve to obtain Figure 5.26.

The curve will be a good approximation of the data in Table 5.2 if it gives an accurate impression of them. The area under the curve in any interval will be approximately the same as the area of the histogram for the same interval. That is, the area under the curve between zero and 125.5 will contain 0.1160 of the total area under the curve, the area under the curve between 125.5 and 150.5 will contain 0.2520 of the total area, and so on. If the curve is approximately a normal curve, we can find the areas under the curve between two points by using Table 4.

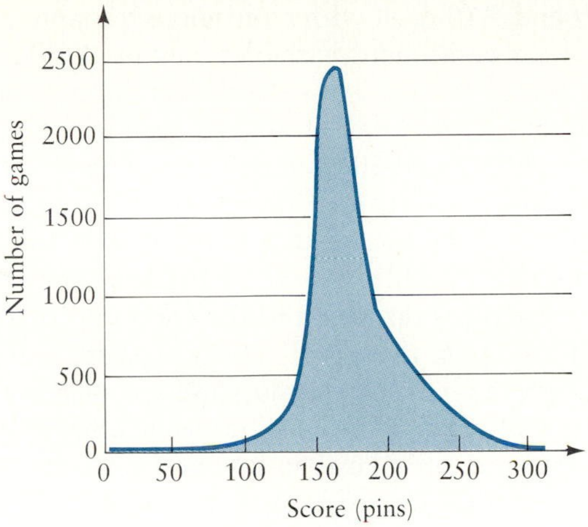

Figure 5.26
Graph of continuous approximation for the bowling game scores of Table 5.2 and Figure 5.25.

5.2.3 Continuity Correction

We can approximate many discrete distributions satisfactorily by a continuous distribution, as we saw in Figure 5.26. If we assume that a smooth curve is a reasonable approximation for the actual probability distribution of the variable, we can estimate probabilities for individual values of the variable or for ranges of values. We note that if a random variable is continuous, we can calculate probabilities only for ranges of the variable. If we use a continuous distribution to approximate the probabilities for a discrete random variable, we must use a **continuity correction.** Each value of the original (discrete) random variable is represented in a histogram by an interval from one-half unit below the number to one-half unit above. Thus 50 is represented on the base by the interval 49.5 to 50.5. When we use a continuous distribution to approximate a discrete distribution, we must maintain this distinction. Thus for the continuous representation of the discrete distribution we still have 50 represented on the base by the interval 49.5 to 50.5, and the area under the curve (see Figure 5.27) from 49.5 to 50.5 would be equal to $P(50)$.

Continuity Correction

When we use a continuous random variable to approximate a discrete random variable, each value of the discrete variable is represented by a range of values of the continuous variable. This **continuity correction** is achieved by adding and subtracting half the unit of measurement to each value of the discrete variable. The resulting range of values gives the values of the continuous variable that correspond to the single value of the discrete variable.

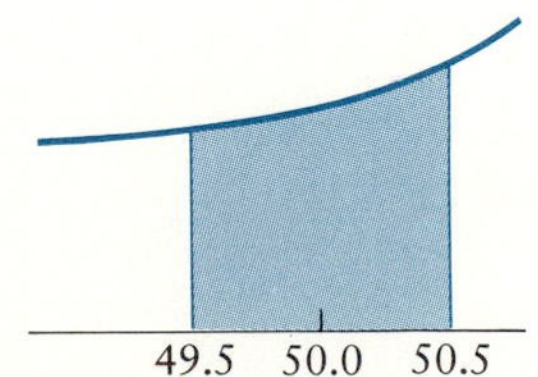

Figure 5.27
The shaded area represents $P(50)$, using the continuity correction.

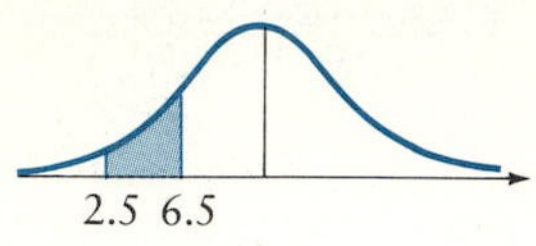

Figure 5.28
Illustration of the continuity correction for the light bulb data of Example 5.11. Here $P(3 < x < 6)$, where x is the number of defective light bulbs.

Examples 5.9 and 5.10 dealt with continuous variables. If variables are discrete, or if they are continuous variables measured to the nearest whole unit (such as the nearest minute), you should use the continuity correction. In Example 5.10, for instance, if the time had been measured to the nearest minute, then "between 48 and 50 minutes" would be 49 minutes only and would have been represented by 48.5 to 49.5—which would have changed the result considerably. The use of such terms as *between* and *inclusive* is critical with a discrete variable. The term *between* does not include the endpoints. If the variable is discrete and the endpoints are to be included, we should say "from a to b, inclusive." This way there is no confusion.

The question of exactly *when* we must use a continuity correction has not been completely resolved. It is clear that if the units in which a discrete variable is measured are fairly large in comparison with the range of the data, we need the continuity correction. It is equally clear that if the units are very small in comparison with the range of the data, the difference between the histogram and a continuous curve is negligible, so the continuity correction is not necessary. The dividing point between the two cases is not at all clear. We recommend that you use the continuity correction whenever the variable is discrete and measured in integers. The prime consideration is whether lack of the continuity correction will distort the results. If there is doubt, use the continuity correction.

Example 5.11

A machine produces light bulbs that are shipped in lots of 1000. On the average, a lot has 10 defective. The distribution of defective bulbs is approximately normal with a standard deviation of about 3.15. What is the probability that a particular lot will have at least 3 but not more than 6 defective bulbs? What is the probability that it will have more than 15 defective bulbs?

Solution

Since the data (number of defective bulbs) are discrete and measured in integers, we need the continuity correction. To find $P(3 \leqslant x \leqslant 6)$ we note that the interval for 3 is 2.5 to 3.5, and the interval for 6 is 5.5 to 6.5. Since we wish to include 3, 6, and everything in between, the interval representing "at least 3 but not more than 6" is 2.5 to 6.5, as shown in Figure 5-28.

The standard normal deviates for 2.5 and 6.5 are $(2.5 - 10)/3.15$ and $(6.5 - 10)/3.15$, or about -2.38 and -1.11 respectively. The associated areas, from Table 4, are 0.4913 and 0.3665. The area between 2.5 and 6.5, then, is about $0.4913 - 0.3665$ or about 0.1248. Since 2.38 and 1.11 are rounded, we conclude that $P(3 \leqslant x \leqslant 6) \doteq 0.125$.

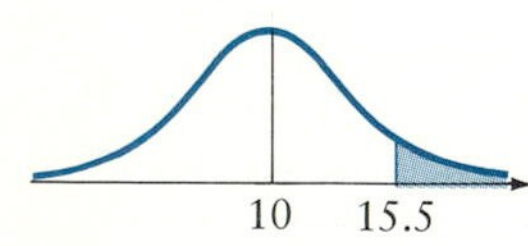

Figure 5.29
$P(x > 15)$, where x represents the number of defective light bulbs in Example 5.11.

To determine $P(x > 15)$, we note that 15 is represented by the interval from 14.5 to 15.5. Since 15 is not to be included, we want the area under the curve *to the right of* 15.5 as shown in Figure 5.29. For $x = 15.5$, we have $z = (15.5 - 10)/3.15 \doteq 1.75$. The area under the standardized normal curve between zero and 1.75 is 0.4599 and $0.5000 - 0.4599 = 0.0401$, so $P(x > 15) \doteq 0.040$.

Example 5.12

In a certain district the monthly rental for apartments is approximately normally distributed with a mean of \$384.22 and a standard deviation of \$76.40. Above what value is the highest 30% of the monthly rentals in this district?

Solution

Although the variable is discrete, the unit (cents) is very small in comparison with the range of the data (6σ or about \$450, using the empirical rule). Further, its values are not given in integers. Thus we do not need the continuity correction. According to Table 4, 20% of the area under a standardized normal curve lies between the mean and $z \doteq 0.52$, so 30% of the area under the curve will be to the right of $z = 0.52$. (See Figure 5.30.) Thus we have $(x - 384.22)/76.40 \doteq 0.52$. Solving for x, we find $x \doteq 423.95$, so about 30% of the rentals are above \$423.95.

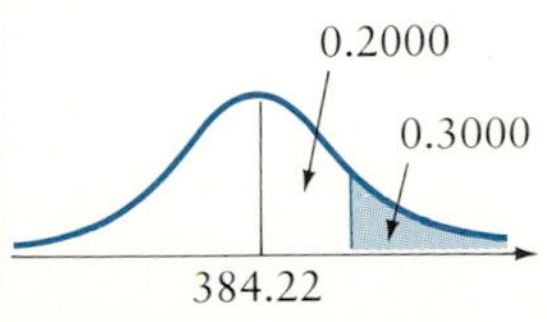

Figure 5.30
Graph of probability distribution for the rent data of Example 5.12.

PROFICIENCY CHECK

13. A botanist has irradiated 100 caladium rhizomes and wants to see what effect the radiation has on the incidence of their nongermination. On the average 30 of 100 rhizomes will not germinate, but the number will vary from sample to sample of 100 with a standard deviation of 5.4. What is the probability that fewer than 20 rhizomes do not germinate?

5.2.4 The Normal Approximation to the Binomial

One of the most important types of random variables in statistics is the binomial random variable. Yet as the number of trials increases, it becomes more and more cumbersome to use. If we wanted to determine the probability of, say, 30 or more heads in 50 tosses of a coin, we might have to make, in the absence of appropriate binomial tables, 21 separate calculations of individual probabilities. Fortunately as n (the number of trials) increases, the normal distribution is a reasonably satisfactory approximation to a binomial distribution—particularly if p (the probability of success on one trial) is close to 0.5. Figure 5.31 shows a graph of a binomial distribution with $n = 10$ and $p = 0.5$; a continuous curve has been superimposed on the graph.

We can see that the continuous curve in Figure 5.31 bears a strong resemblance to a normal curve. As the number of trials becomes greater and greater, the graph of a binomial distribution becomes closer and closer to a normal curve. For very large numbers of trials, the approximation remains valid even if the probability of success differs substantially from 0.5. The

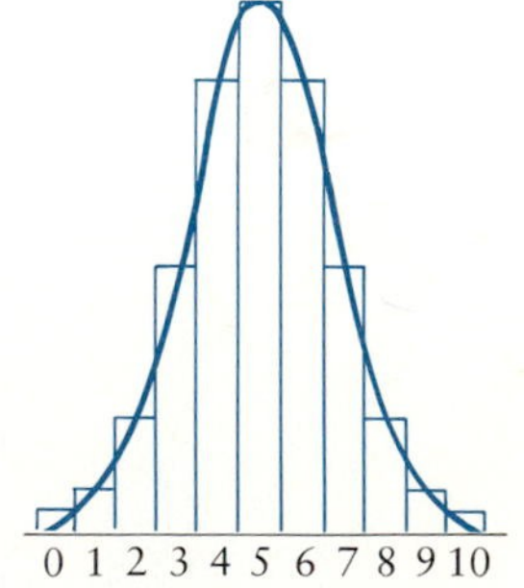

Figure 5.31
Histogram for binomial distribution with $n = 10$ and $p = 0.5$ with a continuous curve superimposed.

next section presents several computer printouts to show the increasing accuracy of the approximation as n increases, even when p is very different from 0.5.

There are a number of methods for deciding whether a normal curve is a sufficiently close approximation to the graph of a binomial distribution.* Among them is the following rule of thumb.

The Normal Approximation to the Binomial Probability Distribution

If a binomial random variable x has a probability p of success on one trial, the probability distribution of x for n trials can be approximated by a normal random variable with mean and standard deviation

$$\mu = np \qquad \text{and} \qquad \sigma = \sqrt{np(1 - p)}$$

respectively, provided that both np (the expected number of successes) and $n(1 - p)$ (the expected number of failures) are at least 5.

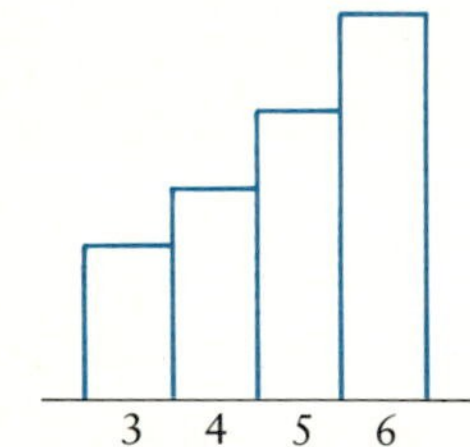

Figure 5.32
Histogram for the binomial distribution from $x = 3$ to $x = 6$ (Example 5.11).

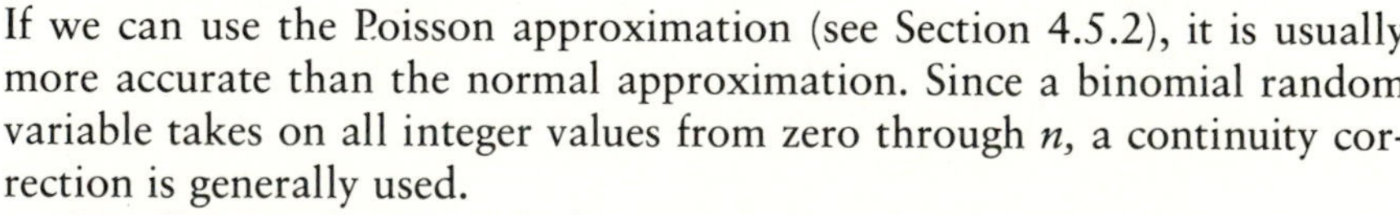

If we can use the Poisson approximation (see Section 4.5.2), it is usually more accurate than the normal approximation. Since a binomial random variable takes on all integer values from zero through n, a continuity correction is generally used.

Recall Example 5.11. The histogram for the graph of the binomial distribution from $x = 3$ to $x = 6$ appears as in Figure 5.32. Since $np = 10$ and $n(1 - p) = 990$, we can use the normal approximation to the binomial. We can approximate the binomial random variable with a normal random variable that has mean 10 ($np = 10$) and standard deviation $\sqrt{1000(0.01)(0.99)}$, or $\sqrt{9.9}$—that is, about 3.15. That portion of the normal curve for the variable with $\mu = 10$ and $\sigma \doteq 3.15$ is superimposed on the graph of Figure 5.32 and shown in Figure 5.33.

The shaded portion in Figure 5.33 is the area under the normal curve from 2.5 to 6.5, a very close approximation to the area in the rectangles. In Example 5.11, the shaded area was found to be about 0.125. Direct calculation using the binomial formula shows us that the area in the rec-

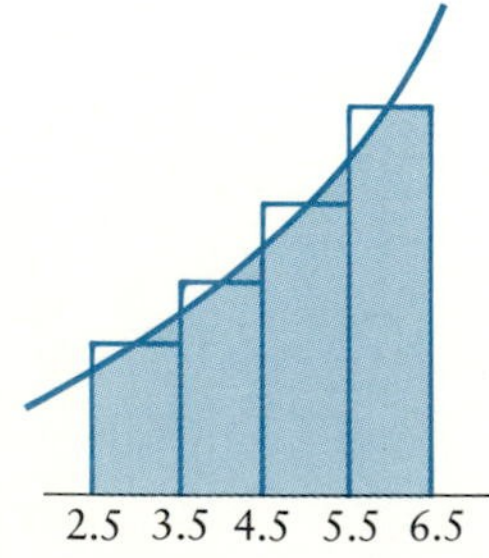

Figure 5.33
Portion of normal curve superimposed on histogram of Figure 5-32.

*Mendenhall (1983, p. 235) suggests that the normal approximation is valid if the interval $np \pm 2\sqrt{np(1 - p)}$ lies within the range of the data—that is, from zero to n. McClave and Dietrich (1985, p. 209) prefer the interval $np \pm 3\sqrt{np(1 - p)}$ to lie within the range of the data. The rule given here appears to be the most widely used and seems to represent a compromise between these two positions.

tangles—hence the exact probability (to three decimal places) of having from 3 to 6 defective bulbs—inclusive, is 0.126. The normal approximation agrees quite well with this value. Using a Poisson approximation with $m = 10$ gives a probability (from Table 3) of 0.127, again agreeing quite well.

Example 5.13

Suppose that a very good basketball player makes, on the average, 50% of his field-goal attempts. Using 0.50 as the probability that he will make a particular attempt, what is the probability that he will make at least 30 of his next 50 attempts?

Solution

This is a binomial experiment with $n = 50$ and $p = 0.5$. The probability distribution for the number of successful attempts can be approximated by the probability distribution for a normal random variable x with $\mu = 50(0.5) = 25$ and $\sigma = \sqrt{50(0.5)(0.5)} = \sqrt{12.5} \doteq 3.54$. Now 30 is represented on the continuous scale by 29.5 to 30.5; since we want 30 or more, we must determine the probability that $x > 29.5$. The standard normal deviate for 29.5 is $(29.5 - 25)/3.54$ or about 1.27. The area under the normal curve between 29.5 and the mean is 0.3980 (from Table 4). Then $P(x > 29.5) \doteq 0.5000 - 0.3980 = 0.102$. Thus the probability that he will make at least 30 of his next 50 field-goal attempts is about 0.102.

Example 5.14

Use the normal approximation to find the probability of obtaining 5 or more successes in 25 trials of a binomial experiment if the probability of success on a given trial is 0.20. Compare your result with the true probability to three decimal places. Repeat for $p = 0.40$.

Solution

Using the normal approximation with $n = 25$ and $p = 0.2$, we have $\mu = np = 25(0.2) = 5$ and $\sigma = \sqrt{np(1-p)} = \sqrt{25(0.2)(0.8)} = \sqrt{4} = 2$. Since 5 is represented by the interval 4.5 to 5.5, we want the probability that the normal random variable is greater than 4.5. The standard normal deviate for 4.5 is $(4.5 - 5)/2 = -0.25$, which corresponds to an area of 0.0987 between -0.25 and the mean. Since the area above 4.5 includes the entire area above the mean, the probability of obtaining at least five successes is about $0.5000 + 0.0987$, or about 0.599. From the table of binomial probabilities, $P(x \geqslant 5) \doteq 0.579$. The approximation is not particularly good, but this is to be expected since $np = 5$, the minimum requirement for use of the normal approximation.

For $n = 25$ and $p = 0.4$, we have $\mu = 25(0.4) = 10$ and $\sigma = \sqrt{25(0.4)(0.6)} = \sqrt{6} \doteq 2.45$. The standard normal deviate for 4.5 is $(4.5 - 10)/2.45$, or about -2.24. The area under the normal curve between $z = 2.24$ and the mean is 0.4875, so $P(x \geqslant 5) \doteq 0.988$. From Table 1, we have $P(x \geqslant 5) \doteq 0.991$, so the approximation is very good.

PROFICIENCY CHECKS

14. Determine the probability of getting from 6 to 10 sevens, inclusive, in 45 rolls of a pair of dice. What is the probability of obtaining exactly 3 sevens?
15. A quarterback completes, on the average, 60 percent of his passes. What is the probability that he will complete no more than 45 of his next 100 passes?
16. There are 1000 fuses in a shipment, and they are rated 98% reliable. Reliability in a fuse is the probability that it will function properly under the conditions for which it was designed. What is the probability that a shipment of 1000 fuses will contain at least 27 defective?

Problems

Practice

5.15 A coin is tossed 40 times. If 25 or more heads appear, a player receives $15. If 25 or more tails appear, the player must pay $10. What is the player's expected profit (or loss) for one game of 40 tosses?

5.16 A test has 225 multiple-choice questions, each with one correct and four incorrect answers. Each question is answered randomly. What is the probability of obtaining more than 60 correct answers?

5.17 A thumbtack is tossed 75 times. It falls point up, on the average, about 30% of the time. What is the probability that it will fall point up at least 10 times but no more than 25 times?

Applications

5.18 Number 10 cans of peaches are supposed to hold 72 ounces. The canner sets the machine to fill each can with 72.15 ounces on the average. The distribution of fill weights is approximately normal with a standard deviation of 0.10 ounce.

a. Approximately how many of 100,000 cans will have less than72 ounces?
b. The manufacturer thinks this result is too high. Assuming that the standard deviation remains unchanged, what mean should the manufacturer set (to two decimal places) so that no more than 100 cans in 100,000 contain less than 72 ounces?

5.19 According to *USA Today* (Jan. 21, 1986) the average annual cost of attending a four-year public college or university for 1985–1986 was $3430. If the standard deviation is $815, what is the probability that the annual cost of attending such a college selected at random will be more than $5000?

5.20 The final exam scores in a statistics class are normally distributed with a mean of 70 and a standard deviation of 10. Assume that the grades are discrete.

a. If the lowest passing mark is 60, what percentage of the class fails?

b. If the highest 80 percent are to pass, what should be the lowest passing score?

5.21 A certain machine lasts on the average 5 years with a standard deviation of 6 months. Assuming that the lives of these machines are normally distributed, find the probability that a given machine will last less than 4.4 years.

5.22 The Bureau of Labor statistics (*Newsweek*, Feb. 17, 1986) showed that 50% of all women with children under the age of 2 are working. Suppose that a random sample of households is interviewed in a door-to-door survey conducted on Sundays only, and there are 115 households containing a woman with children under 2. Assuming independence of outcomes and the Bureau of Labor statistics figure, what is the probability that fewer than 50 of these women are working?

5.23 A poll is taken to learn whether children suffering from AIDS should be allowed to attend school. If 60% of the population believe they should, what is the probability that the majority of a random sample of 85 persons will nevertheless be opposed?

5.24 A basket of peaches in a certain fruit stand has about 84 peaches. This number varies a bit, however, and the number of peaches per basket is normally distributed with a mean of 84.2 and a standard deviation of 2.1. What is the probability that a basket selected at random contains between 80 and 90 peaches?

5.25 Approximately 6% of the male population is color-blind. If a random sample of 250 male adults is tested for color-blindness, what is the probability that at least 10, but no more than 20, will be color-blind?

5.26 Rats are being tested to see if they prefer a square or a triangle. In this experiment, the rat can step on a square or a triangle and will receive food in either case. The figures are the same in area, color, and composition, so the only difference is shape. If there is no preference, it can be assumed that the rat will choose each shape with a probability 0.5. What is the probability that if there really is no preference, a rat will choose the triangle 40 or more times in 60 trials?

5.27 You drive to class every morning. The time can be approximated by a normal random variable with a mean of 20 minutes 24 seconds and a standard deviation of 5 minutes 12 seconds. At what time should you leave home if you want the probability of being late to your 8 AM class to be no more than 0.05? It takes 3 minutes to get from your car to the classroom.

5.28 Dowel rods for a certain product should have a diameter of exactly 4 millimeters. A tolerance of 0.020 millimeter is allowed. Rods with diameters greater than 4.020 millimeters or less than 3.980 millimeters are not usable. Company A guarantees that the standard deviation on a lot with mean diameter 4 millimeters will be no greater than 0.016 millimeter and will charge $400 per lot of 10,000. Company B guarantees that the standard deviation on a lot with mean diameter 4 millimeters will be no more than 0.01 millimeter and will charge \$460 for a lot of 10,000. Which company should get the order if the major criterion is cost per usable dowel rod?

5.29 A test for right and left preference of salmon is being conducted prior to construction of a "ladder" to aid salmon swimming upstream in getting around a dam. Temporary artificial channels are constructed to observe whether salmon swimming upstream prefer the left or right channel. A total of 732 salmon are observed. Of these, 348 swim up the left channel and the remainder use the right channel. Determine the probability that 348 or fewer will choose the left channel by chance if there really is no preference.

5.30 To test whether a process is in control, a reading is taken on a machine. If the reading is greater than 860.0, the machine will have to be retooled. If the process is in control, the daily readings will have a mean of 832.4 with a standard deviation of 10.2. What is the probability of retooling a machine by mistake? That is, what is the probability of getting a reading above 860.0 when the process is in control?

5.31 It has been found that sex ratios in families are conditional. That is, if a couple has a daughter, the probability that the next child will be a daughter is greater than if the first child had been a son. If a run of children of the same sex occurs, the probability of the next child being the same sex increases with each additional child. Suppose that the probability of having a boy if the first two children are girls is 0.44. A hundred families have two girls and are expecting a third child. If all births are single births, what is the probability that at least one-half of the 100 children born will be boys?

5.32 A survey organization regularly sends out questionnaires. The number of replies on a mailing of 1000 is approximately normally distributed with a mean of 585.3 and a standard deviation of 41.2. If 1000 questionnaires are mailed out, what is the probability, on the basis of past experience, of receiving at least 650 replies?

5.33 According to *USA Weekend* (Feb. 14–16, 1985), 74% of Americans would be willing to lend \$100 or more to a friend. A random sample of 300 people is interviewed. Assuming that the binomial model applies with $p = 0.74$, what is the probability that at least 80% of the people will say they would be willing to lend at least \$100 to a friend?

5.34 According to *American Demographics Magazine* (Dec. 1985), 56% of all families have two or more wage earners. Suppose a random sample of 100 families is interviewed. What is the probability that more than half the families will have two or more wage earners?

5.3 Computer Usage

In Section 5.2.4 we noted that we can approximate binomial probabilities by using the normal curve, provided that both np and $n(1 - p)$ are at least 5. To illustrate this point further we can use the computer to graph the binomial probabilities for specific values of n and p and determine visually whether the resulting graph closely resembles a normal curve. This procedure provides us with a visual check of the accuracy of the normal approximation.

In this section we will also discuss using the computer to determine whether or not a particular random variable is a normal random variable. In the remaining chapters of the book we will frequently encounter the assumption that the variable is normal. It is important that we have ways of checking whether this assumption seems justified, and we can use the computer to do this. We will also see how we can obtain probabilities for a normal random variable using an SAS statement.

5.3.1 Minitab and the Normal Approximation

Obtaining the graph of a binomial distribution using Minitab is relatively simple. We can use the PDF command and the BINOMIAL subcommand to generate the binomial probabilities and then store these values in order to plot them. The following commands illustrate how to do this for a binomial random variable with $n = 20$ and $p = 0.5$:

```
MTB > SET the following in C1
DATA> 0:20
DATA> END of data
MTB > PDF for the values in C1 and store them in C2 ;
SUBC> BINOMIAL n = 20 p = .5 .
MTB > PLOT C2 as the vertical axis C1 as the horizontal .
```

In this illustration the data line 0:20 causes the integers from zero through 20 to be stored in C1. The PDF command finds the binomial probabilities for each value in C1 and stores these probabilities in C2. The resulting graph is plotted as usual. Note that use of the PDF command is different here from its use in Chapter 4. In Chapter 4 we used it to print the binomial probabilities; here we use it to compute these values and store them for future use.

To show how well the normal approximation fits a binomial distribution, we will consider two cases, both with $n = 50$. In the first case $p = 0.5$, so that np and $n(1 - p)$ are both equal to 25. According to the statements in Section 5.2.4, the approximation should be quite good. The Minitab printout for the appropriate program is shown in Figure 5.34. We can see that the graph closely resembles a normal curve and, as a consequence, we can expect the normal probabilities to provide good approximations to binomial probabilities.

```
MTB > SET the following in C1
DATA> 0.50
DATA> END of data
MTB > PDF for the values in C1 store them in C2 ;
SUBC> BINOMIAL n = 50 p = .5 .
MTB > PLOT C2 as the vertical axis C1 as the horizontal
```

Figure 5.34
Minitab printout for normal approximation to the binomial distribution where $n = 50$ and $p = 0.5$.

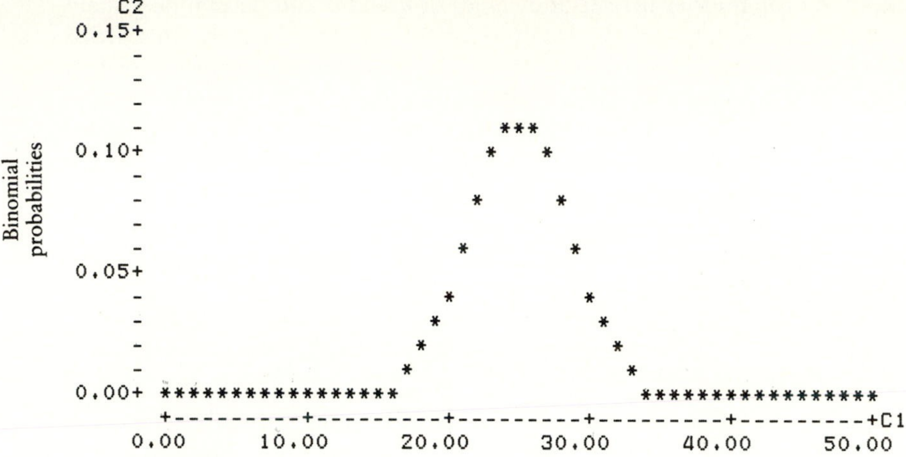

Our second illustration is the case when $n = 50$ and $p = 0.05$. In this case $np = 2.5$; since $2.5 < 5$, the normal approximation procedure may not work well. The Minitab printout for the appropriate program is shown in Figure 5.35. We can see quite clearly that the graph does not resemble a normal curve; consequently the normal probabilities would not be good approximations to the binomial probabilities.

Figure 5.35
Minitab printout for normal approximation to the binomial distribution where $n = 50$ and $p = 0.05$.

```
MTB > PDF for the values in C1 store them in C3 ;
SUBC> BINOMIAL n = 50 p = .05 .
MTB > PLOT C3 as the vertical axis C1 as the horizontal
```

Binomial probabilities

```
        C3
      0.30+
          -
          -   *
          -
          -    *
      0.20+  *
          -
          -
          -     *
          -
      0.10+
          - *
          -      *
          -
          -       *
      0.00+        ********************************************
            +---------+---------+---------+---------+---------+C1
          0.00     10.00     20.00     30.00     40.00     50.00
```

5.3.2 Checking Normality Using Minitab

Since the assumption of normality is frequently important in statistical inference, we may need to investigate the shape of the probability distribution of a random variable if we are unsure whether it is normal. Generally we do this by using a sample of data from the population we are investigating. If the histogram of the data appears to be reasonably close to a bell curve in shape, we can then apply the empirical rule. Moreover, recall from the discussion of the pseudo-standard deviation that approximately 50 percent of the data in a normal distribution should fall between 0.675 standard deviations above the mean and 0.675 standard deviations below the mean. Thus we can obtain the mean and standard deviation of the data and apply the criteria stated here.

Criteria for Normality

It is reasonable to consider a random variable normally distributed if:

1. A graph of the data resembles a bell curve.
2. The data are distributed so that approximately 50% of the data lie between $\bar{x} - 0.675s$ and $\bar{x} + 0.675s$, approximately 68% of the data lie between $\bar{x} - s$ and $\bar{x} + s$, approximately 95% of the data lie between $\bar{x} - 2s$ and $\bar{x} + 2s$, and nearly all (99.7%) of the data lie between $\bar{x} - 3s$ and $\bar{x} + 3s$.

This procedure works especially well with large samples.

Example 5.15

Apply the criteria for normality to the pollution data of Example 2.15 and determine whether or not the data are reasonably normal—that is, whether or not the population of data from which the sample was taken is reasonably close to a normal distribution.

Solution

Figure 5.36 shows the Minitab program and output used to obtain a histogram, stem and leaf plot, and the mean and standard deviation of the data.

```
MTB > SET the following data in column C1
DATA>  47 70 84 46 29 64 43 61 46 40 41 59
DATA>  58 72 88 57 39 60 47 62 58 38 33 54
DATA>  67 59 81 63 44 57 54 54 60 47 42 63
DATA>  72 54 77 69 57 51 59 57 52 62 48 60
DATA>  88 61 54 61 61 61 67 30 54 70 52 69
DATA>  74 50 70 72 60 58 62 44 48 54 41 66
DATA>  67 73 42 52 53 59 68 33 58 83 48 58
DATA>  73 68 41 44 50 54 70 41 54 88 44 42
DATA>  64 79 43 37 44 60 74 49 67 69 42 47
DATA>  61 82 37 33 58 48 66 52 49 57 45 48
DATA> END of data
```

Figure 5.36
Minitab printout for pollution data of Example 5.15: checking for normality.

```
MTB > NAME for C1 is 'P-INDEX'

MTB > HISTOGRAM the data in C1

      HISTOGRAM of P-INDEX N = 120

      MIDPOINT      COUNT
         30            2        **
         35            5        *****
         40           11        ***********
         45           14        **************
         50           14        **************
         55           15        ***************
         60           24        ************************
         65           10        **********
         70           12        ************
         75            5        *****
         80            3        ***
         85            2        **
         90            3        ***
```

```
MTB > DESCRIBE the data in C1

                     N        MEAN      MEDIAN     TRMEAN     STDEV     SEMEAN
      P-INDEX       120       56.7       57.0       56.5       13.1       1.2
                    MIN        MAX         Q1         Q3
      P-INDEX      29.0       88.0       47.0       66.0
```

```
MTB > STEM-AND-LEAF the data in C1

       STEM-AND-LEAF OF P-INDEX
       LEAF UNIT = 1.0

          1     2.  9
          5     3*  0333
          9     3.  7789
         25     4*  0111122223344444
         39     4.  56677778888899
         56     5*  00122223444444444
       (15)     5.  777778888889999
         49     6*  000001111112223344
         31     6.  66777788999
         20     7*  00002223344
          9     7.  79
          7     8*  1234
          3     8.  888
```

We see from the histogram that an assumption of normality is not unreasonable. Since the mean is 56.7 and the standard deviation is 13.1, then,

Table 5.3 SUMMARY OF RESULTS FOR EXAMPLE 5.15.

INTERVAL	NUMBER OF OBSERVATIONS	%	EXPECTED %
47.9–65.5	57	48	50
43.6–69.8	80	67	68
30.5–82.9	113	94	95
17.4–96.0	120	100	99.7

since $0.675(13.1) \doteq 8.8$, 50% of the data should lie between 47.9 and 65.5; further, 68% should lie between 43.6 and 69.8, 95% should lie between 30.5 and 82.9, and almost all of the data should lie between 17.4 and 96. We count the number of observations in each range. The percentage for each observation is reported in Table 5.3. We can see from Table 5.3 that the data set agrees remarkably well with the expectations for normal data, and the weight of the evidence is preponderantly in favor of the data being normal.

Another procedure we may use is the **normal probability plot.** We can obtain such a plot for data stored in C1 by using the following Minitab commands:

```
MTB > NSCORES the data stored in C1 and store the results in C2
MTB > PLOT C1 on the vertical axis and C2 on the horizontal axis
```

The z-score for each observation is computed and stored in C2; Minitab then plots the value of each observation against its z-score. If the data are normally distributed, the plot should be a straight line. If the graph is not a straight line and has some curvature, we assume that the random variable is not normally distributed. This procedure is particularly useful for small samples since there may not be enough observations to use the method of Example 5.15 effectively, but it can also be used for large samples.

Example 5.16

Refer to the spore culture data of Example 1.1. Obtain a normal probability plot for the data, and determine whether or not the distribution of data is reasonably normal.

Solution

The Minitab program for this example is shown in Figure 5.37. Notice that this graph is not a straight line; thus the normality assumption may not be reasonable for this random variable.

```
MTB > SET the following data in C1
DATA> 57 59 58 61 56 63 57 69 59 59 57 63 72 59
DATA> 62 60 59 63 59 65 57 61 58 59 63
DATA> END of data
MTB > NSCORES the data in C1 and store in C2
MTB > PLOT C1 on the vertical axis and C2 on the horizontal
```

Figure 5.37 Minitab printout for normal probability plot for spore culture data of Example 5.16.

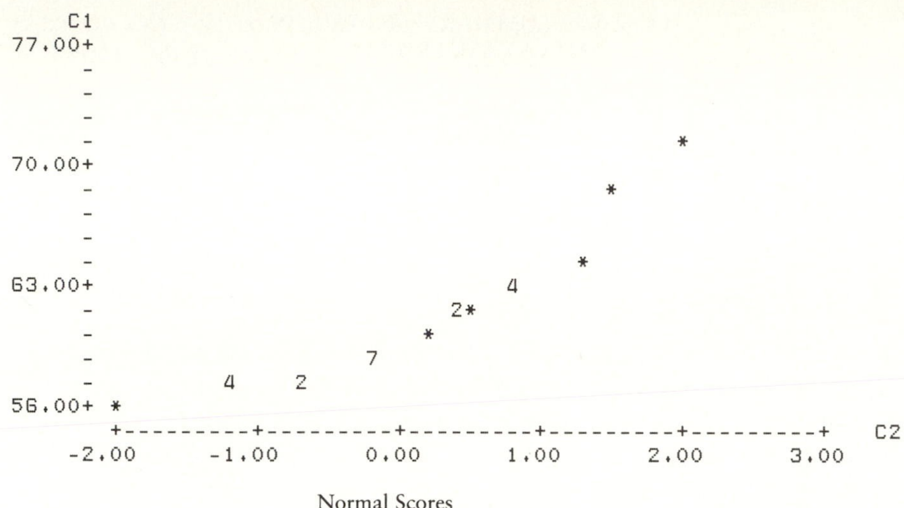

Normal Scores

5.3.3 SAS Statements for This Chapter

We may use SAS statements to determine whether or not a normal approximation to a binomial random variable is justified. We can obtain the probability for each value of a binomial random variable by using the PROBBNML statement. We can then plot these values against the values of the variable by using the PLOT procedure. The statement PROC PLOT; followed by the statement PLOT *variable 1* variable 2;* will give a printout of the graph with variable 1 on the vertical axis and variable 2 on the horizontal axis. Unless otherwise specified, the printout will use 'A' for each point. We will see later that if additional observations have the same value for both variables, 'B' will be used if there are two points with the same observations, 'C' if there are three, and so on. In this case we know that there is only one observation for each value of the variable (its probability) so we choose to use an asterisk (*) for each point. To do so we insert "='*'" after the second variable and before the semicolon. Thus the following program will give the graph of a binomial random variable with $n = 50$ and $p = 0.5$:

```
DATA APPROX;
 DO X = 0 TO 50;
 CPX = PROBBNML(.5, 50, X);
 IF X > 0 THEN CPX1 = PROBBNML(.5, 50, X-1);
 IF X = 0 THEN PROBX = CPX;
 IF X > 0 THEN PROBX = CPX-CPX1;
 OUTPUT;
 END;
PROC PLOT; PLOT PROBX * X = '*';
```

The output of this program is shown in Figure 5.38. Compare this output with Figure 5.34. We see that the approximate normality of the graph is confirmed.

```
DATA APPROX:
   DO X = 0 to 50;
   CPX = PROBBNML(.5,50,X);
   IF X > 0 THEN CPX1 = PROBBNML(.5,50,X-1);
   IF X = 0 THEN PROBX = CPX;
   IF X > 0 THEN PROBX = CPX - CPX1;
   OUTPUT;
   END;
PROC PLOT; PLOT PROBX * X = '*';
```

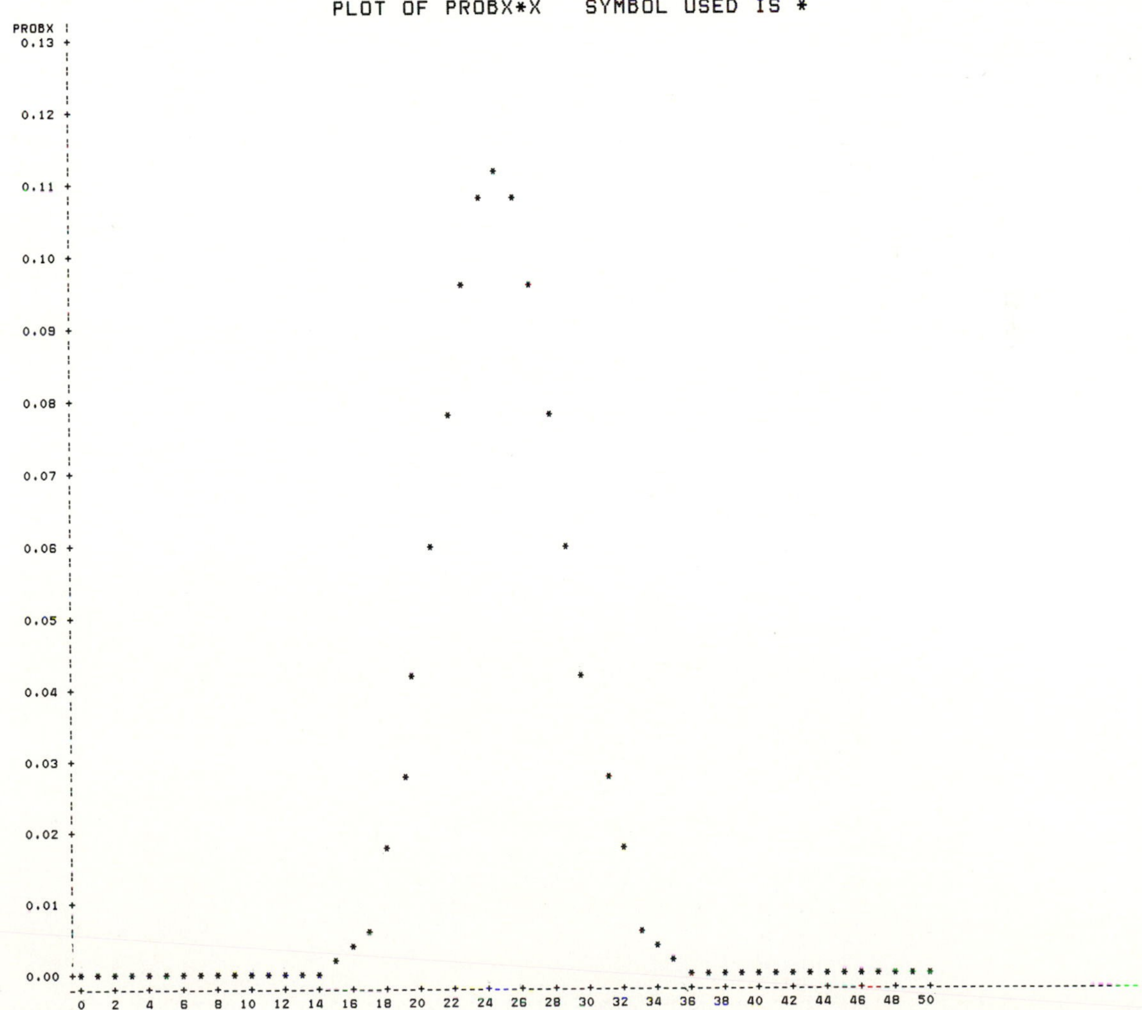

Figure 5.38
SAS printout for normal approximation to the binomial distribution, where $n = 50$ and $p = 0.5$.

We use a similar program to obtain Figure 5.39, with $p = 0.05$ instead of 0.5. We see the same nonnormal appearance that is evident in Figure 5.35.

Figure 5.39
SAS printout for normal approximation to the binomial distribution where $n = 50$ and $p = 0.05$.

```
DATA APPROX2;
   DO X = 0 TO 50;
   CPX = PROBBNML(.05,50,X);
   IF X > 0 THEN CPX1 = PROBBNML(.05,50,X-1);
   IF X = 0 THEN PROBX = CPX;
   IF X > 0 THEN PROBX = CPX - CPX1;
   OUTPUT;
   END;
PROC PLOT; PLOT PROBX * X = '*';
```

SAS

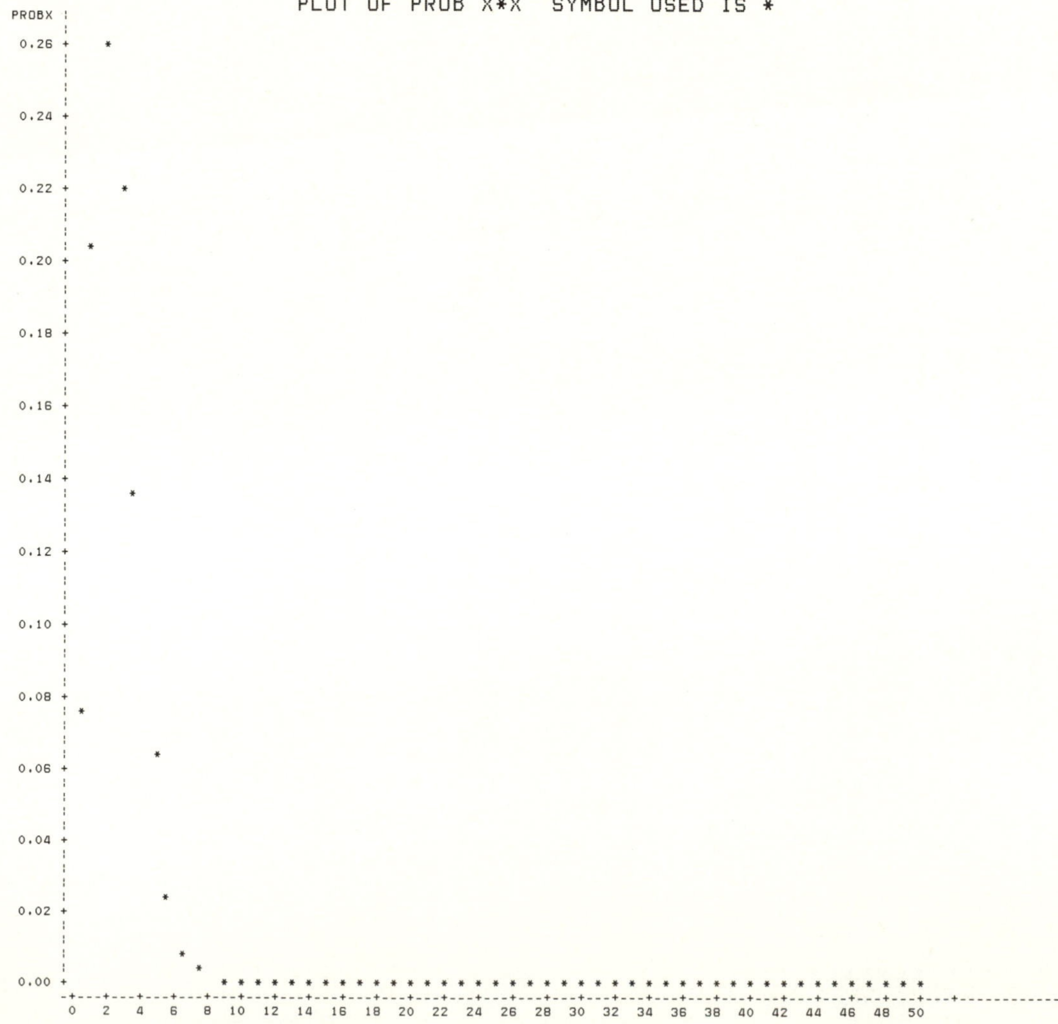

To check on normality, we may use the output of the UNIVARIATE procedure with the option PLOT. In addition to obtaining the mean and standard deviation, this procedure also gives us a stem and leaf diagram. See Figure 2.15 for the output of PROC UNIVARIATE PLOT; for the data of Example 2.16 (and Example 5.15). We can obtain the histogram with the CHART procedure; see Section 1.3.2 for details on this procedure. We can use the mean (56.7167) and standard deviation (13.0526) to determine whether the criteria for normality are met. The procedure also provides us with the stem and leaf plot and box plot that may give us further information about the normality of the data set.

The normal probability plot discussed in Section 5.3.2 is a by-product of the UNIVARIATE PLOT procedure discussed in Section 2.5.2. We use the spore culture data of Example 1.1 (and Example 5.16) to obtain the SAS printout shown in Figure 5.40. On the normal probability plot, asterisks mark the data points in the data set; the plus signs indicate where the points should lie if the data are normal. We can see that the data are probably not normal, confirming the result of Example 5.16.

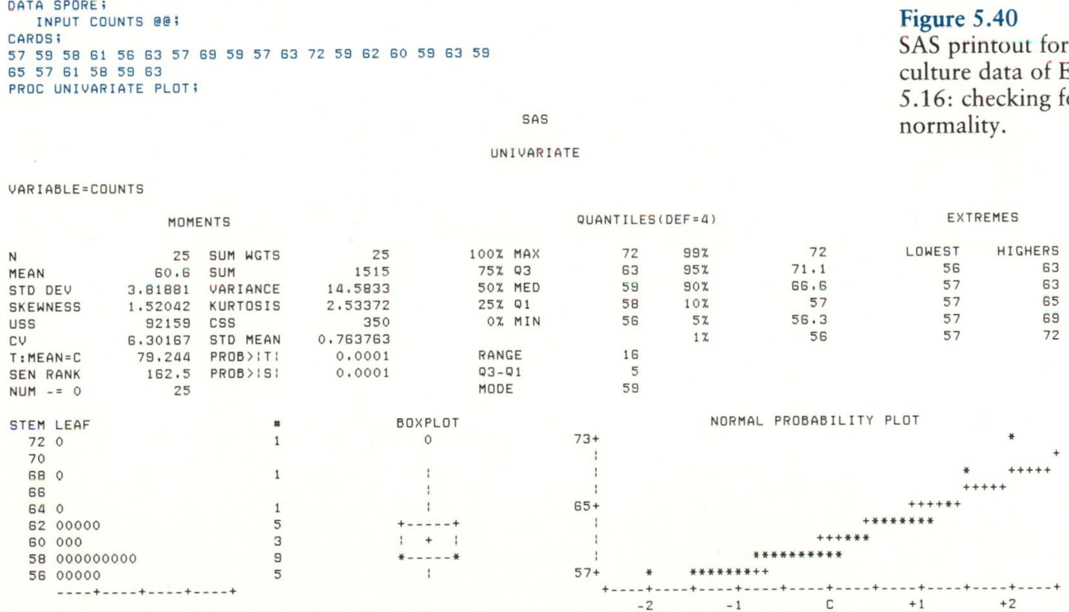

```
DATA SPORE;
   INPUT COUNTS @@;
CARDS;
57 59 58 61 56 63 57 69 59 57 63 72 59 62 60 59 63 59
65 57 61 58 59 63
PROC UNIVARIATE PLOT;

                                          SAS

                                       UNIVARIATE

VARIABLE=COUNTS

              MOMENTS                                          QUANTILES(DEF=4)                              EXTREMES

N                  25  SUM WGTS          25      100% MAX       72     99%          72       LOWEST    HIGHERS
MEAN             60.6  SUM             1515       75% Q3        63     95%        71.1           56         63
STD DEV       3.81881  VARIANCE     14.5833       50% MED       59     90%        66.6           57         63
SKEWNESS      1.52042  KURTOSIS     2.53372       25% Q1        58     10%          57           57         65
USS             92159  CSS              350        0% MIN       56      5%        56.3           57         69
CV            6.30167  STD MEAN    0.763763                              1%          56           57         72
T:MEAN=C       79.244  PROB>|T|      0.0001      RANGE          16
SEN RANK        162.5  PROB>|S|      0.0001      Q3-Q1           5
NUM -= 0           25                            MODE           59

STEM LEAF                     #          BOXPLOT                       NORMAL PROBABILITY PLOT
  72 0                        1             0           73+                                                *
  70                                                      |                                                     +
  68 0                        1             |             |                                           *    +++++
  66                                        |             |                                           +++++
  64 0                        1             |           65+                                     +++++*
  62 00000                    5          +-----+          |                                +*******
  60 000                      3          |  +  |          |                           ++++++
  58 000000000                9          *-----*          |                     **********
  56 00000                    5             |           57+    *     *********++
     ----+----+----+----+                                  +----+----+----+----+----+----+----+----+----+----+
                                                               -2        -1         C        +1        +2
```

Figure 5.40
SAS printout for spore culture data of Example 5.16: checking for normality.

As a final observation we note that the SAS statement PROBNORM computes probabilities for a normal distribution. The statement PROBNORM(Z) computes the probability that a random variable z which is N(0,1) is less than the value specified. We may use the statement to determine $P(x < k)$ if x is $N(\mu,\sigma^2)$ by writing (after the data step)

```
P = PROBNORM((K - μ)/σ);
```

We may obtain $P(k_1 < x < k_2)$ by writing

```
P1 = PROBNORM((K1 - μ)/σ);
P2 = PROBNORM((K2 - μ)/σ);
P = P2 - P1;
```

As an example, suppose we want to determine the probability that a normal random variable with mean 10 and standard deviation 5 is between 8 and 13. We write

```
DATA EXAMPLE;
 P1 = PROBNORM((8 - 10)/5);
 P2 = PROBNORM((13 - 10)/5);
 P = P2 - P1;
 PUT P;
RUN;
```

The result is shown in Figure 5.41. Thus if x is N(10,25), then $P(8 < x < 13) = 0.3811686$.

Figure 5.41
SAS printout for $P(8 < x < 13)$, where x is N(10,25).

```
DATA EXAMPLE;
 P1 = PROBNORM((8-10)/5);
 P2 = PROBNORM((13-10)/5);
 P = P2 - P1;
 PUT P;
RUN;

0.3811686
```

Problems

Practice

5.35 Find graphs of the probability distribution for each of the following binomial random variables. Determine whether the normal approximation is appropriate in each case. Compare with the rule that np and $n(1 - p)$ should both be 5.
a. x is B(30, 0.4)
b. x is B(30, 0.98)
c. x is B(10, 0.3)
d. x is B(50, 0.3)

Applications

5.36 Consider the reaction time data given in Problem 1.8. Use a histogram and the criteria for normality to determine whether the normality assumption is reasonable.

5.37 Consider the Miami Female Study data given in Appendix B.1. For the variable height:

a. Use the histogram and the criteria for normality to determine whether it is reasonable to assume that height is normally distributed.

b. Obtain a normal probability plot for the height data and use it to check the normality assumption. Do the results check with those of part (*a*)?

5.38 Use Minitab to obtain a normal probability plot for the pollution data of Example 5.15. Does the graph resemble a straight line? What does this shape suggest?

5.39 Use the SAS statement PROBNORM to obtain the following probabilities, and compare them with the result using Table 4:

a. $P(210.7 < x < 235.8)$ if x is N(215, 25)

b. $P(34.2 < x < 66.8)$ if x is N(51.5, 60.3)

5.40 We can use the computer to generate a random sample of observations based on a normal random variable with a specified mean and standard deviation. With Minitab the appropriate commands are as follows:

```
MTB > RANDOM K1 normal observations in each column C, ... C ;
SUBC> NORMAL mean = K2 standard deviation = K3.
```

where K1 is the number of observations required in each column.

Using the SAS System, we find that the appropriate statement is RANNOR(*seed*). The statement

```
X = M + S*RANNOR(seed);
```

will produce a normal variate (single value of a normal variable) from a distribution with mean M and standard deviation S. We can use a do loop to obtain as many observations as we need. See previous chapters for information on the do loop (especially Section 4.7).

Use the computer to generate a sample of 100 observations of a normal random variable with mean 25 and standard deviation 5.

a. Obtain a histogram and apply the criteria for normality.

b. Obtain and interpret a normal probability plot.

c. Obtain the sample mean and standard deviation. Are they close to the true values?

d. What do you conclude from all this?

5.4 Case 5: The Dubious Decisions—Solution

The probability of a female name being selected is 0.4, since two of the names are female and three male. If selections are at random and independent of each other, the number of female names selected will be a binomial random variable with mean equal to 50(0.4), or 20, and standard deviation

$\sqrt{50(0.4)(0.6)}$, or about 3.464. Since the number of trials is 50, we can use the normal approximation to the binomial. We wish to determine the probability of 17 or fewer female names being selected by chance. Thus we want the area to the left of 17.5 under a normal curve for which the variable has a mean of 20 and a standard deviation of 3.464. The standard normal deviate is $(17.5 - 20)/3.464$, or about -0.72. From Table 4, we find that the probability of obtaining a value of z less than -0.72 is about 0.2358. Thus a result this extreme would happen, by chance, about one time in four. The result is not particularly unlikely to have happened by chance if no bias is present, and we must conclude that the result of the experiment does not support a charge of sexual bias. Statisticians are reluctant to conclude that an event did not happen by chance unless the probability that it did happen by chance is quite small. This topic is investigated in detail in Chapter 7. On the other hand, since the probability that the event did occur by chance was only one in four, the result does not fully support a contention that there was no bias. When neither viewpoint can be supported, the results of an experiment are often termed "inconclusive" and further investigation is often warranted.

5.5 Summary

A continuous random variable is a random variable that can take on any value in an interval. A continuous random variable x is defined over a sample space by a function $y = f(x)$. If we graph $y = f(x)$ over the interval in which x is defined, the area under the curve is equal to 1. If x_1 and x_2 are in the interval, the probability that x will be between x_1 and x_2, that is, $P(x_1 < x < x_2)$, is the area under the graph of $y = f(x)$ between x_1 and x_2.

When a continuous random variable is used to approximate a discrete random variable, we must use a continuity correction.

Among the most important and useful sets of continuous distributions in statistics is the set of normal distributions. If the probability distribution of a random variable is a normal distribution, we say that the random variable is normally distributed or that it is a normal random variable. The graph of a normal distribution is called a normal curve. A normal distribution is completely determined by the mean and variance of the variable. If x is a normal random variable with mean μ and variance σ^2, we symbolize this by writing "x is $N(\mu,\sigma^2)$." A particularly important normal distribution is the standardized normal distribution—the probability distribution of the standardized normal variable z, where z is $N(0,1)$.

The values of the standard normal deviate are used to enter Table 4. Table 4 gives the areas under the standardized normal curve between $z = 0$ and any value of z from zero to 3.59. These areas can be used to determine the probability that a normal random variable will lie between any two numbers.

An important use of the normal distribution is as an approximation to a binomial distribution.

It is frequently necessary to verify that a set of data comes from a normal distribution. We may investigate this by obtaining a normal probability plot and noting whether or not the graph is a straight line. If it is not, the data are probably not normal. We may also obtain a histogram and the mean and standard deviation and apply the criteria for normality.

Chapter Review Problems

Practice

5.41 A true-false test is answered by tossing a coin for each question and answering true if the result is a head and false if the result is a tail. The test consists of 100 questions. What is the probability of getting at least 60 correct answers?

5.42 A fair die is rolled 180 times and the number of dots on top is observed. What is the probability of getting:

a. exactly 20 sixes?
b. more than 40 threes?
c. at least 50 numbers divisible by 3?

Applications

5.43 On a certain IQ test, scores are approximately normally distributed with a mean of 100 and a standard deviation of 16. Assuming that the variable is continuous, determine:

a. the proportion of scores between 80 and 110
b. the proportion of scores above 130
c. the proportion of scores below 85
d. the score above which lie 25% of the scores
e. the score with a standard normal deviate of 1.12
f. the score with a standard normal deviate of -0.87
g. the probability that a score selected at random will be between 100 and 120
h. the probability that a score selected at random will be less than 110

5.44 According to *USA Today* (Jan. 21, 1986) the average annual expense for out-of-state students to attend a four-year public college or university was $5137 for 1985–1986. If 55% of such schools have annual expenses above $5000 and the distribution of expenses is random, what is the standard deviation of the distribution?

5.45 Scores on a memorization test are normally distributed with a standard deviation of 11.62. If 69.5% of the scores are above 70.34, what is the mean score for the test?

5.46 A machine that fills fruit juice cans can be set to fill the cans with a mean of from 44.80 to 48.20 ounces. The standard deviation of the fills will be 0.08 ounce regardless of the mean. The manufacturer wishes to comply with federal regulations requiring the stated net weight of the contents to be accurate. Since it is nearly impossible to be certain that *all* cans will contain at least 46 ounces, limits must be set on the acceptable number that do not.

a. If the manufacturer wants no more than 100 of every 10,000 cans to contain less than 46 ounces, what should be set as the mean fill?
b. What should be set as the mean fill if no more than 100 of every 100,000 cans should contain less than 46 ounces?

5.47 Chitwood and McBride (1985) reported that 27.3% of students surveyed in grades 10–12 had used marijuana at least once. Of these, 47% stated that they no longer used marijuana. If these percentages are correct as applied to a population, what is the probability that if 443 students reported that they had used marijuana at least once, more than half no longer use it?

5.48 On the average, 60% of the graduates of a certain high school go on to college. This year's graduating class has 150 students in it. Using past records as a criterion, what is the probability that fewer than 75 of the graduates will attend college? One hundred or more? Exactly 100?

5.49 An amateur gardener decides to put in a border of petunias along his driveway. He buys a package of mixed petunia seeds. According to the label, one-fifth of the seeds should grow into pink petunias and four-fifths into red and white variegated petunias. Assume that the seeds were selected at random from a population mixed in this proportion and that the binomial model applies. If 200 seeds germinate, what is the probability that there will be at least 30 but not more than 50 pink petunias?

5.50 A company wishes to test market a higher-priced but more attractively packaged version of a product. It will be deemed a success if 30% of the customers who buy the product buy the higher-priced version. To test the marketability of the new version, both versions of the product are displayed prominently in a store and a careful record is kept of the number of sales of each version. Suppose that 53 of the first 200 sales are for the higher-priced version.
a. What is the probability of such an extreme result (53 or less, since the expected number of sales for $p = 0.30$ would be 60) if the actual proportion is 30%?
b. What is the probability of such an extreme result if actually 20% of the customers who buy the product will buy the higher-priced version?

5.51 A biologist needs a minimum of 10 specimens of a species of annelid with an abnormal alimentary dysfunction. On the average, about 20% of all members of the species possess this characteristic. If she goes on an expedition to collect specimens, it is time consuming to collect more than needed but even more time consuming to collect fewer, because then she would have to make another expedition. It is impractical to test each specimen for the dysfunction in the field. Since 10 is 20% of 50, one of her students suggests that she bring back 50 specimens.
a. If she does, what is the probability that she will not have enough with the specified dysfunction—that is, fewer than 10?
b. If she wants the probability that she will not have enough to be less than 0.01, how many should she bring back? (*Note:* This latter problem requires solution of a quadratic equation.)

5.52 A bacterial culture is subjected repeatedly to a treatment designed to reduce the bacteria count. The average count is 76 with a standard deviation of 16. Twelve percent of the cultures had counts below the viable stage. What is the lowest count possible to sustain the viable stage?

Index of Terms

Glossary of Symbols

N(0,1)	normally distributed variable with mean zero and variance 1
$N(\mu,\sigma^2)$	normally distributed variable with mean μ and variance σ^2
z	standardized normal variable; standard normal deviate
z_a	value of z for which $P(z > z_a) = a$

Answers to Proficiency Checks

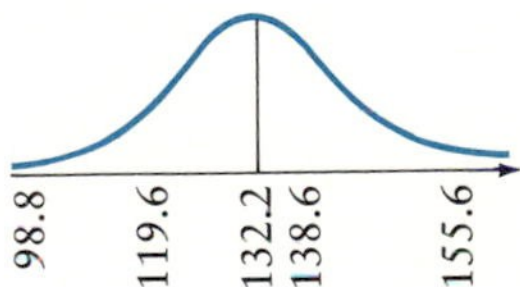

1. a. 2.05 b. −2.93 c. −1.11 d. 0.56
2. a. 0.33 b. 0.89 c. 0.56 d. 1.07
3. $\sigma = (42.1 - 31.7)/1.3 = 8.0$
4. $\mu = 11.3 - (2.1)(11.7) \doteq -13.3$

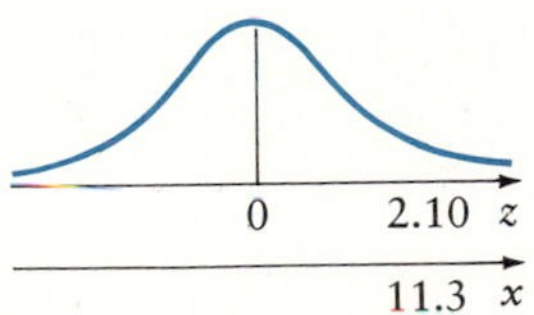

5. a. 0.4082 b. 0.0987 c. 0.4979 d. 0.4732
6. a. 1.36 b. 0.67 c. 2.20 d. 0.46
7. a. 0.4394 b. 0.3238 c. 0.6900 d. 0.1028
8. a. 0.2236 b. 0.0359 c. 0.8944 d. 0.9608

9. a. $P(120 < x < 145) = 0.4938$
   b. $P(100 < x < 130) = 0.4771 + 0.3413 = 0.8184$
   c. $P(x > 125) = 0.5000 - 0.1915 = 0.3085$

10. a. $P(x > 75) = 0.5000 - 0.4987 = 0.0013$
    b. $P(55 < x < 72) = 0.3413 + 0.4918 = 0.8331$
    c. $P(x < 65) = 0.3413 + 0.5000 = 0.8413$

11. a. $z \doteq 1.10$, so $P(x > 17) \doteq 0.5000 - 0.3643 = 0.1357 \doteq 0.136$
    b. $P(x \geqslant 17) = P(x > 17) = 0.136$.
    c. One ohm is 1/1.75—about 0.57 standard deviation from the mean. Thus the question is really asking "what is the area under the standardized normal curve between $z = -0.57$ and $z = 0.57$?" From Table 4, we find the answer to be 2(0.2157) or 0.4314. If x is the resistance of the resistor, we may write $P(|x - 15.08| < 1) \doteq 0.431$.
    d. 0.6293

12. a. $(70 - 50)/10 = 2.00$ and $(35 - 50)/10 = -1.5$. Thus the probability is $0.4332 + 0.4772$, or 0.9104.
    b. About 910.

13. $z = (19.5 - 30)/5.4 \doteq -1.94$. Since $P(x < 20) \doteq 0.0262$, we would conclude that this has less than a 3% chance of happening if the radiation had no effect.

14. $P(6 < x < 10) \doteq 0.2881 + 0.3849 = 0.6730 \doteq 0.673$; $P(3) \doteq 0.4772 - 0.4452 \doteq 0.032$

15. $z \doteq -2.96$; $P(x < 45) \doteq 0.0015$

16. $z \doteq 1.47$; $P(x \geqslant 27) \doteq 0.071$

References

Chitwood, Dale D., and Duane C. McBride. "The Cessation of Marijuana Use." *Florida Journal of Anthropology* 10 (1985): 33–47.

McClave, James T., and Frank H. Dietrich II. *Statistics*. 3rd ed. San Francisco: Dellen Publishing Co., 1985.

Mendenhall, William. *Introduction to Probability and Statistics*. 6th ed. Boston: Duxbury Press, 1983.

6

Estimating Population Parameters

For the past several chapters we have been dealing with the characteristics of a complete statistical population—that is, a set of data consisting of all possible or hypothetically possible observations of a certain phenomenon. Probability, in a broad sense, deals with using known population characteristics to obtain information about a sample or subset of the population. More often—perhaps most often—our data are incomplete. We know the characteristics of a sample but not the population from which it was drawn. The procedures we use to estimate the characteristics of a population from those of a sample drawn from it belong to the field of statistical inference.

In an earlier problem a team of marine biologists took samples in order to estimate the proportion of plankton in bloom. If a sample showed 40% of the plankton in bloom, this would not mean that 40% of all the plankton in the estuary were in bloom. The actual value could be more or less than 40%. It is important to know how close a sample statistic is to the population parameter it estimates. A sample statistic is an estimate for a population parameter, but only an estimate. Of greater use is a range of values—an interval that probably includes the population parameter. In this chapter we will discuss the general procedure for obtaining such an interval and then see how we can use this method to obtain interval estimates for a population mean, the proportion in a dichotomous population, and a population variance, each with a stated degree of certainty.

6.0 Case 6: The Ballpark Estimate

Some major league baseball players have bonus clauses built into their contracts. Bonuses are given for many different reasons, among them attendance at the ballpark. One ballplayer, especially popular with the public, was to receive a bonus based on attendance. Early in the season this athlete was tempted to make an investment on the basis of his expected bonus. He was not sure how much to expect, however. Since his team was a pennant contender this year, he could not use last year's attendance as a guide. He obtained the attendance for the eight home games that had been played thus far. The figures were 15,886, 17,364, 31,137, 23,319, 24,376, 32,192, 19,388, and 14,617. The player could multiply the mean attendance per game, 22,285, by the number of games to obtain an estimate for the season's total attendance. Although there were 81 home dates, some games would be rained out and rescheduled as part of a doubleheader. Last season there were 75 playing dates, so he multiplied the mean by 75 to obtain 1,671,366 as an estimate for the season's attendance.

When he discussed his proposed venture with his accountant, however, the accountant pointed out that there was a 50% chance that the attendance would be lower than this, as well as a 50% chance that it would be higher. He also noted that the attendance figures were highly variable. What was needed, he said, was a rock-bottom figure—a number that could be counted on. He said it was possible, though extremely unlikely, that *no one* would attend the rest of the games, so the player could only count on what had already been collected. The ballplayer agreed that this was so, but he wanted to be "darn sure" of knowing how much he could count on. After much discussion, the accountant finally got the ballplayer to define "darn sure." They agreed they should obtain a figure that was highly likely to be a conservative minimum for the season's attendance—so likely, in fact, that there was no more than 1 chance in 20 that the season's attendance would be lower. The accountant consulted a statistics text and soon had the information.

6.1 Samples and Sampling Distributions

The procedure of taking a sample and analyzing the results to make inferences about the population has certain built-in difficulties. Suppose, for example, a public-opinion research company is interested in determining whether or not the public thinks that Social Security benefits should be limited or reduced. If a researcher asks opinions at the annual meeting of the American Association of Retired Persons, the results are not likely to reflect the views of the public as a whole. Similarly, asking for opinions on certain political issues at a meeting of the Young Republicans or the Young Democrats will not necessarily reflect the opinions of the general public.

If we want to generalize characteristics of a sample to those of a population, our sample must represent the population. Two sources of error may affect the results of a sample and the extent to which it represents the population from which it is drawn. The first of these is called **systematic error.** This type of error is the result of incorrectly obtaining or analyzing data. Examples include weighing with a scale that weighs incorrectly, measuring with a ruler that is not accurate, or, on a questionnaire, asking ill-defined or misleading questions. If the sample does not represent the population it is supposed to represent—as in the examples concerning opinions on Social Security benefits or political issues—we cannot generalize the result to the population. Systematic error may also result from asking the wrong question. Asking "Are you a Republican or Democrat?" when we wish to know whether a person is a political liberal or conservative can yield quite misleading results, particularly in localities where practically everyone has registered as belonging to the same party, at least for the purpose of local elections. We can often avoid systematic error, but first we must be aware of the problem.

The other source of error is called **sampling error.** Sampling error is the difference between the sample statistic being measured and the corresponding population parameter. If we choose a sample in such a way that it represents the population, we can estimate the size of the sampling error by applying statistical theory. Most of the statistical theory developed to estimate population parameters is based on the assumption that the sample statistic was obtained from a random sample (as discussed in Section 3.3).

6.1.1 Sampling Distributions

Our usual purpose in obtaining a random sample is to make some sort of inference about the population. First we must define the variable of interest. For example, we might want to estimate the mean number of cars that pass a certain corner between 7 AM and 11 PM so that we can decide whether or not to build a service station on the corner. The random variable, then, is the number of cars passing the corner during these hours. The mean number of cars per day is the population parameter μ. With the permission of the highway department we install traffic counters and take daily readings for 10 weekdays with the following numbers of cars: 284, 386, 273, 308, 317, 281, 309, 290, 278, 271. These numbers have a mean of 299.7. Since this is a sample of the population we are interested in, we write $\bar{x} = 299.7$ and may then use this number as an estimate for μ, the mean number of cars passing this corner during the specified hours each weekday. On the other hand, if we take another sample of 10 days, or add more days to this sample, it is highly likely that the mean of the new sample will be some number other than 299.7.

Thus it is clear that means of different samples from the same population will vary. If the samples we obtain are random, however, statistical theory provides us with a way of estimating how closely the sample mean estimates the population mean. In fact, the method may be extended so that we can estimate other population parameters such as the standard deviation or the population proportion in a binomial distribution.

For each population parameter—mean, variance, proportion, and others to be defined later—there is at least one corresponding sample statistic. It is customary to denote a general population parameter by θ (theta) and a corresponding sample statistic by $\hat{\theta}$ ("theta-hat"). (We could write $\bar{x} = \hat{\mu}$, for example.) If we were to take all possible samples of the same size and calculate $\hat{\theta}$ for each one, we would be able to construct a probability distribution for $\hat{\theta}$. This probability distribution is called the **sampling distribution** of the statistic $\hat{\theta}$. Since the value of $\hat{\theta}$ differs from sample to sample and the actual value is determined by chance, the statistic is a random variable. The mean of this random variable is denoted $E(\hat{\theta})$ or $\mu_{\hat{\theta}}$, and the standard deviation of $\hat{\theta}$ is denoted $\sigma_{\hat{\theta}}$. The random variable $\hat{\theta}$ is said to be an **estimator** for the population parameter θ, and a single value of the sample statistic $\hat{\theta}$ is called a **point estimate** for θ. Note the distinction: the variable is an estimator, and a value of the statistic is an estimate.

The random variable $\hat{\theta}$ is a good estimator for θ if it possesses two characteristics. First, the expected value of the estimator should be the population parameter; that is, we would like $E(\hat{\theta}) = \theta$. If this is the case, then the random variable $\hat{\theta}$ is called an **unbiased estimator** for θ. A single value of the statistic $\hat{\theta}$ is then called an *unbiased estimate* of θ. It can be shown, for instance, that a sample mean is an unbiased estimator of the population mean; that is, for all samples of size *n*, if $\bar{x}$ is the sample mean and μ is the population mean, then $E(\bar{x}) = \mu$. If the mean of the estimator is not equal to the population parameter, the estimator is said to be **biased.**

Suppose there are two estimators for a population parameter θ. Call them $\hat{\theta}_1$ and $\hat{\theta}_2$. Suppose also that $E(\hat{\theta}_1) = \theta$ and $E(\hat{\theta}_2) < \theta$. The sampling distributions of the two statistics might be as shown in Figure 6.1. As we can see in the figure, most of the values of $\hat{\theta}_1$ are near $E(\hat{\theta}_1)$ and therefore near θ. The mean of the sampling distribution of $\hat{\theta}_1$ actually equals θ. Most of the values of $\hat{\theta}_2$ are near $E(\hat{\theta}_2)$ and therefore *not* so near θ. In fact, most values of $\hat{\theta}_2$ will overestimate the value of θ. Thus $\hat{\theta}_1$ is most likely preferred as an estimator for θ. Note that if the sampling distribution were negatively skewed instead of positively skewed, most of the values of $\hat{\theta}_2$ would underestimate θ.

The second characteristic of a good estimator is relatively small variability of the sampling distribution. (This characteristic is called **efficiency** by statisticians.) If there is more than one unbiased estimator for a population parameter, we naturally want the one for which the point estimates come as close as possible to the parameter. Thus the less variable the estimator, the better it is for purposes of estimation. Consider, for example, three estimators for θ ($\hat{\theta}_1$, $\hat{\theta}_2$, and $\hat{\theta}_3$), all with expected value θ. Their graphs are shown in Figure 6.2.

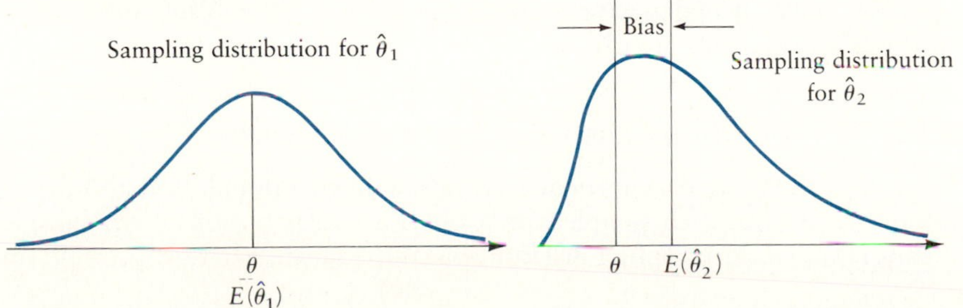

Figure 6.1 Sampling distributions for an unbiased estimator (left) and a biased estimator (right).

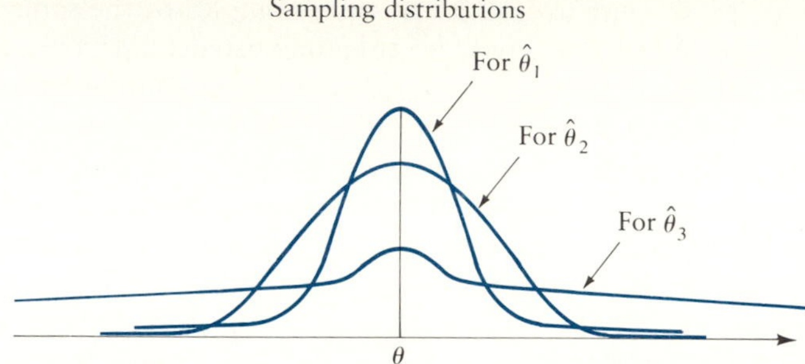

Figure 6.2
Sampling distributions for three unbiased estimators with different variances.

The variance of $\hat{\theta}_1$ is the smallest and the variance of $\hat{\theta}_3$ is the largest. Thus the values of $\hat{\theta}_1$ (the point estimates) will cluster more closely about θ than the values of $\hat{\theta}_2$ or $\hat{\theta}_3$. Since $\hat{\theta}_1$ has the smallest variance of the three estimators for θ, it is the most efficient of the three and therefore the best estimator. It should be mentioned here that the standard deviation of any estimator is called the **standard error** of the estimator and is symbolized $\sigma_{\hat{\theta}}$, where $\hat{\theta}$ is the estimator. Thus the standard error of the mean is $\sigma_{\bar{x}}$, the standard error of standard deviation is σ_s, and so forth.

One of the primary purposes of statistical inference is to estimate population parameters as closely as possible, and good estimators have been determined for most parameters of interest. In the remainder of the book we will use many sampling distributions—all of which provide an unbiased estimate for the parameter of interest and all of which have minimum variability.

Once we have determined a good estimator for a population parameter, we still need to determine the shape of the estimator's sampling distribution so that we can draw inferences from a point estimate. In many cases, as we will see, a sampling distribution is normal or close to normal. In Chapter 5 we investigated the normal distribution extensively. We can use the results of this investigation to make inferences about values of a normal random variable. If estimators for a population parameter are normally distributed, we can make inferences about the value of the parameter from a single point estimate.

Estimators for some population parameters do not possess normal sampling distributions; in many cases, however, they have distributions that have been investigated as extensively as the normal. We will encounter some of these distributions later in the book.

6.1.2 The Sampling Distribution of the Mean

If we take all possible samples of a given size from a population and determine the mean of each sample, the probability distribution of the sample means is called the **sampling distribution of the mean,** where $\bar{x}$ is a random variable with mean $\mu_{\bar{x}}$ and standard deviation $\sigma_{\bar{x}}$. Whether we do the sampling with or without replacement makes a difference in the value of $\sigma_{\bar{x}}$, as we will see. Sampling with replacement is analogous to sampling from a very large population. In most cases we assume that sampling was done from a very large population in order to simplify the calculations. The example presented here uses a small population solely for illustrative purposes.

Suppose we have a "population" consisting of six daily sales reports and the number of sales in the reports are 2, 3, 4, 5, 6, and 7. The variable x is the number of sales in a report. The mean and variance of the variable are given by the formulas $\mu = \Sigma x/N$ and $\sigma^2 = \Sigma(x - \mu)^2/N$. For this population $N = 6$, so $\mu = 27/6 = 4.5$ and $\sigma^2 = 17.5/6$ or about 2.9167.

We may use a sample of two reports to estimate the mean of the population. For simplicity, suppose this sampling was done with replacement. There are 6(6) or 36 different possibilities with means ranging from 2 to 7. These data are listed in Table 6.1 together with the mean of each.

A simple count of the number with each mean yields the probability distribution for $\bar{x}$, as presented in Table 6.2. This table gives us the sampling distribution of the mean of samples of size 2 drawn from this population.

Table 6.1 ALL POSSIBLE SAMPLES FOR $n = 2$, WITH THEIR MEANS.

NUMBER OF SALES			NUMBER OF SALES		
FIRST	SECOND	MEAN	FIRST	SECOND	MEAN
2	2	2.0	5	2	3.5
2	3	2.5	5	3	4.0
2	4	3.0	5	4	4.5
2	5	3.5	5	5	5.0
2	6	4.0	5	6	5.5
2	7	4.5	5	7	6.0
3	2	2.5	6	2	4.0
3	3	3.0	6	3	4.5
3	4	3.5	6	4	5.0
3	5	4.0	6	5	5.5
3	6	4.5	6	6	6.0
3	7	5.0	6	7	6.5
4	2	3.0	7	2	4.5
4	3	3.5	7	3	5.0
4	4	4.0	7	4	5.5
4	5	4.5	7	5	6.0
4	6	5.0	7	6	6.5
4	7	5.5	7	7	7.0

Table 6.2 PROBABILITY DISTRIBUTION OF SAMPLE MEANS FOR $n = 2$.

$\bar{x}$	$P(\bar{x})$
2.0	1/36
2.5	2/36
3.0	3/36
3.5	4/36
4.0	5/36
4.5	6/36
5.0	5/36
5.5	4/36
6.0	3/36
6.5	2/36
7.0	1/36

Table 6.3 PROBABILITY DISTRIBUTION OF SAMPLE MEANS FOR $n = 3$.

$\overline{x}$	$P(\overline{x})$
2.00	1/216
2.33	3/216
2.67	6/216
3.00	10/216
3.33	15/216
3.67	21/216
4.00	25/216
4.33	27/216
4.67	27/216
5.00	25/216
5.33	21/216
5.67	15/216
6.00	10/216
6.33	6/216
6.67	3/216
7.00	1/216

We can calculate the mean and variance of $\overline{x}$ for $n = 2$ from this table and find them to be the following: $\mu_{\overline{x}} = 4.5$ and $\sigma^2_{\overline{x}} \doteq 1.4583$.

If we take a sample of size 3, there are $6 \cdot 6 \cdot 6$ or 216 different samples. We can construct these samples systematically by using the results in Table 6.2. For example, the probability that a sample of two will have mean 2.0 is 1/36, so the probability that a sample of two will have a sum of 4 is 1/36. The probability that the third number will be 2 is 1/6, so the probability that the sum of three numbers will be 6 is 1/216. We may extend this process to determine all sample sums and their probabilities and then divide the sums by 3 to obtain sample means. The probability of a sample mean equal to 2.0 is 1/216, for example. Sample means range from 2.0 to 7.0, as before. The probability distribution of the means of samples of size 3 is given in Table 6.3. We can calculate the mean and variance of $\overline{x}$ for $n = 3$. We obtain $\mu_{\overline{x}} = 4.5$ and $\sigma^2_{\overline{x}} \doteq 0.9722$.

Finally, suppose we take a sample of size 4, again with replacement. Sample means again range from 2.0 to 7.0; the probability distribution is shown in Table 6.4. The mean and and variance of $\overline{x}$ for $n = 4$ turn out to be $\mu_{\overline{x}} = 4.5$ and $\sigma^2_{\overline{x}} \doteq 0.7292$. We summarize the findings in Table 6.5. If $n = 1$, of course, this is the population. Now $\sigma^2 = 2.9167$, $2.9167/2 \doteq 1.4583$, $2.9167/3 \doteq 0.9722$, and $2.9167/4 \doteq 0.7292$. In each case $\mu_{\overline{x}} = \mu$ and $\sigma^2_{\overline{x}}$ is equal to the population variance divided by the sample size; that is, $\sigma^2_{\overline{x}} = (\sigma^2)/n$.

We mentioned before that the standard deviation of the sampling distribution of an estimator is called the standard error of the statistic. It can be shown that for samples of size n drawn from a population with mean μ and variance σ^2, the random variable $\overline{x}$ has mean μ and variance σ^2/n. Thus the **standard error of the mean,** $\sigma_{\overline{x}}$, is equal to $\sigma/\sqrt{n}$. If the population standard deviation σ is not known, we can obtain an estimate for $\sigma_{\overline{x}}$, by using a sample standard deviation in place of the population standard deviation. This estimate is called a **sample standard error of the mean** and is symbolized $s_{\overline{x}}$, where $s_{\overline{x}} = s/\sqrt{n}$.

In the closing paragraphs of Section 6.1.1, we noted that once we have determined a good estimator for a population parameter, we may draw inferences from a point estimate if we can determine the shape of the sampling distribution of the estimator. We have determined that the sample mean is a good estimator for the population mean. Thus any sample mean will give us a good estimate of the population mean. We know that the estimate is not perfect, however, and different samples will usually give different point estimates. Of more use than a single point estimate is an *interval estimate*—a range of values assuring us that the interval contains the parameter. To obtain an interval estimate, we must determine the shape of the distribution for the sample statistics.

Returning to the example of the population 2, 3, 4, 5, 6, 7, each with probability 1/6, Figure 6.3 shows the relative frequency polygon for this population. The mean is 4.5; the standard deviation is $\sqrt{2.9167}$, or about 1.70783.

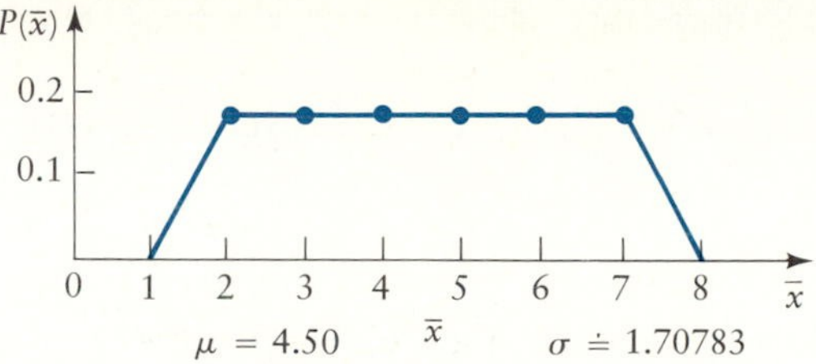

Figure 6.3
Relative frequency polygon for the population 2, 3, 4, 5, 6, 7.

Figures 6.4, 6.5, and 6.6 show the relative frequency polygons for the sampling distributions of $\bar{x}$ for $n = 2$, 3, and 4. Note that as n increases, each curve gets closer in appearance to a normal curve (discussed in Section 5.2), although the curve in Figure 6.6 still is a bit flatter than a normal curve. Figure 6.23 in Section 6.7 shows some computer-generated distributions for sample means of 50 and 1000 random samples of 5 and 50 observations drawn from this population. By the time $n = 50$, the sampling distribution of the mean approaches a normal curve very closely. We can list the characteristics of the sampling distribution of the mean found in this example as three properties of the sampling distribution of any mean.

Table 6.4 PROBABILITY DISTRIBUTION OF SAMPLE MEANS FOR $n = 4$.

$\bar{x}$	$P(\bar{x})$
2.00	1/1296
2.25	4/1296
2.50	10/1296
2.75	20/1296
3.00	35/1296
3.25	56/1296
3.50	80/1296
3.75	104/1296
4.00	125/1296
4.25	140/1296
4.50	146/1296
4.75	140/1296
5.00	125/1296
5.25	104/1296
5.50	80/1296
5.75	56/1296
6.00	35/1296
6.25	20/1296
6.50	10/1296
6.75	4/1296
7.00	1/1296

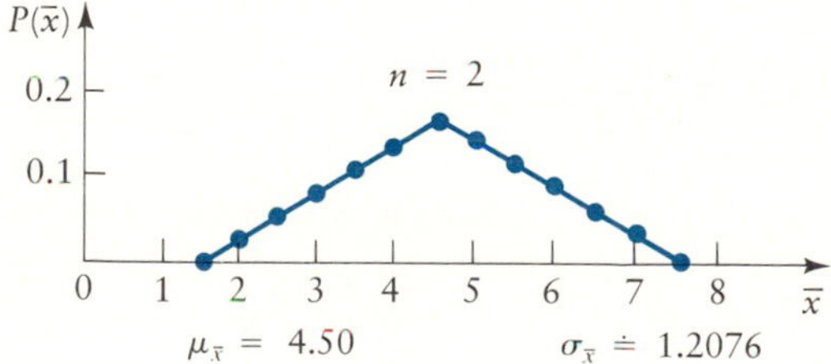

Figure 6.4
Relative frequency polygon of sample means for $n = 2$.

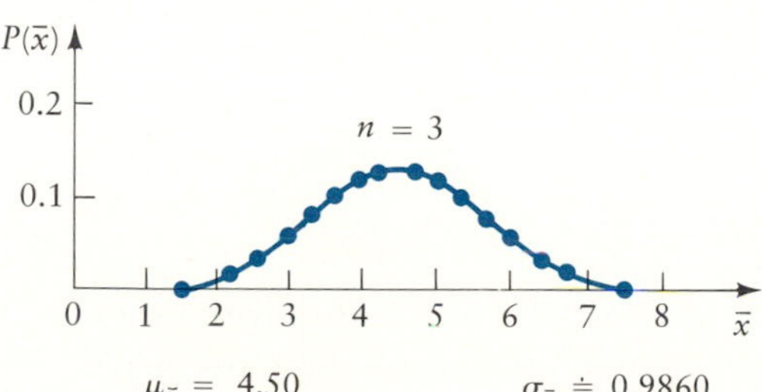

Figure 6.5
Relative frequency polygon of sample means for $n = 3$.

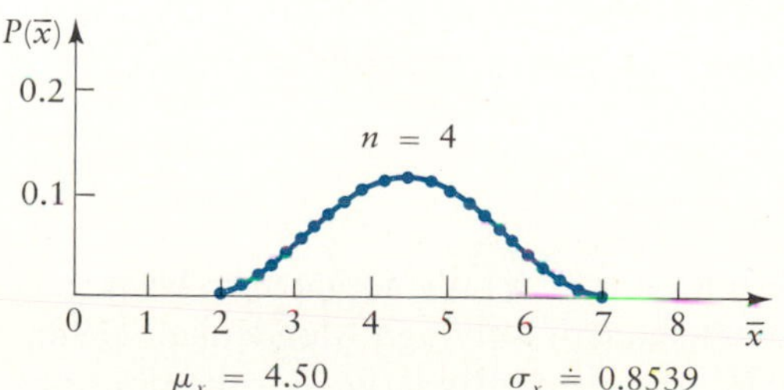

Figure 6.6
Relative frequency polygon of sample means for $n = 4$.

Table 6.5 SUMMARY OF RESULTS FOR $n = 1, 2, 3, 4$.

n	$\mu_{\bar{x}}$	$\sigma^2_{\bar{x}}$
1	4.5	2.9167
2	4.5	1.4583
3	4.5	0.9722
4	4.5	0.7292

Properties of the Sampling Distribution of the Mean

1. The mean of the sampling distribution of the mean is the population mean; that is, $\mu_{\bar{x}} = \mu$.
2. The standard deviation of the sampling distribution of the mean (the standard error of the mean) is equal to the population standard deviation divided by the square root of the sample size; that is, $\sigma_{\bar{x}} = \sigma/\sqrt{n}$.
3. The sampling distribution of the mean approaches the normal distribution for increasingly large samples; that is, as n increases, $\bar{x}$ approaches $N(\mu, \sigma^2/n)$.

Property 3 was illustrated in the previous example, but we will look at its implications in the next subsection.

6.1.3 The Central Limit Theorem

If the random variable x is normal, the random variable $\bar{x}$— the sample mean—will also be normal for any sample size n. If x is not normal, then as n gets larger and larger, the graph of the probability distribution for $\bar{x}$ approaches closer and closer to a normal curve. By the time n gets reasonably large, the agreement is very close. This is an illustration of one of the most powerful results of statistical theory—the **central limit theorem**— presented here as it applies to the sampling distribution of the mean.

The Central Limit Theorem

If we draw random samples of size n from a population with mean μ and standard deviation σ, the sampling distribution of sample means, $\bar{x}$, will be approximately normally distributed, provided that n is sufficiently large. The approximation of the sampling distribution of the mean to the normal distribution will be increasingly accurate as n increases.

"Sufficiently large" in this definition is generally accepted to be at least 30 for distributions that are mound-shaped, although Mendenhall (1983, p. 226) suggests that samples of 25 are sufficiently large. For distributions that are badly skewed, "sufficiently large" is probably closer to 50. If sampling is done without replacement and the sample is relatively large (more than 5% of the population is a common rule of thumb), we must modify the formula for $\sigma_{\bar{x}}$. This *finite population correction factor* is explained in Section 6.1.4.

Note that the central limit theorem makes no mention of the shape of the probability distribution of the population. In Section 6.7, we will see that the sampling distribution of the means of samples from normal, Poisson, exponential, and uniform distributions all obey the central limit theorem. The shape of the population is not important for the central limit theorem to apply, but for populations that vary considerably from mound-shaped, the samples need to be larger in order for n to be sufficiently large.

Now let us see how the central limit theorem applies to even a dichotomous population. Suppose that we draw a sample of eight chips from a large number of poker chips of which half are red and half blue. If we assign the red chips the value 5 and the blue chips the value 10, the theoretical probability distribution is based on a random variable that has the values 5 and 10, each with probability 0.5. The mean of this variable is 7.5 and the standard deviation is 2.5. If we assume sampling with replacement, the sampling distribution for the mean of samples of size 8 is given in Table 6.6 rounded to four decimal places.

Table 6.6
PROBABILITY DISTRIBUTION OF SAMPLE MEANS FOR $n = 8$.

$\bar{x}$	$P(\bar{x})$
5.000	0.0039
5.625	0.0312
6.250	0.1094
6.875	0.2188
7.500	0.2734
8.125	0.2188
8.750	0.1094
9.375	0.0312
10.000	0.0039

We compute the mean and standard deviation of $\bar{x}$, obtaining $\mu_{\bar{x}} = 7.5$ and $\sigma_{\bar{x}} = 2.5/\sqrt{8} \doteq 0.8839$. A graph of the distribution of Table 6.6 is shown in Figure 6.7. Although the graph is not normal, it is mound-shaped, showing that the tendency toward normality is there. In Section 6.7 there are several figures illustrating the validity of the central limit theorem. Figure 6.8 shows a computer-generated graph of the means of 1000 random samples of 50 observations each drawn from the poker-chip population of the last example. This distribution is not the theoretical sampling distribution of the mean, but it approaches a normal curve in appearance. The mean of these sample means is 7.5152, close to the population mean of 7.5, while the standard deviation of the sample means is about 0.3582. Since the population standard deviation is 2.5, we have $\sigma_{\bar{x}} = 2.5/\sqrt{50} \doteq 0.3536$. Thus the results agree quite closely with the theoretical values.

The true importance of the central limit theorem lies in its implications for statistical inference about population means. For sufficiently large sam-

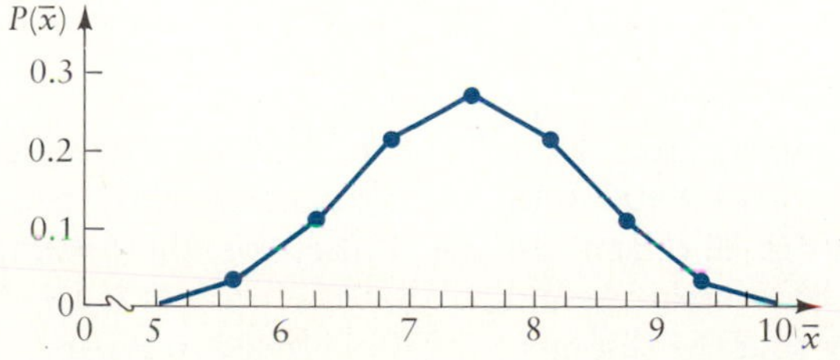

Figure 6.7
Graph of the distribution of Table 6.6.

```
                                                                                                                        FREQ  CUM.  PERCENT     CUM.
Means                                                                                                                         FREQ              PERCENT

6.2 TO 6.4                                                                                                               1     1     0.10     0.10
6.4 TO 6.6   *                                                                                                           3     4     0.30     0.40
6.6 TO 6.8   ****                                                                                                       11    15     1.10     1.50
6.8 TO 7.0   *************                                                                                              33    48     3.30     4.80
7.0 TO 7.2   **********************************************                                                            114   162    11.40    16.20
7.2 TO 7.4   ******************************************************************                                        165   327    16.50    32.70
7.4 TO 7.6   ***********************************************************************************                       207   534    20.70    53.40
7.6 TO 7.8   ************************************************************************************** 215                      749    21.50    74.90
7.8 TO 8.0   ****************************************************                                                      131   880    13.10    88.00
8.0 TO 8.2   **********************************                                                                         84   964     8.40    96.40
8.2 TO 8.4   **********                                                                                                 26   990     2.60    99.00
8.4 TO 8.6   ***                                                                                                         8   998     0.80    99.80
8.6 TO 8.8   *                                                                                                           2  1000     0.20   100.00
             ----+---+---+---+---+---+---+---+---+---+---+---+---+---+---+---+---+---+---+---+---+--
                 1   2   3   4   5   6   7   8   9  10  11  12  13  14  15  16  17  18  19  20  21

                                          PERCENTAGE
```

Figure 6.8
SAS printout of graph of distribution of means of 1000 samples of 50 observations from the binomial poker-chip distribution.

ple sizes, the sampling distribution of the mean tends to be approximately normal with mean $\mu_{\bar{x}} = \mu$ and standard deviation $\sigma_{\bar{x}} = \sigma/\sqrt{n}$. This means that the variable $z = (\bar{x} - \mu)/\sigma_{\bar{x}}$ is a standardized normal variable; that is, if $\bar{x}$ is $N(\mu, \sigma_{\bar{x}}^2)$, then $(\bar{x} - \mu)/\sigma_{\bar{x}}$ is $N(0, 1)$. Thus we can use the areas under the standardized normal curve to find the probability that the population mean lies within a specified range of values. The following two examples apply this principle, and in Section 6.2 we will use it to obtain interval estimates for the population mean.

Example 6.1

A random variable has a mean of 50.0 and a standard deviation of 12.0. A random sample of 36 is selected from the population. What is the probability that the mean of the sample is greater than 52.0?

Solution

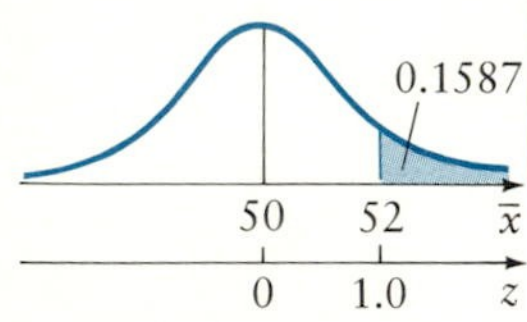

Figure 6.9
Standardized normal curve for Example 6.1.

According to the central limit theorem, the sample means will be approximately normally distributed with mean 50.0 and standard error $\sigma_{\bar{x}} = 12/\sqrt{36} = 2.0$. A drawing of a normal curve with all pertinent information marked is very helpful (see Figure 6.9). The standard normal deviate for $\bar{x} = 52$ is $z = (52 - 50)/2 = 1.00$. The area under the standardized normal curve to the right of $z = 1.00$ (from Table 4) is equal to $0.5000 - 0.3413$, or about 0.1587. Thus the probability that a sample mean will be greater than 52.0 is 0.1587.

Example 6.2

A sample of 100 pieces of copper tubing is examined for defects. If the process is in control, there will be a mean of 3.000 defects per tube with a standard deviation of 0.400. If the sample contains a mean of 3.100 or more defects, the entire shipment will be refused on the assumption that the process is out of control. Assuming that the process is in control, what is the probability that the shipment will be refused in error?

Solution

The shipment will be refused if the mean number of defects in the sample is greater than 3.100. We want the probability that a sample mean will be greater than 3.100 if the process is in control, that is, if $\mu = 3.000$ and $\sigma = 0.400$. Then $\sigma_{\bar{x}} = 0.400/\sqrt{100} = 0.040$ and the standard normal deviate for $\bar{x} = 3.100$ is $z = (3.100 - 3.000)/0.040 = 2.50$. From Table 4, we find $P(\bar{x} > 3.100) = 0.5000 - 0.4938 = 0.0062$, so the chance of rejecting a shipment by mistake is very small, just over 1 in 200.

PROFICIENCY CHECKS

1. A sample of 40 observations is drawn from a large population. The sample standard deviation is 134.8. Use this figure to obtain $s_{\bar{x}}$.
2. A certain brand of tire for a motorcycle has a mean life of 38,000 miles with a standard deviation of 1650 miles. What is the probability that the mean life for a random sample of 64 tires will be less than 37,500 miles?
3. The average IQ of students at a certain university is 125 with a standard deviation of 14. What is the probability that a random sample of 49 students will have a mean IQ greater than 128? What is the probability that the mean will be between 122.5 and 126.0?

6.1.4 The Finite Population Correction Factor

If sampling is done without replacement from a finite population, theoretically, the standard error of the mean is not equal to $\sigma/\sqrt{n}$. If we draw a sample of size n without replacement from a population of size N, the standard error of the mean will be

$$\sigma_{\bar{x}} = \frac{\sigma}{\sqrt{n}}\sqrt{\frac{N-n}{N-1}}$$

The expression $\sqrt{(N-n)/(N-1)}$ is called the **finite population correction factor (FPCF)**. If the sample size n is small in comparison to the population size N, the FPCF is very close to 1 and is usually omitted. As a rule of thumb, the difference is usually considered negligible and we omit the FPCF if n is less than 5% of N.

The finite population correction factor reduces the variance of the sampling distribution of the mean. This result increases the efficiency of the estimator and thus improves its value as an estimator.

Table 6.7 ALL POSSIBLE SAMPLES FOR $n = 4$, WITH THEIR MEANS.

SAMPLE	NUMBER OF SALES	$\bar{x}$
1	2, 3, 4, 5	3.50
2	2, 3, 4, 6	3.75
3	2, 3, 4, 7	4.00
4	2, 3, 5, 6	4.00
5	2, 3, 5, 7	4.25
6	2, 3, 6, 7	4.50
7	2, 4, 5, 6	4.25
8	2, 4, 5, 7	4.50
9	2, 4, 6, 7	4.75
10	2, 5, 6, 7	5.00
11	3, 4, 5, 6	4.50
12	3, 4, 5, 7	4.75
13	3, 4, 6, 7	5.00
14	3, 5, 6, 7	5.25
15	4, 5, 6, 7	5.50

To illustrate use of the FPCF, consider again the population of daily reports with sales of 2, 3, 4, 5, 6, and 7, and suppose we again take a sample of four from the population. This time, however, assume that the sampling is done without replacement. There are 15 different possible samples. We determine the mean of each sample of four as shown in Table 6.7.

The probability distribution of sample means is given in Table 6.8. The mean of $\bar{x}$ in Table 6.8 is 4.50, again equal to the population mean. We can calculate the standard deviation of $\bar{x}$ to be about 0.5401. Using the finite

Table 6.8 PROBABILITY DISTRIBUTION OF SAMPLE MEANS FOR $n = 4$.

$\bar{x}$	$P(x)$
3.50	1/15
3.75	1/15
4.00	2/15
4.25	2/15
4.50	3/15
4.75	2/15
5.00	2/15
5.25	1/15
5.50	1/15

PROFICIENCY CHECKS

4. A random variable x has variance $\sigma^2 = 34.8$. If the population is very large, determine $\sigma_{\bar{x}}$ for the sampling distribution of the mean of a sample of size n, where n is:
   a. 25 b. 50 c. 100 d. 400

5. Suppose that the population in Proficiency Check 4 consists of 5000 observations. Find $\sigma_{\bar{x}}$ for the sample sizes given using the finite population correction factor, and compare the result with the answers to Proficiency Check 4. Judging from the result, should you use the correction factor in most cases?

population correction factor, with $\sigma \doteq 1.7083$, we obtain $\sigma_{\bar{x}} \doteq (1.7083/\sqrt{4})[\sqrt{(6-4)/(6-1)}] \doteq 0.5401$. If we do not use the FPCF, we obtain $\sigma_{\bar{x}} \doteq 0.8542$, which would be the appropriate value of $\sigma_{\bar{x}}$ if we did sampling with replacement. If the sampling is done without replacement, however, the estimator would be more useful if we used the FPCF because the variability of the estimator would be less.

Problems

Practice

6.1 Suppose that a random variable has standard deviation 100.0. Determine the standard error of the mean for samples of size n equal to:

a. 100 b. 1000 c. 10,000 d. 36
e. 144 f. 64 g. 128 h. 1024

6.2 A population has mean 250.0 and standard deviation 40.0. If a sample of size 25 is randomly selected, what is the probability that the sample mean will be:

a. greater than 260.0?
b. less than 230.0?
c. between 248.0 and 255.0?

6.3 A random variable defined on a population of 10,000 observations has a mean of 187.0 with a standard deviation of 31.0. A sample of 100 is randomly obtained. What is the probability that the mean of the sample will be greater than 190.0 if:

a. you make no correction for finite population?
b. you use the finite population correction factor?

6.4 Repeat Problem 6.3 for a population size of 2000 and a sample size of 200.

6.5 A random sample of 100 is taken from a very large population for which the mean of the variable is 72.0 and the standard deviation is 8.0. What is the probability that the difference between the sample mean and the population mean will be less than 0.50? That is, what is the probability that the sample mean will lie between 71.5 and 72.5?

Applications

6.6 In 1985 the federal government required the states to lower the average speed of automobiles to certain limits or face the loss of federal funds. A traffic engineer obtains the following readings for 12 random observations on an interstate highway: 76, 55, 49, 55, 69, 71, 66, 63, 55, 62, 55, 78. Determine the mean, standard deviation, and sample standard error of the mean for this set of data.

6.7 A sample of 100 houseflies is taken from the swarm around a garbage dump. The bacterial count on the feet of the flies is found to have a mean of 10,000 per fly. The sample standard error of the mean is 156. What is the sample standard deviation for the bacterial counts?

6.8 A researcher reports that he has a sample of persons for whom the mean number of jobs held in the last 5 years is 7.6. He says the sample standard deviation is 3.7 and the sample standard error of the mean is 0.925. He neglects to report the sample size, however. From the information given, can you obtain the sample size? If so, what is it?

6.9 Ten years ago students at a university had a mean score of 18.3 on a manual dexterity test; the standard deviation was 2.4. Assume that these parameters are representative of the entire population at that time. A researcher wishes to discover whether there has been any change during the past 10 years. She tests 64 students (of the university's 3752) and obtains a mean score of 19.1 with a standard deviation of 2.2. What is the probability of obtaining a mean this high by chance if the true situation remains as it was 10 years ago?

6.10 A random sample of 40 aluminum stampings is checked each day to ensure that the manufacturing process is in control. A quality index is assigned to each stamping. If the manufacturing process is in control, a day's output of 6000 stampings can be expected to have a mean index of 80.32 with a standard deviation of 3.16. If the process is out of control—that is, if the sample mean is greater than 81.62 or less than 79.02—the manufacturer must begin retooling. What is the probability of retooling by mistake?

6.11 The daily wages of workers in a certain industry are normally distributed with a mean of $97 and a standard deviation of $13.20. If a random sample of size 36 is taken, what is the probability that the mean daily wage of the workers in the sample will be less than $92?

6.12 Suppose the mean weight of all sailors in the U.S. Navy is 172 pounds with a standard deviation of 15 pounds. Determine an interval, symmetric about the mean, within which 90% of all sample means of weights of 100 sailors can be expected to fall.

6.13 The International Coffee Organization (*Pensacola News Journal*, Feb. 13, 1986) reported that the average number of cups consumed per person by people over the age of 10 in the United States was 3.3 per day in 1985. Since some people do not drink coffee, the range can be assumed to be at least 6.6 cups and probably more. Suppose a reasonable estimate for the standard deviation of the number of cups of coffee drunk by a person in one day is 1.8. What is the probability that a random sample of 100 people will drink an average of fewer than three cups per day?

6.14 An elevator bears a plate stating that the load limit is 2640 pounds and no more than 16 persons may occupy the elevator at one time. Assume that the weights of the population riding the elevator are normally distributed with a mean of 156.0 pounds and a standard deviation of 12.0 pounds. What is the probability of overloading the elevator with exactly 16 passengers?

6.15 The Titanic Motor Company runs a quality check weekly on precision parts. The first week in November, a sample of 25 parts is found to have a mean tolerance of 21.00 microns with a sample standard error of the mean of 0.38 micron. The second week in November, a sample of 25 is again taken but 4 are discarded. The remainder have a mean tolerance of 21.80 microns with a sample standard error of the mean equal to 0.25 micron. Determine the mean and standard error of the mean for the combined sample of 46 observations. (This problem is fairly difficult and requires recourse to basic formulas. The time spent with it can be rewarding, though, since research studies are often printed without raw data and it may be necessary to reconstruct some of the intermediate steps to acquire information about a set of data.)

6.2 Estimating the Population Mean

Frequently we wish to know the value of a population parameter but cannot calculate it directly, so we must estimate it. One of the simplest ways to do this is to take a random sample, calculate the statistic that corresponds to the population parameter we wish to estimate, and then use the sample statistic as a point estimate for the population parameter.

In Section 6.1 we attempted to ascertain whether or not to build a service station on a particular corner. We wanted to estimate the mean number of automobiles passing the corner between 7 AM and 11 PM. With the permission of the highway department we install traffic counters and take daily readings for 10 weekdays with the following numbers of cars: 284, 386, 273, 308, 317, 281, 309, 290, 278, 271. These numbers have a mean of 299.7. Since this set is a sample of the population we are interested in, we write $\bar{x} = 299.7$ and may then use this number as a point estimate for μ, the mean number of cars passing this corner during the specified hours each weekday.

Suppose we wish to estimate the average take-home pay of New York secretaries. If we take a survey of 100 secretaries, we may find that the mean take-home pay of this sample is \$285.40 per week with a standard deviation of \$65.40. Both of these numbers are point estimates for the corresponding population parameters. We also know that in both these cases another sample is likely to yield different point estimates.

If we obtain a point estimate $\bar{x}$ for a population mean μ, the difference between $\bar{x}$ and μ is called the **error of estimation**—the sampling error incurred. The absolute difference between $\bar{x}$ and μ is written $|\bar{x} - \mu|$—the absolute value of $(\bar{x} - \mu)$, the positive difference between the two. Recall that this is equal to $(\bar{x} - \mu)$ if $\bar{x}$ is greater than μ and it is equal to $(\mu - \bar{x})$ if $\bar{x}$ is less than μ.

Suppose we obtain the mean IQ score of a random sample of 100 doctors. Suppose that the random variable is known to have a standard deviation of 15. Finally, suppose we wish to know the probability that the error of estimation is less than 3; that is, we want to find the value of $P(|\bar{x} - \mu| < 3)$. The sampling distribution of the mean will have a mean equal to μ and a standard deviation $\sigma_{\bar{x}}$ equal to $15/\sqrt{100} = 1.5$. The graph of the distribution is normal and is shown in Figure 6.10. Since $3 = 2\sigma_{\bar{x}}$, it follows that we are looking for the probability that the sample mean IQ,

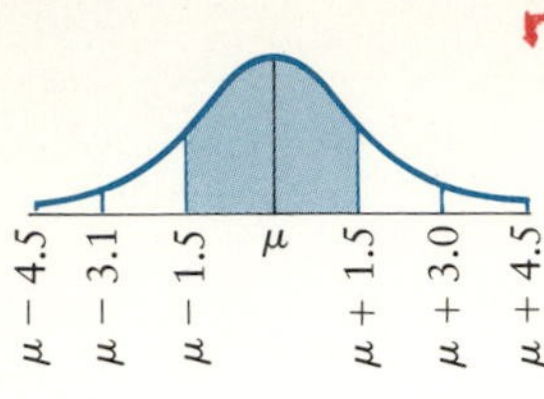

Figure 6.10
Graph of the distribution of IQ scores for 100 students.

$\overline{x}$, will lie between $\mu - 2\sigma_{\overline{x}}$ and $\mu + 2\sigma_{\overline{x}}$. The standard normal deviates for $\mu - 2\sigma_{\overline{x}}$ and $\mu + 2\sigma_{\overline{x}}$ are $z = -2$ and $z = 2$, respectively. From Table 4, we find that 0.4772 of the area under that curve is between zero and $z = 2.00$, so the area under the curve in Figure 6.10 between $\mu - 2\sigma_{\overline{x}}$ and $\mu + 2\sigma_{\overline{x}}$ is 0.9544. Thus the probability that the error is no more than 3 is 0.9544. We can say, then, that if $\overline{x}$ is the mean IQ of any sample of 100 chosen from this population, the probability that $\overline{x}$ is no more than 3 from μ is 0.9544. If the population mean IQ were 108.4, for example, the probability would be 0.9544 that the mean of a sample is between 108.4 − 3 and 108.4 + 3; that is, $P(105.4 < \overline{x} < 111.4) = 0.9544$.

6.2.1 Confidence Intervals

Neyman was a Romanian who taught in the Ukraine (1917–1921) and in Poland (1923–1934). He joined Pearson at University College, London, in 1934. In 1938 he moved to the University of California, Berkeley, where he had a distinguished career. Pearson, son of the famous mathematician Karl Pearson, joined the faculty of University College where his father was the leading figure. In 1933 he took one of the two positions created when his father retired—the other going to Ronald A. Fisher. In addition to the idea of confidence intervals, Neyman and Pearson formulated the hypothesis-testing procedure presented in Chapter 7, an idea suggested to them by William S. Gossett. Neyman and Pearson may be considered the fathers of a great deal of modern statistical methodology.

The ideas presented in the introduction to this section have led to the development of the concept of *confidence intervals*. Confidence intervals are based on a simple idea. Suppose that we sample a population repeatedly and obtain interval estimates for a population parameter from each sample. Some of the intervals will contain the parameter and some will not. The proportion of intervals that do contain the parameter is called the **confidence level** for each interval. This idea is the result of the work of Jerzy Neyman (1894–1981) and Egon S. Pearson (1895–1980).

If, in the example of Figure 6.10, we obtain a sample mean IQ of 102.3, we can add 3 to 102.3 and subtract 3 from 102.3 to obtain **confidence limits** for the population mean; that is, we are 0.9544 confident that the population mean is between 99.3 and 105.3. The interval 99.3 to 105.3 is called a **confidence interval** and 0.9544 is the **confidence level.***

We are saying, then, that we are 95.44% *confident* that the population mean IQ lies between 99.3 and 105.3; this is not a probability statement because μ is not a variable. We can make probability statements only about variables, not about constants.

Generally, we use a confidence level different from 95.44%. It would be more natural to use 95% than 95.44%, for example. If we want a 95% confidence interval, we wish to have 0.9500 of the area under the curve between the confidence limits so that 0.025 is in each tail. Since $z_{.025} = 1.96$, a 95% confidence interval would be found in the preceding example by adding and subtracting $1.96\sigma_{\overline{x}}$ rather than $2\sigma_{\overline{x}}$. Since $1.96(1.5) = 2.94$, if the sample mean were 102.3, our confidence limits would be 102.3 − 2.94, or 99.36, and 102.3 + 2.94, or 105.24. Our 95% confidence interval, then, is 99.36 to 105.24 and we are 95% confident that μ lies in the interval.

*The term *confidence coefficient* is also used interchangeably with confidence level. A few authors express confidence level as a percentage (95.44%) and confidence coefficient as a probability (0.9544).

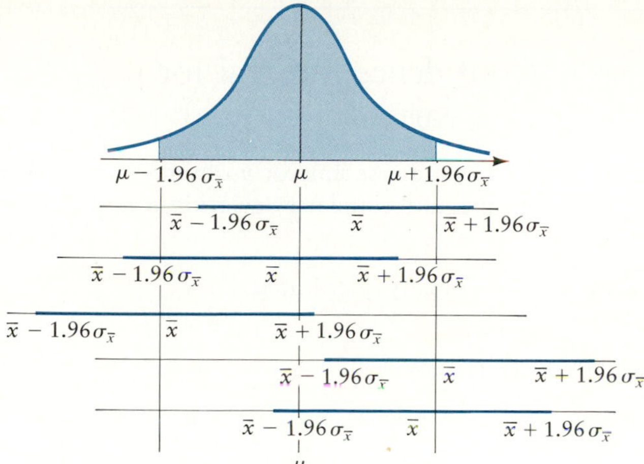

Figure 6.11
Five 95% confidence intervals for the population mean μ.

This interval is an example of a Neyman–Pearson confidence interval. If we take all possible random samples of the same size from a population with mean μ and standard deviation σ and then add and subtract $1.96\sigma_{\bar{x}}$ to each value of $\bar{x}$, we will obtain a very large set of intervals. If the sample mean is within 1.96 standard errors of the population mean, the interval will contain the population mean; if not, the interval will not contain the population mean. A few such intervals are shown in Figure 6.11. We can see from the figure that if $\bar{x}$ lies within $1.96\sigma_{\bar{x}}$ of μ, the confidence interval centered at $\bar{x}$ will contain μ, as in four of the five intervals shown. Since 95% of all sample means lie within 1.96 standard errors of the population mean, it follows that 95% of all confidence intervals constructed in this way will contain the population mean.

A meaningful probability statement, then, is the following: if C is the set of all 95% confidence intervals containing the population mean, and if c is a 95% confidence interval constructed from a random sample, the probability that c is a member of C is 0.95. In other words, 95% of the confidence intervals "capture" μ while 5% do not—providing, of course, all the samples are the same size.

We need not necessarily use 95% for our confidence level. The most commonly used confidence levels are 90%, 95%, 99%, and sometimes 98%. The total area in both tails is usually designated α (alpha), so that $\alpha / 2$ is in each tail. The standard normal deviate for which $\alpha/2$ is in the tail is $z_{\alpha/2}$. Thus the endpoints for the confidence interval (the confidence limits) are $\bar{x} - z_{\alpha/2}\sigma_{\bar{x}}$ and $\bar{x} + z_{\alpha/2}\sigma_{\bar{x}}$. The confidence level for such a confidence interval, then, is the area under the curve between the limits, or $1 - \alpha$. Expressed as a percentage, as confidence levels usually are, we would write $100(1 - \alpha)\%$ or $(1 - \alpha)\ 100\%$. We then have the following characterization of a confidence interval.

Confidence Interval for μ (Variance Known)

Suppose we obtain a random sample of n observations of a random variable x with mean μ and standard deviation σ and that either of the following is true:

1. The variable x is normally distributed.

   or

2. The sample size, n, is at least 30.

Then a $100(1 - \alpha)\%$ confidence interval for μ is the interval

$$\bar{x} - z_{\alpha/2}\sigma_{\bar{x}} \quad \text{to} \quad \bar{x} + z_{\alpha/2}\sigma_{\bar{x}}$$

where

$$\sigma_{\bar{x}} = \frac{\sigma}{\sqrt{n}}$$

or

$$\sigma_{\bar{x}} = \frac{\sigma}{\sqrt{n}}\sqrt{\frac{N-n}{N-1}}$$

if n is more than 5% of the population size.

Note that if the sample is more than 5% of the population and the sampling is done without replacement, we should use the finite population correction factor in computing $\sigma_{\bar{x}}$. Moreover, if the distribution of x is not normal, the confidence interval is approximate. The approximation becomes more accurate as n increases. The endpoints of the confidence interval, $\bar{x} - z_{\alpha/2}\sigma_{\bar{x}}$ and $\bar{x} + z_{\alpha/2}\sigma_{\bar{x}}$, are the confidence limits, and we often write "the confidence limits are $\bar{x} \pm z_{\alpha/2}\sigma_{\bar{x}}$." We can also say that we are $100\,(1 - \alpha)\%$ confident that the statement

$$\bar{x} - z_{\alpha/2}\sigma_{\bar{x}} < \mu < \bar{x} + z_{\alpha/2}\sigma_{\bar{x}}$$

is true.

It can be verified that appropriate standard normal deviates (commonly called **critical values**) for the most widely used confidence levels are

$z_{.05} = 1.645$ for the 90% confidence level
$z_{.025} = 1.96$ for the 95% confidence level
$z_{.01} = 2.33$ for the 98% confidence level
$z_{.005} = 2.58$ for the 99% confidence level

6.2.2 Large-Sample Confidence Intervals (Unknown Population Variance)

In many cases, we do not know the population variance and thus cannot determine $\sigma_{\bar{x}}$ directly. We can obtain a satisfactory approximation to an exact confidence interval by replacing σ in the formula for $\sigma_{\bar{x}}$ by s, the sample standard deviation, thus obtaining the sample standard error $s_{\bar{x}}$ provided that the sample is sufficiently large. For a normally distributed population, "sufficiently large" is generally considered to be at least 30; for a distribution that is not known to be normally distributed, the sample should, as a rule, contain at least 50 in order for the approximation to be satisfactory. We can obtain exact confidence intervals if the population is normally distributed by using the t distribution, which is covered in Section 6.4.

Large-Sample Confidence Interval for μ (Variance Unknown)

Suppose we obtain a random sample of n observations of a random variable x with mean μ and that either of the following is true:

1. The variable x is normally distributed and the sample size, n, is at least 30.

   or

2. The sample size, n, is at least 50.

If $\bar{x}$ and s are the sample mean and standard deviation, an approximate $100(1 - \alpha)\%$ confidence interval for μ is the interval

$$\bar{x} - z_{\alpha/2}s_{\bar{x}} \quad \text{to} \quad \bar{x} + z_{\alpha/2}s_{\bar{x}}$$

where

$$s_{\bar{x}} = \frac{s}{\sqrt{n}}$$

or

$$s_{\bar{x}} = \frac{s}{\sqrt{n}}\sqrt{\frac{N - n}{N - 1}}$$

if n is more than 5% of the population size.

Example 6.3

A metallurgist wishes to determine the melting point of a new alloy. He takes 36 pieces of the alloy and records the melting point of each piece. The mean of the 36 numbers is 2356.0°C, and the standard deviation of the sample is 3.6°C. Obtain a 95% confidence interval for the true melting point of the alloy. What is the probability that he will be off no more than 1.0 degree?

Solution

Since each determination is of the same alloy, it is reasonable to suppose that the melting points of the pieces will be normally distributed. Thus we can find an approximate confidence interval by using $s_{\bar{x}} = 3.6/\sqrt{36} = 0.6$ and $z_{.025} = 1.96$. Since $z_{.025}s_{\bar{x}} = 1.96(0.6) = 1.176 \doteq 1.2$, we can be 95% confident that the melting point is within 1.2° of 2356.0° C—that is, between 2354.8°C and 2357.2°C. Note that we do not need the finite population correction factor since the population is theoretically infinite.

To determine the probability that the maximum error will be no more than 1.0 degree, we need to determine the standard normal deviate of a difference of 1 degree from the mean. Since $s_{\bar{x}} = 0.6$, we have $z = 1/0.6 \doteq 1.67$. From Table 4, the area between $z = 0$ and $z = 1.67$ is 0.4525, so the probability is about 0.9050 that 2356.0°C is no more than 1°C from the population mean—that is, from 1°C above the population mean to 1°C below the population mean.

Example 6.4

A sample of 40 loaves of bread is weighed and found to have a mean weight of 20.24 ounces with a standard deviation of 0.34 ounce. If the sample is a small portion of the daily output, determine 95% and 99% confidence intervals for the mean weight of the entire daily output.

Solution

We can be fairly certain that the distribution of loaf weights is normal since all loaves are baked to the same specification. We do not need the correction factor since the sample is a small portion of the daily output, so $s_{\bar{x}} = 0.34/\sqrt{40} \doteq 0.054$. For a 95% confidence interval we use $z_{.025} = 1.96$ and $1.96(0.054) \doteq 0.11$. The confidence limits are $20.24 - 0.11$ and $20.24 + 0.11$, so the 95% (approximate) confidence interval for the mean daily weight is 20.13 ounces to 20.35 ounces.

For a 99% confidence interval we use $z_{.005} = 2.58$; $2.58(0.054) \doteq 0.14$, so we are 99% certain that the mean weight is between 20.10 ounces and 20.38 ounces. Note that the more confident we wish to be, the wider the interval becomes.

Example 6.5

A sample of sixty-four 30-ampere fuses is found to have a mean peak load of 30.840 amperes with a standard deviation of 0.420 ampere. Give 90% and 98% confidence intervals for the mean peak load of the entire shipment of 1000 fuses from which this sample was randomly selected.

Solution

Since 64 is greater than 50, no assumption of normality is necessary, although it would apply in this case; 64 is greater than 5% of 1000, however, so we must use the finite population correction factor. Then

$$s_{\overline{x}} = \frac{0.420}{\sqrt{64}}\sqrt{\frac{1000-64}{1000-1}} = \frac{0.420}{8}\sqrt{\frac{936}{999}}$$
$$\doteq 0.0525(0.9680) \doteq 0.051$$

Since $z_{.05} = 1.645$, the maximum error with a probability of 90% is about 1.645(0.051), or 0.084, so the 90% confidence interval is 30.756 to 30.924. We have $z_{.01} = 2.33$ and $2.33(0.051) \doteq 0.119$, so the 98% confidence interval is 30.721 to 30.959.

PROFICIENCY CHECKS

6. If $\overline{x} = 100.00$ and $s = 30.00$, determine a 95% confidence interval for μ if n is:
   a. 50 b. 75 c. 100 d. 200

7. If $\overline{x} = 0.560$, $s = 0.122$, and $n = 60$, determine confidence intervals for μ with a confidence level of:
   a. 90% b. 95% c. 98% d. 99%

8. A sample consisting of 100 cultures of bacteria is given a special nutrient designed to stimulate growth. After 3 days the proportion of increase is measured. The mean increase is 1.94 (that is, 2.94 times as many bacteria in each culture on the average) with a standard deviation of 0.30. We wish to use this result to estimate the mean proportionate increase we can expect in all cultures of this type of bacteria with this nutrient.
   a. What is the probability that the mean increase is actually no more than 0.05 from 1.94?
   b. What is the maximum error with a probability of 99%?
   c. Give a 99% confidence interval for the true mean increase.

9. Suppose that the number of cultures of bacteria in Proficiency Check 8 is increased to 250. This time the mean proportionate increase is 1.92 with a standard deviation of 0.32.
   a. What is the probability that the mean increase is actually no more than 0.05 from 1.92?
   b. What is the maximum error with a probability of 99%?
   c. Give a 99% confidence interval for the true mean increase.

Problems

Practice

6.16 A sample of size n is drawn from a population of size N, and the result is used to obtain a 95% confidence interval for the population mean. Determine the width of the confidence interval if the population standard deviation is as given. If not given, N is very large.

a. $n = 100, \sigma = 13.4$
b. $n = 64, N = 500, \sigma = 103.4$
c. $n = 200, \sigma = 0.040$
d. $n = 400, N = 5000, \sigma = 8.63$
e. $n = 36, \sigma = 12.136$

6.17 A sample of size 100 is drawn from a very large population with known standard deviation of 11.60. If the mean of the sample is used to estimate the mean of the population, what is the probability that the estimate will differ from the population mean by no more than:

a. 1.64? b. 0.80? c. 3.81? d. 0.50? e. 2.45?

6.18 A random sample is drawn from a normally distributed population with a known variance of 39.69. Determine 95% confidence intervals for the population mean if the sample mean is 117.6 and the sample size is:

a. 10 b. 25 c. 100 d. 400

6.19 A random sample of size 50 is drawn from a population with known variance of 116.64. If the sample mean is 317.8, determine confidence intervals for the population mean if the confidence level is:

a. 90% b. 95% c. 98% d. 99%

6.20 A random sample of size 75 is drawn from a population. Determine 90% confidence intervals for the population mean if the sample mean and standard deviation are as given.

a. $\bar{x} = 12.3, s = 1.9$ b. $\bar{x} = 12.3, s = 10.6$
c. $\bar{x} = 121.7, s = 1.9$ d. $\bar{x} = 121.7, s = 10.6$

6.21 A random sample of size 60 is drawn from a population. If the sample mean and variance are 0.112 and 0.000169, respectively, determine confidence intervals for the population mean with a confidence level of:

a. 90% b. 95% c. 98% d. 99%

6.22 A sample of 36 observations drawn from a normally distributed population of about 1500 has the following values:

8.6, 11.3, 9.4, 6.3, 8.0, 9.7, 10.8, 9.4, 12.2, 5.4, 7.6, 10.2, 9.4, 7.7, 9.6, 6.9, 9.3, 6.8, 9.1, 11.4, 8.8, 9.3, 12.4, 10.2, 10.5, 8.7, 9.4, 8.3, 9.1, 9.5, 10.4, 7.2, 13.1, 8.8, 6.7, 11.1

What is the maximum error with a confidence level of 90% if we use the mean of the sample to estimate the population mean?

6.23 Using the data of Problem 6.22, what is the probability that the maximum error will be less than 0.8?

Applications

6.24 A random sample of 40 copper tubes is drawn from a shipment and the tubes are found to have a mean interior diameter of 2.080centimeters with a standard deviation of 0.030 centimeter. Determine 95% and 99% confidence intervals for the mean interior diameter of the copper tubes in the shipment:

a. if no finite population correction is made

b. if the shipment contains 500 copper tubes and you use the correction factor.

6.25 A sample of a hundred 10-pound sacks of sugar from a large shipment is weighed and found to have a mean weight of 9.83 pounds with a standard deviation of 0.70 pound. Obtain 90% and 98% confidence intervals for the mean weight of sacks in the shipment.

6.26 A random sample of 50 college seniors has a mean grade-point average (GPA) of 2.80 with a standard deviation of 0.30.

a. Obtain 95% and 99% confidence intervals for the mean GPA of the entire graduating class of 2654 students.

b. If students with GPAs of at least 3.5 graduate with honors, how many students would you estimate will graduate with honors?

c. Fifty members of a certain sorority are graduating. What is the probability that if they constitute a random sample of seniors, the mean GPA of the 50 students will be above 2.90?

6.27 A load of 2500 sacks of grain is inspected by weighing a random sample of 50 sacks and determining a 95% confidence interval for the mean weight of the sacks in the entire load. If the mean of the sample is less than 150 pounds and the confidence interval does not contain 150 pounds, the shipment will be rejected. On a particular day, the mean is 149.0 pounds with a standard deviation of 5.2 pounds. Will the load be rejected?

6.28 According to the American Medical Association (*Newsweek*, February 17, 1986), the average malpractice suit in 1983 was settled for \$72,243. To determine whether this figure is valid for his company, an insurance adjuster takes a random sample of 75 such claims his company has settled in the previous 6 months, obtaining an average of \$102,655 with a standard deviation of \$87,663. Determine a 95% confidence interval for the mean payment of all such claims in his company. Is \$72,243 a reasonable mean figure for his company at the present time? Why or why not?

6.29 A water company wishes to discover the mean water consumption for the month of July in all homes in a certain subdivision. There are 618 homes in the subdivision, and a random sample of 30 homes shows a mean consumption of 11,644 gallons with a standard deviation of 1206 gallons.

a. Without using the correction factor, obtain a 95% confidence interval for the mean water consumption of all homes.

b. Using the correction factor, obtain a 95% confidence interval for the mean water consumption of all homes.

c. What assumption is being made about the population? Do you think this assumption is reasonable? If not, what should be done?

6.30 A total of 400 castings is selected at random from a very large shipment and examined for flaws. The mean number of flaws is found to be 13.40 with a standard deviation of 3.60.

a. What can be said, with a confidence level of 98%, about the size of the error that will be incurred if 13.40 is used to estimate the mean number of flaws in the castings of the entire shipment?

b. Suppose that management does not wish to accept the shipment if there is more than 1 chance in 100 that the mean number of flaws per casting is above 14.0. Should the shipment be rejected?

6.31 A psychologist has doubts about the standardization technique used on a new test, so he decides to make his own determination of the probable mean score on the test. He gives the test to 200 subjects and finds that the mean for the sample is 118.4. He is willing to use the published standard deviation of 22.6 for his computations, but he is willing to take only a 1% chance of being wrong. What is the maximum error in using 118.4 as the mean for the test with a 99% confidence level?

6.32 Income data for a set of 100 incomes, chosen at random from a large population, show a mean family income of $26,443 with a standard deviation of $3762.

a. Using these figures, construct a 95% confidence interval for the population mean.

b. Suppose that the standard deviation of the population is actually $3000. What would be the 95% confidence interval using this value for σ?

c. Which of the two is more likely to be accurate? Why?

6.33 Suppose an environmentalist group claims that the average American family produces 4.8 pounds of organic garbage per day. To test this claim, a public health official constructs a 95% confidence interval for the population mean from a random sample of 50 families for which the mean weight of organic garbage is 4.43 pounds with a standard deviation of 2.31 pounds. What conclusion should be drawn? Is the estimate reasonable, or does the sample result contradict it?

6.3 Estimating the Population Proportion

If sampling is done from a dichotomous population, we generally want to estimate the proportion of the population possessing the attribute of interest. A population proportion is a parameter, since the number of observations possessing the attribute is a constant. In keeping with the usual practice of using Greek letters to represent population parameters, we will represent the population proportion by the Greek letter π (pi). A sample proportion, then, should be represented by the corresponding roman letter, p. This p should not cause any confusion with probability of success in Chapter 4, since the usage is somewhat different.

Examples of population proportion include the proportion of apartments in a city that are vacant, the proportion of males who are color-blind, the proportion of voters who are in favor of an issue, and so on. Proportions are often given as percentages, but we must change them to decimal notation in order to perform the needed arithmetic.

6.3.1 The Standard Error of Proportion

If samples are small, there is no simple method to estimate a population proportion, but we know from Section 5.2.4 that a binomial distribution can be approximated by a normal distribution if both the expected number of successes and the expected number of failures are at least 5. Thus if the actual population proportion is π and the sample size in n, the number of "successes" in a sample (the random variable x, representing the number in the sample that possess the attribute) is approximately normally distributed with mean $n\pi$ and variance $n\pi(1 - \pi)$, provided that both $n\pi$ and $n(1 - \pi)$ are at least 5.

We know that if a variable x is $N(\mu, \sigma^2)$, the variable $(x - \mu)/\sigma$ is $N(0, 1)$. Thus if x is $N(n\pi, n\pi(1 - \pi))$, the variable

$$z = \frac{x - n\pi}{\sqrt{n\pi(1 - \pi)}}$$

is $N(0, 1)$; that is, it has a standardized normal distribution.

If p is the sample proportion, $p = x/n$, so $x = np$. Replacing x by np in the preceding expression for z and then dividing numerator and denominator by n, we have

$$z = \frac{np - n\pi}{\sqrt{n\pi(1 - \pi)}} = \frac{p - \pi}{\sqrt{\pi(1 - \pi)/n}} = \frac{p - \pi}{\sigma_p}$$

Thus sample proportions p from all samples of the same size are approximately normally distributed with mean $\mu_p = \pi$ and standard deviation $\sigma_p = \sqrt{\pi(1 - \pi)/n}$ for samples sufficiently large. That is, if $n\pi$ and $n(1 - \pi)$ are at least 5, the sample statistic p is $N(\pi, \sigma_p^2)$. The quantity σ_p is called the **standard error of proportion.** Since we are estimating π, it is not known, so we estimate σ_p by substituting the sample proportion p for π in the formula for σ_p to obtain the **sample standard error of proportion.** The sample standard error of proportion is symbolized s_p, and $s_p = \sqrt{p(1 - p)/n}$.

If the sample is large in comparison to the population—that is, if $n > 0.05N$—the binomial model does not exactly apply and we must use the hypergeometric model. In effect, we must apply a finite population correction factor. Recall that the variance of a hypergeometric random variable is

$$\sigma^2 = \frac{nAB(A + B - n)}{(A + B)^2(A + B - 1)} = \frac{nAB(N - n)}{N^2(N - 1)}$$

If $\pi = A/N$, then

$$\sigma^2 = \frac{n\pi(1 - \pi)(N - n)}{N - 1}$$

This is the correct variance of x, rather than the $n\pi(1 - \pi)$ we used in the derivation for σ_p above. If we replace $n\pi(1 - \pi)$ by σ^2 as derived above, we obtain

$$z = \frac{np - n\pi}{\sqrt{\dfrac{n\pi(1-\pi)(N-n)}{(N-1)}}} = \frac{p - \pi}{\sqrt{\dfrac{\pi(1-\pi)(N-n)}{n(N-1)}}}$$

so that

$$\sigma_p = \sqrt{\frac{\pi(1-\pi)(N-n)}{n(N-1)}}$$

and

$$s_p = \sqrt{\frac{p(1-p)(N-n)}{n(N-1)}} = \sqrt{\frac{p(1-p)}{n}}\sqrt{\frac{N-n}{N-1}}$$

if the sample is more than 5% of the population. We can see that this finite population correction factor, $\sqrt{(N-n)/(N-1)}$, is the same factor we sometimes use for interval estimation of the population mean.

6.3.2 Confidence Intervals for the Population Proportion

We may use the fact that sample proportions are approximately normally distributed for sufficiently large n to obtain confidence intervals for the population proportion, as we did for the population mean. Since we have to use s_p instead of σ_p, however, all such intervals are approximate. Again, the approximation becomes more nearly accurate as n increases.

The safeguards for the sample size n being greater than 30 or 50, as appropriate, do not apply here since the applicability of the normal approximation to the binomial is assured if np and $n(1 - p)$ are both at least 5.

Thus the confidence limits for a population proportion are $p \pm z_{\alpha/2}s_p$ and we can say that the statement

$$p - z_{\alpha/2}s_p < \pi < p + z_{\alpha/2}s_p$$

is true in approximately $100(1 - \alpha)\%$ of the confidence intervals so obtained. The appropriate values for $z_{\alpha/2}$ remain as discussed in Section 6.2.

Example 6.6

A sample of 100 fuses from a large shipment is found to have 10 defective. Construct 95% and 99% confidence intervals for the proportion of defectives in the shipment. Repeat for a shipment of 500 fuses.

Solution

In the first case, $n = 100$ and $p = 10/100 = 0.10$, so $s_p = \sqrt{(0.1)(0.9)/100} = 0.03$. For a 95% confidence interval, $z_{.025} = 1.96$ and $1.96(0.03) \doteq 0.06$, so the confidence limits will be $0.10 - 0.06$ and 0.10

Confidence Interval for π

Suppose we obtain a random sample of n observations from a dichotomous population with proportion π and that all of the following are true:

1. The sample proportion is p.
2. $np \geqslant 5$.
3. $n(1 - p) \geqslant 5$.

Then a $100(1 - \alpha)\%$ confidence interval for π is the interval

$$p - z_{\alpha/2}s_p \qquad \text{to} \qquad p + z_{\alpha/2}s_p$$

where

$$s_p = \sqrt{\frac{p(1-p)}{n}}$$

or

$$s_p = \sqrt{\frac{p(1-p)}{n}}\sqrt{\frac{N-n}{N-1}}$$

if n is more than 5% of the population size.

$+ 0.06$; thus the 95% confidence interval for π is 0.04 to 0.16. For a 99% confidence interval, $z_{.005} = 2.58$ and $2.58(0.03) \doteq 0.08$, so the confidence limits are 0.02 and 0.18 and the 99% confidence interval is 0.02 to 0.18. That is, we are 99% confident that π lies between 0.02 and 0.18.

If $N = 500$,

$$s_p = \sqrt{(0.1)(0.9)/100}\,\sqrt{(500 - 100)/(500 - 1)} \doteq 0.0269$$

Then $1.96(0.0269) \doteq 0.053$ and $2.58(0.069) \doteq 0.069$. Thus, to the nearest 0.01, the 95% confidence interval is 0.05 to 0.15 and the 99% confidence interval is 0.03 to 0.17. We see that the estimation is more efficient (the confidence intervals are narrower) if we incorporate the finite population correction factor.

Note that π cannot be less than zero or greater than 1.00. If by chance one of the confidence limits is either negative or greater than 1, common sense dictates that we replace it by 0 or 1. If we had a sample proportion equal to 0.35 and for a given confidence level the maximum error of estimation were 0.40, the upper confidence limit for π would be 0.75, but the lower limit would be zero.

A common error made by students is to obtain the correct value of the maximum error of estimate, but then add and subtract to x rather than p.

If we note that the limits must be in the interval from 0 to 1, this mistake can usually be avoided.

PROFICIENCY CHECKS

10. If a dichotomous population proportion is 0.7, determine σ_p to four decimal places if n is:
    a. 30 b. 60 c. 100 d. 400
11. Repeat Proficiency Check 10 for a population size of 5000, and use the finite population correction factor.
12. If 23 of a random sample of 64 workers in a plant support a potential strike, determine the sample standard error of proportion if the number of workers in the plant is very large.
13. Determine, to two decimal places, a 90% confidence interval for the proportion of workers in the plant sampled in Proficiency Check 12 who support a potential strike.
14. It is known that 64% of voters in the county are registered Democrats. A poll is taken among a random sample of voters to determine how many are planning to vote a straight Democratic ticket in the next election. Out of 400 registered Democrats who are polled, 80 plan to vote straight Democratic ticket, 170 plan to vote but not a straight ticket, and 150 do not plan to vote at all.
    a. Construct a 95% confidence interval for the proportion of all registered Democrats who plan on voting in the next election.
    b. Construct a 95% confidence interval for the proportion of all registered Democrats who plan to vote a straight ticket.

Problems

Practice

6.34 A sample of size n is drawn from a population of size N, and the result is used to obtain a 95% confidence interval for the population proportion. Determine the width of the confidence interval if the sample proportion is as given.

a. $n = 100, p = 0.44$
b. $n = 64, N = 500, p = 0.12$
c. $n = 200, p = 0.84$
d. $n = 400, N = 5000, p = 0.231$
e. $n = 36, p = 0.543$

6.35 A sample of size 100 is drawn from a very large population. The sample proportion is 0.60. If the sample proportion is used to estimate the population proportion, what is the probability that the estimate will differ from the population proportion by no more than:
a. 0.02? b. 0.03? c. 0.04? d. 0.06? e. 0.08?

6.36 If a sample proportion is 0.40, determine 95% confidence intervals for the population proportion if the sample size is:
a. 50 b. 100 c. 200 d. 400

6.37 A random sample of size 200 is drawn from a population. Determine 90% confidence intervals for the population proportion if the sample proportion is:
a. 0.15 b. 0.75 c. 0.10 d. 0.37

Applications

6.38 A poll of 408 Floridians over the age of 18 was conducted shortly after the explosion of the space shuttle *Challenger* (Gannett News Service, Feb 5, 1986). When asked whether they felt NASA's reputation had been severely damaged by the tragedy, 74 said yes, 318 said no, and the rest were undecided. Assuming that the sample was random, construct a 98% confidence interval for the true proportion of all Floridians who believed NASA's reputation was severely damaged by the tragedy.

6.39 A survey conducted by *Health* magazine (December 1985) concerned the issue of advertising by doctors. A total of 51% of the respondents thought that doctors who advertise have to charge more. Suppose a survey of 435 adults chosen at random in a city contains 198 who agree with this viewpoint. Obtain a 95% confidence interval for the proportion of all adults in the city who feel this way.

6.40 A sociologist interviews 120 of 650 families in an apartment complex and finds that 48 of them support certain pending legislation. Give 95% and 99% confidence intervals for the proportion of all families in the complex supporting the legislation.

6.41 To determine what proportion of people use brand X cough syrup, 500 people are questioned. It is found that 40 use brand X, 160 use some other brand, and 300 do not use cough syrup. Construct 95% confidence intervals for:
a. the proportion of the population who use brand X
b. the proportion of cough syrup users who use brand X

6.42 In a survey of lottery winners of $1 million or more (Institute for Socioeconomic Studies, February 4, 1986), 52 of 139 surveyed continued working. If there were actually 1213 such winners and the sample was random, construct a 95% confidence interval for the true percentage of all winners who continued working.

6.43 The Lifetime television network has a show entitled "The Dr. Ruth Show" which covers many candid topics some viewers may find offensive. Suppose that the management of a local cable company conducts a survey to gauge customer reaction to the program. A total of 863 people is interviewed. Of these, 244 saw the show and 27 of the 244 were offended by it. Determine a 90% confidence interval for:

a. the proportion of the population who watched the show
b. the proportion of viewers who were offended by the show

6.44 Refer to Problem 6.43. With what probability can we assert that the point estimate for the proportion of the population who viewed the show differs from the true proportion by no more than 0.04? With what probability can we assert that the point estimate for the proportion of viewers who were offended by the show differs from the true proportion by no more than 0.03?

6.4 Confidence Intervals from Small Samples

The preceding sections have shown how to find confidence intervals for the population mean or proportion if all the needed assumptions have been met or the samples are sufficiently large. If the samples are not large enough, we can use alternative methods to obtain confidence intervals provided that other assumptions are met.

If we want a confidence interval for the population proportion but cannot use the normal approximation to the binomial, we can use the binomial distribution itself. This procedure can be complicated, however, so you are advised to consult a mathematical statistician.

In Section 6.2 we learned how to obtain confidence intervals for the population mean. We saw that a $100(1 - \alpha)\%$ confidence interval for a population mean has limits $\overline{x} - z_{\alpha/2}\sigma_{\overline{x}}$ and $\overline{x} + z_{\alpha/2}\sigma_{\overline{x}}$ if the variable x is normally distributed or the sample size is at least 30, regardless of the shape of the distribution for x. We also noted that if the population standard deviation is unknown, we may substitute the sample standard deviation s for σ in the formula for $\sigma_{\overline{x}}$ to obtain $s_{\overline{x}}$. The approximate confidence interval $\overline{x} - z_{\alpha/2}s_{\overline{x}}$ and $\overline{x} + z_{\alpha/2}s_{\overline{x}}$ is generally a close approximation to the true confidence interval for n sufficiently large. The sample size n will be sufficiently large if x is normally distributed and $n \geqslant 30$ or if $n \geqslant 50$ regardless of the shape of the distribution for x. The approximation obtained is not very good for small sample sizes, however, and even for large samples the approximation is often unsatisfactory if the population is badly skewed.

It was known for many years that the approximate confidence intervals tend to be narrower than is justified so that the true confidence level for a given interval is somewhat less than the nominal level. A general method for determining exact confidence intervals using the sample standard deviation in place of σ is not yet known for nonnormal populations, but W. S. Gossett (1876–1937) developed a method for normal populations and published it in 1908.

Thus it becomes important to know whether or not a sample has come from a normal population. Criteria for normality were discussed in Section

5.3. There are also statistical tests for determining whether a data set is normal. We can use the chi-square procedure (discussed in Chapter 9) for this purpose, as well as the Kolmogorov–Smirnov Test, which is found in some texts. (See, for example, Kohler 1985, pp. 479–484.)

6.4.1 The *t* Distribution

William S. Gossett worked for the Guinness Brewery in Dublin and was responsible for quality control. He developed the theory of small-sample estimation (actually applicable to any normal population) so that he could use small samples in his work. His results were published under the pen name "Student" because the firm regarded his work as a trade secret and wished to keep it from competitors. For this reason the probability distribution he developed is sometimes called the *Student t distribution.*

Gossett showed that if the variable $z = (\bar{x} - \mu)/(\sigma/\sqrt{n})$ has a standardized normal distribution, the variable obtained by replacing σ with s, the standard deviation of a sample, has a similarly shaped distribution. This distribution is approximately bell-shaped but has a higher proportion of the area in the tails and is consequently somewhat flatter. He called the variable t, and defined t to be equal to $(\bar{x} - \mu)/(s/\sqrt{n})$. The probability distribution of this variable is called a ***t* distribution.**

The distribution of the variable t depends upon a parameter called **degrees of freedom,** symbolized by the Greek letter ν (nu). If ν is very large, the distribution of t is indistinguishable from a standardized normal distribution. It has different values in different situations and is loosely defined as the number of independent or "freely obtained" observations used to obtain a given quantity. For instance, to obtain the sample standard deviation we must first obtain the numerator $\Sigma(x - \bar{x})^2$. If we use a total of n observations of x to obtain the numerator, the first $n - 1$ of them are "free" to be any numbers we feel like picking, but the last must be the number we need to obtain the fixed value of $\bar{x}$. Thus the number of degrees of freedom associated with the quantity $\Sigma(x - \bar{x})^2$ is $n - 1$. Degrees of freedom for other quantities will be presented as needed.

Now a normal distribution is determined by two parameters—the mean and variance of the variable—whereas a t distribution is determined by a single parameter—the degrees of freedom of the variable. The random variable t has mean zero and variance $\nu/(\nu - 2)$, whereas the random variable z has mean zero and variance 1. Since $\nu/(\nu - 2)$ is always greater than 1, the variance of t is greater than the variance of z, so the t distribution is more variable than the standardized normal distribution (as evidenced by its flatter appearance). For 3 degrees of freedom, the variance is 3; for 4 degrees of freedom, the variance is 2. As ν increases, the variance decreases, approaching 1 as a limit for very large values of ν, so that the t distribution approaches a normal distribution in shape as ν gets larger and larger. For practical purposes, many statisticians consider the two distributions virtually the same for $\nu > 30$.

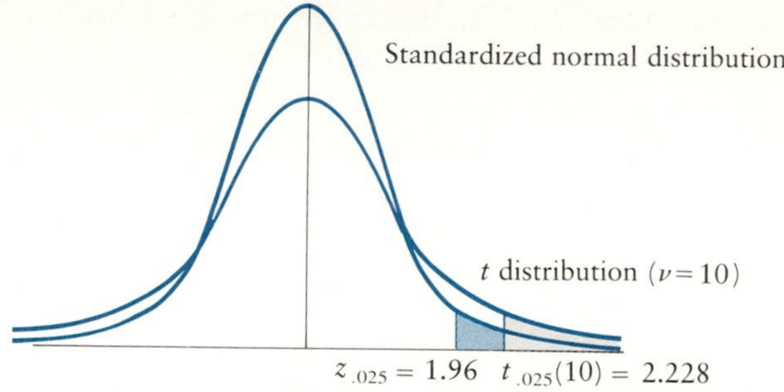

Figure 6.12
Curve of a *t* distribution with 10 degrees of freedom and the standardized normal curve.

Symbolically, we write $t(\nu)$ to indicate a t distribution with ν degrees of freedom. Thus if the variable $(\bar{x} - \mu)/(\sigma/\sqrt{n})$ is N(0, 1), the variable $(\bar{x} - \mu)/(s/\sqrt{n})$ is $t(n - 1)$.

We have one final bit of notation to learn. We use z_α to indicate the value of z, the standard normal deviate, for which the area under the standardized normal curve to the right of z_α is α. Similarly, we use $t_\alpha\{\nu\}$ to indicate the value of t for which the area under the graph of a t distribution with ν degrees of freedom to the right of $t_\alpha\{\nu\}$ is α. These numbers are called **critical values of *t*.** We use the braces for degrees of freedom in critical values rather than parentheses to avoid confusion. Numbers in braces are part of the expression of a critical value and never part of the mathematical term in which they are found. Critical values of t for $\alpha = 0.10, 0.05, 0.025, 0.01$, and 0.005 are listed in Table 5 and on the inside back cover. Critical values are different for each different value of ν and are listed for $\nu = 1$ to 30, 40, 60, 120, and ∞. The symbol ∞ means "infinity." We noted previously that as ν increases, $\nu/(\nu - 2)$ gets closer and closer to 1 so that $t(\infty)$ has mean zero and variance 1; thus $t(\infty)$ is N(0, 1), and $t_\alpha\{\infty\} = z_\alpha$. You can verify this by looking at Table 5. The numbers at the top of Table 5 give the area under the curve to the right of the critical value; the numbers at the side list the degrees of freedom.

The graph of a t distribution with 10 degrees of freedom is shown in Figure 6.12 together with a standardized normal curve. The critical values $z_{.025}$ and $t_{.025}\{10\}$ are shown in the figure. We know that $z_{.025} = 1.96$, and we find from Table 5 that $t_{.025}\{10\} = 2.228$. Thus the shaded area under each curve contains 2.5% of the area under the curve; 95% of the area under a standardized normal curve lies above the interval -1.96 to 1.96, while a wider interval (-2.228 to 2.228) is necessary for the graph of a t distribution with $\nu = 10$.

6.4.2 Confidence Intervals (Normal Population, Unknown Population Variance)

Since we know the sampling distribution of the statistic $t = (\bar{x} - \mu)/(s/\sqrt{n})$ if the distribution of x is normal, we can find exact confidence intervals for the population mean by substituting $t_{\alpha/2}\{n-1\}$ for $z_{\alpha/2}$ in the formula for approximate confidence intervals for μ in Section 6.2. We do not use the finite population correction factor for these intervals.

Confidence Interval for μ (Normal Population, Variance Unknown)

Suppose we obtain a random sample of n observations of a normal random variable x with mean μ. If $\overline{x}$ and s are the sample mean and standard deviation, $100(1 - \alpha)\%$ confidence interval for μ is the interval

$$\overline{x} - t\frac{s}{\sqrt{n}} \quad \text{to} \quad \overline{x} + t\frac{s}{\sqrt{n}}$$

where $t = t_{\alpha/2}\{n - 1\}$.

Note in particular that use of the t distribution to determine confidence intervals is not limited to small samples. The large-sample procedure (Section 6.2) for samples of at least 30 is purely arbitrary and, as noted, yields an approximate interval. As the sample size increases, the approximation improves. But if we desire complete accuracy, we should construct the exact confidence interval using the t distribution. We will follow customary practice in the remainder of this chapter, however, and when a problem asks for a confidence interval from a sample of at least 30, the solutions manual will list the approximate confidence interval unless specified otherwise.

If we need to know degrees of freedom above 30 not listed in Table 5, we can estimate approximate values of $t_{\alpha}\{\nu\}$. For example, $t_{.025}\{50\}$ is approximately halfway between $t_{.025}\{40\}$ and $t_{.025}\{60\}$; that is, $t_{.025}\{50\} \doteq 2.010$. Complete tables are available if needed. (See *Biometrika Tables for Statisticians,* Volume 1, Third Edition, Table 12, by Pearson and Hartley.)

Example 6.7

A sample of 25 observations randomly selected from a normally distributed population has a mean of 32.60 and a standard deviation of 1.30. What are the 95% and 99% confidence intervals for the population mean?

Solution

For a 95% confidence interval, we need $t_{.025}\{24\}$. Referring to Table 5, we look in the column headed 0.025 and in the row opposite 24 in the column headed "degrees of freedom." Here we find 2.064. Thus $t_{.025}\{24\} = 2.064$. Now $s/\sqrt{n} = 1.30/\sqrt{25} = 0.26$, so the maximum error of estimation (the term to be added to and subtracted from 32.60 to obtain the confidence limits) is $(2.064)(0.26) \doteq 0.54$. Thus the 95% confidence limits are

$32.60 - 0.54 = 32.06$ and $32.60 + 0.54 = 33.14$ and the 95% confidence interval is 32.06 to 33.14.

For a 99% confidence interval we need $t_{.005}\{24\} = 2.797$; then $(2.797)(0.26) \doteq 0.73$ and the 99% confidence interval is 31.87 to 33.33. Thus we can say that we are 95% confident that the statement $32.06 < \mu < 33.14$ is true and 99% confident that the statement $31.87 < \mu < 33.33$ is true.

Example 6.8

Archeologists and other scientists often determine the age of artifacts and animal bones by a method known as *carbon dating*. The radioactive element carbon-14 has a half-life of 5730 years. This means that half the carbon-14 originally present in the object will have become a more stable isotope in 5730 years. The proportion of carbon-14 remaining in the item to be dated can be used to estimate the amount of time the object has been in existence. The method is not 100% accurate, however, and several items from a population should be tested whenever possible.

Suppose a researcher wishes to estimate the age of a burial site. He takes six items found in the site and uses the carbon-14 method of dating. He proposes to use the mean of the results to estimate the true age of the site. The mean of the six numbers is found to be 2356, and the standard deviation is 360. Determine a 90% confidence interval for the probable age of the site.

Solution

Since all items came from the same site, we assume they are all the same age, so the distribution of age determinations is probably normal. Thus we may use the t distribution to obtain the confidence interval; $t_{.05}\{5\} = 2.015$, so $(2.015)s/\sqrt{n} = (2.015)(360)/\sqrt{6} \doteq 296.14$, or about 296. Thus the 90% confidence interval is about 2060 to 2652. Given that the method is not 100% accurate, the researcher is 90% certain that the site is about 2000 to 2700 years old.

PROFICIENCY CHECKS

15. If x is normally distributed, $\overline{x} = 100.00$, and $s = 30.00$, determine a 95% confidence interval for μ if n is:
    a. 5 b. 10 c. 20 d. 25

16. If x is normally distributed, $\overline{x} = 0.560$, $s = 0.122$, and $n = 15$, determine confidence intervals for μ with a confidence level of:
    a. 90% b. 95% c. 98% d. 99%

(continued)

PROFICIENCY CHECKS *(continued)*

17. If x is normally distributed, $\overline{x} = 8.66$, $s = 1.18$, and $n = 41$, determine a 95% confidence interval for μ using:
    a. the large-sample procedure (with $z_{.025}$)
    b. the exact procedure (with $t_{.025}\{40\}$)
18. Sixteen cars of the same model are driven by the same driver, as similarly as possible, over the same course using 1 gallon of gasoline. The mean distance driven is 20.3 miles with a standard deviation of 2.7 miles. Establish a 95% confidence interval for the mean miles per gallon for this model of automobile for the course and driver.
19. A pharmacist fills prescriptions for pills from a large bottle which is supposed to contain 1000 tablets. Unexpectedly, she runs short. She checks the next bottle she receives and finds that it contains only 984 pills. She asks several other pharmacists to check the number of pills in an unopened bottle and obtains counts of 991, 999, 1003, 989, and her own 984. Establish a 99% confidence interval for the mean number of pills per "full" bottle.

6.4.3 The Trimmed-Mean Confidence Interval*

If we do not know that the random variable is normally distributed, we cannot use the t distribution to obtain a confidence interval for the population mean. If the population variance is unknown, we can obtain approximate confidence intervals for samples of at least 50, but there is no known general method for obtaining confidence intervals for the population mean from samples less than 50 if the shape of the distribution for the population is not known.

The procedures discussed so far in this chapter are sensitive to deviations from normality. For smaller samples especially, lack of normality in a population reduces the efficiency of the estimators. Samples drawn from badly skewed populations may give rise to interval estimates that do not cover the estimated parameter in even a reasonable proportion of the intervals. Measures that do not give misleading results due to lack of normality are called **robust measures.**

For badly skewed populations, the best robust procedures are the nonparametric methods—those that do not depend upon a normal distribution.

*This section may be omitted without loss of continuity.

A few such procedures are given in Chapter 10, although confidence intervals are not discussed there. We may also use nonparametric procedures for long-tailed distributions, but other procedures are often better (Gross 1976; Koopmans 1981, pp. 233–238). The trimmed-mean interval estimate for the population mean given below is an example of a robust procedure.

If a nonnormal data set is symmetric but has long tails, we can use a method for small samples (less than 50) that is based on the 10% trimmed and Winsorized data sets discussed in Section 2.4. The use of a box plot is recommended to determine whether a symmetric distribution has long tails (signaled by the presence of outliers). We can use the trimmed mean, $\overline{x}(T)$, to estimate the population mean. In most cases the trimmed data set is sufficiently near to normal to allow us to use the methods of Section 6.2.2 (if the trimmed data set contains at least 30 observations) or those of Section 6.4.2 (if not).

Rather than using the standard deviation of the trimmed data set in place of s in the appropriate formulas, however, Koopmans (1981, pp. 235–236) gives a procedure that leads to a more nearly accurate confidence interval. We can illustrate the procedure by using an example.

Example 2.14 presented a data set consisting of 25 observations, and we computed the mean and standard deviation of the resulting trimmed data set. There were 25 pieces of data in the set, and we trimmed three data points off each end. In the following stem and leaf plot, the data points to be trimmed are underlined. We can see that the sample is reasonably symmetric with a slight positive skew.

2 | $\underline{9}$
3 | $\underline{3}$
3 | $\underline{7}$77888899
4 | 0002233
4 | 5557
5 | $\underline{0}$
5 | $\underline{9}$
6 |
6 | $\underline{6}$

When we remove the underlined values, the resulting trimmed data set has the following stem and leaf plot:

3 | 77888899
4 | 0002233
4 | 5557

The trimmed mean is $\overline{x}(T) \doteq 40.48$. Rather than use $s(T)$, however, we use the standard deviation of the Winsorized data set to obtain the appropriate standard error, $s_{\overline{x}(T)}$. Recall that we obtain the Winsorized data set from the trimmed data set by replacing the trimmed values with the value which would be trimmed next if one additional point were trimmed off each end.

In this case, the values would be 37 and 47. The stem and leaf plot for the Winsorized data set is shown here:

```
3|77777888899
4|0002233
4|5557777
```

The underlined values are those that were added.

We now obtain $s(W)$, the standard deviation of the Winsorized data set. In this case

$$s(W) = \sqrt{\frac{(42{,}600 - 1028^2/25)}{24}} \doteq 3.7005$$

We then use the following formula to find $s_{\bar{x}(T)}$, the **standard error of the trimmed mean,** where n is the number of observations in the original (and Winsorized) data set and h is the number of observations in the trimmed data set:

$$s_{\bar{x}(T)} = \frac{s(W)}{\sqrt{h}}\sqrt{\frac{n-1}{h-1}}$$

We use the resulting values of $\bar{x}(T)$ and $s_{\bar{x}(T)}$ in the usual way to obtain confidence intervals for μ.

Trimmed-Mean Confidence Interval for μ

Suppose we obtain a random sample of n observations of a random variable x with mean μ and the distribution of the variable is symmetric.

If $\bar{x}(T)$ is the mean of the trimmed data set consisting of h observations and $s(W)$ is the standard deviation of the Winsorized data set, then a trimmed-mean $100(1 - \alpha)\%$ confidence interval for μ is the interval

$$\bar{x}(T) - z_{\alpha/2}s_{\bar{x}(T)} \quad \text{to} \quad \bar{x}(T) + z_{\alpha/2}s_{\bar{x}(T)} \qquad \text{if } h \geq 30$$

or

$$\bar{x}(T) - t \cdot s_{\bar{x}(T)} \quad \text{to} \quad \bar{x}(T) + t \cdot s_{\bar{x}(T)} \qquad \text{if } h < 30$$

where $t = t_{\alpha/2}\{h - 1\}$ and

$$s_{\bar{x}(T)} = \frac{s(W)}{\sqrt{h}}\sqrt{\frac{n-1}{h-1}}$$

Example 6.9 Determine a 95% confidence interval for the population mean based on the data of Example 2.14 (shown earlier).

Solution

We have $\overline{x}(T) = 40.48$, $s(W) = 3.7005$, $n = 25$, and $h = 19$. Therefore

$$s_{\overline{x}(T)} = \frac{3.7005}{\sqrt{19}}\sqrt{\frac{24}{18}} \doteq 0.9803$$

and $t_{.025}\{18\} = 2.101$. Since $2.101(0.9803) \doteq 2.06$, the confidence interval is 38.42 to 42.54.

A 95% confidence interval for μ, obtained from the original data set, is 38.92 to 45.08. This interval is larger than the trimmed-mean confidence interval, and therefore less efficient, due to the influence of the outliers.

The trimmed-mean confidence interval is applicable only if the data set is reasonably symmetric. When the data set is badly skewed, nonparametric methods are usually preferred. For small nonnormal data sets, the trimmed-mean confidence interval is usually preferred, but if the data set is not reasonably symmetric and the population variance is known, we may use Chebyshev's theorem, first mentioned in Chapter 2. We will not discuss this method here.

PROFICIENCY CHECK

20. Obtain a trimmed-mean 90% confidence interval for the population mean based on the following data:

    32, 27, 28, 44, 29, 31, 33, 28, 61, 26, 27, 34, 5, 31, 28, 11

Problems

Practice

6.45 A sample of size n is obtained and the result is used to construct a 95% confidence interval for the population mean. Determine the width of the confidence interval if the sample standard deviation is as given.

a. $n = 10$, $s = 13.4$
b. $n = 24$, $s = 103.4$
c. $n = 20$, $s = 0.040$
d. $n = 41$, $s = 8.63$, using the exact procedure ($t_{.025}\{40\}$)
e. $n = 41$, $s = 8.63$, using the approximate procedure ($z_{.025}$)

6.46 A random sample of size 25 is drawn from a normally distributed population. Determine 90% confidence intervals for the population mean if the sample mean and standard deviation are as given.

a. $\overline{x} = 12.3, s = 1.9$ b. $\overline{x} = 12.3, s = 10.6$
c. $\overline{x} = 121.7, s = 1.9$ d. $\overline{x} = 121.7, s = 10.6$

6.47 A random sample of size 15 is drawn from a normally distributed population. If the sample mean and variance are 0.112 and 0.000169, respectively, determine confidence intervals for the population mean with a confidence level of:

a. 90% b. 95% c. 98% d. 99%

6.48 A sample of 12 observations drawn from a normally distributed population of about 1500 has the following values: 12.2, 9.4, 7.7, 9.1, 11.4, 8.8, 8.7, 9.4, 9.5, 13.1, 8.8, 11.1. What is the maximum error with a confidence level of 90% if you use the mean of the sample to estimate the population mean?

6.49 A random sample is drawn and the following values are obtained: 2, 14, 23, 26, 27, 28, 28, 31, 33, 35, 37, 58.

a. Make a box plot for the data. Are there any outliers?
b. Obtain a trimmed-mean 90% confidence interval for the population mean.
c. Obtain a 90% confidence interval for μ using the assumption that the population is normal.
d. Which of these two intervals is more efficient? Which do you think is better in this case?

Applications

6.50 A new alloy is subjected to nine determinations of hardness. As a result the mean value on the Moh scale of hardness is 0.630 with a standard deviation of 0.081. Obtain 95% and 99% confidence intervals for the true hardness of the alloy.

6.51 A superball is dropped and the height of the bounce is measured. The proportion of the original height to which the ball returns is called the *coefficient of restitution* for the substance. Four determinations of the coefficient of restitution for this superball yield values of 0.84, 0.78, 0.86, and 0.81. Give 95% and 99% confidence intervals for the true coefficient of restitution for this ball.

6.52 According to a University of Pennsylvania study (*American City & County*, December 1985), meetings that involve overhead graphics are 28% shorter than those in which these graphics are not used. To obtain data about their own company, a group of sales managers use overhead graphics in their meetings and keep records of the lengths of the meetings. The times (in minutes) for similar meetings are 78, 91, 57, 72, 66, 83, 67, 74, 62, and 58. Assuming normality, obtain a 90% confidence interval for the mean length of all such meetings.

6.53 In another company, the manager reads the study cited in Problem 6.52 and conducts a training program to educate personnel in the use of overhead graphics. Records are then kept of the length of time (in minutes) needed to conduct the meetings. They are as follows: 81, 43, 76, 69, 77, 64, 97, 71, 68, 73. Obtain a trimmed-mean 90% confidence interval for the mean length of all such meetings.

6.54 An office manager decides to undertake a study of telephone calls at her business office. For one day she times all incoming calls and outgoing calls. Assuming that the day's calls make up a random sample of calls (an assumption that may be open to argument), determine 95% confidence intervals for the length of incoming calls and for the length of outgoing calls if 17 incoming calls last an average of 5.16 minutes with a standard deviation of 1.12 minutes and 12 outgoing calls last an average of 4.13 minutes with a standard deviation of 2.36 minutes. (*Note:* If we determine more than one confidence interval on the same set of data, the confidence level of the entire collection of confidence intervals is somewhat less than the confidence level of any one interval. If we construct k confidence intervals, each with confidence level α, the confidence level of the set of intervals is no less than $1 - (1 - \alpha)^k$; that is, if we are 100 $(1 - \alpha)$% confident that any one of k confidence intervals is correct, we are $100[1 - (1 - \alpha)^k]$% confident that all of them are correct. This is an application of what is known as *Kimball's Inequality*.)

6.55 *Venture* magazine (November, 1985) featured a story about a device called the Truant. To assess the need for such a device, a school district superintendent obtains a random sample of daily absentee figures for the past semester. The numbers of suspected truancies on each of 28 days are as follows:

132, 117, 143, 114, 125, 133, 197, 134, 113, 143, 121, 108,
131, 109, 117, 116, 84, 102, 153, 116, 98, 122, 127, 113, 111,
65, 122, 114

Assuming that these were all truancies and using a 10% trimmed data set, construct a 95% confidence interval for the true mean daily truancies in the district. Discuss whether or not you needed the trimmed-mean approach.

6.56 Sociologists classify kin groups by number of persons living in the same household. A married couple with three children is a kin group of size 5. A person living alone is a kin group of size 1. Suppose that 20 households are examined and the mean kin-group size for the sample is found to be 3.14 with a standard deviation of 0.83. Construct a 90% confidence interval for the mean kin-group size of the population from which the sample is taken.

6.57 Suppose an environmentalist group claims that the average American family produces 4.8 pounds of organic garbage per day. To test this claim, a public health official constructs a 95% confidence interval for the population mean from a random sample of ten families for which the mean weight of organic garbage is 4.63 pounds with a standard deviation of 2.31 pounds. What conclusion should be drawn? Is the estimate reasonable, or does the sample result contradict it?

6.58 Birth weights of 20 babies born at a local hospital in 1985 are taken at random from the records. The data, measured to the nearest tenth of a pound, are reproduced here:

6.1, 6.6, 7.1, 3.1, 6.9, 7.3, 9.2, 6.4, 7.3, 6.3,
5.5, 1.8, 7.0, 6.9, 8.3, 10.3, 8.0, 6.8, 7.0, 6.9

Assuming normal distribution, estimate the mean weight of all babies born in this hospital in 1985 with a confidence level of 95%.

6.59 Refer to Problem 6.58. An analyst is not sure that the birth weights are normally distributed. Construct a box plot for the data, checking for outliers. Obtain the trimmed-mean 95% confidence interval and compare your result with the interval estimate in Problem 6.58. Discuss the implications of the results.

6.5 Estimating Needed Sample Size

6.5.1 Sample Size Needed to Estimate a Population Mean

Often we find it necessary to estimate a population parameter correctly within a stated maximum error. In Example 6.3, suppose the metallurgist wishes to obtain a 95% confidence interval for the melting point of the alloy that is no more than 2° wide rather than the 2.4° obtained. In other words, suppose he wishes to estimate the melting point correctly to within 1° with a confidence level of 95%. In either case, he is saying that he wants $1.96s_{\bar{x}} = 1$. Since he has an estimate for s from the prior experiment, he may use that. Thus he wants $1.96(3.6/\sqrt{n}) = 1$. Solving this equation for n, he has $\sqrt{n} = 1.96(3.6)/1$ so that $n = [1.96(3.6)]^2 = (7.056)^2 = 49.79$, or about 50. Thus he needs about 14 more determinations, provided that the standard deviation does not change.

In general, if we wish to place limits on the size of the confidence interval, we may use the known value of $z_{\alpha/2}$, the limits we wish to place, and our best estimate for σ to obtain the needed sample size. If E is the maximum desired error of estimation (E is one-half the width of the confidence interval), then we have $E = z_{\alpha/2}\sigma_{\bar{x}} = z_{\alpha/2}\sigma/\sqrt{n}$. Solving this expression for n, we obtain the following rule.

Estimating Needed Sample Size (Population Mean)

If we wish to estimate a population mean to within E with a confidence level of $(1 - \alpha)$ or to obtain a $100(1 - \alpha)\%$ confidence interval no more than $2E$ in width, the sample size we need is estimated to be n, where

$$n = \left(\frac{z_{\alpha/2} \cdot \sigma}{E}\right)^2$$

and σ is the population standard deviation. If σ is not known, we can estimate it from a previous result or conservatively estimate it to be approximately one-fourth of the range.

Example 6.10

A researcher wishes to poll a sample in order to get public opinion on a certain issue. He will ask the people in his sample to rate their confidence in the administration on a scale of 0 to 100. How many people should he poll in order to estimate the population mean within 5 units with a 90% confidence level?

Solution

The maximum allowable error is 5—that is, the 90% confidence interval will be no wider than 10—and $z_{.05} = 1.645$. Although we have no estimate for σ, the range is about 100, so we can estimate that σ will probably be no wider than 25. Then we estimate $n = [(1.645)(25)/5]^2$, which is about 67.65. Since n must be an integer, $n = 68$. Generally we round upward, even if the obtained value is closer to the next lower number, unless the fractional part is very small. This is a slightly conservative safeguard in case our estimate for σ is too small.

If we know the size of the population, we can incorporate the finite population correction factor into the formula to obtain a less conservative (that is, smaller and therefore more efficient) estimate for the needed sample size. For ease of computation we assume the correction factor to be $\sqrt{(N - n)/N}$ instead of $\sqrt{(N - n)/(N - 1)}$. The formula then becomes

$$n = \frac{(z_{\alpha/2} \cdot \sigma)^2}{E^2 + (z_{\alpha/2} \cdot \sigma)^2/N}$$

In Example 6.5, for instance, suppose we want to estimate the mean peak load within 0.05, with a confidence level of 90%. We have 0.420 as our estimate for σ, $z_{.05} = 1.645$, $N = 1000$, and $E = 0.05$, so $n = (1.645 \cdot 0.420)^2/[(0.05)^2 + (1.645 \cdot 0.420)^2/1000]$, or about 160 or 161. Without using the correction factor we would have $n = 191$. If we did take a sample of 160, of course, we would have to use the correction factor in computing the confidence interval.

Example 6.11

Refer to Example 6.8. What size sample should the researcher take if he wishes to be 90% certain that he is not more than 150 years from the true age of the site?

Solution

To be certain of being off by no more than 150 years, the researcher can use the formula for estimating sample size. In this case $E = 150$, $z_{.05} = 1.645$, and we estimate $\sigma = 360$. Then $n = [(1.645)(360)/150]^2 \doteq 16$. Since this is less than 30, we cannot use $z_{.05}$ and would need to use $t_{.05}$. If we use $t_{.05}\{15\}$, however, $n = [(1.753)(360)/150]^2 \doteq 18$, so we would have to use $t_{.05}\{17\}$; this gives us $n = [(1.740)(360)/150]^2 \doteq 17.44$ so that a sample size of 18 would be acceptable. This procedure of trying one solution after another until the result fits is called *iteration*. One problem with this method, however, is that variances of small samples are highly variable. Thus if the sample of 18 had a standard deviation greater than 360, the sample would probably be too small to achieve the desired maximum error. There are two ways of ensuring the accuracy of the result when the formula gives us a sample size less than 30. The first is to use a sample size of at least 30, so that we can obtain an approximate confidence interval using $z_{\alpha/2}$; the second is to construct a confidence interval for the population standard deviation (see Section 6.6) using the upper confidence limit as the estimate for σ. In this way we can be sure that the sample size is adequate.

PROFICIENCY CHECK

21. Suppose a sample standard deviation is 0.30. Determine the sample size you would need to estimate the population mean within 0.05 with a confidence level of 99% if you use the sample result to estimate the population standard deviation.

6.5.2 Sample Size Needed to Estimate a Population Proportion

As in Section 6.5.1, we can estimate the number needed in a sample to satisfy some specified limit for the maximum error or width of a confidence interval for a population proportion. If E is the maximum error allowed with confidence level $100(1 - \alpha)\%$, then $E = z_{\alpha/2}\sigma_p$. We set

$$E = z_{\alpha/2}\sqrt{\frac{\pi(1 - \pi)}{n}}$$

Solving for n, we obtain

$$n = \frac{(z_{\alpha/2})^2(\pi)(1 - \pi)}{E^2}$$

Since π is not known, we can estimate it by using a prior sample result. Even better, we could use a confidence interval (with the desired degree of confidence) from historical data or a pilot study (a preliminary experiment conducted to obtain such estimates and make sure the main experiment will be successful). If both confidence limits are greater than 0.5 or less than 0.5, use the one closer to 0.5; if 0.5 is included in the interval, use 0.5. If no estimate is available, use 0.5 as the estimate. The reason for this choice is that if $\pi = 0.5$, then $\pi(1 - \pi) = 0.25$. For any other value of π, $\pi(1 - \pi)$ will be less than 0.25; for instance $(0.4)(0.6) = 0.24$, $(0.3)(0.7) = 0.21$, and so on. Using 0.5 as the estimate for π yields the largest possible sample size we might require, so our sample would be adequate whatever the true value of π.

Estimating Needed Sample Size (Population Proportion)

If we wish to estimate a dichotomous population proportion to within E with a confidence level of $(1 - \alpha)$ or to obtain a $100(1 - \alpha)\%$ confidence interval no more than $2E$ in width, the sample size we need is estimated to be n, where

$$n = \frac{(z_{\alpha/2})^2\pi(1 - \pi)}{E^2}$$

and π is the population proportion. We can estimate π from a previous result; if no estimate is available, we will obtain the maximum value of n when $\pi = 0.5$.

As before, if we know the population size N we can obtain a smaller estimate for n by using the finite population correction factor. In this case we get

$$n = \frac{(z_{\alpha/2})^2\pi(1 - \pi)}{E^2 + (z_{\alpha/2})^2\pi(1 - \pi)/N}$$

Example 6.12

An agronomist wishes to examine a sample of a certain hybrid strain of corn to see what proportion exhibits a certain genetic characteristic. What sample size should he examine if he wants to estimate the proportion accurate to within 0.01 with a confidence level of 95%? Assume (*a*) that 0.20 is a reasonable estimate for π and (*b*) that nothing is known about π.

Solution

If $E = 0.01$ and $z_{.025} = 1.96$, then $n = (1.96)^2(\pi)(1 - \pi)/(0.01)^2$.

a. Assuming that $\pi = 0.20$, we have $n = (1.96)^2(0.2)(0.8)/(0.01)^2 \doteq 6147$.
b. Knowing nothing about π, we let $\pi = 0.5$, so n will be a maximum. Then $n = (1.96)^2(0.5)(0.5)/(0.01)^2 \doteq 9604$.

PROFICIENCY CHECKS

22. Suppose we wish to determine the proportion of registered Democrats who plan to vote in the next election correct to

(*continued*)

PROFICIENCY CHECKS *(continued)*

within 0.02 with a probability of 98%. How many registered Democrats should we poll if:
a. we use the results of Proficiency Check 14?
b. nothing is known about the value of π?

23. A toothpaste manufacturer would like to know what percentage of people in a locality use her company's brand of toothpaste.
a. Determine the sample size needed to ensure 95% confidence that the error in the estimate will not exceed 0.03.
b. Suppose 1100 people are polled and 212 say they use the toothpaste. Determine a 95% confidence interval for the population proportion.

Problems

Practice

6.60 Using the data of Problem 6.22, what sample size should you take if you want to obtain a 95% confidence interval no more than 1.0 wide?

6.61 What sample size should you take to estimate the population proportion within 0.05 with a confidence level of 95% if:
a. a pilot study shows a sample proportion of 0.2?
b. no pilot study results are available?

6.62 What sample size should you take to estimate the population proportion correct to within 0.01 with a confidence level of:
a. 90%? b. 95%? c. 98%? d. 99%?

Applications

6.63 A sociologist wishes to sample a population to establish the validity of a test given several years ago. The standard deviation of that test, 14.8, is used to estimate the current standard deviation, and she wishes to estimate the current mean correctly to within 2.0 units. What sample size should she take to have a confidence level of:
a. 90%? b. 95%? c. 98%? d. 99%?

6.64 To obtain data for his thesis, a psychology major plans to interview people to determine whether their reaction to a certain situation is positive or negative. He wants to estimate the true proportion to within 0.02 with a confidence level of 95%. If a pilot study shows 60 of 100 people with positive reactions, how many should he interview?

6.65 Refer to Problem 6.30. In the interests of economy, management is considering reducing the number of castings to be examined from shipments. On the other hand, they wish to be able to estimate the mean number of flaws correctly to within 0.5 with a confidence level of 98%. Can they reduce the number inspected? If so, what is the minimum number they should inspect?

6.66 A survey reported in *Electronic Education* (February 1986) that only 20% of secondary school students who use the computer to play games are girls but that 50% of the students using a computer for word processing are girls. A teacher wishes to estimate the proportion of girls among the students in the school who use the computer for each of these activities correctly to within 0.05 with a confidence level of 90%. Using the survey results as reasonable estimates for the true proportions in the school, how large a sample is needed in each case?

6.67 Refer to Problem 6.43. We wish to obtain a 95% confidence interval for the proportion of people who watched the show. If the interval is to be no more than 0.04 wide, how large a sample is needed? How would you determine the sample size necessary to estimate the proportion of viewers who were offended by the show correctly to within 0.02 with a 95% level of confidence? (*Note:* Remember that you can control the size of the sample but not the number of people in the sample who actually watched the show.)

6.68 Suppose that a sample of 40 sales orders drawn from a shipment of 5000 similar sales orders shows an average order of \$154.87 with a standard deviation of \$88.30.

a. Determine a 95% confidence interval for the mean order in the shipment.

b. Determine the necessary sample size to be 95% confident that your estimate is off by no more than \$10.

c. Suppose that the orders range from \$15 to \$722. Would the use of one-fourth the range for the standard deviation appear to be too far off in this case? What sample size would you need in this case in order to be 95% confident that the estimate is off by no more than \$10?

d. If the result in part (*c*) suggests that the population is positively skewed, what does this imply about the accuracy of the confidence interval obtained in part (*a*)?

e. Use the confidence interval in part (*a*) to obtain an interval estimate for the total value of the 5000 orders. What is the probability that the actual dollar value will be less than the minimum value of your interval?

6.6 Estimating Population Variance

Sometimes we need a confidence interval for the population variance or standard deviation. Often, in industrial applications, the variability of a manufacturing process must be kept small; in biological applications, the variability of a genetic trait in a population may be used to indicate the presence or absence of a mutation; or we may need to estimate the size of the population variance in order to obtain the needed sample size for a certain estimation, as in Section 6.5. In such cases we must be able to obtain a confidence interval for the population variance or standard deviation.

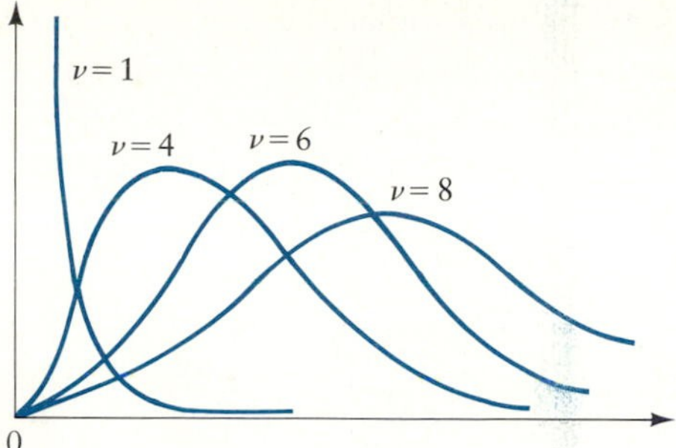

Figure 6.13
Graph of four chi-square distributions with $\nu = 1$, $\nu = 4$, $\nu = 6$, and $\nu = 8$.

6.6.1 The Chi-Square Distribution

If the random variable x has a normal distribution, we can obtain exact confidence intervals for σ^2 and σ by using the **chi-square distribution.** A number of sampling distributions, including sample variances, are distributed as chi square. Suppose we obtain all samples of n observations of a normal random variable and determine the variance s^2 of each sample. If the population variance is σ^2, the statistic

$$\chi^2 = \frac{(n-1)s^2}{\sigma^2}$$

is called **chi square** (where χ is the Greek letter chi, pronounced kī) and has a distribution that has been studied extensively. The distribution of χ^2 is determined completely by the degrees of freedom, ν. Like the t distribution, the shape of a chi-square distribution depends on ν, and as ν increases the curve is very nearly a normal curve (although not a standardized one). Unlike the t distribution, however, the graph of a chi-square distribution changes quite a bit as ν changes. Values of the variable can never be negative, so the graph is bounded below at zero. The distributions are positively skewed; the degree of skewness decreases as ν increases, however, and the graph approaches symmetry. A chi-square distribution with ν degrees of freedom is symbolized $\chi^2(\nu)$. The variable χ^2 has mean $\mu_{\chi^2} = \nu$ and variance $\sigma^2_{\chi^2} = 2\nu$. The graph of chi-square distributions for $\nu = 1, 4, 6$, and 8 is illustrated in Figure 6.13.

6.6.2 Confidence Intervals for the Population Variance

The chi-square distributions have been studied extensively and tables have been prepared for a large number of degrees of freedom. Critical values of χ^2 are symbolized $\chi^2_\alpha\{\nu\}$; the proportion of the area under the graph of $\chi^2(\nu)$ to the right of $\chi^2_\alpha\{\nu\}$ is α. Since chi-square curves are not symmetric, however, $\chi^2_{1-\alpha}\{\nu\}$ is not equal to $-\chi^2_\alpha\{\nu\}$. (Recall that for critical values of t and z, we have $t_{1-\alpha}\{\nu\} = -t_\alpha\{\nu\}$ and $z_{1-\alpha} = -z_\alpha$, respectively.)

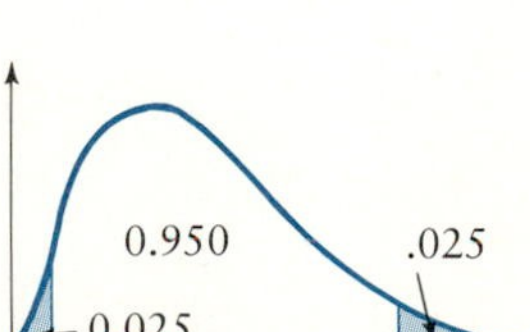

Figure 6.14
Areas under the chi-square curve for $\chi^2_{.975}\{\nu\}$ and $\chi^2_{0.25}\{\nu\}$.

We can be certain that 95% of all values of χ^2 computed from a sampling distribution of s^2 will lie between $\chi^2_{.975}\{\nu\}$ and $\chi^2_{.025}\{\nu\}$, as shown in Figure 6.14. This is true since 0.025 of the area under the curve lies to the right of

$\chi^2_{.025}\{v\}$ and 0.975 of the area under the curve lies to the right of $\chi^2_{.975}\{v\}$. Since s^2 has $n - 1$ degrees of freedom, the statistic $\chi^2 = (n - 1)s^2/\sigma^2$ has $n - 1$ degrees of freedom. Thus for 95% of the sample variances calculated from samples of size n drawn from a normal population with variance σ^2,

$$\chi^2_{.975}\{n-1\} < \frac{(n-1)s^2}{\sigma^2} < \chi^2_{.025}\{n-1\}$$

Solving $\chi^2_{.975}\{n - 1\} < (n - 1)s^2/\sigma^2$ for σ^2, we have

$$\sigma^2 < \frac{(n-1)s^2}{\chi^2_{.975}\{n-1\}}$$

Similarly, solving $(n - 1)s^2/\sigma^2 < \chi^2_{.025}\{n - 1\}$ for σ^2 we have

$$\frac{(n-1)s^2}{\chi^2_{.025}\{n-1\}} < \sigma^2$$

Putting these statements together, we get a 95% confidence interval for the population variance. That is, for such samples the statement

$$\frac{(n-1)s^2}{\chi^2_{.025}\{n-1\}} < \sigma^2 < \frac{(n-1)s^2}{\chi^2_{.975}\{n-1\}}$$

will be true 95% of the time. Similarly, we can obtain the following formula for a $100(1 - \alpha)\%$ confidence interval for the population variance in a normal population.

Confidence Interval for the Population Variance in a Normal Population

Suppose we obtain a random sample of n observations of a normal random variable x with variance σ^2. If s^2 is the sample variance, a $100(1 - \alpha)\%$ confidence interval for the population variance of σ^2 is the interval

$$\frac{(n-1)s^2}{\chi^2_{\alpha/2}} \quad \text{to} \quad \frac{(n-1)s^2}{\chi^2_{1-\alpha/2}}$$

where $\chi^2_{\alpha/2} = \chi^2_{\alpha/2}\{n - 1\}$ and $\chi^2_{1-\alpha/2} = \chi^2_{1-\alpha/2}\{n - 1\}$.

Example 6.13

A sample of 12 castings is taken from a daily production run. To ensure good quality and not too much variation between castings in a run, a confidence interval for the population variance or standard deviation is obtained. If the standard deviation of the number of flaws per casting for the sample is 6.34, determine a 95% confidence interval for the variance of the flaws during the run and a 90% confidence interval for the standard deviation of flaws during the run.

Solution

We can assume normality since presumably the products are all made to the same specifications. Using Table 6, we obtain $\chi^2_{.975}\{11\} = 3.816$ and $\chi^2_{.025}\{11\} = 21.92$. Then $11(6.34)^2/21.92 \doteq 20.17$ and $11(6.34)^2/3.816 \doteq 115.87$, so a 95% confidence interval for the population variance is 20.17 to 115.87. To determine a 90% confidence interval for the standard deviation, we first need to obtain a 90% confidence interval for the variance. We get $\chi^2_{.95}\{11\} = 4.575$ and $\chi^2_{.05}\{11\} = 19.68$; $11(6.34)^2/19.68 \doteq 22.47$ and $11(6.34)^2/4.575 \doteq 96.65$. These are 90% confidence limits for the population variance. The square roots of these numbers are 90% confidence limits for the population standard deviation, so the 90% confidence interval for σ is 4.74 to 9.83.

Example 6.14

A sample of 81 observations has a variance of 114.06. Determine a 90% confidence interval for the population variance assuming that the population is normally distributed.

Solution

From Table 6 we get $\chi^2_{.95}\{80\} = 60.39$ and $\chi^2_{.05}\{80\} = 101.9$. Then $(80)(114.06)/101.9 \doteq 89.55$ and $(80)(114.06)/60.39 \doteq 151.10$, so we can be 90% certain that σ^2 lies between these limits. We are equally 90% certain that σ lies between the square roots of the limits—that is, between 9.46 and 12.29.

PROFICIENCY CHECKS

24. Suppose that x is normally distributed, $\bar{x} = 100.00$, and $s = 30.00$. Determine a 95% confidence interval for σ^2 if n is:
    a. 5 b. 10 c. 20 d. 25
25. Suppose that x is normally distributed, $\bar{x} = 0.560$, $s = 0.122$, and $n = 15$. Determine confidence intervals for σ with a confidence level of:
    a. 90% b. 95% c. 98% d. 99%
26. If x is normally distributed, $\bar{x} = 8.66$, $s = 1.18$, and $n = 41$, determine a 95% confidence interval for σ.

Problems

Practice

6.69 A sample of size n is drawn from a normal population and the result is used to obtain a 95% confidence interval for the population standard deviation. Determine the width of the confidence interval if the sample standard deviation is as given.

a. $n = 10, s = 13.4$
b. $n = 6, s = 103.4$
c. $n = 20, s = 0.040$
d. $n = 24, s = 8.63$
e. $n = 36, s = 12.136$

6.70 A random sample is drawn from a normally distributed population. The sample variance is 39.69. Determine 95% confidence intervals for the population variance if the sample size is:

a. 10 b. 15 c. 20 d. 50

6.71 A random sample of size 25 is drawn from a normal population. If the sample variance is 13.18, determine confidence intervals for the population variance if the confidence level is:

a. 90% b. 95% c. 98% d. 99%

6.72 A random sample of size 30 is drawn from a normal population. Determine 90% confidence intervals for the population variance if the sample standard deviation is:

a. $s = 1.9$ b. $s = 4.6$ c. $s = 8.9$ d. $s = 10.6$

6.73 A random sample of size 30 is drawn from a normal population. If the sample mean and variance are 0.112 and 0.000169, respectively, determine confidence intervals for the population standard deviation with a confidence level of:

a. 90% b. 95% c. 98% d. 99%

Applications

6.74 Variance is an important aspect of quality control in that variability of output is a measure of consistency. If a machine making ball bearings is too variable in its output, much of the production run will be unacceptable—either too large or too small in diameter. Suppose that the first 20 ball bearings produced by a machine have a mean diameter of 6.003 millimeters with a standard deviation of 0.017 millimeter. Determine 90% and 99% confidence intervals for the standard deviation of the diameter of the ball bearings in the run.

6.75 When a new production line is being started, management must get an estimate of the mean and variability of the time required to perform tasks in order to time the movement of the line. A sample of 25 workers performs the same task and requires a mean of 4.11 minutes and a standard deviation of 1.85 minutes to do the task. Obtain 95% confidence intervals for the mean and standard deviation of the time required of all workers to do the task.

6.76 Refer to Problem 6.75. Suppose that management wants to set a time for the product to stay at the station so that they can be 95% certain that the average worker will complete the task within the allotted time. To do this they will take the upper limit of a 95% confidence interval for σ and use that value as an estimate of the sample standard deviation in computing a 95% confidence interval for the population mean. They will then use the upper level of that confidence interval for estimating the maximum time needed to complete the task. Since they will underestimate the maximum standard deviation only about 2.5% of the time and consequently underestimate the maximum mean time 2.5% of the time, the likelihood that they will be wrong in both cases can be shown to be less than the sum of these values, or about 5%. Thus they are 95% confident that the mean time will be enough. Use the data of Problem 6.75 to determine the time allocation they should set for that product.

6.7 Computer Usage

In this section we will use the computer for two purposes. First we can learn how to obtain confidence intervals. Example 6.15 shows how to obtain a confidence interval for the population mean using Minitab and Examples 6.17, 6.18, and 6.19 show how to use SAS statements to obtain confidence intervals for the population mean, population proportion, and population variance, respectively. Example 6.16 and the last (unnumbered) example of Section 6.7.2 are designed to provide additional insight into the central limit theorem. We also use Minitab to clarify the concept of the confidence interval.

6.7.1 Minitab Commands for This Chapter

To find a confidence interval for a population mean we can use the DESCRIBE command to calculate the sample mean and standard deviation and then apply the formulas to obtain the confidence limits. Alternatively we may use the commands TINTERVAL or ZINTERVAL to find confidence intervals for the population mean based on t or z, respectively.

To obtain the confidence interval for the population mean with limits $\overline{x} \pm t_{\alpha/2}s/\sqrt{n}$, we use the following command:

```
MTB > TINTERVAL with K % confidence for the data stored in C
```

where K is the confidence level in percent (such as 90 or 95). If we omit the confidence level, the default value for K is 95.

To obtain a confidence interval for the population mean with limits $\overline{x} \pm z_{\alpha/2}\sigma/\sqrt{n}$, we use the following command:

```
MTB > ZINTERVAL with K1 % confidence sigma = K2 for the data stored in C
```

where K1 is the desired confidence level and K2 is the known population standard deviation. Example 6.16 illustrates the confidence interval using t.

Example 6.15

Consider the pollution indices presented originally in Example 2.15. In Example 5.15 we found that the normality assumption is reasonable for this random variable so we can base a confidence interval for the population mean on the *t* distribution. Use Minitab to obtain a 95% confidence interval for the mean pollution index for the city.

Solution

We enter the data in the same manner as in Figure 2.10. The remainder of the program and the results are shown in Figure 6.15. Thus the 95% confidence interval for the mean pollution index is 54.4 to 59.1.

Our next example illustrates the central limit theorem (CLT). Recall that the CLT tells us that the distribution of the sample mean is approximately normal for large samples obtained from a population. Thus if we take many large samples from the same population and find the sample mean for each sample, the distribution of these sample means should closely resemble a normal curve, no matter how the population distribution appears.

We can take random samples from a population with specified distribution and parameters by using the appropriate commands. The mean for each sample can be computed and stored in a column, and we can construct a bar graph (using the HISTOGRAM command) to obtain information about the shape of the probability distribution of the sample means.

To generate a random sample from a normal population on Minitab, we use the RANDOM command and the NORMAL subcommand. For example, we can generate samples from a normal distribution with $\mu = 30$ and $\sigma = 5$ by using the commands

```
MTB >  RANDOM 100 observations in each of C1 - C50 ;
SUBC>  NORMAL mean = 30 standard deviation = 5 .
```

Each column will be a random sample of size 100; each row will be a random sample of size 50. For our purposes we will consider the numbers in the rows to represent the random samples, so we will have 100 random samples of size 50. We can find the sample means for each row by using the RMEAN command.

We can use a similar procedure for a nonnormal population as well, simply specifying the details for the desired distribution. In the following example we show the effect of the CLT on both a normal population and a skewed population generated by a Poisson random variable.

Figure 6.15
Minitab printout for confidence interval of mean pollution index (Example 6.15).

(The data are entered here in the same manner as in Table 2.28.)

```
MTB > TINTERVAL with 95 % confidence for the data in C1


                N        MEAN       STDEV  SE MEAN        95.0 PERCENT C.I.
P=INDEX       120        56.7        13.1      1.2       (   54.4,     59.1)
```

Example 6.16

Generate 100 random samples of size 50 from a normal population with $\mu = 30$ and $\sigma = 5$, and inspect the histogram of the sample means. Does the bar graph suggest that the distribution of all sample means may be normal? Repeat the experiment using a Poisson random variable with mean $m = 5$.

Solution

The program and printout for the normal population are shown in Figure 6.16. The bar graph for these means does resemble the normal curve closely. This ageement suggests that we can approximate the distribution of these sample means with a normal curve. It can be shown theoretically that the true sampling distribution of the sample mean is exactly normal when the population is normal with mean $\mu_{\bar{x}} = 30$ and standard deviation $\sigma_{\bar{x}} = 5/\sqrt{50} \doteq 0.707$. From the output of the DESCRIBE command we find that the average of these 100 sample means is 29.967 and the standard deviation of the sample means is 0.737, both of which agree reasonably well with the theoretical values. As the number of samples increases, the agreement becomes even better.

The program and printout for the Poisson population are shown in Figure 6.17. Notice again that the sample means have been computed and stored in C51. The histogram for the means again resembles a normal curve, so again we conclude that the distribution of the sample means can be closely approximated by a normal distribution. Recall that the mean and

```
MTB > RANDOM 100 observations in C1 - C50 ;
SUBC> NORMAL mean = 30 standard deviation = 5 .
MTB > RMEAN C1 - C50 store the means in C51
MTB > HISTOGRAM the values in C51
```

Histogram of C51 The 100 sample means

```
Midpoint    Count
   28.5       7        *******
   29.0      12        ************
   29.5      20        ********************
   30.0      24        ************************
   30.5      23        ***********************
   31.0      12        ************
   31.5       1        *
   32.0       0
   32.5       1        *

MTB > DESCRIBE the values in C51

          N      MEAN    MEDIAN    TRMEAN     STDEV    SEMEAN
C51     100    29.967    30.057    29.963     0.737     0.074

        MIN       MAX        Q1        Q3
C51  28.387    32.271    29.411    30.423
```

Figure 6.16
Minitab printout showing the sampling distribution of the mean for a normal population (Example 6.16).

Figure 6.17
Minitab printout showing the sampling distribution of the mean for a Poisson population (Example 6.16).

```
MTB > RANDOM 100 observations in C1 - C50 ;
SUBC> POISSON with mean = 5 .
MTB > RMEAN C1 - 50 store the means in C51
MTB > HISTOGRAM the values in C51


Histogram of C51   ← the 100 sample means

Midpoint    Count
     4.4      2       **
     4.5      4       ****
     4.6      7       *******
     4.7      7       *******
     4.8     13       *************
     4.9     12       ************
     5.0      9       *********
     5.1     20       ********************
     5.2     13       *************
     5.3      1       *
     5.4      6       ******
     5.5      4       ****
     5.6      2       **

MTB > DESCRIBE the values in C51


             N        MEAN      MEDIAN    TRMEAN     STDEV    SEMEAN
C51        100       4.991       5.010     4.988     0.274     0.027

           MIN         MAX          Q1        Q3
C51      4.440       5.620       4.785     5.180
```

variance of a Poisson distribution are both equal to m. In this case, then, we should have $\mu_{\bar{x}} = m = 5$ and $\sigma_{\bar{x}} = \sqrt{m}/\sqrt{n} = \sqrt{5}/\sqrt{50} \doteq 0.316$.

From the output of the DESCRIBE command we find that the average of these 100 sample means is 4.991, which agrees reasonably well with the theoretical value, and the standard deviation of the sample means is 0.274. The latter value is somewhat different from the theoretical value, but as the number of samples increases the agreement becomes considerably better.

We may also use the capacities of the computer to clarify the concept of a confidence interval. Suppose that we have a single population with mean μ and take many samples of the same size from the population. For each sample we can find a 90% confidence interval for μ. Will all these confidence intervals be the same? Of course not. Not all samples will have the same mean and standard deviation, so not all the confidence intervals will be the same. Any two samples with different means will generate different confidence intervals. Further, if two samples have the same mean and our procedure uses the sample standard deviation (either with a t or large-sample procedure using $s_{\bar{x}}$), then the confidence intervals will be different unless the sample standard deviations are the same. Thus most, if not all, of these confidence intervals will be different. How many of these confidence intervals will include the true value of μ? Since the confidence level is 90%, we should expect about 90% of these intervals to include μ. We can illustrate these facts using Minitab.

Figure 6.18
Minitab printout for multiple confidence intervals for a population mean

```
MTB > RANDOM 50 observations in C1 - C50 ;
SUBC> NORMAL mean = 30 standard deviation = 5 .
MTB > TINTERVAL with 90 % confidence for data in C1 - C50

           N       MEAN     STDEV  SE MEAN       90.0 PERCENT C.I.
C1        50      30.59      5.25     0.74     (  29.35,   31.84)
C2        50      30.77      4.73     0.67     (  29.65,   31.89)
C3        50      30.98      6.04     0.85     (  29.55,   32.41)
C4        50      28.63      5.01     0.71     (  27.45,   29.82)*
C5        50      30.01      4.58     0.65     (  28.93,   31.10)
C6        50      29.54      5.13     0.73     (  28.33,   30.76)
C7        50      29.89      5.05     0.71     (  28.69,   31.09)
C8        50      30.69      5.40     0.76     (  29.41,   31.97)
C9        50      29.31      5.38     0.76     (  28.03,   30.58)
C10       50      29.78      5.37     0.76     (  28.50,   31.05)
C11       50      29.40      4.53     0.64     (  28.33,   30.48)
C12       50      30.33      5.59     0.79     (  29.00,   31.66)
C13       50      30.44      5.20     0.73     (  29.20,   31.67)
C14       50      29.79      5.25     0.74     (  28.55,   31.04)
C15       50      29.81      4.92     0.70     (  28.64,   30.98)
C16       50      29.39      4.56     0.64     (  28.31,   30.47)
C17       50      29.43      4.95     0.70     (  28.26,   30.61)
C18       50      29.12      5.02     0.71     (  27.93,   30.31)
C19       50      29.63      4.73     0.67     (  28.51,   30.76)
C20       50      30.64      5.43     0.77     (  29.36,   31.93)
C21       50      29.98      4.56     0.65     (  28.90,   31.07)
C22       50      30.40      5.03     0.71     (  29.21,   31.60)
C23       50      30.08      5.30     0.75     (  28.82,   31.34)
C24       50      29.63      4.66     0.66     (  28.52,   30.74)
C25       50      30.21      5.12     0.72     (  29.00,   31.42)
C26       50      30.08      4.88     0.69     (  28.93,   31.24)
C27       50      31.34      4.53     0.64     (  30.26,   32.41)*
C28       50      30.51      4.33     0.61     (  29.48,   31.53)
C29       50      30.52      6.62     0.94     (  28.95,   32.08)
C30       50      31.75      4.74     0.67     (  30.63,   32.88)*
C31       50      29.81      5.20     0.74     (  28.58,   31.04)
C32       50      30.99      5.59     0.79     (  29.66,   32.32)
C33       50      30.67      4.92     0.70     (  29.50,   31.83)
C34       50      31.05      4.90     0.69     (  29.89,   32.21)
C35       50      29.29      5.15     0.73     (  28.06,   30.51)
C36       50      28.53      4.28     0.60     (  27.52,   29.55)*
C37       50      29.98      4.52     0.64     (  28.91,   31.06)
C38       50      30.24      4.87     0.69     (  29.08,   31.40)
C39       50      29.53      5.26     0.74     (  28.28,   30.78)
C40       50      31.51      5.36     0.76     (  30.24,   32.78)*
C41       50      29.42      5.80     0.82     (  28.04,   30.79)
C42       50      29.94      5.36     0.76     (  28.66,   31.21)
C43       50      29.77      5.28     0.75     (  28.52,   31.02)
C44       50      29.53      5.79     0.82     (  28.16,   30.90)
C45       50      30.62      5.11     0.72     (  29.41,   31.83)
C46       50      29.24      5.07     0.72     (  28.04,   30.44)
C47       50      29.10      4.98     0.70     (  27.92,   30.28)
C48       50      31.42      3.92     0.55     (  30.48,   32.35)*
C49       50      29.77      4.55     0.64     (  28.69,   30.85)
C50       50      30.59      4.57     0.65     (  29.50,   31.67)
```

*Confidence intervals that do not contain 30.

First we obtain a random sample of size 50 from a normal population with mean 30 and standard deviation 5; then we obtain a 90% confidence interval for the population mean by using the TINTERVAL command. We repeat this procedure to obtain a total of 50 such intervals. The program and printout of the 50 intervals are shown in Figure 6.18. Notice that each interval has different upper and lower limits. If we count the number of intervals containing the population mean, we find that forty-four intervals actually contain 30 while six do not. Thus 88% of our sample of confidence intervals do contain the mean—which is satisfactorily close to the theoretical 90%.

6.7.2 SAS Statements for This Chapter

Although the SAS manuals give no functions or procedures for obtaining confidence intervals, the method for doing so is quite simple. The MEANS procedure will give us the mean, standard deviation, and standard error of the mean for a sample, as well as other information. We may use the output of the MEANS procedure with statements that will allow us to obtain confidence intervals for a population mean, proportion, or variance without difficulty. Since we need only the mean and standard error of each sample, we will only calculate those measures. The statement

```
PROC MEANS MEAN STDERR;
```

will cause the mean and standard error of the sample to be calculated and printed. If we wish all the other information—number of observations, standard deviation, minimum value, maximum value, sum, variance, and coefficient of variation—we may simply write PROC MEANS; and these will all be calculated and printed. We may use the OUTPUT statement to save $\overline{x}$ and $s_{\overline{x}}$ and give them variable names as specified. The statement

```
OUTPUT OUT = A MEAN = XBAR STDERR = SE;
```

will result in the mean and standard error being given the variable names XBAR and SE, respectively, added to the original data set; the new data set will be given the name A. If we wish to save the standard deviation as well, we may use STD = *name* as part of the output statement.

We can then use the new data set to construct confidence intervals. To do so we need a new DATA step since we may construct new variables only in a data step. We write

```
DATA CONF;
  SET A;
```

This statement gives the data set A a new name and allows us to create new variables. We create the variables LLIMIT (the lower limit) and ULIMIT (the upper limit) as follows on the next page.

```
LLIMIT = XBAR - (t or z) * STDERR;
ULIMIT = XBAR + (t or z) * STDERR:
```

where t or z is the appropriate multiple for the sample. We may then print the variable we want by using the PRINT procedure. To print the confidence limits only, we write

```
PROC PRINT;
   VAR LLIMIT ULIMIT;
```

We can give the output an appropriate title by writing

```
TITLE '95% CONFIDENCE LIMITS FOR THE POPULATION MEAN';
```

or other suitable title. Note that this procedure requires us to specify the value of t or z to be used.

We illustrate by repeating Example 6.15 for the pollution data of Example 2.15. Since we had already stored the data in a file, we need only to retrieve them and write the program as shown here. Because this is a very large sample ($n = 120$), we use the large-sample procedure. Figure 6.19 shows the statements used and the output giving 90%, 95%, and 99% confidence intervals for the population mean. Although the results are given to four decimal places, if rounded off to one decimal place they agree with those of Example 6.15.

If we need to use the finite population correction factor, we may write

```
SE = STDERR * SQRT ((N - n) / (N - 1));
```

and use SE in place of STDERR in the formulas for ULIMIT and LLIMIT.

```
DATA PCLLUT;
  INPUT INDEX @@;
  CARDS;
PROC MEANS;
  OUTPUT OUT = A MEAN = XBAR STDERR = STDERR;
DATA CONF90;
  SET A;
  LLIMIT = XBAR - 1.645*STDERR;
  ULIMIT = XBAR + 1.645*STDERR;
PROC PRINT;
  VAR LLIMIT ULIMIT;
TITLE '90% CONFIDENCE LIMITS FOR THE POPULATION MEAN';
DATA CONF95;
  SET A;
  LLIMIT = XBAR - 1.96*STDERR;
  ULIMIT = XBAR + 1.96*STDERR;
PROC PRINT;
  VAR LLIMIT ULIMIT;
TITLE '95% CONFIDENCE LIMITS FOR THE POPULATION MEAN';
DATA CONF99;
  SET A;
  LLIMIT = XBAR - 2.576*STDERR;
  ULIMIT = XBAR + 2.576*STDERR;
PROC PRINT;
  VAR LLIMIT ULIMIT;
TITLE '99% CONFIDENCE LIMITS FOR THE POPULATION MEAN';
```

Figure 6.19
SAS printout for large-sample confidence interval for mean pollution index, Example 6.15.

```
                                                    SAS
VARIABLE      N          MEAN           STANDARD       MINIMUM        MAXIMUM      STD ERROR
                                        DEVIATION       VALUE          VALUE        OF MEAN          SUM          VARIANCE       C.V.

INDEX       120    56.71666667      13.05269432   29.00000000    88.00000000     1.19154252   6806.0000000  170.37282913    23.014
                                      90% CONFIDENCE LIMITS FOR THE POPULATION MEAN

                                                  OBS     LLIMIT      ULIMIT

                                                   1      54.7566     58.6768
                                      95% CONFIDENCE LIMITS FOR THE POPULATION MEAN

                                                  OBS     LLIMIT      ULIMIT

                                                   1      54.3812     59.0521
                                      99% CONFIDENCE LIMITS FOR THE POPULATION MEAN

                                                  OBS     LLIMIT      ULIMIT

                                                   1      53.6473     59.7861
```

Example 6.17

Use SAS to find 90%, 95%, and 99% confidence intervals for the population mean if a random sample from a normal population has the following values:

3, 5, 6, 4, 7, 8, 6, 9, 4, 5, 6, 7, 3, 3, 5, 4, 7, 8, 9, 3

Solution

The program and output are shown in Figure 6.20.

Figure 6.20
SAS printout for small-sample confidence intervals for the population mean.

```
INPUT FILE

DATA START;
  INPUT X @@;
  CARDS;
3 5 6 4 7 8 6 9 4 5 6 7 3 3 5 4 7 8 9 3
PROC MEANS;
  OUTPUT OUT = A MEAN = XBAR STDERR = STDERR;
DATA CONF90;
  SET A;
  LLIMIT = XBAR - 1.729 * STDERR;
  ULIMIT = XBAR + 1.729 * STDERR;
PROC PRINT;
  VAR LLIMIT ULIMIT;
TITLE '90% CONFIDENCE LIMITS FOR THE POPULATION MEAN';
DATA CONF95;
  SET A;
  LLIMIT = XBAR - 2.093 * STDERR;
  ULIMIT = XBAR + 2.093 * STDERR;
PROC PRINT;
  VAR LLIMIT ULIMIT;
TITLE '95% CONFIDENCE LIMITS FOR THE POPULATION MEAN';
DATA CONF99;
  SET A;
  LLIMIT = XBAR - 2.861 * STDERR;
  ULIMIT = XBAR + 2.861 * STDERR;
PROC PRINT;
  VAR LLIMIT ULIMIT;
TITLE '99% CONFIDENCE LIMITS FOR THE POPULATION MEAN';
```

```
OUTPUT FILE

                                              SAS

VARIABLE   N      MEAN        STANDARD      MINIMUM      MAXIMUM     STD ERROR       SUM        VARIANCE      C.V.
                              DEVIATION      VALUE        VALUE       OF MEAN

X         20   5.60000000   2.01049876   3.00000000   9.00000000   0.44956119   112.00000000  4.04210526   35.902
                          90% CONFIDENCE LIMITS FOR THE POPULATION VARIANCE

                                    OBS    LLIMIT     ULIMIT

                                     1     2.54811    7.58893
                          95% CONFIDENCE LIMITS FOR THE POPULATION VARIANCE

                                    OBS    LLIMIT     ULIMIT

                                     1     2.3379     8.6234
                          99% CONFIDENCE LIMITS FOR THE POPULATION VARIANCE

                                    OBS    LLIMIT     ULIMIT

                                     1     1.99067    11.2215
```

We may use similar SAS statements to construct confidence intervals for a population proportion as well. We could write

```
DATA EXAMPLE;
      P = X / N;
  SEP = SQRT(P * (1 - P) / N);
  LLIMIT90 = P - 1.645 * SEP;
  ULIMIT90 = P + 1.645 * SEP;
  LLIMIT95 = P - 1.96 * SEP;
  ULIMIT95 = P + 1.96 * SEP;
  LLIMIT99 = P - 2.576 * SEP;
  ULIMIT99 = P + 2.576 * SEP;
  CARDS;
data here--x n x n x n ...
PROC PRINT;
```

If there is only one sample we may wish to omit the CARDS statement and enter the appropriate numbers instead of X and N in the second line. Of course there is no need to obtain confidence limits for 90%, 95%, and 99% confidence intervals unless they are all wanted.

Example 6.18

A random sample of 91 voters contains 43 who say they are happy with the present administration. Obtain 90%, 95%, and 99% confidence intervals for the proportion of voters who are happy with the present administration.

Solution

Figure 6.21 shows the confidence intervals. We can be 90% confident that the proportion of voters who are happy with the present administration is about 0.39 to 0.56, while we can be 95% confident that the proportion of voters who are happy with the present administration is about 0.37 to 0.58. Finally, we can be 99% confident that the proportion of voters who are happy with the present administration is about 0.34 to 0.61.

Figure 6.21
SAS printout for confidence intervals for the population proportion (Example 6.18).

```
INPUT FILE

DATA START;
  INPUT X N;
  P = X/N;
  STDERRP = SQRT(P*(1-P)/N);
CARDS ;
43 91
PROC PRINT;
DATA CONF90;
  SET START;
  LLIMIT = P - 1.645*STDERRP;
  ULIMIT = P + 1.645*STDERRP;
PROC PRINT;
  VAR LLIMIT ULIMIT;
TITLE '90% CONFIDENCE LIMITS FOR THE POPULATION PROPORTION';
DATA CONF95;
  SET START;
  LLIMIT = P - 1.96*STDERRP;
  ULIMIT = P + 1.96*STDERRP;
PROC PRINT;
  VAR LLIMIT ULIMIT;
TITLE '95% CONFIDENCE LIMITS FOR POPULATION PROPORTION';
DATA CONF99;
  SET START;
  LLIMIT = P - 2.576*STDERRP;
  ULIMIT = P + 2.576*STDERRP;
PROC PRINT;
  VAR LLIMIT ULIMIT;
TITLE '99% CONFIDENCE LIMITS FOR POPULATION PROPORTION';

OUTPUT FILE

                         SAS

      OBS     X      N          P           STDERRP

       1      43     91     0.472527     0.0523351
90% CONFIDENCE LIMITS FOR THE POPULATION PROPORTION

             OBS      LLIMIT       ULIMIT

              1      0.386436     0.558619
95% CONFIDENCE LIMITS FOR THE POPULATION PROPORTION

             OBS      LLIMIT       ULIMIT

              1      0.369951     0.575104
99% CONFIDENCE LIMITS FOR THE POPULATION PROPORTION

             OBS      LLIMIT       ULIMIT

              1      0.337712     0.607343
```

We may use statements similar to those of Example 6.18 to obtain confidence limits for a population variance. The MEANS procedure will obtain the variance of a sample for us, and we can name it by using the output statement with VAR = *name*. Then we can create a new data set and determine the lower limit by writing LLIMIT = $(n - 1) * (s^2)/(\chi^2_{\alpha/2})$, where the expressions in parentheses are replaced by, respectively, the sample size minus one, the name chosen for the variance, and the appropriate critical value of chi square, all without parentheses. The upper confidence limit is written similarly.

Example 6.19

Using the data of Example 6.18, obtain a 95% confidence interval for the population variance.

Solution

The program and printout for this example are shown in Figure 6.22.

```
INPUT FILE

DATA START;
  INPUT X @@;
  CARDS;
3 5 6 4 7 8 6 9 4 5 6 7 3 3 5 4 7 8 9 3
PROC MEANS;
  OUTPUT OUT = A MEAN = XBAR STDERR = STDERR;
DATA CONF90;
  SET A;
  LLIMIT = XBAR - 1.729 * STDERR;
  ULIMIT = XBAR + 1.729 * STDERR;
PROC PRINT;
  VAR LLIMIT ULIMIT;
TITLE '90% CONFIDENCE LIMITS FOR THE POPULATION MEAN';
DATA CONF95;
  SET A;
  LLIMIT = XBAR - 2.093 * STDERR;
  ULIMIT = XBAR + 2.093 * STDERR;
PROC PRINT;
  VAR LLIMIT ULIMIT;
TITLE '95% CONFIDENCE LIMITS FOR THE POPULATION MEAN';
DATA CONF99;
  SET A;
  LLIMIT = XBAR - 2.861 * STDERR;
  ULIMIT = XBAR + 2.861 * STDERR;
PROC PRINT;
  VAR LLIMIT ULIMIT;
TITLE '99% CONFIDENCE LIMITS FOR THE POPULATION MEAN';
```

```
OUTPUT FILE

                                              SAS

VARIABLE   N       MEAN          STANDARD       MINIMUM       MAXIMUM      STD ERROR        SUM          VARIANCE      C.V.
                                 DEVIATION      VALUE         VALUE        OF MEAN

X          20   5.60000000     2.01049876    3.00000000    9.00000000   0.44956119   112.00000000   4.04210526    35.902
                                   90% CONFIDENCE LIMITS FOR THE POPULATION MEAN

                                             OBS    LLIMIT      ULIMIT

                                              1     4.82271     6.37729
                                   95% CONFIDENCE LIMITS FOR THE POPULATION MEAN

                                             OBS    LLIMIT      ULIMIT

                                              1     4.65907     6.54093
                                   99% CONFIDENCE LIMITS FOR THE POPULATION MEAN

                                             OBS    LLIMIT      ULIMIT

                                              1     4.31381     6.88619
```

Figure 6.22
SAS printout for small-sample confidence intervals for the population variance, Example 6.19.

We may illustrate the validity of the central limit theorem by using SAS statements as well as Minitab commands. Rather than repeat Example 6.17, we will instead expand the illustration given in Section 6.1, using a uniform distribution with values 2, 3, 4, 5, 6, and 7, each equally likely. We use (*a*) 50 random samples of 5 observations. (*b*) 1000 random samples of 5 observations, (*c*) 50 random samples of 50 observations, and (*d*) 1000 random samples of 50 observations to show that the central limit theorem applies to this distribution. As the SAS statements for this example are rather complicated and will most likely not be repeated by students, we merely show the input file for part (*a*) and all four output files in Figure 6.23a, b, c, d.

Figure 6.23*a*
SAS printout for input and output for illustration of the central limit theorem for 50 random samples of 5 observations each.

```
INPUT FILE

DATA START;
  DO J = 1 TO 50;
    GP = J;
  DO I = 1 to 5;
  X = INT(6*RANUNI(I) + 2);
  OUTPUT;
  END;
  END;
PROC SORT;
  BY GP;
PROC MEANS MEAN NOPRINT;
  VAR X;
  BY GP;
OUTPUT OUT = A MEAN = XBAR;
DATA B;
  SET A;
  IF 2.0 <= XBAR < 2.5 THEN MEANS = '2.0 TO 2.5';
  IF 2.5 <= XBAR < 2.9 THEN MEANS = '2.5 TO 2.9';
  IF 2.9 <= XBAR < 3.3 THEN MEANS = '2.9 TO 3.3';
  IF 3.3 <= XBAR < 3.7 THEN MEANS = '3.3 TO 3.7';
  IF 3.7 <= XBAR < 4.1 THEN MEANS = '3.7 TO 4.1';
  IF 4.1 <= XBAR < 4.5 THEN MEANS = '4.1 TO 4.5';
  IF 4.5 <= XBAR < 4.9 THEN MEANS = '4.5 TO 4.9';
  IF 4.9 <= XBAR < 5.3 THEN MEANS = '4.9 TO 5.3';
  IF 5.3 <= XBAR < 5.7 THEN MEANS = '5.3 TO 5.7';
  IF 5.7 <= XBAR < 6.1 THEN MEANS = '5.7 TO 6.1';
  IF 6.1 <= XBAR < 6.5 THEN MEANS = '6.1 TO 6.5';
  IF 6.5 <= XBAR < 7.0 THEN MEANS = '6.5 TO 7.0';
PROC MEANS;
  VAR XBAR;
PROC FREQ;
  TABLES MEANS;
PROC CHART;
  HBAR MEANS/TYPE = PCT;
```

OUTPUT FILE (A)

SAS

VARIABLE	N	MEAN	STANDARD DEVIATION	MINIMUM VALUE	MAXIMUM VALUE	STD ERROR OF MEAN	SUM	VARIANCE	C.V.
XBAR	50	4.39600000	0.67671475	2.40000000	6.00000000	0.09570192	219.80000000	0.65794286	15.394

SAS

MEANS	FREQUENCY	PERCENT	CUMULATIVE FREQUENCY	CUMULATIVE PERCENT
2.0 TO 2.5	1	2.0	1	2.0
2.9 TO 3.3	1	2.0	2	4.0
3.3 TO 3.7	6	12.0	8	16.0
3.7 TO 4.1	7	14.0	15	30.0
4.1 TO 4.5	14	28.0	29	58.0
4.5 TO 4.9	8	16.0	37	74.0
4.9 TO 5.3	11	22.0	48	96.0
5.3 TO 5.7	1	2.0	49	98.0
5.7 TO 6.1	1	2.0	50	100.0

```
                                   SAS

                          PERCENTAGE BAR CHART

MEANS                                                                      FREQ  CUM.   PERCENT    CUM.
                                                                                 FREQ             PERCENT

2.0 TO 2.5  ****                                                              1     1     2.00      2.00

2.9 TO 3.3  ****                                                              1     2     2.00      4.00

3.3 TO 3.7  ************************                                          6     8    12.00     16.00

3.7 TO 4.1  ****************************                                      7    15    14.00     30.00

4.1 TO 4.5  ********************************************************         14    29    28.00     58.00

4.5 TO 4.9  ********************************                                  8    37    16.00     74.00

4.9 TO 5.3  ********************************************                     11    48    22.00     96.00

5.3 TO 5.7  ****                                                              1    49     2.00     98.00

5.7 TO 6.1  ****                                                              1    50     2.00    100.00

           ----+---+---+---+---+---+---+---+---+---+---+---+---+---+
               2   4   6   8  10  12  14  16  18  20  22  24  26  28

                              PERCENTAGE
```

Figure 6.23*b*
SAS output for illustration of the central limit theorem for 1000 random samples of 5 observations.

OUTPUT FILE (B)

SAS

VARIABLE	N	MEAN	STANDARD DEVIATION	MINIMUM VALUE	MAXIMUM VALUE	STD ERROR OF MEAN	SUM	VARIANCE	C.V.
XEAR	1000	4.49880000	0.74907922	2.40000000	6.40000000	0.02368796	4498.8000000	0.56111968	16.651

SAS

MEANS	FREQUENCY	PERCENT	CUMULATIVE FREQUENCY	CUMULATIVE PERCENT
2.0 TO 2.5	4	0.4	4	0.4
2.5 TO 2.9	10	1.0	14	1.4
2.9 TO 3.3	40	4.0	54	5.4
3.3 TO 3.7	96	9.6	150	15.0
3.7 TO 4.1	152	15.2	302	30.2
4.1 TO 4.5	205	20.5	507	50.7
4.5 TO 4.9	183	18.3	690	69.0
4.9 TO 5.3	170	17.0	860	86.0
5.3 TO 5.7	81	8.1	941	94.1
5.7 TO 6.1	47	4.7	988	98.8
6.1 TO 6.5	12	1.2	1000	100.0

SAS

FREQ	CUM. FREQ	PERCENT	CUM. PERCENT
4	4	0.40	0.40
10	14	1.00	1.40
40	54	4.00	5.40
96	150	9.60	15.00
152	302	15.20	30.20
205	507	20.50	50.70
183	690	18.30	69.00
170	860	17.00	86.00
81	941	8.10	94.10
47	988	4.70	98.80
12	1000	1.20	100.00

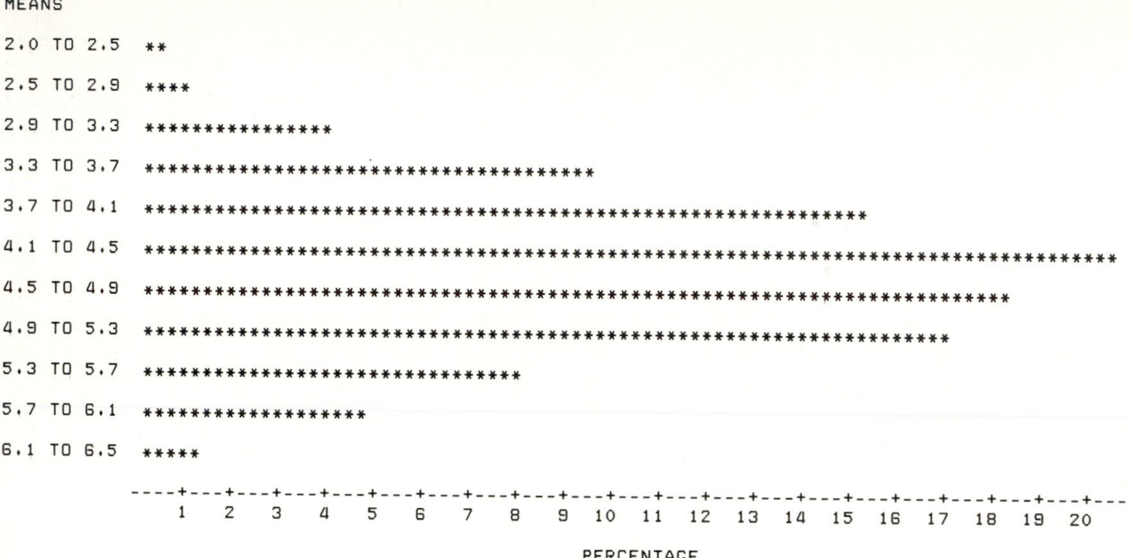

Figure 6.23c
SAS output for illustration of the central limit theorem for 50 random samples of 50 observations.

OUTPUT (C)

SAS

VARIABLE	N	MEAN	STANDARD DEVIATION	MINIMUM VALUE	MAXIMUM VALUE	STD ERROR OF MEAN	SUM	VARIANCE	C.V.
XBAR	50	4.52640000	0.27616736	3.94000000	5.16000000	0.03905596	226.32000000	0.07626841	6.101

SAS

MEANS	FREQUENCY	PERCENT	CUMULATIVE FREQUENCY	CUMULATIVE PERCENT
3.8 TO 4.0	1	2.0	1	2.0
4.0 TO 4.2	5	10.0	6	12.0
4.2 TO 4.4	10	20.0	16	32.0
4.4 TO 4.6	12	24.0	28	56.0
4.6 TO 4.8	13	26.0	41	82.0
4.8 TO 5.0	7	14.0	48	96.0
5.0 TO 5.2	2	4.0	50	100.0

SAS

FREQ	CUM. FREQ	PERCENT	CUM. PERCENT
1	1	2.00	2.00
5	6	10.00	12.00
10	16	20.00	32.00
12	28	24.00	56.00
13	41	26.00	82.00
7	48	14.00	96.00
2	50	4.00	100.00

```
                         PERCENTAGE BAR CHART

MEANS

3.8 TO 4.0  ****

4.0 TO 4.2  ********************

4.2 TO 4.4  ****************************************

4.4 TO 4.6  ************************************************

4.6 TO 4.8  ****************************************************

4.8 TO 5.0  ****************************

5.0 TO 5.2  ********

            ----+---+---+---+---+---+---+---+---+---+---+---+---+
                2   4   6   8  10  12  14  16  18  20  22  24  26

                             PERCENTAGE
```

Figure 6.23*d*
SAS output for illustration of the central limit theorem for 1000 random samples of 50 observations.

OUTPUT FILE (D)

SAS

VARIABLE	N	MEAN	STANDARD DEVIATION	MINIMUM VALUE	MAXIMUM VALUE	STD ERROR OF MEAN	SUM	VARIANCE	C.V.
XBAR	1000	4.50874000	0.24352272	3.76000000	5.20000000	0.00770086	4508.7400000	0.05930332	5.401

SAS

MEANS	FREQUENCY	PERCENT	CUMULATIVE FREQUENCY	CUMULATIVE PERCENT
3.6 TO 3.8	1	0.1	1	0.1
3.8 TO 4.0	17	1.7	18	1.8
4.0 TO 4.2	77	7.7	95	9.5
4.2 TO 4.4	207	20.7	302	30.2
4.4 TO 4.6	326	32.6	628	62.8
4.6 TO 4.8	244	24.4	872	87.2
4.8 TO 5.0	103	10.3	975	97.5
5.0 TO 5.2	24	2.4	999	99.9
5.2 TO 5.4	1	0.1	1000	100.0

SAS

FREQ	CUM. FREQ	PERCENT	CUM. PERCENT
1	1	0.10	0.10
17	18	1.70	1.80
77	95	7.70	9.50
207	302	20.70	30.20
326	628	32.60	62.80
244	872	24.40	87.20
103	975	10.30	97.50
24	999	2.40	99.90
1	1000	0.10	100.00

```
MEANS                        PERCENTAGE BAR CHART

3.6 TO 3.8

3.8 TO 4.0  ***

4.0 TO 4.2  ***************

4.2 TO 4.4  *****************************************

4.4 TO 4.6  *****************************************************************

4.6 TO 4.8  *************************************************

4.8 TO 5.0  *********************

5.0 TO 5.2  *****

5.2 TO 5.4

            ----+---+---+---+---+---+---+---+---+---+---+---+---+---+---+---+-
                2   4   6   8  10  12  14  16  18  20  22  24  26  28  30  32

                                 PERCENTAGE
```

Note that 5 observations are not sufficient to provide an approximately normal distribution for $\overline{x}$ with 50 samples, but with 1000 samples the graph approaches normality. For 50 observations the graph of $\overline{x}$ is definitely mound-shaped with 50 samples, and with 1000 samples the normal distribution is clearly suggested.

As a last example we show how to use SAS statements to obtain a trimmed-mean confidence interval. Figure 6.24 shows the program and results. The confidence interval for the original sample is shown as well. Note that the trimmed-mean confidence interval is narrower and therefore more efficient.

Figure 6.24
Confidence interval for the trimmed mean, compared with confidence interval from original sample.

```
INPUT FILE

DATA FIRST;
  INPUT SCORE @@;
CARDS;
73 62 31 59 94 71 58 73 54 67 91 70 73 71 61 55 67 68 48 73
PROC MEANS MEAN STD STDERR NOPRINT;
  OUTPUT OUT = A MEAN= XBAR STD = S STDERR = SE;
DATA CONF;
  SET A;
  LLIMIT = XBAR - 1.729 * SE;
  ULIMIT = XBAR + 1.729 * SE;
PROC PRINT;
  VAR XBAR S LLIMIT ULIMIT;
TITLE '95% CONFIDENCE INTERVAL FOR POPULATION MEAN FROM ORIGINAL SAMPLE';
DATA TRIM;
  INPUT SCORE @@;
CARDS;
73 62 59 71 58 73 54 67 70 73 71 61 55 67 68 73
PROC MEANS MEAN NOPRINT;
  OUTPUT OUT = B MEAN = XBART;
DATA EXTRA;
  INPUT SCORE @@;
CARDS;
54 54 73 73
DATA WINSOR;
  SET TRIM EXTRA;
PROC MEANS STD NOPRINT;
  OUTPUT OUT = C STD = SW;
DATA CONFINT;
  MERGE B C;
  SE = (SW / SQRT(16)) * SQRT (19/15);
  LLIMIT = XBART - 1.753 * SE;
  ULIMIT = XBART + 1.753 * SE;
PROC PRINT;
  VAR XBART SW LLIMIT ULIMIT;
TITLE 'TRIMMED 95% CONFIDENCE INTERVAL FOR POPULATION MEAN';

OUTPUT FILE
```

```
95% CONFIDENCE INTERVAL FOR POPULATION MEAN FROM ORIGINAL SAMPLE

        OBS     XBAR          S          LLIMIT      ULIMIT

         1      65.95       13.8506      60.5951     71.3049
     TRIMMED 95% CONFIDENCE INTERVAL FOR POPULATION MEAN

        OBS     XBART         SW          LLIMIT      ULIMIT

         1     65.9375      7.49368      62.2414     69.6336
```

Problems

Applications

6.77 Consider the reaction time data given in Problem 1.8.

a. Determine whether it is reasonable to assume that the data set is approximately normal. (See Problem 5.36.)

b. Obtain a 95% confidence interval for the mean time.

c. Using SAS statements, obtain a 90% confidence interval for the population variance.

6.78 Refer to the Miami Female Study data in Appendix B.1.

a. Determine a 95% confidence interval for the mean height of Miami females.

b. Determine a 95% confidence interval for the mean number of credit hours taken by Miami students.

6.79 According to a poll of 1007 adults (R. H. Bruskin Associates, 1985), 4% of the respondents were allergic to dairy products. Suppose instead that actually 42 of the respondents were allergic to dairy products. Use SAS statements to determine a 95% confidence interval for the proportion of the population that is allergic to dairy products, assuming that the sample was random.

6.80 Use Minitab to simulate 100 samples of size 15 from a normal distribution with a mean of 25 and a standard deviation of 4. For each sample obtain a 95% confidence interval for the mean using the TINTERVAL command. How many of your confidence intervals actually contain the mean? Is your answer close to what you expected?

6.81 We may use SAS statements to simulate m samples of size n from a normal distribution with a mean of μ and a standard deviation of σ. We can generate the samples by using statements similar to those in the first 13 lines of the input for Figure 6.23, using appropriate limits for I and J. In line 5 we need the appropriate normal function (see Problem 5.40), and in line 13 we need to add STDERR = SE to the OUTPUT statement. We can then set data set A into another data set and create the variables LLIMIT and ULIMIT, writing LLIMIT = XBAR − t * SE and ULIMIT = XBAR + t * SE for appropriate values of t. For very large samples we may use the appropriate value of z. Then the PRINT procedure will write out all m intervals. Use the statements outlined above to simulate 100 samples of size 15 from a normal distribution with a mean of 25 and a standard deviation of 4. For each sample obtain a 95% confidence interval.

How many of the intervals actually contain the mean? Is your answer close to what you expected?

6.82 Use SAS statements to simulate 100 samples of size 15 from a Poisson distribution with mean 10. For each sample obtain a 95% confidence interval using the correct value of t. (Note that t is not appropriate in this case since the distribution is not normal.) How many intervals contain the mean? Repeat the procedure using samples of size 50 and $z = 1.96$. How many intervals contain the mean? What do you conclude?

6.83 There are many continuous probability distributions besides the normal distribution. For example, an *exponential distribution* is a continuous distribution which is highly skewed. With Minitab we can generate a random sample of size K1 from an exponential distribution with mean K2 by using the following commands:

```
MTB >   RANDOM K1 observations in C ;
SUBC>   EXPONENTIAL with mean = K2 .
```

Using SAS statements, we can generate a random observation from an exponential distribution with mean M with the following statement:

```
X = RANEXP(seed ) / M;
```

a. Generate a random sample of size 50 from an exponential distribution with a mean of 5. Obtain a bar graph for the resulting data set.
b. Generate 100 random samples of size 50 from an exponential distribution with a mean of 5. Obtain a bar graph for the means of these random samples. Does the central limit theorem appear to hold in this case?

6.8 Case 6: The Ballpark Estimate—Solution

The accountant found that the standard deviation of the 8 days' attendance figures was about 6701. Since presumably the distribution of attendances for the season would be normal, he could use the t distribution to obtain an exact confidence interval from the sample. The ballplayer wanted only 1 chance in 20 of estimating too high, so the area under the curve to the right of the critical value would be 0.95. Then $t_{.95}\{7\} = -t_{.05}\{7\} = -1.895$, and $1.895(6701/\sqrt{8}) \doteq 4489$. Since $22{,}285 - 4489 = 17{,}796$, the ballplayer could be 95% confident that the mean attendance would be at least 17,796 per game. Multiplying this figure by 75, they obtained a lower bound (with a confidence level of 95%) of 1,334,700. An upper bound, obtained the same way, would be 75(22,285 + 4489) or 2,008,050, so a 90% confidence interval for the season's attendance would be 1,334,700 to 2,008,050 (since 5% of the area under the curve is in each tail).

6.9 Summary

Under certain conditions we can estimate population parameters from sample statistics. A random variable $\hat{\theta}$ is a good estimator for a population parameter θ if it is unbiased ($E(\hat{\theta}) = \theta$) and efficient (variance is small). A sample mean $\overline{x}$ is called a **point estimate** for the population mean. It is an **unbiased estimator** because $E(\overline{x}) = \mu$. It is also the most efficient estimator. In this chapter we saw how to find interval estimates for the mean, variance, and standard deviation of a random variable x and for the population proportion π of a binomial random variable from a randomly obtained sample.

Since means of different samples are different, it is usual to give a range of values that is likely to contain the population mean—a **confidence interval.** The **confidence level** is equal to the proportion or percentage of all intervals constructed in the same way from samples of the same size that will actually contain the population mean.

The confidence limits for estimating a population mean are summarized in Table 6.9. The actual limits used depend on factors such as size and shape of the sample distribution and on whether the standard deviation of the population is known or unknown. We can obtain confidence intervals for a population proportion π in a similar manner by using the normal approximation to the binomial distribution.

If we can estimate the population variance, we can determine the sample size needed to obtain a desired confidence level estimate for the population mean correctly within certain limits. We can use a similar formula to estimate the sample size

Table 6.9 SUMMARY OF CONFIDENCE LIMITS FOR ESTIMATING A POPULATION MEAN.

POPULATION STANDARD DEVIATION	SHAPE OF DISTRIBUTION	SAMPLE SIZE	CONFIDENCE LIMITS
Known	Normal	Any n	$\overline{x} \pm z_{\alpha/2} \cdot \sigma_{\overline{x}}$
	Nonnormal	$n \geq 30$	$\overline{x} \pm z_{\alpha/2} \cdot \sigma_{\overline{x}}$
Unknown	Normal	$n \geq 30$	$x \pm z_{\alpha/2} \cdot s_{\overline{x}}$
		Any n	$\overline{x} \pm t_{\alpha/2} \cdot s/\sqrt{n}$
	Nonnormal	$n \geq 50$	$\overline{x} \pm z_{\alpha/2} \cdot s_{\overline{x}}$
	Nonnormal	$30 \leq n < 50$	$\overline{x}(T) \pm z_{\alpha/2}\ s_{\overline{x}(T)}$
	Symmetric	$n < 30$	$\overline{x}(T) \pm t_{\alpha/2} \cdot s_{\overline{x}(T)}$
	Nonnormal, skewed	$n < 50$	No general method (use nonparametric methods)

needed to obtain a desired confidence level estimate for the population proportion correctly within certain limits.

We can also obtain confidence intervals for the population variance and standard deviation if the variable is normally distributed. We can get confidence intervals for the population standard deviation from the square root of these limits.

Chapter Review Problems

Applications

6.84 A random sample of 120 Graduate Record Examination scores is taken from those of a university's graduating class of 4182. If the mean of the sample is 1082 with a standard deviation of 108, determine 95% and 99% confidence intervals for:

a. the population mean

b. the population standard deviation

6.85 A newspaper reporter believes that 75% of retired couples prefer apartment living to house living. She wishes to sample retired couples and obtain a 99% confidence interval correctly to within 0.05. What sample size should she take if:

a. the newspaper estimate is used as a basis?

b. no estimate is used?

6.86 Under conditions of stress, the mean heartbeat rate of 64 persons is found to be 134 with a standard deviation of 12. Determine 90% and 95% confidence intervals for the population's mean heartbeat rate under the conditions of the experiment.

6.87 To determine the educational level of a certain ethnic group in a particular locale, a sociologist interviews 50 members of the group. The mean educational level is 11.4 years of school with a standard deviation of 3.2 years. Determine 95% and 98% confidence intervals for the true mean educational level.

6.88 Four subjects are randomly selected from a population and asked to make subjective judgments involving interpersonal relationships. The judgments are entered on a rating sheet with the ratings weighted and an overall score is obtained. These four subjects obtain scores of 173, 217, 143, and 166. Assuming that the population is normal, estimate a 95% confidence interval for the population mean.

6.89 A researcher wishes to obtain the mean income of a certain area correctly to within \$100 with a confidence level of 95%. What size of random sample should she take if \$500 is a reasonable estimate for the population standard deviation?

6.90 According to the *New York Times* (Jan. 5, 1986), the president had a 56% approval rating among blacks. It was later found that the survey was based on interviews with 110 blacks. Construct a 90% confidence interval for the true proportion of blacks who approve of the president.

6.91 A *Washington Post* poll (reported in the *Wall Street Journal,* Jan. 23, 1986) conducted with a sample of 1022 blacks found that 63% of the respondents disapproved of the president. Construct a 90% confidence interval for the true proportion of blacks who disapprove of the president. Compare your result with Problem 6.90. Is it important to know how the sample was obtained and the exact nature of the questions?

6.92 A random sample of 80 insurance salesmen shows 56 in opposition to proposed revisions in their company's pension plan. Construct 95% and 98% confidence intervals for the proportion of all salesmen in the company who oppose the revisions.

6.93 A new strain of bacteria is introduced into the water supply of a group of 84 white mice. As a result, 36 of the mice are affected adversely with muscular spasms and general debilitation. Give 95% and 99% confidence intervals for the proportion of all white mice that will be affected adversely under like conditions.

6.94 Sixteen mentally handicapped patients are given a task requiring a certain degree of mental dexterity. The mean time required to do the task is 11 minutes 33 seconds with a standard deviation of 3 minutes 12 seconds. Construct 90% and 95% confidence intervals for the mean time they require to do this task.

6.95 A random sample of 200 families of 750 in an apartment complex is interviewed, and the mean number of children per family is found to be 3.2 with a standard deviation of 0.8. Calculate the 95% confidence interval for the mean number of children per family in the apartment complex.

6.96 Weights of 20 babies born at a local hospital in 1985 had a mean of 6.34 pounds with a standard deviation of 1.67 pounds. (See Problem 6.60.) Suppose we wish to estimate the mean weight of the babies born at this hospital in 1985 correct to within ½ pound with a 99% confidence level. What sample size should we take?

6.97 In a shipment of 1000 phonograph records, one of the first ten inspected shows defects. Using this result as an estimate for the proportion defective in the shipment, how many should be inspected as a random sample to estimate the true proportion defective correctly to within 0.05 with a 95% confidence level?

6.98 In 1984, many polls were taken to determine which of the two presidential candidates was preferred. In a preliminary poll (June 1984) it was found that 56% of those polled preferred Reagan to Mondale. How many people should have been polled to be correct to within 0.05 with a confidence level of 99% if:

a. the results of the preliminary poll were used as an estimate?
b. the pollsters wished to be absolutely certain they had polled a sufficient number?

6.99 The receiving department of a large electronics manufacturer uses the following rule in deciding whether to accept or reject a shipment of 100,000 identical small parts. Select a random sample of 400 parts from the lot. If 3% or more of these parts are defective, reject the entire lot; otherwise accept the lot. What is the probability that a lot containing 2% defective parts will be rejected?

6.100 To estimate the mean efficiency rating of all machines produced by a company during the same period, efficiency ratings of 12 identical machines are chosen at random from among the company's output of thousands over the past 5 years. The ratings of the 12 machines are 0.820, 0.913, 0.764, 0.881, 0.902, 0.893, 0.663, 0.862, 0.812, 0.778, 0.932, 0.824.

a. If the mean of this sample is used to estimate the population mean, what can be said with a confidence of 95% about the possible size of the error?

b. Determine a 95% confidence interval for the population variance and standard deviation.

Index of Terms

Glossary of Symbols

E	maximum allowable error with a given confidence level; half-width of a confidence interval
$\chi^2(\nu)$	chi-square distribution with ν degrees of freedom
$\chi^2_\alpha\{\nu\}$	critical value of chi square for a chi-square distribution with ν degrees of freedom (α is the area under the curve to the right of the critical value)
μ	population mean
μ_p	mean, sampling distribution of proportion
$\mu_{\hat{\theta}}$	mean, sampling distribution of the statistic $\hat{\theta}$
$\mu_{\bar{x}}$	mean, sampling distribution of the mean

N	population size
n	sample size
ν	number of degrees of freedom
π	population proportion
p	sample proportion
σ	population standard deviation
σ^2	population variance
σ_p	standard error of proportion
$\sigma_{\hat{\theta}}$	standard error of the random variable $\hat{\theta}$
$\sigma_{\overline{x}}$	standard error of the mean
s	sample standard deviation
$s(W)$	standard deviation of a Winsorized data set
s^2	sample variance
s_p	sample standard error of proportion
$s_{\overline{x}}$	sample standard error of the mean
$s_{\overline{x}(T)}$	sample standard error of the trimmed mean
$t(\nu)$	t distribution with ν degrees of freedom
$t_\alpha\{\nu\}$	critical value of t for a t distribution with ν degrees of freedom (α is the area under the curve to the right of the critical value)
θ	population parameter
$\hat{\theta}$	sample statistic; estimator for population parameter θ
$\overline{x}$	sample mean
$\overline{x}(T)$	trimmed mean (mean of a trimmed data set)
$\sqrt{\dfrac{N-n}{N-1}}$	finite population correction factor
z_α	critical value of z for a standardized normal distribution (α is the area under the curve to the right of the critical value)

Answers to Proficiency Checks

1. $134.8/\sqrt{40} \doteq 21.3$

2. $z \doteq -2.42$; $P(\overline{x} < 17{,}500) \doteq 0.008$

3. $P(\overline{x} > 128.0) = 0.0668$; $P(122.5 < \overline{x} < 126.0) = 0.3944 + 0.1915 = 0.5859$

4. **a.** 1.18 **b.** 0.83 **c.** 0.59 **d.** 0.29

5. **a.** 1.18 **b.** 0.83 **c.** 0.58 **d.** 0.28

6. **a.** 91.68 to 108.32 **b.** 93.21 to 106.79 **c.** 94.12 to 105.88
   d. 95.84 to 104.16

7. **a.** 0.534 to 0.586 **b.** 0.529 to 0.591 **c.** 0.523 to 0.597
   d. 0.519 to 0.601

8. **a.** $s_{\overline{x}} = 0.30/\sqrt{100} = 0.03$; $z = 0.05/0.03 \doteq 1.67$; the probability is about 0.9050.

b. $(2.58)(0.03) \doteq 0.08$
c. 1.87 to 2.01

9. a. $s_{\bar{x}} = 0.32/\sqrt{250} \doteq 0.02$; $z = 0.05/0.02 = 2.50$; the probability is about 0.9876.
b. $(2.58)(0.02) \doteq 0.05$
c. 1.87 to 1.97

10. a. 0.0837 b. 0.0592 c. 0.0458 d. 0.0229

11. a. 0.0834 b. 0.0588 c. 0.0454 d. 0.0220

12. 0.060

13. 0.26 to 0.46

14. a. $p = 250/400 = 0.625$; $s_p \doteq 0.24$; the 95% confidence interval for p is 0.578 to 0.672.
b. $p = 80/250 = 0.32$; $s_p \doteq 0.030$; the 95% confidence interval is 0.261 to 0.379.

15. a. 62.76 to 137.24 b. 78.54 to 121.46 c. 85.96 to 114.04
d. 87.62 to 112.38

16. a. 0.505 to 0.615 b. 0.492 to 0.628 c. 0.477 to 0.643
d. 0.466 to 0.654

17. a. 8.30 to 9.02 b. 8.29 to 9.03

18. $t_{.025}\{15\} = 2.131$; we are 95% confident that the mean mileage lies between 18.9 and 21.7 miles per gallon.

19. $\bar{x} = 993.2$, $s \doteq 7.69$, $t_{.005}\{4\} = 4.604$; 977.4 to 1009.0

20. $\bar{x}(T) = 29.50$; $s(W) \doteq 3.0523$; $s_{\bar{x}(T)} \doteq 1.0289$; $t_{.05}\{11\} = 1.796$; confidence limits are 29.50 ± 1.85; confidence interval is 27.65 to 31.35.

21. $n = [(2.58)(0.30)/0.05]^2$ or about 240

22. a. 3181 b. 3393

23. a. 1067 or 1068 b. 0.169 to 0.216

24. a. 323.0641 to 7431.5830 b. 425.8048 to 2999.5667
c. 520.5115 to 1919.9353 d. 548.7233 to 1741.7810

25. a. 0.094 to 0.178 b. 0.089 to 0.192 c. 0.085 to 0.211
d. 0.082 to 0.226

26. 0.97 to 1.51

References

Gross, A. M. "Confidence Interval Robustness with Long-Tailed Symmetric Distributions." *Journal of the American Statistical Association* 71 (1976): 409–416.

Kohler, Heinz. *Statistics for Business and Economics*. Glenview, IL: Scott, Foresman and Company, 1985.

Koopmans, Lambert H. *An Introduction to Contemporary Statistics.* Boston: Duxbury Press, 1981.

Mendenhall, William. *Introduction to Probability and Statistics.* 6th ed. Boston: Duxbury Press, 1983.

Mosteller, Frederick, and John W. Tukey. *Data Analysis and Regression.* Reading, MA: Addison-Wesley, 1977.

7

Hypothesis Testing

Decisions are generally based on a belief regarding the true state of things. People buy stock in the belief that its value will increase rather than decrease, for example. A manufacturer of fertilizer may believe that his new formula will produce greater yields than another type. A team of biologists may believe they have discovered a new species. A motion picture producer wishes to determine whether a new picture will be a success. A politician wants to know if he has a good chance against another politician. In each case, statistical inference provides an excellent systematic method for assessing whether or not the belief is likely to be true.

This chapter presents methods for testing hypotheses, or beliefs, and drawing conclusions about the probability that these hypotheses are true. Often these methods are extensions of the estimation techniques discussed in Chapter 6. First we will discuss the general method of formulating and testing hypotheses and then learn how to apply these methods to hypotheses about a population mean. We will cover methods applicable to both large and small samples. Hypotheses about the population proportion in a dichotomous population are treated next, and the chapter closes with a method to test a hypothesis concerning a population variance.

7.0 Case 7: The Radioactive Residue

After months of searching for the right land, the Ajax Home-builders Corporation bought a large piece of property in order to build a subdivision providing "affordable housing for thousands of families who would otherwise be unable to have a home of their own." Just after the development was started, a group called Citizens for Safety sought an injunction to halt the development on the grounds that there was hazardous radioactivity. The site had been used as a dumping ground for radioactive waste, the group contended, and was unsafe for human habitation. Ajax said they had known about the dumping, but they had been assured by a government agency that the residual radiation was safe and no harm would result if people lived there. The judge issued a temporary injunction and referred the case to an agent of the Environmental Protection Agency for an opinion. The agent suggested that tests should be conducted to determine whether the radioactive residue was potentially harmful. The groups agreed to hire an outside agency to take readings and determine the level of radioactivity present.

The results of the testing merely led to more dispute. A total of 20 readings were taken in randomly selected locations, and the average reading showed an annual rate of 432 rem, well under the recognized tolerance level of 500 rem. People absorb as much as 200 rem per year from natural radiation, so Ajax concluded that the site was perfectly safe. The citizen's group, however, noted that there was considerable variation from site to site and that the readings ranged from 410 to 473 rem with a standard deviation of 20.8. Ajax replied that all the readings were below 500. The citizen's group responded that although this was so, it did not mean that readings from other locations would always be below 500, and they called for a more thorough testing. The agency making the tests protested, however, saying that it would take months to test every site and a random sample was sufficient. They also pointed out that if the radioactive readings were normally distributed from location to location (a reasonable supposition) and the mean and standard deviation were 432 and 20.8, as in the sample, the standard normal deviate of 500 would be 3.27 and the probability of a single reading being above 500 would be less than 0.0005.

After much arguing, the two groups agreed that if the average reading for the entire property were above 450, the project should be canceled; if it were less than 450, the project could be built. The agent stated, however, that unless it could be shown that the mean radioactive level was less than 450 rem beyond a reasonable doubt, she would recommend to the judge that the injunction be made permanent. When pressed, the agent defined "reasonable doubt" to be more likely than 1 chance in 100. She also asked the agency making the tests whether the results could be used to make that determination; the agency said that they could. What will the recommendation be?

7.1 Hypothesis-Testing Techniques

A belief can be formulated as a *hypothesis.* A statistical hypothesis is a statement regarding the value of one or more population parameters. A hypothesis supporting the belief is called the **research hypothesis.**

Suppose that a laboratory has developed a treatment which is believed to be effective against eczema. The belief is that the drug will reduce or eliminate eczema in a sufferer. In order to investigate this belief statistically, we must determine what happens to eczema sufferers if left untreated and show that there is improvement if the new treatment is used. There is often more than one way to test a hypothesis. One possible approach in this case is to investigate the extent of rashes in a sufferer and determine whether the extent diminishes more with the use of this treatment than without it. A second approach is to determine the average time it takes an attack to run its course and show that the average time is reduced by using the treatment. Still another approach is to show that the proportion of cases which remain after a specified period of time is less with the use of the treatment than without it.

Suppose that research has shown that if white mice are infected with a case of eczema, 2/5 of them will be symptom-free naturally within 4 weeks. We can assert that if the new treatment is effective, more than 2/5 of the mice will be symptom-free after 4 weeks. Thus a statement about a population parameter that will support our belief is the hypothesis $\pi > 0.4$, where π is the proportion of all infected mice who will be symptom-free after 4 weeks with the new treatment.

In another situation, a social agency wishes to determine whether the income level of families in a large housing development has remained the same or changed (when adjusted for inflation) over the past 5 years. They determine that the family income level 5 years ago was $16,788 and that, adjusted for inflation, the same income level would be $20,278 today. In the absence of any evidence to the contrary, the agency believes there has been no change. The statement about a population parameter that will support this belief is the hypothesis $\mu = 20{,}278$, where μ is the mean income, in dollars, of the families in the development.

Consider a production line where the foreman believes that a machine produces too many items which do not meet standard specifications. On the average, they are still meeting the specifications, but the variability may be such that an unacceptable proportion fails to meet the requirements. One statement about a population parameter supporting this belief is the hypothesis $\sigma^2 > \sigma_0^2$, where σ_0^2 is the maximum acceptable variance in the product. Another such hypothesis is $\pi > \pi_0$, where π_0 is the maximum acceptable proportion of items not meeting the standard.

7.1.1 Formulating the Hypotheses

Once we have formulated a research hypothesis, we can test it by experimentation. To do so, we first formulate two hypotheses: the *null hypothesis* and the *alternate hypothesis*. The **null hypothesis** is a statement that the population parameter is equal to a specific value, such as $\pi = 0.4$ or $\mu = 20{,}278$ in the preceding cases. The purpose of experimentation is to try to prove or disprove this hypothesis. In the general case, we write $H_0{:}\theta = \theta_0$; the hypothesis to be tested is that the population parameter has the value stated.

Ronald A. Fisher (1890–1962) viewed hypothesis testing as a method for forming an *opinion* as to whether or not a specific value of the parameter is likely to be the true value. Using Fisher's method, we first determine the sampling distribution of an estimator for the parameter based on the

Fisher was one of the most productive statisticians who ever lived. In about 50 years he published nearly 300 papers, most of them introducing new concepts. His book Statistical Methods for Research Workers *(1925) is possibly the most important statistics book ever written. His idea of interval estimates was different from the Neyman–Pearson concept, as was his hypothesis-testing technique. Indeed, these and other differences led to a bitter and long-lasting feud. Although his views on these matters no longer prevail, his introduction of such concepts as unbiasedness and efficiency and his work with experimental design formed the foundation of modern statistical inference.*

hypothesized value of the parameter; then we determine the probability that a sample result as extreme as the one we obtained could occur by chance. If the probability is greater than 0.05, we conclude that the null hypothesis is true. If the probability is less than 0.05 but not less than 0.01, we conclude that the null hypothesis is unlikely to be true. If the probability is less than 0.01, we conclude that the hypothesis is highly unlikely to be true.

In 1933, Jerzy Neyman and Egon S. Pearson (originators of the confidence interval) introduced the idea of an **alternate hypothesis**—a hypothesis opposing the null hypothesis. As in Fisher's method, they determined the sampling distribution of an estimator for the parameter based on the hypothesized value of the parameter. In contrast to the opinion-formation view of Fisher, however, this procedure is used to make a *decision* between two alternatives—whether or not to market a product, for example. Unlike Fisher's method, it allows for the formulation of a one-sided research hypothesis, such as $\pi > 0.4$. If the research hypothesis is not a statement of equality such as $\mu = 20{,}278$, then the alternate hypothesis is a hypothesis that supports the research hypothesis. If the research hypothesis is a statement of equality, it then becomes the null hypothesis. This method of hypothesis testing is the one in use today.

To determine whether the new treatment for eczema is effective, for example, we might infect a random sample of mice with eczema and then use the treatment to see what proportion is symptom-free after 4 weeks. If most are symptom-free, the treatment is probably effective. On the other hand, there is a remote possibility that the treatment is not effective and the result is due to chance. Suppose, for example, we treat 15 mice; if the treatment is not effective, the probability that any one mouse will be symptom-free without treatment is 0.4. The probability that at least 12 mice will be symptom-free by chance can be found (from Table 2) to be 0.002. Thus if 12 of the 15 mice are symptom-free, the probability that the treatment is ineffective and the results are due to chance is 0.002; if we conclude that the treatment is effective, the probability that we are wrong will be no more than 0.002. If all 15 mice are cured, the probability that this might happen by chance if the treatment were ineffective is $(0.4)^{15}$, or about 0.00000107. The probability of being wrong is exceedingly small in this case, but it still exists.

Conversely, suppose the treatment is so effective that it doubles (to 0.8) the proportion of mice who will be symptom-free after 4 weeks. The probability that 6 or fewer of 15 mice will be symptom-free after 4 weeks in this case is only about 0.001—but it could happen. In that case we would probably conclude erroneously that the treatment was not effective. Thus we can never be *completely* certain about the effectiveness of the treatment. We must always take chances. *Determining when to take these chances and controlling the risk involved is the essence of statistical hypothesis testing.*

The statistical hypothesis-testing procedure requires that we determine an appropriate sampling distribution of a sample statistic based on the assumption that the null hypothesis about the population parameter is true. If the hypothesis concerns a population mean, the appropriate sampling distribution is the sampling distribution of the mean; if it concerns a population proportion, we should use the sampling distribution of the sample proportion, and so on. To learn what the sampling distribution looks like, we must make some assumptions. Usually we assume that the null hypothesis is true and decide whether or not an assumption of normality for the population is appropriate.

A list of all the assumptions used in analyzing the results of the experiment constitutes a statistical **model** for the experiment. We then take a sample and observe whether or not the sample result is consistent with the model. In the case of the eczema treatment, our research hypothesis is $\pi > 0.4$; the null hypothesis would then be $H_0{:}\pi = 0.4$ and the model would be a binomial population with $\pi = 0.4$. If our sample result showed a result inconsistent with this model in such a way that we believe $\pi > 0.4$, we would conclude that the treament was effective. If not—that is, if we were to conclude that $\pi \leqslant 0.4$ is reasonable—we would be forced to conclude that the treatment is either not effective or counterproductive.

For a simple experiment involving one sample, the model is more often assumed than explicitly discussed. Nevertheless, we must thoroughly understand the elements of the model. For example, if we wish to test the hypothesis $\mu = \mu_0$ and we believe that the distribution of the variable x is normal, these two elements explicitly determine our model.

If our research hypothesis is about the population parameter θ, we would probably state it in one of the following ways: $\theta = \theta_0$, $\theta > \theta_0$, $\theta < \theta_0$, $\theta \leqslant \theta_0$, or $\theta \geqslant \theta_0$. In each of these cases, θ_0 is some hypothetical reference value of the population parameter. The only statement we can use to define a model is the statement $\theta = \theta_0$. The value θ_0 is called the **reference value** of the parameter θ, and the statement $\theta = \theta_0$ is the null hypothesis. The null hypothesis (designated H_0) is the hypothesis that the true value of the population parameter θ is no different from (hence the word *null*) the reference value θ_0. We use the null hypothesis to determine whether or not we can reasonably conclude that the research hypothesis is true. The alternate hypothesis (designated H_1 or H_a—no consensus exists) is then formulated according to the research hypothesis. If the research hypothesis is $\theta = \theta_0$, the alternate hypothesis is $\theta \neq \theta_0$. If the research hypothesis is $\theta > \theta_0$ or $\theta \geqslant \theta_0$, the alternate hypothesis must be $\theta > \theta_0$. This is because the null and alternate hypotheses must be mutually exclusive, and $\theta = \theta_0$ and $\theta \geqslant \theta_0$ are not. Similarly, if our research hypothesis is $\theta < \theta_0$ or $\theta \leqslant \theta_0$, the alternate hypothesis is $\theta < \theta_0$. If the research hypothesis is a statement of strict inequality (not $\leqslant$ or $\geqslant$), the research hypothesis and the alternate hypothesis are the same.

After we have formulated the hypotheses, we collect data to see whether the sample result is consistent or inconsistent with the null hypothesis. *If the sample result supports the alternate hypothesis, we reject the null hypothesis in favor of the alternate hypothesis.* Note that the null and alternate hypotheses are not necessarily exhaustive. If the null and alternate hypotheses are $H_0{:}\theta = \theta_0$ and $H_1{:}\theta > \theta_0$ and the sample result suggests that θ is probably less than θ_0, this result contradicts the alternate hypothesis. Our conclusion would be that we cannot conclude that H_1 is true, rather than that H_0 is true.

7.1.2 Two Types of Errors

The possibility of error is implicit in every statistical test. When we use a sample result to test the validity of a null hypothesis, we know that sample statistics vary. In the case of the eczema treatment, even if all 15 mice were symptom-free we noted that the result might still be due to chance. It is possible to reject a true hypothesis as well as to accept a false one. If we reject a true null hypothesis, the error is called a **type I error**; if we accept a false null hypothesis, the error is called a **type II error.** (These designations were introduced in 1933 by Neyman and Pearson.) The probability of making a type I error is designated α (alpha), and the probability of making a type II error is designated β (beta). Of course, there are also two ways of making the *correct* decision. Table 7.1 lists the possible decisions and outcomes.

The null hypothesis is generally designated H_0. We can use conditional probability statements to identify the four outcomes and probabilities listed in Table 7.1. Thus $P(\text{reject } H_0 | H_0 \text{ is true}) = \alpha$, $P(\text{accept } H_0 | H_0 \text{ is true}) = 1 - \alpha$, $P(\text{reject } H_0 | H_0 \text{ is false}) = 1 - \beta$, and $P(\text{accept } H_0 | H_0 \text{ is false}) = \beta$. Before we conduct the test, we know nothing about the actual value of the parameter being tested. If we assume that H_0 is actually true, we can usually determine the precise sampling distribution of the statistic used for testing the hypothesis so we can control the value of α. The value of α is usually set before the sample is obtained, and α is equal to the probability of making an error in rejecting the null hypothesis. This probability, called the **level of significance** for the test, designates the willingness of the experimenter to make a type I error. The probability of correctly accepting the null hypothesis $(1 - \alpha)$, is called the **level of confidence.** As you might guess,

Table 7.1 DECISIONS AND OUTCOMES IN HYPOTHESIS TESTING.

	ACTION	
NULL HYPOTHESIS IS	ACCEPT NULL HYPOTHESIS	REJECT NULL HYPOTHESIS
True (probability)	No error $(1 - \alpha)$	Type I error (α)
False (probability)	Type II error (β)	No error $(1 - \beta)$

there is a strong relationship between this level of confidence and the one in a confidence interval. Common preselected levels of significance are 0.10, 0.05, and 0.01, depending on the risk the experimenter is willing to take of making a type I error. Levels of 0.01, sometimes 0.001, are usually reserved for decisions for which the risk of making a mistake must be kept small—such as business decisions involving large amounts of money. If we cannot reject the null hypothesis at the stated level of significance, we usually conclude that H_0 is true or say that the results are inconclusive.

In many cases we decide upon an acceptable value of α and divide the possible outcomes into two groups before we select the sample. One group makes up the **rejection region**; the other group comprises the **acceptance region**. If the sample result falls into the rejection region, we reject the null hypothesis in favor of the alternate hypothesis. If the sample result falls into the acceptance region, we fail to reject the null hypothesis, concluding either that the null hypothesis is probably true or that the results are inconclusive—depending upon the nature of the hypotheses, the actual sample result, and the other possibilities, if any, that are open to us. We will discuss the options available to us later in this section.

There is no other name given to β, but $1 - \beta$ is called the **power** of the test—that is, the probability of correctly rejecting a false null hypothesis. Since a type II error can only occur if H_0 is actually false, the precise nature of the sampling distribution cannot be known. Some analysis is possible, however, and the relationship between α, β, and n is discussed later in this section.

PROFICIENCY CHECKS

1. If you spin a coin, the probability that a head will be showing when the coin comes to rest is not necessarily 0.5. A Lincoln penny, for instance, has more mass on the head side of the coin than the tail side, and this difference affects the probability of the coin showing a head. Suppose that you spin a coin 60 times, obtaining 35 heads. Deciding that this is sufficient evidence, you conclude that the probability of obtaining a head is not equal to 0.5. If the probability is actually 0.5, what kind of error have you made?
2. Suppose you take a different denomination of coin and again spin it 60 times. This time you obtain 27 heads. Since half of 60 is 30, you conclude that the probability of obtaining a head is 0.5. Unknown to you, however, the probability of obtaining a head is 0.4. What type of error have you made?

7.1.3 The Test Statistic

Separating all possible outcomes into acceptance and rejection regions can often be a formidable, even impossible, task. Furthermore, it is the responsibility of the experimenter to report procedures and results in such a way that they can be understood and even duplicated by others. (The term *replication* refers to the repetition of an experiment under the same conditions.) Moreover, it may be that others interested in the result of an experiment have different levels of significance in mind than the experimenter. To accommodate all these considerations, a procedure has been developed in which one of the "standard" random variables is used. So far we have studied three such variables: z, the standardized normal random variable, which is $N(0, 1)$; the t random variable, which is $t(\nu)$; and the χ^2 random variable, which is $\chi^2(\nu)$.

The **test statistic** to be used is determined by the parameter θ appearing in the null hypothesis, the shape of the distribution of the sample statistic $\hat{\theta}$, and the size of the sample. To test $H_0{:}\pi = \pi_0$, for instance, we observe that if H_0 is true and n is sufficiently large, the sampling distribution of sample proportion is normally distributed with mean π_0 and standard error σ_p, where $\sigma_p = \sqrt{\pi_0(1 - \pi_0)/n}$. That is, p is approximately $N(\pi_0, \sigma_p^2)$, so that the statistic $(p - \pi_0)/\sigma_p$ is approximately $N(0, 1)$. Thus our test statistic is z, where $z = (p - \pi_0)/\sigma_p$. Other test statistics are used in other situations. The actual value of the test statistic obtained in the experiment is often designated with a "c" or "calc," for calculated value, or with an asterisk(*). We will use the asterisk; thus if the test statistic is z, the actual value of the test statistic obtained from a sample result is written z^*.

7.1.4 The Rejection Region

In most cases an acceptable value of α is decided upon and the possible outcomes are divided into two groups before the sample is selected. One group makes up the *rejection region* and the other group the *acceptance region*. Once we have identified the test statistic, determining the rejection region is greatly simplified. We relate the sample result to the probability distribution of the test statistic and divide the area under the graph of this distribution into these two regions. The area above the rejection region is α, and the area above the acceptance region is $1 - \alpha$.

If, for example, the null hypothesis is $H_0{:}\theta = \theta_0$, the alternate hypothesis is $H_1{:}\theta \neq \theta_0$, and the test statistic is z, we say that the alternate hypothesis is **two-sided** and the rejection region is **two-tailed.** The alternate hypothesis is two-sided because we wish to reject H_0 if either $\theta > \theta_0$ or $\theta < \theta_0$—that is, if the true parameter is either greater than or less than θ_0. The rejection region is two-tailed because we will reject H_0 if z^* is far enough from zero in either direction so that the combined probability of being that far by chance if H_0 is true is less than α. Figure 7.1 illustrates a two-tailed rejection region for a test statistic z.

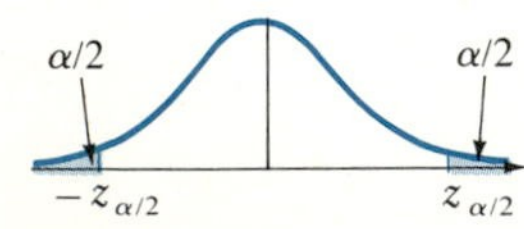

Figure 7.1
A two-tailed rejection region for z.

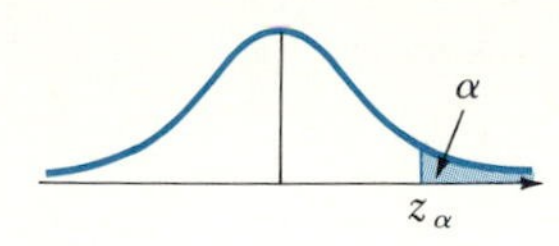

Figure 7.2
A one-tailed rejection region for z for $H_1{:}\theta > \theta_0$.

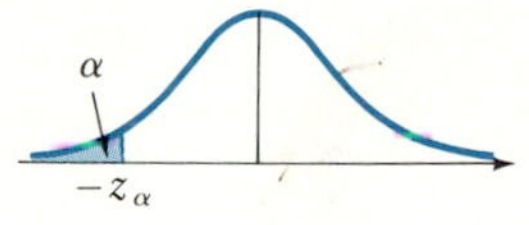

Figure 7.3
A one-tailed rejection region for z for $H_1{:}\theta < \theta_0$.

The sum of the two shaded areas in Figure 7.1 is equal to α; thus if H_0 is true, the probability that z^* is greater than $z_{\alpha/2}$ or less than $-z_{\alpha/2}$ is equal to α. The rejection region, then, is that portion of the z axis below the shaded area and consists of all values of z greater than $z_{\alpha/2}$ or smaller than $-z_{\alpha/2}$. The values $z_{\alpha/2}$ and $-z_{\alpha/2}$ are called the **critical values** of the test statistic. If the alternate hypothesis is $H_1{:}\theta \neq \theta_0$, we have the following **decision rule**: reject H_0 if $|z^*| > z_{\alpha/2}$. If we reject H_0, we can decide whether $\theta > \theta_0$ or $\theta < \theta_0$ on the basis of the sample result. If $|z^*| > z_{\alpha/2}$ and z^* is positive, we conclude that $\theta > \theta_0$; if $|z^*| > z_{\alpha/2}$ and z^* is negative, we conclude that $\theta < \theta_0$.

If the null hypothesis is $H_0{:}\theta = \theta_0$, the alternate hypothesis is $H_1{:}\theta > \theta_0$, and the test statistic is z, we say that the alternate hypothesis is **one-sided** and the rejection region is **one-tailed**. The alternate hypothesis is one-sided because we wish to reject H_0 in favor of H_1 if $\theta > \theta_0$ but not if $\theta \leqslant \theta_0$—that is, only if the true parameter is greater than θ_0. The rejection region is one-tailed because we will reject H_0 only if z^* is so much greater than zero that the probability of being that far by chance if H_0 is true is smaller than α. Figure 7.2 illustrates a one-tailed rejection region for a test statistic z.

The shaded area in Figure 7.2 is equal to α; thus if H_0 is true, the probability that z^* is greater than z_α is equal to α. The rejection region, then, is that portion of the z axis below the shaded area and consists of all values of z greater than z_α. If the alternate hypothesis is $H_1{:}\theta > \theta_0$, we have the following decision rule: reject H_0 if $z^* > z_\alpha$.

In the remaining case, if the null hypothesis is $H_0{:}\theta = \theta_0$, the alternate hypothesis is $H_1{:}\theta < \theta_0$, and the test statistic is z, we again have a one-sided alternate hypothesis and a one-tailed rejection region. The alternate hypothesis is one-sided because we wish to reject H_0 in favor of H_1 if $\theta < \theta_0$ but not if $\theta \geqslant \theta_0$—that is, only if the true parameter is smaller than θ_0. The rejection region is one-tailed because we will reject H_0 only if z^* is so much smaller than zero that the probability of being that far by chance if H_0 is true is smaller than α. Figure 7.3 illustrates a one-tailed rejection region for a test statistic z.

The shaded area in Figure 7.3 is equal to α; thus if H_0 is true, the probability that z^* is smaller than $-z_\alpha$ is equal to α. The rejection region, then, is that portion of the z axis below the shaded area and consists of all values of z less than $-z_\alpha$. If the alternate hypothesis is $H_1{:}\theta < \theta_0$, we have the following decision rule: reject H_0 if $z^* < -z_\alpha$.

7.1.5 The Decision and Conclusions

Once we have formulated the hypotheses and decided on the decision rule, all that remains is to obtain the data and see what the result is. We calculate the sample test statistic, such as z^*, and compare the result to the decision rule. This comparison leads to one of two possible **decisions**: reject the null hypothesis in favor of the alternate hypothesis or fail to reject the null hypothesis.

After the decision has been made, the experimenter must present the result of the decision—these are the **conclusions**. If the decision is to reject the null hypothesis in favor of the alternate hypothesis, the conclusions are generally clear; nonetheless, the meaning of the decision must be expressly discussed. If the decision is to fail to reject the null hypothesis, the conclusions are generally not so clear. If the sample result clearly confirms the null hypothesis, many statisticians say that the null hypothesis should be accepted. In some cases, however, although the test statistic was not in the rejection region, it may have been so close to the rejection region that the experimenter is unwilling to conclude that H_0 is true. In such cases the experimenter will often **reserve judgment**, or report that the results were inconclusive, and recommend a replication of the experiment. Replication is not always possible, of course. If a truck driver brings a load of parts to the loading dock, the quality control expert may take a sample to ensure that the parts are acceptable. If the results are inconclusive, however, a decision must nonetheless be made whether or not to unload the shipment. When the null hypothesis must either be accepted or rejected and it cannot be rejected (because the test statistic is not in the rejection region), it must be accepted. Note further that reserving judgment is appropriate *only* when we fail to reject H_0. *If the decision is to reject H_0, we do not reserve judgment.*

7.1.6 Summary of Hypothesis-Testing Procedures

The foregoing subsections have presented a systematic procedure for formulating and testing a hypothesis designed to verify or deny a belief held by an experimenter. First it is necessary to formulate a research hypothesis, usually a statement about a population parameter that will verify the belief. After we have done this, the following five steps constitute the simple **hypothesis-testing procedure**.

1. *Statement of hypotheses:*
   H_0: Statement of the null hypothesis—a statement about the population parameter that will define a sampling distribution
   H_1: Statement of the alternate hypothesis to be accepted if H_0 is rejected

2. *The decision rule:* This rule is the basis for decisions and is determined entirely by three conditions: the appropriate sampling distribution, the alternate hypothesis (one- or two-sided), and the level of significance. The statement of the decision rule is usually as follows: reject H_0 if the sample value of the test statistic is greater than (or smaller than) the critical value of the test statistic. If we reject H_0, then we conclude that H_1 is true.

3. *Sample results:* We obtain the value of the test statistic from the sample and denote it with an asterisk (*).

4. *Decision:* We compare the sample test statistic to the decision rule and decide either to reject H_0 or to fail to reject H_0.

5. *Conclusions:* We explain the consequences of the decision. The conclusions "reserve judgment" or "the result is inconclusive" may be reached only if the decision was to fail to reject the null hypothesis and the absence of a definitive conclusion is an allowable outcome of the experiment. Moreover, it is a good idea to include the observed level of significance—the actual probability that a type I error has been made. This probability, often called the P value, is discussed in Section 7.1.7.

If the decision is to reject the null hypothesis, the result is said to be **significant at the [α] level**—that is, at the 0.05 level, the 0.01 level, or whatever significance level we chose before we conducted the experiment. Note that this is the level of *statistical significance* for the test. The practical implications (or practical *significance*) of the result are not implicit in the decision to reject or fail to reject a null hypothesis. A further discussion is often necessary to investigate fully the ramifications of the result.

The five-step procedure outlined here, or a variation of it, is applicable to most statistical tests and will safeguard against improper uses or conclusions. Generally, aside from selection of α, the only features that vary from problem to problem are the precise hypotheses that are formulated and the test statistic. The decision rule is not selected—it is *dictated* by the elements of the problem.

Example 7.1

A farmer takes a sample of grapefruit and weighs them. His standard requires a mean weight of at least 12.5 ounces for his crop. If the mean weight is at least 12.5 ounces, the crop should be marketed; if the mean weight is less than 12.5 ounces, the crop should not be marketed. Give suitable null and alternate hypotheses if *(a)* he has high standards and will not market the crop unless he is reasonably sure that the mean weight is greater than 12.5 ounces and if *(b)* he is anxious to market the crop and will do so unless it has a mean weight less than 12.5 ounces.

Solution

In either case we will have $H_0{:}\mu = 12.5$, where μ is the mean weight of the crop. In case *(a)*, however, he wants to be certain that the crop meets the standard if he markets it, so he will market it only if $\mu > 12.5$. Thus the alternate hypothesis is $H_1{:}\mu > 12.5$, and he will market the crop only if the mean weight of the sample yields a sample test statistic falling in the right-tail rejection region.

In case *(b)*, he wants to be certain that the crop does not meet the standard if he does not market it, so he will not market it only if $\mu < 12.5$. In this case the alternate hypothesis is $H_1{:}\mu < 12.5$, and he will market the crop unless the sample test statistic is in the left-tail rejection region. In either case the alternate hypothesis is one-sided and the rejection region is one-tailed. Inconclusive results, however, would have opposite effects. In case *(a)*, an inconclusive result would result in the crop not being marketed; in case *(b)*, the same result would cause the crop to be marketed.

Example 7.2 Two applicants for a job agree to a competition in which each will perform a certain task a number of times in order to have a sample of each applicant's performance on the job. The applicant whose performance is significantly better will be given the job. Set up suitable null and alternate hypotheses for the experiment.

Solution

The key word here is *significantly*. Slight differences may not show up as true differences (not attributable to chance) until after a number of trials. Hence we are searching for the true difference in their long-term performance. First we must determine a parameter measuring job performance. Suppose the job consists of assembling the parts for a musical instrument. We can measure job performance by the number of errors, proportion of defectives, length of time required for the task, or some other such measurement. Then if θ_1 and θ_2 measure the long-term values of the parameters for the two candidates, we have

$$H_0:\theta_1 = \theta_2$$
$$H_1:\theta_1 \neq \theta_2$$

The alternate hypothesis is two-sided, since we are only interested in some difference in the two performances. We do not address ourselves here to the problem of what to do if we cannot reject H_0; perhaps they will both be hired, some other criteria will be devised, or further trials will be conducted.

PROFICIENCY CHECKS

3. A chemist believes that the combination of two chemicals in specified amounts will result in a precipitation of at least 30 cubic centimeters of a compound. To test this belief a number of trial preparations will be conducted and the resulting precipitate measured. What are the null and alternate hypotheses for the experiment?
4. In Proficiency Checks 1 and 2, an experiment was conducted on the probability that a spinning coin will come to rest showing a head. Give the null and alternate hypotheses for the experiment.

7.1.7 Observed Significance Levels—*P* Values

When an experimenter reports the results of an experiment, the conclusions are based on the decision rule. The decision rule is set up according to the experimenter's willingness to risk making a type I error—the level of significance α. Others who may wish to use the results of the experiment, however, may have a different level of significance in mind, so it is useful to know the actual level of significance for an experiment. For example, if a decision rule is to reject H_0 if $z^* > 1.645$, and the value of z^* is 2.02, the experimenter would reject H_0 and conclude H_1 at a 0.05 level of significance (if that was the value of α chosen before the test). The actual level of significance—as opposed to the level of significance *chosen* before the test—is the probability that a value of z is greater than the obtained value of 2.02, that is, $P(z > 2.02)$. From Table 4 we find this probability to be $0.5000 - 0.4783$, or 0.0217. This is the **observed significance level** or ***P* value** (*P* for probability) of the result. It is the smallest α such that the observed sample results in rejecting H_0. Thus people reading the result of the experiment can decide for themselves whether the risk of making a type I error is satisfactory.

Many test reports, especially in the social sciences, do not give a preselected value of α; instead the *P* value for the result is reported so that the reader can make a decision based on a personal selection of α. Thus *P* values represent a different way of reporting the result of a statistical test. Even if a value of α has been selected, a good report usually lists the *P* value as well. In more complex problems where the use of a computer is indicated, many computer programs show the *P* value as part of the output. When the *P* value is given, readers can make an independent judgment based on the following procedure: if $P < \alpha$ (chosen by the reader), then H_0 will be rejected; if $P \geqslant \alpha$, then H_0 will not be rejected. Note also that although an experimenter may have selected a significance level of 0.05, a *P* value of 0.002 for the experiment is much stronger evidence against H_0 than the experimenter called for. If *P* is not given, this extent of the evidence for the alternate hypothesis will not be apparent.

Observed Significance Level (*P* Value)

The observed significance level of an experiment is computed as the probability that a value of the test statistic will be at least as extreme as the sample value of the test statistic in the direction of the alternate hypothesis. Thus it is the smallest α for which the observed sample results in rejecting H_0.

The *P* value for an experiment depends on the location of the rejection region. If the rejection region is two-tailed and the test statistic is z, the *P* value for the experiment is $P(z > |z^*|) + P(z < -|z^*|)$, that is, $2 \cdot P(z > |z^*|)$. If the rejection region is one-tailed and the test statistic is z, the *P* value for the experiment is either $P(z > z^*)$ or $P(z < z^*)$, depending upon whether the rejection region is in the left or right tail, respectively.

Example 7.3

An experiment is conducted in which the null hypothesis is $H_0{:}\mu = 156.0$. Suppose the sample test statistic is $z^* = 1.87$. What is the P value for the action of rejecting H_0 if the alternate hypothesis is: *(a)* $H_1{:}\mu \neq 156.0$; *(b)* $H_1{:}\mu > 156.0$? Suppose the sample test statistic is $t^* = 1.87$ and there are 13 degrees of freedom. What is the P value for the action of rejecting H_0 if the alternate hypothesis is: *(c)* $H_1{:}\mu \neq 156.0$; *(d)* $H_1{:}\mu > 156.0$?

Solution

a. Since this is a two-sided alternate hypothesis that will have a two-tailed rejection region, we want the probability that z would be greater than 1.87 or smaller than -1.87. This probability is equal to $P(|z| > 1.87)$ or $2P(z > 1.87)$. Since $2(0.5000 - 0.4693) = 2(0.0307) = 0.0614$, this is the P value for the experiment. If an experimenter had been conducting the test at a 0.10 level of significance, H_0 would be rejected; at a lower level of significance, 0.05 or smaller, the result would probably have been judged inconclusive.

b. Here we want only $P(z > 1.87)$, which is 0.0307, so this is the P value for the one-sided alternate hypothesis. We would have rejected H_0 at a significance level of 0.05 but not at a level of 0.01.

c. Again we have a two-tailed rejection region and want the probability that $|t| > 1.87$. We do not have extensive tables of areas under the t curves, however (although they are available). In these cases we state *ranges* of values. We find from Table 5 that $t_{.05}\{13\} = 1.771$. Thus $P(t > 1.771) = 0.05$ and $P(t < -1.771) = 0.05$; therefore $P(|t| > 1.771) = 0.10$. Now $1.87 > 1.771$, so $P(|t| > 1.87) < 0.10$. On the other hand, $t_{.025}\{13\} = 2.160$, so $P(|t| > 2.160) = 0.05$. Since $1.87 < 2.16$, we have $P(|t| > 1.87) > 0.05$. We thus conclude that the P value for this result is between 0.05 and 0.10—that is, $0.05 < P < 0.10$. This is because $t_{.025} < t^* < t_{.05}$, and the test is two-sided so that the total rejection region is twice the rejection region in one side. Thus we would reject H_0 if $\alpha = 0.10$ but not if $\alpha = 0.05$.

d. Since this is a one-tailed rejection region and we have found that $t_{.025} < t^* < t_{.05}$, then $0.025 < P < 0.05$. We would reject the null hypothesis in favor of the hypothesis $\mu > 156.0$ if $\alpha = 0.10$ or 0.05 but not if $\alpha = 0.01$ or less.

PROFICIENCY CHECKS

5. A test concerning a population proportion has $z^* = 2.18$. What would the P value be if the alternate hypothesis was:
   a. $\pi > \pi_0$? b. $\pi \neq \pi_0$?

6. A one-sided test concerning a population variance has 25 degrees of freedom. Determine the P value for the test if χ^{2*} is:
   a. 31.65 b. 39.33 c. 55.33

7.1.8 Power of a Test

We observed earlier that the *power of a statistical test* is the probability of rejecting a false null hypothesis. Let us assume for the moment that there are only two possibilities—accepting H_0 or rejecting H_0. If H_0 is false, the probability that it will be accepted (and that a type II error will be committed) is designated β; thus if H_0 is false, the probability that it will be rejected is $1 - \beta$. Therefore the power of a statistical test is $1 - \beta$. To determine the power of a test, we must know the true value of the population parameter. Suppose, for example, the hypotheses being tested are $H_0{:}\mu = \mu_0$ and $H_1{:}\mu > \mu_0$ and that the sampling distribution of the mean is normal. We would divide the possible values of $\bar{x}$ into acceptance and rejection regions by a value $\bar{x}_c$ (c for critical value). The value of $\bar{x}_c$ corresponds to z_α; that is $(\bar{x}_c - \mu_0)/s_{\bar{x}} = z_\alpha$ so that $P(\bar{x} > \bar{x}_c) = \alpha$. Thus the rejection region (see Figure 7.4) is the part of the horizontal axis under the normal curve representing the sampling distribution of the mean to the right of $\bar{x}_c$.

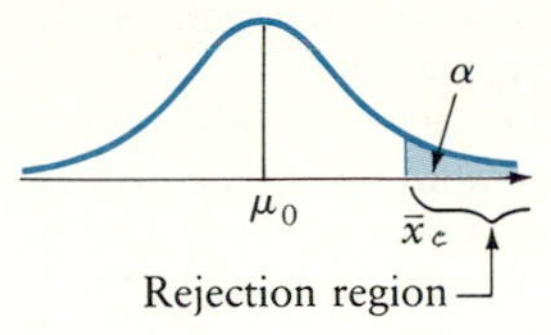

Figure 7.4
Rejection region for $H_1{:}\mu_1 > \mu_0$.

If the population mean is actually some value other than μ_0, say μ_1, we will commit a type II error if we accept the null hypothesis. Now suppose $\mu_1 > \mu_0$. Since we could accept H_0 if $\bar{x} < \bar{x}_c$, the probability of making a type II error (β) is the area under the probability distribution with mean $\mu = \mu_1$ to the left of $\bar{x}_c$, as shown in Figure 7.5. We cannot determine the value of β unless we know that H_0 is false and know the actual value of μ_1. In Figure 7.5, the value of μ_1 is fairly close to μ_0, so β is large; that is, if μ_0 and μ_1 are close, the probability of making a type II error is great. If μ_1 is farther from μ_0, then β will decrease as shown in Figure 7.6.

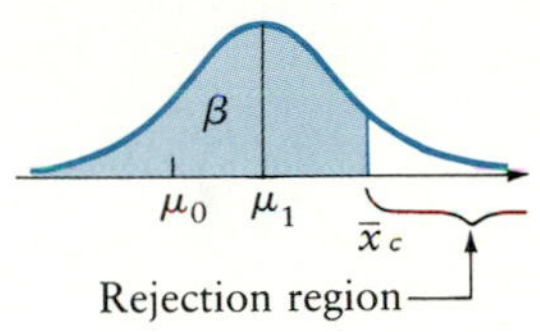

Figure 7.5
Rejection region and β for $H_1{:}\mu_1 > \mu_0$ if H_1 is true.

Suppose that we are to test a null hypothesis $H_0{:}\mu = \mu_0$ against $H_1{:}\mu > \mu_0$. Suppose further that (unknown to us) H_0 is false and that actually $\mu = \mu_1$, where $\mu_1 > \mu_0$. In Figure 7.7, the sampling distribution of the mean used to test the null hypothesis (centered at μ_0) is on the left, and the actual sampling distribution of the mean (centered at μ_1) is on the right. The rejection region, based on the left-hand curve, is to the right of the critical value $\bar{x}_c$, while the acceptance region is to the left of $\bar{x}_c$. If the sample mean is greater than $\bar{x}_c$, the null hypothesis will be rejected; if it is less than $\bar{x}_c$, the null hypothesis will not be rejected. If we accept the null hypothesis, we will commit a type II error since it is false. The probability of this happening is β, as shown in Figure 7.7.

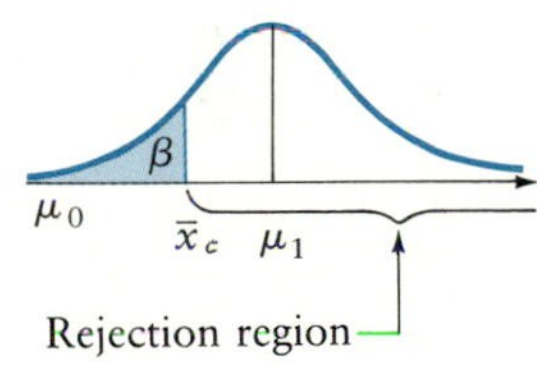

Figure 7.6
Rejection region and β for the case where μ_0 and μ_1 are far apart.

One way of decreasing β is to increase the value of α by moving $\bar{x}_c$, to the left. This step will make β smaller and increase the power of the test,

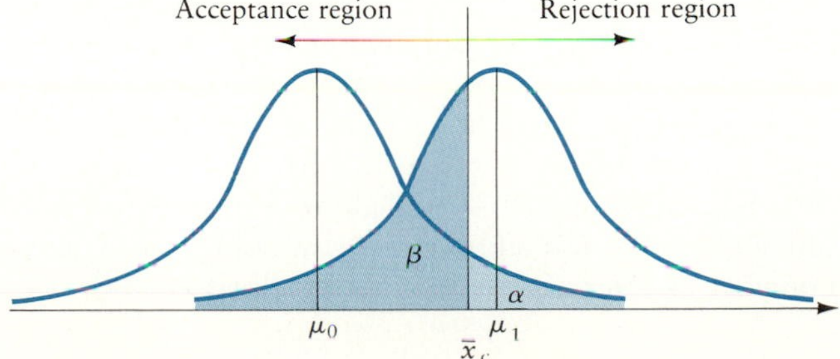

Figure 7.7
Acceptance and rejection regions and α and β for a test of $H_0{:}\mu = \mu_0$ against $H_1{:}\mu_1 > \mu_0$.

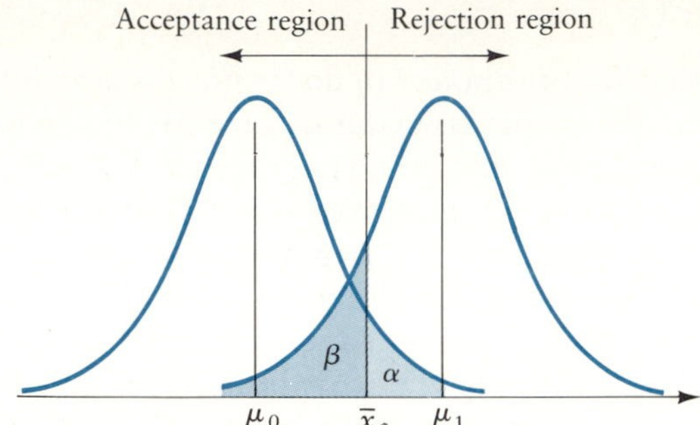

Figure 7.8
The effect of increasing sample size at a constant α: a decrease in β and an increase in the power of the test.

but it will also increase the likelihood of a type I error. If we wish to decrease β without increasing the value of α, we may do so by increasing the sample size. Recall that $s_{\bar{x}} = s/\sqrt{n}$; if n is increased the standard error will be reduced, so that both the curves will be taller and narrower. If we keep α constant and increase the sample size, the dividing line between the regions will move to the left as shown in Figure 7.8, decreasing β and increasing $1 - \beta$, the power of the test.

There are procedures for determining the sample size we need in order to reject H_0 correctly with a minimum specified power. These methods require specification of the level of significance and the minimum difference between μ_0 and μ_1 to be detected. (See, for example, Neter and others [1985], pp. 547–549, 602–605.) A further discussion of power may be found in Section 7.5.

7.1.9 Designing the Experiment

It is important to design the experiment before it is actually performed in order to avoid potential problems. Care should be taken to distinguish between *mensurative* (measurement) experiments and *manipulative* experiments. **Mensurative experiments** simply involve making measurements at one or more points in space and time—measuring the achievement of students in different schools, for example, and then comparing the results. If students in one school score significantly better on a test, we cannot conclude that it is because of the schools they attend. There is an association between schools and achievement, but it may be due to factors other than the school—differences in level of affluence, prior background of students, and many other factors, any combination of which may account for the differences.

In a **manipulative experiment**, different experimental units receive different treatments and assignment of treatments to experimental units is randomized. We would conduct a manipulative experiment to measure differences in achievement from school to school by collecting a group of students and then assigning them at random to the different schools. All factors other than school are thus controlled by random selection. In this case we may attribute differences to school alone. Manipulative experiments let us make inferences in "cause-and-effect" terminology, while mensurative experiments only let us infer association. See Hurlbert (1984).

7.1.10 Confidence Intervals and Hypothesis Testing

When we reject a null hypothesis, we do not know the value of the population parameter in question. All we know is that the population parameter is most likely more than or less than a particular value. We can estimate its true value by constructing a confidence interval for the parameter.

When we cannot reject the null hypothesis, we can construct a confidence interval for the population parameter in order to obtain estimates for the value of β. We may use the extremes of the confidence interval to estimate the probable value of the true parameter and calculate β for these extremes, thus obtaining an interval estimate for the probability of making a type II error. This information can help us decide whether to accept H_0, reserve judgment, or gather additional data.

Some statisticians prefer to avoid the formal hypothesis-testing procedure entirely, instead constructing a confidence interval for the parameter and drawing conclusions from that. Suppose that we wish to test $H_0{:}\mu = \mu_0$ against $H_1{:}\mu \neq \mu_0$. If the level of significance is α, we may use a sample result to construct a $100(1 - \alpha)\%$ confidence interval for μ. *If the value of μ_0 is not included in the interval, we may reject H_0; if the value of μ_0 is included in the interval, we fail to reject H_0.* An advantage of this procedure is that we can test many null hypotheses at once; we may reject $H_0{:}\mu =$ [any value not in the interval].

For a one-sided alternative, the procedure is a bit more complicated. To test $H_0{:}\mu = \mu_0$ against $H_1{:}\mu > \mu_0$ with a level of significance equal to α, we construct a $100(1 - 2\alpha)\%$ confidence interval for μ. If μ_0 is in the right-hand tail (greater than the upper confidence limit), we reject H_0; if μ_0 is not in the right-hand tail, we fail to reject H_0. Thus the hypothesis $H_0{:}\mu =$ [any value greater than the upper confidence limit] will be rejected. A similar procedure applies if we wish to test $H_0{:}\mu = \mu_0$ against $H_1{:}\mu < \mu_0$ with a level of significance equal to α. We construct a $100(1 - 2\alpha)\%$ confidence interval for μ. If μ_0 is in the left-hand tail (smaller than the lower confidence limit), we reject H_0; if μ_0 is not in the left-hand tail, we fail to reject H_0. Thus the hypothesis $H_0{:}\mu =$ [any value smaller than the lower confidence limit] will be rejected.

Problems

Practice

In Problems 7.1 through 7.6, you are given a research hypothesis. Set up null and alternate hypotheses to test this hypothesis.

7.1 $\pi \leq 0.6$

7.2 $\sigma^2 > 132.77$

7.3 $\mu = 1225.0$

7.4 $\mu \geqslant 1.58$

7.5 $\sigma^2 = 0.001765$

7.6 $\pi \neq 0.15$

Applications

7.7 We wish to test the hypothesis that a motel room on Santa Rosa Beach (in season) is priced, on the average, between \$78 and \$125. Under what conditions would we make a type I error? A type II error?

7.8 A firm decides to open a branch office if the volume of business is great enough. It samples its daily orders to estimate the proportion of orders going into that area. The critical amount is 5%. Formulate suitable null and alternative hypotheses for the experiment if:

a. the firm is cautious and does not wish to open a new office unless it must

b. the firm is experiencing a period of expansion and wishes to open a branch office unless it is totally unjustified

7.9 In an earlier example a social agency wished to determine whether the income level of families in a large housing development has remained the same or changed over the past 5 years. They determine that the mean income 5 years ago was \$16,788, which, if adjusted for inflation, would be \$20,278 today. They take a sample of families and determine the mean income for the sample. What are the null and alternate hypotheses for the experiment?

7.10 From her many years of experience, a statistics instructor knows that grades in her classes have averaged about 50 percent A's and B's. The instructor believes that this term's classes are superior to most others in the past and takes a sample of midterm grades to test this belief. What are the null and alternative hypotheses she will use?

7.11 A car manufacturer wishes to test the claim that unleaded gasoline reduces air pollution by decreasing the amount of pollutants discharged into the air by exhaust. Give null and alternate hypotheses for an experiment designed to determine whether this claim is true.

7.12 A grocer carries two products that are identical except for brand name. Since his shelf space is limited, he has decided to eliminate one brand from his inventory. Brand A is advertised more extensively than brand B, and the grocer would like to take advantage of the advertising offered by brand A. On the other hand, he thinks that his customers have recently shown a preference for brand B. If there is no real preference he would prefer to stock brand A, but he would stock brand B if more than half his customers who buy the product prefer brand B. Suppose he polls a sample of customers to determine which brand they prefer. What will be his null and alternate hypotheses?

7.13 A teacher tries two methods of teaching verbs to her French class. She divides the class randomly into two parts and uses method I for one part of the class and method II for the other part. She wants to determine which method is better. What are the null and alternate hypotheses?

7.14 A businessman claims that his business is going to be better this month than last, so he examines a sample of new orders to prove or disprove his contention. What null and alternate hypotheses would be appropriate to his experiment?

7.15 Advertisements for a retirement community say that "over 90% of our residents prefer living here to anyplace else." A newspaperman samples residents of the community to see whether the statement is correct. Give null and alternate hypotheses for the experiment if: *(a)* he wishes to prove that the statement is true; *(b)* he wishes to disprove the statement. (*Hint:* What research hypothesis or belief is appropriate in each case?)

7.16 A secretary is testing a new word processor. The office manager is trying to decide whether to replace the current typewriters with the processor. He will base his decision upon whether or not the average number of correct pages typed by the secretary exceeds the average number with the current typewriter (52.3 per 8-hour day). Give null and alternate hypotheses for the experiment if:

a. the office manager wishes to buy the new processors unless the results indicate that the average number is higher with the current typewriters
b. the office manager does not wish to buy the new processors unless they are significantly better than the current typewriters

7.17 An oil firm is testing a gasoline additive that may or may not increase gasoline mileage. To discover the additive's properties, they test the gasoline with the additive in a large sample of automobiles against the same gasoline without the additive in the same sample. They then draw conclusions. Three different sets of null and alternate hypotheses are possible. List each one and discuss the probable motivation for each.

7.18 An advertisement for aspirin states that "aspirin substitutes have not been shown to be safer than aspirin." How might that assertion have been obtained using the experimental procedure? Can you conclude that aspirin is just as safe as aspirin substitutes? Can you conclude that aspirin substitutes are not safer than aspirin? Can you conclude that aspirin is not safer than aspirin substitutes? Is it important to know who is making the statement? Why? Can you draw any conclusions from the statement? If so, what?

7.19 To ensure quality control in a canning plant, an engineer selects a random sample of 40 cans and takes their mean weight. If the procedure is in control, the population mean weight will be 16.42 ounces. The level of significance used is 0.05, and the shipment will be passed unless the sample mean is significantly below standard. The engineer weighs Monday's sample and tests its mean against the norm.

a. The mean is found to be so far below 16.42 that only 1 sample in 15 would fall so low by chance if the population mean were actually 16.42. Will the shipment be passed? If it were rejected, what would the P value be?
b. Will the sample be passed if the mean of only 1 sample in 25 would fall so low by chance? If the shipment were rejected, what would the P value be?
c. Suppose the sample mean is so much in excess of 16.42 that only 1 sample in 25 would weigh so much by chance if the mean were actually 16.42. Will the sample be passed?

7.20 A researcher tests samples of two models of sewing machine, models A and B, and wants to decide which model actually does the better job. As a result of the sample, he finds that the sample of model A performs better than the sample of model B. His hypotheses have been formulated in accordance with the guidelines set forth here. As a result of the test, he decides that the probability of type II error precludes acceptance of the null hypothesis. There are three possible alternate hypotheses. Which of these is not likely to have been used here? Why?

7.2 Hypotheses About a Population Mean

The hypothesis-testing procedures presented in Section 7.1 are applicable in a great many situations involving the test of a belief. In each case we need to determine the null and alternative hypotheses, set the significance level, and then determine the appropriate sampling distribution to obtain the test statistic.

In this section we will see how the procedures we used to obtain confidence intervals for the population mean in Chapter 6 can be used to test hypotheses about μ. Large-sample procedures will be given first and then the use of the t distribution. Finally, we will learn how to use the trimmed mean.

If we wish to test a hypothesis about a population mean, the null hypothesis will be H_0:$\mu = \mu_0$, where μ_0 is the hypothetical value of the population mean. Recall from Chapter 6 that according to the central limit theorem, the sampling distribution of the mean will be approximately normal if n, the sample size, is sufficiently large. If the random variable is normally distributed, the sampling distribution of the mean will be normal for any sample size; if the variable is not normal, "sufficiently large" is taken to mean $n \geqslant 30$. Thus if our null hypothesis is true so that the population mean is μ_0, and if the central limit theorem applies, then the sampling distribution of the mean will be $N(\mu_0, \sigma_{\bar{x}}^2)$, so the test statistic is z, where $z = (\bar{x} - \mu_0)/\sigma_{\bar{x}}$.

7.2.1 Large-Sample Tests

If we do not know the population variance (and usually we do not) we may still use z as a test statistic provided that the sample is sufficiently large. Recall that "sufficiently large" was defined in Chapter 6 to be at least 50 for any distribution and at least 30 if the distribution of the variable is normal. The results we obtain using z as the test statistic will be approximate and will understate the observed level of significance slightly—that is, the actual probability of type I error will be somewhat larger than what the value of z indicates. In most cases this discrepancy is not a problem. If the obtained value of z^* lies just slightly beyond the critical value of z, however, we cannot reject H_0 with certainty. Exact tests using the t statistic are appropriate for any normally distributed variable and must be used if $n < 30$. These tests are covered in Section 7.2.2. Recall that we can use a normal probability plot (using the computer) to determine whether a sample is reasonably close to normally distributed.

There are three possible alternate hypotheses, each dictating a different decision rule. In each case we assume that the appropriate test statistic is z, where $z = (\bar{x} - \mu_0)/s_{\bar{x}}$. This assumption holds if the random variable is normal and the sample is at least 30; if the variable is not known to be normal, the sample must be at least 50. The selected significance level is α, so that if the alternate hypothesis is one-sided, the critical value of the test statistic is z_α; if the alternate hypothesis is two-sided, the critical values are $z_{\alpha/2}$ and $-z_{\alpha/2}$. If the alternate hypothesis is two-sided, we may write the decision rule in either of the two equivalent forms: reject H_0 if $z^* > z_{\alpha/2}$ or if $z^* < -z_{\alpha/2}$; or reject H_0 if $|z^*| > z_{\alpha/2}$. The three possible cases are given here.

Large Sample Hypothesis Tests for the Population Mean

Suppose we obtain a random sample of n observations of a random variable x and that either of the following is true:

1. The population variance, σ^2, is known and either:
   a. the random variable is normal
   b. the variable is not known to be normal and the sample is at least 30
2. The population variance is unknown and either:
   a. the random variable is normal and the sample is at least 30
   b. the variable is not known to be normal and the sample is at least 50

Then we may test the hypothesis $H_0{:}\mu = \mu_0$ by using the following procedure. The sample value of the test statistic used to test the hypothesis is z^*. If $\bar{x}$ and s are the mean and standard deviation of the sample, then $z^* = (\bar{x} - \mu_0)/\sigma_{\bar{x}}$ if σ^2 is known or $z^* = (\bar{x} - \mu_0)/s_{\bar{x}}$ if σ^2 is unknown. The value of $\sigma_{\bar{x}}$ or $s_{\bar{x}}$ is calculated with the finite population correction factor if it is appropriate.

The decision rule is dictated by the alternate hypothesis and the level of significance α chosen for the test. The three possibilities are as follows:

Case I	**Case II**	**Case III**
	Null and Alternate Hypotheses	
$H_0{:}\mu = \mu_0$ $H_1{:}\mu \neq \mu_0$	$H_0{:}\mu = \mu_0$ $H_1{:}\mu > \mu_0$	$H_0{:}\mu = \mu_0$ $H_1{:}\mu < \mu_0$
	Decision Rule	
Reject H_0 if $\lvert z^*\rvert > z_{\alpha/2}$	Reject H_0 if $z^* > z_\alpha$	Reject H_0 if $z^* < -z_\alpha$

Example 7.4

A firm wants to decide whether or not to purchase a new letter sorter for its office. Since the sorter is too expensive to buy if it will not be used sufficiently, the firm arranges to try the sorter for 6 weeks to make a decision. The firm works on Saturday, so there are 36 working days in this period. The office manager decides that she will order the machine if she can be 95% sure that it will sort at least 4000 letters per day. Taking what she hopes to be a random sample of 36 days' performance of the machine, she performs an experiment with a research hypothesis $\mu \geqslant 4000$ and a significance level of $\alpha = 0.05$. What will her course of action be if this sample has a mean of 4132 letters sorted per day with a standard deviation of 324 letters?

Solution

Since the research hypothesis is $\mu \geqslant 4000$, the reference value of the population mean is 4000. Thus the null hypothesis is $\mu = 4000$. To support the research hypothesis, the alternate hypothesis must be $\mu > 4000$. If the null hypothesis is rejected in favor of the alternate hypothesis at the 0.05 level of significance, the sorter will be purchased. If we can reasonably assume that the number of letters sorted daily is normally distributed, we may use the test statistic z. The rejection region is one-tailed, so the critical value of the test statistic is $z_{.05} = 1.645$. We use the hypothesis-testing procedure and obtain the results given here:

1. $H_0{:}\mu = 4000$
   $H_1{:}\mu > 4000$
2. Decision rule: Reject H_0 if $z^* > 1.645$.
3. Results: $n = 36$, $\bar{x} = 4132$, and $s_{\bar{x}} = 324/\sqrt{36} = 54$, so

$$z^* = \frac{\bar{x} - \mu_0}{s_{\bar{x}}} = \frac{4132 - 4000}{54} = \frac{132}{54} \doteq 2.44$$

4. Decision: Since $2.44 > 1.645$, we reject H_0 and conclude that the mean number of letters sorted per day is probably greater than 4000.
5. Conclusion: Since the mean is probably greater than 4000, the firm will purchase the sorter.

We can calculate the observed significance level from Table 4. Since $P(z > 2.44) = 0.0073$, we have a P value of 0.0073 for this experiment. This value provides a reassuring degree of confirmation that the course of action is correct, since the probability of concluding incorrectly that $\mu > 4000$ is only 0.0073 (if μ actually is 4000) or less (if $\mu < 4000$).

Example 7.5

A sociologist examining a large apartment complex wishes to discover whether the mean number of persons per family can be assumed to be equal to 4.80. He interviews 100 of the 750 families in the complex, and obtains a mean of 4.70 persons per family with a standard deviation of 0.80. If his significance level is set at 0.05, what conclusions does he draw?

Solution

Since the sample is greater than 50, no additional assumptions need to be made. The alternate hypothesis is two-sided, so the rejection region is two-tailed. The critical value of the test statistic is $z_{\alpha/2}$. Here $\alpha = 0.05$, so we use $z_{.025} = 1.96$. We then have the following procedure:

1. $H_0: \mu = 4.80$
   $H_1: \mu \neq 4.80$
2. Decision rule: Reject H_0 if $|z^*| > 1.96$.
3. Results: Since 100 is more than 5% of 750, we should use the finite population correction factor. Thus we have

$$s_{\bar{x}} = \frac{0.80}{\sqrt{100}}\sqrt{\frac{650}{749}} \doteq 0.0745$$

   and

$$z^* = \frac{\bar{x} - \mu_0}{s_{\bar{x}}} = \frac{4.70 - 4.80}{0.0745} = \frac{-0.10}{0.0745} \doteq -1.34$$

4. Decision: Since $|-1.34| = 1.34$, and 1.34 is not greater than 1.96, we fail to reject H_0.
5. Conclusion: The mean number of persons per family unit in this complex does not differ significantly from 4.80. It is reasonable to use 4.80 as the mean where appropriate (on reports, for example).

The observed significance level is equal to $P(z > 1.34) + P(z < -1.34)$, or $2P(z > 1.34)$. From Table 4, we have $2(0.0901) = 0.1802$, so that if H_0 had been rejected, there would have been a substantial chance of being wrong. On the other hand, we must consider the possibility of a type II error. It is probably not possible to reserve judgment in a case like this, so the following comment might be added to the report: "Although we considered the risk of type II error to be fairly high, we decided to use 4.80 since the sample was a substantial fraction of the population and the sample result was not significantly different (at the 0.05 level of significance) from 4.80."

PROFICIENCY CHECKS

7. A research hypothesis is that $\mu \neq 18.7$. The level of significance is 0.10; sample results are $n = 60$, $\bar{x} = 19.3$, and $s = 3.8$. Test the hypothesis using the hypothesis-testing procedure.
8. A research hypothesis is that $\mu \geqslant 150.0$. The level of significance is 0.01; sample results are $n = 40$, $\bar{x} = 156.3$, $s = 15.4$,

(continued)

PROFICIENCY CHECKS *(continued)*

and it is reasonable to believe that the distribution is normal. Test the hypothesis.

9. Compute and interpret the P values for Proficiency Checks 7 and 8.

10. A brochure claims that college students at the university spend an average of \$250 per year on textbooks. To test this hypothesis, a random sample of 144 students is polled and the amount each student spent for textbooks during the year is recorded. The mean amount is \$264 with a standard deviation of \$54. Using a 0.05 level of significance, does this result support the brochure's claim?

7.2.2 Tests in a Normal Population

We first considered the t distribution in Section 6.4. There we noted that if the statistic $(\bar{x} - \mu)/(\sigma/\sqrt{n})$ is N(0, 1), then the statistic $(\bar{x} - \mu)/(s/\sqrt{n})$ is $t(n - 1)$. We can use this fact to test hypotheses about a population mean if the random variable is normal, no matter what size the sample is. The large-sample procedure outlined in the preceding section applies to a normal variable only if the sample size is at least 30, and even then the result is approximate. In Example 7.4 the appropriate test statistic is actually t, not z. The critical value of t in that case would be $t_{.05}\{35\}$, which (from Table 5) is about 1.691. Since the test statistic was 2.44, the conclusions would be the same, although the P value is actually larger than calculated there.

In our examples, proficiency checks, and problems, we will continue the conventional practice of using z as test statistic instead of t for $n \geqslant 30$. In most cases the difference is not great. If the test statistic we obtain lies between the critical values of z and t, however, caution is advised. In this case, use of the exact test is recommended.

Example 7.6

A chemist is testing the hypothesis that the boiling point of a certain substance is 846°C. She makes four determinations and obtains values of 844, 847, 845, and 844°. What conclusions can you draw at a significance level of 0.05?

Solution

Since it is the same substance in each case, a normal distribution is likely. There is a two-tailed rejection region, and $t_{.025}\{3\} = 3.182$. Then we have:

1. $H_0: \mu = 846$
   $H_1: \mu \neq 846$

Hypothesis Tests for the Population Mean (Normal Population, Unknown Population Variance)

Suppose we obtain a random sample of n observations of a random variable x and that both of the following are true:

1. The population variance σ^2 is not known.
2. The random variable is normal.

Then we may test the hypothesis $H_0{:}\mu = \mu_0$ by using the following procedure. The sample value of the test statistic used to test the hypothesis is t^*. If $\bar{x}$ and s are the mean and standard deviation of the sample, then $t^* = (\bar{x} - \mu_0)/(s/\sqrt{n})$.

The decision rule is dictated by the alternate hypothesis and the level of significance α chosen for the test. The three possibilities are as follows:

	Case I	**Case II**	**Case III**
Null and Alternate Hypotheses	$H_0{:}\mu = \mu_0$ $H_1{:}\mu \neq \mu_0$	$H_0{:}\mu = \mu_0$ $H_1{:}\mu > \mu_0$	$H_0{:}\mu = \mu_0$ $H_1{:}\mu < \mu_0$
Decision Rule	Reject H_0 if $\lvert t^*\rvert > t_{\alpha/2}\{n - 1\}$	Reject H_0 if $t^* > t_\alpha\{n - 1\}$	Reject H_0 if $t^* < -t_\alpha\{n - 1\}$

2. Decision rule: Reject H_0 if $|t^*| > 3.182$.
3. Results: $n = 4$, $\bar{x} = 845$, and $s \doteq 1.4$, so

$$t^* = \frac{845 - 846}{1.4/\sqrt{4}} = \frac{-1}{0.7} \doteq -1.43$$

4. Decision: Since $|-1.43| = 1.43$, and 1.43 is not greater than 3.182, we fail to reject H_0.
5. Conclusion: The chemist might well conclude that the true boiling point is 846°C, provided that she has a theoretical basis for this result. It is difficult to calculate P values for the t statistic without extensive tables. We observe from Table 5, however, that $t_{.10}\{3\} = 1.638$, so we know that the P value for this test is at least 0.20, since $1.43 < 1.638$ and the rejection region is two-tailed.

Example 7.7

A psychologist contends that today's young people are more responsible than those of an earlier generation. To prove this belief he shows evidence that nine young people selected at random and given a test designed to measure responsibility showed a mean index of 0.74 as compared to 0.63 for the young people 30 years ago. Sample scores were 0.73, 0.64, 0.93, 0.84, 0.52, 0.80, 0.71, 0.74, and 0.75. High scores on the test indicate responsibility; low scores indicate lack of responsibility. If the nine tests had a standard deviation of 0.117, can his claim be substantiated at a 0.05 significance level?

Solution

His research hypothesis is $\mu > 0.63$, where μ is the mean index of responsibility for today's young people. Assuming a normal distribution, we may use the t statistic; $t_{.05}\{8\} = 1.860$, so we have the following:

1. $H_0: \mu = 0.63$
   $H_1: \mu > 0.63$
2. Decision rule: Reject H_0 if $t^* > 1.860$.
3. Results: $n = 9$, $\bar{x} = 0.74$, and $s \doteq 0.117$, so

$$t^* = \frac{0.74 - 0.63}{0.117/\sqrt{9}} = \frac{0.11}{0.039} \doteq 2.82$$

4. Decision: Since $2.82 > 1.860$, we reject H_0.
5. Conclusion: The psychologist can conclude at a significance level of 0.05 that today's young people are more responsible than those of 30 years ago, providing that the test actually measures responsibility.

Example 7.8

A drug manufacturer claims that his new drug causes faster red-cell buildup in anemic persons than the drug currently used. A team of doctors tests the drug on six persons and compares the results with the current buildup factor of 8.3. The six persons have factors of 6.3, 7.8, 8.1, 8.3, 8.7, and 9.4. Try to substantiate the manufacturer's claim at a 0.01 level of significance.

Solution

Assuming the factors to be normally distributed, the appropriate test statistic is t^*. We are trying to substantiate the claim, so that becomes the alternate hypothesis; $t_{.01}\{5\} = 3.365$, so we have:

1. $H_0: \mu = 8.3$
   $H_1: \mu > 8.3$
2. Decision rule: Reject H_0 if $t^* > 3.365$.
3. Results: $n = 6$, $\bar{x} = 8.1$, and $s \doteq 1.04$, so

$$t^* = \frac{8.1 - 8.3}{1.04/\sqrt{6}} \doteq -0.47$$

4. Decision: Since $-0.47 < 3.365$, we fail to reject H_0.
5. Conclusion: Since t^* is negative, we might examine the claim a little more closely, but certainly the new drug is not better than the old one.

Example 7.9

An experiment is conducted to test the hypothesis that a population mean is a certain value. A random sample of 31 observations is obtained, and the level of significance chosen for the test is 0.05. The alternate hypothesis is two-sided and the researcher obtains a test statistic value $z^* = 2.01$, assuming that the population is normally distributed. The researcher then rejects the null hypothesis. Is this decision valid?

Solution

As discussed earlier, the appropriate test statistic in this case is actually t. Thus the researcher really obtained $t^* = 2.01$. From Table 5 we have $t_{.025}\{30\} = 2.042$, so $t^* = 2.01$ is not actually in the rejection region. The correct decision in this case is probably to reserve judgment. For $n = 31$, values of the test statistic between 2.042 and 1.96 would be in this gray region. Always be careful to allow for such a possibility.

PROFICIENCY CHECKS

11. A research hypothesis is that $\mu \neq 18.7$. The level of significance is 0.10; sample results are $n = 20$, $\bar{x} = 19.3$, and $s = 3.8$. If the variable is normally distributed, test the hypothesis using the hypothesis-testing procedure.
12. A research hypothesis is that $\mu \geqslant 150.0$. The level of significance is 0.01; sample results are $n = 4$, $\bar{x} = 156.3$, and $s = 15.4$, and it is reasonable to believe that the distribution is normal. Test the hypothesis.
13. If it is operating properly, a soft-drink machine should give each customer a drink containing, on the average, 9 ounces. To see if the machine is in adjustment, the service personnel take a sample consisting of six drinks and measure the contents. On one occasion the mean is 9.2 ounces with a standard deviation of 0.18 ounce. Test the hypothesis that the true mean is 9 ounces. Use the 0.05 level of significance and estimate the P value for the result.

7.2.3 Using the Trimmed Mean

For nonnormal populations, we cannot test hypotheses concerning the population mean by using the results of a small sample. In most cases, the best way to test such a hypothesis is to use a robust method, preferably a nonparametric test. The applicable nonparametric tests, discussed in Chapter 10, are the signs test and the Wilcoxon T Test. If the data set is reasonably symmetric, however, suggesting that the population is symmetric, we may use the trimmed mean to test a hypothesis that $\mu = \mu_0$. The test statistic is $t^* = [\bar{x}(T) - \mu_0]/s_{\bar{x}(T)}$, where

$$s_{\bar{x}(T)} = \frac{s(W)}{\sqrt{h}}\sqrt{\frac{n-1}{h-1}}$$

Recall that $s(W)$, the standard deviation of the Winsorized data set, is obtained by replacing the trimmed values by the values that would have been trimmed next. We thus have the following procedure.

Hypothesis Tests for the Population Mean (Nonnormal Symmetric Population)

Suppose we obtain a random sample of n observations of a random variable x and that both of the following are true:

1. The population variance σ^2 is not known.
2. The random variable is nonnormal but symmetric.

Then we may test the hypothesis $H_0{:}\mu = \mu_0$ by using the following procedure. The sample value of the test statistic used to test the hypothesis is t^*. If $\bar{x}(T)$ is the mean of the trimmed sample containing h observations and $s(W)$ is the standard deviation of the Winsorized data set, then $t^* = [\bar{x}(T) - \mu_0]/s_{\bar{x}(T)}$, where

$$s_{\bar{x}(T)} = \frac{s(W)}{\sqrt{h}}\sqrt{\frac{n-1}{h-1}}$$

The decision rule is dictated by the alternate hypothesis and the level of significance α chosen for the test. The three possibilities are as follows:

Case I	**Case II**	**Case III**
	Null and Alternate Hypotheses	
$H_0{:}\mu = \mu_0$ $H_1{:}\mu \neq \mu_0$	$H_0{:}\mu = \mu_0$ $H_1{:}\mu > \mu_0$	$H_0{:}\mu = \mu_0$ $H_1{:}\mu < \mu_0$
	Decision Rule	
Reject H_0 if $\lvert t^*\rvert > t_{\alpha/2}\{h-1\}$	Reject H_0 if $t^* > t_{\alpha}\{h-1\}$	Reject H_0 if $t^* < -t_{\alpha}\{h-1\}$

Example 7.10

Refer to Example 7.7. Suppose the population of responsibility indexes is known to be nonnormal (long-tailed) but symmetric. Use the trimmed mean to test the claim.

Solution

Since $n = 9$, we obtain a 10% trimmed sample by removing one observation from each end. We remove 0.52 and 0.93. The resulting sample has mean $\bar{x}(T) = 0.74$. We obtain the Winsorized data set by adding 0.71 and 0.84 to the trimmed data set. The standard deviation of the Winsorized data set is $s(W) = 0.0747$. Then $s_{\bar{x}(T)} = (0.0747/\sqrt{7})\sqrt{8/6} \doteq 0.0326$, so $t^* = (0.74 - 0.63)/0.0326 \doteq 3.50$. Since $t_{.05}\{6\} = 1.943$, the new rejection rule is to reject H_0 if $t^* > 1.943$. Since $3.50 > 1.943$, we reject H_0. The conclusions are the same.

Problems

Practice

7.21 A sample of size n is drawn from a normally distributed population to test a research hypothesis that the population mean differs from 110.0. Test the hypothesis at the 0.05 level of significance if the sample mean is 112.4, the sample standard deviation is 4.4, and the sample size is:
a. $n = 5$ **b.** $n = 25$ **c.** $n = 40$ **d.** $n = 100$ **e.** $n = 200$

7.22 A random sample of 25 observations is drawn from a normal population to test the research hypothesis that the population mean is greater than 1.60. The sample mean is 1.63 and the sample standard deviation is 0.064. Obtain the P value (range) for the test. Test the hypothesis at a significance level of:
a. 0.10 **b.** 0.05 **c.** 0.025 **d.** 0.01 **e.** 0.005

How can you use the P value to conduct these tests?

7.23 A random sample of 75 observations is drawn from a population to test the research hypothesis that the population mean is greater than 1.60. The sample mean is 1.63 and the sample standard deviation is 0.108. Obtain the P value for the test. Test the hypothesis at a significance level of:
a. 0.10 **b.** 0.05 **c.** 0.025 **d.** 0.01 **e.** 0.005

7.24 A random sample of 60 observations is drawn from a normal population to test whether or not the population mean is 3450.0. The significance level is chosen at 0.05 and the sample mean is 3516.2. Test this hypothesis if the sample standard deviation is:
a. 127.3 **b.** 181.5 **c.** 203.8 **d.** 255.6 **e.** 387.8

7.25 A random sample of 10 observations is drawn from a normal population to test whether or not the population mean is 3450.0. The significance level is chosen at 0.05 and the sample mean is 3516.2. Test this hypothesis if the sample standard deviation is:
a. 57.3 **b.** 81.5 **c.** 103.8 **d.** 155.6 **e.** 227.8

7.26 A random sample of 22 observations is taken from a normal population to test the hypothesis that the population mean is equal to a specified amount. The sample standard deviation is 0.013. Let D represent the difference between the sample mean and hypothesized mean. If $D > D^*$ the null hypothesis will be rejected. What is the value of D^* for:
a. $\alpha = 0.10$? **b.** $\alpha = 0.05$? **c.** $\alpha = 0.01$?

7.27 Repeat Problem 7.26 for a sample of 68 observations and a sample standard deviation of 0.021.

7.28 A random sample of 41 observations is drawn from a normal population to test the hypothesis that $\mu = 35.0$. The alternate hypothesis is $H_1: \mu \neq 35.0$ and $\alpha = 0.05$. Suppose that $\bar{x} = 39.3$ and $s = 13.89$. Test the hypothesis using *(a)* the z test statistic and *(b)* the t test statistic. What is your conclusion in each case? Discuss the results and generalize.

7.29 A random sample of 15 is drawn from a normal population with a known variance of 139.69 to test the hypothesis that $\mu = 120.0$. Test this hypothesis if the sample mean is 116.2 and the significance level $\alpha =$
a. 0.10 **b.** 0.05 **c.** 0.02 **d.** 0.01

7.30 Estimate the P value for the test conducted in:
a. Problem 7.21*b* **b.** Problem 7.21*c* **c.** Problem 7.24*b*
d. Problem 7.25*b* **e.** Problem 7.29

Applications

7.31 A county farm agent wishes to determine the effect of a new fertilizer on the yield of corn per acre compared to that of the most popular brand. The popular brand is known to yield an average of 2.12 tons per acre. He persuades 64 farmers to plant an acre of corn each and treat it with the new fertilizer. These 64 acres have an average yield of 2.18 tons per acre with a standard deviation of 0.30 ton. Assuming that the yield is attributed solely to the fertilizer and the sample is random, test the hypothesis that there is no difference between the two fertilizers. Use the 0.05 level of significance.

7.32 More than half the U.S. population was covered by state seat-belt laws by 1986 according to *Newsweek* (February 3, 1986). Seventeen states had laws mandating use of seat belts, and there was evidence that auto fatalities were being reduced. Suppose an analyst obtains information from 20 randomly selected similar-sized counties in these states and finds that changes in traffic fatalities after the law's enactment range from a 21% decrease to a 5% increase. Fatalities have decreased, on the average, 4.6% in these counties with a standard deviation of 6.2%. At what level of significance does this information allow us to conclude that fatalities have decreased in these states? Can we conclude that the seat-belt law is responsible for the decrease? What further studies might be suggested?

7.33 The price commissioner says that the average profit per new car sale should be no more than \$500. A certain dealer is suspected of violating this policy. To test this suspicion, a sample of one day's sales is taken with the following profit per sale in dollars: 465, 515, 534, 498, 489, 545, 562, 508. The commissioner wants to be fair, so the 0.01 level of significance is used. Use the results to determine whether the dealer is violating the policy.

7.34 According to government estimates, the average amount spent on food per week for a family of four is \$120. A social worker thinks that the average amount, at least in this city, is actually higher. To test this belief a random sample of 40 families of four is asked to keep track of their expenditures for food for 1 week. The mean amount spent per family is \$126.50 with a standard deviation of \$14.40. At what level of significance has the social worker's belief been substantiated?

7.35 In nine tests under prescribed conditions, an automobile averages 27 miles per gallon with a standard deviation of 1.7 miles. The company wishes to advertise that this model will get at least 26 miles per gallon.
a. Using the 0.05 level, does this result substantiate the claim?
b. What sample size is needed to substantiate the claim at the 0.01 level?

7.36 Coffee cans are filled to a "net weight" of 16 ounces, but there is considerable variability. In fact, a random sample of eight cans of a particular brand shows net weights of 15.94, 16.01, 15.98, 16.01, 16.12, 15.95, 16.00, and 15.81 ounces. Test (at $\alpha = 0.05$) the hypothesis that the net weight of a can is actually 16 ounces.

7.37 In Problem 7.36 a quality control expert has reason to believe that the coffee can fills are not normally distributed but the distribution of fills is symmetric. Repeat Problem 7.36 using the trimmed-mean procedure.

7.38 A group of 40 tenth-grade students is given a paragraph and asked to memorize as much as possible. Their retention is then measured on a scale from 1 to 99. The contention being tested is that the average tenth-grade student will score at least 70 on the scale. It is assumed that the distribution of scores is normal. This group scores a mean of 66.93 with a standard deviation of 11.62. Using the 0.05 level of significance, does this result:
a. "prove" the contention?
b. "disprove" the contention?

In one of these cases we need the exact (t) test to make the correct decision. Which test is it, what is the result of the exact test, and what is the conclusion? How does this result differ from the approximate (z) test?

7.39 A shipment of 10,000 copper rods must be accepted or rejected on the basis of a sample of 100. The shipment will be accepted unless there are significantly more than 13.7 blemishes per bar, using a significance level of 0.01. The sample of 100 for a particular shipment has an average of 13.9 blemishes with a standard deviation of 1.4. Will the shipment be accepted or rejected? What is the P value for the result?

7.40 Injections of a drug must be properly maintained. A certain firm manufactures ampoules labeled 400 units and takes quality control samples using a 0.05 level of significance to reject lots which may not be maintaining that level. A sample of 10 ampoules from each of six lots is analyzed daily for content, and each lot is accepted or rejected on the basis of the sample. Variability is uncertain from lot to lot, but it is known that the number of units per ampoule in each lot is normally distributed. Which lots will be accepted and which rejected if the following are the results for a particular day?

Lot A: $\bar{x} = 400.7$, $s = 3.4$
Lot B: $\bar{x} = 401.1$, $s = 1.8$
Lot C: $\bar{x} = 398.7$, $s = 3.2$
Lot D: $\bar{x} = 401.3$, $s = 0.6$
Lot E: $\bar{x} = 401.7$, $s = 2.4$
Lot F: $\bar{x} = 399.2$, $s = 0.8$

7.41 A biologist wishes to determine whether an insect population found only in one location of a forest belongs to a certain species. The only morphological characteristic which appears different from that of the known members of the species is wing length. The mean wing length of the species is 15.4 millimeters. The biologist measures wing length for 25 insects from the forest location and finds that the mean wing length is 17.4 millimeters with a standard deviation of 2.3 millimeters. Can she conclude, at the 0.05 significance level, that the insects are of a different species?

7.42 Sixteen tests are made of the effervescing time of a certain combination of ingredients. The purpose of the tests is to discover whether this combination is the same as that of a rival product, which has a mean effervescing time of 16.2 seconds. The sample mean is 14.8 seconds with a standard deviation of 1.6 seconds. What do the results of the tests show at a 0.01 level of significance? What is the observed level of significance of the test?

7.43 A government agency routinely tests food products to see if they meet federal requirements at a 0.10 level of significance. In order to be labeled "ice cream" the product must contain a minimum of 10% butterfat. A sample of 10 half-gallons of a certain brand shows butterfat percentages of 9.6, 10.1, 9.9, 9.7, 10.0, 9.9, 9.8, 10.1, 9.9, and 9.7. What conclusions will the government agency reach if the hypotheses are formulated to favor:
a. the consumer? **b.** the manufacturer?

7.44 A new drug is hailed by its manufacturer as an excellent treatment for a disease. Using the standard drug, patients take an average of 9.2 days to recover. Using the new drug, 50 patients take an average of 9.3 days with a standard deviation of 1.1 days. The manufacturer claims that this result shows that the old drug is not significantly better than the new, so they will market it. Can their claim be substantiated at a 0.05 level of significance?

7.45 Suppose that a second set of eight coffee cans shows the same mean and standard deviation as the set given in Problem 7.36. Combine the two samples and test the same hypothesis at the same level of significance. What conclusion do you reach this time? Compare both the decision rules and the sample test statistics you obtain.

7.46 A cigarette manufacturer claims that one of their brands contains no more than 12 milligrams of tar. Five cigarettes are smoked via machine and found to contain, respectively, 15, 11, 13, 12, and 15 milligrams of tar. Can the manufacturer's claim be rejected at the 0.05 level of significance?

7.47 A new type of tire is run through an endurance test to see whether or not it can last an average of 25,000 miles under the worst conditions. A sample of 80 tires is tested and lasts an average of 25,226 miles with a standard deviation of 863 miles. Suppose the company does not wish to market the tire unless they can prove it will last, on the average, at least 25,000 miles under these conditions, and they wish to use a significance level of 0.01. Will the new tire be marketed?

7.48 A traffic survey proclaims that the mean length of time required to go through a tunnel at rush hour is 3.5 minutes. A skeptical motorist is certain that the claim is too low. He times his trips through the tunnel for a total of 22 trips during rush hour. He finds that his mean time is 4.4 minutes with a standard deviation of 0.8 minute. At about what level of significance can his belief be substantiated?

7.49 A certain sedative causes physiological changes that last for a period of time. A researcher wishes to test the accepted mean duration of effects of 1.3 hours and claims that the duration is actually greater. She uses 12 subjects and finds the sample to have a mean duration of 1.8 hour with a standard deviation of 0.7 hours. Can she substantiate her claim at the 0.05 level of significance?

7.50 Specifications for a candy-making machine call for it to produce bars weighing a mean of 2.30 ounces with a standard deviation of 0.12 ounce. A daily sample of 100 bars is weighed all together to decide if the machine is in control. If the total weight of the bars is above or below a preset weight, the machine should be retooled. If a significance level of 0.05 is used, above or below what total weight should the machine be retooled?

7.51 Suppose, in Problem 7.50, that the daily sample consists of only ten bars. What are the limits if weights of the bars are normally distributed?

7.3 Hypotheses About a Population Proportion

Public opinion polls play a large part in our everyday lives—particularly during presidential election years, when we are subjected to weekly polls telling us that Jones is preferred to Smith by 54% to 46% and so on. Biologists study the proportion of a species with a certain characteristic. A manufacturer wishes to determine whether a certain proportion of the population will buy a product. In such cases we may wish to test a hypothesis concerning the value of a proportion in a dichotomous population.

7.3.1 Large-Sample Tests

In Section 6.3 we determined confidence intervals for the population proportion. We noted that if π is a population proportion, then for sufficiently large samples drawn from the population, sample proportions p will be approximately normally distributed with mean π and standard error σ_p, where $\sigma_p = \sqrt{\pi(1 - \pi)/n}$ or $\sqrt{\pi(1 - \pi)/n} \cdot \sqrt{(N - n/(N - 1)}$ if we use the finite population correction factor. If a population proportion is π_0, then the sample proportions will be approximately $N(\pi_0, \sigma_p^2)$, so the variable $(p - \pi_0)/\sigma_p$ will be $N(0, 1)$. Recall that "sufficiently large" is defined to be a sample size n such that if π_0 is the population proportion, $n\pi_0$ and $n(1 - \pi_0)$ are each at least 5. Thus, using the hypothesis-testing technique outlined in Section 7.1, we have the following.

Large-Sample Hypothesis Tests for the Population Proportion

Suppose we obtain a random sample of n observations from a dichotomous population. We may test the hypothesis $H_0{:}\pi = \pi_0$ provided that both of the following are true:

1. $n\pi_0 \geq 5$
2. $n(1 - \pi_0) \geq 5$

The sample value of the test statistic used to test the hypothesis is z^*. If p is the sample proportion, $z^* = (p - \pi_0)/\sigma_p$. The value of σ_p is calculated with the finite population correction factor if it is appropriate.

The decision rule is dictated by the alternate hypothesis and the level of significance α chosen for the test. The three possibilities are as follows:

Case I	Case II	Case III
	Null and Alternate Hypotheses	
$H_0{:}\pi = \pi_0$ $H_1{:}\pi \neq \pi_0$	$H_0{:}\pi = \pi_0$ $H_1{:}\pi > \pi_0$	$H_0{:}\pi = \pi_0$ $H_1{:}\pi < \pi_0$
	Decision Rule	
Reject H_0 if $\lvert z^*\rvert > z_{\alpha/2}$	Reject H_0 if $z^* > z_\alpha$	Reject H_0 if $z^* < -z_\alpha$

There is a fundamental difference between hypothesis testing for the population proportion and confidence intervals for the population proportion. When we obtain a confidence interval, the population proportion is not known and we must use s_p. In testing a hypothesis about a population proportion, we calculate the test statistic by assuming that the population proportion is a specific value, π_0, and we may use σ_p based on this value.

If we reject H_0 and subsequently obtain a confidence interval for π, we must use s_p as in Section 6.3, since we still do not know π.

Example 7.11

A real estate promoter claims that at least 75% of the retired couples living in an apartment village prefer apartment living to single-unit living. Try to disprove this assertion at the 0.05 level of significance if a random sample of 100 couples of the 2200 in the village are interviewed and 63 of them indeed prefer apartment living.

Solution

We use the hypothesis-testing procedure outlined in Section 7.1. Since we are trying to disprove the assertion, our research hypothesis is $\pi < 0.75$. The critical value of z is $z_{.05} = 1.645$, and we need not use the correction factor since 100 is less than 5 percent of 2200.

1. $H_0{:}\pi = 0.75$
   $H_1{:}\pi < 0.75$
2. Decision rule: Reject H_0 if $z^* < -1.645$.
3. Results: $\sigma_p = \sqrt{(0.75)(0.25)/100} \doteq 0.043$, so

$$z^* = \frac{p - \pi_0}{\sigma_p} = \frac{0.63 - 0.75}{0.043} \doteq -2.79$$

4. Decision: Reject H_0.
5. Conclusion: The promoter's claim is exaggerated; $\pi < 0.75$.

Example 7.12

Of 400 plants of hybrid corn, 79 are found to have the recessive characteristic under study. By the Mendelian Law of Inheritance, 25% of the parent stock (the variety of corn the stock was taken from) would exhibit the characteristic. A researcher hypothesizes that the hybrid differs from the parent stock and wishes to use this result to test the claim at the 0.01 level of significance. Test this hypothesis.

Solution

If this were the parent stock, we would expect 25% of the 400 to have the characteristic, so a significant deviation from this percentage might indicate a difference. Thus we have:

1. $H_0{:}\pi = 0.25$
   $H_1{:}\pi \neq 0.25$
2. Decision rule: Reject H_0 if $|z^*| > 2.58$.
3. Results: $\sigma_p = \sqrt{(0.25)(0.75)/400} \doteq 0.022$ and $p \doteq 79/400 = 0.198$, so

$$z^* = \frac{p - \pi_0}{\sigma_p} \frac{0.198 - 0.25}{0.022} \doteq -2.36$$

4. Decision: Fail to reject H_0.
5. Conclusion: Since $|-2.36|$ is not greater than 2.58, we cannot reject the null hypothesis. On the other hand, the P value for this two-tailed test is less than 0.02. Thus the choice of α resulted in the failure to reject H_0. The most reasonable conclusion is to reserve judgment and continue testing.

PROFICIENCY CHECKS

14. A research hypothesis is that $\pi \neq 0.30$. The level of significance is 0.05; the sample result for $n = 60$ is $p = 0.4$. Test the hypothesis using the hypothesis-testing procedure.
15. A research hypothesis is that $\pi \geq 0.15$. The level of significance is 0.02; sample results are that 8 of 40 have the attribute. Test the hypothesis.
16. Compute and interpret the P values for Proficiency Checks 14 and 15.
17. A manufacturer claims that at most 1.5 percent of a certain product have any defects. A sample of 1000 of this product reveals that 22 have defects. Based on this sample, can the claim be refuted at the 0.01 level of significance? At what level of significance can the claim be refuted?

7.3.2 Small-Sample Tests

If a sample is too small to meet the requirement that both $n\pi_0$ and $n(1 - \pi_0)$ are at least 5, the best approach is to expand the size of the sample. If this is not practical, we can use the fact that sample proportions from a dichotomous population have approximately a binomial distribution. Thus we can test a hypothesis that $\pi = \pi_0$ by using the binomial model directly.

Instead of obtaining a rejection region, it is simpler to obtain the P value for the result and compare this value of P to the level of significance chosen for the test. Thus our decision rule for a small-sample test of a population proportion will be the following: reject H_0 if $P < \alpha$. We can calculate the P values directly or obtain them from either the Table of Cumulative Binomial Probabilities (Table 2), using $p = \pi_0$, or the Table of Cumulative Poisson Probabilities (Table 3), using $m = n\pi_0$. Recall that if the alternate hypothesis is two-sided, we must double the probability obtained from the table.

In Section 7.1 we discussed a case in which a treatment was believed to be effective against eczema. It was stated that the treatment would be considered effective if it could be shown that more than 2/5 of a group of eczema-infected mice were symptom-free after 4 weeks. A good way to test the effectiveness of the new eczema treatment is to use the treatment on a sample of mice with eczema and then observe the proportion of the sample that are symptom-free after 4 weeks. Thus if π is the population proportion for all symptom-free mice after 4 weeks of treatment, the research hypothesis is $\pi > 0.4$. If the sample size is sufficiently large, the methods presented earlier in this section apply. Suppose, however, that only ten mice actually had a case of eczema and were treated with the treatment. It may be that a sufficiently large sample was infected for $n\pi_0$ to be greater than 5, but only these ten actually contracted a case. Further, suppose that seven of the ten were symptom-free after 4 weeks. What is the probability that as many as seven of the ten would be symptom-free if actually $\pi = 0.4$? We can turn to Table 2 (with $n = 10$, $p = 0.4$) and determine that if the population proportion is actually 0.4, $p = 1 - 0.945$ and 0.055. This is the observed significance level of the test—the P value. If this probability is sufficiently small, less than the level of significance α chosen for the test, we reject H_0 in favor of H_1. If a significance level of, say, 0.10 had been chosen for the test, we would conclude that treatment was effective at the 0.10 level of significance.

We would use the hypothesis-testing procedure as follows:

1. $H_0: \pi = 0.4$
   $H_1: \pi > 0.4$
2. Decision rule: Reject H_0 if $P \leqslant 0.10$.
3. Results: Since seven of the ten mice were symptom-free, we refer to Table 2 with $n = 10$, $p = 0.4$, and find $P(x \geqslant 7) = 1 - P(x \leqslant 6) = 0.055$. Since the alternate hypothesis is one-sided, the rejection region is one-tailed, so $P = 0.055$.
4. Decision: Reject H_0.
5. Conclusion: The contention is supported, and the results support the belief that the treatment is effective.

Example 7.13

A social worker interviews families in a large apartment complex to determine their feelings about some proposed rule changes. He believes that fewer than 30 percent of the families favor the changes. Before he has time to complete his survey, he has to return to his office for a meeting. Nonetheless, he does manage to interview 15 families and finds that only 3 of them favor the changes. Is this sufficient information to support his belief at the 0.05 level of significance?

Solution

1. $H_0: \quad \pi = 0.3$
   $H_1: \quad \pi < 0.3$
2. Decision rule: Reject H_0 if $P < 0.05$.

3. Results: Since three of the sample favor the changes, we go to Table 2, with $p = 0.3$ and $n = 15$, to determine $P(x \leq 3)$. We obtain $P(x \leq 3) = 1 - P(x \geq 4) = 1 - 0.703 = 0.297 = P$.
4. Decision: Fail to reject H_0.
5. Conclusion: The social worker's belief is not supported. The P value is 0.297, clearly indicating that a result such as this could occur quite easily—more than one time in four—by chance.

Example 7.14

Some food that has been contaminated by radiation is inadvertently fed to laboratory rats before the mistake is discovered. One of the psychologists decides to see whether the accidental feeding has any effect on the rats' abilities to solve problems. She trains the rats to solve a maze in order to obtain food. Normally 80% of rats so trained can traverse the maze without error after 75 trials. Of the 18 rats that ate the food, however, only 10 of them can do so. Does this result provide evidence that there may have been some effect on the rats? Use the 0.05 level of significance.

Solution

This situation has a two-sided alternate hypothesis since the psychologist has no way of hypothesizing whether the effect on problem-solving ability, if any, is positive or negative. Since 80% of 18 is 14.4, and 10 is less than 14.4, the P value for the test will be two times $P(x \leq 10)$.

1. H_0: $\pi = 08$
   H_1: $\pi \neq 0.8$
2. Decision rule: Reject H_0 if $P < 0.05$.
3. Results: Since ten of the sample solved the maze, we go to Table 2, with $p = 0.8$ and $n = 18$, and determine $P(x \leq 10)$. We obtain $P(x \leq 10) = 0.016$. Thus the P value is 0.032.
4. Decision: Reject H_0.
5. Conclusion: The psychologist can conclude that this group of rats does not exhibit the same ability to solve the maze as rats in general. Whether this difference is due to the food is another question to be answered; moreover, the sample was not random but consisted of those that happened to eat the food. Nonetheless, the results of the analysis point the way for further research.

PROFICIENCY CHECKS

18. A research hypothesis is that $\pi \geq 0.1$. The level of significance is 0.02; sample results are that 4 of 20 have the attribute. Test the hypothesis.

(continued)

PROFICIENCY CHECKS *(continued)*

19. A manufacturer claims that at most 1.5 percent of a certain product have any defects. Two of a sample of 40 of this product have defects. Based on this sample, can the claim be refuted at the 0.01 level of significance? At what level of significance can the claim be refuted?

Problems

Practice

7.52 A sample of size n is used to test a research hypothesis that the population proportion differs from 0.4. Test the hypothesis at the 0.05 level of significance if the sample proportion is 0.5 and the sample size is:
a. $n = 400$ **b.** $n = 250$ **c.** $n = 40$ **d.** $n = 20$ **e.** $n = 8$

7.53 A random sample of 140 observations is used to test the research hypothesis that the population proportion is greater than 0.75. The sample contains 114 with the attribute. Test this hypothesis at a significance level of:
a. 0.10 **b.** 0.05 **c.** 0.025 **d.** 0.01 **e.** 0.005

7.54 A random sample of 60 observations is used to test whether or not the population proportion is 0.35. The significance level is chosen at 0.05. Test this hypothesis if the sample contains x^* with the attribute, where x^* is:
a. 12 **b.** 18 **c.** 24 **d.** 35 **e.** 48

7.55 A random sample of 100 observations is obtained to test whether or not the population proportion is 0.25. With what level of significance can the hypothesis be rejected if x^* have the attribute, where x^* is:
a. 11? **b.** 20? **c.** 31? **d.** 43? **e.** 72?

7.56 In repeated tests with a pair of dice, 7 is rolled 138 times in 540 tries. Does this result support (at the 0.05 level) a belief that the dice are fair (at least as far as 7's are concerned)?

7.57 A random sample of 15 is used to test the hypothesis that $\pi = 0.3$. Test this hypothesis if $p = 2/15$ and the significance level is:
a. $\alpha = 0.10$ **b.** $\alpha = 0.05$ **c.** $\alpha = 0.02$ **d.** $\alpha = 0.01$

Applications

7.58 A television manufacturer claims that at least 80% of its picture tubes last 2 years or more. A consumer protection agency challenges this claim and obtains evidence that a random sample of 200 tubes sold over 2 years ago has 56 defective. Is this sufficient evidence to reject the manufacturer's claim at a 0.01 level of significance?

7.59 A supply house has ordered a large quantity of items and is willing to accept no more than 3% defective. It is somewhat liberal, however, and will accept the shipment unless there are significantly more than 3% defective in a sample of 200, using a 0.10 level of significance. If the sample contains eight defective items, will the shipment be rejected?

7.60 The Gallup Poll claims that 60% of the electorate supports a certain proposition. The Roper Poll decides to test this contention at the 0.05 level of significance. What are their conclusions if a random sample of 1476 people contains 835 who favor the proposition?

7.61 To test a hypothesis that, on the average, at least 40% of consumers buying soft drinks still prefer bottles to cans, a soft drink distributor uses one day's sales to test the hypothesis at the 0.05 level. If the day's sales represent a random sample, what conclusions can they draw if there are 687 bottle sales and 876 can sales?

7.62 In a study relating student seating (proximity to the teacher) to grades, three classes with a total of 97 students are randomly rearranged with respect to seating. If a student moves closer to the teacher and his grade increases, or the student moves farther from the teacher and his grade decreases, this is considered a success since it supports the experimenter's belief. If there were no relationship between a student's grades and proximity to the teacher, a student would be equally likely to be a success or failure. Thus the null hypothesis would be $\pi = 0.5$. The research hypothesis is $\pi > 0.5$. If 64 students are considered to be successes, at what level of significance can the research hypothesis be substantiated?

7.63 A grocer carries two products that are identical except for brand name. Since her shelf space is limited, she has decided to eliminate one brand from her inventory. Brand A is advertised more extensively than brand B, and the grocer would like to take advantage of the advertising offered by brand A. On the other hand, she thinks that her customers have recently shown a preference for brand B. If there is no real preference she would prefer to stock brand A, but she would stock brand B if more than half her customers who buy the product prefer brand B. She polls a sample of customers to determine which of the brands they prefer. Of 163 customers who say they would buy the product, 101 prefer brand B. Using the 0.05 level of significance, which brand should she stock?

7.64 A psychologist has developed an "irresistible advertising display," which he claims will entice customers to buy the product displayed rather than a competing brand at least 60% of the time. A dog food manufacturer decides to give the display a try and sets it up in the local supermarket. Using a 10% level of significance, at least how many of the first n customers must buy the manufacturer's brand to substantiate the claim if n is:
a. 150? **b.** 75? **c.** 25? **d.** 10?

7.65 A biologist working with a new strain of virus finds that rats injected with the virus develop symptoms similar to those of the common cold. He wishes to compare the duration of these symptoms with that of the cold, so he injects 100 rats with the virus and observes the duration of the symptoms. The symptoms of the common cold last, on the average, for 14 days. He wishes to test the hypothesis that at least half the rats will have symptoms lasting longer than 14 days. What can he conclude at a 0.01 level of significance if 61 rats have symptoms lasting longer than 14 days?

7.66 A television network is offering a new show and expects at least 40% of the viewers to watch the premier performance. When a poll of 600 families is taken by telephone, it is found that 413 of the families were watching television and 196 of those were watching the new show. At what level of significance can the network support its contention?

7.67 An electronic typesetter is in use, but it is erratic—about 30% of all pages set contain errors. An R&D team has developed an attachment that they claim will reduce errors. To test the device, they attach it to a typesetter that has an error rate of 30%. A total of 25 pages is set in the usual way. The number of pages containing at least one error is given by x. The level of significance is 0.05.

a. Using the large-sample procedure, what values of x will lead to the conclusion that the attachment is effective?

b. Using the small-sample procedure, what values of x will lead to the conclusion that the attachment is effective?

7.68 According to a report in *Beverage Digest* (*San Francisco Examiner*, Feb. 2, 1986), the Coca Cola Company leads all soft drink manufacturers with 40% of all sales for its product. To determine whether a certain town differs from this percentage, a local Coca Cola bottler commissions a survey. The survey asks a total of 2137 people if they drink soft drinks, and if so what drink they prefer. Of 1203 who drink soft drinks, 438 prefer a Coca Cola product. At what level of significance can you conclude that the proportion in the town is not 0.40?

7.69 A manufacturer samples the daily output for acceptability. The limit for defectives has been set at 5% of the daily output. The daily output will be accepted only if there is 90% confidence that there are no more than 5% defective. If a sample of 80 units is tested daily, for what number of defectives will the lot be accepted? Rejected?

7.70 Patients who contract a rare disease have only a 1% chance of survival, but a new method of treatment is being tested throughout the country. A total of 400 patients have been treated and 10 have survived. Does this result show that the new treatment is more successful than the old? Use the 0.05 level of significance.

7.71 According to *Business Week* (Jan. 27, 1986), the civilian unemployment rate dropped to 6.9% in December 1985. Suppose that a random sample of 3126 adults nationwide shows 1408 employed, 126 unemployed but seeking employment, and the remainder retired or nonworking. Test the hypothesis that the unemployment rate is actually 6.9%. Use the 0.05 level of significance.

7.4 Hypotheses About the Population Variance

In addition to hypothesis tests concerning a population mean or a population proportion, we often use tests concerning the value of a population variance. In industrial applications, for example, variations in a product's quality must be kept to a minimum. If variability exceeds certain allowable limits, steps must be taken. In Section 6.6 we saw that sample variances have a sampling distribution that can be related to a chi-square distribution. We noted that for a normally distributed population where σ^2 is the population variance and s^2 is the variance of a sample of n observations, the random variable $(n - 1)s^2/\sigma^2$ has a chi-square distribution with $n - 1$ degrees of freedom; that is, $(n - 1)s^2/\sigma^2$ is $\chi^2(n - 1)$. Then for a null hypothesis that $\sigma^2 = \sigma_0^2$, or $\sigma = \sigma_0$, the test statistic will be chi square

Hypothesis Tests for the Population Variance (Normal Population)

Suppose we obtain a random sample of n observations of a normal random variable x for which the population variance is hypothesized to be σ_0^2. Then we may test the hypothesis $H_0{:}\sigma^2 = \sigma_0^2$ by using the following procedure. The sample value of the test statistic used to test the hypothesis is χ^{2*}. If s^2 is the variance of the sample, then $\chi^{2*} = (n - 1)s^2/\sigma_0^2$.

The decision rule is dictated by the alternate hypothesis and the level of significance α chosen for the test. The three possibilities are as follows:

Case I	Case II	Case III
Null and Alternate Hypotheses		
$H_0{:}\sigma^2 = \sigma_0^2$	$H_0{:}\sigma^2 = \sigma_0^2$	$H_0{:}\sigma^2 = \sigma_0^2$
$H_1{:}\sigma^2 \neq \sigma_0^2$	$H_1{:}\sigma^2 > \sigma_0^2$	$H_1{:}\sigma^2 < \sigma_0^2$
Decision Rule		
Reject H_0	**Reject H_0**	**Reject H_0**
if $\chi^{2*} > \chi^2_{\alpha/2}\{n - 1\}$ or if $\chi^{2*} < \chi^2_{1-\alpha/2}\{n - 1\}$	if $\chi^{2*} > \chi^2_{\alpha}\{n - 1\}$	if $\chi^{2*} < \chi^2_{1-\alpha}\{n - 1\}$

(χ^2). We obtain the sample value of the test statistic from a single sample variance s^2 and find that it is equal to χ^{2*}, where $\chi^{2*} = (n - 1)s^2/\sigma_0^2$. We thus have the hypothesis-testing procedures for the population variance shown on page 398. If the test concerns the population standard deviation, the procedures are exactly the same, except that the hypotheses concern σ rather than σ^2.

Example 7.15

To test the hypothesis that $\sigma^2 = 125$, a random sample of 81 is found to have a variance of 114.0624. Use $\alpha = 0.10$ to test this hypothesis, assuming that the population is normally distributed.

Solution

In this case we need $\chi^2_{.95}\{80\}$ and $\chi^2_{.05}\{80\}$. From Table 6, the values are 60.3915 and 101.879. We then have:

1. $H_0{:}\sigma^2 = 125$
   $H_1{:}\sigma^2 \neq 125$
2. Decision rule: Reject H_0 if $\chi^{2*} > 101.879$ or if $\chi^{2*} < 60.879$.
3. Results: The sample standard deviation is 10.68, so the sample variance is 114.0624. Then $\chi^{2*} = 80(114.0624)/125 \doteq 73.00$.
4. Decision: Fail to reject H_0.
5. Conclusion: 125 seems to be a satisfactory value for σ^2.

Example 7.16

A sample of 12 castings is taken from a daily production run. To ensure good quality and minimum variation between castings in a run, the population variance or standard deviation should be held to a minimum. The run will be passed unless the standard deviation of the number of flaws per casting is more than 5; in that case the castings will all be inspected. If the standard deviation of the number of flaws per casting for the sample is 6.34, use a 0.01 level of significance and determine whether the run should be inspected. It is reasonable to assume normality of number of flaws since the castings are made to the same specifications.

Solution

Using Table 6, we obtain $\chi^2_{.01}\{11\} = 24.7250$. Then:

1. $H_0{:}\sigma = 5$
   $H_1{:}\sigma > 5$
2. Decision rule: Reject H_0 if $\chi^{2*} > 24.7250$.
3. Results: The sample variance is $(6.34)^2 = 40.1956$; $\sigma_0^2 = 25$, so $\chi^{2*} = 11(40.1956)/25 = 17.6861$.
4. Decision: Fail to reject H_0.
5. Conclusion: The run will be passed.

PROFICIENCY CHECKS

20. If x is normally distributed, $\bar{x} = 100.00$, and $s = 30.00$, test the hypothesis that $\sigma^2 = 1000.00$ using $\alpha = 0.05$ if n is:
a. 5 **b.** 10 c. 20 **d.** 25

21. If x is normally distributed, $\bar{x} = 0.560$, $s = 0.122$, and $n = 15$, test the hypothesis that $\sigma > 0.085$ if α is:
a. 0.10 **b.** 0.05 c. 0.01

22. If x is normally distributed, $\bar{x} = 8.66$, $s = 1.18$, and $n = 47$, can you conclude, at the 0.05 level, that $\sigma < 1.50$? What is the P value for this test?

Problems

Practice

7.72 A sample of size n is drawn from a normal population and the result is used to test a research hypothesis about a population standard deviation. Formulate null and alternate hypotheses and test at the 0.05 level if the research hypothesis about σ is as given and n and s are as shown.
a. $\sigma = 15.0$; $n = 10$, $s = 13.4$
b. $\sigma < 115$; $n = 6$, $s = 103.4$
c. $\sigma \geqslant 0.035$; $n = 20$, $s = 0.050$
d. $\sigma = 8.50$; $n = 24$, $s = 8.63$
e. $\sigma > 12$; $n = 30$, $s = 12.136$

7.73 A random sample is drawn from a normally distributed population. The sample variance is 39.69. Using the 0.05 level of significance, test the hypothesis that the population variance is 50 if the sample size is:
a. 10 **b.** 15 c. 20 **d.** 50

7.74 A random sample of size 36 is drawn from a normal population. Test the hypothesis that $\sigma^2 = 30.0$ if the sample standard deviation is as given. Use $\alpha = 0.10$.
a. $s = 1.9$ **b.** $s = 4.6$ c. $s = 8.9$ **d.** $s = 10.6$

7.75 A random sample of size 30 is drawn from a normal population. If the sample variance is 0.000169, test the hypothesis that $\sigma^2 = 0.0004$ using the significance level:
a. 0.10 **b.** 0.05 c. 0.02 **d.** 0.01

Applications

7.76 Variance is an important aspect of quality control, in that variability of output is a measure of consistency. If a machine making ball bearings is highly variable in its output, much of the production run will be unacceptable—either too large or too small in diameter. Suppose that the first 20 ball bearings have a mean diameter of 6.003 millimeters with a standard deviation of 0.017 millimeter. Using the 0.05 level of significance, test the hypothesis that the standard deviation of the bearings produced by the machine is greater than 0.015.

7.77 A biologist wishes to determine whether conditions in various parts of the bay are causing fish to mature more slowly. One way to measure this is to determine whether there is a great deal of variability among the size of the young of the species. A sample is taken at random from randomly chosen sections of the bay. The sample consists of 46 pompano young, and the standard deviation of the weights is 15.86 grams. Test the hypothesis that this result is not significantly different from the theoretical value of 12 grams, using the 0.02 level of significance.

7.78 When a new production line is being started, management must get an estimate of the mean and variability of the time required to perform tasks in order to time the movement of the line. The original estimate of the standard deviation for a certain production line was 2.00 minutes. Management now thinks that this was an overestimate and wishes to reduce the time accordingly. A sample of 25 workers performs the same task and requires a mean of 4.11 minutes with a standard deviation of 1.85 minutes to do the task. Using the 0.05 level of significance, should management conclude that $\sigma < 2.00$?

7.79 In Problem 7.40, it was decided to accept or reject each lot of ampoules on the basis of a test that the mean of each lot is 400. Suppose in addition that the standard deviation of each lot cannot be more than 2.0. Lots from which sample standard deviations are significantly greater than 2.0 at the 0.05 level are rejected. Which lots will be accepted and which rejected on the basis of this test if the following are the results for a particular day?

Lot A: $\bar{x} = 400.7$, $s = 3.4$
Lot B: $\bar{x} = 401.1$, $s = 1.8$
Lot C: $\bar{x} = 398.7$, $s = 3.2$
Lot D: $\bar{x} = 401.3$, $s = 0.6$
Lot E: $\bar{x} = 401.7$, $s = 2.4$
Lot F: $\bar{x} = 399.2$, $s = 0.8$

Compare with the results of Problem 7.40. Most likely, each lot would have to pass both tests. In this case, which lots would be acceptable?

7.5 Computer Usage

All the hypothesis tests presented in this chapter may be performed using the computer. Minitab commands are available to perform one-sample t and z tests for the population mean; SAS statements can be used for all the tests.

7.5.1 Minitab Commands for This Chapter

We can use Minitab to test a hypothesis that a population mean has a specified value. When the population variance is known, we use the command ZTEST; and when it is unknown, we use the command TTEST. Since TTEST calculates and uses the sample standard deviation, we can also use this command for a large-sample z test even if the population is not known to be normal.

When we know the population variance, the Minitab command is

```
MTB > ZTEST of mean=K1 with stdev=K2 for the data in column C
```

Suppose, for example, we wish to test the hypothesis H_0:$\mu = 10.8$ against the alternative hypothesis H_1:$\mu \neq 10.8$. If the data are stored in column 4 and the population variance is 16, we would write

```
MTB > ZTEST of mean=10.8 with stdev=4 for the data in column C4
```

If the value for the mean is not specified, the default value is zero. Note also that the two-sided test is the default test; to specify a one-sided alternative we need the ALTERNATIVE subcommand. To specify a lower-tail test—a test of the alternate hypothesis H_1:$\mu < \mu_0$—we use the subcommand ALTERNATIVE $= -1$; for an upper-tail test of the alternate hypothesis H_1:$\mu > \mu_0$, we use the subcommand ALTERNATIVE $= +1$. Thus to test the hypothesis H_0:$\mu = 10.8$ against the alternative H_1:$\mu < 10.8$, we would write

```
MTB > ZTEST of mean=10.8 with stdev=4 for the data in column C4;
SUBC> ALTERNATIVE=-1 .
```

If the population variance is unknown and the sample is sufficiently large, or if the sample is small but the population is approximately normal, we can test the hypothesis H_0:$\mu = \mu_0$ by using the TTEST command. With this procedure Minitab calculates and uses the sample standard deviation in the formula. The command to write is

```
MTB > TTEST of mean=K using the data in column C
```

As before, the default value for the mean is zero and the default alternative is two-sided. We can test one-sided alternatives by using the ALTERNATIVE subcommand as indicated previously. As an example, to test the hypothesis H_0:$\mu = 0.08$ against the alternate hypothesis H_1:$\mu > 0.08$ for the data in column C12 the Minitab commands are

```
MTB > TTEST of mean=0.08 using the data in column C12;
SUBC> ALTERNATIVE=+1 .
```

Note that the comments and cautions concerning the assumption of normality mentioned previously still hold and that we should use the ZTEST and TTEST commands only for appropriate data. Graphic analysis, such as histograms and stem and leaf plots, provide quick checks for determining whether the data are approximately normal. A criterion for normality, as well as methods for checking on normality, were discussed in Section 5.3.

Note also that the ALTERNATIVE subcommand only changes the resulting P value for the test. The value of the test statistic (z^* or t^*) does not change for different alternate hypotheses. If the test statistic is t, for example, for lower-tail tests the P value is defined to be $P(t < t^*)$; for upper-tail tests the P value is $P(t > t^*)$; for two-tailed tests the P value is $2P(t > |t^*|)$. For other test statistics the P value is defined similarly. Note that Minitab does not ask for a specification of α prior to the test, but merely gives the P value for the test. It is up to the user to determine whether or not the result is significant—that is, whether or not to reject H_0.

Example 7.17

The last exam given to an introductory statistics class of 25 students is a standardized test that has been given to 2500 students at other colleges and universities. For the 2500 students, the standard test resulted in a mean of 74.2 and a standard deviation of 10.1. Using the results given here, determine whether this class of 25 students is "standard." That is, is it reasonable to assume that this class's test scores could be a random sample of a population with mean 74.2 and a standard deviation of 10.1? Use the 0.05 level of significance.

88, 92, 74, 60, 77, 73, 82, 75, 65, 80, 94, 68
76, 78, 81, 80, 68, 65, 72, 71, 74, 75, 96, 61, 87

Solution

Since we wish to know simply whether the class is "standard" or not, a two-sided alternative is appropriate. We may test $H_0: \mu = 74.2$ against the alternate hypothesis $H_1: \mu \neq 74.2$, using the z test with known population standard deviation $\sigma = 10.1$. The Minitab commands and output are given in Figure 7.9.

```
MTB > SET the following data into column C1
DATA> 88 92 74 60 77 73 82 75 65 80 94 68
DATA> 76 78 81 80 68 65 72 71 74 75 96 61 87
DATA> ENDOFDATA
MTB > NAME for column C1 is 'SCORE'
MTB > ZTEST of mean=74.2 with stdev=10.1 for the data in C1

TEST OF MU = 74.2 VS MU N.E. 74.2
THE ASSUMED SIGMA = 10.1

              N      MEAN     STDEV    SE MEAN        Z      P VALUE
SCORE        25     76.48      9.68        2.0     1.13         0.26
```

Figure 7.9
Minitab printout of the test data in Example 7.17.

The command ZTEST gives us $z^* = 1.13$ and a P value of 0.26 (that is, $2P(z > 1.13) = 0.26$). Since the obtained significance level (P value) is not smaller than the level of α chosen for the test, we may not reject H_0. For these data we conclude that the mean test score for this class of 25 students is *not* significantly different from the standard. Although the sample mean (76.48) is larger than 74.2, the difference is not statistically significant. Given the variability of the test scores ($\sigma = 10.1$), we are not unlikely to obtain a sample mean this large just by chance. In fact, we would expect a mean this far from 74.2 (that is, at least 76.48 or no more than 71.92) about one time in four on the average (since $P = 0.26$) from the population of "standard" classes.

Example 7.18

In Example 7.8, we tested the hypothesis that a new drug causes faster red-cell buildup in anemic persons than the drug currently in use. We used six observations to test the hypothesis that the mean buildup is greater than 8.3. Use Minitab with the data to test this hypothesis at the 0.05 level of significance.

Solution

Figure 7.10 shows the Minitab commands and the output needed to test the hypothesis $H_0{:}\mu = 8.3$ against the alternative $H_1{:}\mu > 8.3$. Compare with the results of Example 7.8. Note the use of the ALTERNATIVE = +1 subcommand to specify the upper-tail test. Comparing these results to those of Example 7.8, we see that they are identical. Note also that the P value of 0.67 is the probability that a value of t from a t distribution with 5 degrees of freedom will be greater than -0.47. Since the P value is larger than 0.05, we fail to reject H_0 and conclude that the mean red-cell buildup is not significantly greater than 8.3.

Figure 7.10
Minitab printout of the red-cell buildup data (Example 7.18).

```
MTB > SET the following data into column C1
DATA> 6.3 7.8 8.1 8.3 8.7 9.4
DATA> ENDofdata
MTB > NAME for column C1 is 'BLD CELL'
MTB > TTEST of mean=8.3 using the data in C1;
SUBC> ALTERNATIVE=+1.

TEST OF MU = 8.30 VS MU G.T. 8.30

             N      MEAN     STDEV   SE MEAN       T     P VALUE
BLD CELL     6      8.10      1.04      0.43   -0.47        0.67
```

The next example illustrates the interpretation of the level of significance α in hypothesis testing. Its interpretation is very similar to the degree of confidence for confidence intervals discussed in Chapter 6. Recall that a 95% confidence interval is interpreted to mean that 95% of the confidence intervals constructed in the same way from the same size random samples should include the true population mean. Recall also that we illustrated this concept using randomly generated samples. We can use a similar procedure to illustrate the interpretation of α.

We know that α, the level of significance, is the probability of rejecting H_0 when H_0 is true. We will generate 20 random samples of size $n = 20$ from a normal population with mean $\mu = 7.2$ and standard deviation $\sigma = 1.5$. If we test $H_0{:}\mu = 7.2$ against $H_1{:}\mu \neq 7.2$ with $\alpha = 0.10$, we would expect to reject H_0 incorrectly approximately 10% of the time—that is, in 2 of the 20 tests. Figure 7.11 shows the results of such random sampling. Note that we reject H_0 in exactly 2 of the 20 tests. This is precisely what we should have expected. If we were to repeat this experiment, it is unlikely that we would reject exactly 2 in every case, but we should expect to reject approximately twice each time with a long-term average almost exactly equal to 2.

As a final example we illustrate the difference between the misleading conclusion "accept H_0" and the correct conclusion "fail to reject H_0." We

```
MTB > RANDOM 20 observations into C1-C20;
SUBC> NORMAL mean=7.2 stdev=1.5.
MTB > TTEST of mean=7.2 using the data in C1-C20

TEST OF MU = 7.20 VS MU N.E. 7.20

          N      MEAN     STDEV   SE MEAN        T     P VALUE
C1       20      7.51      1.30      0.29     1.08       0.29
C2       20      6.33      1.58      0.35    -2.47      0.023*
C3       20      7.64      1.63      0.37     1.22       0.24
C4       20      6.95      1.58      0.35    -0.70       0.49
C5       20      7.46      1.41      0.31     0.82       0.42
C6       20      7.05      1.49      0.33    -0.44       0.66
C7       20      7.17      1.10      0.25    -0.12       0.91
C8       20      7.57      1.52      0.34     1.09       0.29
C9       20      7.40      1.55      0.35     0.59       0.56
C10      20      7.09      1.47      0.33    -0.35       0.73
C11      20      6.67      1.47      0.33    -1.62       0.12
C12      20      6.58      1.22      0.27    -2.28      0.034*
C13      20      7.41      1.36      0.30     0.68       0.51
C14      20      7.29      1.74      0.39     0.23       0.82
C15      20      7.46      1.09      0.24     1.06       0.30
C16      20      6.78      1.77      0.40    -1.06       0.30
C17      20      6.98      1.67      0.37    -0.60       0.56
C18      20      7.63      1.55      0.35     1.24       0.23
C19      20      7.16      1.43      0.32    -0.12       0.91
C20      20      7.03      1.27      0.28    -0.60       0.56
```

*Denotes those tests where we incorrectly reject H_0.

Figure 7.11 Minitab printout to illustrate the interpretation of α.

generate 20 random samples of size 20 from a normal population with mean $\mu = 7.3$ and standard deviation $\sigma = 1.5$. For each sample we test $H_0{:}\mu = 7.2$ against the alternate hypothesis $H_1{:}\mu \neq 7.2$. If we reject H_0, then, we do so correctly. On the other hand, our hypothesis value of μ is close to the true value (7.3), so we should not expect to reject H_0 very often, even though it is false.

The results of the sampling are shown in Figure 7.12. These results support our expectations, since we reject H_0 correctly in only 2 of the 20 tests. In the other 18 cases, if we accept H_0 we imply that H_0 is true, which is clearly not the case. Rather we should fail to reject H_0 with the understanding that the mean is not significantly different from our hypothesized value but not necessarily equal to it. Our results imply that the *power* of the test used for this test—a sample of size 20, a small difference between μ and μ_0 (0.1 compared to $\sigma = 1.5$), and $\alpha = 0.05$—is quite small, about 0.10 (since we correctly rejected 2 of the 20 tests). The power in this case can be found to be approximately 0.113; thus we would expect to reject H_0 correctly in about 11.3% of the tests—a result borne out by the actual results in which we reject H_0 in 10% of the tests. In Section 7.5.3 we will learn how to extend these calculations to obtain a *power curve*.

Figure 7.12
Minitab printout of 20 tests of $H_0{:}\mu = 7.2$ against $H_1{:}\mu \neq 7.2$.

```
MTB > RANDOM 20 observations into C1-C20;
SUBC> NORMAL mean=7.3 stdev=1.5.
MTB > TTEST of mean=7.2 using the data in C1-C20

TEST OF MU = 7.20 VS MU N.E. 7.20
```

	N	MEAN	STDEV	SE MEAN	T	P VALUE
C1	20	6.85	1.72	0.39	-0.91	0.37
C2	20	7.25	1.56	0.35	0.14	0.89
C3	20	7.44	1.63	0.36	0.66	0.52
C4	20	7.21	1.65	0.37	0.04	0.97
C5	20	7.71	1.46	0.33	1.56	0.13
C6	20	7.61	1.30	0.29	1.40	0.18
C7	20	7.31	1.02	0.23	0.49	0.63
C8	20	7.23	1.34	0.30	0.09	0.93
C9	20	7.13	1.72	0.38	-0.18	0.86
C10	20	7.85	1.64	0.37	1.77	0.093
C11	20	7.40	1.58	0.35	0.57	0.58
C12	20	8.25	1.27	0.28	3.69	0.0016*
C13	20	7.47	1.22	0.27	0.98	0.34
C14	20	7.28	1.15	0.26	0.31	0.76
C15	20	7.70	1.52	0.34	1.48	0.15
C16	20	7.76	1.39	0.31	1.79	0.089
C17	20	7.919	0.862	0.19	3.73	0.0014*
C18	20	7.17	1.55	0.35	-0.09	0.93
C19	20	6.90	1.27	0.28	-1.07	0.30
C20	20	7.64	1.58	0.35	1.23	0.23

*Denotes those tests where we reject H_0 using $\alpha = 0.05$.

7.5.2 SAS Statements for This Chapter

We can use SAS procedures and statements to test any of the hypotheses presented in this chapter. The actual statements are similar to those we used in Chapter 6 to obtain confidence intervals. To test hypotheses about a population mean or variance we first use the MEANS procedure to obtain the mean, variance, and standard deviation of the sample, as needed. Sometimes we use the OUTPUT statement to rename these variables for use in a DATA step to obtain the sample value of the test statistic and the P value. There is one special case. To test the hypothesis $H_0{:}\mu = 0$ against $H_1{:}\mu \neq 0$, we may use the output of the MEANS procedure as follows:

```
PROC MEANS MEAN STD STDERR T PRT;
```

The statements after PROC MEANS indicate that we wish only those statistics to be computed. In addition to the mean, standard deviation, and standard error of the mean, we will obtain t^* (T) and the P value (PRT) for a two-sided test.

In most cases the hypothetical population mean is different from zero. We may create a new variable by subtracting μ_0 from each value of x. We can then use the MEANS procedure to obtain t^* and the P value to test the hypothesis that $x - \mu_0 = 0$. Suppose, for example, we wish to test the hypothesis $H_0{:}\mu = 10.8$ against the alternate hypothesis $H_1{:}\mu \neq 10.8$ using a t test. We write

```
DATA START;
  INPUT variable@@;
  Y = variable - 10.8;
CARDS;
data here
PROC MEANS MEAN STD STDERR T PRT;
```

If the alternate hypothesis is one-sided, we need to divide the obtained P value by 2 and make sure that the sign of t^* agrees with the alternate hypothesis. If it does not, we cannot reject the null hypothesis. If we wish, we may create a new data set and list only $\bar{x}$, t^*, and the correct P value for a one-sided test. See Figure 7.14 for an example.

If we know the population standard deviation, we need to construct the test using the given value of σ. We can use a variation of the preceding set of statements. To obtain the P values we use the PROBNORM function. PROBNORM(Z^*) gives the probability that z will be less than z^*. To test $H_1{:}\mu < \mu_0$ we would write P = PROBNORM(ZSTAR) to obtain the P value. To test $H_1{:}\mu > \mu_0$ we would write P = 1 − PROBNORM(ZSTAR) to obtain the P value. For a two-sided alternative we want $P(z > |z^*|) + P(z < -|z^*|) = 2P(z > |z^*|)$. We can use ABS(ZSTAR) to obtain $|z^*|$ so that P is as shown in the program written here:

```
DATA START;
  INPUT variable @@;
CARDS;
data here
PROC MEANS MEAN NOPRINT;
  OUTPUT OUT = A MEAN = XBAR;
DATA TEST;
  SET A;
  ZSTAR = (XBAR - μ0) / σ;
  P = (1 - PROBNORM(ABS(ZSTAR))) * 2;
PROC PRINT;
```

To test a hypothesis about a population variance, we need to know the sample variance. We can obtain it from the MEANS procedure by using the key word VAR; the test statistic is χ^2. The function PROBCHI(χ^{2*}, *df*) computes the probability that a value of χ^2 will be less than the specified value with degrees of freedom as given—that is, $P(\chi^2 < \chi^{2*})$. To obtain $P(\chi^2 > \chi^{2*})$ we use 1 − PROBCHI(χ^{2*}, *df*). For a two-sided alternative the procedure is a bit more complicated. Since all values of χ^2 are positive and the curve is not symmetric, we cannot use the ABS statement as before. The simple solution is merely to compare the obtained value of χ^{2*} with the critical values, $\chi^2_{\alpha/2}$ and $\chi^2_{1-\alpha/2}$. We can obtain the *P* value by doubling $P(\chi^2 > \chi^{2*})$ if χ^{2*} is large or doubling $P(\chi^2 < \chi^{2*})$ if χ^{2*} is small. We can do this with the computer by calculating both these values and then doubling the smaller of them. If we name χ^{2*} CHISTAR, the SAS statements

```
P1 = PROBCHI(CHISTAR, df);
P2 = 1 - PROBCHI(CHISTAR, df);
PVAL = 2 * (MIN (P1, P2));
```

will obtain the *P* value for the two-sided test. The MIN function selects the smallest value of the variables listed in the parentheses. (The MAX function selects the largest.)

Example 7.19

Repeat Example 7.17 using SAS statements.

Solution

The program and output are shown in Figure 7.13. The results are slightly different from those of Example 7.18 ($z^* = 1.18$ instead of 1.13) because SAS procedures retain more decimal places than Minitab commands. The values of TSTAR and PVAL for X have no meaning, and YBAR is unimportant. We could compute XBAR by adding YBAR to 74.2. Another alternative is to obtain $\bar{x}$ explicitly by creating a new data set. See Figure 7.14 for an example.

Example 7.20

Refer to Problem 7.43. Use SAS statements to test the hypothesis that the butterfat content of the ice cream is 10% if the alternate hypothesis is formulated to favor *(a)* the consumer and *(b)* the manufacturer.

```
INPUT FILE

DATA START:
  INPUT X @@;
  Y = X - 74.2;
CARDS ;
88 92 74 60 77 73 82 75 65 80 94 68
76 78 81 80 68 65 72 71 74 75 96 61 87
PROC MEANS MEAN STDERR T PRT;
PROC PRINT;

OUTPUT FILE

VARIABLE            MEAN          STD ERROR        T        PR>|T|

X            76.48000000         1.93573414    39.51        0.0001
Y             2.28000000         1.93573414     1.18        0.2504
```

Figure 7.13
SAS printout for the statistics test data of Example 7.17.

```
INPUT FILE

DATA START;
  INPUT X @@;
  Y = X - 10;
CARDS;
9.6 10.1 9.9 9.7 10.0 9.9 9.8 10.1 9.9 9.7
PROC MEANS MEAN STDERR T PRT NOPRINT;
  VAR Y;
  OUTPUT OUT = A MEAN = YBAR T = TSTAR PRT = TWOP;
DATA ANSWER:
  SET A;
  XBAR = 10 + YBAR;
  PVAL = TWOP / 2;
PROC PRINT;
  VAR XBAR TSTAR PVAL ;

OUTPUT FILE

                         SAS

        OBS      XBAR      TSTAR        PVAL

         1       9.87      -2.414     0.0194949
```

Figure 7.14
SAS printout for the butterfat content of ice cream data (Example 7.20).

Solution

The printout for this problem is given in Figure 7.14. In each case the null hypothesis is $H_0{:}\mu = 10$. For part *(a)* we have $H_1{:}\mu > 10$, so the product will not be acceptable unless it exceeds government specifications. Since t^* is negative, we cannot reject the null hypothesis in favor of the alternative, so the ice cream would not meet government standards. For part *(b)* we have $H_1:\mu < 10$, so the product will be acceptable unless it is significantly substandard. In this instance the P value (about 0.0195) is less than 0.10, so we reject H_0 and the ice cream is still not acceptable.

Example 7.21

In Example 7.19 (and 7.17) we found that the standard deviation was about 9.68 instead of the theoretical value, $\sigma = 10.1$. Use SAS procedures to determine whether or not we can reasonably assume that the standard deviation of the population from which the sample was selected is 10.1 (that is, the variance is 102.01). Use a 0.10 level of significance.

Solution

The printout for this example is shown in Figure 7.15. Note that this is a two-sided alternative, $H_1:\sigma^2 \neq 102.01$. The statement obtaining the P value is written accordingly. The P value is much larger than 0.10, indicating that the assumption $\sigma = 10.1$ is reasonable.

To test a hypothesis about a population proportion, we need to obtain $z = (p - \pi_0)/\sigma_p$. We can obtain this value easily in the DATA step. We can enter x and n either as data following the CARDS statement or as actual numbers in the DATA step. We may write

```
DATA TEST;
   INPUT x n;
   P = x / n;
   SEP = SQRT(π0 * (1 - π0)/N);
   ZSTAR = (P - π0)/SEP;
   PVAL = (1 - PROBNORM(ZSTAR)) * 2;
PROC PRINT;
```

for a two-sided alternative, where π_0 is the hypothetical population proportion. The italics indicate numbers to be entered. We use SEP for standard error of proportion. For a one-sided alternative, we find the P value by using PROBNORM(ZSTAR) for a lower-tail test or 1 − PROBNORM(ZSTAR) for an upper-tail test.

Figure 7.15
SAS printout for a test of the population variance for Example 7.21.

```
INPUT FILE

DATA START;
  INPUT X @@;
CARDS;
88 92 74 60 77 73 82 75 65 80 94 68
76 78 81 80 68 65 72 71 74 75 96 61 87
PROC MEANS VAR NOPRINT;
  OUTPUT OUT = A VAR = SSQ;
DATA TEST;
  SET A;
  CHISTAR = (24 * SSQ) / 102.01
```

```
  P1 = PROBCHI(CHISTAR, 24);
  P2 = 1 - PROBCHI(CHISTAR, 24);
  P = (MIN (P1, P2)) * 2;
PROC PRINT;

OUTPUT FILE

                              SAS

OBS      SSQ       CHISTAR      P1          P2         P
 1     93.6767     22.0394   0.423085    0.576915   0.846171
```

Example 7.22

A firm is introducing a new version of a product and wishes to determine whether or not it should retain the old version of the item. The accounting department has ascertained that it will be feasible to retain the old version unless fewer than 25 percent of the present customers prefer the old version. A market survey of 1343 consumers finds 232 who use the product. These 232 are given a sample of the new product and then are asked if they would switch to the new version; 188 say that they would. Is this sufficient proof (at the 0.05 level of significance) that the old version should be retained? Use the appropriate SAS statements.

Solution

A printout of the program and results is shown in Figure 7.16. We see that there is less than a 2% chance that the true proportion is not smaller than 0.25. Thus the old version should not be retained.

```
INPUT FILE

DATA TEST;
  P = 44/232;
  SEP = SQRT(.25 * .75 / 232);
  ZSTAR = (P - .25) / SEP;
  PVAL = PROBNORM(ZST AR);
PROC PRINT;

OUTPUT FILE

                          SAS

OBS       P          SEP         ZSTAR       PVAL
 1     0.189655   0.0284287    -2.1227    0.0168906
```

Figure 7.16
SAS printout for the market survey data of Example 7.22.

7.5.3 The Power Curve

The power of a test has frequently been mentioned in this chapter. Recall that the power of a test is the probability that the test will correctly reject H_0. The power is denoted $1 - \beta$. We cannot determine the power unless

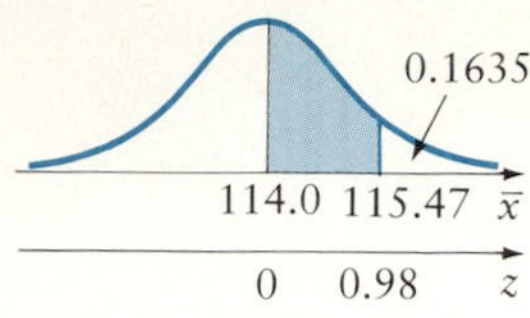

Figure 7.17
The power of the test H_0:μ = 113.0 against H_1:μ_1 > 113.0 if μ = 114.0.

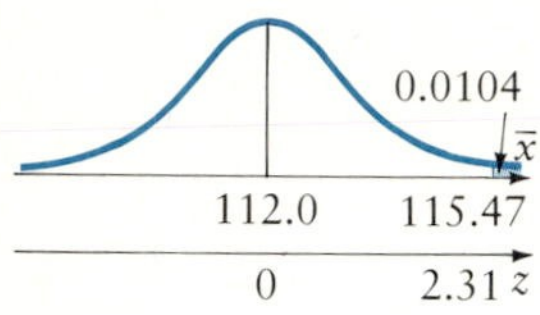

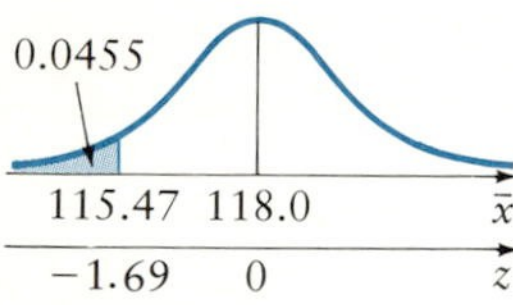

Figure 7.18
The power of the test H_0:μ = 113.0 against H_1:μ > 113.0 if μ = 112.0 (top) and if μ = 118.0 (bottom).

Table 7.2 POWER OF THE ONE-SIDED TEST FOR SEVERAL VALUES OF μ_1.

μ_1	$1 - \beta$
111.0	0.0014
112.0	0.01
113.0	0.05 (= α)
114.0	0.16
115.0	0.38
116.0	0.64
117.0	0.85
118.0	0.95
119.0	0.99
120.0	0.999

we know the true value of the parameter; we can, however, determine what the power will be for an interval of values of the parameter.

Suppose we wish to test the hypothesis H_0:μ = 113.0 against the alternate hypothesis H_1:μ > 113.0. Suppose that μ_1 is the true value of μ. If μ_1 < 113.0, the power of the test is very small; if μ_1 is considerably greater than 113.0, the power of the test will be very large, approaching 1.00 as μ_1 gets larger. We can calculate values of $1 - \beta$ for possible values of μ_1 and plot them on a graph. The resulting curve is called a **power curve** for the test.

To illustrate how this can be done, suppose that μ_1 = 114.0, σ = 15, and a random sample of 100 is obtained to test H_0:μ = 113.0 against H_1:μ_1 > 113.0 at a 0.05 level of significance. We will reject H_0 if $z^* >$ 1.645. Since $z^* = (\bar{x} - 113.0)/(15/\sqrt{100}) = (\bar{x} - 113.0)/1.5$, we have $z^* \geqslant 1.645$ if $(\bar{x} - 113.0)/1.5 > 1.645$. We can solve this latter expression for $\bar{x}$ to obtain $\bar{x} > 115.47$ approximately. Thus if $\bar{x} > 115.47$, then $z^* >$ 1.645, so we will reject H_0 for a test in which the sample mean is greater than 115.47. Now if the population mean is really 114.0, the power of the test will be the probability that $\bar{x} > 115.47$. This situation is shown in Figure 7.17.

We can find the shaded area in Figure 7.17 by using Table 4; z = 1.47/1.5 ≐ 0.98. Thus $1 - \beta \doteq 0.16$. Figure 7.18 illustrates two other possibilities: if μ_1 is actually 118.0, then $1 - \beta \doteq 0.95$; and if μ is actually 112.0, then $1 - \beta \doteq 0.01$. A short table of values $1 - \beta$ for this test is shown in Table 7.2. For purposes of continuity, if H_0 is actually true we define $1 - \beta = \alpha$ even though we cannot commit a type II error if H_0 is true.

A graph of the power curve for this test is shown in Figure 7.19. Values of $1 - \beta$ for values of μ_1 not listed in Table 7.2 may be estimated from the curve. If the alternate hypothesis is H_1:μ < 113.0, the power curve will be a mirror image of the one in Figure 7.19; for H_1:μ > 113.0, the power curve is shown in Figure 7.20.

If the alternate hypothesis is two-sided, such as H_1:$\mu \neq 113.0$, the rejection region is two-tailed. Thus if α = 0.05, we reject H_0:μ = 113.0 if z^* > 1.96 or if $z^* < -1.96$. If σ = 1.5 we can determine the corresponding critical values of $\bar{x}$ to be 110.06 and 115.94. Thus if μ_1 = 115.0, we can determine $1 - \beta$ to be about 0.24 as shown in Figure 7.21.

We can repeat this procedure for various values of μ_1 to obtain the power table shown in Table 7.3. The graph of the power curve for this test is shown in Figure 7.22.

We can use the computer to obtain a simulated power curve that will approximate the true power curve for a test. We saw in Figure 7.12 that the null hypothesis H_0:μ = 7.2 was rejected correctly in favor of H_1:$\mu \neq 7.2$ in 2 of 20 cases for an estimated power of 0.10 when the true mean was 7.3 We illustrate the same procedure using SAS statements for μ_1 = 7.8. Since the use of only 20 tests leads to very rough results, we use the capacities of the SAS system to obtain 1000 confidence intervals.

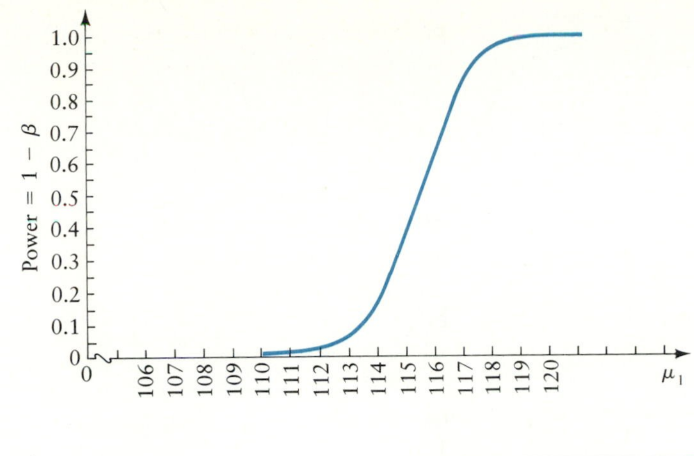

Figure 7.19
Power curve for the test $H_0{:}\mu = 113.0$ against $H_1{:}\mu > 113.0$

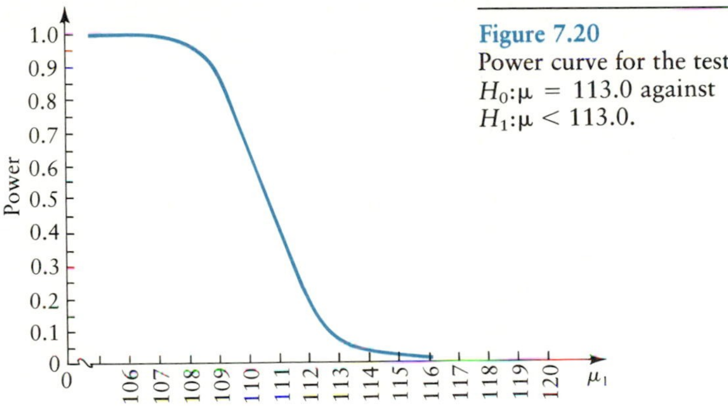

Figure 7.20
Power curve for the test $H_0{:}\mu = 113.0$ against $H_1{:}\mu < 113.0$.

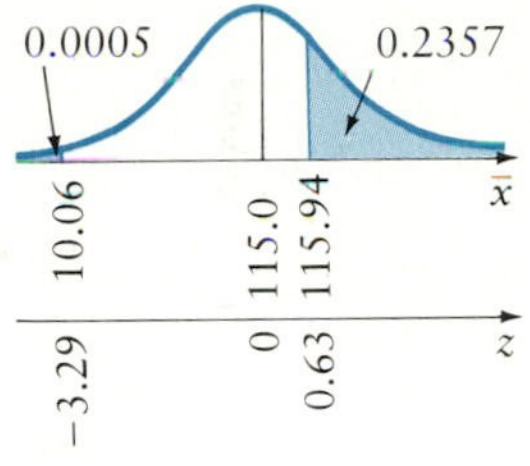

Figure 7.21
Power of the test $H_0{:}\mu = 113.0$ against $H_1{:}\mu \neq 113.0$ if $\mu = 115.0$.

Table 7.3 POWER OF THE TWO-SIDED TEST FOR SEVERAL VALUES OF μ_1.

μ_1	$1 - \beta$
106.0	0.9966
107.0	0.98
108.0	0.91
109.0	0.76
110.0	0.52
111.0	0.24
112.0	0.10
113.0	0.05 (= α)
114.0	0.10
115.0	0.24
116.0	0.52
117.0	0.76
118.0	0.9
119.0	0.98
120.0	0.9966

The function RANNOR (*seed*) generates a standardized random variate. (A *variate* is a single value of a variable.) Multiplying the result of this function by 1.5 and adding it to 7.8 will give us a variate of the normal distribution with mean 7.8 and standard deviation 1.5. We do this 20 times to obtain a sample of size 20 ($I = 1$ to 20) and then repeat this procedure 1000 times to obtain 1000 such samples numbered from 1 to 1000 ($J = 1$ to 1000). We obtain the mean of each sample (with the MEANS procedure) and then create a new data set including $z^* = (\bar{x} - 7.2)/(1.5/\sqrt{n})$ and the P value for each sample. To avoid printing these values and counting those for which $P < 0.05$, we create still another data set with the variable *count*. We let count = 1 if $P < 0.05$ and let count = 0 if $P \geq 0.05$. Using the MEANS procedure we obtain the sum of the count values, and this will be the number of tests in which H_0 was rejected. We could print this value, but in Figure 7.23 we have chosen to create a new data set to define the variable power, where power = count/1000 and is thus an estimate for the power of the test for $\mu_1 = 7.8$. The program and output are shown in Figure 7.23.

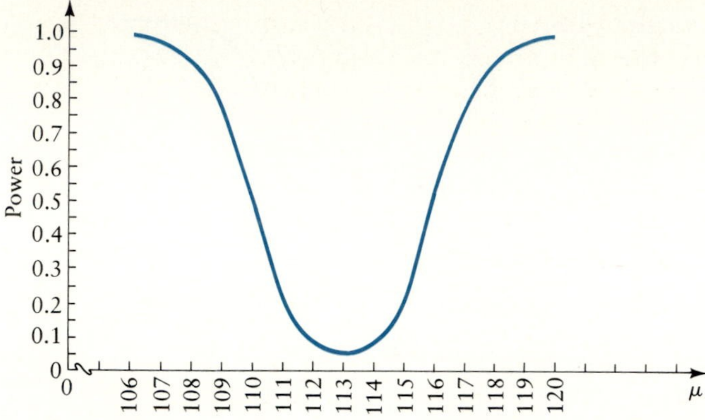

Figure 7.22
Power curve for the test $H_0\colon \mu = 113.0$ against $H_1\colon \mu \neq 113.0$.

Figure 7.23
SAS printout for simulation of the power of the test $H_0{:}\mu = 7.2$ against $H_1{:}\mu = 7.2$ if $\mu = 7.8$.

```
INPUT FILE
DATA START;

  DO J = 1 TO 1000;
     GP = J;
  DO I = 1 TO 20;
  X = 7.8 + 1.5 * RANNOR(I);
  OUTPUT;
  END;
  END;
PROC MEANS MEAN NOPRINT;
  VAR X;
  BY GP;
  OUTPUT OUT = A MEAN = XBAR;
DATA TEST;
  SET A;
  ZSTAR = (XBAR - 7.2) / (1.5 / SQRT(20));
  P = (1 - PROBNORM(ABS (ZSTAR))) * 2;
DATA COUNTS;
  SET TEST;
  IF P < 0.05 THEN COUNT = 1;
  ELSE COUNT = 0;
PROC MEANS SUM NOPRINT ;
  VAR COUNT;
  OUTPUT OUT = B SUM = TOTAL;
DATA LAST;
  SET B;
  POWER = TOTAL / 1000;
PROC PRINT;
  VAR POWER;
TITLE "POWER OF TEST FOR MEAN = 7.8";

OUTPUT FILE

          POWER OF TEST FOR MEAN = 7.8

                   OBS     POWER

                    1      0.422
```

The estimated value of $1 - \beta$ when the true mean is 7.8 is about 0.422. This is very close to the true value, about 0.426. We can repeat the procedure as many times as we need for different values of μ_1 to obtain an approximate power curve for the test.

Problems

Applications

7.80 Refer to Problem 7.36. Use the computer to test the hypothesis that the mean weight of coffee cans of this brand is actually 16 ounces.

7.81 Refer to Problem 7.37. If the coffee-can fills are not normal but symmetric, the trimmed-mean procedure is appropriate. Using SAS procedures and statements, write a program that will use the trimmed-mean procedure. Refer to Figure 6.24 for some suggestions on how you can do this.

7.82 Recall the 50 weights of professional football players listed in Example 1.2. The bar graph and stem and leaf plots for the data were given in Chapter 1. In both diagrams the data look approximately bell-shaped so that normality appears to be a reasonable assumption. Assuming this to be a random sample of all weights of professional football players, use the computer to test whether professional football players weigh, on the average, more than the average American male, who weighs 174.6 pounds. Use the 0.05 level of significance.

7.83 Weights of American males have a mean of 174.6 pounds and range from about 110 pounds to 350 pounds, except for a few extreme cases. A reasonable value for the standard deviation of these weights would be about 60. Use appropriate SAS statements and the data of Example 1.2 to test the hypothesis that the standard deviation of the weights of professional football players is equal to 60. Use $\alpha = 0.05$.

7.84 Use SAS statements to work each of the following problems about a population proportion:
a. Problem 7.62 **b.** Problem 7.63 **c.** Problem 7.65

7.85 We can test a hypothesis about a population proportion by using the SAS function PROBBNML. Write a program to solve Example 7.12.

7.86 Use SAS procedures and statements to repeat the examples shown in Figures 7.11 and 7.12. Refer to the appropriate sections of Chapters 5 and 6 for details on sampling.

7.87 Some statisticians are concerned that hypothesis-testing procedures consider only one specific value in the null hypothesis. As a result these statisticians believe that hypothesis testing is too restrictive, and thus they favor using confidence intervals instead to test their hypotheses. They argue that confidence intervals deal directly with the parameters of interest and hence are more informative about their signs and magnitudes. They also incorporate both the sample size and the sample variability. Another advantage is that once a confidence interval has been obtained, any two-sided alternate hypothesis can be tested as follows. If the hypothetical value of the parameter is *not* contained in the interval, then reject H_0. If the value is contained in the interval, fail to reject H_0.

Suppose that, on 15 tests, the reaction time of a volunteer to a given stimulus has the following values in hundredths of a second: 12, 14, 9, 13, 11, 12, 11, 11, 9, 12, 14, 10, 11, 10, 12. Using the computer, obtain a 95% confidence interval for the true mean reaction time of the volunteer. What would you conclude about the hypotheses $H_0{:}\mu = 10$ and $H_1{:}\mu \neq 10$? Conduct a formal hypothesis test using the computer. Does this test agree with your previous conclusion? Can you conduct a hypothesis test about the hypotheses $H_0{:}\mu = 12$ and $H_1{:}\mu \neq 12$ using the same confidence interval? What do you conclude? Can you use the same hypothesis test for this test as for the previous hypotheses?

7.6 Case 7: The Radioactive Residue—Solution

The result of the random sampling of radiation readings was subjected to a statistical test of the research hypothesis. This hypothesis, we recall, was that the mean for the entire property was less than 450. Since there were 20 readings and the readings for the entire property could be assumed to be normally distributed, the t test statistic was used with the critical value $t_{.01}\{19\} = 2.539$. Thus we have the following formal hypothesis test:

1. $H_0{:}\mu = 450$
   $H_1{:}\mu < 450$
2. Decision rule: Reject H_0 if $t^* < -2.539$.
3. Results: $n = 20$, $\bar{x} = 432$, $s = 20.8$. Then
   $t^* = (432 - 450)/(20.8/\sqrt{20}) = -3.87$.
4. Decision: Reject H_0.
5. Conclusion: The developer's contention has been proved; the housing project may be built.

7.7 Summary

Statistical hypothesis testing and decision making are useful tools in investigating the validity of a belief. We can formulate a belief as a hypothesis. A statistical hypothesis is a statement regarding the value of one or more population parameters.

A hypothesis supporting the belief is called the **research hypothesis.** A hypothesis that a population parameter is equal to a specific value is called a **null hypothesis.**

When we make a decision concerning the validity of a belief, two types of error are possible. A **type I error** occurs when we reject a true null hypothesis; a **type II error** occurs when we accept a false null hypothesis. We can calculate the probability of a type I error; it is designated α and called the **level of significance** for the test. The probability of a type II error is designated β; we cannot calculate β directly, since it depends upon the unknown value of the population parameter. The probability of correctly accepting a true null hypothesis is designated $1 - \alpha$ and called the *level of confidence* for the test. The probability of correctly rejecting a false null hypothesis is designated $1 - \beta$ and called the **power** of the test.

To test the validity of a null hypothesis about a parameter, we assume the null hypothesis to be true. Then we can determine the nature of the sampling distribution of the sample statistic (estimator of the population parameter) and use it to construct a test statistic. The test statistic and the chosen level of significance define the rejection region. If the sample value of the test statistic falls in the rejection region, we reject the null hypothesis.

The observed significance level of a test, or P value, is the smallest value of α that will lead to rejection of H_1 for the given result. The procedures for testing the null hypothesis $H_0{:}\mu = \mu_0$ are summarized in Table 7.4.

The decision rule is defined by the rejection region, which depends upon the alternate hypothesis, the test statistic, and the level of significance α. The appropriate decision rule for the various alternate hypotheses for each test statistic is listed in Table 7.5.

Table 7.4 SUMMARY OF PROCEDURES FOR TESTING THE NULL HYPOTHESIS $H_0{:}\mu = \mu_0$.

POPULATION STANDARD DEVIATION	SHAPE OF DISTRIBUTION	SAMPLE SIZE	TEST STATISTIC
Known	Normal	Any n	$z = (\bar{x} - \mu_0)/\sigma_{\bar{x}}$
	Nonnormal	$n \geq 30$	$z = (\bar{x} - \mu_0)/\sigma_{\bar{x}}$
Unknown	Normal	$n \geq 30$	$z = (\bar{x} - \mu_0)/s_{\bar{x}}$
		Any n	$t = (\bar{x} - \mu_0)/(s/\sqrt{n})$
	Nonnormal	$n \geq 50$	$z = (\bar{x} - \mu_0)/s_{\bar{x}}$
	Nonnormal, symmetric	$n < 50$	$t = [\bar{x}(T) - \mu_0]/s_{\bar{x}(T)}$
	Nonnormal, skewed	$n < 50$	No general method (use nonparametric methods)

Table 7.5 DECISION RULES FOR VARIOUS POSSIBILITIES.

NULL AND ALTERNATE HYPOTHESES		
$H_0{:}\mu = \mu_0$ $H_1{:}\mu \neq \mu_0$	$H_0{:}\mu = \mu_0$ $H_1{:}\mu > \mu_0$	$H_0{:}\mu = \mu_0$ $H_1{:}\mu < \mu_0$
DECISION RULE		
Reject H_0 if	Reject H_0 if	Reject H_0 if
$\lvert z^*\rvert > z_{\alpha/2}$ or $\lvert t^*\rvert > t_{\alpha/2}\{n - 1\}$	$z^* > z_\alpha$ or $t^* > t_\alpha\{n - 1)$	$z^* < -z_\alpha$ or $t^* < -t_\alpha\{n - 1\}$

To test the hypothesis $H_0{:}\pi = \pi_0$ we use the test statistic $z = (p - \pi_0)/\sigma_p$. The decision rule using the sample value z^* of the test statistic z is the same as in Table 7.5, with π and π_0 in the hypotheses instead of μ and μ_0, respectively.

For small samples, we obtain the P value of the result by using the binomial tables or the Poisson approximation. If the P value is less than α, we reject the null hypothesis.

To test the hypothesis $\sigma^2 = \sigma_0^2$ or $\sigma = \sigma_0$ we use the test statistic $\chi^2 = (n - 1)s^2/\sigma_0^2$ provided that the variable is normal.

Chapter Review Problems

Practice

7.88 Can we conclude that $\mu \neq 150$ at the 0.05 level of significance if:
a. $\bar{x} = 135$, $s = 33$, and $n = 60$?
b. $\bar{x} = 135$, $s = 18$, $n = 15$, and x is normal?

7.89 Test the research hypothesis $\pi \geq 0.75$ at the 0.01 level of significance if $p = 0.80$ and n is:
a. 25 **b.** 100 **c.** 400

7.90 At what level of significance can we reject the null hypothesis $H_0{:}\mu = 35$ if $\bar{x} = 37.4$, $s = 6.9$, $n = 60$, and the alternate hypothesis is:
a. $H_1{:}\mu > 35$ **b.** $H_1{:}\mu \neq 35$ **c.** $H_1{:}\mu < 35$

7.91 Test the research hypothesis $\mu < 150$ at the 0.05 level of significance if x is normal,
a. $\bar{x} = 135$, $s = 18$, and $n = 15$
b. $\bar{x} = 135$, $r = 18$, and $n = 15$

7.92 Test the research hypothesis $\sigma < 64$ at the 0.05 level of significance if:
a. $s = 52.6$ and $n = 60$
b. $s = 50.3$, $n = 30$, and the variable is normal in each case

Applications

7.93 A manufacturer of a certain model of car claims that the average mileage of this model is at least 30 miles per gallon using regular unleaded gasoline. *Wheels* magazine tests this claim with a sample of nine cars. Over a prescribed course the mileages are 29.8, 28.1, 26.7, 27.7, 29.4, 28.6, 31.3, 26.8, and 27.1. If they are willing to take no more than 1 chance in 100 of making an error, can they reject the manufacturer's claim?

7.94 The Bureau of Home Relations reports that 30 percent of all marriages reach a divorce court within the first year of marriage. A skeptical statistician finds that in a random sample of 400 couples married more than a year ago, 108 reached a divorce court in the first year of marriage. What can you conclude about the bureau's report at a 0.01 level of significance?

7.95 In a certain area, air conditioning repair workers make $135.50 a day on the average. A sample of 40 repairmen working for ABC Air Conditioning Service has a mean daily wage of $121.95 with a standard deviation of $27.10. At what level of significance can the union use this result to prove their contention that ABC pays substandard wages?

7.96 An advertising agency is trying to prove that at least 18% of all families watching television at the time watch the show "All in a Night's Work." A random sample of 1500 families reveals that 1143 were watching television and that 227 of those were watching the show. The sponsor wants the contention proved at the 0.01 level or the contract will be canceled. What will the sponsor do?

7.97 In a study of 100 patients admitted to a hospital suffering from a ruptured appendix, the mean stay was found to be 9.53 days with a standard deviation of 2.83 days. Does this result support the contention of the hospital administrator that this is significantly less than the national average of 10.17 days? Use the 0.05 level of significance.

7.98 The hospital administrator, while pursuing the question posed in Problem 7.97, notes that nationally the standard deviation of the hospital stay for the same condition is 3.81 days. Can she conclude, at the 0.05 level of significance, that the standard deviation of the sample of 100 persons (that is, 2.83) is significantly different from the national average?

7.99 A random sample of a day's output of a chemical reveals that the ester level for 20 vials is 11.6% with a standard deviation of 0.8%. Specifications require a minimum of 12% but allow output to pass if a sample is not significantly lower ($\alpha = 0.05$). Will the output be passed?

7.100 Testing for tensile strength of a new alloy, a metallurgist obtains readings (in pounds per square inch) of 117.6, 122.4, 119.8, 118.8, 121.6, and 123.4. The hypothetical tensile strength is 120.0 psi. At the 0.05 level of significance, do these tests tend to confirm or deny the theory?

7.101 A firm wishes to determine whether it should expand its offerings and is considering increasing the output of Type 7 widgets. It decides to increase the output if the daily demand exceeds 27,500 on the average. A sample of 40 days during the past 6 months is examined and the mean daily demand for these 40 days is found to be 27,654 with a standard deviation of 384. If the level of significance chosen is 0.05, should the firm increase the output?

7.102 Under the usual treatment, a certain disease produces undesirable side effects in 25% of the victims. A new drug has been developed that is claimed to be effective in reducing the side effects of the disease. A sample of 213 patients with the disease, receiving the usual treatment, is treated with the new drug. Test the hypothesis that the drug is effective in reducing the proportion of patients developing side effects if 41 of the patients develop side effects. Use the 0.025 level of significance.

7.103 A hospital administrator introduces a new accounting system designed to reduce the average amount of unpaid bills. Under the old system, the average unpaid bill was $217.22 with a standard deviation of $34.06. Under the new system, the first 50 unpaid bills show an average of $203.11 with a standard deviation of $47.32. Is this sufficient evidence to conclude that the new system has reduced the amount of the average unpaid bill? Use the 0.01 level of significance.

7.104 Refer to Problem 7.103. While working with the new accounting system, the administrator notices that the standard deviation of $47.32 is more than the $34.06 standard deviation under the old system. Can she conclude, at a 0.01 level of significance, that the variability of the unpaid bills under the new system will be different from the variability of the unpaid bills under the old system?

7.105 A medical study shows that, in a survey of 15,000 men who had a heart attack and then subsequently died, 60% died of a second heart attack. The case histories of 20 men are under study. These men had a heart attack, were treated with a new beta-blocking drug, and then subsequently died. Eight of the 20 died of a subsequent heart attack. Is this sufficient evidence to show that the beta blocker was effective? Use the 0.05 level of significance. What are some sources of problems in attempting to use the result as conclusive?

7.106 Repeat Problem 7.105, but this time suppose the study consists of ten men, of which three died of a subsequent heart attack.

7.107 Rats are being tested to see if they prefer a square or triangle. In an experiment, the rat can step on a square or a triangle and will receive food. The figures are the same in area, color, and composition; the only difference is shape. Suppose a rat chooses the triangle 39 times in 60 trials. Test the hypothesis, at the 0.10 level, that there is really no preference.

7.108 Sixteen mentally handicapped patients are given a task requiring a certain degree of mental dexterity. The mean time required to do the task is 11 minutes 33 seconds with a standard deviation of 3 minutes 12 seconds. The average time required for normal persons is theoretically 10 minutes. Does this result support, at the 0.01 level of significance, a contention that mentally handicapped persons take longer at this task than normal persons?

7.109 Suppose the standard deviation of normal persons doing the task mentioned in Problem 7.108 is theoretically 2 minutes. Use this value to test the hypothesis if the times are known to be normal.

STD DEV

7.110 Test the hypothesis that the standard deviation of the 16 mentally handicapped patients of Problem 7.108 is different from the theoretical standard deviation given in Problem 7.109. Use the 0.01 level of significance.

7.111 A test for right and left preference of salmon is conducted prior to construction of a "ladder" to aid salmon in swimming upstream to get around a dam. Artificial channels are constructed to observe whether salmon swimming upstream prefer the left or right channel. A total of 732 salmon are observed. Of these, 348 swim up the left channel and the remainder use the right channel. Using the 0.10 level of significance, can you conclude that there is no preference?

7.112 A sample of a hundred 10-pound sacks of sugar from a large shipment is found to have a mean weight of 9.83 pounds with a standard deviation of 0.70 pound. Can government officials conclude that the mean weight of the shipment is below 10 pounds if a 0.01 level of significance is used? If a 0.05 level of significance had been used, would this have been more favorable to the manufacturer or the consumer than the 0.01 level?

7.113 A Gallup Poll commissioned by the *New York Times–Mirror* reported that 53 percent of the more than 3000 respondents "harbor attitudes either mildly or highly critical of the news media" (*Wall Street Journal*, Jan. 27, 1986). A local newspaper editor believes that less than half the citizens in the city are critical of the news media and hires a local polling organization to repeat the Gallup Poll but with a smaller sample. The survey obtains a random sample of 112 citizens of whom 47 are critical of the news media. Does this result prove the editor's contention at the 0.05 level of significance? At what level of significance is the result proved?

test proportion

7.114 A sample of 200 light bulbs from a large number shows exactly 4 to be defective. Can this result be used to conclude that $\pi > 0.005$ at the 0.05 level of significance? At what level of significance can it be concluded?

7.115 The International Coffee Organization (*Pensacola News Journal*, Feb. 13, 1986) reported that the percentage of people over the age of 10 who drink coffee dropped from 75% in 1962 to 55% in 1985. Suppose that a grocer wishes to know whether the local community differs from the reported national average. A random sample of 543 people over the age of 10 shows 336 who drink coffee. At what level of significance can the grocer conclude that the proportion of coffee drinkers in the local community is not 0.55?

7.116 A sample of 50 potential customers is shown an advertising presentation and asked to rate it on a scale of 1 to 10 in terms of whether or not it would make them more likely to buy the product. A full-scale test will be scheduled if the result indicates, at a 0.05 level of significance, that the mean rating of the display for all potential customers is at least 5.0. The presentation receives a mean rating of 6.3 with a standard deviation of 1.6. Will a full-scale test be scheduled?

7.117 A machine must maintain a certain level of quality in its output or else it will have to be shut down for repairs. The machine produces 20 electronic components during a particular day's run. The mean impedance for the components is 154 ohms with a standard deviation of 8.3 ohms. The production line will be shut down if either the population mean is not 150 ohms or the standard deviation is greater than 7.5 ohms. Will the production line be shut down if a 0.10 level of significance is used?

Index of Terms

Glossary of Symbols

Symbol	Meaning
α	probability of type I error
β	probability of type II error
χ^{2*}	sample value of a chi-square test statistic
H_0	null hypothesis
H_1	alternate hypothesis
μ_0	hypothetical value of the population mean
P	observed significance level for a statistical test (*P* value)
π_0	hypothetical value of the population proportion
σ_0	hypothetical value of the population standard deviation
σ_0^2	hypothetical value of the population variance
t^*	sample value of a *t* test statistic
z^*	sample value of a *z* test statistic

Answers to Proficiency Checks

1. By mistakenly deciding that the probability is not 0.5, you have rejected a true hypothesis, committing a type I error. The probability of obtaining as many as 35 heads in 60 tries is about 0.38, so this would happen about 38% of the time by chance and is not an unlikely occurrence.

2. In this case you have accepted a false hypothesis, that the probability is 0.5, and have committed a type II error. It can be shown that if the true probability is 0.4, about one-fourth of all samples of size 60 will have 27 or more heads. It would be better simply to fail to reject H_0 and reserve judgment.

3. If μ is the mean amount of precipitate in cubic centimeters, we have $H_0: \mu = 30$ and $H_1: \mu > 30$.

4. $H_0: \pi = 0.5$; $H_1: \pi \neq 0.5$

5. From Table 4, we see that $z = 2.18$ corresponds to an area of 0.4854. Thus *(a)* $P = 0.0146$ and *(b)* $P = 0.0292$.

6. $\chi^2_{.10} = 34.38$, $\chi^2_{.05} = 37.65$, $\chi^2_{.025} = 40.65$, $\chi^2_{.01} = 44.31$, and $\chi^2_{.005} = 46.93$. *(a)* Since $31.65 < 34.38$, $P > 0.10$; *(b)* $37.65 < 39.33 < 40.65$, so $0.025 < P < 0.05$; *(c)* $55.33 > 46.93$, so $P < 0.005$.

7. 1. $H_0: \mu = 18.7$; $H_1: \mu \neq 18.7$

   2. Decision rule: Reject H_0 if $|z^*| > 1.645$.

   3. Results: $n = 60$, $\bar{x} = 19.3$, $s_{\bar{x}} = 3.8/\sqrt{60} \doteq 0.49$, so $z^* = (\bar{x} - \mu_0)/s_{\bar{x}} = (19.3 - 18.7)/0.49 = 0.6/0.49 \doteq 1.22$.

   4. Decision: Since 1.22 is not greater than 1.645, we fail to reject H_0.

   5. Conclusion: Since we do not know the setting of the problem, we cannot reach any conclusion other than that $\mu = 18.7$.

8. 1. $H_0: \mu = 150$; $H_1: \mu > 150$

   2. Decision rule: Reject H_0 if $z^* > 2.33$.

   3. Results: $\bar{x} = 156.3$ and $s_{\bar{x}} = 15.4/\sqrt{40} \doteq 2.43$, so $z^* = (156.3 - 150.0)/2.43 = 6.3/2.43 \doteq 2.59$.

   4. Decision: Since $2.59 > 2.33$, we reject H_0.

   5. Conclusion: Since we do not know the setting of the problem, we cannot reach any conclusion other than $\mu > 150.0$.

9. In Proficiency Check 7, $z = 1.22$ and the rejection region is two-tailed. The P value for rejection of H_0, then, is 0.2224. In Proficiency Check 8, $z = 2.59$ and the rejection region is one-tailed, so the P value is 0.0048.

10. Here we are testing the hypothesis that the annual mean dollar amount spent for textbooks by all students at the university is $250 against an alternative that it is not. Thus:

    1. $H_0: \mu = 250$; $H_1: \mu \neq 250$

    2. Decision rule: Reject H_0 if $|z^*| > 1.96$.

    3. Results: $n = 144$, $\bar{x} = 264$, $s_{\bar{x}} = 54/\sqrt{144} = 4.5$, so $z^* = (264 - 250)/4.5 = 14/4.5 \doteq 3.11$.

    4. Decision: Since $3.11 > 1.96$, we reject H_0.

    5. Conclusion: The mean is not $250. On the basis of the sample, we would say that the mean is likely to be greater than $250. We can estimate the mean by using a confidence interval. The P value for this problem is equal to 0.0018, so we can be very confident about the conclusion.

11. 1. $H_0: \mu = 18.7$; $H_1: \mu \neq 18.7$

2. Decision rule: Reject H_0 if $|t^*| > 1.729$.

3. Results: $n = 20$, $\bar{x} = 19.3$, $s = 3.8$, so
$$t^* = \frac{19.3 - 18.7}{3.8/\sqrt{20}} \doteq \frac{0.6}{0.85} \doteq 0.706.$$

4. Decision: Since 0.706 is not greater than 1.729, we fail to reject H_0.

5. Conclusion: Since we do not know the setting of the problem, we cannot reach any conclusion other than that $\mu = 18.7$ may well be reasonable.

12. 1. $H_1{:}\mu = 150$; $H_1{:}\mu > 150$

2. Decision rule: Reject H_0 if $t^* > 4.541$.

3. Results: $\bar{x} = 156.3$, $s = 15.4$, so $t^* = \dfrac{156.3 - 150.0}{15.4/\sqrt{4}} \doteq 0.82$.

4. Decision: Since 0.82 is not greater than 4.541, we fail to reject H_0.

5. Conclusion: Since we do not know the setting of the problem, we cannot reach any conclusion other than that μ is probably not greater than 150.0.

13. 1. H_0: $\mu = 9.0$; H_1: $\mu \neq 9.0$

2. Decision rule: Reject H_0 if $|t^*| > 2.571$.

3. Results: $n = 6$, $\bar{x} = 9.2$, $s = 0.18$, so $t^* = \dfrac{9.2 - 9.0}{0.18/\sqrt{6}} \doteq 2.722$.

4. Decision: Since $2.722 > 2.571$, we reject H_0.

5. Conclusion: The setting is probably off and needs to be adjusted.

Since $t_{.01}\{5\} = 3.365$, and $2.571 < 2.722 < 3.365$, $P(t > 2.722)$ is between 0.01 and 0.025. The rejection region is two-tailed, so the P value is between 0.02 and 0.05.

14. 1. H_0: $\pi = 0.30$; H_1: $\pi \neq 0.30$

2. Decision rule: Reject H_0 if $|z^*| > 1.96$.

3. Results: $\sigma_p = \sqrt{(0.3)(0.7)/60} \doteq 0.0592$, and $p = 0.4$, so $z^* = (0.4 - 0.3)/0.0592 \doteq 1.69$.

4. Decision: Fail to reject H_0.

5. Conclusion: Although we cannot reject the hypothesis that $\eta = 0.3$, there appears to be a good chance of a type II error, since z^* is fairly large. So, if possible, reserve judgment.

15. 1. H_0: $\pi = 0.15$; H_1: $\pi > 0.15$

2. Decision rule: Reject H_0 *if* $z^* > 2.33$.

3. Results: $\sigma_p = \sqrt{(0.15)(0.85)/40} \doteq 0.0565$, and $p = 8/40 = 0.20$, so $z^* = (0.20 - 0.15)/0.0565 \doteq 0.89$.

4. Decision: Fail to reject H_0.

5. Conclusion: In this case there appears to be no reason not to accept H_0.

16. In Proficiency Check 14, $z^* = 1.69$ and the rejection region is two-tailed, so $P = 2(0.5000 - 0.4545) = 2(0.0455) = 0.0910$; thus H_0 could be rejected at a 0.10 level of significance. In Proficiency Check 15, $z^* = 0.89$ and the rejection region is one-tailed, so $P = 0.5000 - 0.3133 = 0.1867$; thus H_0 could be rejected at a 0.20 level of significance (which is rarely used). That is, there is nearly 1 possibility in 5 that the result is attributable to chance.

17. Since we wish to refute the manufacturer's assertion, we believe that his claim, $\pi \leqslant 0.015$, is false. Thus the research hypothesis for our experiment is $\pi > 0.015$.

1. $H_0: \pi = 0.015$; $H_1: \pi > 0.015$
2. Decision rule: Reject H_0 if $z^* > 2.33$.
3. Results: $\sigma_p = \sqrt{0.015(0.985)/1000} = 0.003844$ and $p = 22/1000 = 0.022$, so $z^* = (0.022 - 0.015)/0.003844 \doteq 1.82$.
4. Decision: Fail to reject H_0.
5. Conclusion: Although we cannot reject the hypothesis that $\pi = 0.015$, there appears to be a good chance of a type II error since z^* is fairly large. So, if possible, we should reserve judgment.

The P value associated with this result—since $z^* = 1.82$ and the rejection region is one-tailed—is 0.0344. Thus the manufacturer's claim may not be true, although we cannot reject it at the 0.01 level. It could have been rejected at the 0.05 level, however.

18.
1. $H_0: \pi = 0.1$; $H_1: \pi > 0.1$
2. Decision rule: Reject H_0 if $P < 0.02$.
3. Results: From Table 2, with $p = 0.1$ and $n = 20$, we obtain $P(x \geqslant 4) = 0.133$.
4. Decision: Fail to reject H_0.
5. Conclusion: $\pi > 0.1$.

19. Since $n\pi_0 = 0.6$, which is much less than 5, we cannot use the procedure for large samples. On the other hand, the value of π_0 (0.015) is not listed in Table 2. To determine whether we can use Table 3, we must see if $n > 1000\pi_0$. Since $1000\pi_0 = 15$ and $n = 40$, we can use Table 3 with $m = 0.6$. To reject the manufacturer's claim that $\pi \leqslant 0.015$, we must show that $\pi > 0.015$. Thus:

1. $H_0: \pi = 0.015$; $H_1: \pi > 0.015$
2. Decision rule: Reject H_0 if $P < 0.01$.
3. Results: From Table 3 with $m = 0.6$ we obtain $P(x \geqslant 2) = 0.122$. Thus $P = 0.122$.
4. Decision: Fail to reject H_0.
5. Conclusion: We cannot refute the claim at the 0.01 level. It can be rejected at a 0.122 level of significance. Clearly we should reserve judgment in this case and take a larger sample if possible.

20. $\chi^2_{.025}\{4\} = 11.14$, $\chi^2_{.025}\{9\} = 19.02$, $\chi^2_{.025}\{19\} = 32.85$, $\chi^2_{.025}\{24\} = 39.36$, $\chi^2_{.975}\{4\} = 0.4844$, $\chi^2_{.975}\{9\} = 2.700$, $\chi^2_{.975}\{19\} = 8.906$, $\chi^2_{.975}\{24\} = 12.40$. Thus we have:

1. $H_0{:}\sigma^2 = 1000.00$; $H_1{:}\sigma^2 \neq 1000.00$
2. Decision rule: Reject H_0 if:
   (a) $\chi^{2*} > 11.14$ or if $\chi^{2*} < 0.4844$
   (b) $\chi^{2*} > 19.02$ or if $\chi^{2*} < 2.700$
   (c) $\chi^{2*} > 32.85$ or if $\chi^{2*} < 8.906$
   (d) $\chi^{2*} > 39.36$ or if $\chi^{2*} < 12.40$
3. Results: $\chi^{2*} = (n - 1)(30)^2/1000 =$
   (a) 3.6 *(b)* 8.1 *(c)* 17.1 *(d)* 21.6
4. Decision: Fail to reject H_0.
5. Conclusion: $\sigma^2 = 1000$ appears to be reasonable.

21. $\chi^2_{.10}\{14\} = 21.06$, $\chi^2_{.05}\{14\} = 23.68$, and $\chi^2_{.01}\{14\} = 29.14$. Thus we have:

1. $H_0{:}\sigma = 0.085$; $H_1{:}\sigma > 0.085$
2. Decision rule: Reject H_0 if: *(a)* $\chi^{2*} > 21.06$; *(b)* $\chi^{2*} > 23.68$; *(c)* $\chi^{2*} > 29.14$.
3. Results: $\chi^{2*} = (14)(0.122)^2/(0.085)^2 \doteq 28.84$.
4. Decision: *(a)* Reject H_0; *(b)* reject H_0; *(c)* fail to reject H_0.
5. Conclusion: We can conclude that $\sigma > 0.085$ at a significance level of 0.10 or 0.05, but not 0.01. Since $\chi^2_{.025}\{14\} = 26.12$, we conclude that the P value is between 0.025 and 0.01; that is, the significance level at which H_0 can be rejected is P, where $0.01 < P < 0.025$.

22. Since this test has a left-hand rejection region, the critical value of χ^2 will be $\chi^2_{.95}\{46\}$. Now $\chi^2_{.95}\{40\} = 26.51$ and $\chi^2_{.95}\{60\} = 43.19$, so $\chi^2_{.95}\{46\}$ will be approximately 6/20 of the way from $\chi^2_{.95}\{40\}$ to $\chi^2_{.95}\{60\}$. Further, 43.19 − 26.51 is 16.68 and $(0.3)(16.68) \doteq 5.00$. Therefore we estimate $\chi^2_{.95}\{46\}$ to be 31.51. Thus:

1. $H_0{:}\sigma = 1.50$; $H_1{:}\sigma < 1.50$
2. Decision rule: Reject H_0 if $\chi^{2*} < 31.51$.
3. Results: $\chi^{2*} = (46)(1.18)^2/(1.50)^2 \doteq 28.47$.
4. Decision: Reject H_0.
5. Conclusion: We can conclude that $\sigma < 1.50$. To determine P we obtain $\chi^2_{.975}\{46\}$ using the same procedure we used to obtain $\chi^2_{.95}\{46\}$. We find $\chi^2_{.975}\{46\} \doteq 29.25$. Since $28.47 < 29.25$, we must next obtain $\chi^2_{.99}\{46\}$. This is estimated to be 26.76. Since $28.47 > 26.76$, we can stop. As $\chi^2_{.99}\{46\} < 28.47 < \chi^2_{.975}\{46\}$, we have $0.01 < P < 0.025$ (since these are the areas to the left of $\chi^2_{.99}\{46\}$ and $\chi^2_{.975}\{46\}$, respectively). If χ^{2*} had been less than $\chi^2_{.99}\{46\}$, we would have continued the procedure, finding $\chi^2_{.995}\{46\}$.

References

Hogg, R. V. and A. J. Craig. *Introduction to Mathematical Statistics,* 4th ed. New York: Macmillan Publishing Company, 1978.

Hurlbert, Stuart H. "Pseudoreplication and the Design of Ecological Field Experiments." *Ecological Monographs* 54(2) (1984): 187–211.

Neter, John, William Wasserman, and Michael H. Kutner. *Applied Linear Statistical Models.* 2nd ed. Homewood, IL: Richard D. Irwin, Inc., 1985.

Winer, B. J. *Statistical Principles in Experimental Design,* 2nd ed. New York: McGraw-Hill Book Company, 1971.

8

Statistical Inferences from Two Samples

In the preceding two chapters we learned how to obtain confidence interval estimates for a population mean, a population variance or standard deviation, and the population proportion in a dichotomous population. We also learned how to test hypotheses concerning these population parameters. In many cases we wish to apply the methods of statistical inference to answer questions of interest based on two or more samples. Specifically we address ourselves to the question: do these samples come from the same or different populations? The answer may tell us whether a vaccine is effective, whether a toothpaste with fluoride is more effective in fighting cavities than a toothpaste without fluoride, which of several advertising displays is likely to sell more of the product, and similar information.

This chapter deals primarily with statistical inference from two samples. We may obtain confidence intervals for and test hypotheses about the difference between two population parameters by using an extension of the methods of Chapters 6 and 7. We first give a general result that often applies when two independent random samples are obtained. Two samples are said to be **independent** if the selection of one sample has no effect on the selection of the second sample. The specific application of this result depends upon how the assumptions underlying these procedures have been met.

In Section 8.2 we apply the general result to obtain confidence intervals for the difference between two population means. We will use a variety of different approaches depending upon whether or not the populations are normal, whether or not the population variances are known and are equal or unequal, and the sample size. In Section 8.3.1 we will extend these procedures to tests of hypotheses about the differences between population means, and in Section 8.3.2 we show how to apply the methods of previous chapters to a special case of nonindependent samples—when observations in one sample are matched with observations in the second sample.

There are also procedures for estimating the difference between proportions in two dichotomous populations and for testing hypotheses concerning the difference between population proportions. We close the chapter by applying certain procedures to sample variances in order to draw inferences about the variances of the population from which the samples were drawn. In all of these instances we increase our knowledge of two important kinds of statistical inference—estimation and hypothesis testing.

8.0 Case 8: The Personality Profiles

A statistician read an article in the April 18, 1985, issue of *USA Today* with great interest. Part of the article read:

> Competitive, harried Type A's are about three times more likely than laid-back Type B's to have painful stomach disorders, reveals a new study.
>
> In a study of 127 men and women, two San Jose (Calif.) State University psychologists found that 11 percent of the 61 Type B's reported stomach trouble—ulcers, gastrointestinal pain, nausea, diarrhea—compared to 30 percent of 66 Type A's.

The remainder of the article speculated on reasons for this difference and stated that 54 percent of Type A's claimed they were in excellent health, while only 36 percent of Type B's rated their health as excellent.

The statistician wondered whether the claimed difference was real or due merely to chance. He also was curious about the statement that "Type A's are about three times more likely than . . . Type B's to have painful stomach disorders." He wondered, as well, how all this related to the fact that 54% of the Type A's claimed they were in excellent health, compared to only 36% of the Type B's. He therefore sat down with his calculator and did some figuring; his results will be available soon.

8.1 Sampling from Two Populations

One important application of statistical inference is to answer research questions that ask whether some experimental technique actually works, or which of two possible methods works better, or whether these samples could have come from the same population. To answer such questions a researcher needs to have at least two samples that are subjected to the techniques under study and be reasonably certain that the only difference between the samples which bears on the results is the variable being investigated. In this section we discuss the design of an experiment that allows such inferences to be drawn and a special case of the central limit theorem that allows us to investigate the questions being asked.

8.1.1 Experimental Design

Although we have been able to test hypotheses and draw inferences about a population parameter, we have not yet been able to conduct a true statistical experiment. In a statistical experiment researchers are interested in one and only one variable. All other variables are controlled so that any change or difference can be attributed solely to the variable under consideration.

Differences between the observed value and the actual value of a parameter are called *experimental error*. There are many sources of experimental error in the design and analysis of an experiment—primarily the existence of uncontrolled variables. One of the best explanations of the sources of error and ways of overcoming them may be found in *Experimental and Quasi-Experimental Designs for Research* by Donald T. Campbell and Julian C. Stanley (1963). Although their booklet was initially designed as a guide to educational research, the principles apply to any experiment.

A true statistical experiment begins with two or more samples randomly selected from a population or randomly assigned to the required number of groups. One or more groups receive an experimental treatment; one group is designated as a **control group** and receives no treatment. A testing instrument or procedure of some sort is used and the results are analyzed to see if there are differences between the groups.

In a study by Piotrowski and Dunham (1983), the researchers focused on the natural maturing of internal control in children. Basically, internal control characterizes those who believe they have control of their lives, while external control characterizes those who believe their lives are under the control of luck, chance, fate, or other people. As children mature, it is natural for their belief in internal control to grow stronger. The researchers wished to determine whether or not children in a city hit by a hurricane were retarded in their maturation of internal control in comparison to children in a city not hit by a hurricane. The study investigated the effects of Hurricane Eloise (September 23, 1975) on control of elementary school students. In order to remove as many sources of experimental error as possible, two coastal cities 50 miles apart on Florida's Panhandle were selected for study. The two cities had received identical hurricane warnings; however, one city was hit by the hurricane while the other was not seriously affected by it. The county hit by the hurricane was devastated and was subsequently declared a federal disaster area. Since internal control scores had been found to be significantly related to socioeconomic status, schools were selected from similar socioeconomic status.

Students in both cities had already been given tests measuring internal control in the fourth grade. As a part of the study, the researchers selected students at random from similar schools in both cities and administered similar tests from 5 to 8 weeks after the hurricane. If μ_1 is the mean increase (negative if a decrease) in internal control score for all students in city 1, and μ_2 is the mean increase in internal control score for all students in city 2, the hypothesis $\mu_1 = \mu_2$ is to be tested—that is, the hypothesis that there is no difference between the populations in the increase of internal control score.

In this experiment it was necessary to obtain a difference between test scores before and after the hurricane. If tests after the hurricane only had been administered, differences might have existed before the hurricane and thus could have been attributable to other factors. The students in the city that was not hit by the hurricane composed the control group (since they were not exposed to the variable of interest). Such a design is called a *pretest–posttest control group design.* The pretest can be eliminated if subjects are randomly assigned to the groups. This randomization assures homogeneity of experience prior to the experiment. That is, we can be reasonably sure that differences among the background of the subjects are minimized by mixing them thoroughly. This kind of experiment without a pretest is called a *completely randomized design.*

If the second city selected had been subjected to a flood rather than a hurricane, the same tests could have been conducted. The results, however, would be attributed to differences between the two disasters; no conclusion could have been drawn as to whether either had an effect that would not have been present had there been no disaster.

In Chapter 11 we will learn how to test hypotheses about three or more population means by using a completely randomized design. If we use a pretest and there are three or more populations, analysis of this design requires *analysis of covariance,* which is beyond the scope of this book. (See Neter and others [1985], chap. 25.) Use of a pretest may also introduce a possible source of error that can be controlled. For further information consult the topic "Solomon 4 design" in the booklet by Campbell and Stanley (1963).

In most experiments the calculations are much simpler if the samples are equal in size. Sometimes it is impossible to achieve this end, even though we begin with equal sample sizes. We may lose subjects through attrition or make errors that result in unequal sample sizes. In other cases equal sample sizes are not desirable. If the populations are considerably different in size, we may want to obtain samples whose relative sizes are proportional to the populations being sampled. Suppose, for instance, we wish to study differences in the length of time a spectator has to stand in line at a football game in order to get a hot dog. If the samples are to be taken from stadiums with different capacities (say 40,000 in one and 15,000 in the other), we may want to obtain samples whose ratio approximates the ratio of the populations. Textbooks on sample survey design often cover this topic. The following example illustrates the difference between a true statistical experiment and a *case study* involving only one sample.

8.1.2 Case Study: Physical Fitness, Personality, and IQ

In a story released by the News America Syndicate on May 1, 1985, the columnist wrote that "people who keep physically fit have different personalities from those who don't, and they're even smarter." Dr. A. H. Ismail, a professor of physical education at Purdue University, released the results of two studies. In one study, 60 men ranging in age from 25 to 65 were psychologically and physiologically tested before and after completing a 4-month physical fitness program. According to Dr. Ismail, "subjects who tested low in emotional stability before the exercise program showed marked improvement in final tests after the program. At the same time, they also demonstrated graphic improvements in terms of a lowering of levels of serum cholesterol, blood sugar, and blood pressure." This is an example of a **case study**—a study of the effects of an experimental treatment on one sample. No proof exists that these changes would not have occurred naturally even if the men had not completed the fitness program. What was needed was a control group—a group that matched the experimental group in all respects except that they did not receive the fitness program.

Dr. Ismail also reported the results of a true **experimental study**—a study of the effectiveness of an experimental treatment conducted by using two samples that differ only in the presence or absence of the treatment. A total of 48 men averaging 48 years of age were divided into two equal groups. The experimental group exercised regularly for about 1½ hours three times a week for 4 months. The control group did not exercise. Both groups were given biochemical, physiological, and psychological tests before and after the program. According to Dr. Ismail, "the tests revealed the men who had engaged in regular exercise did better on mental processes controlled by the left hemispheres of their brains. They did significantly better [no significance level was given] on tests measuring such things as math and logical reasoning. They even showed slight increases in IQ scores." The term "significantly" means that he was able to reject a null hypothesis of no difference between the groups on the results of the "tests measuring such things as math and logical reasoning." Since he did not say that the increase in IQ scores was significant, we can infer that the increase might be attributed to chance. This chapter presents methods for determining whether such experimental results are due to chance or can be attributed to the experimental treatment.

8.1.3 Sampling Distribution of a Difference Between Two Sample Statistics

In order to tell whether or not two samples could have come from the same population, or from populations with the same parameter, we need to obtain the sampling distribution of the difference between two sample statistics.

Suppose that random variable 1 has a parameter θ_1, while random variable 2 has a parameter θ_2. We usually want to know the value of $\theta_1 - \theta_2$. In the preceding example involving internal control scores, if $\mu_1 = \mu_2$ then $\mu_1 - \mu_2 = 0$, while $\mu_1 \neq \mu_2$ is equivalent to $\mu_1 - \mu_2 \neq 0$. Assume that $\hat{\theta}_1$ and $\hat{\theta}_2$ are unbiased estimators of θ_1 and θ_2, respectively. Recall that this means that $E(\hat{\theta}_1) = \theta_1$ and $E(\hat{\theta}_2) = \theta_2$. Suppose that all samples of size n_1 are taken from population 1 and the sample statistic $\hat{\theta}_1$ is calculated for each sample and that all samples of size n_2 are taken from population 2 and the sample statistic $\hat{\theta}_2$ is calculated for each sample. The probability distribution for the set of all possible differences $\hat{\theta}_1 - \hat{\theta}_2$ is called the *sampling distribution of the difference* for the parameter of interest.

Now $E(\hat{\theta}_1 - \hat{\theta}_2) = E(\hat{\theta}_1) - E(\hat{\theta}_2) = \theta_1 - \theta_2$, and it can be shown that if the samples are independent, $\sigma^2_{\hat{\theta}_1 - \hat{\theta}_2} = \sigma^2_{\hat{\theta}_1} + \sigma^2_{\hat{\theta}_2}$. The following result is important and will be used frequently in this and later chapters.

Sampling Distribution of the Difference Between Two Sample Statistics

Suppose we obtain independent random samples of size n_1 and n_2 from populations with parameters θ_1 and θ_2, respectively. If we obtain unbiased estimates $\hat{\theta}_1$ and $\hat{\theta}_2$ for θ_1 and θ_2, respectively, and the sampling distributions of $\hat{\theta}_1$ and $\hat{\theta}_2$ are normal, then the sampling distribution of $\hat{\theta}_1 - \hat{\theta}_2$ will be normal. The mean of the distribution will be $\theta_1 - \theta_2$ and the variance of the distribution will be $\sigma^2_{\hat{\theta}_1 - \hat{\theta}_2} = \sigma^2_{\hat{\theta}_1} + \sigma^2_{\hat{\theta}_2}$, where $\sigma^2_{\hat{\theta}_1}$ and $\sigma^2_{\hat{\theta}_2}$ are the variances of the sampling distributions of $\hat{\theta}_1$ and $\hat{\theta}_2$, respectively.

8.2 Estimating Differences Between Population Means

Quite often we want to compare samples representing possibly different populations. We may wish to determine whether two samples of animals are of the same species, which advertising display will sell more merchandise, whether two chemical compounds are the same, whether a group subjected to some medical treatment differs from a control group, and similar questions. The next section presents a method for determining whether two samples represent populations with the same parameter, but if the parameters do differ, a common question is "by how much?"

Suppose that two different displays of a disposable razor are set up at various stores at different locations in such a way that the assignment of displays is random. Suppose that the average weekly sales of the razor are 26.5 with a standard deviation of 9.1 in 49 stores using display I and 22.4 with a standard deviation of 6.7 in 54 stores using display II. The difference between the sample means is 4.1, so this is an estimate for the difference in true mean weekly sales for the two displays.

We can get a confidence interval for the true difference in mean weekly sales for the two displays by applying the result of the last section and the methods of Chapter 6. If μ_1 and μ_2 are the true population means for two populations with variances σ_1^2 and σ_2^2, respectively, and $\overline{x}_1$ and $\overline{x}_2$ are the means of any two samples drawn from population 1 and population 2, respectively, the sampling distribution of $\overline{x}_1 - \overline{x}_2$ will be normally distributed for normally distributed populations or for n_1 and n_2 sufficiently large, with mean $\mu_1 - \mu_2$ and variance $\sigma^2_{\bar{x}_1 - \bar{x}_2}$, where $\sigma^2_{\bar{x}_1 - \bar{x}_2} = \sigma^2_{\bar{x}_1} + \sigma^2_{\bar{x}_2}$. Since we already know that $\sigma^2_{\bar{x}_1} = \sigma_1^2/n_1$ and $\sigma^2_{\bar{x}_2} = \sigma_2^2/n_2$ (if we do not use the finite population correction factor), we can determine $\sigma^2_{\bar{x}_1 - \bar{x}_2}$. The square root of $\sigma^2_{\bar{x}_1 - \bar{x}_2}$ is $\sigma_{\bar{x}_1 - \bar{x}_2}$ and is called the **standard error of the difference** (or the standard error of the difference for the mean). Thus we have

$$\sigma_{\bar{x}_1 - \bar{x}_2} = \sqrt{\frac{\sigma_1^2}{n_1} + \frac{\sigma_2^2}{n_2}}$$

We should use the finite population correction factor if it is appropriate, but usually it is not necessary.

Thus if the sample statistic $\overline{x}_1 - \overline{x}_2$ is $N(\mu_1 - \mu_2, \sigma^2_{\bar{x}_1 - \bar{x}_2})$, then the statistic $[(\overline{x}_1 - \overline{x}_2) - (\mu_1 - \mu_2)]/\sigma_{\bar{x}_1 - \bar{x}_2}$ is $N(0, 1)$. We can then obtain a confidence interval for $\mu_1 - \mu_2$ as in Chapter 6. The rule is shown here.

Confidence Interval for $\mu_1 - \mu_2$ (Variances Known)

Suppose we obtain two independent random samples—a sample of n_1 observations of a random variable x_1 with mean and standard deviation μ_1 and σ_1 and a sample of n_2 observations of a random variable x_2 with mean and standard deviation μ_2 and σ_2—and that either of the following is true:

1. both variables x_1 and x_2 are normally distributed

OR

2. n_1 and n_2 are both at least 30.

Then a $100(1 - \alpha)\%$ confidence interval for $\mu_1 - \mu_2$ is the interval

$$\overline{x}_1 - \overline{x}_2 - z_{\alpha/2}\sigma_{\bar{x}_1 - \bar{x}_2} \quad \text{to} \quad \overline{x}_1 - \overline{x}_2 + z_{\alpha/2}\sigma_{\bar{x}_1 - \bar{x}_2}.$$

To apply this rule we need the following assumptions: that the samples are randomly obtained from the populations, that the samples are independent of each other, that the populations are normal, and that the population variances are known. One assumption must always be met if we are to obtain valid inferences—the samples must be random. For most tests to be valid, the samples must be independent of each other. We refer to such samples as *independent random samples*.

The assumption of normality of the variables is not necessary if the samples are sufficiently large (at least 30 if the variances are known) so that the distribution of sample means will be approximately normal. Frequently we do not know the population variances. In such cases we may use our best estimate for the population variances—the sample variances—but this procedure introduces a bit of error into the result because the sample variances are variable. If the populations are normally distributed, this variability may also be overcome if the samples are sufficiently large (at least 30). If the populations are not normal and the population variances are not known, we need even larger sample sizes to compensate for these problems—samples should be at least 50. For smaller samples some alternative methods are discussed in Sections 8.2.2 and 8.2.3.

8.2.1 Large-Sample Confidence Intervals for $\mu_1 - \mu_2$ (Unknown Population Variance)

In many cases, if not most, we do not know the population variances and thus cannot determine $\sigma_{\bar{x}_1 - \bar{x}_2}$ directly. As before, however, we can obtain a satisfactory approximation for sufficiently large sample sizes by substituting s_1^2 and s_2^2 for σ_1^2 and σ_2^2 in the formula for the standard error of the difference to obtain a sample standard error of the difference $s_{\bar{x}_1 - \bar{x}_2}$ where

$$s_{\bar{x}_1 - \bar{x}_2} = \sqrt{\frac{s_1^2}{n_1} + \frac{s_2^2}{n_2}}$$

Large-Sample Confidence Interval for $\mu_1 - \mu_2$ (Unknown σ^2)

Suppose we obtain two independent random samples—a sample of n_1 observations of a random variable x_1 with mean μ_1 and a sample of n_2 observations of a random variable x_2 with mean μ_2—and that either of the following is true:

1. both variables x_1 and x_2 are normally distributed with n_1 and n_2 both at least 30

   OR

2. n_1 and n_2 are both at least 50.

If $\bar{x}_1$, $\bar{x}_2$, s_1^2, and s_2^2 are the sample means and variances, respectively, an approximate $100(1 - \alpha)\%$ confidence interval for $\mu_1 - \mu_2$ is the interval

$$\bar{x}_1 - \bar{x}_2 - z_{\alpha/2} s_{\bar{x}_1 - \bar{x}_2} \quad \text{to} \quad \bar{x}_1 - \bar{x}_2 + z_{\alpha/2} s_{\bar{x}_1 - \bar{x}_2}$$

where $s_{\bar{x}_1 - \bar{x}_2}$ is the sample standard error of the difference.

"Sufficiently large" is at least 30 for normal populations and at least 50 for nonnormal populations, as explained above. We can thus obtain an approximate or large-sample confidence interval for $\mu_1 - \mu_2$ as shown on page 436.

Example 8.1

Two different displays of a disposable razor are set up at various stores at different locations in such a way that the assignment of displays is random. The average weekly sales of the razor are 26.5, with a standard deviation of 9.1 in 51 stores using display I, and 22.4 with a standard deviation of 6.7 in 54 stores using display II. Determine a 95% confidence interval for the true difference in mean weekly sales per store with the two displays (that is, the difference in mean weekly sales per store expected of all sales of the razor from type I displays and from type II displays).

Solution

We have $\bar{x}_1 - \bar{x}_2 = 4.1$. The sample standard error of the difference is

$$s_{\bar{x}_1 - \bar{x}_2} = \sqrt{\frac{(9.1)^2}{51} + \frac{(6.7)^2}{54}} \doteq 1.567$$

For a 95% confidence interval, we use $z_{.025} = 1.96$; we will need to add and subtract $(1.96)(1.567) \doteq 3.1$. Then $4.1 - 3.1 = 1.0$, and $4.1 + 3.1 = 7.2$, so a confidence interval for $\mu_1 - \mu_2$ is 1.0 to 7.2. Thus we can be 95% certain that display type I will sell an average of 1.0 to 7.2 more razors per week per store than display type II.

PROFICIENCY CHECKS

1. If $\bar{x}_1 = 110.00$, $s_1 = 35.00$, $\bar{x}_2 = 100.00$, and $s_2 = 30.00$, determine a 95% confidence interval for $\mu_1 - \mu_2$ if n_1 and n_2 are:
   a. 50 b. 75 c. 100 d. 200
2. If $\bar{x}_1 = 0.560$, $s_1 = 0.122$, $\bar{x}_2 = 0.600$, $s_2 = 0.154$, $n_1 = 60$, and $n_2 = 70$, determine confidence intervals for $\mu_1 - \mu_2$ with a confidence level of:
   a. 90% b. 95% c. 98% d. 99%
3. Two sets of elementary school students in the same grade are taught mathematics by two different methods. Group I is taught by method 1 and group II learns by method 2. The results of a learning test are analyzed to estimate the differences in mean scores using the two methods. Group I has 50 students with a mean score of 64.1 and a standard deviation of 8.5; group II has 40 students with a mean score of 61.0 and a standard deviation of 7.8. Construct a 95% confidence interval for the true difference in mean scores with the two methods.

8.2.2 Confidence Intervals from Small Samples

When the populations are normal and we know the population variances, we may obtain confidence intervals for differences between population means for any samples. If the variances are not known, we require large samples when $s_{\bar{x}_1 - \bar{x}_2}$ is substituted for $\sigma_{\bar{x}_1 - \bar{x}_2}$, as we have seen. If the samples are small, and variances unknown, however, we must use other methods.

If the populations from which the small samples are taken are both normal, we can use the *t* distribution to construct a confidence interval for $\mu_1 - \mu_2$, provided that both populations have equal variance. The assumption of equal variance is very important and may be tested by using the *F* test presented in Section 8.6.2. If the populations have a common variance σ^2, we would have

$$\begin{aligned}\sigma^2_{\bar{x}_1 - \bar{x}_2} &= \sigma^2_{\bar{x}_1} + \sigma^2_{\bar{x}_2} \\ &= \frac{\sigma^2}{n_1} + \frac{\sigma^2}{n_2} \\ &= \sigma^2\left(\frac{1}{n_1} + \frac{1}{n_2}\right)\end{aligned}$$

so that

$$\sigma_{\bar{x}_1 - \bar{x}_2} = \sigma\sqrt{\frac{1}{n_1} + \frac{1}{n_2}}$$

We obtain an estimate for σ^2 by obtaining a *weighted* average of the variances of the two samples, using the degrees of freedom as weights. We multiply each variance by its degrees of freedom and divide the sum by the total degrees of freedom. Each variance is then proportionally represented. We indicate the estimate for σ^2 by $\hat{\sigma}^2$, so that

$$\hat{\sigma}^2 = \frac{(n_1 - 1)s_1^2 + (n_2 - 1)s_2^2}{n_1 + n_2 - 2}$$

The term $\hat{\sigma}^2$ is called the **pooled variance** from the two samples. We can also obtain it by pooling (putting together) the sums of squares of the two samples and dividing by the total number of degrees of freedom:

$$\hat{\sigma}^2 = \frac{\Sigma(x_1 - \bar{x}_1)^2 + \Sigma(x_2 - \bar{x}_2)^2}{n_1 + n_2 - 2}$$

Then the statistic

$$\frac{(\bar{x}_1 - \bar{x}_2) - (\mu_1 - \mu_2)}{\hat{\sigma}\sqrt{(1/n_1) + (1/n_2)}}$$

has a *t* distribution with $n_1 + n_2 - 2$ degrees of freedom.

Applying the methods of Chapter 6, we can calculate a confidence interval for $\mu_1 - \mu_2$ by using $t_{\alpha/2}\{n_1 + n_2 - 2\}$ as the critical value.

Confidence Interval for $\mu_1 - \mu_2$ (Normal Populations, Variances Unknown but Equal)

Suppose we obtain two independent random samples—a sample of n_1 observations of a random variable x_1 with mean μ_1 and a sample of n_2 observations of a random variable x_2 with mean μ_2—and that both of the following are true:

1. Both variables x_1 and x_2 are normally distributed.
2. The population variances are equal.

If $\bar{x}_1$, $\bar{x}_2$, s_1^2, and s_2^2 are the sample means and variances, respectively, a $100(1 - \alpha)\%$ confidence interval for $\mu_1 - \mu_2$ is the interval

$$\bar{x}_1 - \bar{x}_2 - t \cdot \hat{\sigma}\sqrt{\frac{1}{n_1} + \frac{1}{n_2}} \quad \text{to} \quad \bar{x}_1 - \bar{x}_2 + t \cdot \hat{\sigma}\sqrt{\frac{1}{n_1} + \frac{1}{n_2}}$$

where $t = t_{\alpha/2}\{n_1 + n_2 - 2\}$ and

$$\hat{\sigma} = \sqrt{\frac{(n_1 - 1)s_1^2 + (n_2 - 1)s_2^2}{n_1 + n_2 - 2}}$$

Example 8.2

A chemical company wants to build a new plant in either East Quincy or New South Hampton. One of their criteria is the index of methane, since the plant will put methane into the atmosphere. Readings of methane proportions are taken at randomly spaced intervals in the two cities. Eight readings in East Quincy show an average of 0.23 parts per million (ppm) with a standard deviation of 0.07 ppm, while 11 readings in New South Hampton show an average of 0.32 ppm with a standard deviation of 0.12 ppm. Determine a 95% confidence interval for the difference in actual mean methane levels for the two cities, and interpret the result.

Solution

In Section 8.6.2 we will learn how to test the assumption that the two population variances are equal. Until then, we will just have to assume that they are. If that is a reasonable assumption in this case (and it is), we first calculate $\hat{\sigma}$:

$$\hat{\sigma} = \sqrt{\frac{7(0.07)^2 + 10(0.12)^2}{17}} \doteq 0.1024$$

The critical value of t for a 95% confidence interval with 17 degrees of freedom is $t_{.025}\{17\}$, or 2.110. Thus the number to be added and subtracted, $t_{\alpha/2}\{\nu\}\hat{\sigma}\sqrt{(1/n_1) + (1/n_2)}$, is $(2.110)(0.1024)\sqrt{1/8 + 1/11} \doteq 0.10$. It is customary (though not essential) to designate the larger sample mean as $\bar{x}_1$; thus in this case we have $\bar{x}_1 - \bar{x}_2 = 0.32 - 0.23 = 0.09$, so the confidence interval is -0.01 to 0.19. This result indicates that New South Hampton's mean methane index may be as much as 0.19 higher than East Quincy's, or East Quincy's mean index may be as much as 0.01 higher than New South Hampton's—all with a 95% level of confidence. In terms of deciding which city has a higher index, the results are inconclusive.

Comment: If $n_1 = n_2$, the calculations may be simplified somewhat. If $n_1 = n_2 = n$, then

$$\hat{\sigma} = \sqrt{\frac{(n_1 - 1)s_1^2 + (n_2 - 1)s_2^2}{n_1 + n_2 - 2}} = \sqrt{\frac{(n - 1)s_1^2 + (n - 1)s_2^2}{n + n - 2}}$$

$$= \sqrt{\frac{(n - 1)s_1^2 + (n - 1)s_2^2}{2n - 2}} = \sqrt{\frac{(n - 1)(s_1^2 + s_2^2)}{2(n - 1)}}$$

$$= \sqrt{\frac{s_1^2 + s_2^2}{2}}$$

$$\text{and} \sqrt{\frac{1}{n_1} + \frac{1}{n_2}} = \sqrt{\frac{1}{n} + \frac{1}{n}} = \sqrt{\frac{2}{n}}$$

so

$$\hat{\sigma}\sqrt{\frac{1}{n_1} + \frac{1}{n_2}} = \sqrt{\frac{s_1^2 + s_2^2}{2}}\sqrt{\frac{2}{n}} = \sqrt{\frac{s_1^2 + s_2^2}{n}}$$

The sample sizes must be equal in order for us to use this formula, and the number of degrees of freedom is $2(n - 1)$.

PROFICIENCY CHECKS

4. If $\bar{x}_1 = 110.00$, $s_1 = 35.00$, $\bar{x}_2 = 100.00$, $s_2 = 30.00$, and both populations are normal, determine a 95% confidence interval for $\mu_1 - \mu_2$ if n_1 and n_2 are:
   a. 8 b. 12 c. 25 d. 50 (see Proficiency Check 1).
5. If $\bar{x}_1 = 0.560$, $s_1 = 0.122$, $\bar{x}_2 = 0.600$, $s_2 = 0.154$, $n_1 = 12$, $n_2 = 15$, and both populations are normal, determine confidence intervals for $\mu_1 - \mu_2$ with a confidence level of: *(continued)*

PROFICIENCY CHECKS *(continued)*

a. 90% b. 95% c. 98% d. 99%

6. Two sets of elementary school students are taught mathematics by two different methods. The results of a learning test are analyzed to estimate the differences in mean scores using the two methods. Group 1 has 18 students with a mean score of 64.1 and a standard deviation of 5.1; group 2 has 24 students with a mean score of 61.0 and a standard deviation of 5.9. Construct a 95% confidence interval for the true difference in mean scores with the two methods.

8.2.3 The Fisher–Behrens t' for Small-Sample Confidence Intervals

If the samples are small, the two populations are normally distributed, and the population variances are unknown but not equal, we can use an alternative method, developed by Sir Ronald A. Fisher and W. U. Behrens, to

Small-Sample Confidence Interval for $\mu_1 - \mu_2$ (Normal Populations, Variances Unknown and Unequal)

Suppose we obtain two independent random samples—a sample of n_1 observations of a random variable x_1 with mean μ_1 and a sample of n_2 observations of a random variable x_2 with mean μ_2—and that both of the following are true:

1. Both variables x_1 and x_2 are normally distributed.
2. The population variances are unknown and unequal.

If $\bar{x}_1$, $\bar{x}_2$, s_1^2, and s_2^2 are the sample means and variances, respectively, a $100(1 - \alpha)\%$ confidence interval for $\mu_1 - \mu_2$ is the interval

$$\bar{x}_1 - \bar{x}_2 - t'\sqrt{\frac{s_1^2}{n_1} + \frac{s_2^2}{n_2}} \quad \text{to} \quad \bar{x}_1 - \bar{x}_2 + t'\sqrt{\frac{s_1^2}{n_1} + \frac{s_2^2}{n_2}}$$

where $t' = t_{\alpha/2}\{\nu\}$ and

$$\text{df} = \nu = \frac{(s_1^2/n_1 + s_2^2/n_2)^2}{\left[\frac{(s_1^2/n_1)^2}{n_1 - 1} + \frac{(s_2^2/n_2)^2}{n_2 - 1}\right]}$$

get an approximate confidence interval. The statistic we use is called t' ("t-prime"), where

$$t' = \frac{(\bar{x}_1 - \bar{x}_2) - (\mu_1 - \mu_2)}{\sqrt{\dfrac{s_1^2}{n_1} + \dfrac{s_2^2}{n_2}}}$$

This statistic is distributed approximately as $t(\nu)$, where

$$\nu = \frac{(s_1^2/n_1 + s_2^2/n_2)^2}{\left[\dfrac{(s_1^2/n_1)^2}{n_1 - 1} + \dfrac{(s_2^2/n_2)^2}{n_2 - 1}\right]}$$

The value of ν we obtain by using this formula is usually not an integer; generally we round to the nearest whole number. Thus we can find a confidence interval for $\mu_1 - \mu_2$ as shown on page 441.

Example 8.3

Repeat Example 8.2 but assume that the standard deviation of the 11 readings for New South Hampton is 0.22 and the population variances are probably unequal.

Solution

We must use t' instead of t, so we obtain the standard error $\sqrt{(0.07)^2/8 + (0.22)^2/11} \doteq 0.0708$. The number of degrees of freedom is ν, where

$$\nu = \frac{[(0.07)^2/8 + (0.22)^2/11]^2}{\left[\dfrac{[(0.07)^2/8]^2}{7} + \dfrac{[(0.22)^2/11]^2}{10}\right]}$$

or about 12.6. Rounding, we may use $\nu = 13$. Then $t_{.025}\{13\} = 2.160$, and $2.160(0.0708) \doteq 0.15$. Since $0.32 - 0.23 = 0.09$, the 95% confidence interval is given by -0.06 to 0.24. Thus New South Hampton's mean methane index may be as much as 0.24 higher than East Quincy's, or East Quincy's mean index may be as much as 0.06 higher than New South Hampton's—all with a 95% level of confidence. In terms of deciding which city has a higher index, the results are inconclusive. Note that the interval is wider than that obtained in Exercise 8.2. This is the result of both the larger variance for sample 2 and the smaller number of degrees of freedom resulting from the Fisher–Behrens method.

8.2.4 Confidence Intervals Using Trimmed Means*

When the populations are not known to be both normal and the sample sizes are small, we may be able to use the trimmed means as in Section 6.4.4.

*Section 8.2.4 may be omitted without loss of continuity.

If a box plot reveals that either data set has outliers but both sets appear to be symmetric, we may remove 10% of the data from each set and obtain the trimmed means for each data set. Then we can use $\bar{x}_1(T) - \bar{x}_2(T)$ to estimate $\mu_1 - \mu_2$. Recall that $s^2_{\bar{x}(T)} = [s^2(W)/h][(n - 1)/(h - 1)]$, where $s(W)$ is the standard deviation of the Winsorized data set, n is the number of observations in the original data set, and h is the number of observations in the trimmed data set. Since, for any sample of size n, $s^2_{\bar{x}} = s^2/n$, it follows that $s^2 = ns^2_{\bar{x}}$. Thus $ns^2_{\bar{x}}$ is an estimate for σ^2. For the trimmed sample, $hs^2_{\bar{x}(T)}$ is an estimate for σ^2. Thus if we have two samples with size n_1 and

Trimmed-Mean Confidence Interval for $\mu_1 - \mu_2$ (Nonnormal Symmetric Populations)

Suppose we obtain two independent random samples—a sample of n_1 observations of a random variable x_1 with mean μ_1 and a sample of n_2 observations of a random variable x_2 with mean μ_2—and that the distributions of both variables are symmetric.

If (for $i = 1$ and 2) $\bar{x}_i(T)$ is the mean of the trimmed data set consisting of h_i observations and $s_i(W)$ is the standard deviation of the Winsorized data set, then we can obtain a trimmed-mean $100(1 - \alpha)\%$ confidence interval for $\mu_1 - \mu_2$.

1. If the population variances are equal, the interval is

$$\bar{x}_1(T) - \bar{x}_2(T) - t\cdot\hat{\sigma}\sqrt{1/h_1 + 1/h_2} \quad \text{to} \quad \bar{x}_1(T) - \bar{x}_2(T) + t\cdot\hat{\sigma}\sqrt{1/h_1 + 1/h_2}$$

where $t = t_{\alpha/2}\{h_1 + h_2 - 2\}$ and

$$\hat{\sigma} = \sqrt{\frac{(n_1 - 1)s_1^2(W) + (n_2 - 1)s_2^2(W)}{h_1 + h_2 - 2}}$$

2. If the population variances are unequal, the interval is

$$\bar{x}_1(T) - \bar{x}_2(T) - t'\cdot\sqrt{s^2_{\bar{x}_1(T)} + s^2_{\bar{x}_2(T)}} \quad \text{to} \quad \bar{x}_1(T) - \bar{x}_2(T) + t'\cdot\sqrt{s^2_{\bar{x}_1(T)} + s^2_{\bar{x}_2(T)}}$$

where $t = t_{\alpha/2}\{\nu\}$, $s^2_{\bar{x}_1(T)}$ and $s^2_{\bar{x}_2(T)}$ are as defined above, and

$$\nu = \frac{(s^2_{\bar{x}_1(T)} + s^2_{\bar{x}_2(T)})^2}{\left[\dfrac{(s^2_{\bar{x}_1(T)})^2}{h_1 - 1} + \dfrac{(s^2_{\bar{x}_2(T)})^2}{h_2 - 2}\right]}$$

n_2 and the trimmed samples are of size h_1 and h_2, we have

$$s^2_{\bar{x}_1(T)} = \frac{s_1^2(W)}{h_1} \cdot \frac{n_1 - 1}{h_1 - 1} \qquad \text{and} \qquad s^2_{\bar{x}_2(T)} = \frac{s_2^2(W)}{h_2} \cdot \frac{n_2 - 1}{h_2 - 1}$$

Then $h_1 s^2_{\bar{x}_1(T)}$ and $h_2 s^2_{\bar{x}_2(T)}$ are estimates for σ_1^2 and σ_2^2, respectively. If we can reasonably conclude that $\sigma_1^2 = \sigma_2^2$ (using the methods of Section 8.6.2), we may use the normal t procedure with the pooled estimate $\hat{\sigma}^2$. If we cannot conclude that $\sigma_1^2 = \sigma_2^2$, we should use the Fisher–Behrens procedure.

Problems

Practice

8.1 Two samples of size n are used to obtain a 95% confidence interval for the difference between the population means. Determine the width of the confidence interval if the population standard deviations are as given:
a. $n = 100, \sigma_1 = 13.4, \sigma_2 = 9.4$
b. $n = 64, \sigma_1 = 115.7, \sigma_2 = 103.4$
c. $n = 200, \sigma_1 = 0.040, \sigma_2 = 0.052$
d. $n = 400, \sigma_1 = 9.97, \sigma_2 = 8.63$
e. $n = 36, \sigma_1 = 12.136, \sigma_2 = 15.432$

8.2 A sample of size 100 is drawn from a very large population with known standard deviation of 11.60. A second sample of size 80 is drawn from a very large population with known standard deviation 14.32. If the means of the samples are used to estimate the difference between the means of the populations, what is the probability that the estimate will differ from the true difference by no more than:
a. 2.32? **b.** 1.13? **c.** 5.39? **d.** 0.50? **e.** 3.45?

8.3 Two random samples are drawn from normally distributed populations. The variance of the first population is 62.41, and the variance of the second population is 39.69. Determine 95% confidence intervals for the difference between the population means if the first sample mean is 123.8, the second sample mean is 117.6, and the sample sizes are both equal to:
a. 10 **b.** 25 **c.** 100 **d.** 400

8.4 A random sample of size 50 is drawn from a population with known variance of 116.64, and the sample mean is 317.8. A random sample of size 60 is drawn from a second population with a known variance of 88.36, and the sample mean is 301.9. Determine confidence intervals for the difference between the population means if the confidence level is:
a. 90% **b.** 95% **c.** 98% **d.** 99%

8.5 Random samples of sizes 75 and 57, respectively, are drawn from two populations. Determine 90% confidence intervals for the difference between the population means if the sample means and standard deviations are as given:
a. $\bar{x}_1 = 12.3, s_1 = 1.9; \bar{x}_2 = 11.4, s_2 = 2.1$
b. $\bar{x}_1 = 12.3, s_1 = 10.6; \bar{x}_2 = 11.4, s_2 = 11.3$

c. $\bar{x}_1 = 121.7, s_1 = 1.9; \bar{x}_2 = 112.8, s_2 = 2.1$
d. $\bar{x}_1 = 121.7, s_1 = 10.6; \bar{x}_2 = 112.8, s_2 = 11.3$

8.6 Random samples of size 60 are drawn from two populations. If the first sample mean and variance are 0.112 and 0.000169, respectively, and the second sample mean and variance are 0.087 and 0.000121, respectively, determine confidence intervals for the difference between the population means with a confidence level of:
a. 90% **b.** 95% **c.** 98% **d.** 99%

8.7 Random samples of sizes 25 and 19, respectively, are drawn from two normal populations. Determine 90% confidence intervals for the difference between the population means if the sample means and standard deviations are as given:
a. $\bar{x}_1 = 12.3, s_1 = 1.9; \bar{x}_2 = 11.4, s_2 = 2.1$
b. $\bar{x}_1 = 12.3, s_1 = 10.6; \bar{x}_2 = 11.4, s_2 = 11.3$
c. $\bar{x}_1 = 121.7, s_1 = 1.9; \bar{x}_2 = 112.8\ s_2 = 2.1$
d. $\bar{x}_1 = 121.7, s_1 = 10.6; \bar{x}_2 = 112.8, s_2 = 11.3$

8.8 Random samples of size 15 are drawn from two normally distributed populations. If the first sample mean and variance are 0.112 and 0.000169, respectively, and the second sample mean and variance are 0.087 and 0.000121, respectively, determine confidence intervals for the difference between the population means with a confidence level of:
a. 90% **b.** 95% **c.** 98% **d.** 99%

8.9 Repeat Problem 8.8 but assume that the variance of the second sample is 0.000036 and the population variances are not equal.

Applications

8.10 A psychologist is investigating the effect of an alkaloid derivative on the reaction times of subjects in an emergency situation. One hundred volunteers are randomly divided into two groups. One group is administered the alkaloid derivative and the other is given a placebo. Then each subject is placed in a driving simulator and given a 10-minute simulated drive with a number of simulated emergencies. The median reaction time is recorded for each subject. For group A, the control group, the mean reaction time for all subjects is 0.78 second with a standard deviation of 0.066 second. Group B, the experimental group that receives the alkaloid derivative, has a mean reaction time of 0.87 second with a standard deviation of 0.072 second. Determine 90% and 95% confidence intervals for the difference in reaction time for the two groups. Interpret the confidence interval.

8.11 To determine the best site for a service station, two likely locations are surveyed and counts are kept for a month of the number of cars passing the location. At location T, the mean daily traffic count is 116.3 with a standard deviation of 22.8 for 31 days. At location S, the mean daily traffic count for 30 days (one day's results are lost) is 138.2 with a standard deviation of 38.8. Determine a 99% confidence interval for the difference in mean daily traffic counts at the two locations, and interpret the result.

8.12 A superball is dropped and the height of the bounce is measured. The proportion of the original height to which the ball returns is called the *coefficient of restitution* for the substance. Four determinations of the coefficient of restitution for this superball yield values of 0.84, 0.78, 0.86, and 0.81. The coefficients of restitution for a second superball are measured five times, yielding 0.88, 0.81, 0.90, 0.84, and 0.85. Give 95% and 99% confidence intervals for the true difference between the coefficients of restitution for these balls, and interpret the result.

8.13 Family incomes are often used to determine a community's eligibility to receive federal aid. A survey to discover which of two communities is in greater need of federal assistance selects 36 families at random from Mariposa and 44 at random from Santa Luna. The mean income for the Mariposa sample is $17,114 with a standard deviation of $3916, while the Santa Luna sample has a mean of $14,544 with a standard deviation of $1944. Determine a 95% confidence interval for the mean difference in family income for the two communities, and interpret your result.

8.14 To study telephone calls at a business office, the office manager times incoming calls and outgoing calls for 1 day. Assuming that the day's calls make up a random sample of calls (which may be open to argument), determine a 95% confidence interval for the mean difference in the length of incoming calls and the length of outgoing calls if 17 incoming calls last an average of 5.16 minutes with a standard deviation of 1.12 minutes and 12 outgoing calls last an average of 4.13 minutes with a standard deviation of 2.36 minutes. Can you draw any conclusion from this result?

8.15 A women's group claims that female college graduates who go to work are paid less than their male counterparts. They cite the results of a study of 1983 college graduates who went to work immediately upon graduation. A total of 32 women graduates are currently making a mean salary of $18,132 per year with a standard deviation of $1409, while 40 male graduates are currently making an average of $19,844 with a standard deviation of $1208. Obtain a 99% confidence interval for the true difference in mean salary for men and women. Does the result substantiate the women's group's claim?

8.16 A company has heard from a small but vocal minority that its already liberal fringe benefit plan is not liberal enough. Since the majority of these employees work the night shift, the company decides to compare night-shift workers and day-shift workers regarding their degree of satisfaction with the fringe benefits. Two independently selected random samples are given questionnaires asking them to rate various aspects of fringe benefit policy. They need not identify themselves on the forms, except for the shift they work, so their answers can be frank. A score of 80 is perfect, while the lowest possible score is 8. A total of 37 day-shift workers rate the fringe benefit package an average of 61.3, with a standard deviation of 11.5, while 52 night-shift workers rate the package an average of 47.7 with a standard deviation of 24.6. Use this information to obtain a 90% confidence interval for the difference between the mean ratings for the two shifts. Can the company conclude anything about the comparative opinions of the two shifts?

8.17 Kenneth DeMeuse (1985) reported a study in which firms provided continuing education for their workers in the workplace. Part of the study dealt with the supervisors' support for the continuing education program as perceived by the employees. The two groups under study were those who participated in the program and those who did not. Suppose that random samples of each group were selected and that the ratings (on a scale of 1 to 7—7 for the most positive attitude perceived by the employees) were as follows:

Participants	4, 6, 6, 5, 7, 4, 6, 4, 7, 5, 6, 4, 4, 4, 5
Nonparticipants	1, 3, 4, 3, 5, 1, 2, 3, 7, 4, 3, 2, 7, 6, 4, 2, 2, 1, 3

We wish to obtain a 95% confidence interval for the difference in mean ratings by the two populations.

a. Construct a box plot for the samples. Do there appear to be outliers? If there are outliers in either sample, the normality assumption may not be satisfied.
b. Obtain the means and variances for the two samples. If the variance of one sample is much more than four times the variance of a second sample, the population variances may not be equal. Does it appear that the population variances are equal?
c. Obtain a 95% confidence interval for the difference in population means, using the appropriate procedure. If you use trimmed-mean procedures, check the Winsorized variances for equal population variance.

8.18 A new alloy is subjected to nine determinations of hardness resulting in a mean value on Moh's scale of 0.630 with a standard deviation of 0.081. When a second measuring instrument is used on the same alloy, eight determinations yield a mean of 0.572 with a standard deviation of 0.074. Determine a 90% confidence interval for the difference in mean hardness of this alloy as measured by the two instruments.

8.19 Repeat Problem 8.17 but assume that the standard deviation of the first group was 1.58, the standard deviation of the second group was 0.35, and the variances of the populations could not be considered equal.

8.3 Testing a Hypothesis About Differences Between Population Means

Since a true statistical experimental design requires at least two samples, sometimes we must determine whether two population parameters differ—an important application of statistical inference. We can apply the methods of Chapter 7 to this end, using the sampling distributions developed in Section 8.2.

Two different applications arise, depending upon whether the samples are independent or not. Two samples are independent if the selection of one sample has no effect on the selection of the other. If a researcher wishes to determine whether one brand of seed corn will produce a better crop than a second brand, he can plant random samples of seed corn of each variety in similar soils and conditions to determine which has the greater yield. To determine which of two advertising displays will sell a product better, stores can be randomly selected and the displays assigned randomly to the selected stores. Stores using display 1 would be considered a random sample of all stores that could use display 1, while stores using display 2 would be considered a random sample of all stores that could use display 2.

On the other hand, samples that are paired in some way are not independent. The classic example of paired samples is the before-and-after experiment in which observations are made on a subject both before and after participation. Paired-sample designs are treated entirely differently from independent-sample designs. We will examine the case of independent samples first.

8.3.1 Tests Using Two Independent Samples

Suppose that two independent random samples are obtained. Sample 1 consisting of n_1 observations has a mean $\bar{x}_1$, while sample 2 consisting of n_2 observations has a mean $\bar{x}_2$. A researcher may wish to test the hypothesis that if μ_1 is the mean of the population from which sample 1 was taken and μ_2 is the mean of the population from which sample 2 was taken, then $\mu_1 = \mu_2$, or else that $\mu_1 - \mu_2 = \delta_0$, where δ_0 (the Greek letter delta) is some specified difference. The first hypothesis is the special case of the second when $\delta_0 = 0$.

Now if the samples are independent, we know from Section 8.2 that, for sufficiently large samples, the sampling distribution of $\bar{x}_1 - \bar{x}_2$ is $N(\mu_1 - \mu_2, \sigma^2_{\bar{x}_1 - \bar{x}_2})$. That is, the random variable z, where $z = [(\bar{x}_1 - \bar{x}_2) - (\mu_1 - \mu_2)]/\sigma_{\bar{x}_1 - \bar{x}_2}$ is N(0, 1). If we wish to test the hypothesis that $\mu_1 - \mu_2 = \delta_0$, our test statistic will be based on the size of the samples and whether or not we know the population variances. Thus if we wish to test the hypothesis $\mu_1 - \mu_2 = \delta_0$, and the population variances are known and the samples are both at least 30, the test statistic will be $z = [(\bar{x}_1 - \bar{x}_2) - \delta_0]/\sigma_{\bar{x}_1 - \bar{x}_2}$. If both populations are normally distributed, the samples are less than 30, and the population variances are unknown but equal, the test statistic will be $t = [(\bar{x}_1 - \bar{x}_2) - \delta_0]/[\hat{\sigma}\sqrt{(1/n_1) + (1/n_2)}]$. We can use the trimmed means in certain cases. In extreme cases, when no other methods apply, we may use nonparametric methods. The nonparametric method most frequently used when we cannot apply the t or z tests is the Mann–Whitney U Test, discussed in Section 10.2. The other cases were discussed in Section 8.2 and are presented in Table 8.1. If $\delta_0 = 0$, the null hypothesis is usually written $\mu_1 = \mu_2$.

The decision rule is defined by the rejection region, which depends upon the alternate hypothesis, the test statistic, and the level of significance α. The

Table 8.1 HYPOTHESIS TESTS FOR THE DIFFERENCE BETWEEN TWO POPULATION MEANS.

POPULATION STANDARD DEVIATIONS	SHAPE OF DISTRIBUTIONS	SAMPLE SIZES	TEST STATISTIC
Both known	Both normal	Any n	$z = [(\bar{x}_1 - \bar{x}_2) - \delta_0]/\sigma_{\bar{x}_1 - \bar{x}_2}$
	One or both not normal	$n_1, n_2 \geq 30$	$z = [(\bar{x}_1 - \bar{x}_2) - \delta_0]/\sigma_{\bar{x}_1 - \bar{x}_2}$
One or both unknown	Both normal	$n_1, n_2 \geq 30$	$z = [(\bar{x}_1 - \bar{x}_2) - \delta_0]/s_{\bar{x}_1 - \bar{x}_2}$
		Any n $\sigma_1^2 = \sigma_2^2$	$t = \dfrac{(\bar{x}_1 - \bar{x}_2) - \delta_0}{\hat{\sigma}\sqrt{(1/n_1) + (1/n_2)}}$
		n_1 or $n_2 < 30$ $\sigma_1^2 \neq \sigma_2^2$	$t' = \dfrac{(\bar{x}_1 - \bar{x}_2) - \delta_0}{\sqrt{(s_1^2/n_1) + (s_2^2/n_2)}}$
	Not both normal	$n_1, n_2 \geq 50$	$z = [(\bar{x}_1 - \bar{x}_2) - \delta_0]/s_{\bar{x}_1 - \bar{x}_2}$
	Both symmetric (use trimmed means)	n_1 or $n_2 < 50$ $\sigma_1^2 = \sigma_2^2$	$t = \dfrac{[\bar{x}_1(T) - \bar{x}_2(T)] - \delta_0}{\hat{\sigma}\sqrt{(1/h_1) + (1/h_2)}}$
		n_1 or $n_2 < 30$ $\sigma_1^2 \neq \sigma_2^2$	$t' = \dfrac{[\bar{x}_1(T) - \bar{x}_2(T)] - \delta_0}{\sqrt{s_{\bar{x}_1(T)}^2 + s_{\bar{x}_2(T)}^2}}$
	Not both symmetric	n_1 or $n_2 < 50$	No general method (use the Mann–Whitney U Test)

value of the test statistic obtained from the sample is designated z^*, t^*, or t'^*. Table 8.2 lists the appropriate decision rule for the various alternate hypotheses for each test statistic.

Example 8.4

Two samples of menhaden are taken, one from Escambia Bay and one from Pensacola Bay. The 52 fish from Pensacola Bay have a mean weight of 12.1 ounces with a standard deviation of 2.1 ounces; the 37 fish from Escambia Bay have a mean weight of 11.6 ounces with a standard deviation of 1.3 ounces. The biologists wish to determine whether the fish are from the same population, in which case the samples would have come from populations with the same mean. Test this hypothesis at a 0.05 level of significance.

Solution

We can probably assume that the weights of the fish are normally distributed, in which case the large-sample procedure is applicable. Therefore we have the following:

1. H_0: $\mu_1 = \mu_2$
   H_1: $\mu_1 \neq \mu_2$

Table 8.2

Null and Alternate Hypotheses

H_0: $\mu_1 - \mu_2 = \delta_0$	H_0: $\mu_1 - \mu_2 = \delta_0$	H_0: $\mu_1 - \mu_2 = \delta_0$
H_1: $\mu_1 - \mu_2 \neq \delta_0$	H_1: $\mu_1 - \mu_2 > \delta_0$	H_1: $\mu_1 - \mu_2 < \delta_0$

Decision Rule

Reject H_0 if $\lvert z^*\rvert > z_{\alpha/2}$	Reject H_0 if $z^* > z_\alpha$	Reject H_0 if $z^* < -z_\alpha$
	OR	
$\lvert t^*\rvert > t_{\alpha/2}\{n_1 + n_2 - 2\}$	$t^* > t_\alpha\{n_1 + n_2 - 2\}$	$t^* < -t_\alpha\{n_1 + n_2 - 2\}$
	OR	
$\lvert t'^*\rvert > t_{\alpha/2}\{\nu\}$	$t'^* > t_\alpha\{\nu\}$	$t'^* < -t_\alpha\{\nu\}$

$$\text{where } \nu = \frac{(s_1^2/n_1 + s_2^2/n_2)^2}{\left[\frac{(s_1^2/n_1)^2}{n_1 - 1} + \frac{(s_2^2/n_2)^2}{n_2 - 1}\right]}$$

OR (using the trimmed means)

$\lvert t^*\rvert > t_{\alpha/2}\{h_1 + h_2 - 2\}$	$t^* > t_\alpha\{h_1 + h_2 - 2\}$	$t^* < -t_\alpha\{h_1 + h_2 - 2\}$
	OR	
$\lvert t'^*\rvert > t_{\alpha/2}\{\nu\}$	$t'^* > t_\alpha\{\nu\}$	$t'^* < -t_\alpha\{\nu\}$

$$\text{where } \nu = \frac{(s^2_{\bar{x}_1(T)} + s^2_{\bar{x}_2(T)})^2}{\left[\frac{(s^2_{\bar{x}_1(T)})^2}{h_1 - 1} + \frac{(s^2_{\bar{x}_2(T)})^2}{h_2 - 1}\right]}$$

2. Decision rule: Reject H_0 if $|z^*| > 1.96$.
3. Results: We are given $n_1 = 52, \bar{x}_1 = 12.1, s_1 = 2.1$ and $n_2 = 37, \bar{x}_2 = 11.6, s_2 = 1.3$, so that $s_{\bar{x}_1 - \bar{x}_2} = \sqrt{(2.1)^2/52 + (1.3)^2/37} \doteq 0.3612$ and $z^* = (12.1 - 11.6)/0.3612 \doteq 1.38$.
4. Decision: Fail to reject H_0.
5. Conclusion: The observed significance level is about 0.17; the probability of making a type II error is probably large, so it would be a good idea to reserve judgment. Another sample is probably easy to obtain. If not, the biologists must conclude that the population means do not differ significantly.

Example 8.5 To test the hypothesis that fifth grade-girls are, on the average, at least an inch taller than fifth-grade boys, two random samples of boys and girls from a large school are measured and their heights recorded. Sixteen boys average 49.8 inches in height with a standard deviation of 5.1 inches, and 15 girls average 53.9 inches with a standard deviation of 4.7 inches. Test the hypothesis at the 0.05 level of significance.

Solution

Since the samples are small but probably from normal populations, we can use the t test statistic. The variances are quite close, so the assumption of equal population variances is probably met. The rejection region is one-tailed, and the critical value is $t_{.05}\{29\} = 1.699$. We can either designate μ_1 as the mean height for girls and μ_2 as the mean height for boys or use the descriptive subscripts μ_g and μ_b.

1. H_0: $\mu_g - \mu_b = 1.0$
   H_1: $\mu_g - \mu_b > 1.0$
2. Decision rule: Reject H_0 if $t^* > 1.699$.
3. Results: $n_g = 15$, $\bar{x}_g = 53.9$, $s_g = 4.7$; $n_b = 16$, $\bar{x}_b = 49.8$, $s_b = 5.1$.
   Thus $\hat{\sigma} = \sqrt{\dfrac{14(4.7)^2 + 15(5.1)^2}{29}} \doteq 4.911$ and $t^* =$
   $(53.9 - 49.8 - 1)/(4.911\sqrt{1/15 + 1/16}) \doteq 1.756$.
4. Decision: Reject H_0.
5. Conclusion: Since we reject H_0, we conclude that the average girl is at least 1 inch taller than the average boy in the fifth grade of this school. We cannot obtain the observed significance level from Table 5. Since $t_{.05}\{29\} = 1.699$, $t_{.025}\{29\} = 2.045$, and $t^* = 1.756$, then $0.025 < P < 0.05$; usually this is simply reported as $P < 0.05$. We assume that if the P value were less than the next lower tabled value, such as 0.025, we would report it as $P < 0.025$. In this case, then, $P < 0.05$.

Example 8.6

Electronic Media (January 6, 1986) reported a study of television stations broken down into network affiliate stations, VHF independent stations (channels 2 to 13), and UHF independent stations (channels 14 to 83). Suppose information regarding cost of a 30-second commercial during prime time Monday through Friday is obtained for two random samples, one of 14 network affiliates and one of 24 independent stations. There is considerable variability in stations from market to market and even within markets, but within the same market network affiliates usually charge more than independents. Suppose the hypothesis being investigated is that network affiliates charge, on the average, at least \$500 more for a 30-second commercial than independents and that the mean charge for the network stations is \$883 with a standard deviation of \$213, while for the independents the mean charge is \$247 with a standard deviation of \$63. Try to "prove" the hypothesis at a 0.05 level of significance.

Solution

The samples are small and probably from normal populations, but the sample variances are far apart. Moreover, we can show that the population variances are not equal (see Section 8.6.2), so the assumption of equal population variances is not met. Thus the proper test statistic is the Fisher–Behrens

t'. The degrees of freedom for this test equal ν, where

$$\nu = \frac{(213^2/14 + 63^2/24)^2}{\left[\dfrac{(213^2/14)^2}{13} + \dfrac{(63^2/24)^2}{23}\right]} \doteq 14$$

The rejection region is one-tailed and the critical value is $t_{.05}\{14\} = 1.761$. We let μ_1 be the mean cost of a 30-second commercial in prime time Monday through Friday for all network affiliate stations and μ_2 be the mean cost for all independent stations. then:

1. H_0: $\mu_1 - \mu_2 = 500$
   H_1: $\mu_1 - \mu_2 > 500$
2. Decision rule: Reject H_0 if $t^* > 1.761$.
3. Results: $n_1 = 14$, $\bar{x}_1 = 883$, $s_1 = 213$; $n_2 = 24$, $\bar{x}_2 = 247$, $s_2 = 63$. Thus $\sqrt{(213)^2/14 + (63)^2/24} \doteq 58.36$ and $t'^* = [(883 - 247) - 500]/58.36 \doteq 2.330$.
4. Decision: Reject H_0.
5. Conclusion: Since we reject H_0, we conclude that the mean charge for network affiliate stations for a 30-second commercial in prime time Monday through Friday is at least $500 more than the comparable charge for independent stations. Since $t_{.025}\{14\} = 2.145$, $t_{.01}\{14\} = 2.624$, and $t^* = 2.330$, then $P < 0.025$.

PROFICIENCY CHECKS

7. A research hypothesis is that $\mu_1 - \mu_2 \neq 10.0$. The level of significance is 0.10; sample results were $n_1 = 60$, $\bar{x}_1 = 50.3$, $s_1 = 3.8$, $n_2 = 50$, $\bar{x}_2 = 38.4$, $s_2 = 1.9$. Test the hypothesis.
8. Repeat Proficiency Check 7 if $n_1 = 18$, $n_2 = 14$, the populations are normal, and $\sigma_1^2 = \sigma_2^2$.
9. Give the observed significance level for Proficiency Checks 7 and 8.
10. Repeat Proficiency Check 8 if $\sigma_1^2 \neq \sigma_2^2$.
11. A research hypothesis is that $\mu_1 - \mu_2 \geqslant 100.0$. The level of significance is 0.01. Sample results are $n_1 = 40$, $\bar{x}_1 = 1356.3$, $s_1 = 125.4$, $n_2 = 34$, $\bar{x}_2 = 1168.9$, $s_2 = 123.7$, and it is reasonable to believe the distribution is normal. Test the hypothesis.
12. Repeat Proficiency Check 11 if $n_1 = 4$ and $n_2 = 6$. *(continued)*

PROFICIENCY CHECKS *(continued)*

13. Obtain the P values for both Proficiency Checks 11 and 12.
14. A psychologist gives a test to eight blue-collar workers and nine white-collar workers to test a hypothesis that white-collar workers should score higher on the test. The blue-collar workers get scores of 23, 18, 22, 21, 16, 17, 19, and 12, and the white-collar workers get scores of 23, 19, 27, 14, 26, 28, 25, 24, and 26. What conclusions can you draw at the 0.05 level of significance? What is the observed level of significance? Assume:
    a. the populations are normal with equal variances
    b. the populations are normal with unequal variances
    c. the populations are nonnormal but symmetric with equal variances
    d. the populations are nonnormal but symmetric with unequal variances

8.3.2 Tests Using Paired Samples

In many cases we can pair the members of two samples according to certain criteria. People may be matched on sex, IQ scores, age, or other factors thought to have a bearing on the outcome of an experiment. Plots of ground may be divided into two equal parts, or two dyes may be tested by applying each dye to halves of several pieces of cloth. Many methods of pairing are possible. In other cases there may be only one sample but two measurements made on the sample—as in before-and-after tests, usually called pretests and posttests. If the subjects are the same, the two tests are called **repeated observations.**

We still wish to test the hypothesis $\mu_1 - \mu_2 = \delta_0$, where μ_1 is the mean of the population represented by sample 1, μ_2 is the mean of the population represented by sample 2, and δ_0 is the hypothesized difference between them. Since the subjects are paired, in this case we can get a more powerful test of the hypothesis than for independent samples.

Paired-difference tests are a simple example of a *randomized block design,* which is discussed in detail in Section 12.1. In this type of design, comparisons are made within homogeneous groupings called *blocks* in order to reduce the amount of variation caused by external sources. Using the same cloth to test two dyes removes the variation that is present in different cloths so that we can concentrate on the effect of the dyes on this type of cloth. Confidence intervals obtained by genuine blocking are more efficient and hypothesis tests in blocked experiments are more powerful than for independent samples, so that we can obtain a more accurate estimation of differences. For small samples, however, we lose sensitivity since the number of degrees of freedom will be half that for independent samples. If we can pair samples on a reasonable criterion, however, paired-difference testing gives more information than independent sample testing.

To test the hypothesis $\mu_1 - \mu_2 = \delta_0$, we obtain the difference score d for each pair—that is, for pair 1, d_1 is the score made by subject 1 of sample 1 minus the score made by subject 1 of sample 2, and so on. The average of the difference scores, $\bar{d}$, is an unbiased estimator for δ, the true difference between μ_1 and μ_2, and we can test the hypothesis $\mu_1 - \mu_2 = \delta_0$ by testing the hypothesis $\delta = \delta_0$ in the same way as for a one-sample test. The standard deviation of all values of d is designated s_d and calculated in the usual way:

$$s_d^2 = \frac{\Sigma(d - \bar{d})^2}{n - 1}$$
$$= \frac{\Sigma d^2 - (\Sigma d)^2/n}{n - 1}$$

and s_d is the square root of s_d^2. Then $s_{\bar{d}}^2 = s_d^2/n$ and we can use the large-sample z test statistic (if $n \geq 50$ or if the population can be assumed normal and $n \geq 30$) or the t test statistic (if the population can be assumed normal, particularly if $n < 30$). We can use the trimmed-mean procedure for the sample of differences if the number of pairs is small and the population of differences can be assumed to be symmetric but not normal. Since the population variance is generally not known and the finite population correction factor is usually not appropriate, there are generally only three appropriate procedures for testing the null hypothesis $\mu_1 - \mu_2 = \delta_0$ if the samples are paired.

Hypothesis Tests for $\mu_1 - \mu_2$ (Paired Samples)

Suppose we randomly obtain n pairs of observations of random variables x_1 and x_2 with means μ_1 and μ_2, respectively, and compute the difference $d = x_1 - x_2$ for each pair. Suppose also that one of the following is true:

1. The number of pairs, n, is at least 50

   OR

2. The population of differences is normal

   OR

3. The population of differences is nonnormal but symmetric.

Then we may test the hypothesis $\mu_1 - \mu_2 = \delta_0$.

The sample value of the test statistic used to test the hypothesis $\mu_1 - \mu_2 = \delta_0$ depends upon the situations numbered above.

1. If $n \geq 50$, the test statistic is z^*, where $z^* = (\bar{d} - \delta_0)/(s_d/\sqrt{n})$.
2. If the population is normal, the test statistic is t^*, where $t^* = (\bar{d} - \delta_0)/(s_d/\sqrt{n})$ with $n - 1$ degrees of freedom; we can use z^* if $n \geq 30$.
3. If the population is nonnormal but symmetric, we may use the test statistic $t^* = (\bar{d} - \delta_0)/s_{\bar{d}(T)}$. This is analogous to the one-sample case; $s_{d(T)} = [s_{\bar{d}}(W)/\sqrt{h}]\sqrt{(n-1)/(h-1)}$, where $s_d(W)$ is the standard deviation of the Winsorized data set of differences. The number of degrees of freedom is $h - 1$.

The decision rule is dictated by the alternate hypothesis and the level of significance α we choose for the test. The three possibilities are as follows:

Case I	Case II	Case III
	Null and Alternate Hypotheses	
H_0: $\mu_1 - \mu_2 = \delta_0$ H_1: $\mu_1 - \mu_2 \neq \delta_0$	H_0: $\mu_1 - \mu_2 = \delta_0$ H_1: $\mu_1 - \mu_2 > \delta_0$	H_0: $\mu_1 - \mu_2 = \delta_0$ H_1: $\mu_1 - \mu_2 < \delta_0$
	Decision Rule	
Reject H_0 if	**Reject H_0 if**	**Reject H_0 if**
1. $\lvert z^*\rvert > z_{\alpha/2}$	$z^* > z_\alpha$ OR	$z^* < -z_\alpha$
2. $\lvert t^*\rvert > t_{\alpha/2}\{n - 1\}$	$t^* > t_\alpha\{n - 1\}$ OR	$t^* < -t_\alpha\{n - 1\}$
3. $\lvert t^*\rvert > t_{\alpha/2}\{h - 1\}$	$t^* > t_\alpha\{h - 1\}$	$t^* < -t_\alpha\{h - 1\}$

If we wish, we can obtain a confidence interval for $\mu_1 - \mu_2$ with limits $\bar{d} - z_{\alpha/2}(s_d/\sqrt{n})$ and $\bar{d} + z_{\alpha/2}(s_d/\sqrt{n})$ or $\bar{d} - t_{\alpha/2}\{n - 1\}(s_d/\sqrt{n})$ and $\bar{d} + t_{\alpha/2}(n - 1)(s_d/\sqrt{n})$ or, using the trimmed mean, $\bar{d} - t_{\alpha/2}\{h - 1\}(s_{d(T)})$ and $\bar{d} + t_{\alpha/2}\{h - 1\}(s_{d(T)})$, as appropriate. If we reject the null hypothesis, a confidence interval is useful for estimating the true value of δ; if we cannot reject the null hypothesis, a confidence interval is useful in estimating a range of values for δ.

Example 8.7

Ten people take part in a diet experiment. Their weights are measured at the beginning and at the end of the diet period. Use the following data to determine whether the diet is effective at the 0.05 level of significance.

SUBJECT	STARTING WEIGHT	ENDING WEIGHT
1	183	177
2	144	145
3	151	145
4	163	162
5	155	151
6	159	163
7	178	173
8	184	185
9	142	139
10	137	138

Solution

The differences are 6, −1, 6, 1, 4, −4, 5, −1, 3, −1; thus $\bar{d} = 1.8$, $s_d \doteq 3.4897$, and if the diet is effective then δ will be positive. Hence the critical value of the test statistic is $t_{.05}\{9\} = 1.833$ and we have:

1. H_0: $\mu_1 = \mu_2$
   H_1: $\mu_1 > \mu_2$
2. Decision rule: Reject H_0 if $t^* > 1.833$.
3. Results: $t^* = 1.8/(3.4897/\sqrt{10}) \doteq 1.8/1.1035 \doteq 1.631$.
4. Decision: Fail to reject H_0.
5. Conclusion: We cannot conclude that the diet is effective. Since $P < 0.10$, however, we would like to continue experimenting.

Problems

Practice

8.20 A random sample of size 50 is drawn from a population with known variance of 116.64; the sample mean is 305.8. A random sample of size 60 is drawn from a second population with a known variance of 88.36, and the sample mean is 301.9. Test the hypothesis that the true difference between the population means is 10.0 using a level of significance of:
a. 0.10 **b.** 0.05 **c.** 0.02 **d.** 0.01

8.21 Random samples of size 75 and 57, respectively, are drawn from two populations. Using a 0.10 level of significance, test the hypothesis that $\mu_1 = \mu_2$ if the sample means and standard deviations are as given:
a. $\bar{x}_1 = 12.3$, $s_1 = 1.9$; $\bar{x}_2 = 11.4$, $s_2 = 2.1$
b. $\bar{x}_1 = 12.3$, $s_1 = 10.6$; $\bar{x}_2 = 11.4$, $s_2 = 11.3$
c. $\bar{x}_1 = 121.7$, $s_1 = 1.9$; $\bar{x}_2 = 112.8$, $s_2 = 2.1$
d. $\bar{x}_1 = 121.7$, $s_1 = 10.6$; $\bar{x}_2 = 112.8$, $s_2 = 11.3$

PROFICIENCY CHECKS

15. Two different toothpastes are tested on two sets of children carefully matched for age, sex, diet, fluoride treatments, and so on. During the test period, the number of new cavities for the subjects is:

SUBJECT PAIR	TOOTHPASTE A	TOOTHPASTE B
1	3	2
2	1	1
3	4	1
4	2	0
5	5	4
6	3	1
7	3	2
8	1	1
9	3	2
10	2	1

Test the hypothesis that there are no differences between the toothpastes in mean number of cavities; use the 0.01 level of significance.

16. A test is designed to compare the wearing qualities of two brands of motorcycle tires. One tire of each brand is placed on each of six motorcycles that are driven for a specified mileage. The two tires are then exchanged (front and rear) and driven for an equal number of miles. At the conclusion of the test, wear is measured (in thousandths of an inch). The data are given here:

	MOTORCYCLE					
	1	2	3	4	5	6
Brand A	98	61	38	117	88	109
Brand B	102	60	46	125	94	111

8.22 Random samples of size 60 are drawn from two populations. If the first sample mean and variance are 0.112 and 0.000169, respectively, and the second sample mean and variance are 0.087 and 0.000121, respectively, test the hypothesis that $\mu_1 > \mu_2$ using:
a. $\alpha = 0.05$ **b.** $\alpha = 0.025$ **c.** $\alpha = 0.01$ **d.** $\alpha = 0.005$

8.23 Random samples of sizes 25 and 19, respectively, are drawn from two populations. Using a 0.10 level of significance, test the hypothesis that $\mu_1 = \mu_2$ if the sample means and standard deviations are as given:

a. $\bar{x}_1 = 12.3$, $s_1 = 1.9$; $\bar{x}_2 = 11.4$, $s_2 = 2.1$
b. $\bar{x}_1 = 12.3$, $s_1 = 10.6$; $\bar{x}_2 = 11.4$, $s_2 = 11.3$
c. $\bar{x}_1 = 121.7$, $s_1 = 1.9$; $\bar{x}_2 = 112.8$, $s_2 = 2.1$
d. $\bar{x}_1 = 121.7$, $s_1 = 10.6$; $\bar{x}_2 = 112.8$, $s_2 = 11.3$

8.24 Random samples of size 15 are drawn from two normally distributed populations. If the first sample mean and variance are 0.112 and 0.000169, respectively, and the second sample mean and variance are 0.087 and 0.000121, respectively, test the hypothesis that $\mu_1 > \mu_2$ using:
a. $\alpha = 0.05$ **b.** $\alpha = 0.025$ **c.** $\alpha = 0.01$ **d.** $\alpha = 0.005$

8.25 Repeat Problem 8.24 but assume that the variance of the second population is 0.000036 and the population variances are not equal.

Applications

8.26 A teacher tries two methods of teaching verbs to his French class. He has divided the class randomly into two samples of 14 students each. Using method 1, the students learn an average of 43.8 verbs each with a standard deviation of 4.6. The other sample, using method 2, learns an average of 38.6 verbs each with a standard deviation of 5.2. Theory suggests that method 1 is superior to method 2. Test the theory, using a 0.01 level of significance.

8.27 The Department of Institutional Research at a state college wishes to determine how their seniors do on the Graduate Record Examination compared to seniors at a private university in the same city. A random sample of 50 state college seniors is found to have an average of 1183 on the GRE with a standard deviation of 137. The private university sample of 100 has a mean of 1168 with a standard deviation of 146. Test (at $\alpha = 0.05$) the hypothesis that there is no difference in the populations. What is the observed significance level?

8.28 A group of 15 subjects is given tests before and after experiencing a learning situation. The results of the tests are shown here. Using the 0.05 level of significance, test the hypothesis that learning actually did take place. Assume that learning is shown by an increase in score.

SUBJECT	BEFORE	AFTER
1	27	29
2	21	32
3	34	29
4	24	27
5	30	31
6	27	26
7	33	35
8	31	30
9	22	29
10	27	28
11	33	36
12	17	15
13	25	28
14	26	26
15	23	26

8.29 To determine which of two brands of golf balls is better for his long game, a professional golfer drives two samples of 50 balls each from the tee, using the same club for each shot and alternating brands at random. For brand 1, the mean yardage is 231 yards with a standard deviation of 24 yards. For brand 2 the mean yardage is 224 with a standard deviation of 20. At what level of significance can he conclude that there is a difference in the yardage of the two brands?

8.30 An amateur chemist discovers some residue left in a test tube and hypothesizes that it is the same as the residue left in a different tube the preceding week. He subjects the residue to four determinations of melting point and obtains 1543°C, 1540°C, 1542°C, and 1543°C. He tests the residue from the previous week three times for its melting point, getting 1545°C, 1544°C, and 1544°C. Do these results tend to confirm or deny, at the 0.10 level of significance, that the residues are the same substance?

8.31 To test the effect of two different fertilizers on corn yield, six 1-acre plots of corn are fertilized with fertilizer 1 and six 1-acre plots are fertilized with fertilizer 2. The plots fertilized with fertilizer 1 show a mean yield of 1.22 tons per plot with a standard deviation of 0.22 ton, while the plots fertilized with fertilizer 2 show a mean yield of 1.51 tons per plot with a standard deviation of 0.11 ton. Do these results support, at the 0.01 level of significance, the manufacturer's claim that fertilizer 2 will produce a yield at least 0.1 ton per acre more than fertilizer l? At what level of significance can you conclude that the mean yield with fertilizer 2 is greater than with fertilizer 1?

8.32 Two varieties of corn are planted in paired plots (one plot split in half) to see if there is a difference in the yield. The same fertilizer is used throughout. Using the 0.05 level of significance, determine whether there is a significant difference between the yields. Yields are given in tons per plot. Paired plots are of the same size, although different pairs are not necessarily of a similar size.

	PLOT				
	1	2	3	4	5
Variety A	12.3	9.1	10.2	6.4	14.4
Variety B	10.8	7.6	10.3	6.0	13.5

8.33 A social worker wishes to test whether or not the average amount spent on food for a family of four is the same in South Fidelia as it is in Punta Grassa. She therefore asks a random sample of 40 families of four in each town to keep track of their expenditures for food for one week. The mean amounts spent for food per family is \$126.50 with a standard deviation of \$14.40 in South Fidelia and \$119.75 with a standard deviation of \$13.66 in Punta Grassa. At what level of significance can she conclude that there is a difference in the mean amount spent on food for a family of four in the two towns?

8.34 A hospital is considering two suppliers for ampoules of a wonder drug. Both suppliers deliver ampoules that are supposed to contain 500,000 units of the drug. An analysis is made of a sample of six ampoules from each company. Brand A has a mean of 510,000 units with a standard deviation of 20,000, while brand B has a mean of 490,000 units with a standard deviation of 15,000.

Can you conclude that there is probably no difference between the brands? Use a 0.05 level of significance. At what level of significance can you conclude that there is a difference?

8.35 Keller and others (1984) conducted a study in which they found that people in the 38–59 age group (middle age) are less anxious about death than older and younger people. A researcher attempting to replicate this study obtains a random sample of people and administers the same test of anxiety. The results are as follows:

GROUP	NUMBER	MEAN SCORE ON TEST	STANDARD DEVIATION
Middle-aged	21	34.7	13.7
Other ages	23	37.1	15.2

Using the 0.05 level of significance, determine whether there is a difference in the mean scores of the two age groups.

8.36 To test the efficiency of a procedure designed to increase the daily output of a machine, five machines are tested with and without the procedure. The following results are obtained:

	MACHINE				
	A	B	C	D	E
With	15.6	18.2	14.3	14.9	16.7
Without	14.4	16.7	13.1	14.4	14.7

Test the hypothesis that the procedure is effective. Use $\alpha = 0.05$.

8.37 Repeat Problem 8.36 but suppose that ten different machines are used, five with the procedure and five without, with the same results as above. Compare the results of the analyses.

8.38 A survey is made to test the hypothesis that college graduates read at least four books per year more than nongraduates. A random sample of 66 college graduates reports reading an average of 9.3 books per year, with a standard deviation of 4.6 books, while 78 nongraduates report an average of 4.9 books per year with a standard deviation of 3.1 books. Test the hypothesis at the 0.05 level of significance. Discuss the probable reason for the large standard deviations, and give the probable shape of the distributions.

8.4 Differences Between Population Proportions

Methods similar to those for comparing two population means, discussed in the preceding two sections, may be used to compare two population proportions as well. Candidates for political office may wish to compare their popularity in one district to that in another, expressed as a proportion of the voters preferring them to their opponent. A factory may wish to compare the proportion of machined items that are acceptable from two assembly lines. Biologists may wish to compare the proportions in two populations bearing the same genetic characteristic. In this section we will discuss the usual method of obtaining a confidence interval for a difference between population proportions and also the method for testing hypotheses about the difference between two population proportions. Recall that a formal hypothesis test allows us to make a decision concerning a certain value of a parameter whereas a confidence interval allows us to make many simultaneous hypothesis tests. A confidence interval also allows us to set limits on the value of β if we cannot reject the null hypothesis.

We learned in Section 6.3 that if we take samples of size n from a binomial population with proportion π, the distribution of the sample proportion, p, is $N(\pi, \sigma_p^2)$, where $\sigma_p^2 = \pi(1 - \pi)/n$, provided that $n\pi$ and $n(1 - \pi)$ are both at least 5. We should use the finite population correction factor when it is appropriate.

If we take samples of size n_1 from a binomial population with proportion π_1 and also take samples of size n_2 from a binomial population with proportion π_2, the differences between the sample proportions, $p_1 - p_2$, will have a normal distribution with mean $\pi_1 - \pi_2$ and variance $\sigma_{p_1-p_2}^2$, where $\sigma_{p_1-p_2}^2 = \sigma_{p_1}^2 + \sigma_{p_2}^2$. The quantity $\sigma_{p_1-p_2}$ is called the **standard error of the difference for proportions.** If we wish to estimate $\pi_1 - \pi_2$, then π_1 and π_2 will not be known, so we must use the sample standard error obtained by using p_1 and p_2 in place of π_1 and π_2. Thus the sample value of the standard error of the difference for proportions is $s_{p_1-p_2}$:

$$s_{p_1-p_2} = \sqrt{\frac{p_1(1 - p_1)}{n_1} + \frac{p_2(1 - p_2)}{n_2}}$$

8.4.1 Confidence Intervals

Since sample proportions are normally distributed for sufficiently large samples, so that $p_1 - p_2$ is $N(\pi_1 - \pi_2, \sigma_{p_1-p_2}^2)$, we may obtain confidence intervals for $\pi_1 - \pi_2$ from sample proportions. Since we must use $s_{p_1-p_2}$ in place of $\sigma_{p_1-p_2}$, however, all such intervals are approximate. The procedure is summarized on page 462. We should use the finite population correction factor in the calculation of $s_{p_1}^2$ and $s_{p_2}^2$ if needed.

Example 8.8

Of 548 persons randomly selected and questioned, 317 identify themselves as political moderates or conservatives; of these, 204 think the administration is doing a good job. The remaining 231 identify themselves as liberals; of these, 109 think the administration is doing a good job. Determine a 99% confidence interval for the difference in proportion of conservatives/moderates and liberals who think the administration is doing a good job.

Confidence Interval for $\pi_1 - \pi_2$

Suppose we obtain two independent random samples from dichotomous populations—a sample of n_1 observations from a population with proportion π_1 and also a sample of n_2 observations from a population with proportion π_2—and that all of the following are true:

1. The sample proportions are p_1 and p_2.
2. n_1p_1 and n_2p_2 are both at least 5.
3. $n_1(1 - p_1)$ and $n_2(1 - p_2)$ are both at least 5.

Then an approximate $100(1 - \alpha)\%$ confidence interval for the population proportion is the interval

$$p_1 - p_2 - z_{\alpha/2}s_{p_1-p_2} \quad \text{to} \quad p_1 - p_2 + z_{\alpha/2}s_{p_1-p_2}$$

where

$$s_{p_1-p_2} = \sqrt{\frac{p_1(1-p_1)}{n_1} + \frac{p_2(1-p_2)}{n_2}}$$

Solution

If sample 1 is the 317-person group and sample 2 is the other group, then $p_1 = 204/317 \doteq 0.64$ and $p_2 = 109/231 \doteq 0.47$, so that $p_1 - p_2 \doteq 0.17$. The standard error is

$$s_{p_1-p_2} \doteq \sqrt{\frac{(0.64)(0.36)}{317} + \frac{(0.47)(0.53)}{231}} \doteq 0.0425$$

(*Note:* When making the calculations for $s_{p_1-p_2}$, we should use $p_1 = 204/317$ and $p_2 = 109/231$ in the formula; any rounding should be done after calculations are completed.) For a 99% confidence interval we use $z_{.005} = 2.58$; since $2.58(0.0425) \doteq 0.11$, our confidence limits are $0.17 - 0.11 = 0.06$ and $0.17 + 0.11 = 0.28$. We are thus 99% confident that the difference in proportions of moderates/conservatives and liberals who think the administration is doing a good job is between 0.06 and 0.28, with a higher proportion of moderates/conservatives than liberals.

PROFICIENCY CHECKS

17. If $p_1 = 0.54$ and $p_2 = 0.42$, determine a 95% confidence interval for $\pi_1 - \pi_2$ if n_1 and n_2 are:
    a. 50 b. 100 c. 200 d. 400

(*continued*)

PROFICIENCY CHECKS *(continued)*

18. If $p_1 = 0.117$, $p_2 = 0.157$, $n_1 = 60$, and $n_2 = 70$, determine confidence intervals for $\pi_1 - \pi_2$ with a confidence level of:
a. 90% b. 95% c. 98% d. 99%

19. A toothpaste manufacturer would like to survey people in two different localities to determine the difference in the proportion of people who use her company's product. Suppose 800 people are polled in locality A and 154 say they use the toothpaste, while 108 of 780 people in locality B say they use the toothpaste. Determine a 95% confidence interval for the difference in the population proportions.

8.4.2 Hypothesis Tests

Suppose that a sample consisting of n_1 observations has a proportion p_1, while a sample consisting of n_2 observations has a proportion p_2. A researcher could test the hypothesis that if π_1 is the proportion of the population from which sample 1 was taken and π_2 is the proportion of the population from which sample 2 was taken, then $\pi_1 = \pi_2$ or else $\pi_1 - \pi_2 = \delta_0$, where δ_0 is some specified difference. The first hypothesis is the special case of the second when $\delta_0 = 0$, but the two hypotheses are tested using slightly different techniques.

We know that, for sufficiently large samples, the sampling distribution of $p_1 - p_2$ is $N(\pi_1 - \pi_2, \sigma^2_{p_1-p_2})$. In other words, the statistic $z = [(p_1 - p_2) - (\pi_1 - \pi_2)]/\sigma_{p_1-p_2}$ is $N(0, 1)$. Since we do not know either π_1 or π_2, we must use $s_{p_1-p_2}$ in place of $\sigma_{p_1-p_2}$. If we wish to test the hypothesis that $\pi_1 - \pi_2 = \delta_0$, our test statistic will be $z = [(p_1 - p_2) - \delta_0]/s_{p_1-p_2}$ and the usual hypothesis-testing procedures apply.

Suppose we wish to test the hypothesis that $\pi_1 = \pi_2$. If this is true, then we can let $\pi_1 = \pi_2 = \pi$ and we have

$$\sigma_{p_1-p_2} = \sqrt{\frac{\pi(1-\pi)}{n_1} + \frac{\pi(1-\pi)}{n_2}}$$
$$= \sqrt{\pi(1-\pi)\left(\frac{1}{n_1} + \frac{1}{n_2}\right)}$$

Now we still do not know π, but if $\pi_1 = \pi_2$ then p_1 and p_2 are both estimators of π, so we can obtain an even better estimate by simply combining the samples and denoting the resulting **pooled proportion** by $\hat{\pi}$ ("pi-hat"). If the sample sizes are n_1 and n_2, $p_1 = x_1/n_1$, and $p_2 = x_2/n_2$, then

Large-Sample Hypothesis Tests for Differences Between Population Proportions

Suppose we obtain two independent random samples from dichotomous populations—a sample of n_1 observations from a population with proportion π_1 and also a sample of n_2 observations from a population with proportion π_2—and that all of the following are true:

1. The sample proportions are p_1 and p_2.
2. n_1p_1 and n_2p_2 are both at least 5.
3. $n_1(1 - p_1)$ and $n_2(1 - p_2)$ are both at least 5.

Then we may test either of the following null hypotheses:

$$H_0: \quad \pi_1 - \pi_2 = \delta_0,$$

or

$$H_0: \quad \pi_1 = \pi_2$$

The sample value of the test statistic used to test either hypothesis in each situation is z^*, where

$$z^* = \frac{p_1 - p_2 - \delta_0}{s_{p_1 - p_2}}$$

or

$$z^* = \frac{p_1 - p_2}{s_{\hat{\pi}}}$$

The decision rule is dictated by the alternate hypothesis and the level of significance α chosen for the test. The three possibilities are as follows:

	Case I	Case II	Case III
	Null and Alternate Hypotheses		
1.	H_0: $\pi_1 - \pi_2 = \delta_0$ H_1: $\pi_1 - \pi_2 \neq \delta_0$	H_0: $\pi_1 - \pi_2 = \delta_0$ H_1: $\pi_1 - \pi_2 > \delta_0$	H_0: $\pi_1 - \pi_2 = \delta_0$ H_1: $\pi_1 - \pi_2 < \delta_0$
		or	
2.	H_0: $\pi_1 = \pi_2$ H_1: $\pi_1 \neq \pi_2$	H_0: $\pi_1 = \pi_2$ H_1: $\pi_1 > \pi_2$	H_0: $\pi_1 = \pi_2$ H_1: $\pi_1 < \pi_2$
	Decision Rule		
	Reject H_0 if $\lvert z^* \rvert > z_{\alpha/2}$	Reject H_0 if $z^* > z_\alpha$	Reject H_0 if $z^* < -z_\alpha$

$$\hat{\pi} = \frac{x_1 + x_2}{n_1 + n_2} = \frac{n_1 p_1 + n_2 p_2}{n_1 + n_2}$$

and the sample standard error of the pooled estimate will be $s_{\hat{\pi}}$, where $s_{\hat{\pi}} = \sqrt{\hat{\pi}(1 - \hat{\pi})(1/n_1 + 1/n_2)}$.

Example 8.9

On a certain campus a poll shows that 45 of 60 underclassmen are in favor of a proposed judicial reform, while 48 of 80 upperclassmen are in favor of it. Do these results support the contention (at $\alpha = 0.05$) that a greater proportion of underclassmen than upperclassmen supports the reform?

Solution

Since the null hypothesis is $\pi_1 = \pi_2$, we use a pooled proportion in determining z^*. We have $\hat{\pi} = (45 + 48)/(60 + 80) \doteq 0.664$ and $s_{\hat{\pi}} = \sqrt{(0.664)(0.336)(1/60 + 1/80)} \doteq 0.0807$. The research hypothesis is that the proportion for underclassmen (π_1) is greater than the proportion for upperclassmen (π_2), so we have:

1. H_0: $\pi_1 = \pi_2$
   H_1: $\pi_1 > \pi_2$
2. Decision rule: Reject H_0 if $z^* > 1.645$.
3. Results: $p_1 = 45/60 = 0.75$, $p_2 = 48/80 = 0.60$, and $s_{\hat{\pi}} \doteq 0.0807$, so $z^* = 0.15/0.0807 \doteq 1.86$.
4. Decision: Reject H_0.
5. Conclusion: A higher proportion of underclassmen than upperclassmen supports the reform.

Example 8.10

A political candidate takes a poll and finds that of 400 voters, only 118 support his candidacy. Alarmed, he immediately hires an advertising agency to take his message to the voters. Six weeks into the campaign he has lunch with a statistics professor who asks him if the advertising agency has been successful. The candidate cites the latest poll, which shows that, of 400 voters, 176 support his candidacy. He asks the statistician if the advertising agency's claim that support for his candidacy has been increased by at least 10 percentage points is borne out by the poll. He is willing to take a 5% chance of being wrong. Is the agency's claim "proved?" At what level of significance has the claim been "proved?"

Solution

If π_1 is the proportion of voters who supported his candidacy before the campaign and π_2 is the proportion now, the agency claims that $\pi_2 - \pi_1 \geqslant 0.10$. This claim can be tested as follows:

1. H_0: $\pi_2 - \pi_1 = 0.10$
   H_1: $\pi_2 - \pi_1 > 0.10$
2. Decision rule: Reject H_0 if $z^* > 1.645$.

3. Results: $p_1 = 118/400 = 0.295$, $p_2 = 176/400 = 0.44$, $s_{p_1-p_2} = \sqrt{(0.295)(0.705)/400 + (0.44)(0.56)/400} \doteq 0.0337$, so $z^* = [(0.44 - 0.295) - 0.10]/0.0337 \doteq 1.34$.
4. Decision: Fail to reject H_0.
5. Conclusion: The agency's claim has not been "proved" at the 0.05 level of significance. The observed level of significance, however, is about 0.09, so the claim may have some merit.

PROFICIENCY CHECKS

20. If $p_1 = 0.54$ and $p_2 = 0.42$, test the hypothesis that $\pi_1 - \pi_2 > 0.05$ at $\alpha = 0.05$ if n_1 and n_2 are:
a. 50 b. 100 c. 200 d. 400
21. If $p_1 = 0.117$, $p_2 = 0.271$, $n_1 = 60$, and $n_2 = 70$, test the hypothesis that $\pi_1 = \pi_2$ at a significance level of:
a. 0.10 b. 0.05 c. 0.02 d. 0.01
22. What are the observed significance levels in Proficiency Checks 20 and 21?
23. A union official surveys people in two different localities to determine whether there is a difference in the proportion of people who own Japanese cars. Suppose 800 people are polled in locality A and 154 say they own a Japanese car, while 108 of 780 people in locality B say they own a Japanese car. Can you conclude, at the 0.01 level of significance, that there is a difference?

Problems

Practice

8.39 Two samples of size n are used to obtain a 95% confidence interval for the difference in population proportions. Determine the width of the confidence interval if the sample proportions are as given:
a. $n = 100$, $p_1 = 0.44$, $p_2 = 0.52$
b. $n = 64$, $p_1 = 0.12$, $p_2 = 0.04$
c. $n = 200$, $p_1 = 0.84$, $p_2 = 0.59$
d. $n = 400$, $p_1 = 0.231$, $p_2 = 0.339$
e. $n = 36$, $p_1 = 0.543$, $p_2 = 0.521$

8.40 Using the data of Problem 8.39, construct a 90% confidence interval for $\pi_1 - \pi_2$ in each case.

8.41 Use the data of Problem 8.39 to test the hypothesis given. Letters are matched with those in Problem 8.39. Use $\alpha = 0.05$.
a. $\pi_1 - \pi_2 = 0.05$ **b.** $\pi_1 > \pi_2$ **c.** $\pi_1 - \pi_2 > 0.15$
d. $\pi_1 - \pi_2 < -0.10$ **e.** $\pi_1 = \pi_2$

8.42 If two sample proportions are 0.40 and 0.30, determine 95% confidence intervals for the difference in the population proportions if the sample sizes are:
a. $n_1 = 40, n_2 = 50$ **b.** $n_1 = 150, n_2 = 100$
c. $n_1 = 180, n_2 = 200$ **d.** $n_1 = 500, n_2 = 400$

8.43 Two samples of size 200 are obtained. Determine 90% confidence intervals for the difference in the population proportions if the sample proportions are:
a. 0.15 and 0.23 **b.** 0.71 and 0.79 **c.** 0.15 and 0.07
d. 0.36 and 0.28

8.44 Use the data of Problem 8.43 to test the hypothesis, in each case, that $\pi_1 = \pi_2$. Use the 0.05 level of significance.

Applications

8.45 A random sample of size 200 has 60 who favor increased government subsidies at the expense of certain social programs. A second sample of size 140 has 70 who favor the subsidies. Determine confidence intervals for the difference in the population proportions and interpret the intervals if the confidence level is:
a. 90% **b.** 95% **c.** 98% **d.** 99%

8.46 In Problem 8.45, the population from which the second sample was taken has always averaged about 10 percentage points more than the first in favor of government subsidies. At what significance level can you conclude that this difference has changed?

8.47 Two companies submit bids and a sample of ball bearings to gain a contract. Firm A has 82 acceptable ball bearings out of 90, while firm B has 96 acceptable out of 110. Firm B's price is slightly lower than firm A's, so unless it can be shown than firm A has a higher percentage of acceptable bearings (with $\alpha = 0.01$), firm B will get the contract. Who will get the contract?

8.48 A biologist wishes to determine whether two insect populations differ. The prime morphological characteristic that differentiates the two populations appears to be the incidence of white eyes. The biologist observes that 42 of 100 from one population and 23 of 100 from the other population have white eyes. Can you conclude that the two populations are different species? Use the 0.01 level of significance.

8.49 Use the results of Problem 8.48 to construct a 99% confidence interval for the difference between the population proportions.

8.50 A class is randomly divided into two sections of 18 each, and one section is given instruction in assembling a puzzle. Each subject is then tested to see if he or she can assemble the puzzle within 1 minute. The result in the test group (given instructions) shows that 12 can do the task, while the result in the control group (given no instructions) shows that only 6 are successful. Theory dictates that the instruction should increase the success rate by at least 15 percentage points. Using the 0.05 level of significance, do these results bear out the theoretical contention?

8.51 Two cities of approximately the same size and type, South Fidelia and Punta Grassa, are interested in the proportion of city employees who actually live within the city limits. A survey of 132 randomly selected employees of South Fidelia discloses that 91 of them live within the city limits, while only 65 of 104 of a random sample of Punta Grassa employees live within the city limits. Use this set of data to determine a 90% confidence interval for the difference in the proportion of city employees who live within the city limits for the two cities.

8.52 *Professional Builder Magazine* reported the results of a 1984 survey of homebuyers. The survey asked the question "What would you pay extra for in a home?" The answers included fireplace (62%), better carpeting (53%), paved driveway (52%), self-cleaning ovens (51%), and extra cabinets (45%). Suppose that prospective homebuyers are asked the question "For the same amount of money, would you prefer a fireplace or better carpeting?" Of 122 men respondents, 81 say they would prefer a fireplace, while 59 of 106 women respondents say they would prefer a fireplace. Assume that the samples are random, independent samples of potential homebuyers in the area.

a. Test the hypothesis that there is no difference in the proportion of men and women potential homebuyers in the area who would prefer a fireplace. Use $\alpha = 0.05$.

b. Construct a 95% confidence interval for the difference in the true proportion of men and women potential homebuyers in the area who would prefer a fireplace. Does this result confirm the result of part (*a*)?

8.53 An apartment owner has two buildings that are identical except that one is painted yellow and the other is painted brown. He wonders whether there is any difference in occupancy rates in the two buildings. (Rents are identical for apartments of the same floor plan and location within the building.) He looks over the records for the past 4 years since the buildings were painted. If any apartment has been vacant a total of more than 6 months, he rates it "poor occupancy rate." Each building contains 26 apartments. In the yellow building, seven apartments are rated "poor," while 11 of the apartments in the brown building are so rated. Can you conclude, at the 0.05 level of significance, that there is a difference in the "poor occupancy rate" of the two buildings? What is the implication?

8.54 Two sites are being considered for a shopping center. The major store in the shopping center will be J. Sawbuck and Company. Sawbuck hires a firm to survey the two areas to see if they would patronize the new store. A sample of 350 residents of East Wadlington Heights shows that 254 say they would patronize the new Sawbuck store if it were built in their area, while 353 of 400 residents of Clearwater Park say they would patronize the new Sawbuck store if it were built in their area. Determine a 95% confidence interval for the difference in the population proportions.

8.55 Refer to Problem 8.54. Sawbuck looks at the survey results and remarks that East Wadlington Heights is somewhat more affluent than Clearwater Park. He thinks it would be better to build the store in East Wadlington Heights unless the proportion of potential customers is at least 8 percentage points higher in Clearwater Park. Using a 0.05 significance level and the results of the survey in Problem 8.54, where should the store be built?

8.56 A survey of 35 cities by Coldwell Banker (January 1986) showed a difference in office vacancy rate by location. The vacancy rate for downtown offices was 16.5% while for suburban offices it was 22%. Assume that the 35 cities represent a random sample of cities in the United States. If there were 5145 downtown offices and 4583 suburban offices in the survey, obtain a 95% confidence interval for the difference in the true proportions nationwide.

8.57 Two moving companies are applying for the job of moving a major concern across the country. The office manager views completion of the move within the promised times as the major criterion for the choice. She feels that there is no difference between the companies since company A has completed 344 out of 388 moves on time during the last year, while company B has completed 217 of 232 on time. Test her belief at the 0.05 level of significance.

8.58 Two machines are being examined for possible replacement. Machine A produced 27 defectives on the last run of 200, while machine B produced 18 defectives on the last run of 200. Using a 98% confidence level, estimate the difference in proportion of defectives for the two machines.

8.5 Estimating Needed Sample Size

In many cases, particularly when small samples are used, hypothesis tests yield inconclusive results or confidence intervals are too wide to be of practical use. In such cases we can determine the sample size required to obtain a confidence interval no wider than a stated maximum. In this section we study the formulas used to obtain the necessary sample sizes.

8.5.1 Sample Size Needed to Estimate Differences Between Population Means

In Example 8.1 we need a confidence interval no more than 5 units wide. Thus $1.96s_{\bar{x}_1-\bar{x}_2}$ should be no more than 2.5 units. If we let both n_1 and n_2 equal n (assume that the samples will be the same size), we can set $1.96s_{\bar{x}_1-\bar{x}_2} = 2.5$ and solve for n. In general, if we assume that $n_1 = n_2 = n$ and $\sigma_1 = \sigma_2 = \sigma$, we can choose a maximum desired error (E) or a maximum confidence interval width ($2E$) and we have

$$E = z_{\alpha/2}\sqrt{\frac{\sigma_1^2}{n_1} + \frac{\sigma_2^2}{n_2}}$$

$$= z_{\alpha/2}\sqrt{\frac{2\sigma^2}{n}}$$

Solving this expression for n, we have the following rule.

Determining Sample Size Needed to Estimate the Difference Between Population Means

To estimate the difference between two population means to within a maximum error E with a confidence level of $1 - \alpha$ or to obtain a $100(1 - \alpha)\%$ confidence interval no more than $2E$ in width, the needed sample sizes are each estimated to be n, where

$$n = 2\left(\frac{z_{\alpha/2} \cdot \sigma}{E}\right)^2$$

and σ is the common population standard deviation. If σ is not known, we can estimate it from a previous result or take it to be approximately one-fourth the range.

If we are estimating σ from a previous result, such as Example 8.1, it is best to use the larger of the two sample standard deviations. In this way we guard against choosing a sample too small to obtain the desired efficiency. If we choose a smaller value for σ and the actual sample values of s_1 or s_2 are larger, the width of the obtained confidence interval will be greater than desired. Using the results of Example 8.1, suppose we want a 95% confidence interval no more than 5.0 razors in width. The needed sample sizes are estimated to be $n = 2[(1.96)(9.1)/2.5]^2 \doteq 102$.

8.5.2 Sample Size Needed to Estimate Differences Between Population Proportions

In Example 8.7, perhaps we desire a confidence interval no more than 0.10 wide. Thus $2.58\sigma_{p_1 - p_2}$ should be no more than 0.05. If we let both n_1 and n_2 equal n (assume that the samples will be the same size), we can set $2.58\sigma_{p_1 - p_2} = 0.05$ and solve for n. In general, if we assume that $n_1 = n_2 = n$, we can choose a maximum desired error (E) or a maximum confidence interval width ($2E$) and we have

$$E = z_{\alpha/2}\sqrt{\frac{\pi_1(1 - \pi_1)}{n_1} + \frac{\pi_2(1 - \pi_2)}{n_2}}$$

$$= z_{\alpha/2}\sqrt{\frac{\pi_1(1 - \pi_1) + \pi_2(1 - \pi_2)}{n}}$$

Solving this expression for n, we have the following rule.

Determining Sample Size Needed to Estimate the Difference Between Population Proportions

To estimate the difference between two population proportions to within E with a confidence level of $1 - \alpha$ or to obtain a $100(1 - \alpha)\%$ confidence interval no more than $2E$ in width, the needed sample sizes are each estimated to be n, where

$$n = \frac{(z\alpha_{/2})^2(\pi_1(1 - \pi_1) + \pi_2(1 - \pi_2)]}{E^2}$$

Since π_1 and π_2 are not known, we can estimate them by using a prior sample result. If no estimates are available, use 0.5 as the estimate for both.

Example 8.11

Determine the sample sizes needed in Example 8.7 to obtain a 99% confidence interval no more than 0.10 in width:

a. using the sample results as estimates for π_1 and π_2
b. using no prior estimates
c. if the samples should be in the approximate ratio 3 to 2 (as in Example 8.7)

Solution

a. $$n = \frac{(2.58)^2[0.64(0.36) + 0.47(0.53)]}{0.05^2}$$

$$\doteq 1277$$

b. $$n = \frac{(2.58)^2[0.5(0.5) + 0.5(0.5)]}{0.05^2}$$

$$\doteq 1331$$

c. In this case we should use

$$E = z_{\alpha/2}\sqrt{\frac{\pi_1(1 - \pi_1)}{n_1} + \frac{\pi_2(1 - \pi_2)}{n_2}}$$

with $n_1 = 3x$ and $n_2 = 2x$. Solving

$$0.05 = 2.58\sqrt{\frac{0.64(0.36)}{3x} + \frac{0.47(0.53)}{2x}}$$

for x, we obtain $x \doteq 536.11$, so $n_1 \doteq 1072$ and $n_2 \doteq 1608$. If we let $\pi_1 = \pi_2 = 0.5$, we obtain $n_1 \doteq 1664$, and $n_2 \doteq 1109$.

PROFICIENCY CHECKS

24. Two samples are to be drawn from two populations to estimate $\mu_1 - \mu_2$. A single sample drawn earlier had a standard deviation of 0.30. What sample sizes will you need to estimate the difference between the population means correct to within 0.05 with a confidence level of 99%?

25. Two samples are to be drawn from two populations to estimate $\pi_1 - \pi_2$. A single sample drawn earlier had a proportion of 0.30. What sample sizes will you need to estimate the difference between the population proportions correct to within 0.05 with a confidence level of 95%?

26. Refer to Proficiency Check 19. Determine the sample sizes needed to ensure 95% confidence that the error in the estimate will not exceed 0.05.

Problems

Practice

8.59 What sample sizes should you take to estimate the difference between population proportions within 0.05 with a confidence level of 95% if:
- **a.** a pilot study shows sample proportions of 0.2 and 0.3?
- **b.** no pilot study results are available?

8.60 What sample sizes should you take to estimate the difference between population proportions correctly to within 0.10 with a confidence level of:
a. 90%? **b.** 95%? **c.** 98%? **d.** 99%?

Applications

8.61 Refer to Problem 8.32. What sample size do you need to be certain of rejecting the null hypothesis at the 0.05 level of significance if the actual difference between yields is at least 1.0 ton per acre?

8.6 Variances of Two Populations

In Sections 6.6 and 7.4 we discussed methods for investigating the value of a population variance. In many practical situations we must determine the relationship between two population variances. A bottling company may wish to know whether the variability of fill levels is the same in each assembly line. A potential employer may wish to examine the variability in job performance of candidates for a position. A researcher may wish to determine whether there is a difference in the variability of certain characteristics in experimental and control groups. Before we can use the t distribution in tests or confidence intervals about means in independent samples, the important assumption of equality of variance in the two populations must be satisfied. In these and other cases we need to be able to test a hypothesis that two population variances are equal. In this section we will see how this may be done.

8.6.1 The *F* Distribution

To test the hypothesis that two population variances are equal, we need the F distribution, named after Sir Ronald Fisher. The **F distribution** is the ratio of two chi-square distributions, $F = (\chi_1^2/\nu_1)/(\chi_2^2/\nu_2)$, and is defined entirely by the two parameters ν_1 and ν_2, the degrees of freedom associated with the numerator and denominator, respectively. The graph of an F distribution resembles that of the chi-square distribution in that it is skewed to the right. Critical values of an F distribution are indicated by $F_\alpha\{\nu_1, \nu_2\}$. For example, $F_{.025}$ is the value of F that has 0.025 of the area under the curve to its right, while $F_{.975}$ is the value of F that has 0.975 of the area under the curve to its right and consequently 0.025 of the area under the curve to its left. These values are shown in Figure 8.1.

The mean of an F distribution is $\nu_2/(\nu_2 - 2)$, provided that $\nu_2 > 2$. The variance is

$$\frac{2\nu_2^2(\nu_1 + \nu_2 - 2)}{\nu_1(\nu_2 - 2)^2(\nu_2 - 4)}$$

provided that $\nu_2 > 4$.

8.6.2 Testing Equality of Population Variances

In Section 6.6 we saw that the variance of a population, σ^2, and the variance of a sample of n observations drawn from it, s^2, can be related by means of the chi-square statistic and that $(n - 1)s^2/\sigma^2 = \chi^2$, provided that the population is normal. The chi-square distribution is defined by its degrees of freedom, in this case $n - 1$.

Now if we obtain two independent random samples—a sample of size n_1 from a normal population with variance σ_1^2 and a sample of size n_2 from a normal population with variance σ_2^2—and the sample variances are s_1^2 and s_2^2, respectively, then the ratio of the two sample variances is

$$\frac{s_1^2}{s_2^2} = \frac{\sigma_1^2[\chi_1^2/(n_1 - 1)]}{\sigma_2^2[\chi_2^2/(n_2 - 1)]}$$

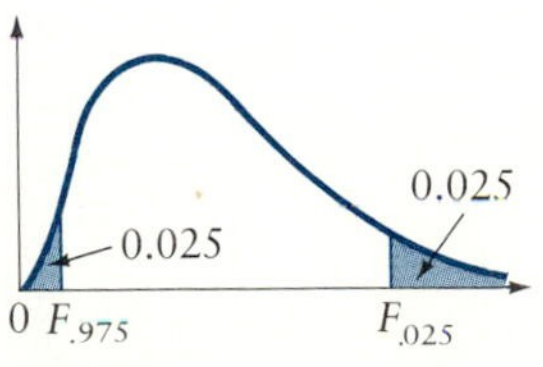

Figure 8.1
Graph of an F distribution with critical values $F_{.975}$ and $F_{.025}$.

If $\sigma_1^2 = \sigma_2^2$, the ratio is an F distribution with $(n_1 - 1)$ and $(n_2 - 1)$ degrees of freedom; that is, s_1^2/s_2^2 is $F(n_1 - 1, n_2 - 1)$. A hypothesis that $\sigma_1^2 = \sigma_2^2$, then, can be tested with an F test statistic. Critical values for the right tail of the F distribution are given in Tables 7 through 10 for $\alpha = 0.05, 0.025, 0.01$, and 0.005. To determine $F_{.05}\{4, 9\}$, for example, we turn to Table 7. The degrees of freedom for the numerator are given in the column headings, and the degrees of freedom for the denominator are given in the row headings on the left side of each row. We find the column headed 4 and the row headed 9 and see that the corresponding critical value of F is 3.63; that is, $F_{.05}\{4, 9\} = 3.63$. A portion of Table 7 is reproduced in Table 8.3. Note that $F_{.05}\{9, 4\} = 6.00$, so it is important not to confuse the order of the degrees of freedom.

There is an interesting relationship between related F distributions. Suppose that we have an F distribution with ν_1 and ν_2 degrees of freedom. Let F^* be any value on the horizontal axis and let α be the area under the graph of $F(\nu_1, \nu_2)$ to the right of F^*. The graph of $F(\nu_2, \nu_1)$ is related to the graph of $F(\nu_1, \nu_2)$ so that if $1/F^*$ is the reciprocal of F^*, the area under the graph of $F(\nu_2, \nu_1)$ to the left of $1/F^*$ is also equal to α. The implication of this is that $F_{1-\alpha}\{\nu_2, \nu_1\} = 1/F_\alpha\{\nu_1, \nu_2\}$. This means that although F distributions are not symmetric, only the upper-tailed critical values are necessary. Thus since $F_{.05}\{4, 9\} = 3.63$, then $F_{.95}\{9, 4\} = 1/3.63$ or about 0.275.

When we test the null hypothesis $\sigma_1^2 = \sigma_2^2$, the possible alternate hypotheses are that $\sigma_1^2 \neq \sigma_2^2$, $\sigma_1^2 > \sigma_2^2$, or $\sigma_1^2 < \sigma_2^2$. If the alternative is one-sided, we can minimize computations if we choose the subscripts so that the alternate hypothesis is $\sigma_1^2 > \sigma_2^2$. Then the rejection region will always consist of values of the F test statistic greater than F_α. If the alternate hypothesis is that $\sigma_1^2 \neq \sigma_2^2$, the rejection region will contain all values of F greater than $F_{\alpha/2}\{n_1 - 1, n_2 - 1\}$ or smaller than $F_{1-\alpha/2}\{n_1 - 1, n_2 - 1\}$, as in Figure 8.1. Calculations can again be minimized if we choose the subscripts so that the

Table 8.3 CRITICAL VALUES OF $F_{.05}$

DEGREES OF FREEDOM FOR DENOMINATOR	DEGREES OF FREEDOM FOR NUMERATOR								
	1	2	3	4	5	6	7	8	9
1	161	200	216	225	230	234	237	239	241
2	18.5	19.0	19.2	19.2	19.3	19.3	19.4	19.4	19.4
3	10.1	9.55	9.28	9.12	9.01	8.94	8.89	8.85	8.81
4	7.71	6.94	6.59	6.39	6.26	6.16	6.09	6.04	6.00
5	6.61	5.79	5.41	5.19	5.05	4.95	4.88	4.82	4.77
6	5.99	5.14	4.76	4.53	4.39	4.28	4.21	4.15	4.10
7	5.59	4.74	4.35	4.12	3.97	3.87	3.79	3.73	3.39
8	5.32	4.46	4.07	3.84	3.69	3.58	3.50	3.44	3.39
9	5.12	4.26	3.86	3.63	3.48	3.37	3.29	3.23	3.18
10	4.96	4.10	3.71	3.48	3.33	3.22	3.14	3.07	3.02

F test statistic is greater than 1, that is, $s_1^2 > s_2^2$, and we need only to determine the right-tailed rejection region. We thus have the following procedures for testing a hypothesis that $\sigma_1^2 = \sigma_2^2$.

Hypothesis Test for Equal Population Variances

Suppose we obtain two independent random samples—a sample of n_1 observations of a random variable with variance σ_1^2 and a sample of n_2 observations of a random variable with variance σ_2^2—and both variables are normally distributed. Then we may test the hypothesis $H_0{:}\sigma_1^2 = \sigma_2^2$.

If s_1^2 and s_2^2 are the sample variances, the sample value of the test statistic used to test the hypothesis is F^*, where $F^* = s_1^2/s_2^2$, provided that we choose the subscripts as discussed above.

The decision rule is dictated by the alternate hypothesis and the level of significance α chosen for the test. There are two possibilities:

Case I	**Case II**
Null and Alternate Hypotheses	
H_0: $\sigma_1^2 = \sigma_2^2$	H_0: $\sigma_1^2 = \sigma_2^2$
H_1: $\sigma_1^2 \neq \sigma_2^2$	H_1: $\sigma_1^2 > \sigma_2^2$
Decision Rule	
Reject H_0 if $F^* > F_{\alpha/2}\{n_1-1, n_2-1\}$	Reject H_0 if $F^* > F_{\alpha}\{n_1-1, n_2-1\}$

Example 8.12

Two samples of production line workers are asked to try new procedures for performing the same task. The first sample, 25 workers from production line 1, requires a mean of 4.11 minutes with a standard deviation of 1.85 minutes to perform the task. The second sample, 24 workers from production line 2, takes a mean of 3.35 minutes with a standard deviation of 1.17 minutes. The experimenter wants to determine whether there are differences in the capabilities of the workers in the two lines. The first hypothesis to be tested is that there is no difference in the variability of the two sets of workers; the second hypothesis is that there is no difference in the mean time required to perform the task. The results of the first hypothesis test will affect the testing procedure used for the second. Test the hypothesis that the population variances are equal, at the 0.05 level of significance, and discuss the implications of the conclusion for the second hypothesis test.

Solution

We assume that the distribution of times for individual workers is normal. Then to test the hypothesis $\sigma_1^2 = \sigma_2^2$ we observe that the alternate hypothesis is two-sided and the critical value of the test statistic is $F_{.025}\{24,23\} = 2.30$ (by interpolation). Then:

1. H_0: $\sigma_1^2 = \sigma_2^2$
   H_1: $\sigma_1^2 \neq \sigma_2^2$
2. Decision rule: Reject H_0 if $F^* > 2.30$.
3. Results: $F^* = (1.85)^2/(1.17)^2 \doteq 2.50$.
4. Decision: Reject H_0.
5. Conclusion: The variances are not equal. We can conclude that $\sigma_1^2 > \sigma_2^2$ with no more than a 0.05 chance of being wrong.

We next wish to test the hypothesis $\mu_1 = \mu_2$. Since the population variances are not equal, the t test is not appropriate and we should use the Fisher–Behrens t' procedure.

8.6.3 Confidence Intervals

Sometimes we want to know the ratio of the population variances—that is, we would like to estimate σ_1^2/σ_2^2. In some applications, as in the production line of Example 8.12, it may be most efficient to replace or retrain workers, or to retool machinery, if the ratio of the variability in two populations exceeds a certain value.

In Section 8.6.2, we mentioned an important relationship between F distributions. The F distribution with ν_1 and ν_2 degrees of freedom is the inverse of the F distribution with ν_2 and ν_1 degrees of freedom. This means that $F(\nu_2, \nu_1) = 1/F(\nu_1, \nu_2)$. In particular, $F_{1-\alpha}\{\nu_2, \nu_1\} = 1/F_\alpha\{\nu_1, \nu_2\}$.

We noted before that

$$\frac{s_1^2}{s_2^2} = \frac{\sigma_1^2[\chi_1^2/(n_1 - 1)]}{\sigma_2^2[\chi_2^2/(n_2 - 1)]}$$

or

$$\frac{\sigma_1^2}{\sigma_2^2} = \frac{s_1^2[\chi_2^2/(n_2 - 1)]}{s_2^2[\chi_1^2(n_1 - 1)]} = \frac{s_1^2}{s_2^2}\cdot F(\nu_2, \nu_1)$$

Thus a $100(1 - \alpha)\%$ confidence interval for σ_1^2/σ_2^2 will have limits $(s_1^2/s_2^2)\ F_{1-\alpha/2}\{\nu_2,\ \nu_1\}$ and $(s_1^2/s_2^2)F_{\alpha/2}\{\nu_2,\ \nu_1\}$. Since $F_{1-\alpha/2}\{\nu_2,\ \nu_1\} = 1/F_{\alpha/2}\{\nu_1, \nu_2\}$, we can write the lower limit as $(s_1^2/s_2^2)[1/F_{\alpha/2}\{\nu_1, \nu_2\}]$ so that we do not need additional tables for the lower tails.

Example 8.13

Using the data of Example 8.12, determine a 95% confidence interval for σ_1^2/σ_2^2 and interpret the result.

Confidence Interval for σ_1^2/σ_2^2

Suppose we obtain two independent random samples—a sample of n_1 observations of a random variable with variance σ_1^2 and a sample of n_2 observations of a random variable with variance σ_2^2—and both variables are normally distributed. If s_1^2 and s_2^2 are the sample variances, a $100(1 - \alpha)\%$ confidence interval for σ_1^2/σ_2^2 will be the interval

$$\frac{s_1^2}{s_2^2}\cdot\frac{1}{F_{\alpha/2}\{\nu_1, \nu_2\}} \quad \text{to} \quad \frac{s_1^2}{s_2^2}\cdot F_{\alpha/2}\{\nu_2, \nu_1\}$$

where $\nu_1 = n_1 - 1$ and $\nu_2 = n_2 - 1$.

Solution

We obtain $F_{.025}\{24,23\} = 2.30$ and $F_{.025}\{23,24\} \doteq 2.28$. The lower limit is $[(1.85)^2/(1.17)^2](1/2.30) \doteq 1.10$ and the upper limit is $[(1.85)^2/(1.17)^2](2.28) \doteq 5.70$. If the ratio is 1.10, this means the larger variance is 1.10 times as large (or 0.10 times larger than) the smaller variance. If the ratio is 5.70, this means the larger variance is 5.70 times as large (or 4.70 times larger than) the smaller variance. Thus, we are 95% confident that the workers on assembly line 1 are more variable in the time needed to perform the task—varying at least 10% more than the workers on assembly line 2 but less than 470% more.

PROFICIENCY CHECKS

27. If $n_1 = 61$ and $n_2 = 51$, use the 0.05 level of significance to test the hypothesis that $\sigma_1^2 = \sigma_2^2$ if:
    a. $s_1 = 3.8, s_2 = 2.8$ b. $s_1 = 10.3, s_2 = 11.2$
    c. $s_1 = 145, s_2 = 157$ d. $s_1 = 0.012, s_2 = 0.009$
28. If $s_1 = 3.58$ and $s_2 = 2.18$, use the 0.05 level of significance to test the hypothesis that $\sigma_1^2 > \sigma_2^2$ if:
    a. $n_1 = 5, n_2 = 8$ b. $n_1 = 8, n_2 = 5$ c. $n_1 = 25, n_2 = 8$
    d. $n_1 = 16, n_2 = 41$
29. Determine a 95% confidence interval for each part of Proficiency Check 28.

Problems

Practice

8.62 Two independent random samples are drawn from normally distributed populations. The variance of the first sample is 132.41, and the variance of the second sample is 43.69. Test, at the 0.01 level, the hypothesis that $\sigma_1^2 > \sigma_2^2$ if sample sizes are both equal to:
a. 6 **b.** 16 **c.** 25 **d.** 40

8.63 Independent random samples of size 15 are drawn from two populations. The variances are 0.000169 and 0.000036. Give confidence intervals for σ_1^2/σ_2^2 using a confidence level of:
a. 90% **b.** 95% **c.** 98% **d.** 99%

8.64 Two random samples are drawn from normally distributed populations. The standard deviation of the first population is 21.6, and the standard deviation of the second population is 38.64. Test, at the 0.01 level, the hypothesis that $\sigma_1^2 = \sigma_2^2$ if:
a. $n_1 = 5, n_2 = 8$ **b.** $n_1 = 8, n_2 = 5$ **c.** $n_1 = 25, n_2 = 8$
d. $n_1 = 16, n_2 = 41$

8.65 Random samples are drawn from two populations. The standard deviations are $s_1 = 1237$ and $s_2 = 1567$. Give 95% confidence intervals for σ_1^2/σ_2^2 if:
a. $n_1 = 15, n_2 = 8$ **b.** $n_1 = 8, n_2 = 15$ **c.** $n_1 = 25, n_2 = 16$
d. $n_1 = 16, n_2 = 35$

Applications

8.66 To study telephone calls at a business office, the office manager times incoming calls and outgoing calls for 1 day. She finds that 17 incoming calls last an average of 5.16 minutes with a standard deviation of 1.12 minutes and 12 outgoing calls last an average of 4.13 minutes with a standard deviation of 2.36 minutes. Test the hypothesis that the variances of the populations are equal. Use the 0.05 level of significance. (See Problem 8.14.)

8.67 A chemical company wants to build a new plant in either East Quincy or New South Hampton. One of the criteria is the index of methane, since the plant will put methane into the atmosphere. Readings of methane proportions are taken at randomly spaced intervals in the two cities. Eight readings in East Quincy show an average of 0.23 parts per million (ppm) with a standard deviation of 0.07 ppm, while 11 readings in New South Hampton show an average of 0.32 ppm with a standard deviation of 0.22 ppm. (See Examples 8.2 and 8.3.) Test, at $\alpha = 0.05$, that the population variances are equal.

8.68 A hospital is considering two suppliers for ampoules of a wonder drug. Both suppliers deliver ampoules that are supposed to contain 500,000 units of the drug. An analysis is made of a sample of six ampoules from each company. Brand A has a mean of 510,000 units with a standard deviation of 20,000, while brand B has a mean of 490,000 units with a standard deviation of 15,000. Can you conclude that there is probably no difference in the variability of the brands? Use a 0.05 level of significance. (See Problem 8.34.)

8.69 A survey is made to test the hypothesis that college graduates read at least two books per year more than nongraduates. A random sample of 66 college graduates reports reading an average of 4.3 books per year, with a standard deviation of 4.6 books, while 78 nongraduates report an average of 1.9 books per year with a standard deviation of 3.1 books. Test the hypothesis that the population variances are equal. Use the 0.05 level of significance. (See Problem 8.38.)

8.7 Computer Usage

We can perform t tests and obtain confidence intervals for the difference between two population means using Minitab. We can also use SAS statements for this purpose and write programs to perform tests or obtain confidence intervals concerning two population proportions.

8.7.1 Minitab Commands for This Chapter

We can obtain the two-sample test and confidence interval by using the TWOSAMPLE-T command in Minitab. By specifying different subcommand options, we can perform most of the two-sample tests. The basic form of the command is

```
MTB > TWOSAMPLE-T with K% confidence first sample in C second in C
```

When using the TWOSAMPLE-T command note that the sample data are stored in two columns. The K% confidence specifies the level of confidence for the confidence interval of the difference in the two population means. The default confidence coefficient for the confidence interval is 95%. We can use the ALTERNATIVE subcommand to specify the alternate hypothesis as before. The default alternate hypothesis is two-sided.

We use the POOLED subcommand when the population variances can be assumed equal and are pooled together. If the POOLED subcommand is specified (see Example 8.14) the "equal, but unknown, variance" test is performed. Otherwise the Fisher–Behrens t' approximation is used. Note that Minitab has no command similar to ZTEST for comparing two population means when the population variances are known.

We can also test the equality of two population means by using the TWOT command for sample data that have been stored in a slightly different and perhaps more standard format. The command is

```
MTB > TWOT with K% confidence, data in C, sample identifier in C
```

(The POOLED and ALTERNATIVE subcommands are also available and work in the same manner as indicated earlier.) The TWOT command is used when the data for both samples are stored in one column and the sample identifier is in another column. For example, we could store the two samples of data (10 8 4 3) and (6 5 4 4 8) in two columns C1 and C2 and use the TWOSAMPLE command. We could also store all nine data points in column C1 and the sample identifier in column C2. The commands for storing the data in this format are

```
MTB > SET the following data into column C1
DATA> 10 8 4 3 6 5 4 4 8
MTB > SET the sample identifier into column C2
DATA> 1 1 1 1 2 2 2 2 2
MTB > ENDOFDATA
```

The resulting Minitab data set looks like this:

OBSERVATION	C1	C2
1	10	1
2	8	1
3	4	1
4	3	1
5	6	2
6	5	2
7	4	2
8	4	2
9	8	2

This is the more common method of storing data in the computer when using Minitab. The statistical results, however, are the same in both instances. Only the data storage format and the command to execute the t test are different.

Recall that the appropriate two-sample test to perform is determined in part by whether the population variances are equal. Therefore you should always check to see if this is a reasonable assumption. You can do so in several ways. First, you can conduct an F test to see whether the two population variances are equal (Section 8.6.2). Second, you can use an approximate rule of thumb: consider two population variances equal unless one of the sample variances is much more than four times the other. Finally, you can use graph aids, such as histograms and stem and leaf diagrams, to compare the spreads of the two samples. This visual check is convenient because you can also inspect for approximate normality of the data. Once the data are in Minitab you can then use the STDEV, DESCRIBE, HISTOGRAM, or STEM-AND-LEAF commands.

For paired data recall that we simply take the difference of the paired observations and use the appropriate one-sample analysis. You can either enter the differences or enter the pairs of data and let Minitab compute the differences by using the LET command. These differences and the appropriate Minitab commands from Chapters 6 and 7 (ZINTERVAL, TINTERVAL, ZTEST, and TTEST) would then be used to test the relevant hypothesis or to obtain a confidence interval. To do a paired t test on the difference of columns 4 and 2 and to obtain the confidence interval for the difference in the means, for example, we would type

```
MTB > LET C5=C4-C2
MTB > TTEST of the differences in C5
MTB > TINTERVAL of the differences in C5
```

Example 8.14

In a study of water quality, the EPA measures the mean bacterial count on several different days at two locations on a certain river. Site 1 is 100 meters downriver from the discharge pipes of a local paper mill; site 2 is 1 kilometer upriver from the mill. The results follow:

Below the mill	30.1	36.2	33.4	28.2	29.8	34.9
Above the mill	29.7	30.3	26.4	27.3	31.7	

The EPA wishes to determine whether the mean bacterial count is greater downriver from the mill discharge than at the site upriver. At $\alpha = 0.05$, use Minitab to conduct this test.

Solution

The Minitab results are presented in Figure 8.2. The results indicate that the mean bacterial count below the mill is not significantly greater than the mean bacterial count upriver. The P value of 0.054 of $t^* = 1.79$ is greater than α for our test. Note that the pooled variance form of the test is appropriate because the larger variance is less than four times the smaller variance. The 95% confidence interval for the difference in the two population means is -0.8 to 6.84. This result confirms the hypothesis test because the interval includes zero. On the other hand, since $P = 0.054$ and is not much greater than 0.05, the risk of a type II error is probably high. It would probably be best to reserve judgment and take further readings.

```
MTB > SET the following data into C1
DATA> 30.1 36.2 33.4 28.2 29.8 34.9
DATA> SET the following data into C2
DATA> 29.7 30.3 26.4 27.3 31.7
DATA> NAME C1 'MILL' C2 'UPRIVER'
MTB > STDEV of C1, store in K1
   ST.DEV. =      3.1887
MTB > STDEV of C2, store in K2
   ST.DEV. =      2.1845
MTB > LET K3 = K1/K2
MTB > PRINT K3                  ← Ratio of the sample standard deviations
K3        1.45971
MTB > TWOSAMPLE-T test using 'MILL' and 'UPRIVER';
SUBC> POOLED;
SUBC> ALTERNATIVE=+1.
```

Figure 8.2
Minitab printout for the bacterial count data of Example 8.14.

```
TWOSAMPLE T FOR MILL VS UPRIVER
                N          MEAN          STDEV          SE MEAN
MILL            6          32.10          3.19             1.3
UPRIVER         5          29.08          2.18            0.98

95 PCT CI FOR MU MILL - MU UPRIVER: (-0.8, 6.84)
TTEST MU MILL = MU UPRIVER (VS GT): T=1.79 P=0.054 DF=9.0
```

Example 8.15

Refer to Example 8.7 in which a diet is being tested to determine its effectiveness. Use Minitab to test the hypothesis that the diet is effective at the 0.05 level of significance.

Solution

Since the two samples of starting and ending weights are not independent, we may use the paired t test. The Minitab commands and output are shown in Figure 8.3.

Comparing the results to those of Example 8.7, note that the same conclusion is reached. Note, also, however, that the P value of 0.069 is near 0.05. While the results are not statistically significant, there is a mean weight loss in the sample of about 2 pounds. For this reason it might be well to reserve judgment and collect more data for analysis.

Figure 8.3
Minitab printout of the diet data of Example 8.7.

```
MTB > READ the following data into column C1 and C2
DATA> 183 177
DATA> 144 145
DATA> 151 145
DATA> 163 162
DATA> 155 151
DATA> 159 163
DATA> 178 173
DATA> 184 185
DATA> 142 139
DATA> 137 138
DATA> ENDOFDATA
     10 ROWS READ
MTB > LET C3=C1-C2
MTB > NAME C1 'START' C2 'END' C3 'DIFFER'
MTB > TTEST of the data in C3;
SUBC> ALTERNATIVE=+1.

TEST OF MU = 0 VS MU G.T. 0

             N      MEAN      STDEV      SE MEAN        T      P VALUE
DIFFER      10      1.80       3.49          1.1     1.63        0.069

MTB > TINTERVAL of the data in C3

             N      MEAN      STDEV      SE MEAN      95.0 PERCENT C.I.
DIFFER      10      1.80       3.49          1.1         (-0.7, 4.3)
```

8.7.2 SAS Statements for This Chapter

To test whether or not there are significant differences between means of two independent samples, we can use the TTEST procedure. When the data are entered, the TTEST procedure will perform both the standard independent samples t test using the pooled variance and the Fisher–Behrens t test assuming that the population variances are unequal. It will also print out an F test determining whether or not the population variances are equal.

Suppose we have two samples of data consisting of numbers of minutes between incoming telephone calls at two different offices, Jones Co. and Smith, Inc.. Suppose the two samples of data are (10 8 4 3) and (6 5 4 4 8), respectively. We may write the following:

```
DATA CALLS;
  INPUT COMPANY $ MINUTES @@;
CARDS;
JONES 10 JONES 8 JONES 4 JONES 3
SMITH 6 SMITH 5 SMITH 4 SMITH 4 SMITH 8
```

We test the hypothesis that there is no difference in mean time between telephone calls for the two companies by writing

```
PROC TTEST;
  CLASS COMPANY;
  VAR MINUTES;
```

The CLASS statement indicates which variable is the independent variable; the VAR statement names the dependent variable. The statement given above says to analyze the differences in minutes between companies. The obtained P value applies for a two-sided alternative; if the alternate hypothesis is one-sided, halve the P value and check the obtained value of t^* to see if the sign agrees with the alternate hypothesis.

To test a hypothesis such as H_0: $\mu_1 - \mu_2 = \delta_0$, we could create a second variable. If $\mu_1 - \mu_2 = \delta_0$ then $\mu_1 = \mu_2 + \delta_0$, so we can create a variable $y = x_2 + \delta_0$ and test the hypothesis that $\mu_1 = \mu_y$. Since the variance of y will be the same as the variance of x_2, the F test will still be valid. If we wanted to test the hypothesis that the mean time between telephone calls for Jones Co. is 1.5 minutes longer than that of Smith, Inc., we could write

```
DATA CALLS;
  INPUT COMPANY $ MIN @@;
  IF COMPANY = 'JONES' THEN MINUTES = MIN;
  IF COMPANY = 'SMITH' THEN MINUTES = MIN + 1.5;
  CARDS;
JONES 10 JONES 8 JONES 4 JONES 3
SMITH 6 SMITH 5 SMITH 4 SMITH 4 SMITH 3
PROC TTEST;
  CLASS COMPANY;
  VAR MINUTES;
```

Example 8.16

Refer to Example 8.14. Use SAS statements to determine whether the difference between the mean bacterial count at the sites is the same. Use the 0.05 test of significance.

Solution

The SAS input and output files are shown in Figure 8.4. Note that the values of the variable SITE are DOWNRIVR and UPRIVER. A SAS name cannot have more than eight letters. We have used a renaming procedure to reduce the amount of writing.

Figure 8.4
SAS printout for the bacterial count data of Examples 8.14 and 8.16.

```
INPUT FILE

DATA COUNT:
   INPUT PLACE $ COUNT @@;
   IF PLACE = 'D' THEN SITE = 'DOWNRIVR';
   IF PLACE = 'U' THEN SITE = 'UPRIVER';
CARDS;
D 30.1 D 36.2 D 33.4 D 28.2 D 29.8 D 34.9
U 29.7 U 30.3 U 26.4 U 27.3 U 31.7
PROC TTEST;
   CLASS SITE;
   VAR COUNT;

OUTPUT FILE
```

```
                                    SAS
                              TTEST PROCEDURE

VARIABLE: COUNT
```

SITE	N	MEAN	STD DEV	STD ERROR	MINIMUM	MAXIMUM	VARIANCES	T	DF	PROB > |T|
DOWNRIVR	6	32.10000000	3.18873015	1.30179363	28.20000000	36.20000000	UNEQUAL	1.8555	8.7	0.0975
UPRIVER	5	29.08000000	2.18449079	0.97693398	26.40000000	31.70000000	EQUAL	1.7892	9.0	0.1072

```
FOR HO: VARIANCES ARE EQUAL, F' = 2.13 WITH 5 AND 4 DF    PROB > F' = 0.4835
```

The F test shows that we may conclude that the population variances are equal, so we use the equal-variance t test. Since $t^* = 1.7892$ and the P value (0.1072) is not less than 0.05, we cannot reject H_0. On the other hand, P is close to 0.05, so we ought to reserve judgment and collect more data. Note in the printout that $t^* = 1.7892$ with 9 degrees of freedom while $t'^* = 1.8555$ with about 8.7 degrees of freedom. Since the variances are not significantly different, t^* is the appropriate test statistic.

To test a hypothesis concerning paired samples we use the one-sample procedure discussed in Section 7.5.2. We enter the data and create a difference score. If the hypothesis is H_0: $\mu_1 - \mu_2 = \delta_0$, we create a variable equal to $x_1 - x_2 - \delta_0$. In either case we use PROC MEANS to test the hypothesis. Suppose, for instance, that the bacterial count in the river is measured on the same days with the following results:

DAY	UPRIVER	DOWNRIVER
1	29.7	30.1
2	30.3	36.2
3	26.4	28.2
4	27.3	29.8
5	31.7	34.9

The following program tests the hypothesis of no difference between the sites:

```
DATA COUNT;
  INPUT UPRIVER DOWNRIVR @@;
  DIFF = UPRIVER - DOWNRIVR;
CARDS;
29.7 30.1 30.3 36.2 26.4 28.2 27.3 29.8 31.7 34.9
PROC MEANS MEAN STDERR T PRT;
  VAR DIFF;
```

The output of this program is shown in Figure 8.5. We see that the pairing does make a difference: $t^* = 3.03$ and the P value (two-tailed) is 0.0388. Thus we reject H_0 and conclude that there is a higher bacterial count downriver than upriver.

```
                                    SAS

VARIABLE            MEAN            STD ERROR          T        PR>|T|
                                     OF MEAN

DIFF            2.76000000          0.91137259        3.03      0.0388
```

Figure 8.5
SAS printout for paired-sample test of bacterial count data.

We can obtain confidence intervals by using a procedure similar to that in Section 6.7.2. To obtain a 95% confidence interval for the difference between the Jones and Smith population means, we may write

```
DATA CALLS;
  INPUT COMPANY MINUTES @@;
CARDS;
JONES 10 JONES 8 JONES 4 JONES 3
SMITH 6 SMITH 5 SMITH 4 SMITH 4 SMITH 8
```

```
PROC MEANS NOPRINT;
  VAR MINUTES;
  BY COMPANY;
OUTPUT OUT = A MEAN = XBARJ XBAR S VAR = VARJ VARS;
DATA CONF;
  SET A;
SEDIFF = SQRT((4*VARJ + 5*VARS)/9) * SQRT(1/5 + 1/6);
LLIMIT = XBARJ - XBARS - 2.262 * SEDIFF;
ULIMIT = XBARJ - XBARS + 2.262 * SEDIFF;
PROC PRINT;
```

When there are two groups (in this case the companies), the output statement needs two names for each variable that is to be saved from the output. Here XBARJ is the mean for the Jones data while XBARS is the mean for the Smith data. If only one name is given, it will be applied to the first group. Thus if we had written VAR = VAR; the Jones variance would be given the name VAR but the Smith variance would not be saved.

To test a hypothesis about a confidence interval for the difference between population proportions or to construct such a confidence interval, we can use procedures similar to those given in Sections 6.7.2 and 7.5.2.

Problems

Applications

8.70 Two different traffic signal settings are being tested to determine which is more efficient in reducing the number of delays along a certain traffic corridor. Using the 0.05 level of significance, the traffic engineer records the mean speed of cars passing through this corridor at random times during several weeks with each traffic light setting. He obtains the following results:

Setting A	30.2	29.9	30.1	30.9	28.6	
Setting B	28.4	31.2	29.5	30.2	30.6	30.1

a. Can we conclude that the population variances are different?
b. Determine whether the two traffic signal settings are related to the mean speed of traffic along the corridor.

8.71 Mutual State Bank is considering a campaign to increase the rate of its customers' savings. Before mounting an expensive campaign, however, the bank conducts a trial campaign on ten randomly selected accounts with the following results:

	MONTHLY SAVINGS DEPOSITS ($)									
BEFORE CAMPAIGN	100	75	50	75	125	30	85	15	40	10
AFTER CAMPAIGN	125	90	50	50	150	40	80	25	40	12

Use the 0.05 level of significance to determine whether the campaign should be conducted. Your criterion for decision is whether there will be an increase in saving.

8.72 In a study investigating the effect of caffeine on heart rate (in beats per minute), two randomly assigned groups of adult males are given different amounts of caffeine. Group 1 is given two cups of coffee and the men's heart rates are measured 15 minutes later. Group 2 is given two cups of coffee as well, but an extra 50 milligrams of caffeine is added to each cup. Their heart rates are then measured 15 minutes later. Using the following data and the appropriate t test, determine whether the heart rates differ for the two populations.

Group 1	82	62	70	73	66	71	
Group 2	100	93	84	120	71	69	50

8.8 Case 8: The Personality Profiles—Solution

To test the hypothesis that there is a difference between the two personality types in incidence of stomach trouble, the statistician tested the hypothesis $\pi_1 = \pi_2$. Since 11 percent of 61 is about 7 and 30 percent of 66 is 20, he concluded that the number of individuals reporting stomach trouble in each case was 7 and 20, respectively. The pooled proportion, $\hat{\pi}$, is $(7 + 20)/(61 + 66)$ or about 0.2126. Then $s_{\hat{\pi}} = \sqrt{(0.2126)(0.7874)(1/61 + 1/66)} \doteq 0.073$. The value of the test statistic z is $z^* = (0.11 - 0.30)/(0.073) \doteq -2.61$. For a two-sided alternative with a two-tailed rejection region, $P \doteq 0.009$. Thus the statistician concluded that there was indeed a significant difference between the two populations in incidence of stomach problems.

A 95% confidence interval for the difference, however, was based on $s_{p_1-p_2}$, since it had been concluded that $\pi_1 \neq \pi_2$. The value of $s_{p_1-p_2}$ was determined to be equal to $\sqrt{(0.11)(0.89)/61 + (0.30)(0.70)/66} \doteq 0.0692$; $1.96(0.0692) \doteq 0.14$, so a 95% confidence interval for the difference is a 0.05 to 0.33 greater proportion of type A's than type B's with stomach problems. The statement "type A's are about three times more likely . . ." has no basis for two reasons. First, the type A's in the sample were only two times *more* likely than (three times *as* likely as) type B's to have stomach problems. Second, according to the confidence interval the difference in proportion could be as small as 0.05, which is only 45% more likely (0.05/0.11).

The difference in the proportion who said they were in excellent health was probably not due to chance ($z^* = 2.07$), but there was no basis laid for any relationship between health and stomach problems. The statistician concluded that the study probably had a great deal of merit, but he decided that he would like to read the study itself and draw his own conclusions rather than rely on the newspaper account.

8.9 Summary

It is often desirable to estimate the difference between two population parameters or to test a hypothesis about the difference between two population parameters. These inferences are usually based on the sampling distribution of the difference between two sample statistics.

In all inferences about the difference between two sample means, the samples must have been randomly picked from their populations, and the samples must be independent of each other. Two additional assumptions—the populations are normal and their variances are known—should preferably be satisfied as well, but if not modified procedures are available.

Hypothesis-testing procedures were summarized earlier in Table 8.2. Confidence limits for each case are summarized here in Table 8.4.

If two samples are paired using appropriate criteria, or if an experiment involves repeated measures on one sample, as in before-and-after tests, difference scores are employed.

If the samples are sufficiently large, we can use the sampling distribution for the difference between sample proportions to obtain interval estimates for $\pi_1 - \pi_2$ or to test hypotheses about the value of $\pi_1 - \pi_2$.

Table 8.4 CONFIDENCE LIMITS FOR DIFFERENCES BETWEEN POPULATION MEANS.

POPULATION STANDARD DEVIATIONS	SHAPE OF DISTRIBUTIONS	SAMPLE SIZES	CONFIDENCE LIMITS
Both known	Both normal	Any n	$(\bar{x}_1 - \bar{x}_2) \pm z_{\alpha/2}\sigma_{\bar{x}_1-\bar{x}_2}$
	One or both not normal	$n_1, n_2 \geqslant 30$	$(\bar{x}_1 - \bar{x}_2) \pm z_{\alpha/2}\sigma_{\bar{x}_1-\bar{x}_2}$
One or both unknown	Both normal	$n_1, n_2 \geqslant 30$	$(\bar{x}_1 - \bar{x}_2) \pm z_{\alpha/2}s_{\bar{x}_1-\bar{x}_2}$
		Any n $\sigma_1^2 = \sigma_2^2$	$(\bar{x}_1 - \bar{x}_2) \pm t_{\alpha/2}\hat{\sigma}\sqrt{(1/n_1 + 1/n_2)}$
		n_1 or $n_2 < 30$ $\sigma_1^2 \neq \sigma_2^2$	$(\bar{x}_1 - \bar{x}_2) \pm t_{\alpha/2}\sqrt{(s_1^2/n_1 + s_2^2/n_2)}$ (modified degrees of freedom)
	Not both normal	$n_1, n_2 \geqslant 50$	$(\bar{x}_1 - \bar{x}_2) \pm z_{\alpha/2}s_{\bar{x}_1-\bar{x}_2}$
	Both symmetric (use trimmed means)	n_1 or $n_2 < 50$ $\sigma_1^2 = \sigma_2^2$	$[(\bar{x}_1(T) - \bar{x}_2(T)] \pm t_{\alpha/2}\hat{\sigma}\sqrt{(1/h_1 + 1/h_2)}$
		n_1 or $n_2 < 30$ $\sigma_1^2 \neq \sigma_2^2$	$[(\bar{x}_1(T) - \bar{x}_2(T)] \pm t_{\alpha/2}\sqrt{(s^2_{\bar{x}_1(T)} + s^2_{\bar{x}_2(T)})}$ (modified degrees of freedom)
	Not both symmetric	n_1 or $n_2 < 50$	No general method

We can also use the sampling distribution of the sample proportions to test a hypothesis about the difference between two population proportions. We may test the null hypothesis $\pi_1 - \pi_2 = \delta_0$ by using the test statistic z.

The sample size needed for constructing a confidence interval of a desired size or for placing an upper bound on the difference between μ_1 and μ_2 or the difference between π_1 and π_2 can be obtained by using the procedures outlined in Section 8.5.

We can also test hypotheses that population variances are equal or obtain interval estimates for the ratio of two population variances. We do not use the sampling distribution of a difference in this case, however; instead we use the F distribution, defined by two parameters—the degrees of freedom for the numerator, ν_1, and the degrees of freedom for the denominator, ν_2. Recall that the samples are numbered so that $s_1^2 > s_2^2$.

Chapter Review Problems

Practice

8.73 Two independent random samples of size 60 and 72 have means and standard deviations, respectively, $\bar{x}_1 = 112.6$, $s_1 = 24.8$, $\bar{x}_2 = 103.9$, $s_2 = 19.7$.
a. Using $\alpha = 0.05$, test the hypothesis that $\mu_1 = \mu_2$.
b. Determine a 95% confidence interval for $\mu_1 - \mu_2$.

8.74 Refer to Problem 8.73. Test the hypothesis that $\sigma_1^2 = \sigma_2^2$, using $\alpha = 0.05$, and determine a 95% confidence interval for σ_1^2/σ_2^2.

8.75 Two independent random samples of size 25 and 21 drawn from normal populations have means and standard deviations, respectively, $\bar{x}_1 = 112.6$, $s_1 = 18.8$, $\bar{x}_2 = 103.9$, $s_2 = 13.9$.
a. Using $\alpha = 0.05$, test the hypothesis that $\mu_1 = \mu_2$.
b. Determine a 95% confidence interval for $\mu_1 - \mu_2$.

8.76 In problem 8.75b, how large should the samples be to obtain confidence intervals no wider than 5.0?

8.77 In Problem 8.75, suppose that $s_1 = 24.6$.
a. Using $\alpha = 0.05$, test the hypothesis that $\sigma_1^2 = \sigma_2^2$.
b. Using the appropriate test, test the hypothesis that $\mu_1 = \mu_2$ for $\alpha = 0.05$.

8.78 If 45 of a random sample of 60 people prefer brand X while 24 of a different random sample of 48 people prefer brand X, test the hypothesis that the two samples come from the same population. Use $\alpha = 0.10$.

8.79 Using the data of Problem 8.78, construct a 95% confidence interval for the difference between the population proportions.

8.80 In problem 8.79, how large should the samples be to construct a confidence interval no wider than 0.10?

Applications

8.81 An antipyretic is being tested as a replacement for aspirin. A total of nine experimental animals are given artificially high temperatures and the drug is administered. Given the before and after temperatures, test the hypothesis that the drug is effective; use the 0.05 level of significance.

Before	107.2	111.4	109.3	106.5	113.7	108.4	107.7	111.9	109.3
After	106.1	111.7	105.4	107.2	109.8	108.8	106.9	109.6	110.5

8.82 Forty percent of the geldings racing at a certain track ran in the money, while only 32 percent of the fillies did so. If the sample is based on the performance of 105 geldings and 75 fillies, test the jockey's hypothesis (at the 0.05 significance level) that geldings run better than fillies.

8.83 Two samples of potential customers are shown an advertising presentation and asked to rate it on a scale of 1 to 10 in terms of whether it would make them more likely to buy the product. A total of 23 people rate presentation A a mean of 5.6 with a standard deviation of 1.9, while presentation B has a mean rating of 6.3 with a standard deviation of 1.6 for its sample of 27. Determine a 99% confidence interval for the difference between the mean ratings of the two presentations and interpret this interval.

8.84 Use the data of problem 8.83 to determine whether the two presentations would probably have different results with the public.

8.85 Determine whether or not the standard deviations of the populations in Problem 8.83 are equal. Use the 0.01 level of significance. What are the implications of this result?

8.86 A sample of 200 light bulbs from company G shows 24 defective. A sample of 180 light bulbs from company S shows 27 defective. Construct a 95% confidence interval for the actual difference in proportion defective for the two companies.

8.87 Refer to Problem 8.86. What size samples would you need to obtain a 95% confidence interval at most 0.10 in width?

8.88 A traffic engineer tries two different settings for traffic signals in an effort to find the most efficient setting. He uses the number of traffic delays per week as his criterion. With setting A he obtains a mean of 26.1 delays per week for 11 weeks, with a standard deviation of 5.6 delays. With setting B he obtains a mean of 22.4 traffic delays per week for 13 weeks with a standard deviation of 6.3 delays. What can he conclude (at $\alpha = 0.05$) about the two settings?

8.89 A study by Ware and Stuck (1985) reported on the frequency of male and female models in pictures in advertisements for or about computers in three randomly selected issues of computer magazines. Three different magazines were selected in three different months. They also reported on whether the models were portrayed as active (using the computer) or passive (looking at the computer or watching someone else use it). A total of 212 men and 59 women were shown as active, while 44 men and 47 women were shown as passive.

a. Test the hypothesis, using $\alpha = 0.01$, that there are no differences in proportion of active poses for the two populations.
b. Obtain a 99% confidence interval for the difference between the population proportions.

8.90 Two different toothpastes are tested on two groups of ten children each. After 6 months of use, the group using toothpaste A has a mean of 2.7 cavities with a standard deviation of 1.25, while the group using toothpaste B has a mean of 1.5 new cavities with a standard deviation of 1.08. Determine a 95% confidence interval for the true mean difference in cavities for users of the two toothpastes. Can any conclusion be reached?

8.91 Is the t test statistic appropriate for Problem 8.90?

8.92 To compare the vitamin E content of two different brands of vitamin capsules, a sample of 12 capsules of each brand is selected. Brand A is found to contain, on the average, 5000 USP units with a standard deviation of 400 units, while brand B has an average of 4700 USP units with a standard deviation of 300 units. Using the 0.05 level of significance, can you conclude that the vitamin E content of the two brands is different?

8.93 A random sample of college personnel records reveals that among 150 students from cities of 25,000 or more, 60 belong to fraternities or sororities, while of 90 students from rural areas or small towns, only 27 belong to these social organizations. A psychologist is studying differences between socialization processes undergone by city dwellers as opposed to those experienced by rural inhabitants. She tests the hypothesis (at $\alpha = 0.05$) that there are differences in the proportion of students in each category belonging to the social organizations. What are the conclusions?

8.94 A random sample of 80 insurance salesmen working for company A shows 56 in opposition to proposed revisions in the current pension plan. A random sample of 75 insurance salesmen working for company B is given company A's plan as if it were being proposed for company B's salesmen. A total of 44 are opposed to it. Determine a 99% confidence interval for the difference in proportion of salesmen of the two companies who favor the plan.

8.95 A medical researcher studies the effect of two drugs on guinea pigs infected with pneumococcus bacillus. One group has a mortality of 17 animals in 44; the second has a mortality of 24 in 52. Can you conclude, at a 0.01 level of significance, that there is a difference in the drugs?

8.96 Two companies submit samples of light bulbs for bids. Company A's sample of 300 bulbs has a mean life of 1065 hours with a standard deviation of 133 hours. Company B's sample of 300 has a mean life of 1047 hours with a standard deviation of 56 hours. If mean bulb life is the primary criterion for awarding the contract, do the samples yield sufficient information to decide (using $\alpha = 0.05$) which company should get the contract? If so, which one?

8.97 In Problem 8.96, can you conclude that the variances of the populations are equal? Use the 0.01 level of significance. Determine a 99% confidence interval for σ_1^2/σ_2^2.

8.98 Use the data of Problem 8.96, but suppose the bulbs are to be placed in the ceiling of a warehouse and changed every 1000 hours. The primary criterion is that the company producing the greatest proportion of bulbs lasting at least 1000 hours will get the contract. Do the samples yield sufficient information to decide (using $\alpha = 0.05$) which company should get the contract? If so, which one?

8.99 In a study of attitudes of military personnel toward women, the Attitude Toward Women Scale (ATWS) is used to assess general attitude. The questionnaire is administered to 172 naval personnel: 48 marines and 124 naval airmen.* The marines have a mean score of 86.4 with a standard deviation of 18.50; the naval airmen have a mean of 99.7 with a standard deviation of 18.32. The hypothesis being tested is that naval airmen, on the average, are more tolerant toward women (higher score) by an average of at least 5 points. At what significance level does the result substantiate that hypothesis?

8.100 A study of 12 juvenile delinquent boys and 12 nondelinquent boys, all from similar backgrounds, is conducted to assess differences in self-concept. The following results are obtained on the Osgood Semantic Differential Test.* The nondelinquent boys have a mean of 96.4 with a standard deviation of 8.1, while the delinquent boys have a mean of 98.5 with a standard deviation of 10.8. Are the obtained differences significant at the 0.05 level of significance?

8.101 A study of personality differences among LSD users and nonusers is conducted with 40 users and 40 nonusers as subjects.* The Minnesota Multiphasic Personality Inventory is administered. Some of the results are given here:

	NONUSERS		USERS	
SCALE	MEAN	SD	MEAN	SD
Depression	55.94	13.60	57.10	11.83
Hysteria	58.05	7.92	62.08	8.31
Paranoia	55.85	10.12	59.35	9.69
Schizophrenia	55.05	12.15	61.25	12.82

Using the 0.025 level of significance, in which cases would the hypothesis that users score higher than nonusers be substantiated? (*Note:* This means that the probability that all four conclusions are correct is at least $1 - 4(0.025) = 0.90$.)

Index of Terms

*Research from unpublished master's theses at the University of West Florida.

Glossary of Symbols

d	difference score (paired sample)
$\bar{d}$	mean of difference scores
δ_0	hypothetical difference between population parameters
$F(\nu_1, \nu_2)$	F distribution with ν_1 and ν_2 degrees of freedom
$F_\alpha\{\nu_1, \nu_2\}$	critical value of F
F^*	sample value of F test statistic
μ_1, μ_2	means of two populations
$\mu_1 - \mu_2$	difference between two population means
p_1, p_2	proportions in two samples
$p_1 - p_2$	difference between two sample proportions
π_1, π_2	proportions in two populations
$\pi_1 - \pi_2$	difference between two population proportions
$\hat{\pi}$	pooled proportion
s_d	standard deviation of difference scores
$s_{\bar{d}}$	standard error of difference scores
$s_{p_1-p_2}$	sample standard error of the difference for proportions
$s_{\hat{\pi}}$	sample standard error of pooled proportion
$s_{\bar{x}_1-\bar{x}_2}$	sample standard error of the difference for means
s_1^2, s_2^2	variances of two samples
s_1^2/s_2^2	ratio of two sample variances
$\sigma_{p_1-p_2}$	standard error of the difference for proportions
$\sigma_{\hat{\theta}_1-\hat{\theta}_2}$	standard error of the difference of two sample statistics
$\sigma_{\bar{x}_1-\bar{x}_2}$	standard error of the difference for means
$\hat{\sigma}^2$	pooled variance of two samples
σ_1^2, σ_2^2	variances of two populations
t'	Fisher–Behrens test statistic
θ_1, θ_2	parameters of two populations
$\theta_1 - \theta_2$	difference of parameters of two populations
$\hat{\theta}_1, \hat{\theta}_2$	sample statistics; estimators for θ_1 and θ_2
$\hat{\theta}_1 - \hat{\theta}_2$	difference between two sample statistics
$\bar{x}_1, \bar{x}_2$	means of two samples
$\bar{x}_1 - \bar{x}_2$	difference between two sample means

Answers to Proficiency Checks

1. **a.** -2.78 to 22.78 **b.** -0.43 to 20.43 **c.** 0.96 to 19.04 **d.** 3.61 to 16.39
2. $s_{\bar{x}_1-\bar{x}_2} \doteq 0.0242$. **a.** -0.080 to 0 **b.** -0.087 to 0.007 **c.** -0.096 to 0.016 **d.** -0.103 to 0.023
3. $s_{\bar{x}_1-\bar{x}_2} \doteq 1.7222$, $1.96s_{\bar{x}_1-\bar{x}_2} \doteq 3.4$; -0.3 to 6.5
4. **a.** -24.96 to 44.96 **b.** -17.60 to 37.60

c. $t_{.025}\{48\} \doteq 2.013$; -9.03 to 29.03
d. $t_{.025}\{98\} \doteq 1.99$; -2.97 to 22.97

5. $\hat{\sigma} \doteq 0.1408$, $\hat{\sigma}\sqrt{1/12 + 1/15} \doteq 0.0545$. **a.** -0.133 to 0.053
b. -0.152 to 0.072 **c.** -0.176 to 0.096 **d.** -0.192 to 0.112

6. $\hat{\alpha} = \sqrt{[17(5.1)^2 + 23(5.9)^2]/40} \doteq 5.5740$; $\hat{\sigma}\sqrt{1/18 + 1/24} \doteq 1.7380$; $t_{.025}\{40\} = 2.021$; -0.4 to 6.6

7. 1. H_0: $\mu_1 - \mu_2 = 10.0$; H_1: $\mu_1 - \mu_2 - 10.0$.
   2. Decision rule: Reject H_0 if $|z^*| > 1.645$.
   3. Results: $s_{\bar{x}_1 - \bar{x}_2} = \sqrt{(3.8)^2/60 + (1.9)^2/50} \doteq 0.5593$, $z^* = (50.3 - 38.4 - 10)/0.5593 \doteq 3.40$.
   4. Decision: Reject H_0.
   5. Conclusion: Since we do not know the situation, no conclusion is possible except that μ_1 and μ_2 differ by at least 10.

8. 1. H_0: $\mu_1 - \mu_2 = 10.0$; H_1: $\mu_1 - \mu_2 \neq 10.0$.
   2. Decision rule: Reject H_0 if $|t^*| > 1.697$.
   3. Results: $\hat{\sigma} = \sqrt{[17(3.8)^2 + 13(1.9)^2]/30} \doteq 3.1220$, $t^* = 1.9/[3.1220\sqrt{1/18 + 1/14}] \doteq 1.9/1.1125 \doteq 1.708$.
   4. Decision: Reject H_0.
   5. Conclusion: Since we do not know the situation, no conclusion is possible except that μ_1 and μ_2 differ by at least 10.

9. Since $P(z > 3.40) = 0.0003$ and the rejection region is two-tailed, $P \doteq 0.0006$. Since $1.708 > t_{.05}\{30\}$, but not greater than $t_{.025}\{30\}$, and the rejection region is two-tailed, $p < 0.10$.

10. We must use the Fisher–Behrens t', so we first need to determine ν. Thus $\nu = [(3.8)^2/18 + (1.9)^2/14]^2/\{[(3.8)^2/18]^2/17 + [(1.9)^2/14]^2/13\} \doteq 26$.
    1. H_0: $\mu_1 - \mu_2 = 10.0$; $H1$: $\mu_1 - \mu_2 \neq 10.0$.
    2. Decision rule: Reject H_0 if $|t'^*| > 1.706$.
    3. Results: $t^* = 1.9/\sqrt{(3.8)^2/18 + (1.9)^2/14} \doteq 1.845$.
    4. Decision: Reject H_0.
    5. Conclusion: Since we do not know the situation, no conclusion is possible except that μ_1 and μ_2 differ by at least 10.

11. 1. H_0: $\mu_1 - \mu_2 = 100.0$; H_1: $\mu_1 - \mu_2 > 100.0$.
    2. Decision rule: Reject H_0 if $z^* > 2.33$.
    3. Results: $s_{\bar{x}_1 - \bar{x}_2} = \sqrt{(125.4)^2/40 + (123.7)^2/50} \doteq 29.0375$, $z^* = (1356.3 - 1168.9 - 100)/29.0375 \doteq 3.01$.
    4. Decision: Reject H_0.
    5. Conclusion: Since we do not know the situation, no conclusion is possible except that μ_1 is at least 100.0 greater than μ_2.

12. 1. H_0: $\mu_1 - \mu_2 = 100.0$; H_1: $\mu_1 - \mu_2 > 100.0$.
    2. Decision rule: Reject H_0 if $t^* > 2.896$.
    3. Results: $\hat{\sigma} = \sqrt{[3(125.4)^2 + 5(123.7)^2]/8} \doteq 124.3402$, $t^* = 87.4/[124.3402\sqrt{1/4 + 1/6}] \doteq 1.089$.
    4. Decision: Fail to reject H_0.
    5. Conclusion: Since we do not know the situation, no conclusion is possible except that μ_1 is not at least 100 more than μ_2.

13. Here the rejection region is one-tailed. $P < 0.0013$; $P > 0.10$.

14. In each case the hypotheses to be tested are:
1. H_0: $\mu_1 = \mu_2$; H_1: $\mu_1 > \mu_2$.
a. Let sample 1 be the white-collar workers (since $\bar{x}_1 > \bar{x}_2$ in that case), so that $n_1 = 9$, $n_2 = 8$, and the standard procedures for the t test apply. The rest of the steps are as follows:
2. Decision rule: Reject H_0 if $t^* > 1.753$.
3. Results: Letting sample 1 be white-collar workers, we have $\bar{x}_1 \doteq 23.56$, $s_1^2 \doteq 19.78$, $\bar{x}_2 = 18.5$, $s_2^2 \doteq 12.8571$; $\hat{\sigma} = \sqrt{[8(19.78) + 7(12.8571)]/15} \doteq 4.0681$, $t^* = (23.56 - 18.5)/(4.0681\sqrt{1/9 + 1/8}) \doteq 2.560$.
4. Decision: Reject H_0.
5. Conclusion: White-collar workers do have a higher mean than blue-collar workers. Since $t_{.01}\{15\} = 2.602$, we may conclude that $P < 0.05$ (but not less than 0.02).
b. In this case we must use the Fisher–Behrens t'. The number of degrees of freedom is found by using the formula, and we have $\nu = (19.78/9 + 12.8571/8)^2/[(19.78/9)^2/8 + (12.8571/8)^2/7] \doteq 14.88$, or about 15.
2. Decision rule: Reject H_0 if $t'^* > 1.753$.
3. Results: $t'^* = (23.56 - 18.5)/\sqrt{19.78/9 + 12.8571/8} \doteq 2.594$.
4. Decision: Reject H_0.
5. Conclusion: White-collar workers do have a higher mean than blue-collar workers. Since $t_{.01}\{15\} = 2.602$, we may conclude that $P < 0.025$ (but not less than 0.01).
c. Here we need to use the trimmed means. We trim one observation from each end. We can use the pooled variance and $h_1 = 7$, $h_2 = 6$. The rest of the steps are as follows:
2. Decision rule: Reject H_0 if $t^* > 1.796$.
3. Results: $\bar{x}_1(T) \doteq 24.29$, $s_1^2(W) \doteq 9.75$, $\bar{x}_2(T) = 18.83$, $s_2^2(W) \doteq 6.4107$; $\hat{\sigma} = \sqrt{[8(9.75) + 7(6.4107)]/11} \doteq 3.3422$ and $t^* = (24.29 - 18.83)/(3.3422\sqrt{1/7 + 1/6}) \doteq 2.94$.
4. Decision: Reject H_0.
5. Conclusion: White-collar workers do have a higher mean than blue-collar workers. Since $t_{.005}\{11\} = 3.106$, we may conclude that $P < 0.005$.
d. In this case we must use the Fisher–Behrens t'. We first find $s^2_{\bar{x}_1(T)}$ and $s^2_{\bar{x}_2(T)}$; $s^2_{\bar{x}_1(T)} = (9.75/7)(8/6) \doteq 1.8571$ and $s^2_{\bar{x}_2(T)} = (6.4107/6)(7/5) \doteq 1.4958$. Then the number of degrees of freedom is found by using the formula, so we have $\nu = (1.8571 + 1.4958)^2/[(1.8571)^2/6 + (1.4958)^2/5] \doteq 11$.
2. Decision rule: Reject H_0 if $t'^* > 1.796$.
3. Results: $t'^* \doteq (24.29 - 18.83)/\sqrt{1.8571 + 1.4958} \doteq 2.982$.
4. Decision: Reject H_0.
5. Conclusion: White-collar workers do have a higher mean than blue-collar workers. Since $t_{.01}\{11\} = 2.718$, and we may conclude that $P < 0.025$ (but not less than 0.01).

15. The differences are 1, 0, 3, 2, 1, 2, 1, 0, 1, 1. Then $\bar{d} = 1.2$, $s_d = 0.9189$, $n = 10$, and $t_{.005}\{9\} = 3.250$. Thus we have:
1. H_0: $\mu_1 = \mu_2$; H_1: $\mu_1 \neq \mu_2$.
2. Decision rule: Reject H_0 if $|t^*| > 3.250$.
3. Results: $t^* = 1.2/(0.9189/\sqrt{10}) \doteq 1.2/0.2906 \doteq 4.130$.
4. Decision: Reject H_0.
5. Conclusion: Subjects using toothpaste B have significantly fewer cavities.

16. In this case we are testing the hypothesis $\mu_A - \mu_B = 0$. This is a paired-sample test since, in each case, the tires are matched for wear and tear by being placed on the same motorcycle. The differences are $-4, 1, -8, -8, -6, -2$. Then, since $n = 6$ and the alternative is two-sided, $t_{.025}\{5\} = 2.571$. The mean of the differences is $\bar{d} \doteq -4.57$ and $s_d \doteq 3.56$. Then:
   1. H_0: $\mu_A - \mu_B = 0$; H_1: $\mu_A - \mu_B \neq 0$.
   2. Decision rule: Reject H_0 if $|t^*| > 2.571$.
   3. Results: $t^* = -4.57/(3.56/\sqrt{6}) \doteq -4.57/1.4534 \doteq 3.096$.
   4. Decision: Reject H_0.
   5. Conclusion: Tire B will wear less than tire A. They should market tire B.

17. **a.** -0.07 to 0.31 **b.** -0.02 to 0.26 **c.** 0.02 to 0.22 **d.** 0.05 to 0.19

18. $s_{p_1-p_2} \doteq 0.0601$. **a.** -0.139 to 0.059 **b.** -0.158 to 0.078
c. -0.180 to 0.100 **d.** -0.195 to 0.115

19. $p_1 = 0.1925$, $p_2 \doteq 0.1385$, $s_{p_1-p_2} \doteq 0.0186$; 0.017 to 0.091.

20. 1. H_0: $\pi_1 - \pi_2 = 0.05$; H_1: $\pi_1 - \pi_2 > 0.05$.
   2. Decision rule: Reject H_0 if $z^* > 1.645$.
   3. Results:
      a. $s_{p_1-p_2} \doteq 0.0992$; $z^* = [(0.54 - 0.42) - 0.05]/0.0992 \doteq 0.71$
      b. $s_{p_1-p_2} \doteq 0.0701$; $z^* \doteq 1.00$
      c. $s_{p_1-p_2} \doteq 0.0496$; $z^* \doteq 1.41$
      d. $s_{p_1-p_2} \doteq 0.0351$; $z^* \doteq 2.00$
   4. Decision: a, b, c: fail to reject H_0; d: reject H_0.
   5. Conclusion: a, b, c: $\pi_1 - \pi_2$ is not greater than 0.05; d: $\pi_1 - \pi_2 > 0.05$.

21. 1. H_0: $\pi_1 = \pi_2$; H_1: $\pi_1 \neq \pi_2$.
   2. Decision rule:
      a. Reject H_0 if $|z^*| > 1.645$.
      b. Reject H_0 if $|z^*| > 1.96$.
      c. Reject H_0 if $|z^*| > 2.33$.
      d. Reject H_0 if $|z^*| > 2.58$.
   3. Results: $\hat{\pi} = [(0.117)(60) + (0.271)(70)]/(60 + 70) = 0.2$ and $s_{\hat{\pi}} \doteq \sqrt{(0.2)(0.8)(1/60 + 1/70)} \doteq 0.0704$, so $z^* = (0.117 - 0.271)/0.0704 \doteq -2.19$.
   4. Decision: a, b: reject H_0; c, d: fail to reject H_0.
   5. Conclusion: a, b: $\pi_1 \neq \pi_2$; c, d: $\pi_1 = \pi_2$.

22. Proficiency Check 20: **a.** 0.2389 **b.** 0.1587 **c.** 0.0793 **d.** 0.0228 (one-tailed). Proficiency Check 21: 0.0286 (two-tailed).

23. 1. H_0: $\pi_1 = \pi_2$; H_1: $\pi_1 \neq \pi_2$.
   2. Decision rule: Reject H_0 if $|z^*| > 2.58$.
   3. Results: $\hat{\pi} = (154 + 108)/(800 + 780) \doteq 0.1658$ and $s_{\hat{\pi}} \doteq \sqrt{(0.1658)(0.8342)(1/800 + 1/780)} \doteq 0.0187$, so $z^* = (0.1925 - 0.1385)/0.0187 \doteq 2.89$.
   4. Decision: Reject H_0.
   5. Conclusion: Yes, there is a difference. Apparently a higher proportion in locality A are vegetarians. $P = 0.0038$.

24. $n = 2[(2.58)(0.30)/0.05]^2 \doteq 480$.

25. About 645.

26. 768 (using no prior estimates for π_1 and π_2) or 362 (if the sample results are used).

27. 1. H_0: $\sigma_1^2 = \sigma_2^2$; H_1: $\sigma_1^2 \neq \sigma_2^2$.
    2. Decision rule: Reject H_0 if $F^* > 1.71 = F_{.025}\{60, 50\}$.
    3. Results: **a.** $F^* = 1.84$; **b.** $F^* = 1.18$; **c.** $F^* = 1.17$; **d.** $F^* = 1.78$.
    4. Decision: a, d: reject H_0; b, c: fail to reject H_0.
    5. Conclusions: a, d: $\sigma_1^2 \neq \sigma_2^2$; b, c: $\sigma_1^2 = \sigma_2^2$.

28. 1. H_0: $\sigma_1^2 = \sigma_2^2$; H_1: $\sigma_1^2 \neq \sigma_2^2$.
    2. Decision rule: Reject H_0 if:
       a. $F^* > 4.12 = F_{.05}\{4, 7\}$
       b. $F^* > 6.09 = F_{.05}\{7, 4\}$
       c. $F^* > 3.41 = F_{.05}\{24, 7\}$
       d. $F^* > 1.92 = F_{.05}\{15, 40\}$
    3. Results: $F^* = 2.70$.
    4. Decision: a, b, c: fail to reject H_0; d: reject H_0.
    5. Conclusions: a, b, c: $\sigma_1^2 \neq \sigma_2^2$; d: $\sigma_1^2 = \sigma_2^2$.

29. Now $(3.58)^2/(2.18)^2 \doteq 2.70$; then:
    a. $F_{.025}\{4, 7\} = 5.52$ and $F_{.025}\{7, 4\} = 9.07$, so the limits are $2.70(1/5.52) \doteq 0.49$ and $2.70(9.07) \doteq 24.49$.
    b. $F_{.025}\{7, 4\} = 9.07$ and $F_{.025}\{4, 7\} = 5.52$, so the limits are $2.70(1/9.07) \doteq 0.30$ and $2.70(5.52) \doteq 14.90$.
    c. $F_{.025}\{24, 7\} = 4.42$ and $F_{.025}\{7, 24\} = 2.87$, so the limits are $2.70(1/4.42) \doteq 0.61$ and $2.70(2.87) \doteq 7.75$.
    d. $F_{.025}\{15, 40\} = 2.18$ and $F_{.025}\{40, 15\} = 2.59$, so the limits are $2.70(1/2.18) \doteq 1.24$ and $2.70(2.59) \doteq 6.99$.

References

Campbell, Donald T., and Julian C. Stanley. *Experimental and Quasi-Experimental Designs for Research.* Chicago: Rand McNally, 1963.

DeMeuse, Kenneth P. "Employees' Responses to Participation in an In-House Continuing Education Program: An Exploratory Study." *Psychological Reports* 57 (1985): 1099–1109.

Keller, John W., Dave Sherry, and Chris Piotrowski. "Perspectives on Death: A Developmental Study." *Journal of Psychology* 116 (1984): 137–142.

Neter, John, William Wasserman, and Michael H. Kutner. *Applied Linear Statistical Models.* 2nd ed. Homewood, IL: Richard D. Irwin, Inc., 1985.

Piotrowski, Chris, and Frances Y. Dunham. "Locus of Control Orientation and Perception of 'Hurricane' in Fifth Graders." *Journal of General Psychology* 109 (1983): 119–127.

SAS Institute, Inc., *SAS® User's Guide: Statistics, Version 5 Edition.* Cary, N.C.: SAS Institute, Inc., 1985.

Ware, Mary Catherine, and Mary Frances Stuck. "Sex-Role Messages vis-à-vis Microcomputer Use: A Look at the Pictures." *Sex Roles* 13 (1985): 205–214.

9

Chi-Square Analysis

One important topic in statistics is the analysis of count or enumerative data. **Count data** give the number of observations in several classifications or categories. Thus count data often result from counting numbers of observations of qualitative variables. In many cases an experimenter has hypothesized the distribution of count data and needs to see how closely the observed numbers agree with the hypothesized numbers. The famous genetic studies of Gregor Mendel were conducted to determine whether the results of his hybridization experiments agreed with his predicted distribution of characteristics in the offspring. His results agreed amazingly well—so well in fact that R. A. Fisher (1890–1962), in a famous paper, showed that it was highly unlikely that Mendel's results were obtained without a little help. Fisher believed that Mendel's gardener helped support his master's ideas out of a desire to please. (See Bennett [1965] for details.)

For many years experimenters who wished to demonstrate a close agreement between theoretical and observed count data merely published the data alongside the theoretical counts and drew conclusions according to how close they seemed to be. In 1900 Karl Pearson (1857–1936) published an important paper (Pearson [1900]) introducing the *chi-square distribution* for determining whether or not discrepancies between observed and theoretical counts were significant. The importance of this discovery was of the first rank and was named by *SCIENCE84* magazine as one of 20 discoveries that have shaped our lives (Hacking [1984]). This distribution is the same one that arose in connection with the sampling distribution of the sample variance (covered in Section 6.6). This application had been published in 1875 by Friedrich Helmert, but Helmert's work was apparently unknown to Pearson.

In this chapter we will study two applications of the chi-square distribution—to count data in a one-way classification and to count data in a two-way classification.

9.0 Case 9: The Controversial Coke

In mid-1985, the hottest topic of continuing discussion was probably the decision by the Coca Cola Bottling Company to change the formula for Coca Cola. Discussions ran pro and con, and occasionally they were quite heated. Many comparative tastings were held, some official and some not so official. In one of the less official ones, a blind taste test of the old Coke versus the new Coke versus Pepsi was conducted at McGuire's Irish Pub in Pensacola, Florida. Twenty-five people tasted all three without knowing which was which. Twelve preferred the old-formula Coke, seven preferred the new formula, and Pepsi was close behind with six people saying they liked it best. Only three people correctly named which cups had which drink. Were these results meaningful, or would such a distribution of results occur by chance if the participants simply chose one of the drinks at random?

9.1 Testing Goodness of Fit

Karl Pearson's discovery of the chi-square distribution made it possible to perform statistical tests on count data. If we can obtain the number of observations that would be expected in each classification, we can compare them to the numbers of actual or observed observations in order to see how well the observed data fit the hypothetical distribution of the data. In this section we use the chi-square (χ^2) distribution to compare the observations of a single categorical variable with theoretical values. In this way we can confirm or deny a hypothesis about the distribution of the variable.

9.1.1 The Chi-Square Distribution

The distribution of the **chi-square statistic,** χ^2, is determined completely by the degrees of freedom, ν. Like the t distribution, the particular shape of a chi-square distribution depends on ν, and as ν increases the curve becomes very nearly a normal curve (although not a standardized one); unlike the t distribution, however, the graph of a chi-square distribution changes quite a bit as ν increases. Values of the variable can never be negative, so the graph is bounded below at zero; it is positively skewed. The degree of skewness decreases as ν increases, however, and the graph then approaches symmetry. A chi-square distribution with ν degrees of freedom is symbolized $\chi^2(\nu)$. The variable χ^2 has mean $\mu_{\chi^2} = \nu$ and variance $\sigma^2_{\chi^2} = 2\nu$. Chi-square distributions for $\nu = 1$, 4, and 10 are graphed in Figure 9.1.

The chi-square distributions have been studied extensively, and tables have been prepared for a large number of degrees of freedom. Critical values of chi square are symbolized $\chi^2_\alpha\{\nu\}$; the proportion of the area under the graph of $\chi^2(\nu)$ to the right of $\chi^2_\alpha\{\nu\}$ is α. Table 6 at the back of the book gives critical values of chi square; for the purposes of this chapter, we will usually need values of $\chi^2_{.05}\{\nu\}$, $\chi^2_{.01}\{\nu\}$, and occasionally $\chi^2_{.10}\{\nu\}$. For 6 degrees of freedom, for example, $\chi^2_{.05}\{6\} = 12.5916$ and $\chi^2_{.01}\{6\} = 16.8119$. These values are shown in Figures 9.2 and 9.3, respectively.

Suppose that a retailer is interested in the consumer preference of four different brands of instant coffee and obtains information on the last 800 sales of instant coffee at a retail outlet. For brands A, B, C, and D, respec-

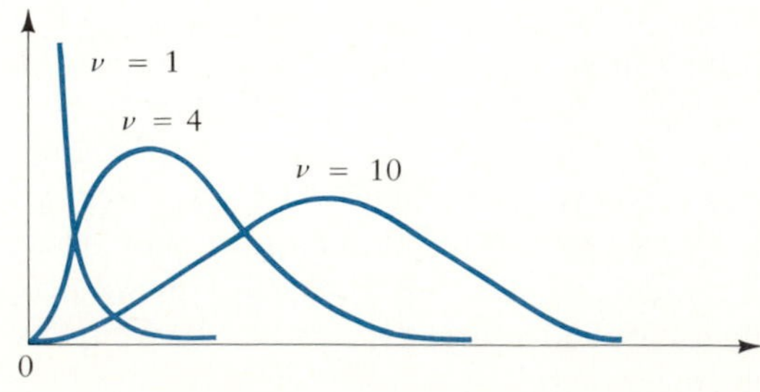

Figure 9.1
Three chi-square distributions.

tively, the numbers buying each brand are 190, 198, 187, and 225. If we hypothesize that all brands are equally popular, we would expect 200 people to buy each one. Since we know the number expected in each category, we can obtain a statistic whose sampling distribution can be compared to the chi-square distribution. Suppose we have a random sample from a population and can obtain the **expected frequency** (f_e) in each of the categories into which count data are classified. Then if f_o is the **observed frequency** in each of the categories, the statistic $\Sigma[(f_o - f_e)^2/f_e]$ has approximately a chi-square distribution for sufficiently large samples. Samples are usually considered to be "sufficiently large" if each expected frequency is at least 5. Some statistics texts use X^2 for an obtained sample value of the test statistic; in keeping with our previous practice, we will use χ^{2*}.

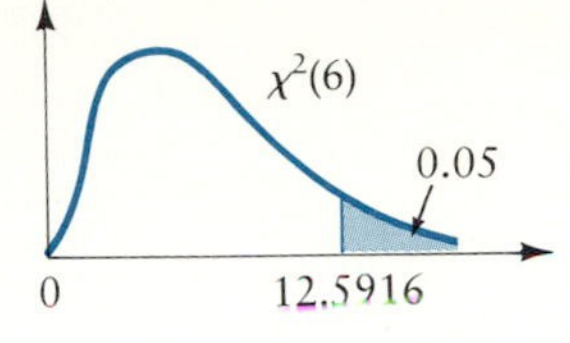

Figure 9.2
Curve showing $\chi^2_{.05}\{6\}$.

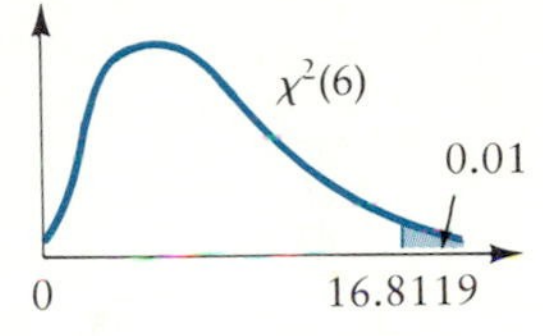

Figure 9.3
Curve showing $\chi^2_{.01}\{6\}$.

The value of $(f_o - f_e)^2/f_e$ for each category gives a measure of the proportionate deviation of the observed values from the expected values. If χ^{2*} is small (close to zero), the agreement between observed and expected values is good; if χ^{2*} is large, the agreement is not good. We can use the chi-square distribution to determine how much disagreement is enough to enable us to decide that the observed data do not fit the hypothetical distribution.

Throughout this chapter we adhere to the requirement that each value of f_e must be at least 5. The primary reason for the restriction is that small differences between f_o and f_e can be magnified if f_e, the denominator of the fraction $(f_o - f_e)^2/f_e$, is small and the difference is fairly large in comparison to f_e. William G. Cochran (1909–1980) made a study of analyses with small values of f_e and wrote a number of papers on the subject. These papers are summarized in an article by Fienberg (1984). Cochran suggested that values of f_e as small as 2 are acceptable if f_o and f_e are fairly close. As a rule of thumb, if there are small values of f_e but χ^{2*} is not greater than the critical value, the procedure is satisfactory. If χ^{2*} is greater than the critical value but categories with small values of f_e are not major contributors to the total, the procedure is again satisfactory. If those categories are a major contributing factor to a significant value of χ^{2*}, however, the results are probably not valid and some modifications (shown later) should take place.

There is a shortcut formula for χ^{2*}. It can be shown that $\chi^{2*} = \Sigma[f_o^2/f_e] - N$, where N is the total number of observations. This formula is often easier to use than the defining formula. (See the note at the end of the chapter.)

As we noted earlier, if the observed and expected frequencies in each category are close together, the value of χ^{2*} will be close to zero, since in each category $(f_o - f_e)^2$ would be close to zero. If the observed and expected frequencies in some categories are not close, then $(f_o - f_e)^2$ will be large for those categories and χ^{2*} will be large. Thus a hypothesis that sample data are drawn from a population possessing the theoretical distribution can be rejected for large values of χ^{2*}. Since the sampling distribution of our test statistic is known, we may apply the hypothesis-testing techniques of previous chapters and obtain a critical value of χ^2—a dividing line between rejection of the null hypothesis and failure to reject it, based on the level of significance (α) decided upon beforehand.

The only constraint on the sample value of the test statistic is that the total number of observations must equal N; thus the number of degrees of freedom for the test statistic is $k - 1$, where k is the number of categories or classifications.

When values of f_e are smaller than 5, some modifications may be made. One way to satisfy this assumption is to combine categories so that each f_e will be at least 5. As mentioned earlier, we may relax this assumption in a few special cases.

Testing Goodness of Fit

If a random sample of data drawn from a population is classified into several categories, we may test the hypothesis H_0: the data conform to a specified distribution. Such a test is a test of "goodness of fit." The sample value of the test statistic used to test this hypothesis is χ^{2*}, where

$$\chi^{2*} = \sum \frac{(f_o - f_e)^2}{f_e} = \sum \frac{f_o^2}{f_e} - N$$

provided each value of f_e is at least 5.

The decision rule will be: Reject H_0 if $\chi^{2*} > \chi^2_\alpha\{k - 1\}$, where α is the level of significance and k is the number of categories.

Example 9.1

Suppose that the retailer mentioned earlier obtained the results given. Test the hypothesis, using the 0.05 level of significance, that the four brands of coffee are equally popular.

Solution

We would expect 200 people to buy each brand. Thus our null hypothesis is H_0: the brands are equally popular in the population; the alternate hypothesis is that the brands are not equally popular. The decision rule is to reject H_0 if $\chi^{2*} > 7.815$, as $\chi^2_{.05}\{3\} = 7.815$. The sample value of the test statistic is χ^{2*}, where

$$\chi^{2*} = \frac{190^2}{200} + \frac{198^2}{200} + \frac{187^2}{200} + \frac{225^2}{200} - 800 = 4.49$$

Since 4.49 does not exceed 7.815, we fail to reject H_0 and may conclude that the differences are due to chance.

Note that although we used the hypothesis-testing technique and format, it was presented somewhat less formally than in previous chapters. We will use this shortcut form from now on.

Example 9.2

A firm wishes to estimate the economic composition of a community where they are considering opening a branch office. They have examined data concerning this and similar communities and have obtained a hypothetical distribution of income that they feel to be approximately accurate. To check this accuracy, they take a random sample of 500 families and family incomes. They have hypothesized the following distribution of income levels.

CATEGORY	INCOME LEVEL	PROPORTION
I	\$30,000 or over	0.40
II	\$25,000–\$29,999	0.20
III	\$20,000–\$24,999	0.20
IV	\$15,000–\$19,999	0.10
V	under \$15,000	0.10

The sample yields the following results:

Category	I	II	III	IV	V	Total
Number	166	97	134	61	42	500

Do these results substantiate the hypothesized distribution at the 0.05 level of significance?

Solution

Here the hypothesis is that the proportions of the population in each category are 0.40, 0.20, 0.20, 0.10, and 0.10, as shown. Since $k = 5$, there are 4 degrees of freedom and $\chi^2_{.05}\{4\} = 9.488$. We reject H_0 if $\chi^{2*} > 9.488$. Since the total number of families is 500, we can calculate the expected number for each category by multiplying each expected proportion by 500. Then from the sample results,

$$\begin{aligned}\chi^{2*} &= \frac{166^2}{200} + \frac{97^2}{100} + \frac{134^2}{100} + \frac{61^2}{50} + \frac{42^2}{50} - 500 \\ &= 21.13\end{aligned}$$

Since $21.13 > 9.488$, reject H_0 and conclude that the population income levels are not described by the hypothetical distribution. We note that $\chi^2_{.005}\{4\} = 14.86$; thus the P value for this test is less than 0.005.

PROFICIENCY CHECKS

1. Observed frequencies in five categories are 22, 44, 31, 19, and 34. Using the 0.05 level of significance, test the hypothesis that the data come from a population with equal proportions in each category.
2. Using the data of Proficiency Check 1, test the hypothesis that the data come from a population in which the categories, in the order given, are in the ratio 2:4:3:2:4. Use the 0.05 level of significance.
3. To determine whether there is any relationship between age and fear of the dark in persons known to be afraid of the dark, a psychologist obtains age data on 200 persons under care for fear of the dark. Ages range from 5 to 33 and are as shown:

Age group	5–9	10–14	15–19	20–24	25–29	30–34
Number	31	53	41	29	24	22

For the local area, the age groups constitute percentages of the population as shown:

Age group	5–9	10–14	15–19	20–24	25–29	30–34
Percentage	6.1	7.3	9.3	8.5	9.4	7.9

Test the hypothesis that there is no relationship between age level and fear of the dark—that is, that the distribution of ages in the sample is not significantly different from that in the population. Use a 0.01 level of significance.

9.1.2 Testing Normality*

Many of the tests in Chapters 6, 7, and 8 are based on the assumption that the distribution from which the sample was taken is normal. We can use chi-square analysis, as well as several other methods, to test this assumption. (The Kolmogorov–Smirnov Test, for example, is widely used to test normality.)

The procedure can be outlined as follows. To determine whether it is reasonable to conclude that a set of sample data is likely to have come from a normal population, first use the mean and standard deviation of the sample to estimate the mean and standard deviation of the population if not known or assumed. Then group the sample data into convenient classes, calculate sample z values for the class boundaries, and determine the area under the standard normal curve between these values to obtain the hypothetical proportion of the sample in each class. Multiply each proportion

*This section may be omitted without loss of continuity.

by the total number of observations to obtain f_e for each class, and calculate χ^{2*} in the usual way. The chi-square test statistic for this result will have $k - 1 - m$ degrees of freedom, where k is the number of classes and m is the number of population parameters estimated. If we use the sample mean and standard deviation to estimate the population mean and standard deviation, then $m = 2$ and the number of degrees of freedom will be $k - 3$. The hypothesis being tested is that the sample came from a population with a normal distribution.

Example 9.3

A frequency distribution of family incomes was given in Table 2.6 and is reproduced in Table 9.1. Use a chi-square test to determine whether it is reasonable to conclude that the population from which the sample is taken is a normal population. Use $\alpha = 0.10$.

Solution

We first obtain the mean and standard deviation of the sample. We find that $\bar{x} \doteq \$10{,}055$ and $s \doteq \$6028$. If the sample is from a normal population, we would probably not have a need for classes more than 3 standard deviations above or below the mean, so we will not have a need for any classes beginning above about 28,000; since the data are bounded below at zero, we would not have any area under the curve below zero. Using the sample z-score transformation, we determine the sample z values for the class boundaries and the areas (from Table 4) under the normal curve between the mean and each class boundary from the upper boundary of the first

Table 9.1 DISTRIBUTION OF FAMILY INCOMES.

INCOME ($)	NUMBER OF FAMILIES
42,000–44,999	3
39,000–41,999	7
36,000–38,999	12
33,000–35,999	17
30,000–32,999	28
27,000–29,999	60
24,000–26,999	216
21,000–23,999	268
18,000–20,999	448
15,000–17,999	621
12,000–14,999	949
9000–11,999	1421
6000– 8999	2844
3000– 5999	2123
0– 2999	294

class to 26,999.5. For the boundary 5999.5, for example, $z = (5999.5 - 10{,}055)/6028 \doteq -0.67$. From Table 4, the area under the standard normal curve between -0.67 and zero is 0.2486. The results are shown in Table 9.2.

We can find the hypothetical proportion in each class by determining the area under the standard normal curve between these boundaries. We determine these areas by taking the differences in the areas for the class boundaries from Table 9.2 (the sum for the class containing the mean). Since $n = 9311$, we can obtain values of f_e by multiplying each hypothetical proportion by 9311. These values are listed in Table 9.3.

Some of the results obtained in Table 9.3 are shown graphically in Figure 9.4, superimposed on a normal curve.

Table 9.2 Z SCORES AND AREAS FOR CLASS BOUNDARIES.

CLASS BOUNDARY	z	AREA
2999.5	−1.17	0.3790
5999.5	−0.67	0.2486
8999.5	−0.18	0.0714
11,999.5	0.32	0.1255
14,999.5	0.82	0.2939
17,999.5	1.32	0.4066
20,999.5	1.82	0.4656
23,999.5	2.31	0.4896
26,999.5	2.81	0.4975

Table 9.3 CALCULATION OF EXPECTED FREQUENCIES FOR EACH CLASS.

CLASS	HYPOTHETICAL PROPORTION	f_e
Below 3000	0.5000 − 0.3790 = 0.1210	1126.63
3000– 5999	0.3790 − 0.2486 = 0.1304	1214.15
6000– 8999	0.2486 − 0.0714 = 0.1772	1649.91
9000–11,999	0.0714 + 0.1255 = 0.1969	1833.34
12,000–14,999	0.2939 − 0.1255 = 0.1684	1567.97
15,000–17,999	0.4066 − 0.2939 = 0.1127	1049.35
18,000–20,999	0.4656 − 0.4066 = 0.0590	549.35
21,000–23,999	0.4896 − 0.4656 = 0.0240	223.46
24,000–26,999	0.4975 − 0.4896 = 0.0079	73.56
27,000 or over	0.5000 − 0.4975 = 0.0025	23.28

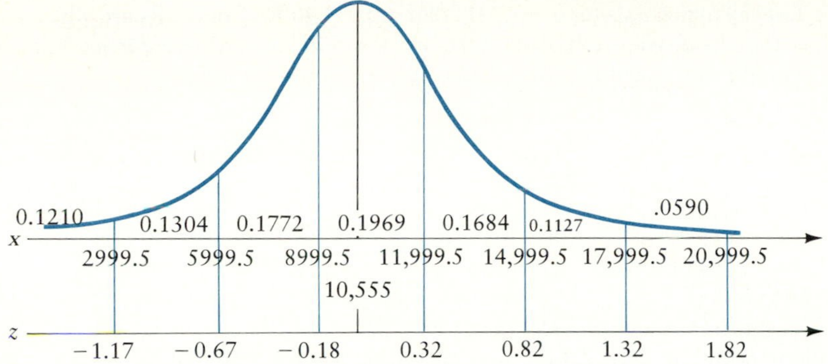

Figure 9.4
Income data of Table 9.3 superimposed on a normal curve for Example 9.3.

Then

$$\begin{aligned}\chi^{2*} &= \frac{294^2}{1126.63} + \frac{2123^2}{1214.15} + \frac{2844^2}{1649.91} + \frac{1421^2}{1833.34} + \frac{949^2}{1567.94} \\ &\quad + \frac{621^2}{1049.35} + \frac{448^2}{549.35} + \frac{268^2}{223.46} + \frac{216^2}{73.56} + \frac{127^2}{23.28} - 9311 \\ &\doteq 3437\end{aligned}$$

There are $10 - 1 - 2 = 7$ degrees of freedom and $\chi^2_{.05}\{7\} = 14.07$ and $\chi^2_{.01}\{7\} = 18.48$. We did not specify a level of significance, but the obtained value of χ^{2*} is significant well beyond the 0.01 level. Thus we can safely conclude that the sample is not from a normal distribution. If we graph the data, we can see strong positive skewing.

Problems

Practice

9.1 We want to test a die to see if it is fair. We roll it 60 times with the following results. Using $\alpha = 0.05$, test the fairness of the die.

Number of dots	1	2	3	4	5	6
Frequency	8	13	10	8	12	9

9.2 Four balls are drawn from an urn, with replacement, and the number of white balls in each set of four is recorded. The experiment is repeated 100 times with the following results.

NUMBER OF WHITE BALLS	FREQUENCY
0	11
1	16
2	38
3	33
4	2

Test the hypothesis (at $\alpha = 0.01$) that exactly 40% of the balls are white—that is, that the distribution of outcomes is not significantly different from that which would be obtained in a binomial distribution with $p = 0.4$.

Applications

9.3 David Phillips (see Tanur [1978]) examined the dates of death of 1251 notable Americans—persons likely to receive more than usual attention on their birthdays. He noticed that the incidence of death in the 6 months preceding the birthdays of these people was quite a bit below what would be expected, indicating that many of these people may have postponed their deaths until after their birthdays. This idea is called the *death-dip hypothesis*. Phillips' data (involving the period 1897–1960) gave the number of months before or after the birthmonth in which these notable people died. Given the number of people dying in each month as shown here, what do you conclude?

6 mo. before	5 mo. before	4 mo. before	3 mo. before	2 mo. before	1 mo. before
90	100	87	96	101	86

Birth month	1 mo. after	2 mo. after	3 mo. after	4 mo. after	5 mo. after
119	118	121	114	113	106

9.4 When a bridge mix (candy) is boxed, the manufacturer claims that the pieces include chocolate-covered peanuts, chocolate-covered raisins, nougat centers, and fudge centers in the ratio 3:3:2:2. A random sample of 200 pieces of the mix contains 43 peanuts, 51 raisins, 54 nougat centers, and 52 fudge centers. Should the manufacturer check on the quality control for the mixture? Use the 0.05 level of significance.

9.5 A scholarship committee is to award 40 scholarships chosen from among a very large number of applicants. Grade-point average (GPA) is one of the pieces of information to be considered, but it is not supposed to carry undue weight. The committee finds that 42% of the applicants have GPAs over 3.5, 24% have GPAs from 3.2 to 3.5, 21% have GPAs from 3.0 to 3.2, and the remainder have GPAs from 2.8 to 3.0. When the scholarships are awarded, 23 of the 40 students have GPAs over 3.5, 10 have GPAs from 3.2 to 3.5, 6 have GPAs from 3.0 to 3.2, and only one has a GPA under 3.0. Using the 0.05 level of significance, determine whether the distribution of GPAs in the group awarded the scholarships differs significantly from that of the population. Interpret this result.

9.6 Tests for goodness of fit are particularly useful in genetics to test whether a set of offspring has bred true to type or is mutant. A biologist examines several litters of guinea pigs with respect to color of coat and length of hair. Genetic probabilities are computed, and a hypothetical ratio for the respective categories shown here is found to be 12:6:4:2:18:12:6:4. If the offspring differ significantly from the expected frequencies, it can be speculated that the litters may contain mutations. Use the 0.05 level of significance to determine whether the results shown here could have come from a population with the hypothetical proportions.

CHARACTERISTIC	NUMBER OF OFFSPRING
long hair, brown-white	61
long hair, black-white	33
long hair, brown-black	23
long hair, white	12
short hair, brown-white	72
short hair, black-white	49
short hair, brown-black	21
short hair, white	12

9.7 In a study by Ware and Stuck (1985), a total of 426 illustrations were examined in three mass market computer magazines for October 1982, January 1983, and April 1983. The illustrations included 727 separate individuals: 465 men, 196 women, 38 boys, and 28 girls. The general population includes 34.25% men, 37.55% women, 14.44% boys, and 13.76% girls. Can we conclude, at the 0.01 level of significance, that this distribution is representative of the population?

9.8 A production line includes a machine that is prone to breakdown. In order to plan more effectively, an efficiency expert wants to test the hypothesis that the number of breakdowns follows a Poisson distribution. The number of breakdowns per day for 90 days is recorded; there were no breakdowns on 48 days, one breakdown on 30 days, two breakdowns on 10 days, and three breakdowns on 2 days. No day had more than three breakdowns. Use the 0.01 level of significance to test the hypothesis.

9.9 The distribution of ages of employees of a large corporation is hypothesized to follow a normal distribution. A sample of 1322 employees has the distribution of ages shown in the table on the right. Use the method of Example 9.3 to determine whether or not the sample could reasonably have come from a normal population.

AGES	NUMBER
18–25	144
26–35	242
36–45	365
46–55	324
56–65	202
66–75	45

9.2 Contingency Tables

In Section 9.1, we used the chi-square distribution to test the hypothesis of goodness of fit when data are arranged in a one-way classification. In such a classification, the observations of just one variable of interest are counted. When we arrange count data in a two-way classification, the resulting table is called a **contingency table.** Suppose, for instance, that the chairman of a university statistics department wishes to assess the effect of prior course work in logic and college algebra on the success of students in statistics. Grades of students in a year's statistics classes at the university are obtained, and each grade is classified by whether or not the student had done prior course work in logic or college algebra. The resulting contingency table is shown in Table 9.4.

Table 9.4 CONTINGENCY TABLE FOR GRADE IN STATISTICS AND PRIOR COURSE WORK.

	GRADE					
PRIOR COURSE WORK	A	B	C	D	F	TOTAL
1. College algebra, but not logic	15	20	40	5	0	80
2. Logic, but not college algebra	10	15	70	20	5	120
3. Both college algebra and logic	10	20	25	5	0	60
4. Neither college algebra nor logic	15	15	75	30	55	190
Total	50	70	210	60	60	450

9.2.1 Testing Independence of Factors

The purpose of analyzing the data presented in Table 9.4 is to determine whether there is a relation between grade in statistics and prior course work in college algebra and logic, or if the two factors are independent. The various categories of the factors are called **levels** of the factors. Thus the factor "prior course work" has four levels, while the factor "grade" has five levels. The intersection in the table of one level of a factor with one level of the other factor is called a **cell.** There are 20 cells in Table 9.4. In general, if a contingency table has r rows and c columns, the number of cells is rc.

If we can determine the number of expected frequencies in each cell, the statistic $\Sigma[(f_o - f_e)^2/f_e]$ will have a chi-square distribution, as in Section 9.1. As will be shown, the number of degrees of freedom is equal to $(r - 1)(c - 1)$.

We may determine expected frequencies in each cell if we assume that the factors are independent. Since 80/450 of the students had prior work in college algebra but not logic, we use this proportion as an estimate of the proportion of all students taking statistics who have had this prior course work. Since 50/450 of the students received an A in the course, we use this ratio as an estimate of the proportion of all students taking statistics who receive an A. If the factors are independent, the joint probability that a statistics student selected at random will have had prior course work in college algebra but not logic *and* will receive an A is (80/450)(50/450). Finally, the expected number of 450 students who will have had prior course work in college algebra but not logic *and* will receive an A is this probability times the number of students, 450, or (80/450)(50/450)(450). This calculation may be simplified to (80)(50)/450.

In general, if $f_e(i, j)$ is the expected frequency in cell (i, j) (the cell in the ith row and the jth column) $f_r(i)$ is the observed frequency in the ith row, $f_c(j)$ is the observed frequency in the jth column, and N is the total number of observations, then $f_e(i, j) = [f_r(i) \cdot f_c(j)]/N$. To illustrate, let us find the value of $f_e(2, 3)$; this number is the expected frequency in the second row

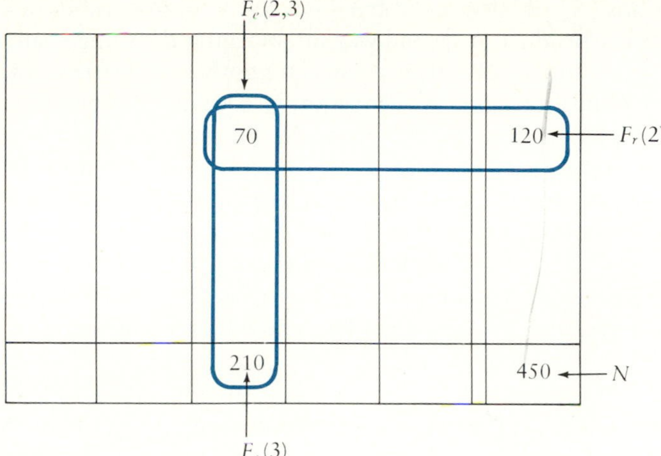

Figure 9.5
Determining the value of $f_e(2, 3)$ in the contingency table.

and third column—students who have had prior work in logic but not college algebra *and* who received a C. Figure 9.5 illustrates the procedure graphically.

Since $f_r(2) = 120$, $f_c(3) = 210$, and $N = 450$, then $f_e(2, 3) = (120 \cdot 210)/450$, or 56. We can calculate the remainder of the expected frequencies in a similar manner, and these are shown in parentheses in Table 9.5. For simplicity, levels of the factor "prior course work" are listed 1, 2, 3, 4, corresponding to those in Table 9.4.

To test the null hypothesis H_0: the factors are independent, we can obtain χ^{2*} and reject H_0 if $\chi^{2*} > \chi^2_\alpha\{\nu\}$, where $\nu = (r - 1)(c - 1)$. We can justify the number of degrees of freedom in either of two ways. First, each column contains r cells, with the constraint that the frequencies in the jth column must total $f_c(j)$. Thus each column has $(r - 1)$ degrees of freedom. Similarly,

Table 9.5 OBSERVED AND EXPECTED VALUES.

	GRADE				
PRIOR COURSES	A	B	C	D	F
1	15	20	40	5	0
	(8.9)	(12.4)	(37.3)	(10.7)	(10.7)
2	10	15	70	20	5
	(13.3)	(18.7)	(56.0)	(16.0)	(16.0)
3	10	20	25	5	0
	(6.7)	(9.3)	(28.0)	(8.0)	(8.0)
4	15	15	75	30	55
	(21.1)	(29.6)	(88.7)	(25.3)	(25.3)

each row has $(c - 1)$ degrees of freedom, so the total number of degrees of freedom is the product. A second way of obtaining the same result is to note that the total number of cells is rc, so the number of degrees of freedom is equal to rc minus the number of constraints. One constraint is that the total number of observations must sum to N. Further, we must estimate the population proportion for each level of each factor. We used 50/450, for instance, to estimate the proportion of statistics students receiving A grades. Since the column factor has r levels, we must estimate the population proportion for all but one of the levels—the last one must be 1.00 minus the sum of the rest—so we estimate $r - 1$ proportions. Similarly, for the row factor we estimate $c - 1$ proportions. The number of degrees of freedom, then, is $rc - (r - 1) - (c - 1) - 1 = rc - r - c + 1 = (r - 1)(c - 1)$.

We proceed to test the hypothesis that a student's grade in statistics is independent of prior course work in algebra and logic, using the 0.01 level of significance. Since all values of f_e are at least 5, we may use the chi-square test statistic.

The number of degrees of freedom is $(4 - 1)(5 - 1) = 12$, so the critical value is $\chi^2_{.01}\{12\} = 26.22$. Thus we are testing the hypothesis H_0: grade achievement in statistics and prior course work in college algebra and/or logic are independent factors, and we will reject H_0 if $\chi^{2*} > 26.22$. Using the shortcut formula we have

$$\begin{aligned}\chi^{2*} &= 15^2/8.9 + 20^2/12.4 + 40^2/37.3 + 5^2/10.7 + 0/10.7 \\ &\quad + 10^2/13.3 + 15^2/18.7 + 70^2/5 + 20^2/16 + 5^2/16 \\ &\quad + 10^2/6.7 + 20^2/9.3 + 25^2/28 + 5^2/8 + 0/8 + 15^2/21.1 \\ &\quad + 15^2/29.6 + 75^2/88.7 + 30^2/25.3 + 55^2/25.3 - 450 \\ &\doteq 106.59\end{aligned}$$

Since $106.59 > 26.22$, we reject H_0 and conclude that there is some relationship between success in statistics and prior course work in college algebra or logic (or both). We summarize the rule for testing independence of factors on the next page.

If the decision is to reject the hypothesis of independence, we must take care in interpreting the results. The precise nature of the dependence is not always easy to determine. Furthermore, statistical dependence, as shown by our analysis of contingency tables, does not imply a cause-and-effect relationship. Sometimes further analysis of the data may be helpful. In the previous example, we could combine lines 1 and 3 and lines 2 and 4 and compare the grades with college algebra against those without college algebra ($\chi^{2*} \doteq 547$); we could combine lines 2 and 3 and lines 1 and 4 and compare the grades with logic against those without logic ($\chi^{2*} \doteq 30.5$); or we could compare line 1 with line 2 ($\chi^{2*} \doteq 15.86$), concluding that there is a difference between the statistics grades of those who had prior course work in algebra and those who had prior course work in logic. Quite often, however, our results are not conclusive. Further, repeated testing increases the probability of at least one type I error. See the discussion in the introduction to Chapter 11 and in Case 11.

Chi-Square Test For Independence

If a random sample of data is classified in a contingency table, we may test the hypothesis H_0: the factors are independent. The sample value of the test statistic used to test this hypothesis is x^{2*}, where

$$\chi^{2*} = \sum \frac{(f_o - f_e)^2}{f_e} = \sum \frac{f_o^2}{f_e} - N$$

and each value of f_e is at least five.

The decision rule will be as follows: Reject H_0 if $\chi^{2*} > \chi^2_\alpha\{(r-1)\cdot(c-1)\}$, where α is the level of significance, r is the number of rows, and c is the number of columns.

Example 9.4

A college collects income data for students 5 years after graduation. Excluding students still in school, the data on income level as compared with undergraduate grade-point average (GPA) is shown in Table 9.6. Test the hypothesis, at the 0.05 level of significance, that there is no relationship between undergraduate GPA and income level 5 years after graduation—that is, that the two factors are independent.

Solution

First we calculate expected frequencies for each cell, as shown in Table 9.7. Since two of the expected frequencies are below 5, we need to combine levels of one or both factors to bring expected frequencies to at least 5. To illustrate the method, we will combine factor levels.

Table 9.6 CONTINGENCY TABLE FOR EXAMPLE 9.4.

	INCOME CATEGORY (DOLLARS PER YEAR)				
	1	2	3	4	5
GPA	UNDER $15,000	$15,000–$19,999	$20,000–$24,999	$25,000–$29,999	$30,000 AND UP
3.50–4.00	1	18	22	9	2
3.00–3.49	5	63	39	16	11
2.50–2.99	10	156	77	34	14
2.00–2.49	11	80	46	38	2

Table 9.7 OBSERVED AND EXPECTED VALUES.

	INCOME CATEGORY					
GPA	1	2	3	4	5	TOTAL
3.50–4.00	1	18	22	9	2	52
	(2.1)	(25.2)	(14.6)	(7.7)	(2.3)	
3.00–3.49	5	63	39	16	11	134
	(5.5)	(65.0)	(37.7)	(19.9)	(5.9)	
2.50–2.99	10	156	77	34	14	291
	(12.0)	(141.1)	(81.9)	(43.2)	(12.9)	
2.00–2.49	11	80	46	38	2	177
	(7.3)	(85.8)	(49.8)	(26.3)	(7.8)	
Total	27	317	184	97	29	654

To compensate for the small expected frequency, 2.1, in cell (1, 1), we could combine columns 1 and 2 or rows 1 and 2. To compensate similarly for cell (1, 5), we could combine columns 4 and 5 or rows 1 and 2. In order to decide which to combine, we use the following rule: *when combining cells in order to compensate for small expected frequencies, combine rows or columns in such a way as to retain the largest possible number of degrees of freedom.* In this way we retain the largest number of cells and thus have a greater degree of sensitivity—a better chance of detecting differences if they exist.

If we combine columns, we will have three columns and four rows, so we will have $(4 - 1)(3 - 1) = 6$ degrees of freedom; if we combine rows, we will have three rows and five columns, so we will have $(3 - 1)(5 - 1) = 8$ degrees of freedom. Thus we should combine rows 1 and 2. Because of possible rounding errors, we should recalculate values of f_e instead of just combining those from the previous table. The revised contingency table, with expected frequencies in parentheses, is given in Table 9.8.

Table 9.8 OBSERVED AND EXPECTED VALUES FOR COMBINED ROWS.

	INCOME CATEGORY					
GPA	1	2	3	4	5	TOTAL
3.00–4.00	6	81	61	25	13	186
	(7.7)	(90.2)	(52.3)	(27.6)	(8.2)	
2.50–2.99	10	156	77	34	14	291
	(12.0)	(141.1)	(81.9)	(43.2)	(12.9)	
2.00–2.49	11	80	46	38	2	177
	(7.3)	(85.8)	(49.8)	(26.3)	(7.8)	
Total	27	317	184	97	29	654

We then obtain $\chi^{2*} = 6^2/7.7 + 81^2/90.2 + \cdots + 2^2/7.8 - 654 \doteq 22.04$. Since $\chi^2_{.05}\{8\} = 15.51$, we can reject H_0—that undergraduate GPA and income level 5 years after graduation are independent—and conclude that there is some relationship between the factors. As in the previous example, however, further analysis is needed to shed more light on the nature of the relationship.

PROFICIENCY CHECKS

4. Determine whether factors A and B are independent for the data in the following table. Use the 0.05 level of significance.

	FACTOR A				
FACTOR B	1	2	3	4	5
1	21	16	29	33	21
2	26	20	32	26	16
3	33	24	29	21	13

5. A study is conducted on the relationship between religious affiliation and attitude toward artificial birth control. The three major religious groups in the area are represented. Use the following data to determine whether the factors are independent. Use $\alpha = 0.01$.

	RELIGIOUS GROUP		
ATTITUDE	A	B	C
For birth control	123	64	43
Against birth control	77	86	57

9.2.2 The 2 × 2 Contingency Table

If each of the factors has only two levels, we may use a special formula to simplify calculations. Instead of calculating f_e for each cell, we use a simple formula making use of row and column totals. Suppose we designate the cell frequencies *a, b, c, d,* as shown here,

a	*b*
c	*d*

and let the total of all observed frequencies be N. Then

$$\chi^{2*} = \frac{N(ad - bc)^2}{(a + b)(c + d)(a + c)(b + d)}$$

where $a + b$ and $c + d$ are the row totals, and $a + c$ and $b + d$ are the column totals.

Note that the chi-square distribution is continuous but the values of a, b, c, d are discrete. For small values of f_e, this discrepancy may cause the obtained value of χ^2 to be slightly larger than is warranted by the data. Thus we are more likely to reject H_0 than the data actually justify. Frank Yates (1934) suggested using a continuity correction as we do when we use the normal distribution to approximate a discrete distribution. He suggested reducing $|f_o - f_e|$ by 0.5 before squaring the term in the basic formula. This continuity correction is called **Yates' correction factor** and should be used if any f_e in a 2×2 table is less than 10. As a result, the obtained value χ^{2*} is decreased slightly, thus making the test a little more conservative. Mosteller (1973, p. 174) observes that this correction factor should not be used for larger tables, where it overcorrects. For very small samples ($N < 20$) an alternative test is Fisher's exact test. See Mosteller and others (1970, p. 317), or Conover (1971, pp. 163ff) for details. If Yates' correction factor is needed, a slight modification of the basic formula yields

$$\chi^{2*} = \frac{N(|ad - bc| - N/2)^2}{(a + b)(c + d)(a + c)(b + d)}$$

Example 9.5

A mathematics instructor wonders whether fear of mathematics is related to sex. An attitude inventory test is administered to 200 students to determine their anxiety level toward mathematics. Each student is classified high or low in "math anxiety." Students are also classified according to sex. The results are shown in Table 9.9. Determine (at the 0.05 level of significance) whether math anxiety and sex are independent factors.

Solution

Here $a = 36$, $b = 52$, $c = 26$, $d = 86$, the row totals are 88 and 112, the column totals are 62 and 138, and $N = 200$. Then

$$\chi^{2*} = \frac{200(36 \cdot 86 - 26 \cdot 52)^2}{(88)(112)(62)(138)} \doteq 7.21$$

Since $\chi^2_{.05}\{1\} = 3.841$, we reject the hypothesis and conclude that there is some relation between sex and math anxiety. In this group a higher proportion of males than females has a high level of anxiety.

Table 9.9 MATH ANXIETY LEVEL BY SEX.

	MATH ANXIETY	
SEX	HIGH	LOW
Male	36	52
Female	26	86

9.2.3 Proportions in Several Samples

Suppose that we draw several samples from dichotomous populations that have the same variable of interest. A hypothesis that the population proportion is the same in each case is often called a test of **homogeneity.** We are hypothesizing that the populations are homogeneous—all the same—

Table 9.10 RESULTS OF CAMPUS POLL.

	FRESHMEN	SOPHOMORES	JUNIORS	SENIORS	TOTAL
Yes	91	91	82	73	337
No	69	76	88	86	319
No opinion	60	47	21	18	146
Total	220	214	191	177	802

in regard to the variable under consideration. In these tests the number of observations to be drawn from each population is fixed in advance. We can test a hypothesis of homogeneity in essentially the same way as a hypothesis of independence by using contingency tables, but the assumptions are a bit different. When we use a contingency table, we are testing a hypothesis concerning the independence of two factors in one population. A test of homogeneity, however, involves two or more populations.

Assume that we have taken a poll on campus asking students the question "Should students sit on the University Council and have a voice in the administration of the university?" We have decided to poll approximately 800 students, with proportions of each class corresponding to the true proportions of freshmen, sophomores, juniors, and seniors in the university. Rounding to whole numbers of people yields the number actually polled. The results are shown in Table 9.10.

We would like to test the hypothesis that the proportions of all four classes with an opinion that would answer yes is the same—that is, that $\pi_1 = \pi_2 = \pi_3 = \pi_4$, where π_i is the proportion in the ith population. The row with no opinion can be dropped and the resulting table (Table 9.11) can be analyzed in the same way as in a test of independence. In fact, the distinction between the two tests when $r = 2$ is often blurred. The primary difference is that the column totals are fixed in a test of homogeneity. We can see that, if $r = 2$, and k is the number of samples (columns), then the number of degrees of freedom is $k - 1$. We can calculate the values of f_e in the same way as with a contingency table. Thus we have the following rule:

Table 9.11 RESULTS OF POLL FOR STUDENTS WITH OPINIONS.

	FRESHMEN	SOPHOMORES	JUNIORS	SENIORS	TOTAL
Yes	91	91	82	73	337
No	69	76	88	86	319
Total	160	167	170	159	656

Hypothesis Tests for Homogeneity

Suppose we obtain independent random samples of data from k dichotomous populations. Then we may test the hypothesis H_0: $\pi_1 = \pi_2 = \cdots = \pi_k$. The sample value of the test statistic used to test this hypothesis is χ^{2*}, where

$$\chi^{2*} = \sum \frac{(f_o - f_e)^2}{f_e} = \sum \frac{f_o^2}{f_e} - N$$

provided that each value of f_e is at least 5.

The decision rule will be: Reject H_0 if $\chi^{2*} > \chi^2_\alpha\{k - 1\}$, where α is the level of significance.

Example 9.6

Using the data of Table 9.11, determine whether the proportions in each population are the same. Use the 0.05 level of significance.

Solution

The values of f_e for each cell can be calculated and are shown in parentheses in Table 9.12. Then $\chi^{2*} = 91^2/82.2 + \cdots + 86^2/77.3 - 656 \doteq 5.157$. Since $\chi^2_{.05}\{3\} = 7.815$, we cannot reject H_0 and we conclude that our sample does not show a significant difference of opinion among the classes.

Table 9.12 OBSERVED AND EXPECTED FREQUENCIES FOR CAMPUS POLL.

	FRESHMEN	SOPHOMORES	JUNIORS	SENIORS	TOTAL
Yes	91	91	82	73	337
	(82.2)	(85.8)	(87.3)	(81.7)	
No	69	76	88	86	319
	(77.8)	(81.2)	(82.7)	(77.3)	
Total	160	167	170	159	656

PROFICIENCY CHECKS

6. Four random samples of voters are obtained and the number in each sample in favor of candidate A or B is determined. Use the 0.10 level of significance to determine whether the population proportions are equal.

	SAMPLE			
PREFERENCE	1	2	3	4
Candidate A	54	38	46	38
Candidate B	27	36	57	46

7. To get an idea of brand loyalty, a merchant asks customers buying a deodorant whether they have bought "Certain" before. He also notes whether or not they are now buying "Certain" or some other brand. At the end of the day he has the following information.

	DEODORANT CUSTOMERS PRIOR EXPERIENCE	
DEODORANT BOUGHT TODAY	HAVE BOUGHT "CERTAIN"	HAVE NOT BOUGHT "CERTAIN"
"Certain"	8	6
Other brand	4	12

Assuming that the sample was random and no customer bought more than one brand of deodorant, determine at the 0.05 level of significance whether there is a difference between the proportion of customers buying "Certain" today who are repeat buyers and those who are not.

Problems

Practice

9.10 Determine whether factors A and B are independent for the data in the following table. Use the 0.05 level of significance.

	FACTOR A			
FACTOR B	1	2	3	4
1	13	19	25	23
2	16	20	22	16
3	23	28	19	11
4	27	28	14	7

9.11 Data are classified in a two-way contingency table as shown here. Test the hypothesis, at the 0.05 level of significance, that the factors are independent.

	FACTOR A	
FACTOR B	1	2
1	48	56
2	32	66

9.12 Using the data of Example 9.4, test the hypothesis by combining rows rather than columns. Note that loss of degrees of freedom causes a marked difference in the result.

Applications

9.13 Random samples of married couples in five different income categories are obtained. Each couple is asked whether they agree on which church to attend. If the couple does not attend church but has agreed not to do so, they are also classified as "agree." Use the 0.05 level of significance to determine whether the population proportions are equal.

	INCOME LEVEL				
	1	2	3	4	5
Agree	83	66	55	48	35
Disagree	45	39	57	56	57

9.14 A survey of business failures during the past year shows that 44 of 110 small businesses failing were in business less than a year. The same survey shows that 24 of 75 medium-size businesses and 14 of 50 large businesses failing were in business less than a year. Use the 0.05 level of significance and determine whether the difference in observed proportions is due to chance.

9.15 The Virginia Slims American Women's Opinion Poll (a 1985 survey of 3000 women and 1000 men) asked the question, "Given the relative advantages in today's society, is it better to be a man or a woman?" The results were as follows:

OPINION	MEN	WOMEN
Better to be a man	501	1,470
Better to be a woman	78	241
No difference	392	1,228
Don't know	29	61

Source: *USA Today,* December 9, 1985.

Test the hypothesis that the respondent's answer to the question is independent of sex. Use the 0.10 level of significance.

9.16 A marriage counselor keeps records of the reasons given by husband and wife for marriage difficulty. (He notes that the stated reasons may not be the true reasons.) He classifies each couple into two factors—husband's reason and wife's reason—with the results shown here. Test the hypothesis at a 0.05 level of significance that husbands' and wives' reasons are independent.

HUSBAND'S REASON FOR MARITAL DISCORD

WIFE'S REASON	MONEY	CHILDREN	CONSIDERATION*	ALL OTHER
Money	86	31	132	19
Children	17	64	43	13
Consideration*	54	39	132	33
All other	30	17	37	54

*Including "doesn't love me any more."

9.17 In a recent poll, 1000 residents in a particular city were asked their opinion about a proposed zoning ordinance. The residents were classified according to educational level, and the responses were recorded within each educational level. Use the 0.05 level of significance to test the hypothesis that there is no difference of opinion due to educational level.

	EDUCATIONAL LEVEL		
OPINION	LESS THAN HIGH SCHOOL	HIGH SCHOOL BUT NO COLLEGE	ATTENDED COLLEGE
Agree	77	159	164
Disagree	131	157	112
No opinion	42	134	24

9.18 Suppose that five samples of fish roe are taken from different species to determine whether the proportion of viable roe is the same for each species. The number of viable and nonviable roe for each species is given here. Use the 0.05 level of significance to test whether the proportion is the same for each species.

	SPECIES				
	I	II	III	IV	V
Viable roe	134	78	104	89	95
Nonviable roe	16	22	46	11	5

9.19 In a study of technological complexity and the relative punitiveness of legal sanctions in 20 West African and North African societies, a researcher finds that eight societies are low in both aspects, eight societies are high in both aspects, while two each are high in one aspect and low in the other.* Use the 0.05 level of significance to determine whether the two factors are related.

9.20 A survey by the R. H. Bruskin Association (*USA Today,* December 10, 1985) reported the answer to the question, "How often do you read Ann Landers?" The results were as follows:

	FREQUENTLY	OCCASIONALLY	NEVER
Men	19%	49%	32%
Women	47%	38%	14%

a. Is it possible to perform a chi-square analysis of the data without knowing the number of people who were polled?

*Data taken from unpublished master's thesis, University of West Florida.

b. The article stated that 740 people were polled. Does this information make it possible to perform the analysis?

c. Suppose that 355 men and 385 women were polled. Is this information needed?

d. Determine whether or not response was independent of the sex of the respondent. Use the 0.01 level of significance.

9.21 A study relating anxiety to success in aviation training was conducted at the Pensacola Naval Air Station using 258 officer candidates who were randomly selected from beginning aviation classes.* Each was given a test of anxiety and classified as very high, high, medium, or low. At the conclusion of the program, 189 students had passed while 69 had dropped out. The number in each anxiety level is given here:

ANXIETY LEVEL	PASS	DROP
Very high	8	11
High	54	25
Medium	87	29
Low	40	4

Test whether the tendency to drop out is independent of anxiety level. Use the 0.01 level of significance.

9.22 A researcher wishes to determine whether there is a relationship between age and opinion on long-term revision of Social Security benefits. Test the hypothesis that there is no relationship between opinion and age. Use the 0.05 level of significance.

	AGE GROUP					
OPINION	15–24	25–34	35–44	45–54	55–64	OVER 64
Agree	23	26	36	20	12	8
Disagree	9	16	38	31	23	18
No opinion	8	8	6	9	5	4

9.23 A study comparing dream recall against need for achievement as measured on a standard test gave the following results.*

	DREAM RECALL CATEGORIES		
NEED TO ACHIEVE	NON-RECALLERS	MODERATE RECALLERS	FREQUENT RECALLERS
Low	4	9	3
Medium	19	17	22
High	7	4	5

Determine whether there is any association between frequency of dream recall and need for achievement as measured on this test. Use the 0.05 level of significance.

9.24 Polls are taken in various locations with regard to the opinion of citizens on a proposed bond issue. Results are as follows:

*Data taken from unpublished master's thesis, University of West Florida.

OPINION	AVON PARK	WILSON CENTER	BAYVIEW	DODD PARK
Favoring	58	115	64	81
Opposing	103	139	88	92
No opinion	137	104	90	58

a. Test the hypothesis that, of those with an opinion, proportions favoring the bond issue are the same from area to area. Use the 0.05 level of significance for each part.

b. Test the hypothesis that the proportions of those with no opinion are the same from area to area.

c. Test the hypothesis that there is no relation between area of residence and response to the poll.

9.3 Computer Usage

9.3.1 Minitab Commands for This Chapter

We can conduct tests for independence and equal proportions by using the Minitab commands CHISQUARE or TABLE. The choice between the two depends on whether the contingency table has been stored in Minitab or must be formed from raw data in Minitab. The CHISQUARE command is

```
MTB > CHISQUARE analysis of the table in C ... C.
```

The CHISQUARE command assumes that the $r \times c$ contingency table of observed frequencies has already been computed and stored in the first r rows of the specified columns. For example, to read the table

25	53
42	98

into columns C1 and C2 we would write

```
MTB > READ the following table into columns C1 and C2
DATA> 25 53
DATA> 42 98
DATA> ENDOFDATA
```

In order to read the contingency table into Minitab, the number of observations in each cell must be known. If the data have not been counted but each observation is classified in two ways, the data are in the raw data form. The Minitab data set consists of n observations, and the columns correspond to the different classification variables of these n observations. For each observation, the cell into which it would fall in the contingency table is given by the value of these classification variables. Suppose, for example, that seven people are asked which candidate they prefer and they are also classified by sex. The results are as follows:

PERSON	SEX	CANDIDATE PREFERRED
1	M	A
2	M	B
3	M	B
4	F	A
5	F	A
6	F	A
7	F	B

C1	C2
1	1
1	2
1	2
2	1
2	1
2	1
2	2

We could store the data in Minitab by storing sex in C1—1 for male and 2 for female—and candidate preference in C2—1 for candidate A and 2 for candidate B. The stored data would look like the figure in the margin. The corresponding contingency table for this data set with C1 (sex) as rows and C2 (candidate preference) as columns would look like this:

		C2	
		1	2
C1	1	1	2
	2	3	1

If the raw data have been stored into two columns as above, then we can use the TABLE command with the CHISQUARE K subcommand to form the contingency table from the raw data and to perform the chi-square test. If $K = 1$, the printed contingency table will contain only the observed frequencies; if $K = 2$, the table will contain both the observed and expected frequencies. The chi-square test is performed in either case. The general form of the command is

```
MTB > TABLE C by C;
SUBC> CHISQUARE K.
```

This command counts the observed frequency in each cell, computes the expected frequencies, and performs the chi-square test. The output from TABLE differs slightly from that of CHISQUARE. The two outputs can be compared in Examples 9.7 and 9.8.

Example 9.7

Refer to Table 9.4, which classifies students by achievement in a statistics course and prior course work in algebra and logic. Use the CHISQUARE command to test the hypothesis that the two factors are independent. Use the 0.05 level of significance.

Solution

The Minitab commands and output are shown in Figure 9.6. Note that you can refer to consecutive columns in Minitab by using the hyphen (-), in which case all columns from the first named to the last named, inclusive, will be used. Thus C1-C5 means that columns C1, C2, C3, C4, and C5 will be used.

The observed value of chi square, $\chi^{2*} = 106.22$, differs slightly from the 106.59 value obtained for the data in Section 9.2.1. This difference is due to rounding off in the manual computation there. Note also that the sum of the numbers listed in the table does not add up to exactly 106.22. This is because all of the intermediate numbers listed in the output are rounded to one or two decimal places, while the actual calculations are carried out to many more. For example, $(20 - 12.4)^2/12.4 \doteq 4.66$, but the corresponding value is listed as 4.59 because the actual value of f_e is $12.44444\cdots$ and the computations are carried out accordingly.

We reach the same conclusions in either case since both values of χ^{2*} are much greater than the critical value, $\chi^2_{.05}\{12\}$, which is equal to 21.03. (Note that the *P* value is not given in the output so that the critical value of chi square must be obtained.) We conclude that there is a significant relation between statistics success and prior course work in algebra or logic; that is, the factors are not independent.

Figure 9.6
Minitab printout of the χ^2 test of independence for the statistics achievement data, Example 9.7.

```
MTB > READ the following data into C1 C2 C3 C4 C5
DATA> 15 20 40 5 0
DATA> 10 15 70 20 5
DATA> 10 20 25 5 0
DATA> 15 15 75 30 55
DATA> ENDOFDATA
     4 ROWS READ
MTB > CHISQUARE analysis of table in columns C1-C5
EXPECTED FREQUENCIES ARE PRINTED BELOW OBSERVED FREQUENCIES
            C1       C2       C3       C4       C5    TOTALS
     1      15       20       40        5        0       80
           8.9     12.4     37.3     10.7     10.7
     2      10       15       70       20        5      120
          13.3     18.7     56.0     16.0     16.0
     3      10       20       25        5        0       60
           6.7      9.3     28.0      8.0      8.0
     4      15       15       75       30       55      190
          21.1     29.6     88.7     25.3     25.3
TOTALS      50       70      210       60       60      450

TOTAL CHI SQUARE =

           4.20 +  4.59 + 0.19 + 3.01 + 10.67 +
           0.83 +  0.72 + 3.50 + 1.00 +  7.56 +
           1.67 + 12.19 + 0.32 + 1.13 +  8.00 +
           1.77 +  7.17 + 2.11 + 0.86 + 34.74 +

                     = 106.22

DEGREES OF FREEDOM = ( 4-1) x ( 5-1) =  12
```

Observed cell frequency (→ 15)

Expected cell frequency (→ 8.9)

Cell chi-square value $= \dfrac{(15 - 8.9)^2}{8.9}$ (→ 4.20)

Example 9.8

When the data are in raw form, as in the Miami Female Study data, we use the TABLE command to perform the chi-square test. Use this command to determine whether Miami females who are not in sororities are more likely to have boyfriends than those females who are not in sororities. Use the 0.05 level of significance.

Solution

The results, using the TABLE command, are shown in Figure 9.7. Note that we have used the RETRIEVE command to access the previously saved Miami Female Study data. If you cannot SAVE and RETRIEVE data sets, you will need to enter the data using the SET or READ command.

The critical value, $\chi^2_{.05}\{1\}$ is 3.841; we compare the obtained value, $\chi^{2*} = 0.051$, and conclude that there is no difference. The proportion of Miami females with steady boyfriends does not depend on whether the female is in a sorority or not.

Although there is no formal goodness-of-fit procedure in Minitab, we can still perform the test. By using the SET or READ and LET commands, Minitab will compute χ^{2*} for the data. Since the computation of this statistic by hand can be quite tedious and is subject to error, having Minitab do the calculations will probably eliminate many mistakes.

Although we could also use this technique to test for normality or for fit to a Poisson, binomial, or other discrete distribution, we would have to compute the expected frequencies by hand and enter them into Minitab. This requirement reduces the usefulness of the computer to perform a chi-square test for normality. The techniques presented in Chapter 5 using bar graphs and normal probability plots to check for normality are sufficient for most situations.

In Example 9.2 recall that the economic composition of a community was investigated to determine whether the hypothetical distribution of incomes was appropriate. Much of the computation can be done with Minitab. The

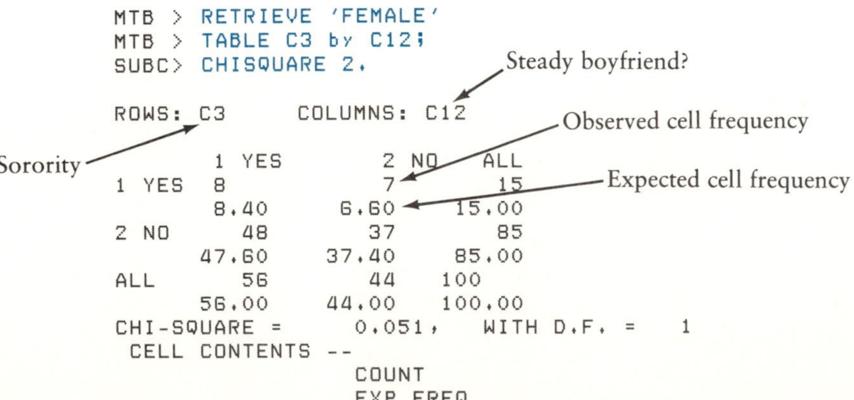

Figure 9.7
Minitab printout for chi-square test of Miami Female Study data of Example 9.8.

```
MTB > SET the following observed frequencies into C1
DATA> 166 97 134 61 42
DATA> SET the following expected frequencies into C2
DATA> 200 100 100 50 50
DATA> NAME C1 'OBS' C2 'EXP'
MTB > LET K1 = SUM(((C1-C2)**2)/C2
MTB > LET K2 = SUM(((ABSOLUTE(C1-C2)-0.5)**2)/C2)
MTB > PRINT K1 K2
K1        21.1300<------------Uncorrected chi-square statistic
K2        20.2262<------------Yates corrected chi-square statistic
```

Figure 9.8
Minitab printout of the chi-square test of goodness of fit for the community economic composition data of Example 9.2.

SUM function obtains the sum of the numbers of a column while the ABSOLUTE function takes the absolute value of the numbers in each column. Figure 9.8 shows the Minitab program and the output for this problem.

Compare the results to those of Example 9.2. The test statistic value is again 21.13, and since $\chi^2_{.05}\{4\} = 9.488$, we reject H_0 and conclude that the model is not appropriate for the population from which the data were obtained. Note that K1 gives the uncorrected chi-square statistic whereas K2 gives the Yates-corrected statistic. The result of the test is the same in both cases.

9.3.2 SAS Statements for This Chapter

We can use SAS procedure FREQ to perform a chi-square test of independence or equality of proportions using the SAS System. If the data have been counted and put into cells, the data are entered and we obtain the contingency table by using the TABLES statement. The option CHISQ causes the value of χ^{2*} and the P value for χ^{2*} to be computed and printed. If the data are in a 2×2 contingency table, Yates' correction factor will be incorporated and reported as CONTINUITY ADJ. CHI-SQUARE along with its P value; Fisher's exact test will be reported as well. Many other statistics we have not discussed are also printed out.

For the Miami Female Study data in Example 9.8 we could write the following program:

```
DATA MIAMI;
 INPUT SORORITY $ BOYFRIEND $ COUNT @@;
CARDS;
YES YES 8 YES NO 7 NO YES 48 YES YES 37
PROC FREQ;
 WEIGHT COUNT;
 TABLES SORORITY * BOYFRIEND / CHISQ;
```

Example 9.9

A study of Florida high school students (grades 10–12) was conducted in the 1970s, and the results were reported by Chitwood and McBride (1985). One of the findings concerned the relationship between parental approval and peer approval of marijuana use among students who initiated the use of marijuana. The data collected are shown here. Determine whether parental approval and peer approval are independent. Use the 0.01 level of significance.

PEER APPROVAL	PARENTAL APPROVAL	
	YES	NO
Yes	52	181
No	11	144

Solution

The SAS program and output table are shown in Figure 9.9. Note that each cell contains the frequency, the percentage of the total, percentage of the row, and percentage of the column; the row totals include percentage of row while the column totals include percentage of column. The value of χ^{2*} is 15.855 and is significant beyond the 0.001 level. We do not need either the CONTINUITY ADJ. CHI-SQUARE or FISHER'S EXACT TEST value since the expected frequencies are sufficiently large. Thus we reject the hypothesis of independence and conclude that there is a relationship between peer and parental approval for students initiating marijuana use.

Figure 9.9
SAS printout for the marijuana use approval data of Example 9.9.

```
INPUT FILE

DATA TEST:
   INPUT PEER $ PARENT $ COUNT @@;
CARDS;
YES YES 52 YES NO 181 NO YES 11 NO NO 144
PROC FREQ;
   WEIGHT COUNT;
   TABLES PEER * PARENT / CHISQ;

OUTPUT FILE
```

```
                    SAS

          TABLE OF PEER BY PARENT
```

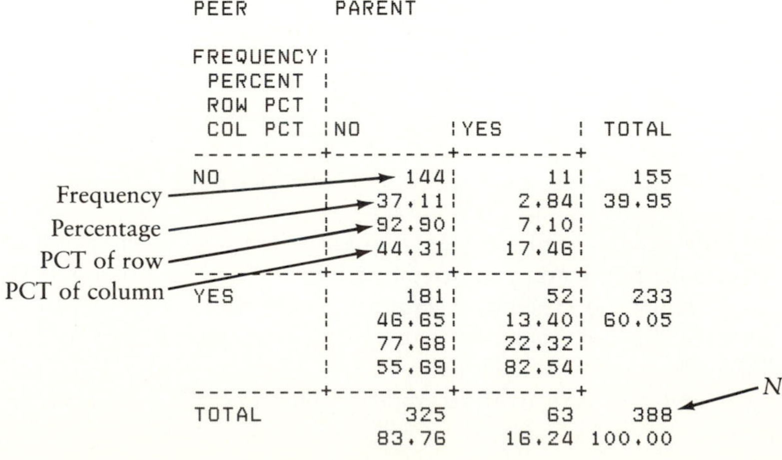

```
PEER       PARENT

FREQUENCY|
 PERCENT |
 ROW PCT |
 COL PCT |NO       |YES      |  TOTAL
---------+---------+---------+
NO       |     144 |      11 |    155
         |   37.11 |    2.84 |  39.95
         |   92.90 |    7.10 |
         |   44.31 |   17.46 |
---------+---------+---------+
YES      |     181 |      52 |    233
         |   46.65 |   13.40 |  60.05
         |   77.68 |   22.32 |
         |   55.69 |   82.54 |
---------+---------+---------+
TOTAL          325        63      388
             83.76     16.24   100.00
```

```
            STATISTICS FOR TABLE OF PEER BY PARENT

STATISTIC                              DF      VALUE      PROB
--------------------------------------------------------------
CHI-SQUARE                              1     15.855     0.000
LIKELIHOOD RATIO CHI-SQUARE             1     17.418     0.000
CONTINUITY ADJ. CHI-SQUARE              1     14.756     0.000
MANTEL-HAENSZEL CHI-SQUARE              1     15.814     0.000
FISHER'S EXACT TEST (1-TAIL)                             0.000
                    (2-TAIL)                             0.000
PHI                                            0.202
CONTINGENCY COEFFICIENT                        0.198
CRAMER'S V                                     0.202

SAMPLE SIZE = 388
```

χ^{2*} (arrow to VALUE 15.855); *P* value (arrow to PROB 0.000)

If the data are not counted prior to entry, each observation is classified in two ways; the WEIGHT statement is omitted from the FREQ procedure. Consider the example that opened Section 9.3.1. Seven people were asked which candidate they prefer; the people were also classified by sex. The results were as follows:

PERSON	SEX	CANDIDATE PREFERRED
1	M	A
2	M	B
3	M	B
4	F	A
5	F	A
6	F	A
7	F	B

We could write the following SAS program:

```
DATA EXAMPLE;
 INPUT SEX $ CANDIDATE $ @@;
CARDS;
M A M B M B F A F A F A F B
PROC FREQ;
 TABLES SEX * CANDIDATE / CHISQ;
```

This program would produce exactly the same output as the following program:

```
DATA EXAMPLE;
 INPUT SEX $ CANDIDATE $ COUNT @@;
CARDS;
M A 1 M B 2 F A 3 F B 1
PROC FREQ;
 WEIGHT COUNT;
 TABLES SEX * CANDIDATE / CHISQ;
```

We can also use SAS statements to produce a goodness-of-fit test; as in Minitab, there is no specific program to do this. Unlike Minitab, the SAS System allows tests for a fit to a Poisson or binomial random variable. Suppose the following data have been collected:

Observation	0	1	2	3	4	5	6
Frequency	11	16	23	18	7	3	1

We could hypothesize that the data come from a Poisson distribution. We can use the mean of this data set to estimate m. We find that the mean is 165/79, or about 2.09. Suppose that we wish to test the hypothesis that the data could have come from a Poisson distribution with $m = 2$. We first enter the data in a SAS data set:

```
DATA OBSERVED;
 INPUT OBS @@;
CARDS;
11 16 23 18 7 3 1
```

We then obtain the expected frequencies using the POISSON function. Recall that POISSON(m, n) gives the probability that an observation from a Poisson distribution is less than or equal to n. We can obtain the probability that a Poisson random variable is equal to n by taking the difference between $P(x \leq n - 1)$ and $P(x \leq n)$. Since the total number of observations is 79, we can multiply this probability by 79 to obtain f_e for each value of the variable. (Recall that $P(x = 0) = P(x \leq 0)$ so that we need to make special provisions for this case.) We may write

```
DATA EXPECTED;
   DO X = 0 to 6;
      P1 = POISSON(2, X);
      IF X > 0 THEN P2 = POISSON(2, X-1);
      IF X = 0 THEN P = P1;
      IF X > 0 THEN P = P1 - P2;
      EXP = 79 * P;
      OUTPUT;
      END;
```

We then put the data sets together by using the statement MERGE, which pairs variables with the same observation number:

```
DATA TOTAL;
   MERGE OBSERVED EXPECTED;
```

Finally we compute the cell totals for each data set and obtain the sum:

```
   CELL = (OBS - EXP)**2 / EXP;
PROC MEANS SUM NOPRINT;
   VAR CELL;
   OUTPUT OUT = A SUM = CHISQ;
DATA TEST;
   SET A;
   P = 1 - PROBCHI(CHISQ, 6);
```

PROBCHI(x, df) computes the probability that a random variable with a chi-square distribution, with df degrees of freedom, will be less than the value given. The P value for a chi-square test is upper-tailed, so we need 1 − PROBCHI(x, df). We may then print out the value by using PROC PRINT. We may also print intermediate data sets as desired.

The printout for this problem is shown in Figure 9.10. We see that $\chi^{2*} \doteq 3.535$ and the P value is about 0.74, so we may conclude that a Poisson distribution with $m = 2$ fits the data very well.

If the expected frequencies are known, we may enter them in the same DATA step in which the observed frequencies are entered. In Example 9.2 the economic composition of a community was investigated to determine whether the hypothetical distribution of incomes was appropriate. We may write

```
DATA FIRST;
 INPUT OBS EXP @@;
 CELL = (OBS - EXP)**2/EXP;
CARDS;
166 200 97 100 134 100 61 50 42 50
PROC MEANS SUM NOPRINT;
 VAR CELL;
 OUTPUT OUT = A SUM = CHISQ;
DATA TEST;
 SET A;
 P = 1 - PROBCHI(CHISQ, 4);
PROC PRINT;
```

The output of this program is shown in Figure 9.11. We note that $\chi^{2*} = 21.13$ and the P value is about 0.0003. Thus we reject the hypothesis and conclude that the hypothesized distribution is not correct for the population from which the sample was obtained.

```
        DATA OBSERVED;
           INPUT OBS @@;
        CARDS;
        DATA EXPECTED;
```

Figure 9.10
SASLOG for testing fit to a Poisson distribution.

```
          DO X = 0 TO 6;
             P1 = POISSON (2,X);
             IF X > 0 THEN P2 = POISSON (2,X - 1);
             IF X = 0 THEN P = P1;
             IF X > 0 THEN P = P1 - P2;
             EXP = 79 * P;
             OUTPUT;
             END;
       DATA TOTAL;
          MERGE OBSERVED EXPECTED;
          CELL = (OBS - EXP) ** 2 / EXP;
       PROC MEANS SUM NOPRINT;
          VAR CELL;
          OUTPUT OUT = A SUM = CHISQ;
       DATA TEST;
          SET A;
          P = 1 - PROBCHI (CHISQ, 6);
       PUT CHISQ P;

3.53488 0.7393235
```

Figure 9.11
Computer printout for Example 9.2.

```
INPUT FILE

DATA FIRST;
   INPUT OBS EXP @@;
   CELL = (OBS - EXP)**2/EXP;
CARDS;
166 200 97 100 134 100 61 50 42 50
PROC MEANS SUM NOPRINT;
   VAR CELL;
   OUTPUT OUT = A SUM = CHISQ;
DATA TEST;
   SET A;
   P = 1 - PROBCHI(CHISQ, 4);
PROC PRINT;

OUTPUT FILE

                              SAS
                OBS     CHISQ          P
                 1      21.13     0.000298418
```

Problems

Applications

9.25 Refer to Example 9.7. If the rows and columns are interchanged, do you think the results would change? Interchange the rows and columns and perform the chi-square test. Do the results change?

9.26 An auto dealer wishes to know why customers buy the model of car that they buy. He collects information and finds that performance and economy are separately named the most important reason by the following numbers of women and men buyers:

	PERFORMANCE	ECONOMY
Women	23	47
Men	27	38

Use the 0.01 level of significance to determine whether sex and reason for buying are independent.

9.27 Use the Miami Female Study data in Appendix B.1 to determine whether females in sororities have higher grade-point averages than females who are not in sororities. Treat the grade-point average as a classification or categorical variable.

9.28 Refer to Problem 9.2. Use SAS statements to solve the problem. (Use the PROBBNML statement instead of POISSON.)

9.29 We can use the SAS function PROBNORM to determine probabilities in a normal distribution (see Section 5.3.3) provided that we know the mean and variance. Using the methods shown here and the suggestions in Example 9.3, solve Problem 9.9.

9.4 Case 9: The Controversial Coke—Solution

If we assume that the three drinks are equally likely to be picked, we can test the hypothesis that each drink will be picked one-third of the time. This would mean that $f_e = 8\frac{1}{3}$ for each drink. Since $f_o = 12$, 7, and 6, we obtain $\chi^{2*} = 2.48$. Since $\chi^2_{.10}\{2\} = 4.605$, we cannot conclude that the difference is other than chance.

Since each participant knew which three drinks were to be tasted, the probability that any given person could correctly identify the three drinks by chance is (1/3)(1/2)(1), or 1/6. Out of 25 people the expected number who would identify the three drinks correctly by chance is 25(1/6), or about 4.17. The probability that no more than 3 of 25 people would do so by chance is $(5/6)^{25} + 25(1/6)(5/6)^{24} + (25 \cdot 24/2)(1/6)^2(5/6)^{23} + (25 \cdot 24 \cdot 23 \cdot 3 \cdot 2)(1/6)^2(5/6)^{23}$, or about 0.3816. Thus we can conclude that this result could have been obtained by chance.

9.5 Summary

The chi-square distribution is often useful when we are comparing expected frequencies to observed frequencies of count data. In this chapter we learned how to use it to determine whether results of a sample conform to a hypothetical distribution, whether two factors into which a sample is classified are independent, and whether proportions among samples from different populations differ significantly.

The sample value of the test statistic used to test these hypotheses is χ^{2*}, where

$$\chi^{2*} = \sum \frac{(f_o - f_e)^2}{f_e}$$

and f_o and f_e are the observed and expected frequencies, respectively.

A shortcut formula, usually simpler to calculate, is

$$\chi^{2*} = \sum \frac{f_o^2}{f_e} - N$$

where N is the total number of observations. To test the hypotheses in this chapter, we required that each value of f_e be at least 5; some exceptions are possible, however, if f_e and f_o are in close agreement.

The critical value of the test statistic is denoted $\chi^2_\alpha\{\nu\}$, where ν is the appropriate number of degrees of freedom. If m population parameters are estimated in order to obtain the values of f_e, the number of degrees of freedom is $k - 1 - m$.

For a 2 × 2 contingency table with entries arranged as

a	b
c	d

the following formula is appropriate without Yates' correction factor

$$\chi^{2*} = \frac{N(ad - bc)^2}{(a + b)(c + d)(a + c)(b + d)}$$

and the following formula incorporates Yates' correction factor:

$$\chi^{2*} = \frac{N(|ad - bc| - N/2)^2}{(a + b)(c + d)(a + c)(b + d)}$$

Finally, we can use the chi-square distribution to test for homogeneity of several populations.

Chapter Review Problems

Practice

9.30 Observed frequencies in six categories are 122, 144, 131, 119, 150, and 134. Using the 0.05 level of significance, test the hypothesis that the data come from a population with equal proportions in each category.

9.31 Determine whether factors A and B are independent for the data in the following table. Use the 0.05 level of significance.

FACTOR B	FACTOR A 1	2	3	4
1	73	19	45	93
2	66	20	52	76
3	83	23	79	51
4	67	28	54	37

9.32 Five random samples are obtained, and the number of conservatives and liberals in each sample is determined. Use the 0.05 level of significance to determine whether the population proportions are equal.

	SAMPLE 1	2	3	4	5
Conservative	28	30	26	28	18
Liberal	20	19	27	22	29

9.33 A coin is tossed three times and the number of heads is recorded. The experiment is repeated 80 times. There were no heads 7 times, one head 24 times, two heads 35 times, and three heads 14 times. Test the hypothesis that the coin is biased (that is, not fair). If it is biased, in what direction is the bias? Use the 0.05 level of significance.

Applications

9.34 Using the data of Table 9.4 (Section 9.2), test the hypothesis that prior course work in college algebra and logic (separately) have equal effects on success in statistics. Use the 0.05 level of significance.

9.35 A sociologist contends that requirements for mortgage approval by lending institutions tend to keep lower-income people from buying a house even if they can afford one. To back up this contention, the following data are cited:

	INCOME CATEGORY (DOLLARS PER YEAR)			
GPA	UNDER $15,000	$15,000–$19,999	$20,000–$24,999	$25,000–AND OVER
Home owners	38	64	31	12
Renters	55	58	15	8

Assuming that the sample was random, do these data support the contention at the 0.05 level of significance?

9.36 In an apartment complex, a sociologist asks a sample of workers if they are satisfied with their work. He then classifies type of work into three categories. Using the following results, test the hypothesis that job satisfaction is not related to type of work. Use the 0.05 level of significance.

	WHITE COLLAR	BLUE COLLAR	MENIAL
Satisfied	81	124	44
Dissatisfied	49	76	56

9.37 In a large firm, salespeople are classified as aggressive, nonaggressive, and shy. Their sales for the month are classified as high, average, or low. Given the following data, test the hypothesis, at $\alpha = 0.01$, that relative aggressiveness and sales are independent.

	SALES		
	HIGH	AVERAGE	LOW
Aggressive	64	28	38
Nonaggressive	45	22	29
Shy	28	29	27

9.38 A medical researcher is testing the hypothesis that there is a relationship between blood type and susceptibility to a certain liver condition. About 40% of the population in his locality has type O blood, 30% has type A, 20% has type B, and 10% has type AB. He obtains data on all patients admitted to the hospitals in the locality during the past 10 years who have had this condition. He finds that there were 1043 with blood type O, 801 with type A, 861 with type B, and 745 with type AB. Can he safely conclude, at $\alpha = 0.01$, that there is a relationship?

9.39 A biologist exposes some culture plates to each of five different, but related, strains of bacillus. She wishes to test the hypothesis that strain makes no difference in the proportion of plates that will attain propagation levels in 36 hours. Given the following data, decide whether or not the hypothesis is substantiated at the 0.05 level of significance.

	STRAIN				
	A	B	C	D	E
Number reaching propagation level	32	23	28	36	31
Number not reaching propagation level	68	77	72	64	69

9.40 To test whether susceptibility to disease A and prior contracting of disease B are related, a sample of 1000 persons is randomly selected during an epidemic of disease A. The severity of each case of disease A, if any, and prior severity of any case of disease B in the sample are recorded. The results are given here. Test, at the 0.01 level of significance, to see if there is any relationship between disease A and prior experience with disease B.

	DISEASE A		
DISEASE B	DID NOT CONTRACT	LIGHT CASE	SEVERE CASE
Did not contract	302	267	31
Light case	86	156	8
Severe case	12	77	61

9.41 A store manager hypothesizes that the number of customers who enter her store during the first 10 minutes after she returns from lunch (that is, from 1:00 to 1:10) follows a Poisson distribution. She observes the occurrences over a period of 20 weeks (120 days) and finds that the distribution of customers was as follows:

Number of customers	0	1	2	3	4	5	6	7	8
Number of occurrences	10	25	32	22	17	9	4	0	1

Test this hypothesis at the 0.05 level of significance.

9.42 Chitwood and McBride (1985) made a study of the use and cessation of use of marijuana among Florida students (grades 10–12) in the 1970s. The number of students in each of five socioeconomic categories is given here (category I is highest).

Category	I	II	III	IV	V
Number of students	173	153	490	582	368

a. The number of students who reported using marijuana, by socioeconomic class, is reported here:

Category	I	II	III	IV	V
Number of students	76	53	146	170	67

Using the 0.05 level of significance, can you conclude that marijuana use is not related to socioeconomic class? If the hypothesis is rejected, in what direction do you think the difference lies?

b. The number of students who once used marijuana but reported that they no longer use it is given here by socioeconomic class:

Category	I	II	III	IV	V
Number of students	28	16	68	83	39

Can you conclude that there is a difference between the socioeconomic classes?

Index of Terms

Glossary of Symbols

χ^{2*}	sample value of chi-square test statistic
$\chi^2(\nu)$	chi-square distribution with ν degrees of freedom
$\chi^2_\alpha\{\nu\}$	critical value of chi square for a chi-square distribution with ν degrees of freedom (α is the area under the curve to the right of the critical value)
f_e	expected frequency in a cell
f_o	observed frequency in a cell
N	total number of observations
$\pi_1, \pi_2, \ldots$	hypothetical proportions for several populations

Answers to Proficiency Checks

1. Expected frequencies are 30 in each category. $\chi^{2*} \doteq 13.267$ and $\chi^2_{.05}\{4\} = 9.488$, so we reject the hypothesis that the categories are equally likely, and conclude that they are not.

2. Expected frequencies in the respective categories will be 20, 40, 30, 20, and 40. $\chi^{2*} = 22^2/20 + 44^2/40 + 31^2/30 + 19^2/20 + 34^2/40 - 150 \doteq 1.5833$, which is less than 9.488. Thus the sample fits the hypothetical distribution quite well.

3. The sum of the percentages is 48.5, so we can determine each expected frequency by taking the percentage for the category, dividing it by 48.5, and then multiplying by 200. We thus obtain the expected frequencies, shown here with the observed frequencies:

AGE GROUP	5–9	10–14	15–19	20–24	25–29	30–34
f_o	31	53	41	29	24	22
f_e	25.15	30.10	38.35	35.05	38.76	32.58

Since $\chi^{2*} = 31^2/25.15 + 53^2/30.10 + 41^2/38.35 + 29^2/35.05 + 24^2/38.76 + 22^2/32.58 - 200 \doteq 29.077$ and $\chi^2_{.01}\{5\} = 15.09$, we reject the hypothesis that the age groups are distributed as the population and conclude that the distribution differs. An examination of the observed versus expected frequencies reveals that there is a higher proportion than expected in the younger categories and a lower proportion than expected in the older categories.

4. The expected frequency in each cell is shown here in parentheses. (Note that f_e of 26.7 is actually $26\frac{2}{3}$ and f_e of 16.7 is actually $16\frac{2}{3}$.)

	FACTOR A					
FACTOR B	1	2	3	4	5	TOTAL
1	21	16	29	33	21	120
	(26.7)	(20.0)	(30.0)	(26.7)	(16.7)	
2	26	20	32	26	16	120
	(26.7)	(20.0)	(30.0)	(26.7)	(16.7)	
3	33	24	29	21	13	120
	(26.7)	(20.0)	(30.0)	(26.7)	(16.7)	
Total	80	60	90	80	50	360

Then $\chi^{2*} = 21^2/26.7 + \cdots + 13^2/16.7 - 360 = 9.21$ (using $26\frac{2}{3}$ and $16\frac{2}{3}$). Since $\chi^2_{.05}\{8\} = 15.51$, we conclude that the factors are independent.

5. The expected frequencies are shown in parentheses:

	RELIGIOUS GROUP			
ATTITUDE	A	B	C	TOTAL
For birth control	123	64	43	230
	(102.2)	(76.7)	(51.1)	
Against birth control	77	86	57	220
	(97.8)	(73.3)	(48.9)	
Total	200	150	100	450

Then $\chi^{2*} \doteq 15.59$ and $\chi^2_{.01}\{2\} = 9.210$, so we conclude that the factors are not independent. Note that this problem could also be approached in terms of equal proportions.

6. The number in each sample expected to favor each candidate if H_0: $\pi_1 = \pi_2 = \pi_3 = \pi_4$ is true is shown in parentheses:

	SAMPLE				
PREFERENCE	1	2	3	4	TOTAL
Candidate A	54	38	46	38	176
	(41.7)	(38.1)	(53.0)	(43.2)	
Candidate B	27	36	57	46	166
	(39.3)	(35.9)	(50.0)	(40.8)	
Total	81	74	103	84	342

Then $\chi^{2*} \doteq 10.67$ (10.39 if the values of f_e are used to more decimal places) and $\chi^2_{.10}\{3\} = 6.251$, so we conclude that the population proportions are not equal.

7. Here we have a 2×2 contingency table. Since all expected frequencies will be below 10, Yates' correction factor is appropriate. Using the special formula, we have

$$\chi^{2*} = \frac{30(|8 \cdot 12 - 4 \cdot 6| - 30/2)^2}{(12)(18)(14)(16)} = \frac{5(72-15)^2}{2(18)(14)(16)} = \frac{5(57)^2}{36(224)}$$
$$\doteq 2.015$$

Since $\chi^2_{.05}\{1\} = 3.841$, we cannot conclude that any differences are not due to chance. What appears to be a real difference may be due to chance.

Derivation of the Shortcut Formula for χ^{2*}

To show that $\Sigma[(f_o - f_e)^2/f_e] = \Sigma[f_o^2/f_e] - N$, first square $(f_o - f_e)$ to obtain $f_o^2 - 2f_of_e + f_e^2$. Then

$$\Sigma(f_o^2 - 2f_of_e + f_e^2) = \Sigma f_o^2 - 2\Sigma f_of_e + \Sigma f_e^2$$

so that

$$\begin{aligned}\Sigma[(f_o - f_e)^2/f_e] &= \Sigma(f_o^2/f_e) - 2\Sigma(f_of_e/f_e) + \Sigma(f_e^2/f_e) \\ &= \Sigma(f_o^2/f_e) - 2\Sigma f_o + \Sigma f_e\end{aligned}$$

Now both Σf_o and Σf_e are equal to N, so

$$\begin{aligned}\Sigma[(f_o - f_e)^2/f_e] &= \Sigma(f_o^2/f_e) - 2N + N \\ &= \Sigma(f_o^2/f_e) - N\end{aligned}$$

References

Bennett, J. H., ed., *Experiments in Plant Hybridisation*. Edinburgh: Oliver & Boyd, 1965.

Chitwood, Dale D., and Duane C. McBride. "The Cessation of Marijuana Use." *Florida Journal of Anthropology* 10 (1985):33–47.

Conover, W. J. *Practical Nonparametric Statistics*. New York: John Wiley & Sons, 1971.

Fienberg, Stephen E. "The Contributions of William Cochran to Categorical Data Analysis." In *W. G. Cochran's Impact on Statistics,* P.S.R.S. Rao and J. Sedransk, eds. New York: John Wiley & Sons, 1984.

Hacking, Ian. "Trial by Number." *SCIENCE84* 5 (9) (November 1984): 69–70.

Mosteller, Frederick and Robert E. K. Rourke. *Sturdy Statistics*. Reading, MA: Addison-Wesley, 1973.

Mosteller, Frederick, Robert E. K. Rourke, and George B. Thomas, Jr. *Probability with Statistical Applications*. 2nd ed. Reading, MA: Addison-Wesley, 1970.

Pearson, Karl. "On the Criterion That a Given System of Deviations from the Probable in the Case of a Correlated System of Variables Is Such That It Can Be Reasonably Supposed to Have Arisen from Random Sampling." *London, Edinburgh and Dublin Philosophical Magazine and Journal of Science,* Series 5, 50 (1900): 157–175. Reprinted in *Karl Pearson's Early Statistical Papers* (Cambridge: Cambridge University Press, 1956).

Tanur, Judith, M., and others. *Statistics: A Guide to the Unknown*. 2nd ed . San Francisco: Holden-Day, 1978.

Ware, Mary Catherine, and Mary Frances Stuck. "Sex Role Messages vis-à-vis Microcomputer Use: A Look at the Pictures." *Sex Roles* 13 (1985): 205–214.

Yates, Frank. "Contingency Tables Involving Small Numbers and the χ^2 Test." *Journal of the Royal Statistical Society,* 1 (1934): 217–235

10

Nonparametric Methods

For most of the statistical tests described in this book, an important assumption must be met if they are to be applied correctly. This assumption is that the distribution of each population from which the sample or samples are drawn is normal. These statistical tests are reasonably *robust,* allowing considerable latitude, and deviations from normality are permissible. The central limit theorem, for instance, allows us to bypass the normality assumption for sufficiently large samples. If the distribution from which a sample is drawn is badly skewed or is otherwise seriously nonnormal, however, these statistical tests will not yield meaningful results for smaller samples. When data are badly skewed, the standard deviation of the data set tends to be large, and in such cases the probability of incorrectly failing to reject the null hypothesis is quite high. The tests presented in this chapter reduce the value of β in such cases and provide much more meaningful results than the usual tests.

A second assumption upon which most of the usual tests rest is that meaningful sample statistics, such as the mean and variance, can be derived from the samples and used to estimate the corresponding population parameters. Data that are purely nominal in nature (such as "increase, decrease, no change"), or which are given in rank order only, do not yield such meaningful results. In the social sciences, for example, much research is done by means of questionnaires. In many cases the responses are given on a five-point or seven-point scale from "strongly agree" to "strongly disagree." Numbers associated with these results are not measured on an interval scale (discussed in Chapter 1), and the mean and variance of such numbers are probably not valid.

The term nonparametric tests *is in fact a misnomer since these tests usually depend upon some parameter—a population proportion or median, for instance. They never depend on a normal distribution, however, so the term* distribution-free tests *is probably more nearly accurate though less widely used.*

Statisticians have devised alternative procedures that can be used to test hypotheses about data which are not normal or are not measured on an interval scale. Since these tests do not depend on the shape of the distribution, they are called **distribution-free tests.** Moreover, they do not depend upon the mean and variance, so they are also called **nonparametric tests.** These tests may be used for data that are nominal or in ranks. In Chapter 6, and particularly in Chapters 7 and 8, we occasionally referred to nonparametric tests. We have already discussed one nonparametric technique—the chi-square tests of Chapter 9. Often we can also use nonparametric tests as quick checks when the usual tests require extensive calculation of means and variances, but they are less powerful than their counterparts that utilize the central limit theorem. When nonparametric tests yield inconclusive results, the usual tests, if appropriate, are often preferred.

Nonparametric tests are quite robust in the sense of the term introduced in Section 6.4.3. A major change in one value has very little effect on these tests.

In most experimental situations, samples are drawn from populations that are reasonably close to normal or the samples are sufficiently large so that the central limit theorem applies. If the data come from a population that is bounded on one end, however, there is a good chance that the population distribution is not normal. For example, income distributions are bounded at the lower end (zero) while for all practical purposes they are unlimited at the upper end. Distributions of incomes tend to bunch up around the lower end. When we obtain sample data from such a bounded distribution, we should check the distribution for skewness. If the samples are small—smaller than 50—the large-sample techniques presented in the preceding chapters are not appropriate when the population variances are unknown, so nonparametric methods should be used. Even if we know the population variances, many of the nonparametric tests are preferred when the data sets are not symmetric.

Many nonparametric tests have been devised. This chapter presents three alternatives to the *t* tests for normal populations and large-sample *z* tests. Other nonparametric methods will be presented in later chapters.

10.0 Case 10: The Sullen Salesman

A salesman entered the office of the president of a large department store chain. He was confident of success. He told the president that he and his company were offering a short intensive course in sales techniques that would raise the sales in each department. To prove his point he produced a study his company had made. When he asked the president if he knew anything about statistics, the president replied that indeed he did. The salesman then explained the study.

In a large department store, a careful check was made of the records of salespeople in all its departments. The purpose was to find pairs of salespeople within a department who had as nearly as possible the same record of sales as far back as possible. Pairs of salespeople in 15 departments were identified as being suitable for the study. One of each pair was randomly selected and assigned to an experimental group. This group was given a short intensive course in sales techniques, while the other group was not. Both groups were informed that they were participating in an experiment, so all knew about the study. It is reasonable to assume that the only difference between the groups was the short course, since the subjects were randomly assigned to the two groups. At the end of 6 months, sales during the final week of the sixth month were compared. The results are shown in Table 10.1.

The salesman went on to say that the results had been subjected to statistical analysis. His company had tested the hypothesis that there was no difference in mean sales for the two populations against the alternate hypothesis that the mean for population 1 (instructed) was greater than that

Table 10.1 SALES DATA FOR CASE STUDY.

PAIR	SALES (UNITS) INSTRUCTED	NOT INSTRUCTED
1	13	10
2	39	39
3	6	4
4	77	79
5	110	108
6	33	27
7	16	12
8	87	74
9	15	11
10	33	37
11	18	11
12	51	41
13	15	15
14	30	17
15	22	25

of population 2 (not instructed). He said that the differences had a mean of 3.667 with a standard deviation of 5.341 and that the sample t value of 2.659 was greater than the critical value, at the 0.01 level of significance, of 2.624. Thus the company was satisfied that the course was effective.

The president looked at the data for a while and then asked why there was such a large difference in sales from department to department. The salesman replied that different items are sold in different departments. Department 3, for example, sold tractors, while department 5 sold hardware. Within each department, however, assured the salesman, sales units were comparable. "Then you are telling me," said the president, "that the mean difference of 3.667 is an average of tractors, paint, and who knows what. Those numbers don't measure the same things at all, so your analysis is worthless. If you can come up with proof that your course works, at the same significance level as you claim, I'll hire your firm. But your analysis of the data is totally meaningless. Good day, sir!"

As the unhappy salesman left the office, he vowed to find out who had done the analysis and to see if the data could be analyzed correctly. When he returned to the home office, he found that the executive who prepared the analysis had been replaced by a recent college graduate who had had a course in statistics. "Of course, you can't take the mean of different kinds of data," she said, "but there are many tests which don't require you to use the mean and standard deviation. I'll see if one of them can be useful. They are called nonparametric tests."

10.1 Nonparametric Tests in One Sample or Paired Samples

A common research problem involves the testing of two paired samples—either before-and-after observations or matched pairs (as discussed in Section 8.3.2). The t test can be used only if the samples come from populations that are normal, and the z test may be used only if the samples are large. Two convenient nonparametric alternatives to these tests are the *sign test* and the *Wilcoxon T Test.* For small samples, either of these alternatives may be used in place of the t test; for large samples, they may replace the z test.

10.1.1 The Paired-Samples Sign Test

The **sign test** is a special application of the binomial distribution to test a null hypothesis which implies that a population proportion is 0.5. Suppose a hypothesis states that there is no difference in the distributions of the populations from which two paired samples are taken. If we can calculate the differences between the values of the variable for each pair, and if this hypothesis is true, each difference that is not zero should be equally likely to be positive or negative. In other words, the distribution of difference scores would come from a population of differences in which the proportion of positive (or negative) scores is 0.5. We can then use Table 2 to obtain the probability of obtaining a result as extreme as the sample result, by chance, if the population proportion of positive scores, π, is 0.5. If this probability is smaller than α, the significance level chosen for the test, we conclude that the distributions of the variable for the two populations are different. (This is actually the procedure employed in Section 7.3.2 for small-sample tests concerning a population proportion.)

The sign test is possibly the oldest nonparametric test in use. Arbuthnott (1710) obtained birth records in London for 82 years and recorded a plus sign (+) for each year in which more males were born than females and a minus sign (−) if more females were born than males. There were no ties, but he might well have recorded zero if the same number of each sex were born. His results showed 82 plus signs! He concluded that more males are likely to be born in a year than females. The probability of obtaining a result this extreme if there is no difference in proportion of males and females being born (that is, $\pi = 0.5$) is $(0.5)^{82}$ or about 0.0000000 000000000000000002068 (that is, 2.068×10^{-25}). Thus his conclusion was probably reasonable.

Sometimes we cannot obtain an actual difference score but we may be able to use some qualitative measure of difference. Suppose, for example, that patients at a hospital suffering from condition X are given an experimental drug. After a period of time the hospital can determine whether the condition of each patient has improved, deteriorated, or remained the same. The sign test can be used to test the hypothesis that the drug is effective.

The test statistic used to test the hypothesis that there is no difference in the distributions of the populations is either x_+, the number of positive differences, or x_-, the number of negative differences. The sample values of the test statistic are designated x_+^* and x_-^*. There are three possible alternate hypotheses, frequently adapted to the situation. We will agree that if the populations are population 1 and population 2 and the corresponding samples are sample 1 and sample 2, we can obtain the difference scores by subtracting each value in sample 2 from the corresponding value in sample 1. For qualitative variables, an increase from sample 1 to sample 2 will be considered positive and a decrease will be considered negative. Then if population 1 has a larger median than population 2, most of the differences or changes should be positive; conversely, if population 1 has a smaller median than population 2, most of the differences or changes will be negative. Thus the null hypothesis may be that the populations have equal medians, and the alternate will be that they do not have equal medians or that one median is greater than the other. The first of these tests has a two-tailed rejection region, while the second has a one-tailed rejection region.

Suppose, for example, that 40 college students are paired on the basis of academic achievement, major, course load, year in school, and other factors. Then the 40 students are randomly assigned to two groups with one student from each matched pair in each group. The groups are called A and B. The students are asked to keep a record of the number of hours they devote to studying their courses in the 2 weeks before midterm examinations. After midterms the students participate in one of two different seminars designed to foster good study habits. All students in a given group attend the same seminar. The purpose of the study is to assess the relative effectiveness of the two seminar presentations. During the 2 weeks before finals, the students record the number of hours they devote to studying their courses. A measure of the effectiveness of either seminar is an increase in

Table 10.2 INCREASES IN STUDY HOURS FOR STUDENTS ATTENDING TWO SEMINARS.

	INCREASE IN HOURS OF STUDY		
PAIR	A	B	DIFFERENCE
1	13	7	+6
2	2	4	−2
3	−1	3	−4
4	4	12	−8
5	11	14	−3
6	7	9	−2
7	12	6	+6
8	−3	2	−5
9	−4	−1	−3
10	7	3	+4
11	5	6	−1
12	2	5	−3
13	1	−1	+2
14	0	7	−7
15	5	1	+4
16	2	1	+1
17	3	9	−6
18	4	5	−1
19	−2	1	−3
20	3	5	−2

the number of hours for a particular student. We may use the sign test to determine whether either seminar is more effective. Table 10.2 shows these increases and the difference between them.

The null hypothesis is H_0: the median increase in study hours is the same for both populations; the alternate hypothesis is that the median increases are not the same. Suppose we are willing to take 1 chance in 20 of rejecting H_0 in error; then $\alpha = 0.05$. Thus if the probability of obtaining a result as extreme as the actual result by chance is less than 0.05, we can reject H_0. The decision rule will then be to reject H_0 if $P(x \geqslant x_+^*) < 0.025$ or if $P(x \geqslant x_-^*) < 0.025$, where the probabilities are based on a population proportion of 0.5. Since there are 14 negative differences and only 6 positive differences, $P(x \geqslant 14)$ will be much smaller than $P(x \geqslant 6)$, so we turn to Table 2 at the back of the book. We want $P(x \geqslant 14) = 1 - P(x \leqslant 13)$, so we have $p = 0.5$, $n = 20$, and $r = 13$. Then $P(x \geqslant 14) = 0.058$ so we cannot reject H_0. Although we cannot conclude that there is a difference in the medians of the populations, the probability of a type II error is probably great, so we reserve judgment. We should use a more powerful test, if possible, or obtain additional data. The procedure used to test this hypothesis and the other possible procedures are summarized here.

Sign Test for Differences Between Paired Samples

Suppose n pairs of observations of variables x_1 and x_2 are randomly obtained from populations with medians Md_1 and Md_2, respectively. We let x_+^* and x_-^* represent the number of increases and decreases from sample 1 to sample 2. The sample value of the test statistic used to test the hypothesis $\text{Md}_1 = \text{Md}_2$ is x_+^* or x_-^*.

The decision rule is dictated by the alternate hypothesis and the level of significance α chosen for the test. It is obtained by assuming that the probability that a particular change is an increase or a decrease is 0.5. The three possibilities are as follows:

Case I	**Case II**	**Case III**
	Null and Alternate Hypotheses	
H_0: $\text{Md}_1 = \text{Md}_2$ H_1: $\text{Md}_1 \neq \text{Md}_2$	H_0: $\text{Md}_1 = \text{Md}_2$ H_1: $\text{Md}_1 > \text{Md}_2$	H_0: $\text{Md}_1 = \text{Md}_2$ H_1: $\text{Md}_1 < \text{Md}_2$
	Decision Rule	
Reject H_0 if either $P(x \geqslant x_+^*) < \alpha/2$ or $P(x \geqslant x_-^*) < \alpha/2$	Reject H_0 if $P(x \geqslant x_+^*) < \alpha$	Reject H_0 if $P(x \geqslant x_-^*) < \alpha$

For large samples (generally considered $n > 25$ for the sign test), we may use the normal approximation to the binomial, correcting for continuity. Since $\pi = 0.5$, the mean will be equal to $0.5n$ and the standard deviation will be equal to $\sqrt{n(0.5)(0.5)} = 0.5\sqrt{n}$. The test statistic will then be z^*, where

$$z^* = \frac{(x_+^* \pm 0.5) - 0.5n}{0.5\sqrt{n}}$$

The ambiguous "± 0.5" is used to indicate that the difference between x_+^* and $0.5n$ is to be reduced by 0.5. If $x_+^* > 0.5n$, we subtract 0.5; if $x_+^* <$

Large-Sample Sign Test for Differences Between Paired Samples

Suppose n pairs of observations of variables x_1 and x_2 are randomly obtained from populations with medians Md_1 and Md_2, respectively. We let x_+^* and x_-^* represent the number of increases and decreases from sample 1 to sample 2. The sample value of the test statistic used to test the hypothesis $\mathrm{Md}_1 = \mathrm{Md}_2$ is z^*, where

$$z^* = \frac{(x_+^* - 0.5) - 0.5n}{0.5\sqrt{n}}$$

if $x_+^* > 0.5n$ or

$$z^* = \frac{(x_+^* + 0.5) - 0.5n}{0.5\sqrt{n}}$$

if $x_+^* < 0.5n$ and x_+^* is the number of positive differences.

The decision rule is dictated by the alternate hypothesis and the level of significance, α, chosen for the test. The three possibilities are as follows:

Case I	**Case II**	**Case III**
	Null and Alternate Hypotheses	
H_0: $\mathrm{Md}_1 = \mathrm{Md}_2$	H_0: $\mathrm{Md}_1 = \mathrm{Md}_2$	H_0: $\mathrm{Md}_1 = \mathrm{Md}_2$
H_1: $\mathrm{Md}_1 \neq \mathrm{Md}_2$	H_1: $\mathrm{Md}_1 > \mathrm{Md}_2$	H_1: $\mathrm{Md}_1 < \mathrm{Md}_2$
	Decision Rule	
Reject H_0 if $\lvert z^*\rvert > z_{\alpha/2}$	Reject H_0 if $z^* > z_\alpha$	Reject H_0 if $z^* < -z_\alpha$

$0.5n$, however, we add 0.5, since $x^*_+ - 0.5n$ will be negative. We could use x^*_- just as easily as x^*_+ in defining the test statistic, but the decision rules would be different. The value of z^* using x^*_- would be exactly the same in absolute value as that given above, but opposite in sign.

10.1.2 Differences of Zero

Since differences of zero are neither positive nor negative, some statisticians prefer to ignore them, reducing the sample size accordingly. Differences of zero tend to substantiate a null hypothesis of no difference, however, so they probably should not be ignored entirely. One way to take zeros into account is to count half the zeros as positive and half as negative. If there is an odd number of zeros, remove one from the sample. If the example of Section 10.1.1 had had 5 differences of zero, 11 negative differences, and 4 positive differences, we would have counted two of the zeros as positive and two as negative, ignoring the remaining one and thus reducing the sample size to 19. We would then have obtained $P(x \geqslant 13) = 1 - P(x \leqslant 12) = 0.084$ for $n = 19$, $p = 0.5$, and $r = 12$.

PROFICIENCY CHECK

1. Two different toothpastes are tested on two sets of children carefully matched for age, sex, diet, fluoride treatments, and so on. During the test period, the number of new cavities for the subjects are as follows:

SUBJECT PAIR	TOOTHPASTE A	TOOTHPASTE B
1	3	2
2	1	1
3	4	1
4	2	0
5	5	4
6	3	1
7	3	2
8	1	1
9	3	2
10	2	1

Use the sign test to test the hypothesis that there are no differences between the toothpastes in median number of cavities. Use the 0.05 level of significance.

10.1.3 The One-Sample Sign Test

We may also use the sign test as a quick test for a hypothesis that the median of a population is a certain value. If we obtain a sample and wish to test the hypothesis that the median of the population is a certain value, say Md_0, we let x_+^* be the number of observations above the median and x_-^* be the number of observations below the median. Since a value equal to the median has a difference of zero from the median, we can count half the values at the median as positive and half as negative, ignoring one of the observations if the number is odd. If the true median is Md_1, and if, for example, $\text{Md}_1 > \text{Md}_0$ (Figure 10.1), then we would expect most of the differences to be positive. For small samples the test statistic is x_+^* or x_-^*, depending on the alternate hypothesis, and for large samples, ($n > 25$) the test statistic is z^* as in the paired-samples sign test.

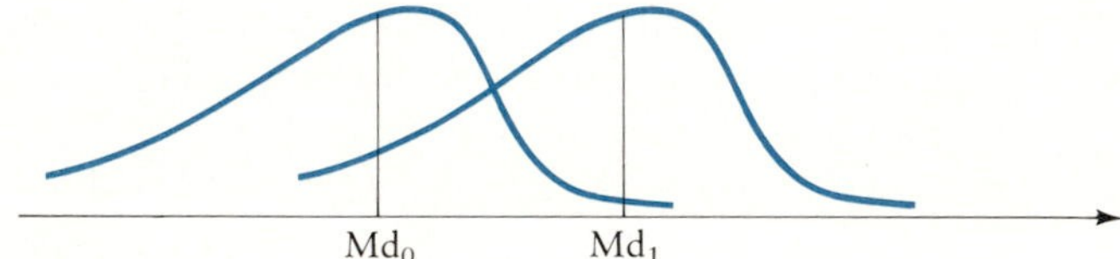

Figure 10.1
Two populations where $\text{Md}_1 > \text{Md}_0$: most of the differences $\text{Md}_1 - \text{Md}_0$ will be greater than zero.

Sign Test for a Population Median

If a random sample of n observations of a variable x is taken from a population with median Md, we may test the hypothesis $\text{Md} = \text{Md}_0$. For small samples, the sample value of the test statistic used to test the hypothesis $\text{Md} = \text{Md}_0$ is x_+^* or x_-^*, where x_+^* is the number of observations greater than Md_0 and x_-^* is the number of observations smaller than Md_0. For large samples ($n > 25$) the sample value of the test statistic is z^*, where

$$z^* = \frac{(x_+^* - 0.5) - 0.5n}{0.5\sqrt{n}}$$

if $x_+^* > 0.5n$ or

$$z^* = \frac{(x_+^* + 0.5) - 0.5n}{0.5\sqrt{n}}$$

if $x_+^* < 0.5n$.

The decision rule is dictated by the alternate hypothesis and the level of significance α chosen for the test. For small samples it is obtained by assuming that the probability an observation is greater than Md_0 is 0.5 or that the probability an observation is smaller than Md_0 is 0.5. The three possibilities are on the next page.

Case I	Case II	Case III
	Null and Alternate Hypotheses	
H_0: Md = Md$_0$ H_1: Md $\neq$ Md$_0$	H_0: Md = Md$_0$ H_1: Md > Md$_0$	H_0: Md = Md$_0$ H_1: Md < Md$_0$
	Decision Rule	
	(Small Samples)	
Reject H_0 if either $P(x \geqslant x_+^*) < \alpha/2$ or $P(x \geqslant x_-^*) < \alpha/2$	Reject H_0 if $P(x \geqslant x_+^*) < \alpha$	Reject H_0 if $P(x \geqslant x_-^*) < \alpha$
	(Large Samples)	
Reject H_0 if $\lvert z^*\rvert > z_{\alpha/2}$	Reject H_0 if $z^* > z_\alpha$	Reject H_0 if $z^* < -z_\alpha$

Example 10.1

A typing school claims that in a 6-week intensive course it can train students to type, on the average, more than 60 words per minute (WPM). A random sample of 12 graduates is given a typing test. Table 10.3 shows the median number of words per minute typed by each student. Test the hypothesis, using $\alpha = 0.05$, that the median typing speed of graduates is greater than 60 words per minute.

Table 10.3 TYPING SPEED OF GRADUATES.

STUDENT	WPM	WPM − 60
A	81	+21
B	76	+16
C	63	+3
D	71	+11
E	66	+6
F	59	−1
G	88	+28
H	73	+13
I	80	+20
J	66	+6
K	58	−2
L	70	+10

Solution

We wish to show that Md > 60. To prove this, we will try to reject H_0: Md = 60 in favor of H_1: Md > 60. The decision rule will be to reject H_0 if $P(x \geq x_+^*) < 0.05$. We have $x_+^* = 10$, so from Table 2 with $n = 12$, $p = 0.5$, and $r = 9$, we find $P(x \geq 10) = 1 - P(x \leq 9) = 0.019$. Thus we reject H_0 and conclude that the contention has been proved.

PROFICIENCY CHECKS

2. Fifth-grade students having difficulty with fractions are taught by a new method. Using traditional methods, it usually takes students 4 weeks to learn the content of these lessons. With the new method, thirteen students take less than 4 weeks to learn it, five take 4 weeks to learn it, while six do not learn it by the fourth week. Test the hypothesis that this method is more effective than the traditional approach. Use the 0.10 level of significance.
3. Encouraged by his apparent success, the fifth-grade teacher of Proficiency Check 2 uses the method with his entire class the following year. Twenty-nine students take less than 4 weeks to learn the material, four take 4 weeks, and fifteen take more than 4 weeks. Test the hypothesis that the new method is more effective than the traditional approach. Use the 0.05 level of significance.

10.1.4 The Wilcoxon *T* Test

Many experiments involve only an indication whether or not there has been a change in some variable. For these experiments the sign test is the preferred test. If we can rank the differences, a somewhat more powerful test than the sign test is the **Wilcoxon *T* test** (or *Wilcoxon signed-ranks test*) introduced by Frank Wilcoxon in 1945. This test is more powerful than the sign test because it makes use of the *magnitudes* of the differences in a paired-samples test or the deviations from an assumed median in a one-sample test. In this test, the absolute values of the differences (or deviations from Md_0) are ranked from smallest to largest. For a particular sample, the sum of the ranks of the positive values is designated T_+^* while the sum of the ranks of the negative values is designated T_-^*. If differences or deviations of zero occur, they are considered half positive and half negative, and one is discarded if the number is odd.

If two or more values are the same, each is assigned the average of all the ranks they occupy. Thus if there are four zeros, they would be assigned ranks 1, 2, 3, 4; the average of these ranks is $(1 + 4)/2 = 2.5$. Two would be positive and two would be negative.

Table 10.4 ALL POSSIBLE RANKINGS, FOUR OBSERVATIONS.

RANKS OF POSITIVE DIFFERENCES	RANKS OF NEGATIVE DIFFERENCES	T_+	T_-
1, 2, 3, 4		10	0
2, 3, 4	1	9	1
1, 3, 4	2	8	2
1, 2, 4	3	7	3
1, 2, 3	4	6	4
1, 2	3, 4	3	7
1, 3	2, 4	4	6
1, 4	2, 3	5	5
2, 3	1, 4	5	5
2, 4	1, 3	6	4
3, 4	1, 2	7	3
1	2, 3, 4	1	9
2	1, 3, 4	2	8
3	1, 2, 4	3	7
4	1, 2, 3	4	6
	1, 2, 3, 4	0	10

The logic behind tests involving ranks is relatively simple. Suppose there are only four observations; then the ranks (assuming there are no ties) are 1, 2, 3, and 4. We can list *all* the possibilities as in Table 10.4.

We can easily determine the probability distribution of T (the smaller of the values of T_+ and T_-) for $n = 4$ as shown in the margin. No matter how large n becomes, we can obtain a probability distribution for T and calculate the exact probability of obtaining a T as small as T^*. We can determine similar distributions for other nonparametric tests.

T	$P(T)$
0	0.125
1	0.125
2	0.125
3	0.250
4	0.250
5	0.125

To illustrate the Wilcoxon T test, let us use the data of Example 10.1 (Table 10.3). When we rank the absolute value of the differences from smallest to largest, the results are as shown in Table 10.5. The signs of the differences are noted for convenience. Then $T^*_+ = 75$ and $T^*_- = 3$. It can be shown that the sum of the first n positive integers is $n(n + 1)/2$; that is $1 + 2 + 3 + \cdots + n = n(n + 1)/2$. We can use this fact as a check—the sum of T^*_+ and T^*_- will always be $n(n + 1)/2$. In this case, $n = 12$ and $12(13)/2 = 78$, so the results check.

Now if there is a considerable difference between the two populations, or if the true median is much different from Md_0, one of the values T^*_+ or T^*_- will be very small. Table 11 in the back of the book lists the critical values of the test statistic T, which differ from sample size to sample size. We designate the critical values as $T_\alpha\{n\}$, where α is the probability of obtaining a value of T smaller than or equal to $T_\alpha\{n\}$ if the null hypothesis is true, for a sample of size n. A portion of Table 11 is shown in Table 10.6.

Table 10.5 DIFFERENCES AND RANKS FOR DATA OF EXAMPLE 10.1.

DIFFERENCE	RANK	COMMON RANK (IF NEEDED)
−1	1	
−2	2	
+3	3	
+6, +6	4, 5	4.5
+10	6	
+11	7	
+13	8	
+16	9	
+20	10	
+21	11	
+28	12	

Table 10.6 PORTION OF TABLE 11 FOR $T_\alpha\{n\}$.

SAMPLE SIZE n	PROBABILITY THAT T IS LESS THAN CRITICAL VALUE			
	0.050	0.025	0.010	0.005
10	11	8	5	3
11	14	11	7	5
12	17	14	10	7
13	21	17	13	10
14	26	21	16	13
15	30	25	20	16
16	36	30	24	19

We find, for example, that $T_{.025}\{15\} = 25$. Thus if the null hypothesis is true, a sample value of T obtained from a sample of size 15 will be as small as 25 only 0.025 of the time. If the alternate hypothesis is two-sided, the sample value of the test statistic is designated T^* and is the smaller of T^*_+ and T^*_-. The decision rule will be to reject H_0 if T^* is smaller than or equal to $T_{\alpha/2}\{n\}$. For a one-sided alternative, if the alternate hypothesis implies that there will be more positive than negative values, reject H_0 if $T^*_- \leqslant T_\alpha\{n\}$; if the alternate hypothesis implies that there will be more negative than positive values, reject H_0 if $T^*_+ \leqslant T_\alpha\{n\}$. In the preceding example, $n = 12$ and the alternate hypothesis is that Md > 60. Since $T_{.05}\{12\} = 17$ (see Table 10.6), the decision rule is to reject H_0 if $T^*_- \leqslant 17$. Since $T^*_- = 3$, we reject H_0 and conclude that indeed Md > 60; that is, the contention has been proved.

For large samples ($n > 25$), it has been shown that if the null hypothesis of no difference is true, the statistic T is approximately normally distributed with mean μ_T and variance σ_T^2, where $\mu_T = n(n + 1)/4$ and $\sigma_T^2 = n(n + 1)(2n + 1)/24$. We can obtain the sample value of the test statistic, z^*, by using either one of the values T_+^* or T_-^*. The statement of the decision rule depends on which one we select. For consistency, we can state it in terms of T_+^*. Thus if $z^* = (T_+^* - \mu_T)/\sigma_T$, for a two-sided alternate hypothesis the decision rule will be to reject H_0 if $|z^*| > z_{\alpha/2}$. For a one-sided alternative, if the alternate hypothesis implies that there will be more positive than negative values, reject H_0 if $z^* > z_\alpha$; if the alternate hypothesis implies that there will be more negative than positive values, reject H_0 if $z^* < -z_\alpha$.

We thus have the following characterization of the Wilcoxon T test, including both the paired-sample and the one-sample version of the test.

Wilcoxon T Test for Differences Between Paired Samples

If n pairs of observations of a variables x are randomly obtained from populations with medians Md_1 and Md_2, respectively, and the difference $d = x_1 - x_2$ is computed for each pair, we may test the hypothesis $\text{Md}_1 = \text{Md}_2$. For small samples, the sample value of the test statistic used to test the hypothesis $\text{Md}_1 = \text{Md}_2$ is T_+^*, T_-^*, or T^*. Here T_+^* is the sum of the ranks of positive values of d, T_-^* is the sum of the ranks of negative values of d, and T^* is the smaller of the values T_+^* and T_-^*. For large samples ($n > 25$) the test statistic is z^*, where $z^* = (T_+^* - \mu_T)/\sigma_T$, $\mu_T = n(n + 1)/4$ and $\sigma_T = \sqrt{n(n + 1)(2n + 1)/24}$.

The decision rule is dictated by the alternate hypothesis and the level of significance α chosen for the test. The three possibilities are as follows:

Case I	Case II	Case III
Null and Alternate Hypotheses		
H_0: $\text{Md}_1 = \text{Md}_2$	H_0: $\text{Md}_1 = \text{Md}_2$	H_0: $\text{Md}_1 = \text{Md}_2$
H_1: $\text{Md}_1 \neq \text{Md}_2$	H_1: $\text{Md}_1 > \text{Md}_2$	H_1: $\text{Md}_1 < \text{Md}_2$
Decision Rule		
(Small Samples)		
Reject H_0 if $T^* \leqslant T_{\alpha/2}\{\text{n}\}$	Reject H_0 if $T_-^* \leqslant T_\alpha\{n\}$	Reject H_0 if $T_+^* \leqslant T_\alpha\{n\}$

(Large Samples)

Reject H_0 if	Reject H_0 if	Reject H_0 if
$\lvert z^*\rvert > z_{\alpha/2}$	$z^* > z_\alpha$	$z^* < -z_\alpha$

Wilcoxon *T* Test for a Population Median

If a random sample of n observations of a variable x is taken from a population with median Md, we may test the hypothesis Md = Md_0. For small samples, the sample value of the test statistic used to test the hypothesis Md = Md_0 is T^*_+, T^*_-, or T^*, where T^*_+ is the sum of the ranks of observations greater than Md_0, T^*_- is the sum of the ranks of observations smaller than Md_0, and T^* is the smaller of these two numbers. For large samples ($n > 25$) the test statistic is z^*, where $z^* = (T^*_+ - \mu_T)/\sigma_T$, $\mu_T = n(n + 1)/4$ and $\sigma_T = \sqrt{n(n + 1)(2n + 1)/24}$.

The decision rule is dictated by the alternate hypothesis and the level of significance α chosen for the test. The three possibilities are as follows:

Case I	Case II	Case III
Null and Alternate Hypotheses		
H_0: Md = Md_0	H_0: Md = Md_0	H_0: Md = Md_0
H_1: Md $\neq$ Md_0	H_1: Md > Md_0	H_1: Md < Md_0
Decision Rule		
(Small Samples)		
Reject H_0 if $T^* \leq T_{\alpha/2}\{n\}$	Reject H_0 if $T^*_- \leq T_\alpha\{n\}$	Reject H_0 if $T^*_+ \leq T_\alpha\{n\}$
(Large Samples)		
Reject H_0 if $\lvert z^*\rvert > z_{\alpha/2}$	Reject H_0 if $z^* > z_\alpha$	Reject H_0 if $z^* < -z_\alpha$

Example 10.2 Thirty people take part in a diet experiment. Their weights are measured at the beginning and at the end of the diet period. Use the data in Table 10.7 to determine whether the diet is effective. Use the 0.05 level of significance.

Table 10.7 WEIGHTS OF SUBJECTS IN DIET EXPERIMENT.

SUBJECT	STARTING WEIGHT	ENDING WEIGHT	DIFFERENCE
1	183	177	+6
2	144	145	−1
3	151	145	+6
4	163	162	+1
5	155	151	+4
6	159	163	−4
7	178	173	+5
8	184	185	−1
9	142	139	+3
10	137	138	−1
11	145	136	+9
12	172	155	+17
13	134	136	−2
14	160	160	0
15	121	123	−2
16	207	196	+11
17	153	149	+4
18	138	145	−7
19	172	171	+1
20	193	185	+8
21	127	127	0
22	118	119	−1
23	175	164	+11
24	166	164	+2
25	183	188	−5
26	191	174	+17
27	174	165	+9
28	153	144	+9
29	163	172	−9
30	129	129	0

Solution

There are three zeros, so we discard one and use the other two, counting one positive and one negative. Ranking the data, we have the results shown in Table 10.8.

Now we wish to test the hypothesis that the diet is effective. If this is true, Md_1 would be greater than Md_2; that is, the dieters would lose weight. Thus we wish to test H_0: $\mathrm{Md}_1 = \mathrm{Md}_2$ against H_1: $\mathrm{Md}_1 > \mathrm{Md}_2$. If the tables were large enough to give $T_\alpha\{29\}$, we would reject H_0 if $T^*_- \leq T_\alpha\{29\}$.

Table 10.8 TABLE OF DIFFERENCES AND THEIR RANKS FOR DIET EXPERIMENT.

DIFFERENCES	RANKS	COMMON RANK
−0, +0	1, 2	1.5
−1, −1, −1, −1, +1, +1	3, 4, 5, 6, 7, 8	5.5
−2, −2, +2	9, 10, 11	10
+3	12	
−4, +4, +4	13, 14, 15	14
−5, +5	16, 17	16.5
+6, +6	18, 19	18.5
−7	20	
+8	21	
−9, +9, +9, +9	22, 23, 24, 25	23.5
+11, +11	26, 27	26.5
+17, +17	28, 29	28.5

Such tables are available, and $T_\alpha\{29\} = 141$. Since we do not have these tables and $n > 25$, we will use the test statistic z and reject H_0 if $z^* > 1.645$, since $z_{.05} = 1.645$. Now $T_-^* = 117.5$ and $T_+^* = 317.5$; $\mu_T = 29(30)/4 = 217.5$ and $\sigma_T = \sqrt{(29)(30)(59)/24} \doteq 46.25$. Then $z^* = (317.5 - 217.5)/46.25 \doteq 2.16$ and we reject H_0, concluding that the diet is effective.

PROFICIENCY CHECKS

4. What is the decision rule if we wish to test the hypothesis H_0: $\text{Md}_1 = \text{Md}_2$ against H_1: $\text{Md}_1 > \text{Md}_2$, given:
   a. $n = 15, \alpha = 0.05$? b. $n = 21, \alpha = 0.01$?
   c. $n = 40, \alpha = 0.05$?

5. What is the decision rule if we wish to test the hypothesis H_0: $\text{Md}_1 = \text{Md}_2$ against H_1: $\text{Md}_1 \neq \text{Md}_2$ given:
   a. $n = 15, \alpha = 0.05$? b. $n = 21, \alpha = 0.01$?
   c. $n = 40, \alpha = 0.05$?

6. Two different toothpastes are tested on two sets of children carefully matched for age, sex, diet, fluoride treatments, and so on. During the test period, the number of new cavities for the subjects are as follows:

(continued)

PROFICIENCY CHECKS *(continued)*

SUBJECT PAIR	TOOTHPASTE A	TOOTHPASTE B
1	3	2
2	1	2
3	4	1
4	2	0
5	5	4
6	3	1
7	3	2
8	1	1
9	3	2
10	2	2

Use the Wilcoxon T test to test the hypothesis that there are no differences between the toothpastes in median number of cavities. Use the 0.05 level of significance.

Problems

Practice

10.1 A random sample of 18 observations is obtained to determine whether the population median is a certain value. What will the conclusion be if 12 observations are above this value, 4 are below the value, and 2 are equal to it and the level of significance is:
a. 0.10? **b.** 0.05? **c.** 0.01?

10.2 A random sample of 18 observations is obtained to test the hypothesis that the population median is greater than a certain value. What will the conclusion be if 12 observations are above this value, 4 are below the value, and 2 are equal to it and the level of significance is:
a. 0.10? **b.** 0.05? **c.** 0.01?

10.3 A random sample of 18 observations is obtained to determine whether the population median is a certain value; 10 of the observations are above this value, 6 are below the value, and 2 are equal to it. What will the conclusion be if the rank sum of those above the value (not including the zeros) is 133 and the level of significance is:
a. 0.10? **b.** 0.05? **c.** 0.01?

10.4 A random sample of 18 observations is obtained to test the hypothesis that the population median is greater than a certain value; 10 of the observations are above this value, 6 are below the value, and 2 are equal to it. What will

the conclusion be if the rank sum of those above the value (not including the zeros) is 133 and the level of significance is:
a. 0.05? **b.** 0.025? **c.** 0.01?

Applications

10.5 In the past, the number of defectives produced by a machine per lot of 1000 has had a median of 8. To test whether the machine is still producing the same amount of defectives, the machine's output for the past week is randomly sampled and 30 lots of 1000 are examined for defects. The number of defectives in the lots, in order of production, are 3, 11, 6, 22, 9, 12, 11, 12, 7, 12, 8, 14, 10, 7, 6, 8, 16, 10, 9, 10, 13, 8, 4, 10, 12, 9, 13, 11, 9, 10. The quality control expert believes that the number of defectives has increased. Using the sign test, test his conjecture at the 0.05 level of significance.

10.6 Repeat Problem 10.5 using the Wilcoxon T test and the large-sample z. Are the results different if you use Table 11?

10.7 An employee for a food processing plant notices a marked reduction in deterioration in unrefrigerated food which has been exposed to gamma radiation. She matches ten food samples by type, quantity, degree of freshness, and source and exposes one of each pair, randomly selected, to the radiation. She then determines the deterioration rate of the foods over a period of 1 week during which all foods are stored as they normally would be. The deterioration rate is defined by an analysis of various factors, including retention of nutritive value, moisture, appealing appearance, and so forth. The results are shown here:

	DETERIORATION RATE	
PAIR	NOT EXPOSED	EXPOSED
A	0.32	0.19
B	0.18	0.21
C	0.28	0.21
D	0.34	0.17
E	0.22	0.24
F	0.17	0.12
G	0.41	0.24
H	0.23	0.28
I	0.30	0.18
J	0.26	0.17

Use the sign test and a 0.05 level of significance to determine whether the radiation has in fact reduced the deterioration rate.

10.8 Repeat Problem 10.7 using the Wilcoxon T test. Do the assumptions for a t test appear to be satisfied?

10.9 To test the effectiveness of an anti-litter campaign, a group of 50 people at a convention is asked to listen to a 10-minute demonstration on the effects of littering. Afterward they are asked whether they are now more or less opposed to littering than they were before the demonstration. Of the 46 who respond, 22 say they are more opposed, 12 say they are less opposed, and 12 say they have not changed their feelings. Ignoring the problems involved in the analysis of such data (nonrandomness, degree of feeling before, desire to please the questioners, and so forth), determine whether the demonstration has had the desired effect. Use the 0.05 level of significance.

10.10 A test is designed to compare the wearing qualities of two brands of motorcycle tires. One tire of each brand is placed on each of six motorcycles that are then driven for a specified number of days. The two tires are then exchanged (front and rear) and driven for an equal number of miles. The motorcycles are not driven for the same number of miles or under (necessarily) the same conditions. At the conclusion of the test, wear is measured (in thousandths of an inch). The data are given here. Management wishes to market the tire that will wear less than the other in similar driving. The analyst decides that the differences are not directly comparable, since the motorcycles have been driven for differing numbers of miles and under differing conditions, so that a nonparametric test is indicated. Use the sign test. Are the results of this test conclusive, or is further testing needed? Use the 0.05 level of significance.

	MOTORCYCLE					
	1	2	3	4	5	6
Brand A	98	61	38	117	88	109
Brand B	102	60	46	125	94	111

10.11 The analyst in Problem 10.10 is not satisfied with the results of the sign test. He wants to use the Wilcoxon T test, but since the difference scores were not obtained over the same mileages they are not directly comparable. To compensate for the differing conditions, he expresses the mileages of brand B as percentages of the mileages of brand A. The differences can then be ranked and the Wilcoxon T test used. Repeat the analysis of Problem 10.10 using the Wilcoxon T test. The percentages are shown here:

	MOTORCYCLE					
	1	2	3	4	5	6
Brand A	100	100	100	100	100	100
Brand B	104.1	98.4	121.1	106.8	106.8	101.8

10.12 A large chain of grocery stores wishes to test the effectiveness of an advertising display on the sales of a certain product. Sixty of their retail stores are paired on pertinent characteristics such as size, weekly volume, usual stock of the product, type of community, and so on. By flipping a coin, one store in each of the 30 pairs is selected to display the product while the other store is to carry it as usual. Given the following data, use the sign test to determine whether the display is effective. Use the 0.05 level of significance.

	WEEKLY SALES			WEEKLY SALES	
PAIR	WITHOUT DISPLAY	WITH DISPLAY	PAIR	WITHOUT DISPLAY	WITH DISPLAY
1	68	74	16	65	73
2	91	90	17	93	96
3	66	69	18	74	69
4	56	64	19	80	73
5	79	84	20	95	108
6	63	60	21	112	114
7	91	84	22	76	73
8	64	72	23	31	47
9	91	99	24	72	81
10	67	73	25	98	100
11	40	46	26	44	43
12	84	93	27	81	79
13	111	95	28	98	114
14	78	83	29	54	61
15	97	109	30	72	90

10.13 Repeat Problem 10.12 using the Wilcoxon T test. Would the t test be appropriate? Compare the results of the sign test with the Wilcoxon T test. Are the P values about the same?

10.14 The clients at a mental health clinic are periodically classified as "condition improved," "condition worsened," or "no change in condition." A routine check on the 15 patients of a new staff member reveals that, after three months, 10 are classified as "condition improved"and 3 as "condition worsened." The other two have shown no change. Can you conclude, at the 0.05 level of significance, that she is effective?

10.15 To test the efficiency of a procedure designed to increase the daily output of a machine, six machines are tested with and without the procedure. The results are as follows:

	MACHINE					
	A	B	C	D	E	F
With	15.6	18.2	14.3	14.9	14.3	16.7
Without	14.4	16.7	13.1	14.4	14.8	14.7

Use the most powerful nonparametric technique to test the hypothesis, at the 0.05 level of significance, that the procedure is effective.

10.2 A Nonparametric Test for Two Independent Samples

Another common research problem involves the testing of two independent samples. The t test can be used only if the samples come from populations that are normal, and the z test may be used only if the samples are large. When one or more assumptions for using the t test or z test are violated, a convenient nonparametric alternative to these tests for two independent samples is the *Mann–Whitney U test*. In this test we rank the data from smallest to largest and determine the rank sums. If one sample has ranks 1–5 and 7, for example, while the other sample has ranks 6 and 8–12, it is clear that the numbers in the first sample are smaller than those in the second sample. For the first sample the rank sum is 22, while for the second sample the rank sum is 56. We can easily conclude that the median of the population from which the first sample is drawn is probably smaller than the median of the population from which the second sample is drawn. The distributions may be as shown in Figure 10.2. For samples in which the differences are not so obvious, we need a statistical test to determine a decision rule. The U test is such a statistical test.

10.2.1 The Mann–Whitney U Test

The **Mann–Whitney *U* test,** like the Wilcoxon T test, makes use of the ranks of the data. The U test is equivalent to another test, the *Wilcoxon rank-sum test,* that is not be covered here but may be found in the booklet by Wilcoxon and Wilcox (1964).

In the Mann–Whitney U test, we rank the two samples in numerical order, smallest to largest, as if they were one set of data. The sum of the ranks in sample 1 is designated R_1 and the sum of the ranks in sample 2 is designated R_2. Then we calculate two values of the U test statistic, one for each sample. These sample values of the test statistic U can be designated U_1^* and U_2^*, where

$$U_1^* = n_1 n_2 + \frac{n_1(n_1 + 1)}{2} - R_1$$

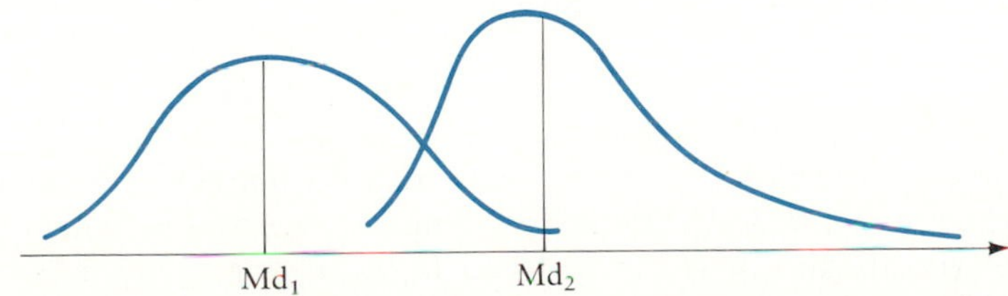

Figure 10.2
Possible distributions for sample 1 with ranks 1–5 and 7 and sample 2 with ranks 6 and 8–12: $Md_2 > Md_1$.

and

$$U_2^* = n_1 n_2 + \frac{n_2(n_2 + 1)}{2} - R_2$$

Since the ranks are 1, 2, 3, . . . , $n_1 + n_2$ and the sum of the first $n_1 + n_2$ integers is $(n_1 + n_2)(n_1 + n_2 + 1)/2$, we must have $R_1 + R_2 = (n_1 + n_2)(n_1 + n_2 + 1)/2$. Moreover, $U_1^* + U_2^* = n_1 n_2$. These requirements serve as a convenient check on arithmetic.

If the median of population 1 is greater than that of population 2, the average rank in sample 1 will be larger than the average rank in sample 2. Thus if $R_1 > R_2$, the hypothesis that $\text{Md}_1 > \text{Md}_2$ is supported. Since R_1 and R_2 are subtracted from $n_1 n_2 + n_1(n_1 + 1)/2$ and $n_1 n_2 + n_2(n_2 + 1)/2$, respectively, small values of U_1^* support an alternate hypothesis that $\text{Md}_1 > \text{Md}_2$, while small values of U_2^* support an alternate hypothesis that $\text{Md}_1 < \text{Md}_2$. An alternate hypothesis H_1: $\text{Md}_1 \neq \text{Md}_2$ is supported by small values of either U_1^* or U_2^*.

Suppose, for example, that 19 applicants for a position are interviewed by three administrators and rated on a scale from 1 to 5 as to suitability for the position, higher ratings indicating a greater suitability for the position. Each applicant is given a score that is the total of the three ratings. Although college education is not a requirement for the position, a personnel director thinks it might have some bearing on suitability for the position. Group A has an educational background of less than 2 years of college, while group B has completed at least 2 years of college. We can use the Mann–Whitney U test to determine whether there is a difference that might be associated with educational background. Scores are shown in Table 10.9.

To rank the data, it is helpful to have a worksheet, particularly when there is a large amount of data. Thus we could list all the scores, from low to high, for the groups in adjacent columns. Then we would list the ranks, computing the common ranks where appropriate, and list the rank sums for each sample. Such a completed worksheet might resemble Table 10.10.

There are 19 observations, so the sum of R_1 and R_2 must be $19(20)/2 = 190$. Since $129.5 + 60.5 = 190$, the results are consistent. We then compute U_1^* and U_2^*. As $n_1 = 10$ and $n_2 = 9$,

$$U_1^* = (10)(9) + (10)(11)/2 - 129.5 = 15.5$$

and

$$U_2^* = (10)(9) + (9)(10)/2 - 60.5 = 74.5$$

Checking, $U_1^* + U_2^*$ must equal $n_1 n_2$; since $15.5 + 74.5 = 90 = (10)(9)$, these results are again consistent.

Table 12 in the back of the book lists critical values of U for $\alpha = 0.05$, 0.025, 0.01, and 0.005. The table lists sample sizes n and n', where $n \geqslant n'$, and the critical value of U can be listed $U_\alpha\{n_1, n_2\}$ or $U_\alpha\{n_2, n_1\}$; the two are interchangeable. A portion of Table 12 is reproduced in Table 10.11. If

Table 10.9

GROUP A	GROUP B
8	7
9	11
13	9
14	4
11	8
10	6
12	11
14	9
13	10
10	

Table 10.10 WORKSHEET FOR RANK SUMS.

SCORES A	SCORES B	RANKS	COMMON RANK	RANK SUMS A	RANK SUMS B
	4	1			1
	6	2			2
	7	3			3
8	8	4, 5	4.5	4.5	4.5
9	9, 9	6, 7, 8	7	7	14
10, 10	10	9, 10, 11	10	20	10
11	11, 11	12, 13, 14	14	13	26
12		15		15	
13, 13		16, 17	16.5	33	
14, 14		18, 19	18.5	37	
			Totals	129.5	60.5

Table 10.11 PORTION OF TABLE 12 FOR $U_\alpha\{n_1, n_2\}$ OR $U_\alpha\{n_2, n_1\}$.

SAMPLE SIZE n	SAMPLE SIZE n'	PROBABILITY THAT U IS LESS THAN OR EQUAL TO CRITICAL VALUE 0.05	0.025	0.01	0.005
3	3	0			
4	3	0			
.	.	.	.	.	.
6	2	0			
	3	2	1		
	4	3	2	1	0
	5	5	3	2	1
	6	7	5	3	2
.	.	.	.	.	.
10	2	1	0		
	3	4	3	1	0
	4	7	5	3	2
	5	11	8	6	4
	6	14	11	8	6
	7	17	14	11	9
	8	20	17	13	11
	9	24	20	16	13
	10	27	23	19	16

the samples are of size 5 and 6, for example, $U_{.05}\{5, 6\} = U_{.05}\{6, 5\} = 5$, so that the probability of obtaining a sample value of U as small as 5 by chance, if H_0 is true, is less than or equal to 0.05. Similarly, if the samples are 4 and 7, then $U_{.025}\{7, 4\} = 3$.

For a one-sided alternate hypothesis, we will reject $H_0\text{:Md}_1 = \text{Md}_2$ if either U_1^* or U_2^*, as appropriate, is smaller than or equal to $U_\alpha\{n, n'\}$; for a two-sided alternate hypothesis, we designate U^* as the smaller of U_1^* and U_2^* and reject $H_0\text{:Md}_1 = \text{Md}_2$ if $U^* \leq U_{\alpha/2}\{n_1, n_2\}$.

For the example with 19 applicants, the null hypothesis is H_0: $\text{Md}_1 = \text{Md}_2$, and the alternate hypothesis is H_1: $\text{Md}_1 \neq \text{Md}_2$. Thus we have a two-sided alternative and the test statistic is U^* (the smaller of 15.5 and 74.5); $U^* = 15.5$. Since $U_{.025}\{10, 9\} = 20$, the decision rule is to reject H_0 if $U^* \leq 20$. Since $15.5 \leq 20$, we reject H_0 and conclude that there was a difference that might be associated with educational background. Since ranks for group A were higher than for group B, we have an indication of the direction of the difference.

For sample sizes of at least 10 each, the sampling distribution of the variable U will be approximately normal with mean μ_U and variance σ_U^2, where

$$\mu_U = \frac{n_1 n_2}{2}$$

and

$$\sigma_U^2 = \frac{n_1 n_2(n_1 + n_2 + 1)}{12}$$

The test statistic for these large samples is z, where $z^* = (U_1^* - \mu_U)/\sigma_U$ or $z^* = (U_2^* - \mu_U)/\sigma_U$, as appropriate. For consistency, we can always use U_2, since large values of U_2 lead to the conclusion that $\text{Md}_1 > \text{Md}_2$, while small values of U_2 support a conclusion that $\text{Md}_1 < \text{Md}_2$. Thus we have the following description of the Mann–Whitney U test.

Mann–Whitney *U* Test for Differences Between Independent Samples

If n_1 and n_2 observations of a variable x are randomly obtained from populations with medians Md_1 and Md_2, respectively, and the samples are independent, we may test the hypothesis $\text{Md}_1 = \text{Md}_2$. For small samples, the sample value of the test statistic used to test the hypothesis $\text{Md}_1 = \text{Md}_2$ is U_1^*, U_2^*, or U^*, where $U_1^* = n_1 n_2 + n_1(n_1 + 1)/2 - R_1$, $U_2^* = n_1 n_2 + n_2(n_2 + 1)/2 - R_2$, and U^* is the smaller of U_1^* and U_2^*. When the data are ranked as in one sample, smallest to largest, R_1 is the sum of the ranks in sample

1 and R_2 is the sum of the ranks in sample 2. For large samples ($n_1 > 10$ and $n_2 > 10$), the test statistic is z^*, where $z^* = (U_2^* - n_1n_2/2)/\sqrt{n_1n_2(n_1 + n_2 + 1)/12}$.

The decision rule is dictated by the alternate hypothesis and the level of significance α chosen for the test. The three possibilities are as follows:

Case I	**Case II**	**Case III**
	Null and Alternate Hypotheses	
H_0: $\text{Md}_1 = \text{Md}_2$ H_1: $\text{Md}_1 \neq \text{Md}_2$	H_0: $\text{Md}_1 = \text{Md}_2$ H_1: $\text{Md}_1 > \text{Md}_2$	H_0: $\text{Md}_1 = \text{Md}_2$ H_1: $\text{Md}_1 < \text{Md}_2$
	Decision Rule	
	(Small Samples)	
Reject H_0 if $U^* \leqslant U_{\alpha/2}\{n_1, n_2\}$	Reject H_0 if $U_1^* \leqslant U_\alpha\{n_1, n_2\}$	Reject H_0 if $U_2^* \leqslant U_\alpha\{n_1, n_2\}$
	(Large Samples)	
Reject H_0 if $\lvert z^*\rvert > z_{\alpha/2}$	Reject H_0 if $z^* > z_\alpha$	Reject H_0 if $z^* < -z_\alpha$

Example 10.3

In the preceding example suppose there are 23 applicants for the position rather than 19. There are two more in group A with scores of 11 and 12 and two more in group B with scores of 9 and 8. Test the hypothesis that group A applicants will have higher scores than group B applicants.

Solution

The rankings for the revised data are given in Table 10.12. Since both n_1 and n_2 are greater than 10, we may use the normal approximation. We have $U_2^* = (12)(11) + (11)(12)/2 - 84.5 = 113.5$, $\mu_U = (12)(11)/2 = 66$, and $\sigma_U = \sqrt{(12)(11)(12 + 11 + 1)/12} \doteq 16.248$, so $z^* = (113.5 - 66)/16.248 \doteq 2.92$. We wish to test the null hypothesis H_0:$\text{Md}_1 = \text{Md}_2$ against the alternate hypothesis H_1:$\text{Md}_1 > \text{Md}_2$, so the decision rule is to reject H_0 if $z^* > 1.645$. Thus we reject H_0 and conclude that applicants with at least 2 years of college tend to receive higher scores than those with less education.

Table 10.12 WORKSHEET FOR RANK SUMS, REVISED DATA.

SCORES			COMMON	RANK SUMS	
A	B	RANKS	RANK	A	B
	4	1			1
	6	2			2
	7	3			3
8	8, 8	4, 5, 6	5	5	10
9	9, 9, 9	7, 8, 9, 10	8.5	8.5	25.5
10, 10	10	11, 12, 13	12	24	12
11, 11	11, 11	14, 15, 16, 17	15.5	31	31
12, 12		18, 19	18.5	37	
13, 13		20, 21	20.5	41	
14, 14		22, 23	22.5	45	
			Totals	191.5	84.5

PROFICIENCY CHECKS

7. What is the decision rule to test the hypothesis H_0: $\text{Md}_1 = \text{Md}_2$ against the alternate hypothesis H_1: $\text{Md}_1 \neq \text{Md}_2$ at the 0.05 level of significance if n_1 and n_2, respectively, are:
   a. 4 and 6? b. 7 and 5? c. 8 and 9? d. 12 and 15?

8. Use the Mann–Whitney U test and the 0.05 level of significance to test the contention that $\text{Md}_1 < \text{Md}_2$ if the sample results are as follows:

Sample 1	12	13	15	17	22	23
Sample 2	14	16	24	26		

9. Use the Mann–Whitney U test and the 0.05 level of significance to test the contention that $\text{Md}_1 \neq \text{Md}_2$ if the sample results are as follows:

Sample 1	2	3	5	7	12	13	18	19	22	31	36	42		
Sample 2	4	6	7	12	14	16	19	19	23	24	26	36	44	48

Problems

Practice

10.16 What is the decision rule if we wish to test the hypothesis H_0: $\text{Md}_1 = \text{Md}_2$ against H_1: $\text{Md}_1 > \text{Md}_2$ given:
a. $n_1 = 5, n_2 = 8, \alpha = 0.05$? **b.** $n_1 = 10, n_2 = 7, \alpha = 0.01$?
c. $n_1 = 40, n_2 = 44, \alpha = 0.05$?

10.17 What is the decision rule if we wish to test the hypothesis H_0: $\text{Md}_1 = \text{Md}_2$ against H_1: $\text{Md}_1 \neq \text{Md}_2$ given:
a. $n_1 = 5, n_2 = 8, \alpha = 0.05$? b. $n_1 = 10, n_2 = 7, \alpha = 0.01$?
c. $n_1 = 40, n_2 = 44, \alpha = 0.05$?

10.18 If $n_1 = 8, n_2 = 7$, and $R_1 = 44.5$, determine U_1^* and U_2^*.

10.19 If $n_1 = 8, n_2 = 7$, and $R_1 = 44.5$, can we conclude, at the 0.05 level, that $\text{Md}_1 \neq \text{Md}_2$?

10.20 If $n_1 = 18, n_2 = 17$, and $R_1 = 234.5$, can we conclude, at the 0.05 level, that $\text{Md}_1 \neq \text{Md}_2$?

Applications

10.21 A biologist investigating the toxicity of a certain substance is concerned with finding a chemical that will retard the substance's toxic properties. He puts equal amounts of the substance into two equal-size tanks containing goldfish and then puts a predetermined amount of a chemical into tank A. He observes how many hours pass before the goldfish succumb to the effects of the toxic substance. The following tabulation gives the number of hours (to the nearest hour) each goldfish survives after the tanks have been treated. Can it be asserted, at $\alpha = 0.05$, that the chemical retards the toxic properties of the substance? Use the U test.

Tank A	55	57	61	63	65	66	66	67	68	68	69	70	70	72	76
Tank B	48	52	54	55	57	60	63	65	68	69	70	70	71	73	75

10.22 A businessman wishes to use the best sales technique to display and sell a product. He displays the product in two different styles at each of six different stores. If the number of sales for the test period at each store is as shown here, can it be determined, at the 0.05 level of significance, using the U test, whether there are differences in the sales levels for the two displays?

Display 1	86	79	83	81	75	79
Display 2	77	80	69	74	71	72

10.23 Residual levels of DDT, in parts per billion, are measured in the blood of fish in two different estuaries of a bay. Use the U test to determine, at the 0.01 level of significance, whether you can conclude that there is no difference between the estuaries in DDT blood level of fish.

Estuary A	15	11	27	9	33	16	22	28	21
Estuary B	6	21	9	13	11	10	15	13	

10.24 A business manager wants to place an order for thermal spirit masters. She wishes to test two different brands, which cost the same, to determine whether there is a difference in the number of legible copies each brand will make. She has a secretary randomly use eight of each brand to copy the same document. The number of easily legible copies each one makes is counted, using the same criterion of legibility on each set of copies. Given the following data, use the U test and the 0.05 level of significance to determine whether there is a difference between the brands in number of easily legible copies per master.

Brand A	66	64	61	57	63	64	68	62
Brand B	64	67	68	65	69	67	71	64

10.25 A large chain of grocery stores wishes to test the effectiveness of an advertising display on the sales of a certain product. Thirty of their retail stores are randomly divided into two groups. In one group, each store uses the display with the product; in the other group, each store carries the product as usual. Given the following data, use the U test to determine whether the display is effective. Use the 0.05 level of significance.

WEEKLY SALES	
WITHOUT DISPLAY	WITH DISPLAY
68	99
91	72
66	84
56	60
79	84
63	64
91	90
64	112
88	90
67	93
40	95
84	61
78	109
31	79
95	59

10.3 Computer Usage

We can use a computer to perform all the tests described in this chapter. Minitab commands will be shown first, then SAS statements.

10.3.1 Minitab Commands for This Chapter

We can perform the sign test for differences between paired samples stored in columns 1 and 2 by using the following Minitab commands:

```
MTB > LET C3 = C1 - C2
MTB > STEST difference in median = 0 for C3
```

With the preceding commands Minitab will perform a two-sided sign test. To have Minitab perform the one-sided tests we can use the ALTERNATIVE subcommand. The ALTERNATIVE = 1 specifies an upper-tailed test while the

ALTERNATIVE = −1 specifies the lower-tailed test. When performing the sign test Minitab deletes the zero differences from the data set and adjusts the sample size accordingly. Consequently, when there are zero differences the sign test will give results that are slightly different from those obtained in this chapter.

To perform the sign test for a population median based on data stored in column 1, we use the following Minitab command:

```
MTB > STEST with median = K for the data in C1
```

where K is the hypothesized value for the median contained in the null hypothesis. As before, this command specifies that a two-sided test be performed. To obtain a one-sided test we use the ALTERNATIVE subcommand. To test H_0: Md = 10 against H_1: Md > 10 for data stored in column 1, for example, we use the following commands:

```
MTB > STEST with median = 10 for the data in C1 ;
SUBC> ALTERNATIVE = 1 .
```

Observations that equal the hypothesized value of the population median (such as 10 in the preceding example) will be deleted and the test will be based on the remaining observations.

We can perform the Wilcoxon T test (the Wilcoxon signed-rank test) for the difference between paired samples stored in columns 1 and 2 on Minitab by using the following commands:

```
MTB > LET C3 = C1 - C2
MTB > WTEST difference in median = 0 for data in C3 .
```

Like the STEST command, the WTEST command performs a two-sided test. We can perform a one-sided test by using the ALTERNATIVE subcommand in the same manner as for the sign test. The zero values will be deleted by this command as they were using the STEST command; consequently, the values obtained by Minitab will differ slightly from those obtained in this book.

Example 10.4

Refer to Example 10.2, which deals with the diet experiment. The data set consists of a set of 30 paired samples. Use the Wilcoxon T test to determine whether the diet is effective.

Solution

The Minitab printout for the T test of the hypothesis H_0: $Md_1 = Md_2$ against the alternate H_1: $Md_1 > Md_2$ is presented in Figure 10.3. From the printout we note that the n used for this test is only 27 since there are three zero differences that are deleted by Minitab. On the printout the P value is 0.014. Since the P value is less than 0.05, we reject H_0 and conclude that the diet is effective. Note that this final conclusion matches the conclusion obtained in Example 10.2.

Figure 10.3
Minitab printout for Wilcoxon T test for paired samples for the diet data of Example 10.4.

```
MTB > SET the following in C1
DATA> 183 144 151 163 155 159 178 184 142 137 145 172 134 160 121
DATA> 207 153 138 172 193 127 118 175 166 183 191 174 153 163 129
DATA> SET the following in C2
DATA> 177 145 145 162 151 163 173 185 139 138 136 155 136 160 123
DATA> 196 149 145 171 185 127 119 164 164 188 174 165 144 172 129
DATA> END of data
MTB > LET C3 = C1 - C2
MTB > WTEST median = 0 for the data in C3 ;
SUBC> ALTERNATIVE = 1 .

TEST OF CENTER = 0 VERSUS CENTER G.T. 0

            N FOR    WILCOXON                  ESTIMATED
       N    TEST     STATISTIC     P-VALUE      CENTER
C3    30     27        280.5        0.014        2.500
```

The Mann–Whitney U test for independent samples stored in columns 1 and 2 is performed on Minitab by using the following commands:

```
MTB>MANN-WHITNEY test and confidence interval for data in C1 and C2 .
```

The procedure performs a two-sided test.

Example 10.5

Refer to Problem 10.25. This data set consists of two independent samples based on weekly sales for a grocery store. Use Minitab to perform the Mann–Whitney U test to solve the problem. Use the 0.05 level of significance.

Solution

The printout is shown in Figure 10.4. The data in C1 are the weekly sales when an advertising display is used; the data in C2 are the weekly sales

Figure 10.4
Minitab printout for the Mann–Whitney U test for the grocery store sales data of Example 10.5.

```
MTB > SET the following data in C1
DATA> 99 72 84 60 84 64 90 112 90 93 95 61 109 79 59
DATA> END of data
MTB > NAME C1 "WITH"
MTB > SET the following data in C2
DATA> 68 91 66 56 79 63 91 64 88 67 40 84 78 31 95
DATA> END of data
MTB > NAME C2 "WITHOUT"
MTB > MANN-WHITNEY test and confidence internal for data in C1 and C2

WITH                 N = 15           MEDIAN =              84.000
WITHOUT              N = 15           MEDIAN =              68.000
A POINT ESTIMATE FOR ETA1-ETA2 IS  11.99
A 95.4 PERCENT C.I. FOR ETA1-ETA2 IS (    -3.1,    26.9)
```

$\underbrace{\text{ETA1-ETA2}}_{Md_1 - Md_2}$

```
W =    269.5
TEST of ETA1 = ETA2 VS. ETA1 N.E. ETA2 IS SIGNIFICANT AT 0.1300
```

without the advertising display. Minitab will provide us with a test for H_0: $\mathrm{Md}_1 = \mathrm{Md}_2$ against H_1: $\mathrm{Md}_1 \neq \mathrm{Md}_2$. The Minitab printout gives the sum of the ranks for the first sample, labeled W = 269.5 on the printout. The medians are labeled ETA1 and ETA2. The MANN-WHITNEY command on Minitab always gives the P value for the two-sided test. In this problem the two-sided P value on the printout (labeled "significant at") is 0.1300. Since this two-sided P value is not less than 0.05, we cannot reject H_0.

There are nonparametric confidence intervals for population medians that can be found by using t. For the one-sample problem we can find a confidence interval for the population median by using the Minitab command

```
MTB > WINTERVAL with confidence near K% for the data stored in C
```

Since this procedure is associated with the Wilcoxon T test, some statisticians call this the confidence interval for the population median based on the Wilcoxon signed-rank test. Because of the manner in which the confidence interval is found, only certain confidence levels are available for this procedure. The user supplies a desired confidence level and Minitab selects the confidence interval with the confidence level nearest the desired level. The attained confidence level for this procedure is specified on the printout.

If there are two independent samples, we can find a confidence interval for the difference between the two population medians by using a procedure based on the Mann–Whitney U test. This confidence interval is given on the printout for the Mann–Whitney test. Again only certain confidence levels are available for the confidence interval procedure. Minitab uses the possible confidence level that is closest to 95%.

Information on the formulas and rationale for these nonparametric confidence intervals can be found in texts on nonparametric statistics. Hollander and Wolfe (1973) present a good discussion of this topic.

Example 10.6

Refer to the Miami Female Study data given in Appendix B.1. Using Minitab, calculate a 95% (approximately) confidence interval for the median number of credit hours taken by the population of all Miami females.

Solution

Figure 10.5 gives the program and results for this problem. Note that the achieved confidence level is 95%, which in this case is precisely what was specified in the command.

```
MTB > RETRIEVE 'FEMALE'
MTB > NAME for C10 is 'NCRHRS'
MTB > WINTERVAL with confidence near 95% for C10

                  ESTIMATED     ACHIEVED
             N     CENTER      CONFIDENCE     CONFIDENCE     INTERVAL
NCRHRS     100     17.00         95.0         (  16.50,       17.50)
```

Figure 10.5 Minitab printout for finding a confidence interval for the median number of credit hours taken by Miami females (Example 10.6).

In the printout of Figure 10.4 there is a confidence interval for the difference between the two population medians. This interval is based on two independent samples taken from the respective populations, For this problem the 95.4% confidence interval for $\text{Md}_1 - \text{Md}_2$ is -3.1 to 26.9.

10.3.2 SAS Statements for This Chapter

We can use SAS statements and procedures to perform any of the tests discussed in this chapter. Although there are no specific procedures to perform the sign test or Wilcoxon *T* test, we may write a program to perform either of these tests. Refer to the data of Table 10.2 where we wished to determine which seminar, if either, was more effective. To perform the sign test we need first to obtain the differences and then sort them into positive, negative, and zero differences. We write

```
DATA INCREASE;
  INPUT A B @@;
  DIFF = A - B;
  IF DIFF > 0 THEN PLUS = 1;
  ELSE PLUS = 0;
  IF DIFF = 0 THEN ZERO = 1;
  ELSE ZERO = 0;
  IF DIFF < 0 THEN MINUS = 1;
  ELSE MINUS = 0;
CARDS;
data entered here
```

In this way each observation will have a value of 1 for one of the variables PLUS, ZERO, and MINUS and 0 for the other two. We then determine how many pluses, zeros, and minuses we have by writing

```
PROC MEANS SUM NOPRINT;
  VAR PLUS ZERO MINUS;
  OUTPUT OUT = A SUM = NUMPLUS NUMZERO NUMMINUS;
```

The variables NUMPLUS, NUMZERO, and NUMMINUS will be the number of plus signs, zeros, and negative signs, as indicated. We then write

```
DATA TEST;
  SET A;
```

If we wish to ignore zeros, we write

```
XSTAR = MIN(NUMPLUS, NUMMINUS);
```

to obtain the smaller of x^*_+ and x^*_-. If we wish to include half the zeros (discarding one if the number of zeros is odd), we write

```
EXTRA = INT(NUMZERO/2);
XPLUS = NUMPLUS + EXTRA;
XMINUS = NUMMINUS + EXTRA;
XSTAR = MIN(XPLUS, XMINUS);
```

Then we obtain n by writing

```
XTOTAL = XPLUS + XMINUS;
```

Finally, we obtain the probability of a value at least as small as x^*_- by chance by writing

```
PROB = PROBBNML(.5,XTOTAL,XSTAR);
```

Since the alternative is two-sided, we double the obtained probability to get the P value:

```
PVALUE = 2 * PROB;
```

The output for this program is shown in Figure 10.6. We see that there are 6 pluses and 14 minuses, and the probability of obtaining no more than 6 pluses is about 0.05766. This confirms the result obtained previously. The P value is about 0.115, so we cannot conclude that there is a significant difference in the increases of study hours for the two seminars.

```
INPUT FILE

DATA INCREASE;
  INPUT A B @@;
  DIFF = A - B;
  IF DIFF > 0 THEN PLUS = 1;
  ELSE PLUS = 0;
  IF DIFF = 0 THEN ZERO = 1;
  ELSE ZERO = 0;
  IF DIFF < 0 THEN MINUS = 1;
  ELSE MINUS = 0;
CARDS;
13 7 2 4 -1 3 4 12 11 14 7 9 12 6 -3 2 -4 -1 7 3 5 6
2 5 1 -1 0 7 5 1 2 1 3 9 4 5 -2 1 3 5
PROC MEANS SUM NOPRINT;
  VAR PLUS ZERO MINUS;
  OUTPUT OUT = A SUM = NUMPLUS NUMZERO NUMMINUS;
DATA TEST;
  SET A;
  EXTRA = INT(NUMZERO/2);
  XPLUS = NUMPLUS + EXTRA;
  XMINUS = NUMMINUS + EXTRA;
  XSTAR = MIN (XPLUS, XMINUS);
  XTOTAL = XPLUS + XMINUS;
  PROB = PROBBNML(.5,XTOTAL,XSTAR);
  PVALUE = 2 * PROB;
PROC PRINT;
```

Figure 10.6
SAS printout for the sign test of the seminar data of Table 10.2.

```
OUTPUT FILE

                                        SAS
OBS   NUMPLUS   NUMZERO   NUMMINUS   EXTRA   XPLUS   XMINUS   XSTAR   XTOTAL      PROB       PVALUE
 1       6         0         14        0       6       14       6       20     0.0576591   0.115318
```

We illustrate the procedure for the one-sample sign test by using the data of Example 10.1. We enter the data and obtain the difference score: $x - 60$. The remainder of the program is the same as for the paired-sample sign test except for the final line. Since the alternate hypothesis is H_1: Md > 60, we want to reject H_0 if $P(x \geqslant x^*_+) < 0.05$. The PROBBNML function gives the probability that $x \leqslant m$, where m is the number entered in the function. Now $P(x \geqslant x^*_+) = P(x > x^*_+ - 1) = 1 - P(x \leqslant x^*_+ - 1)$ or $P(x \geqslant x^*_+) = P(x \leqslant n - x^*_+)$; we can use either method to obtain the P value for this problem. In the program discussed here we chose the latter method. The output is shown in Figure 10.7. We see that the P value is about 0.019—the same that we found in Example 10.1. Thus we conclude that the claim is correct.

Figure 10.7
SAS printout for the one-sample sign test of the median typing speed data of Example 10.1.

```
INPUT FILE

DATA WPM;
  INPUT WPM @@;
  DIFF = WPM - 60;
  IF DIFF > 0 THEN PLUS = 1;
  ELSE PLUS = 0;
  IF DIFF = 0 THEN ZERO = 1;
  ELSE ZERO = 0;
  IF DIFF < 0 THEN MINUS = 1;
  ELSE MINUS = 0;
CARDS;
81 76 63 71 66 59 88 73 80 66 58 70
PROC MEANS SUM NOPRINT;
  VAR PLUS ZERO MINUS;
  OUTPUT OUT = A SUM = NUMPLUS NUMZERO NUMMINUS;
DATA TEST;
  SET A;
  EXTRA = INT(NUMZERO/2);
  XPLUS = NUMPLUS + EXTRA;
  XTOTAL = NUMMINUS + NUMPLUS + EXTRA;
  PVALUE = PROBBNML(.5,XTOTAL,XTOTAL - XPLUS);
PROC PRINT;

OUTPUT FILE

                                 SAS
OBS   NUMPLUS   NUMZERO   NUMMINUS   EXTRA   XPLUS   XTOTAL     PVALUE
 1       10        0          2        0       10      12     0.0192871
```

We can use a similar procedure to perform the Wilcoxon T test. In this case we must rank the absolute values of the data from smallest to largest. To use the Wilcoxon T test on the data of Table 10.2, we first write

```
DATA INCREASE;
  INPUT A B @@;
  DIFF = A - B;
  ABDIFF = ABS(DIFF);
CARDS;
data entered here
```

The ABS(x) function obtains the absolute value of x. We then rank the absolute differences by using PROC RANK, naming the ranks RANKDIFF. The RANK procedure creates an output data set without the necessity of an output statement. We write

```
PROC RANK;
  VAR ABDIFF;
  RANKS RANKDIFF;
```

We retrieve this data set by writing

```
DATA TEST;
  SET;
```

Then we differentiate the ranks of the positive, zero, and negative differences by writing

```
IF DIFF > 0 THEN PLUS = RANKDIFF;
ELSE PLUS = 0;
IF DIFF = 0 THEN ZERO = RANKDIFF;
ELSE ZERO = 0;
IF DIFF < 0 THEN MINUS = RANKDIFF;
ELSE MINUS = 0;
```

We can use the MEANS procedure to obtain the sums of the positive, zero, and negative ranks, named SUMPLUS, SUMZERO, and SUMMINUS. We then add half the sum of the zero ranks to each other rank sum and obtain T^*_+ and T^*_-. Finally, we write

```
TMEAN = 20*21/4;
STD = SQRT(20*21*41/24);
ZSTAR = (TPLUS - TMEAN)/SDT;
PROB = 1 - PROBNORM(ABS(ZSTAR));
PVALUE = 2 * PROB;
```

Note that this set of statements is specific to the problem since we know that $n = 20$. If we preferred, we could rewrite the program to use n.

Note also that we do not remove the extra zero difference if the number of zero differences is odd. For reasonably large n, this is a satisfactory procedure. If necessary we can write the program to remove the extra zero. The output for this program is shown in Figure 10.8. We see that $z^* = -1.4$ and the P value is about 0.16. (This is a two-sided test.) We conclude that the differences between the groups may be attributed to chance.

Figure 10.8
SAS printout of the Wilcoxon T Test on the seminar data of Table 10.2.

```
INPUT FILE

DATA INCREASE;
  INPUT A B @@;
  DIFF = A - B;
  ABDIFF = ABS(DIFF);
CARDS;
13 7 2 4 -1 3 4 12 11 14 7 9 12 6 -3 2 -4 -1 7 3
5 6 2 5 1 -1 0 7 5 1 2 1 3 9 4 5 -2 1 3 5
PROC RANK;
  VAR ABDIFF;
  RANKS RANKDIFF;
DATA TEST;
  SET;
  IF DIFF > 0 THEN PLUS = RANKDIFF;
  ELSE PLUS = 0;
  IF DIFF = 0 THEN ZERO = RANKDIFF;
  ELSE ZERO = 0;
  IF DIFF < 0 THEN MINUS = RANKDIFF;
  ELSE MINUS = 0;
PROC MEANS SUM NOPRINT;
  VAR PLUS ZERO MINUS;
  OUTPUT OUT = A SUM = SUMPLUS SUMZERO SUMMINUS;
DATA TEST;
  SET A;
  EXTRA = SUMZERO/2;
  TPLUS = SUMPLUS + EXTRA;
  TMINUS = SUMMINUS + EXTRA;
  TMEAN = 20*21/4;
  SDT = SQRT(20*21*41/24);
  ZSTAR = (TPLUS - TMEAN)/SDT;
  PROB = 1 - PROBNORM(ABS(ZSTAR));
  PVALUE = 2 * PROB;
PROC PRINT;

OUTPUT FILE

                                             SAS
OBS  SUMPLUS  SUMZERO  SUMMINUS  EXTRA  TPLUS  TMINUS  TMEAN    SDT     ZSTAR    PROB       PVALUE
 1    67.5       0      142.5      0     67.5   142.5   105   26.7862   -1.4   0.0807604  0.161521
```

We can perform the Mann–Whitney U test by using the SAS procedure NPAR1WAY with the option WILCOXON. (The Mann–Whitney test is a version of the Wilcoxon Rank-Sum test.) After the data are entered we write

```
PROC NPAR1WAY WILCOXON;
  VAR x;
  CLASS factor;
```

The italics indicate that the appropriate name should be supplied. The output of this procedure gives the P value for a two-sided alternative; U_{-}^{*} is always used to obtain z^{*}. For a one-sided alternative, we must divide the P value by two.

Figure 10.9 shows the program for the data of Example 10.3. We see that $z^{*} \doteq -2.92$ and the P value is 0.0036. This result confirms that of Example 10.3.

```
INPUT FILE

DATA RATINGS;
  INPUT GROUP $ RATING @@;
CARDS;
A 8 A 9 A 13 A 14 A 11 A 10 A 12 A 14 A 13 A 10 A 11 A 12
B 7 B 11 B 9 B 4 B 8 B 6 B 11 B 9 B 10 B 9 B 8
PROC NPAR1WAY WILCOXON;
  VAR RATING;
  CLASS GROUP;

OUTPUT FILE
```

Figure 10.9
SAS printout of the Mann–Whitney U Test for the applicant data of Example 10.3.

```
                              SAS

   ANALYSIS FOR VARIABLE RATING CLASSIFIED BY VARIABLE GROUP
              AVERAGE SCORES WERE USED FOR TIES

                  WILCOXON SCORES (RANK SUMS)

                     SUM OF        EXPECTED        STD DEV        MEAN
LEVEL        N       SCORES        UNDER HO        UNDER HO       SCORE

  A         12       191.50         144.00          16.12         15.96
  B         11        84.50         132.00          16.12          7.68

                 WILCOXON 2-SAMPLE TEST (NORMAL APPROXIMATION)
                 (WITH CONTINUITY CORRECTION OF .5)
                 S=  84.50    Z=-2.9151    PROB >|Z|=0.0036 <---- P value

                 T-TEST APPROX. SIGNIFICANCE=0.0080

                 KRUSKAL-WALLIS TEST (CHI-SQUARE APPROXIMATION)
                 CHISQ=  8.68    DF=  1    PROB > CHISQ=0.0032
```

Problems

Applications

10.26 Refer to Problem 10.7. Use SAS statements to perform the sign test to solve the problem.

10.27 Refer to Problem 10.8. Use the computer to perform the Wilcoxon *T* Test to solve the problem.

10.28 Refer to the Miami Female Study data in Appendix B.1. Use the Wilcoxon *T* Test to determine whether is it reasonable to use 120 pounds as the median weight for all Miami females. Use the 0.05 level of significance.

10.29 Use the computer to solve Problem 10.21.

10.30 Use the computer to solve Problem 10.23.

10.31 Refer to the Miami Female Study data in Appendix B.1. Use the Mann–Whitney test to determine whether there is a difference in the median number of credit hours taken for those who have a job and for those who do not have a job. Use the 0.05 level of significance.

10.32 Refer to Problem 10.5.

a. Use Minitab to find a confidence interval for the population median with a confidence level near 95%.

b. Use the confidence interval found in part (*a*) to determine whether it is reasonable to conclude that Md = 8. What is the level of significance for the test?

10.33 Refer to Problem 10.23.

a. Use Minitab to find a confidence interval for the difference between the two population medians. Use a confidence level near 95%.

b. Explain why a confidence interval might be more useful in this problem than the result of the hypothesis test.

10.4 Case 10: The Sullen Salesman—Solution

After a while the young executive came back to the salesman and said, "I've found a procedure that will work on the data. It's called the sign test." She then wrote down the results of the study:

	SALES (UNITS)	
PAIR	INSTRUCTED	NOT INSTRUCTED
1	13	10
2	39	39
3	6	4
4	77	79

5	110	108
6	33	27
7	16	12
8	87	74
9	15	11
10	33	37
11	18	11
12	51	41
13	15	15
14	30	17
15	22	25

"Ten of the differences were in favor of the group that had the course," she said, "while only three were in favor of the other group. Splitting the two who were the same, we find that the probability of at least 11 of 15 results favoring the course group, if there is really no difference, is only 0.059. Thus we can be sure, at a 0.059 level of significance, that the course was effective."

"That won't do it," said the salesman. "He said I had to have proof at the 0.01 significance level, and this isn't good enough."

"Don't give up," she said. "We can use the Wilcoxon *T* test. The data have to be rankable, and we can't compare tractors to hammers, but we can compare percentages. Percentages are pure numbers, and a percentage increase in one department can be compared with a percentage increase in another department."

She then revised the figures, expressing the sales of the group that had the course as a percentage of the sales in the other group, with the results shown here.

	PERCENTAGE OF SALES			
PAIR	INSTRUCTED	NOT INSTRUCTED	DIFFERENCE	RANK
1	130	100	+30	10
2	100	100	0	1.5
3	150	100	+50	13
4	97	100	−3	4
5	102	100	+2	3
6	122	100	+22	8
7	133	100	+33	11
8	118	100	+18	7
9	136	100	+36	12
10	89	100	−11	5
11	164	100	+64	14
12	124	100	+24	9
13	100	100	0	1.5
14	176	100	+76	15
15	88	100	−12	6

The sum of the negative ranks, plus the rank of one zero, was 16.5, and the sum of the positive ranks, plus the rank of the other zero, was 103.5. To test the hypothesis that the population from which sample 1 was taken had greater sales than the other, at the 0.01 level of significance, the sum of the negative ranks would have to be less than or equal to the critical value, 20. Since 16.5 is smaller than 20, it could be concluded, at the 0.01 level of significance, that the course was effective.

Armed with this new information, the salesman was no longer sullen and returned to his task with fire and determination.

10.5 Summary

When the data collected to test a hypothesis fail to satisfy one or more of the assumptions needed to perform the usual t test or z test, we can often use tests that do not make assumptions about the shape of the population distributions and do not use the mean and variance. These are called distribution-free or nonparametric tests. Three such tests were discussed in this chapter: the sign test, the Wilcoxon T test, and the Mann–Whitney U test.

One advantage of nonparametric tests is that they are relatively easy to use. Calculations may be done quickly, and if statistically significant results are obtained, significant results would probably have been obtained if the usual parametric tests, such as the t test and z test, had been used. It should be noted, however, that these tests are somewhat less powerful than their counterparts which depend upon normality or parameters. Using nonparametric tests, we are more likely to make a type II error. It is usually wise, therefore, to use these tests only when necessary.

When we cannot use the paired-samples or one-sample t test or z test, we may use the sign test or the Wilcoxon T test. The sign test merely requires that we can assign all data a positive or negative sign or a value of zero. We may use the Wilcoxon T test on data that can be ranked.

If we wish to test for population differences using two independent samples, a nonparametric alternative to the independent-sample t test or z test is the Mann–Whitney U test. This test, like the Wilcoxon T test, requires that the data can be ranked.

Chapter Review Problems

Applications

10.34 A group of 15 accelerated tenth graders is given a paragraph and asked to memorize as much of it as possible. Their retention is measured on a scale of 1 to 99, with a normed mean of 50. If ten of the students score above 50 and five score below 50, test the hypothesis that this group memorizes better than average. Use the 0.05 level of significance.

10.35 Repeat Problem 10.34 for a group of 50 students, of whom 31 score above 50, 2 score 50, and 17 score below 50.

10.36 In Problem 10.34, suppose the rank sum for the students above 50 is 93. Repeat the test.

10.37 In Problem 10.35, suppose the rank sum for the students above 50 is 941.5. Repeat the test.

10.38 Refer to Problem 10.34. Suppose that two different classes are given the retention tests. The first class contains 15 students; the second contains 18 students. The rank sum for the first class is 208. Test the hypothesis that there is no difference between the classes in memorization ability.

10.39 Ten people took part in a diet experiment. Their weights were measured at the beginning and at the end of the diet period. Use the data below to determine if the diet was effective at the 0.05 level of significance.

SUBJECT	STARTING WEIGHT	ENDING WEIGHT
1	183	177
2	144	145
3	151	145
4	163	162
5	155	151
6	159	163
7	178	173
8	184	185
9	142	139
10	137	138

Use the sign test to determine whether the diet was effective. Use the 0.05 level of significance.

10.40 Repeat Problem 10.39, using the Wilcoxon T test.

10.41 A psychologist gives a test to four blue-collar workers and five white-collar workers to see if there is any significant difference (at $\alpha = 0.05$) in their test scores. The blue-collar workers make scores of 23, 18, 22, and 21, and the white-collar workers make scores of 23, 28, 25, 24, and 26. Using a nonparametric technique, what conclusions can you draw from the study?

10.42 A group of 15 subjects was given tests before and after being subjected to a learning situation. The results of the tests are shown here. Test the hypothesis that learning actually took place, using the 0.05 level of significance. Assume that if the subject learned, this would be shown by an increase in score.

SUBJECT	BEFORE	AFTER
1	27	29
2	21	32
3	34	29
4	24	27
5	30	31
6	27	26
7	33	35
8	31	30
9	22	29
10	27	28
11	33	36
12	17	15
13	25	28
14	26	26
15	23	26

Use the sign test to determine if learning took place. Use the 0.05 level of significance.

10.43 Repeat Problem 10.42, using the Wilcoxon T test.

10.44 Two varieties of corn are planted in paired plots (one plot split in half) to see if there is a difference in the yield. The same fertilizer is used throughout. Determine if there is a significant difference between the yields, using the 0.05 level of significance. Yields are given in tons per plot. Paired plots are of the same size, though different pairs are not necessarily of a similar size.

	PLOT				
	1	2	3	4	5
Variety A	12.3	9.1	10.2	6.4	14.4
Variety B	10.8	7.6	10.3	6.0	13.5

Use the sign test to see if there is a difference in the yield. Use the 0.05 level of significance.

10.45 Repeat Problem 10.44, using the Wilcoxon T test.

10.46 Coffee cans are filled to a "net weight" of 16 ounces, but there is considerable variability. In fact, a random sample of 8 cans of a particular brand showed net weights of 15.94, 16.01, 15.98, 16.01, 16.12, 15.95, 16.00, and 15.81 ounces. Test (at $\alpha = 0.05$) the hypothesis that the net weight of the brand of coffee is actually 16 ounces. Use the sign test and a 0.05 level of significance to test the hypothesis that $\text{Md} = 16$.

10.47 Repeat Problem 10.46, using the Wilcoxon T test.

10.48 Suppose a second set of coffee cans had the same distribution above, below, and equal to 16 ounces as in Problem 10.46. Combine the two sets and use the sign test to test the hypothesis that $\text{Md} = 16$, using $\alpha = 0.05$.

10.49 Repeat Problem 10.48, using the Wilcoxon T test.

10.50 A government testing agency routinely tests food products to see if they meet government requirements at a 0.10 level of significance. In order to be labelled "ice cream" the product must contain a minimum of 10% butterfat. A sample of ten half-gallons of a certain brand showed butterfat percents of 9.6, 10.1, 9.9, 9.7, 10.0, 9.9, 9.8, 10.1, 9.9, and 9.7. What conclusions would the government agency reach if the hypotheses were formulated to favor
a. the consumer? **b.** the manufacturer?

Use the sign test and the 0.01 level of significance to test the hypothesis Md = 10.

10.51 Repeat Problem 10.50 using the Wilcoxon T test.

10.52 To test the efficiency of a procedure designed to increase the daily output of a machine, five machines are tested with and without the procedure, with the following results.

	MACHINE				
	A	B	C	D	E
With	15.6	18.2	14.3	14.9	16.7
Without	14.4	16.7	13.1	14.4	14.7

Use a nonparametric test to test the hypothesis that the procedure is effective. Use $\alpha = 0.05$.

10.53 Repeat Problem 10.52, if ten different machines were used, five with the procedure and five without the procedure, with the same results as above. Compare the results of the analyses.

Index of Terms

Glossary of Symbols

Symbol	Meaning
Md_0	hypothetical population median
Md_1, Md_2	medians of two populations
μ_T	mean of distribution of *T* statistic (large sample)
μ_U	mean of distribution of *U* statistic (large sample)
σ_T^2	variance of distribution of *T* statistic (large sample)
σ_U^2	variance of distribution of *U* statistic (large sample)
T^*_+	sum of positive ranks in *T* test
T^*_-	sum of negative ranks in *T* test

T^*	smaller of T^*_+ and T^*_-	$U_\alpha\{n_1, n_2\}$	critical value of U (α is the probability of obtaining a sample value of $U \leqslant U_\alpha\{n_1, n_2\}$ if H_0 is true)
$T_\alpha\{n\}$	critical value of T (α is the probability of obtaining a sample value of $T \leqslant T_\alpha\{n\}$ if H_0 is true)	x^*_+	number of positive signs in sample (sign test)
U^*_1, U^*_2	sample values of U statistic	x^*_-	number of negative signs in sample (sign test)
U^*	smaller of U^*_1 and U^*_2		

Answers to Proficiency Checks

1. When we determine the differences in number of cavities, we find that eight are positive and two are zero. We thus test the hypothesis H_0: $\text{d}_1 = \text{Md}_2$ against the alternate H_1: $\text{d}_1 \neq \text{Md}_2$. The decision rule is to reject H_0 if $P(x \geqslant x^*_+) < 0.025$ or if $P(x \geqslant x^*_-) < 0.025$. Counting the two zeros as one positive and one negative, we have $n = 10$, $p = 0.5$, and $r = 9$, so from Table 1 we have $P(x \geqslant 9) = 0.011$. We therefore reject H_0 and conclude that there is a difference. On the basis of the evidence, it appears that toothpaste B is more effective in fighting cavities.

2. Here we have H_0: Md = 4 weeks with an alternate hypothesis that Md is less than 4 weeks. We will reject H_0 if $P(x \geqslant x^*_+) < 0.10$. Since there are five zeros, we count $x^*_+ = 8$ and $x^*_- = 15$ and discard one zero. Then $n = 23$, $p = 0.5$, and we want $P(x \geqslant 15)$, which is the same as $1 - P(x \leqslant 14)$, or 0.105. Note that this is the same as $P(x \leqslant 8)$, since $1 - p = 0.5$, as well. According to the decision rule, we fail to reject H_0, but it is very close. Since $P(x \leqslant 7) = 0.047$ and $P(x \leqslant 9) = 0.202$, we should probably report the observed significance level, 0.105. We might say that the contention has been proved at *approximately* the 0.10 level of significance.

3. This proficiency check is like the preceding one except that $x^*_+ = 15 + 2 = 17$ and $x^*_- = 29 + 2 = 31$. Then our test statistic will be z^*, where $z^* = [(17.5 - 24)/(0.5\sqrt{48})] \doteq -1.88$. The decision rule is to reject H_0 if $z^* < -1.645$. Thus we reject H_0 and conclude that the method is more effective than traditional ones.

4. **a.** Reject H_0 if $T^*_- \leqslant 30$. **b.** Reject H_0 if $T^*_- \leqslant 49$.
c. Reject H_0 if $z^* > 1.645$.

5. **a.** Reject H_0 if $T^* \leqslant 25$. **b.** Reject H_0 if $T^* \leqslant 42$.
c. Reject H_0 if $|z^*| > 1.96$.

6. The differences are +1, −1, +3, +2, +1, +2, +1, 0, +1, 0. Arranged in rank order, counting one zero as negative and the other positive, we have −0, +0, −1, +1, +1, +1, +1, +2, +2, +3; ranks are 1.5 for 0, 5 for 1, 8.5 for 2, and 10 for 3. Thus $T^*_- = 6.5$ and $T^*_+ = 48.5$. Since the alternate hypothesis is $H_1{:}\text{Md}_1 \neq \text{Md}_2$ and $T_{.025}\{10\} = 8$, we will reject H_0 if $T^* \leqslant 8$. Since T^*, the smaller of T^*_- and T^*_+, is 6.5, we reject H_0 and conclude that there is a difference.

On the basis of the evidence, it appears that toothpaste B is more effective in fighting cavities.

7. **a.** Reject H_0 if $U^* \leq 2$. **b.** Reject H_0 if $U^* \leq 5$.
c. Reject H_0 if $U^* \leq 15$. **d.** Reject H_0 if $|z^*| > 1.96$.

8. Rankings are as follows:

Sample 1	1 2 4 6 7 8
Sample 2	3 5 9 10

Thus $R_1 = 28$ and $R_2 = 27$. Then $U_1^* = (6)(4) + (6)(7)/2 - 28 = 17$, and $U_2^* = (6)(4) + (4)(5)/2 - 27 = 7$. We wish to test the hypothesis H_0: $\text{Md}_1 = \text{Md}_2$ against the alternate hypothesis H_1: $\text{Md}_1 < \text{Md}_2$. Since $U_{.05}\{6, 4\} = 3$, the decision rule is to reject H_0 if $U_2^* \leq 3$. Since $U_2^* = 7$, we fail to reject H_0. The mean ranks are 4⅔ for sample 1 and 6.75 for sample 2. Since these are not the same, we may wish to reserve judgment.

9. Rankings are as follows:

Sample 1	1 3 4 6.5 8.5 10 13 15 17 21 22.5 24
Sample 2	2 5 6.5 8.5 11 12 15 15 18 19 20 22.5 25 26

Thus $R_1 = 145.5$ and $R_2 = 205.5$. Now $n_1 = 12$ and $n_2 = 14$, so we may use the test statistic z. Then $U_2^* = (12)(14) + (14)(15)/2 - 205.5 = 67.5$. We wish to test the hypothesis H_0: $\text{Md}_1 = \text{Md}_2$ against the alternate hypothesis H_1: $\text{Md}_1 \neq \text{Md}_2$. The decision rule is to reject H_0 if $|z^*| > 1.96$; we have $\mu_U = 12(14)/2 = 84$ and $\sigma_U = \sqrt{(12)(14)(27)/12} \doteq 19.442$, so $z^* = (67.5 - 84)/19.442 \doteq -0.85$. Fail to reject H_0—again we may wish to reserve judgment.

References

Arbuthnott, J. "An Argument for Divine Providence, Taken from the Constant Regularity Observed in the Births of Both Sexes." *Philosophical Transactions* 27 (1710): 186–190.

Conover, W. J. *Practical Nonparametric Statistics.* New York: John Wiley & Sons, 1971.

Hollander, M., and D. Wolfe. *Nonparametric Statistical Methods.* New York: John Wiley & Sons, 1973.

Mann, H. B., and D. R. Whitney. "On a Test of Whether One of Two Random Variables Is Stochastically Larger Than the Other." *Annals of Mathematical Statistics* 18 (1947): 50–60.

Mosteller, Frederick, and Robert E. K. Rourke. *Sturdy Statistics.* Reading, MA: Addison-Wesley, 1973.

Noether, Gottfried E. *Elements of Nonparametric Statistics.* New York: John Wiley & Sons, 1967.

Siegel, Sidney. *Nonparametric Statistics for the Behavioral Sciences.* New York: McGraw-Hill, 1956.

Wilcoxon, F. "Individual Comparisons by Ranking Methods." *Biometrics* I (1945): 80–83.

Wilcoxon, Frank, and Roberta A. Wilcox. *Some Rapid Approximate Statistical Procedures.* Rev. ed. Pearl River, NY: Lederle Laboratories, 1964.

11

Single-Factor Analysis of Variance

In Chapter 8 we presented methods for comparing means of two populations—ways to test H_0: $\mu_1 = \mu_2$. Quite often we need to compare means of more than two populations. We may wish to determine which of several treatments for a disease is the best, which of several brands of paint dries the fastest, whether or not different levels of experience influence efficiency of workers, and many other questions involving differences among several populations. The procedure that has been developed to answer such questions is called *analysis of variance*. In this chapter we apply the analysis of variance procedure to determine whether or not several populations have the same mean.

Using the computer to aid in the calculation of test statistics becomes much more attractive for such computationally intense analyses as analysis of variance. Typically only the smallest data sets are analyzed by hand. After doing a few large analyses by hand, your appreciation of the computer will increase immensely. For this reason, computer usage is integrated into the appropriate sections throughout the rest of the book. Rather than include both Minitab and the SAS System, however, we will use only the SAS System for these illustrative purposes; Minitab will be covered in a separate section at the end of the chapter. The Minitab user should be able to integrate the Minitab commands into the text as we proceed. They parallel the SAS procedures and are equally useful.

11.0 Case 11: The Welcome Relief

A pharmaceutical company had developed several different compounds for the treatment of allergic headaches and wished to determine the most effective product for promotion and distribution. Six of these compounds were isolated for testing and possible distribution. Thirty sufferers from allergic headaches were randomly divided into six groups, and each group was randomly assigned one of the compounds. In each group the subjects were asked to take the prescribed dosage of the compound when they had an allergic headache and record the number of minutes it took for them to feel relief. After a week the data were collected by the company's research department. For each subject the median time to relief was recorded. The mean and variance for the six groups are given here:

GROUP	MEAN TIME TO RELIEF (MIN)	VARIANCE
A	29.8	17.7
B	28.8	20.7
C	32.0	25.0
D	24.0	48.5
E	34.8	47.7
F	20.2	27.7

A junior analyst was given the data to analyze for the purpose of deciding which compounds should be studied further. He conducted t tests between each pair of samples. In each case the null hypothesis was that there is no difference in the population means. Each test was conducted at the 0.05 level of significance. The critical value in each case was $t_{.025}\{8\}$, or 2.306. The result of each t test is shown here:

COMPARISON	t^*	COMPARISON	t^*	COMPARISON	t^*
A–B	0.36	B–C	−1.05	C–E	−0.73
A–C	−0.74	B–D	1.29	C–F	3.63
A–D	1.59	B–E	−1.62	D–E	2.46
A–E	−1.38	B–F	2.76	D–F	0.97
A–F	3.19	C–D	2.09	E–F	3.76

He rushed to his supervisor to show her the analysis. "There are many interesting results here," he said. "For instance . . ." "Excuse me for interrupting," said his supervisor, "but you have a few problems here. You performed fifteen t tests, each at the 0.05 level of significance, so you are 95% confident that each of them is correct. How confident are you that all of them are correct?" "Why, 95% confident, of course. Isn't that right?"

"Think about something," said his supervisor. "If the probability that event A will occur is 0.8, while the probability that event B will occur is 0.6, what is the probability that both will occur, if they are independent? That's right . . . 0.6(0.8) or 0.48. If they are not independent, the probability could be higher, but it would never be as high as 0.6 unless event B occurs whenever event A does. So if you are 95% confident that the difference between sample A and sample B is significant, and 95% confident that the difference between sample A and sample C is significant, the confidence that *both* differences are significant is at least (0.95)(0.95) or 0.9025, but almost certainly not 0.95. You did fifteen comparisons, so your confidence that all your results are correct could be as low as $(0.95)^{15}$ or about 0.46, although it's probably higher since the results are not independent. Therefore, although you may be correct in at least some of your conclusions, the analysis certainly does not substantiate it at the significance level you claim."

"Isn't there anything I can do?" asked the analyst. "Yes, there is," she replied. "A procedure called analysis of variance was developed for this very reason. Let me have your raw data, and I'll have the analysis done for you."

11.1 Comparing Means of More Than Two Populations

Suppose that we sample from three populations with means μ_1, μ_2, and μ_3 and obtain sample means $\overline{x}_1$, $\overline{x}_2$, and $\overline{x}_3$. If these sample means are practically the same, we may conclude that the population means are not different.

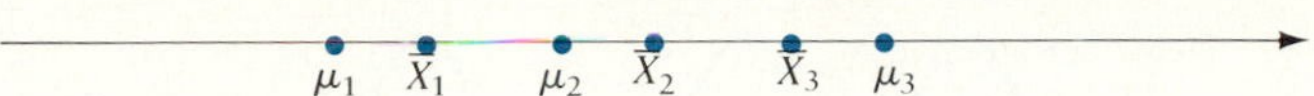

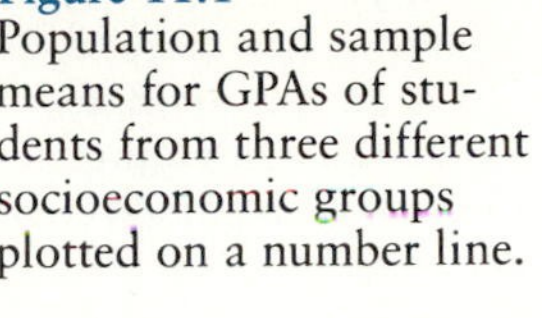

Figure 11.1
Population and sample means for GPAs of students from three different socioeconomic groups plotted on a number line.

Suppose that at least two of the sample means are far apart. Can we then conclude that those population means are different? One answer to this question can be found by using the analysis of variance technique. In Case 11 we noted that repeated t tests will have a lower confidence level, as a group, than each individual test. The analysis of variance technique was developed to test a hypothesis that all population means are equal against an alternative that they are not all equal—that is, at least two of the population means are different.

In Section 7.1.1 we stated that a model for a statistical test consists of the assumptions we make in order to use the test. The assumptions underlying the analysis of variance procedure parallel those used for the t test of two independent samples. Each sample must be randomly and independently obtained from normal populations with equal variances.

Now suppose we let μ_1, μ_2, and μ_3 represent the mean grade-point averages (GPAs) in college for all students from three different socioeconomic groups and that the three distributions have a common variance σ^2. Further, suppose that we obtain a random sample from each group and the sample means are $\overline{x}_1$, $\overline{x}_2$, and $\overline{x}_3$. These numbers may be plotted on a number line as shown in Figure 11.1.

In each case the sample mean is reasonably close to the population mean. Because of the variability of sample means, however, the fact that the population means are different may not be evident. If σ^2 is large, the sampling distribution of the mean will be highly variable for each population. Possible sampling distributions of the means of the three populations are shown in Figure 11.2. We see that $\overline{x}_1$ and $\overline{x}_3$, as well as $\overline{x}_2$, could very possibly have come from the population with mean μ_2.

All three sample means are covered by the sampling distribution of μ_2. In fact, the variability of the population means is no greater than the variability of the observations—as may be seen by the fact that all three population means are covered by the distribution of population 2. On the other hand, if σ^2 is small as in Figure 11.3, then $\overline{x}_1$ clearly comes from a different population than does $\overline{x}_3$ while $\overline{x}_2$ and $\overline{x}_3$ are very unlikely to come from the same population. The population means are more highly variable than the observations within each population.

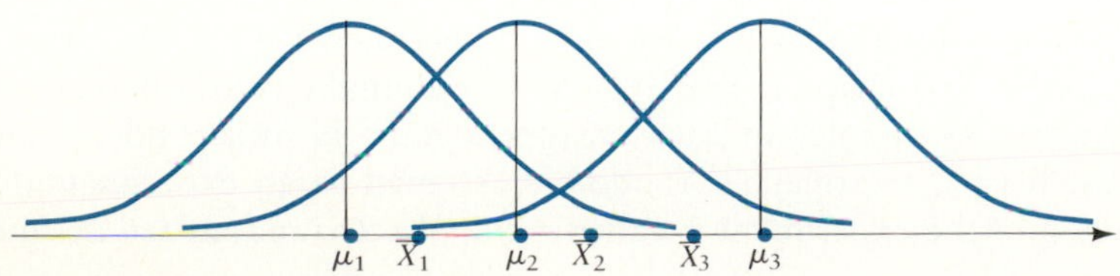

Figure 11.2
Sampling distributions of the means of three populations in Figure 11.1: large common variance.

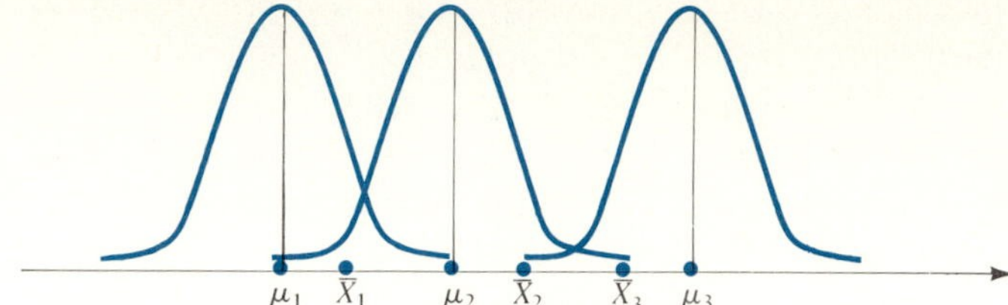

Figure 11.3
Sampling distributions of the means of the three populations in Figure 11.1: small common variance.

Thus if the variation between sample means is large and the common population variance σ^2 is small, we may correctly conclude that the population means are different. The term **analysis of variance,** then, indicates what it attempts to do. Its procedures analyze the total variability of all data of interest and then show how much variability is caused by actual differences between population means and how much is due to variability of sample means within the separate populations. Analysis of variance uses these two sources of variation to obtain estimates for σ^2. One estimate is from the sample variances. Each sample variance is an unbiased estimate for σ^2, so we can obtain an even better estimate by pooling these sample variances. A second estimate is based on the sample means. If in fact the population means are equal, we can use the variance of the sample means to obtain an estimate for $\sigma^2_{\bar{x}}$, which we can then use to estimate σ^2. The extent to which these two estimates differ from each other is the basis for a decision whether or not we can reject the hypothesis that all population means are equal.

We can apply the analysis of variance procedure to several different situations. In each situation different *treatments,* such as the compounds in Case 11, are assigned to different experimental units, such as the people who took the compounds. The overall collection of similar treatments to be studied is called a **factor.** The different forms the factor can take are called **levels** of the factor. In Case 11, the factor was drug and there were six factor levels—the different compounds. Factors are usually categorized as *experimental* or *classification* factors. An **experimental factor** is one in which we assign the factor levels at random to the experimental groups, as in Case 11, while a **classification factor,** such as sex, age, or educational background, is not under the researcher's control but merely pertains to characteristics of the experimental groups. The different socioeconomic levels in the previous example provide an example of a classification factor. In Case 11, drug is an experimental factor. If more than one factor is involved, each combination of different levels of the factors is called a **treatment.** If there is only one factor, each factor level is a treatment. In Case 11, there were six different treatments. The distinction between factor levels and treatments will be apparent when we study experiments with more than one factor in the next chapter. Finally, if we are to make valid inferences from the results, the samples in each treatment must be independent random samples. If each treatment is randomly assigned to an experimental unit, the design of the experiment is called a **completely randomized design.**

With this preliminary discussion in mind, we present the analysis of variance model for a completely randomized design. Suppose we have k independent random samples drawn from populations with means μ_1, μ_2, $\mu_3, \ldots, \mu_k$. The number of observations in the corresponding samples are $n_1, n_2, n_3, \ldots, n_k$. The values of the observations in the samples are denoted by x_{ij}, and each differs from the population mean by ϵ_{ij} (the Greek letter *epsilon*). Then each observation can be written

$$x_{ij} = \mu_i + \epsilon_{ij}$$

where x_{ij} is the value of the jth observation in the ith treatment
μ_i are parameters, the population means
ϵ_{ij} are $N(0, \sigma^2)$
$i = 1, 2, 3, \ldots, k$
$j = 1, 2, 3, \ldots, n_i$

Each of the ϵ_{ij} is called a **random error term;** "random" refers to ϵ being a random variable and "error" simply means the difference between the population mean and the value of the observation. The distribution of each population is $N(\mu_i, \sigma^2)$, and since each $\epsilon_{ij} = x_{ij} - \mu_i$, the distribution of ϵ will be $N(0, \sigma^2)$ for each population and will have the same shape as the distribution of x for each population. That is, the ϵ_{ij} are $N(0, \sigma^2)$ for $i = 1, 2, 3, \ldots, k$.

Ultimately, we wish to test hypotheses that the treatment means are equal. For reasons noted in Section 11.0, use of repeated t tests is not appropriate. The analysis of variance procedure consists essentially of obtaining two estimates for the value of the common variance σ^2 and comparing them. These procedures are outlined in the next section.

11.2 One-Way Analysis of Variance

In this section we define the two estimates for σ^2 mentioned before and give computational procedures for obtaining them. Suppose, for example, that we wish to determine whether or not the mean weight loss is the same for diets A, B, and C. If we have 30 subjects and randomly assign each of them to one of three experimental units and randomly assign one of the diets to each group, we have an example of a completely randomized design. The weight loss for each subject after a suitable period of time is the dependent variable. The factor is diet. Each diet is a level of the factor diet and is also a treatment. If the variation between mean weight losses for the samples is large in relation to the variation of individual weight losses within each sample, we will conclude that the difference is due to differences in the diets—that is, at least one diet is more effective than one of the others. In this section we will see how to obtain estimates for both sources of variation and construct a formal test.

11.2.1 Analysis of Variance Procedures

Suppose we have k samples, each corresponding to a treatment, and the samples have $n_1, n_2, n_3, \ldots, n_i, \ldots, n_k$ observations, respectively. Then the ith sample has n_i observations with a mean $\overline{x}_i$ and variance s_i^2. We assume that the analysis of variance model as stated in Section 11.1 applies, so that each sample mean $\overline{x}_i$ is an estimate for the corresponding treatment mean μ_i.

Since we are assuming that each population has the same variance, one estimate for σ^2 is a weighted average of the variances in the k samples as each is an unbiased estimate of σ^2. The pooled variance $\hat{\sigma}^2$ in the t test for independent samples is obtained in this way. This pooled variance is called the **mean square for error** (MSE) and is defined as

$$\text{MSE} = \frac{(n_1 - 1)s_1^2 + (n_2 - 1)s_2^2 + \cdots + (n_k - 1)s_k^2}{N - k}$$

where $s_1^2, s_2^2, \ldots, s_k^2$ are the k sample variances. Since each sample variance has $n_i - 1$ degrees of freedom, the degrees of freedom for the numerator are equal to $\sum_i (n_i - 1) = N - k$. The numerator is called the **sum of squares for error** (SSE). Each

$$s_i^2 = \sum_j (x_{ij} - \overline{x}_i)^2/(n_i - 1),$$

so that

$$(n_i - 1)s_i^2 = \sum_j (x_{ij} - \overline{x}_i)^2 = \sum_j x_{ij}^2 - (\sum_j x_{ij})^2/n_i.$$

For simplicity we denote

$$\sum_j x_{ij} \text{ by } T_i,$$

the sum of the values of the observations in the ith treatment. Thus

$$\begin{aligned}\text{SSE} &= \sum_j x_{1j}^2 - T_1^2/n_1 + \sum_j x_{2j}^2 - T_2^2/n_2 + \cdots + \sum_j x_{kj}^2 - T_k^2/n_k \\ &= \sum_i \sum_j x_{ij}^2 - \sum_i T_i^2/n_i \\ &= \Sigma x^2 - [T_1^2/n_1 + T_2^2/n_2 + \cdots + T_k^2/n_k]\end{aligned}$$

Note that we replaced

$$\sum_i \sum_j x_{ij}^2$$

by Σx^2. Thus Σx^2 means the sum of the squares of the values of all observations. This formula for SSE is the *shortcut formula*. Since $(n_i - 1)s_i^2 = \sum_j (x_{ij} - \overline{x}_i)^2$, we may also write

$$\text{SSE} = \sum_i \sum_j (x_{ij} - \overline{x}_i)^2$$

This is called the *defining formula* for SSE. Recall that the degrees of freedom associated with a quantity is the number of independent observations used to obtain the quantity. For the quantity $\Sigma(x - \overline{x})^2$, if there are n observations of the random variable x, the number of degrees of freedom is $n - 1$. We may verify that the number of degrees of freedom associated with SSE is $N - k$ by noting that there are a total of N observations and that k population parameters $(\mu_1, \mu_2, \ldots, \mu_k)$ are estimated.

We obtain a second estimate for σ^2 by noting that if we knew σ^2 we would be able to obtain the variance $\sigma^2_{\overline{x}}$ of sample means, $\overline{x}$, since $\sigma^2_{\overline{x}} = \sigma^2/n$ for a sample of size n. Conversely, if we can obtain an estimate for $\sigma^2_{\overline{x}}$, then, since $\sigma^2 = n\sigma^2_{\overline{x}}$, we can estimate σ^2. If we assume that all the treatment means are equal, say to μ, since we know the sample treatment means, $\overline{x}_1, \overline{x}_2, \ldots, \overline{x}_k$, we can estimate μ by taking a weighted average of the sample means. We denote this estimate by $\overline{\overline{x}}$. (As a weighted average, $\overline{\overline{x}} = \sum_i n_i\overline{x}_i/\sum_i n_i = \sum_i[n_i(\sum_j x_{ij}/n_i)]/\sum_i n_i = (\sum_i\sum_j x_{ij})/N$, since $\sum_i n_i = N$.) Then

$$[\sum_i(\overline{x}_i - \overline{\overline{x}})^2]/(k - 1)$$

will be an estimate for $\sigma^2_{\overline{x}}$. Now for a sample of size n_i, we have $\sigma^2 = n_i\sigma^2_{\overline{x}}$, so $\sum_i n_i(\overline{x}_i - \overline{\overline{x}})^2/(k - 1)$ is another estimate for σ^2. This estimate is called the **mean square for treatments (MSTR)**. The defining formula for MSTR is

$$\text{MSTR} = \frac{\sum_i n_i(\overline{x}_i - \overline{\overline{x}})^2}{k - 1} = \frac{n_1(\overline{x}_1 - \overline{\overline{x}})^2 + n_2(\overline{x}_2 - \overline{\overline{x}})^2 + \cdots + n_k(\overline{x}_k - \overline{\overline{x}})^2}{k - 1}$$

Generally we use a somewhat simpler formula for computational purposes. The quantity

$$\sum_i n_i(\overline{x}_i - \overline{\overline{x}})^2$$

has k observations of the random variable $\overline{x}$, and the population mean is estimated by $\overline{\overline{x}}$, so there are $k - 1$ degrees of freedom associated with the numerator. The numerator of MSTR is called the **sum of squares for treatments (SSTR)**; thus MSTR is equal to SSTR divided by the appropriate degrees of freedom. Each $\overline{x}_i = (\sum_j x_{ij})/n_i$, and

$$\overline{\overline{x}} = (\sum_i\sum_j x_{ij})/N, \text{ where } N = n_1 + n_2 + n_3 + \cdots + n_k = \sum_i n_i.$$

We again denote $\sum_j x_{ij}$ by T_i; we also denote $\sum_i\sum_j x_{ij}$ by T, the sum of all the values of all observations.

Then

$$\begin{aligned}\text{SSTR} &= \sum_i n_i(\overline{x}_i - \overline{\overline{x}})^2 \\ &= \sum_i n_i(T_i/n_i - T/N)^2\end{aligned}$$

$$= \sum_i n_i[(T_i/n_i)^2 - 2(T_i/n_i)(T/N) + (T/N)^2]$$
$$= \sum_i n_i[(T_i/n_i)^2)] - 2\sum_i n_i(T_i/n_i)(T/N) + \sum_i n_i(T/N)^2$$
$$= \sum_i (T_i)^2/n_i - 2(T/N)\sum_i T_i + (T/N)^2\sum_i n_i$$

since T/N is a constant. Now $\sum_i T_i = T$ and $\sum_i n_i = N$, so we have

$$\text{SSTR} = \sum_i (T_i)^2/n_i - 2(T/N)T + N(T/N)^2$$
$$= \sum_i (T_i)^2/n_i - 2(T)^2/N + (T)^2/N$$
$$= \sum_i (T_i)^2/n_i - (T)^2/N$$

or

$$\text{SSTR} = \left[\frac{T_1^2}{n_1} + \frac{T_2^2}{n_2} + \frac{T_3^2}{n_3} + \cdots + \frac{T_k^2}{n_k}\right] - \frac{(T)^2}{N}$$

It can be seen easily that

$$\text{SSTR} + \text{SSE} = \Sigma x^2 - \frac{(T)^2}{N}$$

which is called the **total sum of squares (SSTO).**

Example 11.1

Suppose we randomly give 30 subjects one of three experimental diets (10 per diet). After 3 months the weight loss (to the nearest pound) by each subject is recorded. In the following tabulation a negative weight loss indicates a gain:

POUNDS LOST		
DIET A	DIET B	DIET C
8	11	3
4	2	6
5	7	8
7	4	1
3	−1	5
7	8	12
4	3	7
6	3	9
7	6	5
1	7	11

Obtain MSTR and MSE for the data.

Solution

$T_1 = 8 + 4 + \cdots + 1 = 52$, $T_2 = 11 + 2 + \cdots + 7 = 50$, $T_3 = 3 + 6 + \cdots + 11 = 67$; $T = 52 + 50 + 67 = 169$; $\Sigma x^2 = 8^2 + 4^2 + \cdots + 11^2 = 1227$; $n_1 = n_2 = n_3 = 10$; $N = 30$. Then

$$\text{SSTR} = \left[\frac{52^2}{10} + \frac{50^2}{10} + \frac{67^2}{10}\right] - \frac{169^2}{30}$$
$$\doteq 969.3 - 952.0333 \doteq 17.2667$$

and

$$\text{SSE} = 1227 - \left[\frac{52^2}{10} + \frac{50^2}{10} + \frac{67^2}{10}\right]$$
$$= 1227 - 969.3 = 257.7$$

Since $k = 3$, we have MSTR $\doteq 17.2667/(3 - 1) \doteq 8.6333$ and MSE $= 257.7/(30 - 3) \doteq 9.5444$. Thus SSTO $= 1227 - 169^2/30 \doteq 274.9667 = 17.2667 + 257.7$.

11.2.2 Partitioning Total Variation

The total variability of all x values about the overall mean is

$$\sum_i \sum_j (x_{ij} - \bar{\bar{x}})^2 \quad \text{or simply} \quad \Sigma(x - \bar{\bar{x}})^2.$$

This is called the *total sum of squares* of deviations from the overall mean (SSTO). (It can be shown that this is also equal to $\Sigma x^2 - T^2/N$.) Some of this variability is due to differences *between* the sample treatment means, and the remainder is due to variability *within* the samples.

Much of the theory of analysis of variance rests on the concept of **partitioning** the total sum of squares into the various sources of variability. Moreover, use of this concept simplifies calculations. For the single-factor case, these sources are the sum of squares due to treatments (SSTR), which measures variability between samples, and the sum of squares due to error (SSE), which measures variability within the samples (Figure 11.4).

Thus SSTO = SSTR + SSE. Since SSE = SSTO − SSTR, we may obtain SSE by subtraction once we have determined SSTO and SSTR. For single-factor analysis of variance the advantage is slight, but in two-factor analysis

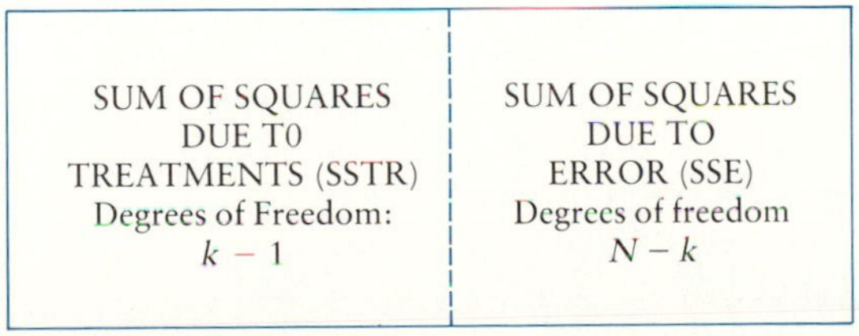

SUM OF SQUARES DUE T0 TREATMENTS (SSTR) Degrees of Freedom: $k - 1$	SUM OF SQUARES DUE TO ERROR (SSE) Degrees of freedom $N - k$

Figure 11.4
Partitioning the total sum of squares and degrees of freedom.

of variance, as we will see in the next chapter, the method provides a great simplification.

Now SSTO $= \Sigma(x - \bar{\bar{x}})^2$, so the number of degrees of freedom for SSTO is $N - 1$. We can partition the degrees of freedom for the total variation into the degrees of freedom for treatments $(k - 1)$ and the degrees of freedom for error $(N - k)$. We obtain the mean squares by dividing each sum of squares by the appropriate degrees of freedom.

If we repeat the experiment many times, we will find that the expected value of all the MSE is the population variance; that is, $E(\text{MSE}) = \sigma^2$ so that the variable MSE is an unbiased estimator of σ^2. It can also be shown that if μ is the mean of the μ_i, then

$$E(\text{MSTR}) = \sigma^2 + \sum_i n_i(\mu_i - \mu)^2/(k - 1).$$

Then if all treatment means are equal

$$\sum_i n_i(\mu_i - \mu)^2 = 0$$

so the variable MSTR is also an unbiased estimator for the common variance σ^2.

We thus have the following formulas for the analysis of variance. The defining formulas are given first in each case, followed by the shortcut formulas that are generally used.

Formulas for Analysis of Variance

Total Sum of Squares

$$\begin{aligned}\text{SSTO} &= \Sigma(x - \bar{\bar{x}})^2 \\ &= \Sigma x^2 - \frac{(T)^2}{N}\end{aligned}$$

Sum of Squares for Treatments

$$\begin{aligned}\text{SSTR} &= \sum_i n_i(\bar{x}_i - \bar{\bar{x}})^2 \\ &= \left[\frac{T_1^2}{n_1} + \frac{T_2^2}{n_2} + \frac{T_3^2}{n_3} + \cdots + \frac{T_k^2}{n_k}\right] - \frac{(T)^2}{N}\end{aligned}$$

Sum of Squares for Error

$$\begin{aligned}\text{SSE} &= \sum_i \sum_j (x_{ij} - \bar{x}_i)^2 = \sum_i (n_i - 1)s_1^2 \\ &= \Sigma x^2 - \left[\frac{T_1^2}{n_1} + \frac{T_2^2}{n_2} + \frac{T_3^2}{n_3} + \cdots + \frac{T_k^2}{n_k}\right]\end{aligned}$$

$$= \text{SSTO} - \text{SSTR}$$

Mean Squares

$$\text{MSTR} = \frac{\text{SSTR}}{k - 1}$$

$$\text{MSE} = \frac{\text{SSE}}{N - k}$$

where $N = n_1 + n_2 + n_3 + \cdots + n_k$, T_i is the sum of the values of the observations in the ith sample treatment, T is the sum of the values of all observations, and $\overline{\overline{x}}$ is the overall mean.

The results of the partitioning of the total sum of squares are usually summarized in an analysis of variance table, frequently called an **ANOVA table,** as shown here.

Analysis of Variance Table

The partition of total sum of squares (SSTO) into SSTR and SSE and the calculation of MSTR and MSE are summarized in an analysis of variance table:

SOURCE OF VARIATION	DF	SS	MS
Treatments	$k - 1$	SSTR	MSTR
Error	$N - k$	SSE	MSE
Total	$N - 1$	SSTO	

Example 11.2

Suppose that three sets each of four different makes of tires are tested on the same car and under the same conditions. The mean number of feet required to stop at 10 miles per hour is then recorded for each set of the four makes:

Make I:	22	24	23
Make II:	23	26	23
Make III:	24	26	25
Make IV:	24	25	23

Assuming that the mean number of feet required for stopping is normally distributed, construct the analysis of variance table for the data.

Solution

$\Sigma x^2 = 22^2 + 24^2 + 23^2 + 23^2 + 26^2 + \cdots + 25^2 + 23^2 = 6930$; each $n_i = 3$, $N = 12$; $T_1 = 69$, $T_2 = 72$, $T_3 = 75$, and $T_4 = 72$; $T = 288$. Therefore

$$\text{SSTO} = 6930 - (288)^2/12 = 6930 - 6912 = 18$$
$$\text{SSTR} = [69^2/3 + 72^2/3 + 75^2/3 + 72^2/3] - (288)^2/12 = 6918 - 6912 = 6$$
$$\text{SSE} = 18 - 6 = 12$$

and $N - 1 = 11$, $k - 1 = 3$, $N - k = 8$, and the ANOVA table is as shown:

SOURCE OF VARIATION	DF	SS	MS
Treatments	3	6	2.000
Error	8	12	1.500
Total	11	18	

PROFICIENCY CHECKS

1. Construct an ANOVA table for the results of Example 11.1.
2. Suppose that four samples have data as shown here. Construct an ANOVA table.

   I: 2 4 3 4
   II: 5 8 5
   III: 8 10 9
   IV: 14 15 13 9

11.2.3 The *F* Test

If the analysis of variance model applies to data from several samples, we may test the hypothesis that all the treatment means are equal (that is, $H_0: \mu_1 = \mu_2 = \cdots = \mu_k$) by using the **F test,** first discussed in Section 8.6.1. If we obtain two independent estimates for the same population variance, the ratio of all possible pairs of such estimates will have a sampling distribution called the ***F* distribution,** developed by Sir Ronald Fisher. This distribution is determined by two parameters: the degrees of freedom for the numerator (ν_1), and the degrees of freedom for the denominator (ν_2). An F distribution with ν_1 and ν_2 degrees of freedom is denoted $F(\nu_1, \nu_2)$.

We know that

$$E(\text{MSTR}) = \sigma^2 + \sum_i n_i(\mu_i - \mu)^2 \text{ and } E(\text{MSE}) = \sigma^2$$

If H_0 is true, $E(\text{MSTR}) = E(\text{MSE}) = \sigma^2$, so both MSTR and MSE will be unbiased estimates for σ^2. In that case the set of all ratios MSTR/MSE will have an F distribution with $\nu_1 = k - 1$ and $\nu_2 = N - k$. Now since

$$\text{MSTR} = [\sum_i n_i(\overline{x}_i - \overline{\overline{x}})^2]/(k-1),$$

if the treatment means are equal the sample means will be close to one another, so MSTR will be small; conversely, if the treatment means are not equal, the sample means will not be so close and MSTR will be relatively large. Thus if F^* is a sample value of F, relatively large values of F^* will lead to the conclusion that the treatment means are not equal, while relatively small values of F^* will lead to the conclusion that the treatment means could well be all equal.

Critical values of F are denoted $F_\alpha\{\nu_1, \nu_2\}$, and are listed in Tables 7, 8, 9, and 10 for $\alpha = 0.05, 0.025, 0.01$, and 0.005, respectively. To determine $F_{.05}\{3, 8\}$, for example, we turn to Table 7. The degrees of freedom for the numerator are given in the column headings, and the degrees of freedom for the denominator are given in the row headings on the left side of each row. We find the column headed 3 and the row headed 8 and determine the corresponding critical value of F to be 4.07; that is, $F_{.05}\{3, 8\} = 4.07$. A portion of Table 7 is reproduced in Table 11.1. Note that $F_{.05}\{8, 3\} = 8.85$, so it is important not to confuse the two.

For the data of Example 11.2, we get $F^* = 2.000/1.500 \doteq 1.33$; $F_{.05}\{3, 8\} = 4.07$ and $F_{.01}\{3, 8\} = 7.59$. These values are shown in Figure 11.5. We would thus be unable to reject the null hypothesis. The F test for analysis of variance is summarized here.

Table 11.1 PORTION OF TABLE 7 FOR CRITICAL VALUES OF $F_{.05}$.

DEGREES OF FREEDOM FOR DENOMINATOR	DEGREES OF FREEDOM FOR NUMERATOR 1	2	3	4	5	6	7	8	9
1	161	200	216	225	230	234	237	239	241
2	18.5	19.0	19.2	19.2	19.3	19.3	19.4	19.4	19.4
3	10.1	9.55	9.28	9.12	9.01	8.94	8.89	8.85	8.81
4	7.71	6.94	6.59	6.39	6.26	6.16	6.09	6.04	6.00
5	6.61	5.79	5.41	5.19	5.05	4.95	4.88	4.82	4.77
6	5.99	5.14	4.76	4.53	4.39	4.28	4.21	4.15	4.10
7	5.59	4.74	4.35	4.12	3.97	3.87	3.79	3.73	3.39
8	5.32	4.46	4.07	3.84	3.69	3.58	3.50	3.44	3.39
9	5.12	4.26	3.86	3.63	3.48	3.37	3.29	3.23	3.18
10	4.96	4.10	3.71	3.48	3.33	3.22	3.14	3.07	3.02

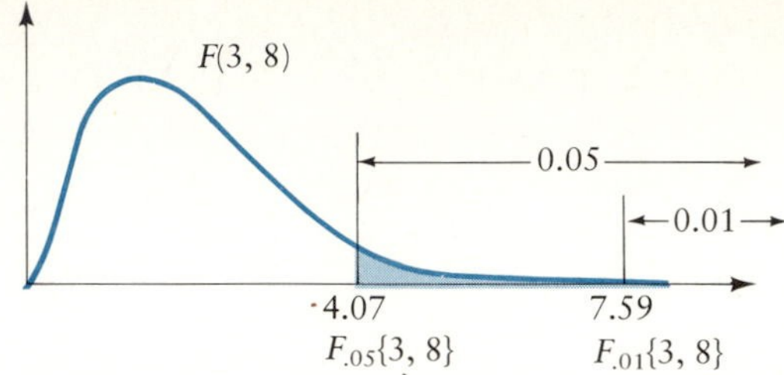

Figure 11.5 Values of $F_{.05}\{3, 8\}$ and $F_{.01}\{3, 8\}$.

Hypothesis Test for Treatment Means (Completely Randomized Design)

Suppose we obtain independent random samples of observations from k populations with means $\mu_1, \mu_2, \ldots, \mu_k$ and that both of the following are true:

1. Each population is normal.
2. The variance of each population is σ^2.

Then we may test the hypothesis H_0: $\mu_1 = \mu_2 = \cdots = \mu_k$.

The sample value of the test statistic used to test the hypothesis is F^*, where $F^* = \text{MSTR/MSE}$.

The decision rule is dictated by the level of significance α chosen for the test.

Null and Alternate Hypotheses

H_0: $\mu_1 = \mu_2 = \cdots = \mu_k$

H_1: not all μ_i are equal

Decision Rule

Reject H_0 if $F^* > F_\alpha\{k-1, N-k\}$

where N is the total number of observations in all samples.

Example 11.3

An English teacher hypothesizes that simplification of paragraphs promotes understanding of content. He divides his class randomly into three groups of ten each and gives a standardized test of paragraph understanding to group A. Groups B and C are given the same test with the paragraphs modified in two different ways, both designed to promote understanding. Each group is given the same questions and scored in the same way. The results follow:

A: 38 54 39 52 63 54 47 52 46 25

B: 58 44 63 94 72 42 89 68 53 47

C: 76 51 83 84 51 67 40 89 76 53

Using the 0.05 level of significance, test the hypothesis that there are differences among the treatments.

Solution

We wish to test the hypothesis that all treatment means are equal against the alternate hypothesis that not all the means are equal. We can use the analysis of variance procedure if all the assumptions are met. Procedures to check on normality have been given previously. A procedure to test the assumption that population variances are equal will be given in Section 11.4. The assumption of random independent samples is met by the random assignment of students into the three levels. We can determine the necessary sums as shown in Table 11.2.

Then $T = 470 + 630 + 670 = 1770$, and

$$\Sigma x^2 = 23{,}124 + 42{,}616 + 47{,}518 = 113{,}258$$

$$\text{SSTO} = 113{,}258 - (1770)^2/30 = 113{,}258 - 104{,}430 = 8828$$

$$\text{SSTR} = [(470)^2/10 + (630)^2/10 + (670)^2/10] - (1770)^2/30$$

$$= 106{,}670 - 104{,}430 = 2240$$

$$\text{SSE} = 8828 - 2240 = 6588$$

The ANOVA table is shown here with the value of F^*:

SOURCE OF VARIATION	DF	SS	MS	F^*
Treatments	2	2240	1120	4.59
Error	27	6588	244	
Total	29	8828		

Table 11.2 COMPUTATIONS FOR EXAMPLE 11.3

x_{1j}	x_{1j}^2	x_{2j}	x_{2j}^2	x_{3j}	x_{3j}^2
38	1444	58	3364	76	5776
54	2916	44	1936	51	2601
39	1521	63	3969	83	6889
52	2704	94	8836	84	7056
63	3969	72	5184	51	2601
54	2916	42	1764	67	4489
47	2209	89	7921	40	1600
52	2704	68	4624	89	7921
46	2116	53	2809	76	5776
25	625	47	2209	53	2809
470	23,124	630	42,616	670	47,518

The critical value of F is $F_{.05}\{2, 27\}$, which is 3.35; thus we reject the hypothesis that all treatment means are equal and conclude that there are differences among the three samples not due to chance. Since $F_{.025}\{2, 27\} = 4.24$ and $F_{.01}\{2, 27\} = 5.49$, we conclude that the P value for this experiment is between 0.025 and 0.01.

PROFICIENCY CHECKS

3. Determine the decision rule to test the hypothesis H_0: $\mu_1 = \mu_2 = \mu_3 = \mu_4 = \mu_5$ if $\alpha = 0.05$ and N is:
   a. 20 b. 35 c. 65 d. 125
4. Determine the decision rule to test the hypothesis H_0: $\mu_1 = \mu_2 = \mu_3 = \mu_4 = \mu_5$ if $\alpha = 0.01$ and N is:
   a. 20 b. 35 c. 65 d. 125
5. Test the hypothesis H_0: $\mu_1 = \mu_2 = \mu_3 = \mu_4$ at the 0.05 level of significance if $N = 34$, SSTO $= 266.5$, and SSTR is:
   a. 91.4 b. 67.4 c. 55.7 d. 43.8
6. Refer to Proficiency Check 1. Using the 0.05 level of significance, determine whether or not there is any significant difference in weight loss among the three diets.
7. Use the results of Proficiency Check 2 to test the hypothesis that all treatment means are equal. Use the 0.01 level of significance. What is the P value for this test?
8. What conclusions may be drawn in Example 11.2?

11.2.4 Confidence Intervals for Treatment Means

We may obtain a confidence interval for an individual treatment mean (μ_i) by using the method of Section 6.4.2 or for the difference between two treatment means (μ_i and μ_j) by using the method of Section 8.2.2. The best estimate for the population variance σ^2 is MSE, so we use MSE in place of the sample variance or variances. Thus $s_{\bar{x}_i} = \sqrt{\text{MSE}/n_i}$ for each treatment and $s_{\bar{x}_i - \bar{x}_j} = \sqrt{\text{MSE}(1/n_i + 1/n_j)}$ for each pair of treatments.

Example 11.4

Refer to Example 11.3. Obtain a 95% confidence interval for the mean of group A and a 95% confidence interval for the difference between the means of group A and group B.

Solution

Since MSE $= 244$ and $n_i = 10$ for each group, the standard error is $\sqrt{244/10} \doteq 4.940$; $t_{.025}\{27\} = 2.052$ and $2.052(4.940) \doteq 10.14$. Since $\bar{x}_a$

Confidence Intervals for Treatment Means

A $100(1 - \alpha)\%$ confidence interval for μ_i is the interval

$$\bar{x}_i - t\sqrt{\frac{\text{MSE}}{n_i}} \quad \text{to} \quad \bar{x}_i + t\sqrt{\frac{\text{MSE}}{n_i}}$$

A $100(1 - \alpha)\%$ confidence interval for $\mu_i - \mu_j$ is the interval

$$\bar{x}_i - \bar{x}_j - t\sqrt{\text{MSE}\left(\frac{1}{n_i} + \frac{1}{n_j}\right)} \text{ to } \bar{x}_i - \bar{x}_j + t\sqrt{\text{MSE}\left(\frac{1}{n_i} + \frac{1}{n_j}\right)}$$

In both cases $t = t_{\alpha/2}\{N-k\}$.

= 47.0, the confidence interval is 47.0 − 10.14 to 47.0 + 10.14—that is, 36.86 to 57.14.

To estimate $\mu_a - \mu_b$ we find $\sqrt{244(1/10 + 1/10)} \doteq 6.986$; then $2.052(6.986) \doteq 14.33$. Thus $\bar{x}_a = 47.0$ and $\bar{x}_b = 63.0$, so $\bar{x}_a - \bar{x}_b = -16.0$ and a 95% confidence interval for $\mu_a - \mu_b$ is -30.33 to -1.67. It might be preferable to express the limits as positive numbers; in this case we obtain a 95% confidence interval for $\mu_b - \mu_a$ from 1.67 to 30.33.

Comment If we obtain multiple $100(1 - \alpha)\%$ confidence intervals, the confidence level for the entire set of intervals is much less than $100(1 - \alpha)\%$. If we obtain several confidence intervals for treatment means, it might be best to use the *Bonferroni procedure* discussed in Section 11.3. Many procedures have been developed for obtaining multiple confidence intervals for the differences between treatment means. The *Tukey procedure* is covered in Section 11.3.1, and other procedures are mentioned there as well.

Another word of caution is in order. We cannot use a single confidence interval to determine whether treatment means differ unless we have selected the particular comparison prior to the experiment. After the analysis has been completed we cannot look at the data and decide which treatments to compare using a single confidence interval. If we select a pair of treatments at random to compare, the sample means, on the average, will not be as far apart as the largest and smallest sample means. When performing comparisons suggested by the data, we must use a multiple-comparison procedure taking into account all possible comparisons.

11.2.5 The SAS Procedure ANOVA

We use the SAS procedure ANOVA to perform a single-factor analysis of variance. We can illustrate the use of the ANOVA procedure with the data of Example 11.3. We first create a SAS data set. We may do this in one of several ways; we use the simplest method here. In this case the two variables are paragraph simplification (none, method 1, or method 2) and test scores. We denote the variable paragraph simplification by TRT (for treatment) and the factor levels as A, B, and C. The test scores are denoted by SCORE. The data step, then, will be

```
DATA TEST;
  INPUT TRT $ SCORE @@;
CARDS;
```

and the data will be listed

```
A 38 A 54 A 39 A 52 A 63 A 54 A 47 A 52 A 46 A 25
B 58 B 44 B 63 B 94 B 72 B 42 B 89 B 68 B 53 B 47
C 76 C 51 C 83 C 84 C 51 C 67 C 40 C 89 C 76 C 53
```

The INPUT statement says that we will be entering two variables, TRT and SCORE, in that order, and the $ sign indicates that the variable TRT is non-numeric. Thus the first entry is A 38, indicating that TRT = A and SCORE = 38. The @@ symbol allows us to continue entering variables on the same line. (Early versions of the SAS System required a separate line with a semicolon after the data lines.)

We then perform the ANOVA procedure as follows:

```
PROC ANOVA;
  CLASS TRT;
  MODEL SCORE = TRT;
```

The CLASS statement identifies the independent variable (TRT) while the model statement identifies the dependent variable (SCORE), which is a function of the independent variable. The output of this program is shown in Figure 11.6.

The important features of the output in Figure 11.6 are numbered. ANOVA first prints a table that includes:

1. The name of each factor in the CLASS statement
2. The number of different levels of the CLASS variables
3. The values of the CLASS variables
4. The number of observations in the data set

An analysis of variance table is then printed for each dependent variable in the MODEL statement. This table gives:

5. The CORRECTED TOTAL sum of squares for the dependent variable (SSTO)
6. The sum of squares attributed to the MODEL. (This is equal to SSTO − SSE. For one-way analysis of variance it is identical to SSTR.)

```
                              SAS

                ANALYSIS OF VARIANCE PROCEDURE

                   CLASS LEVEL INFORMATION

                 CLASS      LEVELS     VALUES

             (1) TRT             3 (2)   A B C (3)

        NUMBER OF OBSERVATIONS IN DATA SET = 30 (4)
                              SAS

                ANALYSIS OF VARIANCE PROCEDURE

DEPENDENT VARIABLE: SCORE

SOURCE            DF    SUM OF SQUARES         MEAN SQUARE     F VALUE        PR > F       R-SQUARE           C.V.

MODEL            2 (8)  2240.00000000 (6)  1120.00000000 (11)   4.59 (13)   0.0192 (14)  0.253738 (15)  26.4754 (16)

ERROR           27 (9)  6588.00000000 (7)   244.00000000 (12)             ROOT MSE                  SCORE MEAN

CORRECTED TOTAL 29 (10) 8828.00000000 (5)                               15.62049935 (17)         59.00000000 (18)

SOURCE            DF        ANOVA SS       F VALUE       PR > F

TRT (19)         2 (20) 2240.00000000 (21)   4.59 (22)   0.0192 (23)
```

Figure 11.6 SAS printout of the ANOVA procedure for the paragraph simplification data of Example 11.3.

7. The sum of squares attributed to ERROR (SSE)
8. The degrees of freedom for the model ($k - 1$)
9. The degrees of freedom for error ($N - k$)
10. The total degrees of freedom ($N - 1$)
11. The mean square for the model (MSTR)
12. MSE
13. A value of F obtained by dividing MSTR by MSE
14. The P value for this test
15. R-SQUARE (symbolized r^2), which is equal to SSTR/SSTO. (It is a measure of association between the variables that is covered in Chapter 13.)
16. C. V., the coefficient of variation for the dependent variable (see Section 2.3.2)
17. ROOT MSE ($\sqrt{\text{MSE}}$)
18. The MEAN of the dependent variable

ANOVA then prints information about each factor. The names of the factors are listed under SOURCE (19). If there is only one factor in the model, (20), (21), (22), and (23) will duplicate (8), (6), (13), and (14), respectively. We see from the printout that the results of Example 11.3 are confirmed. The P value for the experiment is found to be 0.0192.

Example 11.5

Twenty-four employees are hired and given a test designed to predict success on the job. Higher scores theoretically indicate a greater likelihood of success. After 6 months, their progress is evaluated. They are classified as successful, moderately successful, barely adequate, and inadequate. Although 3 of the 24 are no longer with the company, they were employed long enough to have been classified (two as inadequate, one as moderately successful). The scores are given here for each group:

GROUP	TEST SCORES
Successful	64, 49, 61, 57, 54, 44, 49
Moderately successful	55, 60, 38, 44, 41, 45, 41
Barely adequate	33, 19, 64, 38, 37, 31
Inadequate	33, 22, 48, 16

Use a 0.05 level of significance and determine whether there is a difference among the treatments in test scores.

Solution

We enter the data using S M B and I for the four groups. The complete program is shown in Figure 11.9 in Section 11.3.2. The ANOVA table is shown in Figure 11.7.

In Figure 11.7 we see that $F^* = 5.09$. Since $F_{.05}\{3, 20\} = 3.10$ and $5.09 > 3.10$, we reject H_0. The printout also gives us the P value: 0.0088. We may reject H_0 by using the fact that $0.0088 < 0.05$. In either case we can conclude that there are differences among the treatment (population) means. (We may also say that there are *significant differences* among the sample means.)

Figure 11.7
SAS printout of ANOVA table for job success data of Example 11.5.

```
                              SAS

                 ANALYSIS OF VARIANCE PROCEDURE

                    CLASS LEVEL INFORMATION

            CLASS      LEVELS      VALUES

            GROUP         4        B I M S

         NUMBER OF OBSERVATIONS IN DATA SET = 24
                              SAS

                 ANALYSIS OF VARIANCE PROCEDURE

DEPENDENT VARIABLE: SCORE

SOURCE            DF   SUM OF SQUARES     MEAN SQUARE     F VALUE        PR > F   R-SQUARE          C.V.

MODEL              3   1835.77976190    611.92658730        5.09        0.0088   0.433176       25.2182

ERROR             20   2402.17857143    120.10892857                  ROOT MSE               SCORE MEAN

CORRECTED TOTAL   23   4237.95833333                                10.95942191              43.45833333
```

We can use the statement MEANS to obtain the means for each group as part of the ANOVA procedure by writing

```
PROC ANOVA;
  CLASS SUCCESS;
  MODEL SCORE = SUCCESS;
  MEANS SUCCESS;
```

If we want other information about each group, such as the variance, we need to use PROC MEANS with the added statement BY SUCCESS. Prior to these statements we need to write the statements PROC SORT and BY SUCCESS (each followed by ;) so that the data are correctly sorted.

Problems

Practice

11.1 The following analysis of variance table is partially filled in. Complete the table.

SOURCE OF VARIATION	DF	SS	MS
Treatments			303.90
Error	22		12.15
Total		1482.9	

11.2 An analysis of variance table is given here:

SOURCE OF VARIATION	DF	SS	MS
Treatments	2	0.0548	0.0274
Error	21	0.1607	0.0077
Total	23	0.1881	

Using the 0.05 level of significance, test the hypothesis that there are no differences among the treatment means.

11.3 Three samples have the following data:

1: 2, 3, 3, 4, 2, 5, 6

2: 5, 6, 4, 7, 8, 9

3: 3, 7, 9, 5, 8, 9, 11

a. Construct the analysis of variance table for the data.
b. Test the hypothesis, at the 0.05 level of significance, that there are no differences among the treatment means.

11.4 Three samples have the following data:

1: 20, 30, 30, 40, 20, 50, 60

2: 50, 60, 40, 70, 80, 90

3: 30, 70, 90, 50, 80, 90, 110

a. Construct the analysis of variance table for the data.
b. Test the hypothesis, at the 0.05 level of significance, that there are no differences among the treatment means.
c. Note that each observation in this problem is ten times the corresponding observation in Problem 11.3. Compare the analyses. What do you conclude?

Applications

11.5 Students are randomly assigned to four different groups, and four different methods are used to teach the same material to the groups, one method per group. Each group contains 21 students. The means and standard deviations for each group are given below. Use the defining formulas to obtain MSTR and MSE, and test the hypothesis that there are no differences among the treatment means. Use the 0.05 level of significance. Note the effect of different means and different sample standard deviations on the results.

a. $\overline{x}_1 = 78.4, \overline{x}_2 = 88.4, \overline{x}_3 = 71.6, \overline{x}_4 = 70.4$
$s_1 = 14.7, s_2 = 14.4, s_3 = 21.6, s_4 = 19.8$
b. $\overline{x}_1 = 78.4, \overline{x}_2 = 88.4, \overline{x}_3 = 71.6, \overline{x}_4 = 70.4$
$s_1 = 19.7, s_2 = 18.4, s_3 = 24.6, s_4 = 29.8$
c. $\overline{x}_1 = 78.4, \overline{x}_2 = 88.4, \overline{x}_3 = 81.6, \overline{x}_4 = 80.4$
$s_1 = 14.7, s_2 = 14.4, s_3 = 21.6, s_4 = 19.8$

11.6 Anthropologists have uncovered three burial mounds some distance apart. They find several adult skulls in each mound. The skulls measure as follows in each mound (in centimeters):

I: 48.4, 46.2, 47.1, 46.3, 45.9

II: 52.8, 49.6, 50.5, 48.7

III: 44.2, 44.7, 46.1, 45.4

a. What are the factor levels? Is the factor a classification factor or experimental factor?
b. Can you conclude at the 0.05 level that the skulls are representative of the same population?
c. Obtain a 95% confidence interval for the difference in skull measurements for the populations of mounds II and III.

11.7 A psychologist wishes to determine whether prior learning of a maze has an effect on subsequent learning of a different maze in white mice and, if so, whether length of time since the learning experience makes a difference. A number of mice are divided randomly into four groups. Group A is taught the maze immediately, group B learns the maze 3 months later, and group C learns the maze 5 months later. Group D is not taught the maze at all and is designated the control group. Six months after the groupings the psychologist tests how long it takes the four groups of white mice to learn the same maze. The number of trials required for each mouse to learn the current maze is given here:

A	B	C	D
13	9	6	16
11	13	9	10
6	15	13	14
18	11	10	17
7	10	9	9
15	6	7	11
14	7		14
	9		

a. What type of design is employed here?
b. What is the factor under study? Is this factor a classification factor or experimental factor?
c. Can you conclude that there is a difference among the mean trial scores for the various learning groups? Use the 0.05 level of significance.
d. Obtain a 95% confidence interval for the mean number of trials required for all mice with no prior training.

11.8 Cultural anthropologists studying six different cultures did some research into the "waiting time" of an interview—that is, the time spent in amenities before getting down to business. Eleven interviews were timed in each culture. The waiting times (in minutes) are shown here:

CULTURE	WAITING TIMES
A	11, 7, 8, 4, 6, 9, 8, 5, 7, 9, 10
B	6, 4, 8, 2, 7, 6, 3, 5, 7, 6, 3
C	5, 9, 8, 7, 6, 10, 9, 11, 9, 8, 7
D	6, 6, 5, 4, 5, 7, 6, 3, 5, 4, 8
E	13, 14, 16, 12, 17, 11, 14, 16, 13, 11, 12
F	7, 6, 5, 8, 11, 3, 8, 14, 3, 7, 5

a. Decide whether there are differences among the cultures in regard to this aspect. Use the 0.01 level of significance.
b. Obtain a 99% confidence interval for the mean waiting time for culture E.
c. What is the approximate observed significance level (P value) for the test in part (a)?

11.9 Students at five schools are being studied to discover whether there are differences from school to school in increase in ambition as measured by an "ambition level" index. Students included in the study are selected at random from each school and given tests evaluating this index during their freshman and senior years, and the increase or decrease is calculated. Equal numbers are selected initially, but attrition reduces the numbers to the data given here:

SCHOOL	DIFFERENCE SCORES
A	+11, +32, +18, +16, +14, −3, +24, +13, +6
B	−4, +24, +16, −2, +13, +17, +22, +24, +16, +9, +12

C	+13, +31, +34, −7, +18, +14, +34, +28, +19
D	+16, +11, −2, +4, +13, +14, +19, +11, +7, +20
E	+8, −11, +4, +7, +12, −1, −3, +24, +17, +28, +9, +14, + 22

a. Are the differences between the schools significant at the 0.05 level?
b. Which school shows the greatest mean increase? The least?
c. Obtain a 90% confidence interval for the mean increase for all students at school A.

11.10 A researcher wishes to test the mileage of five leading brands of gasoline. She obtains 40 different automobiles of the same make and year and divides them randomly into five groups. Each group uses one and only one brand of gasoline and keeps careful records for 1 month while driving at random over different types of terrain and under varying conditions. (One automobile was wrecked, and the records are unavailable.) The results are given below.

MILES PER GALLON

BRAND E	BRAND G	BRAND P	BRAND S	BRAND T
22.6	18.4	20.8	20.1	24.3
21.9	19.6	20.8	21.2	19.8
22.1	20.6	17.9	18.7	22.3
17.9	22.5	19.2	21.8	23.1
20.9	19.8	18.7	22.3	22.7
21.6	20.8	19.1	19.8	22.3
22.4	20.4	18.8	21.1	19.8
19.9	18.2	18.0		21.5

a. Can you conclude that there is a difference in mean mileage for the five brands? Use the 0.05 level of significance.
b. Obtain a 95% confidence interval for the mean mileage for brand E.
c. Obtain a 90% confidence interval for the difference in mean mileage for brands S and T.

11.11 Students are randomly divided into three groups and taught a lesson using one of three methods—programmed instruction, a tape recorder plus visual aids, and television. A learning test is used to determine the amount of information retained. The scores are given here. Test the hypothesis of no difference between treatment means at the 0.05 level of significance.

METHOD	SCORES
Programmed	2, 3, 1, 9, 3, 6, 9, 1, 15, 4, 1
Tape	2, 9, 15, 6, 9, 12, 13, 9, 12, 9
Television	9, 12, 6, 12, 15, 15, 6, 9, 15

11.12 A survey of businesses that have failed during the past year reveals that the mean length of time in business for 42 large businesses is 6.2 years with a standard deviation of 2.2 years; for 34 medium-size businesses it is 3.9 years with a standard deviation of 1.6 years; and 47 small businesses have a mean life of 3.1 years with a standard deviation of 1.8 years.

a. Assume that the data are collected at random from the various populations. Calculate T_1, T_2, T_3, and T. Use these values to obtain SSTR.

b. Pool the sample variances to obtain SSE.

c. Construct the ANOVA table for the data and test, at the 0.05 level of significance, whether a difference in duration prior to failure is associated with size.

d. Obtain a 95% confidence interval for the mean number of years in business of all failed medium-size businesses.

e. Prior to data collection it was decided to test whether or not there is a significant difference between the time in business for large and small failed businesses. Obtain a 95% confidence interval for this difference and comment.

11.3 A Multiple-Comparison Test

We use the analysis of variance procedure to test whether or not treatment means differ. If we conclude that not all treatment means are equal, the next step is to determine what differences exist between treatment means. In many applications, the question of interest is whether there is a "best" treatment—and if so, which one. For example, we may wish to use a paint that will dry most quickly among several competing brands. A race-car driver may wish to drive the car that has the fastest acceleration among several cars.

A large number of procedures have been developed to make pairwise comparisons among the sample treatment means—that is, comparisons between pairs of means. If only a few comparisons are planned in advance, the most efficient procedure may be the one developed by Carlo Bonferroni (1892–1960). Using the **Bonferroni procedure,** we conduct multiple t tests with a reduced level of significance for each test. If we conduct s tests and desire a family level of significance of α, each test is conducted at an α/s level of significance. If five tests are to be conducted at a family significance level of 0.10, for example, we will conduct each test at a 0.02 level of significance, using a critical value of $t_{.01}\{N - k\}$. In many applications, if not most, the individual significance levels are fractional, so a series of tables has been developed to provide critical values of $t_{\alpha/2s}\{\nu\}$ for a large number of possibilities. (See Bailey [1977].) Section 11.4.4 shows an application of this procedure. As s increases, the comparative efficiency of the Bonferroni procedure decreases.

If most or all pairwise comparisons are intended, or if they are not planned in advance, other procedures are better. Each procedure is based on a different viewpoint. Some are more conservative than others and are therefore more likely to allow a type II error, while others are said to understate the true level of significance. In addition to the Bonferroni, widely used tests include the Duncan, Gabriel, Ryan–Einot–Gabriel–Welsch, Scheffe, Sidak, Student–Newman–Keuls, Fisher's least significant difference (LSD), Tukey, and Waller–Duncan procedures. All these tests are available with the SAS ANOVA procedure; many of them are discussed in Milliken and Johnson (1984, pp. 29–45). For further information consult the references at the end of this chapter.

11.3.1 Tukey's Multiple-Comparison Test

One of the simplest procedures for making multiple comparisons following a significant F test was developed by John Tukey. It uses a modification of the **Studentized range statistic** q, where

$$q = \frac{[\overline{x}_i(\max) - \overline{x}_i(\min)]}{\sqrt{\text{MSE}/n}}$$

and $x_i(\max)$ is the largest sample mean, $x_i(\min)$ is the smallest sample mean, and n is the common sample size. This number gives a bound on the difference between any two means. Critical values of q are determined by two parameters—the number of means being compared and the degrees of freedom for MSE.

To determine whether or not two treatment means differ, say μ_i and μ_j, we determine the **honestly significant difference** (HSD) between $\overline{x}_i$ and $\overline{x}_j$. The standard error of the difference between the samples, $s_{\overline{x}_i - \overline{x}_j}$, is equal to $\sqrt{\text{MSE}(1/n_i + 1/n_j)}$. We use the formula

$$\text{HSD} = \frac{q_\alpha\{k, N-k\}}{\sqrt{2}}\sqrt{\text{MSE}(1/n_i + 1/n_j)}$$

where $q_\alpha\{k, N - k\}$ is listed in Table 13 for $\alpha = 0.05$ and 0.01. If the

Tukey's Multiple-Comparison Test

Suppose the hypothesis H_0: $\mu_1 = \cdots = \mu_k$ can be rejected at an α level of significance. We can conclude that $\mu_i \neq \mu_j$ for any two treatment means if $|\overline{x}_i - \overline{x}_j|$ exceeds HSD, where

$$\text{HSD} = \frac{q_\alpha\{k, N-k\}}{\sqrt{2}}\sqrt{\text{MSE}(1/n_i + 1/n_j)}$$

If all sample sizes are equal to n, then

$$\text{HSD} = q_\alpha\{k, N-k\}\sqrt{\text{MSE}/n}$$

difference between $\overline{x}_i$ and $\overline{x}_j$ exceeds HSD, we conclude that $\mu_i \neq \mu_j$. If all sample sizes are equal to n, then HSD $= q_\alpha\{k, N - k\}\sqrt{\text{MSE}/n}$.

Tukey's test was developed to perform all pairwise comparisons, and it is exact if all samples are equal in size. If they are not equal in size, we must find a different value of HSD for most pairs of samples. A procedure for approximating HSD is given at the end of this section.

Example 11.6

Use the data of Example 11.3 to determine which treatment means are different.

Solution

In this example $F^* = 4.59$, which is greater than $F_{.05}\{2, 27\} = 3.35$, so we conclude that the treatment means are not all equal. The sample means are $\overline{x}_a = 47$, $\overline{x}_b = 63$, and $\overline{x}_c = 67$; MSE $= 244$. To determine $q_{.05}\{3, 27\}$ we refer to Table 13. A portion of Table 13 is given in Table 11.3.

Since $q_{.05}\{3, 24\} = 3.53$ and $q_{.05}\{3, 30\} = 3.49$, $q_{.05}\{3, 27\}$ is about 3.51. Thus HSD $= 3.51\sqrt{244/10} \doteq 17.34$ and the only conclusion we can draw is that the method of simplification used in group C does promote understanding, since $\overline{x}_c - \overline{x}_a = 20$. No other differences are significant. The results are often indicated by listing the groups in order of their sizes from largest to smallest and then drawing a line to connect the factor levels that do not differ significantly. In this case we would have the following configuration:

$$\underline{\text{C} \quad \text{B}} \quad \text{A}$$
$$\quad\quad \underline{\text{B} \quad \text{A}}$$

This diagram indicates that B and C do not differ significantly and B and A do not differ significantly. Since C and A are not connected by a line, they do differ significantly.

Table 11.3 PORTION OF TABLE 13 FOR CRITICAL VALUES OF q_k.

DF FOR MSE	α	k = NUMBER OF MEANS BEING COMPARED 2	3	4	5	6	7	8	9	10
20	.05	2.95	3.58	3.96	4.23	4.45	4.62	4.77	4.90	5.01
	.01	4.02	4.64	5.02	5.29	5.51	5.69	5.84	5.97	6.08
24	.05	2.92	3.53	3.90	4.17	4.37	4.54	4.68	4.81	4.92
	.01	3.96	4.54	4.91	5.17	5.37	5.54	5.69	5.81	5.92
30	.05	2.89	3.49	3.84	4.10	4.30	4.46	4.60	4.72	4.83
	.01	3.89	4.45	4.80	5.05	5.24	5.40	5.54	5.65	5.76

Another common procedure is to obtain confidence intervals for the difference between two treatment means. A confidence interval for $\mu_i - \mu_j$ has confidence limits

$$\bar{x}_i - \bar{x}_j - [q_\alpha\{k, N-k\}/\sqrt{2}]\sqrt{\text{MSE}(1/n_i + 1/n_j)}$$

and

$$\bar{x}_i - \bar{x}_j + [q_\alpha\{k, N-k\}/\sqrt{2}]\sqrt{\text{MSE}(1/n_i + 1/n_j)}$$

that is, $\bar{x}_i - \bar{x}_j \pm \text{HSD}$. Confidence intervals which do not include zero indicate that the difference between the means is significant.

Comment If the sample sizes are unequal but do not differ too greatly, an approximate value of HSD which will apply to all differences can be obtained by using the *harmonic mean* of the sample sizes in the formula for HSD instead of n. The **harmonic mean** of a set of numbers, ñ (n-tilde), is the reciprocal of the mean of the reciprocals. Thus an approximate value of HSD is

$$\text{HSD} = q_\alpha\{k, N-k\}\sqrt{\text{MSE}/\tilde{n}}$$

where

$$\tilde{n} = \frac{1}{(1/n_1 + 1/n_2 + \cdots + 1/n_k)/k} = \frac{1}{(\Sigma 1/n_i)/k}$$

This approximate value of HSD can be used for quick comparisons. In case of doubt, it is best to use the exact value of HSD.

PROFICIENCY CHECKS

9. Determine critical values of q for $\alpha = 0.05$ if $k = 4$ and N is:
   a. 12 b. 20 c. 34 d. 54 e. 100
10. Determine HSD if $\alpha = 0.05$, $k = 4$, $N = 24$, MSE $= 12.56$, and the sample sizes are:
    a. both 6 b. 5 and 7 c. 6 and 7 d. 3 and 8
11. Use the results of Proficiency Checks 2 and 7 to determine which treatment means are different.

11.3.2 Using the SAS Statement MEANS *factor*/TUKEY

We may perform any of the tests listed in the introduction to this section by using SAS statements. We will illustrate the method for using the Tukey test. For the statements necessary to perform other tests refer to the *SAS® User's Guide: Statistics* (1985, pp. 117–118). We can use the MEANS statement of the ANOVA procedure to perform these tests. After the PROC statement and the CLASSES and MODEL statements, we write the statement MEANS/TUKEY factor to indicate that the means are to be computed and the Tukey HSD test is to be performed on all pairwise comparisons of the listed factor.

If the sample sizes are equal, HSD will be computed and the means listed in descending order. Those that are not significantly different are indicated with a line of letters joining them. If we want confidence intervals, we specify CLDIFF after the statement TUKEY and before the semicolon. If no level of α is specified, the tests will be performed at a family confidence level of 95%. We may specify other levels of α in the MEANS statement, as shown here:

```
MEANS factor / TUKEY CLDIFF ALPHA = 0.10;
```

This statement specifies that the sample treatment means are to be computed for the factor specified, all pairwise comparisons are to be made using the Tukey procedure and a family confidence level of 0.10, and the results are to be printed as confidence intervals.

If the sample sizes are unequal, the procedures are the same except that confidence intervals will be obtained as part of the procedure. If we specify the option LINES, the means will be listed in descending order and the nonsignificant differences indicated by a line of letters. The approximate value of HSD will be used.

Figure 11.8 shows how to perform multiple comparisons on the data of

Figure 11.8
SAS printout of the MEANS *factor*/TUKEY statement for the paragraph simplification data.

```
INPUT FILE

DATA ANOVA;
   INPUT TRT SCORE @@;
CARDS;
1 38 1 54 1 39 1 52 1 63 1 54 1 47 1 52 1 46 1 25
2 58 2 44 2 63 2 94 2 72 2 42 2 89 2 68 2 53 2 47
3 76 3 51 3 83 3 84 3 51 3 67 3 40 3 89 3 76 3 53
PROC ANOVA;
   CLASS TRT;
   MODEL SCORE = TRT;
   MEANS TRT / TUKEY;

OUTPUT FILE

                              SAS

                  ANALYSIS OF VARIANCE PROCEDURE

TUKEY'S STUDENTIZED RANGE (HSD) TEST FOR VARIABLE: SCORE
NOTE: THIS TEST CONTROLS THE TYPE I EXPERIMENTWISE ERROR RATE,
      BUT GENERALLY HAS A HIGHER TYPE II ERROR RATE THAN REGWQ
```

```
        ALPHA=0.05   DF=27   MSE=244
        CRITICAL VALUE OF STUDENTIZED RANGE=3.506
        MINIMUM SIGNIFICANT DIFFERENCE=17.32

MEANS WITH THE SAME LETTER ARE NOT SIGNIFICANTLY DIFFERENT.

       TUKEY     GROUPING               MEAN       N   TRT

                        A              67.000     10   3
                        A
                B       A              63.000     10   2
                B
                B                      47.000     10   1
```

Example 11.3. The ANOVA table is omitted. The results corroborate those obtained above. Only treatments C and A differ significantly.

In Figure 11.9 we show the multiple comparisons for the data of Example 11.5 at a family significance level of 0.10. The complete program for Figures 11.7 and 11.9 is shown. The ANOVA table is omitted since it is shown in Figure 11.7. Significant differences are marked '***' as indicated on the printout. This result shows that the group experiencing success scored significantly higher on the test than the groups that were barely adequate or inadequate. No other differences are found.

Figure 11.9
SAS printout of multiple comparisons for job success data of Example 11.5

```
INPUT FILE

DATA ANOVA;
  INPUT GROUP $ SCORE @@;
CARDS;
S 64 S 49 S 61 S 57 S 54 S 44 S 49
M 55 M 60 M 38 M 44 M 41 M 45 M 41
B 33 B 19 B 64 B 38 B 37 B 31
I 33 I 22 I 48 I 16
PROC MEANS;
PROC ANOVA;
  CLASS GROUP;
  MODEL SCORE = GROUP;
  MEANS GROUP / TUKEY ALPHA = 0.10;

OUTPUT FILE (MULTIPLE COMPARISONS ONLY)

                                        SAS

                    ANALYSIS OF VARIANCE PROCEDURE

TUKEY'S STUDENTIZED RANGE (HSD) TEST FOR VARIABLE: SCORE
NOTE: THIS TEST CONTROLS THE TYPE I EXPERIMENTWISE ERROR RATE

     ALPHA=0.1   CONFIDENCE=0.9   DF=20   MSF=120.109
     CRITICAL VALUE OF STUDENTIZED RANGE=3.462
```

```
COMPARISONS SIGNIFICANT AT THE 0.1 LEVEL ARE INDICATED BY '***'

                     SIMULTANEOUS                    SIMULTANEOUS
                         LOWER       DIFFERENCE          UPPER
     GROUP            CONFIDENCE       BETWEEN        CONFIDENCE
   COMPARISON            LIMIT          MEANS            LIMIT

   S     - M            -6.625          7.714           22.053
   S     - B             2.075         17.000           31.925      ***
   S     - I             7.436         24.250           41.064      ***

   M     - S           -22.053         -7.714            6.625
   M     - B            -5.639          9.286           24.210
   M     - I            -0.278         16.536           33.350

   B     - S           -31.925        -17.000           -2.075      ***
   B     - M           -24.210         -9.286            5.639
   B     - I           -10.066          7.250           24.566

   I     - S           -41.064        -24.250           -7.436      ***
   I     - M           -33.350        -16.536            0.278
   I     - B           -24.566         -7.250           10.066
```

Problems

Practice

11.13 Determine HSD if MSE $= 1232.56$, $k = 3$, $\alpha = 0.05$, and each sample size is:

a. 6 **b.** 10 **c.** 12 **d.** 15 **e.** 18

11.14 Determine HSD if MSE $= 0.0435$, $k = 5$, $\alpha = 0.01$, $N = 45$, and the sample sizes are:

a. 7 and 8 **b.** 5 and 9 **c.** 11 and 12 **d.** 6 and 12

11.15 Approximate HSD if MSE $= 0.435$, $\alpha = 0.01$, and the sample sizes are 7, 8, 8, 9, and 13.

11.16 In an analysis of variance, it was determined that the treatment means were not all equal. The sample means were $\overline{x}_1 = 37.3$, $\overline{x}_2 = 42.5$, $\overline{x}_3 = 31.2$, and MSE $= 42.007$. All samples had six observations. Using the 0.05 level of significance, determine which treatment means differed.

Applications

11.17 In Problem 11.5*a* there were significant differences among the groups taught by different methods.

a. Perform all multiple comparisons and determine which differences are significant. Use $\alpha = 0.05$.

b. Use the Tukey procedure to construct confidence intervals (based on all pairwise comparisons) for treatments whose means differ significantly with a family confidence level for all such intervals of 95%.

11.18 In Problem 11.6 skulls were removed from three different burial mounds.
   a. Perform all pairwise comparisons at the 0.05 level of significance.
   b. Interpret your results in terms of whether or not the skulls could have come from the same population.

11.19 In Problem 11.8 an analysis was performed on the "waiting times" for several cultures. Use the 0.01 level of significance and perform all pairwise comparisons. Summarize your conclusions.

11.20 Refer to Problem 11.10. Using the 0.05 level of significance, perform all pairwise comparisons. How do these results clarify the results of Problem 11.10?

11.21 Refer to the teaching method results of Problem 11.11.
   a. Obtain all confidence intervals for the true difference in mean scores for each pair of teaching methods. Use a family confidence level of 95%.
   b. How can you use the confidence intervals to determine which teaching methods differ?
   c. Can you tell which teaching method appears to be best? Worst? Summarize your results.

11.22 Use the data of Problem 11.12 and perform all multiple comparisons. Use the 0.05 level of significance. Obtain a 95% confidence interval, using the Tukey procedure, for the difference between the true mean number of years in business for the sizes that differ.

11.4 Additional Topics in Analysis of Variance*

We have considered several procedures to test the assumption of normality. Another assumption that must be met is the assumption of equal treatment variance. If this assumption is violated, the procedure is not valid. Sample variances that could have come from populations with equal variance are said to be **homogeneous.** In this section we discuss a test to determine whether sample variances are homogeneous and then turn to some tests we can use if the assumptions of equal variance and normality are not satisfied.

11.4.1 Testing Homogeneity of Sample Variances

A number of procedures for testing whether the sample variances are homogeneous have been devised. The most notable tests are those designed by Bartlett, Cochran, and Hartley. Of these, Hartley's test is simplest and easiest to understand.

*This section may be omitted without loss of continuity.

We can test the hypothesis that two population variances are equal by taking the ratio of the sample variances. The set of all ratios of sample variances of sample sizes n_1 and n_2 determines an F distribution with $n_1 - 1$ and $n_2 - 1$ degrees of freedom (see Section 8.5). If there are k samples, however, the number of possible comparisons is equal to the number of combinations of k objects taken two at a time; that is

$$\binom{k}{2} = \frac{k(k-1)}{2}$$

We encounter the same problems with multiple comparisons of sample variances as we do with multiple comparisons of sample means. Thus if we conduct ten comparisons at a significance level of 0.05, the probability that all comparisons are correct could be as low as $(0.95)^{10}$ or about 0.60.

Of all possible ratios of the sample variances, one will be largest; this is denoted $F_{\max}$. We define

$$F_{\max} = \frac{s_i^2(\max)}{s_i^2(\min)}$$

where $s_i^2(\max)$ is the largest sample variance and $s_i^2(\min)$ is the smallest sample variance. Hartley compiled critical values of the distribution of the $F_{\max}$ statistic, so the test is called **Hartley's $F_{\max}$** or **Hartley's test.** Tables of critical values of $F_{\max}$ for $\alpha = 0.05$ and 0.01 are given in Table 14. We denote critical values of $F_{\max}$ by $F_{\max(\alpha)}\{k, \nu_i(\max)\}$, where k is the number of treatments and $\nu_i(\max)$ is the maximum number of degrees of freedom in any sample. The sample value of $F_{\max}$ obtained in an analysis is denoted $F^*_{\max}$. Hartley's test assumes that the samples are independent random samples and that the populations are normal. Since this test is quite sensitive to departures from normality, you should not use it if there is reason to suspect that the assumption of normality is not justified. See Neter and others (1985, pp. 618–623) or Winer (1971, pp. 205–210). Neter and colleagues cover the Bartlett test as well, while Winer also discusses the Bartlett and Cochran tests.

Note that the F test used in the analysis of variance is relatively unaffected by unequal variance when the treatment sample sizes are approximately equal, provided that the differences in variance are not unusually large. Thus if the populations are reasonably normal, we can use a low level of α (such as 0.01) for the Hartley test since only large differences among the variances are important. Thus the possibility of a type II error is not so important with this test as with some other tests. See Neter and others (1985, p. 624) for a discussion.

If we conclude that the population variances are not all equal—that is, that the sample variances are not homogeneous—we cannot use the analysis of variance procedure. Sometimes transformations of data are possible, but more often we must use a nonparametric procedure in such a case. One such procedure, the *Kruskal–Wallis H test,* is detailed in this section.

Hartley's Test for Homogeneity of Sample Variance

Suppose we obtain independent random samples of observations from k populations such that:

1. Each of the populations is normal.
2. The population variances are σ_i^2, $i = 1, 2, \cdots, k$.

Then we may test the hypothesis H_0: $\sigma_1^2 = \sigma_2^2 = \cdots = \sigma_k^2$.

The sample value of the test statistic used to test the hypothesis is F^*_{max}, where $F^*_{max} = s_i^2(\text{max})/s_i^2(\text{min})$.

The decision rule is dictated by the level of significance α chosen for the test.

Null and Alternate Hypotheses

$$H_0: \sigma_1^2 = \sigma_2^2 = \cdots = \sigma_k^2$$
$$H_1: \text{not all } \sigma_1^2 \text{ are equal}$$

Decision Rule

$$\text{Reject } H_0 \text{ if } F^*_{max} > F_{max(\alpha)}\{k, \nu_i(\text{max})\}$$

where k is the number of treatments and $\nu_i\{\text{max}\}$ is the maximum number of degrees of freedom in any sample.

Example 11.7

Use the data of Example 11.3 to determine whether the sample variances are homogeneous.

Solution

Using the data of Table 11.2, we find that $s_a^2 = [23{,}124 - (470)^2/10]/9 \doteq 114.889$, $s_b^2 = [42{,}616 - (630)^2/10]/9 \doteq 325.111$, and $s_c^2 = [47{,}518 - (670)^2/10]/9 \doteq 292.000$. Then $F^*_{max} = 325.111/114.889 \doteq 2.83$. From Table 14, we have $F_{max(.05)}\{3, 9\} = 5.34$, so we fail to reject H_0: σ_i^2 are all equal and conclude that the variances are equal; that is, the sample variances are homogeneous.

We can use the SAS procedure PROC MEANS to obtain the sample variances. Use the BY *factor* statement to determine the variances for each sample. If there is more than one variable, specify the variable of interest. If the data are not sorted correctly, use the PROC SORT statement. In this case we might write

```
PROC SORT;
  BY TRT;
PROC MEANS;
  VAR SCORE;
  BY TRT;
```

PROFICIENCY CHECKS

12. Determine the critical value of F_{max} if $\alpha = 0.05$ and the size of each sample is n, where:
a. $n = 8, k = 4$ b. $n = 16, k = 3$ c. $n = 26, k = 5$

13. Determine whether the sample variances are homogeneous ($\alpha = 0.05$) if $n_1 = 8$, $s_1^2 = 0.0556$, $n_2 = 7$, $s_2^2 = 0.1473$, $n_3 = 9$, $s_3^2 = 0.0865$, and $n_4 = 6$, $s_4^2 = 0.0865$.

14. Use the data of Proficiency Check 2 to determine whether the sample variances are homogeneous. Use the 0.05 level of significance.

11.4.2 Nonparametric Analysis of Variance

In the event that one or more of the treatment populations is badly skewed, or the sample variances are not homogeneous, we must use alternative procedures. If the samples are independent random samples and we can determine the overall median of the data, we may use a chi-square test to compare the number in each sample above or below the overall median with the expected number—half of each sample. Another test uses the Studentized range statistic q. (See Neter and others [1985, pp. 636–638].) If we can rank the data, we can use a somewhat more powerful test: the Kruskal–Wallis H Test.

11.4.3 The Kruskal–Wallis *H* Test

If we are comparing k independent random samples, a nonparametric alternative to the analysis of variance procedure is the *Kruskal–Wallis One-Way Analysis of Variance,* or the **Kruskal–Wallis H test,** suggested by W. H. Kruskal and W. A. Wallis (1952). To use this test, we rank all the data as if they were in one sample, smallest to largest, as in the Mann–Whitney U test. Data with the same values are given the average of the ranks occupied. We then compute the sum of the ranks for each sample; R_i is the sum of the ranks in the ith sample. The statistic used for this test is the H statistic, calculated from the formula

$$H = \frac{12}{N(N+1)}\left(\frac{R_1^2}{n_1} + \frac{R_2^2}{n_2} + \cdots + \frac{R_k^2}{n_k}\right) - 3(N+1)$$

where the n_i are the various sample sizes, R_i are the rank sums in each sample, and N is the total number of observations ($N = \sum_i n_i$). For small sample sizes, special tables are needed. (See Owen 1962 or Neave 1978.) If each sample consists of at least five observations, H is distributed approximately as χ^2 with $k - 1$ degrees of freedom and we can use the critical values of χ^2 from Table 6. We can use the H test to test the hypothesis that the treatment medians do not differ significantly.

The Kruskal–Wallis *H* Test

Suppose we obtain k independent random samples of observations from populations with medians $\text{Md}_1, \text{Md}_2, \cdots, \text{Md}_k$. Then we may test the hypothesis

$$\text{H}_0\text{: } \text{Md}_1 = \text{Md}_2 = \cdots = \text{Md}_k$$

The sample value of the test statistic used to test the hypothesis is H^*. If $n_1, n_2, n_3, \cdots, n_k$ are the sample sizes, $R_1, R_2, R_3, \cdots, R_k$ are the rank sums in each sample, and N is the total number of observations $\left(N = \sum_i n_i\right)$, then

$$H^* = \frac{12}{N(N+1)}\left(\frac{R_1^2}{n_1} + \frac{R_2^2}{n_2} + \cdots + \frac{R_k^2}{n_k}\right) - 3(N+1)$$

The decision rule is dictated by the level of significance α chosen for the test.

Null and Alternate Hypotheses

$H_0\text{: } \text{Md}_1 = \text{Md}_2 = \cdots = \text{Md}_k$
$\text{H}_1\text{: not all } \text{Md}_i \text{ are equal}$

Decision Rule

Reject H_0 if $H^* > \chi^2_\alpha\{k-1\}$

Example 11.8

In Example 10.3, a total of 23 applicants for a position is interviewed by three administrators and rated on a scale of 1 to 5 as to suitability for the position. Each applicant is given a score that is the total of the three ratings. Although college education is not a requirement for the position, a personnel director thinks that it might have some bearing on suitability for the position. The 23 applicants are classified into three groups—group X with no college, group Y with some college but no degree, and group Z with a college degree. Use the H test to determine whether there are significant differences among the groups. Use the 0.05 level of significance. Scores are given here:

Group X: 7, 9, 4, 8, 6, 11, 10, 11
Group Y: 11, 12, 9, 9, 10, 10, 8
Group Z: 8, 13, 14, 11, 12, 14, 13, 9

Solution

The null hypothesis is H_0: treatment medians are all equal. We rank the data as shown in Table 11.4. Then $R_1 = 62.5$, $R_2 = 80.0$, and R3 = 133.5; as a check, the sum of the ranks must equal $N(N + 1)/2$. In this case, $N = 23$, $23(24)/2 = 276$, and $62.5 + 80 + 133.5 = 276$. Then

$$H^* = \frac{12}{23(24)}\left(\frac{(62.5)^2}{8} + \frac{(80)^2}{7} + \frac{(133.5)^2}{8}\right) - 3(24)$$
$$\doteq 6.921$$

There are at least five observations in each sample, so we reject H_0 if $H^* > 5.991$, since $\chi^2_{.05}\{2\} = 5.991$. Since $H^* = 6.921$, we conclude that there are differences among the treatment medians.

Table 11.4

SCORES					RANK SUMS		
X	Y	Z	RANKS	COMMON RANK	X	Y	Z
4			1		1		
6			2		2		
7			3		3		
8	8	8	4, 5, 6	5	5	5	5
9	9, 9	9	7, 8, 9, 10	8.5	8.5	17	8.5
10	10, 10		11, 12, 13	12	12	24	
11, 11	11	11	14, 15, 16, 17	15.5	31	15.5	15.5
	12	12	18, 19	18.5		18.5	18.5
		13, 13	20, 21	20.5			41
		14, 14	22, 23	22.5			45
				Totals	62.5	80.5	133.5

PROFICIENCY CHECK

15. Suppose that four samples yield data as shown here. Use the H test with a 0.05 level of significance and determine whether the treatment medians are equal.

I:	2	4	3	4	6	
II:	5	8	5	9	7	
III:	8	10	9	8	10	11
IV:	14	15	13	9	11	

11.4.4 A Multiple-Comparison Procedure

If the H test shows significant differences among the treatments, we can make multiple comparisons to determine differences between pairs of treatments by comparing the mean ranks of the samples, $\overline{R}_i$, where $\overline{R}_i = \overline{R}_i/n_i$. This method uses the Bonferroni procedure, which was introduced in Section 11.3. The **least significant difference** between sample mean ranks $\overline{R}_i$ and $\overline{R}_j$ is equal to LSD, where

$$\text{LSD} = z_{\alpha/2s}\sqrt{\frac{N(N+1)}{12}\left(\frac{1}{n_i} + \frac{1}{n_j}\right)}$$

Further n_i and n_j are the sample sizes and s is the number of comparisons; $s = k(k - 1)/2$, where k is the number of treatments. Note that we must compute LSD for all pairs of treatments if the sample sizes are different. We can find an approximate value of LSD by using the harmonic mean of sample sizes, $\tilde{n}$. Then

$$\text{LSD} = z_{\alpha/2s}\sqrt{\frac{N(N+1)}{6\tilde{n}}}$$

If all sample sizes are equal ($n_i \equiv n$), then $\tilde{n} = n$.

Example 11.9

Use the results of Example 11.8 to determine differences between treatments.

Solution

The mean ranks are $\overline{R}_x = 62.5/8 \doteq 7.81$, $\overline{R}_y = 80.0/7 \doteq 11.43$, and $\overline{R}_z = 133.5/8 \doteq 16.69$. There are three comparisons, so we use $z_{.05/6} = z_{.0083} \doteq 2.39$. Then

$$\text{LSD} = 2.39\sqrt{\frac{23(24)}{12}\left(\frac{1}{n_i} + \frac{1}{n_j}\right)}$$

A Nonparametric Multiple-Comparison Test

Suppose the hypothesis H_0: $\text{Md}_1 = \cdots = \text{Md}_k$ can be rejected at an α level of significance by use of the Kruskal–Wallis H test. We can conclude that $\text{Md}_i \neq \text{Md}_j$ for any two treatments if $|\overline{R}_i - \overline{R}_j|$ exceeds LSD, where

$$\text{LSD} = z_{\alpha/2s}\sqrt{\frac{N(N+1)}{12}\left(\frac{1}{n_i}+\frac{1}{n_j}\right)}$$

In this test n_i and n_j are the sample sizes and $s = k(k-1)/2$, where k is the number of treatments. We can find an approximate value of LSD by using the harmonic mean of sample sizes, $\tilde{n}$. Then

$$\text{LSD} = z_{\alpha/2s}\sqrt{\frac{N(N+1)}{6\tilde{n}}}$$

for any two treatments. If all sample sizes are equal to n, then $\tilde{n} \equiv n$.

Now $|\overline{R}_x - \overline{R}_z| = 8.88$ and both samples are of size 8, so LSD $\doteq$ 8.10. Thus the difference between treatments X and Z is significant. For the other two comparisons, the sample sizes are 7 and 8, so LSD = 8.39; $|\overline{R}_x - \overline{R}_y| = 3.62$, and $|\overline{R}_y - \overline{R}_z| = 5.26$, so these are not significant. The only significant difference is between treatment X and treatment Z; that is, the college graduates appear to receive significantly higher scores than the applicants who did not attend college. For comparison, the approximate value of LSD is found by using $\tilde{n} = 1/[(1/8 + 1/7 + 1/8)/3] \doteq 7.6364$; then LSD $\doteq$ 8.30.

PROFICIENCY CHECK

16. Use the results of Proficiency Check 15 to determine differences between treatments.

11.4.5 Using NPAR1WAY for the *H* Test

We first encountered the SAS procedure NPAR1WAY in Section 10.4 to perform the Mann–Whitney U test. The Kruskal–Wallis H test is a generalization of this test (also known as the Wilcoxon Rank-Sum test), and the statement WILCOXON is used to select the appropriate procedure. The

Figure 11.10
SAS printout of the NPAR1WAY WILCOXON procedure for the job suitability data of Example 11.8.

```
INPUT FILE

DATA KRUSWAL;
  INPUT GRP $ SCORE @@;
CARDS;
A 7 A 9 A 4 A 8 A 6 A 11 A 10 A 11
B 11 B 12 B 9 B 9 B 10 B 10 B 8
C 8 C 13 C 14 C 11 C 12 C 14 C 13 C 9
PROC NPAR1WAY WILCOXON;
  CLASS GRP;
  VAR SCORE;

OUTPUT FILE


                                 SAS

  ANALYSIS FOR VARIABLE SCORE CLASSIFIED BY VARIABLE GRP
              AVERAGE SCORES WERE USED FOR TIES


                WILCOXON SCORES (RANK SUMS)


                         SUM OF    EXPECTED     STD DEV        MEAN
LEVEL            N       SCORES    UNDER HO    UNDER HO       SCORE

A                8        62.50       96.00       15.37        7.81
B                7        80.00       84.00       14.85       11.43
C                8       133.50       96.00       15.37       16.69


KRUSKAL-WALLIS TEST (CHI-SQUARE APPROXIMATION)
CHISQ=    7.03      DF=  2     PROB > CHISQ=0.0298
```

NPAR1WAY procedure uses a modification of the formula for H^* that corrects for ties. This correction results in a small increase in H^*.

Figure 11.10 shows the program and result of applying the NPAR1WAY WILCOXON procedure to the data of Example 11.8. The value of H^* is 7.03, obtained by using the modified procedure, and the P value is given as 0.0298. Thus we reject H_0 and conclude that there are differences in ratings among education levels.

Problems

Practice

11.23 Determine whether the sample variances are homogeneous ($\alpha = 0.05$) if $n_1 = 12$, $s_1^2 = 43.86$, $n_2 = 13$, $s_2^2 = 39.92$, $n_3 = 11$, $s_3^2 = 12.83$, and $n_4 = 12$, $s_4^2 = 49.92$.

11.24 Determine whether the sample variances are homogeneous ($\alpha = 0.05$) if $n_1 = 18$, $s_1^2 = 5432$, $n_2 = 22$, $s_2^2 = 10{,}877$, and $n_3 = 19$, $s_3^2 = 1815$.

11.25 Determine whether the sample variances are homogeneous ($\alpha = 0.01$) if $n_1 = 11$, $s_1^2 = 0.0117$, $n_2 = 11$, $s_2^2 = 0.0036$, $n_3 = 11$, $s_3^2 = 0.0279$, $n_4 = 11$, $s_4^2 = 0.0093$, and $n_5 = 11$, $s_5^2 = 0.0062$.

11.26 Three samples have the following data:

1: 2, 3, 3, 4, 2, 5, 6

2: 5, 6, 4, 7, 8, 9

3: 3, 7, 9, 5, 8, 9, 11

Use the 0.05 level of significance.

a. Determine whether the variances are homogeneous.
b. Test the hypothesis that there are no differences among the treatment medians.
c. If there are differences among the treatment medians, make all multiple comparisons and draw conclusions.

Applications

11.27 Refer to Problem 11.5. In each case determine whether the variances are homogeneous. If they are not, can you perform any other tests without the actual data?

11.28 In Problem 11.6 skulls were obtained from several burial mounds. Determine whether the variances are homogeneous. If they are not, can any other tests be performed?

11.29 In Problem 11.7 the analysis of the data resulted in a conclusion that none of the groups of mice differed significantly in their ability to learn the maze. Would the Kruskal–Wallis Test be likely to result in a significant result? Why?

11.30 Refer to the culture waiting-time results of Problem 11.8.

a. Are the variances homogeneous?
b. Rank the data and perform a nonparametric ANOVA on the data. Use the 0.01 level of significance.
c. What is the approximate observed significance level (P value) for this test?
d. Use the Bonferroni procedure to perform all multiple comparisons at the 0.05 level of significance.
e. Repeat part (*d*) at the 0.01 level of significance.
f. Compare the results of parts (*d*) and (*e*) with the results of Problem 11.19. What do you conclude about the relative power of the Kruskal–Wallis Test compared to the usual ANOVA?

11.31 Refer to Problem 11.11. Perform the Kruskal–Wallis ANOVA on the data, including pairwise comparisons if appropriate. Use the 0.05 level of significance. What do you conclude?

11.5 Minitab Commands for This Chapter

There are two Minitab commands that perform one-way analysis of variance. Their use depends on how the samples of data are stored. The AOVONEWAY command is used when the samples of data are stored in separate columns. When all the sample data are stored in one column and a sample identifier is stored in another column, the ONEWAY command is used. The format for these two commands is similar to that for the two-sample T test in Chapter 8. The AOVONEWAY command is given by

```
MTB > AOVONEWAY for the data in columns C · · · C
```

The ONEWAY command is

```
MTB > ONEWAY using the data in C, sample identifier in C
```

Both of these commands produce the same output.

To form the nonparametric Kruskal–Wallis H test using Minitab, the command is

```
MTB >  KRUSKAL-WALLIS using the data in C, sample identi-
       fier in C
```

Note that this command is similar to the ONEWAY command in that the data are stored in one column and the sample identifier is stored in another column. The following examples illustrate the use of these commands.

Example 11.10

Repeat Example 11.3 using the Minitab command AOVONEWAY.

Solution

The necessary Minitab commands and output are shown in Figure 11.11. The analysis of variance table is the same one we obtained previously. The value of F^* is 4.59; the critical value, $F_{.05}\{2, 27\}$, is 3.35. Since $4.59 > 3.35$, we conclude that the three exams give significantly different results.

The lower part of the output gives a summary of the samples. The sample sizes, sample means, and standard deviations are listed for each sample. This information can be used to perform comparison tests, such as the Tukey multiple-comparison test, that are not available in Minitab. For instance, we may perform Hartley's F_{max} test. For this set of data we have $F^{*}\text{max} = (18.03)^2/(10.72)^2 \doteq 2.83$. Since $F_{max(.05)}\{3, 9\} = 5.34$, and $2.83 < 5.34$, we may conclude that the sample variances are homogeneous.

The pooled standard deviation ($\sqrt{\text{MSE}}$) is also shown in Figure 11.11. Here it is equal to 15.62. Individual 95% confidence intervals using the confidence limits

$$\bar{x}_i \pm t_{.025}\{N - k\}\sqrt{\text{MSE}/n_i}$$

```
MTB > SET the first sample of data into column C1
DATA> 38 54 39 52 63 54 47 52 46 25
DATA> SET the second sample of data into column C2
DATA> 58 44 63 94 72 42 89 68 53 47
DATA> SET the third sample of data into column C3
DATA> 76 51 83 84 51 67 40 89 76 53
DATA> NAME C1 'TEST A' C2 'TEST B' C3 'TEST C'
MTB > AOVONEWAY using the data in columns C1 C2 C3

ANALYSIS OF VARIANCE
SOURCE      DF        S        MS        F
FACTOR       2     2240      1120     4.59
ERROR       27     6588       244
TOTAL       29     8828
                                     INDIVIDUAL 95 PCT CI'S FOR MEAN
                                     BASED ON POOLED STDEV
LEVEL        N     MEAN     STDEV   ----------+---------+---------+------
TEST A      10    47.00     10.72    (-------*--------)
TEST B      10    63.00     18.03              (--------*-------)
TEST C      10    67.00     17.09                 (--------*-------)
                                    ----------+---------+---------+------
POOLED STDEV = 15.62 <-- √MSE                48        60        72
```


Figure 11.11 Minitab commands and output for the paragraph simplification data of Example 11.3.

are also shown. Note that the family confidence level for all three intervals is at least 0.857 (that is, 0.95^3). If all three confidence intervals were desired at a family confidence level of 0.95, we could apply the Bonferroni procedure. We would have to use $t_{.025/3}$ to obtain the intervals. These critical values of t are given in Bailey (1977).

Example 11.11

Use the Minitab command ONEWAY to analyze the job success data of Example 11.5. Recall that the ONEWAY command assumes that the data are in one column and a sample identifier is in another.

Solution

We use the SET command to put the data in column C1. In column C2, again using the SET command, we give all data from the "successful"group a value of 1, the "moderately successful" group a value of 2, and so on. Note that any numbers could be used so long as they are different for each of the four groups. The commands and output are shown in Figure 11.12.

The observed value of F^* is 5.09; the critical value, $F_{.05}\{3, 20\}$, is 3.10. We therefore conclude that at least two of the population means are different. The results agree with those of Example 11.5.

We next illustrate how to use the Minitab output to perform Tukey's multiple-comparison test. Using $\alpha = 0.05$, we obtain

$$\begin{aligned} \text{HSD} &= (q_{.05}\{3, 20\}/\sqrt{2})\sqrt{\text{MSE}(1/n_i + 1/n_j)} \\ &= (3.58/\sqrt{2})\sqrt{120(1/n_i + 1/n_j)} \\ &= 27.73\sqrt{(1/n_i + 1/n_j)} \end{aligned}$$

```
MTB > SET the sample data in column C1
DATA> 64 49 61 57 54 44 49
DATA> 55 60 38 44 41 45 41
DATA> 33 19 64 38 37 31
DATA> 33 22 48 16
DATA> SET the following sample identifier into column C2
DATA> 1 1 1 1 1 1 1 2 2 2 2 2 2 2 3 3 3 3 3 3 4 4 4 4
DATA> NAME C1 'SCORE' C2 'GROUP'
MTB > ONEWAY using the data in C1, with the sample identifier in C2

ANALYSIS OF VARIANCE ON SCORE
SOURCE     DF        SS        MS        F
GROUP       3      1836       612     5.09 ◄— Treatments
ERROR      20      2402       120
TOTAL      23      4238
                                    INDIVIDUAL 95 PCT CI'S FOR MEAN
                                    BASED ON POOLED STDEV
LEVEL       N      MEAN     STDEV   -----+---------+---------+---------+-
    1       7     54.00      7.16                          (------*------)
    2       7     46.29      8.12                   (-------*------)
    3       6     37.00     14.87           (-------*-------)
    4       4     29.75     14.06   (---------*--------)
                                    -----+---------+---------+---------+-
POOLED STDEV = 10.96 ◄— √MSE           24        36        48        60
```

Figure 11.12
Minitab commands and output for the job success data of Example 11.5.

Note that the value depends on the sample sizes of the two samples being compared. Table 11.5 presents the differences in the sample mean and, for each comparison, the sample sizes as well as the HSD value. We conclude that the "successful" group mean is significantly different from the "barely adequate" and "inadequate" group means.

Example 11.12

Use the Minitab command KRUSKAL–WALLIS to analyze the data in Example 11.8.

Solution

The commands and output are given in Figure 11.13. Note that we have obtained bar graphs to illustrate that the data in groups X and Z are skewed

Table 11.5

GROUP	$\lvert x_i - x_j \rvert$	n_i, n_j	HSD
1, 2	$\lvert 54.00 - 46.29 \rvert = 7.71$	7, 7	14.82
1, 3	$\lvert 54.00 - 37.00 \rvert = 17.00$*	7, 6	15.43
1, 4	$\lvert 54.00 - 29.75 \rvert = 24.25$*	7, 4	17.38
2, 3	$\lvert 46.29 - 37.00 \rvert = 9.29$	7, 6	15.43
2, 4	$\lvert 46.29 - 29.75 \rvert = 16.54$	7, 4	17.38
3, 4	$\lvert 37.00 - 29.75 \rvert = 7.25$	6, 4	17.90

*Significant differences.

```
MTB > SET the sample data in column C1
DATA> 7 9 4 8 6 11 10 11
DATA> 11 12 9 9 10 10 8
DATA> 8 13 14 11 12 14 13 9
DATA> SET the sample identifier into column C2
DATA> 1 1 1 1 1 1 1 1 2 2 2 2 2 2 2 3 3 3 3 3 3 3 3
DATA> NAME C1 'SCORE' C2 'SAMPLE'
MTB > HISTOGRAM the data in 'SCORE';
SUBC> BY the sample identifier in 'SAMPLE'.

HISTOGRAM OF SCORE     SAMPLE = 1    N = 8

MIDPOINT     COUNT
       4         1     *
       5         0
       6         1     *
       7         1     *
       8         1     *
       9         1     *
      10         1     *
      11         2     **

HISTOGRAM OF SCORE     SAMPLE = 2    N = 7

MIDPOINT     COUNT
       8         1     *
       9         2     **
      10         2     **
      11         1     *
      12         1     *

HISTOGRAM OF SCORE     SAMPLE = 3    N = 8

MIDPOINT     COUNT
       8         1     *
       9         1     *
      10         0
      11         1     *
      12         1     *
      13         2     **
      14         2     **

MTB > KRUSKAL-WALLIS test of the data in 'SCORE', with samples in 'SAMPLE'

LEVEL       NOBS    MEDIAN   AVE. RANK    Z VALUE
    1          8     8.500         7.8      -2.16
    2          7    10.000        11.4      -0.27
    3          8    12.500        16.7       2.42
OVERALL       23                  12.0

H = 6.921 <------------------------ Kruskal–Wallis test statistic
H(ADJ. FOR TIES) = 7.028
```

Figure 11.13 Minitab commands and output for the applicant score data of Example 11.8.

and not normal. Thus the usual analysis of variance procedure should not be used. Instead we use the nonparametric Kruskal–Wallis Test. Note that since the sample data are contained in one column we use the BY subcommand with the HISTOGRAM command to obtain separate bar graphs for the three samples.

The results of the KRUSKAL–WALLIS command agree with those in Example 11.8. As we noted in Section 11.4.5, there is a modification of the formula for H^* that we should use if the number of ties is large. Both the uncorrected and corrected values of H^* are given in the printout. The critical value is $\chi^2_{.05}\{2\}$, or 5.991. Both of the obtained values of H^*, 6.921 and 7.028, are greater than the critical value, so we conclude that at least two of the population medians are different.

The mean ranks (AVE. RANK) are shown on the printout, and these can be used for multiple comparisons. We obtain the LSD value for the Bonferroni procedure by using the harmonic mean, since the three sample sizes are very close. Then

$$\begin{aligned} \text{LSD} &= z_{\alpha/2s}\sqrt{N(N+1)/6\tilde{n}} \\ &\doteq 2.39\sqrt{23(24)/(6\{7.64\})} \\ &\doteq 2.39(3.4701) \\ &\doteq 8.30 \end{aligned}$$

Since only the mean ranks of groups X and Z differ by more than 8.3, we conclude that the medians for these groups are significantly different. This conclusion confirms the result of Example 11.9.

Problems

Applications

11.32 Refer to Problem 11.12. Use the computer to determine whether the mean mileages are equal for the five brands of gas. Are the variances significantly different? Perform all pairwise comparisons at a 0.05 family level of significance and state your conclusions.

11.33 A traffic checkpoint is established to assess the deterrent effect of four different approaches. The speed limit in the vicinity is 55 mph. For four different days, the following approaches are employed. In each case the speed limit sign is prominently displayed. Method I: speed limit sign only; method II: a sign "speed limits strictly enforced"; method III: a sign "speed radar controlled"; method IV: an empty highway patrol car sitting by the side of the road. Speeds of cars going past are checked by radar. Only out-of-state cars are included in the results to eliminate bias due to local drivers noticing the differences. The speeds of out-of-state cars are recorded for the four days as given below.

METHOD	SPEEDS OF CARS
I	72, 55, 63, 61, 74, 52, 56, 67, 55, 68, 78, 71, 66
II	61, 60, 54, 55, 69, 58, 66, 73, 68, 61, 53
III	73, 60, 55, 62, 53, 49, 66, 62, 55, 53, 59, 61
IV	55, 57, 52, 62, 54, 55, 50, 58, 55, 62, 67

a. Using the appropriate Minitab or SAS method, obtain bar graphs to compare visually the distributions of the speeds for the four different methods. Are the data symmetric or skewed?

b. Conduct Hartley's Test to determine whether the variances are homogeneous. Use $\alpha = 0.05$.

c. Use the appropriate procedure to determine whether the mean speed of cars differs for the four different deterrent methods. Use $\alpha = 0.05$.

d. If you conclude that there are differences, perform the appropriate pairwise comparison tests. What are your conclusions? Use $\alpha = 0.05$.

11.34 Refer to the Miami Female Study data in Appendix B.1. Determine whether the mean number of credit hours differs significantly for the five groups of females with different grade-point averages. To ensure adequate sample sizes in each group, combine the 2.00 and under GPA group with the next higher GPA group. Use $\alpha = 0.05$. Perform all pairwise comparisons and state your conclusions.

11.6 Case 11: The Welcome Relief—Solution

The results of the analysis of variance are shown in Table 11.6. The 0.05 critical value, $F_{.05}\{5, 24\}$, is 2.71; $F^* \doteq 4.57$, so we can conclude that there is a difference in time of relief for at least two of the compounds. For Tukey's multiple-comparison test, since all samples are equal in size, HSD $= q_{.05}\{6, 24\}\sqrt{31.2167/6} \doteq 4.37(2.2810) \doteq 9.97$. Differences between sample means greater than 9.97 then indicate that the treatment means for those populations are not equal. Differences ($|\overline{x}_i - \overline{x}_j|$) are as follows:

GROUP	DIFF.	GROUP	DIFF.	GROUP	DIFF.
A, B	1.0	B, C	3.2	C, E	2.8
A, C	2.2	B, D	4.8	C, F	11.8*
A, D	5.8	B, E	6.0	D, E	10.8*
A, E	5.0	B, F	8.6	D, F	3.8
A, F	9.6	C, D	8.0	E, F	14.6*

*Significant difference.

Table 11.6 ANOVA TABLE FOR CASE II.

SOURCE OF VARIATION	DF	SS	MS
Treatments	5	712.6667	142.5333
Error	24	749.2000	31.2167
Total	29	1461.8667	

The only conclusions we can reach are that relief times are longer for C and E than for F and longer for E than for D. The difference between A and F is almost significant. Since F is clearly better than C and E, it was decided to drop compounds C and E from the program and continue research on the remaining four preparations.

11.7 Summary

To test for differences between means of several treatments, we use the analysis of variance procedure. There are several important assumptions for the use of this procedure: Independent random samples must be obtained from the treatment populations, the populations must be normally distributed, and all population variances must be equal. Analysis of variance provides methods to analyze the total variability of all data in the samples and to determine how much variability is due to actual differences between sample means and how much is caused by variability within the samples. The total variability of all values about the overall mean is measured by the total sum of squares (SSTO) of deviations from that mean. SSTO is partitioned into the sum of squares due to treatments (SSTR), which measures the variability between sample means, and the sum of squares due to error (SSE), which measures the variability within the samples.

In this chapter we showed how to obtain the values of SSTO, SSTR, and SSE and how to use these values to construct F tests and confidence intervals. If we may reject H_0 and conclude that there are differences among the treatment means, we may compare treatments, two at a time, by using Tukey's multiple-comparison test.

We can test the assumption of equality of treatment variances by means of Hartley's F_{max} test. If the treatment populations are not normal, or the treatment variances are not equal, we cannot use the usual analysis of variance procedure. An alternative procedure is the Kruskal-Wallis H test. If we reject H_0 with the H test and conclude that there are differences among the treatment medians, treatments may be compared, two at a time, by using a multiple comparison test.

Chapter Review Problems

Practice

11.35 The ANOVA table given here is partially filled in. Complete the table.

SOURCE OF VARIATION	DF	SS	MS
Treatments	3		0.4589
Error			
Total	35	1.9774	

11.36 An ANOVA table is given here:

SOURCE OF VARIATION	DF	SS	MS
Treatments	4	66.844	16.711
Error	25	116.070	4.643
Total	29	182.941	

Using the 0.05 level of significance, test the hypothesis that there are no differences among the treatment means.

11.37 Four samples have the following data:

1: 8, 13, 9, 12, 7, 9, 11, 14

2: 15, 16, 14, 17, 8, 9, 11

3: 13, 7, 19, 5, 8, 9, 11, 12

4: 11, 16, 16, 14, 21, 17, 18, 22

a. Construct the ANOVA table for the data.
b. Test the hypothesis, at the 0.05 level of significance, that there are no differences among the treatment means.
c. If there are differences among the treatment means, use Tukey's Multiple-Comparison Test to determine which differences are significant.

11.38 Determine whether the sample variances are homogeneous ($\alpha = 0.05$) if $n_1 = 18$, $s_1^2 = 4.83$, $n_2 = 16$, $s_2^2 = 19.75$, $n_3 = 18$, $s_3^2 = 18.25$, and $n_4 = 17$, $s_4^2 = 34.91$.

11.39 Use the Kruskal–Wallis H test to analyze the data of Problem 11.37. Determine which treatments, if any, differ significantly.

Applications

11.40 Five companies submit samples of paint to a firm that is considering the purchase of a large quantity. Six samples of paint are tested to find the paint with the shortest drying time. The drying times are:

COMPANY	DRYING TIMES (MIN)
I	34, 36, 29, 38, 35, 32
II	30, 34, 32, 31, 28, 30
III	27, 32, 31, 30, 34, 30
IV	28, 35, 29, 29, 37, 33
V	34, 31, 36, 38, 40, 37

a. Perform an analysis of variance on the data. Are there significant differences in the mean drying times for the paints? Use the 0.05 level of significance.
b. Perform all pairwise comparisons using the 0.05 level of significance.
c. The firm wants to buy the paint with the shortest mean drying time. Can they determine, from this analysis, which paint to buy, or should they select certain paints for further testing? If so, which ones?

11.41 Stenographers trained in three different systems of stenography are tested to measure their maximum dictation rate in words per minute. Assume that the stenographers are sampled at random.

SYSTEM	MAXIMUM DICTATION RATES
A	147, 188, 162, 144, 157, 179, 165, 180
B	143, 161, 167, 145, 173, 160, 154
C	173, 152, 194, 186, 166, 194, 178, 192, 186

a. Use analysis of variance to determine whether there are differences among the systems. Use the 0.05 level of significance.
b. Use multiple pairwise comparisons to determine which systems are different.
c. Obtain a 90% confidence interval for the mean dictation rate for system C.
d. Do the results provide conclusive evidence as to which system is best? Discuss.

11.42 A medical researcher is testing the effect of a drug in inhibiting the release of adrenalin. She wishes to use the 0.05 level of significance. She randomly selects five samples from volunteers, each sample composed of ten male individuals, and injects the men of the samples with 0, 10, 20, 30, and 40 cc, respectively, of the drug. Each man is then subjected to a stress situation, and the stress level at which adrenalin is released is measured. The data are shown here:

AMOUNT OF DRUG INJECTED (CC)	STRESS LEVEL AT WHICH ADRENALIN IS RELEASED
0	14, 21, 16, 18, 23, 15, 19, 22, 26, 17
10	13, 17, 19, 18, 21, 16, 25, 18, 22, 23
20	16, 22, 18, 23, 22, 25, 19, 17, 21, 25
30	21, 19, 24, 22, 28, 23, 22, 28, 24, 20
40	26, 22, 23, 25, 27, 29, 22, 24, 27, 26

a. There is considerable variation among the variances within each sample. Can you conclude, using the 0.05 level of significance, that the variances are homogeneous?
b. Using the appropriate analysis of variance technique, determine whether there are significant differences among the groups. Use the 0.01 level of significance.
c. Obtain the means for each level of drug and plot them against cubic centimeters of drug injected. Do you notice a downward trend? Chapter 13 presents a method for using the fact that the factor levels are meaningful numbers.

11.43 A comparison of recovery rates of patients suffering from a disease and given three different treatments is shown here:

TREATMENT	DAYS TO RECOVERY
A	3, 8, 6, 9, 7, 4, 9
B	7, 6, 9, 5, 5, 6, 5
C	4, 3, 5, 2, 6, 3, 2

a. Use the 0.05 level of significance to decide whether there is a difference in recovery rates for the treatments.
b. Perform all pairwise comparisons using the 0.05 level of significance.
c. Does the analysis indicate which treatment requires the fewest days to recovery? Justify your answer.

11.44 A researcher wishes to test the mileage of three different makes of automobile. She obtains six of each make, using the same gasoline in each car, and randomly assigns 18 different drivers to the cars. Each car is then driven repeatedly over the same course. The mean mileage for each car is shown here:

GASOLINE	MILES PER GALLON
A	17.6, 13.4, 19.2, 15.7, 13.8, 18.8
B	20.8, 19.7, 21.3, 20.6, 20.1, 22,2
C	21.9, 24.6, 16.7, 20.2, 17.8, 23.4

a. She notices that mileages for gasoline B are much less variable than the other two. Using the 0.05 level of significance, show that the sample variances are not homogeneous.
b. Using the Kruskal–Wallis H test and the 0.05 level of significance, test whether the median mileages for the gasolines are the same.
c. Using the Bonferroni technique, perform all pairwise comparisons. What are your conclusions?

11.45 A teacher wishes to compare the effectiveness of three different types of mathematics instruction. Thirty tenth-grade students of approximately equal ability and background are selected and divided randomly into three groups of ten. Each group is taught by the same teacher but by a different teaching method. After one semester of instruction, a final examination is administered with the results shown here. Does this evidence support a contention that the three methods do not yield equal results? Use the 0.05 level of significance.

METHOD	FINAL EXAMINATION SCORES
A	71, 74, 83, 64, 95, 74, 88, 96, 57, 66
B	73, 59, 84, 91, 75, 68, 75, 83, 94, 87
C	91, 82, 88, 68, 76, 98, 85, 81, 79, 87

11.46 An attitude test is given to 40 participants, ten in each of four educational groups, to determine intensity of commitment toward a hot social issue. Unfortunately, one participant fails to complete the test, and one makes a mistake about educational status and has to be reclassified. Thus results are available for only 39 subjects, as shown here. Are the differences among the groups attributable to chance? Use the 0.05 level of significance.

EDUCATIONAL ATTAINMENT (GRADE LEVEL)	ATTITUDE SCORE
Less than 8	12, 9, 13, 11, 8, 7, 14, 9, 11, 8, 6
8–11	12, 11, 13, 7, 16, 10, 9, 12, 10
12–15	14, 8, 11, 15, 12, 11, 13, 16, 14, 15
16 or more	9, 13, 9, 16, 17, 15, 8, 15, 17

11.47 Four samples of 30 soldiers selected at random from different companies are given sharpshooter tests. The mean number of points per ten shots for each soldier are given here:

COMPANY	RESULTS (MEAN SCORED)	STANDARD DEVIATION
A	$\bar{x}_a = 57.8$	$s_a = 14.3$
B	$\bar{x}_b = 62.3$	$s_b = 19.4$
C	$\bar{x}_c = 58.4$	$s_c = 16.5$
D	$\bar{x}_d = 48.9$	$s_d = 12.6$

a. Do the data substantiate, at the 0.01 level of significance, the contention that the companies are equal in marksmanship?
b. If not, can you determine which company is the best or the worst? Summarize the results.
c. Are the variances homogeneous? Use the 0.05 level of significance.

11.48 Residual levels of DDT, in parts per billion (PPB), are measured in the blood of fish in four different estuaries of a bay. Results are shown here:

ESTUARY	DDT LEVELS (PPB IN THE BLOOD)
A	15, 11, 27, 9, 33, 16, 22, 28, 11, 21, 17, 22
B	6, 21, 9, 13, 11, 10, 15, 13, 17, 12, 8, 13
C	26, 11, 9, 17, 7, 24, 18, 14, 13, 17, 15, 19
D	16, 28, 41, 27, 16, 22, 18, 37, 26, 19, 32, 17

a. There is a considerable difference among the variances of the samples. Can you conclude that the sample variances are homogeneous? Use the 0.05 level of significance.
b. Perform an analysis of variance on the data. At about what significance level can you conclude that there are differences among the DDT levels in the estuaries?
c. Using $\alpha = 0.05$, determine which differences among the sample means are significant. Summarize your results.

11.49 A businessman wishes to use the best technique to display and sell a product. He displays the product in three different ways at each of three comparable stores, changing the display weekly. He obtains six weekly sales reports on each display (two at each store, staggering the weeks) with the results shown here:

DISPLAY	SALES (TOTAL UNITS)
1	86, 79, 83, 81, 75, 79
2	77, 80, 69, 74, 71, 72
3	81, 75, 73, 84, 76, 85

a. Do the data indicate that there are differences among the displays in mean sales? Use $\alpha = 0.05$.
b. Obtain confidence intervals for the differences between each pair of population means. Use the 95% family confidence level. What do you conclude?

11.50 Two independently obtained random samples consisting of ten subjects each are given different training programs designed to improve mental awareness. A third group, the control group, receives no training but is given the test. The test groups have means of 16.8 and 19.6, with variances of 8.61 and 11.72, respectively. The control group, also consisting of ten subjects, has a mean of 14.6 and a variance of 9.44.

a. Are the differences among the samples significant at the 0.01 level?
b. Determine whether either of the experimental groups is significantly different from the control group or from the other experimental group.
c. Are the sample variances homogeneous? Use the 0.05 level of significance.

Index of Terms

Glossary of Symbols

Symbol	Meaning
ANOVA	analysis of variance
ϵ	random error variable
ϵ_{ij}	random error term
F^*	sample value of *F* statistic
$F(\nu_1, \nu_2)$	*F* distribution with ν_1 and ν_2 degrees of freedom
$F_\alpha\{\nu_1, \nu_2\}$	critical value of *F*
F_{max}	Hartley's F_{max} statistic
F^*_{max}	sample value of F_{max}
$F_{max(\alpha)}\{k, \nu_i(\max)\}$	critical value of F_{max}
H^*	sample value in the Kruskal–Wallis test
HSD	Tukey's honestly significant difference
LSD	least significant difference
μ_i	*i*th treatment mean
MSE	mean square for error

MSTR	mean square for treatments	R_i	sum of ranks, *i*th sample
N	total number of observations in all samples	$\overline{R}_i$	mean rank, *i*th sample
n_i	size of *i*th sample	SSE	sum of squares for error
$\tilde{n}$	harmonic mean of sample sizes	SSTR	sum of squares for treatments
q	Studentized range variable	SSTO	total sum of squares
$q_\alpha\{k, N - k\}$	critical value of Studentized range statistic	$\overline{\overline{x}}$	mean of all N observations

Answers to Proficiency Checks

1.

SOURCE OF VARIATION	DF	SS	MS
Diets	2	17.2667	8.6333
Error	27	257.7000	9.5444
Total	29	274.9667	

2. $\Sigma x^2 = 2^2 + 4^2 + 3^2 + 4^2 + 5^2 + \cdots + 13^2 + 9^2 = 1075$; $n_1 = n_4 = 4$, $n_2 = n_3 = 3$, $N = 14$; $T_1 = 13$, $T_2 = 18$, $T_3 = 27$, and $T_4 = 51$; $T = 109$. Therefore

$$\text{SSTO} = 1075 - (109)^2/14 \doteq 1075 - 848.643 = 226.357$$
$$\text{SSTR} = [13^2/4 + 18^2/3 + 27^2/3 + 51^2/4] - (109)^2/14$$
$$\doteq 1043.5 - 848.643 = 194.857$$
$$\text{SSE} \doteq 226.357 - 194.857 = 31.5$$

Thus $N - 1 = 13$, $k - 1 = 3$, $N - k = 10$, and the ANOVA table is as shown:

SOURCE OF VARIATION	DF	SS	MS
Treatments	3	194.857	64.952
Error	10	31.500	3.150
Total	13	226.357	

3. Reject H_0 if $F^* > F_{.05}\{4, N - 5\}$, where: **a.** $F_{.05}\{4, 15\} = 3.06$; **b.** $F_{.05}\{4, 30\} = 2.69$; **c.** $F_{.05}\{4, 60\} = 2.53$; **d.** $F_{.05}\{4, 120\} = 2.45$.

4. Reject H_0 if $F^* > F_{.01}\{4, N - 5\}$, where: **a.** $F_{.01}\{4, 15\} = 4.89$; **b.** $F_{.01}\{4, 30\} = 4.02$; **c.** $F_{.01}\{4, 60\} = 3.65$; **d.** $F_{.01}\{4, 120\} = 3.48$.

5. Since $F_{.05}\{3, 30\} = 2.92$, we reject H_0 if $F^* > 2.92$.
   a. SSTR = 91.4, SSE = 175.1; MSTR = 91.4/3 ≐ 30.467, MSE = 175.1/30 ≐ 5.837, F^* = 30.467/5.837 ≐ 5.22; reject H_0
   b. SSTR = 67.4, SSE = 199.1; MSTR = 67.4/3 ≐ 22.467, MSE = 199.1/30 ≐ 6.637, F^* = 22.467/6.637 ≐ 3.39; reject H_0
   c. SSTR = 55.7, SSE = 210.8; MSTR = 55.7/3 ≐ 18.567, MSE = 210.8/30 ≐ 7.027, F^* = 18.567/7.027 ≐ 2.64; fail to reject H_0
   d. SSTR = 43.8, SSE = 222.7; MSTR = 43.8/3 = 14.60, MSE = 222.7/30 ≐ 7.423, F* = 14.60/7.423 ≐ 1.97; fail to reject H_0

6. F* = 8.633/9.5444 ≐ 0.90; $F_{.05}\{2, 27\} \doteq 3.36$. Thus the differences among the diets are probably due to chance.

7. We wish to test the hypothesis $H_0{:}\mu_1 = \mu_2 = \mu_3 = \mu_4$ at the 0.01 level of significance. Since $F_{.01}\{3, 10\} = 6.55$, we will reject H_0 if $F^* > 6.55$. From the ANOVA table, F* = 64.952/3.15 ≐ 20.62, so we reject H_0 and conclude that the treatment means are not all equal; $P < 0.005$ since $F_{.005}\{3, 10\} = 8.08$.

8. F^* = 2.000/1.500 ≐ 1.33 and $F_{.05}\{3, 8\} = 4.07$. Differences are not significant.

9. a. $q_{.05}\{4, 8\} = 4.53$ b. $q_{.05}\{4, 16\} = 4.05$ c. $q_{.05}\{4, 30\} = 3.84$
   d. $q_{.05}\{4, 50\}$ is halfway between $q_{.05}\{4, 40\}$ and $q_{.05}\{4, 60\}$, or about 3.77.
   e. $q_{.05}\{4, 96\}$ is between $q_{.05}\{4, 60\} = 3.74$ and $q_{.05}\{4, 120\} = 3.69$; since 96 is 36/60, or 0.60 of the way from 60 to 120, and 0.6{3.74 − 3.69} = 0.6{0.05} = 0.03, we have $q_{.05}\{4, 96\} = 3.74 - 0.03$, or 3.71.

10. $q_{.05}\{4, 20\} = 3.96$, so HSD = $(3.96/\sqrt{2})\sqrt{12.56(1/n_i + 1/n_j)} = 2.800\sqrt{12.56(1/n_i + 1/n_j)}$.
   a. HSD = $2.800\sqrt{12.56(1/6 + 1/6)} = 2.800\sqrt{12.56(1/3)} \doteq 5.73$; or HSD = $3.96\sqrt{12.56(1/6)} \doteq 5.73$
   b. HSD = $2.800\sqrt{12.56(1/5 + 1/7)} \doteq 5.81$
   c. HSD = $2.800\sqrt{12.56(1/6 + 1/7)} \doteq 5.52$
   d. HSD = $2.800\sqrt{12.56(1/3 + 1/8)} \doteq 6.72$

11. MSE = 3.15, $q_{.01}\{4, 10\} = 4.33$, so HSD = $(4.33/\sqrt{2})\sqrt{3.15(1/n_i + 1/n_j)} = 3.0618\sqrt{3.15(1/n_i + 1/n_j)}$. The sample sizes are 4 for samples I and IV and 3 for samples II and III. Then $\bar{x}_1 = 3.25$, $\bar{x}_2 = 6.0$, $\bar{x}_3 = 9.0$, and $\bar{x}_4 = 12.75$. The comparisons and values of HSD are shown here:

COMPARISON	DIFFERENCE	HSD
$\bar{x}_1 - \bar{x}_2$	−2.75	4.15
$\bar{x}_1 - \bar{x}_3$	−5.75*	4.15
$\bar{x}_1 - \bar{x}_4$	−9.50*	3.84
$\bar{x}_2 - \bar{x}_3$	−3.00	4.44
$\bar{x}_2 - \bar{x}_4$	−6.75*	4.15
$\bar{x}_3 - \bar{x}_4$	−3.75	4.15

*Significant difference.

In the following diagram, differences that are *not* significant are indicated by drawing a line under them. In this case we list them as follows:

12.75 9.00 6.00 3.75

12. **a.** $F_{\max(.05)}\{4, 7\} = 8.44$ **b.** $F_{\max(.05)}\{3, 15\} = 3.54$
c. $F_{\max(.05)}\{5, 25\} = 3.16$ by interpolation between 3.54 and 2.78

13. $F_{\max(.05)}\{4, 8\} = 7.18$ and $F^*_{\max} = 0.1473/0.0556 \doteq 2.65$, so the variances are homogeneous.

14. The variances are 0.9167, 3.0000, 1.000, and 6.9167; $F^*_{\max} = 6.9167/0.9167 \doteq 7.55$ and $F_{\max(.05)}\{4, 3\} = 39.2$, so the variances are homogeneous.

15. The ranks within each sample are as shown here:

I:	1	3.5	2	3.5	7	
II:	5.5	10	5.5	13	8	
III:	10	15.5	13	10	15.5	17.5
IV:	20	21	19	13	17.5	

Then the rank sums are 17, 42, 81.5, and 90.5. We will reject H_0 if $H^* > 5.991$, and

$$H^* = \frac{12}{21(22)}\left(\frac{(17)^2}{5} + \frac{(42)^2}{5} + \frac{(81.5)^2}{6} + \frac{(90.5)^2}{5}\right) - 3(22)$$
$$\doteq 15.966$$

Thus we reject H_0 and conclude that there are differences among the treatments.

16. The mean ranks are $\overline{R}_1 = 17/5 = 3.4$, $\overline{R}_2 = 42/5 = 8.4$, $\overline{R}_3 = 81.5/6 \doteq 13.58$, and $\overline{R}_4 = 90.5/5 = 18.1$. Since $4(3)/2 = 6$, there are six comparisons, so we use $z_{.05/12} = z_{.0042} \doteq 2.64$. Then

$$\text{LSD} = 2.64\sqrt{\frac{21(22)}{12}\left(\frac{1}{n_i} + \frac{1}{n_j}\right)}$$

For the comparison between treatment 3 and any other treatment, the samples are of sizes 6 and 5, so LSD $\doteq$ 9.92; for the other comparisons, the sample sizes are each 5, so LSD = 10.36. For comparison, the approximate value of LSD is found by using $\tilde{n} = 1/[(1/5 + 1/5 + 1/6 + 1/5)/4] \doteq 5.217$; then LSD $\doteq$ 10.14. Some of the mean rank differences are so close to the approximate LSD that it would be wiser to use the exact values. The only significant differences are between treatment 1 and treatment 3 (10.18) and between treatment 1 and treatment 4 (14.7). Clearly treatment 1 has a lower mean rank than treatments 3 and 4, but there are no other differences.

References

Bailey, J. R. "Tables of the Bonferroni t Statistic." *Journal of the American Statistical Association* 72 (1977): 469–478.

Conover, W. J. *Practical Nonparametric Statistics*. 2nd ed. San Francisco: Holden-Day, 1980.

Duncan, D. B. "*T*-Tests and Intervals for Comparisons Suggested by the Data." *Biometrics* 31 (1975): 339–359.

Dunnett, C. W. "Pairwise Multiple Comparisons in the Homogeneous Variance Unequal Sample Size Case." *Journal of the American Statistical Association* 75 (1980): 372.

Fisher, Ronald A. *The Design of Experiments*. 3rd ed. Edinburgh: Oliver & Boyd, 1942.

Gabriel, K. R. "A Simple Method of Multiple Comparisons of Means." *Journal of the American Statistical Association* 73 (1978): 364.

Kruskal, W. H., and W. A. Wallis. "Use of Ranks in One-Criterion Variance Analysis." *Journal of the American Statistical Association* 47 (1952): 583–621.

Mendenhall, William. *Introduction to Probability and Statistics*. 6th ed. Boston: Duxbury Press, 1982.

Milliken, George A., and Dallas E. Johnson. *Analysis of Messy Data*. Vol. I: *Designed Experiments*. Belmont, CA: Lifetime Learning Publications, 1984.

Neave, H. R. *Statistical Tables*. London: George Allen & Unwin, 1978.

Neter, John, William Wasserman, and Michael H. Kutner. *Applied Linear Statistical Models*. Homewood, IL: Richard D. Irwin, 1985.

Owen, Donald B. *Handbook of Statistical Tables*. Reading, MA: Addison-Wesley, 1962.

Ryan, T. A. "Multiple Comparisons in Psychological Research."*Psychological Bulletin* 56 (1959): 26–47.

Ryan, T. A. "Significance Tests for Multiple Comparison of Proportions, Variances, and Other Statistics." *Psychological Bulletin* 57 (1960): 318–328.

SAS Institute Inc. *SAS® User's Guide: Statistics, Version 5 Edition*. Cary, NC: SAS Institute, 1985.

Scheffe, Henry. *The Analysis of Variance*. New York: John Wiley & Sons, 1959.

Snedecor, George W., and William G. Cochran. *Statistical Methods*. 6th ed. Ames: Iowa State University Press, 1967.

Walpole, Ronald E. *Introduction to Statistics*. 3rd ed. New York: Macmillan, 1983.

Winer, B. J. *Statistical Principles in Experimental Design*. New York: McGraw-Hill, 1971.

12

Two-Factor Analysis of Variance

In Section 8.3 we tested hypotheses concerning differences between two population means. There were two cases: the method used to test the hypothesis when samples are independent is different from the method used for samples that are paired, or matched. Similarly, there are two fundamentally different designs for the analysis of variance procedure. The completely randomized design of Chapter 11 involved independent samples, so we used analysis of variance to test the hypothesis that the treatment means are equal. If the data are classified in two ways and we assign experimental units randomly to cells, the appropriate model is still the completely randomized design. If there are three or more samples and the observations can be matched, however, the design is called a *randomized block* design, and we can use a more sensitive analysis of variance.

We begin by presenting this new design with full analysis techniques, including multiple comparisons and SAS techniques, and a nonparametric alternative. We then show how to analyze data in a completely randomized design in which the observations are classified in two ways (called a *factorial arrangement* of data). Full analysis of this type of experiment includes multiple-comparison procedures and the use of SAS statements. As in Chapter 11, we use SAS procedures throughout the chapter to illustrate how we may use the computer to perform most of these tests; there is a separate section on the use of Minitab.

12.0 Case 12: The Delighted Dietitian

Fred Smith, a dietitian at the Nutrition Research Laboratory, had been working on a new weight reduction diet and was hoping for statistical proof that this new diet was not only very effective but better than the leading diet. He had obtained six random samples of 12 adults each. Groups A and B had been given the leading diet; group B had also been given an exercise program. Groups C and D had been given his diet, and group D was also given the exercise program. There had been two control groups, E and F. Group E had simply been weighed at the start and the conclusion of the experiment, while group F had been given an exercise program but no diet. At the end of 6 weeks, all subjects were weighed and the weight loss for each person was recorded. Four people had gained weight—three in group E and one in group F—and the "weight loss" for these persons had been recorded as negative. The results were as follows:

GROUP	MEAN WEIGHT LOSS (LB)	VARIANCE
A	4.075	1.2220
B	5.375	3.0020
C	5.233	1.8933
D	7.067	2.6533
E	0.408	4.1190
F	1.725	2.9333

The dietitian obtained an analysis of variance table and was able to reject the hypothesis that all treatments are equal at more than the 0.01 level of significance.

SOURCE OF VARIATION	DF	SS	MS	F^*
Treatments	5	360.3161	72.0632	27.33
Error	66	174.0500	2.6371	
Total	71	534.3661		

Since $F_{.01}\{5, 66\}$ is less than 3.34 (about 3.32 by interpolation), it is clear that the treatment means are not all equal. To compare the sample treatment means, Tukey's HSD was computed. It was decided to use the 0.01 level of significance; $q_{.01}\{6, 66\} \doteq 4.98$, so HSD $= 4.98\sqrt{2.6371/12} \doteq 2.33$. Signed differences between sample means are listed here:

COMPARISON	DIFF.	COMPARISON	DIFF.	COMPARISON	DIFF.
D–C	1.834	C–B	−0.142	B–F	3.650*
D–B	1.692	C–A	1.158	B–E	4.967*
D–A	2.992*	C–F	3.508*	A–F	2.350*
D–F	5.142*	C–E	4.825*	A–E	3.667*
D–E	6.659*	B–A	1.300	F–E	1.317

*Significant difference.

When the sample means were arranged in numerical order, a line was used to indicate which differences are not significant:

D B C A F E

(Lines: D–B–C underlined; B–C–A underlined; F–E underlined.)

The conclusions were that his diet with exercise (group D) was more effective than the leading diet without exercise (group A) and no diet, either with exercise (group F) or without exercise (group E), and that either diet was more effective than no diet, both with and without exercise. There was no clear indication, however, that his diet was more effective than the leading diet under the same conditions.

The dietitian looked at his figures. His diet plus exercise was significantly better than the leading diet without exercise, but he wanted to know whether it was the diet or the exercise that made the difference. He consulted a university statistician and explained his problem. After looking over the data, the statistician told him that he actually had two factors—diet and exercise—and not only could they be tested separately, but he could also determine whether there was any relationship between either diet and exercise. He said he would call the dietitian as soon as the results were available.

12.1 ANOVA for the Randomized Block Design

Suppose that a researcher wishes to test the mileage obtained by four different brands of gasoline. If, for instance, 20 automobiles are used, 5 are assigned at random to each gasoline, and the mileages are analyzed, the design used is the *completely randomized design.* Suppose, however, that only 5 automobiles are used, each gasoline is used in each car, and the order of use determined at random. In this design, differences between automobiles are controlled by using each gasoline in each car. Since each automobile is used with each gasoline, automobile is a factor as well as gasoline. The levels of the factor of interest are considered to be the treatments (different gasolines in this case), while the levels of the factor that is used to match the observations are called *blocks.* (In this case, the blocks are the cars.) A **block** is the set of matched observations. If the treatments are assigned at random within each block, the design is called a **randomized block design.**

12.1.1 The Randomized Complete Block

If each treatment appears exactly once within each block, the design is called a **randomized complete block.** Incomplete block designs occur when some of the treatments are not present in every block. We use this strategy when smaller block sizes (such as twins or litter mates) provide more homogeneous experimental units than we would find in larger blocks each with all treatments. Many plans for incomplete block designs are available; see, for example, Cochran and Cox (1957) or Klatworthy (1973). Since analysis of the incomplete block design is not covered here, whenever we refer to the "randomized block design" we mean the randomized *complete* block design.

The model for the randomized block is shown here. Each observation is x_{ij}:

$$x_{ij} = \mu + \beta_i + \tau_j + \epsilon_{ij}$$

where x_{ij} is the value of the observation in the ith block and jth treatment

μ is the overall mean
β_i is the deviation from μ of the ith block mean
τ_j is the deviation from μ of the jth treatment mean
ϵ_{ij} are $N(0, \sigma^2)$
$i = 1, 2, 3, \ldots, n; j = 1, 2, 3, \ldots, k$

Thus each observation x_{ij} is equal to an overall constant μ plus the effect of the ith block, β_i, plus the effect of the jth treatment, τ_j, plus a random error term, ϵ_{ij}. Another important assumption is that the different blocks do not themselves have an effect on the results that differs from treatment to treatment. Such an effect is called *interaction* and is discussed fully in Section 12.2.1. Tests for interaction and situations in which its presence is unimportant are covered in Neter and others (1985, 773–782 and 911–920).

In the analysis of a randomized block experiment we usually wish to determine whether the treatment means are equal (equivalent to testing the hypothesis that all the τ_j are equal to zero). Occasionally we also wish to determine whether the block means are equal (equivalent to testing the hypothesis that all the β_i are equal to zero). Usually we are investigating only the treatment means, so the blocks are used primarily to reduce the variability of experimental error.

In the example with gasoline and automobiles, we assume that we wish to determine whether there are differences between these gasolines and that the choice of automobiles is arbitrary, so that any other automobiles could have been used as well.

There are many uses for a randomized block design. In agriculture, we may want to test the effects of various crop-improvement methods, such as fertilizer or tillage. Different plots of ground may have soil conditions that vary from plot to plot, so it would be difficult to tell what differences were due to fertilizer and what differences were due to soil conditions. If we used each fertilizer on each plot of ground, randomly arranged in equal subplots, each plot would be a randomized complete block and we could analyze differences between fertilizers by the ANOVA method presented here.

12.1.2 Partitioning Total Sum of Squares

In Chapter 11 we analyzed the total variation among the observations to determine how much variation was due to differences between the treatment means and how much was due to variability within the treatments. In the randomized block design there is an additional source of variation: the differences between block means. The analysis in this case consists of partitioning the total sum of squares (SSTO) for the values of the observations into a sum of squares due to treatments (SSTR), a sum of squares due to blocks (SSB), and a sum of squares due to error (SSE). The method is to obtain the SSTR in the same way as with one-way analysis of variance and then to obtain SSB by using the same procedure, considering the blocks as treatments. Then SSE = SSTO − SSTR − SSB. If there are k treatments and n blocks, the degrees of freedom for treatments and blocks are $k - 1$ and $n - 1$, respectively. Total degrees of freedom are $kn - 1$, so degrees of freedom for error are $kn - 1 - (k - 1) - (n - 1) = kn - k - n + 1 = (k - 1)(n - 1)$. The procedure for partitioning the total sum of squares is summarized below. Symbols refer to those shown in Table 12.1. Each x_{ij} is the value of the observation for treatment j, block i; T_j is the sum of the values of the observations for treatment j, B_i is the sum of the values for block i, $T = \sum_i \sum_j x_{ij}$ is the sum of all the values, and

$$\Sigma x^2 = \sum_i \sum_j x_{ij}^2$$

Table 12.1 SYMBOLS USED IN ANOVA FORMULAS.

	TREATMENTS					
BLOCKS	T_1	T_2	T_3	$\cdots$	T_k	B_i
b_1	x_{11}	x_{12}	x_{13}	$\cdots$	x_{1k}	B_1
b_2	x_{21}	x_{22}	x_{23}	$\cdots$	x_{2k}	B_2
b_3	x_{31}	x_{32}	x_{33}	$\cdots$	x_{3k}	B_3
.	.	.	.		.	.
.	.	.	.		.	.
.	.	.	.		.	.
b_n	x_{n1}	x_{n2}	x_{n3}	$\cdots$	x_{nk}	B_n
T_j	T_1	T_2	T_3	$\cdots$	T_k	T

Formulas for Analysis of Variance (Randomized Complete Block)

Total Sum of Squares

$$\text{SSTO} = \Sigma(x - \bar{\bar{x}})^2$$
$$= \Sigma x^2 - \frac{T^2}{N}$$

Sum of Squares for Treatments

$$\text{SSTR} = \sum_j n(\overline{T}_j - \bar{\bar{x}})^2$$
$$= \frac{T_1^2 + T_2^2 + T_3^2 + \cdots + T_k^2}{n} - \frac{T^2}{N}$$

where $N = kn$ and T_j and $\overline{T}_j$ are the total and mean, respectively, of the values of the observations in the jth sample treatment.

Sum of Squares for Blocks

$$\text{SSB} = \sum_i k(\overline{B}_i - \bar{\bar{x}})^2$$
$$= \frac{B_1^2 + B_2^2 + B_3^2 + \cdots + B_n^2}{k} - \frac{T^2}{N}$$

where $N = kn$ and B_i and $\overline{B}_i$ are the total and mean, respectively, of the values of the observations in the ith block.

Sum of Squares for Error

$$\text{SSE} = \text{SSTO} - \text{SSTR} - \text{SSB}$$

Mean Squares

$$\text{MSTR} = \frac{\text{SSTR}}{k - 1}$$

$$\text{MSB} = \frac{\text{SSB}}{n - 1}$$

$$\text{MSE} = \frac{\text{SSE}}{(k - 1)(n - 1)}$$

is the sum of the squares of all values. Further $\overline{T}_j$ is the mean for the treatment j, $\overline{B}_i$ is the mean for block i, and $\overline{\overline{x}}$ is the overall mean.

12.1.3 *F* Tests

As in the analysis of the completely randomized design, we can now test the hypothesis that all *treatment* means are equal by using the test statistic F. The sample value of the test statistic is $F^* = \text{MSTR/MSE}$. We reject the hypothesis if $F^* > F_\alpha\{k - 1, (k - 1) \cdot (n - 1)\}$. We may also test the hypothesis that all *block* means are equal by using $F^* = \text{MSB/MSE}$. We reject the hypothesis if $F^* > F_\alpha\{n - 1, (k - 1) \cdot (n - 1)\}$.

Hypothesis Tests for Randomized Complete Block Design

Suppose we obtain random samples from n blocks and randomly assign one unit from each block to each of k treatments. Suppose further that the following are true:

1. The response to the jth treatment in the ith block is from a normal distribution.
2. The variance of each of the n_k populations is σ^2.
3. There is no interaction between the treatments and the blocks (see Section 12.2.4).

Then we may test the hypotheses that the treatment means are equal and the block means are equal.

The sample value of the test statistic used to test the hypothesis that the *treatment* means are equal is F^*, where $F^* = \text{MSTR/MSE}$; the sample value of the test statistic used to test the hypothesis that the *block* means are equal is F^*, where $F^* = \text{MSB/MSE}$. The decision rules are dictated by the level of significance α chosen for the test.

Null and Alternate Hypotheses

H_0: treatment means are equal	H_0: block means are equal
H_1: not all treatment means are equal	H_1: not all block means are equal

Decision Rules

Reject H_0 if

$F^* > F_\alpha\{k-1, (k-1)\cdot(n-1)\}$	$F^* > F_\alpha\{n-1, (k-1)\cdot(n-1)\}$

The results of the analysis and the F tests are summarized in an ANOVA table such as Table 12.2.

Example 12.1

Four different brands of gasoline, all of the same grade, are used in five cars. Each gasoline is used in each of the cars, all in random order. The mileage for each gasoline for each car is shown in Table 12.3. Using the 0.01 level of significance, determine whether there is a difference in mileage due to gasolines.

Table 12.2 TYPICAL ANOVA TABLE FOR RANDOMIZED BLOCK RESULTS.

SOURCE OF VARIATION	df	SS	MS	F^*
Treatments	$k-1$	SSTR	SSTR/$(k-1)$	MSTR/MSE
Blocks	$n-1$	SSB	SSB/$(n-1)$	MSB/MSE
Error	$(k-1)(n-1)$	SSE	SSE/$(k-1)(n-1)$	
Total	$N-1$	SSTO		

Table 12.3 MILEAGE DATA FOR EXAMPLE 12.1.

	GASOLINE			
CAR	A	B	C	D
1	21.8	22.4	20.6	23.1
2	24.6	24.9	25.6	26.4
3	31.3	34.2	30.6	33.7
4	24.1	25.3	22.4	26.8
5	23.1	27.3	26.1	28.6

Solution

Table 12.4 shows the data of Table 12.3 along with treatment and block totals. Since $\Sigma x^2 = 13{,}952.81$,

$$\text{SSTO} = 13{,}952.81 - (522.9)^2/20 \doteq 281.59$$
$$\text{SSTR} = (124.9^2 + 134.1^2 + 125.3^2 + 138.6^2)/5 - (522.9)^2/20 \doteq 27.35$$
$$\text{SSB} = (87.9^2 + 101.5^2 + 129.8^2 + 98.6^2 + 105.1^2)/4 - (522.9)^2/20 \doteq 239.95$$

Then SSE $\doteq 281.59 - 27.35 - 239.95 = 14.29$. The ANOVA table is shown in Table 12.5.

Now $F_{.01}\{3, 12\} = 5.95$ and $F_{.01}\{4, 12\} = 5.41$; since $7.66 > 5.95$ and $50.41 > 5.41$, we reject the null hypotheses and conclude that there are significant differences among the treatments and significant differences among the blocks. This means that there are nonchance variations among the mileages of the gasolines (treatments) and also nonchance variations among the mileages of the cars. The latter finding was to be expected, since the reason for blocking is to remove from SSE the variation due to differences between cars. Nonetheless, this result confirms the wisdom of the decision to use the randomized block design.

Table 12.4 TOTALS FOR MILEAGE DATA.

	GASOLINE				
CAR	A	B	C	D	B_i
1	21.8	22.4	20.6	23.1	87.9
2	24.6	24.9	25.6	26.4	101.5
3	31.3	34.2	30.6	33.7	129.8
4	24.1	25.3	22.4	26.8	98.6
5	23.1	27.3	26.1	28.6	105.1
T_j	124.9	134.1	125.3	138.6	522.9

Table 12.5 ANOVA TABLE FOR MILEAGE DATA.

SOURCE OF VARIATION	df	SS	MS	F^*
Treatments	3	27.35	9.12	7.66
Blocks	4	239.95	59.99	50.41
Error	12	14.29	1.19	
Total	19	281.59		

PROFICIENCY CHECKS

1. The values of observations for a randomized block are shown here:

	TREATMENT				
BLOCK	1	2	3	4	5
1	4	6	9	2	7
2	15	16	18	9	14
3	8	7	11	5	8
4	19	23	17	12	16

   Obtain the ANOVA table and determine whether the treatment means are different. Use the 0.01 level of significance.

2. An office manager wishes to assess the extent to which experience reduces errors in the handling of paperwork. Twelve clerks are selected at random and classified by experience: I, 0–3 months; II, 3–12 months; III, 1–3 years; IV, over 3 years. There are three clerks in each experience category, and they are randomly reassigned to three departments (A, B, and C) in which none of the twelve has had any experience. Each clerk within a department has about the same number and complexity of forms to process, but the number and complexity differ from department to department. The clerks are monitored for 6 weeks and the total number of errors is recorded. This information is shown here:

	EXPERIENCE CATEGORY			
DEPARTMENT	I	II	III	IV
A	21	12	8	7
B	38	24	11	12
C	16	12	7	4

   Use the 0.05 level of significance to determine whether there are differences in errors due to experience.

12.1.4 Multiple-Comparison Procedures

We may use Tukey's multiple-comparison procedure to perform all pairwise comparisons between treatment means. Since all sample treatments contain n observations, we have $\text{HSD} = q_\alpha\{k, (k-1)\cdot(n-1)\}\sqrt{\text{MSE}/n}$ for differences between sample treatment means.

Example 12.2

Use the results of Example 12.1 to perform all pairwise comparisons of treatment means at the 0.01 level of significance.

Solution

Since $q_{.01}\{4, 12\} = 5.50$, we have HSD $= 5.50\sqrt{1.19/5} \doteq 2.68$. We can compute the differences, and significant differences are shown below. Means connected by a line are not significantly different.

D	B	C	A
27.72	26.82	25.06	24.98

(Line under D, B, C; line under B, C, A.)

We can conclude that gasoline D gets higher mileage than gasoline A. No other differences can be seen.

PROFICIENCY CHECKS

3. Use the results of Proficiency Check 1 to perform all pairwise comparisons at the 0.01 level of significance.
4. Use the results of Proficiency Check 2 to perform all pairwise comparisons at the 0.05 level of significance. Interpret the results of the analysis.

12.1.5 Friedman's Nonparametric ANOVA

If the data are in ranks or the assumptions of equal variances or normality of treatments are not met, we need to use nonparametric analysis of variance. Milton Friedman (1937) has devised a nonparametric analysis of variance for a randomized block: **Friedman's F_r test** or **Friedman's two-way ANOVA.**

We can use Friedman's test only to test differences among the treatments. It can be adapted to test differences among the blocks, but unlike the usual ANOVA it cannot test both with one analysis. Friedman's test is especially useful when a panel of reviewers has ranked a set of objects according to some criterion. Tasters may be asked to rank soft drinks in order of their preference, for instance, and the results can be analyzed to see if there is a difference in preference for the different soft drinks. Tasters are the blocks and the soft drinks are the treatments.

To use the Friedman test, we rank values of the observations, usually from low to high, within each block. (If rankings are from high to low, the resulting value of F_r will be the same.) We give tied observations the mean of the tied ranks as in other nonparametric tests. After we determine the rank sum R_j for each sample, we compute the F_r statistic from the formula

$$F_r = \frac{12}{nk(k+1)} \sum_j R_j^2 - 3n(k+1)$$

where the number of blocks is n and there are k treatments. If there are at least five blocks, F_r is distributed approximately as chi square with $k - 1$ degrees of freedom and we can use the chi-square table of critical values. For smaller numbers of blocks, we need special tables (see Neave 1978).

The Friedman F_r Test

Suppose we obtain random samples from n blocks and randomly assign one unit from each block to each of k treatments with medians Md_1, $\text{Md}_2, \ldots, \text{Md}_k$. Then we may test the hypothesis that the treatment medians are equal. The sample value of the test statistic used to test the hypothesis is F_r^*, where

$$F_r^* = \frac{12}{nk(k+1)} \sum_j R_j^2 - 3n(k+1)$$

and the R_j are the rank sums in each sample.

The decision rule is dictated by the level of significance α chosen for the test.

Null and Alternate Hypotheses

H_0: $\text{Md}_1 = \text{Md}_2 = \cdots = \text{Md}_k$
H_1: not all Md_j are equal

Decision Rule

Reject H_0 if $F_r^* > \chi^2_\alpha\{k - 1\}$, provided $n \geq 5$.

As with the Kruskal–Wallis H test, we can make multiple comparisons using the Bonferroni approach. The least significant difference between mean ranks of two samples, $\overline{R}_i$ and $\overline{R}_j$, is equal to LSD, where

$$\text{LSD} = z_{\alpha/2s}\sqrt{\frac{k(k+1)}{6n}}$$

All samples are of size n, and s is the number of comparisons; $s = k(k-1)/2$, where k is the number of treatments.

Multiple Comparison Following the F_r Test

If we can reject the hypothesis H_0: $\text{Md}_1 = \cdots = \text{Md}_k$ at an α level of significance by use of the Friedman F_r Test, then we can conclude that $\text{Md}_i \neq \text{Md}_j$ for any two treatments if $|\overline{R}_i - \overline{R}_j|$ exceeds LSD. In this case

$$\text{LSD} = z_{\alpha/2s}\sqrt{\frac{k(k+1)}{6n}}$$

where n is the common sample size, $s = k(k-1)/2$, and k is the number of treatments.

Example 12.3

Use the data of Example 12.1 to perform analysis of variance using the Friedman F_r test.

Solution

We reproduce the data of Example 12.1 here:

	GASOLINE			
CAR	A	B	C	D
1	21.8	22.4	20.6	23.1
2	24.6	24.9	25.6	26.4
3	31.3	34.2	30.6	33.7
4	24.1	25.3	22.4	26.8
5	23.1	27.3	26.1	28.6

Within each block (car) we rank the data from low to high, as shown here:

	GASOLINE			
CAR	A	B	C	D
1	2	3	1	4
2	1	2	3	4
3	2	4	1	3
4	2	3	1	4
5	1	3	2	4
R_j	8	15	8	19

Then

$$F_r^* = \frac{12}{5(4)(5)}(8^2 + 15^2 + 8^2 + 19^2) - 3(5)(5)$$

$$= \frac{12}{100}(714) - 75 = 10.68$$

Since $\chi^2_{.01}\{3\} = 11.35$, we conclude that there are no differences among the treatments. Using ANOVA, we did obtain significant differences. This result points out that nonparametric techniques are less powerful than those using parameters. In ranking data on an interval scale we ignore the magnitude of the differences between consecutively ranked observations. Thus we do not generally use nonparametric techniques unless assumptions underlying the usual model are violated or the data are presented in ranks, as in the next example.

Example 12.4

Six tasters are asked to taste five different colas and rank them in order of preference. The rankings are given in Table 12.6. Using the 0.05 level of significance, determine whether there are differences in the preferences.

Solution

Rank sums, respectively, are 25, 15, 9, 14, and 27. Then

$$F_r^* = (12/6\{5\}\{6\})(25^2 + 15^2 + 9^2 + 14^2 + 27^2) - 3(6)(6)$$
$$= (12/180)(1856) - 108 \doteq 15.733$$

Since $\chi^2_{.05}\{4\} = 9.488$, the treatments do differ. There are $5(4)/2 = 10$ pairwise comparisons; $z_{.05/20} = z_{.0025} = 2.81$, so

$$\text{LSD} = 2.81\sqrt{5(6)/6(6)} \doteq 2.57$$

Treatment means are, respectively, 4.17, 2.50, 1.50, 2.33, and 4.50. The only significant differences are between cola 1 and cola 3, and between cola 3 and cola 5. Cola 3 is ranked significantly higher than cola 1 and cola 5.

PROFICIENCY CHECK

5. Five reviewers rate three entertainers by ability to perform before a live audience. The ratings are given in the following table. Determine, at the 0.05 level of significance, whether there are differences in the performers' abilities.

	PERFORMER		
REVIEWER	1	2	3
1	2	1	3
2	3	1	2
3	2	1	3
4	2	1	3
5	1	2	3

Table 12.6 RANKINGS OF COLAS.

	COLA				
RATER	1	2	3	4	5
1	5	3	2	1	4
2	4	2	1	3	5
3	4	1	2	4	5
4	5	2	1	3	4
5	4	3	1	2	5
6	3	5	2	1	4

12.1.6 Using SAS Procedures for the Randomized Complete Block

We may use the SAS procedure ANOVA to analyze the randomized block. The procedures for entering data are similar to those we used previously except that each observation is classified in two ways. Suppose, for example, we wish to enter the data of Table 12.3. Using the most straightforward entry method, we may write

```
DATA TEST;
  INPUT CAR GAS $ MPG @@;
CARDS;
1 A 21.8 1 B 22.4 1 C 20.6 1 D 23.1
2 A 24.6 2 B 24.9 2 C 25.6 2 D 26.2
3 A 31.3 3 B 34.2 3 C 30.6 3 D 33.7
4 A 24.1 4 B 25.3 4 C 22.4 4 D 26.8
5 A 23.1 5 B 27.3 5 C 26.1 5 D 28.6
```

Recall that we use the $ symbol to indicate a nonnumerical variable. There are several other ways to enter the data; see the *SAS User's Guide* for details.

To perform the analysis of variance we write

```
PROC ANOVA;
  CLASSES CAR GAS;
  MODEL MPG = CAR GAS;
```

Using the Tukey procedure, we may obtain all pairwise comparisons on gasolines by writing

```
MEANS GAS / TUKEY;
```

If we want to use different pairwise comparison procedures, the appropriate name may be written instead of TUKEY. See Section 11.3.3. for more details.

The output of the complete program is shown in Figure 12.1. The primary difference between this table and those shown in Chapter 11 is that

Figure 12.1
SAS printout for ANOVA for the gasoline mileage data of Table 12.3.

```
INPUT FILE

DATA  TEST;
  INPUT CAR GAS $ MPG @@;
CARDS;
1 A 21.8 1 B 22.4 1 C 20.6 1 D 23.1
2 A 24.6 2 B 24.9 2 C 25.6 2 D 26.2
3 A 31.3 3 B 34.2 3 C 30.6 3 D 33.7
4 A 24.1 4 B 25.3 4 C 22.4 4 D 26.8
5 A 23.1 5 B 27.3 5 C 26.1 5 D 28.6
PROC ANOVA;
  CLASSES CAR GAS;
  MODEL MPG = CAR GAS;
  MEANS GAS / TUKEY;

OUTPUT FILE

                    SAS

    ANALYSIS OF VARIANCE PROCEDURE
       CLASS LEVEL INFORMATION
    CLASS      LEVELS       VALUES
    CAR           5        1 2 3 4 5
    GAS           4        A B C D

NUMBER OF OBSERVATIONS IN DATA SET = 20

                    SAS

    ANALYSIS OF VARIANCE PROCEDURE


DEPENDENT VARIABLE: MPG

SOURCE               DF   SUM OF SQUARES      MEAN SQUARE

MODEL                 7    266.99250000       38.14178571
ERROR                12     14.53300000        1.21108333
CORRECTED TOTAL      19    281.52550000

SOURCE               DF       ANOVA SS       F VALUE   PR > F

CAR                   4    240.26300000       49.60    0.0001
GAS                   3     26.72950000        7.36    0.0047
```

DF for cars — SS (car)

DF for gas — SS (gas)

```
F VALUE     PR > F     R-SQUARE        C.V.

 31.49        0.0001   0.948378          6.8205
           ROOT MSE                    MPG MEAN
         1.10049231                 16.13500000
```

```
                                SAS

                 ANALYSIS OF VARIANCE PROCEDURE

    TUKEY'S STUDENTIZED RANGE (HSD) TEST FOR VARIABLE: MPG
NOTE: THIS TEST CONTROLS THE TYPE I EXPERIMENTWISE ERROR RATE,
   BUT GENERALLY HAS A HIGHER TYPE II ERROR RATE THAN REGWQ

         ALPHA=0.05 DF=12 MSE=1.21108
         CRITICAL VALUE OF STUDENTIZED RANGE=4.199
         MINIMUM SIGNIFICANT DIFFERENCE=2.0664

 MEANS WITH THE SAME LETTER ARE NOT SIGNIFICANTLY DIFFERENT.

         TUKEY     GROUPING        MEAN       N      GAS

                       A          17.6800     5       D
                       A
            B          A          16.8200     5       B
            B
            B                     15.0600     5       C
            B
            B                     14.9800     5       A
```

SSTR, SSB, and the appropriate values of F^* are shown in the lower portion of the ANOVA table. Note that MS(car) and MS(gas) are not given. We see that F^* for the factor gas is 7.36 with $P = 0.0047$. Thus we reject H_0: mean mileages are the same for all gasolines and conclude that there are differences. The Tukey multiple-comparison procedure is shown, indicating that D and B do not differ significantly and B, C, and A do not differ significantly.

To perform Friedman's F_r Test we use the RANK procedure to rank the observations within each block. For the data of Example 12.3 we write

```
PROC RANK;
  BY CAR;
  RANKS RGAS;
  VAR MPG;
```

The new variable RGAS has the values shown in Example 12.3. We may then group the ranks into different gasolines and obtain the means and sums of the ranks for each gasoline by writing

```
PROC SORT;
  BY GAS;
PROC MEANS N MEAN SUM;
  VAR RGAS;
  BY GAS;
```

The SAS input and output for this program are given in Figure 12.2. We may then use the rank sums (8, 15, 8, 19) to obtain F_r^* and the rank means (1.6, 3.0, 1.6, 3.8) for the multiple-comparison tests. We could write a program to do this, but it is probably just as easy to do it by hand.

NOTE: If one or more blocks are incomplete, the SAS procedure GLM may be used in place of ANOVA.

Figure 12.2
SAS printout to obtain Friedman's F_r Test for the gasoline mileage data of Example 12.3.

```
INPUT FILE

DATA  TEST;
  INPUT CAR GAS $ MPG @@;
CARDS;
1 A 21.8 1 B 22.4 1 C 20.6 1 D 23.1
2 A 24.6 2 B 24.9 2 C 25.6 2 D 26.2
3 A 31.3 3 B 34.2 3 C 30.6 3 D 33.7
4 A 24.1 4 B 25.3 4 C 22.4 4 D 26.8
5 A 23.1 5 B 27.3 5 C 26.1 5 D 28.6
PROC RANK;
  BY CAR;
  RANKS RGAS;
  VAR MPG;
PROC SORT;
  BY GAS;
PROC MEANS N MEAN SUM;
  VAR RGAS;
  BY GAS;

OUTPUT FILE

                                         SAS

VARIABLE     LABEL                                          N        MEAN               SUM
-----------------------------------------GAS=A ------------------------------------------------
RGAS           RANK FOR VARIABLE MPG                        5     1.60000000        8.00000000
-----------------------------------------GAS=B ------------------------------------------------
RGAS           RANK FOR VARIABLE MPG                        5     3.00000000       15.00000000
-----------------------------------------GAS=C ------------------------------------------------
RGAS           RANK FOR VARIABLE MPG                        5     1.60000000        8.00000000
-----------------------------------------GAS=D ------------------------------------------------
RGAS           RANK FOR VARIABLE MPG                        5     3.80000000       19.00000000
```

Problems

Practice

12.1 The following analysis of variance table is partially filled in. Complete the table.

SOURCE OF VARIATION	DF	SS	MS
Treatments			316.40
Blocks	3		
Error	12		12.15
Total		2482.9	

12.2 An analysis of variance table is given here:

SOURCE OF VARIATION	DF	SS	MS
Treatments	2	0.0548	0.0274
Blocks	7	0.0920	0.0131
Error	14	0.0413	0.0030
Total	23	0.1881	

Using the 0.05 level of significance, test the hypothesis that there are no differences among the treatment means.

12.3 A randomized block design leads to the following data:

	TREATMENT			
BLOCK	I	II	III	IV
A	11	17	16	13
B	24	33	34	19
C	9	13	15	11

a. Obtain SSTO, SSA, and SSB.
b. What is SSE?
c. Construct the analysis of variance table for the data.
d. Test the hypothesis, at the 0.05 level of significance, that there are no differences among the treatment means.

12.4 Using Friedman's F_r test, analyze the data of Problem 12.3

Applications

12.5 A manufacturer decides to test the efficiency of three procedures designed to increase the daily output of a machine. Five machines are tested for one day each with each of the three procedures and then without a procedure. For each machine, the procedures are randomly assigned. Outputs for one day for each machine with each procedure are shown here:

	PROCEDURE			
MACHINE	A	B	C	NONE
A	15.6	15.8	14.8	14.4
B	18.2	18.6	17.8	16.7
C	14.3	14.4	12.9	13.1
D	14.9	15.3	14.8	14.4
E	16.7	16.4	15.1	14.7

a. What design is employed here?
b. What is the factor of interest?
c. Is this factor an experimental or classification factor?
d. Obtain SSTO, SS(machines), and SS(procedures).
e. Use ANOVA to determine whether there are differences due to procedure. Use the 0.05 level of significance.
f. Can you determine which procedure yields the highest output? Use the 0.05 family level of significance.

12.6 Several instructors are arguing whether the hour of the class has an effect on the performance of students. Each instructor who has class at 8:00, 10:00, 1:00, and 2:00 (there are three such instructors) gives a quiz to these classes and obtains the median grade on the quiz. The results are given below. Although the classes are not randomly assigned to the time periods and students may not be enrolled in the classes at random, we can use the methods of this section to provide information (although not conclusive evidence) about the effects of different time periods. Bearing these reservations in mind, analyze the results at the 0.05 level of significance.

	HOUR OF CLASS			
INSTRUCTOR	8:00	10:00	1:00	3:00
Jones	17	20	13	16
Smith	14	15	15	17
White	16	18	11	14

a. What is the factor under study? Is this factor an experimental or classification factor?
b. Can you conclude that there is a difference in quiz grades due to hour?
c. Are there differences due to instructor?
d. The variability among median grades in Smith's class is less than in the other instructors' classes. Thus perhaps a nonparametric analysis is appropriate. Repeat parts (*b*) and (*c*) using the appropriate nonparametric analysis.

12.7 Several different advertising displays are developed to market a product. To discover which is the most effective, each of five different displays is used in four different stores (the order determined randomly) and the number of unit sales in one (30-day) month is recorded as shown here:

	DISPLAY				
STORE	A	B	C	D	E
I	325	417	229	356	319
II	154	231	181	203	167
III	567	819	488	592	578
IV	144	207	113	216	149

a. Decide whether any of the displays will yield more sales. Use the 0.05 level of significance.
b. What is the observed significance level (P value) for the test?

12.8 Five companies submit samples of paint for consideration. The prime criterion is drying time. Suppose that a sample of each paint is tested on each of six different surfaces—masonry, smooth drywall, wood, and others. Given the drying times presented below, can you determine, at the 0.05 level of significance, which paint to buy or which paints to test further? (Note that blocks and treatments are interchanged from the previous tables.)

	SURFACE					
COMPANY	A	B	C	D	E	F
I	34	36	29	38	35	32
II	30	32	28	34	31	30
III	30	32	27	34	31	30
IV	29	33	29	37	35	28
V	36	37	31	38	40	34

12.9 Four different candidates for a position are interviewed by seven top executives and rated for several traits on a scale of 1 to 5. The scores are added together to get a rating for each individual by each executive. The candidates are interviewed in random order by each executive.

	CANDIDATES			
RATERS	LEE	JACOBS	WILKES	DELAP
Moore	42	25	29	33
Gaston	28	31	24	29
Heinrich	44	38	40	39
Seldon	33	30	31	28
TerHand	48	44	46	47
Waters	26	28	22	25
Pierce	42	41	37	45

a. If higher scores indicate a better rating, can you determine which candidate should get the job? Use the 0.05 level of significance.
b. Do there appear to be differences among ratings by the executives? Use the 0.05 level of significance.
c. What is the maximum level of significance for both tests?

12.10 A study was conducted in 1985 (unpublished) to investigate whether or not there was a significant difference in grades given by different departments in a university. The investigators proposed to use students as blocks, comparing grades in different departments for the same students. One of the analyses concerned grades earned by students in three different departments in the social science division. A total of 14 students were found who had taken at least two courses in each of these departments, with grade-point averages (GPA) as shown here.

STUDENT	DEPT. A	DEPT. B	DEPT. C
1	3.25	3.60	3.34
2	2.66	3.02	3.14
3	2.54	2.33	2.67
4	3.16	3.00	3.25
5	2.00	2.00	2.17
6	2.40	2.25	3.00
7	2.88	2.75	3.00
8	2.75	1.88	2.50
9	3.00	2.75	3.50
10	4.00	3.70	3.85
11	3.63	3.75	3.70
12	3.50	2.70	3.30
13	2.15	2.50	2.85
14	2.85	2.50	2.30

a. Using a 0.05 level of significance, determine whether there are differences between the departments in GPA.
b. Determine which differences between mean GPAs are significant.

12.2 Two-Way Analysis of Variance

In the randomized block design, we assign experimental units randomly within blocks. If we classify data in several ways and randomly assign experimental units to combinations of factor levels, we again have a completely randomized design. In Chapter 9, we classified *count* data in two ways and analyzed them using the chi-square distribution. A study in which we classify *continuous* data in two or more ways is called a **factorial experiment** (or factorial arrangement of treatments). A **complete factorial experiment** is an experiment in which there are observations for all possible combinations of factor levels.

Suppose, for example, that we wish to compare the effectiveness of two types of advertising displays and we also vary the sale price of an item from location to location, using three different sale prices. This design would be termed a 2×3 factorial experiment. We may select a number of similar stores and assign them randomly to the combinations of display and sale price. If we assign at least one store to each combination, the factorial experiment would be complete and we could obtain information about the effectiveness of the displays and also about the effect of the different sale prices. If there are at least two observations per combination of display and sale price, we may also obtain information about any possible interrelationship between the two factors. In this section we will discuss analysis of variance procedures for a two-factor factorial experiment using a completely randomized design.

12.2.1 ANOVA for a Factorial Experiment

We analyze a two-factor factorial experiment, or simply two-factor experiment, by using two-way analysis of variance. If there is only one observation for each combination of factors, the analysis methods of the previous section apply. For computational purposes we usually designate the factors A and B. Suppose that factor A has a levels and factor B has b levels; then there are ab combinations of factor levels. Each combination of factor levels is called a *treatment,* and each set of sample values for a treatment is called a **cell.** The methods of this section apply only to analyses in which the number of observations is the same in each cell. For analyses in which the number of observations per cell differs, see Neter and others (1985, pp. 746–772) or Winer (1971, pp. 402–425).

We use the methods of one-way analysis of variance to obtain the sum of squares due to factor A (SSA) and the sum of squares due to factor B (SSB). If s_{ij}^2 is the variance of the observations in a cell, then $(n - 1)s_{ij}^2$ is the sum of squares for each cell. We add these sums of squares to obtain the sum of squares due to error (SSE). The remainder of the total sum of squares (SSTO) is called the **sum of squares due to interaction (SSAB).** Two factors are said to **interact** if the difference between treatment means for two given levels of one factor is not the same for all levels of the other factor. Thus factor A does not have the same effect at all levels of factor B, nor does factor B have the same effect at all levels of factor A.

Suppose, for example, an experiment is conducted in which there are two factors, A and B, each with two levels. The means for the sample treatments are shown in Table 12.7. The change from level 1 to 2 of treatment B is 2.6 at level 1 of treatment A and -0.5 (a decrease) at level 2 of treatment A. Similarly the change from level 1 to 2 of treatment A is -4.6 at level 1 of treatment B and -7.5 (both decreases) at level 2 of treatment B. Since the differences are not the same, the factors *interact.* If the analysis shows that this interaction is between the factors themselves (at the population level), the interaction is significant.

The model for the complete two-factor experiment is shown here. Each observation is x_{ijk}:

$$x_{ijk} = \mu + \alpha_i + \beta_j + \alpha\beta_{ij} + \epsilon_{ijk}$$

where x_{ijk} is the kth observation in the ith level of factor A and jth level of factor B

μ is the overall mean
α_i is the deviation from μ of the ith mean of factor A
β_j is the deviation from μ of the jth mean of factor B
$\alpha\beta_{ij}$ is the deviation from $(\mu + \alpha_i + \beta_j)$ of the ijth treatment mean
ϵ_{ijk}, are $N(0, \sigma^2)$
$i = 1, 2, 3, \ldots, a$
$j = 1, 2, 3, \ldots, b$
$k = 1, 2, 3, \ldots, n$

Thus each observation x_{ijk} is equal to an overall constant μ, plus the effect of the ith level of factor A, α_i, plus the effect of the jth level of factor B, β_j, plus the interaction effect $\alpha\beta_{ij}$ when factor A is at the ith level and factor B is at the jth level, plus a random error term, ϵ_{ijk}. We assume for the purposes of this book that only the observed levels of factors A and B are of interest. That is, we do not wish to generalize our conclusions in either case to a larger population of factor levels. This assumption implies that all factors are *fixed.* If this assumption is not true, the analysis of the data will be affected. See Neter and others (1985, pp. 782–791) for a discussion.

Table 12.7 SAMPLE TREATMENT MEANS.

	TREATMENT B	
TREATMENT A	b_1	b_2
a_1	15.7	18.3
a_2	11.3	10.8

The analysis consists of two parts. We first wish to determine whether there are interaction effects—that is, if all $\alpha\beta_{ij} = 0$. If there are no interaction effects, we then proceed to test whether or not the population means for each factor level are equal. The first step in the analysis is to partition the total sum of squares into the sums of squares due to each source of variation.

12.2.2 Partitioning Total Sum of Squares

Analysis of the complete two-factor design consists of partitioning the total sum of squares (SSTO) for the values of the observations into a sum of squares due to factor A (SSA), a sum of squares due to factor B (SSB), a sum of squares due to interaction (SSAB), and a sum of squares due to error (SSE). The method is to obtain the SSA and SSB in the same way as with one-way analysis of variance, treating each factor in turn as the treatment, and then to obtain SSE as the sum of $(n - 1)s_{ij}^2$ for the cell observations. The interaction sum of squares (SSAB) is obtained as SSAB = SSTO − SSA − SSB − SSE. If factor A has a levels and factor B has b levels, the degrees of freedom for factors A and B are $a - 1$ and $b - 1$, respectively. Since each cell will contain $n - 1$ degrees of freedom and there are ab cells, the number of degrees of freedom for error is $ab(n - 1) = abn - ab = N - ab$, where N is the total number of observations. Total degrees of freedom are $N - 1$, so degrees of freedom for interaction are $N - 1 - (a - 1) - (b - 1) - (N - ab) = ab - a - b + 1 = (a - 1) \cdot (b - 1)$.

The procedure for partitioning the total sum of squares is summarized below. Symbols refer to those shown in Table 12.8. Table 12.8 is called an ***AB* Summary Table,** since the actual values of the x_{ijk} are not shown. Each AB_{ij} is the sum of the values of the observations in cell ij; A_i is the sum of the values of the observations for level i of factor A, B_j is the sum of the values for level j of factor B, $T = \sum_i \sum_j \sum_k x_{ijk}$ is the sum of all the values,

Table 12.8 AB SUMMARY TABLE.

	FACTOR *B*					
FACTOR *A*	b_1	b_2	b_3	$\cdots$	b_b	A_i
a_1	AB_{11}	AB_{12}	AB_{13}	$\cdots$	AB_{1b}	A_1
a_2	AB_{21}	AB_{22}	AB_{23}	$\cdots$	AB_{2b}	A_2
a_3	AB_{31}	AB_{32}	AB_{33}	$\cdots$	AB_{3b}	A_3
.	.	.	.		.	.
.	.	.	.		.	.
.	.	.	.		.	.
a_a	AB_{a1}	AB_{a2}	AB_{a3}	$\cdots$	AB_{ab}	A_a
B_j	B_1	B_2	B_3	$\cdots$	B_b	T

and

$$\Sigma x^2 = \sum_i \sum_j \sum_k x_{ijk}^2$$

is the sum of the squares of all values. Further $\overline{A}_i$ is the mean for the factor A level i, $\overline{B}_j$ is the mean for factor B level j, $\overline{AB}_{ij}$ is the mean for treatment ij, and $\overline{\overline{x}}$ is the overall mean.

Formulas for Analysis of Variance (Complete Two-Factor Experiment)

Total Sum of Squares

$$\text{SSTO} = \Sigma(x - \overline{\overline{x}})^2 = \Sigma x^2 - \frac{T^2}{N}$$

Sum of Squares for Factor *A*

$$\text{SSA} = \sum_i bn(\overline{A}_i - \overline{\overline{x}})^2 = \frac{(A_1^2 + A_2^2 + A_3^2 + \cdots + A_a^2)}{bn} - \frac{T^2}{N}$$

where $N = abn$; A_i and $\overline{A}_i$ are the total and mean, respectively, of the values of the observations in the ith level of factor A.

Sum of Squares for Factor *B*

$$\text{SSB} = \sum_j an(\overline{B}_j - \overline{\overline{x}})^2 = \frac{(B_1^2 + B_2^2 + B_3^2 + \cdots + B_b^2)}{an} - \frac{T^2}{N}$$

where $N = abn$; B_j and $\overline{B}_j$ are the total and mean, respectively, of the values of the observations in the jth level of factor B.

Sum of Squares for Error

$$\text{SSE} = \sum_i \sum_j \sum_k (x_{ijk} - \overline{\text{AB}}_{ij})^2 = \Sigma x^2 - \frac{(AB_{11}^2 + AB_{12}^2 + AB_{13}^2 + \cdots + AB_{ab}^2)}{n}$$

where AB_{ij} and $\overline{AB}_{ij}$ are the total and mean, respectively, of the values of the observations in cell ij.

Sum of Squares for Interaction

$$\text{SSAB} = \text{SSTO} - \text{SSA} - \text{SSB} - \text{SSE}$$

Mean Squares

$$\text{MSA} = \frac{\text{SSA}}{a - 1}$$

$$\text{MSB} = \frac{\text{SSB}}{b - 1}$$

$$\text{MSAB} = \frac{\text{SSAB}}{(a - 1)(b - 1)}$$

$$\text{MSE} = \frac{\text{SSE}}{N - ab}$$

12.2.3 *F* Tests

We can test the hypothesis that all interaction effects $\alpha\beta_{ij}$ are equal to zero (that is, that there is no interaction between the factors) by using the test statistic F. The sample value of the test statistic is $F^* = \text{MSAB/MSE}$. We reject the hypothesis if $F^* > F_\alpha\{(a - 1)(b - 1), N - ab\}$. If interaction effects are significant, tests on the main factors do not yield meaningful results. If interaction effects are present, the differences between factor level A means will not be the same for each level of factor B and conversely. Suppose, for example, that we are testing the drying times of several types of paint on several different surfaces. If there is no interaction, this means that the paint which dries the fastest on any one surface will dry the fastest on all surfaces, so we may reach an overall conclusion about the paints. If there is significant interaction, however, it may mean that the same paint may not dry the fastest on the different surfaces. Thus no overall conclusion is possible, and we must investigate the effect of paints by surface.

If there are no interaction effects, we may also test the hypotheses that factor level means are equal. We test the hypothesis that all factor A means are equal by using $F^* = \text{MSA/MSE}$, rejecting the hypothesis if $F^* > F_\alpha\{a - 1, N - ab\}$. We test the hypothesis that all factor B means are equal by using $F^* = \text{MSB/MSE}$, rejecting the hypothesis if $F^* > F_\alpha\{b - 1, N - ab\}$.

The results of the analysis and the F tests are summarized in an ANOVA table such as Table 12.9.

Table 12.9 ANOVA TABLE FOR TWO-FACTOR ANALYSIS.

SOURCE OF VARIATION	DF	SS	MS	F^*
Factor A	$a-1$	SSA	SSA/$(a-1)$	MSA/MSE
Factor B	$b-1$	SSB	SSB/$(b-1)$	MSB/MSE
Interaction (AB)	$(a-1)(b-1)$	SSAB	SSAB/$(a-1)(b-1)$	MSAB/MSE
Error	$N-ab$	SSE	SSE/$(N-ab)$	
Total	$N-1$	SSTO		

Hypothesis Tests for Interaction and Main Effects (Complete Two-Factor Experiment)

Suppose we obtain independent random samples of n observations from ab populations classified in a levels of factor A and b levels of factor B and that both of the following are true:

1. Each population is normal.
2. The variance of each population is σ^2.

Then we may test the hypotheses that the factor level means are equal for each factor and there is no interaction between the factors.

The sample value of the test statistic used to test the hypothesis that there is no interaction between the factors is F^*, where F^* = MSAB/MSE. If there are no interactions we can test whether factor level means are equal for each factor. The sample value of the test statistic used to test the hypothesis that the means of the levels of factor A are equal is F^*, where F^* = MSA/MSE. The sample value of the test statistic used to test the hypothesis that the means of the levels of factor B are equal is F^*, where F^* = MSB/MSE. The decision rules are dictated by the level of significance α chosen for the test.

Null and Alternate Hypotheses

H_0: there is no interaction between the factors
H_1: there is interaction between the factors

H_0: means are the same for all levels of factor A (B)
H_1: not all factor level means are equal

Decision Rules

Reject H_0 if
$F^* > F_\alpha\{(a-1)\cdot(b-1)\ N-ab\}$

Reject H_0 if
$F^* > F_\alpha\{a-1, N-ab\}$
(for factor A)
$F^* > F_\alpha\{b-1, N-ab\}$
(for factor B)

Example 12.5

Cancer patients often receive chemotherapy, radiation treatments, or both. To determine the effect of these treatments, either separately or in combination, a hospital obtains the records of patients who are undergoing one or both treatments. Four levels of chemotherapy treatments (none, light, moderate, heavy) are identified as well as the number of radiation treatments received during the past year (0 through 4). Three patients are randomly selected from those having received each treatment (combination of factor levels). The number of days each patient is hospitalized is recorded (Table 12.10). Use the 0.01 level of significance and analyze the results to determine whether there is a difference in days hospitalized for the main factors and whether there is any interaction between the factors.

Solution

We first write the *AB* summary table (Table 12.11), listing the cell totals, and obtain Σx^2. We have $\Sigma x^2 = 0^2 + 2^2 + 0^2 + 1^2 + 0^2 + 3^2 + \cdots + 11^2 + 9^2 + 12^2 = 1508$. Then

$$\text{SS(chemotherapy level)} = \frac{33^2 + 45^2 + 58^2 + 112^2}{15} - \frac{(248)^2}{60}$$

$$\doteq 1268.1333 - 1025.0667 = 243.0667$$

$$\text{SS(number of treatments)} = \frac{24^2 + 37^2 + 44^2 + 63^2 + 80^2}{12} - \frac{(248)^2}{60}$$

$$\doteq 1187.5 - 1025.0667 = 162.4333$$

$$\text{SSE} = 1508 - \frac{2^2 + 4^2 + 6^2 + 9^2 + 12^2 + \cdots + 22^2 + 27^2 + 32^2}{3}$$

$$= 1508 - 1446 = 62$$

Table 12.10 DATA FOR EXAMPLE 12.5.

	DAYS HOSPITALIZED				
	NUMBER OF TREATMENTS				
CHEMOTHERAPY LEVEL	0	1	2	3	4
None	0	1	3	4	3
	2	0	1	2	5
	0	3	2	3	4
Light	1	1	2	4	6
	1	3	3	3	7
	2	2	3	5	2
Moderate	2	3	1	5	6
	1	2	4	6	8
	3	3	3	4	7
Heavy	4	5	6	8	11
	5	8	7	10	9
	3	6	9	9	12

Table 12.11 AB SUMMARY TABLE FOR EXAMPLE 12.5.

	TREATMENTS					
CHEMOTHERAPY	0	1	2	3	4	TOTALS
None	2	4	6	9	12	33
Light	4	6	8	12	15	45
Moderate	6	8	8	15	21	58
Heavy	12	19	22	27	32	112
Totals	24	37	44	63	80	248

Since SSTO $= 1508 - (248)^2/60 \doteq 1508 - 1025.066 = 482.9333$, then

$$\begin{aligned}\text{SS(interaction)} &= 482.9333 - 243.0667 - 162.4333 - 62 \\ &= 15.4333\end{aligned}$$

The analysis of variance table is given in Table 12.12.

Now $F_{.01}\{12, 40\} = 2.66$, and 0.830 is considerably less than 2.66, so there is no interaction between the factors. Since $F_{.01}\{3, 40\} = 4.31$ and $F_{.01}\{4, 40\} = 3.83$, both factors are significant. That is, there are differences among the factor level means for each factor. Further comparisons will be made in Section 12.3 (Example 12.9); the implications of no interaction are discussed further in Example 12.7.

Example 12.6 A large firm has many stores and wants to determine whether education or experience is a better indication of potential success in a store manager. From its large number of stores, eight managers are selected at random from each of three categories: H (completed high school or less), C (attended college but did not graduate), and B (received a bachelor's degree). Within each category the managers are further divided according to total experience in sales. Equal cell sizes are obtained by having categories as follows: I (less than 5 years sales experience), II (5 but less than 10 years sales experience), III (10 but less than 15 years sales experience), IV (at least 15 years sales experience). To assess degree of success, a formula based on such factors as

Table 12.12 ANOVA TABLE FOR EXAMPLE 12.5.

SOURCE OF VARIATION	DF	SS	MS	F^*
Chemotherapy	3	243.0667	81.0222	52.272
No. of treatments	4	162.4333	40.6083	26.199
Interaction	12	15.4333	1.2861	0.830
Error	40	62.0000	1.5500	
Total	59	482.9333		

Table 12.13 SUCCESS RATING FOR EXAMPLE 12.6.

EXPERIENCE	EDUCATION LEVEL H	C	B
I	0.84	0.92	1.08
	0.89	0.86	1.12
II	0.94	0.96	1.12
	0.97	0.97	1.08
III	1.02	1.08	1.03
	0.99	1.03	1.06
IV	1.13	1.00	1.04
	1.01	1.04	0.98

stock, location, and number of personnel (a *regression* equation) is developed to predict what sales should have been at that store during the preceding year. For each store manager the ratio of actual sales to predicted sales is determined (a success ratio); the results are presented in Table 12.13. Determine, using the 0.05 level of significance, whether any relationship exists between education, experience, and success as a store manager.

Solution

We first write the *AB* summary table (Table 12.14), listing the cell totals, and obtain Σx^2. We have $\Sigma x^2 = 0.84^2 + 0.89^2 + 0.92^2 + \cdots + 1.04^2 + 1.04^2 + 0.98^2 = 24.4652$. Then

$$\text{SS(experience)} = \frac{5.71^2 + 6.04^2 + 6.21^2 + 6.20^2}{6} - \frac{(24.16)^2}{24}$$

$$\doteq 24.3483 - 24.3211 = 0.0272$$

$$\text{SS(education)} = \frac{7.79^2 + 7.86^2 + 8.51^2}{8} - \frac{(24.16)^2}{24}$$

$$\doteq 24.3605 - 24.3211 = 0.0394$$

Table 12.14 AB SUMMARY TABLE FOR EXAMPLE 12.6.

EXPERIENCE	EDUCATION LEVEL H	C	B	TOTALS
I	1.73	1.78	2.20	5.71
II	1.91	1.93	2.20	6.04
III	2.01	2.11	2.09	6.21
IV	2.14	2.04	2.02	6.20
Totals	7.79	7.86	8.51	24.16

Table 12.15 ANOVA TABLE FOR EXAMPLE 12.6.

SOURCE OF VARIATION	DF	SS	MS	F^*
Experience	3	0.0272	0.0091	6.370
Education	2	0.0394	0.0197	13.827
Interaction	6	0.0604	0.0101	7.063
Error	12	0.0171	0.0014	
Total	23	0.1441		

$$\text{SSE} = 24.4652 - \frac{1.73^2 + 1.78^2 + 2.20^2 + 1.91^2 + 1.93^2 + \cdots + 2.14^2 + 2.04^2 + 2.02^2}{2}$$

$$\doteq 24.4652 - 24.4481 = 0.0171.$$

Since SSTO $= 24.4652 - (24.16)^2/24 \doteq 24.4652 - 24.3211 = 0.1441$, then

$$\text{SS(interaction)} = 0.1441 - 0.0272 - 0.0394 - 0.0171 = 0.0604$$

The analysis of variance table is given in Table 12.15. Now $F_{.01}\{6, 12\} = 4.82$, and 7.063 is considerably greater than 4.82, so we conclude that there is significant interaction between the factors. Tests on main effects are not appropriate.

PROFICIENCY CHECKS

6. Obtain an analysis of variance table for the data shown below. Test the hypothesis that there is no interaction between the factors and, if appropriate, the hypotheses that the factor level means are equal. Use the 0.05 level of significance.

	FACTOR *B*	
FACTOR *A*	*A*	*B*
1	1	2
	2	3
	2	4
	2	2
	1	3
2	4	7
	3	8
	1	7
	3	6
	2	8

(*continued*)

PROFICIENCY CHECKS *(continued)*

7. Obtain an analysis of variance table for the data shown below. Test the hypothesis that there is no interaction between the factors and, if appropriate, the hypotheses that the factor level means are equal. Use the 0.01 level of significance.

	FACTOR *B*		
FACTOR *A*	*A*	*B*	*C*
I	2	5	8
	4	6	9
	3	8	11
II	6	8	11
	7	10	15
	6	9	16

12.2.4 Interaction Diagrams

Diagrams depicting the interaction between factors can be quite helpful in interpreting the results. Consider the sample treatment means of Table 12.7 (reproduced here in Table 12.16).

Two interaction diagrams are possible. We could graph the means by using factor A as the horizontal axis or factor B as the horizontal axis. In either case, the vertical axis is the variable x. The two diagrams for Table 12.16 are shown in Figure 12.3. If there were no interaction, the lines in each case would be nearly parallel. The graph on the right shows that there

Table 12.16 SAMPLE TREATMENT MEANS.

	TREATMENT *B*	
TREATMENT *A*	b_1	b_2
a_1	15.7	18.3
a_2	11.3	10.8

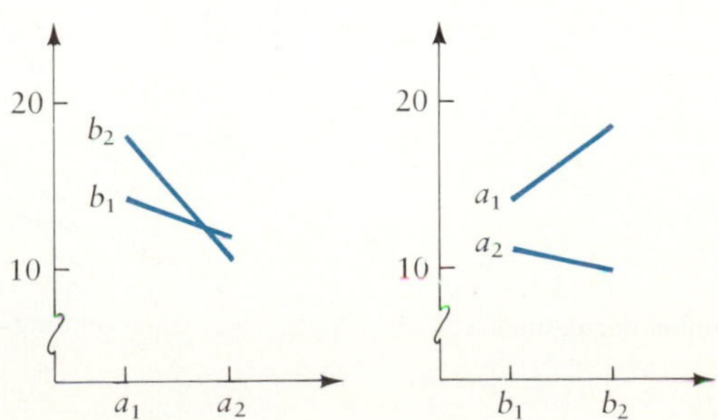

Figure 12.3 Interaction diagrams for the same data set: factor levels of a are along the horizontal axis (left); factor levels of b are along the horizontal axis (right).

may be differences between levels of factor A, although the differences are not the same for each level of factor B. This graph shows differences due to changes in *magnitude* —the size of each mean. The graph on the left clearly shows that there can be no differences between the levels of factor B, since the lines intersect. The mean for b_2 is greater than the mean for b_1 at level a_1, while at level a_2 the opposite is true. This is a result of changes in *rank*—the numerical order of the means of the levels of factor B changes for different levels of factor A.

Example 12.7

Draw the interaction diagrams for the data of Example 12.5. Interpret the results.

Solution

The cell means are shown in Table 12.17. The interaction diagrams are shown in Figure 12.4.

The diagram on the left shows, in general, a steady increase in days in the hospital as number of treatments increases, as well as nearly the same differences between chemotherapy levels at every number of treatments and a greater difference between M and H in each case. The diagram on the right shows a steady increase in days in the hospital as chemotherapy level increases, with a sharp increase from level M to level H; differences between treatment levels are approximately the same at every level of chemotherapy. A minor difference is that two radiation treatments and chemotherapy level

Table 12.17 CELL MEANS FOR EXAMPLE 12.5 AND 12.7.

	TREATMENTS				
CHEMOTHERAPY	0	1	2	3	4
None	0.67	1.33	2.00	3.00	4.00
Slight	1.33	2.00	2.67	4.00	5.00
Moderate	2.00	2.67	2.67	5.00	7.00
Heavy	4.00	6.33	7.33	9.00	10.67

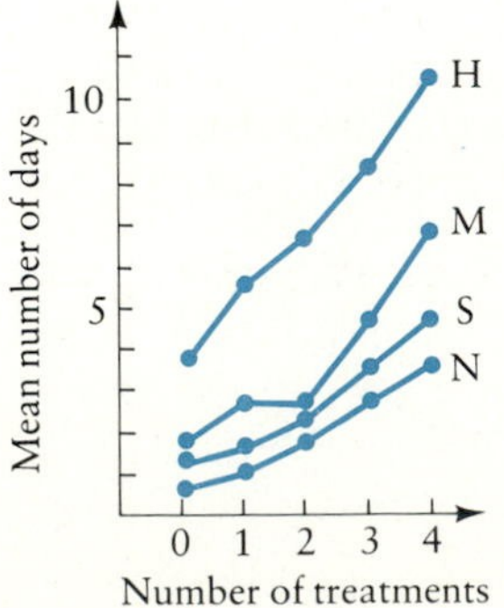

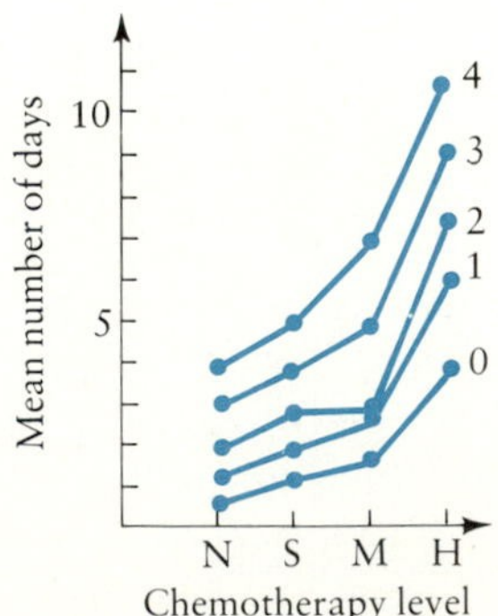

Figure 12.4 Interaction diagrams for the chemotherapy and radiation data of Examples 12.5 and 12.7.

Table 12.18 CELL MEANS FOR EXAMPLES 12.6 AND 12.8.

	EDUCATION LEVEL		
EXPERIENCE	H	C	B
I	0.865	0.890	1.100
II	0.955	0.965	1.100
III	1.015	1.055	1.045
IV	1.070	1.020	1.010

M appear to be slightly out of step with the rest, but overall there is no interaction effect.

When there is no interaction effect, the factor effects are said to be **additive.** That is, the value of each observation is equal to the sum of the overall mean, the effect of factor A, the effect of factor B, and the random error term.

Example 12.8

Draw the interaction diagrams for the data of Example 12.6. Interpret the results.

Solution

The cell means are shown in Table 12.18. The interaction diagrams are shown in Figure 12.5. (Recall that the mean success ratio on the vertical axis is the mean ratio of actual sales to predicted sales). On the left, high school graduates (groups H and C) seem to profit from experience, while college graduates (group B) do not. On the right, experience groups I and II appear to profit from college, while experience groups III and IV do not. One interpretation may be that a college degree can be a substitute initially for experience.

A word of caution in a study such as this one is appropriate, however. This is not a *longitudinal* study—that is, a study of the same people over a period of time. It is rather a *cross-sectional* study—a study of different people at different points in their lives and careers, but at the same point in time. Thus there are many other factors that may account, in part, for

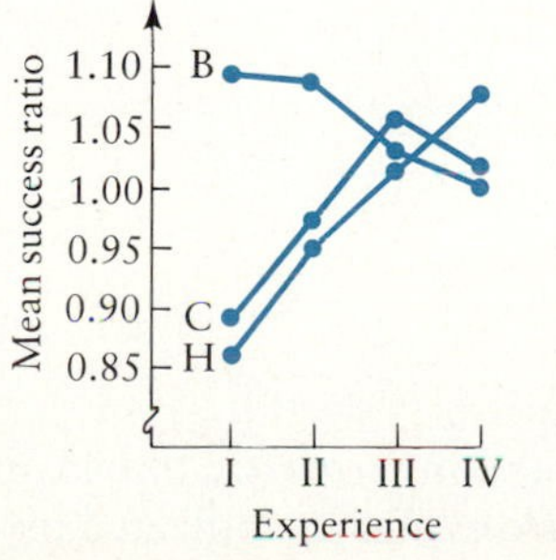

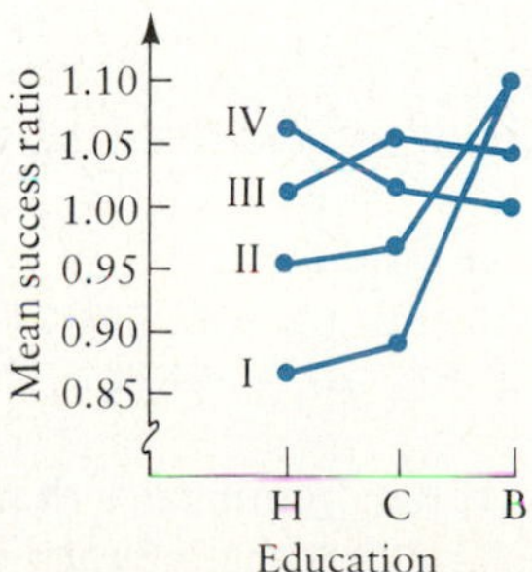

Figure 12.5
Interaction diagrams for the job success ratios of Examples 12.6 and 12.8.

some of the differences. Those with recent college experience may well have studied different subjects than those whose college days are long past. Moreover, those with more experience may have entered their careers from other fields. It is possible that college graduates had success early on and perhaps did not stay in the sales force but moved up to management positions, whereas the less educated salesmen probably stayed on and only the survivors who were good salesmen initially were represented later on. Essentially the "good" college graduates were culled out of the sales force contrasted to the "bad" salespeople who were culled out for the two categories with less education.

PROFICIENCY CHECKS

8. Draw the interaction diagrams for the data of Proficiency Check 6, and interpret the results.
9. Draw the interaction diagrams for the data of Proficiency Check 7, and interpret the results.

12.2.5 Using SAS Procedures for Two-Way Analysis of Variance

The SAS statements for two-way analysis of variance differ very little from those in Section 12.1.6. There are two primary differences. First, since each cell contains more than one observation, we can use the double-input statement illustrated earlier to good effect; second, provisions must be made for the interaction terms. The data of Example 12.6, for instance, may be entered as follows:

```
DATA TWOWAY;
  INPUT EXPER $ EDUC $ RATIO @@;
CARDS;
I H 0.84 I H 0.89 I C 0.92 I C 0.86 I B 1.08 I B 1.12
II H 0.94 II H 0.97 II C 0.96 II C 0.97 II C 1.12 II C 1.08
. . .
```

Once we have entered the data, we write

```
PROC ANOVA;
  CLASSES EXPER EDUC;
MODEL RATIO = EXPER | EDUC;
```

The vertical bar (|) indicates that all interaction terms are to be computed. We can ensure that the interaction terms are explicitly indicated by writing

```
MODEL RATIO = EXPER EDUC EXPER * EDUC;
```

Either method is satisfactory.

We may test the assumption of equality of error variance by obtaining the cell variances. To ensure that the data are sorted properly, we write

```
PROC SORT;
  BY EXPER EDUC;
PROC MEANS;
  VAR RATIO;
  BY EXPER EDUC;
```

We then look at each of the variances and obtain the value of F_{max} for the experiment. Then we compare with the critical value to determine whether the variances are homogeneous.

We illustrate use of the SAS statements by obtaining the printout for the data of Example 12.6 (see Figure 12.6). Since F^* for interaction is 6.24, with a P value of 0.0046, we conclude that interaction is significant. We may draw the interaction diagrams, then, to determine whether comparisons on main effects are reasonable (changes are of magnitude only) or whether comparisons should be made on treatment effects (interactions are due to changes in rank order). We may obtain the cell means by using the MEANS option and writing

```
MEANS EXPER * EDUC;
```

NOTE: If cell sizes are not all equal, the SAS procedure GLM may be used instead of ANOVA.

Figure 12.6
SAS printout for two-factor ANOVA for the manager success data of Example 12.6.

```
INPUT FILE

DATA TWOWAY;
  INPUT EXPER $ EDUC $ RATIO @@;
CARDS;
I H 0.84 I H 0.89 I C 0.92 I C 0.86 I B 1.08 I B 1.12
II H 0.94 II H 0.97 II C 0.96 II C 0.97 II B 1.12 II B 1.08
III H 1.02 III H 0.99 III C 1.08 III C 1.03 III B 1.03 III B 1.06
IV H 1.13 IV H 1.01 IV C 1.00 IV C 1.04 IV B 1.04 IV B 0.98
PROC ANOVA;
  CLASSES EXPER EDUC;
  MODEL RATIO = EXPER | EDUC;

OUTPUT FILE

                              SAS
              ANALYSIS OF VARIANCE PROCEDURE
                  CLASS LEVEL INFORMATION
              CLASS      LEVELS     VALUES
              EXPER         4       I II III IV
              EDUC          3       B C H

            NUMBER OF OBSERVATIONS IN DATA SET = 24
```

```
                                    SAS

                     ANALYSIS OF VARIANCE PROCEDURE

DEPENDENT VARIABLE: RATIO

SOURCE            DF   SUM OF SQUARES   MEAN SQUARE   F VALUE      PR > F   R-SQUARE          C.V.

MODEL             11       0.12267174    0.01115198      7.74      0.0010   0.885578        3.7828

ERROR             11       0.01585000    0.00144091              ROOT MSE               RATIO MEAN

CORRECTED TOTAL   22       0.13852174                          0.03795931               1.00347826

SOURCE          DF       ANOVA SS     F VALUE     PR > F

EXPER            3     0.02405174        5.56     0.0143
EDUC             2     0.04466102       15.50     0.0006
EXPER*EDUC       6     0.05395898        6.24     0.0046
```

DF for interaction — SS (interaction)

Problems

Practice

12.11 The following analysis of variance table is partially filled in. Complete the table.

SOURCE OF VARIATION	DF	SS	MS
Factor A			831.70
Factor B	3		
Interaction		2,884.50	
Error	144	103,258.17	
Total	167	112,876.25	

12.12 An analysis of variance table is given here:

SOURCE OF VARIATION	DF	SS	MS
Factor A	3	0.1288	0.0429
Factor B	7	0.0821	0.0117
Interaction	21	0.1166	0.0056
Error	64	0.3914	0.0061
Total	95	0.7189	

Use the 0.05 level of significance for each test.

a. Is there interaction between the variables?

b. Can you conclude that there are differences among the levels of factor A?

c. Can you conclude that there are differences among the levels of factor B?

d. What is the upper level of significance for all three tests?

12.13 Means for the cells in a complete two-factor study are shown here. Construct the interaction diagrams.

FACTOR A	FACTOR B	
	A	B
1	36.9	28.7
2	29.9	31.4

a. Does there appear to be an interaction between the variables?
b. Is the interaction due to differences in rank or differences in magnitude?
c. If the interaction is significant, would it be reasonable to conduct tests on factors A and B? Why?

12.14 Means for the cells in a complete two-factor study are shown here. Construct the interaction diagrams.

FACTOR A	FACTOR B		
	A	B	C
1	1.12	1.18	1.13
2	1.65	1.45	1.38
3	1.34	1.28	1.15
4	1.35	1.30	1.19

a. Does there appear to be an interaction between the factors?
b. If there is significant interaction, would it be due to differences in rank or differences in magnitude?
c. If there is significant interaction, would it be reasonable to conduct F tests for differences between factor levels? For factor A, factor B, or both?

12.15 The data in a complete two-factor study are shown here:

FACTOR B	FACTOR A			
	I	II	III	IV
A	5	10	9	5
	6	7	7	8
B	13	16	18	6
	11	17	16	13
C	3	7	8	6
	6	6	7	5

a. Construct the analysis of variance table for the data.
b. Test the hypothesis, at the 0.05 level of significance, that there is no interaction between the factors.
c. If interaction is not significant, test the hypothesis that there are no differences among the factor levels. Use the 0.05 level of significance.
d. Draw the interaction diagrams and interpret the results.

Applications

12.16 A petroleum engineer wants to know whether there is any difference in mileage for the three leading brands of gasoline: P, S, and T. Each brand is used in each of four different makes and models of car. The cars of each make and model are matched as to body type, engine, equipment, and so forth to make them comparable. The same driver makes all the test runs over the same test courses, so there is no difference due to driver or course. There are 6 of each type of car (24 in all), so that we have two observations per treatment—that is, for each type of gasoline with each type of car, a total of 24 observations, 2 per cell. The mileages are shown in the table.

AUTOMOBILE	GASOLINE P	S	T
J	21.8	20.9	25.6
	23.4	20.2	27.1
G	29.5	28.6	28.3
	27.8	28.1	28.9
H	25.4	24.2	23.3
	25.9	27.3	22.9
E	33.8	38.1	32.5
	35.9	39.7	35.6

a. Determine SSTO, SS(gasolines), and SS(cars).
b. Obtain SSE and SS(interaction).
c. Construct the ANOVA table.
d. Are there any significant differences due to interaction? Use the 0.05 level of significance.
e. Draw the interaction diagrams. Are tests on main effects indicated? What do you recommend?

12.17 A company is attempting to increase sales by using an incentive program. Three different incentive programs have been proposed and tested in the field. The company has four different branches, and nine salesmen in each branch are randomly selected and divided randomly into three groups of three each. Group A is offered bonuses for sales increasing beyond a certain amount. No competition is involved. Each salesman whose sales increase beyond that amount will receive a \$500 bonus. In group B, a \$1000 bonus will be paid to the salesman of the three whose sales increase by the greatest amount. Group C is a combination of the two approaches. A \$500 bonus will be given to the salesmen who exceed the same increase as in group A, and a further \$500 bonus will be given to the salesman of the three whose sales increase the most. Midway in the campaign the sales are checked. The increase in sales units for each salesman is given in the table:

BRANCH OFFICE	BONUS INCENTIVE PLAN A	B	C	BRANCH OFFICE	BONUS INCENTIVE PLAN A	B	C
	16	18	17		44	31	28
I	22	27	34	III	27	33	54
	9	15	13		22	19	39
	8	13	14		17	19	24
II	7	5	12	IV	17	23	28
	6	11	10		14	17	31

a. Obtain the ANOVA table for the data. Is interaction significant?
b. Are there differences among the plans? Use the 0.05 level of significance.

12.18 A psychologist has designed a test to assess the effect of color and intensity on the working ability of individuals. Fifteen identical offices are painted. In each case the ceiling is white. Five different colors are used for the walls: blue, green, brown, orange, and yellow. In each case, three different intensities are selected—pastel, primary, and highly intense. The cooperating company randomly places four employees in each of the offices for 1 month's time. Each employee does the same type of work, and all are judged to be approximately equally effective at their work. Near the end of the month, each employee is given the identical project. The number of days each worker requires to complete the project is given in the table. Use the data to assess the effect of color and intensity on the work output of employees. All these employees usually work in offices painted white.

	INTENSITY OF COLOR		
COLOR	PASTEL	PRIMARY	INTENSE
Blue	4	6	7
	5	5	6
	5	6	6
Green	3	6	6
	5	4	5
	5	5	6
Brown	7	6	6
	7	5	6
	6	6	5
Orange	5	4	6
	6	5	7
	5	4	7
Yellow	3	4	6
	6	4	6
	5	4	5

a. Construct the ANOVA table for the data.
b. Is there significant interaction? Use the 0.05 level of significance.
c. Draw interaction diagrams. What do you conclude?

12.19 Forty tenth-grade students in a French class take part in an experiment designed to compare the relative effectiveness of four methods for learning French verbs. Twenty males and twenty females are divided randomly into four groups of five each and taught by the different methods. Each student is then given a list of 30 verbs to memorize within a time limit. The number correctly learned is shown in the following table. Analyze the data at the 0.05 level of significance.

METHOD	SEX OF STUDENT MALE	FEMALE
A	22	17
	19	15
	17	22
	19	17
	19	19
B	25	12
	13	16
	20	15
	22	17
	27	19
C	18	27
	20	24
	20	25
	21	18
	16	29
D	20	25
	21	27
	18	26
	19	22
	19	29

a. Construct the ANOVA table for the data.
b. Is the interaction significant?
c. Are there differences between methods?
d. Are there differences due to sex?
e. Summarize your conclusions? Are further tests desirable?

12.20 Two types of advertising display are being considered for a national campaign. To test which is more effective, 24 stores are selected for an experiment—12 large stores and 12 small ones. Within each category, stores have approximately equal sales of the item being displayed. Each set of 12 stores is divided randomly into three groups. In group A, display A is set up; display B is set up in group B; in group C, the control group, no display is set up. Sales data for 6 weeks are shown in the table:

STORE SIZE	DISPLAY GROUP A	B	C
Large	152	172	134
	164	163	136
	161	177	130
	165	174	144
Small	128	149	121
	127	157	108
	143	151	117
	139	160	113

a. Obtain the ANOVA table. Is there significant interaction?
b. Using the 0.01 level of significance, are there significant differences between the store sizes?
c. Are there significant differences among the displays? (Use $\alpha = 0.01$.)
d. Do we need further tests to determine which display, if any, has the greatest sales? Explain.
e. Are further tests needed on store sizes? Explain.

12.3 Multiple-Comparison Tests

If at least one of the F tests in a two-factor analysis of variance proves significant, we usually perform multiple comparisons to determine exactly what differences are significant. If interaction effects are not significant, we are concerned with the **main effects**—differences between the levels of the factor or factors that prove significant. If we have tested several pigments on different surfaces and there is a significant difference between brightness of the pigments but no interaction, we may test further to determine which pigment or pigments are the brightest.

If different pigments are tested on different surfaces and there are significant interaction effects, we need to draw the interaction diagrams before proceeding further. If interaction is due to differences in magnitude only, we may still try to determine which pigment or pigments are the brightest. When the interaction effects are due to changes in rank, however, the same pigments may not be brightest on all surfaces. In such cases we examine **treatment effects**—comparing cell means to determine whether there are differences among the populations from which the observations in each cell are obtained. In each case we may use Tukey's multiple-comparison test, as outlined in Section 11.3, with certain modifications shown in this section.

12.3.1 Analysis of Main Effects

If interactions are not significant but one or both factors show significant differences among the level means, we can use Tukey's multiple-comparison test to assess differences between either, or both, sets of pairs of sample means. In each case we compute the honestly significant difference (HSD) as in one-way analysis of variance. The only difference is that the number of means being compared and the sample sizes will usually be different for the two comparisons.

To test for differences between means of factor A levels, the number of factor levels will be a, the degrees of freedom for error will be $N - ab$, and the number of observations per sample will be bn. Thus we can find HSD by the formula

$$\text{HSD} = q_\alpha\{a, N - ab\}\sqrt{\text{MSE}/bn}$$

To test for differences between means of factor B levels, the number of factor levels will be b, the degrees of freedom for error will be $N - ab$, and the number of observations per sample will be an. Thus we can find HSD by the formula

$$\text{HSD} = q_\alpha\{b, N - ab\}\sqrt{\text{MSE}/an}$$

Now if we make only one set of comparisons, the significance level for the set of comparisons is α; if we perform both sets of multiple comparisons, the significance level for the combined sets of comparisons will probably be higher than α; the same observations that have been made previously still apply. This is because if we are willing to take a chance (α) that not all pairwise comparisons are correct for each set, by performing two sets of comparisons the probability of at least one incorrect comparison may double. Thus if the level of significance for the factor A comparisons is 0.05 and that for the level B comparisons is also 0.05, the level of significance for the full set of comparisons is considered to be 0.10. Actually, it will be somewhat less, but the exact value is difficult to compute. Another way of saying this is that if we are 95% confident that all pairwise comparisons are correct in each set, we are at least 90% confident that all pairwise comparisons are correct in both sets.

Tukey's Multiple-Comparison Test for Main Effects (Two-Factor Experiment)

If we can reject the hypothesis of equal factor level means for factor A at an α level of significance, then we can conclude that means differ for any two levels i and j if $|\overline{A}_i - \overline{A}_j|$ exceeds

$$\text{HSD} = q_\alpha\{a, N-ab\}\sqrt{\text{MSE}/bn}$$

If we can reject the hypothesis of equal factor level means for factor B at an α level of significance, then we can conclude that means differ for any two levels i and j if $|\overline{B}_i - \overline{B}_j|$ exceeds

$$\text{HSD} = q_\alpha\{b, N-ab\}\sqrt{\text{MSE}/an}$$

If both sets of comparisons are made, the level of significance for the entire set of comparisons is no more than 2α.

Example 12.9

Using the results of Example 12.5, conduct all pairwise comparisons at a combined significance level of 0.10.

Solution

MSE = 1.55; for chemotherapy level, there are four levels, each containing 15 observations. Degrees of freedom for error are 40, and each set of comparisons should be conducted at a 0.05 level of significance. For chemotherapy level, $q_{.05}\{4, 40\} = 3.79$, so HSD $= 3.79\sqrt{1.55/15} \doteq 1.22$. Level means are as shown here, with nonsignificant differences marked with a line:

N	L	M	H
2.20	3.00	3.87	7.47

Thus chemotherapy level H has significantly more days in the hospital than any other level, and chemotherapy level M has significantly more days than level N.

For number of radiation treatments, there are five levels, each containing 12 observations. Then $q_{.05}\{5, 40\} = 4.04$, so HSD $= 4.04\sqrt{1.55/12} \doteq 1.45$. Level means are as shown here, with nonsignificant differences marked with a line:

0	1	2	3	4
2.00	3.08	3.67	5.25	6.67

Thus patients with three and four treatments have significantly more days in the hospital than any other number of treatments. Moreover, patients with two treatments have significantly more days in the hospital than patients with no treatments.

A word of caution is in order. No causal relationship between the variables should be inferred. Rather than more radiation or chemotherapy treatments causing patients to spend more time in the hospital, it is possible that these are sicker patients who need more time in the hospital. We cannot draw any causal inferences unless all patients are equally sick at the beginning of treatment and are randomly assigned to the 12 different treatments.

PROFICIENCY CHECK

10. Use the results of Proficiency Check 7 to determine which factor levels differ significantly. Use the 0.01 level of significance.

12.3.2 Analysis of Treatment Effects

If the interaction is significant, we need to examine the interaction diagrams to determine the type of interaction. If interaction diagrams show that differences are due merely to changes in magnitude as in Figure 12.2 (right side), we may perform Tukey's test on that main effect to determine what differences exist between the levels of that factor. If differences are due to changes in rank as in Figure 12.2 (left side) and Figure 12.4, we are usually interested in the treatment means—the means of the various combinations of levels. The total number of treatments is ab, and each cell contains n observations, so HSD $= q_\alpha\{ab, N-ab\}\sqrt{\text{MSE}/n}$.

Tukey's Multiple-Comparison Test for Treatments (Two-Factor Experiment)

If we can reject the hypothesis of no interaction of factors A and B at an α level of significance, then we can conclude that means differ for any two treatments (factor level combinations) ij and $i'j'$ if $|\overline{AB}_{ij} - \overline{AB}_{i'j'}|$ exceeds

$$\text{HSD} = q_\alpha\{ab, N-ab\}\sqrt{\text{MSE}/n}$$

Example 12.10

Using the results of Example 12.6, determine what differences exist between treatment means. Use the 0.05 level of significance.

Solution

Cell means are shown in Table 12.19. Here MSE = 0.0014; there are 12 treatments, each containing two observations, and $q_{.05}\{12, 12\} = 5.62$. Then HSD $= 5.62\sqrt{0.0014/2} \doteq 0.149$. Means are arranged in order, and nonsignificant differences are connected by a line:

1.100 1.100 1.070 1.055 1.045 1.020 1.005 1.010 0.965 0.955 0.890 0.865

The significant differences are those between the treatment with cell mean 0.865 and those with cell means of 1.020 and above and between the treatment with cell mean 0.890 and those with cell means of 1.045 and above. Thus managers with less than 5 years experience and no college do more poorly than all managers with a college degree except those with at least 15 years of experience and all managers with at least 10 years of experience

Table 12.19 CELL MEANS FOR EXAMPLE 12.10.

EXPERIENCE	EDUCATION LEVEL H	C	B
I	0.865	0.890	1.100
II	0.955	0.965	1.100
III	1.005	1.055	1.045
IV	1.070	1.020	1.010

and some college, and managers with no college and at least 15 years of experience. Managers with less than 5 years experience and some college perform similarly except that managers with some college and at least 15 years of experience do not do significantly better. Perhaps these managers who attended college were unable to rise to higher positions or find better jobs, but have not done so poorly that they have been fired and thus are somewhat mediocre.

PROFICIENCY CHECK

11. Use the results of Proficiency Check 6 to determine differences among the treatment means. Use the 0.05 level of significance.

12.3.3 Multiple-Comparison Tests Using SAS Statements

We can obtain multiple comparisons of main effects by using the MEANS statement with PROC ANOVA as in Section 12.1.6 and also in Section 11.3.3. Using the Tukey procedure, we can obtain all multiple pairwise comparisons for each main effect by writing

```
MEANS variable variable / TUKEY;
```

The 0.05 level of significance is the default value for α. We may specify ALPHA = 0.10 or ALPHA = 0.01 instead if desired. Note that if we obtain both sets of comparisons the significance level is increased, possibly to nearly 2α. Figure 12.7 gives the multiple comparisons for the results of Example 12.7.

Figure 12.7 SAS printout of the multiple comparisons for the chemotherapy radiation data of Example 12.7.

```
INPUT FILE

DATA TWOWAY;
  INPUT CHEM $ TRT DAYS @@;
CARDS;
NONE 0 0 NONE 0 2 NONE 0 0 NONE 1 1 NONE 1 0 NONE 1 3
NONE 2 3 NONE 2 1 NONE 2 2 NONE 3 4 NONE 3 2 NONE 3 3
NONE 4 3 NONE 4 5 NONE 4 4 LIGHT 0 1 LIGHT 0 1 LIGHT 0 2
LIGHT 2 1 LIGHT 2 3 LIGHT 2 2 LIGHT 3 2 LIGHT 3 3 LIGHT 3 3
LIGHT 4 4 LIGHT 4 3 LIGHT 4 5 LIGHT 5 6 LIGHT 5 7 LIGHT 5 2
MOD 0 2 MOD 0 1 MOD 0 3 MOD 1 3 MOD 1 2 MOD 1 3
MOD 2 1 MOD 2 4 MOD 2 3 MOD 3 5 MOD 3 6 MOD 3 4
MOD 5 6 MOD 5 8 MOD 5 7 HEAVY 0 4 HEAVY 0 5 HEAVY 0 3
HEAVY 1 5 HEAVY 1 8 HEAVY 1 6 HEAVY 2 6 HEAVY 2 7 HEAVY 2 9
HEAVY 3 8 HEAVY 3 10 HEAVY 3 9 HEAVY 4 11 HEAVY 4 9 HEAVY 4 12

PROC ANOVA;
  CLASSES CHEM TRT;
  MODEL DAYS = CHEM|TRT;
  MEANS CHEM TRT / TUKEY

OUTPUT FILE
```

```
                              SAS

               ANALYSIS OF VARIANCE PROCEDURE

                  CLASS LEVEL INFORMATION

         CLASS      LEVELS     VALUES

         CHEM          4       HEAVY LIGHT MOD NONE

         TRT           5       0 1 2 3 4

          NUMBER OF OBSERVATIONS IN DATA SET = 60

                              SAS

               ANALYSIS OF VARIANCE PROCEDURE

DEPENDENT VARIABLE: DAYS

SOURCE           DF  SUM OF SQUARES  MEAN SQUARE  F VALUE   PR > F  R-SQUARE      C.V.

MODEL            19    420.93333333  22.15438596    14.29   0.0001  0.871618   30.1207

ERROR            40     62.00000000   1.55000000          ROOT MSE           DAYS MEAN

CORRECTED TOTAL  59    482.93333333                     1.24498996          4.13333333


SOURCE           DF      ANOVA SS    F VALUE  PR > F

CHEM              3  243.06666667     52.27   0.0001

TRT               4  162.43333333     26.20   0.0001

CHEM*TRT         12   15.43333333      0.83   0.6201
                              SAS

                      ANALYSIS OF VARIANCE PROCEDURE

                TUKEY'S STUDENTIZED RANGE (HSD) TEST FOR VARIABLE: DAYS
             NOTE: THIS TEST CONTROLS THE TYPE I EXPERIMENTWISE ERROR RATE,
               BUT GENERALLY HAS A HIGHER TYPE II ERROR RATE THAN REGWQ

                    ALPHA=0.05 DF=40 MSE=1.55
                    CRITICAL VALUE OF STUDENTIZED RANGE=3.791
                    MINIMUM SIGNIFICANT DIFFERENCE=1.2185

           MEANS WITH THE SAME LETTER ARE NOT SIGNIFICANTLY DIFFERENT.

                    TUKEY    GROUPING        MEAN      N CHEM

                                A          7.4667     15 HEAVY

                                B          3.8667     15 MOD
                                B
                      C
                      C         B          3.0000     15 LIGHT
                      C         B          2.2000     15 NONE
```

```
                         SAS

           ANALYSIS OF VARIANCE PROCEDURE

    TUKEY'S STUDENTIZED RANGE (HSD) TEST FOR VARIABLE: DAYS
NOTE: THIS TEST CONTROLS THE TYPE I EXPERIMENTWISE ERROR RATE,
   BUT GENERALLY HAS A HIGHER TYPE II ERROR RATE THAN REGWQ

          ALPHA=0.05 DF=40 MSE=1.55
          CRITICAL VALUE OF STUDENTIZED RANGE=4.039
          MINIMUM SIGNIFICANT DIFFERENCE=1.4517

 MEANS WITH THE SAME LETTER ARE NOT SIGNIFICANTLY DIFFERENT.

          TUKEY     GROUPING        MEAN      N  TRT
                        A          6.6667    12   4
                        A
                        A          5.2500    12   3
                        B          3.6667    12   2
                        B
             C          B          3.0833    12   1
             C
             C                     2.0000    12   0
```

If we need multiple comparisons on treatment means, it is probably most efficient to obtain the treatment means by writing A*B in the MEANS option and to determine the value of HSD (or the proper multiple if another procedure is used) by using a hand calculator.

Problems

Practice

The following table shows cell means in a two-factor study. Interaction is not significant. Each cell contains three observations. Use these results for Problems 12.21 through 12.24.

	FACTOR *B*		
FACTOR *A*	*A*	*B*	*C*
1	1.22	1.18	1.13
2	1.55	1.45	1.38
3	1.34	1.28	1.19
4	1.35	1.30	1.19

12.21 Suppose that factor *B* levels are significantly different. Conduct all pairwise comparisons at a 0.05 level of significance if MSE = 0.0272.

12.22 Suppose that factor *B* levels are significantly different. Conduct all pairwise comparisons if MSE = 0.0112. Use a 0.05 level of significance.

12.23 Suppose that factor *A* levels are significantly different. Conduct all pairwise comparisons if MSE = 0.0112. Use the 0.05 level of significance.

12.24 At what level of significance is the set of all comparisons in both Problems 12.22 and 12.23 carried out?

12.25 The following table shows cell means in a complete two-factor study. Interaction is significant. Conduct pairwise comparisons among the treatments at the 0.05 level of significance. Each cell contains six observations; MSE = 9.134.

	FACTOR *B*	
FACTOR *A*	*A*	*B*
1	36.9	28.7
2	29.9	31.4

Applications

12.26 Refer to Problem 12.16. Interactions are significant and due to differences in rank. Suppose we wish to determine which gasoline gets the highest mileage for each automobile. In that case we need to make three comparisons for each car.

a. What value of HSD should you use for the comparisons if $\alpha = 0.05$?

b. Conduct all pairwise comparisons for each automobile. What do you conclude?

12.27 In Problem 12.17 we found significant differences among the incentive programs. Using the 0.05 level of significance, can you determine which plan is most successful? Least successful? What should the company do next?

12.28 Refer to Problem 12.18. The most desirable result is a small number of days.

a. Do the interaction diagrams show interaction due to changes in magnitude or rank? What does this imply?

b. The psychologist noted that interaction was significant and decided to determine whether or not there were significant differences among the combinations of color and intensity at the 0.05 level of significance. Conduct all pairwise comparisons and give your conclusions.

c. Why are all pairwise comparisons on treatments needed in this problem but not in Problem 12.26?

12.29 Refer to Problem 12.19. Can you determine which method is most effective? Least effective? What should the next step be? Use $\alpha = 0.05$.

12.30 In Problem 12.20 we concluded that there were significant differences due to display. Using the 0.01 level of significance, conduct all pairwise comparisons. Can you decide which display yields the greatest number of sales, or is further testing necessary?

12.4 Minitab Commands for This Chapter

We can analyze both the randomized complete block and the two-factor ANOVA by using Minitab. Both analyses are performed using the Minitab command TWOWAY. The number of observations in each cell and the subcommand ADDITIVE determine which analysis is performed. Note that we can use the TWOWAY command only when the number of observations is the same in each cell.

To analyze a randomized block design, the data must be stored in one column with the treatment and block identifiers in two other columns. The Minitab command is

```
MTB > TWOWAY analysis of the data in C, blocks in C,
treatments in C
```

Again the block and treatment identifiers are integers that designate each observation's block and treatment values.

If there are multiple observations (n in each cell), we can perform the two-way analysis of variance either with or without the two-factor interaction term. (If there is only one observation per cell, TWOWAY will automatically exclude the interaction term.) The basic form of the command for a two-way ANOVA is

```
MTB > TWOWAY analysis of the data in C, factor a in C,
factor b in C;
SUBC> ADDITIVE model.
```

To include the interaction term, leave out the ADDITIVE subcommand. The only difference in the output is that the interaction effect row in the ANOVA table is not included when the ADDITIVE subcommand is used.

Note also that the TWOWAY command does not provide multiple comparisons of the means; we must compute multiple comparisons by hand. To determine the sample means we use the TABLE command. We can obtain the cell (treatment sample) means and row and column (factor-level sample) means as follows:

```
MTB > TABLE factor a in C by factor b in C;
SUBC> MEANS of the data in C.
```

Example 12.11

In the gasoline mileage data of Example 12.1 our goal was to determine whether the four brands of gasoline used in the study differed significantly in miles per gallon. The effects of the five different cars are adjusted by using the factor car as a block. Use Minitab commands to repeat the analysis.

Solution

The necessary Minitab commands and the output are shown in Figure 12.8. Compare the Minitab output to the results of Example 12.1 (Table 12.5). Note that some values have been rounded off.

For the factor gasoline, the critical value is $F_{.01}\{3, 12\} = 5.95$; $F^* = 7.36$, so we conclude that after adjusting for the effect of the different cars, the average mileage differs significantly for at least one pair of the four

Figure 12.8
Minitab printout for the gasoline mileage data of Example 12.11.

```
MTB > SET the following data into column C1
DATA> 11.8 12.4 10.6 13.1
DATA> 14.6 14.9 15.6 16.2
DATA> 21.3 24.2 20.6 23.7
DATA> 14.1 15.3 12.4 16.8
DATA> 13.1 17.3 16.1 18.6
DATA> SET the block identifier into column C2
DATA> 1 1 1 1 2 2 2 2 3 3 3 3 4 4 4 4 5 5 5 5
DATA> SET the treatment identifier into column C3
DATA> 1 2 3 4 1 2 3 4 1 2 3 4 1 2 3 4 1 2 3 4
DATA> NAME C1 'MILEAGE' C2 'CAR' C3 'GASOLINE'
MTB > TWOWAY of 'MILEAGE', blocks in 'CAR', treatments in 'GASOLINE'

ANALYSIS OF VARIANCE ON MILEAGE

SOURCE        DF        SS        MS
CAR            4    240.26     60.07
GASOLINE       3     26.73      8.91
ERROR         12     14.53      1.21
TOTAL         19    281.52
MTB > TABLE the data in 'CAR' and 'GASOLINE';
SUBC> MEANS of 'MILEAGE'.

ROWS: CAR   COLUMNS: GASOLINE

              1          2          3          4        ALL

  1      11.800     12.400     10.600     13.100     11.975  <-- Average mileage for car 1
  2      14.600     14.900     15.600     16.200     15.325
  3      21.300     24.200     20.600     23.700     22.450
  4      14.100     15.300     12.400     16.800     14.650
  5      13.100     17.300     16.100     18.600     16.275
ALL      14.980     16.820     15.060     17.680     16.135
           ^
           Average mileage for gasoline brand 1

CELL CONTENTS --
     MILEAGE:MEAN
```

brands of gasoline. Note that if we perform a one-way analysis of variance on the factor gasoline without blocking cars, the gasolines do not have significantly different mileages. The obtained value of F^* is 0.56 in that case, compared to $F_{.01}\{3, 16\} = 5.40$.

Using the Tukey procedure to perform the multiple comparisons on the means of the different brands of gasoline, we obtain

$$\begin{aligned} \text{HSD} &= q_{.05}\{k, (k-1)(n-1)\}\sqrt{\text{MSE}/n} \\ &= q_{.05}\{4, (4-1)(5-1)\}\sqrt{1.21/5} \\ &\doteq 4.20(0.4919) \\ &\doteq 2.07 \end{aligned}$$

Using the table of means in the Minitab output, we observe that the average mileage for brands A and C is significantly different from that of brand D.

We next illustrate the two-way analysis of variance procedure. First we note that there is a simpler way to enter the sample identifiers than by typing the long list of numbers. The command

```
MTB > SET the following into C1
DATA> 2(1 2 4)3
```

produces a column with the following numbers:

1 1 1 2 2 2 4 4 4 1 1 1 2 2 2 4 4 4

The 3 following the parentheses indicates the number of times each number within the parentheses is to be written; the 2 in front of the parentheses indicates the number of times the procedure is to be repeated. When we are using an identifier variable, this use of the SET command eliminates typing in a long list of numbers.

Example 12.12

Use Minitab to analyze the data of Example 12.5.

Solution

The Minitab commands and output are presented in Figure 12.9. Comparing our results to those in Table 12.12, we note that they are identical. The

```
MTB > SET the following data into column C1
DATA> 0 2 0 1 1 2 2 1 3 4 5 3
DATA> 1 0 3 1 3 2 3 2 3 5 8 6
DATA> 3 1 2 2 3 3 1 4 3 6 7 9
DATA> 4 2 3 4 3 5 5 6 4 8 10 9
DATA> 3 5 4 6 7 2 6 8 7 11 9 12
DATA> SET the chemotherapy level identifier into column C2
DATA> 5(1 2 3 4)3
DATA> SET the number of treatments identifier into column C3
DATA> (0 1 2 3 4) 12
DATA> NAME C1 'DAYS' C2 'LEVEL' C3 'NUMBER'
MTB > TWOWAY of data in C1, factor a in C2, factor b in C3

ANALYSIS OF VARIANCE ON ERRORS

SOURCE          DF        SS        MS
LEVEL            3    243.07     81.02
NUMBER           4    162.43     40.61
INTERACTION     12     15.43      1.29
ERROR           40     62.00      1.55
TOTAL           59    482.93
MTB > TABLE the data in 'LEVEL' and 'NUMBER';
SUBC> MEANS of 'DAYS'.

ROWS: LEVEL   COLUMNS: NUMBER

              0         1         2         3          4        ALL

  1      0.6667    1.3333    2.0000    3.0000     4.0000     2.2000
  2      1.3333    2.0000    2.6667    4.0000     5.0000     3.0000
  3      2.0000    2.6667    2.6667    5.0000     7.0000     3.8667
  4      4.0000    6.3333    7.3333    9.0000    10.6667     7.4667
 ALL     2.0000    3.0833    3.6667    5.2500     6.6667     4.1333

CELL CONTENTS --
     ERRORS:MEAN
```

Average number of days for chemotherapy level 1 and 0 treatments

Figure 12.9 Minitab printout for the chemotherapy radiation data of Example 12.12.

F test for interaction has $F^* = 0.8323$ and is not significant ($F_{.01}\{12, 40\} = 2.66$). Both main effects are significant, however; $F^* = 52.27$ with $F_{.01}\{3, 40\} = 4.31$ for chemotherapy level and $F^* = 26.20$ with $F_{.01}\{4, 40\} = 3.83$. We can perform multiple comparisons by hand, as shown before, using the output from the TWOWAY and TABLE commands.

Although Minitab does not have a formal command to perform the Friedman F_r Test, we may use the Minitab commands RANK and TWOWAY to obtain the test statistic.

Example 12.13 Refer to the gasoline/car experiment of Example 12.3. Use Minitab commands to perform the Friedman analysis of variance. Use the 0.01 level of significance.

Solution

We need the following basic steps to perform the analysis:

1. Put each block of data into a separate column by using a SET command.
2. Rank the data within each block by using the RANK command, and store the data into new columns.
3. Combine the rank data into one column by using the STACK command.
4. Enter the block and treatment identifiers by using the SET command.
5. Use TWOWAY to do an analysis of variance on the data.

Figure 12.10 shows the results of performing these commands on the data of Example 12.3.

Figure 12.10 Minitab printout to obtain Friedman's F_r Test for the gasoline mileage data of Example 12.13.

```
MTB  > SET the block one data into C1
DATA> 11.8 12.4 10.6 13.1
DATA> SET the block two data into C2
DATA> 14.6 14.9 15.6 16.2
DATA> SET the block three data into C3
DATA> 21.3 24.2 20.6 23.7
DATA> SET the block four data into C4
DATA> 14.1 15.3 12.4 16.8
DATA> SET the block five data into C5
DATA> 13.1 17.3 16.1 18.6
DATA> RANK the data in C1 store the ranks in C6
MTB  > RANK the data in C2 store the ranks in C7
MTB  > RANK the data in C3 store the ranks in C8
MTB  > RANK the data in C4 store the ranks in C9
MTB  > RANK the data in C5 store the ranks in C10
MTB  > STACK the data in C6, C7, C8, C9, C10, into column C11
MTB  > SET the block identifier into column C12
DATA> (1 2 3 4 5)4
DATA> SET the treatment identifier into column C13
DATA> 5(1 2 3 4)
DATA> NAME C11 'RANKS' C12 'CAR' C13 'GASOLINE'
MTB  > TWOWAY on 'RANKS', blocks in 'CAR', treatments in 'GASOLINE'
```

```
ANALYSIS OF VARIANCE ON RANKS
```

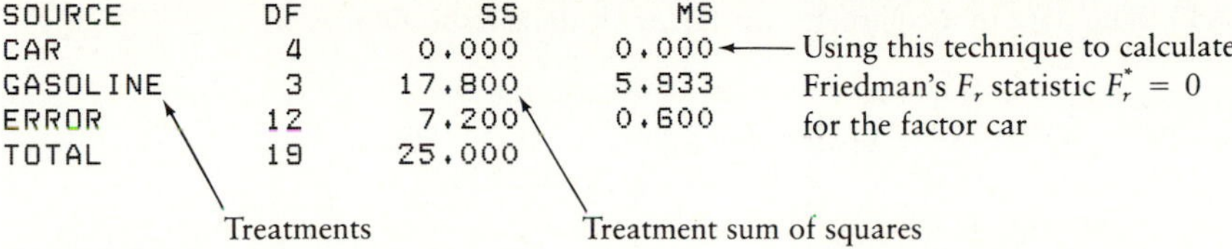

```
SOURCE         DF        SS         MS
CAR             4      0.000      0.000
GASOLINE        3     17.800      5.933
ERROR          12      7.200      0.600
TOTAL          19     25.000
```


The value of SSB—in this case, SS(cars)—will always be zero, since the sum of the block ranks will always be the same for each rank. We may obtain F_r^* from SSTR—here SS(gasoline)—by using the formula

$$F_r^* = \frac{12(\text{SSTR})}{k(k+1)}$$

where k is the number of treatments. Using the results from the output shown in Figure 12.10, we have

$$F_r^* = \frac{12(17.80)}{4(4+1)} = 10.68$$

Since the critical value is $\chi^2_{.01}\{3\} = 11.34$, we conclude that the median mileages are not significantly different for the four gasolines at the 0.01 level of significance. Note, however, that since $\chi^2_{.025}\{3\} = 9.348$, we might wish to reserve judgment and repeat the experiment because of the possibility of a type II error.

Problems

Practice

12.31 Three samples in a complete randomized block design have the following data:

BLOCK	TREATMENT I	II	III
A	23	22	16
B	24	33	19
C	14	18	8
D	13	15	11

a. Construct the analysis of variance table for the data.
b. Test the hypothesis, at the 0.05 level of significance, that there are no differences among the treatment means.
c. Perform all pairwise comparisons, if appropriate.

12.32 Repeat Problem 12.31 using Friedman's F_r Test.

12.33 The data in a complete two-factor study are shown here.

	FACTOR *A*			
FACTOR *B*	I	II	III	IV
A	15	20	19	25
	16	17	17	28
B	23	26	28	36
	21	27	26	33
C	13	17	18	26
	16	16	17	25

a. Construct the analysis of variance table for the data.
b. Test the hypothesis, at the 0.05 level of significance, that there is no interaction between the factors.
c. If interaction is not significant, test the hypothesis that there are no differences among the factor levels. Use the 0.05 level of significance.
d. Draw the interaction diagrams and interpret the results.

Application

12.34

A paint manufacturer is trying to decide which of three pigments will give the brightest color in his paint. Since there might be a difference in brightness due to different wall surfaces, he paints each of four different surfaces with three samples each of the pigment and then tests the surfaces with a light meter. The light meter readings are given here. Use the 0.05 level of significance to determine whether he can decide which pigment to use. The highest readings indicate the brightest pigments.

	PIGMENT		
SURFACE	A	B	C
Smooth	78	74	68
	88	80	63
	74	73	71
Grainy	67	62	58
	68	61	64
	70	66	59
Rough	47	44	38
	42	40	39
	46	43	36
Very rough	29	27	23
	28	30	24
	33	28	24

12.5 Case 12: The Delighted Dietitian—Solution

Before long the dietitian received a call from the university statistician. The statistician said he had performed a two-way analysis of variance on the data with the results shown here:

SOURCE OF VARIATION	DF	SS	MS	F^*
Diet	2	316.2078	158.1039	59.95
Exercise	1	43.2450	43.2450	16.40
Interaction	2	0.8633	0.4317	0.16
Error	66	174.0500	2.6371	
Total	71	534.3661		

Like all the prior analyses, these analyses were carried out at the 0.05 level of significance. Interaction was not significant, but both factor effects were. Since there were only two levels of exercise—some or none—no further analyses were needed. The mean weight loss with exercise was 4.79 pounds, while the mean weight loss without exercise was 3.24 pounds. Thus it can be concluded that exercise is a factor in inducing weight loss. The factor diet had three levels—the leading diet (L), the dietitian's diet (D), and no diet (N)—so Tukey's multiple-comparison test was used. As $q_{.05}\{3, 66\} \doteq 4.27$, we have HSD $= 4.27\sqrt{2.6371/24} \doteq 1.415$. The means for each diet group were as follows:

Diet group	L	D	N
Mean weight loss	4.73	6.15	1.17

All comparisons were significant; thus dieters who use the dietitian's diet can be expected to show, on the average, a greater weight loss than dieters who use the leading diet, and dieters who use either diet can be expected to show a greater weight loss than people who do not diet. Therefore the long-awaited final results were exactly what the dietitian wanted to hear.

12.6 Summary

In this chapter we have examined some analysis of variance methods for continuous data arranged in a two-way classification. If the observations for one factor are matched, as an extension of the matched-pair t test of Section 8.3, each set of matched observations is called a *block* and the design is called a *randomized complete block*. The total sum of squares (SSTO) for this design is partitioned into the sum of squares for treatments (SSTR), the sum of squares for blocks (SSB), and the sum of squares for error (SSE). We can test the hypothesis that treatment means are equal by using $F^* = \text{MSTR/MSE}$; we can test the hypothesis that block means are equal by using $F^* = \text{MSB/MSE}$.

We can perform multiple pairwise comparisons among the treatment means or among the block means by using Tukey's multiple-comparison test. If both sets of multiple comparisons are performed, the significance level for the combined sets of comparisons is no greater than 2α.

If the data are in ranks or either of the assumptions of equal variance or normality of treatments is not met, we can analyze differences between treatment means by using the nonparametric Friedman's F_r Test. We can make multiple comparisons between sample mean ranks by using the LSD statistic.

If there are two different factors, a levels of factor A and b levels of factor B, with n observations per treatment (a treatment is one level of A with one level of B), where $n > 1$, we have a two-factor factorial experiment, or simply a two-factor experiment. We analyze it by means of two-way analysis of variance. Each set of n observations is called a *cell*. The total sum of squares (SSTO) for such an experiment is partitioned into the sum of squares for factor A (SSA), the sum of squares for factor B (SSB), the sum of squares for interaction (SSAB), and the sum of squares for error (SSE). The hypothesis that there is no interaction between the factors can be tested with $F^* = \text{MSAB/MSE}$. If there is no interaction, we can test the hypothesis that the means for the levels of factor A are equal by using $F^* = \text{MSA/MSE}$ and the hypothesis that the means of the levels of factor B are equal by using $F^* = \text{MSB/MSE}$. The presence or absence of interaction between factors can also be determined visually with interaction diagrams.

Multiple comparisons following significant F tests can be carried out for factor A or factor B by using Tukey's multiple-comparison test. If both sets of comparisons are made, the level of significance for the entire set of comparisons is no greater than 2α.

Chapter Review Problems

Practice

12.35 The following analysis of variance table is partially filled in. Complete the table.

SOURCE OF VARIATION	DF	SS	MS
Treatments			0.0039
Blocks	5		
Error		0.1044	
Total	29	0.1456	

12.36 The following analysis of variance table is partially filled in. Complete the table.

SOURCE OF VARIATION	DF	SS	MS
Factor A	1		
Factor B	2		1914.3
Interaction		648.9	
Error	120		27.9
Total		8415.8	

12.37 An analysis of variance table is given here:

SOURCE OF VARIATION	DF	SS	MS
Treatments	3	113.76	37.92
Blocks	4	1237.92	309.48
Error	12	110.16	9.18
Total	19	1461.84	

Using the 0.01 level of significance, test the hypothesis that there are no differences among the treatment means.

12.38 An analysis of variance table is given here:

SOURCE OF VARIATION	DF	SS	MS
Factor A	2	7,188	3,594
Factor B	3	31,842	10,614
Interaction	6	918	153
Error	36	4,356	126
Total	47	44,304	

Using the 0.05 level of significance, test the hypotheses that there are no differences among the factor level means and that there is no interaction.

Applications

12.39 A rustproof coating has been developed to compete with the leading brand. A comparison between the leading brand (A), the second brand (B), and the newly developed product (C) is carried out by coating the metals, exposing them to the usual conditions, and then measuring the extent of corrosion by determining the depth of pitting. In the following table, depth of pitting is given in millimeters. Six different metals are used. Analyze the results at the 0.05 level.

	COATING		
METAL	A	B	C
Iron	1.8	1.6	1.9
Brass	1.2	1.3	1.2
Copper	2.4	2.5	2.0
Steel	1.4	1.4	1.3
Magnesium alloy	2.6	2.5	2.1
Aluminum	1.7	1.6	1.7

a. Construct the ANOVA table for the data.
b. Are there significant differences between the coatings?
c. Draw interaction diagrams. The assumptions for the randomized block include no interaction. Does this assumption seem to be met?
d. Does there appear to be a difference in pitting for the various metals? What steps would you take to investigate these apparent differences further?

12.40 In a variation of the previous experiment, some subjects were to press the button with their preferred hand and others with their nonpreferred hand. Stimuli were limited to a light (visual) or a buzzer (auditory), but not both. Median reaction times are again given in hundredths of a second. Analyze the results at a 0.05 level of significance.

	STIMULUS	
HAND USED	AUDITORY	VISUAL
Preferred	56	72
	61	78
	54	73
	59	78
Nonpreferred	65	80
	64	69
	66	79
	63	77

12.41 An experiment was conducted to test reaction time to auditory and visual stimuli. Each group was to press a button as soon as they saw a light flash or a hand signal (visual stimuli) or heard a buzzer or a bell (auditory stimuli). Group I was given both visual stimuli, group II was given both auditory stimuli, while group III was given mixed visual (light) and auditory (buzzer) stimuli. Each group was further divided into two equal parts; part A received the two signals alternately (regular), while part B received the two signals in random order (random). The following data give the median reaction time for each subject in hundredths of a second:

		STIMULUS	
ORDER	AUDITORY	VISUAL	MIXED
Regular	61	69	58
	57	74	64
	55	67	58
	60	68	63
	54	74	66
Random	70	72	78
	67	74	82
	73	71	79
	66	64	77
	68	72	78

a. Construct the ANOVA table.

b. Determine whether there are significant differences due to order, stimulus, or interaction. Use the 0.05 level in each case.
c. Construct the interaction diagrams. Do they confirm your result in part (*b*)? Are interactions due to changes in rank or to changes in magnitude?
d. Do the interaction diagrams indicate that differences between factor levels should be investigated? Which factor? Are multiple comparisons necessary? What is your conclusion?
e. Which combination of order and stimulus yields the smallest mean for the median reaction times? Use the Tukey procedure to determine whether this is significantly less than the next smallest mean.

Index of Terms

Glossary of Symbols

A_i	sum of values of observations, level *i*, factor *A*
$\overline{A}_i$	mean value of observations, level *i*, factor *A*
α_i	effect of level *i*, factor *A*
AB_{ij}	sum of values of observations, treatment *ij*
$\overline{AB}_{ij}$	mean value of observations, treatment *ij*
$\alpha\beta_{ij}$	interaction effect of level *i*, factor *A*, with level *j*, factor *B*
B_i	sum of values of observations, block *i*
$\overline{B}_i$	mean value of observations, block *i*
β_i	effect of block *i*
B_j	sum of values of observations, level *j*, factor *B*
$\overline{B}_j$	mean value of observations, level *j*, factor *B*
β_j	effect of level *j*, factor *B*
F_r^*	sample value of test statistic, Friedman F_r Test
MSA	mean square due to factor *A*
MSAB	mean square due to interaction
MSB	mean square due to factor *B;* mean square due to blocks
MSE	mean square due to error
MSTR	mean square due to treatments
SSA	sum of squares due to factor *A*
SSAB	sum of squares due to interaction
SSB	sum of squares due to factor *B;* sum of squares due to blocks
SSE	sum of squares due to error
SSTR	sum of squares due to treatments

T	sum of all observations in all samples
T_j	sum of values of observations, treatment j
$\overline{T}_j$	mean value of observations, treatment j
τ_j	effect of treatment j

Answers to Proficiency Checks

1. The totals for treatments and blocks are shown here:

	TREATMENT					
BLOCK	1	2	3	4	5	TOTAL
1	4	6	9	2	7	28
2	15	16	18	9	14	72
3	8	7	11	5	8	39
4	19	23	17	12	16	87
Total	46	52	55	28	45	226

The sum of the squares of the entries is 3,170; SSTO $= 3170 - 226^2/20 = 3170 - 2553.8 = 616.2$. Then SSTR $= (46^2 + 52^2 + 55^2 + 28^2 + 45^2)/4 - 226^2/20 = 2663.5 - 2553.8 = 109.7$; SSB $= (28^2 + 72^2 + 39^2 + 87^2)/5 - 226^2/20 = 3011.6 - 2553.8 = 457.8$; SSE $= 616.2 - 109.7 - 457.8 = 48.7$. The ANOVA table is shown here:

SOURCE OF VARIATION	DF	SS	MS	F^*
Treatments	4	109.7	27.425	6.76
Blocks	3	457.8	152.600	37.60
Error	12	48.7	4.058	
Total	19	616.2		

Since $F_{.01}\{4, 12\} = 5.41$, we can conclude that there are differences among the treatment means.

2. The totals for treatments (experience level) and blocks (department) are shown here:

DEPARTMENT	I	II	III	IV	TOTAL
A	21	12	8	7	48
B	38	24	11	12	85
C	16	12	7	4	39
Total	75	48	26	23	172

In this case $\Sigma x^2 = 3448$; SSTO $= 3448 - (172)^2/12 \doteq 982.667$. SSTR $= (75^2 + 48^2 + 26^2 + 23^2)/3 - (172)^2/12 \doteq 579.333$; SSB $= (48^2 + 85^2 + 39^2)/4 - (172)/12 \doteq 297.167$; SSE $\doteq 982.667 - 579.333 - 297.167 = 106.167$. The results are summarized here:

SOURCE OF VARIATION	DF	SS	MS	F^*
Treatments	3	579.333	193.111	10.91
Blocks	2	297.167	148.583	8.40
Error	6	106.167	17.694	
Total	11	982.667		

Now $F_{.05}\{3, 6\} = 4.76$ and $F_{.05}\{2, 6\} = 5.14$; since $10.91 > 4.75$, there are significant differences in errors among the experience categories. We may also conclude, as expected, that there are significant differences in errors among the departments.

3. Since $q_{.01}\{5, 12\} = 5.84$, we have $\text{HSD} = 5.84\sqrt{4.058/4} \doteq 5.88$. The differences can be computed, and significant differences are shown below. Means connected by a line are not significantly different.

3	2	1	5	4
13.75	13.00	11.50	11.25	7.00

Thus treatments 3 and 2 have higher means than treatment 4. No other differences are found.

4. Since $q_{.05}\{4, 6\} = 4.90$, we have $\text{HSD} = 4.90\sqrt{17.694/3} \doteq 11.90$. The differences can be computed, and significant differences are shown below. Means connected by a line are not significantly different.

I	II	III	IV
25.00	16.00	8.67	7.67

We can conclude that employees with less than 3 months' experience make more errors than those with at least 1 year of experience. No other differences were shown.

5. Rank sums are 10, 6, 14, so $F_r^* = [12/5(3)(4)](10^2 + 6^2 + 14^2) - 3(5)(4) = (12/60)(332) - 60 \doteq 6.4$. As $\chi^2_{.05}\{2\} = 5.991$, the treatments do differ. There are $3(2)/2 = 3$ pairwise comparisons; $z_{.05/6} \doteq z_{.0083} \doteq 2.40$. Then $\text{LSD} = 2.40\sqrt{3(4)/6(5)} \doteq 1.52$. Treatment means are 2.0, 1.2, and 2.8, respectively. The only significant difference is between performer 2 and performer 3. Performer 2 is rated significantly higher than performer 3.

6. The AB summary table is shown here:

	FACTOR *B*		
FACTOR *A*	*A*	*B*	TOTALS
1	8	14	22
2	13	36	49
Totals	21	50	71

Now $\Sigma x^2 = 1^2 + 2^2 + 2^2 + \cdots + 7^2 + 6^2 + 8^2 = 357$. $\text{SSA} = (22^2 + 49^2)/10 - 71^2/20 = 288.5 - 252.05 = 36.45$; $\text{SSB} = (21^2 + 50^2)/10 - 71^2/20$

$= 294.1 - 252.05 = 42.05$; SSE $= 357 - (8^2 + 14^2 + 13^2 + 36^2)/5 = 357 - 345 = 12$. SSTO $= 357 - 71^2/20 = 357 - 252.05 = 104.95$. Then SSAB $= 104.95 - 36.45 - 42.05 - 12 = 14.45$. Also, $a = b = 2$, $n = 5$, $N = 20$. Then we have:

SOURCE OF VARIATION	DF	SS	MS	F^*
Factor A	1	36.45	36.45	48.60
Factor B	1	42.05	42.05	56.07
Interaction	1	14.45	14.45	19.27
Error	16	12.00	0.75	
Total	19	104.95		

Since $F_{.05}\{1, 16\} = 4.49$, it can be seen that there is significant interaction and there are significant differences among the factor level means.

7. The AB summary table is shown here:

	FACTOR B			
FACTOR A	A	B	C	TOTALS
I	9	19	28	56
II	19	27	42	88
Totals	28	46	70	144

Now $\Sigma x^2 = 2^2 + 4^2 + 3^2 + \cdots + 11^2 + 15^2 + 16^2 = 1388$. SSA $= (56^2 + 88^2)/9 - 144^2/18 \doteq 1208.89 - 1152 = 56.89$; SSB $= (28^2 + 46^2 + 70^2)/6 - 144^2/18 = 1300 - 1152 = 148$; SSE $= 1388 - (9^2 + 19^2 + 28^2 + 19^2 + 27^2 + 42^2)/3 = 1388 - 1360 = 28$. SSTO $= 1388 - 144^2/18 = 1388 - 1152 = 236$. Then SSAB $= 236 - 56.89 - 148 - 28 = 3.11$. Also, $a = 2$, $b = 3$, $n = 3$, $N = 18$, so the ANOVA table is:

SOURCE OF VARIATION	DF	SS	MS	F^*
Factor A	1	56.89	56.89	24.42
Factor B	2	148.00	74.00	31.76
Interaction	2	3.11	1.56	0.67
Error	12	28.00	2.33	
Total	17	236.00		

Since $F_{.01}\{2, 12\} = 6.93$, it can be seen that there is no significant interaction. $F_{.01}\{1, 12\} = 9.33$; $24.42 > 9.33$, so there are significant differences among the means for factor level A. Since $31.76 > 6.93$, there are significant differences among the means for factor level B.

8. Here are the cell means:

	FACTOR B	
FACTOR A	A	B
1	1.6	2.8
2	2.6	7.2

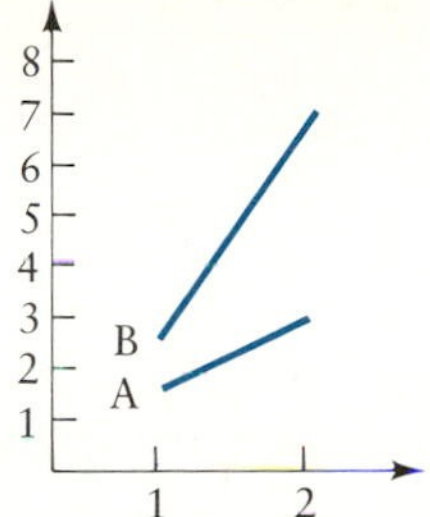

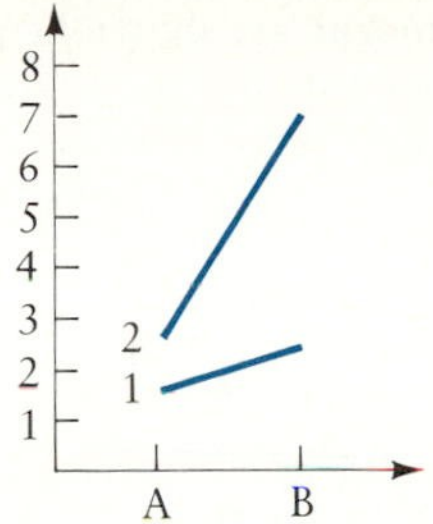

The interaction diagrams are shown above. Both the table and the diagrams show essentially the same thing—that there is interaction. The interaction between the factors results in a large increase at the second level of the two factors. We note that there may be significant differences between the factor levels in both cases, but they are not the same for all levels of the other factor in each case.

9. The cell means are shown in the table:

	FACTOR *B*		
FACTOR *A*	A	B	C
I	3.00	6.33	9.33
II	6.33	9.00	14.00

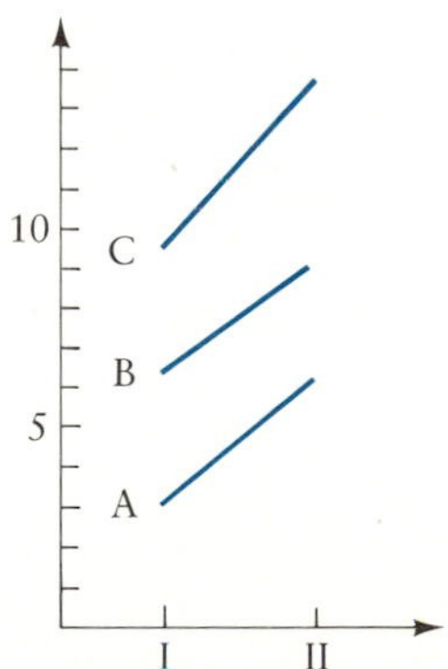

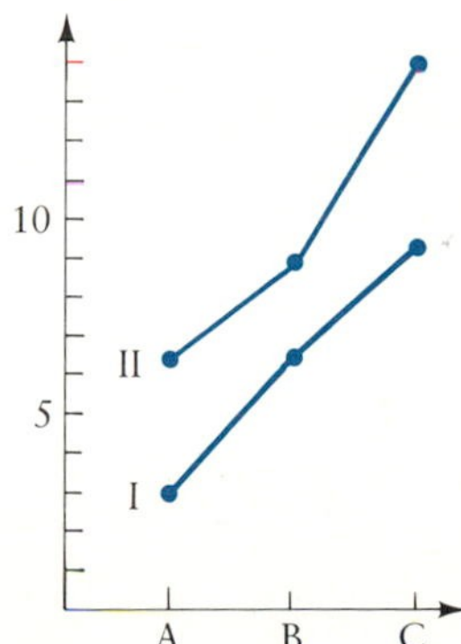

The interaction diagrams are as shown above. In each case the lines are nearly parallel, showing that there is no interaction between the factors.

10. Since factor *A* has only two levels, no comparisons are necessary. The *F* test shows that there are significant differences among the levels of factor *A*, and since there are only two levels, we conclude that level II has a higher mean (sample mean 9.78) than level I (sample mean 6.22). Factor *B* sample means are 4.67, 7.67, and 11.67. Each level contains six observations, and $q_{.01}\{3, 12\} = 5.04$, so HSD $= 5.04\sqrt{2.33/6} \doteq 3.14$. Thus level C is significantly higher than the other two levels.

11. The cell means are shown here:

	FACTOR	
FACTOR *A*	A	B
1	1.6	2.8
2	2.6	7.2

Here $n = 10$, so HSD $= q_{.05}\{4, 16\}\sqrt{0.75/10} = 4.05\sqrt{0.075} \doteq 1.1$. All differences are significant except those between A_1 and A_2 and those between B_1 and A_2; this is shown here:

$$\underline{1.6 \quad 2.6} \quad 2.8 \quad 7.2$$

(with a second line under 2.6 and 2.8)

References

Bailey, J. R. "Tables of the Bonferroni *t* statistic." *Journal of the American Statistical Association* 72(1977): 469–478.

Cochran, W. J., and G. M. Cox. *Experimental Designs.* 2nd ed. New York: John Wiley & Sons, 1957.

Friedman, Milton. "The Use of Ranks to Avoid the Assumption of Normality Implicit in the Analysis of Variance." *Journal of the American Statistical Association* 32 (1937): 675–701.

Klatworthy, Willard H. *Tables of Two Associate-Class Partially Balanced Designs.* NBS Applied Math Series 63. Washington: U.S. Department of Commerce, 1973.

Milliken, George A., and Dallas E. Johnson. *Analysis of Messy Data.* Vol. I: *Designed Experiments.* Belmont, CA: Lifetime Learning Publications, 1984.

Neave, H. R. *Statistical Tables.* London: George Allen & Unwin, 1978.

Neter, John, William Wasserman, and Michael H. Kutner. *Applied Linear Statistical Models.* 2nd ed. Homewood, IL: Richard D. Irwin, 1985.

Winer, B. J. *Statistical Principles in Experimental Design.* 2nd ed. New York: McGraw-Hill, 1971.

13

Linear Regression Analysis

Sir Francis Galton (1822–1911), English scientist, explorer, and anthropologist, was one of the first to investigate the relationship between two variables in a systematic fashion. In one of his studies he studied the connection between the heights of fathers and their sons. He found that men who were taller than average tended to have sons who were taller than average, but not as tall as their fathers, *while men who were shorter than average tended to have sons who were shorter than average,* but not as short as their fathers. *In other words, the sons tended to be closer to the mean height than their fathers; their heights* regress *(fall back) toward the mean. This* regression effect *gave rise to the term* regression analysis.

With a few exceptions, such as contingency tables and analysis of variance, we have been dealing with data determined by one variable. In this chapter we consider data in which the observations consist of values of two variables. Athletes may be classified by height and weight, for example, or we may wish to investigate the relationship between the amount of money spent on a research program and the return from this investment. Two procedures often used to study the relationship between two variables are *regression analysis* and *correlation analysis*. These procedures are related, but they differ in their application. Regression analysis is most useful when values of one variable can be predicted from changes in a second variable, although the analysis itself merely shows an association (rather than a causal relationship) between two variables. When no predictive relationship is presumed between two variables, we can use correlation analysis, but it is not as informative as regression analysis. Correlation analysis is primarily useful in cases where the relationship between the variables is not predictive but merely associative; that is, the variables change together, but no influence of one on the other is necessarily implied.

We cannot infer a cause-and-effect relationship simply as a result of a statistical analysis, either regression or correlation. There may be other variables, not under investigation, that influence either or both variables. For example, teachers' salaries and per capita consumption of alcoholic beverages have both shown a steady increase over the past 40 years. Both are a result of a long-term improvement in economic conditions. Although an analysis investigating the relationship between teachers' salaries and per capita consumption of alcoholic beverages might show a strong association between the two, we cannot conclude that as teachers get more money they consume more alcohol.

In this chapter we first take up regression analysis and learn how to use the SAS System to help in the analysis. We then turn to correlation analysis in Section 13.4 and learn how to use Minitab commands in Section 13.5.

13.0 Case 13: Mail Order Madness

Don Jones, owner of a mail-order business, decided that he had a sure-fire way to increase his sales. He advertised that every order would be shipped on the same day it was received. A friend asked him how he could do this without bankrupting himself hiring extra personnel. Jones explained that he could hire additional clerical and shipping help on a daily basis, provided that he hired them prior to 9:30 AM. The mail was delivered daily by 9:00 AM, and the switchboard closed at 5:00 PM, so no phone orders could be received after then. Clerical and shipping personnel would work until all orders were processed and ready for shipping, and the parcel delivery service would pick up as late as 8:00 PM. All he had to do was make sure he had enough people hired to do the work.

Unfortunately, the total number of daily orders varied widely, from as few as 3500 to as many as 8000, so he could not be certain how many people to hire. He had consulted a statistician, who said it might be possible to estimate the number of orders daily by weighing the mail when it arrives and then predicting the number of orders from the weight. Telephone lines were always busy, so the number of telephone orders was about the same daily. All he had to do was keep track of the weight of the mail and the number of orders daily for a few weeks. From this information the statistician could obtain a *regression equation* that, if valid, could be used to predict the daily orders. For 30 days he collected the information and then took it to the statistician to get it analyzed. "Do you see any sort of pattern?" asked the friend. "You might have bitten off more than you can chew." Together they looked at the data, which are shown here:

DAY	WEIGHT	ORDERS	DAY	WEIGHT	ORDERS
1	7.22	5611	16	5.17	4427
2	10.41	7388	17	6.94	5437
3	4.67	4411	18	4.38	3966
4	6.63	5273	19	11.07	7883
5	3.84	3408	20	7.34	5873
6	4.86	4248	21	8.17	6431
7	8.48	6429	22	6.88	5197
8	8.13	6325	23	9.45	6688
9	11.16	7761	24	5.73	4758
10	9.57	6966	25	3.46	3807
11	6.61	5293	26	10.54	7537
12	4.07	3887	27	5.88	4896
13	3.94	3915	28	5.68	4704
14	9.41	6869	29	6.60	5476
15	7.32	5444	30	9.38	6839

"I can see that the more the mail weighs, the more orders I have," Jones said, "so maybe I'll be able to get what I need. We'll just have to see."

13.1 Fitting the Regression Line

In the past two chapters we have used analysis of variance to determine whether there are differences in a variable for various levels of a factor or factors. If we can express the levels of the factor as interval data (a numerical variable), a different type of analysis may be most helpful. For instance, we can use analysis of variance to determine whether there are differences in number of hours to relief for different dosages of a drug. If the dosages are given as numerical data, we may obtain an even better analysis based on these numbers that may help us to determine the best dosage—even if that dosage was not one of the levels being investigated.

Recall that we can write the equation of a straight line as $y = mx + b$, where m is the slope of the line and b is the y intercept. We can also write $y = a + bx$, where b represents the slope and a is the y intercept; the letters are arbitrary. The important point is that, for the linear equation $y = a + bx$, if $x = 0$ then $y = a$, and for each increase of one unit in x, y will increase by b units. This is called a linear **functional relationship** between x and y. In the ideal case, we could be certain that, for a particular dosage of the drug (x), the number of hours to relief is an exact number (y). The points (x, y) then lie on a straight line.

In most situations, the association between two variables is not exact; that is, the relationship is not functional. In many cases, however, there is certainly a relationship between the two variables in that increases in x are associated with increases in y (or with decreases in y in some cases), but the points (x, y) do not lie exactly on a straight line (see Figure 13.1).

Such a relationship is called a linear **statistical relationship** between x and y. Using statistical techniques we may use such a relationship to obtain estimates for y from known values of x. (In some textbooks a functional relationship between two variables is called a *deterministic model,* since values of y may be *determined* from values of x, while a statistical relationship may be called a *probabilistic model,* since it contains a random error component.) In this section we show how to obtain a linear equation estimating the statistical relationship between two variables and give a method for testing whether or not the regression equation has predictive value (assuming a predictive relationship between the variables).

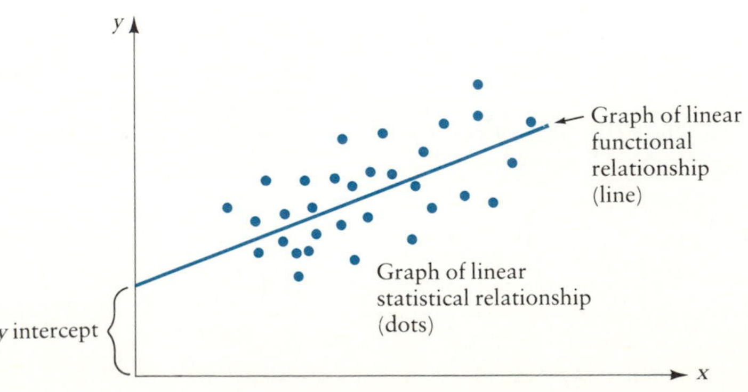

Figure 13.1 Graphs of a linear functional relationship and a linear statistical relationship.

Table 13.1 WATERFRONT FOOTAGE AND SELLING PRICE OF LOTS.

LOT	WATERFRONT FOOTAGE	SELLING PRICE
1	188	140
2	231	160
3	176	130
4	194	130
5	244	180
6	207	160
7	198	140
8	217	150
9	181	140
10	194	150

13.1.1 Statistical Relationship Between Two Variables

A realtor has several waterfront lots to list for sale. In order to set a fair price for the lots, he studies the selling prices of ten similar lots that have sold recently. The most important difference between the lots, as related to selling price, appears to be the amount of waterfront footage. He therefore obtains information relating selling price (in thousands of dollars) to waterfront footage. (See Table 13.1).

Since waterfront footage is presumed to influence selling price, we call waterfront footage (or simply frontage) the independent or **predictor variable** and call selling price (or simply price) the dependent or **response variable.** The term independent does not refer to statistical independence in the sense of Chapter 3; rather, it indicates that this variable does not depend on values of the other variable. We usually denote the predictor variable by x and the response variable by y.

If we plot the data of Table 13.1 on a graph (Figure 13.2), we notice a general upward trend in price as waterfront footage increases. We examine the encircled data more closely in Figure 13.3, where the points are labeled by lot for identification purposes only. Now the upward trend becomes even more apparent. If we could determine the price of a lot exactly from the frontage, there would be a functional relationship between the variables. In this case it appears that greater frontage and higher price go together, if not exactly, so there may well be a statistical relationship between the variables.

The graphs in Figures 13.2 and 13.3 are called **scatterplots** (or *scattergrams*). A scatterplot is useful for determining whether there is a relationship between two variables and, if so, what form the relationship takes. Since a straight line appears to fit the relationship between the variables fairly well, we say that there is a **linear trend** in the graph.

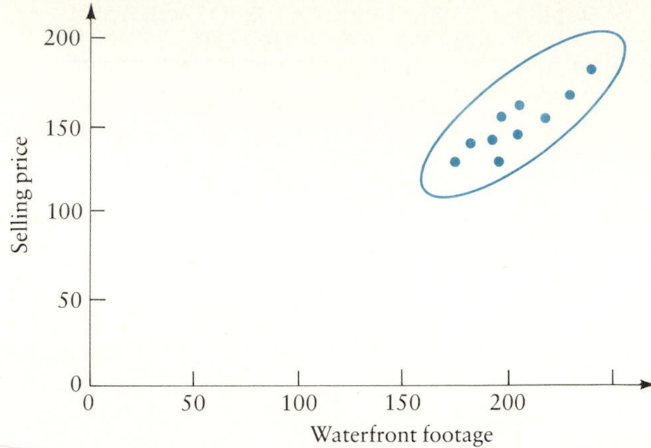

Figure 13.2
Scatterplot for the real estate data.

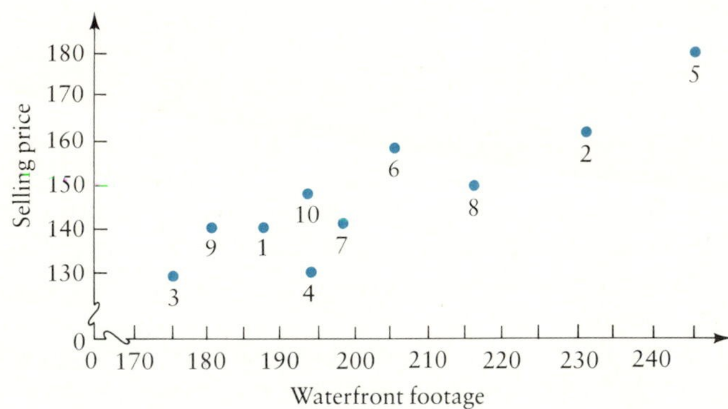

Figure 13.3
Scatterplot of the real estate data of Figure 13.2 (enlarged).

Other examples of scatterplots are shown in Figure 13.4. Plots (*a*), (*c*), and (*d*) show a linear trend. Plot (*b*) shows no trend at all, while plot (*e*) is somewhat linear, somewhat curved. Plot (*f*) is an example of a **curvilinear trend.**

13.1.2 The Linear Regression Model

If two variables appear to have a statistical relationship, regression analysis can determine the function that best fits the data. Each of n observations consists of an ordered pair (x_i, y_i), where $i = 1, 2, 3, \ldots, n$, for which $y_i = f(x_i) + \epsilon_i$, where $y = f(x)$ is the function that best fits the data and ϵ_i is a random error term. (Thus the variable ϵ is a random variable.) If the association between the variables appears to be linear, then the best-fitting function is the equation of a straight line and $f(x_i) = \beta_0 + \beta_1 x_i$, where β_1 is the slope of the line and β_0 is the y intercept. If there is a linear statistical

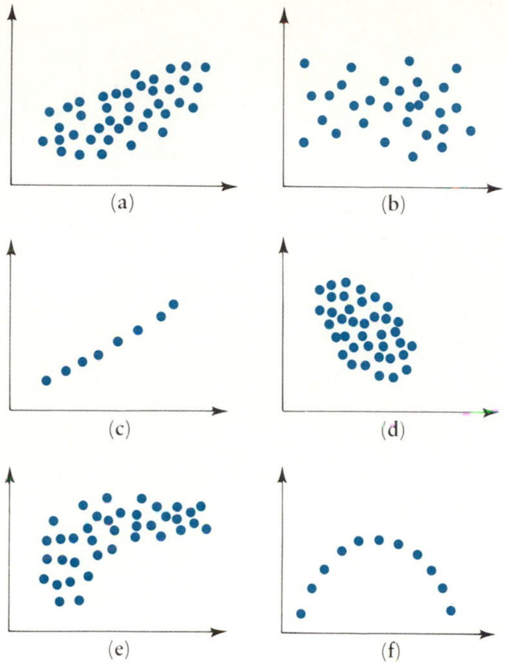

Figure 13.4
Six hypothetical scatterplots.

relationship between variables x and y, then each observation y_i can be defined

$$y_i = \beta_0 + \beta_1 x_i + \epsilon_i$$

where x_i is the ith observation of the predictor variable
y_i is the ith observation of the response variable
β_0 and β_1 are parameters
ϵ_i are $N(0, \sigma^2)$ for each x_i and are independent of each other
$i = 1, 2, \ldots, n$

In order for us to make valid inferences from the model, the assumptions about the ϵ_i must be true. That is, the random error terms are independent of each other and the distribution of the random variable ϵ is the same for each value of x—normal, centered on the regression line, and with the same variance.

Then for any value of x the expected value of the variable y is equal to $E(y)$, where

$$\begin{aligned} E(y) &= E(\beta_0 + \beta_1 x + \epsilon) \\ &= E(\beta_0 + \beta_1 x) + E(\epsilon) \\ &= \beta_0 + \beta_1 x \end{aligned}$$

since β_0 and β_1 are constants, x is fixed, and $E(\epsilon) = 0$. The function

$$E(y) = \beta_0 + \beta_1 x$$

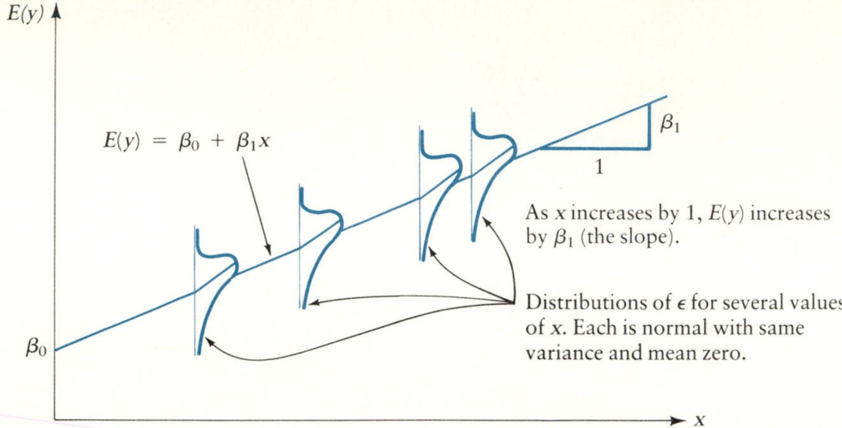

Figure 13.5
Graphs of the hypothetical distributions of the random variable ϵ for several values of x, imposed on the graph of $E(y) = \beta_0 + \beta_1 x$.

is called the **regression function** relating the variables. Figure 13.5 shows a few examples of the distributions of error terms. Note that if $x = 0$ then $E(y) = \beta_0$ and that for each increase of one unit in x, $E(y)$ increases by β_1 units.

Comment Do not confuse $E(y)$ with the usual notation in which the expected value of a variable is its mean; in this case we symbolize the mean of y as μ_y. We might write $E(y)$ more descriptively as $E(y|x)$—that is, the expected value of y given x or the mean of y for a given x. Although both symbols are used, our notation here is common and should cause no confusion.

13.1.3 The Principle of Least Squares

Since we are usually dealing with a sample rather than a population, we cannot determine the regression function. Instead, we obtain estimates for β_0 and β_1 from the sample. These estimates are usually denoted b_0 and b_1. Then for a particular value of x, we have $b_0 + b_1 x$ as an estimate for $E(y)$. We denote this estimate by $\hat{y}$ ("y-hat"). (Recall that in statistics the circumflex symbol (ˆ) denotes an estimator or a point estimate.) Some textbooks use $\hat{\beta}_0$ and $\hat{\beta}_1$ instead of b_0 and b_1, but the latter symbolism is more widely used. The equation $\hat{y} = b_0 + b_1 x$ is called the **sample regression equation** or the **estimated regression function** (and sometimes the *fitted regression equation*). Values of $\hat{y}$ for a particular value of x are called **predicted values** or *fitted values*.

To obtain values of b_0 and b_1 from the sample data, we will use the procedure known as the *principle of least squares*. This principle provides a method for determining the best-fitting equation of any type for a set of data. In this chapter we use the procedure to obtain the best-fitting linear equation. In Figure 13.6, the graph of a sample regression equation is drawn on a scatterplot.

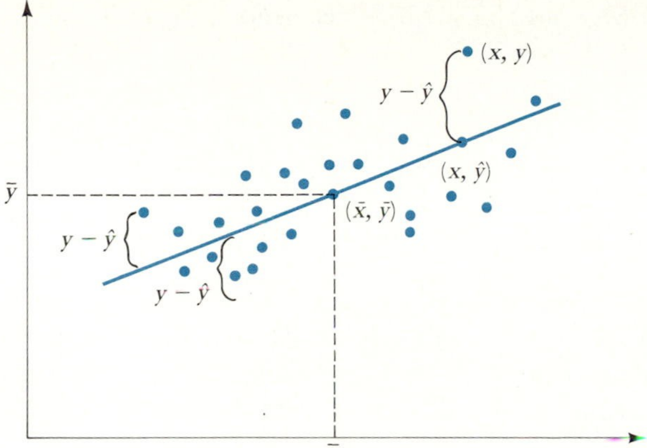

Figure 13.6
Graph of a sample regression equation superimposed on a scatterplot.

For each value of x, the predicted value $\hat{y}$ lies on the line. The difference between the observed value y and the predicted value $\hat{y}$ is $y - \hat{y}$. We define the line of best fit as the one for which the total of vertical distances between each point and the line is minimized. This line is the one for which the sum of the squares (or the absolute values) of these differences is as small as possible. The procedure used to obtain the equation of this line is called the **principle of least squares.** The principle of least squares uses calculus to determine the line for which $\Sigma(y - \hat{y})^2$ is a minimum.

Since $\hat{y} = b_0 + b_1x$, we use the principle to minimize $\Sigma[y - (b_0 + b_1x)]^2$. The values of b_0 and b_1 for which this quantity is a minimum are called **least-squares estimates** for β_0 and β_1. It can be shown (proof omitted) that b_1 and b_0 are equal to the following:

$$b_1 = \frac{\Sigma(x - \overline{x})(y - \overline{y})}{\Sigma(x - \overline{x})^2}$$

and

$$b_0 = \frac{\Sigma y - b_1 \Sigma x}{n}$$

The quantity $\Sigma(x - \overline{x})^2$ is called the **sum of squares for *x*** (SSX); that is, it is the sum of squares of deviations from the mean for the variable x. The quantity $\Sigma(x - \overline{x})(y - \overline{y})$ is the **sum of *xy* terms** (SXY); that is, it is the sum of the products of deviations from the means for the variables x and y (for the same observation). Therefore $b_1 = \text{SXY/SSX}$. By using algebra, we can show that

$$\Sigma(x - \overline{x})(y - \overline{y}) = \Sigma xy - \frac{(\Sigma x)(\Sigma y)}{n}$$

Thus $\text{SXY} = \Sigma xy - (\Sigma x)(\Sigma y)/n$. Recall from Chapter 3 that $\Sigma(x - \overline{x})^2 = \Sigma x^2 - (\Sigma x)^2/n$ so that $\text{SSX} = \Sigma x^2 - (\Sigma x)^2/n$; we can use these shortcut formulas for SXY and SSX. Moreover, since $\overline{y} = \Sigma y/n$ and $\overline{x} = \Sigma x/n$, we can write $b_0 = \overline{y} - b_1\overline{x}$.

Formulas for Fitting a Regression Line

Suppose we obtain a set of n ordered pairs (x_i, y_i) from a population for which the relationship between the variables is described by the regression function $E(y) = \beta_0 + \beta_1 x$. Then we can obtain an estimated regression function $\hat{y} = b_0 + b_1 x$ from the sample. We have

$$b_1 = \frac{\text{SXY}}{\text{SSX}} \quad \text{and} \quad b_0 = \frac{\Sigma y - b_1 \Sigma x}{n} = \overline{y} - b_1\overline{x}$$

where

$$\text{SXY} = \Sigma(x - \overline{x})(y - \overline{y}) = \Sigma xy - \frac{(\Sigma x)(\Sigma y)}{n}$$

and

$$\text{SSX} = \Sigma(x - \overline{x})^2 = \Sigma x^2 - \frac{(\Sigma x)^2}{n}$$

To check for consistency, note that the regression equation must satisfy $\overline{y} = b_0 + b_1\overline{x}$; that is, $(\overline{x}, \overline{y})$ must lie on the sample regression line.

Example 13.1 Use the real estate data of Table 13.1 to determine the sample regression equation, and plot it on the graph.

Solution

The predictor variable (x) is frontage, and the response variable (y) is price. Table 13.2 shows the intermediate sums.

We do not need Σy^2 now, but we will need it later (Example 13.2). Then SXY = 303,130 − (1480)(2030)/10 = 2690 and SSX = 416,392 − $(2{,}030)^2/10$ = 4302. Thus

$$b_1 = \frac{\text{SXY}}{\text{SSX}} = \frac{2690}{4302} \doteq 0.625290563 \doteq 0.6253$$

and

$$b_0 = \frac{1480 - 0.625290563(2030)}{10} \doteq \frac{210.66016}{10} \doteq 21.0660$$

Table 13.2 COMPUTATIONS FOR EXAMPLE 13.1

SUBJECT	x	y	x^2	y^2	xy
1	188	140	35,344	19,600	26,320
2	231	160	53,361	25,600	36,960
3	176	130	30,976	16,900	22,880
4	194	130	37,636	16,900	25,220
5	244	180	59,536	32,400	43,920
6	207	160	42,849	25,600	33,120
7	198	140	39,204	19,600	27,720
8	217	150	47,089	22,500	32,550
9	181	140	32,761	19,600	25,340
10	194	150	37,636	22,500	29,100
Totals	2,030	1,480	416,392	221,200	303,130

Comment When you are calculating the value of b_0, use as many decimal points for b_1 as your calculator can handle, rounding to the number of desired decimal places after completing your computations. There is no rule governing the number of decimal places you should use for b_0 and b_1, other than those consistent with rules of rounding, but since many of the uses for regression analysis involve additional calculations, it is a good idea to maintain a reasonably large number of significant figures. As a rule of thumb, we will give most computations to four decimal places. If the values of b_0 or b_1 are between -1 and $+1$, we will use at least four significant figures.

Thus the sample regression equation for the real estate data is

$$\hat{y} = 21.0660 + 0.6253x$$

For each additional foot of frontage, the estimated increase in mean sales price is 0.6253 thousand dollars, or about \$625; the value of b_0, 21.0660, has no particular meaning here. In other cases, it may have an important meaning such as fixed costs, or start-up time for some activity. As a check that $(\overline{x}, \overline{y})$ is on the regression line, we note that $\overline{y} = 148$, $\overline{x} = 203$, and $21.0660 + 0.6253(203) = 148.0019 \doteq 148$. The difference is due to rounding error.

The graph of this equation is plotted in Figures 13.7 and 13.8 together with the sample points. Figure 13.7 shows the line plotted within the *scope* of the data—the values of x and y that were found in the sample. Figure 13.8 shows the line plotted on the original graph. The dotted portion of the line is outside the scope of the model, and predictions for values of y are not warranted there, since there are no data in the sample to support such projections.

We may use the regression equation to obtain predicted values of y for values of x within the scope of the data. A particular value of x is denoted by x_p, and the corresponding values of y, $\hat{y}$, and $E(y)$ are denoted by y_p, $\hat{y}_p$,

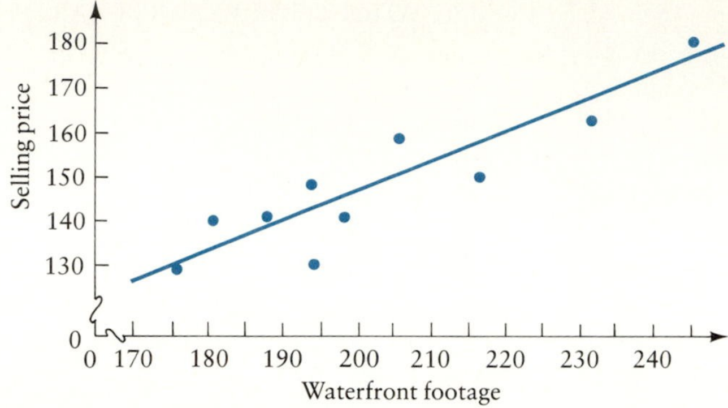

Figure 13.7
Graph of the regression equation $\hat{y} = 21.0660 + 0.6253x$ for the real estate data.

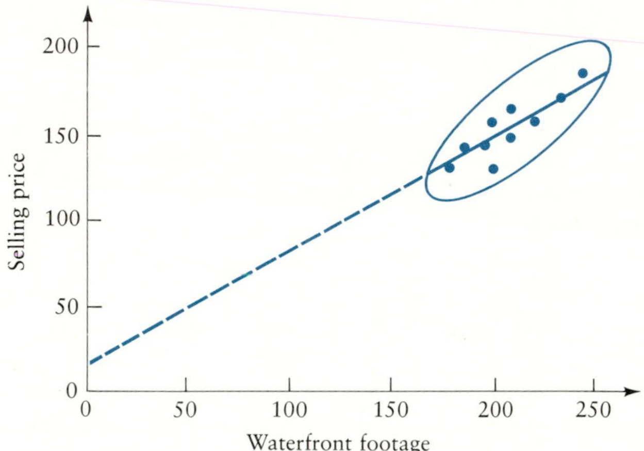

Figure 13.8
Regression line for the real estate data plotted within and outside the scope of the data.

PROFICIENCY CHECKS

1. If $n = 20$ and the intermediate sums in a linear regression problem are as shown here, determine SXY and SSX.

   $\Sigma x = 133.7 \quad \Sigma x^2 = 992.84 \quad \Sigma y = 1475.9 \quad \Sigma xy = 8766.12$

2. Consider the following data:

x	1	1	2	3	4	4	7	9
y	5	6	9	12	14	13	20	24

   a. Draw a scatterplot and determine whether there is a linear trend.
   b. Determine the linear regression equation for the data.
   c. Plot the regression line on the scatterplot.

and $E(y_p)$. If $x_p = 160$, then $\hat{y}_p = 21.0660 + 0.6253(160) \doteq 121.114$, so the predicted mean price for lots with 160 feet of frontage is about \$121,114. For 200 feet of frontage, the predicted mean price is again found by using the regression equation; $21.0660 + 0.6253(200) \doteq 146.126$. The predicted price of lots with 200 feet of waterfront footage is about \$146,126. This is the estimated mean price for all such lots.

13.1.4 Partitioning Total Variation

One of the primary purposes of regression analysis is to reduce the amount of error in prediction. In the real estate data of Example 13.1, the mean frontage ($\bar{y}$) is 148. If we make predictions of price using only $\bar{y}$ for an estimate, the estimated price of each lot would be \$148,000, regardless of frontage. The standard deviation of the predictions, s_y, is equal to $\sqrt{(221{,}200 - (1480)^2/10)/9}$, or about \$15,492. Thus, by the empirical rule, we would expect our predictions to be within about \$30,000 (near $2s_y$) approximately 95% of the time. The variability of our estimates would be quite high. If, on the other hand, we make predictions of price taking frontage into account, we would use the regression equation to obtain $\hat{y}$ for each observation. In this case the variance of the actual numbers as deviations from the predicted values would be close to $\Sigma(y - \hat{y})^2/(n - 1)$ rather than $\Sigma(y - \bar{y})^2/(n - 1)$. Actually, as we will see later, the variance is $\Sigma(y - \hat{y})^2/(n - 2)$. For the data given above, this latter value is about 59.746. Then $\sqrt{59.746} \doteq 7.73$, so we would expect our predictions to be within about \$15,500 (that is, about two times \$7730) approximately 95% of the time.

The sample measure of variability of the actual observations about the predicted values is called the *standard error of estimate* (s_e). It is an estimate for σ, where σ^2 is the common variance of the error terms in the regression model. Here $s_e = 7.73$ and is much smaller than $s_y = 15.13$. This reduction in the variability of our estimates is one of the primary purposes of regression analysis. The standard error of estimate is discussed in Section 13.2.1.

The total amount of variation of the observations of the y variable about $\bar{y}$ is equal to $\Sigma(y - \bar{y})^2$ and is called the total sum of squares for y (SSTO). The total amount of variation of the observations of y about the predicted value $\hat{y}$ for each observation is equal to $\Sigma(y - \hat{y})^2$ and is called the **sum of squares due to error** (SSE). The amount of reduction in variation of the predicted values is called the **sum of squares due to regression** (SSR) and is equal to $\Sigma(\hat{y} - \bar{y})^2$. SSR gives the total variation of the predicted values about the mean. Figure 13.9 shows this relationship.

Clearly the greater the reduction in error variation, the better the regression equation is for predicting future values of y. One of the ways we can test the validity of the regression equation is to use an F test as in Chapters 11 and 12. To do so we must obtain the mean squares for regression and error. These are equal to the corresponding sums of squares divided by the degrees of freedom associated with each term. If there are n observations in

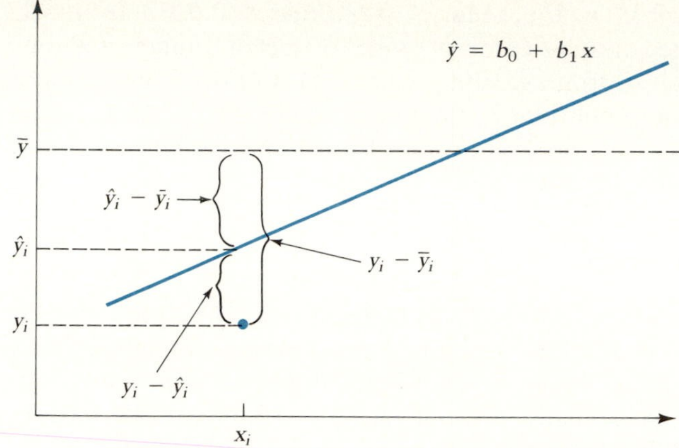

Figure 13.9
Sources of variation in a regression model.

the sample, there are $n - 1$ degrees of freedom for SSTO, since $\Sigma(y - \bar{y})^2$ is based on n observations and one population parameter (μ_y) is estimated. Since $\hat{y} = b_0 + b_1x$, $\Sigma(y - \hat{y})^2 = \Sigma[y - (b_0 + b_1x)]^2$. Thus SSE is also based on n observations, but two population parameters, β_0 and β_1, are estimated so there are $n - 2$ degrees of freedom. Finally SSR, that is, $\Sigma(\hat{y} - \bar{y})^2$, is based on only two observations, the values of b_0 and b_1, and one population parameter (μ_y) is estimated, so the number of degrees of freedom is $2 - 1 = 1$.

We usually compute SSTO, SSE, and SSR using shortcut formulas. The shortcut formulas for partitioning SSTO are summarized here.

Formulas for Partitioning Total Sum of Squares (Regression Analysis)

Total Sum of Squares

$$\text{SSTO} = \Sigma y^2 - \frac{(\Sigma y)^2}{n}$$

Sum of Squares Due to Regression

$$\text{SSR} = \frac{\left[\Sigma xy - \dfrac{\Sigma x \Sigma y}{n}\right]^2}{\left[\Sigma x^2 - \dfrac{(\Sigma x)^2}{n}\right]}$$

$$= \frac{(\text{SXY})^2}{\text{SSX}}$$

Sum of Squares Due to Error

$$SSE = SSTO - SSR$$

Mean Squares

$$MSR = \frac{SSR}{1}$$

$$MSE = \frac{SSE}{n-2}$$

The results of partitioning the total sum of squares are usually summarized in an analysis of variance table such as Table 13.3.

Example 13.2

Use the data and results of Example 13.1 to obtain an analysis of variance table for the regression analysis.

Solution

From Example 13.1, we have $n = 10$, SXY = 2690, and SSX = 4302; SSTO = 221,200 − $(1480)^2/10$ = 2160. Then

$$SSR = \frac{(2690)^2}{4302} \doteq 1682.0316$$

so SSE $\doteq$ 2160 − 1682.0316 = 477.9684. The analysis of variance table is shown in Table 13.4.

Table 13.3 ANOVA TABLE FOR LINEAR REGRESSION ANALYSIS.

SOURCE OF VARIATION	DF	SS	MS
Regression	1	SSR	MSR
Error	$n-2$	SSE	MSE
Total	$n-1$	SSTO	

Table 13.4 ANOVA TABLE FOR EXAMPLE 13.2.

SOURCE OF VARIATION	DF	SS	MS
Regression	1	1682.0316	1682.0316
Error	8	477.9684	59.7460
Total	9	2160.0000	

PROFICIENCY CHECKS

3. Use the results of Proficiency Check 1 and the fact that $\Sigma y^2 =$ 125,251.14 to obtain an analysis of variance table for the regression analysis.

4. Use the results of Proficiency Check 2 to obtain an analysis of variance table for the regression analysis.

13.1.5 Testing the Association Between the Variables

If two variables are statistically related, we say that there is an **association** between the variables. If there is a linear relationship between the variables, then the slope of the regression function, β_1, is not equal to zero. One way to test the association between the variables is to use an F test, as in analysis of variance. This test is useful whether the association between the variables is linear or nonlinear.

Suppose that we obtain all possible samples with n observations from the same population of ordered pairs and calculate MSR and MSE for each sample. We can determine the expected values of MSR and MSE: $E(\text{MSE}) = \sigma^2$ and $E(\text{MSR}) = \sigma^2 + \beta_1^2\text{SSX}$. If there is a linear association between the variables, then $\beta_1 \neq 0$ because as x increases, $E(y)$ changes by an amount equal to β_1 times the change in x. If there is no linear association, then $\beta_1 = 0$ and changes in x have no effect on y. In that case $E(\text{MSR}) = \sigma^2$ so that MSR and MSE are both estimates for σ^2. Thus the ratio MSR/MSE has an F distribution with 1 and $n - 2$ degrees of freedom. If $\beta_1 \neq 0$, MSR will be greater than MSE, so the sample value of the test statistic, F^*, will be large. If F^* is greater than the appropriate critical value of F, we may conclude that $\beta_1 \neq 0$ and there is a linear relationship between the variables. (We may conclude that $\beta_1 > 0$ or $\beta_1 < 0$ according to whether $b_1 > 0$ or $b_1 < 0$.) We thus have the following procedure for determining whether there is a linear relationship between the variables.

F Test for Linear Association Between Two Variables

Suppose we take a random sample of n ordered pairs of observations from a population for which the regression function is $E(y) = \beta_0 + \beta_1 x$ and that for each value of x the deviations between y and $E(y)$ are independent and normally distributed with variance σ^2. Then we may test the hypothesis $\beta_1 = 0$. The sample value of the test statistic used to test the hypothesis is F^*, where $F^* = \text{MSR/MSE}$. The decision rule is dictated by the level of significance α chosen for the test.

Null and Alternate Hypotheses

$H_0: \beta_1 = 0$

$H_1: \beta_1 \neq 0$

Decision Rule

Reject H_0 if $F^* > F_\alpha\{1, n - 2\}$

Table 13.5 ANOVA TABLE FOR EXAMPLE 13.3.

SOURCE OF VARIATION	DF	SS	MS
Regression	1	1682.0316	1682.0316
Error	8	477.9684	59.7460
Total	9	2160.0000	

Example 13.3

Use the real estate data of Example 13.1 and the results of Example 13.2 to determine whether there is a linear association between frontage and price. Use the 0.05 level of significance.

Solution

The analysis of variance table is reproduced in Table 13.5. Then $F^* = 1682.0316/59.7460 \doteq 28.15$; $F_{.05}\{1, 8\} = 5.32$. Thus we can reject the hypothesis that $\beta_1 = 0$ and conclude that $\beta_1 \neq 0$; there is a linear association between the variables.

PROFICIENCY CHECKS

5. Use the data of Proficiency Check 1 and the results of Proficiency Check 3 to determine whether there is a linear association between the variables. Use the 0.05 level of significance.
6. Use the data of Proficiency Check 2 and the results of Proficiency Check 4 to determine whether there is a linear association between the variables. Use the 0.01 level of significance.

13.1.6 The SAS Procedure REG

We can obtain the sample regression equation and analysis of variance tables for regression by using the SAS procedures REG and GLM. We will use the REG procedure here, as it was designed specifically for regression analysis. To obtain the printout we enter the data as shown in previous chapters. Then we use

```
PROC REG;
  MODEL yvariable = xvariable;
```

The printout of PROC REG for the real estate data is given in Figure 13.10. The important values have been marked. We see that $F^* = 28.153$ and the P value is 0.0007. Thus we can conclude that there is a significant association between the variables.

Figure 13.10
SAS printout of the regression analysis for the real estate data.

```
INPUT FILE
DATA LOTS;
  INPUT FRONTAGE PRICE @@;
CARDS;
188 140 231 160 176 130 194 130 244 180 207 160 198 140
217 150 181 140 194 150
PROC REG;
  MODEL PRICE = FRONTAGE;

OUTPUT FILE

DEP VARIABLE: PRICE
```

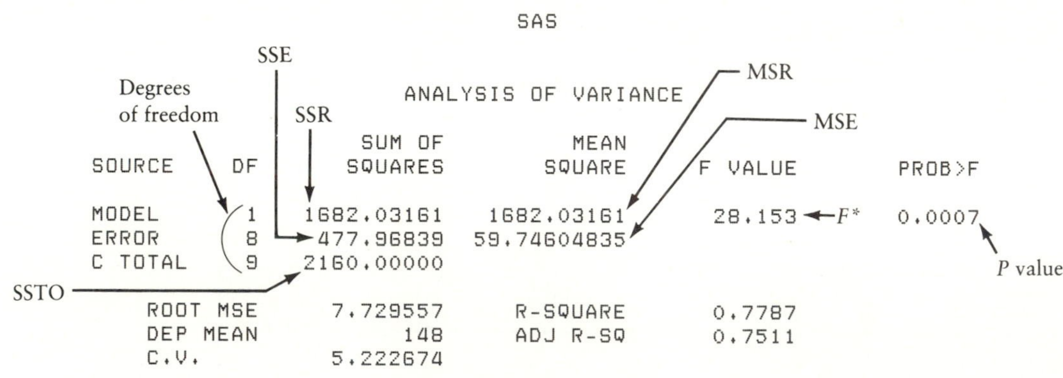

```
                                   SAS

                          ANALYSIS OF VARIANCE

                        SUM OF           MEAN
SOURCE       DF        SQUARES         SQUARE       F VALUE       PROB>F

MODEL         1     1682.03161     1682.03161        28.153       0.0007
ERROR         8      477.96839    59.74604835
C TOTAL       9     2160.00000

     ROOT MSE         7.729557     R-SQUARE          0.7787
     DEP MEAN              148     ADJ R-SQ          0.7511
     C.V.             5.222674
```



```
                         PARAMETER ESTIMATES

                   PARAMETER        STANDARD       T FOR H0:
VARIABLE   DF       ESTIMATE           ERROR     PARAMETER=0      PROB > |T|

INTERCEP    1    21.06601581     24.04753572           0.876          0.4066
FRONTAGE    1     0.62529056      0.11784723           5.306          0.0007
```

b_0 b_1 t^* P value

Problems

Practice

13.1 Consider the following pairs of variables. Is the relationship between them functional or statistical?

a. The edge of a cube and the volume of the cube
b. The number of sales per day in a department store and the daily profit
c. The price of gasoline and the number of gallons of gasoline used by American motorists
d. The depth to which a diver has descended in the ocean and the pressure in pounds per square inch

13.2 A neighborhood car wash has a membership plan. Annual membership is \$20 and members pay 50¢ per wash.

a. Is the relationship between number of washes and the amount a member pays in a year functional or statistical?
b. Obtain an equation for the annual cost (y) to a member who has x car washes during the year.
c. Does the fact that you can obtain an equation to describe this relationship confirm your answer to part (*a*)? Why?

13.3 Consider the following data points:

x	10	13	14	17	16	14
y	15	16	19	23	22	20

a. Draw a scatterplot for the data. Does there appear to be a linear trend?
b. Draw the line $y = 3 + x$ on the graph. Does it appear to fit the data reasonably well?
c. Obtain the squares of the vertical deviations of the observed values from the line $y = 3 + x$. That is, obtain $\Sigma[y - (3 + x)]^2$ for the data.
d. Obtain the least-squares regression line for the data.

13.4 A regression equation for predicting y from x is $\hat{y} = 15.776 + 1.228x$.

a. What is the estimated change in the mean value of y when x increases by 3?
b. If $\bar{y} = 34.43$, determine $\bar{x}$.

13.5 Use the following data for this problem:

x	23	26	33	28	34	39
y	112.6	117.9	123.8	119.7	127.8	133.4

a. Draw a scatterplot of the data. Draw a line of best fit using a freehand estimate.
b. Obtain the sample regression equation for predicting y from x.
c. Graph the regression line with the original data. How well does your freehand estimate compare with the least-squares line?
d. Verify that $\bar{y} = b_0 + b_1\bar{x}$.

13.6 Use the following data for this problem:

x	4.2	3.6	2.8	3.7	1.1	0.8	3.4	1.1
y	0.125	0.176	0.255	0.188	0.317	0.446	0.162	0.312

a. Obtain the sample regression equation for predicting y from x.
b. Graph the regression line with the original data.
c. Obtain the analysis of variance table.
d. Test the hypothesis, at $\alpha = 0.01$, that there is no linear association between the variables.
e. What are the predicted values of y for $x_p = 2.5$ and $x_p = 1.5$?

Applications

13.7 Several farmers in the same county used varying amounts of fertilizer and obtained varying yields of corn. The results are shown here (in hundredweight of fertilizer and tons of corn per acre):

Fertilizer	8.3	9.2	7.7	8.4	8.8	9.6	10.3	8.7	9.1	9.4
Yield	13.6	15.4	12.8	13.4	14.6	15.8	15.5	14.1	14.9	15.6

a. Determine the regression equation for predicting yield from amount of fertilizer.
b. What is the meaning of b_1 in this equation?
c. How would you interpret the value of b_0?
d. Obtain the analysis of variance table for the data. Does this equation have predictive value? Use $\alpha = 0.05$.
e. Predict the yield if 10 hundredweight of fertilizer is used per acre.

13.8 The duration of a certain illness is apparently related to the bacterial count of the infecting organism. Ten patients with the illness have a bacterial count taken upon admission, and the duration of their symptoms is observed. The data are shown here:

PATIENT	COUNT (1000S)	DURATION (DAYS)
A	8	11
B	7	10
C	4	9
D	6	8
E	9	12
F	7	9
G	8	13
H	3	7
I	8	10
J	7	11

a. Construct a scattergram for the data.
b. Derive the regression equation for predicting duration of illness from bacterial count. Plot it on the scattergram as a check of your calculations.
c. Using $\alpha = 0.05$, can you conclude that bacterial count has an effect on duration of illness?
d. Predict duration of illness from bacterial counts of 5 and 7.
e. Why is it not valid to predict duration of illness if the bacterial count is 15?

13.9 In a reforestation project, all the trees are planted at the same time and watered by means of irrigation ditches. After 10 years a random sample of the trees is examined to find their height and distance from the nearest irrigation ditch. The data are given here:

HT (FT)	DISTANCE (FT)	HT (FT)	DISTANCE (FT)	HT (FT)	DISTANCE (FT)
23	80	21	100	27	72
18	113	17	120	23	88
22	108	20	110	19	84
24	92	22	94	18	108
21	87	23	103	24	92
19	110	26	76	23	84
23	90	24	82	21	93

a. Plot the data and determine whether there appears to be a linear trend. What can you say about the sign of β_1?
b. Is distance from the ditch linearly associated with height? Use $\alpha = 0.05$.
c. What is the effect on height of trees for each additional foot from the irrigation ditch?
d. What is the predicted height of trees 90 feet from the irrigation ditch?

13.10 A corporation uses a screening test to discover potential sales ability of its applicants. A linear regression equation is used for the purpose. Since situations vary, the regression equation is refigured monthly according to the latest data for salespeople who have been hired within the last 6 months. Adjusted gross sales for each person are shown here:

SALESPERSON	TEST SCORE	ADJUSTED GROSS SALES ($1000s)
1	97	141
2	132	113
3	88	94
4	154	157
5	143	118
6	119	131
7	157	148
8	89	107
9	134	158
10	135	136
11	162	159
12	155	146
13	113	122
14	124	131
15	108	113
16	136	94
17	117	124
18	182	237
19	130	118
20	122	145

a. Draw a scatterplot for the data. Does there appear to be a linear trend?
b. Obtain the analysis of variance table and determine, at $\alpha = 0.01$, whether the test appears to have predictive value.
c. Obtain an estimate for the mean change in sales (in thousands of dollars) for each increase of one point in test score.
d. Determine predicted sales for applicants who score as follows: 83; 97; 112; 124; 146.

13.2 Analysis of the Linear Regression Model

Once we have obtained a regression equation, we may use it for a variety of purposes if there is an association between the variables. We may make additional inferences and analyses if the sample is random and the following four assumptions about the model are satisfied:

1. The relationship between the variables is linear and $E(\epsilon) = 0$.
2. The error terms (ϵ_i) are independent of each other.
3. For each value of x, the error terms are normally distributed.
4. For each value of x, the variance of the error terms is the same (σ^2).

Two types of additional analysis are usually of interest. We may obtain confidence intervals for population parameters or test hypotheses about them, or we may make predictions about future values of the response variable. In this section we discuss methods for checking on the assumptions for the regression model and making inferences about β_0 and β_1.

13.2.1 The Standard Error of Estimate

The linear regression model is based on the assumption that each observation y_i differs from the expected value, $\beta_0 + \beta_1 x_i$, by a random error term ϵ_i. When we obtain the sample regression equation $\hat{y} = b_0 + b_1 x$, each y_i differs from $b_0 + b_1 x_i$ by a number that is an estimate for the random error term. In keeping with our practice of using Greek letters for population values and Roman letters for sample values, we denote this difference by e_i. Thus for any observation (x_i, y_i), $y_i - \hat{y} = e_i$. Each value of e_i is called a **residual.** The variance of the random variable ϵ is denoted by σ^2. The estimator of σ^2 is the variance of the residuals, s_e^2. This is equal to $\Sigma(e - \overline{e})^2$ divided by the appropriate degrees of freedom for each sample. A characteristic of the least-squares approach is that $\Sigma e = 0$, so $\overline{e} = 0$ and $\Sigma(e - \overline{e})^2 = \Sigma e^2$. Further, $\Sigma e^2 = \Sigma(y - \hat{y})^2$, so the term has $n - 2$ degrees of freedom. Thus the estimator for σ^2 is $\Sigma(y - \overline{y})^2/(n - 2)$, which is equal to MSE, so that MSE $= s_e^2$, the variance of the residuals.

The square root of s_e^2, s_e, is called the **standard error of estimate.** The standard error of estimate is quite important because it estimates the variability of the error terms. Moreover, it is used in the formulas for estimating β_0, β_1, and values of y. If s_e is large, s_{b_0}, s_{b_1}, $s_{\hat{y}_p}$, and s_f will also be large, so the estimates will be highly variable. Thus the smaller the value of s_e, the greater the predictive value of the regression equation. We need not calculate MSE to obtain s_e. By applying the definitions of MSE and SSE, we can derive additional formulas that can be used if the ANOVA table is not determined.

The Standard Error of Estimate

If $\hat{y} = b_0 + b_1x$ is a regression equation obtained from a sample of n observations, then the standard deviation of the residuals is called the standard error of estimate, s_e, where

$$s_e = \sqrt{\text{MSE}}$$

If MSE has not been calculated, we can obtain s_e by using one of the alternative formulas:

$$s_e = \sqrt{\frac{\text{SSTO} - b_1\text{SXY}}{n - 2}}$$

$$s_e = \sqrt{\frac{\Sigma y^2 - b_0\Sigma y - b_1\Sigma xy}{n - 2}}$$

Example 13.4

Determine the standard error of estimate for the data of Example 13.1.

Solution

Since MSE $= 59.7460$, we have $s_e = \sqrt{59.7460} \doteq 7.7296$.

13.2.2 Residual Analysis

The assumptions underlying the linear regression model must be met if inferences and predictions are to be valid. Some of these assumptions may be tested by using a residual plot. The **residual plot** for the linear regression model is a graph of the residuals plotted against $\hat{y}$. When we use a computer for the analysis, the residual plot may be printed out as part of the output. When we are doing the computations by hand, we must compute each residual separately. For a linear regression equation, we could just as easily plot the residuals against x rather than $\hat{y}$, and the same information would be obtained. Using the residual plot, we may verify that the following assumptions are met and that the linear regression model is the best fit to the data:

1. *The association between the variables is linear.* If this is true, the points will scatter at random above and below the zero line.

2. *The variance of the error terms is constant for all values of x.* If this is true, the spread of the residuals will be about the same for all values of x. This assumption may be verified or denied only if n is reasonably large.

3. *The distribution of the error terms is normal.* If this is true, then by the empirical rule about two-thirds of the residuals will lie within s_e of zero and about 95% will lie within $2s_e$ of zero. More precise measurements are possible, but as a rough check this result will be satisfactory. Again n must be reasonably large for this assumption to be verified or denied.

4. *The error terms are independent.* This assumption can often be verified by plotting the residuals against time order—that is, the order in which we obtained the observations. If there is a random pattern, we conclude that the error terms are independent, at least with respect to time. If they are not independent, in most cases successive residuals will be fairly close together. In this case, the response variable is said to be *positively autocorrelated.* Problems of autocorrelation are often dealt with through time-series analysis, which is not covered here. (If successive residuals are far apart—positive, negative, positive, negative, and so on—the response variable is *negatively autocorrelated.* Such autocorrelation is rare.)

We may also use a residual plot to check for the presence of **outliers**—observations that do not appear to belong with the rest of the data. If a residual lies more than $3s_e$ from zero, you should check it to see if it belongs with the data. If the residual is a result of an error of measurement or can be shown not to belong for one reason or another, it should be discarded. *Outliers should not be discarded unless they can be shown to have been included erroneously.*

Regression analysis is reasonably robust with respect to assumptions 2, 3, and 4. Thus some deviation from normality, constant variance, or independence of the error terms is permitted. If the association between the variables is not linear, a different model should be used.

Example 13.5

Use the real estate data of Table 13.1 to construct a residual plot, test the assumptions, and test for the presence of outliers.

Solution

The regression equation is $\hat{y} = 21.0660 + 0.6253x$; we reproduce the data in Table 13.6 together with predicted values ($\hat{y}$) and residuals (e). Waterfront footage is denoted by x, and selling price (in thousands of dollars) is denoted by y. The residuals sum to 0.01, which is satisfactory, allowing for rounding error. The residual plot is shown in Figure 13.11.

Table 13.6 REAL ESTATE DATA FOR EXAMPLE 13.5.

LOT	x	y	$\hat{y}$	e
1	188	140	138.62	1.38
2	231	160	165.51	−5.51
3	176	130	131.12	−1.12
4	194	130	142.37	−12.37
5	244	180	173.64	6.36
6	207	160	150.50	9.50
7	198	140	144.87	−4.87
8	217	150	156.75	−6.75
9	181	140	134.24	5.76
10	194	150	142.37	7.63

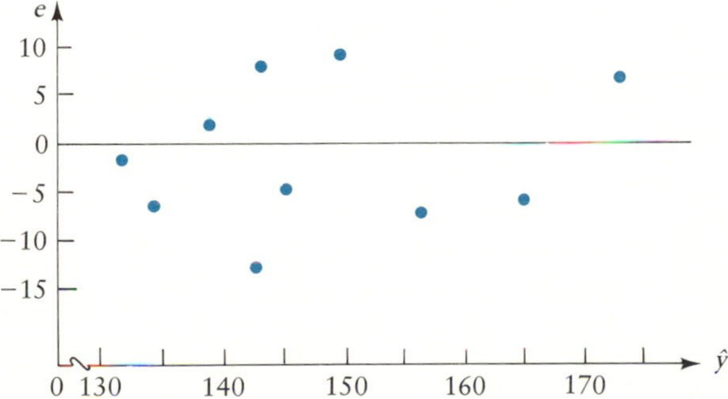

Figure 13.11
Residual plot for the real estate data of Example 13.5.

In Figure 13.11, points are scattered randomly above and below the horizontal axis, so the assumption of linearity appears to be met. There are really too few points to provide a test of equal variance or normality, but we note that—except for one residual, that for lot 4—the points appear equally spread. Further, since $s_e \doteq 7.7296$, eight of the ten residuals have absolute values less than s_e while all have absolute values less than $2s_e$. Thus there appears to be no deviation from normality. The residual for lot 4 is within $2s_e$, so it is not considered an outlier.

PROFICIENCY CHECK

7. Use the data of Proficiency Check 2 and the results of Proficiency Check 4 to construct a residual plot for the data. Test assumptions and summarize your conclusions. Do the residuals sum to zero?

13.2.3 Estimating β_1

If we can conclude that the association between two variables is best described by a linear regression function, $E(y) = \beta_0 + \beta_1 x$, it is usually important to be able to estimate the value of β_1 closely. Clearly each value of b_1 is an estimate for β_1, but we obtain different values of b_1 from different samples. If all possible samples with n observations are obtained from the same population of ordered pairs, we may determine a sampling distribution of the variable b_1. It can be shown that this variable has a normal distribution with mean $E(b_1) = \beta_1$ and variance $\sigma_{b_1}^2 = \sigma^2/\text{SSX}$; that is, b_1 is $N(\beta_1, \sigma_{b_1}^2)$. Then the variable $(b_1 - \beta_1)/\sigma_{b_1}$ is $N(0, 1)$, so that $(b_1 - \beta_1)/s_{b_1}$ has a t distribution. The best estimate for σ^2 is MSE, so the best estimate for $\sigma_{b_1}^2$ is MSE/SSX. Then $s_{b_1} = \sqrt{\text{MSE/SSX}}$ or $s_e/\sqrt{\text{SSX}}$, where s_e is the standard error of estimate. Since MSE has $n - 2$ degrees of freedom, $(b_1 - \beta_1)/s_{b_1}$ is $t(n-2)$. A $100(1 - \alpha)\%$ confidence interval for β_1 then has limits $b_1 - ts_{b_1}$ and $b_1 + ts_{b_1}$, where $t = t_{\alpha/2}\{n-2\}$.

Confidence Interval for β_1

Suppose we take a random sample of n ordered pairs of observations from a population for which the regression function is $E(y) = \beta_0 + \beta_1 x$ and that, for each value of x, the error terms are independent and $N(0, \sigma^2)$. Then a $100(1 - \alpha)\%$ confidence interval for β_1 is the interval

$$b_1 - ts_{b_1} \quad \text{to} \quad b_1 + ts_{b_1}$$

where $t = t_{\alpha/2}\{n-2\}$ and $s_{b_1} = \sqrt{\text{MSE/SSX}}$.

Example 13.6

Use the real estate data of Examples 13.1 through 13.4 to obtain a 95% confidence interval for the increase in mean price of a lot for each increase of 1 foot of waterfront.

Solution

The estimated increase in mean price of a lot for each increase of 1 foot in frontage is \$625.30 (from Example 13.2). Now SSX = 4302 and, from Example 13.4, $s_e \doteq 7.7296$. Then $s_{b_1} = s_e/\sqrt{\text{SSX}} = 7.7296/\sqrt{4302} \doteq 0.1178$. There were ten observations, so we use $t_{.025}\{8\} = 2.306$; $2.306(0.1178) \doteq 0.2718$, so the confidence limits for β_1 are $0.6253 - 0.2718$ and $0.6253 + 0.2718$. We are thus 95% confident that β_1 lies between 0.3535 and 0.8971. That is, we are 95% confident that for each increase of 1 foot of waterfront, the price of a lot will increase between about \$353 and \$897.

Example 13.7

Use the following data to obtain a 99% confidence interval for β_1:

x	11	12	22	14	17	22	18	16
y	44	41	35	42	39	32	37	42

Solution

We have $\Sigma x = 132$, $\Sigma x^2 = 2298$, $\Sigma y = 312$, $\Sigma y^2 = 12{,}284$, $\Sigma xy = 5039$, and $n = 8$. Then

$$\text{SSTO} = 12{,}284 - (312)^2/8 = 116$$
$$\text{SSX} = 2298 - (132)^2/8 = 120$$
$$\text{SXY} = 5039 - (132)(312)/8 = -109$$

Furthermore,

$$b_1 = -109/120 \doteq -0.9083333 \doteq -0.9083$$

and

$$\begin{aligned} s_e &= \sqrt{(\text{SSTO} - b_1\text{SXY})/(n-2)} \\ &= \sqrt{[116 - (-0.9083333)(-109)]/6} \\ &= \sqrt{2.8319} \\ &\doteq 1.6828 \end{aligned}$$

Then $s_{b_1} = s_e/\sqrt{\text{SSX}} \doteq 1.6828/\sqrt{120} \doteq 0.1536$. Finally, $t_{.005}\{6\} = 3.707$, so the confidence limits are $-0.9083 - 3.707(0.1536)$ and $-0.9083 + 3.707(0.1536)$—that is, $-0.9083 - 0.5694$ and $-0.9083 + 0.5694$—so the confidence interval is from -1.4777 to -0.3389. We are 99% certain that β_1 lies between -1.4777 to -0.3389; that is, 99% of the confidence intervals obtained in this way will contain β_1. The data thus indicate, at the 99% level of confidence, that the slope is negative.

PROFICIENCY CHECKS

8. Use the data of Proficiency Checks 1 and 3 to obtain a 95% confidence interval for β_1.
9. Use the data of Proficiency Checks 2 and 4 to obtain a 99% confidence interval for β_1.

13.2.4 Hypothesis Tests Concerning β_1

If the assumptions about the model are met, we know that b_1 is $N(\beta_1, \sigma^2_{b_1})$ so that $(b_1 - \beta_1)/s_{b_1}$ is $t(n-2)$. We can use a t test to test the hypothesis that β_1 has a certain value, say β_1^*, and the test statistic will be $t = (b_1 - \beta_1^*)/s_{b_1}$.

Hypothesis Tests Concerning β_1

Suppose we take a random sample of n ordered pairs of observations from a population for which the regression function is $E(y) = \beta_0 + \beta_1 x$ and the assumptions for regression analysis are met. Then the sample value of the test statistic used to test the hypothesis H_0: $\beta_1 = \beta_1^*$ is t^*, where

$$t^* = \frac{b_1 - \beta_1^*}{s_{b_1}}$$

and $s_{b_1} = s_e/\sqrt{\text{SSX}}$.

The decision rule is determined by the alternate hypothesis and the level of significance α chosen for the test.

Null and Alternate Hypotheses

H_0: $\beta_1 = \beta_1^*$	H_0: $\beta_1 = \beta_1^*$	H_0: $\beta_1 = \beta_1^*$
H_1: $\beta_1 > \beta_1^*$	H_1: $\beta_1 \neq \beta_1^*$	H_1: $\beta_1 < \beta_1^*$

Decision Rules

Reject H_0 if	Reject H_0 if	Reject H_0 if
$t^* > t_\alpha\{n-2\}$	$\lvert t^*\rvert > t_{\alpha/2}\{n-2\}$	$t^* < -t_\alpha\{n-2\}$

Note that we may test the hypothesis H_0: $\beta_1 = 0$ by using either a t test or an F test (Section 13.2). The two tests are equivalent for linear regression (but only for linear regression). We can show that $F^* = (t^*)^2$ as follows:

$$F^* = \frac{\text{MSR}}{\text{MSE}} = \frac{\text{SSR}}{\text{MSE}}$$

Now

$$\text{SSR} = \frac{\text{SXY}^2}{\text{SSX}} = \frac{\text{SXY}^2}{\text{SSX}} \cdot \frac{\text{SSX}}{\text{SSX}} = \frac{\text{SXY}^2}{\text{SSX}^2} \cdot \text{SSX} = b_1^2\text{SSX}$$

and

$$F^* = \frac{b_1^2\text{SSX}}{\text{MSE}} = \frac{b_1^2}{\text{MSE/SSX}} = \frac{b_1^2}{s_{b_1}^2} = (t^*)^2$$

It can be shown that the F distribution with 1 and ν degrees of freedom is the square of the t distribution with ν degrees of freedom; $F_\alpha\{1, \nu\} = (t_{\alpha/2}\{\nu\})^2$. Thus the two tests are equivalent for H_0: $\beta_1 = 0$ with the alternate hypothesis H_1: $\beta_1 \neq 0$. If the hypothetical value of β_1 is not zero or if the alternate hypothesis is not two-sided, we must use the t test.

If the alternate hypothesis is two-sided, we can use a confidence interval to test the hypothesis H_0: $\beta_1 = \beta_1^*$. If the level of significance is α and a $100(1 - \alpha)\%$ confidence interval for β_1 does not include β_1^*, we may reject H_0 and conclude that H_1: $\beta_1 \neq \beta_1^*$.

Example 13.8

Use the results of the real estate example (Example 13.6) to test the conjecture that there is a positive association between the variables—that is, as frontage increases, price increases. Use the 0.05 level of significance.

Solution

We wish to test the hypothesis H_0: $\beta_1 = 0$ against the alternate hypothesis H_1: $\beta_1 > 0$. As $t_{.05}\{8\} = 1.860$, we reject H_0 if $t^* > 1.860$. Now $b_1 = 0.6253$ and $s_{b_1} = 0.1178$, so $t^* = (0.6253 - 0)/0.1178 \doteq 5.308$ and we reject H_0, concluding that there is a positive association between the variables.

Example 13.9

Use the data of Example 13.7 to test the hypothesis that $\beta_1 = -1$. Use the 0.05 level of significance.

Solution

The alternate hypothesis is two-sided and $t_{.025}\{6\} = 2.447$, so we will reject H_0 if $|t^*| > 2.447$. Now $\beta_1 \doteq -0.9083$ and $s_{b_1} \doteq 0.1536$, so $t^* = [-0.9083 - (-1)]/0.1536 \doteq 0.597$. Since 0.597 is not greater than 2.447, we conclude that $\beta_1 = -1$; that is, we do not have enough evidence to conclude that β_1 differs from -1.

PROFICIENCY CHECKS

10. Use the results of Proficiency Check 8 to test the conjecture that $\beta_1 < -10$. Use the 0.05 level of significance.
11. Use the results of Proficiency Check 9 to determine whether $\beta_1 = 1.8$. Use the 0.01 level of significance.

13.2.5 Estimating β_0

In cases where β_0 is meaningful, it is useful to obtain a confidence interval for β_0. We can use a method similar to that we used to obtain a confidence interval for β_1. If we obtain sample values of b_0 from samples of n observations of ordered pairs and the assumptions for the model apply, the sampling distribution of b_0 will have a normal distribution with $E(b_0) = \beta_0$ and $\sigma_{b_0}^2 = \sigma^2(1/n + \bar{x}^2/\text{SSX})$. Since b_0 is $N(\beta_0, \sigma_{b_0}^2)$, then $(b_0 - \beta_0)/\sigma_{b_0}$ is

$N(0, 1)$. Finally, $(b_0 - \beta_0)/s_{b_0}$ is $t(n - 2)$, where $s_{b_0} = s_e\sqrt{1/n + \bar{x}^2/\text{SSX}}$. Thus we have the following procedure for obtaining a confidence interval for β_0.

Confidence Interval for β_0

If we take a random sample of n ordered pairs of observations from a population for which the regression function is $E(y) = \beta_0 + \beta_1 x$, a $100(1 - \alpha)\%$ confidence interval for β_0 is the interval

$$b_0 - ts_{b_0} \quad \text{to} \quad b_0 + ts_{b_0}$$

where $t = t_{\alpha/2}\{n - 2\}$ and

$$s_{b_0} = s_e\sqrt{\frac{1}{n} + \frac{\bar{x}^2}{\text{SSX}}}$$

provided the error terms are independent and $N(0, \sigma^2)$ for each value of x.

Sometimes β_0 should be zero. For example, if a mail-order company does not take telephone orders and the weight of the mail is zero, the number of orders is also zero. If, theoretically, $\beta_0 = 0$ and the sample result confirms this hypothesis, we can use a regression procedure that computes the best-fitting regression line through the origin. This procedure may be found in Neter and others (1985, pp. 160–164).

Example 13.10 Use the data of Example 13.7 to obtain a 95% confidence interval for β_0.

Solution

Since $b_1 \doteq -0.9083$, $\Sigma x = 132$, and $\Sigma y = 312$, we have $b_0 = [312 - b_1(132]/8 = 53.9875$. Since $s_e \doteq 1.6828$, $\bar{x} = 16.5$, and $\text{SSX} = 120$, we have $s_{b_0} \doteq 1.6828\sqrt{\frac{1}{8} + (16.5)^2/120} \doteq 2.6036$. Now $t_{.025}\{6\} = 2.447$ and $2.447(2.6036) \doteq 6.371$, so the 95% confidence interval for β_0 is 47.6165 to 60.3585.

PROFICIENCY CHECK

12. Use the data of Proficiency Checks 2 and 4 to obtain a 99% confidence interval for β_0.

13.2.6 Using SAS Statements for This Section

The printout for PROC REG includes the value of s_e as a part of the procedure; s_e is listed as ROOTMSE. In Figure 13.10, $s_e = 7.729557$. It also provides s_{b_0} and s_{b_1}. In Figure 13.10 we see that $s_{b_0} = 24.04753572$ and $s_{b_1} = 0.11784273$. The value of t^* for testing the hypothesis H_0: $\beta_1 = 0$ is also given on the printout, as is the P value for the two-sided test. (For a one-sided alternative, you should divide this value by 2 and check the sign of t^* to make sure it agrees with the alternate hypothesis.) In Figure 13.10, we see that $t^* = 5.306$ and $P = 0.0007$.

We may obtain a residual plot by using the OUTPUT statement and the PLOT procedure. We may also use the PLOT procedure to obtain both a scatterplot of the data and a plot of the points of the regression line corresponding to values of x in the sample. We write

```
PROG REG;
  MODEL yvariable = xvariable;
  OUTPUT P = YHAT R = RESID;
PROC PLOT;
  PLOT yvariable * xvariable YHAT * xvariable = '*' / OVERLAY;
  PLOT RESID * YHAT / VREF = 0;
```

The names YHAT and RESID in the output statement for the predicted values and residuals, respectively, are arbitrary. The first PLOT statement will give a scatterplot with y on the vertical axis and x on the horizontal axis (*yvariable* * *xvariable*) and also plot the predicted values against the x values (YHAT * *xvariable*). The sample values will be plotted using letters, usually *A*. If more than one sample point has the same x and y values, the points are indicated with *B* (for 2), *C* (for 3), and so forth. We illustrate the use of SAS statements with an example.

Example 13.11

A business conducts a study to determine the effectiveness of training in sales techniques. Eighteen new salespeople are given a training program. They are randomly divided into six groups of three persons and each group is given 5, 10, 15, 20, 25, or 30 hours of training. To assess the effectiveness of the different amounts of training, the number of sales for each salesperson is recorded for the first month after the training program. The results are shown in Table 13.7.

Determine a linear regression equation for predicting sales from hours of training. Test the contention that there is a positive association between the variables; use $\alpha = 0.05$. Construct a scatterplot, plot the regression line, and construct a residual plot.

Solution

Figure 13.12 shows the program and the output. We see that $\hat{y} = 2.4444 + 0.90476x$ is the regression equation. Since $F^* \doteq 120.04$ and the P value is 0.0001, we conclude that there is an association between the variables.

If we connect the points denoted by '*' we will obtain the regression line. The residual plot shows no departure from the assumptions of normality and equal error variance.

Table 13.7 SALES RESULTS FOR EXAMPLE 13.11.

SALESPERSON	HOURS OF TRAINING	SALES IN FIRST MONTH
1	5	8
2	5	11
3	5	6
4	10	12
5	10	9
6	10	7
7	15	15
8	15	13
9	15	18
10	20	23
11	20	27
12	20	19
13	25	28
14	25	21
15	25	23
16	30	28
17	30	31
18	30	30

Figure 13.12
SAS printout of the regression analysis of the sales figures of Example 13.10

```
INPUT FILE

DATA PROGRAM;
  INPUT HOURS SALES @@;
CARDS;
5 8 5 11 5 6 10 12 10 9 10 7 15 15 15 13 15 18 20 23 20 27 20 19
25 28 25 21 25 23 30 28 30 31 30 30
PROC REG;
  MODEL SALES = HOURS;
  OUTPUT OUT = A P = YHAT R = RESID;
PROC PLOT;
  PLOT SALES * HOURS YHAT * HOURS = '*' / OVERLAY;
  PLOT RESID * YHAT / VREF = 0;

OUTPUT FILE


DEP VARIABLE: SALES
```

SAS

ANALYSIS OF VARIANCE

SOURCE	DF	SUM OF SQUARES	MEAN SQUARE	F VALUE	PROB>F
MODEL	1	1074.40476	1074.40476	120.040	0.0001
ERROR	16	143.20635	8.95039683		
C TOTAL	17	1217.61111			

ROOT MSE	2.991721	R-SQUARE	0.8824
DEP MEAN	18.27778	ADJ R-SQ	0.8750
C.V.	16.36808		

PARAMETER ESTIMATES

VARIABLE	DF	PARAMETER ESTIMATE	STANDARD ERROR	T FOR H0: PARAMETER=0	PROB > |T|
INTERCEP	1	2.44444444	1.60800193	1.520	0.1480
HOURS	1	0.90476190	0.08257936	10.956	0.0001

SAS

PLOT OF SALES*HOURS   LEGEND: A = 1 OBS, B = 2 OBS, ETC.
PLOT OF YHAT*HOURS   SYMBOL USED IS *

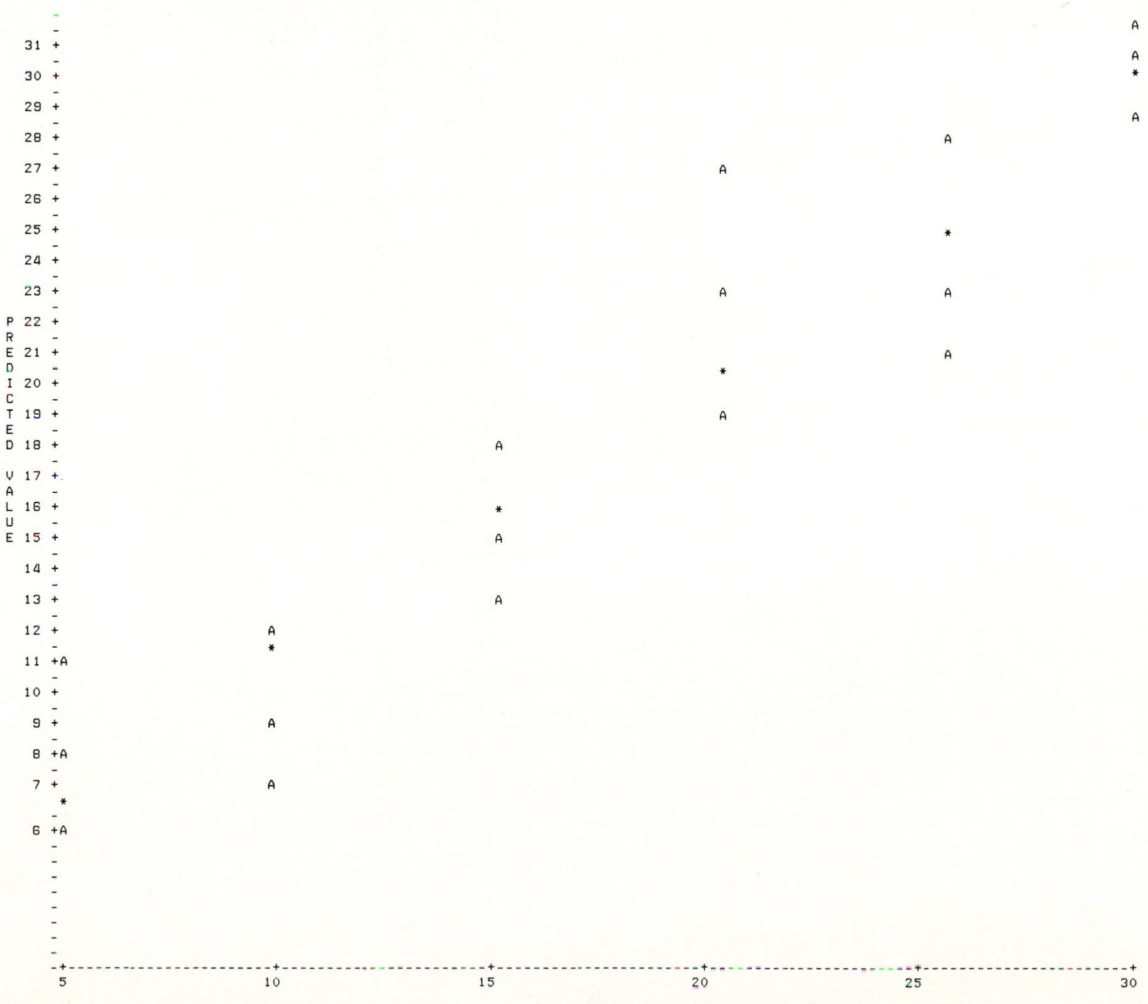

```
                                                                             SAS

                                             PLOT OF RESID*YHAT    LEGEND: A = 1 OBS, B = 2 OBS, ETC.

      -
    8 +
      -
      -
      -
      -
    7 +
      -
      -
      -                                                                                       A
      -
    6 +
      -
      -
      -
      -
    5 +
      -
      -
      -
      -
    4 +A
      -
      -
      -
      -
    3 +                                                                                                                     A
      -
      -
R     -                                                                                       A
E     -
S   2 +                                                         A
I     -
D     -
U     -
A     -
L   1 +A
S     -
      -                           A
      -
      -
    0 +-------------------------------------------------------------------------------------------------------------------------------------------------
      -
      -
      -
      -
   -1 +A                                                        A
      -
      -
      -                                                                                       A
      -
   -2 +                                                                                                                     A
      -
      -                           A
      -
      -
   -3 +                                                         A
      -
      -
      -
      -
   -4 +                                                                                                                     A
      -
      -                           A
      -
      -
   -5 +
      -
      -+-----------------------------+-----------------------------+-----------------------------+-----------------------------+-----------------------------+
  6.968254                      11.492063                     16.015873                     20.539683                     25.063492                     29.587302

                                                                         PREDICTED VALUE
```

NOTE: Another test for normality is the normal probability plot, mentioned in sections 2.5. We may write

```
PROC UNIVARIATE PLOT NORMAL;
VAR RESID;
```

to obtain a normal probability plot and other results.

Problems

Practice

13.11 A regression analysis is conducted on 15 data points and SSE is determined to be 134.51.

a. What is the value of s_e?

b. For $x = 100$, $\hat{y} = 125.4$. If the point (100, 113) is part of the original data set, should this point be considered an outlier? Why?

c. Would the point (100, 132) be considered an outlier? Why?

13.12 A regression analysis of 12 data points has SSE $= 1297.5$, SSX $= 128.6$, and $\bar{x} = 74.6$.

a. Determine s_{b_1}. Interpret its meaning.

b. What is the value of s_{b_0}?

13.13 Suppose an analysis of 21 data points yields the regression equation $\hat{y} = 0.1654 - 0.00237x$.

a. If $s_{b_1} = 0.0000146$, obtain a 95% confidence interval for β_1.

b. Test the hypothesis that there is no linear association between the variables. Use $\alpha = 0.05$.

c. Suppose it has been hypothesized that the slope of the line is -0.002. Test this hypothesis at the 0.05 level of significance.

d. Another hypothesis is that $\beta_1 = -0.0025$. Test this hypothesis at the 0.05 level of significance.

e. Explain how the confidence interval in part (*a*) can be used to test all three hypotheses. Are the results the same? Should you expect them to be?

f. If $s_{b_0} = 0.0844$, obtain a 95% confidence interval for β_0. Would this result tend to confirm or deny the hypothesis that $\beta_0 = 0$?

13.14 An analysis of the cost of producing textbooks is found to have a sample regression equation $\hat{y} = 84{,}355 + 13.88x$, where y measures the total cost in dollars of producing x textbooks. Interpret the meaning of the two numbers in the equation.

13.15 Refer to the data of Problem 13.5.

a. Construct a residual plot.

b. Do any assumptions appear to be violated?

c. Are there any outliers?

d. Determine, at a 0.05 level of significance, whether there is a positive association between the variables (that is, that $\beta_1 > 0$).

13.16 Refer to the data of Problem 13.6.

a. Obtain a 90% confidence interval for β_1.

b. Obtain a 90% confidence interval for β_0.

Applications

13.17 Refer to Problem 13.7.

a. Construct a residual plot.

b. Do any assumptions appear to be violated?

c. It has been hypothesized that each additional hundredweight of fertilizer per acre will cause an increase of 1 ton in yield. Test this hypothesis at the 0.05 level of significance.
d. Obtain a 95% confidence interval for the mean yield per acre if no fertilizer is used. Discuss the validity of this estimate.

13.18 Refer to Problem 13.8.

a. Obtain a 95% confidence interval for the mean change in duration of stay for each increase of 1000 bacteria in the count.
b. Test the hypothesis that there is no association between the variables. Use the 0.05 level of significance.

13.19 Refer to the data of Problem 13.9.

a. Obtain the standard error of estimate.
b. Construct a residual plot.
c. Do any assumptions appear to be violated?
d. Two of the data points appear to be outliers; one tree 84 feet from the ditch grew only to a height of 19 feet, while one 108 feet from the ditch grew to 22 feet. Examine the residual plot to determine whether these extreme values can be attributed to chance or if they should be investigated further. For instance, there may be a hidden source of water.
e. Test the conjecture that trees further from the ditch will be shorter than trees closer to the ditch. Use the 0.05 level of significance.

13.20 Refer to Problem 13.10.

a. Use a *t* test to test the hypothesis that the equation has no predictive value. Use the 0.01 level of significance. Compare the results with those of Problem 13.10.
b. Obtain a 99% confidence interval for β_1. Explain the meaning of this interval.

13.21 A medical researcher is testing the effect of a drug in inhibiting the release of adrenalin. She wishes to use the 0.05 level of significance. She randomly selects five samples from volunteers, each sample composed of ten male individuals, and injects the men of the samples with 0, 10, 20, 30, and 40 cc, respectively, of the drug. Each man is then subjected to a stress situation, and the stress level at which adrenalin is released is measured for each individual. Data are given here:

AMOUNT OF DRUG INJECTED (CC)	STRESS LEVEL AT WHICH ADRENALIN IS RELEASED
0	14, 21, 16, 18, 23, 15, 19, 22, 26, 17
10	13, 17, 19, 18, 21, 16, 25, 18, 22, 23
20	16, 22, 18, 23, 22, 25, 19, 17, 21, 25
30	21, 19, 24, 22, 28, 23, 22, 28, 24, 20
40	26, 22, 23, 25, 27, 29, 22, 24, 27, 26

a. Determine the linear regression equation for predicting the response from amount of drug injected.
b. Obtain the analysis of variance table for the regression analysis.
c. Using the 0.05 level of significance, determine whether there is a linear association between the variables.

d. Construct a residual plot.
e. Do the variables appear to be linearly related?
f. Are there any outliers?
g. Do the responses appear to be normal?
h. Apart from examining the residual plot, is there any method by which you can test the assumption of equal error variances? (See Problem 11.42.)
i. Test the hypothesis that there is a positive association between the variables. Use the 0.05 level of significance.
j. Obtain a 95% confidence interval for the mean change in stress level for each increase of 1 cc in the drug.
k. Obtain a 95% confidence interval for the mean stress level when no drug is injected.

13.3 Interval Estimation of Particular Values of y

A primary purpose for using regression analysis is to estimate a value or values of the response variable from the predictor variable. Two different kinds of estimation are usually important. For a particular value $x = x_p$, we can estimate the overall mean value of y for x_p—that is, $E(y_p)$. We can estimate the mean recovery time for *all* patients receiving x_p cubic centimeters of a drug, for example. If we are interested in the recovery time for Mrs. Jones rather than for all patients, we would want to estimate the recovery time for that individual patient. Thus we can estimate a single value y_p. In this section we show how to obtain interval estimates for both of these cases.

13.3.1 Confidence Intervals for $E(y_p)$

Often we wish to estimate the mean value of y for a particular x—that is, $E(y)$ for a particular x. It might be important, as noted, to estimate the mean recovery time for all patients receiving a certain dosage of a drug. Recall that we denote a particular value of x by x_p. The value of x_p need not be a value that was included in the sample, but it should be within the scope of the data—that is, it should not be greater than the largest value of x in the sample nor less than the smallest value of the sample. The reason for this is that the validity of the sample regression equation outside the scope of the sample is questionable. In Figure 13.13, for example, we see that the sample regression equation fits the data within the scope of the sample data quite well—but outside this range (where sample data were not obtained) additional data could give us an entirely different regression equation.

Within the scope of the data, the best estimate for $E(y)$ for any value of x is clearly $\hat{y}$, since $\hat{y} = b_0 + b_1x$ is the best estimate for the regression

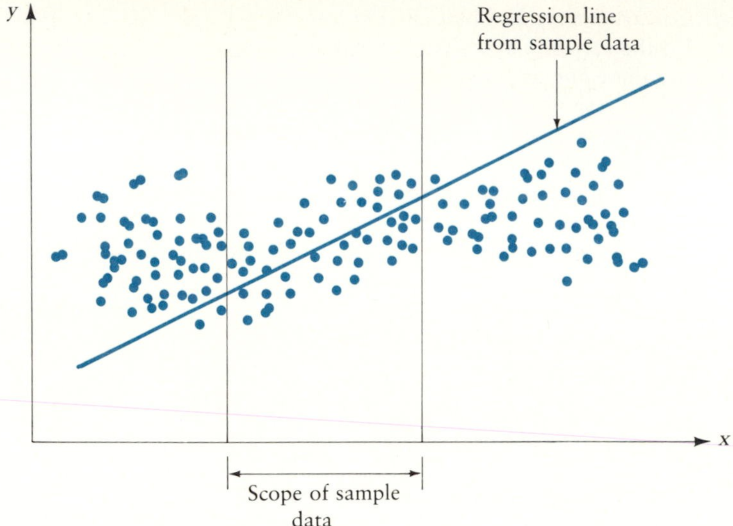

Figure 13.13
Fit of a regression line within and outside the scope of the sample data.

function $E(y) = \beta_0 + \beta_1 x$. Since $E(y_p) = \beta_0 + \beta_1 x_p$, we have $\hat{y}_p = b_0 + b_1 x_p$. Thus we may estimate the mean value of y for a given x, x_p, by using $\hat{y}_p$. On the other hand, it is quite likely that the true value of $E(y_p)$ is different from $\hat{y}_p$, since the sample regression equation is probably not the same as the true regression function. Figure 13.14 shows the error of estimation graphically.

If all possible sample regression equations are obtained from the same population and the assumptions for the model are satisfied, we may use statistical theory to obtain an interval estimate for $E(y_p)$. If we obtain $\hat{y}_p$ for all such samples, we can find the sampling distribution of $\hat{y}_p$. The variance of the errors of estimation shown in Figure 13.14 is denoted by $\sigma^2_{\hat{y}_p}$ and the distribution of $\hat{y}_p$ is normal. It can be shown that

$$\sigma^2_{\hat{y}_p} = \sigma^2\left(\frac{1}{n} + \frac{(x_p - \overline{x})^2}{\text{SSX}}\right)$$

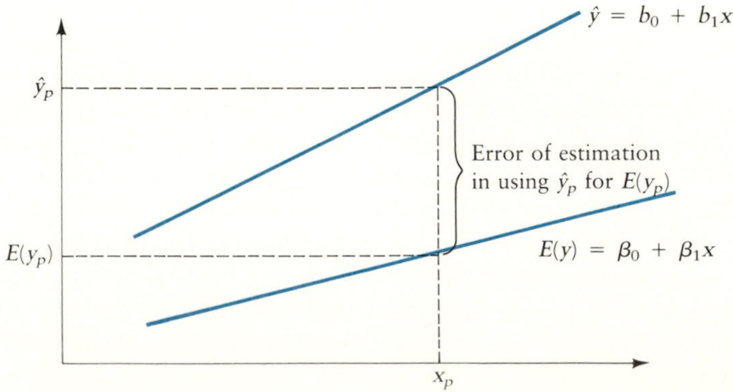

Figure 13.14
The error of estimation incurred when using $\hat{y}_p$ as an estimate for $E(y_p)$.

For each x_p, $\hat{y}_p$ is $N(E(y_p), \sigma^2_{\hat{y}_p})$; then the variable $[\hat{y}_p - E(y_p)]/\sigma_{\hat{y}_p}$ is $N(0, 1)$, so that $[\hat{y}_p - E(y_p)]/s_{\hat{y}_p}$ has a t distribution, where $s_{\hat{y}_p}$ is the best estimate for $\sigma_{\hat{y}_p}$. Since MSE is the best estimate for σ^2 and $s_e = \sqrt{\text{MSE}}$, the best estimate for $\sigma_{\hat{y}_p}$ is

$$s_{\hat{y}_p} = s_e\sqrt{\frac{1}{n} + \frac{(x_p - \overline{x})^2}{\text{SSX}}}$$

Since there are $n - 2$ degrees of freedom associated with s_e, the t distribution has $n - 2$ degrees of freedom.

We can determine a $100(1 - \alpha)\%$ confidence interval for $E(y_p)$ as shown here.

Confidence Interval for $E(y_p)$

Suppose we take a random sample of n ordered pairs of observations from a population for which the regression function is $E(y) = \beta_0 + \beta_1 x$, the sample regression equation is $\hat{y} = b_0 + b_1 x$, and the error terms are independent and $N(0, \sigma^2)$ for each value of x. Then a $100(1 - \alpha)\%$ confidence interval for $E(y_p) = \beta_0 + \beta_1 x_p$ is the interval

$$\hat{y}_p - ts_{\hat{y}_p} \quad \text{to} \quad \hat{y}_p + ts_{\hat{y}_p}$$

where $\hat{y}_p = b_0 + b_1 x_p$, $t = t_{\alpha/2}\{n-2\}$ and

$$s_{\hat{y}_p} = s_e\sqrt{\frac{1}{n} + \frac{(x_p - \overline{x})^2}{\text{SSX}}}$$

The formula for $s_{\hat{y}_p}$ implies that the variability of the estimate increases as x_p gets farther from $\overline{x}$. Estimates are least variable for $x_p = \overline{x}$. The reason for this can be shown from an examination of several sample regression lines. Figure 13.15 shows the graph of a regression function, $E(y) = \beta_0 + \beta_1 x$, and several sample regression lines.

Since each sample line in Figure 13.15 is an estimate for the true regression line, the sample lines tend to cluster about the true line. As the slope (b_1) increases, the y intercept (b_0) decreases, so the sample lines are closest, as a group, to the true regression line near the center and fan out as the differences between values of x and the mean increase. If we obtain upper and lower 95% confidence limits for the confidence intervals for each value of x and plot them with the graph of the sample regression equation, we obtain a 95% **confidence band** for $E(y)$. Figure 13.16 shows such a confidence band. Note that 95% confidence intervals for $E(y_p)$ widen as x_p gets further from $\overline{x}$.

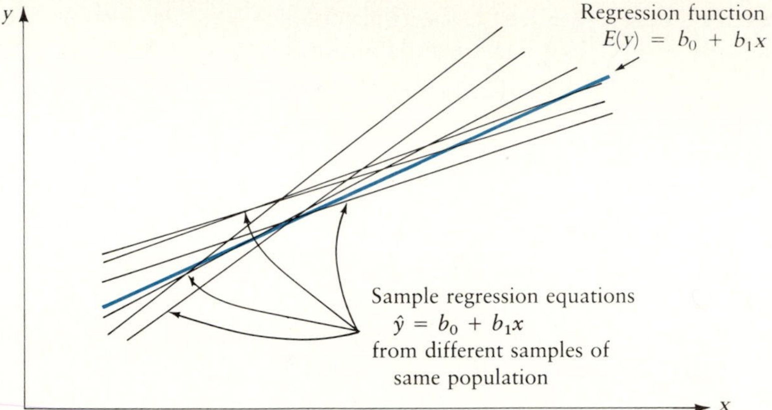

Figure 13.15
Graph of the regression function $E(y) = \beta_0 + \beta_1 x$ and a few sample regression equations.

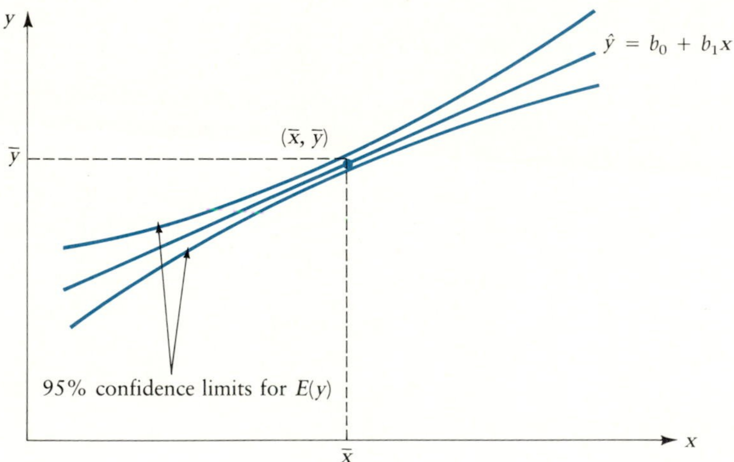

Figure 13.16
A 95% confidence band for $E(y)$.

Example 13.12

Use the real estate data of Table 13.1 to obtain a 95% confidence interval for the mean price of all lots with 200 feet of waterfront footage.

Solution

Since $\hat{y} = 21.0660 + 0.6253x$, if $x_p = 200$ then $\hat{y}_p = 21.0660 + 0.6253(200) = 146.1260$. Thus the predicted price for all lots with 200 feet of frontage is about \$146,126. To obtain a 95% confidence interval for the mean price of all lots with 200 feet of frontage, we must obtain $s_{\hat{y}_p}$. Now SSX = 4302, $s_e \doteq 7.7296$, $n = 10$, and $\bar{x} = 203$, so

$$s_{\hat{y}_p} = s_e\sqrt{\frac{1}{n} + \frac{(x_p - \bar{x})^2}{\text{SSX}}}$$

$$= 7.7296\sqrt{\frac{1}{10} + \frac{(200 - 203)^2}{4302}}$$

$$\doteq 2.4697$$

Now $t_{.025}\{8\} = 2.306$ and $2.306(2.4697) \doteq 5.6952$, so the confidence limits are $146.1260 - 5.6952$ and $146.1260 + 5.6952$. Thus the confidence interval is 140.4308 to 151.8212, and we are 95% confident that the mean price of all lots with 200 feet of frontage is between \$140,431 and \$151,821.

13.3.2 Prediction Intervals for y_p

Often we have to estimate a value of y for a particular x—that is, a value of y_p for x_p. Just as before, it is clear that the best estimate for an individual y_p is $\hat{y}_p = b_0 + b_1 x_p$. For a particular x_p, the sampling distribution of y_p will be normal with mean $E(y_p)$. The error involved in using $\hat{y}_p$ as an estimate for y_p is called the error of prediction (Figure 13.17).

The use of $\hat{y}_p$ to predict y_p incurs two sources of variation—the individual observations themselves are variable, with variance σ^2, and the values of $\hat{y}_p$ are variable with variance $\sigma^2_{\hat{y}_p}$. Recall from Section 8.1 that the variance of the difference between two estimators is the sum of the variances; that is, the variance of $\hat{\theta}_1 - \hat{\theta}_2$, $(\sigma^2_{\hat{\theta}_1 - \hat{\theta}_2})$ is equal to $\sigma^2_{\hat{\theta}_1} + \sigma^2_{\hat{\theta}_2}$. Similarly, the variance of $y_p - \hat{y}_p(\sigma^2_{y_p - \hat{y}_p})$ is equal to $\sigma^2 + \sigma^2_{\hat{y}_p}$, and we can write

$$\sigma^2_{y_p - \hat{y}_p} = \sigma^2(1 + 1/n + (x_p - \overline{x})^2/\text{SSX}).$$

Since the notation $\sigma^2_{y_p - \hat{y}_p}$ is rather cumbersome, we can use a number of alternate notations. One of the terms for $\sigma_{y_p - \hat{y}_p}$ is the **standard error of forecast,** so the notation σ_f is frequently used. We will use this notation; that is, $\sigma^2_{y_p - \hat{y}_p} = \sigma^2_f$.

Then for a particular value $x = x_p$, y_p is $N(E(y_p), \sigma^2_f)$, so $[y_p - E(y_p)]/\sigma_f$ is $N(0, 1)$. Therefore $[y_p - E(y_p)]/s_f$ has a t distribution if s_f is the best estimate for σ_f. Since s_e is the best estimate for σ, the t distribution has 2 degrees of freedom and

$$s_f = s_e\sqrt{1 + \frac{1}{n} + \frac{(x_p - \overline{x})^2}{\text{SSX}}}$$

The term s_f is sometimes symbolized $s_{y-\hat{y}}$.

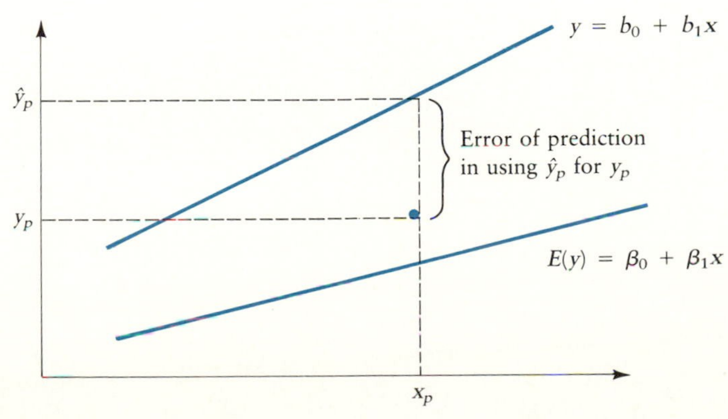

Figure 13.17
The error incurred in using $\hat{y}_p$ as an estimate for y_p.

An interval estimate for y_p is called a **prediction interval** rather than a confidence interval. The term *confidence interval* is used for an interval estimate for a population parameter, a constant. Since values of y_p vary and are therefore not constant, we use the term *prediction interval* to distinguish between the two types of intervals. We can determine a $100(1 - \alpha)\%$ prediction interval for y_p as shown here.

Prediction Interval for y_p

Suppose we take a random sample of n ordered pairs of observations from a population for which the regression function is $E(y) = \beta_0 + \beta_1 x$, the sample regression equation is $\hat{y} = b_0 + b_1 x$, and the error terms are independent and $N(0, \sigma^2)$ for each value of x. Then a $100(1 - \alpha)\%$ prediction interval for the next value of y, y_p, for a given value $x = x_p$ is the interval

$$\hat{y}_p - ts_f \quad \text{to} \quad \hat{y}_p + ts_f$$

where $\hat{y}_p = b_0 + b_1 x_p$, $t = t_{\alpha/2}\{n - 2\}$, and

$$s_f = s_e\sqrt{1 + \frac{1}{n} + \frac{(x_p - \overline{x})^2}{\text{SSX}}}$$

Since there are two sources of variability, s_f is larger than $s_{\hat{y}_p}$; thus a prediction interval for y_p for a particular value of x_p is wider than the corresponding confidence interval for $E(y_p)$, as shown in Figure 13.18. Moreover, we can see from the formula for s_f that s_f increases as $|x_p - \overline{x}|$ increases—that is, as x_p gets farther from $\overline{x}$. If we obtain upper and lower

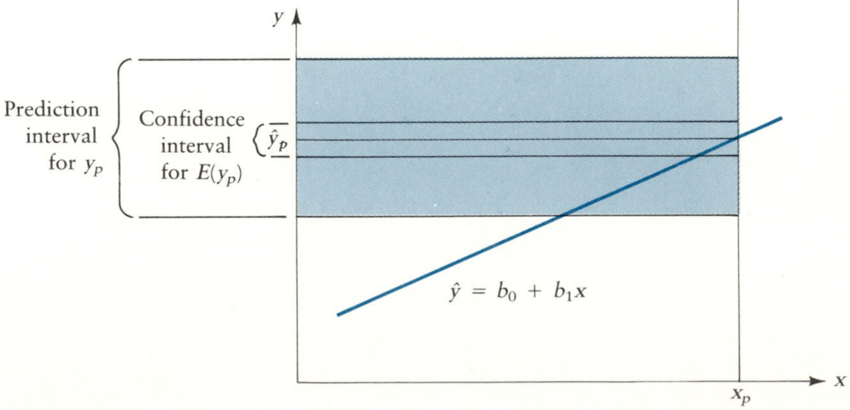

Figure 13.18
Prediction intervals for y_p and confidence intervals for $E(y_p)$ compared.

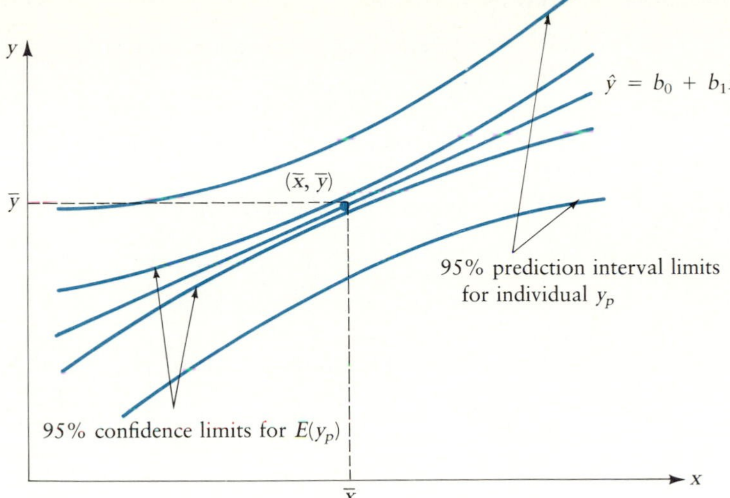

Figure 13.19
95% confidence and prediction bands for $E(y_p)$ and y_p, respectively.

95% confidence limits for prediction intervals for each value of x, we obtain a prediction band for individual values of y. Figure 13.19 shows a 95% confidence band for $E(y_p)$ and a 95% prediction band for y_p for a particular regression equation.

Example 13.13

Use the real estate data of Table 13.1 to obtain the selling price of a lot with 200 feet of frontage.

Solution

Since $\hat{y} = 21.0660 + 0.6253$, if $x_p = 200$ then $\hat{y}_p = 21.0660 + 0.6253(200) = 146.1260$. Thus the predicted price for all lots with 200 feet of frontage is about \$146,126. To obtain a 95% prediction interval for the price of an individual lot with 200 feet of frontage, we must obtain s_f. Now SSX = 4302, $s_e \doteq 7.7296$, $n = 10$, and $\bar{x} = 203$, so

$$s_f = s_e\sqrt{1 + \frac{1}{n} + \frac{(x_p - \bar{x})^2}{\text{SSX}}}$$

$$= 7.7296\sqrt{1 + \frac{1}{10} + \frac{(200 - 203)^2}{4302}}$$

$$\doteq 8.1146$$

Since $t_{.025}\{8\} = 2.306$ and $2.306(8.1146) \doteq 18.7122$, the prediction limits are $146.1260 - 18.7122$ and $146.1260 + 18.7122$. Thus the confidence interval is 127.4138 to 164.8382, and our 95% prediction interval for the price of an individual lot with 200 feet of frontage is \$127,414 to \$164,838.

Example 13.14

Refer to Example 13.11.

a. Obtain a 95% confidence interval for the mean change in sales for each increase of 1 hour in training.
b. Obtain a 95% confidence interval for the mean sales if there is no training.
c. Obtain a 95% confidence interval for the mean sales of all salespeople with 20 hours of training.
d. Obtain a 95% prediction interval for the sales of an individual with 20 hours of training.

Solution

Assume that the computer results of Example 13.11 are not available. First we obtain the intermediate sums: $n = 18$, $\Sigma x = 315$, $\Sigma x^2 = 6825$, $\Sigma y = 329$, $\Sigma y^2 = 7231$, and $\Sigma xy = 6945$. Then SSX = 1312.5, SXY = 1187.5, SSTO $\doteq$ 1217.6111, and

$$b_1 = \frac{1187.5}{1312.5} \doteq 0.9048$$

$$b_0 = \frac{329 - 315b_1}{18} \doteq 2.4444$$

Thus the regression equation is $\hat{y} = 2.4444 + 0.9048x$. We need not obtain the analysis of variance table, but we do need s_e:

$$s_e = \sqrt{\frac{1217.6111 - 1187.5b_1}{16}} \doteq 2.9917$$

a. $s_{b_1} \doteq 2.9917/\sqrt{1312.5} \doteq 0.08258$. For a 95% confidence interval we will use $t_{.025}\{16\} = 2.120$. Since the mean change in sales for each increase of 1 hour in training is β_1, we want a 95% confidence interval for β_1. Since $2.120(0.08258) \doteq 0.1751$, the 95% confidence interval for β_1 is 0.7297 to 1.0799. Each additional hour of training is worth an estimated average of from 0.7297 to 1.0799 sales.

b. Similarly, $s_{b_0} \doteq 2.9917\sqrt{(1/18) + 17.5^2)/1312.5} \doteq 1.6080$; $.120(1.6080) \doteq 3.4090$, so a 95% confidence interval for β_0, the mean sales if there is no training, is -0.9646 to 5.8534. Thus we are 95% confident that salespeople with no training would sell, on the average, fewer than 5.8534 units. (*Caution:* This interval should be interpreted with care, since $x = 0$ is outside the scope of the data.)

c. If $x_p = 20$, then $\hat{y}_p \doteq 2.4444 + 0.9048(20) \doteq 20.5404$. To estimate mean sales of all salespeople with 20 hours of training, we use

$$s_{\hat{y}_p} = 2.9917\sqrt{\frac{1}{18} + \frac{(20 - 17.5)^2}{1312.5}} \doteq 0.7347$$

Then $2.120(0.7347) \doteq 1.5577$, so the 95% confidence interval will be 18.9827 to 22.0981.

d. To predict the sales of an individual with 20 hours of training, we use

$$s_f = 2.9917\sqrt{1 + \frac{1}{18} + \frac{(20 - 17.5)^2}{1312.5}} \doteq 3.0806$$

Since $2.120(3.0806) \doteq 6.5309$, the 95% prediction interval is 14.0095 to 27.0713.

PROFICIENCY CHECKS

13. Use the data of Proficiency Checks 1 and 3 to obtain:
    a. a 95% confidence interval for $E(y_p)$ when $x_p = 6$
    b. a 95% prediction interval for y_p when $x_p = 6$
14. Use the data of Proficiency Checks 2 and 4 to obtain:
    a. a 99% confidence interval for $E(y_p)$ when $x_p = 4.5$
    b. a 99% prediction interval for y when $x_p = 4.5$

13.3.3 SAS Statements for Interval Estimation

We can obtain confidence limits for confidence intervals for $E(y_p)$ and prediction intervals for y_p by using the SAS procedure REG. We write

```
PROC REG;
  MODEL y = x / P CLM CLI;
```

to obtain 95% confidence limits for the confidence intervals for $E(y_p)$ (CLM) and the prediction intervals for y_p (CLI) for x_p = every value of x in the sample. (CLM stands for confidence limits—means—and CLI stands for confidence limits—individual.) If we want 99% or 90% limits, we must use the GLM procedure. See the *SAS® User's Guide: Statistics* for details.

There are several ways to obtain confidence or prediction intervals for a value of x not included in the sample. One way is to include the values of x_p in the data set and use a period (.) for the y variable to indicate that no value of y is to be entered. For the data of Example 13.1, for instance, suppose we wish to obtain 95% interval estimates for the mean price of all lots with 200 feet of frontage and the price of a lot with 224 feet of frontage. We write the input data set as before, including these two values for x:

```
DATA LOTS;
  INPUT FRONTAGE PRICE @@;
CARDS;
188 140 231 160 176 130 194 130 244 180 207 160 198 140
217 150 181 140 194 150 200 . 224 .
PROC REG;
  MODEL PRICE = FRONTAGE / P CLM CLI;
  ID FRONTAGE;
```

The output of the program is shown in Figure 13.20.

The P option causes the predicted values and residuals to be computed for each value of x in the data set. The ID statement causes the specified variable to be printed with the output data set. We see that a 95% confidence interval for $E(y_p)$ when $x_p = 200$ is from about \$140,400 to \$151,800 and a 95% prediction interval for y_p when $x_p = 224$ is from about \$141,600 to \$180,700.

If we want to estimate more than one interval, we encounter the same problems as with multiple estimates in previous chapters. Specific procedures have been developed to deal with this problem. See Neter and others (1985, pp. 154–164) for a discussion of procedures to be used in these situations.

Figure 13.20 SAS printout for confidence and prediction intervals for the real estate data.

INPUT FILE

```
DATA LOTS;
  INPUT FRONTAGE PRICE @@;
CARDS;
188 140 231 160 176 130 194 130 244 180 207 160 198 140
217 150 181 140 194 150 200 . 224 .
PROC REG;
  MODEL PRICE = FRONTAGE / P CLM CLT;
  ID FRONTAGE;
```

```
DEP VARIABLE: PRICE

                                  SAS

                         ANALYSIS OF VARIANCE

                           SUM OF           MEAN
SOURCE        DF          SQUARES         SQUARE        F VALUE        PROB>F

MODEL          1       1682.03161     1682.03161         28.153        0.0007
ERROR          8        477.96839    59.74604835
C TOTAL        9       2160.00000

      ROOT MSE           7.729557       R-SQUARE         0.7787
      DEP MEAN                148       ADJ R-SQ         0.7511
      C.V.               5.222674

                         PARAMETER ESTIMATES

                       PARAMETER           STANDARD      T FOR H0:
VARIABLE    DF          ESTIMATE              ERROR    PARAMETER=0      PROB > |T|

INTERCEP     1       21.06601581        24.04753572          0.876          0.4066
FRONTAGE     1        0.62529056         0.11784723          5.306          0.0007
```

OBS	ID	ACTUAL	PREDICT VALUE	STD ERR PREDICT	LOWER95% MEAN	UPPER95% MEAN	LOWER95% PREDICT	UPPER95% PREDICT	RESIDUAL
1	188	140.0	138.6	3.0165	131.7	145.6	119.5	157.8	1.3794
2	231	160.0	165.5	4.1064	156.0	175.0	145.3	185.7	-5.5081
3	176	130.0	131.1	4.0123	121.9	140.4	111.0	151.2	-1.1172
4	194	130.0	142.4	2.6645	136.2	148.5	123.5	161.2	-12.3724
5	244	180.0	173.6	5.4148	161.2	186.1	151.9	195.4	6.3631
6	207	160.0	150.5	2.4893	144.8	156.2	131.8	169.2	9.4988
7	198	140.0	144.9	2.5143	139.1	150.7	126.1	163.6	-4.8735
8	217	150.0	156.8	2.9490	150.0	163.6	137.7	175.8	-6.7541
9	181	140.0	134.2	3.5632	126.0	142.5	114.6	153.9	5.7564
10	194	150.0	142.4	2.6645	136.2	148.5	123.5	161.2	7.6276
11	200	.	146.1	2.4697	140.4	151.8	127.4	164.8	.
12	224	.	161.1	3.4784	153.1	169.2	141.6	180.7	.

```
SUM OF RESIDUALS                3.55271E-14
SUM OF SQUARED RESIDUALS           477.9684
PREDICTED RESID SS (PRESS)         750.0178
```

Problems

Practice

13.22 A regression analysis of 18 data points yields the regression equation $\hat{y} = 2168.5 + 114.8x$; SSX $= 412.5$, $\bar{x} = 65.8$, and $s_e = 213.4$.
  a. What is $s_{\hat{y}_p}$ for $x_p = 60$? For $x_p = 70$?
  b. Determine a 90% confidence interval for $E(\hat{y}_p)$ for $x_p = 70$.

13.23 A regression analysis of 24 data points yields the regression equation $\hat{y} = 761.665 - 2.544x$; SSX $= 1096.5$, $\bar{x} = 188.6$ and $s_e = 21.87$.
  a. Calculate $s_{\hat{y}_p}$ for $x_p = 200$.
  b. Determine s_f for $x = 200$ by using the formula $s_f = s_e\sqrt{1 + 1/n + (x_p - \bar{x})^2/\text{SSX}}$.
  c. If we use the formula $s_f^2 = s_e^2 + s_{\hat{y}_p}^2$, do we get the same value for s_f?
  d. Obtain a 95% prediction interval for y_p for $x_p = 200$.
  e. If we obtain a 95% confidence interval for $E(y_p)$ for $x_p = 200$, would the interval be wider or narrower than the one in part (*d*)? Why?

13.24 Under what conditions would a confidence interval for $E(y_p)$ be useful? When would a prediction interval be of more use? Explain.

13.25 Refer to the results of Problem 13.5 (and 13.15) and suppose that $x_p = 30$.
  a. Obtain a 95% confidence interval for $E(y_p)$.
  b. Obtain a 95% prediction interval for an individual y_p.

13.26 Refer to the results of Problem 13.6 (and 13.16) and suppose that $x_p = 2.5$.
  a. Obtain a 99% confidence interval for $E(y_p)$.
  b. Obtain a 99% prediction interval for an individual y_p.

Applications

13.27 Refer to the fertilizer-yield data of Problem 13.7 (and 13.17).
  a. Confidence intervals for $E(y_p)$ and prediction intervals for y_p have different interpretations. Which interval would be more useful from the standpoint of Farmer Brown, who will use 9.2 hundredweight per acre?
  b. Which interval would be more useful for the county farm agent, who recommends that farmers use 9.2 hundredweight per acre?

13.28 Refer to Problem 13.8 (and 13.18).

a. Obtain a 95% confidence interval for the mean duration of symptoms for all patients with a bacterial count of 7000.
b. Obtain a 95% prediction interval for the duration of symptoms for a patient who was just admitted and has a bacterial count of 6000.
c. Suppose a patient is admitted with a bacterial count of 6000 and still has not recovered after 15 days in the hospital. Should the doctor and staff suspect additional problems or complications, or is this occurrence within reasonable limits?

13.29 Refer to Problem 13.9 in which we investigated the relationship between height of trees and distance from the irrigation ditch. Which is probably of more use to the grower: a prediction interval or a confidence interval?

13.30 A department of a university randomly selects 20 files of students who graduated with a major in the department within the past year and took the graduate record examination (GRE). The accompanying table gives the undergraduate grade-point average (GPA) and composite score on the GRE verbal and quantitative sections:

STUDENT	GPA	GRE SCORE
1	2.34	910
2	3.61	1340
3	3.08	1160
4	2.77	1420
5	3.13	960
6	2.54	830
7	2.47	940
8	2.38	1060
9	2.91	1230
10	3.17	1280
11	3.28	1140
12	2.08	760
13	3.14	1110
14	3.03	1080
15	2.86	1040
16	2.89	1320
17	3.13	940
18	2.07	680
19	2.71	780
20	2.64	970

a. Obtain a regression equation to predict GRE score from GPA.
b. Give the analysis of variance table for the data.
c. Test the hypothesis that there is no relationship between the variables. Use $\alpha = 0.05$.
d. Obtain a 95% confidence interval for the mean increase in GRE score for each increase of 1.00 in GPA.
e. Give the 95% confidence intervals for the GRE score of all students with a 3.00 GPA.

f. The highest and lowest GPAs for the current graduates are 3.59 and 2.12, respectively. Give a 95% prediction interval for the GRE score of each one.

g. Would the relationship between GPA and GRE be likely to be as strong if students were drawn from different departments rather than just one? Why?

13.31 Refer to the data of Problem 13.21.

a. Obtain a 95% confidence interval for the mean stress level at which adrenalin will be released for all men receiving 20 cc of the drug.

b. Obtain a 95% prediction interval for the stress level at which adrenalin will be released for a patient who is to receive an injection of 20 cc of the drug.

c. In order to monitor deviations from the physiologically normal, which kind of interval estimate should be used? Why?

13.32 Refer to the data of Problem 13.10 (and 13.20).

a. Obtain a 99% confidence interval for the mean sales of all salespeople scoring 110 on the test. Interpret this interval.

b. Obtain a 95% prediction interval for the sales of a person who scored 142 on the test. Interpret this interval.

13.4 Correlation Analysis

In the preceding sections we examined methods of deriving an equation for predicting values of one variable from another. In many other cases there may be a statistical relationship between two variables, but neither is a predictor for the other. If two variables have a statistical relationship we say that the variables are *correlated*. **Correlation** is a measure of the strength of the linear association between the variables. Clearly if there is a predictive relationship between the variables, they are correlated. Thus we can use correlation analysis to assess the association between two variables whether there is a predictive relationship between the variables or not. In an earlier example we noted an association between teachers' salaries and per capita consumption of alcohol. The fact that both are reflections of improved economic conditions causes a positive association to be found between the two variables. Since there is no predictive relationship, regression analysis is not really useful here, but correlation analysis can provide information about the association between the variables.

Sometimes correlation between two variables is a result of chance and chance alone. One study, for example, showed a significant correlation between the number of storks sighted and the number of human births in the Netherlands.

13.4.1 The Coefficient of Determination

Suppose that we obtain a regression line from a set of data relating two variables. If there is a strong association between the variables, the data will fit the regression line closely and the sum of the deviations of observed values from the line, $\Sigma(y - \hat{y})^2$, will be small—that is, SSE will be small. The total amount of variability of the y variable, SSTO, is $\Sigma(y - \bar{y})^2$. If we use $\hat{y}$ as the predicted value for each value of x, the total amount of variation in the predicted values is $\Sigma(y - \hat{y})^2$, or SSE. The amount of *reduction* in error variation is equal to SSTO $-$ SSE $=$ SSR. The value SSR is sometimes called the *explained* variation of y—that portion of SSTO which is explained or predicted by the regression equation. Thus if there is a strong association between the variables so that SSE is small, SSR will be proportionately large.

The *proportion* of reduction in error variation is equal to SSR/SSTO, which measures the proportion of total variation that is explained by the regression model. The value SSR/SSTO is called the **coefficient of determination;** if the regression function is linear, SSR/SSTO is called the coefficient of *linear* determination. The coefficient of determination derived from a sample is symbolized r^2, so that

$$r^2 = \frac{\text{SSR}}{\text{SSTO}}$$

The coefficient of determination for a population is symbolized ρ^2, where ρ is the Greek letter rho. We can see from this discussion that $\rho^2 = \Sigma[y - E(y)]^2/\Sigma(y - \mu_y)^2$, where μ_y is the mean of all values of y in the population and $E(y) = \beta_0 + \beta_1 x$ for each value of x.

Since both numerator and denominator of r^2 are positive, r^2 cannot be negative. On the other hand, SSR cannot be greater than SSTO, so we know that

$$0 \leqslant r^2 \leqslant 1$$

If there is a strong association between the variables, SSE is quite small, so SSR is large and r^2 is close to 1. Conversely, if there is no association between the variables then r^2 is close to zero.

The value of r^2 is the same no matter which variable we choose for x and which for y. Thus r^2 measures the strength of the association between two variables, whether one can be considered as a predictor of the other or not. Further, r^2 is unitless—a pure number, no matter what units x and y are measured in. We can obtain a formula for calculating r^2 by rewriting SSR. This formula is particularly useful when we do not want to perform a regression analysis but wish instead merely to measure the strength of the association between two variables.

Now SSTO is the total variation of the response variable in a linear regression analysis. We note that SSTO $= \Sigma(y - \bar{y})^2$, so that it is the sum of squares for y (SSY). If we do not consider one variable as a predictor and the other as the response, we use SSY instead of SSTO as the total variation of the y variable. Since SSR = $(\text{SXY})^2/\text{SSX}$, we have the following characterization of the coefficient of determination of a sample.

Coefficient of Determination for a Sample

Suppose that we take a sample of n ordered pairs of observations from a population of ordered pairs (x, y). Then the strength of the linear association between the variables can be measured by the coefficient of linear determination, r^2, where

$$r^2 = \frac{(\text{SXY})^2}{\text{SSX} \cdot \text{SSY}}$$

The quantity r^2 is always a pure number.

Since $\text{SXY} = \Sigma xy - (\Sigma x \Sigma y)/n$, $\text{SSX} = \Sigma x^2 - (\Sigma x)^2/n$, and $\text{SSY} = \Sigma y^2 - (\Sigma y)^2/n$, we may substitute into the preceding formula to obtain a raw-data formula. Usually we multiply the numerator and denominator by n^2 to remove the complex fractions, so we have the following raw-data formula for r^2:

$$r^2 = \frac{(n\Sigma xy - \Sigma x \Sigma y)^2}{[n\Sigma x^2 - (\Sigma x)^2][n\Sigma y^2 - (\Sigma y)^2]}$$

Although the operational definition of r^2 is clear—it is the proportion of total variation that is due to the association between the variables—it is not as widely used as another measure of the strength of the association: the coefficient of correlation.

13.4.2 The Coefficient of Correlation

Another measure of the strength of the association between two variables is the **coefficient of correlation** (usually taken to mean *linear* correlation), or **correlation coefficient,** which is symbolized ρ for a population and r for a sample. As these symbols imply, the coefficient of correlation is the square root of the coefficient of determination. It can be either positive or negative. A positive correlation coefficient indicates that as one variable increases the other increases, while a negative correlation coefficient indicates that as one variable increases the other decreases. Thus it has the same sign as β_1 or b_1. Since the sign of b_1 is determined by the sign of SXY, the coefficient of correlation is defined as follows.

Coefficient of Correlation for a Sample

Suppose we take a sample of n ordered pairs of observations from a population of ordered pairs (x, y). Then the strength of the linear association between the variables may be measured by the coefficient of linear correlation, r, where

$$r = \frac{\text{SXY}}{\sqrt{\text{SSX}\cdot\text{SSY}}}$$

$$= \frac{(n\Sigma xy - \Sigma x\Sigma y)}{\sqrt{[n\Sigma x^2 - (\Sigma x)^2][n\Sigma y^2 - (\Sigma y)^2]}}$$

The second formula, called the raw-data formula, was developed by Karl Pearson, and r is often called the *Pearson product-moment correlation coefficient.*

Some scatterplots are shown in Figure 13.21, together with values of r for each set of data. Notice in part *(f)* that although there is no linear association between the variables, the variables are certainly related. The association is curvilinear. Thus even when $r = 0$ we cannot automatically conclude that there is no relationship between the variables. Note also that

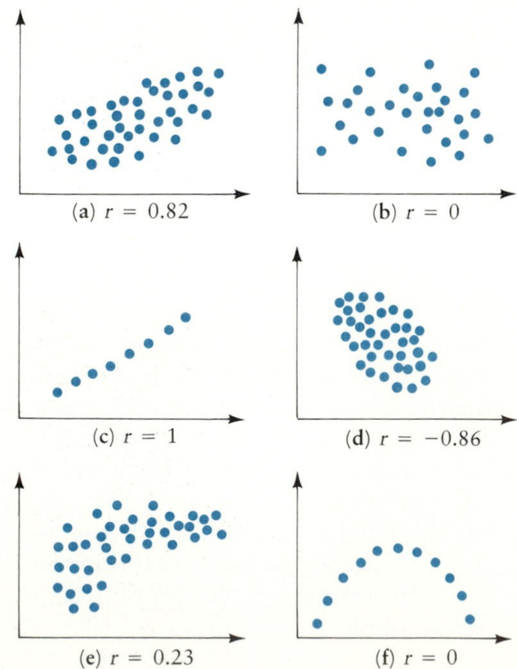

Figure 13.21 Scatterplots with values of r indicated.

association does not imply causation. A high correlation between two variables implies only that they vary linearly in the same way (or opposite way, if r is negative). Note also that high correlation coefficients do not necessarily imply a strong relationship between the variables. If $r = 0.7$ then $r^2 = 0.49$, so 51% of the variation of either variable is still unexplained in terms of the other. Moreover, the degree of association between two variables for a certain value of r depends on the size of the sample.

Example 13.15

Obtain the correlation coefficient for the following set of data:

x	2	3	3	4	5	5	6	6	7	8	8
y	10	14	15	17	18	19	18	22	26	29	27

Solution

We get $\Sigma x = 57$, $\Sigma x^2 = 337$, $\Sigma y = 215$, $\Sigma y^2 = 4549$, $\Sigma xy = 1230$, and $n = 11$. Then SXY $\doteq 115.9091$, SSX $\doteq 41.6364$, and SSY $\doteq 346.7273$, so $r \doteq 115.9091/\sqrt{(41.6364)(346.7273)} \doteq 115.9091/120.1519 \doteq 0.9647$.

Example 13.16

Obtain the coefficient of correlation for the real estate data of Table 13.1.

Solution

From the analysis of variance table (Example 13.2), SSR = 1682.0316 and SSTO = 2160, so $r^2 = 1682.0316/2160 \doteq 0.7787$ and $r \doteq 0.8825$. Thus nearly 78% of the variability in prices is explained by the number of waterfront feet, and we can consider the correlation coefficient to be fairly high. Therefore the price of a lot and the number of waterfront feet are clearly associated.

PROFICIENCY CHECKS

15. Obtain r^2 and r for the data of Proficiency Checks 1 and 3.
16. Obtain r^2 and r for the data of Proficiency Checks 2 and 4.
17. Determine r for the data given here:

x	2	3	6	7	6	4	6	9	11	7
y	8	5	8	3	5	7	5	2	4	4

13.4.3 Testing the Hypothesis $\rho = 0$

If there is no linear association between the variables, $\rho = 0$. On the other hand, if $\rho \neq 0$, the variables are linearly related. We can test the hypothesis that $\rho = 0$ by using a t test. If $\rho = 0$, the sampling distribution of r will be normally distributed with mean zero and a standard error that can be approximated by s_r, where

$$s_r = \sqrt{\frac{1 - r^2}{n - 2}}$$

Then the statistic r/s_r has a t distribution with $n - 2$ degrees of freedom. We cannot obtain confidence intervals for ρ since the foregoing statistic is based on the assumption that $\rho = 0$, but we can test the null hypothesis that $\rho = 0$. We can obtain confidence intervals for ρ and test the hypothesis that ρ is equal to some value other than zero by using *Fisher's z statistic*, z_r. (See, for example, Huntsberger and Billingsley [1973], or Neter and others, [1985].)

Since $r/s_r = r/\sqrt{(1 - r^2)/(n - 2)} = r\sqrt{(n - 2)/(1 - r^2)}$, we can write the sample value of the test statistic either as $t^* = r/s_r$ or $t^* = r\sqrt{(n - 2)/(1 - r^2)}$. Although the two are equivalent, $t^* = r\sqrt{(n - 2)/(1 - r^2)}$ is more widely used.

The hypothesis test shown here is equivalent to both the F test and the t test for $H_0\colon \beta_1 = 0$. Thus we can use it to determine whether there is a

Testing the Hypothesis $\rho = 0$

Suppose we take a random sample of n observations from a population for which the coefficient of correlation is ρ. Then we may test the hypothesis $H_0\colon \rho = 0$. The sample value of the test statistic used to test the hypothesis is t^*, where

$$t^* = r\sqrt{\frac{n - 2}{1 - r^2}}$$

or

$$t^* = \frac{r}{s_r}$$

with $s_r = \sqrt{(1 - r^2)/(n - 2)}$.

The decision rule is determined by the alternate hypothesis and the level of significance α chosen for the test.

Null and Alternate Hypotheses

$H_0\colon \rho = 0$	$H_0\colon \rho = 0$	$H_0\colon \rho = 0$
$H_1\colon \rho > 0$	$H_1\colon \rho \neq 0$	$H_1\colon \rho < 0$

Decision Rules

Reject H_0 if	Reject H_0 if	Reject H_0 if
$t^* > t_\alpha\{n - 2\}$	$\lvert t^*\rvert > t_{\alpha/2}\{n - 2\}$	$t^* < -t_\alpha\{n - 2\}$

linear association between the variables without doing all the calculations needed for the regression analysis. Then if we cannot reject the hypothesis there is no need to perform the regression analysis. Again, however, you should use caution before concluding that $\rho = 0$ and there is no association between the variables. The probability of a type II error may be high, particularly if t^* is near the rejection region. Moreover, you should examine the scattergram to determine whether there may be a nonlinear relationship between the variables.

Example 13.17

Use the data of Example 13.15 to test the hypothesis that there is no correlation between the variables. Use the 0.01 level of significance.

Solution

We have $r = 0.9647$ and $n = 11$, so $t^* = 0.9647\sqrt{9/(1 - 0.9647^2)} \doteq 10.990$. Since $t_{.005}\{9\} = 3.250$ and $r > 0$, we conclude that $\rho > 0$, so there is a linear association between the variables. If we suspect there is a predictive relationship between the variables, regression analysis would be useful.

Example 13.18

A graduate student in psychology hypothesizes that success in graduate school as measured by grade-point average (GPA) is correlated positively with scores on the Graduate Record Examination (GRE)—that is, he believes that high GRE scores and success in graduate school are positively associated. To test his hypothesis he takes a random sample of 75 graduate students' records and finds that the correlation coefficient for the two variables is 0.37. Does this result support his hypothesis at a 0.05 level of significance?

Solution

Here the null hypothesis is H_0: $\rho = 0$ and the alternate hypothesis is H_1: $\rho > 0$. Since $t_{.05}\{73\} \doteq 1.67$, we reject H_0 if $t^* > 1.67$. Now $s_r = \sqrt{(1 - .37^2)/73} \doteq 0.1087$, so $t^* = 0.37/0.1087 \doteq 3.40$. Thus we reject H_0 and conclude that $\rho > 0$ and there is a positive association between the variables.

Note that this result shows there is *some* positive association between the variables, but not how much. Since $r^2 = (0.37)^2 \doteq 0.14$, about 14% of the variation is explained while the remainder is due to chance or other factors not in evidence. Thus although higher GRE scores and graduate GPA tend to vary in the same way, we cannot conclude that the relationship is very strong.

PROFICIENCY CHECKS

18. Using the result of Proficiency Check 15, test the conjecture that ρ is negative. Use the 0.05 level of significance.

(*continued*)

PROFICIENCY CHECKS *(continued)*

19. Using the result of Proficiency Check 16, test the conjecture that ρ is positive. Use the 0.01 level of significance.
20. Using the data of Proficiency Check 17, test the hypothesis that there is no correlation between the variables. Use the 0.05 level of significance.

13.4.4 The Spearman Rank-Difference Correlation Coefficient

If a set of data is classified in two ways and at least one variable is in ranks, we should use a nonparametric test of association. We can also use this procedure if some of the assumptions for linear regression analysis are not met. If the data are arranged in ranks, with each variable ranked in order, we can obtain the **Spearman rank-difference correlation coefficient,** symbolized r_s for a sample. The population Spearman correlation coefficient is symbolized ρ_s.

The order of ranking is not important so long as the low ranks of one variable are associated with low ranks of the other variable. We then compute the difference d in rank for each pair. We find the correlation between the ranked data sets by using the Spearman formula:

$$r_s = 1 - \frac{6(\Sigma d^2)}{n(n^2 - 1)}$$

Note that the differences are squared so that all we need is the absolute value of each difference. No sign is necessary.

The number obtained for r_s is exactly the same as if we had applied the Pearson formula to the ranks. Therefore the Spearman formula is a shortcut way of obtaining the correlation coefficient when the data are in ranks. We may then test the hypothesis H_0: $\rho_s = 0$ by using the t test as before, substituting r_s for r.

Spearman Rank-Difference Correlation Coefficient

Suppose we obtain n ordered pairs of observations (x, y) and each variable is ranked in order. Then we may obtain the Spearman rank-difference correlation coefficient, r_s. The Spearman formula is

$$r_s = 1 - \frac{6(\Sigma d^2)}{n(n^2 - 1)}$$

where each value of d is the difference between the ranks of the two variables for an observation. The sample value of the test statistic used to test the hypothesis H_0: $\rho_s = 0$ is t^*, where

$$t^* = r_s\sqrt{\frac{n-2}{1-r_s^2}}$$

The decision rule is determined by the alternate hypothesis and the level of significance α chosen for the test.

Null and Alternate Hypotheses

H_0: $\rho_s = 0$	H_0: $\rho_s = 0$	H_0: $\rho_s = 0$
H_1: $\rho_s > 0$	H_1: $\rho_s \neq 0$	H_1: $\rho_s < 0$

Decision Rules

Reject H_0 if	Reject H_0 if	Reject H_0 if
$t^* > t_\alpha\{n-2\}$	$\lvert t^*\rvert > t_{\alpha/2}\{n-2\}$	$t^* < -t_\alpha\{n-2\}$

In the case of ties, the rank of each tied value is the mean of all positions the tied values occupy.

Example 13.19

A husband and wife are each asked, independently, to rank six well-known public figures 1 through 6 in order of the degree to which they admire them. Using the data given here, test the hypothesis that husband and wife tend to rate public figures the same way. Use $\alpha = 0.10$. What is the P value for the test?

	PUBLIC FIGURE					
	A	B	C	D	E	F
Husband	6	5	3	1	2	4
Wife	6	3	2	1	4	5

Solution

Since the data are already in ranks, we need only compute the differences and square them. The sum of the squared differences is

$$\Sigma d^2 = 0^2 + 2^2 + 1^2 + 0^2 + 2^2 + 1^2 = 10$$

Then

$$r_s = 1 - \frac{6(10)}{6(6^2-1)} \doteq 0.7143$$

If husband and wife tend to rate public figures the same way, $\rho_s > 0$. Since $t_{.10}\{4\} = 1.533$, we can reject $H_0{:}\rho_s = 0$ and conclude $H_1{:}\rho_s > 0$ if $t^* > 1.533$. Now $t^* = 0.7143\sqrt{4/(1 - 0.7143^2)} \doteq 2.041$ and $2.041 > 1.533$, so we reject H_0 and conclude that the husband and wife do tend to rate public figures the same way. Since the test is one-sided, $t_{.05}\{4\} = 2.132$ and $1.533 < 2.041 < 2.132$, the P value is between 0.05 and 0.10.

The Spearman correlation coefficient is particularly useful in determining whether or not two raters tend to agree on ratings. If ρ_s is positive, there is agreement between the raters; if ρ_s is negative, there is disagreement between the raters; if $\rho_s = 0$, the raters neither agree nor disagree—there is no relationship between the ratings. If there are more than two raters, we can use the Friedman F_r test (Section 12.1) to determine whether or not the raters are in agreement. Other procedures for testing correlation between two variables include gamma (γ) and Kendall's tau (τ). See Goodman and Kruskal (1963 and 1972).

Example 13.20

Using the real estate data of Table 13.1, obtain r_s and test the hypothesis that there is no association between the variables. Use the 0.01 level of significance.

Solution

We reproduce the data in Table 13.8, ranking each variable from low to high. The sum of the d^2 is 27, so $r_s = 1 - 6(27)/(10 \cdot 99) \doteq 0.8364$. To test the hypothesis H_0: $\rho_s = 0$, we use $t^* = 0.8364\sqrt{8/(1 - 0.8364^2)} \doteq 4.315$; $t_{.01}\{8\} = 2.896$, so we reject H_0 and conclude that there is a positive association between the variables.

Table 13.8 WORKSHEET FOR DATA OF EXAMPLE 13.20.

LOT	WATERFRONT FOOTAGE (x)	SELLING PRICE (y)	RANK OF x	RANK OF y	d	d^2
1	188	140	8	7	1	1
2	231	160	2	2.5	−0.5	0.25
3	176	130	10	9.5	0.5	0.25
4	194	130	6.5	9.5	−3	9
5	244	180	1	1	0	0
6	207	160	4	2.5	1.5	2.25
7	198	140	5	7	2	4
8	217	150	3	4.5	−1.5	2.25
9	181	140	9	7	2	4
10	194	150	6.5	4.5	2	4

PROFICIENCY CHECKS

21. Union and management in a labor dispute have agreed on a list of eight issues that should be discussed. Each side has ranked the issues in order of importance. Test the hypothesis that there is neither agreement nor disagreement between the sides on the order in which issues should be discussed.

	ISSUE							
	1	2	3	4	5	6	7	8
Union	7	6	1	3	4	8	5	2
Management	4	5	1	6	7	8	2	3

22. Determine r_s for the data given here. Test the conjecture that there is a negative linear association between the variables. Use the 0.05 level of significance.

x	2	3	6	7	5	4	6	9	11	7
y	8	5	8	3	6	7	5	2	4	4

13.4.5 The SAS Procedure CORR

We can determine the correlation coefficient between two variables by using the SAS procedure CORR. When the variables have been entered in a data set, the statement

```
PROC CORR;
```

will create a correlation matrix for the two variables. Numerous refinements are available (see the *SAS® User's Guide: Basics*, pp. 861–874), but for most purposes the basic procedure is satisfactory. The output includes mean, standard deviation, median, and minimum and maximum for each variable, as well as the correlation coefficient between each pair of variables. The P value for the hypothesis test H_0: $\rho = 0$ is included in each entry of the matrix. Figure 13.22 shows the program and output of PROC CORR for the data of Example 13.15.

The correlation between x and y is 0.96469, and the probability of a type I error if we reject the hypothesis H_0: $\rho = 0$ is 0.0001 (or less). Thus we may reject H_0 and conclude that there is a positive association between the variables.

We can also use the CORR procedure to obtain the Spearman correlation coefficient between two variables. We write

```
PROC CORR SPEARMAN;
```

to obtain the value of r_s for a sample.

Figure 13.22
SAS printout of the correlation analysis of the data of Example 13.15.

```
INPUT FILE

DATA CORR;
  INPUT X Y @@;
CARDS;
2 10 3 14 3 15 4 17 5 18 5 19 6 18 6 22 7 26 8 29 8 27
PROC CORR;

OUTPUT FILE

                              SAS

VARIABLE    N        MEAN        STD DEV          SUM       MINIMUM       MAXIMUM

X          11    5.18181818   2.04049905    57.00000000   2.00000000    8.00000000
Y          11   19.54545455   5.88835523   215.00000000  10.00000000   29.00000000

     PEARSON CORRELATION COEFFICIENTS / PROB > |R| UNDER HO:RHO=0 / N = 11

                                  X         Y

                    X       1.00000   0.96469
                             0.0000    0.0001

                    Y       0.96469   1.00000
                             0.0001    0.0000
```

Problems

Practice

13.33 A correlation analysis is conducted with the following results: $n = 15$, $\Sigma x = 122.5$, $\Sigma x^2 = 1322.88$, $\Sigma y = 1.66$, $\Sigma y^2 = 0.2313$, and $\Sigma xy = 12.916$.

a. Obtain the coefficient of corrrelation for the data.
b. Obtain the coefficient of determination for the data.
c. Which coefficient has a clearer operational definition?

13.34 A correlation coefficient between two variables is found to be 0.58.

a. If $n = 17$, determine whether there is significant correlation between the variables. Use $\alpha = 0.05$.
b. What is the P value for the test?

13.35 Refer to Problem 13.5.

a. Calculate the correlation coefficient.
b. Determine whether there is a positive association between the variables. Use the 0.05 level of significance. Compare your answer with the results of Problem 13.15.

13.36 Refer to Problem 13.6.

a. Calculate the correlation coefficient.
b. Is there a linear association between the variables? Use the 0.01 level of significance. Compare your answer with the results of Problem 13.16.

13.37 Consider the following data:

x	4	8	11	5	9	10
y	10	12	25	11	15	15

a. Determine the coefficient of correlation for the data.
b. Determine the Spearman correlation coefficient for the data.
c. How do you account for the difference between the two? Draw a scatterplot to illustrate your conclusion.

Applications

13.38 A horse owner is investigating the relationship between weight carried and finish position of several horses in his stable. (See the following table.) Calculate r and determine whether it is significant at the 0.05 level. What do you conclude?

WEIGHT CARRIED	FINISH POSITION
110	2
113	6
120	3
115	4
110	6
115	5
117	4
123	2
106	1
108	4
110	1
110	3
120	5
105	7
110	2
115	4
103	1
118	3
115	2
110	7
115	6
105	2
110	3

13.39 A medical researcher measures the blood sugar level of cross-country runners immediately before and after the race. He wishes to discover whether there is any correlation between the decrease in blood sugar level and order of finish in the race. The race has 18 entrants. The following are the net decreases in blood sugar of the 18 entrants in order of finish from first to last (there are no ties): 10.8; 11.7; 9.7; 9.4; 10.3; 11.2; 8.8; 10.2; 7.4; 8.1; 7.7; 9.2; 6.4; 8.3; 9.4; 7.3; 6.1; 8.3.

a. Determine the correlation coefficient between decrease in blood sugar level and finish in the race.
b. Test the hypothesis, at a 0.01 level of significance, that there is no relationship between the variables, and interpret the result.

13.40 An old adage is "You get what you pay for!" Is this true for major league baseball? The average salaries paid by major league baseball teams in 1985 and 1984 are shown in the following table along with each team's winning percentage for the year. Do the results confirm this adage, or is there no relationship between salaries and winning percentage? Support your answers.

	1985		1984	
TEAM	AVERAGE SALARY	WINNING PCT.	AVERAGE SALARY	WINNING PCT.
	American League			
Yankees	$546,364	0.602	$458,544	0.537
Baltimore	438,256	0.516	360,204	0.525
California	433,818	0.556	431,431	0.500
Milwaukee	430,843	0.441	385,215	0.416
Detroit	406,755	0.522	371,332	0.642
Boston	386,597	0.500	297,878	0.531
Toronto	385,995	0.615	295,563	0.549
Kansas City	368,469	0.562	291,160	0.519
Oakland	352,004	0.475	384,027	0.475
White Sox	348,488	0.525	447,281	0.457
Minnesota	258,039	0.475	172,024	0.500
Texas	257,573	0.385	247,081	0.429
Cleveland	219,579	0.370	159,774	0.463
Seattle	169,694	0.457	168,505	0.457
	National League			
Atlanta	$540,988	0.407	$402,689	0.494
Los Angeles	424,273	0.586	316,250	0.488
Cubs	413,765	0.478	422,193	0.596
San Diego	400,497	0.512	311,199	0.568
Philadelphia	399,728	0.463	401,476	0.500
Pittsburgh	392,271	0.354	330,661	0.463
Mets	389,365	0.605	282,952	0.556
St. Louis	386,505	0.623	290,886	0.519
Houston	366,250	0.512	382,991	0.494
Cincinnati	336,786	0.553	269,019	0.432
San Francisco	320,370	0.383	282,132	0.407
Montreal	315,328	0.522	368,557	0.484

Source: Major League Players Association

13.41 A college student is asked to rank a list of ten prominent people in order of his preference that they become president. Two months later he is asked to repeat his ordering of the same list. He has several ties on the list, giving each tied value the same rank. If three are ties for fourth place, he gives them all a 4 but omits numbers 5 and 6. Using the following data, determine whether or not his second ordering is in agreement with his original ranking. Use the 0.05 level of significance. Be sure to use the correct procedure for scoring tied ranks.

Original order	1	2	3	4	4	4	7	8	8	10
Second order	2	2	1	6	8	9	4	7	10	4

13.42 Two professors have rated 11 students in terms of ability as part of a preliminary procedure to determine the recipients of several scholarships. Use the following data to test the hypothesis that there is no relationship between the way the professors rate the students. Let $\alpha = 0.05$.

STUDENT	PROF. X	PROF. Y
A	1	4
B	7	8
C	8	10
D	3	1
E	6	5
F	10	9
G	9	11
H	2	3
I	11	7
J	4	2
K	5	6

13.5 Minitab Commands for This Chapter

Most of the regression techniques presented in this chapter can be done easily and quickly with Minitab. The REGRESS and CORRELATION commands are the two Minitab commands we will need. The REGRESS command will also be useful in the next chapter for multiple regression analysis. The term *regress* is used as follows: when we say "y is regressed on x" we mean that y is to be used as a response variable and x as a predictor variable. When we say "regress y on x," this is the same as "determine a regression equation predicting y from x."

It is a good practice always to plot the data to determine whether the assumption of linear relationship between the two variables is appropriate. The PLOT command, introduced earlier, has the form

```
MTB > PLOT C on the vertical axis against C on the horizontal
      axis.
```

The basic command for obtaining a regression equation is the REGRESS command:

```
MTB >   REGRESS the y's in C on the K predictor variables in C
        ...C;
SUBC>   RESIDUALS store in C;
SUBC>   FITS store in C;
SUBC>   PREDICT using E.
```

Note that the REGRESS command requires you to specify three quantities:

1. The column in which the y data are stored
2. The number of predictor variables in the model
3. The columns in which the predictor variables are stored

For simple linear regression with one predictor variable, you must still indicate that there is one predictor variable. The version of the command for simple linear regression is

```
MTB >   REGRESS the y's in C on the 1 predictor variable in C;
SUBC>   RESIDUALS store in C;
SUBC>   FITS store in C;
SUBC>   PREDICT using E.
```

The RESIDUALS subcommand computes and stores the residuals for future use. The FITS subcommand computes and stores the fitted or predicted values ($\hat{y}$). The PREDICT subcommand computes and lists (but does not store) the predicted values and 95% confidence and prediction intervals for each element specified in E. E may be a number, a Minitab constant (such as K5), or a column of numbers. If E = 10, for example, the predicted value and the 95% confidence and prediction intervals for $x_p = 10$ are listed. If E = C (a column), then $\hat{y}$ and the 95% confidence and prediction intervals for each value in C are listed. Any combination of subcommands may be specified and in any order.

You can specify the amount of output to be written after using the REGRESS command by using the BRIEF command before typing the REGRESS command. Specifying BRIEF 1 requests the minimum output—the regression equation, the table of coefficients, the adjusted and unadjusted values of R^2, the standard error of estimate (listed as S), and the ANOVA table. BRIEF 2 is the default for the amount of output if the brief command is not given. When BRIEF 2 is specified in addition to BRIEF 1, any suspected outliers are identified. If BRIEF 3 is specified, the output will list, in addition to all the foregoing information, each x value, each y value, each fitted (predicted) value, the residual, and the standardized residual for each observation. Note that if the data set is at all large, the BRIEF 3 command will result in a large amount of output.

Example 13.21

Refer to the real estate data of Example 13.1. Use Minitab to obtain the regression equation and analyze the results completely.

Solution

We use the READ command rather than the SET command to ensure that the data pairs are matched correctly. Figure 13.23 shows the Minitab commands and results. We use the SAVE command to save the data set created in the program under the name specified—in this case REALEST. We can retrieve the data and use them in subsequent programs by using the RETRIEVE command. (Precise commands may vary by installation; check to make certain.)

Most of the REGRESS output is self-explanatory, and the results agree with those computed by hand in Example 13.1. The plot of the data suggests that a linear relationship between the variables is certainly appropriate, and the regression results indicate that the relationship is significant. The test of H_0: $\beta_1 = 0$ is significant ($t^* = 5.31$, exceeding the critical value,

```
MTB > NAME C1 'FRONTAGE' C2 'PRICE'
MTB > READ the following data into 'FRONTAGE' and 'PRICE'
DATA> 188 140
DATA> 231 160
DATA> 176 130
DATA> 194 130
DATA> 244 180
DATA> 207 160
DATA> 198 140
DATA> 217 150
DATA> 181 140
DATA> 194 150
DATA> ENDOFDATA
    10 ROWS READ
MTB > PLOT the y value 'PRICE' against the predictor 'FRONTAGE'

  PRICE             Graph of the data
180.00+                                                      *
      -
      -
      -
      -
160.00+                        *                   *
      -
      -        *                        *
      -
      -
140.00+    *     *     *
      -
      - *        *
      -
      -
120.00+
       +---------+---------+---------+---------+---------+FRONTAGE
     175.00    190.00    205.00    220.00    235.00    250.00
```

Figure 13.23
Minitab printout of the regression analysis of the real estate data, Example 13.22.

```
MTB > REGRESS the y variable C2 on 1 predictor in C1;
SUBC> RESIDUALS store in C3;
SUBC> FITS store in C4;
SUBC> PREDICT C1.

The regression equation is
PRICE = 21.1 + 0.625 FRONTAGE

Predictor        Coef  <- b0    Stdev          t-ratio
Constant        21.07          24.05 <- s_b0      0.88
FOOTAGE         0.6253 <- b1   0.1178 <- s_b1     5.31

s = 7.730 <- √MSE     R-sq = 77.9%     R-sq(adj) = 75.1%

Analysis of Variance

SOURCE        DF          SS          MS
Regression     1      1682.0      1682.0
Error          8       478.0        59.7
Total          9      2160.0

     Fit   Stdev.Fit        95% C.I.              95% P.I.
  138.62        3.02   ( 131.66, 145.58)    ( 119.48, 157.76)
  165.51        4.11   ( 156.04, 174.98)    ( 145.32, 185.70)
  131.12        4.01   ( 121.86, 140.37)    ( 111.03, 151.21)
  142.37        2.66   ( 136.23, 148.52)    ( 123.51, 161.23)
  173.64        5.41   ( 161.15, 186.13)    ( 151.87, 195.41)
  150.50        2.49   ( 144.76, 156.24)    ( 131.77, 169.23)
  144.87        2.51   ( 139.07, 150.67)    ( 126.12, 163.62)
  156.75        2.95   ( 149.95, 163.56)    ( 137.67, 175.84)
  134.24        3.56   ( 126.02, 142.46)    ( 114.61, 153.88)
  142.37        2.66   ( 136.23, 148.52)    ( 123.51, 161.23)

MTB > NAME C3 'RESID' C4 'FITS'
MTB > SAVE the data in 'REALEST'   <- Data set (plus residuals
                                      and fitted values) saved for
                                      future use
```

$t_{.025}\{8\} = 2.306$). Since the P values are not given, we must use critical values to test hypotheses.

Note that, except for the critical value of t, all the necessary information is given to compute confidence intervals for β_0 and β_1. The 95% confidence interval for the slope (β_1) has limits

$$b_1 \pm t_{\alpha/2}\{n-2\}s_{b_1}$$

$$0.625 \pm t_{\alpha/2}\{8\}(0.1178)$$

$$0.625 \pm 2.306 \cdot 0.1178$$

$$0.625 \pm 0.2716$$

or 0.3534 and 0.8966. Thus a 95% confidence interval for β_1 is

$$0.3534 \text{ to } 0.8966$$

We can compute a confidence interval for the intercept (β_0) in a similar manner.

The PREDICT subcommand was also used with the column of x data, C1. Note that the output includes the predicted values and the 95% confidence and prediction intervals for all eight x values in C1. The adjusted R^2 and the Durbin–Watson statistic are advanced regression topics that have been mentioned previously. Both are covered in Neter and others (1985).

We can find the correlation coefficient for the data in any pair of columns by using the CORRELATE command. If more than two colums are specified, the correlation between each pair of columns is printed in a table. The general format of the command is

```
MTB >  CORRELATE the data in columns C...C
```

The correlation coefficients that are printed are the usual Pearson product-moment correlation coefficients. To obtain the nonparametric Spearman rank correlation coefficient, we must first rank the data by using the RANK command and then obtain the correlation coefficient between these ranks. To obtain the rank correlation coefficient for the data in columns C6 and C8, for example, we would write

```
MTB >  RANK the data in C6 and store the results in C7
MTB >  RANK the data in C8 and store the results in C9
MTB >  CORRELATE the ranked data in C7 and C9
```

Example 13.22

Refer to the real estate data of Example 13.1. Use Minitab to obtain both the Pearson and Spearman correlation coefficients. Test the significance at the 0.05 level of significance.

Solution

The output of the Minitab program is shown in Figure 13.24. The RETRIEVE command has been used in this program. If you cannot SAVE data sets at your installation, you will need to enter the data as in the previous example. Comparing our results to those of Examples 13.16 and 13.10, we note that the results are identical. Thus the selling price and waterfront footage are positively linearly related to a strong degree.

```
MTB > RETRIEVE the data set in 'REALEST'
MTB > CORRELATE 'FRONTAGE' with 'PRICE'

   CORRELATION OF  FRONTAGE AND PRICE   = 0.882   <- Pearson correlation coefficient

MTB > RANK the 'FRONTAGE' data and store in C5
MTB > RANK the 'PRICE' data and store in C6
MTB > CORRELATE C5 with C6

   CORRELATION OF       C5 AND C6       = 0.832   <- Spearman rank correlation coefficient
```

Figure 13.24 Minitab printout of the correlation analysis of the real estate data, Example 13.22.

Recall that we test H_0: $\rho = 0$ by using $t^* = r\sqrt{(n-2)/(1-r^2)}$. For the real estate data we have

$$t^* = 0.8825\sqrt{\frac{10-2}{1-0.8825^2}}$$
$$\doteq 5.31$$

which is significant at the 0.05 level ($t_{.025}\{8\} = 2.306$). In fact, the for H_0: $\rho = 0$ and for $H_0 = \beta_1 = 0$ tests are equivalent (they give the same results) and the t values should be identical. This result is intuitively correct because the correlation coefficient measures the degree to which two variables are linearly related.

We may also use Minitab to perform a residual analysis and compare the results to those of Example 13.5. Recall that we had previously saved the x values, the y values, the residuals, and the $\hat{y}$ values in Example 13.21. We now simply RETRIEVE the data. (If you cannot save Minitab data sets, you will have to enter the data, use the REGRESS command, and store the residuals and fitted values as in Example 13.21.) Then you can execute the commands given here. Figure 13.25 shows the residual plots that may be obtained, as well as the normal probability plot for testing the normality of the error terms.

The three plots obtained are:

1. The residuals plotted against the predicted values
2. The residuals plotted against the predictor (x)
3. The residuals plotted against the normal scores of the residuals

Figure 13.25
Minitab printout of the residual plots for the real estate data.

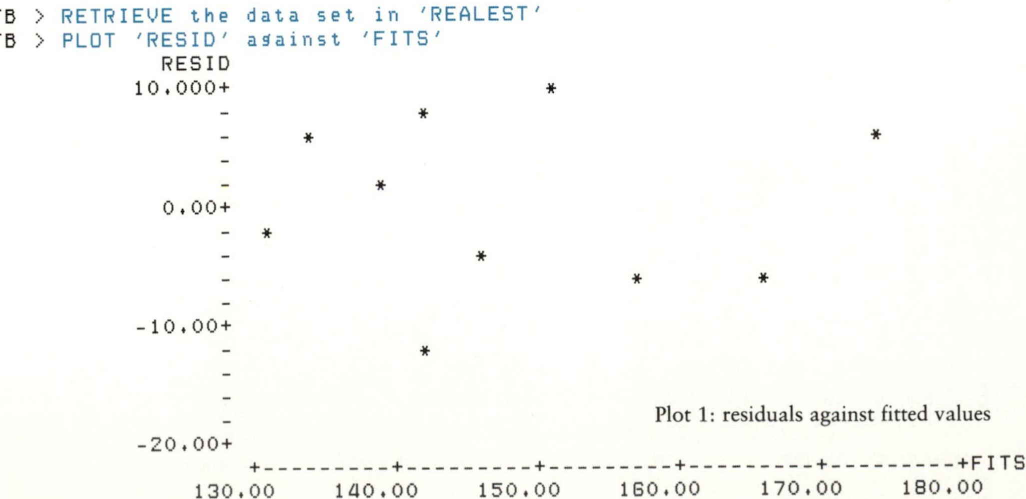

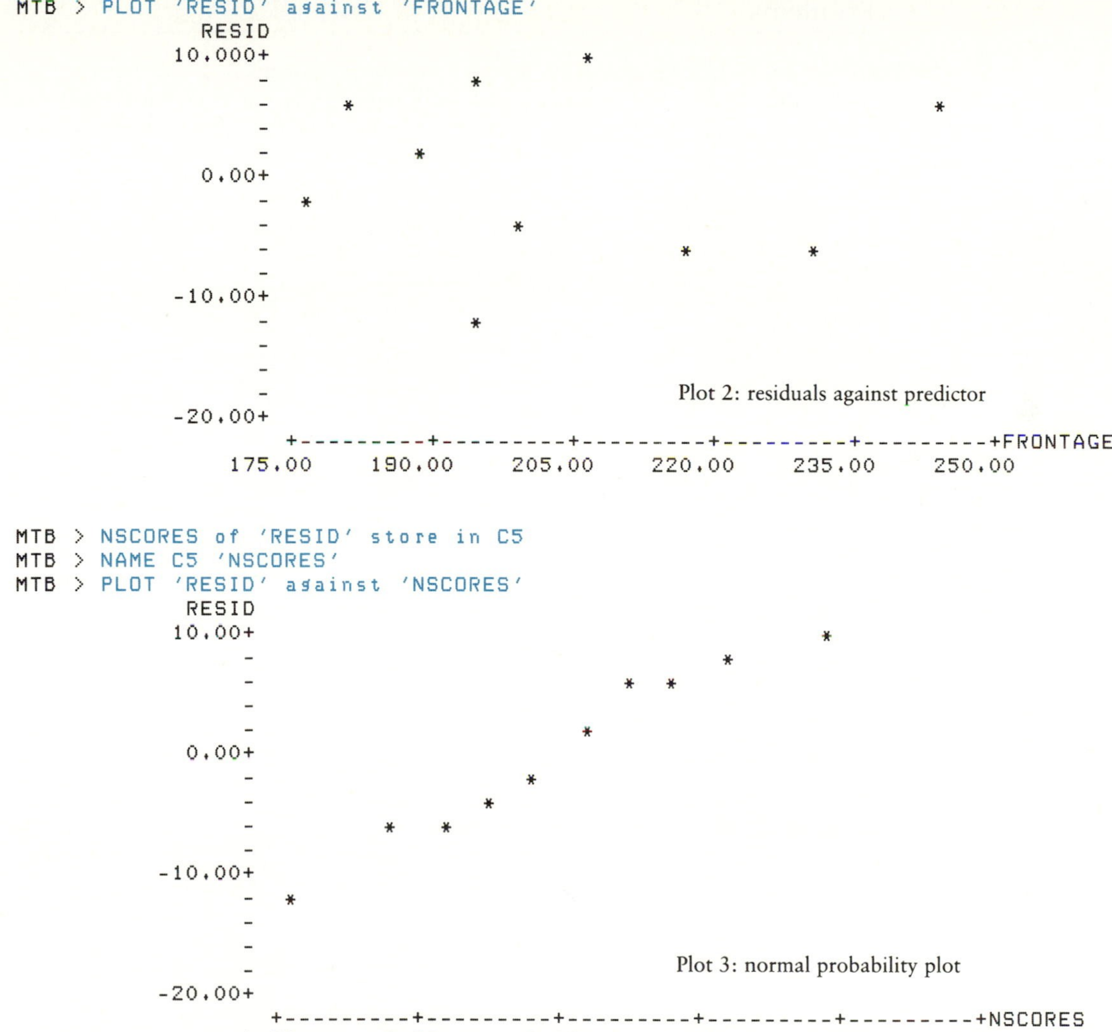

Plot 2: residuals against predictor

Plot 3: normal probability plot

There is no obvious pattern in plot 1 or 2, so we conclude that a linear relationship is appropriate. The relatively constant distances between the maximum and minimum values of the residuals in plots 1 and 2 suggest that the error variances are constant so that the assumption of constant error variance is satisfied. Further, since the normal probability plot (plot 3) is close to a straight line, the normality assumption is not unreasonable.

We emphasize here, as in the analysis of variance chapters, that the assumptions for a statistical test must be met and should always be checked. We need not be overly cautious in interpreting these plots, however. We wish only to identify those cases in which the assumptions are seriously violated. Regression analysis, like many of the statistical techniques we have used thus far, is robust against slight departures from the assumptions.

Problems

13.43 The following table gives the scores made by students in a statistics class on the midterm and final examinations. For predictive purposes, we assume that this is a random sample of all statistics students (although it probably is not).

STUDENT	MIDTERM	FINAL
1	98	90
2	68	82
3	100	97
4	74	78
5	88	77
6	98	93
7	45	82
8	64	80
9	85	77
10	87	99
11	91	98
12	94	77
13	96	95
14	80	95
15	89	92
16	70	80
17	64	75
18	75	65
19	99	88
20	67	78
21	75	63
22	96	93
23	49	53
24	100	90
25	76	53
26	71	88
27	77	84
28	73	58
29	55	88
30	65	63

Obtain the value of r for the data and determine, at the 0.05 level of significance, whether the variables are linearly related.

13.44 Nine advertising promotions are rated independently by a number of executives for clarity of presentation and aesthetic appeal. The median ratings are shown here:

Clarity	1.5	4.3	0.9	5.2	4.4	5.6	0.8	1.7	3.4
Appeal	3.2	1.7	1.6	6.3	3.1	2.8	1.5	3.3	2.7

a. Find the correlation coefficient and determine whether there is a linear association between the variables. Use the 0.05 level of significance.

b. A statistician questions the validity of using the Pearson correlation coefficient for the data, pointing out that the ratings are subjective and, although the rank order may be used, it is doubtful that the numbers themselves will yield meaningful results. Use the Spearman r_s and the 0.05 level of significance to determine whether there is a linear association between the variables. Compare with the result from part (*a*).

13.45 Refer to the USDA survey data set in Appendix B.2.

a. Is there a significant linear relationship between monthly income of household and years of education of head of household?

b. Would your answer to part (*a*) change if you used monthly income of household in thousands of dollars rather than dollars? Would the sample regression equation change?

13.6 Case 13: Mail Order Madness—Solution

Don Jones, owner of the mail-order business, brought in his data. The statistician sent the program off and waited a few minutes for it to come back. The result of the analysis is shown in Figure 13.26.

Figure 13.26
Computer output for Case 14.

```
                                        SAS

DEP VARIABLE: ORDERS
                                 ANALYSIS OF VARIANCE

                             SUM OF             MEAN
          SOURCE     DF     SQUARES           SQUARE        F VALUE        PROB>F

          MODEL       1  47823114.92     47823114.92       1965.850        0.0001
          ERROR      28    681154.45     24326.94457
          C TOTAL    29  48504269.37

              ROOT MSE      155.971       R-SQUARE        0.9860
              DEP MEAN     5581.567       ADJ R-SQ        0.9855
              C.V.         2.794394

                                 PARAMETER ESTIMATES

                         PARAMETER          STANDARD         T FOR H0:
    VARIABLE   DF         ESTIMATE             ERROR       PARAMETER=0        PROB > |T|

    INTERCEP    1       1608.12005       94.03282758            17.102            0.0001
    WT          1        559.66664       12.62275859            44.338            0.0001
```

Studying the printout for the SAS REG procedure, he told Jones that there was a definite linear association between the weight of mail and number of orders and that for each additional pound of mail the number of orders would increase, on the average, by about 560. He showed Jones how to get a 90% prediction interval for the day's orders by using the weight of the mail. For a weight of x_p pounds the predicted number of orders is $\hat{y}_p = 1608.12005 + 559.66664x_p$. It can be shown that $\text{SSX} = \text{SSR}/b_1^2$; then a 90% prediction interval for the actual number of orders has limits

$$\hat{y}_p - 1.701\sqrt{1 + \tfrac{1}{30} + (x_p - \bar{x})^2/152.6789}$$

and

$$\hat{y}_p + 1.701\sqrt{1 + \tfrac{1}{30} + (x_p - \bar{x})^2/152.6789}$$

"If you hire enough people to process and ship the number of orders indicated by the upper limit of the prediction interval," he said, "there is only a 5% chance that you will hire too few people on that day. In the long run, of course, you will probably have to pay overtime one day in twenty on the average. You might analyze the cost effectiveness of using wider prediction intervals, thus hiring more people each day but having to pay less overtime. In addition, you need to keep records and bring the data over for periodic revision." Jones was satisfied that he would be able to keep his promise. When he saw his friend several months later, he told him that the predictions were working extremely well and his business had picked up.

13.7 Summary

If there is a statistical relationship between two variables and one variable can be used to predict values of the other, we use regression analysis to obtain a function for estimating values of the response variable (y) from the predictor variable (x). If the relationship between the variables is linear, we use linear regression analysis. This association can be tested by using an F test.

If there is a linear association between the variables, further analysis and inferences are warranted. In order for us to draw inferences, the following four assumptions about the model must be satisfied (in addition to the assumption that the sample is a random sample from the population):

1. The statistical relationship between the variables is linear, and the expected value of the error terms is zero.
2. For each value of x, the error terms are normally distributed.
3. For each value of x, the variance of the error terms is constant.
4. The error terms are independent of each other.

We can obtain confidence intervals for β_1, β_0, and $E(y_p)$, the expected value of y for a given $x = x_p$, as well as prediction intervals for y_p, an individual future value of y for $x = x_p$. In addition to the F test, we can use a t test to test hypotheses concerning β_1.

The final topic covered in this chapter was correlation analysis. We can use correlation analysis to test the association between two variables even if neither is considered a predictor for the other. We first define the coefficient of determination to be the proportion of total sum of squares for one variable that is explained by the relationship between the variables. The square root of the coefficient of determination is called the coefficient of correlation, symbolized ρ for a population and r for a sample. If at least one variable is in ranks or the assumptions are not all satisfied, the nonparametric Spearman rank-difference correlation coefficient should be used.

Chapter Review Problems

Practice

13.46 Use the data given here:

x	3	3	5	4	5	6	7	5	4	7	3	6
y	11	10	16	13	18	22	23	17	19	26	9	23

a. Derive the regression equation for predicting y from x.
b. Obtain the coefficient of correlation for the data.
c. Obtain r_s for the data.

13.47 A regression analysis includes the following intermediate calculations: $n = 30$, SSX $= 32{,}133$, SSTO $= 188{,}355$, and SXY $= 71{,}302$.
a. Obtain the analysis of variance table for the data.
b. Test the hypothesis that there is no association between the variables. Use the 0.05 level of significance.
c. Obtain r^2 for the data. Interpret its meaning.
d. Calculate r and repeat part (*b*). How do the answers compare?

13.48 A regression analysis has the regression equation $\hat{y} = 41.4564 + 11.8876x$; $n = 15$, SSX $= 45.3687$, $s_e = 44.9960$, $s_{b_1} = 6.6803$, $s_{b_0} = 36.2487$, and $\bar{x} = 5.14$.
a. Obtain a 95% confidence interval for β_1.
b. Obtain a 95% confidence interval for β_0.
c. Give a 95% confidence interval for $E(y_p)$ for $x_p = 5.25$.
d. Give a 95% prediction interval for y_p for $x_p = 5.00$.

13.49 Suppose that, for a set of data, $n = 20$, $\Sigma x = 146.88$, $\Sigma x^2 = 1269.6143$, $\Sigma y = 14.53$, $\Sigma y^2 = 12.1901$, and $\Sigma xy = 122.3048$.
a. Determine SSX, SSTO (SSY), and SXY.
b. Obtain a regression equation to predict y from x.
c. Show the analysis of variance table.
d. Using a 0.01 level of significance, test the association between the variables.
e. Obtain a 99% confidence interval for β_1.
f. Using $\alpha = 0.01$, test the hypothesis that $\beta_1 < 0.10$.

g. Obtain a 99% confidence interval for β_0.
h. Obtain a 99% confidence interval for $E(y_p)$ when $x_p = 7.5$.
i. Obtain a 99% prediction interval for y_p when $x_p = 7.5$.

Applications

13.50 A sociologist is studying the relationship between absenteeism among city employees during the year and their relative "job pride" as measured by an index. She has obtained the following data, where high "job pride" values ostensibly indicate a greal deal of job pride.

Days absent	8	1	3	11	6	7	3	8	10	24	6	2
"Job pride" index	63	82	59	73	84	67	81	73	63	94	81	90

a. Obtain a regression equation for predicting days absent from the index.
b. Plot the data on a scatterplot and draw the regression line. Does it appear to fit the data well?
c. Determine the correlation coefficient for the data. Is there a significant association between the variables at the 0.05 level of significance?
d. Obtain a residual plot for the data.
e. Determine s_e and check for the presence of outliers.

13.51 The employee with the highest job-pride score in Problem 13.50 had been absent the most days. After investigation, the sociologist found that the absenteeism was unavoidable, so she eliminated the data point.
a. Determine the revised regression equation.
b. Plot the data on a scatterplot and draw the regression line. Does it appear to fit the data well?
c. Calculate r and determine whether there is a linear association between the variables. Use $\alpha = 0.05$.
d. Obtain a residual plot. Do any assumptions appear to have been violated?
e. Determine s_e and check for the presence of outliers.

13.52 A stable owner is training his horse, Whirlalong. Trying several different weights on Whirlalong, he obtains the following data for the horse's time for a mile:

Weight (lb)	100	103	105	108	110	112	115	116	119
Time (sec)	99.2	98.2	101.1	102.3	100.9	102.8	105.2	105.3	106.1

a. Determine a regression equation for predicting Whirlalong's time for a mile from the weight carried.
b. Is there a linear relationship between the variables? Use the 0.01 level of significance.
c. How many seconds would you expect each additional pound to add, on the average, to Whirlalong's time for the mile?
d. Give a 99% confidence interval for Whirlalong's mean time for the mile when he carries 110 pounds.
e. Whirlalong is racing in a mile race today and carrying 108 pounds. Give a 99% prediction interval for his time today.
f. From a bettor's standpoint, which of the three would be most useful for betting purposes?

13.53 Refer to Problem 13.43.

Obtain the value of r for the data and determine, at the 0.05 level of significance, whether the variables are linearly related.

a. Determine a regression equation for predicting a score on the final from the midterm score.
b. Determine a 95% confidence interval for β_1. Interpret your result.
c. Give a 95% confidence interval for the mean final exam grade for all students scoring 90 on the midterm.
d. Give a 95% prediction interval for the final exam score for a student who scores 80 on the midterm next semester.

13.54 The management of a large corporation decides to reduce the price of its leading product in order to increase sales and therefore profits. Careful records of sales and profits are kept for 6 months preceding the experiment. Fifty widely separated retail outlets are then selected at random. They are further randomly divided into five groups. Reductions of from \$2 to \$10 per unit are randomly assigned to the groups in increments of \$2, and sales are observed for 3 months. The mean net change in proceeds per month, compared with the mean for the preceding 6 months, is expressed as a percentage. Changes are adjusted to reflect an average 30-day month. The data are given here:

REDUCTION (\$)	% CHANGE IN NET PROCEEDS
2	+5.1 −1.3 +4.4 +0.6 −0.4 +4.7 +3.2 +0.6 +1.9 +6.3
4	−1.4 −8.6 +1.8 +2.1 −1.7 +7.7 +3.6 −2.5 +5.6 +3.8
6	−3.2 +1.3 +2.1 −3.2 −0.4 −11.2 +3.1 −2.4 +5.8 +0.3
8	−7.2 +0.1 −1.6 −2.1 −6.8 −7.2 −2.2 +4.2 −3.6 −1.5
10	+12.3 −3.1 −3.9 −8.6 −5.3 −9.3 −1.5 −3.2 −4.1 −0.9

a. Obtain a regression equation for the data.
b. Write the analysis of variance table for the data.
c. Test the hypothesis that there is no association between the variables. Use the 0.05 level of significance.
d. Obtain a 95% confidence interval for the mean increase or decrease of change in net proceeds in all outlets for each reduction of \$1.
e. Obtain a 95% confidence interval for the mean change in net proceeds if there is no reduction.
f. Obtain a residual plot. Do any assumptions appear to be violated?

Index of Terms

Glossary of Symbols

b_0, b_1	estimated regression parameters, regression coefficients
β_0, β_1	regression parameters
$E(y)$	expected value of y for a given x
$E(y_p)$	expected value of y for a particular $x = x_p$
r	coefficient of correlation for a sample
r_s	Spearman correlation coefficient for a sample
r^2	coefficient of determination for a sample
ρ	coefficient of correlation for a population
ρ_s	Spearman correlation coefficient for a population
ρ^2	coefficient of determination for a population
SSE	sum of squares for error
SSR	sum of squares for regression
$SSTO$	total sum of squares
SSX	sum of squares for x
SSY	sum of squares for y
SXY	sum of xy terms
s_{b_0}	standard error of b_0
s_{b_1}	standard error of b_1
s_e	standard error of estimate
s_f	standard error of forecast
$s_{\hat{y}_p}$	standard error of predicted value of y for $x = x_p$
y_p	individual future value of y for a particular $x = x_p$
$\hat{y}_p$	predicted value of $E(y_p)$ for a particular $x = x_p$

Answers to Proficiency Checks

1. SXY = 8766.12 − (133.7)(1475.9)/20 = 8766.12 − 9866.3915 = −1100.2715; SSX = 992.84 − $(133.7)^2$/20 = 992.84 − 893.7845 = 99.0555. The sample regression equation is $\hat{y} = 148.0495 - 11.1076x$.

2. $\Sigma x = 31$, $\Sigma x^2 = 177$, $\Sigma y = 103$, $\Sigma xy = 529$, $n = 8$; SSX $= 177 - 31^2/8 = 56.875$; SXY $= 529 - 31(103)/8 = 129.875$. Then $b_1 = 129.875/56.875 \doteq 2.2835$ and $b_0 = [103 - b_1(31)]/8 \doteq 4.0264$. Thus $\hat{y} = 4.0264 + 2.2835x$. (For more decimal places, if desired for a check, $b_1 \doteq 2.2835164845$ and $b_0 \doteq 4.026373629$.) The graph is shown below:

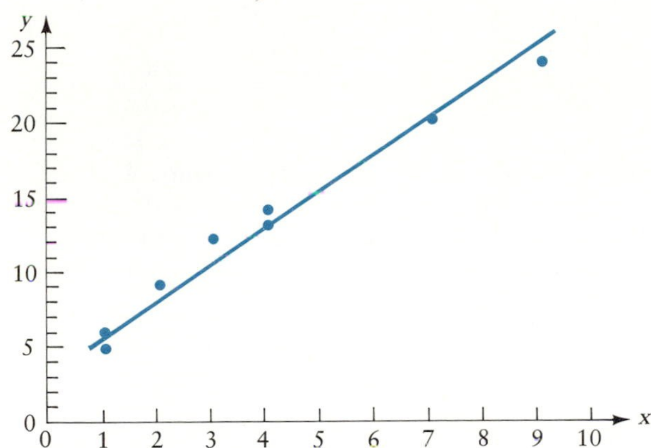

3. SSTO $= 125{,}251.14 - (1475.9)^2/20 = 125{,}251.14 - 108{,}914.0405 = 16{,}337.0995$. SXY $= -1100.2715$, SSX $= 99.0555$, and $n = 20$. Then SSR $= (-1100.2715)^2/99.0555 \doteq 12{,}221.4049$ and SSE $\doteq 16{,}337.0995 - 12{,}221.4049 = 4{,}115.6946$.

SOURCE OF VARIATION	DF	SS	MS
Regression	1	12,221.4049	12,221.4049
Error	18	4,115.6946	228.6497
Total	19	16,337.0995	

4. $\Sigma y^2 = 1627$, so SSTO $= 1627 - (103)^2/8 = 300.875$. SXY $= 129.875$, SSX $= 56.875$, $n = 8$. Then SSR $= (129.875)^2/56.875 \doteq 296.5717$ and SSE $\doteq 300.875 - 296.5717 = 4.3033$.

SOURCE OF VARIATION	DF	SS	MS
Regression	1	296.5717	296.5717
Error	6	4.3033	0.7172
Total	7	300.875	

5. $F^* = 12{,}221.4049/228.6497 \doteq 53.45$; $F_{.05}\{1, 18\} = 4.41$, so we conclude that $\beta_1 \neq 0$. There is a linear association between the variables.

6. $F^* = 296.5717/0.7172 \doteq 413.35$; $F_{.01}\{1, 6\} = 13.7$. Thus $\beta_1 \neq 0$, and there is a linear association between the variables.

7. The residuals are displayed in the following table; $\hat{y} = 4.0264 + 2.2835x$. They do sum to zero.

x	y	$\hat{y}$	e
1	5	6.31	−1.31
1	6	6.31	−0.31
2	9	8.59	0.41
3	12	10.88	1.12
4	14	13.16	0.84
4	13	13.16	−0.16
7	20	20.01	−0.01
9	24	24.58	−0.58
			0.00

The residual plot is shown in the accompanying figure. With so few points it is difficult to tell, but we note that the points lie below the axis on the two ends and above the axis in the center. This pattern appears to violate the assumption of linearity, so perhaps the association between the variables is curvilinear.

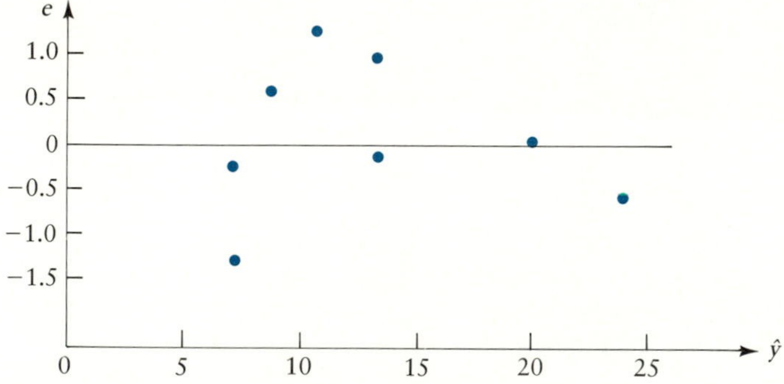

8. $b_1 = -11.1076$, $s_{b_1} = \sqrt{228.6497/99.0555} \doteq 1.5193$, and $t_{.025}\{18\} = 2.101$. Then $2.101(1.5193) \doteq 3.1921$; the confidence limits are $-11.1076 - 3.1921$ and $-11.1076 + 3.1921$, so we are 95% certain that β_1 lies between -14.2997 and -7.9155.

9. $b_1 = 2.2835$, $s_{b_1} = \sqrt{0.7172/56.875} \doteq 0.1123$, and $t_{.005}\{6\} = 3.707$. Then $3.707(0.1123) \doteq 0.4163$, and the 99% confidence interval is 1.8672 to 2.6998.

10. $t_{.05}\{18\} = 1.734$, so we reject H_0 if $t^* < -1.734$. Since $b_1 = -11.1076$ and $s_{b_1} = 1.5193$, we have $t^* = (-11.1076 - \{-10\})/1.5193 \doteq -0.729$; fail to reject H_0. We cannot conclude that $\beta_1 < -10$.

11. $t_{.005}\{6\} = 3.707$, so we reject H_0 if $|t^*| > 3.707$. Since $b_1 = 2.2835$ and $s_{b_1} = 0.1123$, we have $t^* = (2.2835 - 1.8)/0.1123 \doteq 4.305$; reject H_0. We conclude that $\beta_1 \neq 1.8$. For an estimate of the value of β_1, we use the confidence interval shown in Proficiency Check 9.

12. $b_0 \doteq 4.0264$, $s_e = \sqrt{0.7172} \doteq 0.8469$, and $s_{b_0} = s_e\sqrt{1/n + \bar{x}^2/\text{SSX}} \doteq 0.8469\sqrt{1/8 + (3.875)^2/56.875} \doteq 0.5282$; $t_{.005}\{6\} = 3.707$ and $3.707(0.5282) \doteq 1.9581$. Thus the confidence limits for β_0 are $4.0264 - 1.9581$ to $4.0264 + 1.9581$, and the 99% confidence interval is 2.0683 to 5.9845.

13. When $x_p = 6$, $\hat{y}_p \doteq 81.4039$; $t_{.05}\{18\} = 2.101$. Since $\bar{x} = 6.685$

a. $s_{\hat{y}_p} \doteq \sqrt{228.6497(1/20 + (6 - 6.686)^2/99.0555)} \doteq 3.5377$; $2.101(3.5377) \doteq 7.4328$. Then the confidence interval is 73.9711 to 88.8367.

b. $s_f \doteq \sqrt{228.6497(1 + 1/20 + (6 - 6.686)^2/99.0555)} \doteq 15.5295$; $2.101(15.5295) \doteq 32.6275$. Then the prediction interval is 48.7764 to 114.0314.

14. When $x_p = 4.5$ $\hat{y}_p \doteq 14.3022$; $t_{.005}\{8)\} = 3.355$. Since $\bar{x} = 3.875$,

a. $s_{\hat{y}_p} \doteq \sqrt{0.7172(1/8 + (4.5 - 3.875)^2/56.875)} \doteq 0.3075$; $3.355(0.3075) \doteq 1.0318$. Then the confidence interval is 13.2704 to 15.3340.

b. $s_f \doteq \sqrt{0.7172(1 + 1/8 + (4.5 - 3.875)^2/56.875)} \doteq 0.9010$; $3.355(0.9010) \doteq 3.023$. Then the prediction interval is 11.2793 to 17.3250.

15. $r^2 = 12{,}221.4049/16{,}337.0995 \doteq 0.7481$; $r \doteq -0.8649$ (r will be negative since b_1 is negative). Alternatively, $r = -1100.2715/\sqrt{(99.0555)(16{,}337.0995)} \doteq -1100.2715/1272.1162 \doteq -0.8649$.

16. $r^2 = 296.5717/300.875 \doteq 0.9857$; $r \doteq 0.9928$ (r will be positive since b_1 is positive). Alternatively, $r = 129.875/\sqrt{(56.875)(300.875)} \doteq 129.875/130.8139 \doteq 0.9928$.

17. $\Sigma x = 61$, $\Sigma x^2 = 437$, $\Sigma y = 51$, $\Sigma y^2 = 297$, $\Sigma xy = 278$, $n = 10$. Using the raw-data formula,

$$r = \frac{10(278) - (61)(51)}{\sqrt{[10(437) - (61)^2][10(297) - (51)^2]}}$$

$$= \frac{-331}{\sqrt{(649)(369)}} = \frac{-331}{\sqrt{239{,}841}} \doteq \frac{-331}{489.3680}$$

$$\doteq -0.6764$$

Using sums of squares, SSX = 64.9, SSY = 36.9, SXY = −33.1, and the same result is obtained.

18. We wish to test H_0: $\rho = 0$ against H_1: $\rho < 0$. Since $n = 20$ and $t_{.05}\{18\} = 1.734$, we reject H_0 if $t^* < -1.734$. Now $r \doteq -0.8649$, so $t^* = -0.8649\sqrt{18/[1 - (-0.8649)^2]} \doteq -7.310$. Thus we reject H_0 and conclude that $\rho < 0$.

19. We wish to test H_0: $\rho = 0$ against H_1: $\rho > 0$. Since $n = 8$ and $t_{.01}\{6\} = 3.143$, we reject H_0 if $t^* > 3.143$. Now $r \doteq 0.9928$, so $t^* = 0.9928\sqrt{6/[1 - (0.9928)^2]} \doteq 20.302$. Thus we reject H_0 and conclude that $\rho > 0$.

20. We wish to test H_0: $\rho = 0$ against H_1: $\rho \neq 0$. Since $n = 10$ and $t_{.025}\{8\} = 2.306$, we reject H_0 if $|t^*| > 2.306$. Now $r = -0.6764$, so $t^* = -0.6764\sqrt{8/[1 - (-0.6764)^2]} \doteq -2.597$. Thus we reject H_0 and conclude that $\rho \neq 0$. There appears to be a negative association between the variables.

21. $\Sigma d^2 = 3^2 + 1^2 + 0^2 + 3^2 + 3^2 + 0^2 + 3^2 + 1^2 = 38$; $r_s = 1 - 6(38)/8(63) \doteq 0.5476$. $t^* = 0.5476\sqrt{6/(1 - 0.5476^2)} \doteq 1.6031$ and $t_{.025}\{6\} = 2.447$, so we cannot reject H_0: $r_s = 0$. We conclude that there is no significant association between the orders of importance for the two sides; that is, there is neither agreement nor disagreement.

22. The data are ranked as shown here:

x	1	2	5.5	7.5	4	3	5.5	9	10	7.5
y	9.5	5.5	9.5	2	7	8	5.5	1	3.5	3.5

Then $\Sigma d^2 = 8.5^2 + 3.5^2 + 4^2 + 5.5^2 + 3^2 + 5^2 + 0^2 + 8^2 + 6.5^2 + 4^2 = 287$, and $r_s = 1 - 6(287)/10(99) \doteq -0.7394$. To test the hypothesis H_0: $\rho_s = 0$ against the alternate hypothesis H_1: $\rho_s < 0$, we will reject H_0 if $t^* < -t_{.05}\{8\}$—that is, if $t^* < -1.860$. Now $t^* = -0.7394\sqrt{8/[1 - (-0.7394)^2]} \doteq -3.106$. Therefore we reject H_0 and conclude that $\rho_s < 0$.

References

Goodman, L. A., and W. H. Kruskal. "Measures of Association for Cross-Classification III." *Journal of the American Statistical Association* 58 (1963): 310–364.

Goodman, L. A., and W. H. Kruskal. "Measures of Association for Cross-Classification IV." *Journal of the American Statistical Association* 67 (1972): 415–421.

Huntsberger, David V., and Patrick Billingsley. *Elements of Statistical Inference.* 3rd ed. Boston: Allyn & Bacon, 1973.

Neter, John, William Wasserman, and Michael H. Kutner. *Applied Linear Statistical Models.* 2nd ed. Homewood, IL: Richard D. Irwin, 1985.

SAS Institute Inc. *SAS® User's Guide: Basics, Version 5 Edition.* Cary, NC: SAS Institute Inc., 1985.

SAS Institute Inc. *SAS® User's Guide: Statistics, Version 5 Edition.* Cary, NC: SAS Institute Inc., 1985.

14

Multiple Regression

In Chapter 13 we discussed methods of detecting the extent of a linear relationship between one predictor variable x and a response variable y. Often a response variable may be considered to depend on two or more predictor variables. The value of property in a given location may depend upon area, the number of trees per acre, the slope of the land, the view (rated on a numerical scale from poor to excellent), and perhaps the elevation, as well as other factors. Sometimes the relationship between one predictor variable and a response variable is not linear but is described by a polynomial function. The cost per unit of a manufactured commodity usually depends on the number manufactured. As the lot size (x) increases, the cost per unit may decrease to a certain point and then begin to rise again because of overtime, machine breakdown, bottlenecks, and other factors. The relationship between the two variables could be quadratic, depending upon x^2 as well as x. In such cases we can extend the methods of Chapter 13 to **multiple regression analysis.**

In this chapter we consider the principles used to obtain multiple regression equations, learn how to determine whether there is a relationship between the variables, and find out how to test some of the assumptions. We then learn how to obtain confidence intervals for the population parameters and prediction intervals for a specific value of the response variable and discuss ways of determining the best regression equation from several variables. We finally apply multiple regression procedures to obtain quadratic regression equations and make inferences.

We illustrate throughout by using SAS statements and printouts; applications of Minitab are given in the last section. If you are using Minitab, refer to Section 14.4 and use the appropriate commands for each section.

14.0 Case 14: How Much to Send a Letter?

In late 1981, just after the cost of a first-class letter had been raised to 20¢ (the second raise in a year), the author and Dr. Robert E. Lee were discussing the cost of a postage stamp and speculating on the next raise. When would it be, and for how much? Dr. Lee obtained the following information on postage costs:

DATE	FIRST-CLASS POSTAGE	DATE	FIRST-CLASS POSTAGE
1885–1917	2¢	May 16, 1971	8¢
1917–1919	3¢ (Wartime increase)	March 2, 1974	10¢
1919	2¢ (Restored by Congress)	December 31, 1975	13¢ (Temporary)
July 6, 1932	3¢	July 18, 1976	13¢ (Permanent)
August 1, 1958	4¢	May, 1978	15¢
January 7, 1963	5¢	March 22, 1981	18¢
January 7, 1968	6¢	November 1, 1981	20¢

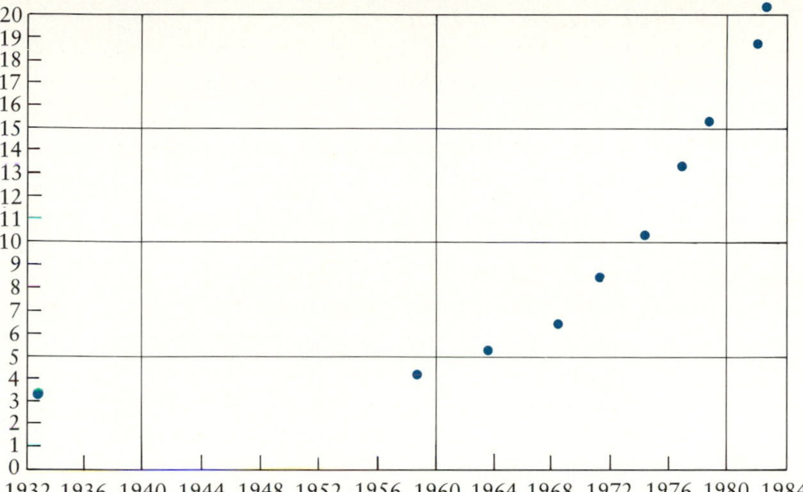

Figure 14.1
Graph of the prices of first-class postage stamps plotted against the dates of the price increases.

We thought it might be interesting to obtain a regression equation for predicting the price of a postage stamp as a function of time. Although we could make no inferences (since most of the assumptions for regression analysis were not met), our procedure would be valid for predictive purposes of the single variable. The graph of the costs, plotted against the date of each increase (Figure 14.1), shows a slight increase at first, followed by ever sharper increases. Clearly we could not use a linear regression equation for prediction, so we turned to multiple regression analysis. After trying several approaches, we eventually came up with an equation that fit the data well.

14.1 Multiple Regression Models

Suppose we have reason to believe that a response variable (y) is influenced by $k - 1$ predictor variables $x_1, x_2, \ldots, x_{k-1}$. A single observation consists of one value of each predictor variable and the response variable. (We denote the number of predictor variables by $k - 1$ rather than k so that the number of population parameters is k rather than $k + 1$.) If there is a linear statistical relationship between each of the predictor variables and the response variable, each observation of y can be defined as follows:

$$y_i = \beta_0 + \beta_1 x_{i1} + \beta_2 x_{i2} + \cdots + \beta_{k-1} x_{i,k-1} + \epsilon_i$$

where x_{ij} is the ith observation of the jth predictor variable
y_i is the ith observation of the response variable
β_0, β_j are parameters
ϵ_i are $N(0, \sigma^2)$
$i = 1, 2, 3, \ldots, n$
$j = 1, 2, 3, \ldots, k - 1$

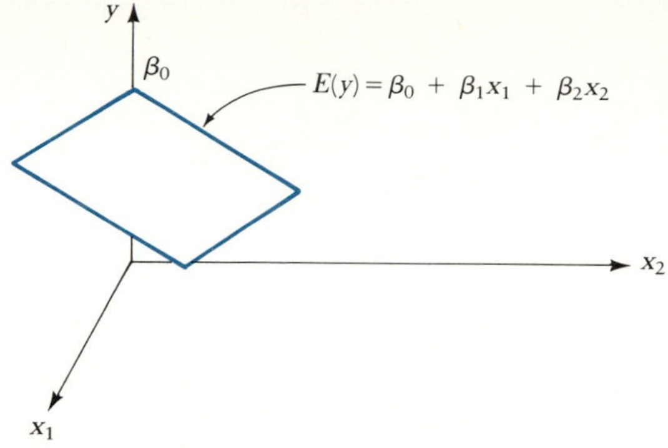

Figure 14.2 Graph of the regression function $E(y) = \beta_0 + \beta_1 x_1 + \beta_2 x_2$.

We write the regression function

$$E(y) = \beta_0 + \beta_1 x_1 + \beta_2 x_2 + \beta_3 x_3 + \cdots + \beta_{k-1} x_{k-1}$$

We can use the method of least squares to obtain the sample regression equation

$$\hat{y} = b_0 + b_1 x_1 + b_2 x_2 + b_3 x_3 + \cdots + b_{k-1} x_{k-1}$$

Note that β_0 denotes the value of y when each predictor variable has a value of zero; $\beta_1, \beta_2, \beta_3, \ldots, \beta_{k-1}$ are the increases in $E(y)$ for an increase of one unit for each predictor variable provided that all other predictor variables are held constant. That is, if x_j is increased by one unit and all other variables are held at the same values, β_j is the increase in $E(y)$.

If $k = 3$ then $k - 1 = 2$, so that $E(y) = \beta_0 + \beta_1 x_1 + \beta_2 x_2$ and there are two predictor variables but three population parameters (β_0, β_1, and β_2). The graph of the regression function is a plane, as shown in Figure 14.2. Note that β_0 is the elevation of the plane when x_1 and x_2 are both equal to zero. The random error terms (ϵ) are the vertical distances from each point (x_{i1}, x_{i2}, y_i) to the regression plane; the point on the regression plane has coordinates $(x_{i1}, x_{i2}, E(y_i))$. Thus each value of ϵ_i is $y_i - E(y_i)$, as shown in Figure 14.3.

14.1.1 The Multiple Regression Equation

Again we use the principle of least squares to find the sample regression equation, but the results are not so simple as when there is only one predictor variable. The least-squares approach in the case of multiple regression leads to a system of k equations in k unknowns, called **normal equations.** (The word *normal* is used here in the sense of regular or standard.) For two predictor variables (x_1, and x_2) and one response variable (y), the normal equations are

$$\Sigma y = nb_0 \quad + b_1 \Sigma x_1 \quad + b_2 \Sigma x_2$$
$$\Sigma x_1 y = b_0 \Sigma x_1 + b_1 \Sigma x_1^2 \quad + b_2 \Sigma x_1 x_2$$

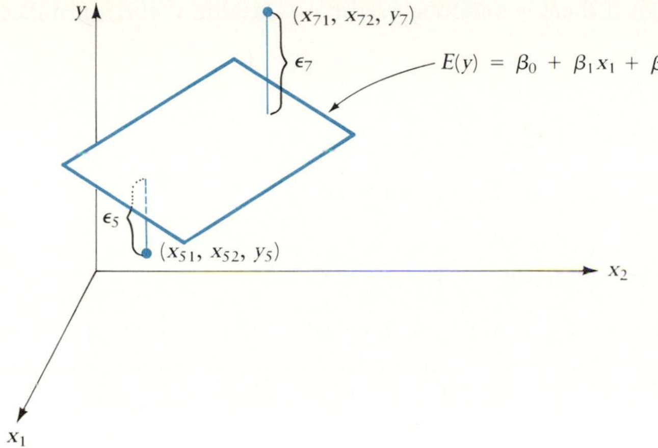

Figure 14.3
Graph of ϵ_5 and ϵ_7 for the regression function $E(y) = \beta_0 + \beta_1 x_1 + \beta_2 x_2$.

$$\Sigma x_2 y = b_0 \Sigma x_2 + b_1 \Sigma x_1 x_2 + b_2 \Sigma x_2^2$$

We can solve these equations for b_0, b_1, and b_2 by using the method of linear combinations discussed in high school and college algebra or by using a matrix inversion approach. (The matrix approach is the most widely used. See Neter and others [1985, pp. 185–225] for a discussion.) Because both these approaches require a considerable amount of arithmetic, we generally use the computer in multiple regression. For the same reason we do not solve the normal equations by hand in this chapter. Instead we illustrate the computer approach by using SAS statements throughout; Minitab commands for these procedures are shown in Section 14.4.

If there are $k - 1$ predictor variables, there are k normal equations; these may be solved for the values of $b_0, b_1, \ldots, b_{k-1}$ to obtain the sample regression equation (or estimated regression function) $\hat{y} = b_0 + b_1 x_1 + b_2 x_2 + b_3 x_3 + \cdots + b_{k-1} x_{k-1}$.

The Multiple Regression Equation

Suppose we obtain a set of n observations from a population for which the relationship between the variables is described by the regression function

$$E(y) = \beta_0 + \beta_1 x_1 + \beta_2 x_2 + \beta_3 x_3 + \cdots + \beta_{k-1} x_{k-1}$$

If the error terms are independent and $N(0, \sigma^2)$, we can obtain a regression equation

$$\hat{y} = b_0 + b_1 x_1 + b_2 x_2 + b_3 x_3 + \cdots + b_{k-1} x_{k-1}$$

from the sample data by simultaneously solving the normal equations.

Note that if there is only one predictor variable x, the normal equations become

$$\Sigma y = nb_0 \quad + b_1\Sigma x$$

$$\Sigma xy = b_0\Sigma x + b_1\Sigma x^2$$

Solving these equations for b_1, we obtain

$$b_1 = \frac{\Sigma xy - (\Sigma x)(\Sigma y)/n}{\Sigma x^2 - (\Sigma x)/n}$$

and since $\Sigma y = nb_0 + b_1\Sigma x$, we have $b_0 = (\Sigma y - b_1\Sigma x)/n$. These are the same formulas we used to obtain b_0 and b_1 in simple linear regression (Chapter 13).

Example 14.1

A doctor believes that two drugs are both helpful in the treatment of a disease. The correct amount of each drug to use has not been determined, however, and she also believes that a combination of the drugs would be most helpful. Injections are started as soon as possible after admission to the hospital and continued until the patient is out of danger—that is, until the patient has reached a satisfactory condition in the judgment of the attending physician. In order to minimize problems with other variables, all conditions are kept as similar as possible: same physician, same nursing staff, and so on. Some factors, such as time of day and the weather, are impossible to control, of course. Given the results shown in Table 14.1, construct a multiple regression equation for predicting time to satisfactory condition from amount of drugs administered.

Solution

Letting x_1 represent the amount of drug 1 and x_2 the amount of drug 2, the normal equations are

Table 14.1 PATIENT RESPONSE TO DRUG DOSAGES.

PATIENT	DRUG 1 (ML)	DRUG 2 (ML)	HOURS OF SATIS-FACTORY CONDITION	PATIENT	DRUG 1 (ML)	DRUG 2 (ML)	HOURS OF SATIS-FACTORY CONDITION
1	1	2	46	9	3	2	39
2	1	4	40	10	3	4	34
3	1	6	38	11	3	6	30
4	1	8	35	12	3	8	33
5	2	2	41	13	4	2	42
6	2	4	36	14	4	4	33
7	2	6	35	15	4	6	34
8	2	8	33	16	4	7	35

Note: Patient 16 was scheduled to receive 8 milliliters of drug 2, but there was a shortage.

$$584 = 16b_0 + 40b_1 + 79b_2$$
$$1433 = 40b_0 + 120b_1 + 196b_2$$
$$2783 = 79b_0 + 196b_1 + 465b_2$$

We can solve these equations by hand using the method of linear combinations or the matrix inversion method. More often we simply enter the data into a computer and obtain the computer printout. (The SAS statements needed to perform the multiple regression are given in Section 14.1.4; the corresponding Minitab commands are shown in Section 14.4). Figure 14.4 presents the computer printout using the SAS procedure REG. The values of b_0, b_1, and b_2 are under PARAMETER ESTIMATE and identified by the corresponding term under the heading VARIABLE. The upper portion of the printout is an analysis of variance table showing the partitioning of the total sum of squares for the response variable. The ANOVA table is discussed further in Section 14.1.2.

Thus the sample regression equation for the data (arbitrarily rounding to four decimal places) is

$$\hat{y} = 46.8973 - 1.4528x_1 - 1.3702x_2$$

We can say that, for each additional milliliter of drug 1 injected, the mean change in hours to satisfactory condition is estimated to be a decrease of about 1.4528 hours if the amount of drug 2 is held constant. For each additional milliliter of drug 2 injected, the mean change in hours to satis-

Figure 14.4 SAS printout for obtaining the multiple regression equation for the drug data of Example 14.1.

SAS

DEP VARIABLE: HOURS

ANALYSIS OF VARIANCE

SOURCE	DF	SUM OF SQUARES	MEAN SQUARE	F VALUE	PROB>F
MODEL	2	176.92946	88.46473020	13.844	0.0006
ERROR	13	83.07053959	6.39004151		
C TOTAL	15	260.00000			

ROOT MSE	2.527853	R-SQUARE	0.6805
DEP MEAN	36.5	ADJ R-SQ	0.6313
C.V.	6.925625		

PARAMETER ESTIMATES

VARIABLE	DF	PARAMETER ESTIMATE	STANDARD ERROR	T FOR H0: PARAMETER=0	PROB > |T|
INTERCEP	1	46.89726027	2.15391393	21.773	0.0001
DRUG1	1	-1.45276478	0.56566991	-2.568	0.0234
DRUG2	1	-1.37019713	0.29223248	-4.689	0.0004

b_0 b_1 b_2

factory condition is estimated to be a decrease of about 1.3702 hours if the amount of drug 1 is held constant. The constant, 46.8973, could be interpreted as an estimate for the mean hours to satisfactory condition for a patient who receives neither drug, but this is not necessarily true, since all patients receive both drugs and an untreated patient would very likely react quite differently.

Note also that we should not use this regression equation to predict hours to satisfactory condition for dosages outside the scope of the data. This pitfall was discussed in Section 13.4.1. If we let $x_1 = 6$ and $x_2 = 12$, for example, we obtain $\hat{y} \doteq 21.74$. Since x_1 in the sample has values from 1 to 4, we cannot generalize above $x_1 = 4$ because we do not know that the regression equation will still be a valid predictor. If two doses of a drug are more effective than one, would it be better to take three, four, five, six, or maybe ten? Doctors know that drugs taken in large quantities may be harmful, so they limit the amount taken by patients. On the other hand, we may investigate the effects of intermediate dosages, such as 2.5 milliliters of drug 1 and 4.75 milliliters of drug 2, since both these values are within the scope of the data.

14.1.2 Analysis of Variance for Multiple Regression

Most of the inferences we make from multiple regression analysis are simply extensions of the same procedures used in simple linear regression analysis. We can test the usefulness of the multiple regression equation as a predictor by performing an F test, and we can use MSE to estimate σ^2. We can determine confidence and prediction intervals and test the assumptions by means of residual plots.

In order to achieve these goals, we need to partition the total sum of squares (SSTO) of the response variable into the sum of squares due to regression (SSR) and the sum of squares due to error (SSE). These sums of squares are defined as they were for linear regression:

$$\text{SSTO} = \Sigma(y - \bar{y})^2$$
$$\text{SSR} = \Sigma(\hat{y} - \bar{y})^2$$
$$\text{SSE} = \Sigma(y - \hat{y})^2$$

In simple linear regression analysis there are two population parameters, so SSE has $n - 2$ degrees of freedom. If there are k population parameters, $\beta_0, \beta_1, \ldots, \beta_{k-1}$, SSE has $n - k$ degrees of freedom. Similarly, SSR has $k - 1$ degrees of freedom instead of $2 - 1$. As usual, SSTO has $n - 1$ degrees of freedom. Formulas for SSR and SSE are given here for reference, but they are not essential since a computer is used for most calculations. The results are usually presented in an ANOVA table, as shown in Table 14.2.

Partitioning Total Sum of Squares (Multiple Regression)

Total Sum of Squares

$$\text{SSTO} = \Sigma y^2 - \frac{(\Sigma y)^2}{n}$$

Sum of Squares Due to Regression

$$\text{SSR} = b_0\Sigma y + b_1\Sigma x_1 y + b_2\Sigma x_2 y + \cdots + b_{k-1}\Sigma x_{k-1} y - \frac{(\Sigma y)^2}{n}$$

Sums of Squares Due to Error

$$\begin{aligned}\text{SSE} &= \text{SSTO} - \text{SSR} \\ &= \Sigma y^2 - b_0\Sigma y - b_1\Sigma x_1 y - b_2\Sigma x_2 y - \cdots - b_{k-1}\Sigma x_{k-1} y\end{aligned}$$

Mean Squares

$$\text{MSR} = \frac{\text{SSR}}{k-1}$$

$$\text{MSE} = \frac{\text{SSE}}{n-k}$$

Table 14.2 ANOVA TABLE FOR MULTIPLE REGRESSION ANALYSIS.

SOURCE OF VARIATION	DF	SS	MS
Regression	$k-1$	SSR	SSR/$(k-1)$
Error	$n-k$	SSE	SSE/$(n-k)$
Total	$n-1$	SSTO	

We can use the F test, where $F^* = \text{MSR/MSE}$, to test the hypothesis that there is an association between *at least one* of the predictor variables and the response variable—that is, that at least one of the β_j $(j > 0)$ is not equal to zero. If all β_j are zero, changes in predictor variables do not have an effect on the response variable. The test is frequently called an *overall F test* , since it simultaneously tests the hypotheses that each $\beta_j = 0$ $(j > 0)$.

F Test for Multiple Regression

Suppose we obtain a set of n observations $(x_1, x_2, \ldots, x_{k-1}, y)$ from a population for which the relationship between the variables is described by the regression function $E(y) = \beta_0 + \beta_1 x_1 + \beta_2 x_2 + \beta_3 x_3 + \cdots + \beta_{k-1} x_{k-1}$. Then we may test the hypothesis H_0: $\beta_1 = \beta_2 = \beta_3 = \cdots = \beta_{k-1} = 0$. The sample value of the test statistic used to test the hypothesis is F^*, where $F^* = \text{MSR/MSE}$. The decision rule is dictated by the level of significance α chosen for the test.

Null and Alternate Hypothesis

H_0: $\beta_1 = \beta_2 = \beta_3 = \cdots = \beta_{k-1} = 0$
H_1: not all β_j are equal to zero

Decision Rule

Reject H_0 if $F^* > F_\alpha\{k-1, n-k\}$

Example 14.2

Obtain the analysis of variance table for the data of Example 14.1, and test the hypothesis $\beta_1 = \beta_2 = 0$. Use the 0.05 level of significance.

Solution

The ANOVA table produced by the SAS procedure REG is shown in Figure 14.5. The P value for the obtained value of F is listed under the heading

Figure 14.5 SAS printout of the ANOVA table for the drug data of Example 14.2.

```
                                         SAS
DEP VARIABLE: HOURS
                               ANALYSIS OF VARIANCE

                             SUM OF              MEAN
    SOURCE        DF         SQUARES             SQUARE      F VALUE       PROB>F

    MODEL          2      176.92946←SSR  88.46473020←      13.844←F*    0.0006←P value
    ERROR         13      83.07053959←SSE  6.39004151 MSR
    C TOTAL       15      260.00000←SSTO            ↖MSE

          ROOT MSE   √MSE→2.527853          R-SQUARE      0.6805
          DEP MEAN       ȳ → 36.5           ADJ R-SQ      0.6313
          C.V.            6.925625

                               PARAMETER ESTIMATES

                          PARAMETER        STANDARD       T FOR H0:
VARIABLE        DF         ESTIMATE           ERROR     PARAMETER=0        PROB > |T|

INTERCEP         1      46.89726027      2.15391393          21.773            0.0001
DRUG1            1      -1.45276478      0.56566991          -2.568            0.0234
DRUG2            1      -1.37019713      0.29223248          -4.689            0.0004
```

Table 14.3.

SOURCE OF VARIATION	DF	SS	MS
Regression	2	176.9295	88.4647
Error	13	83.0705	6.3900
Total	15	260.0000	

PROB > F. Here it is equal to 0.0006; thus we can reject the null hypothesis at any level of significance greater than 0.0006. The value listed beside ROOT MSE is the standard error of estimate s_e; here $s_e \doteq 2.528$. DEP MEAN is the mean of the response variable; here $\bar{y} = 36.5$. C. V., the coefficient of variation, is not discussed here. The term listed beside R-SQUARE is the **coefficient of multiple determination, R^2**, and measures the reduction in error variance due to using the model. Like the coefficient of linear determination (see Section 13.4.1) it is equal to SSR/SSTO. The adjusted R^2 is discussed in Section 14.2.3. We summarize the ANOVA table in Table 14.3.

To test the hypothesis H_0: $\beta_1 = \beta_2 = 0$, we use $F^* = 88.4647/6.3900 \doteq 13.84$. Since $F_{.05}\{2, 13\} = 3.81$, we conclude that there is an association between at least one of the drugs and hours to satisfactory condition.

PROFICIENCY CHECKS

1. Consider the following set of data:

	OBSERVATION						
	1	2	3	4	5	6	7
x_1	1	1	2	3	4	5	7
x_2	2	4	3	4	8	8	11
y	4	6	5	8	12	13	19

For this set of data SSTO $\doteq$ 173.7143, SSR $\doteq$ 172.0532, and the regression equation for predicting y from x_1 and x_2 is

$\hat{y} = 0.7379 + 1.0123x_1 + 0.9638\, x_2$

a. Obtain the ANOVA table and test the hypothesis that there is no association between either predictor variable and the response variable. Use the 0.01 level of significance.
b. Using a computer or a hand calculator, verify the regression equation and the sums of squares.

14.1.3 Residual Analysis

As with simple linear regression, residual plots in which residuals are plotted against $\hat{y}$ may be used to test the assumptions that the distribution of ϵ is normal and that the variance σ^2 is constant throughout the scope of the data. We can test the assumption that the association between y and each x_j is linear by using the plots of the residuals against each x_j. The computer printout of the residuals for the data of Example 14.1 is shown in Figure 14.6. Each observation of y is shown together with the predicted values ($\hat{y}$) and the residuals.

We can graph the residual plots by hand, using the data of Figure 14.6, or obtain them from the computer printout. The computer-generated residual plot is shown in Figure 14.7. (See Section 14.1.5 for the statements you need to obtain these plots.) Note that the points scatter at random about the horizontal axis. Since $s_e \doteq 2.578$, we observe that all the data points lie within $2s_e$ of the horizontal axis and 11 of the 16 lie within s_e of the axis. The residuals appear to be normally distributed. No serious deviations of error variances are noticeable. Thus no assumptions appear to be violated. A normal probability plot may also be used to test normality of the residuals.

Computer-generated plots of the residuals against each predictor variable are shown in Figure 14.8. In both cases there appears to be a curvilinear effect. Observations at both ends are above the axis, while those in the center are below the axis. Thus it might be a good idea to investigate this relationship further by using a quadratic term (see Section 14.3).

PROFICIENCY CHECK

2. Figure 14.9 shows the residual plots for the data of Proficiency Check 1. What do you conclude?

Figure 14.6
SAS printout of the predicted values and the residuals for the drug data of Example 14.1.

OBS	ACTUAL	PREDICT VALUE	RESIDUAL
1	46.0000	42.7041	3.2959
2	40.0000	39.9637	0.0363
3	38.0000	37.2233	0.7767
4	35.0000	34.4829	0.5171
5	41.0000	41.2513	-0.2513
6	36.0000	38.5109	-2.5109
7	35.0000	35.7705	-0.7705
8	33.0000	33.0302	-0.0302
9	39.0000	39.7986	-0.7986
10	34.0000	37.0582	-3.0582
11	30.0000	34.3178	-4.3178
12	33.0000	31.5774	1.4226
13	42.0000	38.3458	3.6542
14	33.0000	35.6054	-2.6054
15	34.0000	32.8650	1.1350
16	35.0000	31.4948	3.5052

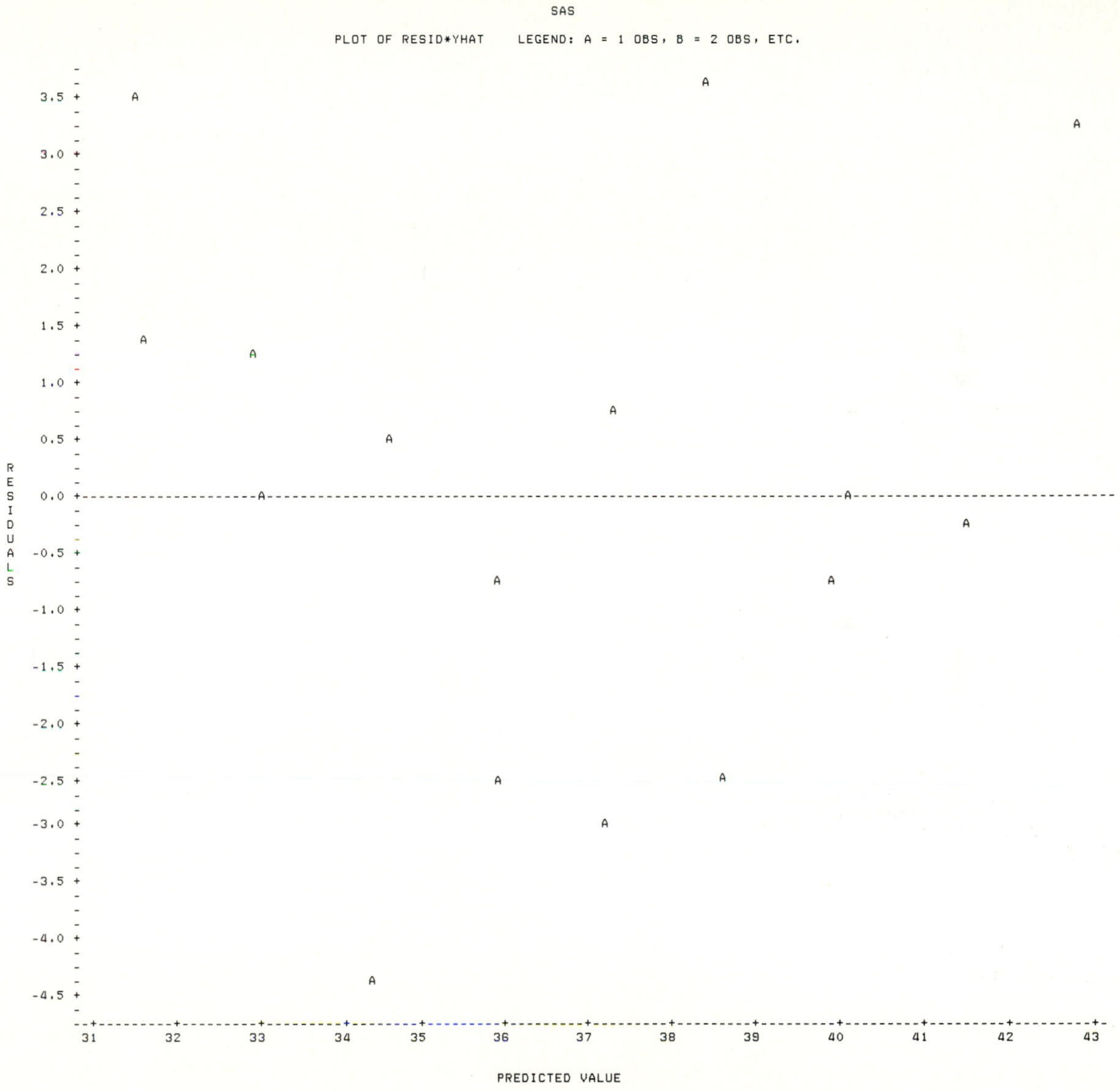

```
                                   SAS

              PLOT OF RESID*YHAT    LEGEND: A = 1 OBS, B = 2 OBS, ETC.

R      3.5 +
E      3.0 +
S      2.5 +
I      2.0 +
D      1.5 +
U      1.0 +
A      0.5 +
L      0.0 +--------------------------------------------------------------
S     -0.5 +
      -1.0 +
      -1.5 +
      -2.0 +
      -2.5 +
      -3.0 +
      -3.5 +
      -4.0 +
      -4.5 +
           --+------+------+------+------+------+------+------+------+------+------+------+------+-
             31     32     33     34     35     36     37     38     39     40     41     42     43

                                   PREDICTED VALUE
```

Figure 14.7
SAS printout of the plot of the residuals against $\hat{y}$ for the drug data of Example 14.1.

Figure 14.8
SAS printouts of the residuals plotted against each predictor variable for the drug data of Example 14.1.

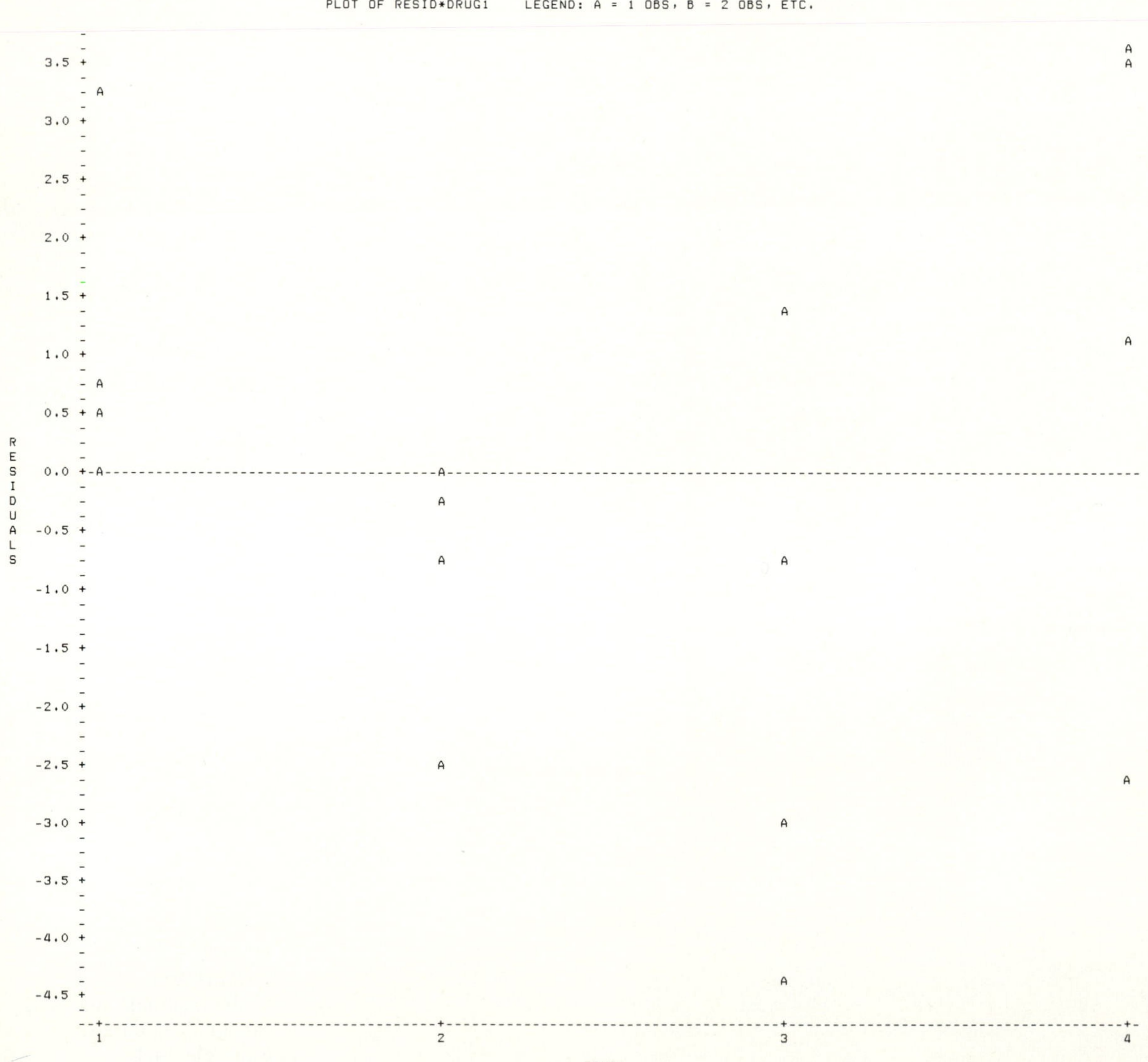

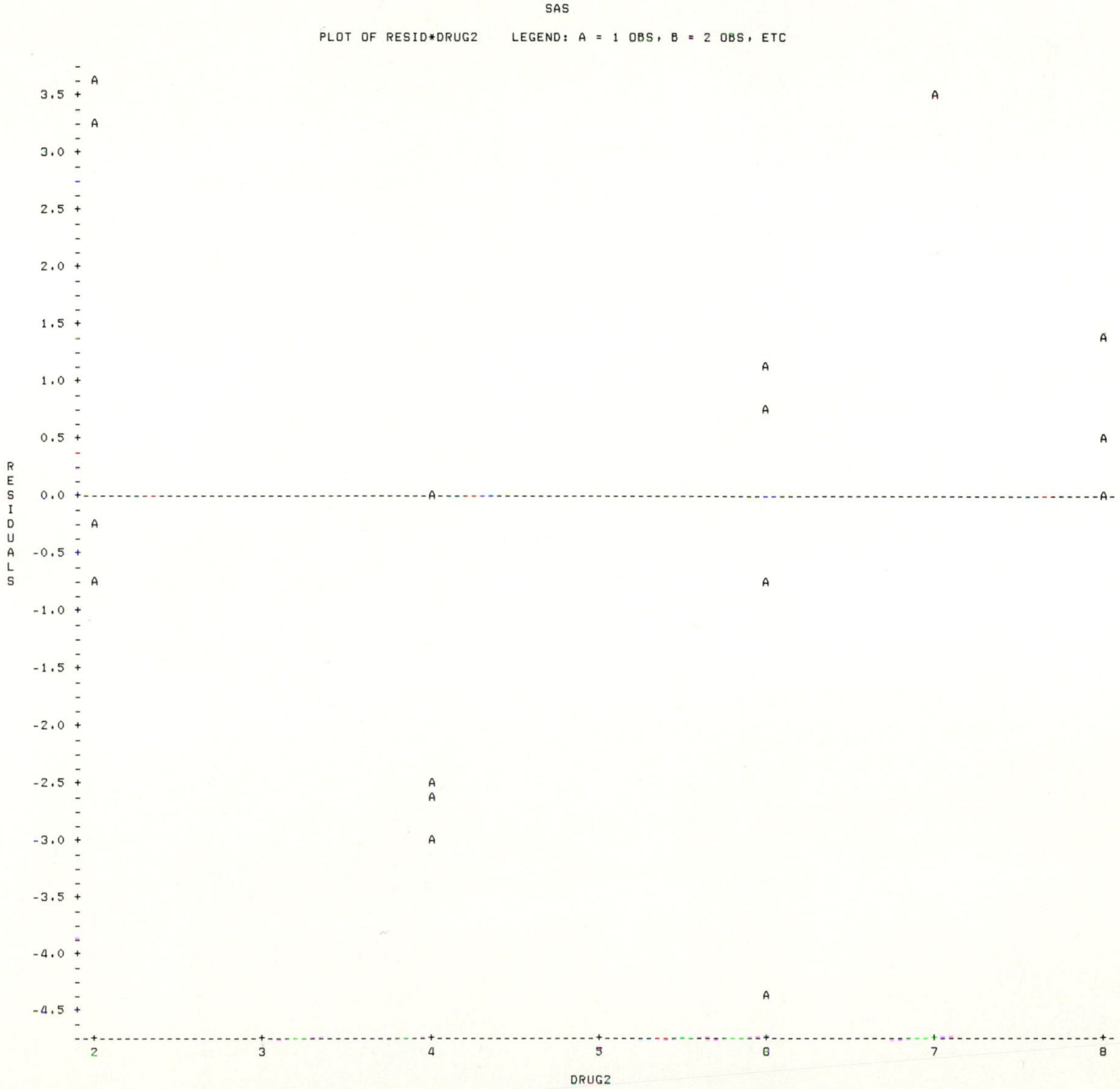

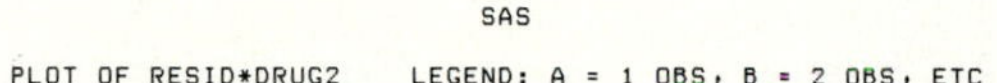

```
                                                       SAS

                               PLOT OF RESID*DRUG2     LEGEND: A = 1 OBS, B = 2 OBS, ETC

         -
         - A
    3.5 +                                                                                           A
         -
         - A
         -
    3.0 +
         -
         -
         -
    2.5 +
         -
         -
         -
    2.0 +
         -
         -
         -
    1.5 +
         -                                                                                                             A
         -
         -                                                                         A
    1.0 +
         -
         -                                                                         A
         -
    0.5 +                                                                                                             A
R        -
E        -
S   0.0 +-------------------------------------A-----------------------------------------------------------------------A-
I        -
D        - A
U        -
A  -0.5 +
L        -
S        - A                                                                       A
   -1.0 +
         -
         -
         -
   -1.5 +
         -
         -
         -
   -2.0 +
         -
         -
         -
   -2.5 +                                     A
         -                                     A
         -
         -
   -3.0 +                                     A
         -
         -
         -
   -3.5 +
         -
         -
         -
   -4.0 +
         -
         -
         -                                                                         A
   -4.5 +
         -
         --+-----------------+-----------------+-----------------+-----------------+-----------------+-----------------+-
           2                 3                 4                 5                 6                 7                 8

                                                         DRUG2
```

Figure 14.9
SAS printout of the residual plots for the data of Proficiency Check 1.

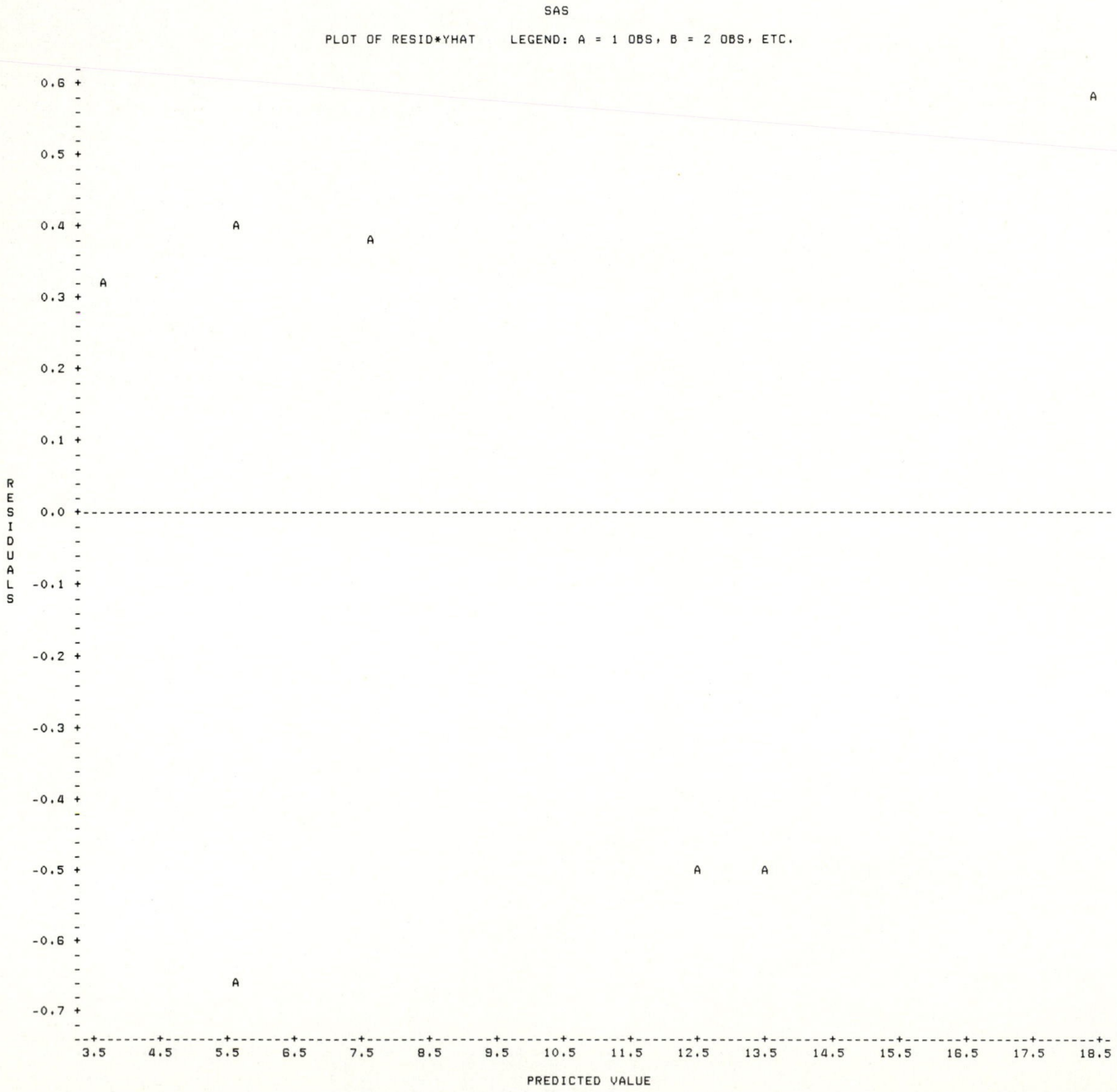

```
                                   SAS

              PLOT OF RESID*YHAT    LEGEND: A = 1 OBS, B = 2 OBS, ETC.

      0.6 +
          -                                                                          A
          -
          -
          -
      0.5 +
          -
          -
          -
          -
      0.4 +          A
          -                   A
          -
          -
          - A
      0.3 +
          -
          -
          -
          -
      0.2 +
          -
          -
          -
          -
      0.1 +
          -
R         -
E         -
S     0.0 +-----------------------------------------------------------------------------
I         -
D         -
U         -
A         -
L    -0.1 +
S         -
          -
          -
          -
     -0.2 +
          -
          -
          -
          -
     -0.3 +
          -
          -
          -
          -
     -0.4 +
          -
          -
          -
          -
     -0.5 +                                                  A    A
          -
          -
          -
          -
     -0.6 +
          -
          -
          -          A
     -0.7 +
          --+----+----+----+----+----+----+----+----+----+----+----+----+----+----+----+-
          3.5  4.5  5.5  6.5  7.5  8.5  9.5 10.5 11.5 12.5 13.5 14.5 15.5 16.5 17.5 18.5

                                   PREDICTED VALUE
```

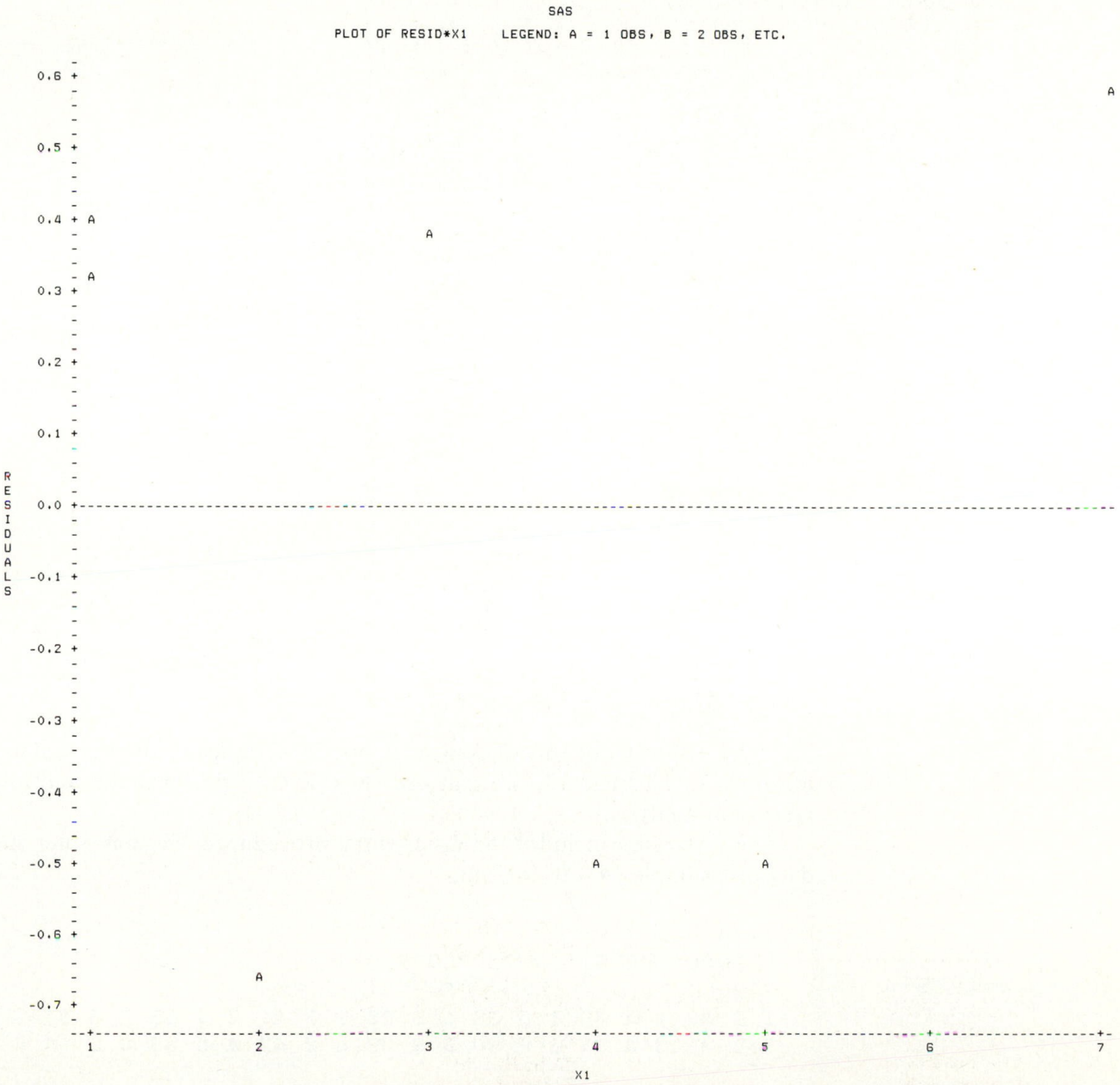

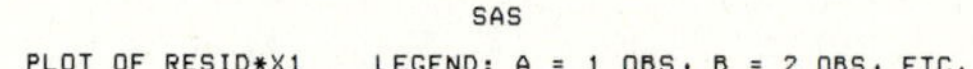


```
 0.6 +
     -                                                                    A
 0.5 +
 0.4 + A
     -                     A
     - A
 0.3 +
 0.2 +
 0.1 +
R
E
S  0.0 +------------------------------------------------------------------------
I
D
U
A
L -0.1 +
S
-0.2 +
-0.3 +
-0.4 +
-0.5 +                                A          A
-0.6 +
     -          A
-0.7 +
     --+----------+----------+----------+----------+----------+----------+-
       1          2          3          4          5          6          7
                                       X1
```

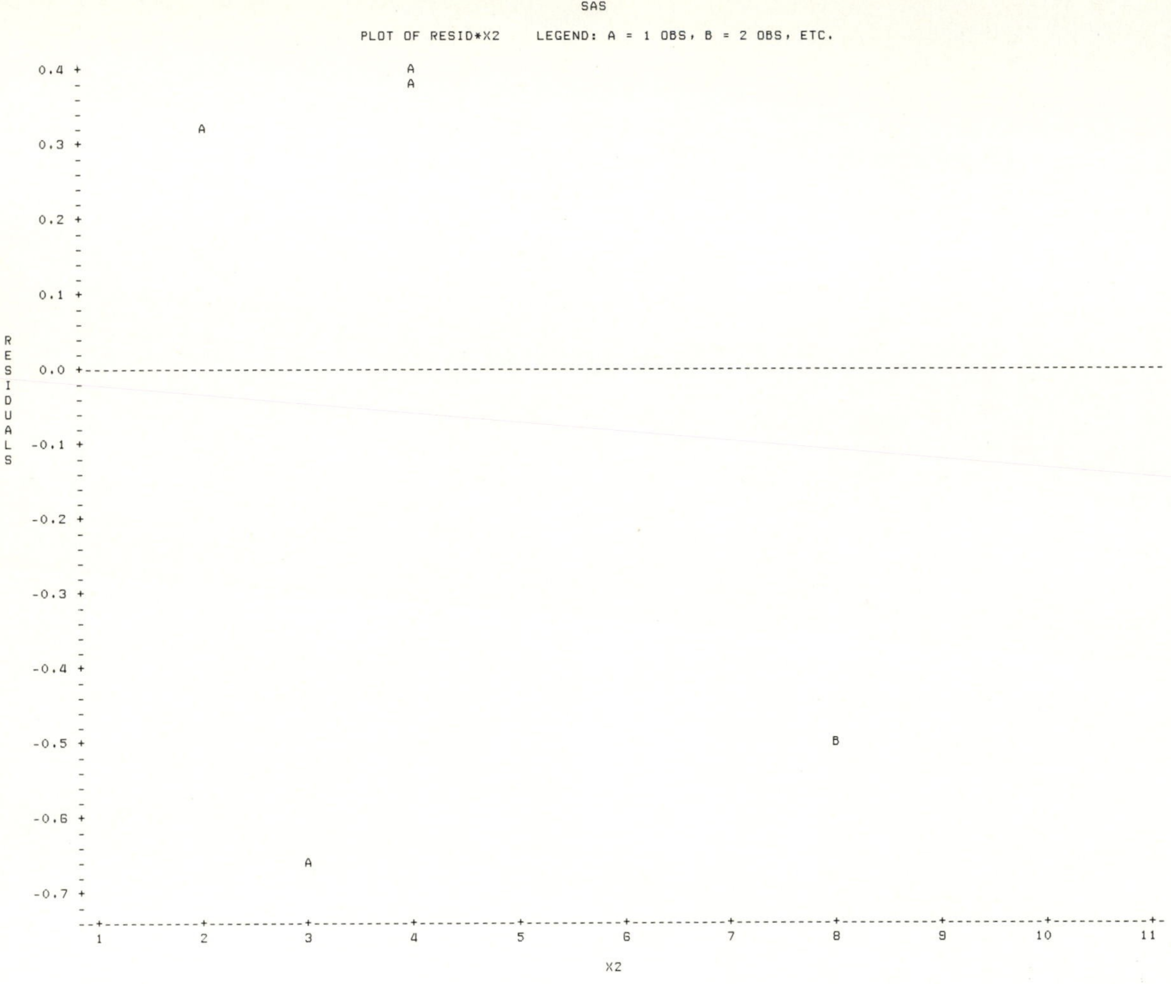

14.1.4 Using SAS Statements

The SAS printouts in this section were obtained by using the same statements as in Chapter 13. We can use PROC REG to perform the multiple regression analysis.

The DATA step includes the usual entry procedures. We may enter the data of Example 14.1 by writing

```
DATA MULT;
  INPUT DRUG1 DRUG2 HOURS @@;
CARDS;
1 2 46 1 4 40 1 6 38 1 8 35 2 2 41 2 4 36 2 6 35 2 8 33
3 2 39 3 4 34 3 6 30 3 8 33 4 2 42 4 4 33 4 6 34 4 7 35
```

We can obtain the ANOVA table by writing

```
PROC REG;
  MODEL HOURS = DRUG1 DRUG2;
```

We can find the predicted values and residuals by adding /P to the model statement. To obtain the residual plots, we need the output statement creating the variables YHAT ($\hat{y}$) and RESID (e). We can get the residual plots by using the PLOT procedure. The computer outputs shown in this section were obtained by using the following statements after entry of the data:

```
PROC REG;
  MODEL HOURS = DRUG1 DRUG2 / P;
  OUTPUT P = YHAT R = RESID;
PROC PLOT;
  PLOT RESID * YHAT / VREF = 0;
  PLOT RESID * DRUG1 / VREF = 0;
  PLOT RESID * DRUG2 / VREF = 0;
```

Problems

Practice

14.1 An analysis of variance table, partially filled in, is given here. There were three predictor variables, x_1, x_2, x_3. Complete the table.

SOURCE OF VARIATION	DF	SS	MS
Regression			33.8144
Error			
Total	23	153.8876	

14.2 An analysis of variance table is given here. Test the hypothesis that there is no association between the response variable and any of the predictor variables. Use the 0.05 level of significance.

SOURCE OF VARIATION	DF	SS	MS
Regression	4	3.3364	0.8341
Error	22	4.1602	0.1891
Total	26	7.4966	

14.3 Using the following data, obtain a multiple regression equation and ANOVA table, and test the hypothesis that there is no association between the variables. Use $\alpha = 0.05$.

	OBSERVATION					
	1	2	3	4	5	6
x_1	0.8	1.4	1.6	1.7	2.0	2.5
x_2	10	12	13	14	16	20
y	36	41	44	47	52	64

Applications

These problems should be solved by using a computer. Problems 14.4 to 14.7 can be solved with a hand calculator.

14.4 The following data were obtained on usage of power at a manufacturing plant. In the table, y represents number of thousands of kilowatt-hours of electricity used in a particular day, x_1 represents average temperature (in degrees Fahrenheit) during the 16-hour working day, and x_2 represents the plant's output in thousands of units.

DAY	y	x_1	x_2
1	11.76	81.8	117.7
2	10.84	80.6	118.2
3	12.74	84.4	123.4
4	9.88	79.7	102.7
5	13.66	85.9	106.2
6	10.39	81.5	105.7
7	12.44	85.8	107.6
8	10.89	78.9	108.1
9	12.16	83.7	110.6
10	11.43	85.2	117.9

a. Develop a regression equation for usage of electricity based on temperature and output.
b. Obtain the ANOVA table, and test the significance of the association between the response and predictor variables.
c. Examine the residual plot. Do the assumptions of constant error variance and normality appear to be satisfied?
d. Plot the residuals against each of the predictor variables. Does the relationship appear to be linear in each case?

14.5 The duration of a particular illness is apparently related to the bacterial count of the infecting organism and the temperature of the patient upon admission. Ten patients with the illness have their bacterial counts (in thousands) determined and their temperatures (in degrees Celsius) taken upon admission. The duration of symptoms is observed. The data are shown here:

PATIENT	COUNT	TEMP.	DURATION (DAYS)
A	8	37.6	11
B	7	39.2	10
C	4	38.5	11
D	6	37.4	8
E	9	38.1	12
F	7	39.1	11
G	8	39.0	13
H	3	37.8	7
I	8	38.2	10
J	7	39.1	11

a. Derive the regression equation for predicting duration of illness from bacterial count and temperature at admission.
b. Interpret the meaning of b_1 and b_2.
c. Obtain the analysis of variance table for the analysis.
d. Using the 0.05 level of significance, test the hypothesis that there is no association between either bacterial count or temperature and length of stay.
e. Construct a residual plot. Do any assumptions appear to be violated? Are there any outliers?

14.6 A study was conducted by Neslin, Henderson, and Quelch (1985) to determine how the price of a product and the number of ads for the product placed in local newspapers during a given month affect the market share of the product. In the study, house brands were excluded since they were not the same in each store. A total of 2293 consumers took part in the study. Six brands of bathroom tissue dominated the market. The data are shown here:

	BRAND					
	A	B	C	D	E	F
Price (¢)	26	26	20	39	25	25
Number of ads	4	1	5	6	2	3
Market share (%)	18.6	5.8	28.3	15.0	12.5	19.8

a. Is there a significant relationship between the predictor variables and the market share? Use $\alpha = 0.05$.
b. What further analyses seem to be indicated?

14.7 Refer to the Neslin study of Problem 14.6. Seven brands of instant coffee dominated the market. The data are shown here:

	BRAND						
	A	B	C	D	E	F	G
Price per oz. (¢)	42	35	37	59	40	41	61
Number of ads	3	2	2	5	4	10	4
Market share (%)	10.3	1.9	2.6	10.1	26.1	44.9	4.2

a. Obtain a regression equation for predicting market share from price and number of ads.
b. Which predictor variable seems to have the most effect on market share? Why?

14.8 A real estate agent found four factors that may have a bearing on the price of a lot in an exclusive subdivision. The factors are area in thousands of square feet (x_1), mean elevation of the lot in feet above sea level (x_2), slope of the lot in degrees (x_3), and view, rated on an ascending scale of 1 to 9 (x_4). Selling price of the lot is given in thousands of dollars (y):

a. Develop a regression equation for lot price, obtain the ANOVA table, and test the significance of the association between the response and predictor variables.
b. Examine the residual plots. What do you conclude?

LOT	x_1	x_2	x_3	x_4	y
1	24	132	6.8	7	59.9
2	18	118	7.9	4	42.5
3	21	141	6.6	6	55.0
4	22	125	11.6	8	66.5
5	28	108	12.7	4	44.9
6	19	128	3.2	7	41.7
7	22	143	1.9	9	72.5
8	25	132	9.6	9	70.0

14.2 Inferences in Multiple Regression

Even when the F test shows that there is some association between at least one predictor variable and the response variable, there is not necessarily an association between each of the predictor variables and the response variable. Perhaps only one or two predictor variables are needed in the model and the others can be discarded. The first stage in further analysis is to determine which of the β_j are equal to zero—that is, which predictor variables can be eliminated. If the assumptions about the model hold, we may draw inferences similar to those in simple linear regression. In Problem 14.8, for instance, a real estate agent isolated four factors that may have a bearing on the price of a lot in an exclusive subdivision. These factors were area, elevation, slope, and view. Suppose, however, that the only factors which really have a bearing on price are area and view. Since the overall F test will show an association between price and *at least one* variable, we need a procedure to separate the factors that do have a bearing on the response variable from those that do not.

14.2.1 Inferences About β_j

As in simple linear regression, each β_j is estimated by b_j. Each b_j is normally distributed with mean β_j and standard error σ_{b_j}. Recall from simple linear regression that b_1 is normally distributed with mean β_1 and standard error σ_{b_1} so that $(b_1 - \beta_1)/s_{b_1}$ has a t distribution with $n - 2$ degrees of freedom, where s_{b_1} is the best estimate for β_1 obtained from the sample. In multiple regression, $(b_j - \beta_j)/s_{b_j}$ has a t distribution with $n - k$ degrees of freedom for each value of j. Thus we can obtain confidence intervals for the β_j and test hypotheses concerning specific values of β_j in ways similar to those of the preceding chapters. Sample standard errors for each b_j (the s_{b_j}) are listed in the computer printouts along with other information. Since the s_{b_j} are difficult to calculate, we use the computer to obtain these values.

Figure 14.10 is the computer printout for the data of Example 14.1. Under the heading STANDARD ERROR are the standard errors of the variables listed under the heading VARIABLE. They are marked as shown; in particular, $s_{b_1} = 0.56566991$ and $s_{b_2} = 0.29223248$. We can test an individual hypoth-

```
                                   SAS
DEP VARIABLE: HOURS
                            ANALYSIS OF VARIANCE

                           SUM OF              MEAN
     SOURCE      DF       SQUARES            SQUARE     F VALUE      PROB>F

     MODEL        2     176.92946       88.46473020      13.844      0.0006
     ERROR       13   83.07053959        6.39004151
     C TOTAL     15     260.00000

          ROOT MSE       2.527853          R-SQUARE      0.6805
          DEP MEAN           36.5          ADJ R-SQ      0.6313
          C.V.           6.925625

                             PARAMETER ESTIMATES

                          PARAMETER       STANDARD     T FOR H0:
VARIABLE      DF           ESTIMATE          ERROR   PARAMETER=0     PROB > |T|

INTERCEP       1        46.89726027     2.15391393        21.773         0.0001
DRUG1          1        -1.45276478     0.56566991        -2.568         0.0234
DRUG2          1        -1.37019713     0.29223248        -4.689         0.0004
```

s_{b_0}, s_{b_1}, s_{b_2} (standard errors)

t^* for H_0: $\beta_1 = 0$

t^* for H_0: $\beta_2 = 0$

P value for H_0: $\beta_1 = 0$

P value for H_0: $\beta_2 = 0$

Figure 14.10
SAS printout for inferences about β_j in the regression model for the drug data of Example 14.1.

esis that $\beta_j = 0$ by using the usual t test. The P values for each of these tests are listed under the heading PROB > |T|.

Hypothesis Tests Concerning β

Suppose we obtain a random sample of n observations from a population for which the relationship between the variables is described by the regression function $E(y) = \beta_0 + \beta_1 x_1 + \beta_2 x_2 + \beta_3 x_3 + \cdots + \beta_{k-1} x_{k-1}$. If the error terms are independent and $N(0, \sigma^2)$, we may test the hypotheses H_0: $\beta_j = 0$ $(j = 1, 2, 3, \ldots, k - 1)$. The sample value of the test statistic used to test each hypothesis is t^*, where $t^* = b_j/s_{b_j}$. The decision rule is dictated by the alternate hypothesis and the level of significance α chosen for the test.

Null and Alternate Hypotheses

H_0: $\beta_j = 0$	H_0: $\beta_j = 0$	H_0: $\beta_j = 0$
H_1: $\beta_j > 0$	H_1: $\beta_j \neq 0$	H_1: $\beta_j < 0$

Decision Rules

Reject H_0 if	Reject H_0 if	Reject H_0 if
$t^* > t_\alpha\{n - k\}$	$\lvert t^* \rvert > t_{\alpha/2}\{n - k\}$	$t^* < -t_\alpha\{n - k\}$

The values of $t^* = b_j/s_{b_j}$ for each j are also given in the computer printout under the heading T FOR H0: PARAMETER=0; in Figure 14.10, note that $b_1/s_{b_1} \doteq -2.57$ and $b_2/s_{b_2} \doteq -4.69$. The P values are given in each case. For drug 1, the level of significance is 0.0234; for drug 2, the level of significance is 0.0004. Thus both drugs have an effect on recovery time.

Comment The tests just performed are known as *marginal t tests;* each test is performed on the assumption that all other variables are in the model. It is possible, for example, that none of the predictor variables has a t test that indicates an association with the response variable, yet the F test indicates significant association within the model. If two predictor variables are themselves strongly associated, it is possible that they are not both needed in the model. In such a case the t test would probably indicate that neither is needed—*provided that the other is in the model*—and more extensive analysis is necessary. See, for example, Neter and others (1985, pp. 282–296).

We can obtain a confidence interval for an individual value of β_0 or β_j in the usual manner.

Confidence Intervals for β_0 and β_j

Suppose we obtain a random sample of n observations from a population for which the relationship between the variables is described by the regression function $E(y) = \beta_0 + \beta_1 x_1 + \beta_2 x_2 + \beta_3 x_3 + \cdots + \beta_{k-1} x_{k-1}$ and the error terms are independent and $N(0, \sigma^2)$. Then a $100(1 - \alpha)\%$ confidence interval for β_j ($j = 0, 1, 2, 3, \ldots, k$) is the interval

$$b_j - ts_{b_j} \quad \text{to} \quad b_j + ts_{b_j}$$

where $t = t_{\alpha/2}\{n-k\}$.

Example 14.3

Use the results of Example 14.1 as shown in Figure 14.10 to obtain a 95% confidence interval for β_0, β_1, and β_2.

Solution

We have $t_{.025}\{13\} = 2.160$ and $s_{b_0} = 2.15391393$, so $ts_{b_0} \doteq 4.652454089$. The confidence limits are

$$46.89726027 - 4.652454089 \doteq 42.2448$$

and

$$46.89726027 + 4.652454089 \doteq 51.5497$$

so a 95% confidence interval for β_0 is about 42.2448 to 51.5497. Similarly, $s_{b_1} = 0.56566991$ and $ts_{b_1} \doteq 1.221847006$. The confidence limits are

$$-1.45276478 - 1.221847006 \doteq -2.6746$$

and

$$-1.45276478 + 1.221847006 \doteq -0.2309$$

so a 95% confidence interval for β_1 is about -2.6746 to -0.2309. Finally, $s_{b_2} = 0.29223248$ and $ts_{b_2} \doteq 0.631222157$. The confidence limits are

$$-1.37019713 - 0.631222157 \doteq -2.0014$$

and

$$-1.37019713 + 0.631222157 \doteq -0.7390$$

so a 95% confidence interval for β_2 is about -2.0014 to -0.7390.

Comment If we obtain more than one confidence interval, the probability that all are correct is much less than 95%. In this case, since there are three intervals the probability that all are correct could be as small as $(0.95)^3$ or 86%. The same caution applies to further confidence and prediction intervals.

14.2.2 Confidence and Prediction Intervals

There are formulas for obtaining confidence intervals for $E(y_p)$ and prediction intervals for y_p, but they are not simple. Fortunately, many computer programs give confidence and prediction intervals as part of the output. In Figure 14.11, the 95% confidence limits for $E(y_p)$ are given for each observation in the sample for the data of Example 14.1. We can find confidence intervals for $E(y_p)$ for values of x that are not in the sample by entering these x_p with the data and indicating the corresponding y value as missing.

Figure 14.11 SAS printout of confidence intervals for $E(y_p)$ and prediction intervals for individual y_p for the drug data of Example 14.1.

OBS	ACTUAL	PREDICT VALUE	STD ERR PREDICT	LOWER 95% MEAN	UPPER 95% MEAN	LOWER 95% PREDICT	UPPER 95% PREDICT	RESIDUAL
1	46.0000	42.7041	1.3830	39.7163	45.6919	36.4791	48.9291	3.2959
2	40.0000	39.9637	1.1011	37.5849	42.3425	34.0070	45.9204	0.0363
3	38.0000	37.2233	1.0933	34.8614	39.5853	31.2733	43.1733	0.7757
4	35.0000	34.4829	1.3643	31.5354	37.4304	28.2772	40.6887	0.5171
5	41.0000	41.2513	1.1113	38.8504	43.6523	35.2858	47.2169	-0.2513
6	36.0000	38.5109	0.7486	36.8936	40.1282	32.8154	44.2065	-2.5109
7	35.0000	35.7705	0.7543	34.1410	37.4001	30.0715	41.4696	-0.7705
8	33.0000	33.0302	1.1228	30.6045	35.4558	27.0546	39.0057	-0.0302
9	39.0000	39.7986	1.0943	37.4345	42.1626	33.8477	45.7494	-0.7986
10	34.0000	37.0582	0.7406	35.4583	38.6581	31.3676	42.7488	-3.0582
11	30.0000	34.3178	0.7633	32.6688	35.9667	28.6132	40.0224	-4.3178
12	33.0000	31.5774	1.1402	29.1142	34.0405	25.5865	37.5683	1.4226
13	42.0000	38.3458	1.3416	35.4475	41.2441	32.1633	44.5283	3.6542
14	33.0000	35.6054	1.0846	33.2623	37.9486	29.6629	41.5480	-2.6054
15	34.0000	32.8650	1.1118	30.4631	35.2670	26.8990	38.8310	1.1350
16	35.0000	31.4948	1.2338	28.8294	34.1603	25.4180	37.5717	3.5052

SUM OF RESIDUALS 1.52767E-13
SUM OF SQUARED RESIDUALS 83.07054
PREDICTED RESID SS (PRESS) 129.1073

As before, prediction intervals for individual values of y_p are wider than confidence intervals for $E(y_p)$ since the standard error for the individual values (σ_f) is related to the standard error used for confidence intervals ($\sigma_{\hat{y}_p}$) by the formula $\sigma_f^2 = \sigma^2 + \sigma_{\hat{y}_p}^2$.

PROFICIENCY CHECKS

3. A regression equation is obtained from a sample of 20 observations for predicting y from x_1 and x_2. Now $\hat{y} = 133.512 + 44.775x_1 - 0.03456x_2$. Suppose that $s_{b_1} = 31.881$ and $s_{b_2} = 0.00589$.
   a. Test the hypotheses H_0: $\beta_j = 0$ ($j = 1, 2$) against the alternate hypotheses H_1: $\beta_j \neq 0$. Use the 0.05 level of significance. What are your conclusions?
   b. Which of the predictor variables appears to have the most effect on y? Why?
   c. Would you say that this equation appears to be a satisfactory predictor, or should an additional analysis be performed? If so, what should be done?
4. Refer to Proficiency Check 3. Obtain 95% confidence intervals for β_1 and β_2. Do these results confirm your conclusions?
5. Refer to Proficiency Check 1. Using the computer, obtain:
   a. a 95% confidence interval for $E(y_p)$ when $x_1 = 5$ and $x_2 = 11$
   b. a 95% prediction interval for y_p when $x_1 = 5$ and $x_2 = 11$

14.2.3 Determining the "Best" Model

Clearly if there are several predictor variables, many different models are possible using linear terms only. If there are $k - 1$ predictor variables, there are two possibilities for each predictor variable—it can be included in the model or left out. Thus the number of possible ways to include or leave out each variable is $2 \cdot 2 \cdots 2 = 2^{k-1}$. This also includes the possibility that all variables are left out, which is not a possible model, so the number of possible models using $k - 1$ predictor variables is equal to $2^{k-1} - 1$.

There are several different criteria for selecting the "best"model for a given set of data. The one that appears to be in keeping with the least-squares procedure is that the best-fitting model among several possible ones is the one with the smallest MSE. Although this criterion is not always the best choice because of cost considerations and other practical factors, from a theoretical viewpoint it is the best fit among those tried. An example of prohibitive cost is the analysis conducted by an air force officer who wanted to determine the optimal replacement time for engine oil in jet aircraft. He used a number of measurement factors, such as viscosity of the oil and a chemical analysis of a number of components of the oil. In this study the "best" regression equation for predicting when the oil should be replaced involved a very costly analysis. If a different analysis was used, however, the regression equation was almost as good and the data needed to make the decision could be obtained at a much reduced cost.

A special technique called *stepwise regression* is one of the better approaches to determining the best-fitting model from among several. We can perform this procedure by using Minitab or the SAS System. Stepwise regression performed with SAS statements is illustrated in Section 14.2.6; the Minitab commands for performing stepwise regression are shown in Section 14.4.

Short of using the stepwise procedure, the simplest way to obtain the best-fitting model is probably the following. Use all variables and select a level of significance (usually 0.05). Perform the regression analysis and check the output. If all variables are significant at the chosen level of significance, this is probably the best model. If one or more variables are not significant at the chosen level, remove the variable with the highest level of significance (the one for which $|t^*|$ is the smallest) and perform the regression procedure again. If the MSE of the resulting analysis is not smaller than that of the prior model, the first model is the best fit. If the MSE is smaller and all variables are significant, this is the best-fitting model. If one or more variables are not significant, repeat the procedure until you obtain the model with the smallest MSE. Thus if there are four predictor variables, there are $2^4 - 1 = 15$ possible models. Using the procedure outlined here, you would need to perform a maximum of four regression procedures to determine the best model.

Another widely used criterion is the *adjusted R^2 criterion*. The coefficient of multiple determination, R^2, measures the improvement in error variance obtained by using the model for predictive purposes. Each time a variable is added, R^2 increases since SSE can never become larger with more predictor variables and SSTO is always the same. Thus we cannot use the largest value of R^2 to determine the best fit. We can, however, adjust for the number of predictor variables in the model and obtain an adjusted coefficient of multiple determination, R_a^2, defined by

$$R_a^2 = 1 - \left(\frac{n-1}{n-k}\right)\frac{\text{SSE}}{\text{SSTO}}$$

Those who use this criterion choose the model for which R_a^2 is the largest.

14.2.4 Interaction

When there are two or more predictor variables in a regression model, an **interaction** effect may be present, similar to the effect we investigated in two-factor analysis of variance. Two factors are said to interact if the rate of change of y for one predictor variable is not the same for all values of

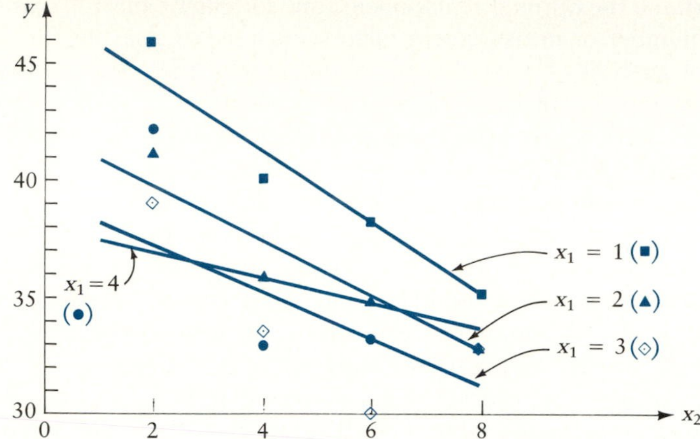

Figure 14.12
Interaction diagram of y plotted against x_2 for values of x_1 for the drug data of Example 14.1.

the other predictor variable. Consider the data of Example 14.1. If we graph the values of y against x_2 for each value of x_1 and sketch a line that approximately fits the data, we obtain the graph shown in Figure 14.12.

Since the lines in Figure 14.12 are not parallel, there may be some interaction. That is, the change in the response variable is not the same for each value of x_1 when x_2 is held constant. We may test for interaction by using the term x_1x_2 in the model. The model will have the following regression function:

$$E(y) = \beta_0 + \beta_1 x_1 + \beta_2 x_2 + \beta_3 x_1 x_2$$

The printout of the ANOVA table for this model is shown in Figure 14.13.

We can test the hypothesis that $\beta_3 = 0$ (there is no interaction) by means of the usual t test. Here we see that $b_3 = 0.14364601$ and the P value for the test of $H_0{:}\beta_3 = 0$ is 0.6149, so we cannot conclude that there is interaction between the variables. Note that the P value for the test of $H_0{:}\beta_1 = 0$ is 0.1704. We cannot reject both these hypotheses however, since, as noted before, the tests are marginal. The test on each coefficient rests on the assumption that the other variables will remain in the model.

Figure 14.13
SAS printout of the ANOVA table for the model $E(y) = \beta_0 + \beta_1 x_1 + \beta_2 x_2 + \beta_3 x_1 x_2$ for the drug data of Example 14.1.

```
INPUT FILE

DATA DRUGS;
  INPUT DRUG1 DRUG2 HOURS @@;
  DRUG12 = DRUG1 * DRUG2;
CARDS;
1 2 46 1 4 40 1 6 38 1 8 35 2 2 41 2 4 36 2 6 35 2 8 33
3 2 39 3 4 34 3 6 30 3 8 33 4 2 42 4 4 33 4 6 34 4 7 35
PROC REG;
  MODEL HOURS = DRUG1 DRUG2 DRUG12;

OUTPUT FILE
```

```
                                   SAS
DEP VARIABLE: HOURS
                           ANALYSIS OF VARIANCE

                             SUM OF            MEAN
      SOURCE      DF        SQUARES          SQUARE      F VALUE     PROB>F

      MODEL        3      178.73600     59.57866738        8.798     0.0023
      ERROR       12    81.26399786      6.77199982
      C TOTAL     15      260.00000

          ROOT MSE          2.602307        R-SQUARE      0.6874
          DEP MEAN              36.5        ADJ R-SQ      0.6093
          C.V.              7.129607

                          PARAMETER ESTIMATES

                         PARAMETER       STANDARD     T FOR H0:
   VARIABLE      DF       ESTIMATE          ERROR   PARAMETER=0     PROB > |T|

   INTERCEP       1    48.58952832     3.95623380        12.282         0.0001
   DRUG1          1    -2.15371699     1.47679484        -1.458         0.1704
   DRUG2          1    -1.71441001     0.73119620        -2.345         0.0371
   DRUG12         1     0.14364601     0.27811732         0.516         0.6149
```

b_3 → 0.14364601; P value for H_0: $\beta_1 = 0$ → 0.1704; P value for H_0: $\beta_3 = 0$ → 0.6149

Thus if we decide to remove one variable, all other variables will be left in the model. When we remove the interaction variable, the P value for the test of H_0: $\beta_1 = 0$ is 0.0234 (see Figure 14.10) so that drug 1 should be included in the model. For a further discussion of interaction terms see Neter and others (1985, pp. 232–237) or Mendenhall and Sincich (1986, pp. 170–176 and 313–319).

14.2.5 Multicollinearity

In the discussion of marginal t tests (Section 14.2.1) we noted that two or more predictor variables may themselves be strongly associated. When some of the predictor variables are associated, we say that **multicollinearity** or *intercorrelation* exists among them. A small amount of multicollinearity is common and poses no particular problems. When there is a large amount of multicollinearity, however, the results may be misleading or in error. It is even possible that small changes in the values of the observations will cause radical changes in the estimated values of the parameters. A regression equation that included the price (per barrel) of crude oil and the price (per gallon) of gasoline as predictor variables would clearly be subject to the problem of multicollinearity. There are several ways to detect multicollinearity:

1. We can obtain correlation coefficients for each pair of variables by using the SAS procedure CORR. Significant correlation coefficients between predictor variables should be investigated.
2. Marginal t tests for the hypotheses that $\beta_j = 0$ may not be significant for any β_j, even though the overall F test is significant. As discussed in

the comment in Section 14.2.1, this result may indicate that we can remove one of the predictor variables if we leave the rest of the variables in the model.

3. We can obtain radically different values for one or more of the b_j when a few observations are removed or added.

To compensate for multicollinearity, several options are available. The simplest option is to use only one of the highly correlated predictor variables. For a further discussion of multicollinearity see Neter and others (1985, pp. 271–282) or Mendenhall and Sincich (1986, pp. 222–228).

14.2.6 SAS Statements for This Section

We can obtain printouts for interaction models by using the REG procedure and simply including the appropriate term in the model statement. To include x_1x_2 in the model, we need to create a new variable $x_{12} = x_1x_2$. We write

```
DATA INTERACT;
  INPUT X1 X2 Y @@;
  X12 = X1*X2;
CARDS;
Data go here
PROC REG;
  MODEL Y = X1 X2 X12;
```

and proceed as usual.

We may obtain 95% confidence intervals for $E(y_p)$ and prediction intervals for y_p by writing

```
PROC REG;
  MODEL Y = · · · / P CLM CLI;
```

We can use the SAS procedure STEPWISE to determine the best model. Like REG, the STEPWISE procedure requires that we define any second-order variables in the data step. Several options are available, but these will not be explored here. The default option is the forward selection procedure; we add variables with significant F statistics one at a time until no further variable has a significant F. The backward procedure used by Minitab (see Section 14.4) is also available. All variables are included in the model and then removed, one at a time, until all the variables remaining produce significant F statistics. The stepwise procedure does not necessarily produce the model with the smallest MSE or the largest adjusted R^2. Other options producing more models to examine are available. The output of the STEPWISE procedure using the BACKWARD option for the real estate data of Problem 14.8 is shown in Figure 14.14.

```
                                              SAS

                        BACKWARD ELIMINATION PROCEDURE FOR DEPENDENT VARIABLE Y

STEP 0    ALL VARIABLES ENTERED         R SQUARE = 0.86726553        C(P) =        5.00000000

                        DF          SUM OF SQUARES         MEAN SQUARE               F        PROB>F

REGRESSION               4           955.32333860        238.83083465             4.90        0.1113
ERROR                    3           146.21166140         48.73722047
TOTAL                    7          1101.53500000

                          B VALUE        STD ERROR          TYPE II SS               F        PROB>F

INTERCEPT          -65.75825159
X1                   0.63474495        1.03413685         18.36129197             0.38        0.5828
X2                   0.55173128        0.42888551         80.65535873             1.65        0.2886
X3                   1.37447066        1.14331845         70.43642123             1.45        0.3155
X4                   3.99885301        2.00503540        193.85970675             3.98        0.1401

BOUNDS ON CONDITION NUMBER:      3.524607,       80.67426
------------------------------------------------------------------------------------------------------

STEP 1    VARIABLES X1 REMOVED          R SQUARE = 0.85059671        C(P) =        3.37674065

                        DF          SUM OF SQUARES         MEAN SQUARE               F        PROB>F

REGRESSION               3           936.96204663        312.32068221             7.59        0.0397
ERROR                    4           164.57295337         41.14323834
TOTAL                    7          1101.53500000

                          B VALUE        STD ERROR          TYPE II SS               F        PROB>F

INTERCEPT          -54.71829597
X2                   0.53924185        0.39361433         77.21894427             1.88        0.2426
X3                   1.71628874        0.91743668        143.98822569             3.50        0.1347
X4                   4.32319957        1.77708634        243.49608593             5.92        0.0718

BOUNDS ON CONDITION NUMBER:      3.516674,       45.98534
------------------------------------------------------------------------------------------------------

STEP 2    VARIABLE X2 REMOVED           R SQUARE = 0.78049549        C(P) =        2.96113433

                        DF          SUM OF SQUARES         MEAN SQUARE               F        PROB>F

REGRESSION               2           859.74310236        429.87155118             8.89        0.0226
ERROR                    5           241.79189764         48.35837953
TOTAL                    7          1101.53500000

                          B VALUE        STD ERROR          TYPE II SS               F        PROB>F

INTERCEPT            9.66363599
X3                   0.88551738        0.74637000         68.07028418             1.41        0.2887
X4                   5.96841138        1.42012818        854.15000275            17.66        0.0085

BOUNDS ON CONDITION NUMBER:      1.146875,       9.174990
------------------------------------------------------------------------------------------------------

STEP 3    VARIABLES X3 REMOVED          R SQUARE = 0.71869965        C(P) =        2.35781398
                        DF          SUM OF SQUARES         MEAN SQUARE               F        PROB>F

REGRESSION               1           791.67281818        791.67281818            15.33        0.0078
ERROR                    6           309.86218182         51.64369697
TOTAL                    7          1101.53500000

                          B VALUE        STD ERROR          TYPE II SS               F        PROB>F

INTERCEPT           20.40818182
X4                   5.36545455        1.37038411        791.67281818            15.33        0.0078

BOUNDS ON CONDITION NUMBER:               1,              2
------------------------------------------------------------------------------------------------------

ALL VARIABLES IN THE MODEL ARE SIGNIFICANT AT THE 0.1000 LEVEL.

                                              SAS

                   SUMMARY OF BACKWARD ELIMINATION PROCEDURE FOR DEPENDENT VARIABLE Y

                     VARIABLE    NUMBER     PARTIAL      MODEL
             STEP    REMOVED         IN        R**2       R**2          C(P)             F       PROB>F

                1    X1               3      0.0167     0.8506       3.37674        0.3767       0.5828
                2    X2               2      0.0701     0.7805       2.96113        1.8768       0.2426
                3    X3               1      0.0618     0.7187       2.35781        1.4076       0.2887
```

Figure 14.14
SAS printout for stepwise regression performed on the real estate data of Problem 14.8.

Problems

Practice

14.9 A regression equation obtained from a sample of 25 observations for predicting y from x_1, x_2, and x_3 is $\hat{y} = 13.6621 + 1.4567x_1 + 0.0004432x_2 - 3.9877x_3$. Suppose that $s_{b_1} = 0.8849$, $s_{b_2} = 0.00009471$, and $s_{b_3} = 1.1538$.

a. Test the hypotheses H_0: $\beta_j = 0$ ($j = 1, 2, 3$) against the alternate hypotheses H_1: $\beta_j \neq 0$. Use the 0.05 level of significance. What are your conclusions?

b. Which of the predictor variables appears to have the greatest effect on y? Why?

c. Would you say that this equation appears to be a satisfactory predictor, or should an additional analysis be performed? If so, what should be done?

d. What are the implications of performing three t tests, each at the 0.05 level of significance?

14.10 Refer to Problem 14.9. Obtain 95% confidence intervals for β_1, β_2, and β_3. Do these results confirm your conclusions in part (*a*)?

14.11 Refer to Problem 14.2. Obtain R^2. Does this value suggest that the model is a good fit to the data?

14.12 Refer to Problem 14.3. Use the computer to fit the model $y = \beta_0 + \beta_1 x_1 + \beta_2 x_2 + \beta_3 x_1 x_2 + \epsilon$ to the data. Is there significant interaction? Use $\alpha = 0.05$.

14.13 Refer to Problem 14.3.

a. Obtain a 95% confidence interval for the expected value of y when $x_1 = 1.5$ and $x_2 = 15$. Use a computer.

b. Obtain a 95% prediction interval for the next value of y when $x_1 = 1.9$ and $x_2 = 17$. Use a computer.

c. Does multicollinearity appear to be a potential problem here? Discuss.

Applications

A computer is required for most of the following problems.

14.14 Refer to Problem 14.4.

a. Test the hypotheses that $\beta_1 = 0$ and $\beta_2 = 0$. Use the 0.05 level of significance in each case.

b. Obtain a 95% confidence interval for the mean change in electricity for an increase of 1 degree in average temperature if output is held constant.

c. Obtain a 95% confidence interval for the mean change in electricity for an increase of 1000 units of output if temperature is held constant.

d. Obtain a 95% confidence interval for the expected usage of electricity for all days in which the mean temperature is 80°F and the output is 110,000 units.

e. Obtain a 95% prediction interval for the usage of electricity on a day in which the mean temperature is 82.3°F and the output is 113,234 units.

14.15 Refer to the duration of illness problem (Problem 14.5).

a. Test the hypothesis that each variable is significant. Use the 0.05 level of significance.

b. Obtain linear regression equations for predicting days in the hospital for each predictor variable. Compare MSE for each with the MSE for the full model (Problem 14.5). Which is the best predictor?
c. Using the best predictor, give a 95% confidence interval for the stay of all patients with a bacterial count of 7000 and a temperature of 38°C.
d. Using the best predictor, give a 95% prediction interval for the stay of the next patient with a bacterial count of 7000 and a temperature of 38°C.

14.16 Refer to the Neslin bathroom tissue study (Problem 14.6).
a. Do both variables contribute to the market share? Use the 0.05 level of significance.
b. What is the best regression equation? Why?
c. Using the best predictor, obtain a 95% confidence interval for the mean market share of a tissue for all months in which the price is 39¢ and four ads are placed.
d. Obtain a 95% prediction interval for the market share of a tissue for the next month in which the price is 39¢ and four ads are placed.

14.17 Refer to the Neslin instant coffee study (Problem 14.7).
a. Do both variables contribute to the market share? Use the 0.05 level of significance.
b. What is the best regression equation? Why?
c. Using the best predictor, obtain a 95% confidence interval for the mean market share of a coffee for all months in which the price is 35¢ per ounce and five ads are placed.
d. Obtain a 95% prediction interval for the market share of a coffee for the next month in which the price is 35¢ per ounce and five ads are placed.

14.18 Refer to Problem 14.8. Using the procedure outlined in Section 14.2.3, determine the regression model with the smallest MSE.
a. Determine a 95% confidence interval for the expected mean price of lots with the same specifications as lot 1 using the full model (all four predictor variables).
b. Determine a 95% confidence interval for the expected mean price of lots with the same specifications as lot 1 using the best model as determined above.
c. Compare the confidence interval widths. Which has the smallest width? Is this what you would expect?
d. Determine a 95% prediction interval for the price of a lot with the same specifications as lot 1. Compare the width of this interval with the width of the confidence interval found in part (*b*). Is this what you would expect? Explain.

14.3 Quadratic Regression

In the past two chapters we have often mentioned the assumption that the association between the variables is linear. We have seen that it is possible,

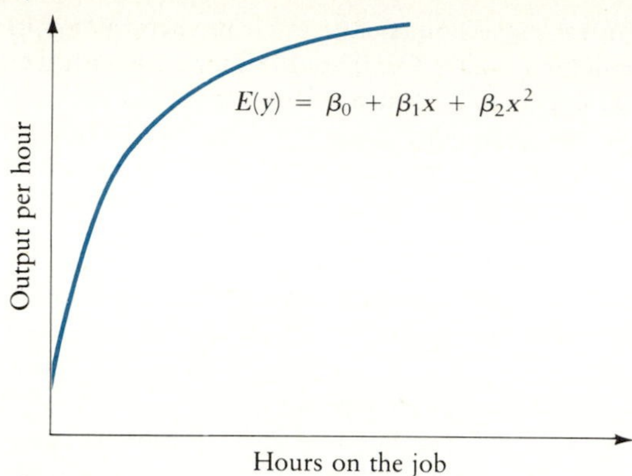

Figure 14.15
Graph of a nonlinear relationship between number of hours on the job and work efficiency.

by means of a residual plot, to ascertain whether this assumption is reasonable. In certain cases we can conduct a test for goodness of fit.

If the association between two variables is not linear, a *quadratic regression function* may be the best-fitting regression function. A **quadratic regression function** is a function of the form $E(y) = \beta_0 + \beta_1 x + \beta_2 x^2$. We can see that this is a special case of a multiple linear regression function in which $x = x_1$ and $x^2 = x_2$; therefore, exactly the same multiple regression methods apply. The multiple regression function in this case is a quadratic function, the graph of which is a parabola. The parameter β_2 measures the curvature of the graph. If $\beta_2 = 0$ and $\beta_1 \neq 0$, then the relationship between the variables is linear. If $\beta_2 \neq 0$, then the relationship is nonlinear.

Suppose, for example, that a group of apprentice workmen is assigned to an assembly line to manufacture a product. If we measure efficiency as a function of time spent on the assembly line, efficiency is likely to increase rapidly as the workers learn their jobs, but as they become more proficient the improvements in efficiency will decrease. The graph of efficiency over a period of time may look like that in Figure 14.15.

The efficiency with no on-the-job training is represented by β_0. When x is small the function increases rapidly, showing that β_1 is positive. As x increases, the increases in efficiency are less rapid, showing that β_2 is negative. If the regression function were quadratic, as x continued to increase the curve would eventually turn down. In actuality, the efficiency would probably level off to a nearly constant value as x becomes sufficiently large. This is another illustration of the danger of predicting beyond the scope of the data.

14.3.1 The Quadratic Regression Equation

A quadratic regression function is an example of a *second-order equation*—an equation in which one or more of the terms contains a variable raised to the second power or a product of two variables. Equations containing an interaction term, such as x_1x_2, are also of the second order. Equations in which all terms other than the constant term contain only one variable to the first degree are called *first-order equations*. Equations in which the sum of the exponents of the variables of one or more terms is greater than 2 are called *higher-order equations*.

If the response variable has a quadratic statistical relationship with the predictor variable, then each observation of y can be defined by

$$y_i = \beta_0 + \beta_1 x_i + \beta_2 x_i^2 + \epsilon_i$$

where x_i is the ith observation of the predictor variable
y_i is the ith observation of the response variable
$\beta_0, \beta_1, \beta_2$ are parameters
ϵ_i are $N(0, \sigma^2)$
$i = 1, 2, 3, \ldots, n$

The quadratic regression function is written

$$E(y) = \beta_0 + \beta_1 x + \beta_2 x^2$$

and the sample quadratic regression equation is written

$$\hat{y} = b_0 + b_1 x + b_2 x^2$$

The Quadratic Regression Equation

Suppose we obtain a set of n ordered pairs (x, y) from a population for which the relationship between the variables is described by the regression function $E(y) = \beta_0 + \beta_1 x + \beta_2 x^2$. Then we can obtain a sample regression equation $\hat{y} = b_0 + b_1 x + \beta_2 x^2$ from the sample data by solving the following normal equations:

$$\Sigma y = nb_0 + b_1\Sigma x + b_2\Sigma x^2$$
$$\Sigma xy = b_0\Sigma x + b_1\Sigma x^2 + b_2\Sigma x^3$$
$$\Sigma x^2 y = b_0\Sigma x^2 + b_1\Sigma x^3 + b_2\Sigma x^4$$

We obtain these equations by replacing x_1 by x and x_2 by x^2 in the normal equations of Section 14.1. The one unfamiliar term is $\Sigma x^2 y$. This is the sum of the products of each y value and the square of the corresponding x value. We obtain the analysis of variance table in the same way as the one for multiple regression. The results are usually presented in an ANOVA table, as shown in Table 14.4.

Partitioning Total Sum of Squares (Quadratic Regression)

Total Sum of Squares

$$\text{SSTO} = \Sigma y^2 - \frac{(\Sigma y)^2}{n}$$

Sum of Squares Due to Regression

$$\text{SSR} = \frac{b_0\Sigma y + b_1\Sigma xy + b_2\Sigma x^2 y - (\Sigma y)^2}{n}$$

Sums of Squares Due to Error

$$\begin{aligned}\text{SSE} &= \text{SSTO} - \text{SSR} \\ &= \Sigma y^2 - b_0\Sigma y - b_1\Sigma xy - b_2\Sigma x^2 y\end{aligned}$$

Mean Squares

$$\text{MSR} = \frac{\text{SSR}}{2}$$

$$\text{MSE} = \frac{\text{SSE}}{n-3}$$

Table 14.4 ANOVA TABLE FOR QUADRATIC REGRESSION ANALYSIS.

SOURCE OF VARIATION	DF	SS	MS
Regression	2	SSR	SSR/2
Error	$n - 3$	SSE	SSE/(n − 3)
Total	$n - 1$	SSTO	

14.3.2 Inferences in Quadratic Regression

We can perform an F test to determine whether β_1 and β_2 are both zero. If both are zero, there is probably no relationship between the variables. Another test usually performed is a t test to determine whether $\beta_2 = 0$. If $\beta_2 \neq 0$, then there is a curvature effect between the two variables. If the F test indicates a relationship between the variables but the t test indicates there is no curvature effect, the relationship between the variables is linear. We may use the residual plot, residuals plotted against $\hat{y}$, to test all the assumptions, since there is only one predictor variable.

No new problems are encountered in constructing confidence and prediction intervals. The same procedures apply as for multiple regression.

PROFICIENCY CHECKS

Use the following data for Proficiency Checks 6 to 8:

x	3	3	5	4	5	6	7	5	4	7	3	6
y	14	12	23	21	22	25	25	23	19	26	13	23

6. a. Using the computer or a hand calculator, verify that the quadratic regression equation for the data is

$$\hat{y} = -14.2919 + 11.8964x - 0.8963x^2$$

   b. If $s_{b_1} \doteq 1.8802$ and $s_{b_2} \doteq 0.1898$, test the curvature effect using the 0.05 level of significance. Is there significant curvature?
   c. Is there a significant linear effect?

7. Obtain the linear regression equation for the data. If MSE for the quadratic model is 1.2724, which equation is the better fit? Does this result agree with your test for significant curvature in Proficiency Check 6?

8. You will need a computer for this problem.
   a. Using the best-fitting equation from Proficiency Check 6 or 7, obtain a 95% confidence interval for $E(y_p)$ when $x = 6$.
   b. Using the best-fitting equation from Proficiency Check 6 or 7, obtain a 95% prediction interval for y_p when $x = 6$.

14.3.3 Using SAS Statements for Quadratic Regression

We can perform quadratic regression analysis by using PROC REG. The x^2 variable must be created in the data step. We may write

```
DATA QUAD;
  INPUT X Y @@;
  XSQ = X*X;
CARDS;
Data go here
PROC REG;
  MODEL Y = X XSQ;
```

Table 14.5.

LOT SIZE (100'S)	UNIT COST (CENTS)
5	5.2 4.9 5.4 5.2
10	3.8 4.1 4.3 4.0
15	3.5 3.3 3.2 3.4
20	2.7 2.3 3.1 2.8
25	1.9 2.0 2.2 2.1
30	2.6 2.5 2.7 2.6

We can obtain 95% confidence and prediction intervals by using the options P CLM and CLI.

The residual plot, plotting RESID against YHAT, provides a method for checking the fit as well as the other assumptions. If the regression equation fits the data well, the residual plot will appear to be random.

14.3.4 A Comprehensive Example

A manufacturer wishes to determine the optimum lot size for manufacturing a product and wants to find a regression equation for predicting unit cost from lot size. He produces four lots each of 500, 1000, 1500, 2000, 2500, and 3000 units, obtaining the unit cost for each lot (Table 14.5). We will show, using a residual plot, that a linear regression equation is not appropriate, and perform a complete quadratic regression analysis of the data.

Figure 14.16 shows the printout for the linear regression analysis and the residual plot. For reference, note that the MSE for the linear regression analysis is about 0.2189. In the residual plot each A represents an observation; the B's represent two identical observations. The graph shows a curvilinear pattern—first above the fitted values ($\hat{y}$), then below, then above. Thus a quadratic model may be the best fit.

Figure 14.16
SAS printout for linear regression analysis and residual plot for the manufacturing cost data.

```
                                  SAS

                          ANALYSIS OF VARIANCE

                         SUM OF          MEAN
SOURCE        DF        SQUARES        SQUARE        F VALUE        PROB>F

MODEL          1    21.72857143   21.72857143         99.250        0.0001
ERROR         22     4.81642857    0.21892857
C TOTAL       23    26.54500000

     ROOT MSE        0.467898      R-SQUARE           0.8186
     DEP MEAN           3.325      ADJ R-SQ           0.8103
     C.V.            14.07212

                          PARAMETER ESTIMATES

                    PARAMETER         STANDARD      T FOR H0:
VARIABLE   DF        ESTIMATE            ERROR    PARAMETER=0     PROB > |T|

INTERCEP    1      5.27500000       0.21779468         24.220         0.0001
LOTSIZE     1     -0.11142857       0.01118490         -9.962         0.0001
```

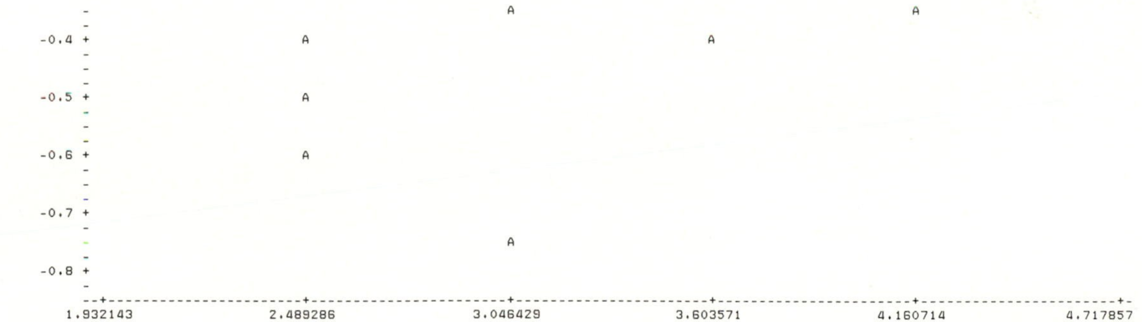

The results of the quadratic regression analysis are shown in Figure 14.17. The sample regression equation is

$$\hat{y} = 6.6875 - 0.3233x + 0.006054x^2$$

The F^* value of 189.135 is significant well beyond the 0.01 level of significance ($P = 0.0001$), so we may conclude that there is an association between the variables.

Figure 14.17
SAS printout of the quadratic regression analysis of the manufacturing cost data.

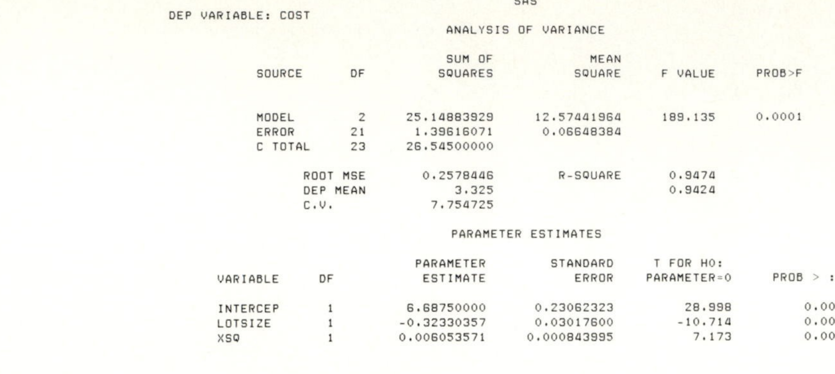

SAS

DEP VARIABLE: COST

ANALYSIS OF VARIANCE

SOURCE	DF	SUM OF SQUARES	MEAN SQUARE	F VALUE	PROB>F
MODEL	2	25.14883929	12.57441964	189.135	0.0001
ERROR	21	1.39616071	0.06648384		
C TOTAL	23	26.54500000			

ROOT MSE	0.2578446	R-SQUARE	0.9474
DEP MEAN	3.325	ADJ R-SQ	0.9424
C.V.	7.754725		

PARAMETER ESTIMATES

VARIABLE	DF	PARAMETER ESTIMATE	STANDARD ERROR	T FOR H0: PARAMETER=0	PROB > :T:
INTERCEP	1	6.68750000	0.23062323	28.998	0.0001
LOTSIZE	1	-0.32330357	0.03017600	-10.714	0.0001
XSQ	1	0.006053571	0.000843995	7.173	0.0001

Figure 14.18
SAS printout of the residual plot for the gradratic regression analysis of the manufacturing cost data.

```
                                        SAS
PLOT OF RESIDUAL*YHAT     LEGEND: A = 1 OBS, B = 2 OBS, ETC.

      0.8 +
      0.7 +
      0.6 +
      0.5 +
          -               A
      0.4 +
      0.3 +                                    A
          -       A
R         -                                                      A
E         -
S     0.2 +                                    A
I         -                                                                          A
D         -       B       A
U         -
A         -
L     0.1 +                                    A
S         -
          -       A       A
          -                                                      A
      0.0 +--------------------------------------A---------------------------------------------
          -                                                                          B
          -                                                      A
     -0.1 +
          -      A
     -0.2 +
          -                                                      A
          -      A
     -0.3 +
          -               A                                                          A
          -      A
     -0.4 +
          -      A
     -0.5 +
        --+--------+--------+--------+--------+--------+--------+--------+--------+--------+--------+--------+--------+----
         2.2      2.4      2.6      2.8      3.0      3.2      3.4      3.6      3.8      4.0      4.2      4.4      4.6

                                          PREDICTED VALUE
```

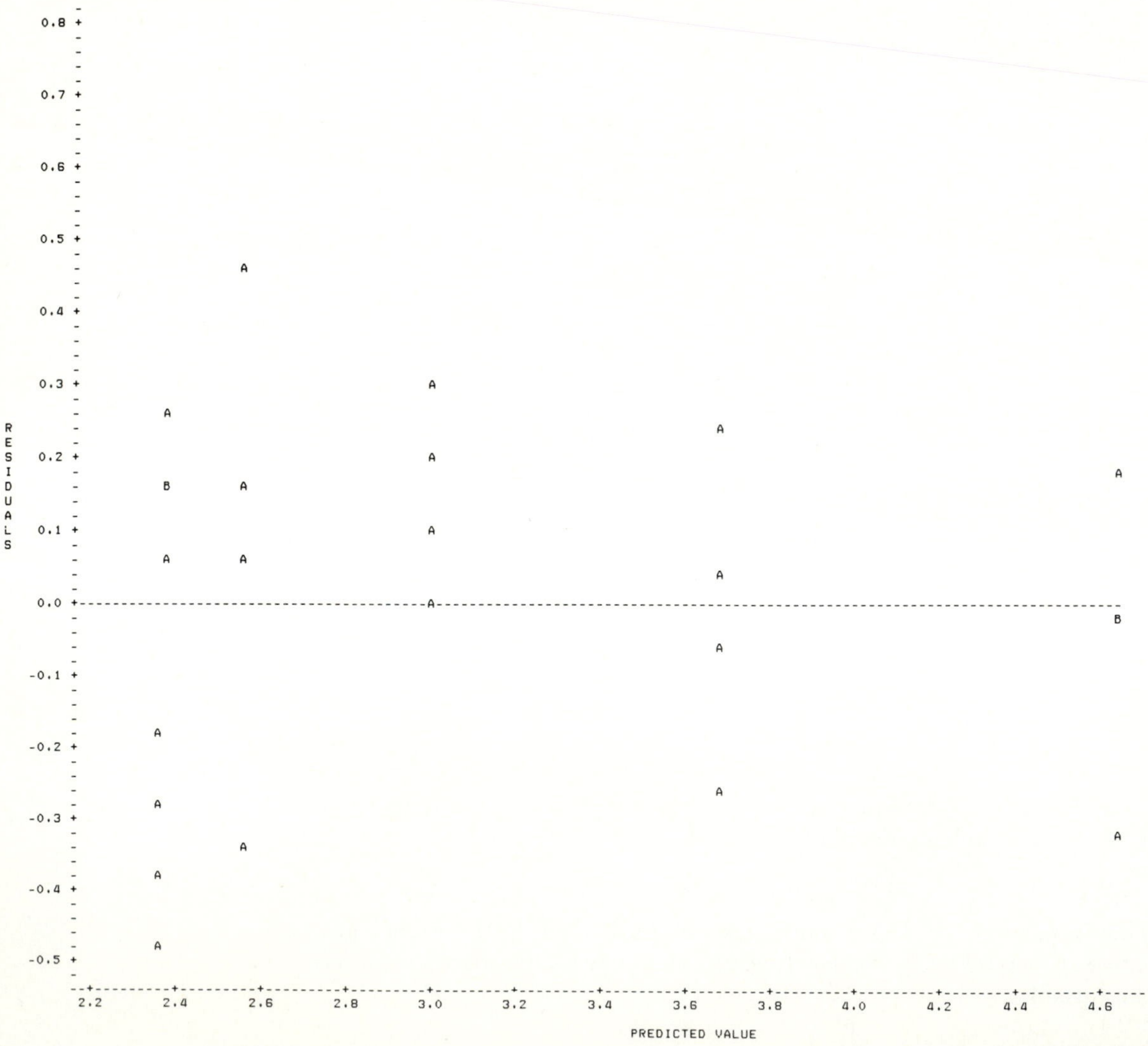

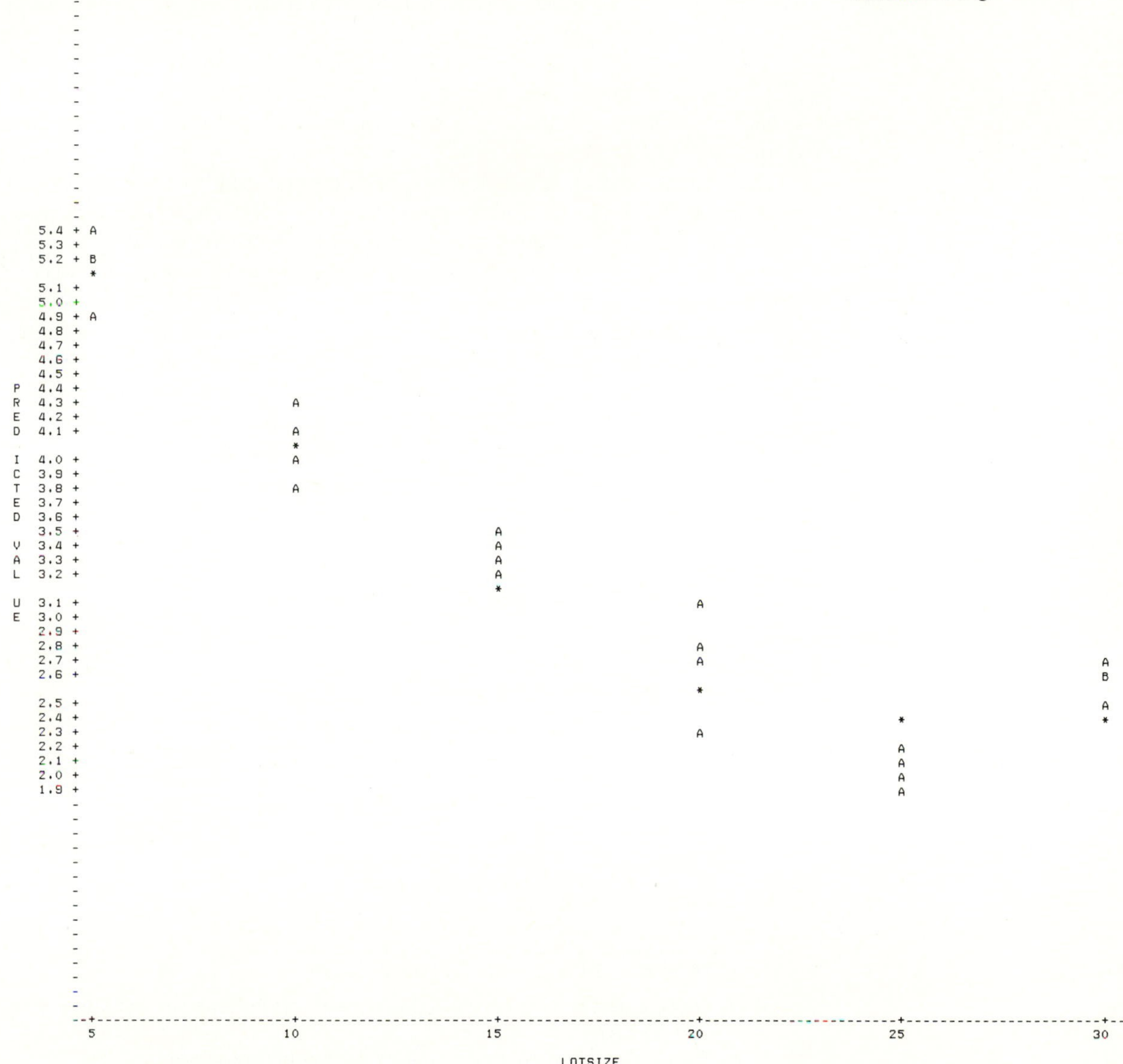

Figure 14.19 SAS printout of the graph of the quadratic regression equation for the manufacturing cost data.

To test for curvature, we wish to test the hypothesis $\beta_2 = 0$; t^* for this test is 7.17, which is significant well beyond the 0.01 level of significance ($P = 0.0001$). Thus the quadratic regression equation seems to be a good fit. Comparing the MSE for the quadratic model (about 0.0665) with the MSE for the linear model (about 0.2189), we see that the quadratic model is indeed a much better fit.

Finally, we obtain a residual plot (Figure 14.18). There may be slight deviations from fit, but they do not appear to be serious. Other assumptions appear to be safe; $s_e \doteq 0.2578$ and all residuals are within $2s_e$ of the fitted values.

To estimate the optimum lot size, we plot the graph of the regression equation and estimate the minimum cost. Figure 14.19 shows the computer plot of the sample data (letters) and predicted values for each x (asterisks). Joining the asterisks freehand, we see that the minimum cost value occurs approximately when lot size is 25 (2500).

The graph of a quadratic regression equation is called a *parabola.* It can be shown that the maximum or minimum value of $\hat{y}$ in the regression equation $\hat{y} = b_0 + b_1x + b_2x^2$ will occur when $x = -b_1/2b_2$. In this example, this would be when $x \doteq 0.3233/0.01210$, or about 26.7. As a result of this analysis we conclude that the optimum lot size, producing the minimum cost, is 2670 units. Of course, this result is subject to sampling error and must be viewed as an approximation.

Problems

Practice

14.19 A quadratic regression equation is

$$\hat{y} = 1143.66 + 122.4x + 3.54x^2$$

a. As x increases, does y increase at an increasing or decreasing rate?
b. If $s_{b_2} = 1.03$ and $n = 18$, is the curvature effect significant? At about what level of significance?
c. Obtain a 95% confidence interval for β_2.

14.20 Use the data given here:

x	7	1	4	4	5	3	9	6	5	7	2	9	8
y	28	3	6	7	13	2	55	20	14	26	1	58	44

a. Draw a scatterplot for the data. Does the trend appear to be linear or curvilinear?
b. Derive a linear regression equation for predicting y from x.
c. Derive a quadratic regression equation for predicting y from x.
d. Determine MSE for both models. Which is the better fit?
e. Conduct a test of the significance of b_2. Use the 0.05 level of significance. Does your result agree with part (*d*)?
f. Use the computer to obtain a 95% confidence interval for $E(y_p)$ when $x = 6.5$. Use both linear and quadratic models. Which is the narrower interval? How does this finding confirm your previous results?

Applications

You should use a computer for these problems.

14.21 In Problem 13.21 a medical researcher tested the effect of a drug inhibiting the release of adrenalin. The linear regression equation indicated that there was a significant positive association between the variables. She now wonders whether the association will remain linear throughout the safe dosage of the drug, up to 60 cubic centimeters per injection. To investigate this possibility she randomly selects ten samples from volunteers, each sample composed of five male individuals, and injects the men of the samples with 15, 20, 25, 30, 35, 40, 45, 50, 55, and 60 cubic centimeters, respectively, of the drug. Each man is then subjected to a stress situation and the stress level at which adrenalin is released is measured for each individual. Data are given here:

AMOUNT OF DRUG INJECTED (CC)	STRESS LEVEL AT WHICH ADRENALIN IS RELEASED
15	14, 19, 19, 20, 18
20	19, 20, 24, 18, 23
25	19, 21, 21, 25, 24
30	20, 23, 26, 25, 22
35	23, 21, 26, 26, 23
40	23, 24, 28, 23, 27
45	26, 25, 28, 26, 27
50	27, 28, 30, 26, 27
55	28, 27, 31, 26, 27
60	27, 28, 26, 28, 30

a. Obtain a linear regression equation and obtain a residual plot. Does the plot suggest nonlinearity?
b. Determine the quadratic regression equation for predicting the response from amount of drug injected.
c. Compare the MSE for the quadratic model with that of the linear model. Does this comparison confirm the nonlinearity suggested by the residual plot?
d. Test the hypothesis that $\beta_2 = 0$. Use the 0.05 level of significance. Do all these results seem to confirm each other?

14.22 According to the *International Civil Aviation Organization Report* (January 1986), the number of global air passenger miles for the past few years is as listed here:

Year	1981	1982	1983	1984	1985
Miles (millions)	752	764	795	841	892

a. Draw a scattergram for the data. Does a linear or quadratic regression equation appear to be appropriate?
b. Obtain a quadratic regression equation, using $x = 1, 2, 3, 4, 5$, indicating the number of years after 1980.
c. Test the curvature effect. Is a quadratic or linear regression equation needed?

d. This is an example of a time series. Prediction intervals are not appropriate, but extrapolation to future time periods is possible, taking into account the possibility of error. Using the regression equation, forecast the number of millions of global air passenger miles for 1990.

14.23 The cost of a 30-second television ad during the Super Bowl game for the past several years is given here in thousands of dollars (*Sports Industry News*, January 24, 1986):

Year	1977	1978	1979	1980	1981	1982	1983	1984	1985	1986
Cost	163	185	222	225	275	355	400	445	525	550

a. Obtain linear and quadratic regression equations for predicting future costs of the ads. For x, use year minus 1976.
b. Which is the better fit?
c. Using the best-fitting equation, predict the cost of an ad in 1990.

14.4 Minitab Commands for This Chapter

To perform a multiple regression analysis using Minitab, we use the REGRESS command discussed in Chapter 13. In multiple regression, however, we need to specify the number of predictor variables and the columns in which they are stored. The general form of the multiple regression command is

```
MTB > REGRESS the y values in C on the K predictors in C ... C;
SUBC> RESIDUALS store in C;
SUBC> FITS store in C;
SUBC> PREDICT using E.
```

These subcommands were presented in the previous chapter and operate in the same manner in multiple regression. Note, however, that in the multiple regression case E must contain the same number of constants, Minitab constants, or columns as there are predictor variables. Once the residuals and fitted values have been stored, we can perform the residual analysis as illustrated in the last chapter. Note that you should also PLOT the residuals against each predictor variable.

Example 14.4

Use Minitab commands to perform multiple regression analysis on the drug data of Example 14.1.

Solution

The Minitab results are presented in Figure 14.20. The Minitab results give the equation $\hat{y} = 46.9 - 1.45x_1 - 1.37x_2$, where x_1 is the amount of drug 1 and x_2 is the amount of drug 2. This equation is the same as that found

```
MTB > READ the following data into columns C1-C3
DATA> 1 2 46
DATA> 1 4 40
DATA> 1 6 38
DATA> 1 8 35
DATA> 2 2 41
DATA> 2 4 36
DATA> 2 6 35
DATA> 2 8 33
DATA> 3 2 39
DATA> 3 4 34
DATA> 3 6 30
DATA> 3 8 33
DATA> 4 2 42
DATA> 4 4 33
DATA> 4 6 34
DATA> 4 7 35
DATA> ENDOF DATA
      16 ROWS READ
MTB > NAME C1 'DRUG1' C2 'DRUG2' C3 'HOURS'
MTB > REGRESS C3 on 2 predictors in C1 C2;
SUBC> RESIDUALS store in C4;
SUBC> FITS store in C5;
SUBC> PREDICT using C1 and C2.

The regression equation is
HOURS = 46.9 - 1.45 DRUG1 - 1.37 DRUG2

Predictor         Coef        Stdev      t-ratio
Constant        46.897        2.154        21.77
DRUG1          -1.4528       0.5657        -2.57
DRUG2          -1.3702       0.2922        -4.69

s = 2.528         R-sq = 68.0%     R-sq(adj) = 63.1%

Analysis of Variance

SOURCE         DF        SS        MS
Regression      2    176.929    88.465
Error          13     83.071     6.390
Total          15    260.000

SOURCE         DF     SEQ SS
DRUG1           1     36.450
DRUG2           1    140.479

     Fit    Stdev.Fit       95% C.I.              95% P.I.
  42.704       1.383   ( 39.716, 45.693)    ( 36.478, 48.931)
  39.964       1.101   ( 37.584, 42.343)    ( 34.006, 45.922)
  37.223       1.093   ( 34.861, 39.586)    ( 31.272, 43.175)
  34.483       1.364   ( 31.535, 37.431)    ( 28.276, 40.690)
  41.251       1.111   ( 38.850, 43.653)    ( 35.284, 47.218)
  38.511       0.749   ( 36.893, 40.129)    ( 32.814, 44.208)
  35.771       0.754   ( 34.141, 37.401)    ( 30.070, 41.471)
  33.030       1.123   ( 30.604, 35.456)    ( 27.053, 39.007)
  39.799       1.094   ( 37.434, 42.163)    ( 33.846, 45.751)
  37.058       0.741   ( 35.458, 38.658)    ( 31.366, 42.750)
  34.318       0.763   ( 32.668, 35.967)    ( 28.612, 40.024)
  31.577       1.140   ( 29.114, 34.041)    ( 25.585, 37.570)
  38.346       1.342   ( 35.447, 41.245)    ( 32.162, 44.530)
  35.605       1.085   ( 33.262, 37.949)    ( 29.661, 41.549)
  32.865       1.112   ( 30.462, 35.268)    ( 26.898, 38.833)
  31.495       1.234   ( 28.829, 34.161)    ( 25.416, 37.573)
```

Figure 14.20
Minitab printout of the multiple regression analysis of the drug data of Example 14.4.

previously. The coefficient table has an entry for each predictor variable in the model, with the estimated coefficient, standard error for each b_j, and t^* for each marginal t test given.

The residual plots do not indicate any problems with the assumptions (Figure 14.21). The plot of the residuals against drug 1 shows a slight curve, so we might include an x_1^2 term in a further analysis. The normal probability plot is shown in Figure 14.22. Although the plot wiggles a bit, it is sufficiently close to a straight line that no problem with the normality assumption is indicated.

Figure 14.21
Minitab printout of the residual plots for the drug data of Example 14.4.

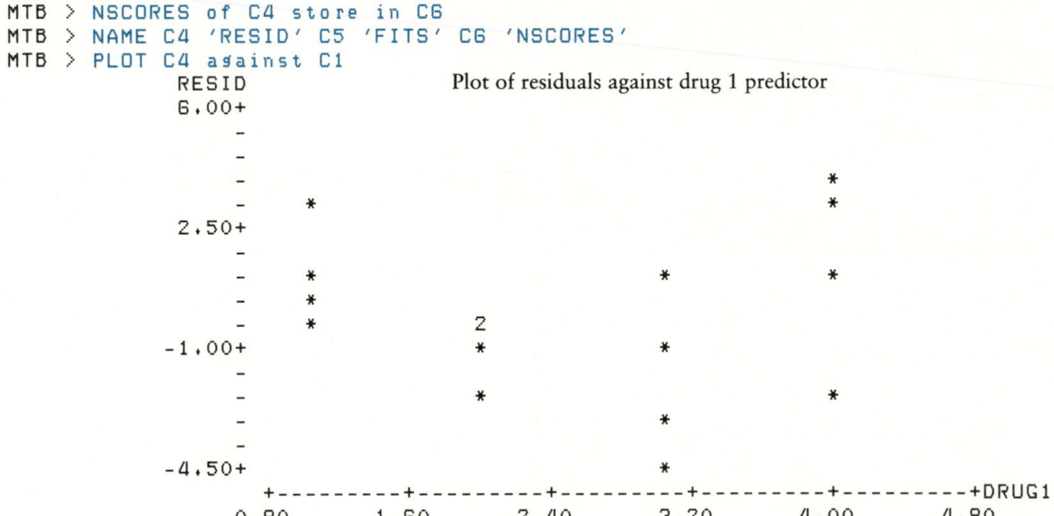

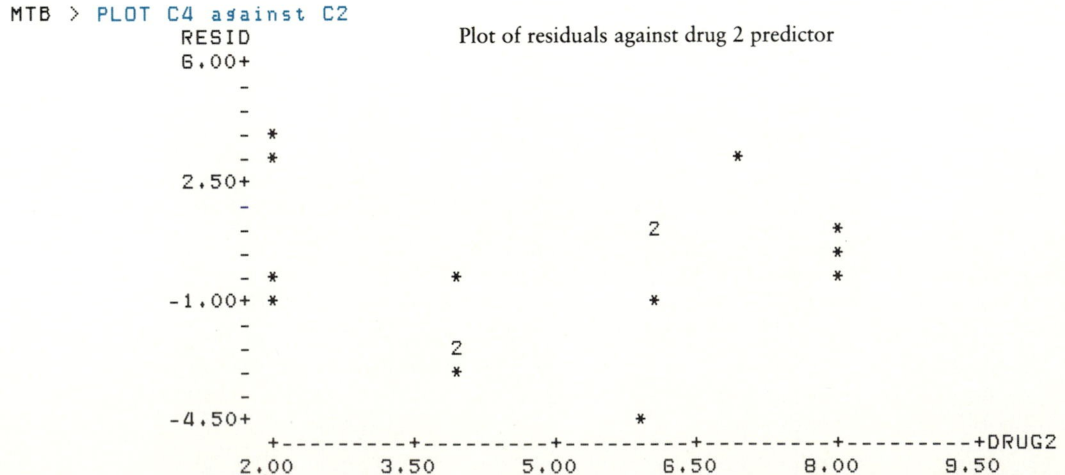

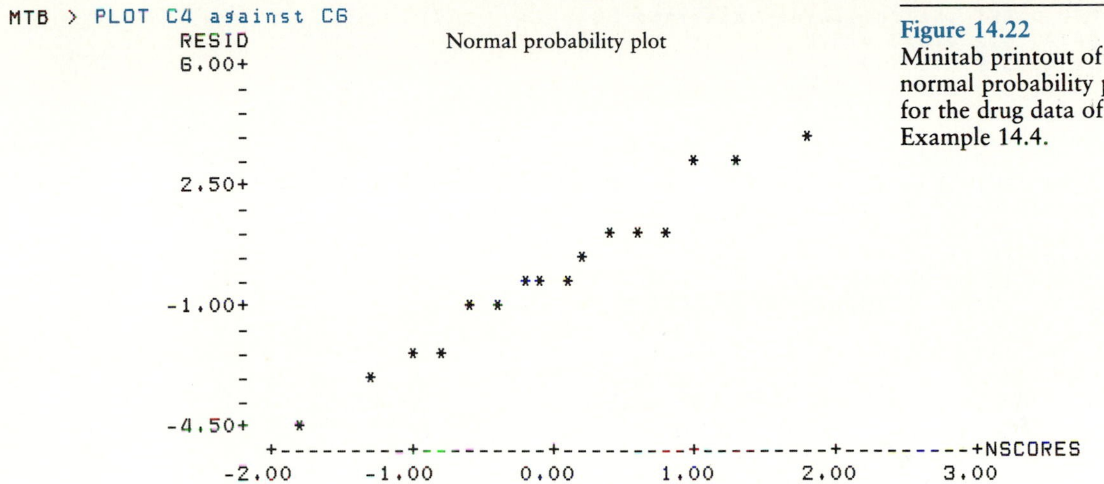

Figure 14.22 Minitab printout of the normal probability plot for the drug data of Example 14.4.

We can also perform quadratic regression with Minitab. The procedure uses the multiple regression analysis commands, creating a new variable in C3 for the x^2 term.

Example 14.5

Use Minitab to perform a quadratic regression analysis on the data of Section 14.3.4.

Solution

The commands and output are shown in Figure 14.23. The printout includes a scatterplot of the data, residual plots, and the normal probability plot. Note that we enter the x and y data into columns C2 and C1 and have Minitab compute and store the quadratic (squared) values of x in C3. (This is done by using the command LET C3 = C2**2.)

The plot of y versus x certainly suggests a nonlinear (perhaps quadratic) relationship between lot size and cost. Compare the regression results to those in Section 14.3.4. The test of the quadratic coefficient, β_2, is highly significant ($t^* = 7.17$; $t_{.025}\{21\} = 2.080$). Thus the inclusion of the quadratic predictor variable is appropriate. The linear coefficient is also significant.

As stated in the example, the MSE of 0.066 is substantially smaller than the MSE for the linear model using only lot size. The inclusion of the quadratic term results in a much better-fitting model. Finally, the residual analysis and normal probability plots do not indicate any severe problem.

The stepwise procedure for finding the best regression equation, outlined in Section 14.2.3, can be done manually by performing the regression with all predictor variables included. Predictors are removed in subsequent runs, based on the results of the previous run, by changing the predictor variable list in the REGRESS command. To find the best regression equation

```
MTB > SET the following data into C1
DATA> 5.2 4.9 5.4 5.2
DATA> 3.8 4.1 4.3 4.0
DATA> 3.5 3.3 3.2 3.4
DATA> 2.7 2.3 3.1 2.8
DATA> 1.9 2.0 2.2 2.1
DATA> 2.6 2.5 2.7 2.6
DATA> SET the following data into C2
DATA>  5   5   5   5
DATA> 10  10  10  10
DATA> 15  15  15  15
DATA> 20  20  20  20
DATA> 25  25  25  25
DATA> 30  30  30  30
DATA> LET C3=C2**2
MTB > NAME C1 'COST' C2 'LOT SIZE' C3 'QUAD'
MTB > PLOT 'COST' against 'LOT SIZE'
```

Plot of the data

```
    COST
   6.00+
       -
       -          *
       -          2
       -          *
   4.50+
       -                  2
       -                  2
       -                          *
       -                          3
   3.00+                                    *
       -                                    2                   3
       -                                    *                   *
       -                                              3
       -                                              *
   1.50+
         +---------+---------+---------+---------+---------+LOT SIZE
       0.00      7.00     14.00     21.00     28.00     35.00
```

```
MTB > REGRESS 'COST' on the 2 predictors in 'LOT SIZE' 'QUAD';
SUBC> RESIDUALS store in C4;
SUBC> FITS store in C5.

THE REGRESSION EQUATION IS
COST = 6.69 - 0.323 LOT SIZE + 0.00605 QUAD

                                 ST. DEV.     T-RATIO =
COLUMN         COEFFICIENT       OF COEF.     COEF/S.D.
                    6.6875         0.2306         29.00
LOT SIZE          -0.32330        0.03018        -10.71
QUAD             0.0060536      0.0008440          7.17

S = 0.2578

R-SQUARED = 94.7 PERCENT
R-SQUARED = 94.2 PERCENT, ADJUSTED FOR D.F.

ANALYSIS OF VARIANCE

  DUE TO        DF          SS     MS=SS/DF
REGRESSION       2      25.149       12.574
RESIDUAL        21       1.396        0.066
TOTAL           23      26.545
```

Figure 14.23
Minitab printout of the quadratic regression analysis of the manufacturing cost data of Example 14.5.

```
FURTHER ANALYSIS OF VARIANCE
SS EXPLAINED BY EACH VARIABLE WHEN ENTERED IN THE ORDER GIVEN
   DUE TO        DF           SS
REGRESSION        2       25.149
LOT SIZE          1       21.729
QUAD              1        3.420

DURBIN-WATSON STATISTIC = 1.81
```

```
MTB > NSCORES of C4 store in C6
MTB > NAME C4 'RESID' C5 'FITS' C6 'NSCORES'
MTB > PLOT 'RESID' against 'FITS'
```

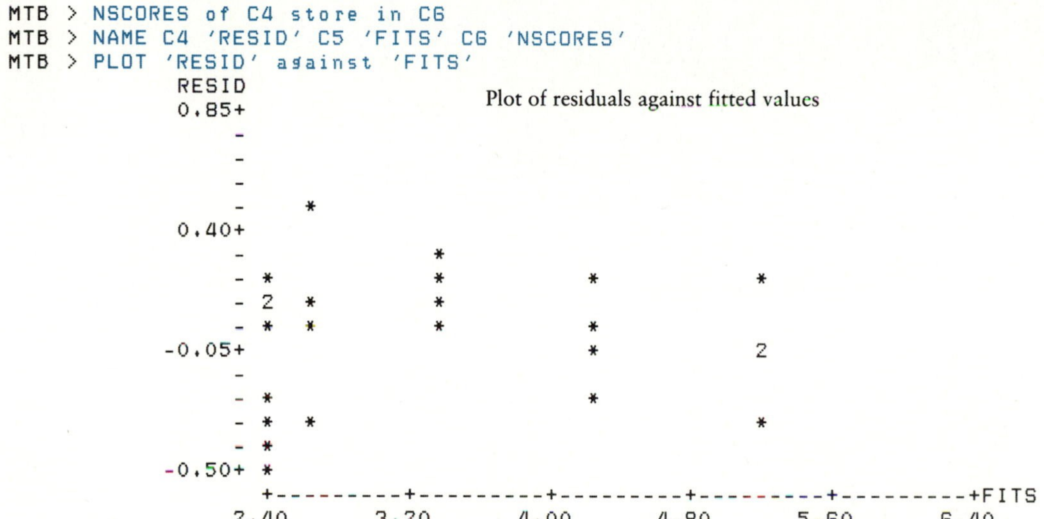

Plot of residuals against fitted values

```
MTB > PLOT 'RESID' against 'LOT SIZE'
```

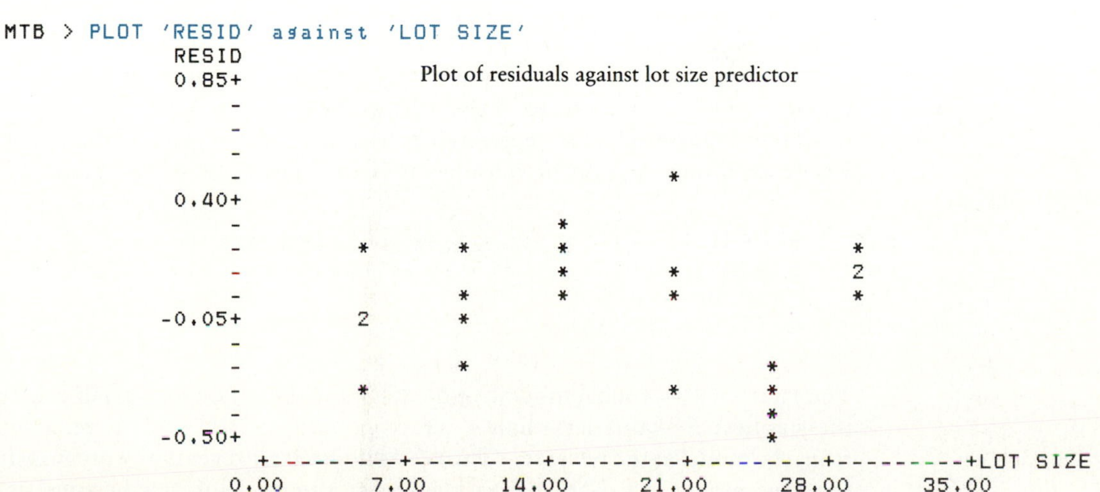

Plot of residuals against lot size predictor

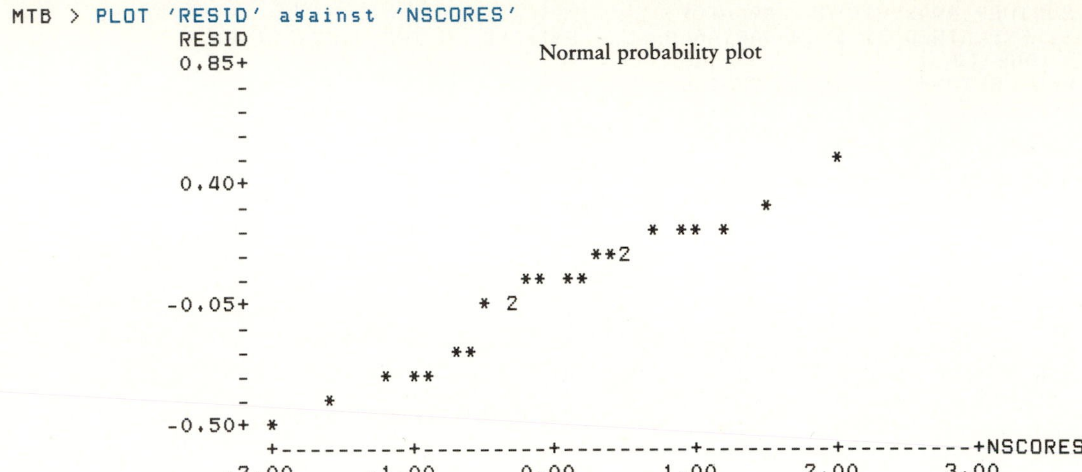

using the y values in C8 and the predictor variables in C1 through C7, for example, the first command is

```
MTB > REGRESS the y's in C8 on the 7 predictors in C1-C7
```

Suppose that the predictor in C5 yields a t^* of 0.02 and is the smallest $|t^*|$ value. We would remove C5 and the next regression command would be

```
MTB > REGRESS the y's in C8 on the 6 predictors in C1 C2 C3 C4
C6 C7
```

We would continue until none of the predictors in our model was nonsignificant (using the chosen level of significance). (Using $|t^*| < 2$ for nonsignificance provides a reasonable procedure.)

You can also have Minitab eliminate the variables automatically. The STEPWISE command can be used to perform the necessary steps of the backward elimination stepwise regression procedure, removing variables one at a time until only significant variables remain. The STEPWISE command is

```
MTB > STEPWISE regression, y in C, predictors in C ... C;
SUBC> ENTER, initially, all predictors in C ... C;
SUBC> FREMOVE = 4;
SUBC> FENTER = 10000.
```

The FREMOVE = 4 subcommand indicates that the predictor variable with the smallest F^* value less than 4 (corresponding to $|t^*| < 2$) is removed from the model at the next step. The FENTER = 10000 prevents any previously removed predictor variables from reentering the model if they become significant. The resulting stepwise analysis uses a level of significance of approximately 0.05 at each step. See Neter and others (1985, pp. 430–435) for a discussion of the stepwise procedure.

Example 14.6

Refer to the real estate data of Problem 14.8. Use the STEPWISE procedure to find the best regression model.

Solution

The Minitab commands and output are shown in Figure 14.24. The output is read as follows. The different models or runs that were done by Minitab are indicated by reading down a column under the appropriate step. For example, at STEP 1 we read down the first column and note that the regression model is

$$\text{Predicted PRICE} = -65.758 + 0.6(\text{SQFT}) + 0.55(\text{ELEV}) + 1.37(\text{SLOPE}) + 4.0(\text{VIEW})$$

That is, $\hat{y} = -65.758 + 0.6x_1 + 0.55x_2 + 1.37x_3 + 4.0x_4$. At this first step we note that the SQFT predictor has $t^* = 0.61$, which is smaller than

```
MTB > NAME C1 'SQFT' C2 'ELEV' C3 'SLOPE' C4 'VIEW' C5 'PRICE'
MTB > READ the following data into C1-C5
DATA> 24 132  6.8 7 59.9
DATA> 18 118  7.9 4 42.5
DATA> 21 141  6.6 6 55.0
DATA> 22 125 11.6 8 66.5
DATA> 28 108 12.7 4 44.9
DATA> 19 128  3.2 7 41.7
DATA> 22 143  1.9 9 72.5
DATA> 25 132  9.6 9 70.0
DATA> ENDOFDATA
      8 ROWS READ
MTB > STEPWISE regression, y in C5, predictors in C1-C4;
SUBC> ENTER initially all predictors C1-C4;
SUBC> FREMOVE=4;
SUBC> FENTER=100000.
STEPWISE REGRESSION OF PRICE ON 4 PREDICTORS, WITH N = 8

     STEP  b0 -->     1          2         3         4
 CONSTANT <--   -65.758    -54.718     9.664    20.408
                               Intercept
SQFT     b1 ------> 0.6
T-RATIO            0.61 <-- t* for SQFT at step 1

ELEV     b2 ------> 0.55      0.54
T-RATIO            1.29       1.37

SLOPE    b3 ------> 1.37      1.72      0.89
T-RATIO            1.20       1.87      1.19

VIEW     b4 ------> 4.0       4.3       6.0       5.4
T-RATIO            1.99       2.43      4.20      3.92

S                  6.98       6.41      6.95      7.19
R-SQ              86.73      85.06     78.05     71.87
```

Figure 14.24 Minitab printout for the stepwise regression analysis of the real estate data of Example 14.6.

the other predictor variable t^* values and has an absolute value less than 2. Thus Minitab removed SQFT from the model. The resulting analysis is summarized in the column labeled STEP 2. The process is repeated, and at the fourth step we note that the only remaining predictor, VIEW, is still significant ($|t^*| \geq 2$). The final Minitab STEPWISE model is

$$\text{Predicted PRICE} = 20.408 + 5.4(\text{VIEW})$$

That is, $\hat{y} = 20.408 + 5.4x_4$.

In terms of fit, this model fits very well when compared to the other models that were considered. In terms of lowest MSE, the best-fitting model is the one given at step 2; S, at the bottom, is $\sqrt{\text{MSE}}$. Thus the model using elevation, slope, and view has the lowest MSE. The model to choose depends on the criteria we are using. We might also investigate the possibility of multicollinearity and interaction between the variables. It is evident that finding the best predictors among several candidates is not a simple task.

Problems

Applications

14.24 Population data (in millions) for the United States for the past seven census reports are given here:

Year	1920	1930	1940	1950	1960	1970	1980
US pop.	106.02	123.20	132.16	151.33	179.32	203.21	226.50

a. Plot the population data against the year. Use $x = 1920, 1930, 1940$, and so on.

b. Fit a linear model to this set of data. How well do the data fit a linear model?

c. Now fit a quadratic model to the data. Do the data fit better than in part (*b*)? How can you tell?

d. Repeat parts (*a*) to (*c*) using $x = 0, 10, 20$, and so on. What, if anything, changes? Why? Which set of x values is the easiest to use?

e. Noting that this is a time series and that prediction is subject to different sources of error than with the usual regression analysis, use the best predictor to forecast US population in 1990, 2000, and 2100.

14.25 We have discussed how we can use regression techniques to analyze the relationship between continuous numerical variables. In some situations, however, we would like to include categorical (nonnumerical) variables in the analysis. In the USDA survey data set (Appendix B.2) we might be interested in knowing whether the sex of the head of household affects monthly income. We may assign numerical values to a categorical variable; this variable is then called an **indicator variable** or **dummy variable** because the coding or assignment of numbers to values is not important. For convenience, indicator variables are usually coded 0 and 1. Dummy variables may have only two values.

Suppose we want to use sex as a predictor variable in a regression model in which the response variable is monthly income and the other predictor variable is years of education of the head of household.

a. Write out the model with sex included.

b. Obtain the multiple regression equation using both education and sex predictors. Is sex of the head of household important in income of elderly households?

14.5 Case 14: How Much to Send a Letter?—Solution

We used several approaches to obtain the best-fitting equation from among those tried. We hasten to add that this analysis is a time series and subject to different sources of error. As mentioned in Problem 14.22, prediction intervals are not appropriate, but extrapolation to future time periods is possible, so long as we note the possibility of error. The article on our findings was submitted to *Mathematics and Computer Education* since we thought it had a pedagogical slant in addition to illustrating a valuable regression technique. The article was accepted for publication and entitled "The Cost of a Postage Stamp, or Up, Up, and Away!"—the cover of the issue in which it was published consisted of a blowup of the US postage stamp featuring hot-air balloons. The results outlined in the article are summarized here.

We used an exponential model—a model including the term e^x, a version of a compound interest formula—but found that the cost was rising faster than an exponential model would explain. We finally found two equations that had nearly the same MSE. One was the equation

$$\hat{y} = -17.01505 - 135.2867(x/10) + 6.73303(x/10)^2 + 329.5963 \ln(x/10)$$

where x is the number of years after 1900 and "ln" is the natural logarithm function. We used $x/10$ instead of x in the solution to reduce the size of the numbers when raised to powers as high as the sixth. The second equation was

$$\hat{y} = 3.2981 - 0.00091902(x/10)^5 + 0.00016674(x/10)^6$$

The first equation had a slightly smaller MSE than the second, but outside the interval 32 to 82 the second equation seemed to be a better predictor. (The first equation predicted negative values for years prior to 1930, for example.) Thus we settled on the second equation for prediction.

Since we did not try powers higher than the sixth, trigonometric functions, or other possibilities such as a constant times a logarithm, we cannot guarantee that this is the best-fitting equation, but it was the best-fitting equation of those we tried. Using this equation, we predicted that the postal rate would be 25¢ in October 1984 and 30¢ in February 1987.

The predicted increase to 25¢ did not take place, although the increase to 22¢ on February 17, 1985 was not far from the predicted date. The sudden abatement of inflation probably had a great deal to do with that outcome. The interested reader is urged to repeat the study, taking into account the increase to 22¢ in 1985.

14.6 Summary

If a response variable depends on two or more predictor variables, we must use multiple regression techniques. The analysis is generally carried out by means of computer programs.

After we obtain the regression equation, we often begin an analysis of the regression function by partitioning the total sum of squares of the response variable into sums of squares due to regression (SSR) and to error (SSE). Again, this procedure is usually part of the computer output.

We use an F test to determine whether or not all the β_j are equal to zero. Confidence intervals and hypothesis tests concerning the β_j require values of s_{b_j} most often taken from computer printouts. We find confidence intervals for $E(y_p)$ and prediction intervals for y_p by computer techniques.

The "best" of several models is often defined to be the one with the smallest MSE. We can use a systematic approach or a computer approach utilizing the stepwise procedure to determine this best-fitting model.

We perform quadratic regression by using the multiple regression approach, substituting x for x_1 and x^2 for x_2.

Chapter Review Problems

Practice

14.26 The following analysis of variance table is partially filled in. There were four predictor variables. Complete the table.

SOURCE OF VARIATION	DF	SS	MS
Regression			312.6654
Error			
Total	33	1406.1215	

a. Determine R^2. Does your result indicate that the model is a good fit?
b. Conduct an F test for the model. Use the 0.05 level of significance. What do you conclude?

14.27 An analysis of variance table is shown here. Test the hypothesis that there is no association between the response variable and any of the predictor variables. Use the 0.05 level of significance. Does the value of R^2 support your results?

SOURCE OF VARIATION	DF	SS	MS
Regression	2	23.8876	11.9438
Error	29	10.6654	0.3678
Total	31	34.5530	

14.28 Using the data given here, conduct a multiple linear regression analysis.

	OBSERVATION								
	1	2	3	4	5	6	7	8	9
x_1	2.4	5.3	3.8	3.7	4.8	5.1	2.8	1.9	5.6
x_2	0.16	0.14	0.32	0.16	0.33	0.24	0.45	0.32	0.27
y	66.7	76.3	70.4	56.8	88.7	83.8	74.3	67.4	87.9

a. Obtain an ANOVA table, and test the hypothesis that there is no association between the variables. Use $\alpha = 0.05$.

b. Using the computer, obtain standard errors for b_1 and b_2 and test the hypothesis H_0: $\beta_1 = 0$ and H_0: $\beta_2 = 0$. Use the 0.05 level of significance.

c. Obtain 95% confidence intervals for β_0, β_1, and β_2.

d. Obtain a 95% confidence interval for $E(y_p)$ when $x_1 = 2.5$ and $x_2 = 0.20$.

e. Obtain a 95% prediction interval for y_p when $x_1 = 2.5$ and $x_2 = 0.20$.

14.29 A multiple regression equation was obtained as follows:

$$\hat{y} = 21.4 + 0.32x_1 - 0.044x_2 + 113.2x_3 + 81.44x_4 - 13.1x_5$$

The intercorrelation matrix is given below. The correlation between pairs of variables is indicated by the intersection of the row and column so numbered; y is variable 0. For example, the correlation between variables x_2 and x_4 is denoted r_{24} and $r_{24} = -0.1320$.

VARIABLE	0	1	2	3	4	5
0	1.0000	0.4123	−0.7143	0.6731	0.0234	−0.5628
1		1.0000	−0.0143	0.8230	0.0451	−0.0387
2			1.0000	−0.3013	−0.1320	0.1349
3				1.0000	0.2041	−0.3011
4					1.0000	−0.0043
5						1.0000

a. Which variable seems to have the greatest association with y?

b. Which variable could most likely be eliminated simply by dropping it from consideration?

c. Two variables appear to contain redundant information (that is, they are not both needed). Which two are they, and which one should probably be dropped?

Applications

You should use a computer for these problems.

14.30 The average price of used cars sold by new car dealers is given here for the past 6 years. (Source: *USA Today,* December 9, 1985.) Fit a quadratic regression line to the time series and forecast the average price of these cars in 1990.

Year	1980	1981	1982	1983	1984	1985
Price ($)	3380	3850	4300	4610	4950	5200

14.31 Refer to the Miami Female Study data (Appendix B.1). Determine the best-fitting multiple linear regression equation for predicting grade-point average (variable 19) from the numerical variables (6, 7, 10, 13, 17, and 18).

14.32 Refer to the USDA survey data (Appendix B.2). Determine the best-fitting multiple linear regression equation for predicting monthly income from the numerical variables (AGEYRS, NUMINHSE, and EDUC).

14.33 (Indicator variables are needed here; refer to Problem 14.25.) Refer to the Miami Female Study data. Repeat Problem 14.31, adding the categorical variables with two classifications (3, 8, 12, 14, 15, and 16). What are your conclusions?

14.34 (Indicator variables are needed here.) Repeat Problem 14.32, adding HHSEH as an indicator variable. What do you conclude?

Index of Terms

Glossary of Symbols

R^2 coefficient of multiple determination

Answers to Proficiency Checks

1. **a.** The ANOVA table is shown here.

SOURCE OF VARIATION	DF	SS	MS
Regression	2	172.0532	86.0266
Error	4	1.6611	0.4153
Total	6	173.7143	

To test the hypothesis $H_0: \beta_1 = \beta_2 = 0$, we use $F^* = 86.0266/0.4153 \doteq 207.14$. Since $F_{.01}\{2, 4\} = 18.0$, we conclude that there is an association between at least one of the predictor variables and the response variable.

b. Using a hand calculator, we find the normal equations to be

$$67 = 7b_0 + 23b_1 + 40b_2$$
$$290 = 23b_0 + 105b_1 + 173b_2$$
$$488 = 40b_0 + 173b_1 + 294b_2$$

Solving these equations simultaneously, we obtain

$$b_0 \doteq 0.737871108 \qquad b_1 \doteq 1.012309920 \qquad b_2 \doteq 0.963794352$$

Sums of squares can be found as follows:

$$\text{SSTO} = 815 - \frac{(67)^2}{7} = 173.7143$$

$$\text{SSR} \doteq b_0(67) + b_1(290) + b_2(488) - \frac{(67)^2}{7}$$
$$\doteq 172.0532$$

so

$$\text{SSE} \doteq 173.7143 - 172.0532 = 1.6611$$

Since $n = 7$ and $k = 3$, the analysis of variance table is as shown above. The ANOVA table produced by SAS is shown here:

```
                                     SAS

DEP VARIABLE: Y
                               ANALYSIS OF VARIANCE

                               SUM OF              MEAN
          SOURCE      DF      SQUARES            SQUARE      F VALUE      PROB>F

          MODEL        2    172.05317       86.02658529      207.154      0.0001
          ERROR        4   1.66111513        0.41527878
          C TOTAL      6    173.71429

                ROOT MSE    0.6444213          R-SQUARE       0.9904
                DEP MEAN     9.571429          ADJ R-SQ       0.9857
                C.V.          6.73276

                               PARAMETER ESTIMATES

                           PARAMETER        STANDARD      T FOR H0:
      VARIABLE      DF      ESTIMATE           ERROR    PARAMETER=0      PROB > |T|

      INTERCEP       1    0.73787111      0.53194636          1.387          0.2377
      X1             1    1.01230992      0.37111266          2.728          0.0526
      X2             1    0.96379435      0.24888954          3.872          0.0180
```

The P value for the obtained value of F is equal to 0.0001; thus we can reject the null hypothesis at a 0.0001 level of significance.

2. The residual plot of residuals against predicted values (first part of Figure 14.9) shows a somewhat random scatter. There are too few points to draw any conclusions, but the dispersion appears to be about the same everywhere. The plot of the residuals against x_1 also appears random; the relationship is linear. The plot of the residuals against x_2 is not so clear. If it were not for the point (3, −0.6538) in the lower left portion of the graph, it would appear to be curvilinear. More observations are needed.

3. a. In each case the critical value is $t_{.025}\{17\} = 2.110$; we reject H_0 if $|t^*| > 2.11$. Since $b_1/s_{b_1} \doteq 1.40$, we cannot reject H_0: $\beta_1 = 0$; $\beta_2/s_{b_2} \doteq -5.87$ and $5.87 > 2.11$, so we reject H_0: $\beta_2 = 0$ and conclude that there is a negative linear association between x_2 and y.
   b. x_2 has the most effect on y since $|t^*|$ is larger for the test on that variable than on the test for x_1.
   c. No; we should eliminate variable x_1 and obtain a linear regression equation using only x_2 as a predictor. If the MSE is smaller for this equation, it will be better than the one using both predictor variables.

4. Since $2.11(31.881) \doteq 67.269$, a 95% confidence interval for β_1 is

   $$-22.494 \quad \text{to} \quad 112.044$$

   Similarly $2.11(0.00589) \doteq 0.01243$, so a 95% confidence interval for β_2 is

   $$-0.04699 \quad \text{to} \quad -0.02213$$

   Since the confidence interval for β_1 includes zero, we cannot reject H_0: $\beta_1 = 0$, but since the confidence interval for β_2 does not include zero, we may conclude that $\beta_2 < 0$. These results confirm the conclusions in Proficiency Check 3.

5. The computer printout of the confidence and prediction intervals for the data of Proficiency Check 1 are shown here. The observation for which $x_1 = 5$ and $x_2 = 11$ was entered as 5 11. and is shown as observation 8. (Recall that the dot is used to show a missing observation.)

OBS	ACTUAL	PREDICT VALUE	STD ERR PREDICT	LOWER95% MEAN	UPPER95% MEAN	LOWER95% PREDICT	UPPER95% PREDICT	RESIDUAL
1	4.0000	3.6778	0.3842	2.6109	4.7446	1.5947	5.7609	0.3222
2	6.0000	5.6054	0.5245	4.1490	7.0617	3.2984	7.9123	0.3946
3	5.0000	5.6539	0.3642	4.6428	6.6649	3.5988	7.7090	-0.6539
4	8.0000	7.6300	0.4085	6.4957	8.7642	5.5116	9.7484	0.3700
5	12.0000	12.4975	0.4093	11.3612	13.6337	10.3780	14.6170	-0.4975
6	13.0000	13.5098	0.3193	12.6233	14.3962	11.5130	15.5065	-0.5098
7	19.0000	18.4258	0.5041	17.0262	19.8253	16.1542	20.6973	0.5742
8	.	16.4012	0.7803	14.2346	18.5677	13.5913	19.2110	.

   a. The 95% confidence interval for $E(y_p)$ is 14.2346 to 18.5677.
   b. The 95% prediction interval for y_p is 13.5913 to 19.2110.

6. The computer printout is shown here:

```
                                      SAS

DEP VARIABLE: Y
                              ANALYSIS OF VARIANCE

                              SUM OF           MEAN
        SOURCE       DF      SQUARES         SQUARE      F VALUE     PROB>F

        MODEL         2    253.54864      126.77432       99.636     0.0001
        ERROR         9  11.45135566     1.27237285
        C TOTAL      11    265.00000

            ROOT MSE        1.127995       R-SQUARE       0.9568
            DEP MEAN            20.5       ADJ R-SQ       0.9472
            C.V.            5.502415

                              PARAMETER ESTIMATES

                          PARAMETER        STANDARD      T FOR H0:
    VARIABLE      DF       ESTIMATE           ERROR    PARAMETER=0      PROB > |T|

    INTERCEP       1   -14.29186603      4.36888026         -3.271          0.0097
    X              1    11.89633174      1.88017166          6.327          0.0001
    XSQ            1    -0.89633174      0.18978969         -4.723          0.0011
```

a. We see that the quadratic regression equation is correct to four decimal places.
b. To test the curvature effect, we must test H_0: $\beta_2 = 0$. The value of t^* for this test is -4.723. Since $4.723 > t_{.025}\{9\} = 2.262$, we conclude that $\beta_2 < 0$ and there is a significant curvature effect. Moreover, the P value is 0.0011, which is less than 0.05.
c. To test the linear effect, we must test H_0: $\beta_1 = 0$. The value of t^* for this test is 6.327. Since $6.327 > 2.262$, we conclude that $\beta_1 > 0$ and there is a significant linear effect. Moreover, the P value is 0.0001, which is less than 0.05.

7. The printout for the linear regression analysis is shown here:

```
                                      SAS

DEP VARIABLE: Y
                              ANALYSIS OF VARIANCE

                              SUM OF           MEAN
        SOURCE       DF      SQUARES         SQUARE      F VALUE     PROB>F

        MODEL         1    225.16901      225.16901       56.531     0.0001
        ERROR        10  39.83098592     3.98309859
        C TOTAL      11    265.00000

            ROOT MSE         1.99577       R-SQUARE       0.8497
            DEP MEAN            20.5       ADJ R-SQ       0.8347
            C.V.            9.735464
```

PARAMETER ESTIMATES

VARIABLE	DF	PARAMETER ESTIMATE	STANDARD ERROR	T FOR H0: PARAMETER=0	PROB > ITI
INTERCEP	1	5.59154930	2.06484830	2.708	0.0220
X	1	3.08450704	0.41024376	7.519	0.0001

The linear regression equation is $\hat{y} = 5.5915 + 3.0845x$. The MSE is 3.9831, so we conclude that the multiple regression equation is a better fit.

8. The printout for the confidence and prediction intervals is shown here; $x = 6$ in observations 6 and 12.

OBS	ACTUAL	PREDICT VALUE	STD ERR PREDICT	LOWER95% MEAN	UPPER95% MEAN	LOWER95% PREDICT	UPPER95% PREDICT	RESIDUAL
1	14.0000	13.3301	0.6242	11.9181	14.7422	10.4138	16.2465	0.6699
2	12.0000	13.3301	0.6242	11.9181	14.7422	10.4138	16.2465	-1.3301
3	23.0000	22.7815	0.4976	21.6559	23.9071	19.9925	25.5704	0.2185
4	21.0000	18.9522	0.4362	17.9655	19.9389	16.2163	21.6880	2.0478
5	22.0000	22.7815	0.4976	21.6559	23.9071	19.9925	25.5704	-0.7815
6	25.0000	24.8182	0.4499	23.8004	25.8360	22.0710	27.5654	0.1818
7	25.0000	25.0622	0.7484	23.3692	26.7552	21.9999	28.1245	-0.0622
8	23.0000	22.7815	0.4976	21.6559	23.9071	19.9925	25.5704	0.2185
9	19.0000	18.9522	0.4362	17.9655	19.9389	16.2163	21.6880	0.0478
10	26.0000	25.0622	0.7484	23.3692	26.7552	21.9999	28.1245	0.9378
11	13.0000	13.3301	0.6242	11.9181	14.7422	10.4138	16.2465	-0.3301
12	23.0000	24.8182	0.4499	23.8004	25.8360	22.0710	27.5654	-1.8182

a. 23.8004 to 25.8360
b. 22.0710 to 27.5654

References

Byrkit, Donald R., and Robert E. Lee. "The Cost of a Postage Stamp, or Up, Up, and Away!" *Mathematics and Computer Education* 6 (1983): 15–24.

Draper, Norman R., and Harry Smith. *Applied Regression Analysis.* 2nd ed. New York: John Wiley & Sons, 1981.

Fraser, D. A. S. *The Structure of Inference.* New York: John Wiley & Sons, 1968.

Fraser, D. A. S. *Inference and Linear Models.* New York: McGraw-Hill, 1979.

Mendenhall, William. *Introduction to Linear Models and the Design and Analysis of Experiments.* Belmont, CA: Wadsworth Publishing Company, 1968.

Mendenhall, William, and Terry Sincich. *A Second Course in Business Statistics: Regression Analysis.* 2nd ed. San Francisco: Dellen, 1986.

Neslin, Scott A., Caroline Henderson, and John Quelch. "Consumer Promotions and the Acceleration of Product Purchases." *Marketing Science* 4 (1985): 147–165.

Neter, John, William Wasserman, and Michael H. Kutner. *Applied Linear Statistical Models.* 2nd ed. Homewood, IL: Richard D. Irwin, 1985.

SAS® User's Guide: Statistics, Version 5 Edition. Cary, NC: SAS Institute, 1985.

Appendix A.1

List of Minitab Commands

In this appendix we provide a summary of the available commands in the 1985 version of Minitab. We thank Minitab, Inc., for letting us reproduce this list from the *Minitab Handbook,* second edition, by B. F. Ryan, B. L. Joiner, and T. A. Ryan, Jr.

Notation

K denotes a constant, which can be either a number, such as 8.3, or a stored constant, such as K14.

C denotes a column, which must be typed with a C directly in front, such as C12. Columns may be named.

E denotes either a constant or a column.

M denotes a matrix, which must be typed with an M directly in front, such as M5.

[ ] denotes an optional argument.

Subcommands are shown indented under the main command.

1. General Information

HELP explains Minitab commands
INFORMATION gives status of worksheet
STOP ends the current session

2. Entering Numbers

READ data [from 'FILENAME'] into C. . .C
SET data [from 'FILENAME'] into C
INSERT data [from 'FILENAME'] between rows K and K of C. . .C
INSERT data [from 'FILENAME'] at the end of C. . .C

READ, SET, and INSERT all have subcommands

  FORMAT (FORTRAN format)
  NOBS = K

END of data
NAME for C is 'NAME', for C is 'NAME' . . . for C is 'NAME'
RETRIEVE the worksheet saved [in 'FILENAME']

3. Outputting Numbers

PRINT the data in C. . .C
PRINT the data in K. . .K

WRITE [to 'FILENAME'] the data in C. . .C

PRINT and WRITE have the subcommand

FORMAT (FORTRAN format)

SAVE [in 'FILENAME'] a copy of the worksheet

4. Editing and Manipulating Data

DELETE rows K. . .K of C. . .C
INSERT (see Section 2)
COPY C. . .C into C. . .C
  USE rows K . . .K
  USE rows where C = K. . .K
  OMIT rows K. . .K
  OMIT rows where C = K. . .K
COPY C into K. . .K
COPY K. . .K into C
CODE (K. . .K) to K. . . (K. . .K) to K for C. . .C store in C. . .C
STACK (E. . .E) on . . . on (E. . .E) store in (C. . .C)
  SUBSCRIPTS into C
UNSTACK (C. . .C) into (E. . .E) . . . (E. . .E)
  SUBSCRIPTS are in C
CONVERT, using table in C,C, the data in C, put in C

You can use LET (see next section) to correct a number in the worksheet.

Examples. LET C2(7) = 12.8
LET C3(5) = '*'

5. Arithmetic

ADD	E to E . . .	to E, put into E
SUBTRACT	E from	E, put into E
MULTIPLY	E by E . . .	by E, put into E
DIVIDE	E by	E, put into E
RAISE	E to the power	E, put into E
ABSOLUTE	value	of E, put into E
SQRT		of E, put into E
LOGTEN		of E, put into E
LOGE		of E, put into E
ANTILOG		of E, put into E
EXPONENTIATE		E, put into E
ROUND	to integer	E, put into E
SIN		of E, put into E
COS		of E, put into E
TAN		of E, put into E
ASIN		of E, put into E
ACOS		of E, put into E
ATAN		of E, put into E
SIGNS		of E, put into E
NSCORE	normal scores	of E, put into E
PARSUMS		of E, put into E
PARPRODUCTS		of E, put into E

LET = expression

Expressions may use the arithmetic operators + − * / and ** (exponentiation) and any of the following: ABSOLUTE, SQRT, LOGTEN, LOGE, ANTILOG, EXPO, ROUND, SIN, COS, TAN, ASIN, ACOS, ATAN, SIGNS, NSCORE, PARSUMS, PARPRODUCTS, COUNT, N, NMISS, SUM, MEAN, STDEV, MEDIAN, MIN, MAX, SSQ, SORT, RANK, LAG.

You can use subscripts to access individual numbers.

Examples. LET C2 = SQRT(C1 − MIN(C1))
LET C3(5) = 4.5

6. Column and Row Operations

COUNT	# of values	in C [put into K]
N	(# of nonmissing values)	in C [put into K]
NMISS	(# of missing values)	in C [put into K]
SUM	of the values	in C [put into K]
MEAN	of the values	in C [put into K]
STDEV	of the values	in C [put into K]
MEDIAN	of the values	in C [put into K]
MINIMUM	of the values	in C [put into K]
MAXIMUM	of the values	in C [put into K]
SSQ	(uncorrected sum of sq.)	for C [put into K]

The following are all done rowwise:

RCOUNT of C. . .C put into C
RN of C. . .C put into C

RNMISS of C. . .C put into C
RSUM of C. . .C put into C
RMEAN of C. . .C put into C
RSTDEV of C. . .C put into C
RMEDIAN of C. . .C put into C
RMINIMUM of C. . .C put into C
RMAXIMUM of C. . .C put into C
RSSQ of C. . .C put into C

7. Plots and Histograms

HISTOGRAM C. . .C
DOTPLOT C. . .C

HISTOGRAM and DOTPLOT both have the subcommands

INCREMENT = K
START at K [end at K]
BY C
SAME scales for all columns

PLOT C versus C
MPLOT C versus C. . .C versus C
LPLOT C versus C, using plotting symbols given by C
TPLOT C versus C versus C (three-dimensional plot)

PLOT, MPLOT, LPLOT, and TPLOT have the subcommands

YINCREMENT = K
YSTART at K [end at K]
XINCREMENT = K
XSTART at K [end at K]

TSPLOT [period = K] data in C
INCREMENT = K
START at K [end at K]
ORIGIN = K
TSTART at K [end at K]

WIDTH of all plots that follow is K spaces
HEIGHT of all plots that follow is K lines

8. Basic Statistics

DESCRIBE C. . .C
BY C

ZINTERVAL [with K% confidence] sigma = K for C. . .C
ZTEST [of mu = K] sigma = K on data in C. . .C
ALTERNATIVE = K
TINTERVAL [with K% confidence] for C. . .C
TTEST [of mu = K] on data in C. . .C
ALTERNATIVE = K
TWOSAMPLE test and c.i. [K% confidence] on C,C
ALTERNATIVE = K
POOLED procedure
TWOT test and c.i. [K% confidence] data in C, groups in C
ALTERNATIVE = K
POOLED procedure
CORRELATION between C. . .C [put into M]
or
CORRELATE C with C [put into M]
COVARIANCE for C. . .C [put into M]
CENTER the data in C. . .C put into C. . .C
LOCATION [subtracting K. . .K]
SCALE [dividing by K. . .K]
MINMAX [with K as min and K as max]

9. Regression

REGRESS C on K predictors C. . .C [store standardized residuals in C [fits in C]]
NOCONSTANT in equation
WEIGHTS are in C
MSE put into K
COEFFICIENTS put into C
XPXINV put into M
HI put into C (leverage)
RESIDUALS put into C (observed—fit)
TRESID put into C (Studentized, or deleted standardized residuals)
COOKD put into C (Cook's distance)
DFITS put into C
VIF (variance inflation factors)
DW (Durbin–Watson statistic)
PURE (pure error lack-of-fit test)
XLOF (experimental lack-of-fit test)
STEPWISE regression of C on predictors C. . .C
FENTER = K (default is 4)

FREMOVE = K (default is 4)
FORCE C. . .C
ENTER C. . .C
REMOVE C. . .C
BEST K alternative predictors (default is 0)
STEPS = K (default depends on output width)

NOCONSTANT in all STEPWISE and REGRESS that follow
CONSTANT return to fitting a constant in STEPWISE and REGRESS
BRIEF output [using print code = K] from REGRESS and ARIMA
NOBRIEF return to default amount of output

10. Analysis of Variance

AOVONEWAY analysis of variance for samples in C. . .C
ONEWAY data in C, subscripts in C [store residuals in C [fits in C]]
TWOWAY data in C, subscripts in C,C [store residuals in C [fits in C]]
ADDITIVE model
INDICATOR variables for subscripts in C, put into C. . .C

11. Nonparametrics

RUNS test above and below K for data in C. . .C
STEST sign test [median = K] data in C. . .C
ALTERNATIVE = K
SINT sign c.i. [K% confidence] data in C. . .C
WTEST Wilcoxon one-sample rank test [center = K] data in C. . .C
ALTERNATIVE = K
WINT Wilcoxon c.i. [K% confidence] data in C. . .C
MANN-WHITNEY test and c.i. [alternative = K] [K% confidence] first sample in C, second sample in C
KRUSKAL-WALLIS test data in C, subscripts in C

12. Tables

TALLY the data in C. . .C
COUNTS
PERCENTS
CUMCNTS cumulative counts
CUMPCTS cumulative percents
ALL four statistics above
CHISQUARE test on table stored in C. . .C
TABLE the data classified by C. . .C
MEANS for C. . .C
MEDIANS for C. . .C
SUMS for C. . .C
MINIMUMS for C. . .C
MAXIMUMS for C. . .C
STDEV for C. . .C
STATS for C. . .C
DATA for C. . .C
N for C. . .C
NMISS for C. . .C
PROPORTION of cases = K [thru K] in C. . .C
COUNTS
ROWPERCENTS
COLPERCENTS
TOTPERCENTS
CHISQUARE analysis [output code = K]
MISSING level [for each classification variable C. . .C]
NOALL in margins
ALL for C. . .C
FREQUENCIES are in C
LAYOUT K by K

13. Time Series

ACF [with up to K lags] for series in C [put into C]
PACF [with up to K lags] for series in C [put into C]
CCF [with up to K lags] between series in C and C
DIFFERENCES [of lag K] for data in C, put into C
LAG [by K] data in C, put into C
ARIMA p=K, d=K, q=K, P=K, D=K, Q=K,

S = K, data in C [put residuals in C [put predicteds in C [put coefficients in C]]]
CONSTANT term in model
NOCONSTANT term in model
STARTING values are in C
FORECAST [forecast origin = K] up to K leads ahead [store forecasts in C [confidence limits in C,C]]
BRIEF output [using print code = K] for ARIMA and REGRESS

See TSPLOT in Section 7 above.

14. Exploratory Data Analysis

STEM-AND-LEAF display of C. . .C
INCREMENT = K
TRIM "outliers"
BY C
SAME increment on all displays
BOXPLOTS for C [levels in C]
LINES = K
NOTCHES
LEVELS K. . .K
MPOLISH C, levels in C,C [put residuals in [fits in C]]
COLUMNS (start iteration with column median)
ITERATIONS = K
EFFECTS put common into K, rows into C, columns into C
COMPARISON values, put into C
RLINE y in C, x in C [put residuals into C [fits into C [coefficients into C]]]
MAXITER = K (maximum number of iterations)
RSMOOTH C, put rough into C, smooth into C
SMOOTH by 3RSSH, twice
CPLOT (condensed plot) y in C versus x in C
Subcommands: LINES, CHARACTERS, XBOUNDS, YBOUNDS
CTABLE (coded table) data in C, row levels in C, column levels in C
Subcommands: MINIMUM, MAXIMUM, EXTREME
ROOTOGRAM data in C [bin boundaries in C]
Subcommands: BOUNDARIES, DRRS, FITTED, COUNTS, FREQUENCIES, MEAN, STDEV
LVALUES of C [put letter values in C [mids in C [spreads in C]]]

15. Sorting

SORT the values in C carry along corresponding rows of C. . .C put into put corresponding rows into C. . .C
RANK the values in C put ranks into C

16. Distributions and Random Data

RANDOM K observations into each of C. . .C
PDF for values in E [store results in E]
CDF for values in E [store results in E]
INVCDF for values in E [store results in E]

RANDOM, PDF, CDF, and INVCDF have the subcommands

BINOMIAL n = K p = K
POISSON mean = K
INTEGERS uniform on K to K
DISCRETE values in C, probabilities in C
NORMAL mu = K sigma = K
UNIFORM continuous uniform on K to K
T degrees of freedom = K
F df numerator = K, denominator = K
Additional subcommands are BERNOULLI, CAUCHY, LAPLACE, LOGISTIC, LOGNORMAL, CHISQUARE, EXPONENTIAL, GAMMA, WEIBUIL, BETA.
SAMPLE K rows from C. . .C put into C. . .C
BASE for random number generator = K

17. Matrices

READ the following data into a K by K matrix M
READ data from 'FILENAME' into a K by K matrix M

PRINT	M. . .M
COPY	C. . .C into M
COPY	M into C. . .C
COPY	M into M
TRANSPOSE	M into M
INVERT	M into M
DIAGONAL	is C, form into M
DIAGONAL	of M, put into C
EIGEN	for M, put values into C [vectors into M]

In the following commands E can be C, K, or M:

ADD	E to E,	put into E
SUBTRACT	E from E,	put into E
MULTIPLY	E by E,	put into E

18. Miscellaneous

NOTE	comments may be put here
ERASE	E. . .E
RESTART	begin fresh Minitab session
NEWPAGE	start next output on a new page
OW	output width = K spaces
OH	output height = K lines
PAPER	(put terminal output on paper)
NOPAPER	
OUTPUT 'FILENAME'	(put output in this file)
NOOUTFILE	(put output just to the terminal
BATCH	batch mode
TSHARE	interactive or timesharing mode

The symbol # anywhere on a line tells Minitab to ignore everything after that on a line.

To continue a command onto another line, end the first line with the symbol &.

19. Stored Commands and Loops

The commands STORE and EXECUTE provide both a simple (missing) (or stored command file) capability and a simple looping capability.

STORE	[in 'FILENAME'] following commands
	(Minitab commands go here)
END	of stored commands
EXECUTE	commands [in 'FILENAME'] [K times]
NOECHO	the commands that follow
ECHO	the commands that follow

The integer part of a column number may be replaced by a stored constant. This is useful in loops.

Example.

```
LET K1 = 5
PRINT C1 - CK1
```

Since K1 = 5, this PRINTS C1 through C5.

Appendix A.2

Selected List of SAS Software Statements

This list has been taken from the Base SAS Software Reference Card. For complete descriptions and operating system notes, see the *SAS User's Guide: Basics, Version 5 Edition* and *SAS User's Guide: Statistics, Version 5 Edition*. In interactive installations, you can use the HELP statement for a brief description of SAS features.

Statement Descriptions

The form of a SAS statement is specified with these conventions:

KEYWORD *parameter* . . .
[*item*|*item*|*item*] *options;*

where

BOLD	indicates that you use exactly the same spelling and form as shown
italics	means that you supply your own information
[*bracketed information*]	is optional
parameters not in brackets	are not optional
three periods (. . .)	means that more than one of the parameters preceding . . . may be optionally specified
vertical bar (|)	separating keywords and options means to choose only one of the terms separated by vertical bars

options are keyword options specific to a particular SAS statement

SAS Statements Used in the DATA Step

BY [DESCENDING] *variable* . . . [NOTSORTED];

CARDS;
data lines
[;]

DATA [[*SASdataset*[(*dsoptions*)]]. . .];

DO;
more SAS statements
END;

DROP *variables;*

END;

IF *expression* THEN *statement;*
ELSE *statement;*

KEEP *variables;*

MERGE *SASdataset*[(*dsoptions* IN = *name*)] *SASdataset*[(*dsoptions* IN = *name*)] . . . [END = *name*];

PUT *variable;*

RENAME *oldname* = *newname* . . . ;

SET[[*SASdataset*[(*dsoptions* IN = *name*)] . . .];

SAS Statements Used in the PROC Step

BY [DESCENDING] *variable* . . . [NOTSORTED];

CLASS *variables;*

FREQ *variable;*

MODEL *dependents* = *independenteffects*/ [*option*];

OUTPUT [*SASdataset*] . . . ;

VAR *variables;*

WEIGHT *variable;*

Functions

ABS(*argument*)

MAX(*argument, argument,* . . .)

MIN(*argument, argument,* . . .)

NORMAL(*seed*)

POISSON(*mean, n*)

PROBBNML(*p, n, m*)

PROBCHI(*x, df*)

PROBHYPR(*N, k, n, x*)

PROBNORM(*x*)

RANBIN(*seed, n, p*)

RANEXP(*seed*)

RANNOR(*seed*)

RANPOI(*seed, mean*)

RANUNI(*seed*)

Procedures (PROC) in *SAS User's Guide: Basics, Version 5 Edition*

PROC CHART [DATA = *SASdataset* LPI = *p*];
[VBAR *variables*[/ any options in the option list, below, and LEVELS = *n* SYMBOL = *'char'* GROUP = *variable* SUBGROUP = *variable* NOSYMBOL NOZEROS G100 ASCENDING

```
    DESCENDING REF = value DATA = SASdataset LPI = p];
    [VBAR variables [/ any options in the option list,
      below, and LEVELS = n SYMBOL = 'char'
      GROUP = variable SUBGROUP = variable
      NOSYMBOL NOZEROS G100 ASCENDING
     DESCENDING REF = value NOSPACE];]
    [HBAR variables [/any options in the option list,
      below, and LEVELS = n SYMBOL = 'char'
      GROUP = variable SUBGROUP = variable
      NOSYMBOL NOZEROS G100 ASCENDING
      DESCENDING REF = value NOSTAT FREQ CFREQ
      PERCENT CPERCENT SUM MEAN];]
    [BLOCK variables [/ any options in the option list,
      below, and LEVELS = n SYMBOL = 'char'
      GROUP = variable SUBGROUP = variable
      NOSYMBOL NOZEROS G100];]
    [PIE variables [/ any option in the option list, below];]
    [STAR variables [/ any option in the option list,
      below];]
    [BY variables;]
option list: MISSING DISCRETE TYPE = FREQ
      TYPE = PERCENT|PCT TYPE = CFREQ
      TYPE = CPERCENT|CPCT TYPE = SUM TYPE = MEAN
      SUMVAR = variable MIDPOINTS = values
      FREQ = variable AXIS = value

PROC CORR [DATA = SASdataset OUTP = SASdataset
      OUTS = SASdataset OUTK = SASdataset
      OUTH = SASdataset PEARSON SPEARMAN
      KENDALL HOEFFDING RANK BEST = n
      VARDEF = DF|WGT|WEIGHT|WDF
      NOSIMPLE NOPRINT NOPROB NOMISS SSCP
      COV NOCORR];

PROC FREQ [DATA = SASdataset ORDER = FREQ|
      DATA|INTERNAL|FORMATTED
      FORMCHAR(1,2,7) = 'string'];
    TABLES requests [/ MISSING LIST OUT = SASdataset
      CHISQ EXPECTED DEVIATION CELLCHI12
      CUMCOL MISSPRINT SPARSE NOFREQ
      NOPERCENT NOROW NOCOL NOCUM NOPRINT];
    [WEIGHT variable;]
    [BY variables;]
```

PROC MEANS [DATA = *SASdataset* NOPRINT MAXDEC = N
VARDEF = DF|WEIGHT|WGT|N|WDF N NMISS
MEAN STD MIN MAX RANGE SUM VAR USS CSS
STDERR CV SKEWNESS KURTOSIS T PRT SUMWEIGHT];
[VAR *variables;*]
[BY *variables;*]
[FREQ *variable;*]
[WEIGHT *variable;*]
[ID *variables;*]
[OUTPUT [OUT = *SASdataset keyword* = *names*];]
where *keyword* is chosen from N NMISS STD MIN
MAX RANGE SUM VAR USS CSS STDERR CV
SKEWNESS KURTOSIS T PRT SUMWGT

PROC PLOT [DATA = *SASdataset* UNIFORM NOLEGEND];
PLOT *vertical*horizontal*|*vertical*horizontal* = *'character'*|
*vertical*horizontal* = *'variable'* [/VAXIS = *values*
HAXIS = *values* VZERO HZERO VREVERSE
VREF = *values* VREFCHAR = *'c'* HREF = *values*
HREFCHAR = *'c'* VPOS = *n* HPOS = *n*
VSPACE = *n* HSPACE = *n* OVERLAY
CONTOUR = *value* $2 = *value* $2 = *value*];
[BY *variables;*]

PROC PRINT [DATA = *SASdataset* N UNIFORM|U
DOUBLE|D ROUND LABEL SPLIT = *'splitchar'*
NOOBS];
[VAR *variables;*]
[ID *variables;*]
[PAGEBY *byvariable;*]
[SUM *variables;*]
[SUMBY *byvariable;*]
[BY *variables;*]

PROC SORT [DATA = *SASdataset* OUT = *SASdataset*
EQUALS NODUPLICATES|NODUP NATIONAL
REVERSE DANISH NORWEGIAN FINNISH
SWEDISH MESSAGE|M LIST|L LEAVE = *n*
TECHNIQUE|T = *xxxx* SORTWKNO = *number*
DIAG SORTSIZE|SIZE = *parameter*];
BY [DESCENDING] *variable. . .* ;

PROC SUMMARY [DATA = *SASdataset* MISSING
NWAY IDMIN DESCENDING ORDER = FREQ|

```
      DATA|INTERNAL|EXTERNAL|FORMATTED
      VARDEF = DF|WEIGHT|WGT|N|WDF];
   [CLASS|CLASSES variables;]
   VAR variables;
   [BY variables;]
   [FREQ variable;]
   [WEIGHT variables,]
   [ID variables;]
   OUTPUT [OUT = SASdataset]keyword
      [(variables)] = [names]. . . ;
where keyword is chosen from N NMISS MEAN
      STD MIN MAX RANGE SUM VAR USS
      CSS CV STDERR T PRT SUMWGT

PROC TABULATE [DATA = SASdataset MISSING
      FORMAT = format ORDER = FREQ|DATA|INTERNAL|
      FORMATTED FORMCHAR [(indexlist)] = 'string'];
   [CLASS variables;]
   [VAR variables;]
   [FREQ variable;]
   [WEIGHT variables,]
   [FORMAT variables format;]
   [LABEL variable = 'label'. . . ;]
   [BY variables;]
   TABLE [expression,] [expression,] expression
      [/MISSTEXT = 'text' FUZZ = nnn RTSPACE|
      RTS = n BOX = PAGE|variablename|'string'
      ROW = FLOAT|CONSTANT|CONST CONDENSE];
   [KEYLABEL keyword = 'text'. . . ;]

PROC UNIVARIATE [DATA = SASdataset NOPRINT
      PLOT FREQ NORMAL PCTLDEF = value
      VARDEF = DF|WGT|WEIGHT|N|WDF];
   [VAR variables];
   [BY variables;]
   [FREQ variables;]
   [WEIGHT variables;]
   [ID variables;]
   [OUTPUT OUT = SASdataset keyword = names. . . ;]
where keyword is chosen from N NMISS MEAN
      SUM STD VAR SKEWNESS KURTOSIS SUMWGT
      MAX MIN RANGE Q3 MEDIAN Q1 QRANGE P1 P5
      P10 P90 P95 P99 MODE SIGNRANK NORMAL
```

Procedures (PROC) in *SAS User's Guide: Statistics, Version 5 Edition*

PROC ANOVA [DATA = *SASdataset*];
[CLASS *variables*];
[MODEL *dependents* = *effects* [/NOUNI INT|
INTERCEPT;]
[MEANS *effects* [/BON DUNCAN GABRIEL
REGWF REGWQ SCHEFFE SIDAK SMM|GT2 SNK
T|LSD TUKEY ALPHA = *p* WALLER
KRATIO = *value* LINES CLDIFF E = *effect*];]
[ABSORB *variables,*]
[FREQ *variable;*]
[TEST H = *effects* E = *effect;*]
[MANOVA [H = *effects* E = *effect* M = *equation*, . . .
[MNAMES = *names* PREFIX = *name*]] [/PRINTER
PRINTE ORTH SHORT CANONICAL SUMMARY];]
[REPEATED *factorname levels* (*levelvalues*)
[CONTRAST [(*ordinalreferencelevel*)]|
POLYNOMIAL|HELMERT|MEAN
[(*ordinalreferencelevel*)] |PROFILE][, . . .]
[/NOM NOU PRINTM PRINTH PRINTE
PRINTRV SHORT SUMMARY CANONICAL];]
[BY *variables;*]

PROC FREQ [DATA = *SASdataset*
ORDER = FREQ|DATA|INTERNAL|FORMATTED
FORMCHAR(1,2,7) = '*string*'];
TABLES *requests* [/MISSING LIST
OUT = *SASdataset* CHISQ MEASURES CMH ALL
SCORES = RANK|TABLE|RIDIT|MODRIDIT
ALPHA = *p* EXPECTED DEVIATION CELLCHI12
CUMCOL MISSPRINT SPARSE NOFREQ
NOPERCENT NOROW NOCOL NOCUM NOPRINT];
[WEIGHT *variable;*]
[BY *variables,*]

PROC GLM [DATA = *SASdataset*
ORDER = FREQ|DATA|INTERNAL|FORMATTED];
[CLASS|CLASSES *variables*];
MODEL *dependents* = *independents* [/NOINT INT|
INTERCEPT NOUNI SOLUTION TOLERANCE
E E1 E2 E3 E4 SS1 SS2 SS3 SS4 P CLM

```
    CLI ALPHA = p XPX INVERSE|I
    SINGULAR = value ZETA = value;]
  [CONTRAST 'label' [INTERCEPT value] effect values . . .
    [/E E= effect ETYPE = n SINGULAR = number];]
  [ESTIMATE 'label' [INTERCEPT value] effect values . . .
    [/E DIVISOR = number SINGULAR = number];]
  [LSMEANS effects [/E STDERR PDIFF E = effect
    ETYPE = n SINGULAR = number];]
  [MANOVA [H = effects E = effect
    M = equation, . . .[MNAMES = names
    PREFIX = name]] [/PRINTH PRINTE
    HTYPE = n ETYPE = n ORTH SHORT
    CANONICAL SUMMARY];]
  [OUTPUT [OUT = SASdataset PREDICTED|P = variables
    RESIDUAL|R = variables];]
  [RANDOM effects[/Q];]
  [REPEATED factorname levels (levelvalues)
    [CONTRAST [(ordinalreferencelevel)]|
    POLYNOMIAL|HELMERT|MEAN
    [(ordinalreferencelevel)]|
    PROFILE][, . . .][/NOM NOU PRINTM PRINTH
    PRINTE PRINTRV SHORT SUMMARY CANONICAL
    HTYPE = n];]
  [TEST H = effects E = effect[/HTYPE = n ETYPE = n];]
  [ABSORB variables,]
  [BY variables;]
  [FREQ variable;]
  [MEANS effects[/DEPONLY BON DUNCAN
    GABRIEL REGWF REGWQ SCHEFFE SIDAK SMM|
    GT2 SNK T|LSD TUKEY ALPHA = p
    WALLER KRATIO = value LINES CLDIFF
    NOSORT E = effect ETYPE = n HTYPE = n];]
  [WEIGHT variable;]

PROC NPAR1WAY [DATA = SASdataset ANOVA
    WILCOXON MEDIAN VW SAVAGE];
  [VAR variables;]
  CLASS variable;
  [BY variables;]

PROC RANK [DATA = SASdataset OUT = SASdataset
    TIES = MEAN|HIGH|LOW DESCENDING
```

```
      GROUPS = n FRACTION|F PERCENT|
      P NORMAL = BLOM|TUKEY|VW SAVAGE];
   [VAR variables;]
   [RANKS names;]
   [BY variables;]

PROC REG [DATA = SASdataset OUTEST = SASdataset
      OUTSSCP = SASdataset NOPRINT SIMPLE
      USSCP ALL COVOUT SINGULAR = n];
   [label:] MODEL dependents = predictors [/NOPRINT
      NOINT ALL XPX I SS1 SS2 STB TOL VIF COVB
      CORRB SEQB COLLIN COLLINOINT ACOV
      SPEC PCORR1 PCORR2 SCORR1 SCORR2
      P CLM CLI DW INFLUENCE PARTIAL];
   [VAR variables;]
   [FREQ variable;]
   [WEIGHT variable;]
   [ID variable;]
   [OUTPUT [OUT = SASdataset PREDICTED|P = names
      RESIDUAL|R = names L95M= names U95M = names
      L95 = names U95 = names STDP = names
      STDR = names STDI = names STUDENT = names
      COOKD = names H = names PRESS = names
      RSTUDENT = names DFFITS = names
      COVRATIO = names];]
   [RESTRICT equation, . . .[/PRINT];]
   [label: TEST equation, . . .[/PRINT];]
   [label: MTEST equation, . . .[/PRINT CANPRINT
      DETAILS];]
   [BY variables;]

PROC RSQUARE [DATA = SASdataset SIMPLE|
      S CORR|C NOINT NOPRINT OUTEST = SASdataset];
   [label:] MODEL response = independents
      [/SELECT = n INCLUDE = i START = n
      STOP = n SIGMA = n ADJRSQ AIC BIC
      CP GMSEP JP MSE PC RMSE SBC SP SSE B];
   [FREQ variable;]
   [WEIGHT variable;]
   [BY variables;]

PROC STEPWISE [DATA = SASdataset];
   MODEL dependents = independents [/NOINT
```

```
    FORWARD|F BACKWARD|B STEPWISE MAXR
    MINR SLENTRY|SLE = value SLSTAY|SLS = value
    INCLUDE = n START = s STOP = s DETAILS];
  [WEIGHT variable;]
  [BY variables;]

PROC TTEST [DATA = SASdataset];
  CLASS variable;
  [VAR variables;]
  [BY variables;]
```

Appendix B.1

Miami Female Study Data

A sample of $n = 100$ females was taken from the population of female students at Miami University in Oxford, Ohio. For each female surveyed the following 19 variables were recorded:

1. Hair color (1 = brown, 2 = black, 3 = blonde, 4 = red)
2. Eye color (1 = brown, 2 = blue, 3 = green)
3. Belong to sorority (1 = yes, 2 = no)
4. Month of birth (1 = January, . . . , 12 = December)
5. Religious preference (1 = Protestant, 2 = Catholic, 3 = Jewish, 4 = Other)
6. Number of brothers
7. Number of sisters
8. Wear glasses (1 = yes, 2 = no)
9. Length of hair (1 = short, 2 = medium, 3 = long)
10. Number of credit hours taking
11. Type of residence (1 = dorm, 2 = apartment, 3 = commute)
12. Have steady boyfriend (1 = yes, 2 = no)
13. Typical number of hours sleep per night
14. Watch soap operas regularly (1 = yes, 2 = no)

15. Currently on diet (1 = yes, 2 = no)

16. Have job (1 = yes, 2 = no)

17. Height in inches

18. Weight in pounds

19. Grade-point average on a four-point scale (2.0 = between 0 and 2.00, 2.5 = between 2.01 and 2.50, 3.0 = between 2.51 and 3.00, 3.5 = between 3.01 and 3.50, 4.0 = between 3.51 and 4.00)

HAIR COLOR	EYE COLOR	SORORITY	MONTH OF BIRTH	RELIGION	# BROTHERS	# SISTERS	WEAR GLASSES	HAIR LENGTH	# CREDIT HOURS	TYPE RESIDENCE	BOYFRIEND	HOURS OF SLEEP	WATCH SOAPS	DIET	JOB	HEIGHT	WEIGHT	GPA
2	1	1	10	1	0	1	1	1	16	1	1	7	2	2	1	60	98	2.0
1	1	2	06	1	1	1	1	2	19	1	2	6	2	1	1	66	140	3.0
1	3	2	08	1	1	1	2	1	17	1	2	8	2	2	2	64	120	4.0
3	3	2	05	1	3	0	2	2	20	1	1	9	1	2	2	63	98	3.0
1	1	2	03	1	1	2	2	2	19	1	1	8	2	2	1	67	125	3.5
1	2	2	04	2	0	1	1	1	18	1	2	7	2	1	2	67	145	3.0
1	1	2	08	1	1	0	1	3	16	1	1	9	2	2	2	64	125	3.0
1	1	2	07	2	0	2	1	1	16	1	2	6	2	2	1	66	123	4.0
1	3	2	07	1	0	3	2	3	19	1	2	6	2	1	1	63	125	3.0
1	1	2	01	2	1	1	1	1	16	1	2	8	1	1	2	69	155	2.5
1	3	2	01	1	2	0	2	2	18	1	1	7	1	1	2	68	130	3.0
4	2	2	09	1	1	1	1	2	17	1	2	7	1	2	2	64	120	2.5
3	2	2	05	1	1	0	2	2	20	1	1	6	2	2	1	66	120	3.0
1	1	2	04	1	1	4	1	2	16	1	1	6	2	2	1	63	117	3.0
1	3	2	05	1	1	1	1	1	19	1	2	7	2	2	2	67	135	3.5
1	2	2	08	2	0	1	2	3	18	1	1	7	2	2	1	66	118	4.0
1	2	2	03	1	2	0	1	1	16	1	2	7	2	2	2	68	175	4.0
1	2	2	11	1	2	0	1	1	16	1	2	8	1	2	1	63	123	3.0
1	3	2	11	1	1	2	1	1	16	1	1	5	2	2	2	64	120	3.0
1	3	1	07	4	0	2	1	1	13	2	2	9	2	2	2	64	125	4.0
1	1	2	10	4	1	1	2	2	15	1	1	8	2	2	1	65	125	3.5
3	2	2	03	1	1	2	1	2	17	1	1	6	2	1	2	63	112	4.0
1	3	2	06	2	2	5	1	3	18	1	2	7	2	1	1	67	147	3.0
2	1	2	11	1	1	1	1	2	18	1	2	7	2	2	1	62	115	3.0
1	1	2	03	1	0	2	1	2	15	1	1	6	2	1	1	62	100	4.0
4	1	2	10	1	0	2	2	3	17	1	1	7	2	1	1	68	128	3.5
3	3	2	04	1	1	1	1	3	16	1	2	6	1	1	1	68	135	3.0
3	2	2	04	1	0	1	1	2	18	1	1	6	2	2	2	68	140	4.0

(continued)

HAIR COLOR	EYE COLOR	SORORITY	MONTH OF BIRTH	RELIGION	# BROTHERS	# SISTERS	WEAR GLASSES	HAIR LENGTH	# CREDIT HOURS	TYPE RESIDENCE	BOYFRIEND	HOURS OF SLEEP	WATCH SOAPS	DIET	JOB	HEIGHT	WEIGHT	GPA
1	2	2	02	1	0	2	1	2	17	1	1	6	2	2	2	70	135	3.0
1	1	2	03	2	4	3	1	2	15	1	1	6	2	2	1	69	125	3.5
3	2	2	01	1	1	2	1	1	15	1	2	7	1	1	2	67	140	3.5
2	1	2	05	3	0	2	1	2	19	1	1	7	2	1	1	62	118	3.5
1	3	2	07	2	2	0	2	2	18	1	2	8	1	1	2	66	124	3.5
1	2	2	12	1	0	3	1	1	18	1	1	8	2	2	2	66	125	3.0
1	3	2	02	4	2	1	1	1	16	1	2	7	1	2	1	68	126	3.0
1	2	2	05	1	0	1	1	1	18	1	1	6	2	1	1	68	143	3.5
3	2	2	06	1	2	0	1	2	21	1	2	7	2	1	1	68	155	2.5
1	1	2	07	1	0	1	1	1	16	1	1	7	1	2	2	66	130	2.5
1	1	2	04	1	2	0	2	2	17	1	2	6	2	2	2	66	145	3.5
1	2	2	02	1	0	2	1	3	18	1	1	7	2	1	1	64	135	2.5
1	2	2	04	1	1	2	1	2	18	1	1	6	2	2	2	64	123	3.5
1	1	2	09	1	0	2	2	1	16	1	2	7	2	2	1	68	136	2.5
4	1	2	04	1	4	2	1	1	15	3	1	7	2	2	2	67	117	4.0
1	3	2	09	2	3	1	2	1	16	1	1	8	1	2	2	66	140	3.0
3	3	2	07	1	1	1	1	1	17	1	1	7	1	2	2	64	122	3.0
1	1	1	05	1	2	2	1	1	15	1	2	6	1	2	1	62	110	3.0
1	1	2	09	2	2	1	1	3	18	1	1	7	2	2	2	66	127	3.0
3	3	2	08	2	3	0	1	2	17	2	1	7	2	2	2	66	125	4.0
1	1	1	10	1	1	1	2	1	17	1	2	7	2	2	2	70	117	3.5
1	3	1	11	2	0	0	1	1	20	1	2	7	2	2	2	63	115	3.0
1	2	2	08	1	4	6	2	2	18	2	1	7	1	2	2	66	120	3.0
1	1	2	10	2	2	1	1	1	17	2	1	7	1	1	2	58	105	4.0
3	2	1	11	1	2	2	1	1	16	1	1	7	1	1	1	61	105	3.5
1	2	2	08	2	1	3	2	2	18	1	1	7	2	2	1	64	110	4.0
1	2	2	07	1	0	2	1	2	20	1	2	7	2	1	2	68	175	3.0
1	3	2	09	1	0	4	1	1	17	2	1	7	2	1	1	65	140	3.0
1	2	2	04	1	1	3	1	1	17	1	1	8	2	2	1	62	115	3.5
1	2	2	12	1	1	1	1	2	17	1	1	8	2	1	1	62	150	3.5
1	1	2	12	2	0	1	1	1	16	1	1	7	2	2	2	65	126	4.0
3	3	2	04	2	3	2	1	1	17	1	1	7	2	2	2	68	150	2.5
1	1	2	10	2	1	0	2	2	19	1	2	8	2	2	1	70	140	3.5
3	1	2	12	1	0	0	1	1	18	1	1	7	2	1	2	65	137	3.5
1	1	2	10	2	1	2	2	3	13	1	1	9	2	2	1	69	138	3.0
4	3	2	11	1	1	0	1	2	15	1	1	7	1	2	1	68	122	3.0

HAIR COLOR	EYE COLOR	SORORITY	MONTH OF BIRTH	RELIGION	# BROTHERS	# SISTERS	WEAR GLASSES	HAIR LENGTH	# CREDIT HOURS	TYPE RESIDENCE	BOYFRIEND	HOURS OF SLEEP	WATCH SOAPS	DIET	JOB	HEIGHT	WEIGHT	GPA
1	2	2	06	1	2	0	1	2	23	1	2	7	1	2	2	64	125	4.0
1	1	2	08	2	0	1	1	1	19	1	1	8	2	2	1	71	120	4.0
1	2	1	03	1	3	0	1	2	12	1	1	6	2	2	2	62	108	3.5
3	3	1	07	4	1	1	1	2	13	1	2	8	1	1	1	64	115	3.0
1	2	2	07	4	0	2	2	1	18	1	2	6	2	2	2	64	120	3.5
1	2	2	07	1	1	0	1	2	16	2	1	8	1	2	1	67	141	2.5
1	2	2	02	4	4	0	2	2	14	2	1	6	2	2	1	67	120	2.5
3	2	2	12	2	0	3	2	3	17	1	1	5	2	1	2	64	120	3.0
3	2	2	06	1	0	4	2	2	15	1	2	6	1	1	1	67	121	2.5
1	3	1	02	1	1	0	2	1	18	1	1	7	1	2	1	68	135	3.5
1	2	2	02	2	2	1	1	2	18	1	2	7	2	2	1	67	120	4.0
1	1	1	12	2	1	1	1	2	15	2	2	5	2	2	2	60	98	3.0
1	3	2	08	3	3	0	2	2	16	1	2	7	1	1	2	63	115	3.0
1	3	1	02	1	0	1	1	1	16	2	1	7	2	1	1	67	127	3.5
3	2	1	06	2	1	2	1	2	16	2	1	6	2	1	2	67	125	3.0
3	3	2	07	1	1	0	2	2	18	1	2	8	2	1	2	64	133	4.0
1	1	2	06	4	2	2	1	2	15	2	2	8	2	1	1	64	123	3.5
1	1	2	04	4	2	1	1	2	16	1	1	8	2	2	1	63	120	3.5
1	2	2	11	2	2	4	2	1	15	1	2	6	2	2	2	67	120	2.5
1	1	2	02	1	2	0	2	1	17	1	2	8	2	2	2	62	112	2.5
1	1	2	05	3	1	2	1	3	14	2	1	5	2	2	2	60	110	4.0
1	2	2	12	1	0	1	1	2	16	2	2	6	2	2	2	66	132	4.0
3	2	2	07	2	0	2	1	1	19	1	2	6	2	2	2	64	130	3.5
3	1	1	10	1	0	0	1	1	24	1	2	4	2	2	2	64	115	3.5
1	1	2	10	1	1	1	2	2	26	1	2	5	2	2	2	62	125	4.0
3	3	2	08	1	0	0	1	3	18	1	1	7	2	1	2	64	130	3.0
3	2	2	09	2	0	1	2	3	18	1	1	6	2	1	2	70	175	3.5
1	2	2	11	1	2	1	1	1	17	1	2	6	2	1	1	63	130	3.5
1	2	2	08	1	2	1	1	2	16	1	2	5	2	2	2	66	145	4.0
1	1	2	02	1	2	1	1	1	18	1	1	6	2	2	2	65	105	4.0
1	1	2	04	1	3	0	1	1	17	1	2	8	2	1	2	67	135	3.5
1	3	2	07	2	5	1	1	1	16	1	2	7	2	2	2	63	115	4.0
1	3	1	12	2	1	2	1	2	17	2	1	7	2	2	2	68	138	3.0
1	1	2	08	4	3	1	1	1	16	2	1	7	2	2	2	67	118	3.0
1	2	2	03	2	0	3	2	1	12	2	1	8	2	2	2	65	135	2.5
3	3	1	11	1	1	0	2	1	18	1	1	6	2	2	2	63	94	3.5

Appendix B.2

Department of Agriculture Survey

The following data for 166 people were taken from a survey of food consumption for elderly people performed by the U.S. Department of Agriculture.

REGION section of the U.S. (1 = northeast, 2 = central, 3 = south, 4 = west)
AGEYRS age in years of head of household
NUMINHSE number of persons living in household
INCMON monthly income of household in dollars
HHSEX sex of head of household (1 = male, 0 = female)
EDUC years of education of head of household (17 = 17 or more years)

OBS	REGION	AGEYRS	NUMINHSE	INCMON	HHSEX	EDUC
1	1	73	2	608	1	8
2	1	64	2	1369	1	9
3	3	64	1	508	0	7
4	2	73	1	218	0	9
5	3	77	1	164	0	4
6	1	86	1	200	0	4
7	1	60	2	4034	1	12
8	1	62	3	400	1	8
9	2	64	1	187	0	10
10	2	63	1	238	0	12

OBS	REGION	AGEYRS	NUMINHSE	INCMON	HHSEX	EDUC
11	4	71	1	648	0	12
12	3	77	1	117	0	7
13	3	77	1	390	0	13
14	3	77	1	264	0	14
15	2	68	2	801	1	16
16	1	67	1	272	0	8
17	3	57	3	600	1	8
18	3	67	2	858	1	12
19	1	59	2	1325	1	12
20	3	56	2	1465	1	12
21	2	65	5	1910	1	12
22	3	79	2	555	1	3
23	4	57	5	5400	1	12
24	2	79	1	187	0	8
25	2	57	2	667	0	13
26	3	78	1	304	1	7
27	4	56	2	1070	1	12
28	3	73	1	344	0	8
29	2	68	1	762	1	12
30	2	60	2	474	0	10
31	1	56	1	456	0	6
32	2	59	2	320	0	10
33	2	68	2	725	1	8
34	1	56	2	445	0	14
35	1	64	2	1429	0	12
36	3	59	3	1412	1	12
37	1	84	1	245	0	12
38	1	67	1	319	0	8
39	3	55	3	302	0	11
40	2	56	1	2042	0	16
41	4	62	2	556	1	12
42	3	74	2	300	1	7
43	1	79	2	387	1	7
44	3	69	1	204	0	12
45	1	56	3	825	1	12
46	2	69	2	318	1	17
47	3	72	2	394	1	8
48	4	74	2	970	1	9
49	3	68	1	270	1	8
50	1	72	2	606	1	12
51	3	63	1	185	0	8
52	2	70	1	179	1	8
53	3	72	2	500	0	10

(*continued*)

OBS	REGION	AGEYRS	NUMINHSE	INCMON	HHSEX	EDUC
54	4	76	1	302	0	14
55	2	64	4	1618	1	8
56	1	68	1	248	1	14
57	3	62	2	2500	1	17
58	1	64	2	462	0	8
59	3	60	2	1376	1	17
60	3	68	2	668	1	11
61	2	55	1	933	0	12
62	3	72	2	675	1	4
63	3	56	2	1332	0	5
64	1	63	1	220	0	10
65	2	68	2	830	1	12
66	3	56	2	466	0	8
67	4	61	2	1908	1	14
68	4	71	1	370	1	14
69	1	88	2	1280	1	12
70	2	64	2	225	1	12
71	3	73	1	359	0	12
72	4	60	2	542	1	12
73	3	74	2	600	1	6
74	2	65	3	628	1	7
75	1	75	2	490	0	9
76	1	83	1	468	0	16
77	1	69	2	546	1	6
78	1	70	1	355	0	10
79	1	69	1	412	1	8
80	2	56	1	9900	0	12
81	2	58	4	2167	1	12
82	2	55	3	1695	1	12
83	2	63	3	375	1	8
84	2	69	1	404	0	7
85	3	64	2	1011	1	9
86	3	66	1	184	1	8
87	3	73	3	905	1	9
88	1	85	1	276	0	6
89	3	62	2	833	1	12
90	2	61	2	1000	1	8
91	1	77	1	316	0	3
92	2	68	1	2300	1	17
93	2	70	1	238	0	10
94	3	56	1	159	0	4
95	4	55	3	141	1	17
96	2	82	2	353	1	12

OBS	REGION	AGEYRS	NUMINHSE	INCMON	HHSEX	EDUC
97	4	75	2	554	1	14
98	2	64	2	517	1	7
99	2	75	1	223	0	8
100	2	56	2	1700	1	12
101	2	55	2	1218	1	12
102	4	63	1	850	0	12
103	1	59	4	1110	1	12
104	3	79	2	207	1	6
105	4	65	1	575	0	11
106	4	65	2	1344	0	17
107	1	67	7	942	1	8
108	1	59	2	1100	1	12
109	2	66	1	481	1	8
110	2	58	2	252	1	8
111	3	70	2	344	1	8
112	3	66	1	150	1	8
113	3	90	2	200	1	6
114	2	62	2	505	1	10
115	3	64	2	349	1	4
116	1	61	1	192	0	9
117	4	65	2	797	1	9
118	3	57	3	3400	1	17
119	1	63	2	902	1	12
120	1	63	2	801	1	12
121	1	83	1	198	0	8
122	1	72	1	427	1	6
123	2	63	2	1330	1	12
124	3	85	2	1539	1	14
125	4	57	4	3035	1	12
126	3	59	2	320	1	7
127	1	56	1	1250	1	12
128	2	58	1	575	0	12
129	4	73	2	359	1	12
130	1	56	1	3186	1	14
131	1	59	4	2004	0	9
132	1	56	2	2100	1	12
133	3	55	2	260	1	12
134	2	56	1	521	0	12
135	2	61	2	750	1	8
136	4	62	1	0	0	14
137	1	56	5	1310	1	17
138	2	63	1	311	1	12

(continued)

OBS	REGION	AGEYRS	NUMINHSE	INCMON	HHSEX	EDUC
139	2	75	2	211	0	8
140	2	75	3	814	0	12
141	1	80	2	419	1	8
142	2	75	2	1043	1	8
143	4	68	2	549	1	10
144	1	79	2	422	1	7
145	1	58	2	1541	1	10
146	1	86	1	268	0	8
147	3	69	2	697	1	12
148	3	56	8	672	1	6
149	2	59	2	336	1	11
150	1	71	1	427	0	6
151	2	77	2	882	0	6
152	2	57	2	2085	1	16
153	4	73	1	265	0	11
154	3	67	3	973	1	12
155	2	73	2	478	1	8
156	4	74	1	677	0	11
157	3	74	2	367	0	6
158	1	83	2	370	1	0
159	1	70	2	271	1	12
160	2	62	2	1905	1	11
161	2	55	3	616	1	3
162	2	80	2	450	1	9
163	2	55	5	2225	1	12
164	3	57	2	537	0	12
165	2	58	3	1200	1	12
166	2	65	2	720	1	12

Summary Measures
Number in Household

NUMINHSE	FREQUENCY	CUM FREQ	PERCENT	CUM PERCENT
1	58	58	34.940	34.940
2	82	140	49.398	84.337
3	15	155	9.036	93.373
4	5	160	3.012	96.386
5	4	164	2.410	98.795
7	1	165	0.602	99.398
8	1	166	0.602	100.000

Years of Education

EDUC	FREQUENCY	CUM FREQ	PERCENT	CUM PERCENT
0	1	1	0.602	0.602
3	3	4	1.807	2.410
4	5	9	3.012	5.422
5	1	10	0.602	6.024
6	11	21	6.627	12.651
7	11	32	6.627	19.277
8	32	64	19.277	38.554
9	10	74	6.024	44.578
10	10	84	6.024	50.602
11	7	91	4.217	54.819
12	52	143	31.325	86.145
13	2	145	1.205	87.349
14	10	155	6.024	93.373
16	4	159	2.410	95.783
17	7	166	4.217	100.000

Sex of Head of Household

HHSEX	FREQUENCY	CUM FREQ	PERCENT	CUM PERCENT
0	61	61	36.747	36.747
1	105	166	63.253	100.000

Region Code

REGION	FREQUENCY	CUM FREQ	PERCENT	CUM PERCENT
1	46	46	27.711	27.711
2	53	99	31.928	59.639
3	46	145	27.711	87.349
4	21	166	12.651	100.000

VARIABLE	N	MEAN	STANDARD DEVIATION	MINIMUM VALUE	MAXIMUM VALUE	STD ERROR OF MEAN
REGION	166	2.2530120	1.001131	1.00000000	4.000000	0.0777028
AGEYRS	166	66.6265060	8.608961	55.00000000	90.000000	0.6681849
NUMINHSE	166	1.9397590	1.071384	1.00000000	8.000000	0.0831555
INCMON	166	851.8433735	1042.908613	0.00000000	9900.000000	80.9454068
HHSEX	166	0.6325301	0.483575	0.00000000	1.000000	0.0375327
EDUC	166	10.0783133	3.251090	0.00000000	17.000000	0.2523335

Appendix C

Answers to Odd-Numbered Problems

Chapter 1

1.1

Limit	*Width*	*Mark*	*Boundary*
20–26	7	23	19.5–26.5
27–33	7	30	26.5–33.5
34–40	7	37	33.5–40.5
41–47	7	44	40.5–47.5
48–54	7	51	47.5–54.5

1.3

Data	*Freq*	*Cum Freq*	*Rel Freq*	*Rel Cum Freq*
95	1	1	0.03	0.03
98	2	3	0.07	0.10
100	1	4	0.03	0.13
102	3	7	0.10	0.23
103	1	8	0.03	0.27
106	1	9	0.03	0.30
107	2	11	0.07	0.37
110	4	15	0.13	0.50
111	3	18	0.10	0.60
112	2	20	0.07	0.67
114	1	21	0.03	0.70
116	2	23	0.07	0.77
120	1	24	0.03	0.80
121	1	25	0.03	0.83

1.3

Data	*Freq*	*Cum Freq*	*Rel Freq*	*Rel Cum Freq*
122	2	27	0.07	0.90
124	2	29	0.07	0.97
125	1	30	0.03	1.00

1.5 **a.** 147. **b.** 160–179, 180–199, 200–219, 220–239, 240–259, 260–279, 280–299, 300–319. **c.** 15.

1.7 **a.**

Data	*Freq*	*Data*	*Freq*	*Data*	*Freq*	*Data*	*Freq*
47.6	1	51.0	1	54.4	1	56.8	1
48.4	1	51.5	1	54.8	1	57.3	1
49.0	1	52.1	1	54.9	1	58.1	2
49.7	1	52.8	2	55.4	1	58.5	1
50.2	1	52.9	1	55.7	1	59.8	1
50.4	1	53.4	1	56.1	1	62.5	1
50.9	1	54.3	2	56.5	1		

b. 1.9; 47.0–48.9, 49.0–50.9, 51.0–52.9, 53.0–54.9, 55.0–56.9, 57.0–58.9, 59.0–60.9, 61.0–62.9

c.

Data	*Freq*	*Cum Freq*	*Data*	*Freq*	*Cum Freq*
47.0–48.9	2	2	55.0–56.9	5	24
49.0–50.9	5	7	57.0–58.9	4	28
51.0–52.9	6	13	59.0–60.9	1	29
53.0–54.9	6	19	61.0–62.9	1	30

d. 5/6

1.9 **a.**

Data	*Rel Freq*	*Data*	*Rel Freq*	*Data*	*Rel Freq*
0.0–4.9	.075	20.0–24.9	.100	40.0–44.9	.050
5.0–9.9	.250	25.0–29.9	.100	45.0–49.9	.075
10.0–14.9	.125	30.0–34.9	.075	50.0–54.9	.025
15.0–19.9	.100	35.0–39.9	.025		

b. .45. **c.** .75. **d.** .20

1.11

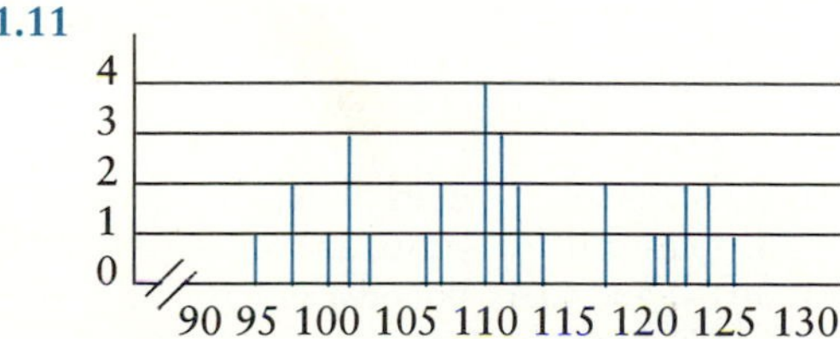

1.13 *See* Solutions Manual

1.15 The total expenses are $1250, so each expenditure must be represented as a fraction of $1250. If we wish, we may calculate the number of degrees for each one as well. Thus, housing is 0.4224 (152°), food is 0.1448 (52°), utilities 0.1552 (56°), transportation 0.1496 (54°), medical expenses 0.0168 (6°), clothing 0.0280 (10°), savings 0.0280 (10°), and miscellaneous 0.0552 (20°).

1.17 *See* Solutions Manual

1.19

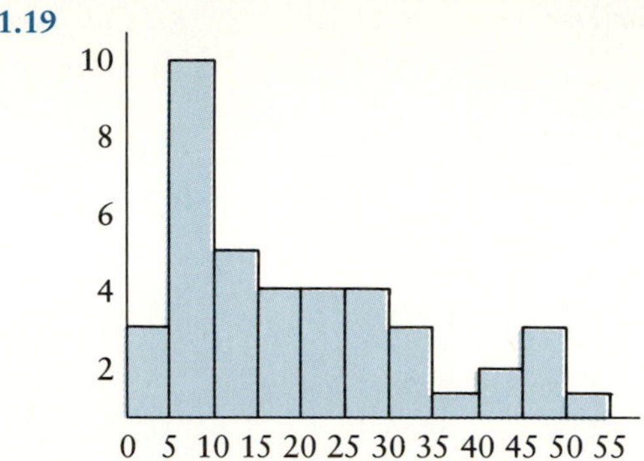

1.21 Mostly round-shaped, with a dip in the center.

1.23 Frequencies are 71, 3, 22, and 4.

1.25 **a.**

Stem	*Leaves*
0	346678999
1	0001111233333444566677777889
2	0223

b.

Data	*Freq*	*Data*	*Freq*	*Data*	*Freq*	*Data*	*Freq*
3	1	9	3	14	3	19	1
4	1	10	3	15	1	20	1
6	2	11	4	16	3	22	2
7	1	12	1	17	4	23	1
8	1	13	5	18	2		

c.

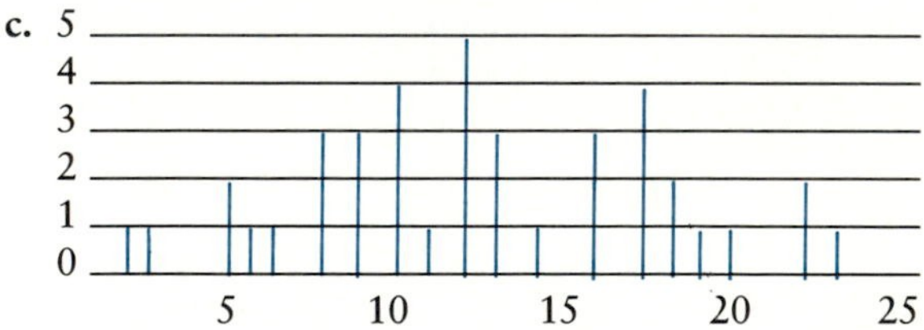

1.27 The class interval is 50, so each class extends 25 units above and below the class mark. Class limits: 101–150, 151–200, 201–250, 251–300, 301–350, 351–400, 401–450, 451–500, 501–550. Class boundaries: 100.5, 150.5, 200.5, 250.5, 300.5, 350.5, 400.5, 450.5, 500.5, 550.5.

1.29 **a.** 16.995 and 18.995. **b.** 0.01. **c.** 11, 12, 13, 14, etc. **d.** 13.995. **e.** 2.00 **f.** 10.94.

1.31

a.

Stem	*Leaves*
2	2223444456677888899 9
3	00112222233444444567777
4	0011111122233678

b.

Data	*Freq*	*Data*	*Freq*
21–23	4	36–38	5
24–26	7	39–41	8
27–29	9	42–44	5
30–32	9	45–47	2
33–35	10	48–50	1

c.

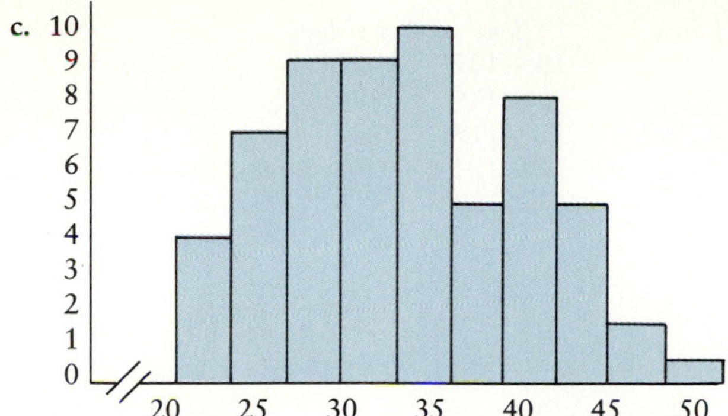

1.33

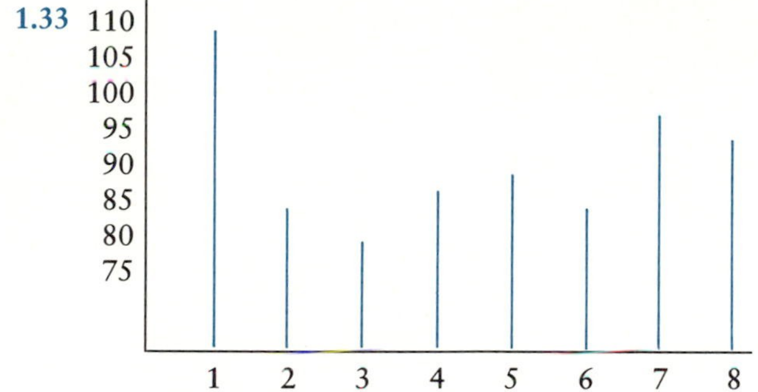

1 = Pacific
2 = Mountains
3 = Plains
4 = South Central
5 = Great Lakes
6 = Southeast
7 = Middle Atlantic
8 = Northeast

1 = \$109.00. 2 = \$84.59. 3 = \$79.39. 4 = \$87.29.
5 = \$88.76. 6 = \$85.19. 7 = \$96.56. 8 = \$94.66.

1.35 **a.**

Stem	*Leaves*	*Stem*	*Leaves*
8	01235579	12	0122222233556777888
9	01133446677889	13	13345
10	01112223333344455667788888999	14	7
11	00011122233333445555566677788999	15	014

b.

Class	*Freq*	*Class*	*Freq*	*Class*	*Freq*
80–84	4	105–109	14	130–134	4
85–89	4	110–114	16	135–139	1
90–94	7	115–119	16	140–144	0
95–99	7	120–124	9	145–149	1
100–104	15	125–129	9	150–154	3

c.

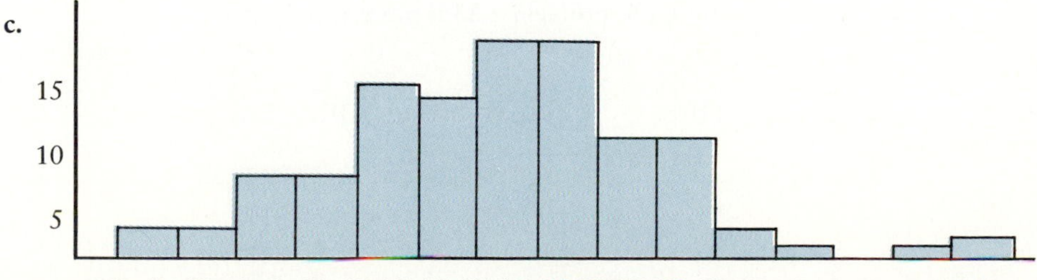

d.

Class	*Rel Freq*	*Class*	*Rel Freq*	*Class*	*Rel Freq*
80–84	.036	105–109	.127	130–134	.036
85–89	.036	110–114	.145	135–139	.009
90–94	.064	115–119	.145	140–144	.000
95–99	.064	120–124	.082	145–149	.009
100–104	.136	125–129	.082	150–154	.027

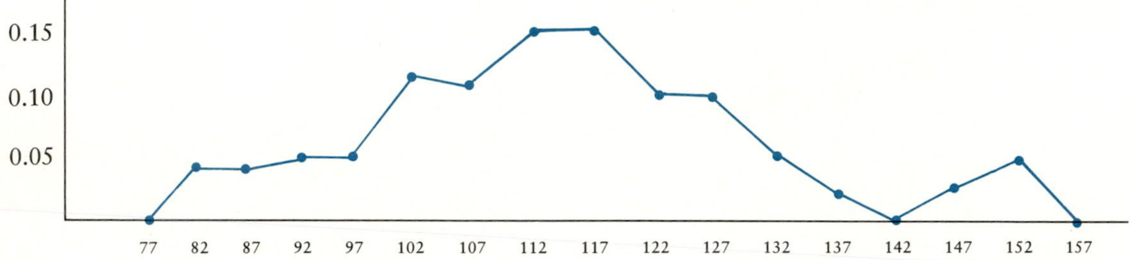

Chapter 2

2.1 mean = 25; md = 24.5; there is no mode.

2.3 mean ≐ 4.31; md = 3.02; there is no mode.

2.5 mean = 70.52; modal class is 70–72.

2.7 mean ≐ 3.71; md = 4.25; mo = 5.

2.9 mean ≐ 20.725.

2.11 mean = 25; $s^2 = 16$; $s = 4$.

2.13 mean ≐ 20.31; $s^2 \doteq 2.22$; $s \doteq 1.49$.

2.15 **a.** 3/4. **b.** 8/9. **c.** 15/16.

2.17 **a.** 68%. **b.** 95%. **c.** 100%. **d.** 97.5%, **e.** 16%. **f.** 16%. **g.** 97.5%,

2.19 mean = 36.1; $s = 1.76$.

2.21 mean ≐ \$95,127; $s \doteq$ \$35,681.

2.23 **a.** $z = -2.5$. **b.** $z \doteq -1.92$. **c.** $z \doteq -1.17$. **d.** $z = -0.25$. **e.** $z = 0.5$. **f.** $z \doteq 1.67$. **g.** $z \doteq 2.83$.

2.25 The distribution of problem 2.23 is more variable.

2.27 IQR = 4.24.

2.29 **a.** md = 110.5; $Q_1 = 102$; $Q_3 = 119$; IQR = 17.

b.

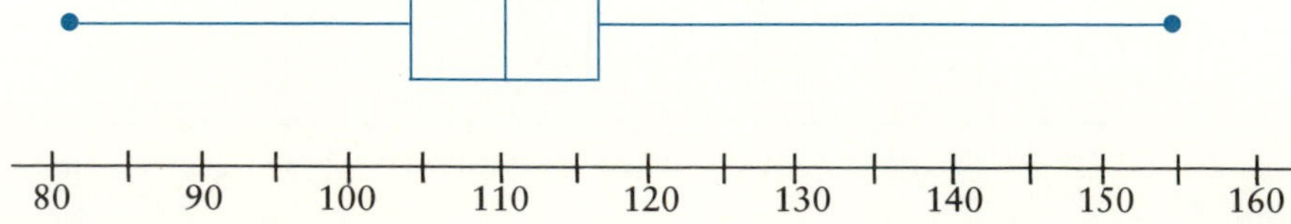

2.31 **a.**

		1985	*1984*
American League	μ	357,319.57	319,287.07
	σ	97,057.39	99,342.09
National League	μ	390,510.50	338,417.08
	σ	56,698.25	52,148.03

b.

	1985	*1984*
American League	27.16	31.11
National League	14.52	15.41

Salaries more variable in American League and in 1984 for both leagues.

c.

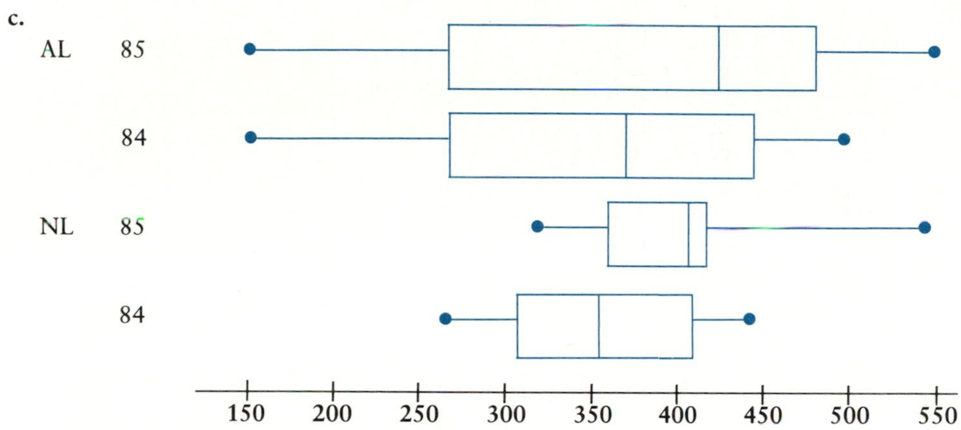

2.33 Production line A's high coefficient of variation makes it the more likely source of the problem.

2.35

Class Mark	*Freq*	*Class Mark*	*Freq*
123	1	138	14
128	11	143	7
133	9	148	4

$x(T) \doteq 135.9$
$s(T) \doteq 6.547$

2.37

Stem	*Leaves*
.3	4677899
.4	2347899
.5	222244455567788889
.6	044466677788889
.7	14447
.8	113
.9	0146
1.0	2

No; skewed positively.

2.39 var = 0.0248; s.d. = 0.1576; PSD = 0.1185; a better estimate of s would use the trimmed data set.

2.41 PSD = 0.107; $s(T) = 0.102$; yes.

2.43 **a.**

Stem	*Leaves*
1	15
3	9
6	479
7	037
8	3469
10	3
11	09

$Q_1 = 8.35$; Md = 17; $Q_3 = 29.9$.

$\bar{x} = 20.725$ $s \doteq 14.55$; PSD $\doteq 15.96$

Stem	*Leaves*
14	46
15	9
16	3
17	7
19	0
20	5
21	78
24	1
26	08
27	47
32	1
34	77
38	4
40	6
43	4
47	08
49	9
50	6

17
1.1
8.35
29.9
50.6
−23.9
62.2

no outliers, slightly positively skewed.

2.45 **a.** mean = 17.03, s.d. ≐ 2.14, variance ≐ 4.6153, c.u. ≐ 12.615.
b. *See* Solutions Manual.
c. 12, 23, 24 are mild outliers; 26 is an extreme outlier.

2.47 mean = 152.68.

2.49 −0.282; 1.297.

2.51 Md ≐ 153; $I \doteq -.101$; data is not significantly skewed.

2.53 mean = 48.5; md = 47.5.

2.55 0.191.

2.57 yes.

2.59 $s \doteq 9.870$.

2.61

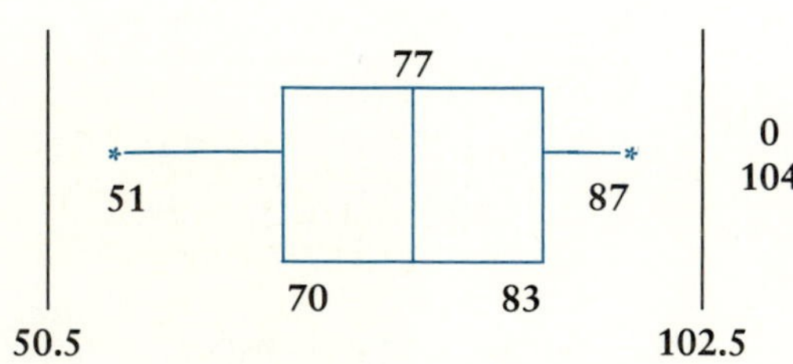

104 is a mild outlier.

2.63 $s(T) \doteq 5.52$, PSD ≐ 5.19

2.65 All are equally valid.

2.67 Test 2 is more highly variable.

2.69 29 to 62.

2.71 $I \doteq 0.36$; $s \doteq 8.17$, PSD $\doteq 8.89$; Md $= 44.5$, $\bar{x} \doteq 45.5$. Yes.

2.73 mean = 12.19; s.d. = 5.79.

2.75 **a.** 5%. **b.** not more than 25%.

2.77 Mr. Jones; yes, Mr. Jones; no.

Chapter 3

3.1 **a.** 0.4. **b.** 0.7. **c.** 0.7. **d.** 0.0.

3.3 **a.** 0.16. **b.** 0.24. **c.** 0.01. **d.** 0.36. **e.** 0.64. **f.** 0.12. **g.** 0.75.

3.5 0.7

3.7 Since all the possible outcomes are stated and they are mutually exclusive, the sum must be 1. But 0.44 + 0.29 + 0.17 = 0.90, which is not possible. He is mistaken.

3.9 **a.** 1280. **b.** 160.

3.11 0.111

3.13 2/15 + 0.9 = 1.033 (or 31/30), which is greater than 1. Since these are mutually exclusive, the statement is false. However, 2/15 + 0.8 = 14/15. Since there are other possibilities (tie, cancellation), the second fan's statement is not inconsistent with the rules of probability. This does not, however, increase his likelihood of being correct.

3.15 **a.** 0.209. **b.** 0.605.

3.17 P(cognitive dissonance) = 0.35; P(distorted reality) = 0.40; $P(cd \text{ and } dr)$ = 32/200 = 0.16. Thus, $P(cd \text{ or } dr)$ = 0.35 + 0.40 − 0.16 = 0.59. The probability of suffering from neither is 1 − 0.59 = 0.41, so out of 200, (0.41) × (200) = 82 would be free from both.

3.19 **a.** 0.554. **b.** 0.27.

3.21 Theoretically the 1 ace divides the 51 remaining cards into 2 parts which, on the average, will be equal. Therefore each part will contain 51/2 or about 25.5 cards, on the average. The part before the first ace will average this many and, adding the first ace, we obtain 26.5.

3.25 **a.** 1/50. **b.** 1/15. **c.** 2/15. **d.** 2/15.

3.27 125/216.

3.29 Since 10 take neither, 90 take one or the other or both. Since 80 + 60 = 140, and 140 − 90 = 50, then 50 take both. Thus $P(E \text{ and } M) = 0.50$. Now $P(M) = 0.60$, but $P(M|E) = 0.50/0.80 = 0.625$. Since they are not equal, the events are not independent. Other approaches are possible.

3.31 **a.** 0.28. **b.** 0.1165. **b.** 0.5714.

3.33 0.214.

3.35 **a.** 0.434. **b.** 0.358.

3.37 **a.** 0.06. **b.** 0.08. **c.** 0.14. **d.** 0.3. **e.** 0.88. **f.** 0.56.

3.39 0.36; 0.91.

3.41 0.97489.

3.43 2,598,960.

3.45 12.

3.47 150.

3.49 **a.** 30. **b.** 45. **c.** 55. **d.** 46. **e.** 56.

3.51 302,400.

3.53 **a.** 0.01.

3.55 **a.** Probability is about 0.31. **b.** about 0.213.

3.57 **a.** 0.40. **b.** 0.28. **c.** $P(A|B) = 0.40$. **d.** Yes. $P(A|B) = P(A)$. **e.** Yes. The events have no points in common.

3.59 **a.** 0.019. **b.** 0.25. **c.** 0.308.

3.61 **a.** 8. **b.** HHH,HHT,HTH,THH. **c.** HHH,HHT,HTH,HTT. **d.** HTH,HTT.

3.63 **a.** 256. **b.** 24.

3.65 **a.** 0.057. **b.** 0.0083. **c.** 0.692. **d.** 0.621.

3.67 **a.** 0.76. **b.** 0.86. **c.** 0.226. **d.** 0.50.

3.69 **a.** 210. **b.** 50. **c.** 80.

3.71 **a.** 210. **b.** 140.

3.73 10000.

3.75 The probability that a certain machine and its "back-up" will both fail is $(0.01)^2$ or .0001. Since there are five machines and the events are mutually exclusive (if a pair breaks down, no other pair has a chance), the probability that the assembly line will shut down is 0.0005.

3.77 **a.** 1/26. **b.** 1/156.

3.79 **a.** 0.74. **b.** $20/37 \doteq 0.5405$

3.81 1/4.

3.83 **a.** 1/225. **b.** 1/225. **c.** 11/312.

3.85 No; probability is $(.545)(.730) + (.012)(.922) + (.443)(.390) \doteq 0.582$.

3.87 No. To make valid inferences about a population from a sample, we must be reasonably certain that the sample is representative of the population.

Chapter 4

4.1 Yes. *See* Solutions Manual for graph.

4.3 Yes

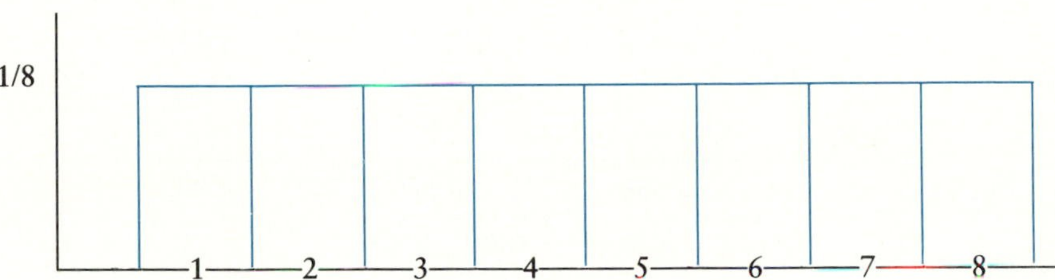

4.5 Yes

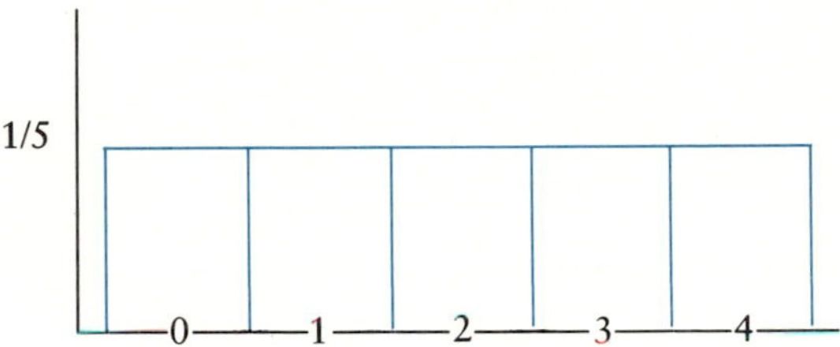

4.7 $P(1) = 2/n$; $P(2) = 5/n$; $P(3) = 10/n$; $P(4) = 17/n$.

See Solutions Manual for histogram.

The sum of the probabilities is $34/n$. Since it is a probability distribution, the sum is 1, so $n = 34$.

4.9 $P(2) = 2/5$. $P(3) = 2/5$. $P(4) = 1/5$.
Note that order is interchangeable up to, but not including, the last bulb attempted.

4.11 $P(0) = 16/81$; $P(4) = 1/81$; the sum of the probabilities must be 1, so that leaves 24/81 for $P(2)$.

4.13 4/3.

4.15 $\mu = 119.46$; $\sigma \doteq 9.58$.

4.17 $\mu = 1200$; $\sigma = 100$.

4.19 The probability of drawing 2 black cards is 25/102. The probability of drawing 1 black, 1 red is 26/51. Thus $E(x) = (25/102)(10) + (26/51)(5) = 5$, and we should pay $5.00 to play.

4.21 $\mu = 5.3$; $\sigma^2 = 0.81$; $\sigma = 0.9$.

4.23 If there are no other possibilities, the probability that he will lose $120,000 is $1 - (0.20 + 0.35 + 0.10 + 0.15 + 0.15) = 0.05$. Thus $E(x) = (0.20)(200{,}000) + (0.35)(120{,}000) + (0.10)(40{,}000) + (0.15)(0) + (0.15)(-60{,}000) + (0.05)(-120{,}000) = 71{,}000$.

4.25 The probability of winning is 1/38, the probability of losing is 37/38. If you win you gain \$35.00, if you lose you lose \$1.00. So $E(x) = (1/38)(35) + (37/38)(-1) = -1/19$.

4.27 $\mu = 1.25$; $\sigma^2 = 0.9375$.

4.29 $\mu = 1.82$; $\sigma^2 \doteq 1.75$.

4.31 **a.** $\mu = 0.4$; $\sigma \doteq 1.11$. **b.** $\mu = 10.4$; $\sigma \doteq 1.11$. **c.** $\mu = 1.2$; $\sigma \doteq 3.33 = 3(1.11)$.

4.33

x	0	1	2	3	4
$P(x)$	0.366	0.418	0.179	0.034	0.002

Thus $\mu = 8/9$; $\sigma^2 = 56/81$

4.35

x	0	1	2	3	4	5
$P(x)$	0.001	0.010	0.044	0.117	0.205	0.246

x	6	7	8	9	10
$P(x)$	0.205	0.117	0.044	0.010	0.001

$\mu = 5$, $\sigma^2 = 2.5$.

4.37 $P(x \leq 5) = 0.981$; $P(5) = 0.054$.

4.39 $P(x > 14) = 0.034$; $P(x < 14) = 0.922$.

4.41 $P(x < 12) = 0.006$; $P(6 \leq x \leq 15) = 0.189$.

4.43 $P(0 < x < 5) = 0.688$.

4.45 $r = 15$.

4.47 $\mu = 7{,}500$; $\sigma = 43.3$.

4.49 $P(2 \leq x \leq 5) \doteq 0.203$.

4.51 $P(x \geq 3) \doteq 0.896$.

4.53 $P(5) = 0.174$. $P(x \geq 8) = 0.109$.

4.55 **a.** $P(x \geq 4) = 0.303$. **b.** $P(2) = 0.249$. **c.** $P(x > 1) = 0.897$. **d.** $P(1 \leq x \leq 4) = 0.921$.

4.57 $P(2\text{ girls}) = 0.325$; $P(2\text{ boys}) = 0.414$.

4.59 **a.** 0.896. **b.** 0.648.

4.61 **a.** 0.983. **b.** 0.967. **c.** 0.990. **d.** 0.996. **e.** 0.999. **f.** 1 −.

4.63 0.633; 0.656.

4.65 $P(x > 10) = 0.952$; $P(10) = 0.031$.

4.67 $P(x \leq 5) = 0.15$.

4.69 mean = 2.2; var = 2.2. The table on the left (p. C-12) lists the values using the defining formula. The table on the right (p. C-12) lists the values using Table 3. Differences are due to the cumulative effect and rounding.

x	$P(x)$	x	$P(x)$
0	.111	0	.111
1	.244	1	.244
2	.268	2	.268
3	.197	3	.196
4	.108	4	.109
5	.048	5	.047
6	.017	6	.018
7	.005	7	.005
8	.002	8	.002

4.71 $m = 1.8$. $P(x \geq 2) = 0.537$.

4.73 $m = 8.0$. 0.184.

4.75 0.109; 16.35.

4.77 0.030.

4.79 $P(6 \leq x \leq 10) = 0.590$.

4.81 $112.

4.83. **a.** 0.04. **b.** 0.30 **c.** 0.66.

4.85 mean = 1.2; $\sigma^2 = 0.56$.

4.87 mean = 0.4; $\sigma \doteq 0.599$.

4.89 $\mu = 2.8$, $\sigma^2 = 0.56$.

4.91 **a.** 0.082. **b.** 0.328.

4.93 **a.** 0.00937. **b.** 0.01116. **c.** 0.015.

4.95 **a.** .0000223. **b.** .0000298. **c.** 0 +.

4.97 **a.** 0.736. **b.** Plan ii ($P \doteq 0.397$).

4.103 $\mu \doteq 3.33$; $\sigma \doteq 0.83$.

4.105 **a.** $n = 50$. **c.** 4.20. **d.** 0.92.

4.107 $1.60.

4.109 **a.** $\mu = 2.5$, $\sigma \doteq 1.37$. **b.** 0.94. **c.** 0.53.

4.111 6.

4.113 5/33.

4.115 **a.** 0.52. **b.** 0.90. **c.** 0.48.

4.117 **a.** A. **b.** A. **c.** B.

4.119 $P(x \geq 9) \doteq 0.046$; investigate.

4.121 0.488.

4.123 0.522; $P(x \geq 3) = 0.026$.

4.125 $P(x \geq 6) = 0.254$; not a strong case.

4.127 $1416/1716 \doteq 0.825$.

Chapter 5

5.1 **a.** −1.60. **b.** 1.30. **c.** −1.76. **d.** −0.29. **e.** 0.63. **f.** 2.50.

5.3 **a.** 0.4854. **b.** 0.3508 + 0.4382 = 0.7890. **c.** 0.4983 − 0.4406 = 0.0577. **d.** 0.1879 − 0.0478 = 0.1401. **e.** 0.4988 + 0.4484 = 0.9472. **f.** 0.1664 + 0.4817 = 0.6481.

5.5 **a.** 0.6826. **b.** 0.9544. **c.** 0.9974. **d.** 0.7994. **e.** 0.9000. **f.** 0.9500. **g.** 0.9802. **h.** 0.9902.

5.7 If 0.1230 of the area under the normal curve is to the right of z, 0.8770 is to the left of z, and 0.3770 is between 0 and z. Therefore $z = 1.16$. If 0.8770 of the area under the normal curve is to the right of z, 0.1230 is to the left of z, and 0.3770 is between 0 and z. Therefore $z = -1.16$.

5.9 If 0.1446 of the area of a normal distribution lies above a score, then 0.3554 lies between the score and the mean. Thus $z = 1.06$. Since $1.06 = (121 - 100)/s$, then $s^2 \doteq 392.5$.

5.11 About 10.4% of the population spends more than 45 minutes.

5.13 About 61% of the insurance policies are between \$600 and \$250.

5.15 Since $p = 1/2$, $np = 20$, $n(1-p) = 20$, $np(1-p) = 10$, so we can use a normal distribution with mean = 20, $s = 3.16$. Using a correction factor and including 25, $z = (24.5 - 20)/3.16 = 1.42$, so $P(X \geq 25) = P(x > 24.5) = 0.5000 - 0.4222 = 0.0778$. Similarly if y represents the number of tails, $P(y > 25) = P(y > 24.5) = 0.0778$. Assuming that there is no exchange of money in the other outcomes, the expected value is $15(0.0778) + (-10)(0.0778) = 0.389$ dollars, or about \$0.39.

5.17 Here $p = 0.30$, $np = 22.5$, $n(1-p) = 52.5$, and $np(1-p) = 15.75$; thus we can use a normal curve with mean = 22.5, $s = 3.97$. Since we want to include 10 and 25, we have $P(10 \leq x \leq 25) = P(9.5 < x < 25.5)$. For the lower limit, $z = (9.5 - 22.5)/3.97 \doteq -3.27$ and virtually all the area is to the right of 9.5. For the upper limit $z = (25.5 - 22.5)/3.97 = 0.76$ and 0.2764 of the area is between 22.5 and 25.5. Then $P(10 \leq x \leq 25) = 0.7764$.

5.19 About 2.7%.

5.21 11.51%.

5.23 Here we have $n = 85$. Now we are interested in those opposing school attendance for these children, so $p = 0.40$. Since $np = 34.0$, $n(1-p) = 51.0$, $np(1-p) = 20.4$, and $s = 4.52$, we can use the normal approximation. A majority of 85 is 43, so we want $P(x > 42.5)$. Hence $z = (42.5 - 34.0)/4.52 = 1.88$. The area between 34.0 and 42.5 is 0.4699, so $P(x \geq 43) = 0.0301$.

5.25 About 0.86.

5.27 You must leave before 7:28.

5.29 0.0985.

5.31 0.1335.

5.33 1.07%.

5.35 **a.** and **d.** appropriate; **b.** and **c.** not appropriate.

5.37 Considering the heights are measured to the nearest inch, the data appear to be normal.

5.39 **a.** 0.8050. **b.** 0.9627.

5.41 Now $p = 1/2$, and we have a binomial experiment. Since $np = 50$, $np(1-p) = 25$, we can use a normal curve with mean = 50, $s = 5$. Then $P(x \geq 60) = P(x > 59.5)$ and $z = (59.5 - 50)/5 = 1.90$. The area under the curve between 50 and 59.5 is 0.4713, so $P(x \geq 60) = 0.0287$.

5.43 **a.** About 0.63. **b.** 0.0301. **c.** 0.1736. **d.** 110.8. **e.** 117.9. **f.** 86.1. **g.** 0.3944. **h.** about 0.73.

5.45 76.27.

5.47 About 0.10.

5.49 Since $p = 1/5$, mean $= np = 40$, $np(1-p) = 32$, $s = 5.66$. We want $P(30<x<50) = P(29.5<x<50.5)$. $z = (29.5 - 40)/5.66 = -1.86$ for the lower limit, and $z = 1.86$ for the upper limit. Thus $P(30<x<50) = 2(0.4686) = 0.9372$.

5.51 **a.** About 0.43. **b.** 93.

Chapter 6

6.1 **a.** 10. **b.** 3.16. **c.** 1. **d.** 16.67. **e.** 8.33. **f.** 12.5. **g.** 8.84. **h.** 3.125.

6.3 **a.** 0.1660. **b.** 0.1660.

6.5 About 0.47.

6.7 1,560.

6.9 0.0038.

6.11 0.0116.

6.13 0.0475.

6.15 mean $\doteq 21.4$; $s_{\bar{x}} = 0.24$.

6.17 **a.** 0.8414. **b.** 0.5098. **c.** 0.9990. **d.** 0.3328. **e.** 0.9652.

6.19 **a.** 315.288 to 320.312. **b.** 314.806 to 320.794. **c.** 314.241 to 321.359. **d.** 313.859 to 321.741.

6.21 **a.** 0.1092 to 0.1148. **b.** 0.1087 to 0.1153. **c.** 0.1081 to 0.1159. **d.** 0.1077 to 0.1163.

6.23 0.994.

6.25 9.71 to 9.95; 9.67 to 9.99.

6.27 A 0.95 confidence interval is 147.6 to 150.4. Since 150 is in the confidence interval, the shipment will not be rejected.

6.29 **a.** 11,212 to 12,076. **b.** 11,223 to 12,065.
The differences are very slight; only about 1/12 of one percent.

6.31 $E = (2.58)(22.6/14.14) \doteq 4.12$.

6.33 3.79 to 5.07 pounds; claim is reasonable.

6.35 **a.** 0.3182. **b.** 0.4582. **c.** 0.5878. **d.** 0.7776. **e.** 0.8968.

6.37 **a.** 0.11 to 0.19. **b.** 0.70 to 0.80. **c.** 0.07 to 0.13. **d.** 0.31 to 0.43.

6.39 0.41 to 0.50.

6.41 **a.** 0.0562 to 0.1038. **b.** 0.1446 to 0.2554.

6.43 For those who saw "The Dr. Ruth Show" the proportion is $p = 244/863 \doteq 0.283$. Thus $s_p = 0.015$. $(1.645)(0.015) \doteq 0.025$, so the 90% confidence interval is 0.258 to 0.308. For the proportion offended, $p = 27/244 \doteq 0.111$, $s_p \doteq 0.020$. $(1.645)(0.020) \doteq 0.033$. Thus the interval would be 0.078 to 0.144.

6.45 **a.** 19.2. **b.** 87.3. **c.** 0.037. **d.** 5.45. **e.** 5.28.

6.47 **a.** 0.106 to 0.118. **b.** 0.105 to 0.119. **c.** 0.103 to 0.121. **d.** 0.102 to 0.122.

6.49

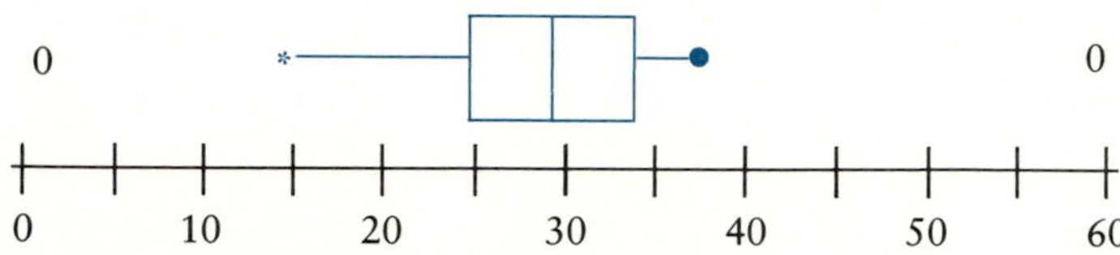

a. 2 and 58 are mild outliers.
b. $\overline{x}(T) = 28.875$; $s(W) \doteq 4.776$.

$$s_{\overline{x}(T)} = \frac{s(W)}{\sqrt{8}}\sqrt{\frac{11}{7}} \doteq 2.1167.$$

24.86 to 32.89.
c. 21.57 to 35.43.
d. The trimmed-mean interval.

6.51 $\overline{x} \doteq 0.82$; $s = 0.035$, $s_{\overline{x}} = 0.0175$. For 0.95 confidence, $\alpha = 0.05$, $t = 3.182$ (for 3 degrees of freedom), so $E = (0.0175)(3.182) = 0.06$. Thus a 0.95 confidence interval is 0.76 to 0.88. For 0.99 confidence, $\alpha = 0.01$, $t = 5.841$, and $E = (0.0175)(5.841) = 0.10$, and a 0.99 confidence interval is 0.72 to 0.92.

6.53 67.6 to 77.2

6.55 113.38 to 126.63. Yes, three mild outliers.

6.57 2.98 to 6.28; estimate is reasonable.

6.59

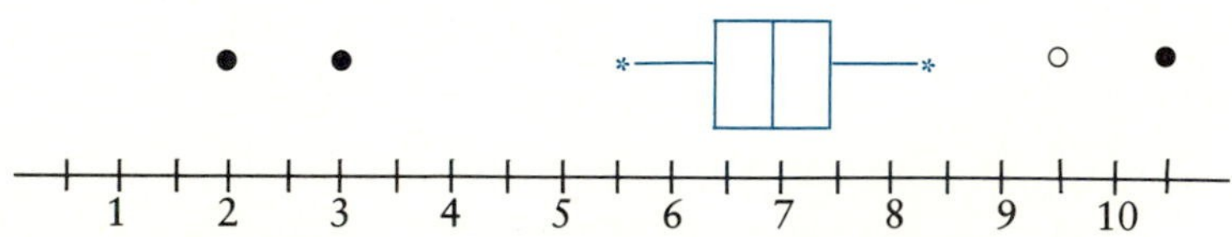

6.37 to 7.43.

6.61 **a.** 246. **b.** 384.

6.63 **a.** 148. **b.** 210. **c.** 297. **d.** 365.

6.65 Yes; 281 or 282.

6.67 1949; need 945 who watched the show. Using the sample result we know we are 95% confident that at least 25.8% of the population watched the show. Since 945 is 25.8% of 3663, we could use a sample of 3663.

6.69 **a.** 9.218 to 24.465. **b.** 64.55 to 253.60. **c.** 0.030 to 0.058. **d.** 6.707 to 12.105. **e.** 9.847 to 15.815.

6.71 **a.** 8.69 to 22.84. **b.** 8.04 to 25.51. **c.** 7.36 to 29.13. **d.** 6.94 to 32.00.

6.73 **a.** 0.0107 to 0.0166. **b.** 0.0104 to 0.0175. **c.** 0.0099 to 0.0185. **d.** 0.0097 to 0.0193.

6.75 3.35 to 4.87 minutes; 1.44 to 2.57 minutes.

6.77 **a.** 057 to 0.65. **b.** 0.0105 to 0.0192. **c.** 0.0188 to 0.0346.

6.79 0.029 to 0.054.

6.85 **a.** 499. **b.** 666.

6.87 10.5 to 12.3 years; 10.3 to 12.5 years.

6.89 96.

6.91 0.605 to 0.655.

6.93 0.323 to 0.534; 0.289 to 0.568.

6.95 3.10 to 3.30.

6.97 138 or 139.

6.99 0.0764.

Chapter 7

7.1 $H_0 : \pi = 0.6, H_1 : \pi < 0.6$.

7.3 $H_0 : \mu = 1225.0, H_1 : \mu \neq 1225.0$.

7.5 $H_0 : \sigma^2 = 0.001765, H_1 : \sigma^2 \neq 0.001765$.

7.7 If we reject the hypothesis and the true mean is between \$78 and \$125. If we accept the hypothesis and the mean is not between \$78 and \$125.

7.9 $H_0 : \mu = 20{,}278, H_1 : \mu \neq 20{,}278$.

7.11 If μ_0 is the mean amount of pollutants discharged by regular gasoline we have $H_0 : \mu = \mu_0, H_1 : \mu < \mu_0$.

7.13 If μ_1 and μ_2 are the mean number of verbs learned in a given time period with the two methods, we may test $H_0 : \mu_1 = \mu_2$ against $H_1 : \mu_1 \neq \mu_2$.

7.15 **a.** $H_0 : \pi = 0.9, H_1 : \pi > 0.9$. **b.** $H_0 : \pi = 0.9, H_1 : \pi < 0.9$.

7.17 Let μ_1 = mileage with the additive and μ_2 = mileage without the additive. In each case we have $H_0 : \mu_1 = \mu_2$. If H_1 is $\mu_1 \neq \mu_2$, they are simply trying to determine if the additive affects mileage. If they want to use the additive only if it increases mileage, they would have $H_1 : \mu_1 > \mu_2$. If they want to use the additive unless it is proved ineffective, they would have $H_1 : \mu_1 < \mu_2$.

7.19 **a.** Yes; $P \doteq 0.0667$. **b.** No; $P = 0.04$. **c.** No; $\bar{x} > 16.42$

7.21 **a.** $t^* \doteq 1.22$; $t_{.025}\{4\} = 2.776$. Fail to reject H_0. **b.** $t^* \doteq 2.73$; $t_{.025}\{24\} \doteq 2.064$. Reject H_0. **c.** $z^* \doteq 3.45$; $z_{.025} = 1.96$. Reject H_0. **d.** $z^* \doteq 5.45$; reject H_0. **e.** $z^* \doteq 7.71$; reject H_0.

7.23 $z^* \doteq 2.41$; $P \doteq 0.008$. **a., b., c., d.** reject H_0; $\mu > 1.60$. **e.** Fail to reject H_0.

7.25 Conclude $\mu \neq 3450$ if $|t^*| > 2.262$.
a. $t^* = 3.653$. **b.** $t^* = 2.569$. **c.** $t^* = 2.017$. **d.** $t^* = 1.345$.
e. $t^* = 0.919$. Reject H_0 in **a., b.**; fail to reject H_0 in **c., d., e.**

7.27 About **a.** 0.0042. **b.** 0.0050. **c.** 0.0066.

7.29 $z^* \doteq -2.42$. **a., b., c.** Reject H_0. **d.** Fail to reject H_0.

7.31 $z^* = 1.60$; no significant difference.

7.33 $t^* \doteq 1.299$; $t_{.01}\{7\} = 2.998$. Policy is not violated.

7.35 **a.** $t^* \doteq 1.765$; $t_{.05}\{8\} = 1.860$. Claim is not substantiated. **b.** For $n = 19$, $t^* \doteq 2.564$ and $t_{.01}\{18\} = 2.552$.

7.37 $\overline{x}(T) \doteq 15.98$, $s(W) \doteq 0.03207$; $t^* \doteq -1.29$; $t_{.025}\{5\} = 2.571$. Conclude $\mu = 16$.

7.39 $z^* \doteq 1.43$. Accept the shipment. $P \doteq .0764$.

7.41 $t^* \doteq 4.348$; $t_{.025}\{24\} = 2.064$. Species is different.

7.43 $t^* \doteq -2.41$. **a.** Product does not meet standard; do not accept. **b.** Product is substandard; do not accept it.

7.45 $t^* \doteq -1.069$; $t_{.025}\{15\} = 2.131$. $\mu = 16$. $|t^*|$ is larger and the critical value is smaller.

7.47 $z^* \doteq 2.34$; market the tires.

7.49 $t^* \doteq 2.47$; $t_{.05}\{11\} = 1.796$. Claim is substantiated.

7.51 About 22.256 to 23.744 ounces for the ten.

7.53 $z^* \doteq 1.76$; **a., b.** Reject H_0. **c., d., e.** Fail to reject H_0.

7.55 **a.** 0.0012. **b.** 0.25. **c.** 0.165. **d., e.** $P < .0001$.

7.57 $P = 0.254$. Fail to reject H_0 in each case.

7.59 $z^* \doteq 0.83$; no.

7.61 $z^* \doteq 3.19$; $\pi > 0.40$.

7.63 $z^* \doteq 3.05$; brand B.

7.65 $z^* = 2.20$; $z_{.01} = 2.33$. Probably should reserve judgment.

7.67 **a.** $x \leqslant 3$. **b.** $x \leqslant 3$.

7.69 Lot will be accepted if there are no defectives, or if there is one defective. Otherwise it will be rejected. This assumes lot will be accepted only if $\pi < 0.05$.

7.71 $n = 1534$, $p \doteq 0.082$; $z^* \doteq 2.03$. Rate is higher than 6.9%.

7.73 **a.** $\chi^{2*} \doteq 7.14$; $\chi^2_{.975}\{9\} = 2.70$. Fail to reject H_0. **b.** $\chi^{2*} \doteq 11.11$; $\chi^2_{.975}\{14\} = 5.629$. Fail to reject H_0. **c.** $\chi^{2*} \doteq 15.08$; $\chi^2_{.975}\{19\} = 8.906$. Fail to rejects H_0. **d.** $\chi^{2*} \doteq 38.90$; $\chi^2_{.975}\{49\} \doteq 31.7$. Fail to reject H_0.

7.75 $\chi^{2*} \doteq 12.25$. $\chi^2 < \chi^2_{1-\alpha/2}\{29\}$ for α in **a.**, **b.**, **c.**, **d**; reject H_0.

7.77 $\chi^{2*} \doteq 78.606$; $\chi^2_{.01}\{45\} \doteq 69.9$. $\sigma > 12$.

7.79 Reject A and C. Only lot B passes both tests.

7.83 $\chi^{2*} \doteq 6.539$; $\sigma < 60$.

7.87 10.54 to 12.26.

7.89 **a.** $z^* \doteq 0.58$. **b.** $z^* \doteq 1.15$. **c.** $z^* \doteq 2.31$. $z_{.01} = 2.33$. Fail to reject H_0 in each case.

7.91 **a.** $t^* \doteq -3.227$; $t_{.05}\{14\} = 1.761$. **b.** $z^* \doteq -3.23$; $z_{.05} = 1.645$. Reject H_0 in each case; $\mu < 150$.

7.93 $t^* \doteq -3.126$; $t_{.01}\{8\} = 2.896$. Yes; $\mu < 30$.

7.95 $z^* \doteq -3.16$; $P \doteq 0.0008$.

7.97 $z^* \doteq -2.26$; yes.

7.99 $t^* \doteq -2.236$; no.

7.101 $z^* \doteq 2.54$; yes.

7.103 $z^* = -2.11$; no.

7.105 $z^* = -1.83$; yes.

7.107 $z^* \doteq 2.32$; $z_{.05} = 1.645$. Reject H_0.

7.109 $z^* = 3.10$; contention is supported.

7.111 $z^* \doteq -1.33$; no preference.

7.113 $z^* = -1.70$; yes. $P \doteq 0.045$.

7.115 $z^* \doteq 3.22$; $P \doteq 0.012$

7.117 Mean is not 150 ($t^* = 2.155$) but variance is 7.5. Note that it would be reasonable to use $\sigma = 7.5$ and obtain $z^* \doteq 2.39$. The result is the same.

Chapter 8

8.1 **a.** 6.42. **b.** 76.0. **c.** 0.018. **d.** 2.58. **e.** 12.826.

8.3 **a.** 0.06 to 12.46. **b.** 2.24 to 10.16. **c.** 4.22 to 8.18. **d.** 5.21 to 7.19.

8.5 **a.** 0.32 to 1.48. **b.** −2.28 to 4.08. **c.** 8.32 to 9.48. **d.** 5.72 to 12.08.

8.7 **a.** −0.12 to 1.92. **b.** −4.69 to 6.49. **c.** 7.88 to 9.92. **d.** 3.31 to 14.49.

8.9 **a.** 0.019 to 0.031. **b.** 0.017 to 0.033. **c.** 0.016 to 0.034. **d.** 0.014 to 0.036. (Note that t' must be used; $\nu \doteq 20$.)

8.11 -43.0 to -0.8.

8.13 \$1168 to \$3972.

8.15 \$902 to \$2522 higher mean salary for men.

8.17 **a.** No outliers. **b.** $\bar{x} \doteq 5.133$, $s_1 \doteq 1.1255$; $\bar{x}_2 \doteq 3.316$, $s_2 \doteq 1.857$. **c.** 0.70 to 2.93.

8.19 -0.93 to 1.05.

8.21 **a.** $z^* \doteq 2.54$; $\mu_1 \neq \mu_2$. **b.** $z^* \doteq 0.47$; differences not significant. **c.** $z^* \doteq 25.12$; $\mu_1 \neq \mu_2$ **d.** $z^* \doteq 4.60$; $\mu_1 \neq \mu_2$.

8.23 $t_{.05}\{42\} \doteq 1.683$. **a.** $t^* \doteq 1.49$. **b.** $t^* \doteq 0.27$. **c.** $t^* \doteq 14.71$. **d.** $t^* \doteq 2.68$. Differences are significant in c and d.

8.25 $\nu = 20$ (see Problem 8.9), $t' \doteq 6.76$. All tests are significant.

8.27 $z^* \doteq 0.62$; $P \doteq 0.5352$.

8.29 $z^* \doteq 1.58$; $P \doteq 0.1142$.

8.31 $t^* \doteq 1.89$; $P < 0.05$.

8.33 $z^* \doteq 2.15$; $P \doteq 0.0316$.

8.35 $t^* \doteq -0.548$; no significant difference.

8.37 $t^* \doteq 1.418$; procedure is not effective.

8.39 **a.** 0.28. **b.** 0.19. **c.** 0.17. **d.** 0.12. **e.** 0.46.

8.41 **a.** $z^* \doteq -1.85$; reserve judgment. **b.** $z^* \doteq 1.69$; $\pi_1 > \pi_2$. **c.** $z^* \doteq 2.31$; $\pi_1 - \pi_2 > 0.15$. **d.** $z^* \doteq -0.25$; $\pi_1 - \pi_2 \not< -0.10$. **e.** $z^* \doteq 0.19$; $\pi_1 = \pi_2$.

8.43 **a.** -0.14 to -0.02. **b.** -0.15 to -0.01. **c.** 0.03 to 0.13. **d.** 0.004 to 0.16.

8.45 **a.** 0.11 to 0.29. **b.** 0.10 to 0.30. **c.** 0.08 to 0.32. **d.** 0.06 to 0.34.

8.47 $z^* \doteq 0.88$; firm B.

8.49 0.023 to 0.357.

8.51 -0.038 to 0.167.

8.53 $z^* \doteq -1.18$; no.

8.55 $z^* \doteq 2.67$; Clearwater Park.

8.57 $z^* \doteq -2.14$; use company B.

8.59 **a.** 569. **b.** 768.

8.61 Using $\sigma^2 \doteq 9.33$, we have $n \doteq 72$.

8.63 **a.** Using $F_{.05}\{14,14\} \doteq 2.483$, 1.891 to 11.66. **b.** Using $F_{.025}\{14,14\} \doteq 2.983$, 1.57 to 14.00. **c.** Using $F_{.01}\{14,14\} \doteq 3.71$, 1.265 to 17.42. **d.** Using $F_{.005}\{14,14\} \doteq 4.31$, 1.089 to 20.23.

8.65 **a.** 0.135 to 2.106. **b.** 0.184 to 2.868. **c.** 0.231 to 1.521. **d.** 0.276 to 1.633.

8.67 $F^* \doteq 9.88 > 4.76 = F_{.025}\{10,7\}$; not equal.

8.69 $F^* \doteq 2.20 > 1.62 \doteq F_{.025}\{65,77\}$; not equal.

8.71 $t^* \doteq 1.208$; do not continue campaign.

8.73 **a.** $z^* \doteq 2.20$; $\mu_1 \neq \mu_2$. **b.** 0.95 to 16.45.

8.75 **a.** $t^* \doteq 1.76$; $\mu_1 = \mu_2$. **b.** -1.31 to 18.71.

8.77 **a.** $F^* \doteq 3.13$; $\sigma_1^2 \neq \sigma_2^2$. **b.** $t'^* = 1.51$; $t_{.025}\{39\} \doteq 2.023$.

8.79 0.07 to 0.43

8.81 $t^* \doteq -1.62$; $t_{.05}\{8\} = 1.86$.

8.83 -2.03 to 0.63.

8.85 $F^* = 1.41$; $\sigma_1^2 = \sigma_2^2$. The t test is appropriate.

8.87 358.

8.89 **a.** $z^* \doteq 5.42$, difference is significant. **b.** 0.13 to 0.41.

8.91 $F^* \doteq 1.34$; yes.

8.93 $z^* \doteq 1.56$; difference is not significant.

8.95 $z^* \doteq -0.75$; no difference.

8.97 $F^* \doteq 5.64$; population variances are not equal. Since $F_{.005}\{299,299\} \doteq 1$, no confidence interval may be determined.

8.99 $z^* \doteq 2.65$; $P \doteq 0.004$.

8.101 $z^* \doteq 0.41, 2.22, 1.58, 2.22$. Hysteria and schizophrenia.

Chapter 9

9.1 $\chi^{2*} = 2.20$; die is fair.

9.3 $\chi^{2*} \doteq 17.192$; $\chi^2_{.10}\{11\} = 17.28$. Hypothesis is not supported.

9.5 $\chi^{2*} \doteq 6.383$; $\chi^2_{.05}\{3\} = 7.815$. Difference is not significant.

9.7 $\chi^{2*} \doteq 303.7$; $\chi^2_{.01}\{3\} = 11.34$. Not at all representative.

9.9 $\chi^{2*} \doteq 19.26$; $\chi^2_{.01}\{3\} = 11.34$. Not normal. Distribution is flat in the middle, and has a preponderant proportion in the 18–25 group.

9.11 $\chi^{2*} \doteq 3.845$; $\chi^2_{.05}\{1\} = 3.841$. Factors are not independent.

9.13 $\chi^{2*} \doteq 22.206$; $\chi^2_{.05}\{4\} = 9.488$. Proportions are not all equal.

9.15 $\chi^{2*} \doteq 3.29$; $\chi^2_{.10}\{3\} = 6.251$. Responses are independent of sex.

9.17 $\chi^{2*} \doteq 81.347$; $\chi^2_{.05}\{4\} = 9.488$. Opinions differ by educational level.

9.19 $\chi^{2*} = 5.00$ (Yates' correction factor required); $\chi^{2*}\{1\} = 3.841$. Factors are related.

9.21 $\chi^{2*} \doteq 17.558$; $\chi^2_{.01}\{3\} = 11.34$. Factors are not independent.

9.23 $\chi^{2*} \doteq 5.405$; $\chi^2_{.05}\{4\} = 9.488$. No association.

9.25 No difference.

9.27 $\chi^{2*} \doteq 0.013$; no differences.

9.31 $\chi^{2*} \doteq 38.356$; $\chi^2_{.05}\{9\} = 16.92$. Factors are not independent.

9.33 (Binomial model needed.) $\chi^{2*} \doteq 4.533$; $\chi^2_{.05}\{3\} = 7.815$. The coin is not biased.

9.35 $\chi^{2*} \doteq 9.489$; $\chi^2_{.05}\{3\} = 7.815$. The contention is supported.

9.37 $\chi^{2*} \doteq 7.024$; $\chi^2_{.01}\{4\} = 13.28$. The factors are independent.

9.39 $\chi^{2*} \doteq 4.476$; $\chi^2_{.05}\{4\} = 9.488$. Strain makes no difference.

9.41 $m = 2.5$. We combine number of customers $= 6, 7, 8$. Then $\chi^{2*} \doteq 0.763$. Agreement is excellent.

Chapter 10

10.1 $P = 0.048$; reject $H_0 : \text{Md} =$ the value in each case.

10.3 Zeroes each are 1.5; $T^* = 36.5$. $T_{.025}\{18\} = 40$, $T_{.01}\{18\} = 33$. Thus $.02 < P < .05$. **a., b.** Reject H_0. **c.** Fail to reject H_0.

10.5 $n = 29$, $x^*_+ = 22$; $z^* = 2.60$, $z_{.05} = 1.645$. The conjecture is correct; $\text{Md} > 8$.

10.7 $P = 0.172$; conjecture is not substantiated.

10.9 $z^* = 1.33$; differences are not significant.

10.11 $T^* = 1$; $T_{.05}\{6\} = 2$. Reject H_0; brand A wears less.

10.13 $T^*_+ = 370$; $z^* \doteq 2.83$. $P = .002$. No, since pairs may differ in many ways. $P \doteq 0.02$ for problem 10.12.

10.15 $T^* = 1.5$, $T_{.05}\{6\} = 2$. Procedure is effective.

10.17 **a.** Reject H_0 if $U^* < 6$. **b.** Reject H_0 if $U^* < 14$. **c.** Reject H_0 if $|z^*| > 1.96$.

10.19 $U^* = 8.5$; $U_{.025}\{7, 8\} = 10$.

10.21 $|z^*| \doteq 0.66$; differences are not significant.

10.23 $U^* = 14$; $U_{.005}\{8,9\} = 9$. Fail to reject H_0; probably need to continue sampling.

10.25 $U^* = 75.5$; $z^* = -1.54$. Display is not effective.

10.31 $z^* = -0.80$; no significant difference.

10.33 **a.** $0 - 16(95.1\%)$. **b.** *See* Solutions Manual.

10.35 $z^* \doteq 1.84 > 1.645$. Group memorizes better than average.

10.37 $z^* \doteq (943 - 637.5)/103.59 \doteq 2.95$. Group memorizes better than average.

10.39 $x_+^* = 4$; $P = 0.377$. Not effective.

10.41 $U^* = 0.5$; $U_{.025}\{4, 5\} = 1$. Therefore there is a difference.

10.43 $T^* = 23$; $T_{.05}\{14\} = 26$. Yes, learning took place.

10.45 $T^* = 1$. Difference is not significant.

10.47 $T^* = 9$; $T_{.025}\{7\} = 2$. Md = 16.

10.49 $T^* = 46.5$; $T_{.025}\{16\} = 30$. Md = 16.

10.51 **a.** $T^* = 39$; $T_{.01}\{9\} = 3$. Fail to reject H_0 : Md = 10 in favor of H_1 : Md > 10. Product will not pass.

b. $T^* = 6$. Fail to reject H_0 : Md = 10 in favor of H_1 : Md < 10. Product will pass.

10.53 $U^* = 6.5 : U_{.05}\{5, 5\} = 3$. The procedure is not efficient.

Chapter 11

11.1 DfTR = 4, dfTO = 26; SSTR = 1215.6, SSE = 267.3.

11.3 **a.** DfTR = 2, dfE = 17; SSTR = 56.2714, SSE = 74.9286. **b.** $F^* = 6.38$; differences are significant.

11.5 $F_{.05}\{3,80\} \doteq 2.73$ for each test. Then **a.** MSTR = 1431.36, MSE = 320.5125, $F^* = 4.47$; differences are significant. **b.** MSTR = 1431.36, MSE = 554.9625, $F^* = 2.58$; differences are not significant. **c.** MSTR = 395.36, MSE = 320.5125, $F^* = 1.23$; differences are not significant.

11.7 **a.** The completely randomized design. **b.** Prior learning of the maze; experimental, since rats are assigned to the groups at random. **c.** $F^* = 2.06$, $P = 0.1322$; differences are not significant. **d.** 10.43 to 15.57.

11.9 **a.** $F^* = 1.53$; differences are not significant. **b.** School C (20.44); school E (10.00). **c.** 8.78 to 20.34.

11.11 $F^* = 6.54$, $P = 0.0048$, $F_{.05}\{2,27\} \doteq 3.35$. Differences are significant.

11.13 **a.** 52.67. **b.** 38.97. **c.** 35.22. **d.** 31.18. **e.** 28.30.

11.15 1.108.

11.17 **a.** HSD $\doteq$ 14.53; groups 2 and 3 differ significantly and groups 2 and 4 differ significantly. **b.** A 95% confidence interval for $\mu_2 - \mu_3$ is 2.3 to 31.3; a 95% confidence interval for $\mu_2 - \mu_4$ is 3.5 to 32.5.

11.19 HSD = 3.2763; culture E takes significantly longer than the others. No other differences are significant.

11.21 **a.** If program = 1, tape = 2, television = 3, the confidence intervals are as

follows: for $\mu_2 - \mu_1$, 0.36 to 9.02; for $\mu_3 - \mu_1$, 1.63 to 10.55; for $\mu_3 - \mu_2$, -3.16 to 5.96.
b. If confidence intervals do not contain zero, the difference is significant.
c. We can tell that using the program seems to be worst, but not which of the other methods is better.

11.23 $F^*_{max} \doteq 3.89$; $F_{max(.05)}\{4,12\} = 4.79$. They are homogeneous.

11.25 $F^*_{max} \doteq 7.75$; $F_{max(.01)}\{5,10\} \doteq 9.6$. They are homogeneous.

11.27 $F_{max(.05)}\{4,20\} = 3.29$. **a.** $F^*_{max} = 2.25$. **b.** $F^*_{max} = 2.62$. **c.** $F^*_{max} = 2.25$. Variances are homogeneous in each case.

11.29 No; differences are not significant with the usual ANOVA, and the Kruskal-Wallis test is less powerful.

11.31 $H^* \doteq 8.80$; $\chi^2_{.05}\{2\} = 5.991$; differences are significant. We use $z_{.0083} \doteq 2.39$. Using the designations as in problem 11.21, the confidence intervals are as follows: for $\overline{R}_2 - \overline{R}_1$, -0.66 to 17.73; for $\overline{R}_3 - \overline{R}_1$, 1.51 to 20.43; for $\overline{R}_3 - \overline{R}_2$, -7.23 to 12.10. The only conclusion is that the group with television had a significantly higher rank average than the group with the programmed text only.

11.33 **a.** Data sets are small and bunched, but appear reasonably symmetric. Comparison of PSD with s for each sample shows reasonably close agreement except for method I; here $s \doteq 8.24$, PSD $\doteq 11.11$, but overall, deviations from normality do not appear to be serious.
b. $F^*_{max} = 2.76$; $F_{max(.05)}\{4,12\} = 4.79$. The variances are homogeneous.
c. Using the usual ANOVA, we find $F^* = 2.76$; $P = 0.0539$; $F_{.05}\{3,43\} \doteq 2.83$. Differences are not significant. Perhaps continued study would be worthwhile.

11.35 DfE = 32; SSTR = 1.3767, SSE = 0.6007; MSE $\doteq$ 0.0188.

11.37 **a.** DfTR = 3, dfE = 27; SSTR $\doteq$ 221.4896, SSE $\doteq$ 339.60714; MSTR $\doteq$ 73.8299, MSE $\doteq$ 12.5780. **b.** $F^* \doteq 5.87$, differences are significant. **c.** Differences between groups 3 and 4 and groups 1 and 4 are significant.

11.39 $H^* = 11.263 > 7.815 = \chi^2_{.05}\{3\}$. Differences are significant. There are six comparisons, so we use $z_{.0042} \doteq 2.64$. Group 4 has a significantly higher rank average than either group 1 or group 3.

11.41 **a.** $F^* = 5.39$; differences are significant. **b.** Means are significantly different for systems B and C. **c.** 172.04 to 188.18. **d.** No; system A could be better than system B or worse than system C.

11.43 **a.** $F^* = 5.5$; differences are significant. **b.** Treatment C has a significantly lower mean than treatment A or B. **c.** Yes—treatment C.

11.45 $F^* = 0.97$; differences are not significant.

11.47 **a.** $F^* \doteq 2.85$; $F_{.01}\{3,116\} \doteq 3.80$. Differences are not significant. **c.** $F^*_{max} \doteq 2.37$; yes.

11.49 **a.** $F^* \doteq 3.93$; $F_{.05}\{2,15\} \doteq 3.68$. Differences are significant.
b. Confidence interval for μ_1—μ_3 is -4.99 to 7.99; confidence interval for

$\mu_1 - \mu_2$ is 0.18 to 13.15; confidence interval for $\mu_2 - \mu_3$ is -11.65 to 1.32. The difference between the means of groups 1 and 2 is significant. Since we know that display 1 is better than display 2, display 2 can be eliminated and the testing continue with the other two displays.

Chapter 12

12.1 DfTR = 4, dfTO = 19; SSTR = 1265.6, SSB = 1071.5, SSE = 145.8; MSB ≑ 357.17.

12.3 **a.** SSTO ≑ 760.9167, SSA ≑ 140.9167m, SSB ≑ 561.1667. **b.** SSE = 58.8333. **c.** MSTR ≑ 46.9722, MSB ≑ 280.5833, MSE ≑ 9.8056. **d.** $F^* = 4.79$; $F_{.05}\{3,6\} = 4.76$. Differences are significant.

12.5 **a.** The randomized block design. **b.** The procedures. **c.** Experimental. **d.** SSTO ≑ 45.6895, SS(Machines) ≑ 32.267, SS(Procedures) ≑ 7.1175. **e.** $F^* = 21.82$; differences are significant. **f.** No; A and B differ significantly from C and D, but not from each other.

12.7 **a.** $F^* \doteq 6.17$, $F_{.05}\{4,12\} = 3.26$. There are differences due to displays. **b.** $P < 0.01$.

12.9 **a.** $F^* = 2.92$; differences are not significant. **b.** $F^* = 23.35$; differences are significant. **c.** 0.10, using Bonferroni's inequality; 0.0975, using Kimball's inequality.

12.11 DfA = 5, dfAB = 15; SSA = 4158.50, SSB = 2575.08; MSB = 858.36, MSAB = 192.30; MSE = 717.07.

12.13 **a.** Yes. **b.** Rank. **c.** No, because of the differences in rank.

12.15 **a.** DfA = 3, dfB = 2, dfAB = 6, dfE = 12; SSA ≑ 70.4583, SSB ≑ 280.5833, SSAB ≑ 29.4167, SSE = 46.5, SSTO ≑ 426.9583; MSA ≑ 23.4861, MSB ≑ 140.2917, MSAB ≑ 4.9028, MSE = 3.875.
b. $F^* = 1.27$; interaction is not significant.
c. $F^* = 6.06$ (factor A), $F^* = 36.20$ (factor B). Both are significant.
d. See Solutions Manual. The diagrams show little interaction.

12.17 **a.** Let plan be factor A and branch be factor B. Then dfA = 2, dfB = 3, dfAB = 6, dfE = 24; SSA ≑ 412.1167, SSB ≑ 2506.8889, SSAB ≑ 150.2778, SSE = 1266.6667, SSTO = 4296; MSA ≑ 206.0833, MSB ≑ 835.6296, MSAB ≑ 25.0463, MSE ≑ 51.1111.
b. $F^* = 4.03$; differences are significant.

12.19 **a.** Sums of squares for method, sex, and interaction are about 152.3, 25.6 and 274.4, respectively. SSE = 251.6, and SSTO ≑ 703.9. MSE ≑ 7.8625. **b.** Interaction is significant ($F^* \doteq 11.63$). **c.** There are significant differences between methods ($F^* \doteq 6.46$). **d.** There are no differences due to sex ($F^* \doteq 3.26$). **e.** There are differences in mean number of verbs memorized for the four groups. Further tests are desirable to analyze the interactions.

12.21 HSD ≑ 0.168; no significant differences.

12.23 HSD ≐ 0.138; level 2 mean differs significantly from each of the others.

12.25 HSD ≐ 4.886; treatment 1A mean differs significantly from all others.

12.27 Plan C has a significantly higher mean sales increase than plan A but not than plan B. Further tests should be conducted on plans B and C.

12.29 No overall test on methods is appropriate, but methods C and D have higher means than methods A and B for females. No other differences are significant.

12.31 **a.** SSTR = 146, SSB = 318, SSE = 42, SSTO = 506; MSE = 7. **b.** $F^* = 10.43$; differences are significant. **c.** The means of treatments 2 and 3 differ significantly (HSD ≐ 5.74).

12.33 **a.** SSA = 432.125, SSB = 384.75, SSAB = 5.25, SSE = 26.5, SSTO = 848.625; MSE ≐ 2.2083. **b.** $F^* = 0.40$; interaction is not significant. **c.** $F^* = 65.23$ (factor A); $F^* = 87.11$ (factor B). Both are significant. **d.** See Solutions Manual. Results show no interaction.

12.35 DfTR = 4, dfE = 20; SSTR = 0.0156, SSB = 0.0256; MSB ≐ 0.0051, MSE ≐ 0.0052.

12.37 $F^* \doteq 4.13$; $F_{.01}\{3,12\} = 5.95$. No significant differences.

12.39 **a.** SS (Metal) ≐ 3.4111; SS (Coating) ≐ 0.0744; SSE ≐ 0.2722; MSE ≐ 0.0272. **b.** $F^* \doteq 1.37$; not significant. **c.** See Solutions Manual. There may be significant interaction. **d.** $F^* = 25.06$; there are differences among the metals. Use a multiple comparison procedure.

12.41 **a.** SS (Order) ≐ 681.6333, SS (Stimulus) ≐ 355.4667, SS (Interaction) ≐ 365.8667, SSE = 240.0; MS (Order) ≐ 681.6333. MS (Stimulus) ≐ 177.7333, MS (Interaction) ≐ 182.9333, MSE = 10.0. **b.** All F tests are significant. **c.** See Solutions Manual. Yes. **d.** Differences between levels of the factor order may be investigated since the interactions here are due to changes in magnitude. Multiple comparisons are not necessary since there is only one degree of freedom. Reaction times are lower for regular order than random, although the difference is negligible for visual stimulus. **e.** Regular order, auditory stimulus. Using $\alpha = 0.05$, HSD ≐ 6.18; the difference between the means is 4.4 (61.8 − 57.4), so the difference is not significant.

Chapter 13

13.1 **a.** functional. **b.** statistical. **c.** statistical. **d.** functional.

13.3 **a.** See Solutions Manual. Yes, upward to the right. **b.** Yes, reasonably well. **c.** 35. **d.** $\hat{y} = 1.9 + 1.2333x$.

13.5 **b.** $\hat{y} = 84.4918 + 1.2473x$.

13.7 **a.** $\hat{y} = 2.9377 + 1.2997x$. **b.** An estimate for the mean increase in yield for each increase of 100 pounds of fertilizer. **c.** It might be an estimate for the mean yield if no fertilizer were used. **d.** SSR ≐ 8.2856, SSE ≐ 1.6155; MSE ≐ 0.2019. **e.** 15.93.

13.9 **a.** $\beta_1 < 0$. **b.** $F^* = 29.935$; yes. **c.** the mean height of trees is estimated to decrease 0.1558 feet for each additional foot from the irrigation ditch. **d.** mean height is predicted to be about 22.52 feet.

13.11 **a.** About 3.2167. **b.** Yes; it is $3.85s_e$ from the regression line. **c.** No, it is only $2.05s_e$ from the regression line.

13.13 **a.** -0.0024006 to -0.0023394. **b.** $t^* \doteq -162.329$; $t_{.025}\{19\} = 2.093$, so it can be concluded that $\beta_1 < 0$. **c.** $t^* \doteq -25.342$; reject H_0; $\beta_1 < -0.002$. **d.** $t^* \doteq 8.904$; reject H_0; $\beta_1 > -0.0025$. **e.** Since neither -0.002 nor -0.0025 is in the interval, we may reject a hypothesis that β_1 is either value. **f.** -0.01125 to 0.3420; confirm.

13.15 **b.** no. **c.** no. **d.** $t^* = 13.780$; reject H_0: $\beta_1 = 0$ in favor of H_1: $\beta_1 > 0$.

13.17 **b.** no. **c.** To test H_0: $\beta_1 = 1$ we obtain $t^* = (1.2997-1)/0.2029 \doteq 1.477$. Fail to reject H_0; the contention is reasonable. **d.** -1.2627 to 7.1381. Since zero is far beyond the scope of the data, the validity is questionable.

13.19 **a.** $s_e \doteq 1.6868$. **c.** No. **d.** Difference of 4.4565 feet is $2.64s_e$ from the line, while the difference of 2.2826 feet is $1.35s_e$ from the line. The tree 84 feet from the ditch should probably be investigated to see if something is interfering with its water supply.

13.21 **a.** $\hat{y} = 18.28 + 0.159x$. **b.** SSR $\doteq 252.81$, SSE $\doteq 489.61$; MSE $\doteq 10.2$. **c.** $F^* = 24.875$; yes, there is. **e.** yes. **f.** The points (0,26), (10,13) might bear investigation. **g.** Close enough. **h.** Hartley's test may be used. **i.** $t^* = 4.978$; conclude $\beta_1 > 0$. **j.** 0.095 to 0.223. **k.** 16.71 to 19.85.

13.23 **a.** $s_{\hat{y}_p} \doteq 8.7532$. **b.** $s_f \doteq 23.5566$. **c.** yes. **d.** 204.01 to 301.72. **e.** Narrower, since $s_{\hat{y}_p} < s_f$.

13.25 **a.** 120.6 to 123.3. **b.** 118.3 to 125.5.

13.27 **a.** Prediction interval. **b.** confidence interval.

13.29 A confidence interval, since the grower will be growing many trees.

13.31 **a.** 20.552 to 22.368. **b.** 14.975 to 27.945. **c.** prediction interval, since individuals will be monitored.

13.33 **a.** $r \doteq -0.1635$. **b.** $r^2 \doteq 0.0267$.

13.35 **a.** $r \doteq 0.9896$. **b.** H_0: $\rho = 0$; H_1: $\rho > 0$; Reject H_0 if $t^* > 2.132$; $t^* \doteq 13.780$; $\rho > 0$.

13.37 **a.** $r \doteq 0.8098$. **b.** $r_s \doteq 0.9857$.

13.39 $r_s \doteq -0.744$; $t^* \doteq -4.45$; there is a significant relationship between the variables.

13.41 $r_s \doteq 0.5449$; $t^* \doteq 1.838$; there is no relationship between the variables at the 0.05 level of significance.

13.43 $r \doteq 0.5345$; $t^* \doteq 3.346$. Yes, they are linearly related.

13.45 **a.** $F^* = 19.916$; yes. **b.** No; yes.

13.47 **a.** SSR $\doteq$ 158,216.6372, SSE $\doteq$ 30,138.3628; MSE $\doteq$ 1076.3701. **b.** $F^* \doteq$ 146.991; they are associated. **c.** $r^2 \doteq 0.8400$. **d.** $r \doteq 0.9165$.

13.49 **a.** SSX = 190.92758, SSTO = 1.634055; SXY = 15.59648. **b.** $\hat{y} = 0.1266 + 0.0817x$. **c.** SSR $\doteq$ 1.2740, SSE $\doteq$ 0.3600; MSE $\doteq$ 0.0200. **d.** $F^* \doteq 63.70$; there is a linear association between the variables. **e.** 0.0522 to 0.1111. **f.** $t^* \doteq -1.789$; fail to reject H_0. **g.** -0.1081 to 0.3613. **h.** 0.6481 to 0.8304. **i.** 0.3222 to 1.1564.

13.51 **a.** $\hat{y} = 17.3567 - 0.1543x$. **c.** $r \doteq -0.4759$; $t^* = -1.623$. No. **d.** No assumptions violated. **e.** $s_e \doteq 3.0595$; no outliers.

13.53 **a.** $\hat{y} = 45.0949 + 0.4551x$. **b.** 0.1766 to 0.7336. **c.** 80.7984 to 91.3110. **d.** 57.7621 to 105.2; if we know that the maximum score on the final is 100, then the upper confidence limit will be 100.

Chapter 14

14.1 DfA = 3, dfE = 20; SSR = 101.4432, SSE = 52.4444; MSE $\doteq$ 2.6222.

14.3 $\hat{y} = 6.4838 - 1.8421x_1 + 3.1002x_2$; DfR = 2, dfE = 3; SSR $\doteq$ 479.0597, SSE $\doteq$ 0.2736; MSR $\doteq$ 239.5299, MSE $\doteq$ 0.0912; $F^* = 2626.308$. There is an association between the variables.

14.5 **a.** $\hat{y} = -35.5514 + 0.5574x_1 + 1.0994x_2$, where count $= x_1$ and temperature $= x_2$.
b. For each additional 1000 bacteria, the mean stay increases about 0.5574 days; for each increase of one degree in temperature, the mean stay increases about 1.0994 days.
c. DfR = 2, dfE = 7; SSR $\doteq$ 17.5253, SSE $\doteq$ 10.8747; MSR $\doteq$ 8.7626, MSE $\doteq$ 1.5535. **d.** $F^* = 5.64$; there is an association. **e.** See solutions manual.

14.7 **a.** $\hat{y} = 13.3416 - 0.4943x_1 + 5.4136x_2$, where price $= x_1$ and number of ads $= x_2$.
b. From the printout, the P value for testing that $\beta_1 = 0$ is 0.1201, and for testing that $\beta_2 = 0$ is 0.0049. Thus, the number of ads has a greater effect on market share than price.

14.9 **a.** $t_{.025}\{22\} = 2.074$; $t^* \doteq 1.65, 4.68, -3.46$ for $i = 1, 2, 3$, respectively. Thus, we can conclude that β_2 is positive and β_3 is negative; we cannot conclude that $\beta_1 \neq 0$.
b. Since $4.68 > 3.46$, we conclude that x_2 has the greatest effect on y.

14.11 $R^2 \doteq 0.445$. Although that is not especially large, since nearly 55% of the variation in y is unexplained, there is probably some association between the predictor variables and the response variable.

14.13 **a.** 49.1021 to 51.3457. **b.** 54.3503 to 57.0246. **c.** The correlation between x_1 and x_2 is about 0.9765. Since they are highly correlated, multicollinearity is likely to be a problem. On the other hand, since β_1 is not significantly different from zero, we should eliminate x_1 and rework the problem.

14.15 **a.** We may conclude that count is significant at the 0.05 level of significance ($t^* = 2.491$, $P = 0.0415$) and temperature is not ($t^* = 1.760$ and $P = 0.1218$). Temperature is significant at a 0.1218 level of significance, however, so it should not be summarily discarded.
b. $\hat{y} = 6.1838 + 0.6293x_1$, MSE $\doteq$ 1.9611, for count; $\hat{y} = -42.726 + 1.3835x_2$, MSE $\doteq$ 2.5643, for temperature. Since the full model MSE (1.5535) is smaller than these, the model using both predictor variables is the best predictor.
c. 8.9974 to 11.2575. **d.** 6.9709 to 13.2840.

14.17 **a.** We may conclude that the number of ads is significant at the 0.05 level of significance ($t^* = 5.636$, $P = 0.0049$) and price is not ($t^* = -1.970$ and $P = 0.1201$). Price is significant at a 0.1201 level of significance, however, so it cannot be completely ignored.
b. $\hat{y} = 25.9351 - 0.2586x_1$, MSE $\doteq$ 291.4752, for price; $\hat{y} = -7.5491 + 5.0981x_2$, MSE $\doteq$ 64.2354, for temperature. Since the full model MSE (40.7456) is smaller than these, the model using both predictor variables is the best predictor.
c. 13.0385 to 33.1810 **d.** 2.7256 to 43.4939.

14.19 **a.** increasing, since $b_2 > 0$. **b.** $t^* \doteq 3.437$; $t_{.005}\{15\} = 2.947$. We may conclude that $\beta_2 \neq 0$ at better than a 0.01 level of significance. **c.** 1.35 to 5.73.

14.21 **a.** $\hat{y} = 16.24 + 0.2133x$. See Solutions Manual for the residual plot; nonlinearity is suggested, but is not pronounced.
b. $\hat{y} = 12.3127 + 0.4588x - 0.0003273x^2$.
c. MSE (linear) $\doteq$ 4.1622, MSE (quadratic) $\doteq$ 3.8748; quadratic is a better fit.
d. $t^* = -2.136$; $t_{.025}\{47\} \doteq 2.014$, $P = 0.0379$; $\beta_2 \neq 0$. Yes.

14.23 $\hat{y} = 81.7333 + 45.9576x$, MSE $= 669.0439$ (linear); $\hat{y} = 134.7333 + 19.4576x + 2.4091x^2$, MSE $= 326.8554$ (quadratic). Cost will be about \$879.32 in 1990.

14.25 $\hat{y} = -362.8402 + 102.4635$ (EDUC) $+ 287.7716$ (HHSEX). No; $t^* = 1.817$, P = 0.071.

14.27 $F^* \doteq 32.47$, $F_{.05}\{2,29\} \doteq 3.31$. There is association. $R^2 = 0.6913$; yes.

14.29 **a.** Variable 2. **b.** Variable 4. **c.** Variables 1 and 3; drop 1.

14.31 Minimum MSE is 0.253011 for $\hat{y} = 3.2956 - 0.0435$ (NOSIS) $+ 0.0341$ (CRED HRS) $- 0.0041$ (WEIGHT).

14.33 Minimum MSE is 0.236036 for $\hat{y} = 3.0442 + 0.1727$ (SOR) $+ 0.0677$ (NOSIS) $+ 0.3374$ (SOAPS) $+ 0.1063$ (JOB) $- 0.0058$ (WEIGHT).

Appendix D

Tables

Table 1 RANDOM NUMBERS

	1	2	3	4	5	6	7	8	9	10	
1	48461	14952	72619	73689	52059	37086	60050	86192	67049	64739	1
2	76534	38149	49692	31366	52093	15422	20498	33901	10319	43397	2
3	70437	25861	38504	14752	23757	59660	67844	78815	23758	86814	3
4	59584	03370	42806	11393	71722	93804	09095	07856	55589	46020	4
5	04285	59554	16085	51555	27501	73883	33427	33343	45507	50063	5
6	77340	10412	69189	85171	29082	44785	83638	02583	96483	76553	6
7	59183	62687	91778	80354	23512	97219	65921	02035	59847	91403	7
8	91800	04218	39979	03927	82564	28777	59049	97532	54540	79472	8
9	12066	24817	81099	48940	69554	55925	48379	12866	51232	21580	9
10	69907	91751	53512	23748	65906	91385	84983	27915	48491	91068	10
11	80467	04873	54053	25955	48518	13815	37707	68687	15570	08890	11
12	78057	67835	28302	45048	56761	97725	58438	91528	24645	18544	12
13	05648	39387	78191	88415	60269	94880	58812	42931	71898	61534	13
14	22304	39246	01350	99451	61862	78688	30339	60222	74052	25740	14
15	61346	50269	67005	40442	33100	16742	61640	21046	31909	72641	15

(continued)

Table 1 RANDOM NUMBERS *(continued)*

	1	2	3	4	5	6	7	8	9	10	
16	66793	37696	27965	30459	91011	51426	31006	77468	61029	57108	16
17	86411	48809	36698	42453	83061	43769	39948	87031	30767	13953	17
18	62098	12825	81744	28882	27369	88183	65846	92545	09065	22655	18
19	68775	06261	54265	16203	23340	84750	16317	88686	86842	00879	19
20	52679	19595	13687	74872	89181	01939	18447	10787	76246	80072	20
21	84096	87152	20719	25215	04349	54434	72344	93008	83282	31670	21
22	63964	55937	21417	49944	38356	98404	14850	17994	17161	98981	22
23	31191	75131	72386	11689	95727	05414	88727	45583	22568	77700	23
24	30545	68523	29850	67833	05622	89975	79042	27142	99257	32349	24
25	52573	91001	52315	26430	54175	30122	31796	98842	37600	26025	25
26	16586	81842	01076	99414	31574	94719	34656	80018	86988	79234	26
27	81841	88481	61191	25013	30272	23388	22463	65774	10029	58376	27
28	43563	66829	72838	08074	57080	15446	11034	98143	74989	26885	28
29	19945	84193	57581	77252	85604	45412	43556	27518	90572	00563	29
30	79374	23796	16919	99691	80276	32818	62953	78831	54395	30705	30
31	48503	26615	43980	09810	38289	66679	73799	48418	12647	40044	31
32	32049	65541	37937	41105	70106	89706	40829	40789	59547	00783	32
33	18547	71562	95493	34112	76895	46766	96395	31718	48302	45893	33
34	03180	96742	61486	43305	34183	99605	67803	13491	09243	29557	34
35	94822	24738	67749	83748	59799	25210	31093	62925	72061	69991	35
36	34330	60599	85828	19152	68499	27977	35611	96240	62747	89529	36
37	43770	81537	59527	95674	76692	86420	69930	10020	72881	12532	37
38	56908	77192	50623	41215	14311	42834	80651	93750	59957	31211	38
39	32787	07189	80539	75927	75475	73965	11796	72140	48944	74156	39
40	52441	78392	11733	57703	29133	71164	55355	31006	25526	55790	40
41	22377	54723	18227	28449	04570	18882	00023	67101	06895	08915	41
42	18376	73460	88841	39602	34049	20589	05701	08249	74213	25220	42
43	53201	28610	87957	21497	64729	64983	71551	99016	87903	63875	43
44	34919	78901	59710	27396	02593	05665	11964	44134	00273	76358	44
45	33617	92159	21971	16901	57383	34262	41744	60891	57624	06962	45
46	70010	40964	98780	72418	52571	18415	64352	90636	38034	04909	46
47	19282	68447	35665	31530	59832	49181	21914	65742	89815	39231	47
48	91429	73328	13266	54898	68795	40948	80808	63887	89939	47938	48
49	97637	78393	33021	05867	86520	45363	43066	00988	64040	09803	49
50	95150	07625	05255	83254	93943	52325	93230	62668	79529	65964	50

Source: F. James Rohlf and Robert R. Sokal, *Statistical Tables* (San Francisco: W. H. Freeman and Company). Copyright © 1969.

Table 2 CUMULATIVE BINOMIAL PROBABILITIES

								p							
n	*r*	.01	.05	.10	.20	.30	.40	.50	.60	.70	.80	.90	.95	.99	*r*
2	0	980	902	810	640	490	360	250	160	090	040	010	002	0+	0
	1	1−	998	99	960	910	840	750	640	510	360	190	098	020	1
	2	1	1	1	1	1	1	1	1	1	1	1	1	1	2
3	0	970	857	729	512	343	216	125	064	027	008	001	0+	0+	0
	1	1−	993	972	896	784	648	500	352	216	104	028	007	0+	1
	2	1−	1−	999	992	973	936	875	784	657	488	271	143	030	2
	3	1	1	1	1	1	1	1	1	1	1	1	1	1	3
4	0	961	815	656	410	240	130	062	026	008	002	0+	0+	0+	0
	1	999	986	948	819	652	475	312	179	084	027	004	0+	0+	1
	2	1−	1−	996	973	916	821	688	525	348	181	052	014	001	2
	3	1−	1−	1−	998	992	974	938	870	760	590	344	185	039	3
	4	1	1	1	1	1	1	1	1	1	1	1	1	1	
5	0	951	774	590	328	168	078	031	010	002	0+	0+	0+	0+	0
	1	999	977	919	737	528	337	188	087	031	007	0+	0+	0+	1
	2	1−	999	991	942	837	683	500	317	163	058	009	001	0+	2
	3	1−	1−	1−	993	969	913	812	663	472	263	081	023	001	3
	4	1−	1−	1−	1−	998	990	969	922	832	672	410	226	049	4
	5	1	1	1	1	1	1	1	1	1	1	1	1	1	5
6	0	941	735	531	262	118	047	016	004	001	0+	0+	0+	0+	0
	1	999	967	886	655	420	233	109	041	011	002	0+	0+	0+	1
	2	1−	998	984	901	744	544	344	179	070	017	001	0+	0+	2
	3	1−	1−	999	983	930	821	656	456	256	099	016	002	0+	3
	4	1−	1−	1−	998	989	959	891	767	580	345	114	033	001	4
	5	1−	1−	1−	1−	999	996	984	953	882	738	469	265	059	5
	6	1	1	1	1	1	1	1	1	1	1	1	1	1	6
7	0	932	698	478	210	082	028	008	002	0+	0+	0+	0+	0+	0
	1	998	956	850	577	329	159	062	019	004	0+	0+	0+	0+	1
	2	1−	996	974	852	647	420	227	096	029	005	0+	0+	0+	2
	3	1−	1−	997	967	874	710	500	290	126	033	003	0+	0+	3
	4	1−	1−	1−	995	971	904	773	580	353	148	026	004	0+	4
	5	1−	1−	1−	1−	996	981	938	841	671	423	150	044	002	5
	6	1−	1−	1−	1−	1−	998	992	972	918	790	522	302	068	6
	7	1	1	1	1	1	1	1	1	1	1	1	1	1	7
8	0	923	663	430	168	058	019	004	001	0+	0+	0+	0+	0+	0
	1	997	943	813	503	255	106	035	009	001	0+	0+	0+	0+	1
	2	1−	994	962	797	552	315	145	050	011	001	0+	0+	0+	2
	3	1−	1−	995	944	806	594	363	174	058	010	0+	0+	0+	3
	4	1−	1−	1−	990	942	826	637	406	194	056	005	0+	0+	4

(continued)

Table 2 CUMULATIVE BINOMIAL PROBABILITIES *(continued)*

								p							
n	*r*	.01	.05	.10	.20	.30	.40	.50	.60	.70	.80	.90	.95	.99	*r*
	5	1−	1−	1−	999	989	950	855	685	448	203	038	006	0+	5
	6	1−	1−	1−	1−	999	991	965	894	745	497	187	057	003	6
	7	1−	1−	1−	1−	1−	999	996	983	942	832	570	337	077	7
	8	1	1	1	1	1	1	1	1	1	1	1	1	1	8
9	0	914	630	387	134	040	010	002	0+	0+	0+	0+	0+	0+	0
	1	997	929	775	436	196	071	020	004	0+	0+	0+	0+	0+	1
	2	1−	992	947	738	463	232	090	025	004	0+	0+	0+	0+	2
	3	1−	999	992	914	730	483	254	099	025	003	0+	0+	0+	3
	4	1−	1−	999	980	901	733	500	267	099	020	001	0+	0+	4
	5	1−	1−	1−	997	975	901	746	517	270	086	008	001	0+	5
	6	1−	1−	1−	1−	996	975	910	768	537	262	053	008	0+	6
	7	1−	1−	1−	1−	1−	996	980	929	804	564	225	071	003	7
	8	1−	1−	1−	1−	1−	1−	998	990	960	866	613	370	086	8
	9	1	1	1	1	1	1	1	1	1	1	1	1	1	9
10	0	904	599	349	107	028	006	001	0+	0+	0+	0+	0+	0+	0
	1	996	914	736	376	149	046	011	002	0+	0+	0+	0+	0+	1
	2	1−	988	930	678	383	167	055	012	002	0+	0+	0+	0+	2
	3	1−	999	987	879	650	382	172	055	011	001	0+	0+	0+	3
	4	1−	1−	998	967	850	633	377	166	047	006	0+	0+	0+	4
	5	1−	1−	1−	994	953	834	623	367	150	033	002	0+	0+	5
	6	1−	1−	1−	999	989	945	828	618	350	121	013	001	0+	6
	7	1−	1−	1−	1−	998	988	945	833	617	322	070	012	0+	7
	8	1−	1−	1−	1−	1−	998	989	954	851	624	264	086	004	8
	9	1−	1−	1−	1−	1−	1−	999	994	972	893	651	401	096	9
	10	1	1	1	1	1	1	1	1	1	1	1	1	1	10
11	0	895	569	314	086	020	004	0+	0+	0+	0+	0+	0+	0+	0
	1	995	898	697	322	113	030	006	001	0+	0+	0+	0+	0+	1
	2	1−	985	910	617	313	119	033	006	001	0+	0+	0+	0+	2
	3	1−	998	981	839	570	296	113	029	004	0+	0+	0+	0+	3
	4	1−	1−	997	950	790	533	274	099	022	002	0+	0+	0+	4
	5	1−	1−	1−	988	922	753	500	247	078	012	0+	0+	0+	5
	6	1−	1−	1−	998	978	901	726	467	210	050	003	0+	0+	6
	7	1−	1−	1−	1−	996	971	887	704	430	161	019	002	0+	7
	8	1−	1−	1−	1−	999	994	967	881	687	383	090	015	0+	8
	9	1−	1−	1−	1−	1−	999	994	970	887	678	303	102	005	9
	10	1−	1−	1−	1−	1−	1−	1−	996	980	914	686	431	105	10
	11	1	1	1	1	1	1	1	1	1	1	1	1	1	11
12	0	886	540	282	069	014	002	0+	0+	0+	0+	0+	0+	0+	0
	1	994	882	659	275	085	020	003	0+	0+	0+	0+	0+	0+	1

(continued)

Table 2 CUMULATIVE BINOMIAL PROBABILITIES *(continued)*

								p							
n	*r*	.01	.05	.10	.20	.30	.40	.50	.60	.70	.80	.90	.95	.99	*r*
	2	1−	980	889	558	253	083	019	003	0+	0+	0+	0+	0+	2
	3	1−	998	974	795	493	225	073	015	002	0+	0+	0+	0+	3
	4	1−	1−	996	927	724	438	194	057	009	001	0+	0+	0+	4
	5	1−	1−	999	981	882	665	387	158	039	004	0+	0+	0+	5
	6	1−	1−	1−	996	961	842	613	335	118	019	001	0+	0+	6
	7	1−	1−	1−	999	991	943	806	562	276	073	004	0+	0+	7
	8	1−	1−	1−	1−	998	985	927	775	507	205	026	002	0+	8
	9	1−	1−	1−	1−	1−	997	981	917	747	442	111	020	0+	9
	10	1−	1−	1−	1−	1−	1−	997	980	915	725	341	118	006	10
	11	1−	1−	1−	1−	1−	1−	1−	998	986	931	718	460	114	11
	12	1	1	1	1	1	1	1	1	1	1	1	1	1	12
13	0	878	513	254	055	010	001	0+	0+	0+	0+	0+	0+	0+	0
	1	993	865	621	234	064	013	002	0+	0+	0+	0+	0+	0+	1
	2	1−	975	866	502	202	058	011	001	0+	0+	0+	0+	0+	2
	3	1−	997	966	747	421	169	046	008	001	0+	0+	0+	0+	3
	4	1−	1−	994	901	654	353	133	032	004	0+	0+	0+	0+	4
	5	1−	1−	999	970	835	574	291	098	018	001	0+	0+	0+	5
	6	1−	1−	1−	993	938	771	500	229	062	007	0+	0+	0+	6
	7	1−	1−	1−	999	982	902	709	426	165	030	001	0+	0+	7
	8	1−	1−	1−	1−	996	968	867	647	346	099	006	0+	0+	8
	9	1−	1−	1−	1−	999	992	954	831	579	253	034	003	0+	9
	10	1−	1−	1−	1−	1−	999	989	942	798	498	134	025	0+	10
	11	1−	1−	1−	1−	1−	1−	998	987	936	766	379	135	007	11
	12	1−	1−	1−	1−	1−	1−	1−	999	990	945	746	487	122	12
	13	1	1	1	1	1	1	1	1	1	1	1	1	1	13
14	0	869	488	229	044	007	001	0+	0+	0+	0+	0+	0+	0+	0
	1	992	847	585	198	047	008	001	0+	0+	0+	0+	0+	0+	1
	2	1−	970	842	448	161	040	006	001	0+	0+	0+	0+	0+	2
	3	1−	996	956	698	355	124	029	004	0+	0+	0+	0+	0+	3
	4	1−	1−	991	870	584	279	090	018	002	0+	0+	0+	0+	4
	5	1−	1−	999	956	781	486	212	058	008	0+	0+	0+	0+	5
	6	1−	1−	1−	988	907	692	395	150	031	002	0+	0+	0+	6
	7	1−	1−	1−	998	969	850	605	308	093	012	0+	0+	0+	7
	8	1−	1−	1−	1−	992	942	788	514	219	044	001	0+	0+	8
	9	1−	1−	1−	1−	998	982	910	721	416	130	009	0+	0+	9
	10	1−	1−	1−	1−	1−	996	971	876	645	302	044	004	0+	10
	11	1−	1−	1−	1−	1−	999	994	960	839	552	158	030	0+	11
	12	1−	1−	1−	1−	1−	1−	999	992	953	802	415	153	008	12
	13	1−	1−	1−	1−	1−	1−	1−	999	993	956	771	512	131	13
	14	1	1	1	1	1	1	1	1	1	1	1	1	1	14

(continued)

Table 2 CUMULATIVE BINOMIAL PROBABILITIES *(continued)*

							p								
n	r	.01	.05	.10	.20	.30	.40	.50	.60	.70	.80	.90	.95	.99	r
15	0	860	463	206	035	005	0+	0+	0+	0+	0+	0+	0+	0+	0
	1	990	829	549	167	035	005	0+	0+	0+	0+	0+	0+	0+	1
	2	1−	964	816	398	127	027	004	0+	0+	0+	0+	0+	0+	2
	3	1−	995	944	648	297	091	018	002	0+	0+	0+	0+	0+	3
	4	1−	999	987	836	515	217	059	009	001	0+	0+	0+	0+	4
	5	1−	1−	998	939	722	403	151	034	004	0+	0+	0+	0+	5
	6	1−	1−	1−	982	869	610	304	095	015	001	0+	0+	0+	6
	7	1−	1−	1−	996	950	787	500	213	050	004	0+	0+	0+	7
	8	1−	1−	1−	999	985	905	696	390	131	018	0+	0+	0+	8
	9	1−	1−	1−	1−	996	966	849	597	278	061	002	0+	0+	9
	10	1−	1−	1−	1−	999	991	941	783	485	164	013	001	0+	10
	11	1−	1−	1−	1−	1−	998	982	909	703	352	056	005	0+	11
	12	1−	1−	1−	1−	1−	1−	996	973	873	602	184	036	0+	12
	13	1−	1−	1−	1−	1−	1−	1−	995	965	833	451	171	010	13
	14	1−	1−	1−	1−	1−	1−	1−	1−	995	965	794	537	140	14
	15	1	1	1	1	1	1	1	1	1	1	1	1	1	15
16	0	851	440	185	028	003	0+	0+	0+	0+	0+	0+	0+	0+	0
	1	989	811	515	141	026	003	0+	0+	0+	0+	0+	0+	0+	1
	2	999	957	789	352	099	018	002	0+	0+	0+	0+	0+	0+	2
	3	1−	993	932	598	246	065	011	001	0+	0+	0+	0+	0+	3
	4	1−	999	983	798	450	167	038	005	0+	0+	0+	0+	0+	4
	5	1−	1−	997	918	660	329	105	019	002	0+	0+	0+	0+	5
	6	1−	1−	999	973	825	527	227	058	007	0+	0+	0+	0+	6
	7	1−	1−	1−	993	926	716	402	142	026	001	0+	0+	0+	7
	8	1−	1−	1−	999	974	858	598	284	074	007	0+	0+	0+	8
	9	1−	1−	1−	1−	993	942	773	473	175	027	001	0+	0+	9
	10	1−	1−	1−	1−	998	981	895	671	340	082	003	0+	0+	10
	11	1−	1−	1−	1−	1−	995	962	833	550	202	017	001	0+	11
	12	1−	1−	1−	1−	1−	999	989	935	754	402	068	007	0+	12
	13	1−	1−	1−	1−	1−	1−	998	982	901	648	211	043	001	13
	14	1−	1−	1−	1−	1−	1−	1−	997	974	859	485	189	011	14
	15	1−	1−	1−	1−	1−	1−	1−	1−	997	972	815	560	149	15
	16	1	1	1	1	1	1	1	1	1	1	1	1	1	16
17	0	843	418	167	023	002	0+	0+	0+	0+	0+	0+	0+	0+	0
	1	988	792	482	118	019	002	0+	0+	0+	0+	0+	0+	0+	1
	2	999	950	762	310	077	012	001	0+	0+	0+	0+	0+	0+	2
	3	1−	991	917	549	202	046	006	0+	0+	0+	0+	0+	0+	3
	4	1−	999	978	758	389	126	025	003	0+	0+	0+	0+	0+	4
	5	1−	1−	995	894	597	264	072	011	001	0+	0+	0+	0+	5
	6	1−	1−	999	962	775	448	166	035	003	0+	0+	0+	0+	6

(continued)

Table 2 CUMULATIVE BINOMIAL PROBABILITIES *(continued)*

							p							
n	*r*	.01	.05	.10	.20	.30	.40	.50	.60	.70	.80	.90	.95	.99	*r*
	7	1−	1−	1−	989	895	641	315	092	013	0+	0+	0+	0+	7
	8	1−	1−	1−	997	960	801	500	199	040	003	0+	0+	0+	8
	9	1−	1−	1−	1−	987	908	685	359	105	011	0+	0+	0+	9
	10	1−	1−	1−	1−	997	965	834	552	225	038	001	0+	0+	10
	11	1−	1−	1−	1−	999	989	928	736	403	106	105	0+	0+	11
	12	1−	1−	1−	1−	1−	997	975	874	611	242	022	001	0+	12
	13	1−	1−	1−	1−	1−	1−	994	954	798	451	083	009	0+	13
	14	1−	1−	1−	1−	1−	1−	999	988	923	690	238	050	001	14
	15	1−	1−	1−	1−	1−	1−	1−	998	981	882	518	208	012	15
	16	1−	1−	1−	1−	1−	1−	1−	1−	998	977	833	582	157	16
	17	1	1	1	1	1	1	1	1	1	1	1	1	1	17
18	0	835	397	150	018	002	0+	0+	0+	0+	0+	0+	0+	0+	0
	1	986	774	450	099	014	001	0+	0+	0+	0+	0+	0+	0+	1
	2	999	942	734	271	060	008	001	0+	0+	0+	0+	0+	0+	2
	3	1−	989	902	501	165	033	004	0+	0+	0+	0+	0+	0+	3
	4	1−	998	972	716	333	094	015	001	0+	0+	0+	0+	0+	4
	5	1−	1−	994	867	534	209	048	006	0+	0+	0+	0+	0+	5
	6	1−	1−	999	949	722	374	119	020	001	0+	0+	0+	0+	6
	7	1−	1−	1−	984	859	563	240	058	006	0+	0+	0+	0+	7
	8	1−	1−	1−	996	940	737	407	135	021	001	0+	0+	0+	8
	9	1−	1−	1−	999	979	865	593	263	060	004	0+	0+	0+	9
	10	1−	1−	1−	1−	994	942	760	437	141	016	0+	0+	0+	10
	11	1−	1−	1−	1−	999	980	881	626	278	051	001	0+	0+	11
	12	1−	1−	1−	1−	1−	994	952	791	466	133	006	0+	0+	12
	13	1−	1−	1−	1−	1−	999	985	906	667	284	028	002	0+	13
	14	1−	1−	1−	1−	1−	1−	996	967	835	499	098	011	0+	14
	15	1−	1−	1−	1−	1−	1−	999	992	940	729	266	058	001	15
	16	1−	1−	1−	1−	1−	1−	1−	999	986	901	550	226	014	16
	17	1−	1−	1−	1−	1−	1−	1−	1−	998	982	850	603	165	17
	18	1	1	1	1	1	1	1	1	1	1	1	1	1	18
19	0	826	377	135	014	001	0+	0+	0+	0+	0+	0+	0+	0+	0
	1	985	755	420	083	010	001	0+	0+	0+	0+	0+	0+	0+	1
	2	999	933	705	237	046	005	0+	0+	0+	0+	0+	0+	0+	2
	3	1−	987	885	455	133	023	002	0+	0+	0+	0+	0+	0+	3
	4	1−	998	965	673	282	070	010	001	0+	0+	0+	0+	0+	4
	5	1−	1−	991	837	474	163	032	003	0+	0+	0+	0+	0+	5
	6	1−	1−	998	932	666	308	084	012	001	0+	0+	0+	0+	6
	7	1−	1−	1−	977	818	488	180	035	003	0+	0+	0+	0+	7
	8	1−	1−	1−	993	916	667	324	088	011	0+	0+	0+	0+	8
	9	1−	1−	1−	998	967	814	500	186	033	002	0+	0+	0+	9

(continued)

Table 2 CUMULATIVE BINOMIAL PROBABILITIES *(continued)*

								p							
n	*r*	.01	.05	.10	.20	.30	.40	.50	.60	.70	.80	.90	.95	.99	*r*
	10	1−	1−	1−	1−	989	912	676	333	084	007	0+	0+	0+	10
	11	1−	1−	1−	1−	997	965	820	512	182	023	0+	0+	0+	11
	12	1−	1−	1−	1−	999	988	916	692	334	068	002	0+	0+	12
	13	1−	1−	1−	1−	1−	997	968	837	526	163	009	0+	0+	13
	14	1−	1−	1−	1−	1−	999	990	930	718	327	035	002	0+	14
	15	1−	1−	1−	1−	1−	1−	998	977	867	545	115	013	0+	15
	16	1−	1−	1−	1−	1−	1−	1−	995	954	763	295	067	001	16
	17	1−	1−	1−	1−	1−	1−	1−	999	990	917	580	245	015	17
	18	1−	1−	1−	1−	1−	1−	1−	1−	999	986	865	623	174	18
	19	1	1	1	1	1	1	1	1	1	1	1	1	1	19
20	0	818	358	122	012	001	0+	0+	0+	0+	0+	0+	0+	0+	0
	1	983	736	392	069	008	001	0+	0+	0+	0+	0+	0+	0+	1
	2	999	925	677	206	035	004	0+	0+	0+	0+	0+	0+	0+	2
	3	1−	984	867	411	107	016	001	0+	0+	0+	0+	0+	0+	3
	4	1−	997	957	630	238	051	006	0+	0+	0+	0+	0+	0+	4
	5	1−	1−	989	804	416	132	021	001	0+	0+	0+	0+	0+	5
	6	1−	1−	998	913	608	250	058	006	0+	0+	0+	0+	0+	6
	7	1−	1−	1−	968	772	416	126	021	002	0+	0+	0+	0+	7
	8	1−	1−	1−	990	887	596	252	057	005	0+	0+	0+	0+	8
	9	1−	1−	1−	997	952	755	412	128	017	001	0+	0+	0+	9
	10	1−	1−	1−	999	983	872	588	245	048	003	0+	0+	0+	10
	11	1−	1−	1−	1−	995	943	748	404	113	010	0+	0+	0+	11
	12	1−	1−	1−	1−	999	979	868	584	228	032	0+	0+	0+	12
	13	1−	1−	1−	1−	1−	994	942	750	392	087	002	0+	0+	13
	14	1−	1−	1−	1−	1−	998	979	874	584	196	011	0+	0+	14
	15	1−	1−	1−	1−	1−	1−	994	949	762	370	043	003	0+	15
	16	1−	1−	1−	1−	1−	1−	999	984	893	589	133	016	0+	16
	17	1−	1−	1−	1−	1−	1−	1−	996	965	794	323	075	001	17
	18	1−	1−	1−	1−	1−	1−	1−	999	992	931	608	264	017	18
	19	1−	1−	1−	1−	1−	1−	1−	1−	999	988	878	642	182	19
	20	1	1	1	1	1	1	1	1	1	1	1	1	1	20
21	0	810	341	109	009	001	0+	0+	0+	0+	0+	0+	0+	0+	0
	1	981	717	365	058	006	0+	0+	0+	0+	0+	0+	0+	0+	1
	2	999	915	648	179	027	002	0+	0+	0+	0+	0+	0+	0+	2
	3	1−	981	848	370	086	011	001	0+	0+	0+	0+	0+	0+	3
	4	1−	997	948	586	198	037	004	0+	0+	0+	0+	0+	0+	4
	5	1−	1−	986	769	363	096	013	001	0+	0+	0+	0+	0+	5
	6	1−	1−	997	891	551	200	039	004	0+	0+	0+	0+	0+	6
	7	1−	1−	999	957	723	350	095	012	001	0+	0+	0+	0+	7
	8	1−	1−	1−	986	852	524	192	035	002	0+	0+	0+	0+	8

(continued)

Table 2 CUMULATIVE BINOMIAL PROBABILITIES *(continued)*

		p													
n	*r*	.01	.05	.10	.20	.30	.40	.50	.60	.70	.80	.90	.95	.99	*r*
	9	1−	1−	1−	996	932	691	332	085	009	0+	0+	0+	0+	9
	10	1−	1−	1−	999	974	826	500	174	026	001	0+	0+	0+	10
	11	1−	1−	1−	1−	991	915	668	309	068	004	0+	0+	0+	11
	12	1−	1−	1−	1−	998	965	808	476	148	014	0+	0+	0+	12
	13	1−	1−	1−	1−	999	988	905	650	277	043	001	0+	0+	13
	14	1−	1−	1−	1−	1−	996	961	800	449	109	003	0+	0+	14
	15	1−	1−	1−	1−	1−	999	987	904	637	231	014	0+	0+	15
	16	1−	1−	1−	1−	1−	1−	996	963	802	414	052	003	0+	16
	17	1−	1−	1−	1−	1−	1−	999	989	914	630	152	019	0+	17
	18	1−	1−	1−	1−	1−	1−	1−	998	973	821	352	085	001	18
	19	1−	1−	1−	1−	1−	1−	1−	1−	994	942	635	283	019	19
	20	1−	1−	1−	1−	1−	1−	1−	1−	999	991	891	659	190	20
	21	1	1	1	1	1	1	1	1	1	1	1	1	1	21
22	0	802	324	098	007	0+	0+	0+	0+	0+	0+	0+	0+	0+	0
	1	980	698	339	048	004	0+	0+	0+	0+	0+	0+	0+	0+	1
	2	999	905	620	154	021	002	0+	0+	0+	0+	0+	0+	0+	2
	3	1−	978	828	332	068	008	0+	0+	0+	0+	0+	0+	0+	3
	4	1−	996	938	543	165	027	002	0+	0+	0+	0+	0+	0+	4
	5	1−	999	982	733	313	072	008	0+	0+	0+	0+	0+	0+	5
	6	1−	1−	996	867	494	158	026	002	0+	0+	0+	0+	0+	6
	7	1−	1−	999	944	671	290	067	007	0+	0+	0+	0+	0+	7
	8	1−	1−	1−	980	814	454	143	021	001	0+	0+	0+	0+	8
	9	1−	1−	1−	994	908	624	262	055	004	0+	0+	0+	0+	9
	10	1−	1−	1−	998	961	772	416	121	014	0+	0+	0+	0+	10
	11	1−	1−	1−	1−	986	879	584	228	039	002	0+	0+	0+	11
	12	1−	1−	1−	1−	996	945	738	376	092	006	0+	0+	0+	12
	13	1−	1−	1−	1−	999	979	857	546	186	020	0+	0+	0+	13
	14	1−	1−	1−	1−	1−	993	933	710	329	056	001	0+	0+	14
	15	1−	1−	1−	1−	1−	998	974	842	506	133	004	0+	0+	15
	16	1−	1−	1−	1−	1−	1−	992	928	687	267	018	001	0+	16
	17	1−	1−	1−	1−	1−	1−	998	973	835	457	062	004	0+	17
	18	1−	1−	1−	1−	1−	1−	1−	992	932	668	172	022	0+	18
	19	1−	1−	1−	1−	1−	1−	1−	998	979	846	380	095	001	19
	20	1−	1−	1−	1−	1−	1−	1−	1−	996	952	661	302	020	20
	21	1−	1−	1−	1−	1−	1−	1−	1−	1−	993	902	676	198	21
	22	1	1	1	1	1	1	1	1	1	1	1	1	1	
23	0	794	307	089	006	0+	0+	0+	0+	0+	0+	0+	0+	0+	0
	1	978	679	315	040	003	0+	0+	0+	0+	0+	0+	0+	0+	1
	2	998	895	592	133	016	001	0+	0+	0+	0+	0+	0+	0+	2
	3	1−	974	807	297	054	005	0+	0+	0+	0+	0+	0+	0+	3

(continued)

Table 2 CUMULATIVE BINOMIAL PROBABILITIES *(continued)*

		p													
n	*r*	.01	.05	.10	.20	.30	.40	.50	.60	.70	.80	.90	.95	.99	*r*
	4	1−	995	927	501	136	019	001	0+	0+	0+	0+	0+	0+	4
	5	1−	999	977	695	269	054	005	0+	0+	0+	0+	0+	0+	5
	6	1−	1−	994	840	440	124	017	001	0+	0+	0+	0+	0+	6
	7	1−	1−	999	928	618	237	047	004	0+	0+	0+	0+	0+	7
	8	1−	1−	1−	973	771	388	105	013	001	0+	0+	0+	0+	8
	9	1−	1−	1−	991	880	556	202	035	002	0+	0+	0+	0+	9
	10	1−	1−	1−	997	945	713	339	081	007	0+	0+	0+	0+	10
	11	1−	1−	1−	999	979	836	500	164	021	001	0+	0+	0+	11
	12	1−	1−	1−	1−	993	919	661	287	055	003	0+	0+	0+	12
	13	1−	1−	1−	1−	998	965	798	444	120	009	0+	0+	0+	13
	14	1−	1−	1−	1−	999	987	895	612	229	027	0+	0+	0+	14
	15	1−	1−	1−	1−	1−	996	953	763	382	072	001	0+	0+	15
	16	1−	1−	1−	1−	1−	999	983	876	560	160	006	0+	0+	16
	17	1−	1−	1−	1−	1−	1−	995	946	731	305	023	001	0+	17
	18	1−	1−	1−	1−	1−	1−	999	981	864	499	073	005	0+	18
	19	1−	1−	1−	1−	1−	1−	1−	995	946	703	193	026	0+	19
	20	1−	1−	1−	1−	1−	1−	1−	999	984	867	408	105	002	20
	21	1−	1−	1−	1−	1−	1−	1−	1−	997	960	685	321	022	21
	22	1−	1−	1−	1−	1−	1−	1−	1−	1−	994	911	693	206	22
	23	1	1	1	1	1	1	1	1	1	1	1	1	1	23
24	0	786	292	080	005	0+	0+	0+	0+	0+	0+	0+	0+	0+	0
	1	976	661	292	033	002	0+	0+	0+	0+	0+	0+	0+	0+	1
	2	998	884	564	115	012	001	0+	0+	0+	0+	0+	0+	0+	2
	3	1−	970	786	264	042	004	0+	0+	0+	0+	0+	0+	0+	3
	4	1−	994	915	460	111	013	001	0+	0+	0+	0+	0+	0+	4
	5	1−	999	972	656	229	040	003	0+	0+	0+	0+	0+	0+	5
	6	1−	1−	993	811	389	096	011	001	0+	0+	0+	0+	0+	6
	7	1−	1−	998	911	565	192	032	002	0+	0+	0+	0+	0+	7
	8	1−	1−	1−	964	725	328	076	008	0+	0+	0+	0+	0+	8
	9	1−	1−	1−	987	847	489	154	022	001	0+	0+	0+	0+	9
	10	1−	1−	1−	996	926	650	271	053	004	0+	0+	0+	0+	10
	11	1−	1−	1−	999	969	787	419	114	012	0+	0+	0+	0+	11
	12	1−	1−	1−	1−	988	886	581	213	031	001	0+	0+	0+	12
	13	1−	1−	1−	1−	996	947	729	350	074	004	0+	0+	0+	13
	14	1−	1−	1−	1−	999	978	846	511	153	013	0+	0+	0+	14
	15	1−	1−	1−	1−	1−	992	924	672	275	036	0+	0+	0+	15
	16	1−	1−	1−	1−	1−	998	968	808	435	089	002	0+	0+	16
	17	1−	1−	1−	1−	1−	999	989	904	611	189	007	0+	0+	17
	18	1−	1−	1−	1−	1−	1−	997	960	771	344	028	001	0+	18
	19	1−	1−	1−	1−	1−	1−	999	987	889	540	085	006	0+	19

(continued)

Table 2 CUMULATIVE BINOMIAL PROBABILITIES *(continued)*

								p							
n	r	.01	.05	.10	.20	.30	.40	.50	.60	.70	.80	.90	.95	.99	r
	20	1−	1−	1−	1−	1−	1−	1−	996	958	736	214	030	0+	20
	21	1−	1−	1−	1−	1−	1−	1−	999	988	885	436	116	002	21
	22	1−	1−	1−	1−	1−	1−	1−	1−	998	967	708	339	024	22
	23	1−	1−	1−	1−	1−	1−	1−	1−	1−	995	920	708	214	23
	24	1	1	1	1	1	1	1	1	1	1	1	1	1	24
25	0	778	277	072	004	0+	0+	0+	0+	0+	0+	0+	0+	0+	0
	1	974	642	271	027	002	0+	0+	0+	0+	0+	0+	0+	0+	1
	2	998	873	537	098	009	0+	0+	0+	0+	0+	0+	0+	0+	2
	3	1−	966	764	234	033	002	0+	0+	0+	0+	0+	0+	0+	3
	4	1−	993	902	421	090	009	0+	0+	0+	0+	0+	0+	0+	4
	5	1−	999	967	617	193	029	002	0+	0+	0+	0+	0+	0+	5
	6	1−	1−	991	780	341	074	007	0+	0+	0+	0+	0+	0+	6
	7	1−	1−	998	891	512	154	022	001	0+	0+	0+	0+	0+	7
	8	1−	1−	1−	953	677	274	054	004	0+	0+	0+	0+	0+	8
	9	1−	1−	1−	983	811	425	115	013	0+	0+	0+	0+	0+	9
	10	1−	1−	1−	994	902	586	212	034	002	0+	0+	0+	0+	10
	11	1−	1−	1−	998	956	732	345	078	006	0+	0+	0+	0+	11
	12	1−	1−	1−	1−	983	846	500	154	017	0+	0+	0+	0+	12
	13	1−	1−	1−	1−	994	922	655	268	044	002	0+	0+	0+	13
	14	1−	1−	1−	1−	998	966	788	414	098	006	0+	0+	0+	14
	15	1−	1−	1−	1−	1−	987	885	575	189	017	0+	0+	0+	15
	16	1−	1−	1−	1−	1−	996	946	726	323	047	0+	0+	0+	16
	17	1−	1−	1−	1−	1−	999	978	846	488	109	002	0+	0+	17
	18	1−	1−	1−	1−	1−	1−	993	926	659	220	009	0+	0+	18
	19	1−	1−	1−	1−	1−	1−	998	971	807	383	033	001	0+	19
	20	1−	1−	1−	1−	1−	1−	1−	991	910	579	098	007	0+	20
	21	1−	1−	1−	1−	1−	1−	1−	998	967	766	236	034	0+	21
	22	1−	1−	1−	1−	1−	1−	1−	1−	991	902	463	127	002	22
	23	1−	1−	1−	1−	1−	1−	1−	1−	998	973	729	358	026	23
	24	1−	1−	1−	1−	1−	1−	1−	1−	1−	996	928	723	222	24
	25	1	1	1	1	1	1	1	1	1	1	1	1	1	25

Table 3 CUMULATIVE POISSON PROBABILITIES

x	.001	.002	.003	.004	.005	.006	.007	.008	.009	.01	.02	.03	.04	.05	.06	.07	.08	.09	.10	.15	x
0	999	998	997	996	995	994	993	992	991	990	980	970	961	951	942	932	923	914	905	861	0
1	1	1	1	1	1	1	1	1	1	1	1	1	999	999	998	998	997	996	995	990	1
2	1	1	1	1	1	1	1	1	1	1	1	1	1	1	1	1	1	1	1	999	2
3	1	1	1	1	1	1	1	1	1	1	1	1	1	1	1	1	1	1	1	1	3

x	.20	.25	.30	.40	.50	.60	.70	.80	.90	1.0	1.1	1.2	1.3	1.4	1.5	1.6	1.7	1.8	1.9	2.0	x
0	819	779	741	670	607	549	497	449	407	368	333	301	273	247	223	202	183	165	150	135	0
1	982	974	963	938	910	878	844	809	772	736	699	663	627	592	558	525	493	463	434	406	1
2	999	998	996	992	986	977	966	953	937	920	900	879	857	833	809	783	757	731	704	677	2
3	1	1	1	999	998	997	994	991	987	981	974	966	957	946	934	921	907	891	875	857	3
4	1	1	1	1	1	1	999	999	998	996	995	992	989	986	981	976	970	964	956	947	4
5	1	1	1	1	1	1	1	1	1	999	999	998	998	997	996	994	992	990	987	983	5
6	1	1	1	1	1	1	1	1	1	1	1	1	1	999	999	999	998	997	997	995	6
7	1	1	1	1	1	1	1	1	1	1	1	1	1	1	1	1	1	999	999	999	7
8	1	1	1	1	1	1	1	1	1	1	1	1	1	1	1	1	1	1	1	1	8

(continued)

Table 3 CUMULATIVE POISSON PROBABILITIES *(continued)*

x	m = 2.1	2.2	2.3	2.4	2.5	2.6	2.7	2.8	2.9	3.0	3.1	3.2	3.3	3.4	3.5	3.6	3.7	3.8	3.9	4.0	x
0	122	111	100	091	082	074	067	061	055	050	045	041	037	033	030	027	025	022	020	018	0
1	380	355	331	308	287	267	249	231	215	199	185	171	159	147	136	126	116	107	099	092	1
2	650	623	596	570	544	518	494	469	446	423	401	380	359	340	321	303	285	269	253	238	2
3	839	819	799	779	758	736	714	692	670	647	625	603	580	558	537	515	494	473	453	433	3
4	938	928	916	904	891	877	863	848	832	815	798	781	763	744	725	706	687	668	648	629	4
5	980	975	970	964	958	951	943	935	926	916	906	895	883	871	858	844	830	816	801	785	5
6	994	993	991	988	986	983	979	976	971	966	961	955	949	942	935	927	918	909	899	889	6
7	999	998	997	997	996	995	993	992	990	988	986	983	980	977	973	969	965	960	955	949	7
8	1	1	999	999	999	999	998	998	997	996	995	994	993	992	990	988	986	984	981	979	8
9	1	1	1	1	1	1	999	999	999	999	999	998	998	997	997	996	995	994	993	992	9
10	1	1	1	1	1	1	1	1	1	1	1	1	999	999	999	999	998	998	998	997	10
11	1	1	1	1	1	1	1	1	1	1	1	1	1	1	1	1	1	999	999	999	11
12	1	1	1	1	1	1	1	1	1	1	1	1	1	1	1	1	1	1	1	1	12

x	m = 4.1	4.2	4.3	4.4	4.5	4.6	4.7	4.8	4.9	5.0	5.1	5.2	5.3	5.4	5.5	5.6	5.7	5.8	5.9	6.0	x
0	017	015	014	012	011	010	009	008	007	007	006	006	005	005	004	003	003	003	003	002	0
1	085	078	072	066	061	056	052	048	044	040	037	034	031	029	027	024	022	021	019	017	1
2	224	210	197	185	174	163	152	143	133	125	116	109	102	095	088	082	077	072	067	062	2
3	414	395	377	359	342	326	310	294	279	265	251	238	225	213	202	191	180	170	160	151	3
4	609	590	570	551	532	513	495	476	458	440	423	406	390	373	358	342	327	313	299	285	4
5	769	753	737	720	703	686	668	651	634	616	598	581	563	546	529	512	495	478	462	446	5
6	879	867	856	844	831	818	805	791	777	762	747	732	717	702	686	670	654	638	622	606	6
7	943	936	929	921	913	905	896	887	877	867	856	845	833	822	809	797	784	771	758	744	7
8	976	972	968	964	960	955	950	944	938	932	925	918	911	903	894	886	877	867	857	847	8
9	990	989	987	985	983	980	978	975	972	968	964	960	956	951	946	941	935	929	923	916	9
10	997	996	995	994	993	992	991	990	988	986	984	982	980	977	975	972	969	965	961	958	10
11	999	999	998	998	998	997	997	996	995	995	994	993	992	990	989	988	986	984	982	980	11
12	1	1	999	999	999	999	999	999	998	998	998	997	997	996	996	995	994	993	992	991	12
13	1	1	1	1	1	1	1	1	999	999	999	999	999	999	998	998	998	997	997	996	13
14	1	1	1	1	1	1	1	1	1	1	1	1	1	1	999	999	999	999	999	999	14
15	1	1	1	1	1	1	1	1	1	1	1	1	1	1	1	1	1	1	1	999	15
16	1	1	1	1	1	1	1	1	1	1	1	1	1	1	1	1	1	1	1	1	16

(continued)

Table 3 CUMULATIVE POISSON PROBABILITIES *(continued)*

	m																				
x	6.1	6.2	6.3	6.4	6.5	6.6	6.7	6.8	6.9	7.0	7.1	7.2	7.3	7.4	7.5	8.0	8.5	9.0	9.5	10.0	x
0	002	002	002	002	002	001	001	001	001	001	001	001	001	001	001	0+	0+	0+	0+	0+	0
1	016	015	013	012	011	010	009	009	008	007	007	006	006	005	005	003	002	001	001	0+	1
2	058	054	050	046	043	040	037	034	032	030	027	025	024	022	020	014	009	006	004	003	2
3	143	134	126	119	112	105	099	093	087	082	077	072	067	063	059	042	030	021	015	010	3
4	272	159	247	235	224	213	202	192	182	173	164	156	147	140	132	100	074	055	040	029	4
5	430	414	399	384	369	355	341	327	314	301	288	276	264	253	241	191	150	116	089	067	5
6	590	574	558	542	527	511	495	480	465	450	435	420	406	392	378	313	256	207	165	130	6
7	730	716	702	687	673	658	643	628	614	599	584	569	554	539	525	453	386	324	269	220	7
8	837	826	815	803	792	780	767	755	742	729	716	703	689	676	662	593	523	456	392	333	8
9	909	902	894	886	877	869	860	850	839	830	820	810	799	788	776	717	653	587	522	458	9
10	953	949	944	939	933	927	921	915	908	901	894	887	879	871	862	816	763	706	645	583	10
11	978	975	972	969	966	963	959	955	951	947	942	937	932	926	921	888	849	803	752	697	11
12	990	989	987	986	984	982	980	978	976	973	970	967	964	961	957	936	909	876	836	792	12
13	996	995	995	994	993	992	991	990	989	987	986	984	982	980	978	966	949	926	898	870	13
14	998	998	998	997	997	997	996	996	995	994	994	993	992	991	990	983	973	959	940	917	14
15	999	999	999	999	999	999	998	998	998	998	997	997	996	996	995	992	986	978	967	951	15
16	1	1	1	1	1	999	999	999	999	999	999	999	999	998	998	996	993	989	982	973	16
17	1	1	1	1	1	1	1	1	1	1	1	999	999	999	999	998	997	995	991	986	17
18	1	1	1	1	1	1	1	1	1	1	1	1	1	1	1	999	999	998	996	993	18
19	1	1	1	1	1	1	1	1	1	1	1	1	1	1	1	1	999	999	998	997	19
20	1	1	1	1	1	1	1	1	1	1	1	1	1	1	1	1	1	1	999	998	20
21	1	1	1	1	1	1	1	1	1	1	1	1	1	1	1	1	1	1	1	999	21
22	1	1	1	1	1	1	1	1	1	1	1	1	1	1	1	1	1	1	1	1	22

Table 4 AREAS UNDER THE NORMAL CURVE.

z	.00	.01	.02	.03	.04	.05	.06	.07	.08	.09
0.0	.0000	.0040	.0080	.0120	.0160	.0199	.0239	.0279	.0319	.0359
0.1	.0398	.0438	.0478	.0517	.0557	.0596	.0636	.0675	.0714	.0753
0.2	.0793	.0832	.0871	.0910	.0948	.0987	.1026	.1064	.1103	.1141
0.3	.1179	.1217	.1255	.1293	.1331	.1368	.1406	.1443	.1480	.1517
0.4	.1554	.1591	.1628	.1664	.1700	.1736	.1772	.1808	.1844	.1879
0.5	.1915	.1950	.1985	.2019	.2054	.2088	.2123	.2157	.2190	.2224
0.6	.2257	.2291	.2324	.2357	.2389	.2422	.2454	.2486	.2517	.2549
0.7	.2580	.2611	.2642	.2673	.2704	.2734	.2764	.2794	.2823	.2852
0.8	.2881	.2910	.2939	.2967	.2995	.3023	.3051	.3078	.3106	.3133
0.9	.3159	.3186	.3212	.3238	.3264	.3289	.3315	.3340	.3365	.3389
1.0	.3413	.3438	.3461	.3485	.3508	.3531	.3554	.3577	.3599	.3621
1.1	.3643	.3665	.3686	.3708	.3729	.3749	.3770	.3790	.3810	.3930
1.2	.3849	.3869	.3888	.3907	.3925	.3944	.3962	.3980	.3997	.4015
1.3	.4032	.4049	.4066	.4082	.4099	.4115	.4131	.4147	.4162	.4177
1.4	.4192	.4207	.4222	.4236	.4251	.4265	.4279	.4292	.4306	.4319
1.5	.4332	.4345	.4357	.4370	.4382	.4394	.4406	.4418	.4429	.4441
1.6	.4452	.4463	.4474	.4484	.4495	.4505	.4515	.4525	.4535	.4545
1.7	.4554	.4564	.4573	.4582	.4591	.4599	.4608	.4616	.4625	.4633
1.8	.4641	.4649	.4656	.4664	.4671	.4678	.4686	.4693	.4699	.4706
1.9	.4713	.4719	.4726	.4732	.4738	.4744	.4750	.4756	.4761	.4767
2.0	.4772	.4778	.4783	.4788	.4793	.4798	.4803	.4808	.4812	.4817
2.1	.4821	.4826	.4830	.4834	.4838	.4842	.4846	.4850	.4854	.4857
2.2	.4861	.4864	.4868	.4871	.4875	.4878	.4881	.4884	.4887	.4890
2.3	.4893	.4896	.4898	.4901	.4904	.4906	.4909	.4911	.4913	.4916
2.4	.4918	.4920	.4922	.4925	.4927	.4929	.4931	.4932	.4934	.4936
2.5	.4938	.4940	.4941	.4943	.4945	.4946	.4948	.4949	.4951	.4952
2.6	.4953	.4955	.4956	.4957	.4959	.4960	.4961	.4962	.4963	.4964
2.7	.4965	.4966	.4967	.4968	.4969	.4970	.4971	.4972	.4973	.4974
2.8	.4974	.4975	.4976	.4977	.4977	.4978	.4979	.4979	.4980	.4981
2.9	.4981	.4982	.4982	.4983	.4984	.4984	.4985	.4985	.4986	.4987
3.0	.4987	.4987	.4987	.4988	.4988	.4989	.4989	.4989	.4990	.4990
3.1	.4990	.4991	.4991	.4991	.4992	.4992	.4992	.4992	.4993	.4993
3.2	.4993	.4993	.4994	.4994	.4994	.4994	.4994	.4995	.4995	.4955
3.3	.4995	.4995	.4996	.4996	.4996	.4996	.4996	.4996	.4996	.4997
3.4	.4997	.4997	.4997	.4997	.4997	.4997	.4997	.4997	.4998	.4998
3.5	.4998	.4998	.4998	.4998	.4998	.4998	.4998	.4998	.4998	.4998

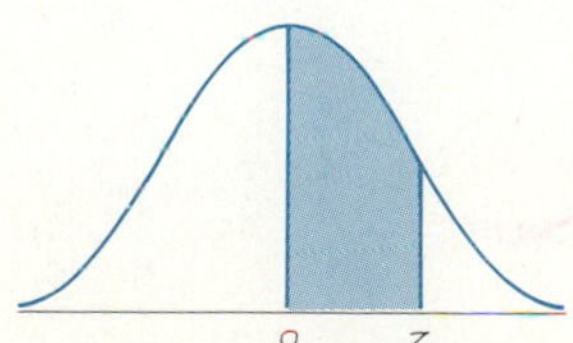

Table 5 CRITICAL VALUES OF t.

DF	$t_{.10}$	$t_{.05}$	$t_{.025}$	$t_{.01}$	$t_{.005}$	DF	$t_{.10}$	$t_{.05}$	$t_{.025}$	$t_{.01}$	$t_{.005}$
1	3.078	6.314	12.706	31.821	63.657	18	1.330	1.734	2.101	2.552	2.878
2	1.886	2.920	4.303	6.965	9.925	19	1.328	1.729	2.093	2.539	2.861
3	1.638	2.353	3.182	4.541	5.841	20	1.325	1.725	2.086	2.528	2.845
4	1.533	2.132	2.776	3.747	4.604	21	1.323	1.721	2.080	2.518	2.831
5	1.476	2.015	2.571	3.365	4.032	22	1.321	1.717	2.074	2.508	2.819
6	1.440	1.943	2.447	3.143	3.707	23	1.319	1.714	2.069	2.500	2.807
7	1.415	1.895	2.365	2.998	3.499	24	1.318	1.711	2.064	2.492	2.797
8	1.397	1.860	2.306	2.896	3.355	25	1.316	1.708	2.060	2.485	2.787
9	1.383	1.833	2.262	2.821	3.250	26	1.315	1.706	2.056	2.479	2.779
10	1.372	1.812	2.228	2.764	3.169	27	1.314	1.703	2.052	2.473	2.771
11	1.363	1.796	2.201	2.718	3.106	28	1.313	1.701	2.048	2.467	2.763
12	1.356	1.782	2.179	2.681	3.055	29	1.311	1.699	2.045	2.462	2.756
13	1.350	1.771	2.160	2.650	3.012	30	1.310	1.697	2.042	2.457	2.750
14	1.345	1.761	2.145	2.624	2.977	40	1.303	1.684	2.021	2.423	2.704
15	1.341	1.753	2.131	2.602	2.947	60	1.296	1.671	2.000	2.390	2.660
16	1.337	1.746	2.120	2.583	2.921	120	1.289	1.658	1.980	2.358	2.617
17	1.333	1.740	2.110	2.567	2.898	∞	1.282	1.645	1.960	2.326	2.576

Source: E. S. Pearson and H. O. Hartley, *Biometrika Tables for Statisticians,* vol. 1, 3rd ed. (Cambridge: The University Press, 1966), Table 12.

t DISTRIBUTION

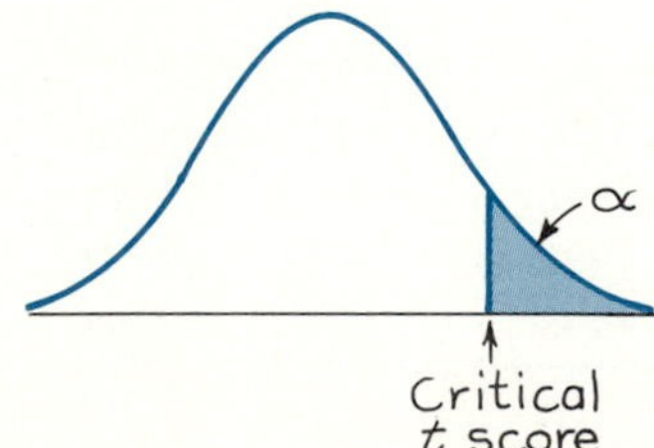

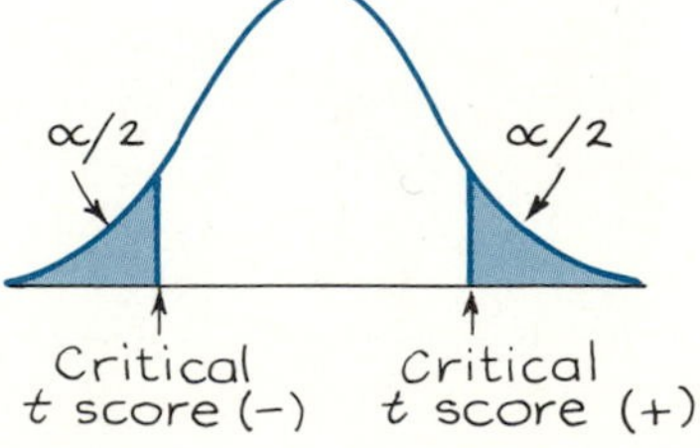

Table 6 CRITICAL VALUES OF χ^2

DF	$\chi^2_{.995}$	$\chi^2_{.99}$	$\chi^2_{.975}$	$\chi^2_{.95}$	$\chi^2_{.90}$	$\chi^2_{.10}$	$\chi^2_{.05}$	$\chi^2_{.025}$	$\chi^2_{.01}$	$\chi^2_{.005}$
1	0.0000393	0.000157	0.000982	0.00393	0.0158	2.706	3.841	5.024	6.635	7.879
2	0.01003	0.02010	0.05064	0.1026	0.2107	4.605	5.991	7.378	9.210	10.60
3	0.0717	0.1148	0.2158	0.3518	0.5844	6.251	7.815	9.348	11.34	12.84
4	0.2070	0.2971	0.4844	0.7107	1.065	7.779	9.488	11.14	13.28	14.86
5	0.4117	0.5543	0.8312	1.145	1.161	9.236	11.07	12.83	15.09	16.75
6	0.6757	0.8721	1.237	1.635	2.204	10.64	12.59	14.45	16.81	18.55
7	0.9893	1.239	1.690	2.167	2.833	12.02	14.07	16.01	18.48	20.28
8	1.344	1.646	2.180	2.733	3.490	13.36	15.51	17.53	20.09	21.95
9	1.735	2.088	2.700	3.325	4.168	14.68	16.92	19.02	21.67	23.59
10	2.156	2.558	3.247	3.940	4.865	15.99	18.31	20.48	23.21	25.19
11	2.603	3.053	3.816	4.575	5.578	17.28	19.68	21.92	24.72	26.76
12	3.074	3.571	4.404	5.226	6.304	18.55	21.03	23.34	26.22	28.30
13	3.565	4.107	5.009	5.892	7.042	19.81	22.36	24.74	27.69	29.82
14	4.075	4.660	5.629	6.571	7.790	21.06	23.68	26.12	29.14	31.32
15	4.601	5.229	6.262	7.261	8.547	22.31	25.00	27.49	30.58	32.80
16	5.142	5.812	6.910	7.962	9.312	23.54	26.30	28.84	32.00	34.27
17	5.697	6.408	7.564	8.672	10.08	24.77	27.59	30.19	33.41	35.72
18	6.265	7.015	8.231	9.390	10.86	25.99	28.87	31.53	34.80	37.16
19	6.844	7.633	8.906	10.12	11.65	27.20	30.14	32.85	36.19	38.58
20	7.434	8.260	9.591	10.85	12.44	28.41	31.41	34.17	37.57	40.00
21	8.034	8.897	10.28	11.59	13.24	29.62	32.67	35.48	38.93	41.40
22	8.643	9.542	10.98	12.34	14.04	30.81	33.92	36.78	40.29	42.80
23	9.260	10.20	11.69	13.03	14.85	32.01	35.17	38.08	41.64	44.18
24	9.886	10.86	12.40	13.85	15.66	33.20	36.42	39.36	42.98	45.56
25	10.52	11.52	13.12	14.61	16.47	34.38	37.65	40.65	44.31	46.93
26	11.16	12.20	13.84	15.38	17.29	35.56	38.88	41.92	45.64	48.29
27	11.81	12.88	14.57	16.15	18.11	36.74	40.11	43.19	46.96	49.64
28	12.46	13.56	15.31	16.93	18.94	37.92	41.34	44.46	48.28	50.99
29	13.12	14.26	16.05	17.71	19.77	39.09	42.56	45.72	49.59	52.34
30	13.79	14.95	16.79	18.49	20.60	40.26	43.77	46.98	50.89	53.67
40	20.71	22.16	24.43	26.51	29.05	51.80	55.76	59.34	63.69	66.77
60	35.53	37.48	40.48	43.19	46.46	74.40	79.08	83.30	88.38	91.95
80	51.17	53.54	57.15	60.39	64.28	96.58	101.9	106.6	112.3	116.3
100	67.33	70.06	74.22	77.93	82.36	118.5	124.3	129.6	135.8	140.2
120	83.85	86.92	91.57	95.70	100.6	140.2	146.6	152.2	159.0	163.6
150	109.1	112.7	118.0	122.7	128.3	172.6	179.6	185.8	193.2	198.4
200	152.2	156.4	162.7	168.3	174.8	226.0	234.0	241.1	249.4	255.3
	0.005	0.01	0.025	0.05	0.10	0.10	0.05	0.025	0.01	0.005
	Area in Left Tail					Area in Right Tail				

Source: E. S. Pearson and H. O. Hartley, *Biometrika Tables for Statisticians,* vol. 1, 3rd ed. (Cambridge: The University Press, 1966), Table 8.

0.05

F

Table 7 CRITICAL VALUES OF $F_{.05}$

DEGREES OF FREEDOM FOR DENOMINATOR	DEGREES OF FREEDOM FOR NUMERATOR 1	2	3	4	5	6	7	8	9
1	161	200	216	225	230	234	237	239	241
2	18.5	19.0	19.2	19.2	19.3	19.3	19.4	19.4	19.4
3	10.1	9.55	9.28	9.12	9.01	8.94	8.89	8.85	8.81
4	7.71	6.94	6.59	6.39	6.26	6.16	6.09	6.04	6.00
5	6.61	5.79	5.41	5.19	5.05	4.95	4.88	4.82	4.77
6	5.99	5.14	4.76	4.53	4.39	4.28	4.21	4.15	4.10
7	5.59	4.74	4.35	4.12	3.97	3.87	3.79	3.73	3.68
8	5.32	4.46	4.07	3.84	3.69	3.58	3.50	3.44	3.39
9	5.12	4.26	3.86	3.63	3.48	3.37	3.29	3.23	3.18
10	4.96	4.10	3.71	3.48	3.33	3.22	3.14	3.07	3.02
11	4.84	3.98	3.59	3.36	3.20	3.09	3.01	2.95	2.90
12	4.75	3.89	3.49	3.26	3.11	3.00	2.91	2.85	2.80
13	4.67	3.81	3.41	3.18	3.03	2.92	2.83	2.77	2.71
14	4.60	3.74	3.34	3.11	2.96	2.85	2.76	2.70	2.65
15	4.54	3.68	3.29	3.06	2.90	2.79	2.71	2.64	2.59
16	4.49	3.63	3.24	3.01	2.85	2.74	2.66	2.59	2.54
17	4.45	3.59	3.20	2.96	2.81	2.70	2.61	2.55	2.49
18	4.41	3.55	3.16	2.93	2.77	2.66	2.58	2.51	2.46
19	4.38	3.52	3.13	2.90	2.74	2.63	2.54	2.48	2.42
20	4.35	3.49	3.10	2.87	2.71	2.60	2.51	2.45	2.39
21	4.32	3.47	3.07	2.84	2.68	2.57	2.49	2.42	2.37
22	4.30	3.44	3.05	2.82	2.66	2.55	2.46	2.40	2.34
23	4.28	3.42	3.03	2.80	2.64	2.53	2.44	2.37	2.32
24	4.26	3.40	3.01	2.78	2.62	2.51	2.42	2.36	2.30
25	4.24	3.39	2.99	2.76	2.60	2.49	2.40	2.34	2.28
30	4.17	3.32	2.92	2.69	2.53	2.42	2.33	2.27	2.21
40	4.08	3.23	2.84	2.61	2.45	2.34	2.25	2.18	2.12
60	4.00	3.15	2.76	2.53	2.37	2.25	2.17	2.10	2.04
120	3.92	3.07	2.68	2.45	2.29	2.18	2.09	2.02	1.96
∞	3.84	3.00	2.60	2.37	2.21	2.10	2.01	1.94	1.88

(continued)

Source: E. S. Pearson and H. O. Hartley, *Biometrika Tables for Statisticians,* vol. 1, 3rd ed. *(Cambridge: The University Press, 1966), Table 18.*

Table 7 CRITICAL VALUES OF $F_{.05}$ (*continued*)

DEGREES OF FREEDOM FOR NUMERATOR

DEGREES OF FREEDOM FOR DENOMINATOR

10	12	15	20	24	30	40	60	120	∞
242	244	246	248	249	250	251	252	253	254
19.4	19.4	19.4	19.4	19.5	19.5	19.5	19.5	19.5	19.5
8.79	8.74	8.70	8.66	8.64	8.62	8.59	8.57	8.55	8.53
5.96	5.91	5.86	5.80	5.77	5.75	5.72	5.69	5.66	5.63
4.74	4.68	4.62	4.56	4.53	4.50	4.46	4.43	4.40	4.37
4.06	4.00	3.94	3.87	3.84	3.81	3.77	3.74	3.70	3.67
3.64	3.57	3.51	3.44	3.41	3.38	3.34	3.30	3.27	3.23
3.35	3.28	3.22	3.15	3.12	3.08	3.04	3.01	2.97	2.93
3.14	3.07	3.01	2.94	2.90	2.86	2.83	2.79	2.75	2.71
2.98	2.91	2.85	2.77	2.74	2.70	2.66	2.62	2.58	2.54
2.85	2.79	2.72	2.65	2.61	2.57	2.53	2.49	2.45	2.40
2.75	2.69	2.62	2.54	2.51	2.47	2.43	2.38	2.34	2.30
2.67	2.60	2.53	2.46	2.42	2.38	2.34	2.30	2.25	2.21
2.60	2.53	2.46	2.39	2.35	2.31	2.27	2.22	2.18	2.13
2.54	2.48	2.40	2.33	2.29	2.25	2.20	2.16	2.11	2.07
2.49	2.42	2.35	2.28	2.24	2.19	2.15	2.11	2.06	2.01
2.45	2.38	2.31	2.23	2.19	2.15	2.10	2.06	2.01	1.96
2.41	2.34	2.27	2.19	2.15	2.11	2.06	2.02	1.97	1.92
2.38	2.31	2.23	2.16	2.11	2.07	2.03	1.98	1.93	1.88
2.35	2.28	2.20	2.12	2.08	2.04	1.99	1.95	1.90	1.84
2.32	2.25	2.18	2.10	2.05	2.01	1.96	1.92	1.87	1.81
2.30	2.23	2.15	2.07	2.03	1.98	1.94	1.89	1.84	1.78
2.27	2.20	2.13	2.05	2.01	1.96	1.91	1.86	1.81	1.76
2.25	2.18	2.11	2.03	1.98	1.94	1.89	1.84	1.79	1.73
2.24	2.16	2.09	2.01	1.96	1.92	1.87	1.82	1.77	1.71
2.16	2.09	2.01	1.93	1.89	1.84	1.79	1.74	1.68	1.62
2.08	2.00	1.92	1.84	1.79	1.74	1.69	1.64	1.58	1.51
1.99	1.92	1.84	1.75	1.70	1.65	1.59	1.53	1.47	1.39
1.91	1.83	1.75	1.66	1.61	1.55	1.50	1.43	1.35	1.25
1.83	1.75	1.67	1.57	1.52	1.46	1.39	1.32	1.22	1.00

Table 8 CRITICAL VALUES OF $F_{.025}$.

DEGREES OF FREEDOM FOR NUMERATOR (columns); DEGREES OF FREEDOM FOR DENOMINATOR (rows)

	1	2	3	4	5	6	7	8	9
1	647.8	799.5	864.2	899.6	921.8	937.1	948.2	956.7	963.3
2	38.51	39.00	39.17	39.25	39.30	39.33	39.36	39.37	39.39
3	17.44	16.04	15.44	15.10	14.88	14.73	14.62	14.54	14.47
4	12.22	10.65	9.98	9.60	9.36	9.20	9.07	8.98	8.90
5	10.01	8.43	7.76	7.39	7.15	6.98	6.85	6.76	6.68
6	8.81	7.26	6.60	6.23	5.99	5.82	5.70	5.60	5.52
7	8.07	6.54	5.89	5.52	5.29	5.12	4.99	4.90	4.82
8	7.57	6.06	5.42	5.05	4.82	4.65	4.53	4.43	4.36
9	7.21	5.71	5.08	4.72	4.48	4.32	4.20	4.10	4.03
10	6.94	5.46	4.83	4.47	4.24	4.07	3.95	3.85	3.78
11	6.72	5.26	4.63	4.28	4.04	3.88	3.76	3.66	3.59
12	6.55	5.10	4.47	4.12	3.89	3.73	3.61	3.51	3.44
13	6.41	4.97	4.35	4.00	3.77	3.60	3.48	3.39	3.31
14	6.30	4.86	4.24	3.89	3.66	3.50	3.38	3.29	3.21
15	6.20	4.77	4.15	3.80	3.58	3.41	3.29	3.20	3.12
16	6.12	4.69	4.08	3.73	3.50	3.34	3.22	3.12	3.05
17	6.04	4.62	4.01	3.66	3.44	3.28	3.16	3.06	2.98
18	5.98	4.56	3.95	3.61	3.38	3.22	3.10	3.01	2.93
19	5.92	4.51	3.90	3.56	3.33	3.17	3.05	2.96	2.88
20	5.87	4.46	3.86	3.51	3.29	3.13	3.01	2.91	2.84
21	5.83	4.42	3.82	3.48	3.25	3.09	2.97	2.87	2.80
22	5.79	4.38	3.78	3.44	3.22	3.05	2.93	2.84	2.76
23	5.75	4.35	3.75	3.41	3.18	3.02	2.90	2.81	2.73
24	5.72	4.32	3.72	3.38	3.15	2.99	2.87	2.78	2.70
25	5.69	4.29	3.69	3.35	3.13	2.97	2.85	2.75	2.68
30	5.57	4.18	3.59	3.25	3.03	2.87	2.75	2.65	2.57
40	5.42	4.05	3.46	3.13	2.90	2.74	2.62	2.53	2.45
60	5.29	3.93	3.34	3.01	2.79	2.63	2.51	2.41	2.33
120	5.15	3.80	3.23	2.89	2.67	2.52	2.39	2.30	2.22
∞	5.02	3.69	3.12	2.79	2.57	2.41	2.29	2.19	2.11

(continued)

Source: E. S. Pearson and H. O. Hartley, *Biometrika Tables for Statisticians,* vol. 1, 3rd ed. (Cambridge: The University Press, 1966), Table 18.

Table 8 CRITICAL VALUES OF $F_{.025}$ *(continued)*

DEGREES OF FREEDOM FOR NUMERATOR

DEGREES OF FREEDOM FOR DENOMINATOR

10	12	15	20	24	30	40	60	120	∞
968.6	976.7	984.9	993.1	997.2	1001	1006	1010	1014	1018
39.40	39.41	39.43	39.45	39.46	39.46	39.47	39.48	39.49	39.50
14.42	14.34	14.25	14.17	14.12	14.08	14.04	13.99	13.95	13.90
8.84	8.75	8.66	8.56	8.51	8.46	8.41	8.36	8.31	8.26
6.62	6.52	6.43	6.33	6.28	6.23	6.18	6.12	6.07	6.02
5.46	5.37	5.27	5.17	5.12	5.07	5.01	4.96	4.90	4.85
4.76	4.67	4.57	4.47	4.42	4.36	4.31	4.25	4.20	4.14
4.30	4.20	4.10	4.00	3.95	3.89	3.84	3.78	3.73	3.67
3.96	3.87	3.77	3.67	3.61	3.56	3.51	3.45	3.39	3.33
3.72	3.62	3.52	3.42	3.37	3.31	3.26	3.20	3.14	3.08
3.53	3.43	3.33	3.23	3.17	3.12	3.06	3.00	2.94	2.88
3.37	3.28	3.18	3.07	3.02	2.96	2.91	2.85	2.79	2.72
3.25	3.15	3.05	2.95	2.89	2.84	2.78	2.72	2.66	2.60
3.15	3.05	2.95	2.84	2.79	2.73	2.67	2.61	2.55	2.49
3.06	2.96	2.86	2.76	2.70	2.64	2.59	2.52	2.46	2.40
2.99	2.89	2.79	2.68	2.63	2.57	2.51	2.45	2.38	2.32
2.92	2.82	2.72	2.62	2.56	2.50	2.44	2.38	2.32	2.25
2.87	2.77	2.67	2.56	2.50	2.44	2.38	2.32	2.26	2.19
2.82	2.72	2.62	2.51	2.45	2.39	2.33	2.27	2.20	2.13
2.77	2.68	2.57	2.46	2.41	2.35	2.29	2.22	2.16	2.09
2.73	2.64	2.53	2.42	2.37	2.31	2.25	2.18	2.11	2.04
2.70	2.60	2.50	2.39	2.33	2.27	2.21	2.14	2.08	2.00
2.67	2.57	2.47	2.36	2.30	2.24	2.18	2.11	2.04	1.97
2.64	2.54	2.44	2.33	2.27	2.21	2.15	2.08	2.01	1.94
2.61	2.51	2.41	2.30	2.24	2.18	2.12	2.05	1.98	1.91
2.51	2.41	2.31	2.20	2.14	2.07	2.01	1.94	1.87	1.79
2.39	2.29	2.18	2.07	2.01	1.94	1.88	1.80	1.72	1.64
2.27	2.17	2.06	1.94	1.88	1.82	1.74	1.67	1.58	1.48
2.16	2.05	1.94	1.82	1.76	1.69	1.61	1.53	1.43	1.31
2.05	1.94	1.83	1.71	1.64	1.57	1.48	1.39	1.27	1.00

0.01

F

Table 9 CRITICAL VALUES OF $F_{.01}$.

DEGREES OF FREEDOM FOR DENOMINATOR	DEGREES OF FREEDOM FOR NUMERATOR								
	1	2	3	4	5	6	7	8	9
1	4052	5000	5403	5625	5764	5859	5928	5982	6023
2	98.5	99.0	99.2	99.2	99.3	99.3	99.4	99.4	99.4
3	34.1	30.8	29.5	28.7	28.2	27.9	27.7	27.5	27.3
4	21.2	18.0	16.7	16.0	15.5	15.2	15.0	14.8	14.7
5	16.3	13.3	12.1	11.4	11.0	10.7	10.5	10.3	10.2
6	13.7	10.9	9.78	9.15	8.75	8.47	8.26	8.10	7.98
7	12.2	9.55	8.45	7.85	7.46	7.19	6.99	6.84	6.72
8	11.3	8.65	7.59	7.01	6.63	6.37	6.18	6.03	5.91
9	10.6	8.02	6.99	6.42	6.06	5.80	5.61	5.47	5.35
10	10.0	7.56	6.55	5.99	5.64	5.39	5.20	5.06	4.94
11	9.65	7.21	6.22	5.67	5.32	5.07	4.89	4.74	4.63
12	9.33	6.93	5.95	5.41	5.06	4.82	4.64	4.50	4.39
13	9.07	6.70	5.74	5.21	4.86	4.62	4.44	4.30	4.19
14	8.86	6.51	5.56	5.04	4.70	4.46	4.28	4.14	4.03
15	8.68	6.36	5.42	4.89	4.56	4.32	4.14	4.00	3.89
16	8.53	6.23	5.29	4.77	4.44	4.20	4.03	3.89	3.78
17	8.40	6.11	5.19	4.67	4.34	4.10	3.93	3.79	3.68
18	8.29	6.01	5.09	4.58	4.25	4.01	3.84	3.71	3.60
19	8.19	5.93	5.01	4.50	4.17	3.94	3.77	3.63	3.52
20	8.10	5.85	4.94	4.43	4.10	3.87	3.70	3.56	3.46
21	8.02	5.78	4.87	4.37	4.04	3.81	3.64	3.51	3.40
22	7.95	5.72	4.82	4.31	3.99	3.76	3.59	3.45	3.35
23	7.88	5.66	4.76	4.26	3.94	3.71	3.54	3.41	3.30
24	7.82	5.61	4.72	4.22	3.90	3.67	3.50	3.36	3.26
25	7.77	5.57	4.68	4.18	3.86	3.63	3.46	3.32	3.22
30	7.56	5.39	4.51	4.02	3.70	3.47	3.30	3.17	3.07
40	7.31	5.18	4.31	3.83	3.51	3.29	3.12	2.99	2.89
60	7.08	4.98	4.13	3.65	3.34	3.12	2.95	2.82	2.72
120	6.85	4.79	3.95	3.48	3.17	2.96	2.79	2.66	2.56
∞	6.63	4.61	3.78	3.32	3.02	2.80	2.64	2.51	2.41

(continued)

Source: E. S. Pearson and H. O. Hartley, *Biometrika Tables for Statisticians,* vol. 1, 3rd ed. (Cambridge: The University Press, 1966), Table 18.

Table 9 CRITICAL VALUES OF $F_{.01}$ *(continued)*

DEGREES OF FREEDOM FOR NUMERATOR

DEGREES OF FREEDOM FOR DENOMINATOR

10	12	15	20	24	30	40	60	120	∞
6056	6106	6157	6209	6235	6261	6287	6313	6339	6366
99.4	99.4	99.4	99.4	99.5	99.5	99.5	99.5	99.5	99.5
27.2	27.1	26.9	26.7	26.6	26.5	26.4	26.3	26.2	26.1
14.5	14.4	14.2	14.0	13.9	13.8	13.7	13.7	13.6	13.5
10.1	9.89	9.72	9.55	9.47	9.38	9.29	9.20	9.11	9.02
7.87	7.72	7.56	7.40	7.31	7.23	7.14	7.06	6.97	6.88
6.62	6.47	6.31	6.16	6.07	5.99	5.91	5.82	5.74	5.65
5.81	5.67	5.52	5.36	5.28	5.20	5.12	5.03	4.95	4.86
5.26	5.11	4.96	4.81	4.73	4.65	4.57	4.48	4.40	4.31
4.85	4.71	4.56	4.41	4.33	4.25	4.17	4.08	4.00	3.91
4.54	4.40	4.25	4.10	4.02	3.94	3.86	3.78	3.69	3.60
4.30	4.16	4.01	3.86	3.78	3.70	3.62	3.54	3.45	3.36
4.10	3.96	3.82	3.66	3.59	3.51	3.43	3.34	3.25	3.17
3.94	3.80	3.66	3.51	3.43	3.35	3.27	3.18	3.09	3.00
3.80	3.67	3.52	3.37	3.29	3.21	3.13	3.05	2.96	2.87
3.69	3.55	3.41	3.26	3.18	3.10	3.02	2.93	2.84	2.75
3.59	3.46	3.31	3.16	3.08	3.00	2.92	2.83	2.75	2.65
3.51	3.37	3.23	3.08	3.00	2.92	2.84	2.75	2.66	2.57
3.43	3.30	3.15	3.00	2.92	2.84	2.76	2.67	2.58	2.49
3.37	3.23	3.09	2.94	2.86	2.78	2.69	2.61	2.52	2.42
3.31	3.17	3.03	2.88	2.80	2.72	2.64	2.55	2.46	2.36
3.26	3.12	2.98	2.83	2.75	2.67	2.58	2.50	2.40	2.31
3.21	3.07	2.93	2.78	2.70	2.62	2.54	2.45	2.35	2.26
3.17	3.03	2.89	2.74	2.66	2.58	2.49	2.40	2.31	2.21
3.13	2.99	2.85	2.70	2.62	2.53	2.45	2.36	2.27	2.17
2.98	2.84	2.70	2.55	2.47	2.39	2.30	2.21	2.11	2.01
2.80	2.66	2.52	2.37	2.29	2.20	2.11	2.02	1.92	1.80
2.63	2.50	2.35	2.20	2.12	2.03	1.94	1.84	1.73	1.60
2.47	2.34	2.19	2.03	1.95	1.86	1.76	1.66	1.53	1.38
2.32	2.18	2.04	1.88	1.79	1.70	1.59	1.47	1.32	1.00

0.005

F

Table 10 CRITICAL VALUES OF $F_{.005}$.

DEGREES OF FREEDOM FOR NUMERATOR

DEGREES OF FREEDOM FOR DENOMINATOR	1	2	3	4	5	6	7	8	9
1	16,211	20,000	21,615	22,500	23,056	23,437	23,715	23,925	24,091
2	198.5	199.0	199.2	199.2	199.3	199.3	199.4	199.4	199.4
3	55.55	49.80	47.47	46.19	45.39	44.84	44.43	44.13	43.88
4	31.33	26.28	24.26	23.15	22.46	21.97	21.62	21.35	21.14
5	22.78	18.31	16.53	15.56	14.94	14.51	14.20	13.96	13.77
6	18.63	14.54	12.92	12.03	11.46	11.07	10.79	10.57	10.39
7	16.24	12.40	10.88	10.05	9.52	9.16	8.89	8.68	8.51
8	14.69	11.04	9.60	8.81	8.30	7.95	7.69	7.50	7.34
9	13.61	10.11	8.72	7.96	7.47	7.13	6.88	6.69	6.54
10	12.83	9.43	8.08	7.34	6.87	6.54	6.30	6.12	5.97
11	12.23	8.91	7.60	6.88	6.42	6.10	5.86	5.68	5.54
12	11.75	8.51	7.23	6.52	6.07	5.76	5.52	5.35	5.20
13	11.37	8.19	6.93	6.23	5.79	5.48	5.25	5.08	4.94
14	11.06	7.92	6.68	6.00	5.56	5.26	5.03	4.86	4.72
15	10.80	7.70	6.48	5.80	5.37	5.07	4.85	4.67	4.54
16	10.58	7.51	6.30	5.64	5.21	4.91	4.69	4.52	4.38
17	10.38	7.35	6.16	5.50	5.07	4.78	4.56	4.39	4.25
18	10.22	7.21	6.03	5.37	4.96	4.66	4.44	4.28	4.14
19	10.07	7.09	5.92	5.27	4.85	4.56	4.34	4.18	4.04
20	9.94	6.99	5.82	5.17	4.76	4.47	4.26	4.09	3.96
21	9.83	6.89	5.73	5.09	4.68	4.39	4.18	4.01	3.88
22	9.73	6.81	5.65	5.02	4.61	4.32	4.11	3.94	3.81
23	9.63	6.73	5.58	4.95	4.54	4.26	4.05	3.88	3.75
24	9.55	6.66	5.52	4.89	4.49	4.20	3.99	3.83	3.69
25	9.48	6.60	5.46	4.84	4.43	4.15	3.94	3.78	3.64
30	9.18	6.35	5.24	4.62	4.23	3.95	3.74	3.58	3.45
40	8.83	6.07	4.98	4.37	3.99	3.71	3.51	3.35	3.22
60	8.49	5.79	4.73	4.14	3.76	3.49	3.29	3.13	3.01
120	8.18	5.54	4.50	3.92	3.55	3.28	3.09	2.93	2.81
∞	7.88	5.30	4.28	3.72	3.35	3.09	2.90	2.74	2.62

(continued)

Source: E. S. Pearson and H. O. Hartley, *Biometrika Tables for Statisticians,* vol. 1, 3rd ed. (Cambridge: The University Press, 1966), Table 18.

Table 10 CRITICAL VALUES OF $F_{.005}$ *(continued)*

DEGREES OF FREEDOM FOR NUMERATOR

DEGREES OF FREEDOM FOR DENOMINATOR

10	12	15	20	24	30	40	60	120	∞
24,224	24,426	24,630	24,836	24,940	25,044	25,148	25,253	25,359	25,465
199.4	199.4	199.4	199.4	199.5	199.5	199.5	199.5	199.5	199.5
43.69	43.39	43.08	42.78	42.62	42.47	42.31	42.15	41.99	41.83
20.97	20.70	20.44	20.17	20.03	19.89	19.75	19.61	19.47	19.32
13.62	13.38	13.15	12.90	12.78	12.66	12.53	12.40	12.27	12.14
10.25	10.03	9.81	9.59	9.47	9.36	9.24	9.12	9.00	8.88
8.38	8.18	7.97	7.75	7.65	7.53	7.42	7.31	7.19	7.08
7.21	7.01	6.81	6.61	6.50	6.40	6.29	6.18	6.06	5.95
6.42	6.23	6.03	5.83	5.73	5.62	5.52	5.41	5.30	5.19
5.85	5.66	5.47	5.27	5.17	5.07	4.97	4.86	4.75	4.64
5.42	5.24	5.05	4.86	4.76	4.65	4.55	4.44	4.34	4.23
5.09	4.91	4.72	4.53	4.43	4.33	4.23	4.12	4.01	3.90
4.82	4.64	4.46	4.27	4.17	4.07	3.97	3.87	3.76	3.65
4.60	4.43	4.25	4.06	3.96	3.86	3.76	3.66	3.55	3.44
4.42	4.25	4.07	3.88	3.79	3.69	3.58	3.48	3.37	3.26
4.27	4.10	3.92	3.73	3.64	3.54	3.44	3.33	3.22	3.11
4.14	3.97	3.79	3.61	3.51	3.41	3.31	3.21	3.10	2.98
4.03	3.86	3.68	3.50	3.40	3.30	3.20	3.10	2.99	2.87
3.93	3.76	3.59	3.40	3.31	3.21	3.11	3.00	2.89	2.78
3.85	3.68	3.50	3.32	3.22	3.12	3.02	2.92	2.81	2.69
3.77	3.60	3.43	3.24	3.15	3.05	2.95	2.84	2.73	2.61
3.70	3.54	3.36	3.18	3.08	2.98	2.88	2.77	2.66	2.55
3.64	3.47	3.30	3.12	3.02	2.92	2.82	2.71	2.60	2.48
3.59	3.42	3.25	3.06	2.97	2.87	2.77	2.66	2.55	2.43
3.54	3.37	3.20	3.01	2.92	2.82	2.72	2.61	2.50	2.38
3.34	3.18	3.01	2.82	2.73	2.63	2.52	2.42	2.30	2.18
3.12	2.95	2.78	2.60	2.50	2.40	2.30	2.18	2.06	1.93
2.90	2.74	2.57	2.39	2.29	2.19	2.08	1.96	1.83	1.69
2.71	2.54	2.37	2.19	2.09	1.98	1.87	1.75	1.61	1.43
2.52	2.36	2.19	2.00	1.90	1.79	1.67	1.53	1.36	1.00

Table 11 CRITICAL VALUES OF T.

SAMPLE SIZE n	$T_{.05}$	$T_{.025}$	$T_{.01}$	$T_{.005}$	SAMPLE SIZE n	$T_{.05}$	$T_{.025}$	$T_{.01}$	$T_{.005}$
5	1				16	36	30	24	19
6	2	1			17	41	35	28	23
7	4	2	0		18	47	40	33	28
8	6	4	2	0	19	54	46	38	32
9	8	6	3	2	20	60	52	43	37
10	11	8	5	3	21	68	59	49	43
11	14	11	7	5	22	75	66	56	49
12	17	14	10	7	23	83	73	62	55
13	21	17	13	10	24	92	81	69	61
14	26	21	16	13	25	101	90	77	68
15	30	25	20	16					

Source: F. Wilcoxon and R. A. Wilcox, *Some Rapid Approximate Statistical Procedures* (Stamford, CT: American Cyanamid Co., 1964), Table 2. Reproduced with the permission of American Cyanamid.

Table 12 CRITICAL VALUES OF U.

SAMPLE SIZES n	n	$U_{.05}$	$U_{.025}$	$U_{.01}$	$U_{.005}$	SAMPLE SIZES n	n	$U_{.05}$	$U_{.025}$	$U_{.01}$	$U_{.005}$
3	3	0					5	8	6	4	2
4	3	0					6	10	8	6	4
	4	1	0				7	13	10	7	6
5	2	0					8	15	13	9	7
	3	1	0			9	2	1	0		
	4	2	1	0			3	4	2	1	0
	5	3	2	1	0		4	6	4	3	1
6	2	0					5	9	7	5	3
	3	2	1				6	12	10	7	5
	4	3	2	1	0		7	15	12	9	7
	5	5	3	2	1		8	18	15	11	9
	6	7	5	3	2		9	21	17	14	11
7	2	0				10	2	1	0		
	3	2	1	0			3	4	3	1	0
	4	4	3	1	0		4	7	5	3	2
	5	6	5	3	1		5	11	8	6	4
	6	8	6	4	3		6	14	11	8	6
	7	11	8	6	4		7	17	14	11	9
8	2	1	0				8	20	17	13	11
	3	3	2	0			9	24	20	16	13
	4	5	4	2	1		10	27	23	19	16

Source: F. Wilcoxon and R. A. Wilcox, *Some Rapid Approximate Statistical Procedures* (Stamford, CT: American Cyanamid Co., 1964), Table 1. Reproduced with the permission of American Cyanamid.

Table 13 CRITICAL VALUES OF q_k.

DF FOR MSE	α	k = NUMBER OF TREATMENTS 2	3	4	5	6	7	8	9	10	11	12	13	14	15
1	.05	18.0	27.0	32.8	37.1	40.4	43.1	45.4	47.4	49.1	50.6	52.0	53.2	54.3	55.4
	.01	90.0	135	164	186	202	216	227	237	246	253	260	266	272	277
2	.05	6.09	8.3	9.8	10.9	11.7	12.4	13.0	13.5	14.0	14.4	14.7	15.1	15.4	15.7
	.01	14.0	19.0	22.3	24.7	26.6	28.2	29.5	30.7	31.7	32.6	33.4	34.1	34.8	35.4
3	.05	4.50	5.91	6.82	7.50	8.04	8.48	8.85	9.18	9.46	9.72	9.95	10.2	10.4	10.5
	.01	8.26	10.6	12.2	13.3	14.2	15.0	15.6	16.2	16.7	17.1	17.5	17.9	18.2	18.5
4	.05	3.93	5.04	5.76	6.29	6.71	7.05	7.35	7.60	7.83	8.03	8.21	8.37	8.52	8.66
	.01	6.51	8.12	9.17	9.96	10.6	11.1	11.5	11.9	12.3	12.6	12.8	13.1	13.3	13.5
5	.05	3.64	4.60	5.22	5.67	6.03	6.33	6.58	6.80	6.99	7.17	7.32	7.47	7.60	7.72
	.01	5.70	6.97	7.80	8.42	8.91	9.32	9.67	9.97	10.2	10.5	10.7	10.9	11.1	11.2
6	.05	3.46	4.34	4.90	5.31	5.63	5.89	6.12	6.32	6.49	6.65	6.79	6.92	7.03	7.14
	.01	5.24	6.33	7.03	7.56	7.97	8.32	8.61	8.87	9.10	9.30	9.49	9.65	9.81	9.95
7	.05	3.34	4.16	4.69	5.06	5.36	5.61	5.82	6.00	6.16	6.30	6.43	6.55	6.66	6.76
	.01	4.95	5.92	6.54	7.01	7.37	7.68	7.94	8.17	8.37	8.55	8.71	8.86	9.00	9.12
8	.05	3.26	4.04	4.53	4.89	5.17	5.40	5.60	5.77	5.92	6.05	6.18	6.29	6.39	6.48
	.01	4.74	5.63	6.20	6.63	6.96	7.24	7.47	7.68	7.87	8.03	8.18	8.31	8.44	8.55
9	.05	3.20	3.95	4.42	4.76	5.02	5.24	5.43	5.60	5.74	5.87	5.98	6.09	6.19	6.28
	.01	4.60	5.43	5.96	6.35	6.66	6.91	7.13	7.32	7.49	7.65	7.78	7.91	8.03	8.13
10	.05	3.15	3.88	4.33	4.65	4.91	5.12	5.30	5.46	5.60	5.72	5.83	5.93	6.03	6.11
	.01	4.48	5.27	5.77	6.14	6.43	6.67	6.87	7.05	7.21	7.36	7.48	7.60	7.71	7.81
11	.05	3.11	3.82	4.26	4.57	4.82	5.03	5.20	5.35	5.49	5.61	5.71	5.81	5.90	5.99
	.01	4.39	5.14	5.62	5.97	6.25	6.48	6.67	6.84	6.99	7.13	7.26	7.36	7.46	7.56
12	.05	3.08	3.77	4.20	4.51	4.75	4.95	5.12	5.27	5.40	5.51	5.62	5.71	5.80	5.88
	.01	4.32	5.04	5.50	5.84	6.10	6.32	6.51	6.67	6.81	6.94	7.06	7.17	7.26	7.36
13	.05	3.06	3.73	4.15	4.45	4.69	4.88	5.05	5.19	5.32	5.43	5.53	5.63	5.71	5.79
	.01	4.26	4.96	5.40	5.73	5.98	6.19	6.37	6.53	6.67	6.79	6.90	7.01	7.10	7.19
14	.05	3.03	3.70	4.11	4.41	4.64	4.83	4.99	5.13	5.25	5.36	5.46	5.55	6.64	5.72
	.01	4.21	4.89	5.32	5.63	5.88	6.08	6.26	6.41	6.54	6.66	6.77	6.87	6.96	7.05
16	.05	3.00	3.65	4.05	4.33	4.56	4.74	4.90	5.03	5.15	5.26	5.35	5.44	5.52	5.59
	.01	4.13	4.78	5.19	5.49	5.72	5.92	6.08	6.22	6.35	6.46	6.56	6.66	6.74	6.82
18	.05	2.97	3.61	4.00	4.28	4.49	4.67	4.82	4.96	5.07	5.17	5.27	5.35	5.43	5.50
	.01	4.07	4.70	5.09	5.38	5.60	5.79	5.94	6.08	6.20	6.31	6.41	6.50	6.58	6.65

(continued)

Table 13 CRITICAL VALUES OF q_k *(continued)*

DF FOR MSE	α	k = NUMBER OF TREATMENTS 2	3	4	5	6	7	8	9	10	11	12	13	14	15
20	.05	2.95	3.58	3.96	4.23	4.45	4.62	4.77	4.90	5.01	5.11	5.20	5.28	5.36	5.43
	.01	4.02	4.64	5.02	5.29	5.51	5.69	5.84	5.97	6.09	6.19	6.29	6.37	6.45	6.52
24	.05	2.92	3.53	3.90	4.17	4.37	4.54	4.68	4.81	4.92	5.01	5.10	5.18	5.25	5.32
	.01	3.96	4.54	4.91	5.17	5.37	5.54	5.69	5.81	5.92	6.02	6.11	6.19	6.26	6.33
30	.05	2.89	3.49	3.84	4.10	4.30	4.46	4.60	4.72	4.83	4.92	5.00	5.08	5.15	5.21
	.01	3.89	4.45	4.80	5.05	5.24	5.40	5.54	5.65	5.76	5.85	5.93	6.01	6.08	6.14
40	.05	2.86	3.44	3.79	4.04	4.23	4.39	4.52	4.63	4.74	4.82	4.91	4.98	5.05	5.11
	.01	3.82	4.37	4.70	4.93	5.11	5.27	5.39	5.50	5.60	5.69	5.77	5.84	5.90	5.96
60	.05	2.83	3.40	3.74	3.98	4.16	4.31	4.44	4.55	4.65	4.73	4.81	4.88	4.94	5.00
	.01	3.76	4.28	4.60	4.82	4.99	5.13	5.25	5.36	5.45	5.53	5.60	5.67	5.73	5.79
120	.05	2.80	3.36	3.69	3.92	4.10	4.24	4.36	4.48	4.56	4.64	4.72	4.78	4.84	4.90
	.01	3.70	4.20	4.50	4.71	4.87	5.01	5.12	5.21	5.30	5.38	5.44	5.51	5.56	5.61
∞	.05	2.77	3.31	3.63	3.86	4.03	4.17	4.29	4.39	4.47	4.55	4.62	4.68	4.74	4.80
	.01	3.64	4.12	4.40	4.60	4.76	4.88	4.99	5.08	5.16	5.23	5.29	5.35	5.40	5.45

Source: E. S. Pearson and H. O. Hartley, *Biometrika Tables for Statisticians,* vol. 1, 3rd ed. (Cambridge: The University Press, 1966), Table 29.

Table 14 CRITICAL VALUES OF F_{max}.

		k = NUMBER OF VARIANCES								
DF	α	2	3	4	5	6	7	8	9	10
2	.05	39.0	87.5	142	202	266	333	403	475	550
	.01	199	448	729	1036	1362	1705	2063	2432	2813
3	.05	15.4	27.8	39.2	50.7	62.0	72.9	83.5	93.9	104
	.01	47.5	85	120	151	184	216	249	281	310
4	.05	9.60	15.5	20.6	25.2	29.5	33.6	37.5	41.4	44.6
	.01	23.2	37	49	59	69	79	89	97	106
5	.05	7.15	10.8	13.7	16.3	18.7	20.8	22.9	24.7	26.5
	.01	14.9	22	28	33	38	42	46	50	54
6	.05	5.82	8.38	10.4	12.1	13.7	15.0	16.3	17.5	18.6
	.01	11.1	15.5	19.1	22	25	27	30	32	34
7	.05	4.99	6.94	8.44	9.70	10.8	11.8	12.7	13.5	14.3
	.01	8.89	12.1	14.5	16.5	18.4	20	22	23	24
8	.05	4.43	6.00	7.18	8.12	9.03	9.78	10.5	11.1	11.7
	.01	7.50	9.9	11.7	13.2	14.5	15.8	16.9	17.9	18.9
9	.05	4.03	5.34	6.31	7.11	7.80	8.41	8.95	9.45	9.91
	.01	6.54	8.5	9.9	11.1	12.1	13.1	13.9	14.7	15.3
10	.05	3.72	4.85	5.67	6.34	6.92	7.42	7.87	8.28	8.66
	.01	5.85	7.4	8.6	9.6	10.4	11.1	11.8	12.4	12.9
12	.05	3.28	4.16	4.79	5.30	5.72	6.09	6.42	6.72	7.00
	.01	4.91	6.1	6.9	7.6	8.2	8.7	9.1	9.5	9.9
15	.05	2.86	3.54	4.01	4.37	4.68	4.95	5.19	5.40	5.59
	.01	4.07	4.9	5.5	6.0	6.4	6.7	7.1	7.3	7.5
20	.05	2.46	2.95	3.29	3.54	3.76	3.94	4.10	4.24	4.37
	.01	3.32	3.8	4.3	4.6	4.9	5.1	5.3	5.5	5.6
30	.05	2.07	2.40	2.61	2.78	2.91	3.02	3.12	3.21	3.29
	.01	2.63	3.0	3.3	3.4	3.6	3.7	3.8	3.9	4.0
60	.05	1.67	1.85	1.96	2.04	2.11	2.17	2.22	2.26	2.30
	.01	1.96	2.2	2.3	2.4	2.4	2.5	2.5	2.6	2.6
∞	.05	1.00	1.00	1.00	1.00	1.00	1.00	1.00	1.00	1.00
	.01	1.00	1.00	1.00	1.00	1.00	1.00	1.00	1.00	1.00

Source: E. S. Pearson and H. O. Hartley, *Biometrika Tables for Statisticians,* vol. 1, 3rd ed. (Cambridge: The University Press, 1966), Table 31.

Index

Table 4 AREAS UNDER THE NORMAL CURVE

z	.00	.01	.02	.03	.04	.05	.06	.07	.08	.09
0.0	.0000	.0040	.0080	.0120	.0160	.0199	.0239	.0279	.0319	.0359
0.1	.0398	.0438	.0478	.0517	.0557	.0596	.0636	.0675	.0714	.0753
0.2	.0793	.0832	.0871	.0910	.0948	.0987	.1026	.1064	.1103	.1141
0.3	.1179	.1217	.1255	.1293	.1331	.1368	.1406	.1443	.1480	.1517
0.4	.1554	.1591	.1628	.1664	.1700	.1736	.1772	.1808	.1844	.1879
0.5	.1915	.1950	.1985	.2019	.2054	.2088	.2123	.2157	.2190	.2224
0.6	.2257	.2291	.2324	.2357	.2389	.2422	.2454	.2486	.2517	.2549
0.7	.2580	.2611	.2642	.2673	.2704	.2734	.2764	.2794	.2823	.2852
0.8	.2881	.2910	.2939	.2967	.2995	.3023	.3051	.3078	.3106	.3133
0.9	.3159	.3186	.3212	.3238	.3264	.3289	.3315	.3340	.3365	.3389
1.0	.3413	.3438	.3461	.3485	.3508	.3531	.3554	.3577	.3599	.3621
1.1	.3643	.3665	.3686	.3708	.3729	.3749	.3770	.3790	.3810	.3930
1.2	.3849	.3869	.3888	.3907	.3925	.3944	.3962	.3980	.3997	.4015
1.3	.4032	.4049	.4066	.4082	.4099	.4115	.4131	.4147	.4162	.4177
1.4	.4192	.4207	.4222	.4236	.4251	.4265	.4279	.4292	.4306	.4319
1.5	.4332	.4345	.4357	.4370	.4382	.4394	.4406	.4418	.4429	.4441
1.6	.4452	.4463	.4474	.4484	.4495	.4505	.4515	.4525	.4535	.4545
1.7	.4554	.4564	.4573	.4582	.4591	.4599	.4608	.4616	.4625	.4633
1.8	.4641	.4649	.4656	.4664	.4671	.4678	.4686	.4693	.4699	.4706
1.9	.4713	.4719	.4726	.4732	.4738	.4744	.4750	.4756	.4761	.4767
2.0	.4772	.4778	.4783	.4788	.4793	.4798	.4803	.4808	.4812	.4817
2.1	.4821	.4826	.4830	.4834	.4838	.4842	.4846	.4850	.4854	.4857
2.2	.4861	.4864	.4868	.4871	.4875	.4878	.4881	.4884	.4887	.4890
2.3	.4893	.4896	.4898	.4901	.4904	.4906	.4909	.4911	.4913	.4916
2.4	.4918	.4920	.4922	.4925	.4927	.4929	.4931	.4932	.4934	.4936
2.5	.4938	.4940	.4941	.4943	.4945	.4946	.4948	.4949	.4951	.4952
2.6	.4953	.4955	.4956	.4957	.4959	.4960	.4961	.4962	.4963	.4964
2.7	.4965	.4966	.4967	.4968	.4969	.4970	.4971	.4972	.4973	.4974
2.8	.4974	.4975	.4976	.4977	.4977	.4978	.4979	.4979	.4980	.4981
2.9	.4981	.4982	.4982	.4983	.4984	.4984	.4985	.4985	.4986	.4987
3.0	.4987	.4987	.4987	.4988	.4988	.4989	.4989	.4989	.4990	.4990
3.1	.4990	.4991	.4991	.4991	.4992	.4992	.4992	.4992	.4993	.4993
3.2	.4993	.4993	.4994	.4994	.4994	.4994	.4994	.4995	.4995	.4955
3.3	.4995	.4995	.4996	.4996	.4996	.4996	.4996	.4996	.4996	.4997
3.4	.4997	.4997	.4997	.4997	.4997	.4997	.4997	.4997	.4998	.4998
3.5	.4998	.4998	.4998	.4998	.4998	.4998	.4998	.4998	.4998	.4998

Table 5 CRITICAL VALUES OF t

DF	$t_{.10}$	$t_{.05}$	$t_{.025}$	$t_{.01}$	$t_{.005}$
1	3.078	6.314	12.706	31.821	63.657
2	1.886	2.920	4.303	6.965	9.925
3	1.638	2.353	3.182	4.541	5.841
4	1.533	2.132	2.776	3.747	4.604
5	1.476	2.015	2.571	3.365	4.032
6	1.440	1.943	2.447	3.143	3.707
7	1.415	1.895	2.365	2.998	3.499
8	1.397	1.860	2.306	2.896	3.355
9	1.383	1.833	2.262	2.821	3.250
10	1.372	1.812	2.228	2.764	3.169
11	1.363	1.796	2.201	2.718	3.106
12	1.356	1.782	2.179	2.681	3.055
13	1.350	1.771	2.160	2.650	3.012
14	1.345	1.761	2.145	2.624	2.977
15	1.341	1.753	2.131	2.602	2.947
16	1.337	1.746	2.120	2.583	2.921
17	1.333	1.740	2.110	2.567	2.898
18	1.330	1.734	2.101	2.552	2.878
19	1.328	1.729	2.093	2.539	2.861
20	1.325	1.725	2.086	2.528	2.845
21	1.323	1.721	2.080	2.518	2.831
22	1.321	1.717	2.074	2.508	2.819
23	1.319	1.714	2.069	2.500	2.807
24	1.318	1.711	2.064	2.492	2.797
25	1.316	1.708	2.060	2.485	2.787
26	1.315	1.706	2.056	2.479	2.779
27	1.314	1.703	2.052	2.473	2.771
28	1.313	1.701	2.048	2.467	2.763
29	1.311	1.699	2.045	2.462	2.756
30	1.310	1.697	2.042	2.457	2.750
40	1.303	1.684	2.021	2.423	2.704
60	1.296	1.671	2.000	2.390	2.660
120	1.289	1.658	1.980	2.358	2.617
∞	1.282	1.645	1.960	2.326	2.576

Source: E. S. Pearson and H. O. Hartley, *Biometrika Tables for Statisticians,* vol. 1, 3rd ed. (Cambridge: The University Press, 1966), Table 12.